ORIGIN AND EVOLUTION OF VIRUSES

ORIGIN AND EVOLUTION OF VIRUSES

SECOND EDITION

Edited by

ESTEBAN DOMINGO
Centro de Biología Molecular "Severo Ochoa" (CSIC-UAM), Campus de Cantoblanco, 28049 Madrid, Spain
and CIBERehd

COLIN R. PARRISH
Cornell University, College of Veterinary Medicine, Baker Institute for Animal Health, Ithaca, NY 14853, USA

JOHN J. HOLLAND
Institute for Molecular Genetics and Division of Biology, University of California, San Diego, CA 92093, USA

ELSEVIER

AMSTERDAM • BOSTON • HEIDELBERG • LONDON
NEW YORK • OXFORD • PARIS • SAN DIEGO
SAN FRANCISCO • SINGAPORE • SYDNEY • TOKYO
Academic Press is an imprint of Elsevier

Academic Press is an imprint of Elsevier
32 Jamestown Road, London NW1 7BY, UK
30 Corporate Drive, Suite 400, Burlington, MA 01803, USA
525 B Street, Suite 1900, San Diego, CA 92101-4495, USA
360 Park Avenue South, New York, NY 10010-1710, USA

First edition 1999
Second edition 2008

British Library Cataloguing in Publication Data
A catalogue record for this book is available from the British Library

Library of Congress Cataloguing-in-Publication Data
A catalogue record for this book is available from the Library of Congress

ISBN: 978-0-12-374153-0

For information on all Elsevier publications
visit our web site at www.elsevierdirect.com

Typeset by Charon Tec Ltd., A Macmillan Company.
(www.macmillansolutions.com)

Printed and bounded in Hungary

08 09 10 11 12 10 9 8 7 6 5 4 3 2 1

Contents

Contributors

Raul Andino
Department of Microbiology and Immunology, University of California, Mission Bay, Genentech Hall, 600 16th Street, Suite S572E, Box 2280, San Francisco, CA 94143-2280, USA

Jamie J. Arnold
Department of Biochemistry and Molecular Biology, The Pennsylvania State University, 201 Althouse Laboratory, University Park, PA 16802, USA

John W. Barrett
Biotherapeutics Research Group, Robarts Research Institute, London, ON N6G, Canada

Hans-Ulrich Bernard
Department of Molecular Biology and Biochemsitry, University of California, Irvine, Sprague Hall, Irvine, CA 92697, USA

Christof K. Biebricher
Max-Planck Institute for Biophysical Chemistry D-37070, Gottingen, Germany

Sebastian Bonhoeffer
Institute of Integrative Biology, ETH Zürich, ETH Zentrum, CHN K12.1, Universitaetsstr. 16, CH-8092 Zurich, Switzerland

A. Bosch
Enteric Virus Laboratory, Department of Microbiology, University of Barcelona, UB Barcelona, Spain

J.J. Bull
Section of Integrative Biology, Institute for Cellular and Molecular Biology, Center for Computational Biology and Bioinformatics, University of Texas, 1 University Station c0930, Austin, TX 78712, USA

M. Buti
Liver Unit, Hospital Universitari, Vall d'Hebron, Barcelona, Spain

Craig E. Cameron
Department of Biochemistry and Molecular Biology, The Pennsylvania State University, 201 Althouse Laboratory, University Park, PA 16802, USA

José-Antonio Daròs
Instituto de Biología Molecular y Celular de Plantas (CSIC-UPV), Campus UPV, 46022 Valencia, Spain

Andrew J. Davison
MRC Virology Unit, Institute of Virology, University of Glasgow, Church Street, Glasgow G11 5JR, UK

Aidan Dolan
MRC Virology Unit, Institute of Virology, University of Glasgow, Church Street, Glasgow G11 5JR, UK

Esteban Domingo
Centro de Biología Molecular "Severo Ochoa" (CSIC-UAM), Universidad Autónoma de Madrid, Cantoblanco, 28049 Madrid, Spain
and Centro de Investigación Biomedica en Red de Enfermedades Hepáticas y Digestivas (CIBERehd)

Núria Duran-Vila
Institut Valenciano de Investigaciones Agrarias (IVIA), 46113 Moncada, Valencia, Spain

Santiago F. Elena
Instituto de Biología Molecular y Celular de Plantas (CSIC-UPV), Campus UPV, CPI, 46022 Valencia, Spain

Cristina Escarmís
Centro de Biología Molecular "Severo Ochoa" (CSIC-UAM), Universidad Autónoma de Madrid, Cantoblanco, 28049 Madrid, Spain

J.I. Esteban
Liver Unit, Department of Medicine, Hospital Universitari Vall d' Hebron, Pg Vall D' Hebron 119–129, 08035 Barcelona, Spain
and Centro de Investigación Biomedica en Red de Enfermedades Hepáticas y Digestiuas (CIBERehd)

Richard Flores
Instituto de Biología Molecular y Celular de Plantas (CSIC-UPV), Campus UPV, CPI, 46022 Valencia, Spain

Fernando García-Arenal
Centro de Biotecnologia y Genómica de Plantas, E.T.S.I. Agrónomos, Universidad Politécnica de Madrid, 28040 Madrid, Spain

Derek Gatherer
MRC Virology Unit, Institute of Virology, University of Glasgow, Church Street, Glasgow G11 5JR, UK

Adrian Gibbs
7 Hutt Street, Yarralumla, ACT 2600, Australia

Mark Gibbs
School of Botany and Zoology, Faculty of Science, Australian National University, Canberra, A.C.T. 0200, Australia

Warner C. Greene
Gladstone Institute of Virology and Immunology, 1650 Owens Street, San Francisco, CA 94158, USA

Kathryn A. Hanley
Department of Biology, New Mexico State University, Las Cruces, NM 88003, USA

Roger W. Hendrix
Department of Biological Sciences and Pittsburgh Bacteriophage Institute, University of Pittsburgh, Pittsburgh, PA 15260, USA

Mónica Herrera
Centro de Biología Molecular " Severo Ochoa" (CSIC-UAM), Universidad Autónoma de Madrid, Cantoblanco, 28049 Madrid, Spain

Karin Hoelzer
Baker Institute for Animal Health, Department of Microbiology and Immunology, College of Veterinary Medicine, Cornell University, Ithaca, NY 14853, USA

John J. Holland
Division of Biology and Institute for Molecular Genetics, University of California at San Diego, 9500 Gilman Drive, La Jolla, California 92093-0116, USA

Edward C. Holmes
Center for Infectious Disease Dynamics, Department of Biology, The Pennsylvania State University, University Park, PA 16802, USA

R. Jardí
Biochemistry Department, Hospital Universitari Vall d'Hebron, Barcelona, Spain and Centro de Investigación Biomedica en Red de Enfermedades Hepáticas y Digestivas (CIBERehd)

M. Martell
Liver Unit, Hospital Universitari, Vall d' Hebron, Barcelona, Spain

Grant McFadden
Department of Molecular Genetics and Microbiology, College of Medicine, University of Florida, 1600 SW Archer Road, ARB Rm R4-295, POB 100266, Gainesville, FL 32610, USA

Duncan J. McGeoch
MRC Virology Unit, Institute of Virology, University of Glasgow, Church Street, Glasgow G11 5JR, UK

Luis Menéndez-Arias
Centro de Biología Molecular "Severo Ochoa" (CSIC-UAM), Universidad Autónoma de Madrid, Cantoblanco, 28049 Madrid, Spain

Viktor Müller
Institute of Biology, Eötvös Loránd University, Pázmány P.s. 1/C, 1117 Budapest, Hungary

Isabel S. Novella
Department of Medical Microbiology and Immunology, University of Toledo, College of Medicine, Toledo 43614, Spain

Kazusato Ohshima
Laboratory of Plant Virology, Faculty of Agriculture, Saga University, 1-banchi, Honjo-machi, Saga 840-8502, Japan

Colin R. Parrish
Baker Institute for Animal Health, Department of Microbiology and Immunology, College of Veterinary Medicine, Cornell University, Ithaca, NY 14853, USA

Celia Perales
Centro de Biología Molecular "Severo Ochoa" (CSIC-UAM), Universidad Autónoma de Madrid, Cantoblanco, 28049 Madrid, Spain and Centro de Investigación Biomedica en Red de Enfermedades Hepáticas y Digestivas (CIBERehd)

J. Quer
Liver Unit, Department of Medicine, Hospital Universitari Vall d'Hebron, Barcelona, Spain and Centro de Investigación Biomedica en Red de Enfermedades Hepáticas y Digestivas (CIBERehd)

F. Rodriguez
Biochemistry Department, Hospital Universitari Vall d'Hebron, Barcelona, Spain and Centro de Investigación Biomedica en Red de Enfermedades Hepáticas y Digestivas (CIBERehd)

Marilyn J. Roossinck
The Samuel Roberts Noble Foundation, Plant Biology Division, 2510 Sam Noble Parkway, P.O. Box 2180, Ardmore, OK 73402, USA

R. Sannjuán
Institut Cavanilles de Biodiversitat i Biologia Evolutiva, Universitat de València, Valencia, Spain

Mario L. Santiago
Gladstone Institute of Virology and Immunology, 1650 Owens Street, San Francisco, CA 94158, USA

Peter Schuster
Institute of Theoretical Chemistry, University of Vienna, Währingstrasse 17, A-1090 Vienna, Austria

Edgar E. Sevilla-Reyes
MRC Virology Unit, Institute of Virology, University of Glasgow, Church Street, Glasgow G11 5JR, UK

Eric Smidansky
Department of Biochemistry and Molecular Biology, The Pennsylvania State University, 201 Althouse Laboratory, University Park, PA 16802, USA

Peter F. Stadler
Bioinformatics Group, Department of Computer Science and Interdisciplinary Center of Bioinformatics, University of Leipzig, Härtelstrasse 16–18, D-04107 Leipzig, Germany

Ronald Van Rij
Hubrecht Institute, Uppsalalaan 8, 3584 CT Utrecht, The Netherlands

Luis P. Villareal
Center for Virus Research, Irvine Research Unit on Animal Viruses, Department of Molecular Biology and Biochemistry, University of California, Irvine, 2332 McGaugh Hall, Irvine, CA 92697, USA

Simon Wain-Hobson
Pasteur Institute, 28 rue du Dr. Roux, 75724 Paris Cedex 15, France

Scott C. Weaver
Center for Tropical Diseases and Department of Pathology, University of Texas Medical Branch, 301 University Boulevard, Galveston, TX 77555-0609, USA

C.O. Wilke
Section of Integrative Biology, Institute for Cellular and Molecular Biology, Center for Computational Biology and Bioinformatics, University of Texas, Austin,TX 78712, USA

Preface to the Second Edition

A second, updated edition of a scientific book is a good sign that the first edition was well accepted, and that the book contents respond to the demands of a dynamic field of activity. The second edition of *Origin and Evolution of Viruses* tries to reflect, as the first edition did, the different molecular strategies that viruses use to evolve within and between hosts, and to provide a view of the complexities of short-term and long-term evolution, with their implications for viral disease. To this aim, the book includes several chapters related to general concepts and tools for the study of virus diversity, evolution, and pathogenesis, and a number of chapters covering specific viral systems including animal, plant, and bacterial viruses. We have broadened the scope of the first edition by including new subjects such as phylogenetic analysis of viral genomes, the enzymological bases of error-prone replication, RNA interference, and cellular functions involved in hypermutagenesis. We have maintained and expanded the chapters needed to provide a historical view of how viruses might have originated, with updated accounts of the primitive RNA world, quasispecies dynamics, and how virus evolution relates to the general evolution of our biosphere. We hope that the reader will find the series of chapters to be both informative and a stimulus for further research.

It is now well established, and it becomes even more transparent upon reading the book, that the diversity of viruses, and their population numbers are astonishing. On top of this, individual viral populations consist of clouds of related variants rather than defined genomic nucleotide sequences in the classical sense. This is carefully explored in the book from the viewpoints of biochemistry and of evolution. It will not escape the reader that the extreme dynamics of viral populations, with all its implications for virus evolution and pathogenesis, represents a translation into virology of the quasispecies concept, proposed initially by Manfred Eigen in an influential paper published 36 years ago. Quasispecies theory was further developed by several of Eigen's colleagues, and some of them have contributed chapters to the book. Quasispecies has represented the introduction of concepts of complexity to virology. The editors wish to highlight the relevance of quasispecies theory for virology, and gratefully dedicate this volume to Manfred Eigen on the occasion of his 80th birthday.

Such a diversity of topics and viral systems would not have been possible without the commitment and hard work of the many authors to whom we are deeply indebted. Our thanks go also to Elsevier, in particular to Lisa Tickner and Maureen Twaig, first for taking the initiative to propose a second edition of the book, and, second, for their careful involvement in the various matters that require attention before a book can be printed.

Esteban Domingo, Colin Parrish, John Holland

'Theory cannot remove complexity, but it shows what kind of "regular" behavior can be expected and what experiments have to be done to get a grasp on the irregularities.'

Manfred Eigen

Preface to the First Edition

Viruses differ greatly in their molecular strategies of adaptation to the organisms they infect. RNA viruses utilize continuous genetic change as they explore sequence space to improve their fitness, and thereby to adapt to the changing environment of their hosts. Variation is intimately linked to their disease-causing potential. Paramount to the understanding of RNA viruses is the concept of quasispecies, first developed to describe the early replicons thought to be components of a primitive RNA world devoid of DNA or proteins. The first chapter of the book deal with theoretical concepts of self-organization, RNA-mediated catalysis and the adaptive exploration of sequence space by RNA replicons. Likely descendants of the RNA world that we can study today are the plant-infecting viroids, and the δ agent (hepatitis D), a unique RNA genome associated with some cases of hepatitis B infection. δ provides an example of a simple, bifunctional molecule that contains a viroid-like replication domain, and a minimal protein-coding domain. It may be a relic of the type of recombinant molecules that may have participated in the transition to the DNA world from the RNA world. The impact of genetic variability of pathogenic RNA viruses is addressed in several chapters that cover specific viruses of animals and plants.

Retroid agents probably had an essential role in early evolution. Not only are they widely distributed and capable of copying RNA into DNA, but they may also have provided regulatory elements, and promoted genetic modifications for adaptation of DNA genomes. Among the retroelements, retroviruses are transmitted as RNA-containing particles, prior to intracellular copying of their RNA genomes into DNA, which can be stably maintained as an insert into the DNA of their hosts. The book discusses retroid agents and retroviruses, with emphasis on human immunodeficiency virus, the most thoroughly scrutinized retrovirus of all. Experiments and modeling meet to try to understand how variation and adaption of this dreaded pathogen lead to a collapse of the human immune system.

DNA viruses are likely to have coevolved with their hosts while the DNA world was developing. The last chapters of the book deal with the interplay between host evolution and DNA virus evolution, including chapters on the simplest and the most complex of the DNA viral genomes known. This broad coverage of topics would not have been possible without the contributions of many experts. We express our most sincere gratitude to all of these authors for having joined in the effort. The strong interdisciplinary flavor of the book is due to their different points of view. We expect the book to take the reader on a long journey (in time and in concepts) from the primitive and basic to the modern and complex.

While this book was in press, Professor Eladio Viñuela passed away on March 9, 1999. Eladio was an outstanding scientist, a pioneer of Virology in Spain, and a friend. The editors dedicate this volume to his memory.

E. Domingo, R.G. Webster, J.J. Holland

CHAPTER

1

Early Replicons: Origin and Evolution*

Peter Schuster and Peter F. Stadler

ABSTRACT

RNA and protein molecules have been found to be both templates for replication and specific catalysts for biochemical reactions. RNA molecules, although very difficult to obtain via plausible synthetic pathways under prebiotic conditions, are the only candidates for early replicons. Only they are obligatory templates for replication, which can conserve mutations and propagate them to forthcoming generations. RNA-based catalysts, called ribozymes, act with high efficiency and specificity for all classes of reactions involved in the interconversion of RNA molecules such as cleavage and template-assisted ligation. The idea of an RNA world was conceived for a plausible prebiotic scenario of RNA molecules operating upon each other and constituting thereby a functional molecular organization. A theoretical account of molecular replication making precise the conditions under which one observes parabolic, exponential, or hyperbolic growth is presented. Exponential growth is observed in a protein-assisted RNA world where plus–minus (±) duplex formation is avoided by the action of an RNA replicase. Error propagation to forthcoming generations is analyzed in the absence of selective by neutral mutants as well as for predefined degrees of neutrality. The concept of an error threshold for sufficiently precise replication and survival of populations derived from the theory of molecular quasispecies is discussed. Computer simulations are used to model the interplay between adaptive evolution and random drift. A model of evolution is proposed that allows for explicit handling of phenotypes.

WHAT IS A REPLICON?

Biology, and evolution in particular, are based on reproduction or multiplication and on variation. Reproduction pure has the property of self-enhancement and leads to exponential growth. Self-enhancement in chemical reactions under isothermal conditions is tantamount to autocatalysis that, in its simplest form, corresponds to a reaction mechanism of the kind:

$$A + Y \xrightarrow{\quad k \quad} 2Y, \qquad (1)$$

*Dedicated to Manfred Eigen, the pioneer of molecular evolution and intellectual father of quasispecies theory, on the occasion of his 80th birthday.

ISBN 978-0-12-374153-0

where **A** is the substrate and **Y** the autocatalyst. Being just an autocatalyst is certainly not enough for playing a role at the origin of life or in evolution. An additional *conditio sine qua non* is the property to act as an encoded instruction for the reproduction process. It is useful to remain rather vague as far as the nature of this instruction is concerned, because there are many possible solutions for template action at the molecular level. In reality the most straightforward candidates for useful templates are heteropolymers built from a few classes of monomers with specific interactions. The proper physical basis for such interactions are charge patterns, patterns of hydrogen bonds, space-filling hydrophobic interactions, and others. We may summarize the first paragraph by saying: "A replicon is an entity that carries the instruction for its own replication in some encoded form."

Precise asexual reproduction gives rise to perfect inheritance. This is essentially true for prokaryotes: bacteria, archaebacteria, and viruses. In sexually reproducing eukaryotes, recombination introduces variation already into the error-free reproduction process.[1] Mutation in the form of unprecise or error-prone reproduction represents the universal kind of variation, which occurs in all organisms and can be sketched by a single overall reaction step:

$$\mathrm{A} + \mathrm{Y} \xrightarrow{k'} \mathrm{Y} + \mathrm{Y}'. \qquad (2)$$

Here, the mutant is denoted by $\mathbf{Y}'$. The rate parameters k and k' refer to two parallel reaction channels. This can be indicated by replacing the two parameters with a single rate constant and reaction (channel) frequencies:

$$\begin{aligned} k &\Rightarrow f \cdot Q \text{ and} \\ k' &\Rightarrow f \cdot Q'. \end{aligned} \qquad (3)$$

In the (improbable) case that $\mathbf{Y}'$ is the only mutant of $\mathbf{Y}$, the two channel frequencies add up to unity: $Q + Q' = 1$. In general, there will be many mutations, $\mathbf{Y}_j \rightarrow \mathbf{Y}_i$, that give rise to variants and conservation of probabilities then leads to the conservation relation:

$$\sum_{i=1}^{n} Q_{ij} = 1 \;\forall\; j = 1, \ldots n, \qquad (4)$$

which expresses that a copy is either error free or contains errors.

It is useful further to distinguish two classes of replicons: (i) obligatory replicons and (ii) optional replicons. All error copies of obligatory replicons can be replicated and thus are replicons themselves. Examples of obligatory replicons are nucleic acid molecules under suitable conditions (Figure 1.1). In Nature practically no restrictions on the initiation and chain propagation of replication are known apart from recognition sites at replication origins and a few other general requirements for replication. An example of a laboratory system is the polymerase chain reaction (PCR), which allows for amplification of DNA templates with (almost) any sequence. Optional replicons are, for example, autocatalytically growing oligonucleotides (von Kiedrowski, 1986) and oligopeptides (Lee *et al.*, 1996) (Figure 1.2). These oligomers lose their capability to act as template (almost always) when a particular nucleotide or amino acid residue is exchanged for any other one. In other words, the property to be a replicon is not common feature of the entire class of molecules but a specific property of certain selected molecules only.

Simple replicons certainly lack the complexity of present-day organisms and are defined best as molecular entities that are capable of replication by means of some mechanism based on interaction with a template. Almost all known replicons are oligomers or polymers composed from a few classes of monomers. Two extreme types of replicons are distinguished: *obligatory replicons*, for which exchange of individual monomeric units yields other replicons with different monomer sequences, and

[1] Sexual reproduction introduces obligatory recombination into the mechanism of inheritance. Recombination in eukaryotes occurs during meiosis and is a highly complex process. In this chapter we are discussing primitive replication systems only and therefore we can dispense with any detailed discussion of recombination.

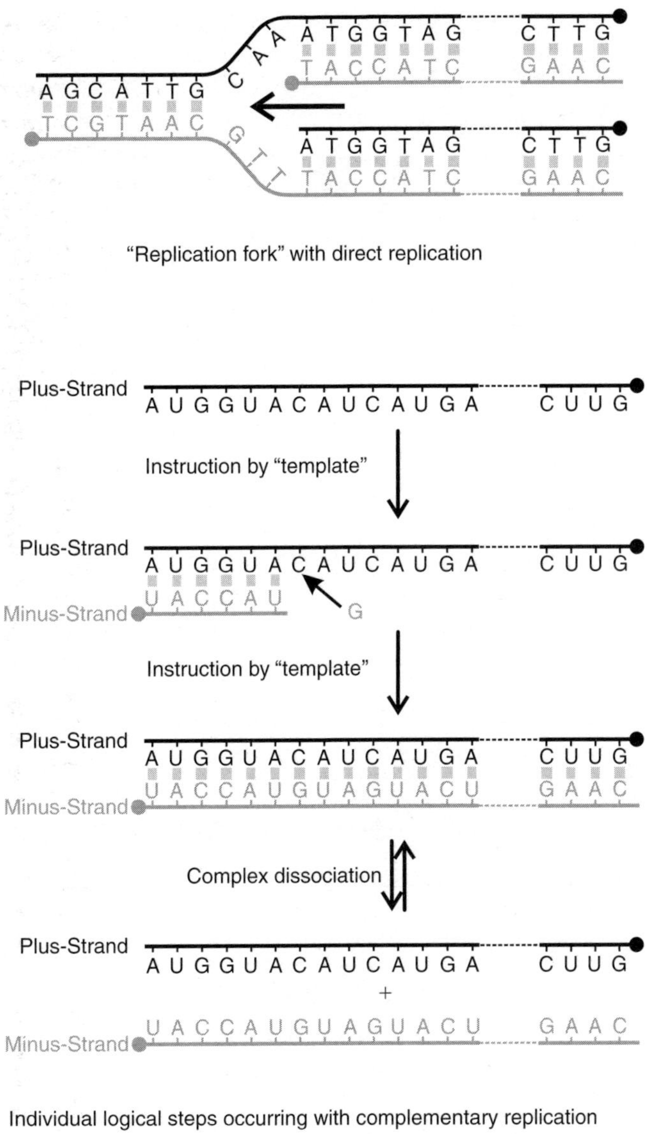

FIGURE 1.1 Template-induced replication of nucleic acids molecules. Direct replication (upper part) is primarily occurring with DNA. It represents a highly sophisticated process involving some 20 enzymes. Template-induced DNA synthesis occurs at the "replication fork," both daughter molecules carry one DNA strand of the parent molecule. Complementary replication (lower part) occurs in Nature with single-stranded RNA molecules. The problem in uncatalyzed complementary replication is complex dissociation. A single enzyme is sufficient for complementary replication of simple RNA bacteriophages, since it causes the separation of plus and minus strands during replication. The two strands separate and form their own single-strand structures before the double helix is completed. Polymerase chain reaction (PCR) follows essentially the same mechanism of complementary replication as shown here. The separation of the two strands of the double helix is accomplished by heating: the complex dissociates spontaneously at higher temperature.

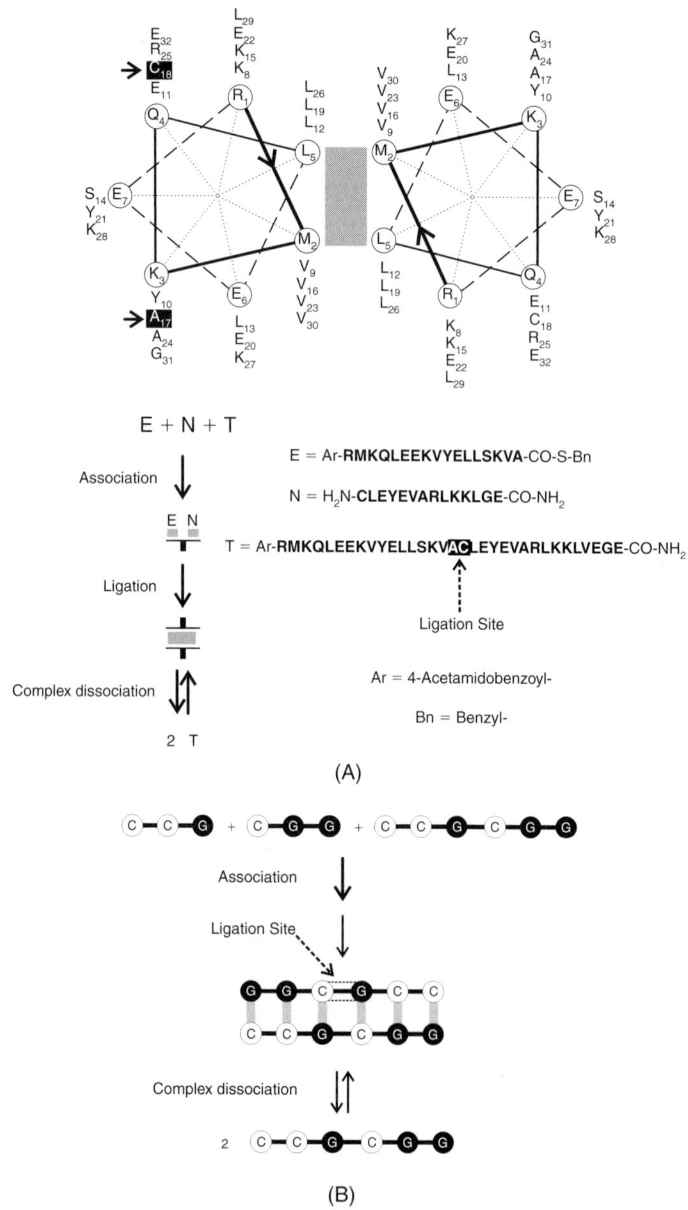

FIGURE 1.2 Oligopeptide and oligonucleotide replicons. (A) An autocatalytic oligopeptide that makes use of the leucine zipper for template action. The upper part illustrates the stereochemistry of oligopeptide template–substrate interaction by means of the helix wheel. The ligation site is indicated by arrows. The lower part shows the mechanism (Lee *et al.*, 1996; Severin *et al.*, 1997). (B) Template-induced self-replication of oligonucleotides (von Kiedrowski (1986)) follows essentially the same reaction mechanism. The critical step is the dissociation of the dimer after bond formation which commonly prevents these systems from exponential growth and Darwinian behavior.

optional replicons where the capability of replication is restricted to certain specific sequences.

More complex replicons (not discussed in detail here) including DNA and protein, compartment structure, and metabolism have been considered as well (Eigen and Schuster, 1982; Gánti, 1997; Szathmáry and Maynard Smith, 1997; Rasmussen *et al.*, 2003; Luisi,

2004). A successful experimental approach to self-reproduction of micelles and vesicles highlights one of the many steps on the way towards a primitive cell: prebiotic formation of vesicle structures (Bachmann *et al.*, 1992). The basic reaction leading to autocatalytic production of amphiphilic materials is the hydrolysis of ethyl caprate. The combination of vesicle formation with RNA replication represents a particularly important step towards the construction of a kind of minimal synthetic cell (Luisi *et al.*, 1994). Primitive forms of metabolism were considered for minimal cells as well (see, e.g., Rasmussen *et al.*, 2004b).

SIMPLE REPLICONS AND THE ORIGIN OF REPLICATION

A large number of successful experimental studies have been conducted to work out plausible chemical scenarios for the origin of early replicons being molecules capable of replication (Mason, 1991). A sketch of such a possible sequence of events in prebiotic evolution is shown in Figure 1.1. Most of the building blocks of present-day biomolecules are available from different prebiotic sources, from extraterrestrial origins, as well as from processes taking place in the primordial atmosphere or near hot vents in deep oceans. Condensation reactions and polymerization reactions formed non-instructed polymers, for example random oligopeptides of the protenoid type (Fox and Dose, 1977).

Template catalysis opens up the door to molecular copying and self-replication. Several small templates were designed by Rebek and co-workers and these molecules do indeed show complementarity and undergo complementary replication under suitable conditions (see, e.g., Tjivikua *et al.*, 1990; Nowick *et al.*, 1991). Like nucleic acids they consist of a backbone whose role is to bring "molecular digits" into stereochemically appropriate positions, so that they can be read by their complements. Complementarity is also based on essentially the same principle as in nucleic acids: Specific patterns of hydrogen bonds allow for recognition of complementary digits and discriminate the non-complementary "letters" of an alphabet. The hydrogen bonding pattern in these model replicons may be assisted by opposite electric charges carried by the complements. We shall encounter the same principle later in the discussion of Ghadiri's replicons based on stable coiled-coils of oligopeptide α-helices (Lee *et al.*, 1996). Autocatalysis in small model systems is certainly interesting because it reveals some mechanistic details of molecular recognition. These systems are, however, highly unlikely to be the basis of biologically significant replicons because they cannot be extended to large polymers in a simple way and hence they are unsuitable for storing a sizeable amount of (sequence) information. Ligation of small pieces to larger units, on the other hand, is a source of combinatorial complexity providing sufficient capacity for information storage and evolution. Heteropolymer formation thus seems inevitable and we shall therefore focus here only on replicons that have this property: nucleic acids and proteins.

A first major transition leads from a world of simple chemical reaction networks to autocatalytic processes that are able to form self-organized systems capable of replication and mutation as required for Darwinian evolution. This transition can be seen as the interface between chemistry and biology since an early Darwinian scenario is tantamount to the onset of biological evolution. Two suggestions were made in this context: (i) autocatalysis arose in a network of reactions catalyzed by oligopeptides (Kauffman, 1993) and (ii) the first autocatalyst was a representative of a class of molecules with obligatory template function in the sense discussed above (Eigen, 1971; Orgel, 1987). The first suggestion works with molecules that are easily available under prebiotic conditions but lacks plausibility because the desired properties—conservation and propagation of mutants—are unlikely to occur with oligopeptides. The second concept suffers from the opposite: it is very hard to derive a plausible scenario for the appearance of the first nucleic acid-like molecules. Once formed, however, they would fulfill most functional requirements for evolutionary optimization.

Until the 1980s biochemists had an empirically well established but nevertheless prejudiced view on the natural and artificial functions of proteins and nucleic acids. Proteins were thought to be Nature's unbeatable universal catalysts, highly efficient as well as ultimately specific, and as in the case of immunoglobulins even tunable to recognize previously unseen molecules. After Watson and Crick's famous discovery of the double helix, DNA was considered to be the molecule of inheritance, capable of encoding genetic information and sufficiently stable to allow for essential conservation of nucleotide sequences over many replication rounds. RNA's role in the molecular concert of Nature was reduced to the transfer of sequence information from DNA to protein, either as mRNA or as tRNA. Ribosomal RNA and some rare RNA molecules did not fit well into this picture: Some sort of scaffolding functions were attributed to them, such as holding supramolecular complexes together or bringing protein molecules into the correct spatial positions required for their functions.

This conventional picture was based on the idea of a complete "division of labor." Nucleic acids, DNA, as well as RNA were the templates, ready for replication and read-out of genetic information but not to do catalysis. Proteins were the catalysts and thus not capable of template function. In both cases these rather dogmatic views turned out to be wrong. In the 1980s Cech and Altman discovered RNA molecules with catalytic functions (Cech, 1983, 1986, 1990; Guerrier-Takada *et al.*, 1983). The name "ribozyme" was created for this new class of biocatalysts because they combine the properties of ribonucleotides and enzymes. Their examples dealt with RNA cleavage reactions catalyzed by RNA; without the help of a protein catalyst a non-coding region of an RNA transcript, a group I intron, cuts itself out during mRNA maturation. The second example concerns the enzymatic reaction of RNaseP, which catalyzes tRNA formation from the precursor poly-tRNA. For a long time biochemists had known that this enzyme consists of a protein and an RNA moiety. It was tacitly assumed that the protein was the catalyst while the RNA component had only a backbone function. The converse, however, is true: the RNA acts as catalyst and the protein provides merely a scaffold required to enhance the efficiency. Even more spectacular was the result from the structure of the ribosome at atomic resolution (Ban *et al.*, 2000; Nissen *et al.*, 2000; Steitz and Moore, 2003): polypeptide synthesis at the ribosome is catalyzed by rRNA and not by ribosomal proteins.

The second prejudice was disproved only about ten years ago by the demonstration that oligopeptides can act as templates for their own synthesis and thus show autocatalysis (Lee *et al.*, 1996; Severin *et al.*, 1997; Lee *et al.*, 1997). In this very elegant work, Ghadiri and co-workers have demonstrated that template action does not necessarily require hydrogen bond formation. Two smaller oligopeptides of chain lengths 17 (E) and 15 (N) are aligned on the template (T) by means of the hydrophobic interaction in a coiled-coil of the leucine zipper type and the 32-mer is produced by spontaneous peptide bond formation between the activated carboxygroup and the free amino residue (Figure 1.2A). The hydrophobic cores of template and ligands consist of alternating valine and leucine residues and show a kind of knobs-into-holes type packing in the complex. The ability of proteins to act as templates is a consequence of the three-dimensional structure of the protein α-helix, which allows the formation of coiled-coils. It requires that the residues making the contacts between the helices fulfill the condition of space filling and thus stable packing. Modification of the oligopeptide sequences alters the interaction in the complex and thereby modifies the specificity and efficiency of catalysis.

A highly relevant feature of oligopeptide self-replication concerns the easy formation of higher replication complexes: Coiled-coil formation is not restricted to two interacting helices; triple helices and higher complexes are known to be very stable as well. Autocatalytic oligopeptide formation may thus involve not only a template and two substrates but, for example, a template and a catalyst that form a triple helix together with the substrates (Severin *et al.*, 1997). Only a very small fraction of all possible peptide sequences fold into three-dimensional structures that are suitable for leucine zipper formation and hence a given autocatalytic oligopeptide is very unlikely to retain the capability of template action on mutation. Peptides are thus optional templates and replicons on a peptide basis are rare.

In contrast to the volume-filling principle of protein packing, the specificity of catalytic RNAs is provided by base pairing and to a lesser extent by tertiary interactions. Both are the results of hydrogen bond specificity. Metal ions, in particular Mg^{2+}, are often involved in RNA structure formation and catalysis, too. The catalytic action of RNA on RNA is exercised in the co-folded complexes of ribozyme and substrate. Since the formation of a catalytic center of a ribozyme that operates on another RNA molecule requires sequence complementarity in parts of the substrate, ribozyme specificity is thus predominantly reflected by the sequence and not by the three-dimensional structure of the isolated substrate. Template action of nucleic acid molecules—being the basis for replication—is a direct consequence of the structure of the double helix. It requires an appropriate backbone provided by the antiparallel ribose-phosphate or 2′-deoxyribose-phosphate chains and a suitable geometry of the complementary purine–pyrimidine pairs. All RNA (and DNA) molecules, however, share these features which, accordingly, are independent of sequence. Every RNA molecule has a uniquely defined complement. Nucleic acid molecules, in contrast to proteins, are therefore obligatory templates. This implies that mutations are conserved and readily propagated into future generations.

Enzyme-free template-induced synthesis of longer RNA molecules from monomers, however, has not been successfully achieved so far (see, e.g., Orgel, 1986). A major problem, among others, is the dissociation of double-stranded molecules at the temperature of efficient replication. If monomers bind with sufficiently high binding constants to the template in order to guarantee the desired accuracy of replication, the new molecules are too sticky to dissociate after the synthesis has been completed. Autocatalytic template-induced synthesis of oligonucleotides from smaller oligonucleotide precursors was nevertheless successful: a hexanucleotide through ligation of two trinucleotide precursors was carried out by von Kiedrowski (1986). His system is the oligonucleotide analogue of the autocatalytic template-induced ligation of oligopeptides discussed above (Figure 1.2). In contrast to the latter system, the oligonucleotides do not form triple-helical complexes. Isothermal autocatalytic template-induced synthesis, however, cannot be used to prepare longer oligonucleotides because of the duplex dissociation problem as mentioned for the template-induced polymerization of monomers.

RNA CATALYSIS AND THE RNA WORLD (FIGURE 1.3)

The first natural ribozymes to be discovered were all RNA-cleaving molecules: the RNA moiety of RNase P (Guerrier-Takada *et al.*, 1983), the class I introns (Cech, 1983), as well as the first small ribozyme called "hammerhead" (Figure 1.4) because of its characteristic secondary structure shape (Uhlenbeck, 1987). Three-dimensional structures are now available for three classes of RNA-cleaving ribozymes (Pley *et al.*, 1994; Scott *et al.*, 1995; Cate *et al.*, 1996; Ferré-D'Amaré *et al.*, 1998) and these data revealed the mechanism of RNA-catalyzed cleavage reactions in full molecular detail. Additional catalytic RNA molecules were obtained through selection from random or partially random RNA libraries and subsequent evolutionary optimization. RNA catalysis in non-natural ribozymes is not only restricted to RNA cleavage: some ribozymes show ligase activity (Bartel and Szostak, 1993; Ekland *et al.*, 1995) and many efforts were undertaken to prepare a ribozyme with full RNA replicase activity. The attempt that comes closest to the

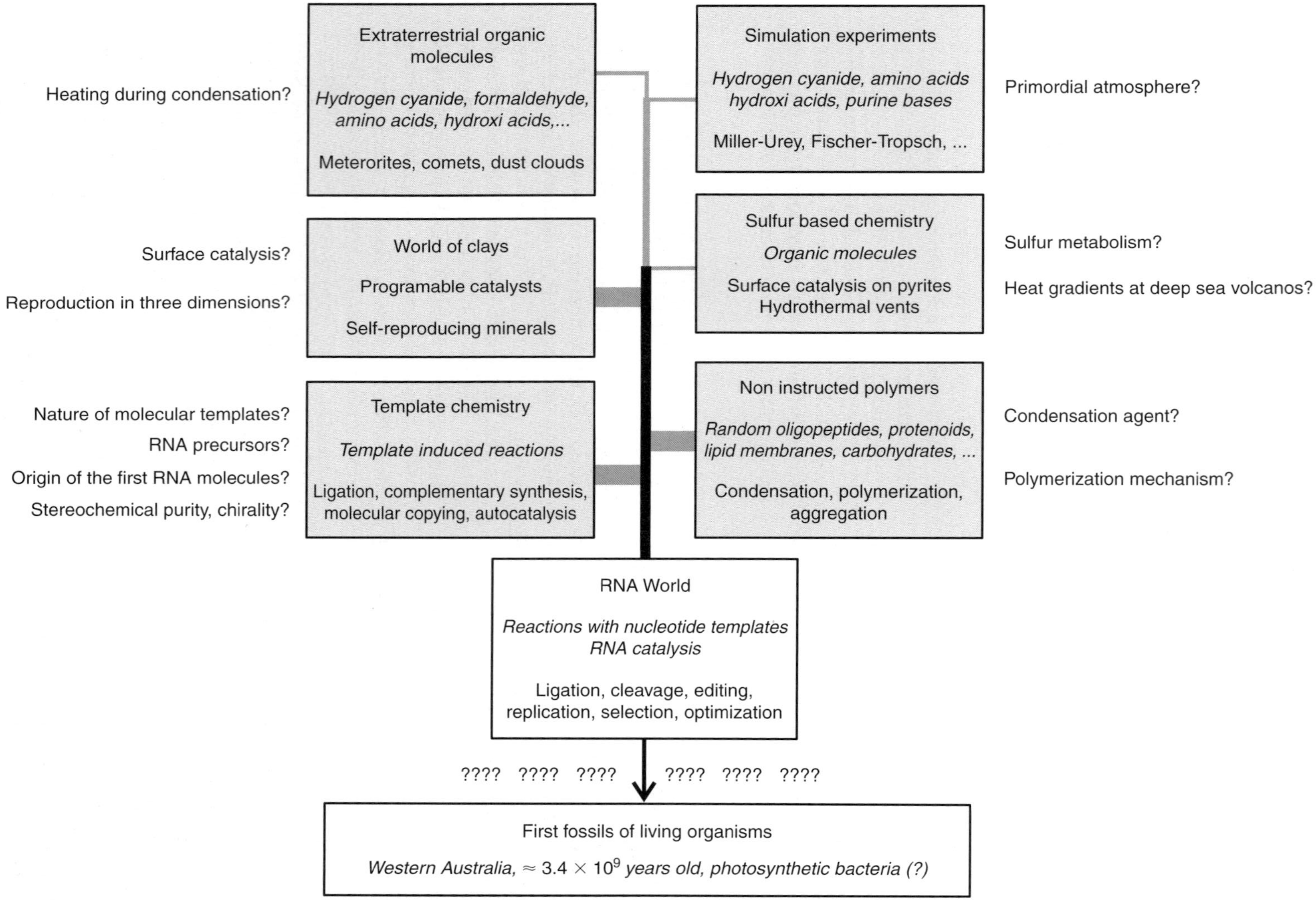

FIGURE 1.3 The RNA world. The concept of a precursor world preceding present-day genetics based on DNA, RNA, and protein is based on the idea that RNA can act as both storage of genetic information and specific catalyst for biochemical reactions. An RNA world is the first scenario on the route to present-day organisms that allows for Darwinian selection and evolution. The question marks along this road to early life indicate important problems. Little is known about further steps (not shown here explicitly) from early replicons to the first cells (Eigen and Schuster, 1982; Maynard Smith and Szathmáry, 1995).

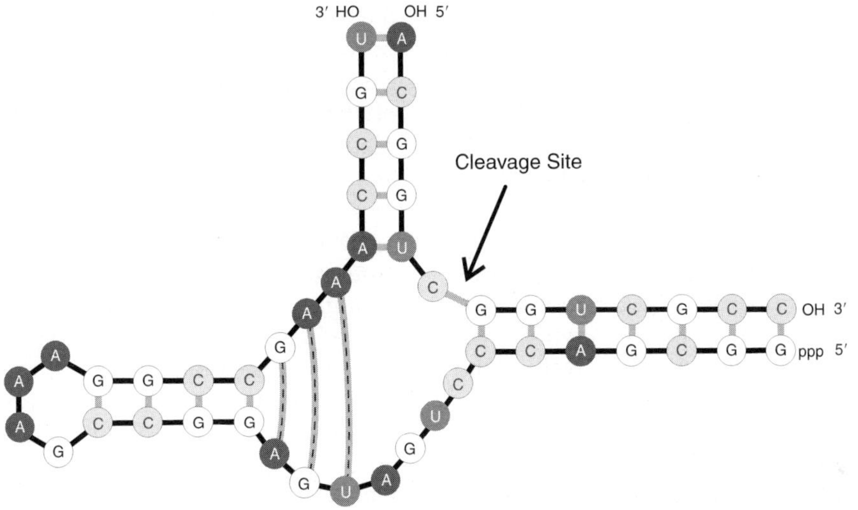

FIGURE 1.4 The hammerhead ribozyme. The substrate is a tridecanucleotide forming two double-helical stacks together with the ribozyme ($n = 34$) in the co-folded complex (Pley *et al.*, 1994). Some tertiary interactions indicated by broken lines in the drawing determine the detailed structure of the hammerhead ribozyme complex and are important for the enzymatic reaction cleaving one of the two linkages between the two stacks. Substrate specificity of ribozyme catalysis is caused by the secondary structure in the co-folded complex between substrate and catalyst.

goal yielded a ribozyme that catalyzes RNA polymerization in short stretches (Ekland and Bartel, 1996). RNA catalysis is not restricted to operating on RNA, nor do nucleic acid catalysts require the ribose backbone. Ribozymes were trained by evolutionary techniques to process DNA rather than their natural RNA substrate (Beaudry and Joyce, 1992), and catalytically active DNA molecules were evolved as well (Breaker and Joyce, 1994; Cuenoud and Szostak, 1995). Polynucleotide kinase activity of ribozymes has been reported (Lorsch and Szostak, 1994, 1995) as well as self-alkylation of RNA on nitrogen (Wilson and Szostak, 1995).

Systematic studies have also revealed examples of RNA catalysis on non-nucleic acid substrates. RNA catalyzes ester, amino acid, and peptidyltransferase reactions (Lohse and Szostak, 1996; Zhang and Cech, 1997; Jenne and Famulok, 1998). The latter examples are particularly interesting because they revealed close similarities between the RNA catalysis of peptide bond formation and ribosomal peptidyltransfer (Zhang and Cech, 1998). A spectacular finding in this respect was that oligopeptide bond cleavage and formation is catalyzed by ribosomal RNA and not by protein: More than 90% of the protein fraction can be removed from ribosomes without losing the catalytic effect on peptide bond formation (Noller *et al.*, 1992; Green and Noller, 1997). These experiments found a straightforward interpretation in the atomic structure of the ribosome (Ban *et al.*, 2000; Nissen *et al.*, 2000). In addition, ribozymes were prepared that catalyze alkylation on sulfur atoms (Wecker *et al.*, 1996) and, finally, RNA molecules were designed that are catalysts for typical reactions of organic chemistry, for example an isomerization of biphenyl derivatives (Prudent *et al.*, 1994). A ribozyme with $Zn^{\oplus}$ and NADH as coenzyme was active in a redox reaction with an aldehyde substrate (Tsukiji *et al.*, 2004). A particularly interesting case is a ribozyme catalyzing the Diels-Alder reaction (Seelig and Jäschke, 1999; Serganov *et al.*, 2005), an organic reaction during which two new carbon–carbon bonds are formed.

For two obvious reasons RNA was chosen to be the preferred candidate for the leading molecule in a scenario at the interface between chemistry and biology: (1) RNA is capable of storing retrievable information, because it is an obligatory replicon, and (2) it has widespread catalytic properties. Although the catalytic properties of RNA are more modest than those of proteins, they are apparently sufficient for processing RNA. RNA molecules operating on RNA molecules form a self-organizing system that can develop molecular organizations with emerging properties and functions. This scenario has been termed the RNA world (see, e.g., Gilbert, 1986; Joyce, 1991, as well as the collective volume by Gesteland and Atkins, 1993, and the recent update, Gesteland *et al.*, 2006). The idea of an RNA world turned out to be fruitful in a different aspect too—it initiated the search for molecular templates and created an entirely new field that may be characterized as template chemistry (Orgel, 1992). Series of systematic studies were performed, for example, on the properties of nucleic acids with modified sugar moieties (Eschenmoser, 1993). These studies revealed the special role of ribose and provided explanations why this molecule is basic to all information-based processes in life.

Chemists working on origin of life problems envisage a number of difficulties for an RNA world being a plausible direct successor of the functionally unorganized prebiotic chemistry (see Figure 1.1 and the reviews by Orgel, 1987, 1992, 2003; Joyce, 1991; Schwartz, 1997): (1) no convincing prebiotic synthesis for all RNA building blocks under the same conditions has been demonstrated, (2) materials for successful RNA synthesis require a high degree of purity that can hardly be achieved under prebiotic conditions, (3) RNA is a highly complex molecule whose stereochemically correct synthesis (3′-5′ linkage) requires an elaborate chemical machinery, and (4) enzyme-free template-induced synthesis of RNA molecules from monomers has not been achieved so far. In particular, the dissociation of duplexes into single strands and the optical asymmetry problem are of major concern. Template-induced synthesis of RNA molecules requires pure optical antipodes. Enantiomeric monomers (containing L-ribose instead of the natural D-ribose) are "poisons" for the polycondensation reaction on the template since their incorporation causes termination of the polymerization process. Currently no plausible conditions are known that could lead to a source of sufficiently pure chiral molecules.[2]

Several suggestions postulating other "intermediate worlds" between chemistry and biology preceding the RNA world have been made. Most of the intermediate information carriers were thought to be more primitive and easier to synthesize than RNA but nevertheless still have the capability of template action (Schwartz, 1997). Glycerol, for example, was suggested as a substitute for ribose because it is structurally simpler and it lacks chirality. However, no successful attempts to use such less sophisticated backbone molecules together with the natural purine and pyrimidine bases for template reactions have been reported so far.

Starting from an RNA world with replicating and catalytically active molecules, it took a long series of many not yet understood steps to arrive at the first cellular organisms with organized cell division and metabolism (see Eigen and Schuster, 1982; Maynard Smith and Szathmáry, 1995). These first precursors of our present-day bacteria and archaea presumably formed the earliest identified fossils (Warrawoona, Western Australia, 3.4×10^9 years old; Schopf, 1993; see Figure 1.1) and/or eventually also the even older kerogen found in the Isua formation (Greenland, 3.8×10^9 years old; Pflug and Jaeschke-Boyer, 1979; Schidlowski, 1988). The correct interpretation of these microfossils as remnants of early forms of life has been questioned (Brasier *et al.*, 2002), although a recent careful consideration of all available information seems to justify the original interpretation (Schopf, 2006).

[2] It is worth noting in this context that an organic reaction has been discovered (Soai *et al.*, 1995) that follows a mechanism for autocatalytic production of optically almost pure chiral material (Frank, 1953); this had been predicted almost 40 years earlier.

REPLICATION AND COUPLING TO ENVIRONMENT

Autocatalysis in Closed and Open Systems

The simple autocatalytic replication reaction according to the overall mechanism 1 is presented here first, because it allows for the derivation of analytical solutions or for complete qualitative analysis. It serves as a simple model for correct replication. First, we consider replication in a closed system (Figure 1.5),[3] where a uniquely defined equilibrium state is approached after a sufficiently long time.[4] Open chemical systems are required to prevent reaction from the approach towards thermodynamic equilibrium. We consider here two examples: (1) a flow reactor (Figure 1.6) and (2) a reaction vessel called a photocell, which allows for coupling of replicon kinetics to a photochemical reaction (Figure 1.7).

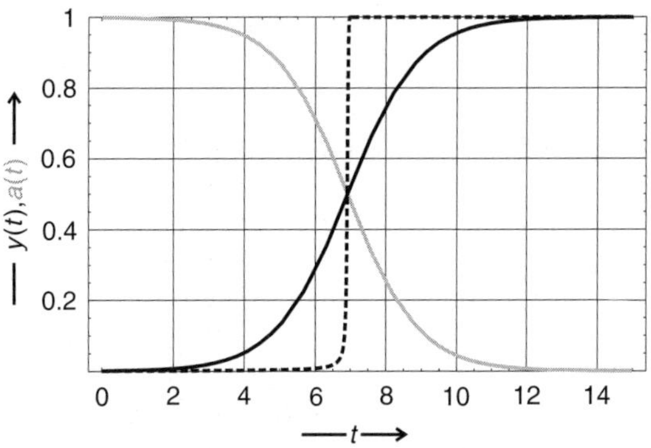

FIGURE 1.5 Replication in a closed system. The figure shows plots of the concentration of the replicator Y (full black line) and the substrate A (gray) as functions of time, $y(t)$ and $a(t)$, respectively, for simple (first order) autocatalysis according to equations (1,1b). Second order autocatalysis (27) leads to the steep curve (broken black line). The curves were adjusted to yield $y = 0.5$ for $t = 6.907$. Choice of parameters: $a(0) = a_0 = 0.999$, $x(0) = x_0 = 0.001$ in arbitrary concentration units (m), $k = 1$ ($m^{-1}t^{-1}$) and $k = 145.35$ ($m^{-2}t^{-1}$) for simple and second order catalysis.

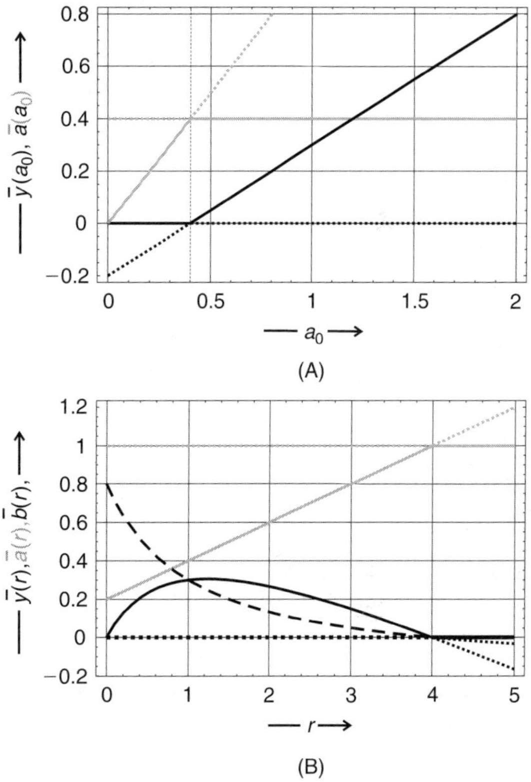

FIGURE 1.6 Replication in a flow reactor. (A) The stationary concentrations $\bar{y}$. (black lines) and $\bar{a}$ (gray lines) as functions of the influx concentration of A, a_0. For the parameter choice applied here we have $\bar{b} = \bar{y}$. Unstable stationary states are shown as dotted lines. A transcritical bifurcation is observed at $a_0 = 0.4$ (m). (B) The stationary concentrations $\bar{y}$. (full black curve), $\bar{a}$ (gray line) and $\bar{b}$ (broken black line) as functions of the flow rate r. Choice of parameters: $k = 5$ ($m^{-1}t^{-1}$), $d = 1$ (t^{-1}).

[3] A closed system exchanges heat but no materials with the environment. A typical example is an isothermal reaction at constant pressure in a closed reaction vessel.

[4] Equation (1) is not correct in the strict sense of thermodynamics, because the reverse reaction, $2Y \rightarrow A + Y$, is not considered explicitly. In order to make the mechanism formally correct the reverse reaction needs to be added, commonly with a (negligibly) small rate constant that makes the analysis a bit more involved but does not change any result or conclusion derived here.

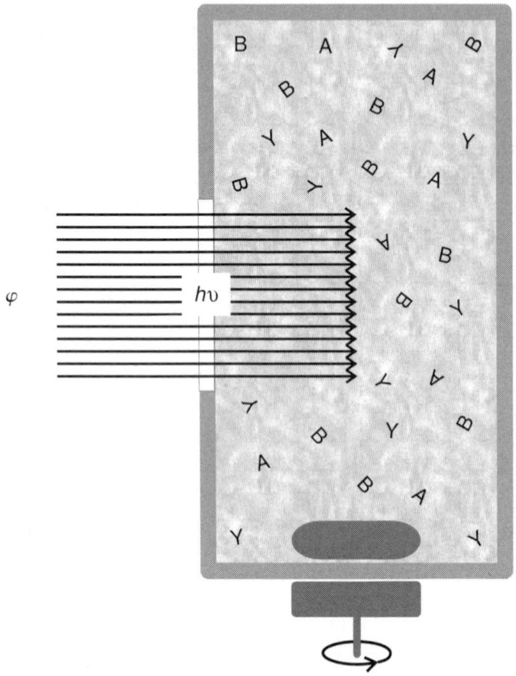

FIGURE 1.7 Photocell as an open system. The autocatalytic reaction A + Y → 2Y is prevented from approaching thermodynamic equilibrium by radiation from a suitable light source. The replicon Y is degraded to yield some low free energy material B, which is activated by means of a photochemical reaction, B + $h\nu$ → A. The reactions inside the photocell are thus driven by a flux of radiation, φ. The solution in the reaction vessel is mixed by magnetic stirring.

Autocatalysis in the closed system is described by the rate equation (concentrations are denoted by lower case letters a = (A) and y = (Y)):

$$-\frac{da}{dt} = -\dot{a} = \frac{dy}{dt} = \dot{y} = k\,a \cdot y, \qquad \text{(1a)}$$

mass conservation, $a(t) + y(t) = a(0) + y(0) = c_0$ (where c_0 is the total concentration), and initial conditions, $a(0) = a_0$ and $y(0) = y_0$. An analytical solution is computed straightforwardly,

$$y(t) = \frac{y_0 c_0}{y_0 + a_0 e^{-kc_0 t}}, \qquad \text{(1b)}$$

and shows the expected behavior in the limits

$$\lim_{t\to o} y(t) = y_0 \text{ and } \lim_{t\to\infty} y(t) = c_0.$$

In other words, all material **A** is converted into **Y** in the long time limit.[5] For $a_0 \gg y_0$ and small t we obtain for the time dependence of the concentration of Y,

$$y(t) \approx y_0 \cdot e^{kc_0 \cdot t} \text{ for small } t,$$

corresponding to exponential growth of the replicon:

$$y(t) \approx c_0 \left(1 - \frac{a_0}{y_0} e^{-kc_0 \cdot t}\right) \text{ for large } t.$$

As shown in Figure 1.5 by means of a numerical example, the initial phase of exponential growth is turned into an exponential approach towards the final state that has a negative exponent with the same (absolute) value, kc_0. Addition of an irreversible decomposition reaction for the replicon Y,

$$\mathrm{Y} \xrightarrow{d} \mathrm{B}, \qquad (5)$$

changes the final state in trivial manner: **Y** is then an intermediate and all material is converted into the decomposition product **B** after sufficiently long time: $\lim_{t\to\infty} b(t) = c_0$. In case of template-induced replication of nucleic acids, for example, A would be the activated monomers, the trinucleotides, whereas B stands for the mononucleotides.

Autocatalysis in the flow reactor considering replication and degradation follows the mechanism

$$\begin{aligned} * &\xrightarrow{a_0 \cdot r} \mathrm{A} \\ \mathrm{A} + \mathrm{Y} &\xrightarrow{k} 2\mathrm{Y} \\ \mathrm{Y} &\xrightarrow{d} \mathrm{B} \\ \mathrm{A, B, Y} &\xrightarrow{r} \varnothing, \end{aligned} \qquad (6)$$

[5] This is a consequence of the assumption that reaction (1) is irreversible.[4] In case the inverse reaction of (1) would be included with a non-zero rate constant the system would approach an equilibrium state at infinite time, which is defined by $\overline{x}/\overline{a} = K$, where K is the equilibrium parameter of the reaction (1).

and is described by the following kinetic differential equation

$$\dot{a} = -kay + r(a_0 - a),$$
$$\dot{y} = \big(ka - (d + r)\big)\, y \text{ and} \qquad (7)$$
$$\dot{b} = dx - rb.$$

The reaction sustains two stationary states: (i) the state of extinction $\bar{a} = a_0, \bar{x} = 0, \bar{b} = 0$, and (ii) the active state:

$$\bar{a} = \frac{d + r}{k}, \; \bar{x} = \frac{ka_0 - (d + r)}{k(d + r)} r, \qquad (7a)$$
$$\bar{b} = \frac{ka_0 - (d + r)}{k(d + r)} d.$$

The two scenarios are separated by a transcritical bifurcation: The active state is stable at $r < ka_0 - d$ and this implies at sufficiently low flow rates r or large enough influx concentrations a_0. In Figure 1.6 the dependence of the stationary concentrations on a_0 and r is shown for a typical example. It is worth noticing that the curve $\bar{y}(r)$ goes through a maximum at $r(\bar{y}_{\max}) = \sqrt{d}(\sqrt{ka_0} - \sqrt{d})$. The value at this flow rate is: $\bar{y}_{\max} = (\sqrt{ka_0} - d)^2/k$. In other words, there exists a flow rate r for every influx concentration a_0 that allows for optimal exploitation of the resources.

Autocatalysis in the photocell is driven by a flux of photons, which are consumed in a (recycling) photoreaction according to the mechanism

$$\mathbf{A} + \mathbf{Y} \xrightarrow{k} 2\mathbf{Y}$$
$$\mathbf{Y} \xrightarrow{d} \mathbf{B} \qquad (8)$$
$$h\upsilon + \mathbf{B} \xrightarrow{\varphi} \mathbf{A},$$

which gives rise to the differential equation

$$\dot{a} = -kay + \varphi b,$$
$$\dot{y} = (ka - d)y \text{ and} \qquad (9)$$
$$\dot{b} = dx - \varphi b.$$

Therefore the system shows mass conservation, $a(t) + y(t) + b(t) = a_0$ and one variable can be eliminated: $b(t) = a_0 - a(t) - y(t)$. There are two steady states: (i) extinction, $\bar{a} = a_0, \bar{y} = b = 0$, and (ii) the active state:

$$\bar{a} = \frac{d}{k}, \; \bar{y} = \frac{ka_0}{k(d + \varphi)} \varphi, \; \bar{b} = \frac{ka_0}{k(d + \varphi)} d.$$

The dependence of the stationary concentration on the total concentration is in full analogy to the plot in Figure 1.6A. Extinction occurs when the total concentrations is too small, $a_0 < d/k$. Plotting the steady state (ii) as a function of the radiation flux φ is different from Figure 1.6B: The curve $\bar{y}(\varphi)$, does not go through a maximum but reaches its highest value in the large flux limit, $\lim_{\varphi \to \infty} \bar{y}(\varphi) = (ka_0 - d)/k$ (Figure 1.8). If a_0 is above threshold, an increase in the flux of photons leads always to an increase in $\bar{y}$.

REPLICATION IN OPEN SYSTEMS

Replicating chemical species are a special class of autocatalysts. In the most general setting, we are dealing with a collection of molecular species called replicators $\{\mathbf{I}_1, \mathbf{I}_2, \ldots\}$, which are capable of replication, $\mathbf{I}_k \to 2\mathbf{I}_k$, and mutation, $\mathbf{I}_j \to \mathbf{I}_k + \mathbf{I}_j$. Template-induced replication requires a source of (energy-rich) building material conveniently subsumed under **A**. In general, waste products **B** are

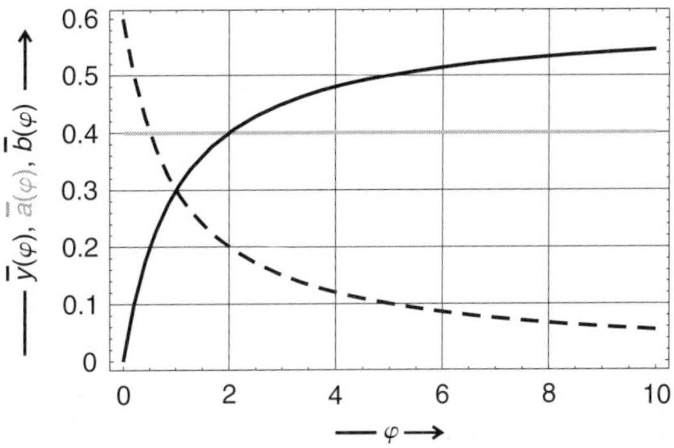

FIGURE 1.8 Steady state in the photocell. The concentrations in the steady state, $\bar{y}$ (black, full line), $\bar{a}$ (gray), and $\bar{b}$ (black, broken line), are plotted as functions of the radiation flux φ. Choice of parameters: $a_0 = 1(m)$, $k = 1(m^{-1}t^{-1})$, and $d = 1(t^{-1})$.

produced through a degradation process. They can be neglected unless they interact further with the replicators or they are recycled. We shall discuss two examples of open systems, the flow reactor (Figure 1.9), where degradation products can be neglected, and the photocell (Figure 1.7), which recycles the degradation products through a photochemical reaction (8). The state of the system and its evolution are conveniently described by time-dependent concentrations of replicators $\mathbf{c}(t) = (c_1(t), c_2(t), \ldots)$ and building blocks $a(t)$, which are determined by initial conditions and kinetic differential equations.

In the flow reactor the ordinary differential equation is of the form:

$$\begin{aligned} \dot{c}_k &= G_k(a, \mathbf{c}) - r\, c_k,\ k = 1, 2, \ldots \\ \dot{a} &= r\,(a_0 - a) - \sum_j G_j(a, \mathbf{c})) \end{aligned} \tag{10}$$

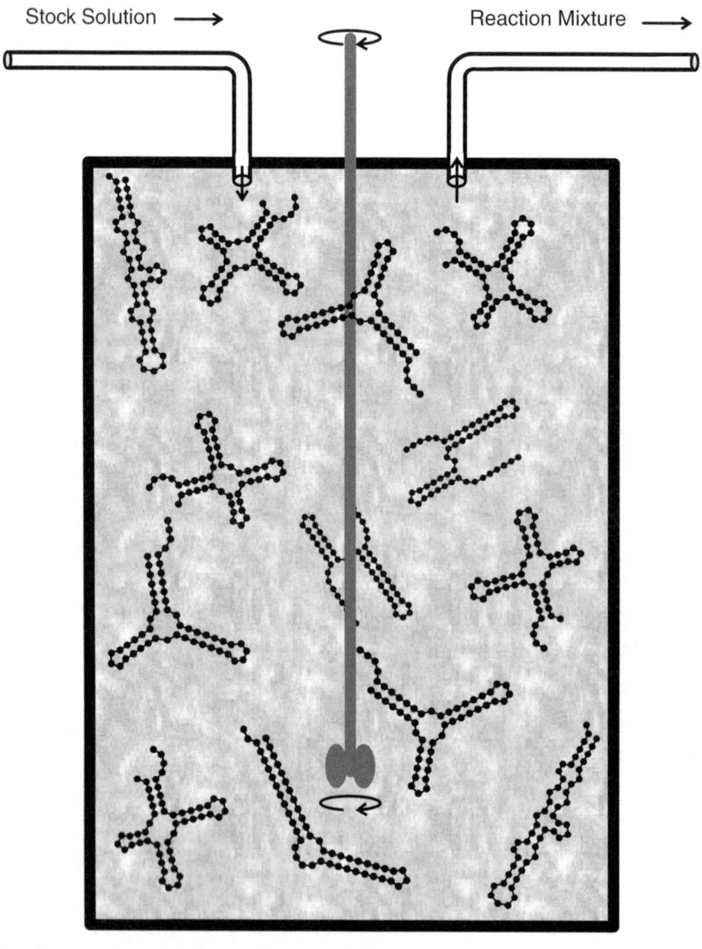

FIGURE 1.9 The flow reactor for the evolution of RNA molecules. A stock solution containing all materials for RNA replication including an RNA polymerase flows continuously into a well-stirred tank reactor and an equal volume containing a fraction of the reaction mixture leaves the reactor. (For different experimental set-ups see Watts and Schwarz, 1997.) The population in the reactor fluctuates around a mean value, $N \pm \sqrt{N}$. RNA molecules replicate and mutate in the reactor, and the fastest replicators are selected. The RNA flow reactor has been used also as an appropriate model for computer simulations (Fontana and Schuster, 1987; Huynen *et al.*, 1996; Fontana and Schuster, 1998a). There, other criteria than fast replication can be used for selection. For example, fitness functions are defined that measure the distance to a predefined target structure and fitness increases during the approach towards the target (Huynen *et al.*, 1996; Fontana and Schuster, 1998a).

The replication functions G_k reflect the kinetics of the mechanism of reproduction and may be highly complex. In case degradation according to (6) is important, the term $-d_k \cdot c_k$ is properly included in the replication function. A differential equation for the total concentration $c = \sum_k c_k$ is derived by summation,

$$\sum_k \dot{c}_k = \dot{c} = \sum_k G_k - rc = \sum_k G_k - \phi(t), \quad (11)$$

where $\phi(t)$ is a concentration weighted generalized flux representing the material flowing out of the reactor. For constant total concentration denoted as constant organization we have $\dot{c} = 0$ and obtain a condition for this flux: $\phi(t) = \Sigma_k G_k$, which implies an adjustable flux $r(t) = \Sigma_k G_k / c$.

Equation (11) has the formal solution

$$c(t) = c(0) + \int_0^t \left(\sum_k G_k - \phi(\tau) \right) d\tau.$$

emphasizing the time-dependence of the total concentration $c(t)$ in the general case. Introducing normalized concentrations for the replicators, $x_k = c_k/c$ and computing their time derivatives

$$\dot{x}_k = \frac{1}{c}(\dot{c}_k - x_k \dot{c}),$$

results in a system of equations for internal equilibration that does not depend explicitly on the flow rate r:

$$\dot{x}_k = \frac{1}{c(t)} \left(G_k(c\mathbf{x}) - x_k \sum_j G_j(c\mathbf{x}) \right). \quad (12)$$

The expression becomes particularly handy if the replication functions G_k are homogeneous in the concentrations c_k, for example—in the simplest case—polynomials of degree λ, $G_k(\mathbf{c}) = c\lambda \cdot G_k(\mathbf{x})$:[6]

$$\dot{x}_k = c(t)^{\lambda-1} \left(G_k(\mathbf{x}) - x_k \sum_j G_j(\mathbf{x}) \right).$$

As long as the total concentration does not vanish (and stays finite), the function $c(t)$ can be absorbed in the time axis. In other words, the survival of the entire system requires that c stays bounded away from 0 for all times. According to equation (11) the balance of the intrinsic net production $\Sigma_k G_k$ and the external dilution flux $r(t)$ determines the survival of the entire system. The internal equilibrium is approached independently of the setup of the particular open system applied. If the reactions of interest are modeled by one-step template-induced replication reactions, the functions G_k are of the form $G_k(a, \mathbf{c}) = c_k f_k(a)$, $\lambda = 1$, and equation (12) is exact in real time, i.e. without the time transforming factor involving c.

In a more general setting, incorrect replication is allowed. This can be described by specifying the probabilities Q_{kj} that a copy of type $\mathbf{I}_k$ is produced from a template of type $\mathbf{I}_j$: $G_k = \Sigma_j Q_{kj} f_j(a) c_j$. In this case, the first line of equation (10) can be rewritten in the form

$$\dot{c}_k = \sum_j Q_{kj} c_j f_j(a, c\mathbf{x}) - rc_k \quad (13)$$

where f_j is a growth rate that depends on the chemical environment. The (quadratic) matrix of replication probabilities $Q = \{Q_{kj}\}$ is a stochastic matrix since every replication has to yield either a correct or an incorrect copy of the template, $\Sigma_k Q_{kj} = 1$. Hence we have,

$$\dot{c} = \sum_k \dot{c}_k = \sum_j c_j f_j(a, c\mathbf{x}) - rc, \quad (14)$$

the mutation terms vanished and the expression for $\dot{c}$ is the same as in case of error-free replication.

For relative concentrations, x_k, a short computation shows that mutual relationship of

[6] The condition of homogeneous replication functions is very often fulfilled when the mechanism of replication is the same for all replicators.

the replicators is described by a differential equation of the form

$$\dot{x}_k = \underbrace{x_k \left[f_k(a, c\mathbf{x}) - \sum_j x_j f_j(a, c\mathbf{x}) \right]}_{\text{selection}} + \underbrace{\sum_j \{Q_{kj} x_j f_j(a, c\mathbf{x}) - Q_{jk} x_k f_k(a, c\mathbf{x})\}}_{\text{mutation}} \quad (15)$$

In the special case in which $r(t)$ is adjusted such that c stays constant, it can be absorbed into the definition of f_j and it is sufficient to consider the internal competition of the replicons.

For replication in the photocell the flow rate r is replaced by the degradation rate parameter d_k in equation (10) and the production term in the equation for $\dot{a}$, $r(a_0 - a)$ is exchanged for $\varphi \cdot [\mathbf{B}] = \varphi \cdot b = \varphi \cdot (a_0 - a - c)$:

$$\begin{aligned} \dot{c}_k &= G_k(a, \mathbf{c}) - d_k c_k, \; k = 1, 2, \ldots \\ \dot{a} &= \varphi(a_0 - a - c) - \sum_j G_j(a, \mathbf{c})). \end{aligned} \quad (16)$$

Defining $\Gamma_k(a, \mathbf{c}) = G_k(a, \mathbf{c}) - d_k c_k$ we obtain for the internal equilibration an expression that is identical with equation (12) except G is replaced by Γ. For simple replication, $\lambda = 1$, we have $\dot{c}_k = c_k(f_k(a) - d_k) = c_k \cdot \gamma_k$ and internal equilibration is described by

$$\dot{x}_k = \gamma_k x_k - x_k \sum_j \gamma_j x_j \text{ with } \gamma_k = f_k(a) - d_k.$$

The introduction of mutation following exactly the same derivation as before is straightforward.

The mathematical derivations above can be summarized as follows:

- The competition of replicators for common resources can be formulated in terms of relative concentrations. Both their total concentration c and the concentration a of the building material enter only as "parameters" into the associated growth rate functions f_k. In particular, if the vector field f is a homogeneous function in c and a, i.e., if $f_k(a, c\mathbf{x}) = a^p c^q f_k(1, \vec{x})$ for all k, then one can absorb the common prefactor $a^p c^q$ into a rescaling of the time axis (Schuster and Sigmund, 1985). In this case, the internal dynamics of the replicators becomes completely independent of the environment. In the limit of small fluxes, the flow reactor and constant organization yield essentially the same results even for non-homogeneous interaction functions (Happel and Stadler, 1999).
- Selection acting of correct copies and the effects of miscopying can be separated into additive contributions. Indeed, the term in the curly brackets disappears when the matrix **Q** is diagonal. Since **Q** is a stochastic matrix by definition, the time dependence of the total concentration, $\dot{c}$, is independent of mutation terms. In other words, the internal production does not depend on the mutations matrix **Q**.
- The overall survival of the system in the flow reactor is governed by the balance between the external dilution flux r and the internal production ϕ. In case of the photocell a minimum amount of material is required for survival according to the condition for the active state (ii) derived from equation (9).

REPLICATION IN LIPID AGGREGATES

These observations remain valid in even more general settings. We consider here an example. Cavalier-Smith (2001) discussed a model for the origin of life in which membranes initially functioned as supramolecular structures to which different replicators attached. In this picture, the membranes are selected as a higher level reproductive unit. From a biophysical point of view, this model is simpler than micellar or vesicular protocells since it avoids the difficulties of modeling the regulation of both growth and fission. More precisely, the "pre-protocell" in Figure 1.10, consists of a lipid aggregate that can grow by inclusion of amphiphilic molecules from the environment. Attached to its surface is a suitable nucleic acid analogue that undergoes uncatalyzed replication in the spirit of the membrane linked replication cycle of the "Los Alamos Bug" (Rasmussen *et al.*, 2003, 2004a).

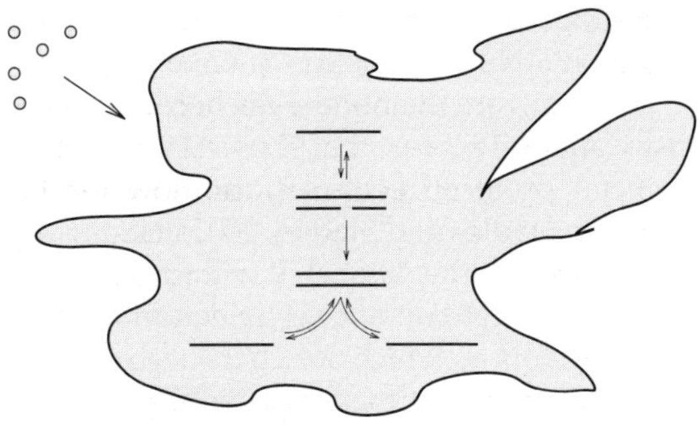

FIGURE 1.10 Model of a protocell precursor. Replicating polymers are attached to the surface of a lipid aggregate which can grow by incorporating amphiphilic molecules from the environment.

The dynamical properties of this model are discussed in some detail in Stadler and Stadler (2007). Suppose the number n_{ka} of replicators of type k embedded in membrane fragment a grows according to the $\dot{n}_{ka} = n_{ka} f_k(\mathbf{c}_a)$ where $\mathbf{c}_a$ is the vector of replicator concentrations in membrane a. Denoting the surface area of the membrane by Ω_a we can write $c_{ka} = n_{ka}/\Omega_a$. A short computation again leads to an equation of the same form as equation (15) for the relative concentrations $x_{ka} = c_{ka}/c_a$ of replicators within each piece of membrane. Furthermore we obtain a set of equations describing to total concentrations c_a of replicators within a given membrane.

$$\dot{c}_a = c_a \sum_j x_j f_j(c_a \cdot \mathbf{x}_a) - c_a \frac{\dot{\Omega}_a}{\Omega_a} \qquad (17)$$

Note that c_a now explicitly depends on the growth law for the membrane itself, i.e. to complete the model we now need to explicitly describe the membrane growth $\dot{\Omega}_a$.

PARABOLIC AND EXPONENTIAL GROWTH

It is relatively easy to derive a kinetic rate equation displaying the elementary behavior of replicons if one assumes (i) that catalysis proceeds through the complementary binding of reactant(s) to free template and (ii) that autocatalysis is limited by the tendency of the template to bind to itself forming an inactive dimer in the manner of product inhibition (von Kiedrowski, 1993). However, in order to achieve an understanding of what is likely to happen in systems where there is a diverse mixture of reactants and catalytic templates, it is desirable to develop a comprehensive kinetic description of as many individual steps in the reaction mechanism of template synthesis as is feasible and tractable from the mathematical point of view.

Szathmáry and Gladkih (1989) over-simplified the resulting dynamics to a simple parabolic growth law $\dot{x}_k \propto x_k^p, 0 < p < 1$ for the concentrations of the interacting template species. This model suffers from a conceptual and a technical problem: (i) under no circumstances does one observe extinction of a species in any parabolic growth model, and (ii) the vector fields are not Lipschitz-continuous on the boundary of the concentration simplex, indicating that we cannot expect uniqueness of solutions, and thus that we cannot take for granted that the system behaves physically reasonable in this area.

In Wills *et al.* (1998), we have derived the kinetic equations for a system of coupled template-instructed ligation reactions of the form

$$\mathbf{A}_i + \mathbf{B}_j + \mathbf{C}_{kl} \underset{\bar{a}_{ijkl}}{\overset{a_{ijkl}}{\rightleftharpoons}} \mathbf{A}_i\mathbf{B}_j\mathbf{C}_{kl} \xrightarrow{b_{ijkl}} \mathbf{C}_{ij}\mathbf{C}_{kl} \underset{\bar{d}_{ijkl}}{\overset{d_{ijkl}}{\rightleftharpoons}} \mathbf{C}_{ij} + \mathbf{C}_{kl} \qquad (18)$$

Here **A**. and **B**. denote the two substrate molecules which are ligated on the template **C**.., for example, the electrophilic, **E**, and the nucleophilic, **N**, oligopeptide in peptide template reactions or the two different trinucleotides, **GGC** and **GCC**, in the autocatalytic hexanucleotide formation (Figure 1.2). This scheme thus encapsulates the experimental results on both peptide and nucleic acid replicons (von Kiedrowski, 1986; Lee *et al.*, 1996).

The following assumptions are straightforward and allow for a detailed mathematical analysis:

(i) the concentrations of the intermediates are stationary in agreement with the "quasi-steady state" approximation (Segel and Slemrod, 1989),
(ii) the total concentration c_0 of all replicating species is constant in the sense of constant organization (Eigen, 1971),

(iii) the formation of heteroduplices of the form $\mathbf{C}_{ij}\mathbf{C}_{kl}$, $ij \neq kl$ is neglected, and
(iv) only reaction complexes of the form $\mathbf{A}_k\mathbf{B}_l\mathbf{C}_{kl}$ lead to ligation.

Assumptions (iii) and (iv) are closely related. They make immediate sense for hypothetical macromolecules for which the template instruction is direct instead of complementary. It has been shown, however, that the dynamics of complementary replicating polymers is very similar to direct replication dynamics if one considers the two complementary strands as "single species" by simply adding their concentrations (Eigen, 1971; Stadler, 1991).

Assumptions (iii) and (iv) suggest a simplified notation of the reaction scheme:

$$\begin{aligned} \mathbf{A}_k + \mathbf{B}_k + \mathbf{C}_k &\underset{\bar{a}_k}{\overset{a_k}{\rightleftharpoons}} \mathbf{A}_k\mathbf{B}_k\mathbf{C}_k \\ &\xrightarrow{b_k} \mathbf{C}_k\mathbf{C}_k \underset{\bar{d}_k}{\overset{d_k}{\rightleftharpoons}} 2\mathbf{C}_k \end{aligned} \tag{19}$$

It can be shown that equation (19) together with the assumptions (i) and (ii) leads to the following system of differential equations for the frequencies or relative total concentrations x_k, i.e. $\sum_k^m x_k = 1$ of the template molecules $\mathbf{C}_k$ in the system (note that x_k accounts not only for the free template molecules but also for those bound in the complexes $\mathbf{C}_k\mathbf{C}_k$ and $\mathbf{A}_k\mathbf{B}_k\mathbf{C}_k$):

$$\dot{x}_k = x_k\left(\alpha_k\varphi(c\beta_k x_k) - \sum_j^m \alpha_j x_j \varphi(c\beta_j x_j)\right), \quad k = 1, \ldots, m, \tag{20}$$

where

$$\varphi(z) = \frac{2}{z}\left(\sqrt{z+1} - 1\right) \quad \varphi(0) = 1, \tag{21}$$

and the effective kinetic constants α_k and β_k can be expressed in terms of the physical parameters a_k, $\bar{a}_k$, etc. This special form of the growth rate function,

$$f_k(c, \vec{x}) = \alpha_k \varphi(c\beta_k x_k) \tag{22}$$

is also obtained from a wide range of alternative template-directed ligation mechanisms, including experimentally studied systems based on DNA triplehelices (Li and Nicolaou, 1994) and the membrane-anchored mechanism suggested for the "Los Alamos Bug" artificial protocell project (Rasmussen *et al.*, 2003; see Stadler and Stadler, 2003; Rasmussen *et al.*, 2004a for the details). It will turn out that survival of replicon species is determined by the constants α_k which we characterize therefore as Darwinian fitness parameters.

Equation (20) is a special form of a replicator equation with the non-linear response functions $f_k(x) := \alpha_k\varphi(\beta_k x_k)$. Its behavior depends strongly on the values of β_k: For large values of z we have $\varphi(z) \sim 2/\sqrt{z}$. Hence equation (20) approaches Szathmáry's expression (Szathmáry and Gladkih, 1989):

$$\dot{x}_k = h_l\sqrt{x_k} - x_k\sum_j^M h_j\sqrt{x_j} \tag{23}$$

with suitable constants h_k. This equation exhibits a very simple dynamics: the mean fitness $\Phi(x) = \sum_j^M h_j\sqrt{x_j}$ is a Ljapunov function, i.e. it increases along all trajectories, and the system approaches a globally stable equilibrium at which all species are present (Wills *et al.*, 1998; Varga and Szathmáry, 1997). Szathmáry's parabolic growth model thus does not lead to selection.

On the other hand, if z remains small, that is, if β_k is small, then $\varphi(\beta_k x_k)$ is almost constant 1 (since the relative concentration x_k is of course a number between 0 and 1). Thus we obtain

$$\dot{x}_k = x_k\left(\alpha_k - \sum_j^M \alpha_j x_j\right) \tag{24}$$

which is the "no-mutation" limit of Eigen's kinetic equation for replication (Eigen, 1971) (see equation (33a); if condition (iv) above is relaxed, we in fact arrive at Eigen's model with a mutation term). Equation (24) leads to survival of the fittest: The species with the largest value of α_k will eventually be the only survivor in the system. It is worth noting that the mean fitness also increases along all orbits of equation (24) in agreement with the no-mutation case (Schuster and Swetina, 1988). The constants β_k that determine whether the system shows Darwinian selection or unconditional coexistence is proportional to the total concentration c_0 of the templates. For small total concentration we obtain equation (24), while for large concentrations, when the formation of the dimers $\mathbf{C}_k\mathbf{C}_k$ becomes dominant, we enter the regime of parabolic growth.

Equation (20) is a special case of a class of replicator equations studied by Hofbauer *et al.* (1981). Restating their main result yields the following: All orbits or trajectories starting from physically meaningful points (these are points in the interior of the simplex S_M with $x_j > 0$ for all $j = 1, 2, \ldots, M$) converge to a unique equilibrium point $\bar{\mathbf{x}} = (\bar{x}_1, \bar{x}_2, \ldots, \bar{x}_M)$ with $\bar{x}_i \geqslant 0$, which is called the ω-limit of the orbits. This means that species may go extinct in the limit $t \to \infty$. If $\bar{x}$ lies on the surface of S_M (which is tantamount to saying that at least one component $\bar{x}_j = 0$) then it is also the ω-limit for all orbits on this surface. If we label the replicon species according to decreasing values of the Darwinian fitness parameters, $\alpha_1 \geq \alpha_2 \geq \ldots \geq \alpha_M$, then there is an index $\ell \geq 1$ such that $\bar{\mathbf{x}}$ is of the form $\bar{x}_i > 0$ if $i \leq \ell$ and $\bar{x}_i = 0$ for $i > \ell$. In other words, ℓ replicon species survive and the $M - \ell$ least efficient replicators die out. This behavior is in complete analogy to the reversible exponential competition case (Schuster and Sigmund, 1985) where the Darwinian fitness parameters α_k are simply the rate constants a_k. If the smallest concentration dependent value $\beta_s(c_0) = \min\{\beta_j(c_0)\}$ is sufficiently large, we find $\ell = M$ and no replicon goes extinct ($\bar{\mathbf{x}}$ is an interior equilibrium point).

The condition for survival of species k is explicitly given by:

$$\alpha_k > \Phi(\bar{\mathbf{x}}) \tag{25}$$

It is interesting to note that the Darwinian fitness parameters α_k determine the order in which species go extinct whereas the concentration-dependent values $\beta_k(c_0)$ collectively influence the flux term and hence set the "extinction threshold." In contrast to Szathmáry's model equation, the extended replicon kinetics leads to both competitive selection and coexistence of replicons depending on total concentration and kinetic constants.

HYPERBOLIC GROWTH

In this section we consider second order autocatalysis which is distinguished from simple (or first order) autocatalysis by the stoichiometry 1 : 2 for substrate **A** and autocatalyst **Y**:

$$\mathbf{A} + 2\mathbf{Y} \xrightarrow{k} 3\mathbf{Y}. \tag{26}$$

Although such a reaction step is often used in simple models for chemical oscillators and pattern formation (Turing, 1952; Nicolis and Prigogine, 1977) as well as non-equilibrium phase transitions (Schlögl, 1972), it occurs in reality only in overall kinetics of many step reactions. The notion of hyperbolic growth is derived from the solution curve of the unconstrained system, $\dot{x} = f \cdot x^2$, the solution curve $x(t) = x_0/(1 - x_0 ft)$ is a hyperbola with the time axis as a horizontal asymptote and a vertical asymptote at $t = 2/(x_0 a)$. The kinetic differential equation for (26) in the closed system can be solved exactly but no explicit expression $x(t)$ is available:

$$t = \frac{1}{k\,a_0}\left(\frac{x - x_0}{x x_0} + \frac{1}{a_0}\ln\frac{x(a_0 - x_0)}{x_0(a_0 - x)}\right). \tag{26b}$$

In Figure 1.5 the solutions curves for first and second order autocatalysis are compared. Second order autocatalysis leads to a comparatively long lag phase and an extremely steep increase in concentration. Precisely such a behavior was observed in the early phase of the infection cycle of a bacteriophage in *Escherichia coli* (Eigen *et al.*, 1991).

In contrast to the weakly coupled networks of replicons considered in previous sections, hypercycles (Eigen, 1971; Eigen and Schuster, 1978a) involve specific catalysis beyond mere template instruction (see Figure 1.11). In the

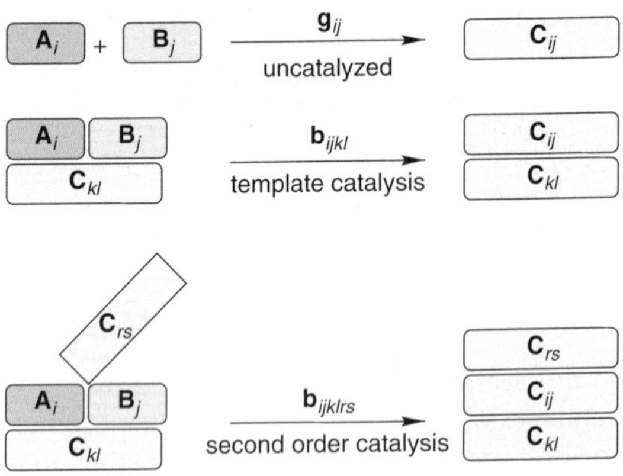

FIGURE 1.11 Modes of template formation. In complex systems of mixed templates and depending on the underlying mechanism of template synthesis, different modes of dynamic behavior are possible. Uncatalyzed synthesis generally corresponds to linear growth. Template-instructed synthesis gives parabolic or exponential growth. The coupling of systems involving second order autocatalysis can also give rise to hyperbolic growth, as has been predicted for hypercycles (Eigen and Schuster, 1978a).

simplest case, where we consider catalyzed replication reactions explicitly, the reaction equations are of the form:

$$(\mathbf{A}) + \mathbf{I}_k + \mathbf{I}_l \rightarrow 2\mathbf{I}_k + \mathbf{I}_l. \tag{27}$$

Here a copy of $\mathbf{I}_k$ is produced using another macromolecular species $\mathbf{I}_l$ as a specific catalyst for the replication reaction. This corresponds to growth rate functions of the form

$$f_k(a, c\mathbf{x}) = \sum_l a_{kl}(a, c)x_l \tag{28}$$

where the matrix $A = \{a_{kl}\}$ describes the network of catalytic interactions. The corresponding kinetic differential equation

$$\dot{x}_k = x_k\left(\sum_l a_{kl}x_l - \phi(\mathbf{x})\right) \tag{29}$$

corresponding to the mechanism (27) has been termed second order replicator equation (Schuster and Sigmund, 1983). These systems can display enormous diversity of dynamic behavior (Hofbauer and Sigmund, 1998).

In case matrix A is diagonal we have $f_k(x) = a_{kk}x_k$, the corresponding dynamical system

$$\dot{x}_k = x_k\left(a_{kk}x_k - \sum_l a_{jj}x_j^2\right) \tag{29a}$$

is known as generalized Schlögl model (Schlögl, 1972; Schuster and Sigmund, 1985): Each replicator considered in isolation shows hyperbolic growth. In the competitive ensemble described by equation (29a) every replicator can be selected, since all pure states corresponding to the corners of the concentration simplex $P_k(S_n) = (x_k = 1,\ x_j = 0\ \forall\ j \neq k)$ are point attractors. Which one is selected depends on the initial conditions. The sizes of the basins of attraction correspond strictly to the values of the replication parameters, i.e. the replicator with the largest a_{kk}-value has the largest basin, the one with the next largest value the next largest basin, etc.

A more realistic version of (27) that might be experimentally feasible is

$$\begin{aligned} &\mathbf{A}_i + \mathbf{B}_j + \mathbf{C}_{kl} + \mathbf{C}_{rs} \underset{\bar{a}_{ijkl}}{\overset{a_{ijkl}}{\rightleftharpoons}} \mathbf{A}_i\mathbf{B}_j\mathbf{C}_{kl} \\ &+ \mathbf{C}_{rs} \underset{\bar{e}_{ijklrs}}{\overset{e_{ijklrs}}{\rightleftharpoons}} \mathbf{A}_i\mathbf{B}_j\mathbf{C}_{kl}\mathbf{C}_{rs} \xrightarrow{b_{ijklrs}} \\ &\mathbf{C}_{ij}\mathbf{C}_{kl}\mathbf{C}_{rs} \xrightarrow{f_{ijklrs}} \mathbf{C}_{ij} \\ &+ \mathbf{C}_{kl}\mathbf{C}_{rs} \underset{\bar{d}_{ijkl}}{\overset{d_{ijkl}}{\rightleftharpoons}} \mathbf{C}_{ij} + \mathbf{C}_{kl} + \mathbf{C}_{rs} \end{aligned} \tag{30}$$

Here the template $\mathbf{C}_{rs}$ plays the role of a ligase for the template-directed replication step. Dynamically, it again leads to replicator equations with non-linear growth functions (Stadler *et al.*, 2000). Depending on the total concentration of replicons, they interpolate between a parabolic growth regime, $f_k \sim x_k^{-1/3}$, and hyperbolic growth $f_k \sim x_k$.

Second order replicator equations, equation (29), are mathematically equivalent to Lotka-Volterra equations used in mathematical ecology (Hofbauer, 1981). Indeed, research in the group of John McCaskill (Wlotzka and McCaskill, 1997; McCaskill, 1997) is dealing with molecular ecologies of strongly interacting replicons.

MOLECULAR EVOLUTION EXPERIMENTS

In the first half of the twentieth century it was apparently out of the question to do conclusive and interpretable experiments on evolving populations on account of two severe problems: (i) time-scales of evolutionary processes are prohibitive for laboratory investigations and (ii) the numbers of possible genotypes are outrageously large and thus only a negligibly small fraction of all possible sequences can be realized and evaluated by selection. If generation times could be reduced to a minute or less, thousands of generations, numbers sufficient for the observation of optimization and adaptation, could be recorded in the laboratory. Experiments with RNA molecules in the test-tube do indeed fulfill this time-scale criterion for observability. With respect to the "combinatorial explosion" of the numbers of possible genotypes the situation is less clear. Population sizes of nucleic acid molecules of 10^{15}–10^{16} individuals can be produced by random synthesis in conventional automata. These numbers cover roughly all sequences up to chain lengths of $n = 27$ nucleotides. These are only short RNA molecules but their length is already sufficient for specific binding to predefined target molecules, for example antibiotics (Jiang *et al.*, 1997) and molecules of similar size, the siRNAs, were found to play an important role in regulation of gene expression (McManus and Sharp, 2002; Mattick, 2004; Marques *et al.*, 2006). Moreover, sequence to structure to function mappings of RNA were found to be highly redundant (Fontana *et al.*, 1993; Schuster *et al.*, 1994) and thus only a small fraction of all sequences has to be searched in order to find solutions to given evolutionary optimization problems.

The first successful attempts to study RNA evolution *in vitro* were carried out in the late 1960s by Sol Spiegelman and his group (Mills *et al.*, 1967; Spiegelman, 1971). They created a "protein assisted RNA replication medium" by adding an RNA replicase isolated from *E. coli* cells infected by the RNA bacteriophage Qβ to a medium for replication that also contains the four ribonucleoside triphosphates (GTP, ATP, CTP, and UTP) in a suitable buffer solution. Qβ RNA and some of its smaller variants start instantaneously to replicate when transferred into this medium. Evolution experiments were carried out by means of the serial transfer technique: Materials consumed in RNA replication are replenished by transfer of small samples of the current solution into fresh stock medium. The transfers were made after equal time steps. In series of up to 100 transfers the rate of RNA synthesis increased by orders of magnitude. The increase in the replication rate occurs in steps and not continuously as one might have expected. Analysis of the molecular weights of the replicating species showed a drastic reduction of the RNA chain lengths during the series of transfers: The initially applied Qβ RNA was 4220 nucleotides long and the finally isolated species contained little more than 200 bases. What happened during the serial transfer experiments was a kind of degradation due to suspended constraints on the RNA molecule. In addition to perform well in replication the viral RNA has to code for four different proteins in the host cell and needs also a proper structure in order to enable packing into the virion. In test-tube evolution these constraints are released and the only remaining requirement for survival are recognition of the RNA by Qβ replicase and fast replication.

Evidence for a non-trivial evolutionary process came a few years later when the Spiegelman group published the results of another serial transfer experiment that gave evidence for adaptation of an RNA population to environmental change. The replication of an optimized RNA population was challenged by the addition of ethidium bromide to the replication medium (Kramer *et al.*, 1974). This dye intercalates into DNA and RNA double helices and thus reduces replication rates. Further serial transfers in the presence of the intercalating substance led to an increase in the replication rate until an optimum was reached. A mutant was isolated from the optimized population which differed from the original variant by three-point mutations. Extensive studies on the reaction kinetics of RNA replication in the Qβ replication assay were performed by Biebricher (Biebricher and Eigen, 1988). These studies revealed consistency of the kinetic data with many-step reaction mechanism. Depending on concentration the growth of template molecules allows to distinguish three phases of the replication process: (i) at low concentration all free template molecules are instantaneously bound by the replicase which is present in excess and therefore the template concentration grows exponentially, (ii) excess of template molecules leads to saturation of enzyme molecules, then the rate of RNA synthesis becomes constant and the concentration of the template grows linearly, and (iii) very high template concentrations impede dissociation of the complexes between template and replicase, and the template concentration approaches a constant in the sense of product inhibition. We neglect plus–minus complementarity in replication by assuming stationarity in relative concentrations of plus and minus strand (Eigen, 1971) and consider the plus–minus ensemble as a single species. Then, RNA replication may be described by the overall mechanism:

$$\begin{aligned} &\mathbf{A} + \mathbf{I}_i + \mathbf{E} \underset{\bar{k}_i}{\overset{k_i}{\rightleftharpoons}} \mathbf{A} + \mathbf{I}_i \cdot \mathbf{E} \xrightarrow{\;a_i\;} \\ &\mathbf{I}_i \cdot \mathbf{E} \cdot \mathbf{I}_i \underset{\bar{k}_i'}{\overset{k_i'}{\rightleftharpoons}} \mathbf{I}_i \cdot \mathbf{E} + \mathbf{I}_i . \end{aligned} \qquad (31)$$

Here **E** represents the replicase and **A** stands for the low-molecular-weight material consumed in the replication process. This simplified reaction scheme reproduces all three characteristic phases of the detailed mechanism (Figure 1.12) and can be readily extended to complementary replication and mutation.

Despite the apparent complexity of RNA replication kinetics the mechanism at the same time fulfills an even simpler overall rate law provided the activated monomers, ATP, UTP, GTP, and CTP, as well as Qβ replicase are present in excess. Then, the rate of increase for the concentration x_i of RNA species $\mathbf{I}_i$

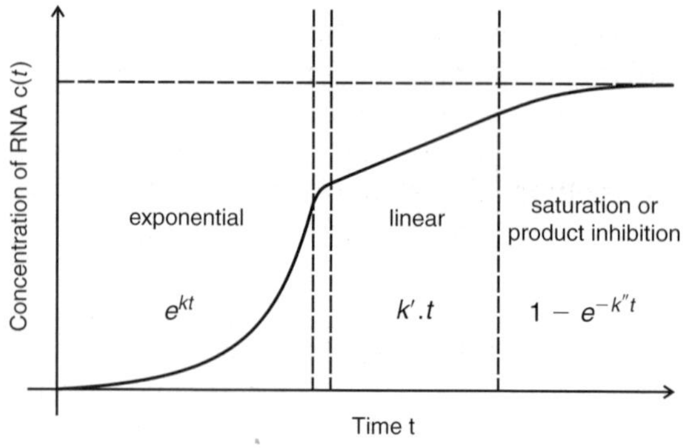

FIGURE 1.12 Replication kinetics of RNA with Qβ replicase. In essence, three different phases of growth are distinguished: (i) exponential growth under conditions with excess of replicase, (ii) linear growth when all enzyme molecules are loaded with RNA, and (iii) a saturation phase that is caused by product inhibition.

follows the simple relation, $\dot{x}_i \propto x_i$, which in absence of constraints ($\phi = 0$) leads to exponential growth. This growth law is identical to that found for asexually reproducing organisms and hence replication of molecules in the test-tube leads to the same principal phenomena that are found with evolution proper. RNA replication in the Qβ system requires specific recognition by the enzyme which implies sequence and structure restrictions. Accordingly only RNA sequences that fulfill these criteria can be replicated. In order to be able to amplify RNA free of such constraints many-step replication assays have been developed. The discovery of the DNA polymerase chain reaction (PCR) (Mullis, 1990) was a milestone towards sequence independent amplification of DNA sequences. It has one limitation: double helix separation requires higher temperatures and therefore conventional PCR works with a temperature program. PCR is combined with reverse transcription and transcription by means of bacteriophage T7 RNA polymerase in order to yield a sequence-independent amplification procedure for RNA. This assay contains two possible amplification steps: PCR and transcription.

Another frequently used assay makes use of the isothermal self-sustained sequence replication reaction of RNA (3SR) (Fahy *et al.*,

1991). In this system the RNA–DNA hybrid obtained through reverse transcription is converted into single-stranded DNA by RNase digestion of the RNA strand, instead of melting the double strand. DNA double strand synthesis and transcription complete the cycle. Here, transcription by T7 polymerase represents the amplification step. Artificially enhanced error rates needed for the creation of sequence diversity in population can be achieved readily with PCR. Reverse transcription and transcription are also susceptible to increase of mutation rates. These two and other new techniques for RNA amplification provided universal and efficient tools for the study of molecular evolution under laboratory conditions and made the use of viral replicases with their undesirable sequence specificities obsolete. Since the 1990s RNA selection experiments have given rise to a new kind of biotechnology making use of evolutionary techniques to create molecules for predefined properties (Klussmann, 2006).

FITNESS LANDSCAPES

So far, we have treated the growth functions f_k as externally given parameters. Only the population dynamics of the replicators $\{\mathbf{I}_1, \mathbf{I}_2, \ldots\}$ has been considered. The function f_k, however, is the mathematical description of the behavior and interactions of a particular chemical entity, the replicator $\mathbf{I}_k$ in a particular environment. In natural evolution, as well as in evolution experiments *in vitro*, mutation (and possibly other mechanisms such as recombination) will cause the emergence of new type of replicons, while existing ones may be driven to extinction by the population dynamics. Thus it is imperative to gain an understanding for the dependence of f_k on the underlying replicons $\mathbf{I}_k$ and to relate this knowledge to the mutual accessibility of variants.

Although the concepts can be generalized further, we restrict ourselves here to the simplest case of constant functions $f_k(\mathbf{x}) = f_k$—we call these fixed values the fitness of $\mathbf{I}_k$—and we assume that our replicons $\mathbf{I}_k$ are sequences of a fixed length n. Sequences can be interconverted by point mutations, hence adjacent sequences differ by a mutation in a single position (it is easy to relax the restriction to point mutations and to include insertions, deletions, and rearrangements into the framework). Let us denote the set of all possible replicon types by Ξ. Given an adjacency relation on Ξ, we can visualize Ξ as a graph, with a adjacent sequences (interrelated by single point mutants) connected by edges.

Fitness can now be seen as a function $f : \Xi \to \mathbb{R}$. Together with the graph structure on Ξ, we speak of a fitness landscape, a concept introduced by Sewall Wright (Wright, 1932) to explain the effect of selection. In the crudest approximation, a population will move in Ξ so as to maximize f. An elaborate mathematical theory has been developed to analyze the structure of fitness landscapes in terms of various measures of ruggedness, i.e. the local variability of fitness values (see Reidys and Stadler, 2001).

Realistic biological fitness landscapes,[7] however, are not just arbitrary functions $f : \Xi \to \mathbb{R}$. In fact, they are naturally decomposed into two steps because it is never the nucleic acid or peptide sequence itself that is subject to selection, but rather the three-dimensional structure that if forms, or the "organism" that it encodes. Hence there is first the map $\Psi : \Xi \to \mathbb{S}$ that connects a sequence with its phenotype, $\mathbf{I}_k \to \Psi\ (k)$. This phenotype is then "evaluated" by its environment. Hence $f_k = \text{eval}(\Psi(k))$ is a composite of the genotype map Ψ and the fitness evaluation function. In biophysically realistic settings, such as the RNA folding model where the phenotype is by the molecular structure and its properties, one observes substantial redundancy in the genotype-phenotype map, i.e. many genotypes give rise to phenotypes that are indistinguishable. As a consequence, there are many sequences $\mathbf{I}_k \in \Xi$ that have the same fitness. Since in particular closely related sequences are often selectively indistinguishable, there is a certain fraction of neutral mutations

[7] Realistic is used here in order to distinguish these landscapes from oversimplified landscape models often used in population genetics.

with the property that $f_k = f_l$. We shall see below that these neutral mutations play a crucial role in molecular evolution.

Many proposals for simple model landscapes have been made, among them the so-called Nk-landscape of Kauffman (1993) has become very popular. In the simplest realistic case that is based on molecular data, the genotype–phenotype map Ψ is defined by folding the biopolymer sequences (RNA, DNA, or peptide) into its three-dimensional structure. In case of RNA and a simplified notion of structure, the so-called secondary structure the map Ψ is sufficiently simple in order to allow for systematic analysis (Schuster, 2006).

Time-dependent fitness landscapes have been discussed some time ago (see, e.g., Kauffman, 1993; Levitan and Kauffman, 1995). Two major effects introduce dynamics into landscapes: (i) fluctuating environments and (ii) co-evolution. More recently these ideas were extended to a comprehensive treatment of dynamic fitness landscapes (Wilke *et al.*, 2001; Wilke and Ronnewinkel, 2001). Successful application of dynamic landscapes requires that the adaptive process on the landscape occurs on a substantially shorter time-scale than the changes of the landscapes, otherwise strong coupling between adaptation and landscape dynamics makes the landscape concept obscure. In case of co-evolution the separation of time-scales is at least questionable.

QUASISPECIES AND ERROR PROPAGATION

Evolution of molecules based on replication and mutation has been discussed above. Here we consider in detail the internal equilibration in populations as formulated in terms of normalized concentrations (15) and extensively discussed before (Eigen, 1971; Eigen and Schuster, 1977; Eigen *et al.*, 1989). Error-free replication and mutation are seen as parallel chemical reactions,

$$\mathbf{A} + \mathbf{I}_j \xrightarrow{f_j Q_{kj}} \mathbf{I}_k + \mathbf{I}_j, \tag{32}$$

and constitute a network, which in principle allows for the formation of every RNA genotype as a mutant of any other genotype, $\mathbf{I}_j \rightarrow \mathbf{I}_k$, eventually through a series of consecutive point mutations, $\mathbf{I}_j \rightarrow \mathbf{I}_l \rightarrow \ldots \rightarrow \mathbf{I}_k$. The materials required for or consumed by RNA synthesis, again denoted by A, are kept constant by adjusting flow and influx material in a kind of chemostat (Figure 1.9). The object of interest is now the distribution of genotypes in the population and its dependence on the mutation rate. We shall be dealing here exclusively with single-strand replication but mention a recent approach that considers semi-conservative DNA replication (Tannenbaum *et al.*, 2004, 2006). Spatially resolved reaction-diffusion dynamics of quasispecies has been studied as well (Altmeyer and McCaskill, 2001; Pastor-Satorras and Solé, 2001).

Quasispecies Equation

The time-dependence of the genotype distribution is described by the kinetic equation

$$\dot{x}_k = x_k(f_k Q_{kk} - \phi(t)) + \sum_{j=1, j\neq k}^{m} f_j Q_{kj} x_j, \quad k = 1,\ldots, m. \tag{33}$$

The replication functions of the molecular species, f_k, are constants under these conditions. The frequencies of the individual reaction channels are contained in the mutation matrix $\mathbf{Q} = \{Q_{kj};\, k, j = 1, \ldots, m\}$. Recall that $\mathbf{Q}$ is a stochastic matrix, $\Sigma_k Q_{kj} = 1$ since every copy is either correct or incorrect.

In the no-mutation limit the mutation matrix $\mathbf{Q}$ is the unit matrix, the kinetic equation has the form

$$\dot{x}_k = x_k(f_k - \phi(t)), \quad i = 1,\ldots, m \text{ with } \phi(t) = \sum_{j=1}^{m} f_j x_j, \tag{33a}$$

and an analytical solution of (33a) is available

$$x_k(t) = \frac{x_k(0)\cdot \exp(f_k t)}{\sum_{j=1}^{m} x_j(0)\cdot \exp(f_j t)}. \tag{33b}$$

The interpretation of the result is straightforward: After sufficiently long time the exponential function with the largest value of the replication rate parameter, $f_M = \max\{f_j;\ j = 1, 2, \ldots, m\}$, dominates the sum in the denominator, and hence $\lim_{t\to\infty} x_M = 1$ and $\lim_{t\to\infty} x_j = 0\ \forall\, j = 1, 2, \ldots, m;\ j \neq M$. The replicator that replicates fastest is selected.

The quantities determining the outcome of selection in the replication–mutation scenario are the products of replication rate constants and mutation frequencies subsumed in the value matrix:[8] $W \doteq \{w_{kj} = f_j Q_{kj};\, k, j = 1, \ldots, m\}$, its diagonal elements, w_{kk}, are called the selective values of the individual genotypes (Eigen, 1971).

The selective value of a genotype is tantamount to its fitness in the case of vanishing mutational backflow and hence the genotype with maximal selective value, $\mathbf{I}_M$,

$$w_{MM} = \max\{w_{kk} \mid i = 1,\ldots,m\} \qquad (34)$$

dominates a population after it has reached the selection equilibrium and is called the master sequence. The notion quasispecies was introduced for the stationary genotype distribution in order to point at its role as the genetic reservoir of an asexual population.

Error Threshold

A simple expression for the stationary frequency can be found if the master sequence is derived from the single peak model landscape that assigns a higher replication rate to the master and identical values to all others, for example $f_M = \sigma_M \cdot f$ and $f_i = f$ for all $i \neq M$ (Swetina and Schuster, 1982; Tarazona, 1992; Alves and Fontanari, 1996). The (dimensionless) factor σ_M is called the superiority of the master sequence. The assumption of a single peak landscape is tantamount to lumping all mutants together into a mutant cloud with average fitness and reminds of a mean field approximation. The probability of being in the cloud is simply $x_c = \sum_{j=1, j\neq M}^{m} x_j = 1 - x_M$ and the replication–mutation problem boils down to an exercise in a single variable, x_M, the frequency of the master. In the sense of a mean field approximation, for example, we define a mean-except-the-master replication rate constant $\bar{f} = \sum_{j\neq M} f_j x_j/(1 - x_M)$. The superiority then reads: $\sigma_M = f_M/\bar{f}$. Neglecting mutational backflow we can readily compute the stationary frequency of the master sequence,

$$\bar{x}_M = \frac{f_M Q_{MM} - \bar{f}}{f_M - \bar{f}} = \frac{\sigma_M Q_{MM} - 1}{\sigma_M - 1}, \qquad (35)$$

which vanishes at some finite replication accuracy, $Q_{MM}|_{\bar{x}_M=0} = Q_{\min} = \sigma_M^{-1}$. Non-zero frequency of the master requires $Q_{MM} > Q_{\min}$. Within the uniform error rate approximation, which assumes that the mutation rate per site and replication event, p, is independent of the nature of the nucleotide and the position in the sequence (Eigen and Schuster, 1977). Then, the single digit accuracy $q = 1-p$ is the mean fraction of correctly incorporated nucleotides and the elements of the mutation matrix for a polynucleotide of chain length n are of the form:

$$Q_{ij} = q^n \left(\frac{1-q}{q}\right)^{dij},$$

with d_{ij} being the Hamming distance between two sequences $\mathbf{I}_i$ and $\mathbf{I}_j$. The critical condition, called the error threshold, $\bar{x}_M = 0$, occurs at a minimum single digit accuracy of

$$q_{\min} = 1 - p_{\max} = \sqrt[n]{Q_{\min}} = \sigma_M^{-1/n}. \qquad (36)$$

Figure 1.13 shows the stationary frequency of the master sequence, $\bar{x}_M$, as a function of the error rate. The "no mutational backflow approximation" cannot describe how populations behave at mutation rates above the error threshold.

[8] In case degradation rates d_k are important they are readily absorbed in the diagonal terms of the value matrix (Eigen, 1971): $w_{kk} = f_k Q_{kk} - d_k$; see also (16) and the definition of $\Gamma(a, \mathbf{c})$.

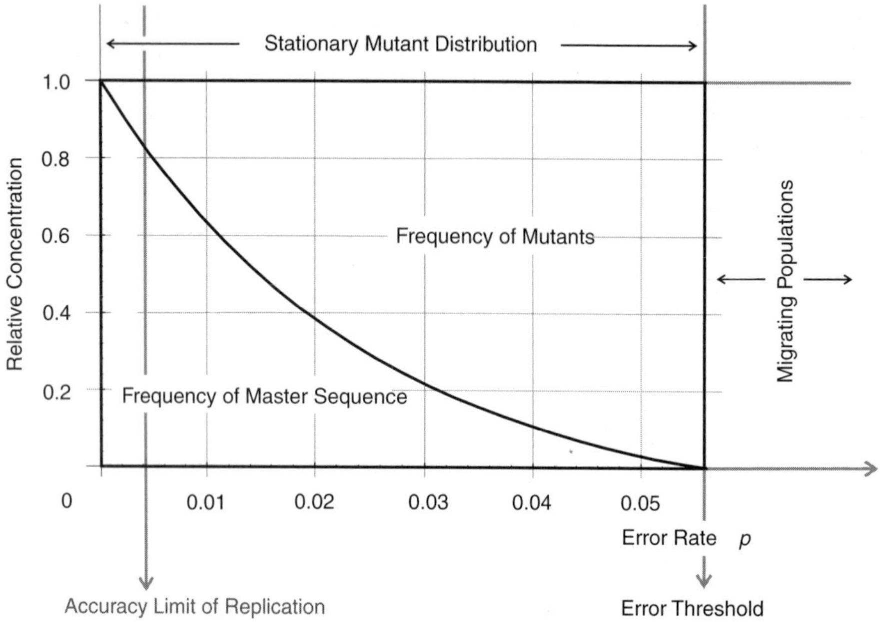

FIGURE 1.13 The genotypic error threshold. The fraction of mutants in stationary populations increases with the error rate p. The formation of a stable stationary mutant distributions, the quasispecies, requires sufficient accuracy of replication: The error rate p has to be below a maximal value known as error threshold, $p < p_{\max}$, tantamount to a minimal replication accuracy, $q > q_{\min}$. Above threshold, populations migrate through sequence space in random walk-like manner (Huynen *et al.*, 1996; Fontana and Schuster, 1998a). There is also a lower limit to replication accuracy which is given by the maximum accuracy of the replication machinery.

Exact Solution of the Quasispecies Equation

Exact solutions of the kinetic equation (33) can be obtained by different techniques (Thompson and McBride, 1974; Jones *et al.*, 1976; Baake and Wagner, 2001; Saakian and Hu, 2006). A straightforward approach starts with a transformation of variables

$$z_k(t) = x_k(t) \cdot \exp\left(\int_0^t \phi(\tau)d\tau\right),$$

that leads to a linear first order differential equation, $\dot{\mathbf{z}} = \mathrm{W} \cdot \mathbf{z}$, which can be solved in terms of the eigenvalue problem

$$\mathrm{W} \cdot \zeta_k = \lambda_k \cdot \zeta_k \text{ with } \zeta_k = \sum_{j=1}^{m} h_{kj} z_j \text{ and}$$

$$\Lambda = \mathrm{H} \cdot \mathrm{W} \cdot \mathrm{H}^{-1}.$$

The eigenvectors ζ_k are linear combinations of the variables z and represent the normal modes of the replication-mutation network, $\Lambda = \{\lambda_1, \lambda_2, \ldots, \lambda_m\}$ is a diagonal matrix, and the transformation matrix **H** contains the coefficients for the eigenvectors. The replication–mutation equation written in terms of eigenvectors of W is of the simple form: $\dot{\zeta}_k = \lambda_k \cdot \zeta_k$ and the solutions after re-introduction of constant population size through the constraint $\phi(t)$ are the same as in equation (33b).

In cases where all genotypes have non-zero fitness and **Q** is a primitive matrix,[9] Perron–Frobenius theorem (Seneta, 1981) applies: The largest eigenvalue λ_0 is real, positive, and non-degenerate.[10] The eigenvector ζ_0

[9] A square matrix **A** with non-negative entries is a primitive matrix, if and only if there exists a positive integer k such that A^k has only strictly positive entries.

[10] A non-degenerate eigenvalue has only one unique eigenvector. Twofold degeneracy, for example, means that two eigenvectors are associated with the eigenvalue and all linear combinations of the two eigenvectors are also solutions of the eigenvalue problem associated with the (twofold) degenerate eigenvalue.

belonging to the largest eigenvalue λ_0 is therefore unique, in addition it has strictly positive components. This purely mathematical result has important implications for the replication–mutation system:

(i) Since $\lambda_0 > \lambda_k \forall\ k = 1, 2, \ldots, m-1$ the eigenvector ζ_0 outgrows all other eigenvectors ζ_k and determines the distribution of genotypes in the population after sufficiently long time: ζ_0 is the stationary distribution of genotypes called the quasispecies.

(ii) All genotypes of the population, $\{\mathbf{I}_1, \mathbf{I}_2, \ldots, \mathbf{I}_m\}$ are present in the quasispecies although the concentration may be extremely small.

It is important to note that quasispecies can also exist in cases where the Perron–Frobenius theorem is not fulfilled. As an example we consider an extreme case of lethal mutants: Only genotype $\mathbf{I}_1$ has a positive fitness value, $f_1 > 0$ and $f_2 = \ldots = f_m = 0$, only the entries $w_{k1} = f_1 Q_{k1}$ are non-zero and hence

$$W = \begin{pmatrix} w_{11} & 0 & \cdots & 0 \\ w_{21} & 0 & \cdots & 0 \\ \vdots & \vdots & \ddots & \vdots \\ w_{m1} & 0 & \cdots & 0 \end{pmatrix} \quad \text{and}$$

$$W^k = w_{11}^k \begin{pmatrix} 1 & 0 & \cdots & 0 \\ \frac{w_{21}}{w_{11}} & 0 & \cdots & 0 \\ \vdots & \vdots & \ddots & \vdots \\ \frac{w_{m1}}{w_{11}} & 0 & \cdots & 0 \end{pmatrix}$$

Clearly, W is not primitive in this example, but $\bar{\mathbf{x}} = (Q_{11}, Q_{21}, \ldots, Q_{m1})$ is a stable stationary mutant distribution and for $Q_{11} > Q_{j1}$ $j = 2, \ldots, m$ (correct replication occurs more frequently than a particular mutation) genotype $\mathbf{I}_1$ is the master sequence. On the basis of a rather idiosyncratic mutation model consisting of a one-dimensional chain of mutations Wagner and Krall (1993) raised the claim that no error thresholds can occur in presence of lethal mutants. In a recent paper Takeuchi and Hogeweg (2007) used a realistic highdimensional mutation model and presented numerically computed examples of perfect error thresholds in the presence of lethal mutants.

Several authors (Leuthäusser, 1987; Tarazona, 1992; Franz *et al.*, 1993; Franz and Peliti, 1997) pointed out an equivalence between the quasispecies model and spin systems. Applying methods of statistical mechanics Franz and Peliti (1997) were able to show that for both models, the single peak fitness landscape and a random fitness model the error threshold corresponds to a first order phase transition. Valandro *et al.* (2000) demonstrated an isomorphism between the quasispecies and percolation models. Earlier work by Haken showed an analogy between selection of laser modes and quasispecies (Haken, 1983a, 1983b).

It is important to note that the appearance of a sharp error threshold depends on the distribution of fitness values in genotype space. The single-peak fitness landscape (Swetina and Schuster, 1982; Franz and Peliti, 1997), the multiple-peak fitness landscape (Saakian *et al.*, 2006), the random fitness landscape (Franz and Peliti, 1997; Campos, 2002), and realistic rugged landscapes (see below) give rise to sharp transitions whereas artificially smooth landscapes, which are often used in population genetics (Wiehe, 1997; Baake and Wagner, 2001), lead to gradual transitions from the replication–mutation ordered quasispecies to the uniform distribution of genotypes.

Random Drift and Truncation of Quasispecies

In contrast to the no-mutational-backflow approximation (35) the concentration of the master sequence does not drop to zero but converges to some small value beyond the error threshold. Nevertheless, the stationary solution of equation (33) changes abruptly within a narrow range of the error rate p. The cause of this change is an avoided crossing of the first two eigenvalues around $p_{\max}$ (Nowak and Schuster, 1989):[11] Below threshold the

[11] The notion of avoided crossing is used in quantum physics for a situation in which two eigenvalues that are coupled by a small off-diagonal element do not cross but approach each other very closely (Figure 1.14).

ζ_0 representing the quasispecies is associated with λ_0 the largest eigenvalue. Above threshold the previous eigenvector ζ_1 is associated with the largest eigenvalue. With further increasing error rates, p, this eigenvector approaches the uniform distribution of genotypes. A uniform distribution of genotypes, however, is no realistic object: Population sizes are almost always below 10^{15} molecules, a value that can be achieved in evolution experiments with molecules. The numbers of viruses in a host hardly exceed 10^{12}. The numbers of possible genotypes exceed these numbers by many orders of magnitude. There are, for example, about 6×10^{45} genotypes of tRNA sequence length $n = 76$. All the matter in the universe would not be sufficient to produce a uniform distribution of these molecules and, accordingly, no stationary distribution of sequences can be formed. Instead, the population drifts randomly through sequence space. This implies that all genotypes have only finite life times, inheritance breaks down and evolution becomes impossible unless there is a high degree of neutrality that can counteract this drastic imbalance (see below).

A similar situation occurs with rare mutations within individual quasispecies. Since

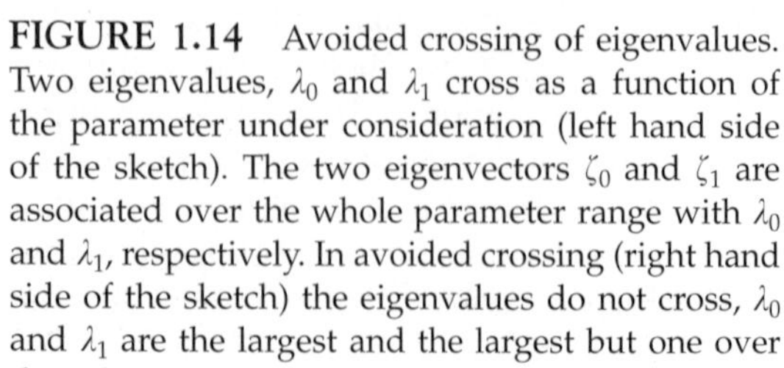

FIGURE 1.14 Avoided crossing of eigenvalues. Two eigenvalues, λ_0 and λ_1 cross as a function of the parameter under consideration (left hand side of the sketch). The two eigenvectors ζ_0 and ζ_1 are associated over the whole parameter range with λ_0 and λ_1, respectively. In avoided crossing (right hand side of the sketch) the eigenvalues do not cross, λ_0 and λ_1 are the largest and the largest but one over the whole range. The two eigenvectors, however, behave roughly as in case of crossing. Before the avoided crossing zone ζ_0 is associated with λ_0 and ζ_1 with λ_1, after crossing, the assignment is inverse: ζ_0 is associated with λ_1 and ζ_1 with λ_0.

every genotype can be reached from any other genotype by a sequel of individual mutations, all genotypes are present in the quasispecies no matter how small their concentrations might be. This, again is contradicting the discreteness at the molecular level. The solution of the problem distinguishes two classes of mutants: (i) frequent mutants, which are almost always present in realistic quasispecies, and (ii) rare mutants that are stochastic elements at the periphery of the deterministic mutant cloud.

In order to be able to study stochastic features of population dynamics around the error threshold in rigorous terms, the replication–mutation system was modeled by a multitype branching process (Demetrius *et al.*, 1985). The main result of this study is the derivation of an expression for the probability of survival to infinite time for the master sequence and its mutants. In the regime of sufficiently accurate replication, i.e. in the quasispecies regime, the survival probability is non-zero and decreases with increasing error rate p. At the critical accuracy pmax this probability becomes zero. This implies that all molecular species which are currently in the populations, master and mutants, will die out in finite times and new variants will appear. This scenario is tantamount to migration of the population through sequence space (Huynen *et al.*, 1996; Huynen, 1996). The critical accuracy q_{min}, commonly seen as an error threshold for replication, can as well be understood as the localization threshold of the population in sequence space (McCaskill, 1984). Later investigations aimed directly at a derivation of the error threshold in finite populations (Nowak and Schuster, 1989; Alves and Fontanari, 1998).

Error Thresholds in Reality

Variations in the accuracy of *in vitro* replication can indeed be easily achieved because error rates can be tuned over many orders of magnitude (Leung *et al.*, 1989; Martinez *et al.*, 1994). The range of replication accuracies which are suitable for evolution is limited by the maximal accuracy that can be achieved by the replication machinery and the minimum accuracy determined by the error threshold (Figure 1.13). Populations in constant environments have an advantage when they operate near the maximal accuracy because then they lose as few copies through mutation as possible. In highly variable environments the opposite is true: It pays to produce as many mutants as possible to maximize the chance of coping successfully with change. RNA viruses live in very variable environments since they have to cope with the highly effective defense mechanisms of the host cells.

The key parameter in testing error thresholds in real populations is the rate of spontaneous mutation, p. The experimental determination of mutation rates per replication and site, which is different from the observed frequency of mutations, is tricky mainly for two reasons: (i) deleterious and most neutral mutations will not be observed on the population level, because they are eliminated earlier by selection, and (ii) in the case of virus replication more than one replication take place in the infected cell (Drake, 1993; Drake and Holland, 1999). Careful evaluated results reveal a rate of roughly $\mu_g = 0.76$ per genome and replication, although the genome lengths vary from $n = 4200$ to $n = 13600$. This finding implies that the mutation rate per replication and nucleotide site is adjusted to the chain length. For a given error rate p the minimum accuracy of replication can be transformed into a maximum chain length $n_{\max}$.[12] Then the condition for the quasispecies error threshold provides a limit for the lengths of genotypes:

$$n < n_{\max} = -\frac{\ln \sigma}{\ln q} \approx \frac{\ln \sigma}{1-q} = \frac{\ln \sigma}{p}. \quad (37)$$

RNA viruses mutate much more frequently than all other known organisms and this is presumably the consequence of two factors: (i) the defense mechanisms of the host provide a highly variable environment, which requires fast adaptation, and (ii) the small genome size is prohibitive for coding enzymes that replicate with high accuracy. The high mutation rate and the vast sequence heterogeneity of RNA viruses (Domingo *et al.*, 1998) suggest that most RNA viruses live indeed near the above mentioned critical value of replication accuracy (Domingo, 1996; Domingo and Holland, 1997) in good agreement with the relation between chain length n and error rate p mentioned above. For a review on medical application of the error threshold in antiviral therapies see, for example, Domingo and Holland (1997), Eigen (2002), Anderson *et al.* (2004), and the special issue of *Virus Research* (Domingo, ed., 2005). In a recent paper, Bull *et al.* (2007) present a theory of lethal mutagenesis that distinguishes crossing the error threshold from the decline of the population, $\lim c(t) \to 0$, which by construction cannot be seen in the quasispecies equation (33). The experimental verification of which of the two effects is the cause of lethal mutagenesis, however, seems to be very subtle.

The justification of the quasispecies concept in the description of RNA virus evolution has been challenged by Edward Holmes and co-workers (Jenkins *et al.*, 2001; Holmes and Moya, 2002; Comas *et al.*, 2005) (see also the reply by Domingo, 2002). They propagate the application of conventional population genetics to RNA virus evolution (Moya *et al.*, 2000, 2004) and raised several arguments against the application of the quasispecies concept to RNA virus evolution. Wilke (2005) performed a careful analysis of both approaches by means of thoughtfully chosen examples and showed the equivalence of both models that apparently has escaped the attention of the quasispecies opponents.[13] Indeed, it is only a matter

[12] The accuracy of replication is determined by the RNA replicase. Fine-tuning of the enzyme allows for an adjustment of the error rate within certain limits.

[13] On the basis of the paper by Wagner and Krall (1993), Wilke concluded erroneously that an error threshold cannot occur in the presence of lethal mutants. Wagner's result was an artifact of the assumption of an unrealistic one-dimensional sequence space. Takeuchi and Hogeweg (2007) have shown the existence of error thresholds on landscapes with lethal variants.

of model economy and taste whether one prefers the top-down approach of population genetics with the plethora of often unclear effects or the sometimes deeply confusing molecular bottom-up approach of biochemical kinetics with the enormous wealth of detail. To address issues of conventional evolutionary biology the language of population genetics provides an advantage; molecular biology and its results, however, are much more easily translated into the formalism of biochemical kinetics as the fast development of systems biology shows (Klipp *et al.*, 2005; Palsson, 2006).

Finally, we relate the concept of error threshold to the evolution of small prebiotic replicons. Uncatalyzed template-induced RNA replication can hardly be more accurate than $q = 0.99$ and this implies that the chain lengths of correctly replicated polynucleotides are limited to molecules with $n < 100$. RNA molecules of this size are neither in a position to code for efficiently replicating ribozymes nor can they develop a genetic code that allows for the evolution of protein enzymes. A solution for this dilemma, often called the Eigen paradox, was seen in functional coupling of replicons in the form of hypercycle (Eigen and Schuster, 1978a, 1978b).

EVOLUTION OF PHENOTYPES AND COMPUTER SIMULATION

The quasispecies concept discussed so far is unable to handle cases where many molecular species have the same maximal fitness.[14] In this section we deal with this case of neutrality first introduced by Kimura (1983) in order to interpret the data of molecular phylogenies. If we had only neutral genotypes the superiority of the master sequence becomes $\sigma_M = 1$ and the localization threshold of the quasispecies converges to the limit of absolute replication accuracy, $q_{\min} = 1$. Clearly, the deterministic model fails, and we have to modify the kinetic equations. For example, there is ample evidence that RNA structures are precisely conserved despite vast sequence variation. Neutrality of RNA sequences with respect to secondary structure is particularly widespread and has been investigated in great detail (Fontana *et al.*, 1993; Schuster *et al.*, 1994; Reidys *et al.*, 1997; Reidys and Stadler, 2001). Here we sketch an approach to handle neutrality within the quasispecies approach (Reidys *et al.*, 2001) and then present computer simulations for a stochastic model based on the quasispecies equations (33) (Fontana and Schuster, 1987, 1998a, 1998b; Fontana *et al.*, 1989; Schuster, 2003).

A Model for Phenotype Evolution

Genotypes are ordered with respect to non-increasing selective values. The first k_1 different genotypes have maximal selective value: $w_1 = w_2 = \ldots = w_{k_1} = w_{\max} = \tilde{w}_1$ (where ~ indicates properties of groups of neutral phenotypes). The second group of neutral genotypes has the highest but one selective value: $w_{k1+1} = w_{k1+2} = \ldots = w_{k1+k2} = \tilde{w}_2 < \tilde{w}_1$, etc. Replication rate constants are assigned in the same way: $f_1 = f_2 = \ldots = f_{k1} = \tilde{f}_1$, etc. In addition, we define new variables, y_j ($j = 1, \ldots, \ell$), that lump together all genotypes folding into the same phenotype:

$$y_j = \sum_{i=k_{j-1}+1}^{k_j} x_i \text{ with } \sum_{j=1}^{\ell} y_j = \sum_{i=1}^{m} x_i = 1. \quad (38)$$

The phenotype with maximal fitness, the master phenotype, is denoted by "M." Since we are heading again for a kind of zeroth-order solution, we consider only the master phenotype and put $k_1 = k$. With $y_M = \sum_{i=1}^{k} x_i$ we obtain the following kinetic differential equation for the set of sequences forming the neutral network of the master phenotype:

$$\dot{y}_M = \sum_{i=1}^{k} \dot{x}_i = y_M(\tilde{f}_M Q_{kk} - \bar{E}) + \sum_{i=1}^{k} \sum_{j \neq i} f_j Q_{ji} x_j. \quad (39)$$

[14] Different examples of fitness landscapes with two highest peaks were analyzed and discussed by Schuster and Swetina (1988). This approach, however, cannot be extended to a substantially larger number of master genotypes.

The mean excess productivity of the population is, of course, independent of the choice of variables: $\bar{E} = \sum_{j=1}^{\ell} \tilde{f}_j y_j = \sum_{i=1}^{m} f_i x_i$.

The mutational backflow is split into two contributions, (i) mutational backflow on the neutral network and (ii) mutational backflow from genotypes not on the network.

$$\dot{y}_M = (\tilde{f}_M \tilde{Q}_{MM} - \bar{E}) y_M + \text{mutational backflow} \tag{40}$$

The next task is to compute the effective replication accuracy $\tilde{Q}_{MM}$.

Phenotypic Error Thresholds

An assumption for the distribution of neutral genotypes in sequence space is required for the calculation of the effective replication accuracy $\tilde{Q}_{MM}$ of the master phenotype. Two assumptions were made (i) uniform distribution of neutral sequences (Reidys *et al.*, 2001) and (ii) a binomial distribution for neutral substitutions as a function of the Hamming distance from the reference sequence (Takeuchi *et al.*, 2005). Both assumptions lead to an expression of the form

$$\tilde{Q}_{MM} = Q_{MM} + \Lambda(1 - Q_{MM}) = q^n \cdot F(q, \lambda, n)$$

where Λ is the fraction of neutral mutants in sequence space and λ is the degree of neutrality, the fraction of neutral mutants in the one-error neighborhood of the reference sequence. The functions $F(q, \lambda, n)$ are of the form

$$F^{(i)}(q, \lambda, n) = 1 + \lambda \frac{1 - q^n}{q^n} \quad \text{for assumption (i) and}$$

$$F^{(ii)}(q, \lambda, n) = \left(1 + \lambda \frac{1 - q}{q}\right)^n \quad \text{for assumption (ii).}$$

The second function was also used in a different version with a tunable parameter ν instead of λ (Wilke, 2001). The calculation of expressions for phenotypic error thresholds is now straight-forward and leads to the following two expressions for the minimal replication accuracy $q_{\min}$:

$$q_{\min}^{(i)} = (1 - p_{\max}^{(i)}) = \left(\frac{\sigma_M^{-1} - \lambda_M}{1 - \lambda_M}\right)^{1/n} \quad \text{and} \tag{41}$$

$$q_{\min}^{(ii)} = (1 - p_{\max}^{(ii)}) = \frac{\sigma_M^{-1/n} - \lambda_M}{1 - \lambda_M} \tag{42}$$

Both equations converge to the expression for genotypic error threshold (36) in the limit $\lambda \to 0$. Both approaches predict a decrease of the minimum accuracy with increasing neutrality but the assumption (ii) leads to a much smaller effect that becomes dominant only close to complete neutrality $\lambda \to 1$. The conclusion of Takeuchi *et al.* (2005) is therefore that neutrality has a very limited influence on the minimum replication accuracy.

Between the genotypic and the phenotypic error threshold the population migrates in sequence space but the phenotype is still conserved. Precisely this behavior is postulated in the observed phylogenies of RNA molecules and RNA viruses. Because of the deterministic nature of the quasispecies equation (33) random drift on neutral spaces or subspaces cannot be described. Such a behavior, however, can be directly observed and analyzed in computer simulations of RNA evolution, which will be the subject of the next subsection.

Computer Simulations

The concept of the phenotypic error threshold allows for an extension of the kinetic equations to the regime of random drift without, however, providing insights into the stochastic process itself. Since a sufficiently high degree of neutrality is required to observe random drift, the RNA sequence-structure map was chosen for the computer simulations because it was known to give rise to vast neutrality and to support random drift (Fontana *et al.*, 1993; Schuster *et al.*, 1994; Huynen *et al.*, 1996). The flow reactor shown in Figure 1.9 was chosen as a proper chemical environment for the simulation of RNA evolution (Fontana and Schuster, 1998a, 1998b). We present only the result that is relevant here (for more details see Schuster, 2003). Solutions of the master equation (Gardiner, 2004) corresponding to the reaction network of the quasispecies equation (33) are approximated by sampling numerically computed trajectories according to a procedure proposed by Gillespie (1976, 1977a, 1977b). In order to be able to evaluate the progress in the individual simulations a fixed target that happens to be the secondary structure of $tRNA^{phe}$, S_τ, was chosen. The fitness function, $f_k = (\alpha + d_S(S_k, S_\tau)/n)^{-1}$, increases with decreasing distance to the target structure S_τ.[15] The trajectories end after the target structure has been reached. Thus the stochastic process has two absorbing barriers: (i) extinction of the population and (ii) reaching the target. The question is whether or not the populations become extinct and whether the trajectories of surviving populations reach the target in reasonable or astronomic times.

A typical trajectory is shown in Figure 1.15. The stochastic process occurs on two timescales: (i) fast adaptive phases during which the population approaches the target are interrupted by (ii) slow epochs of random drift at constant distance from the target, and this gives trajectories the typical stepwise appearance. At the beginning of an adaptive phase the genotype distribution in sequence space is very narrow, typical are widths below Hamming distance 5 for population sizes of $N = 3000$. Then, along the plateau, the width of the population increases substantially up to values of 30 in Hamming distance. At first the population broadens but still occupies a coherent region in sequence space, later it is split into individual clones that continue to diverge in sequence space.[16] Interestingly, the consensus sequence of the population tantamount to the position of the population center in sequence space stays almost invariant during the quasistationary epochs. Eventually, the spreading population finds a genotype of higher fitness and a new adaptive phase is initiated. This is mirrored in sequence space by a jump of the population center and a dramatic narrowing of the population width. In other words, the beginning of a new adaptive period represents a bottleneck in sequence space through which the population has to pass in order to continue the adaptation process. Thus the evolutionary process is characterized by a succession of optimization periods in sequence space, where quasispecies-like behavior is observed, and random drift epochs, during which the population spreads until it finds a genotype that is suitable for further optimization. Two types of processes were observed in the random drift domain: (i) changing RNA sequences at conservation of the secondary structure and (ii) changing sequences overlaid by a random walk in the subspace of structures with equal distance to target. Population sizes were varied between $N = 100$ and $N = 100\,000$ but no significant change was observed in the qualitative behavior of the system except the trivial effect that larger populations can cover greater areas in sequence space.

Systematic studies on the parameter dependence of RNA evolution were reported in a recent simulation (Kupczok and Dittrich, 2006). Increase in mutation rate leads to an error threshold phenomenon that is close to one observed with quasispecies on a single-peak landscape as described above (Swetina and Schuster, 1982; Eigen *et al.*, 1989). Evolutionary optimization becomes more efficient[17] with increasing error rate until the error threshold is reached. Further increase in the error rate leads to an abrupt breakdown of the optimization process. As expected, the distribution of replication rates or fitness values f_k in sequence space is highly relevant too: steep

[15] For the definition of a distance between two structures, $d_S(S_k, S_j)$, see the footnote of Table 1.1.

[16] The same phenomenon has been observed in the evolution of populations on flat landscapes (Derrida and Peliti, 1991).

[17] Efficiency of evolutionary optimization is measured by average and best fitness values obtained in populations after a predefined number of generations.

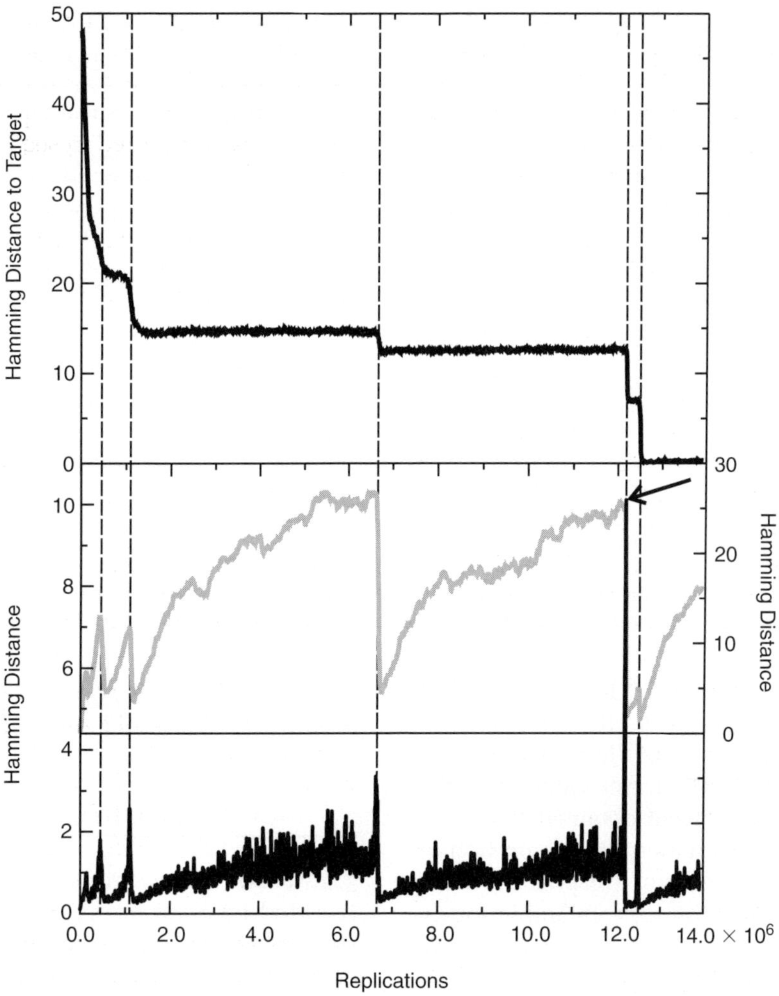

Figure 1.15 Evolutionary optimization of RNA structure. Shown is a single trajectory of a simulation of RNA optimization towards a tRNA^{phe} target with population size $N = 3000$, the fitness function $f_k = (\alpha + d_S(S_k, S_\tau)/n)^{-1}$ with $\alpha = 0.01$ and $n = 76$, and mutation rate $p = 0.001$ per site and replication. The figure shows as functions of time: (i) the distance to target averaged over the whole population, $d_S(S_i, S_\tau)(t)$ (upper black curve), (ii) the mean Hamming distance within the population, $d_P(t)$ (gray, right ordinate), and (iii) the mean Hamming distance between the populations at time t and $t + \Delta t$, $d_C(t, \Delta t)$ (lower black curve) with a time increment of $\Delta t = 8000$. The end of plateaus (vertical lines) are characterized by a collapse in the width of the population and a peak in the migration velocity corresponding to a bottleneck in diversity and a jump in sequence space. The arrow indicates a remarkably sharp peak of Hamming distance 10 at the end of the second long plateau ($t \approx 12.2 \times 10^6$ replications). On the plateaus the center of the cloud stays practically constant (the speed of migration is Hamming distance 0.125 per 1000 replications) corresponding to a constant consensus sequence. Each adaptive phase is preceded by a drastic reduction in genetic diversity, $d_P(t)$, then the diversity increases during the quasistationary epochs and reaches a width of Hamming distance more than 25 on long plateaus.

and rugged fitness functions lead to the sharp threshold behavior as observed with single-peak landscapes, whereas smooth and flat landscapes give rise to a broad maximum of optimization efficiency without an indication of a threshold-like behavior.

Table 1.1 collects some numerical data obtained from repeated evolutionary trajectories

TABLE 1.1 Statistics of the optimization trajectories showing the results of sampled evolutionary trajectories leading from a random initial structure S_0 to the structure of $tRNA^{phe}$, $S\tau$ as target.[a]

Population size N	Number of runs n_t	Real time from start to target		Number of replications (10^7)	
		Mean value	$\sqrt{\text{var}}$	Mean value	$\sqrt{\text{var}}$
1000	120	900	$+1380-542$	1.2	$+3.1-0.9$
2000	120	530	$+880-330$	1.4	$+3.6-1.0$
3000	1199	400	$+670-250$	1.6	$+4.4-1.2$
10 000	120	190	$+230-100$	2.3	$+5.3-1.6$
30 000	63	110	$+97-52$	3.6	$+6.7-2.3$
100 000	18	62	$+50-28$	–	–

Simulations were performed with an algorithm introduced by Gillespie (1976, 1977a, 1977b). The time unit is here undefined. A mutation rate of $p = 0.001$ per site and replication was used. The mean and standard deviation were calculated under the assumption if a log-normal distribution that fits well the data of the simulations.
[a]The structures S_0 and S_τ were used in the optimization:

S_0: ((·(((((((((((((............ (((....)))......))))))·)))))))·)) ... (((......)))

S_τ: ((((((... ((((........))))·(((((.......))))).....(((((.......)))))·)))))).…

The secondary structures are shown in parentheses representation (see, e.g., Schuster, 2006). Every unpaired nucleotide is denoted by a dot, every base pair corresponds to an opening and a closing parenthesis in mathematical notation. The distance between two structures, $d_S(S_k, S_j)$, is computed as the Hamming distance between the two parentheses notation.

under identical conditions.[18] Individual trajectories show enormous scatter in the real time or the number of replications required to reach the target. The mean values and the standard deviations were obtained from statistics of trajectories under the assumption of a log-normal distribution. Despite the scatter three features are seen unambiguously detectable:

(i) A recognizable fraction of trajectories leads to extinction only at very small population sizes, $N < 25$. In larger populations the target is reached with probabilities of measure 1.
(ii) The time to target decreases with increasing population size.
(iii) The number of replications required to reach target increases with population size.

Combining items (ii) and (iii) allows for a clear conclusion concerning time and material requirements of the optimization process: Fast optimization requires large populations whereas economic use of material suggests working with small population sizes just sufficiently large to avoid extinction.

[18] Identical means here that everything was kept constant except the seeds for the random number generators.

CONCLUDING REMARKS

The results on replicons and their evolution reported here are recapitulated in terms of a comprehensive model for evolution considered at the molecular level, which was introduced ten years ago (Schuster, 1997a, 1997b). In most previous models phenotypes were considered only in terms of parameters contained in the kinetic equations and therefore an attempt to include phenotypes as integral parts of the model was made. Mutation and recombination act on genotypes whereas the target of selection, the fitness, is a property of phenotypes. The relations between genotypes and phenotypes are thus an intrinsic part of evolution and no theory can be complete without considering them. The complex process of evolution is partitioned into three simpler phenomena (Figure 1.16): (i) biochemical kinetics, (ii) migration of populations, and (iii) genotype–phenotype mapping. Conventional biochemical kinetics as well as replicator dynamics including quasispecies

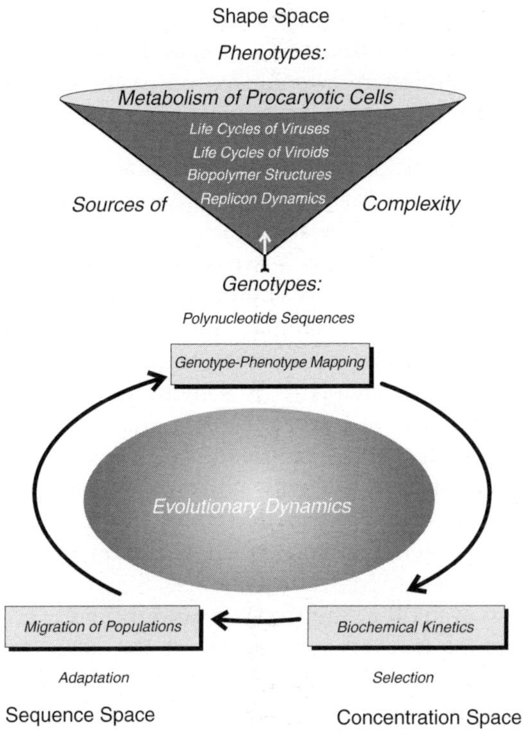

FIGURE 1.16 A comprehensive model of molecular evolution. The highly complex process of biological evolution is partitioned into three simpler phenomena: (i) biochemical kinetics, (ii) migration of populations, and (iii) genotype–phenotype mapping. Biochemical kinetics describes how optimal genotypes with optimal genes are chosen from a given reservoir by natural (or artificial) selection. The basis of population genetics is replication, mutation, and recombination mostly modeled by kinetic differential equations. In essence, kinetics is concerned with selection and other evolutionary phenomena occurring on short time-scales. Population support dynamics describes how the genetic reservoirs change when populations migrate in the huge space of all possible genotypes. Issues are the internal structure of populations and the mechanisms by which the regions of high fitness are found in sequence or genotype space. Support dynamics is dealing with the long-time phenomena of evolution, for example, with optimization and adaptation to changes in the environment. Genotype–phenotype mapping represents a core problem of evolutionary thinking since the dichotomy between genotypes and phenotypes is the basis of Darwin's principle of variation and selection: Variations and their results are uncorrelated in the sense that a mutation yielding a fitter phenotype does not occur more frequently because of the increase in fitness.

theory are modeled by differential equation and therefore miss all stochastic aspects. In the current model kinetics is extended by two more aspects: (i) population support dynamics describing the migration of populations through sequence space and (ii) genotype–phenotype mapping providing the source of the parameters for biochemical kinetics. In general, phenotypes and their formation from genotypes are so complex that they cannot be handled appropriately. In reactions of simple replicons and test-tube evolution of RNA, however, the phenotypes are molecular structures. Then, genotype and phenotype are two features of the same molecule. In these simplest known cases the relations between genotypes and phenotypes boil down to the mapping of RNA sequences onto structures. Folding RNA sequences into structures can be considered explicitly provided a coarse-grained version of structure, the secondary structure, is used (Schuster, 2006). This RNA model is self-contained in the sense that it is based on the rules of RNA secondary structure formation, the kinetics of replication and mutation as well as the structure of sequence space, and it needs no further inputs. The three processes shown in Figure 1.16 are indeed connected by a cyclic mutual dependence in which each process is driven by the previous one in the cycle and provides the input for the next one: (i) folding sequences into structures yields the input for biochemical kinetics, (ii) biochemical kinetics describes the arrival of new genotypes through mutation and the disappearance of old ones through selection, and determines thereby how and where the population migrates, and (iii) migration of the population in sequence space eventually creates the new genotypes that are to be mapped into phenotypes thereby completing the cycle. The model of evolutionary dynamics has been applied to interpret the experimental data on molecular evolution and it was implemented for computer simulations of neutral evolution and RNA optimization in the flow reactor (Huynen *et al.*, 1996; Fontana and Schuster, 1998a, 1998b). Computer simulations allow to follow the optimization process at the molecular level in full detail. What is still needed is a comprehensive mathematical description combining the three processes.

The work with RNA replicons has had a pioneering character. Both the experimental approach to evolution in the laboratory and the development of a theory of evolution are much simpler for RNA than in case of proteins or viruses. On the other hand, genotype and phenotype are more closely linked in RNA than in any other system. The next logical step in theory and experiment consists of the development of a coupled RNA–protein system that makes use of both replication and translation. This achieves the effective decoupling of genotype and phenotype that is characteristic for all living organisms: RNA is the genotype, protein the phenotype and thus, genotype and phenotype are no longer housed in the same molecule. The development of a theory of evolution in the RNA–protein world requires, in addition, an understanding of the notoriously difficult sequence–structure relations in proteins. Issues that are becoming an integral part of research on early replicons are (i) primitive forms of metabolism that can provide the material required for replication (and translation) and (ii) spatial isolation in vesicles or some amphiphilic material that forms compartments.

Molecular evolution experiments with RNA molecules and the accompanying theoretical descriptions made three important contributions to evolutionary biology:

1. The role of replicative units in the evolutionary process has been clarified, the conditions for the occurrence of error thresholds have been laid down, and the role of neutrality has been elucidated.
2. The Darwinian principle of (natural) selection has shown to be no privilege of cellular life, since it is valid also in serial transfer experiments, flow-reactors, and other laboratory assays such as SELEX.
3. Evolution in molecular systems is faster than organismic evolution by many orders of magnitude and thus enables researchers to observe optimization, adaptation, and other evolutionary phenomena on easily accessible time-scales, i.e. within days or weeks.

The third issue made selection and adaptation subjects of laboratory investigations. In all these model systems the coupling between different replicons is weak: in the simplest case there is merely competition for common resources, for example, the raw materials for replication. With more realistic chemical reaction mechanisms a sometimes substantial fraction of the replicons is unavailable as long as templates are contained in complexes. None of these systems, however, comes close to the strong interactions and interdependencies characteristic for co-evolution or real ecosystems. Molecular models for co-evolution are still in their infancy and more experimental work is needed to set the stage for testing the theoretical models available at present.

Virus life cycles represent the next logical step in increasing complexity of genotype–phenotype interactions. The pioneering paper by Weissmann (1974) has shown the way to proceed in the case of an RNA phage that is among the most simple candidates, and indeed the development of phages in bacterial cells can be modeled with sufficient accuracy. A lot of elegant work has been done since then and a wealth of data and models is available but many more experiments and more detailed theories are necessary to decipher the complex interactions of host–pathogen systems on the molecular level.

ACKNOWLEDGMENTS

The work reported here was supported financially by the Austrian Fonds zur Förderung der Wissenschaftlichen Forschung, (Projects No. 11065-CHE, 12591-INF, 13093-GEN, and 14898-MAT), by the European Commission (Project No.PL970189), and by the Santa Fe Institute.

REFERENCES

Altmeyer, S. and McCaskill, J.S. (2001) Error threshold for spatially resolved evolution in the quasispecies model. *Phys. Rev. Lett.* **86**, 5819–5822.

Alves, D. and Fontanari, J.F. (1996) Population genetics approach to the quasispecies model. *Phys. Rev. E* **54**, 4048–4053.

Alves, D. and Fontanari, J.F. (1998) Error threshold in finite populations. *Phys. Rev. E* **57**, 7008–7013.

Anderson, J.P., Daifuku, R. and Loeb, L.A. (2004) Viral error catastrophe by mutagenic nucleosides. *Annu. Rev. Microbiol.* **58**, 183–205.

Baake, E. and Wagner, H. (2001) Mutation-selection models solved exactly with methods of statistical mechanics. *Genet. Res. Camb.* **78**, 93–117.

Bachmann, P.A., Luisi, P.L. and Lang, J. (1992) Autocatalytic self-replicating micelles as models for prebiotic structures. *Nature* **357**, 57–59.

Ban, N., Nissen, P., Hansen, J., Moore, P.B. and Steitz, T.A. (2000) The complete atomic structure of the large ribosomal subunit at 2.4Å resolution. *Science* **289**, 905–920.

Bartel, D.P. and Szostak, J.W. (1993) Isolation of new ribozymes from a large pool of random sequences. *Science* **261**, 1411–1418.

Beaudry, A.A. and Joyce, G.F. (1992) Directed evolution of an RNA enzyme. *Science* **257**, 635–641.

Biebricher, C.K. and Eigen, M. (1988) Kinetics of RNA replication by Qβ replicase. In: *RNA Genetics. RNA Directed Virus Replication* (E. Domingo, J.J. Holland and P. Ahlquist, eds), Vol. I, pp. 1–21. Boca Raton, FL: CRC Press.

Brasier, M.D., Green, O.R., Jephcoat, A.B., Kleppe, A.K., Van Kranendonk, M.J., Lindsay, J.F. *et al.* (2002) Questioning the evidence for Earth's oldest fossils. *Nature* **416**, 76–81.

Breaker, R.R. and Joyce, G.F. (1994) Emergence of a replicating species from an *in vitro* RNA evolution reaction. *Proc. Natl Acad. Sci. USA* **91**, 6093–6097.

Bull, J.J., Sanjuán, R. and Wilke, C.O. (2007) Theory of lethal mutagensis for viruses. *J. Virol.* **81**, 2930–2939.

Campos, P.R.A. (2002) Error threshold transition in the random-energy model. *Phys. Rev. E* **66**, 062904.

Cate, J.H., Gooding, A.R., Podell, E., Zhou, K., Golden, B.L., Kundrot, C.E. *et al.* (1996) Crystal structure of a group I ribozyme domain: Principles of RNA packing. *Science* **273**, 1678–1685.

Cavalier-Smith, T. (2001) Obcells as proto-organisms: Membrane heredity, lithophosphorylation, and the origins of the genetic code, the first cells, and photosynthesis. *J. Mol. Evol.* **53**, 555–595.

Cech, T.R. (1983) RNA splicing: Three themes with variations. *Cell* **34**, 713–716.

Cech, T.R. (1986) RNA as an enzyme. *Sci. Am.* **255**, 76–84.

Cech, T.R. (1990) Self-splicing of group I introns. *Annu. Rev. Biochem.* **59**, 543–568.

Comas, I., Moya, A. and González-Candelas, F. (2005) Validating viral quasispecies with digital organisms: A re-examiniation of the critical mutation rate. *BMC Evol. Biol.* **5**, 1–10.

Cuenoud, B. and Szostak, J.W. (1995) A DNA metalloenzyme with DNA ligase activity. *Nature* **375**, 611–614.

Demetrius, L., Schuster, P. and Sigmund, K. (1985) Polynucleotide evolution and branching processes. *Bull. Math. Biol.* **47**, 239–262.

Derrida, B. and Peliti, L. (1991) Evolution in a flat fitness landscape. *Bull. Math. Biol.* **53**, 355–382.

Domingo, E. (1996) Biological significance of viral quasispecies. *Viral Hepatitis Rev.* **2**, 247–261.

Domingo, E. (2002) Quasispecies theory in virology. *J. Virol.* **76**, 463–465.

Domingo, E. (2005) Virus entry into error catastrophe as a new antiviral strategy. *Virus Res.* **107**, 115–228.

Domingo, E. and Holland, J.J. (1997) RNA virus mutations and fitness for survival. *Annu. Rev. Microbiol.* **51**, 151–178.

Domingo, E., Sabo, D., Taniguchi, T. and Weissmann, C. (1998) Nucleotide sequence heterogeneity of an RNA phage. *Cell* **13**, 735–744.

Drake, J.W. (1993) Rates of spontaneous mutation among RNA viruses. *Proc. Natl Acad. Sci. USA* **90**, 4171–4175.

Drake, J.W. and Holland, J.J. (1999) Mutation rates among RNA viruses. *Proc. Natl Acad. Sci. USA* **96**, 13910–13913.

Eigen, M. (1971) Selforganization of matter and the evolution of macromolecules. *Naturwissenschaften* **58**, 465–523.

Eigen, M. (2002) Error catastrophe and antiviral strategy. *Proc. Natl Acad. Sci. USA* **99**, 13374–13376.

Eigen, M. and Schuster, P. (1977) The hypercycle. A principle of natural self-organization. Part A: Emergence of the hypercycle. *Naturwissenschaften* **64**, 541–565.

Eigen, M. and Schuster, P. (1978a) The hypercycle. A principle of natural self-organization. Part B: The abstract hypercycle. *Naturwissenschaften* **65**, 7–41.

Eigen, M. and Schuster, P. (1978b) The hypercycle. A principle of natural self-organization. Part C: The realistic hypercycle. *Naturwissenschaften* **65**, 341–369.

Eigen, M. and Schuster, P. (1982) Stages of emerging life—Five principles of early organization. *J. Mol. Evol.* **19**, 47–61.

Eigen, M., McCaskill, J. and Schuster, P. (1989) The molecular quasispecies. *Adv. Chem. Phys.* **75**, 149–263.

Eigen, M., Biebricher, C.K., Gebinoga, M. and Gardiner, W.C., Jr. (1991) The hypercycle. Coupling of RNA and protein biosynthesis in the infection cycle of an RNA bacteriophage. *Biochemistry* **30**, 11005–11018.

Ekland, E.H. and Bartel, D.P. (1996) RNA-catalysed RNA polymerization 54 using nucleoside triphosphates. *Nature* **382**, 373–376.

Ekland, E.H., Szostak, J.W. and Bartel, D.P. (1995) Structurally complex and highly active RNA ligases derived from random RNA sequences. *Science* **269**, 364–370.

Eschenmoser, A. (1993) Hexose nucleic acids. *Pure Appl. Chem.* **65**, 1179–1188.

Fahy, E., Kwoh, D.Y. and Gingeras, T.R. (1991) Self-sustained sequence replication (3SR): An isothermal transcription-based amplification system alternative to PCR. *PCR Methods Appl.* **1**, 25–33.

Ferré-D'Amaré, A.R., Zhou, K. and Doudna, J.A. (1998) Crystal structure of a hepatitis delta virus ribozyme. *Nature* **395**, 567–574.

Fontana, W. and Schuster, P. (1987) A computer model of evolutionary optimization. *Biophys. Chem.* **26**, 123–147.

Fontana, W. and Schuster, P. (1998a) Continuity in evolution. On the nature of transitions. *Science* **280**, 1451–1455.

Fontana, W. and Schuster, P. (1998b) Shaping space. The possible and the attainable in RNA genotype-phenotype mapping. *J. Theor. Biol.* **194**, 491–515.

Fontana, W., Schnabl, W. and Schuster, P. (1989) Physical aspects of evolutionary optimization and adaptation. *Phys. Rev. A* **40**, 3301–3321.

Fontana, W., Konings, D.A.M., Stadler, P.F. and Schuster, P. (1993) Statistics of RNA secondary structures. *Biopolymers* **33**, 1389–1404.

Fox, S.W. and Dose, H. (1977) *Molecular Evolution and the Origin of Life*. New York: Academic Press.

Frank, F.C. (1953) On spontaneous asymmetric synthesis. *Biochim. Biophys. Acta* **11**, 459–463.

Franz, S. and Peliti, L. (1997) Error threshold in simple landscapes. *J. Phys. A: Math. Gen.* **30**, 4481–4487.

Franz, S., Peliti, L. and Sellitto, M. (1993) Error threshold in simple landscapes. *J. Phys. A: Math. Gen.* **26**, L1195–L1199.

Gánti, T. (1997) Biogenesis itself. *J. Theor. Biol.* **187**, 583–593.

Gardiner, C.W. (2004) *Handbook of Stochastic Methods for Physics, Chemistry and the Natural Sciences, Springer Series in Synergetics*, 3rd edn. Berlin: Springer-Verlag.

Gesteland, R.F. and Atkins, J.F. (eds) (1993). *The RNA World*. Plainview, NY: Cold Spring Harbor Laboratory Press.

Gesteland, R.F., Thomas, R.C. and Atkins, J.F. (2006). *The RNA World*, 3rd edn. Plainview, NY: Cold Spring Harbor Laboratory Press.

Gilbert, W. (1986) The RNA world. *Nature* **319**, 618.

Gillespie, D.T. (1976) A general method for numerically simulating the stochastic time evolution of coupled chemical reactions. *J. Comp. Phys.* **22**, 403–434.

Gillespie, D.T. (1977a) Concerning the validity of the stochastic approach to chemical kinetics. *J. Stat. Phys.* **16**, 311–318.

Gillespie, D.T. (1977b) Exact stochastic simulation of coupled chemical reactions. *J. Phys. Chem.* **81**, 2340–2361.

Green, R. and Noller, H.F. (1997) Ribosomes and translation. *Annu. Rev. Biochem.* **66**, 679–716.

Guerrier-Takada, C., Gardiner, K., Marsh, T., Pace, N. and Altman, S. (1983) The RNA moiety of ribonuclease P is the catalytic subunit of the enzyme. *Cell* **35**, 849–857.

Haken, H. (1983a) *Laser Theory*. New York: Springer-Verlag.

Haken, H. (1983b) *Synergetics. An Introduction. Springer Series in Synergetics*, 3rd edn. Berlin: Springer-Verlag.

Happel, R. and Stadler, P.F. (1999) Autocatalytic replication in a CSTR and constant organization. *J. Math. Biol.* **38**, 422–434.

Hofbauer, J. (1981) On the occurrence of limit cycles in the Volterra-Lotka differential equation. *Nonlin. Anal.* **5**, 1003–1007.

Hofbauer, J. and Sigmund, K. (1998) *Dynamical Systems and the Theory of Evolution*. Cambridge: Cambridge University Press.

Hofbauer, J., Schuster, P. and Sigmund, K. (1981) Competition and cooperation in catalytic selfreplication. *J. Math. Biol.* **11**, 155–168.

Holmes, E.C. and Moya, A. (2002) Is the quasispecies concept relevant to RNA viruses? *J. Virol.* **76**, 460–462.

Huynen, M.A. (1996) Exploring phenotype space through neutral evolution. *J. Mol. Evol.* **43**, 165–169.

Huynen, M.A., Stadler, P.F. and Fontana, W. (1996) Smoothness within ruggedness. The role of neutrality in adaptation. *Proc. Natl Acad. Sci. USA* **93**, 397–401.

Jenkins, G.M., Worobey, M., Woelk, C.H. and Holmes, E.C. (2001) Evidence for the non-quasispecies evolution of RNA viruses. *Mol. Biol. Evol.* **18**, 987–994.

Jenne, A. and Famulok, M. (1998) A novel ribozyme with ester transferase activity. *Chem. Biol.* **5**, 23–34.

Jiang, L., Suri, A.K., Fiala, R. and Patel, D.J. (1997) Saccharide-RNA recognition in an aminoglycoside antibiotic-RNA aptamer complex. *Chem. Biol.* **4**, 35–50.

Jones, B.L., Enns, R.H. and Rangnekar, S.S. (1976) On the theory of selection of coupled macromolecular systems. *Bull. Math. Biol.* **38**, 15–28.

Joyce, G.F. (1991) The rise and fall of the RNA world. *The New Biologist* **3**, 399–407.

Kauffman, S.A. (1993) *The Origins of Order. Self-Organization and Selection in Evolution*. New York: Oxford University Press.

Kimura, M. (1983) *The Neutral Theory of Molecular Evolution*. Cambridge: Cambridge University Press.

Klipp, E., Herwig, R., Kowald, A., Wieling, C. and Lehrach, H. (2005) *Systems Biology in Practice. Concepts, Implementation, and Application*. Weinheim: Wiley-VCH.

Klussmann, S. (ed.) (2006) The Aptamer Handbook. *Functional Oligonucleotides and Their Applications*. Weinheim: Wiley-VCH.

Kramer, F.R., Mills, D.R., Cole, P.E., Nishihara, T. and Spiegelman, S. (1974) Evolution in vitro: Sequence and phenotype of a mutant RNA resitant to ethidium bromide. *J. Mol. Biol.* **89**, 719–736.

Kupczok, A. and Dittrich, P. (2006) Determinants of simulated RNA evolution. *J. Theor. Biol.* **238**, 726–735.

Lee, D.H., Granja, J.R., Martinez, J.A., Severin, K. and Ghadiri, M.R. (1996) A self-replicating peptide. *Nature* **382**, 525–528.

Lee, D.H., Severin, K., Yokobayashi, Y. and Ghadiri, M.R. (1997) Emergence of symbiosis in peptide self-replication through a hypercyclic network. *Nature* **390**, 591–594.

Leung, D.W., Chen, E. and Goeddel, D.V. (1989) A method for random mutagenesis of a defined DNA segment using a modified polymerase chain reaction. *Technique* **1**, 11–15.

Leuthäusser, I. (1987) Statistical mechanics of Eigen's evolution model. *J. Statist. Phys.* **48**, 343–360.

Levitan, B. and Kauffman, S.A. (1995) Adaptive walks with noisy fitness measurements. *Mol. Diversity* **1**, 53–68.

Li, T. and Nicolaou, K.C. (1994) Chemical self-replication of palindromic duplex DNA. *Nature* **369**, 218–221.

Lohse, P.A. and Szostak, J.W. (1996) Ribozyme-catalyzed amino-acid transfer reactions. *Nature* **381**, 442–444.

Lorsch, J.R. and Szostak, J.W. (1994) In vitro evolution of new ribozymes with polynucleotide kinase activity. *Nature* **371**, 31–36.

Lorsch, J.R. and Szostak, J.W. (1995) Kinetic and thermodynamic characterization of the reaction catalyzed by a polynucleotide kinase ribozyme. *Biochemistry* **33**, 15315–15327.

Luisi, P.L. (ed.) (2004) Prebiotic chemistry and early evolution. *Orig. Life Evol. Biosph.* **34**.

Luisi, P.L., Walde, P. and Oberholzer, T. (1994) Enzymatic RNA synthesis in self-reproducing vesicles: An approach to the construction of a minimal synthetic cell. *Ber. Bunsenges. Phys. Chem.* **98**, 1160–1165.

Marques, J.T., Devosse, T., Wang, D., Zamanian-Daryoush, M., Serbinowski, P., Hartmann, R. *et al.* (2006) A structural basis for discriminating between self and nonself double-stranded RNAs in mammalian cells. *Nat. Biotechnol.* **24**, 559–565.

Martinez, M.A., Vartanian, J.P. and Wain-Hobson, S. (1994) Hypermutagenesis of RNA using human immunodeficiency virus type 1 reverse transcriptase and biased dNTP concentrations. *Proc. Natl Acad. Sci. USA* **91**, 11787–11791.

Mason, S.F. (1991) *Chemical Evolution. Origin of the Elements, Molecules, and Living Systems*. Oxford: Clarendon Press.

Mattick, J.S. (2004) RNA regulation: A new genetics? *Nat. Rev.Genet.* **5**, 316–323.

Maynard Smith, J. and Szathmáry, E. (1995) *The Major Transitions in Evolution*. Oxford: W.H. Freeman.

McCaskill, J. (1984) A localization threshold for macromolecular quasispecies from continuously distributed replication rates. *J. Chem. Phys.* **80**, 5194–5202.

McCaskill, J.S. (1997) Spatially resolved in vitro molecular ecology. *Biophys. Chem.* **66**, 145–158.

McManus, M.T. and Sharp, P.A. (2002) Gene silencing in mammals by small interfering RNAS. *Nat. Rev.Genet.* **3**, 737–747.

Mills, D.R., Peterson, R.L. and Spiegelman, S. (1967) An extracellular Darwinian experiment with a self-duplicating nucleic acid molecule. *Proc. Natl Acad. Sci. USA* **58**, 217–224.

Moya, A., Elena, S.F., Brancho, A., Miralles, R. and Barrio, E. (2000) The evolution of RNA viruses: A population genetics view. *Proc. Natl Acad. Sci. USA,* **97**, 6967–6973.

Moya, A., Holmes, E.C. and González-Candelas, F. (2004) The population genetics and evolutionary epidemiology RNA viruses. *Nat. Rev. Microbiol.* **2**, 279–288.

Mullis, K.B. (1990) The unusual origin of the polymerase chain reaction. *Sci. Am.* **262**, 36–43.

Nicolis, G. and Prigogine, I. (1977) *Self-Organization in Nonequilibrium Systems. From Dissipative Structures to Order through Fluctuations*. New York: John Wiley & Sons.

Nissen, P., Hansen, J., Ban, N., Moore, P.B. and Steitz, T.A. (2000) The structural basis of ribosome activity in peptide bond synthesis. *Science* **289**, 920–930.

Noller, H.F., Hoffarth, V. and Zimniak, L. (1992) Unusual resistance of peptidyl transferase to protein extraction procedures. *Science* **256**, 1416–1419.

Nowak, M. and Schuster, P. (1989) Error thresholds of replication in finite populations. Mutation frequencies and the onset of Muller's ratchet. *J. Theor. Biol.* **137**, 375–395.

Nowick, J.S., Feng, Q., Ballester, T. and Rebek, J., Jr. (1991) Kinetic studies and modeling of a self-replicating system. *J. Am. Chem. Soc.* **113**, 8831–8839.

Orgel, L.E. (1986) RNA catalysis and the origin of life. *J. Theor. Biol.* **123**, 127–149.

Orgel, L.E. (1987) Evolution of the genetic apparatus. A review. *Cold Spring Harbor Symp. Quant. Biol.* **52**, 9–16.

Orgel, L.E. (1992) Molecular replication. *Nature* **358**, 203–209.

Orgel, L.E. (2003) Some consequences of the RNA world hypothesis. *Orig. Life Evol. Biosph.* **33**, 211–218.

Palsson, B.Ø. (2006) *Systems Biology. Properties of Reconstructed Networks*. New York: Cambridge University Press.

Pastor-Satorras, R. and Solé, R.V. (2001) Reaction-diffusion model of quasispecies dynamics. *Phys. Rev. E* **64**, 051909.

Pflug, H.D. and Jaeschke-Boyer, H. (1979) Combined structural and chemical analysis of 3.800-Myr-old microfossils. *Nature* **280**, 483–486.

Pley, H., Flaherty, K. and McKay, D. (1994) Three-dimensional structures of a hammerhead ribozyme. *Nature* **372**, 68–74.

Prudent, J.R., Uno, T. and Schultz, P.G. (1994) Expanding the scope of RNA catalysis. *Science* **264**, 1924–1927.

Rasmussen, S., Chen, L., Nilsson, M. and Abe, S. (2003) Bridging nonliving to living matter. *Artif. Life* **9**, 269–316.

Rasmussen, S., Chen, L., Stadler, B.M.R. and Stadler, P.F. (2004a) Proto-organism kinetics: Evolutionary dynamics of lipid aggregates with genes and metabolism. *Orig. Life Evol. Biosph.* **34**, 171–180.

Rasmussen, S., Chen, L., Deamer, D., Krakauer, D.C., Packard, N.H., Stadler, P.F. and Bedau, M.A. (2004b) Transitions from nonliving to living matter. *Science* **303**, 963–965.

Reidys, C.M. and Stadler, P.F. (2001) Neutrality in fitness landscapes. *Appl. Math. Comput.* **117**, 321–350.

Reidys, C.M., Stadler, P.F. and Schuster, P. (1997) Generic properties of combinatory maps: Neural networks of RNA secondary structures. *Bull. Math. Biol.* **59**, 339–397.

Reidys, C., Forst, C. and Schuster, P. (2001) Replication and mutation on neutral networks. *Bull. Math. Biol.* **63**, 57–94.

Saakian, D.B. and Hu, C.-K. (2006) Exact solution of the Eigen model with general fitness functions and degradation rates. *Proc. Natl Acad. Sci. USA,* **103**, 4935–4939.

Saakian, D.B., Muños, E., Hu, C.-K. and Deem, M.W. (2006) Quasispecies theory for multiple-peak fitness landscapes. *Phys. Rev. E* **73**, 041913.

Schidlowski, M. (1988) A 3.800-million-year isotope record of life from carbon in sedimentary rocks. *Nature* **333**, 313–318.

Schlögl, F. (1972) Chemical reaction models for non-equilibrium phase transitions. *Z. f. Physik A* **253**, 147–161.

Schopf, J.W. (1993) Microfossils of the early archean apex chert: New evidence of the antiquity of life. *Science* **260**, 640–646.

Schopf, J.W. (2006) Fossil evidence of Archaean life. *Philos. Trans. R. Soc. B,* **361**, 869–885.

Schuster, P. (1997a) Genotypes with phenotypes: Adventures in an RNA toy world. *Biophys. Chem.* **66**, 75–110.

Schuster, P. (1997b) Landscapes and molecular evolution. *Physica D* **107**, 351–365.

Schuster, P. (2003) Molecular insight into the evolution of phenotypes. In: *Evolutionary Dynamics—Exploring the Interplay of Accident, Selection, Neutrality, and Function* (J.P. Crutchfield and P. Schuster, eds), pp. 163–215. New York: Oxford University Press.

Schuster, P. (2006) Prediction of RNA secondary structures: From theory to models and real molecules. *Rep. Prog. Phys.* **69**, 1419–1477.

Schuster, P. and Sigmund, K. (1983) Replicator dynamics. *J. Theor. Biol.* **100**, 533–538.

Schuster, P. and Sigmund, K. (1985) Dynamics of evolutionary optimization. *Ber. Bunsenges. Phys.Chem.* **89**, 668–682.

Schuster, P. and Swetina, J. (1988) Stationary mutant distribution and evolutionary optimization. *Bull. Math. Biol.* **50**, 635–660.

Schuster, P., Fontana, W., Stadler, P.F. and Hofacker, I.L. (1994) From sequences to shapes and back: A case study in RNA secondary structures. *Proc. R. Soc. Lond. B* **255**, 279–284.

Schwartz, A.W. (1997) Speculation on the RNA precursor problem. *J. Theor. Biol.* **187**, 523–527.

Scott, W.G., Finch, J.T. and Klug, A. (1995) The crystal structure of an all-RNA hammerhead ribozyme: A proposed mechanism for RNA catalytic cleavage. *Cell* **81**, 991–1002.

Seelig, B. and Jäschke, A. (1999) A small catalytic RNA motif with Diels-Alder activity. *Chem. Biol.* **6**, 167–176.

Segel, L.A. and Slemrod, M. (1989) The quasi-steady state assumption: A case study in perturbation. *SIAM Rev.* **31**, 446–477.

Seneta, E. (1981) *Non-negative Matrices and Markov Chains*, 2nd edn. New York: Springer-Verlag.

Serganov, A., Keiper, S., Malinina, L., Terechko, V., Skripkin, E., Höbartner, C. *et al.* (2005) Structural basis for Diels-Alder ribozyme-catalyzed carbon-carbon bond formation. *Nat. Struct. Mol. Biol.* **12**, 218–224.

Severin, K., Lee, D.H., Granja, J.R., Martinez, J.A. and Ghadiri, M.R. (1997) Peptide self-replication via template directed ligation. *Chemistry* **3**, 1017–1024.

Soai, K., Shibata, T., Morioka, H. and Choji, K. (1995) Asymmetric autocatalysis and amplification of enantiomeric excess of a chiral molecule. *Nature* **378**, 767–768.

Spiegelman, S. (1971) An approach to the experimental analysis of precellular evolution. *Q. Rev. Biophys.* **4**, 213–253.

Stadler, B.M.R. and Stadler, P.F. (2003) Molecular replicator dynamics. *Adv. Complex Syst.* **6**, 47–77.

Stadler, B.M.R., Stadler, P.F. and Schuster, P. (2000) Dynamics of autocatalytic replicator networks based on higher order ligation reactions. *Bull. Math. Biol.* **62**, 1061–1086.

Stadler, P.F. (1991) Complementary replication. *Math. Biosci.* **107**, 83–109.

Stadler, P.F. and Stadler, B.M.R. (2007) Replicator dynamics in protocells. In: *Protocells: Bridging Nonliving and Living Matter* (S. Rasmussen, M. Bedau, L. Chen, D. Deamer, D.C. Krakauer, N.H. Packard and P.F. Stadler, eds) MIT Press. in press.

Steitz, T.A. and Moore, P.B. (2003) RNA, the first molecular catalyst: The ribosome is a ribozyme. *Trends Biochem. Sci.* **28**, 411–418.

Swetina, J. and Schuster, P. (1982) Self-replication with errors—A model for polynucleotide replication. *Biophys. Chem.* **16**, 329–345.

Szathmáry, E. and Gladkih, I. (1989) Sub-exponential growth and coexistence of non-enzymatically replicating templates. *J. Theor. Biol.* **138**, 55–58.

Szathmáry, E. and Maynard Smith, J. (1997) From replicators to reproducers: The first major transition leading to life. *J. Theor. Biol.* **187**, 555–571.

Takeuchi, N. and Hogeweg, P. (2007) Error-thresholds exist in fitness landscapes with lethal mutants. *BMC Evol. Biol.* **7**, 1–11.

Takeuchi, N., Poorthuis, P.H. and Hogeweg, P. (2005) Phenotypic error-threshold: Additivity and epistasis in RNA evolution. *BMC Evol. Biol.* **5**, 1–9.

Tannenbaum, E., Deeds, E.J. and Shakhnovich, E.I. (2004) Semiconservative replication in the quasispecies model. *Phys. Rev. E* **69**, 061916.

Tannenbaum, E., Shirley, J.L. and Shakhnovich, E.I. (2006) Semiconservative quasispecies equations for polysomic genomes: The haploid case. *J. Theor. Biol.* **241**, 791–805.

Tarazona, P. (1992) Error-tresholds for molecular quasispecies as phase transitions: From simple landscapes to spinglass models. *Phys. Rev. A* **45**, 6038–6050.

Thompson, C.J. and McBride, J.L. (1974) On Eigen's theory of the self-organization of matter and the evolution of biological macromolecules. *Math. Biosci.* **21**, 127–142.

Tjivikua, T., Ballester, P. and Rebek, J., Jr. (1990) A self-replicating system. *J. Am. Chem. Soc.* **112**, 1249–1250.

Tsukiji, S., Pattnaik, S.B. and Suga, H. (2004) Reduction of an aldehyde by a NADH/Zn^2+ -dependent redox active ribozyme. *J.Am.Chem. Soc.* **126**, 5044–5045.

Turing, A.M. (1952) A chemical basis of morphogenesis. *Philos. Trans. R. Soc. Lond. B* **337**, 37–72.

Uhlenbeck, O.C. (1987) A small catalytic oligoribonucleotide. *Nature* **328**, 596–600.

Valandro, L., Salvato, B., Caimmi, R. and Galzigna, L. (2000) Isomorphism of quasispecies and peroclation models. *J. Theor. Biol.* **202**, 187–194.

Varga, S. and Szathmáry, E. (1997) An extremum principle for parabolic competition. *Bull. Math. Biol.* **59**, 1145–1154.

von Kiedrowski, G. (1986) A self-replicating hexadeoxynucleotide. *Angew. Chem. Int. Ed. Engl.* **25**, 932–935.

von Kiedrowski, G. (1993) Minimal replicator theory I: Parabolic versus exponential growth. In: *Bioorganic Chemistry Frontiers*, Vol. 3, pp. 115–146. Berlin and Heidelberg: Springer-Verlag.

Wagner, G.P. and Krall, P. (1993) What is the difference between models of error thresholds and Muller's ratchet? *J. Math. Biol.* **32**, 33–44.

Watts, A. and Schwarz, G. (eds) (1997) Evolutionary Biotechnology—From Theory to Experiment. Special *Issue of Biophys. Chem.* **66**, 67–284.

Wecker, M., Smith, D. and Gold, L. (1996) In vitro selection of a novel catalytic RNA: Characterization of a sulfur alkylation reaction and interaction with a small peptide. *RNA* **2**, 982–994.

Weissmann, C. (1974) The making of a phage. *FEBS Lett.* (*Suppl.*) **40**, S10–sS12.

Wiehe, T. (1997) Model dependency of error thresholds: The role of fitness functions and contrasts between the finite and the infinite sites models. *Genet. Res. Camb.* **69**, 127–136.

Wilke, C.O. (2001) Selection for fitness versus selection for robustness in RNA secondary structure folding. *Evolution* **55**, 2412–2420.

Wilke, C.O. (2005) Quasispecies theory in the context of population genetics. *BMC Evol. Biol.* **5**, 1–8.

Wilke, C.O. and Ronnewinkel, C. (2001) Dynamic fitness lansdscapes: Expansions for small mutation rates. *Physica A* **290**, 475–490.

Wilke, C.O., Ronnewinkel, C. and Martinetz, T. (2001) Dynamic fitness lansdscapes in molecular evolution. *Phys. Rep.* **349**, 395–446.

Wills, P.R., Kauffman, S.A., Stadler, B.M. and Stadler, P.F. (1998) Selection dynamics in autocatalytic systems: Templates replicating through binary ligation. *Bull. Math. Biol.* **60**, 1073–1098.

Wilson, C. and Szostak, J.W. (1995) In vitro evolution of a self-alkylating ribozyme. *Nature* **374**, 777–782.

Wlotzka, B. and McCaskill, J.S. (1997) A molecular predator and its prey: Coupled isothermal amplification of nucleic acids. *Chem. Biol.* **4**, 25–33.

Wright, S. (1932). The roles of mutation, inbreeding, crossbreeeding and selection in evolution. In: D.F. Jones (ed.), *International Proceedings of the Sixth International Congress on Genetics*, Vol. 1, pp. 356–366.

Zhang, B. and Cech, T.R. (1997) Peptide bond formation by in vitro selected ribozymes. *Nature* **390**, 96–100.

Zhang, B. and Cech, T.R. (1998) Peptidyl-transferase ribozymes: Trans reactions, structural characterization and ribosomal RNA-like features. *Chem. Biol.* **5**, 539–553.

CHAPTER

2

Structure and Evolution of Viroids

Núria Duran-Vila, Santiago F. Elena, José-Antonio Daròs, and Ricardo Flores

ABSTRACT

Viroids are minimal RNA replicons composed by a single-stranded and highly structured circular small RNA able to infect plants and induce diseases. Viroids lack protein-coding capacity and are therefore parasites of their host transcription machinery. The small size, circularity, high G+C content, and presence of hammerhead ribozymes in some viroids, suggest their evolutionary origin in the RNA world. Phylogenetic reconstructions and structural and biological properties support a classification into two families—*Pospiviroidae* and *Avsunviroidae*—whose members replicate in the nucleus and chloroplast, respectively. Viroids may have a common origin with a class of small satellite RNAs with which they share structural similarities and a rolling circle replication mechanism involving hammerhead ribozymes. Inoculation with infectious viroid-cDNAs results in progenies readily accumulating genetic variation, with *Avsunviroidae* populations being more diverse than *Pospiviroidae* populations. Moreover, assuming that the fitness of a haplotype is determined by its ability to fold into the secondary structure of minimum free energy, the estimated per site deleterious mutation rate in the *Avsunviroidae* is 10-fold higher than in the *Pospiviroidae*. The dissimilar nuclear and chloroplastic RNA polymerases mediating replication in both families may influence their mutation rates, particularly when transcribing atypical RNA templates. Both families also differ in their structural robustness against mutation, with the *Pospiviroidae* rod-like structures being more robust than the *Avsunviroidae* branched structures (and the redundant variants of a specific viroid being more robust than their non-redundant counterparts). Chimeric viroids might have emerged from recombination between co-infecting viroids during transcription by a "jumping" RNA polymerase. Viroids polymorphic populations can be described by the quasispecies model of molecular evolution, and one of its main tenets (that a slow replicator outcompetes a faster one provided that the former is more robust against mutation) has been experimentally proven. Hosts play an important role in shaping the structure of viroid populations. Specific domains of the viroid secondary structure are responsible for symptom expression. Depending on their phylogenetic proximity, interactions between

Origin and Evolution of Viruses
ISBN 978-0-12-374153-0

co-infecting viroids may result in interference (cross-protection) or synergism, both of which may be governed by RNA silencing mechanisms that may have shaped viroid structure and evolution as well. Wild plant species serve as symptomless reservoirs of certain viroids, while their spread and persistence in cultivated species is associated with agricultural practices.

INTRODUCTION

Viroids are unique systems for the study of RNA structure, function, and evolution. They are the minimal RNA replicons characterized so far—their genome is 10-fold smaller than that of the smallest known viral RNA—and in a certain sense are at the frontier of life. Despite being exclusively composed by a single-stranded and highly structured circular RNA of only 246–401 nt (Figure 2.1), viroids contain sufficient information to infect some host plants, to manipulate their gene expression for producing progeny, and, as a consequence, to incite in most cases specific diseases (Diener, 2003). In striking contrast to viruses, which encode proteins that mediate their own replication and movement, viroids depend essentially on host factors for these purposes and can therefore be regarded as parasites of their host transcription machinery (Flores *et al.*, 2005; Daròs *et al.*, 2006).

THE ORIGIN OF VIROIDS: MOLECULAR FOSSILS FROM THE RNA WORLD

Characteristics of the Molecule that Make Them Good Candidates

Certain viroid properties, prominent among which are their small size (a requisite of primordial replicons), circularity (making unnecessary genomic tags for initiating or terminating replication), high G+C content (increasing the fidelity of primitive RNA polymerases) and, specially, the presence of hammerhead ribozymes in members of one of the two families (see below), support the notion of a very ancient evolutionary origin for viroids that may be independent from that of RNA viruses, with which no significant sequence similarity has been detected. In this scenario, viroid origin would go back to the RNA world postulated to have preceded the DNA- and protein-based world now dominant on Earth (Diener, 1989) (Chapter 1).

Hammerhead Ribozymes

Viroids replicate through an RNA-based rolling circle mechanism with three steps: (i) synthesis of longer-than-unit strands catalyzed by a host nuclear or chloroplastic RNA polymerase that reiteratively transcribes the infectious circular template, (ii) cleavage to unit-length mediated by a processing activity, and (iii) circularization resulting from the action of an RNA ligase or from self-ligation. Remarkably, the second step is mediated in some viroids by host enzymes and in others by hammerhead ribozymes embedded in their strands of both polarities.

The discovery 20 years ago of the hammerhead ribozyme in avocado sunblotch viroid (ASBVd) (Hutchins *et al.*, 1986) and in a viroid-like satellite RNA (see below) (Prody *et al.*, 1986), is a landmark in virology with major derivations on the replication and evolutionary origin of these subviral RNAs. The hammerhead ribozyme is a small RNA motif that at room temperature, neutral pH, and in the presence of a divalent metal ion (usually magnesium), self-cleaves *in vitro* at a specific phophodiester bond producing 2′, 3′-cyclic phosphodiester and 5′-hydroxyl termini (Figure 2.1B). Natural hammerhead structures (Flores *et al.*, 2001) have a central core of strictly conserved nucleotides flanked by three helices (I, II, and III) with loose sequence constrains that in most cases are closed by short loops (1, 2, and 3). X-ray crystallography has revealed a complex array of non-Watson–Crick interactions between the nucleotides of the central core that form the

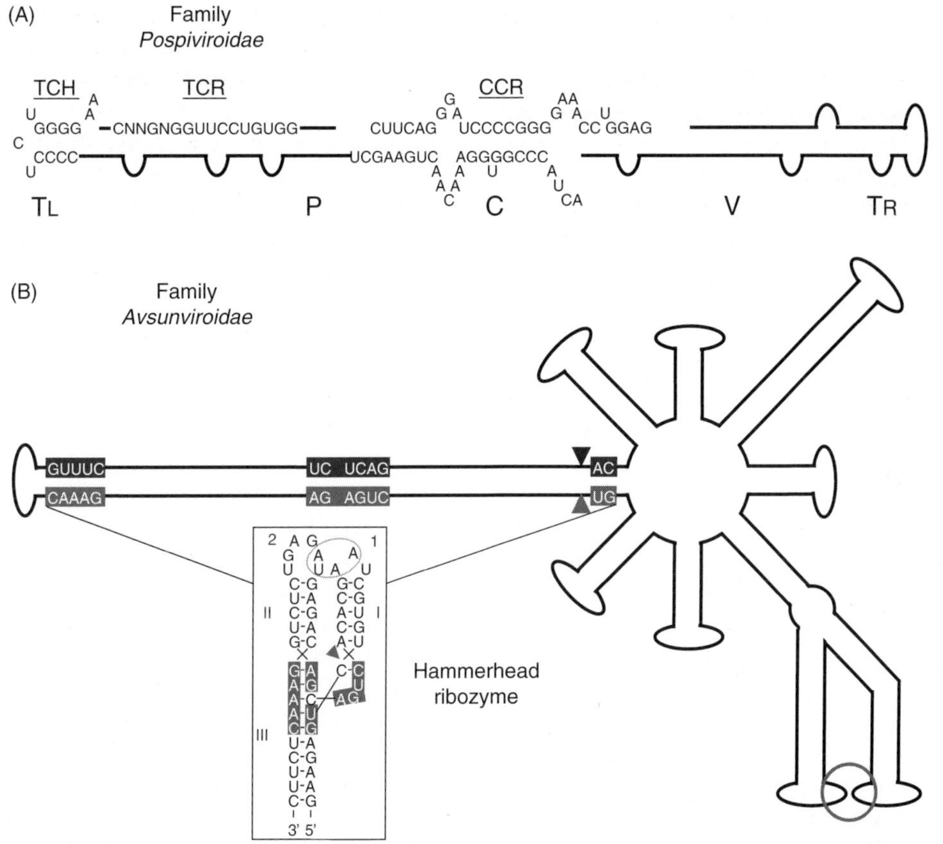

FIGURE 2.1 Viroid structure. (A) Scheme of the rod-like genomic RNA characteristic of the family *Pospiviroidae* with the central (C), pathogenic (P), variable (V) and terminal left and right (T_L and T_R, respectively) domains. The central conserved region (CCR) (genus *Pospiviroid*), the terminal conserved region (TCR) (genera *Pospi-*, *Apsca-*, and part of *Coleviroid*) and the terminal conserved hairpin (TCH) (genera *Hostu-* and *Cocadviroid*), are displayed. (B) Scheme of the branched genomic RNA of PLMVd (family *Avsunviroidae*) in which the sequences conserved in most natural hammerhead ribozymes are displayed on a red and blue background for (+) and (−) polarities, respectively, and the self-cleavage sites by arrowheads. Red circle denotes a kissing-loop interaction. The structure of the (+) hammerhead ribozyme is displayed in the boxed inset, with Roman and Arabic numerals depicting helices I, II, and III, and loops 1 and 2, respectively, and the arrowhead the self-cleavage site. Short black and red lines indicate canonical and non-canonical base pairs, respectively, and the green oval a tertiary interaction between loops 1 and 2 that enhances the catalytic activity. (See plate 1 for the color version of this figure.)

catalytic pocket embracing the cleavage site (Figure 2.1B). There is solid experimental support for the *in vivo* functional role of hammerhead ribozymes in processing the oligomeric viroid RNAs: (i) linear unit-length viroid strands of one or both polarities with 5′ termini identical to those produced in the *in vitro* self-cleavage reactions have been identified in distinct viroid-infected tissues, and (ii) covariations preserving the stability of the hammerhead structures have been found in variants of different viroids (Hernández and Flores, 1992; Daròs *et al.*, 1994; Navarro and Flores, 1997; Fadda *et al.*, 2003).

Recent data have shown that natural *cis*-acting hammerheads self-cleave much faster the RNAs wherein they are contained than their *trans*-acting derivatives in which the peripheral loops 1 and 2 have been removed (De la Peña *et al.*, 2003; Khvorova *et al.*, 2003). These data indicate that regions external to the central conserved core of natural hammerheads play a key catalytic role, most likely because these peripheral loops form tertiary interactions that facilitate the positioning and rigidity of the active site at the low magnesium concentration existing in most physiological conditions (Figure 2.1B). The tertiary interactions, which have been confirmed by X-ray crystallography of a natural hammerhead (Martick and Scott, 2006), could be additionally stabilized *in vivo* by proteins (Daròs and Flores, 2002). Besides being operative at low magnesium concentrations, hammerheads must be finely tuned during viroid replication to catalyze self-cleavage of the oligomeric RNA intermediates but not of the monomeric circular RNAs serving as templates for the successive replication rounds. This regulation is achieved in some viroids by adopting alternative stable foldings that do not promote self-cleavage of the monomeric RNAs, with the active hammerheads being only formed transiently during transcription (Forster and Symons, 1987). In other viroids, active hammerheads can be only formed in their corresponding dimeric or oligomeric replicative intermediates through the adoption of double-hammerhead structures, but not in the monomeric RNAs resulting from their self-cleavage in which the single-hammerhead structures are thermodynamically unstable (Forster *et al.*, 1988). Hammerhead ribozymes might also mediate ligation of the unit-length viroid strands resulting from self-cleavage, although the involvement of a host RNA ligase cannot be excluded.

TAXONOMIC RELATIONSHIPS AMONG VIROIDS AND THEIR RELATIONSHIP WITH OTHER RNAS

Phylogenetic Tree and Taxons (Families, Genera, and Species)

As already indicated, the available evidence suggests that viroids and viruses have an independent evolutionary origin. Viroids, however, may have a common evolutionary origin with a class of small satellite RNAs (which are functionally dependent on a helper virus), the viroid-like satellite RNAs, with which they share structural similarities and the rolling circle replication mechanism involving hammerhead ribozymes. Indeed, application of the likelihood-mapping method to a sequence alignment accounting the local similarities and the insertions/deletions and duplications/rearrangements described for viroids and viroid-like satellite RNA, leads to a reliable phylogenetic reconstruction that is consistent with the biological properties of these RNAs (Elena *et al.*, 2001) (Figure 2.2). From the phylogenetic tree, the approximately 30 known viroid species can be grouped into two families: *Pospiviroidae,* whose type species is potato spindle tuber viroid (PSTVd), and *Avsunviroidae*, with ASBVd as the type species. Each family contains several genera, with an arbitrary level of less than 90% sequence similarity and distinct biological properties (particularly host range, see below) separating species within genera. Viroid-like satellite RNAs appear also grouped according to the type of helper virus they are dependent on.

Biological Properties of Each Family

This classification scheme is supported by other criteria (Flores *et al.*, 2005). PSTVd replicates in the nucleus through an asymmetric rolling-circle mechanism, and ASBVd in the chloroplast through a symmetric variant of this mechanism, with the available evidence indicating that other members of both families behave as their respective type species. Moreover, members of the family *Avsunviroidae* are catalytic RNAs (they can form hammerhead ribozymes in both polarity strands), while members of the family *Pospiviroidae* lack catalytic domains, being characterized for the presence of a central conserved region (CCR) (Figure 2.1). The type

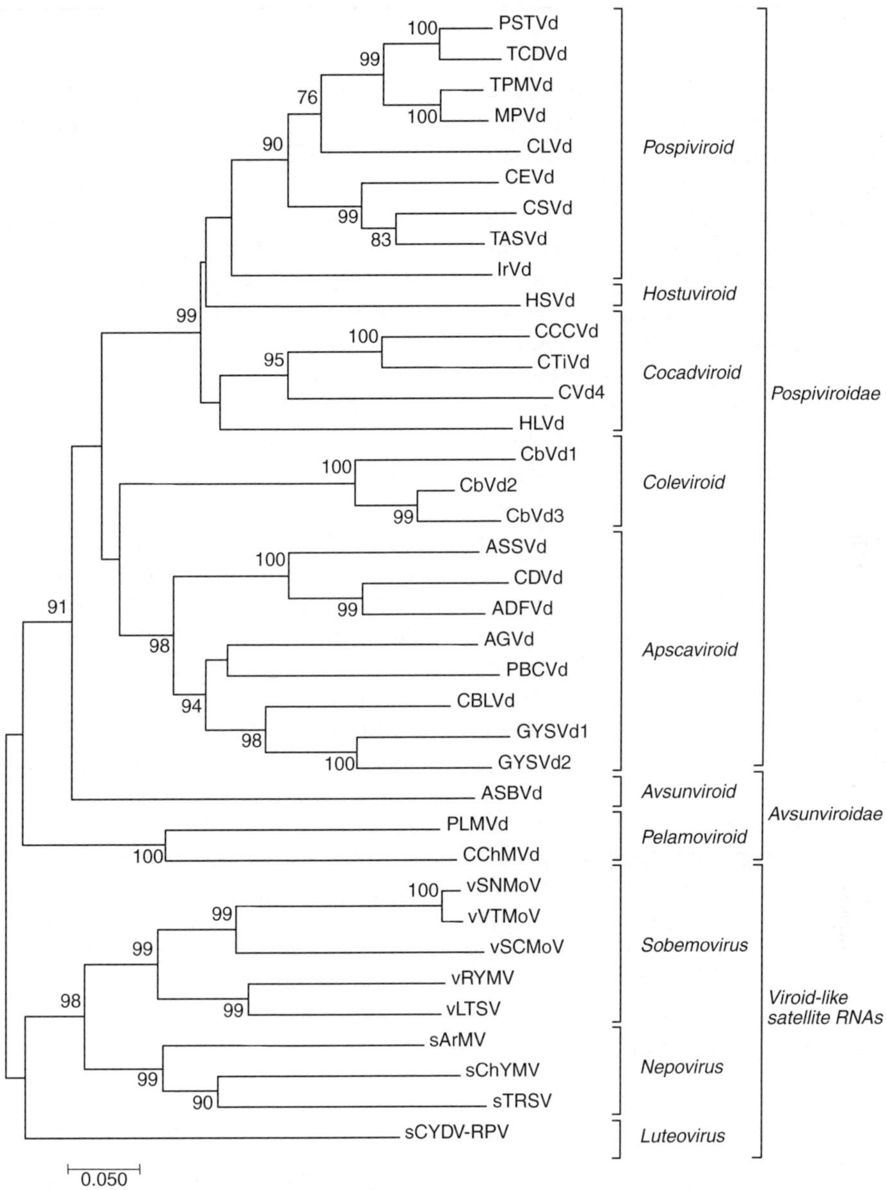

FIGURE 2.2 Neighbor-joining phylogenetic tree obtained from an alignment manually adjusted to take into account local similarities, insertions/deletions, and duplications/rearrangements described in the literature for viroid and viroid-like satellite RNAs. Bootstrap values were based on 1000 random replicates (only values >70% are reported). Viroids: PSTVd (potato spindle tuber); TCDVd (tomato chlorotic dwarf); TPMVd (tomato planta macho); MPVd (Mexican papita); CLVd (columnea latent); CEVd (citrus exocortis); CSVd (chrysanthemum stunt); TASVd (tomato apical stunt); IrVd (iresine 1); HSVd (hop stunt); CCCVd (coconut cadang-cadang); CTiVd (coconut tinangaja); CVd-IV (citrus IV); HLVd (hop latent); CbVd1 (*Coleus blumei* 1); CbVd2 (*Coleus blumei* 2); CbVd3 (*Coleus blumei* 3); ASSVd (apple scar skin); CDVd (citrus dwarf); ADFVd (apple dimple fruit); AGVd (Australian grapevine); PBCVd (pear blister canker); CBLVd (citrus bent leaf); GYSVd1 (grapevine yellow speckle 1); GYSVd2 (grapevine yellow speckle 2); ASBVd (avocado sunblotch); PLMVd (peach latent mosaic); CChMVd (chrysanthemum chlorotic mottle). Viroid-like satellite RNAs: sSNMoV (*Solanum nodiflorum* mottle virus); sVTMoV (velvet tobacco mottle virus); sSCMoV (subterranean clover mottle virus); sRYMV (rice yellow mottle virus); sLTSV (lucerne transient streak virus); sArMV (*Arabis* mosaic virus); sChYMV (chicory yellow mottle virus); sTRSV (tobacco ringspot virus); sCYDV-RPV (cereal yellow dwarf virus-RPV). Adapted from Elena *et al.* (2001).

of CCR and the morphology of the hammerhead structures serve, together with other criteria, to demarcate genera within each family (Figure 2.3).

The host range of members of the family *Avsunviroidae* is restricted to the plants (and closely related species) in which they were initially reported. This is also the case of certain members of the family *Pospiviroidae*, whereas others have a broad host range. Some viroids replicate without inducing phenotypic alterations in their hosts but others cause symptoms in leaves, stems, bark, flowers, fruits, seeds, and reserve organs (tubers). Perhaps the only family-specific symptom is the extreme chlorosis incited by certain variants of ASBVd and peach latent mosaic viroid (PLMVd) (Semancik and Szychowski, 1994; Malfitano *et al.*, 2003), which is most likely related to their ability to invade the shoot apical meristem and block the chloroplast developmental program (Rodio *et al.*, 2007).

MECHANISMS OF GENETIC VARIABILITY

Mutation

The rate and frequency of mutation are often used as synonymous, although in evolutionary genetics have different meanings. The mutation rate is an unavoidable consequence of the reduced fidelity associated with polymerases, which in viroids are DNA-dependent RNA polymerases (DDR) using RNA as a non-natural template, and of the thermodynamic noise. In contrast, the frequency of mutation measures the standing genetic variation in a population, and is the consequence of both mutation rate and natural selection. From an evolutionary standpoint, the most relevant parameter is the rate of mutation.

After inoculating plants with an infectious viroid-cDNA, or its transcripts the resulting

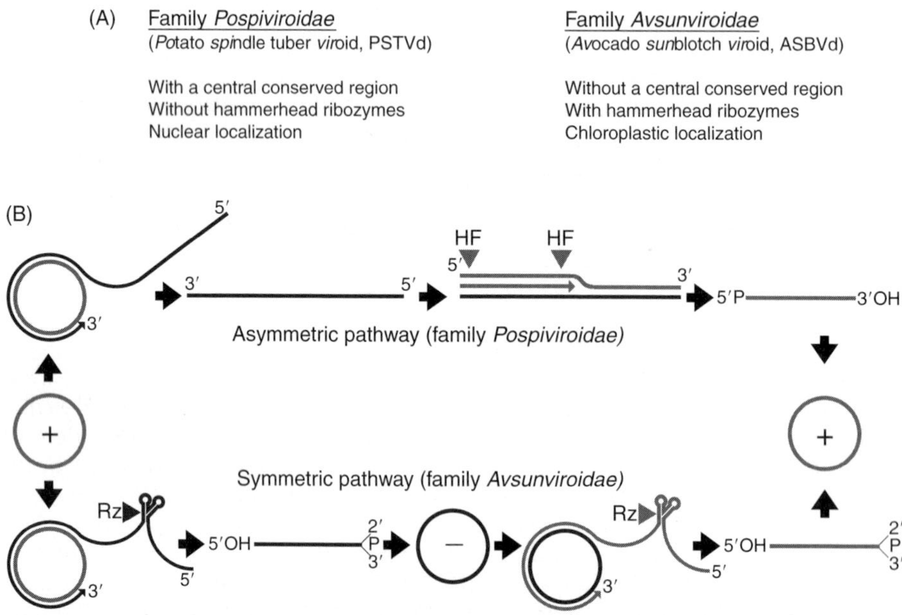

FIGURE 2.3 (A) Distinctive properties of the *Pospiviroidae* and *Avsunviroidae* families. (B) Asymmetric and symmetric pathways of the rolling-circle replication mechanism followed by members of the families *Pospiviroidae* and *Avsunviroidae*, respectively. Red and blue lines refer to (+) and (−) strands, respectively. Arrowheads point to cleavage sites of a host factor (HF) or ribozymes (Rz), and the resulting 5′ and 3′ groups are indicated. (See plate 2 for the color version of this figure.)

populations readily accumulate considerable amounts of genetic variation, as shown by a handful of studies with viroid species from both families. However, the degree of genetic diversity attained by the members of each family is different, as illustrated in Figure 2.4, where two different measures of genetic diversity are compared. The *Avsunviroidae* included in this study are chrysanthemum chlorotic mottle viroid (CChMVd) (Codoñer *et al.*, 2006) and PLMVd (Ambrós *et al.*, 1999) and the *Pospiviroidae* PSTVd (Góra-Sochacka *et al.*, 1997) and citrus bent leaf viroid (CBLVd) (Gandía and Duran-Vila, 2004). Abscises show haplotype diversity, which represents the probability that two molecules randomly taken from the viroid population inhabiting an individual plant would be different. Ordinates correspond to the average number of nucleotide differences between these two randomly sampled molecules. On average, the *Avsunviroidae* have 53.8% more haplotypes than the *Pospiviroidae*. Furthermore, two randomly picked *Avsunviroidae* haplotypes differ in ~7 point mutations, whereas two randomly picked *Pospiviroidae* haplotypes differ, on average, only in ~1 point mutation. Therefore, *Avsunviroidae* populations are more diverse both in terms of abundance of haplotypes and in differences among haplotypes than *Pospiviroidae* populations, being the difference statistically significant (MANOVA, Hotelling's trace $P < 0.0001$) despite the limited data set.

The above discussion is relevant to understand the amount of genetic diversity present on viroid populations after the action of selection and whether differences among viroid families exist. However, can viroid mutation rate be experimentally quantified? The classical genetic method for estimating the per locus mutation rates is the fluctuation test (Luria and Delbrück, 1943). However, this simple approach is impracticable in viroids mostly due to the following limitations: (i)

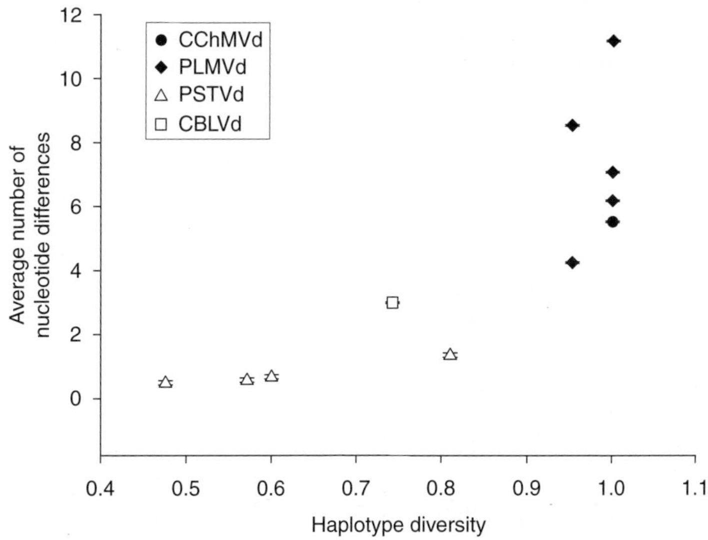

FIGURE 2.4 Diversity indexes for representative viroid species belonging to the families *Pospiviroidae* and *Avsunviroidae*. On average, members of the *Avsunviroidae* (solid symbols) have 53.8% more haplotypes than those of the *Pospiviroidae* (open symbols). Furthermore, on average, the number of nucleotide differences between two randomly taken haplotypes of chrysanthemum chlorotic mottle viroid (CChMVd) or peach latent mosaic viroid (PLMVd) is 5.88 times larger than when two haplotypes of potato spindle tuber viroid (PSTVd) or citrus bent leaf viroid (CBLVd) are compared. Error bars represent one standard error.

it is impossible to identify each and every newly synthesized viroid molecule within a plant inoculated with an infectious cDNA, and (ii) the effect of natural selection at different stages during the infectious cycle, which may indeed affect differentially the accumulation of each mutant genotype, cannot be ruled out, in particular if lethal mutations exist. A possible hybrid alternative relies on the use of experimental data and a classic result from population genetics, namely the relationship between population average fitness and mutational load. Assuming that a population has reached the mutation-selection balance, and that there are no fitness interactions among loci, the average population fitness, $\bar{W}$, only depends on the magnitude of the deleterious genomic mutation rate, U, as $\bar{W} = e^{-U}$ (Kimura and Maruyama, 1966). Therefore, a simple experimental design to estimate U would be as follows: after inoculating plants with infectious cDNAs (with zero initial genetic variability), the replicating and evolving viroid population will reach a genetic equilibrium at which each mutant genotype will be present at a frequency, p_i, that depends on its fitness, W_i. Then, the population will be thoroughly sampled by reverse transcriptase polymerase chain reaction (RT-PCR) followed by sequencing of multiple clones with two goals, first to quantify haplotype frequencies and, second, to determine *in silico* the fitness of each haplotype. Henceforth, the deleterious genomic mutation rate can be estimated as $U = -\log\Sigma_i p_i W_i$. Determining viroids fitness *in vivo* can be troublesome since it would depend on: (i) the existence of appropriate stable genetic markers that would allow the differential quantification of the target genome from a reference strain, and (ii) competition experiments must be run for very short time before polymorphic populations would be created and evolution may become an issue. Both problems seem insurmountable, specially the second one. However, the *in silico* folding of RNA molecules into minimum free energy secondary structures (MFESS) is a simple yet biophysically well-grounded and powerful model for studying the mapping relationships between genotype and phenotype (Fontana, 2002). Thus, assuming that the fitness of a viroid haplotype would be determined by its ability to fold into the right MFESS, then Hamming distance between the mutant's predicted MFESS and the optimal one (here assumed to be the one obtained for the wild-type sequence) can be used as an *in silico* proxy to viroid fitness.

In this study we have proceeded as follows. First, for each viroid genome characterized in the four studies mentioned in the previous paragraph (i.e. CChMVd, PLMVd, PSTVd, and CBLVd), we obtained the MFESS folding using the RNAfold program from the Vienna RNA package version 1.6.4 (Hofacker, 2003). Second, these folds were compared to the one obtained for the corresponding reference sequence using the RNAdistance program (also from the Vienna RNA package). This program provides the Hamming distance, d_H, between the tree-representation of the two folds. Third, following Ancel and Fontana (2000), d_H was converted into a measure of fitness using a hyperbolic function that maps d_H into the range $W_i \in (0, 1)$. A value $W_i = 1$ means that the ith mutant sequence folds exactly into the wild-type structure. A value $W_i = 0$ means that the ith sequence folds in an infinitely different structure. Fourth, combining the frequency in which each haplotype appears in the population sample with its *in silico* fitness, an estimate of the average population fitness, $\bar{W}$, was computed and from this value, U was estimated for each viroid and inoculation experiment. Finally, this estimate can be transformed into a per nucleotide deleterious mutation rate just by dividing U by the actual genome length of each viroid. These data are presented in Figure 2.5. Overall, the per site deleterious mutation rate for the *Avsunviroidae* is ~10-fold higher than the value estimated for *Pospiviroidae* (ANOVA, $P = 0.0002$).

These estimates of deleterious mutation rate are as good as the three assumptions on which they hold. First, after inoculation of a single plant with an infectious cDNA, the viroid populations may still be away from the mutation-selection balance. The currently available data

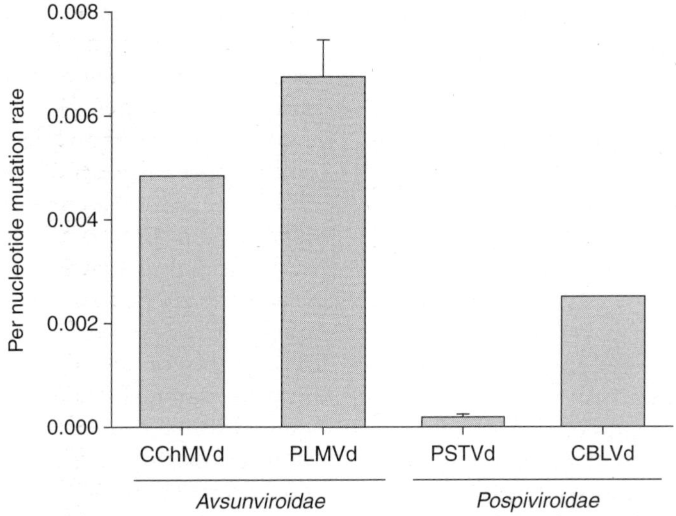

FIGURE 2.5 Estimated per nucleotide mutation rate for representative viroid species belonging to the families *Pospiviroidae* and *Avsunviroidae*. On average, the mutation rate for members of the family *Avsunviroidae* is 10-fold larger than that for members of the *Pospiviroidae* (ANOVA, $P = 0.0002$). For each viroid species, the fitness of each observed haplotype was estimated as a hyperbolic function of the Hamming distance between the predicted minimum free energy folding and the corresponding optimum folding. Folding and comparison among structures was done using the Vienna RNA Package version 1.6.4 (www.tbi.univie.ac.at/~ivo/RNA). Error bars represent one standard error. CChMVd, chrysanthemum chlorotic mottle viroid; PLMVd, peach latent mosaic viroid; PSTVd, potato spindle tuber viroid; CBLVd, citrus bent leaf viroid.

do not inform on this question. Second, epistasis in RNA genomes exists as a consequence of the highly compacted RNA folding into the rod-like and branched structures characteristic of the two viroid families. Indeed, Sanjuán *et al.* (2006b) explored *in silico* the interaction between pairs of mutations for all viroid species and found an overall predominance of antagonistic epistasis (see below). The assumption that the predicted MFESS has any biological relevance is not a solved issue. However, it is informative that UV cross-link experiments have detected interactions among nucleotides that were nearby in the *in silico* predicted structures (Eiras *et al.*, 2007; Wang *et al.*, 2007). Finally, it must be stressed that these estimates must be taken as a lower bound for the mutation rate, since they only consider viable mutations and, therefore, lethal mutations are ignored.

What does generate a 10-fold increase in mutation rate for the *Avsunviroidae*? Replication of nuclear and chloroplastic viroids is mediated by different DDRs. While the *Pospiviroidae* are transcribed by the RNA polymerase II composed of multiple subunits (Mühlbach and Sänger, 1979), the *Avsunviroidae* are presumably transcribed by a nuclear-encoded chloroplastic DDRs structurally similar to the RNA polymerases of certain bacteriophages (i.e. formed by a single subunit) (Navarro *et al.*, 2000). The dissimilar structural complexity of both RNA polymerases may influence their mutation rates, particularly when transcribing RNA templates instead of their physiological DNA templates. These atypical RNA templates may also differentially affect the processivity of RNA polymerases, promoting their jumping during transcription and facilitating the frequent emergence of novel recombinant variants (see below). Furthermore, as a side-effect of electron transduction during photosynthesis, mutagenic free radicals may be more abundant in the chloroplast than in the nucleus, contributing to increase the overall mutation rate.

Mutational Effects on Structure and Differences in Structural Robustness among Families

One topic that has attracted the attention of evolutionary biologists in recent years is the evolution of mutational robustness mechanisms (De Visser *et al.*, 2003). That is, the constancy of phenotypes under perturbations in the genome. Two are the hallmarks of robust systems: (i) the average mutational effect of point mutations on fitness shall be small, and (ii) the dominant type of epistasis among deleterious mutations shall be synergistic, which means that mutations can be accumulated without noticeably affecting fitness until the number reaches a boundary beyond which mutational effects become evident. Robustness may arise as a consequence of genetic redundancy, parallel metabolic networks, or buffering proteins such as chaperons (De Visser *et al.*, 2003). In contrast, sensitive (i.e. non-robust) organisms present the opposite properties: large mutational effects and antagonistic epistasis. Due to their compacted, non-redundant and frequent overlapping reading frames, RNA viruses, and likely viroids, are the prototypes of sensitive genomes (Elena *et al.*, 2006). Viroids provide a unique opportunity to study the evolution of robustness in an oversimplified biological system in which the genotype directly expresses a phenotype (RNA folding).

In a recent series of studies, Sanjuán *et al.* (2006a, 2006b) have explored *in silico* the robustness of viroids genomes. To do so, all possible one-error and two-error mutants were generated for every viroid species. Then, as described above, the structural distances between mutant and wild-type structures were computed. Several conclusions relevant for viroid evolution can be drawn from these studies. First, as expected for simple genomes with no redundancy, the average effect of mutations was large, although the distribution of effects was skewed towards mild effects (Sanjuán *et al.*, 2006a). Also, as expected for non-redundant genomes, antagonistic interactions among pairs of mutations were more abundant than synergistic ones (except for PSTVd) (Sanjuán *et al.*, 2006b). Second, the nature of the folding dramatically affected the result of mutational effects and epistatic interactions (Figure 2.6). For rod-like folds, most mutations have moderate effects, with few sites showing large effects (Figure 2.6A). For branched structures, however, sites located at the right half of the molecule (i.e. the branched part) have strong structural effects (first row in Figure 2.6). Epistasis was more common among pairs of mutations hitting the same structural domain than among pairs hitting different domains, with mutation pairs hitting nearby positions in the secondary structure being more prone to be engaged in antagonistic epistasis (second row in Figure 2.6). In rod-like folds, antagonistic pairs tended to group along the diagonals. Pairs falling along the direct diagonal reveal the multiple hitting effect of the same structure, whereas cases falling along the reverse diagonal correspond to the complementary part of the rod and, hence, indicate multiple hitting or compensatory mutations. A similar pattern is only visible for the left long stem of PLMVd. Altogether these findings suggest that these two viroid families differ in their degree of structural robustness against mutation, with *Pospiviroidae* rod-like structures being, on average, more robust than *Avsunviroidae* branched structures. Third, as has been mentioned above, robustness may be increased through genetic redundancy. The existence of partially redundant variants of several viroids (see below) allows testing this prediction. For example, in coconut cadang-cadang viroid (CCCVd), the mutational effects for the fast non-redundant variant were significantly stronger than for the redundant variant CCCVd-slow. Furthermore, the number of interactions with antagonistic epistasis was smaller for the redundant variant. Similar results were observed for the other two viroids (citrus exocortis viroid, CEVd, versus CEVd-D104), and *Coleus blumei* viroid 1 (CbVd-1) versus CbVd-3 (Sanjuán *et al.*, 2006a, 2006b), providing generality to this conclusion. Therefore, it can be speculated

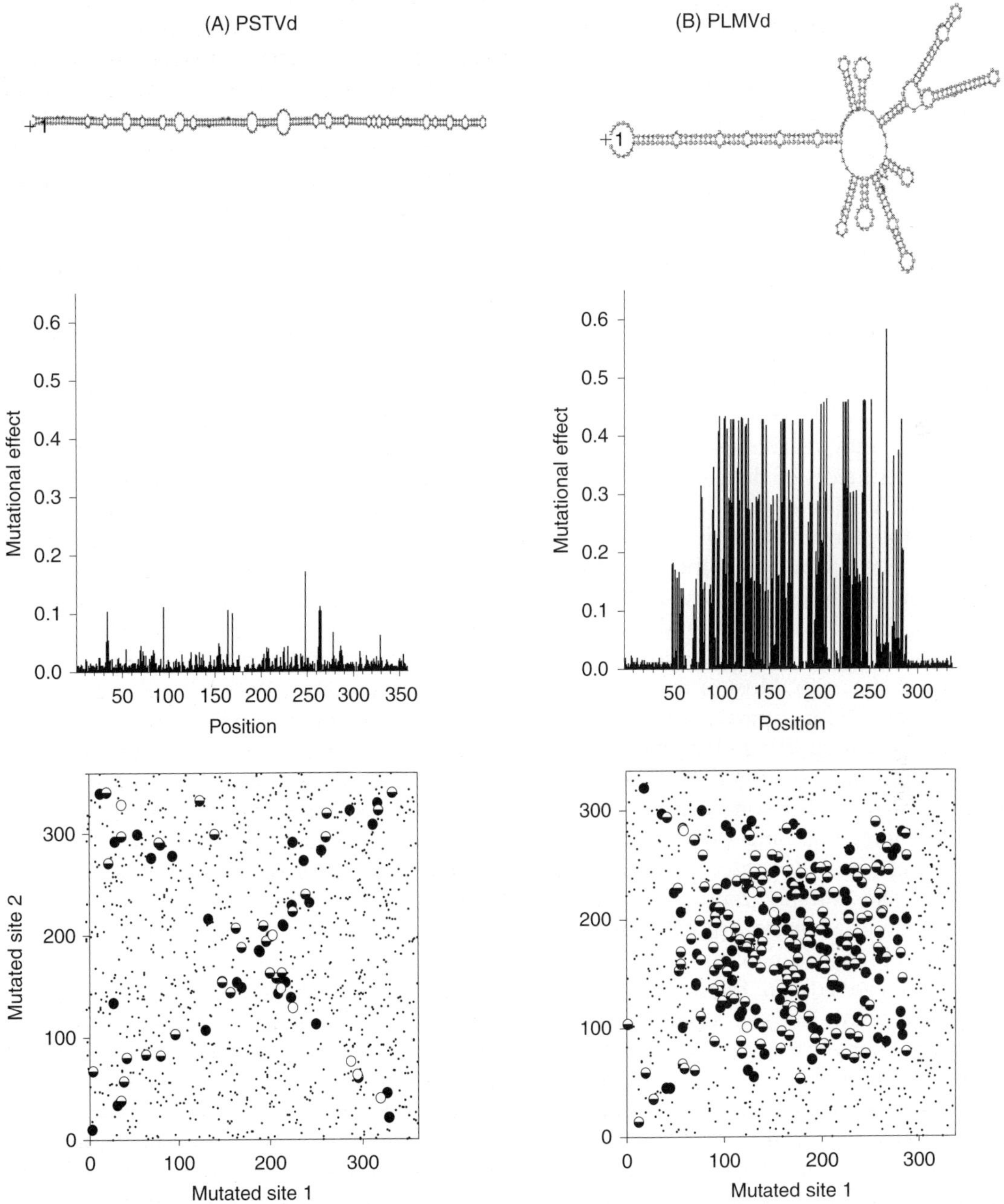

FIGURE 2.6 The magnitude of mutational effects and epistatic interactions do not distribute evenly along viroid molecules and depends on whether the molecule folds into rod-like or branched structures. (A) Potato spindle tuber viroid (PSTVd), the prototypic viroid with rod-like structure and (B) peach latent mosaic viroid (PLMVd), a pelamoviroid with a highly branched secondary structure. The lower panels show the map of all mutation pairs (small dots) and those showing antagonistic epistasis (large dots). White and half-filled dots correspond to compensatory mutations and mostly fall on the diagonals. Solid dots are non-compensatory antagonistic pairs. From Sanjuán *et al.*, (2006a, 2006b).

that redundant variants would be favored when environmental conditions may impose an increase in mutation rate that will provide a fitness advantage due to their increased structural robustness to mutations.

Recombination

Sequence similarities among different viroids have been viewed as evidence that recombination has played a role in the generation of new viroids. The first indirect evidence resulted from the identification of viroids such as tomato apical stunt viroid (TASVd) and tomato planta macho viroid (TPMVd), which share segments of their sequence with PSTVd and CEVd (Keese and Symons, 1985). Since then, the term "chimeric viroid" has been widely used to describe these similarities that were extended to chrysanthemum stunt viroid (CSVd) (Haseloff and Symons, 1981) and columnea latent viroid (CLVd) (Hammond *et al.*, 1989). Most of these viroids have common hosts, particularly within the *Solanaceae*, supporting the hypothesis that they might have emerged in these hosts as a consequence of recombination events between co-infecting ancestors. Although there are exceptions, illustrated by CSVd and CLVd that were initially reported in chrysanthemum and *Columnea erytrophae* respectively, the finding of a new viroid, Mexican papita viroid (MPVd), in the wild *Solanum cardiophyllum* points to solanaceous species remaining in their center of origin and diversification as reservoirs of putative ancestors of the pospiviroids (Martínez-Soriano *et al.*, 1996). Supporting this view, recent surveys have revealed the presence of pospiviroids in the ornamental *Solanum jasminoides* (Verhoeven *et al.*, 2006, 2007; Di Serio, 2007).

Certain pospiviroids, like PSTVd and CEVd, seem to have also contributed to the emergence of members of other genera (Keese and Symons, 1985; Puchta *et al.*, 1991; Di Serio *et al.*, 1996). The vegetative propagation of crops like grapevines and citrus that are naturally infected with several viroids, may have favored the occurrence of recombination events. In this regard, citrus viroid IV (CVd-IV) has been proposed to be a chimeric viroid between CEVd and hop stunt viroid (HSVd), both widespread in citrus, whereas Australian grapevine viroid (AGVd) shares similarities with grapevine yellow speckle viroid (GYSVd-1) and CEVd, which both infect grapevines.

Despite the general acceptance that new viroids can emerge as a result of recombination, there is no experimental evidence showing the generation of a chimeric viroid after co-inoculating a single host with several viroids. However, the identification in *Coleus blumei* plants of CbVd-2 composed of two unchanged parental sequences with sharp boundaries from the right-hand part of CbVd-1 and the left-hand part of CbVd-3, together with the observation that CbVd-2 and its two putative progenitors were found co-infecting the same plants, provides the best available evidence for a true recombination event (Spieker, 1996; Sänger and Spieker, 1997). The infectivity and stability of chimeric viroids constructed *in vitro* further sustains this possibility. Examples include chimeric constructs between strains of CEVd, PSTVd or HSVd (Visvader and Symons, 1986; Góra *et al.*, 1996; Reanwarakorn and Semancik, 1998), as well as between the closely related CEVd and TASVd (Owens *et al.*, 1990; Sano *et al.*, 1992). However, attempts to construct viable chimeras between species of different genera have proved more difficult. In fact, chimeric constructs containing the right-half part of the rod-like secondary structure of CEVd and the left half-part of HSVd and vice versa were not viable, whereas the terminal right domain of both viroids was exchangeable (Sano and Ishiguro, 1998). This illustrates that the structural constrains limiting the viability of recombinant viroids become more difficult to overcome as the phylogenetic distance between species increases.

Discontinuous transcription of a "jumping" RNA polymerase has been proposed as the most likely mechanism accounting for both intramolecular rearrangements and intermolecular recombination between viroids co-infecting a common host (Keese and Symons, 1985). The identification of several CCCVd forms containing duplicated segments of 41, 50, 55, and 100 nucleotides of the left terminal domain and part of the adjacent variable domain provides the best evidence for the existence of discontinuous transcription (Haseloff *et al.*, 1982). This first observation was further supported by the identification of enlarged CEVd variants with duplicated segments of the same domains in a hybrid tomato (Semancik *et al.*, 1994; Szychowski *et al.*, 2005) and in eggplant (Fadda *et al.*, 2003b). These enlarged variants displayed different levels of adaptation: (i) most were infectious in certain hosts but not in others, (ii) some reverted to the original CEVd in some hosts but remained as stable enlarged variants in others, and (iii) those that were infectious induced distinct symptoms in sensitive hosts (Szychowski *et al.*, 2005). This study also showed that only the enlarged variants that were able to fold into a rod-like secondary structure were stable, stressing the structural constraints imposed on the variability of members of the family *Pospiviroidae*. In PLMVd, of the family *Avsunviroidae*, an insertion of 12–13 nucleotides that can be acquired or lost during infection and that is responsible for specific symptoms, always occurs in a defined position and folds into a hairpin (Malfitano *et al.*, 2003; Rodio *et al.*, 2006).

The identification of enlarged viroid variants also exemplifies mechanisms for genome expansion crucial to increase the size and complexity of the small RNAs that initially populated the hypothetical RNA world (Diener, 1989; Chela-Flores, 1994). Classical experiments with the Qβ replicase (Spiegelman, 1971) illustrate that a minimum size is required to achieve autonomous replication of certain RNAs, which like MDV1, share some properties (high G+C content and a compact secondary structure) with viroids (Nishihara *et al.*, 1983). The sequence periodicity found in some viroids also supports the view that their evolution from small RNA precursors occurred through duplication of discrete fragments (Juhasz *et al.*, 1988).

VIROID QUASISPECIES

High Sequence Heterogeneity does not Necessarily Mean a Quasispecies Population Structure

The use of the word quasispecies has become standard among virologists to describe the population structure of RNA viruses. However, this use is inappropriate in many instances because it cannot be simply considered as a synonym for a heterogeneous population. This misuse of the word was already noticed by Eigen (1996).

Perhaps the two most debated concepts in Eigen's theory are the quasispecies effect, that is, the population behaves in such a way that can only be explained through strong mutational coupling between genotypes, and the error threshold, namely, the limit value for mutation rate beyond which the population structure breaks down and the population disperses over genotypic space. However, the existence of an error threshold depends on the assumption of the single-peak fitness landscape in which the wild-type genome has fitness 1 and every mutant has fitness $1 - s > 0$, regardless of whether they have one or many mutations. If lethal mutations exist, or mutational effects are variable, then the error threshold does not exist (Wagner and Krall, 1993; Wilke, 2005; but see Takeuchi and Hogeweg, 2007) (Chapter 1). Regarding the validity of the quasispecies effect, would a quasispecies of lower fitness outcompete a fitter one provided the former has a better support from its mutational neighbors? According to the quasispecies model, selection would maximize the average replication rate of the swarm of genotypes interconnected by mutation rather than favor the genotype with the highest replication rate. Hence, mutation acts as a selective agent and shapes the genome in such a manner that causes the entire quasispecies to become robust against mutations. Thus, in a highly mutagenic environment, a quasispecies occupying a low but flat region in the fitness landscape should outcompete a quasispecies located at a higher but narrower peak when most of its surrounding mutants are unfit. The quasispecies effect has recently been also rebaptized as the survival-of-the-flattest effect (Wilke *et al.*, 2001). Can viroids shed light on this question?

The Quasispecies Effect (or the Survival-of-the-Flattest) in Viroid Populations

To tackle this issue, Codoñer *et al.* (2006) competed populations of CSVd and CChMVd under two environmental conditions that differed in the environment-induced mutation rate. These two viroids were chosen because: (i) in plants inoculated with equivalent amounts of each viroid and for which a systemic infection was established for a long period, the concentration of CChMVd per gram of fresh plant tissue was 20 times lower than for CSVd, suggesting a lower net population growth rate for CChMVd, (ii) the genetic variability of CChMVd quasispecies was much larger than the CSVd quasispecies (for CSVd, values were similar to those found for PSTVd and reported in Figure 2.4), (iii) both viroids invade the whole plant, and the presence of several *Avsunviroidae* and *Pospiviroidae* in the same cell types of infected leaves has been confirmed by *in situ* hybridization (Bonfiglioli *et al.*, 1994, 1996; Bussière *et al.*, 1999), (iv) *in silico*, the size of the neutral neighborhood for the CChMVd molecule is two-fold larger than for CSVd, and (v) the mutation rate for *Avsunviroidae* is 10 times larger than that for *Pospiviroidae* (see above), a factor that creates a stronger selective pressure for the evolution of mutational robustness (van Nimwegen *et al.*, 1999). Therefore, CChMVd quasispecies grow slowly, are genetically heterogeneous, and exhibit a remarkable diversity of RNA secondary structures. In contrast, CSVd quasispecies grow rapidly and are genetically and structurally homogeneous. According to the survival-of-the-flattest effect, CSVd should outcompete CChMVd at low mutation rate. However, at high mutation rate CChMVd should be a superior competitor.

Chrysanthemum plants were co-infected with the two viroids and placed either into control environmental conditions or into mutagenic conditions (UV-irradiated 10 minutes/day with 2 J/cm^2, a dose known to induce intramolecular cross-linking in viroid molecules while still minimizing the impact on plant growth). The ratio CChMVd:CSVd was estimated at different time points after initiated the treatment. The slope of the log regression of this ratio on time is a direct measure of the fitness of CChMVd relative to CSVd. Figure 2.7 shows that an increase in environmental mutagenicity affects the outcome of the competition between the two viroids, in perfect agreement with the predictions of the quasispecies effect. At normal growth conditions, CSVd is 8.6% fitter than CChMVd, as a consequence of its faster replication. In contrast, when the mutation rate was artificially increased, the outcome of the competition was different, and the slower but mutationally more robust CChMVd was not outcompeted anymore, and it was 1.7% fitter than CSVd (Figure 2.7). This reversal of the fortune was due to the two-fold larger neutral neighborhood of CChMVd quasispecies.

These results are the first empirical observation in any biological system of a slower replicating RNA genome outcompeting a faster replicating one, thus clarifying the role of the mutational swarm, as described by the quasispecies theory, for the evolution of RNA replicons.

Selection of Variants

Host-Specific Variants

Viroids have been found in virtually all plant tissues (reviewed by Singh *et al.*, 2003) except in the apical meristem (but see Radio *et al.*, 2007). The exclusion of viruses (and viroids) from this plant compartment was initially

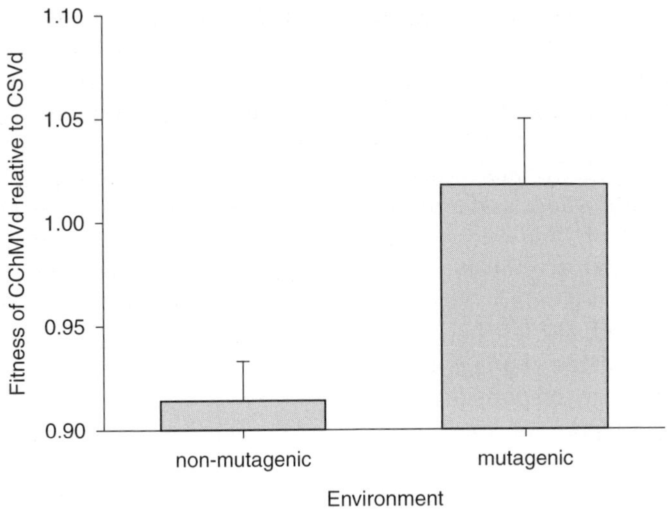

FIGURE 2.7 Fitness of a robust viroid, chrysanthemum chlorotic mottle viroid (CChMVd), relative to a fit one, chrysanthemum stunt viroid (CSVd), in non-mutagenic and in mutagenic environmental conditions. The fitness of CChMVd was significantly higher under mutagenic conditions (t-test, $P = 0.002$). Error bars represent standard errors. From Codoñer *et al.* (2006).

presumed to be due to the absence of vascular tissues but today is interpreted as a result of RNA-silencing phenomena. Hosts play an important role in shaping the composition and structure of viroid populations. The first evidence in this respect was described for CEVd and showed that after serial transmission to different hosts, the populations recovered presented differences in nucleotide composition, titer, and biological properties (Semancik *et al.*, 1993). The role of the host on the population structure was further studied in a natural CEVd isolate from symptomless *Vicia faba* that presented an unusual heterogeneous population, which became more homogeneous after transmission to tomato. When nucleic acid preparations from the infected tomato plants were back inoculated to new *V. faba* plants, the population did not revert to an heterogeneous population like that found originally in *V. faba* but displayed low nucleotide diversity like in tomato (Gandía *et al.*, 2007).

Phylogenetic analysis of variants of HSVd, a viroid with a wide natural host range, showed that they could be separated into several groups corresponding to specific hosts. However, a more detailed analysis revealed that even if a bias for the presence of certain sequences and/or structures in certain hosts was observed, no conclusive host determinants were found and that a number of HSVd isolates probably derived from recombination events (Kofalvi *et al.*, 1997; Amari *et al.*, 2001).

Disease-Specific Variants and Pathogenesis

The characterization of PSTVd, CEVd, and HSVd isolates inducing different symptoms has shown that specific regions of the viroid secondary structure are responsible for the pathogenic response. In PSTVd, mild, intermediate, severe, and lethal strains differ in only a few specific changes located in the "virulence modulating (VM) region" (Schnölzer *et al.*, 1985; Herold *et al.*, 1992) within the pathogenicity domain (Figure 2.1A). Although the virulence of the strains was initially correlated with the instability of the VM region (Schnölzer *et al.*, 1985; Lakshman and Tavantzis, 1993), further attempts to verify this correlation failed. An alternative view proposes that differential bending of the rod-like secondary structure is responsible for the virulence of the strains (Owens *et al.*, 1996).

Similarly, the characterization of severe and mild CEVd strains led to the identification of as many as 26 nucleotide changes located in the pathogenicity and variable domains (Figure 2.1A) (Visvader and Symons, 1985). The inoculation of chimeric *in vitro* constructs containing the sequence of the right-hand part of the secondary structure of one of the strains and the left-hand part of the other, proved that the changes in the pathogenicity domain are responsible, like in PSTVd, for the virulence of the strains (Visvader and Symons, 1986). Even though the stability of the chimeric viroids suggests an apparent lack of interdependence between the two domains of the molecule, the numerous CEVd sequences available in databases show that the specific changes in the pathogenicity domain characteristic of severe and mild strains are always associated with another set of specific changes located in the variable domain. This suggests the existence of some type of tertiary interactions favoring a concerted evolution between distal parts of the molecule.

In contrast with PSTVd and CEVd, the pathogenicity of HSVd in citrus is determined by a set of 5–6 specific changes located in the variable domain of the rod-like secondary structure (Reanwarakorn and Semancik, 1998; Palacio-Bielsa *et al.*, 2004). All the strains characterized so far present a very strict nucleotide composition in these positions affecting the organization of a short helical region and two flanking loops, which apparently do not admit additional changes (Palacio-Bielsa *et al.*, 2004).

Members of the family *Avsunviroidae* offer examples of variants associated with specific symptoms even within a single infected plant. The avocado sunblotch disease has been characterized by the occurrence of a complex syndrome that includes stem streaks, fruit discoloration, and a variety of foliar symptoms. Distinct leaf symptoms such as severe chlorosis associated with vascular tissues, variegations expressed throughout the whole blade, and lack of any visible symptom have been associated with specific ASBVd variants (Semancik and Szychowski, 1994). Other examples of disease-specific variants are found in CChMVd, wherein the pathogenicity determinant for mottling has been mapped at a tetraloop of the branched conformation of the viroid RNA (De la Peña *et al.*, 1999), and in PLMVd, in which an insertion of 12–13 nucleotides that folds into a hairpin accounts for an extreme chlorosis (Malfitano *et al.*, 2003; Rodio *et al.*, 2006).

Interaction between Viroids

Cross-protection

When a plant is infected with a latent strain of a viroid it becomes protected, at least temporarily, against subsequent infection by a severe strain of the same or a closely related viroid: the plant does not develop the symptoms characteristic of the challenging viroid and its accumulation is abolished or attenuated. Similar phenomena were previously reported in viruses. Before the viroid concept was formulated, and as a result of the resembling symptoms induced by viruses and viroids, some diseases now known to have a viroid etiology were presumed to be virus-induced and the corresponding cross-protection phenomena as additional examples of viral cross-protection. Thus, cross-protection between viroids was reported before their fundamental differences with viruses were uncovered. The intrinsic specificity of cross-protection has served for establishing relationships between viroids and for detection bioassays. In fact, the lack of cross-protection between the agent of chrysanthemum chlorotic mottle disease (not yet known to be induced by CChMVd) and PSTVd, CEVd, and CSVd established a first demarcating criterion between members of both viroid families co-infecting a common host (Niblett *et al.*, 1978).

Recently, there has been a renewed interest in cross-protection derived from observations that protection against a virus can be elicited in plants by transgenically expressing non-protein-coding viral RNA sequences (for a review see Hull, 2002). This RNA-mediated cross-protection is mechanistically equivalent to post-transcriptional gene silencing (Ratcliff *et al.*, 1999) (see below). Although the different structural and biological properties separating members of both viroid families led to assume that the mechanisms underlying cross-protection (and pathogenesis) should also differ, a common RNA-silencing mechanism might operate in both instances.

Synergistic Effects

Co-inoculation with two unrelated viroids may result in a synergistic interaction, with symptoms being more severe than those expected for purely additive effects of the two viroids (Serra *et al.*, 2007). Synergistic interactions between distantly related viruses have been also known for a long time and recently have been interpreted as resulting from the concerted action of their viral-encoded suppressors of gene silencing (see below). Because gene silencing also regulates plant development, and because the defensive and the developmental pathways share common components, co-infection by two distinct viruses may lead to enhanced symptom expression as a result of their silencing suppressors impairing different steps of the RNA silencing pathways (Pruss *et al.*, 1997; MacDiarmid, 2005). A parallel interpretation cannot be extrapolated to explain synergism between viroids because, lacking any messenger RNA activity, they do not encode silencing suppressors. However, new data indicate that a plant RNA virus suppresses RNA silencing as a consequence of sequestering its replication enzymes involved in the biogenesis of the small RNAs (siRNAs and miRNAs, see below), the final effectors of silencing (Takeda *et al.*, 2005). Similarly, viroids may also interfere with the RNA-silencing machinery of their hosts, and the synergistic effects of two unrelated co-infecting viroids could result from affecting more than one component of this machinery.

HAS RNA SILENCING SHAPED VIROID STRUCTURE AND EVOLUTION?

RNA silencing is a sequence-specific gene inactivation system present in most eukaryotes, with essential roles in development, chromosome structure, and virus resistance (Matzke and Matzke, 2004). As an antiviral defense, RNA silencing works as an adaptive immune system that recognizes and cleaves the invading RNA. Virus-induced RNA silencing is triggered by double-stranded RNA (dsRNA) sequences that include structured regions of the genomic RNA, replicative intermediates, and products of a cytoplasmic RNA-dependent RNA polymerase (RDR). RNase III-type Dicer enzymes (Dicer-like, DCL, in plants) act on virus dsRNA to produce 21- to 25-nt small interfering RNAs (siRNAs) that are incorporated into the RNA-induced silencing complex (RISC) which include Argonaute (AGO) proteins that mediate sequence-specific cleavage of target RNA. To oppose antiviral RNA silencing, most plant and some animal viruses have evolved silencing suppressor proteins that target key steps in the siRNA pathway by sequestering siRNAs, inhibiting their production or preventing their short- and long-distance spread (Li and Ding, 2006). Plants seem to heavily rely on the antiviral RNA-silencing pathway to fight viruses. Viroids are potential targets of this pathway because they are naked and highly structured RNAs, and also because not encoding any protein, they cannot suppress RNA silencing as viruses do. However, viroids can circumvent the RNA silencing defensive system, as revealed by their ability to infect plants.

Viroids resemble in their secondary structure, with alternate double-stranded stretches separated by single-stranded loops, the precursors of microRNA (miRNA), a class of endogenous small RNAs that are processed in the nucleus by DCL1. This enzyme could also target viroid dsRNAs resulting from replication in the nucleus of members of the family *Pospiviroidae*. In contrast, members of the family *Avsunviroidae* replicate and accumulate in the chloroplast where DCL activity has not been reported so far. Nonetheless, for cell-to-cell and long-distance trafficking, members of both families must move through the cytoplasm of the cells initially infected where DCL isoenzymes are located. Indeed, viroid-derived small RNAs (vd-sRNAs) with the structural properties of siRNAs have been detected in tomato infected with PSTVd (Itaya *et al.*, 2001; Papaefthimiou *et al.*, 2001), and in peach, chrysanthemum, and avocado infected with PLMVd, CChMVd, and ASBVd, respectively (Martínez de Alba *et al.*, 2002; Markarian *et al.*, 2004). Moreover, CEVd sRNAs in infected tomato are phosphorylated and methylated at their 5′ and 3′ termini, respectively, as endogenous plant siRNAs also are, further supporting the view that they are actually DCL products (Martín *et al.*, 2007). Cloning of PSTVd and CEVd sRNAs from infected tomato has shown that most are of (+) polarity and correspond to certain domains of the viroid structure, suggesting that they mainly derive from the action of DCL on preferred regions of the viroid (+) genomic RNA (Itaya *et al.*, 2007; Martín *et al.*, 2007). Furthermore, PSTVd and HSVd sRNAs are biologically active in guiding RISC to cleave viroid RNAs when fused to mRNA reporters (Vogt *et al.*, 2004; Gómez and Pallás, 2007), although the mature viroid RNAs transfected to *Nicotiana benthamiana* protoplasts (Itaya *et al.*, 2007) or transgenically expressed in this same species (Gómez and Pallás, 2007) are resistant to RISC-mediated degradation. Other *in planta* experiments, however, have shown that co-delivery of representative members of both families with their homologous dsRNAs or vd-sRNAs has a negative effect on infectivity, suggesting that, at least in some instances, viroids are targeted by RISC (Carbonell *et al.*, 2008).

These interactions with the RNA-silencing machinery of their hosts might have been the principal driving force shaping viroid structure and evolution (Wang *et al.*, 2004). More specifically, viroids could have evolved their secondary structure as a compromise between resistance to DCL and RISC, which act preferentially against RNAs with compact and relaxed secondary structures, respectively, although subcellular compartmentation, association with host proteins, or active replication could also help viroids to elude their host RNA-silencing machinery.

VIROID EPIDEMIOLOGY

The present understanding of viroid epidemiology in crop plants indicates that spread and persistence of these pathogens is mainly associated with the international exchange of germplasm, vegetative propagation of infected material, and agricultural practices that favor mechanical transmission. However, the hypothesis of the emergence of viroid-like molecules from an early RNA world must be compatible both with their survival until the appearance of suitable host plants and with their perpetuation into such hosts. Perpetuation and evolution of viroids within their initial hosts can only be explained through seed transmission, despite most viroids having been identified in hosts grown as agricultural commodities in which seed transmission is infrequent. However, there are instances (PSTVd, CbVd-1, and ASBVd) in which seed transmission has been demonstrated (reviewed by Singh *et al.*, 2003). In this context, ASBVd offers the most interesting example because symptomless carrier trees present an unusually high rate (95%) of seed transmission (Wallace and Drake, 1962), a situation that clearly favors viroid survival and spread. Therefore, viroids might be more widespread that initially thought and the identification of symptomless wild species carrying known and/or unknown viroids would help to better understand their origin and evolution.

Unfortunately, with a few exceptions, most viroids known today have been identified as disease-causing agents in agricultural crops and, therefore, our knowledge about their origin, variability, and evolution is limited because of specific selection pressures imposed by agricultural practices. Since these practices tend to eliminate low-performing plants, the identification of viroids in vegetatively propagated woody species such as citrus, fruit trees, and grapevine, which underwent clonal selection, indicate that at least in some instances viroid infection might have been linked to a desirable character. In this sense, the general defense responses induced by viroids (Conejero *et al.*, 1990; Vera *et al.*, 1993; Gadea *et al.*, 1996) could explain early observations indicating that CEVd-infected citrus were more resistant to damage by the fungus *Phytophthora* (Rossetti *et al.*, 1980; Solel *et al.*, 1995).

CONCLUDING REMARKS AND PERSPECTIVES

The small size of viroids makes them ideal systems for studying the evolution of a self-replicating RNA because changes in the whole genome, rather than in a fragment thereof as is the usual case in viruses, can be followed. Moreover, the extreme functional simplicity of viroids—there are no interferences between replication and transcription or between transcription and translation as in viruses, with the viroid genotype being essentially expressed in the RNA secondary structure—facilitates interpretation of constraints limiting their evolution. Finally, the presence of hammerhead structures in members of the family *Avsunviroidae* provides a unique opportunity for dissecting the function of these ribozymes in their natural habitat and for getting insights on how primitive replicons might replicate.

REFERENCES

Amari, K., Gómez, G., Myrta, A., Di Terlizzi, B. and Pallás, V. (2001) The molecular characterization of 16 new sequence variants of Hop stunt viroid reveals the existence of invariable regions and a conserved hammerhead-like structure on the viroid molecule. *J. Gen. Virol.* **82**, 953–962.

Ambrós, S., Hernández, C. and Flores, R. (1999) Rapid generation of genetic heterogeneity in progenies from individual cDNA clones of *Peach latent mosaic viroid* in its natural host. *J. Gen. Virol.* **80**, 2239–2252.

Ancel, L.W. and Fontana, W. (2000) Plasticity, evolvability, and modularity in RNA. *J. Exp. Zool.* **288**, 242–283.

Bonfiglioli, R.G., McFadden, G.I. and Symons, R.H. (1994) *In situ* hybridization localizes *Avocado sunblotch viroid* on chloroplast thylakoid membranes and *Coconut cadang cadang viroid* in the nucleus. *Plant J.* **6**, 99–103.

Bonfiglioli, R.G., Webb, D.R. and Symons, R.H. (1996) Tissue and intra-cellular distribution of *Coconut cadang cadang viroid* and *Citrus exocortis viroid* determined by *in situ* hybridization and confocal laser scanning and transmission electron microscopy. *Plant J.* **9**, 457–465.

Bussière, F., Lehous, J., Thompson, D.A., Skrzeckowski, L.J. and Perreault, J. (1999) Subcellular location and rolling circle replication of *Peach latent mosaic viroid*: Hallmarks of group A viroids. *J. Virol.* **73**, 6353–6360.

Carbonell, A., Martínez de Alba, A.E., Flores, R. and Gago, S. (2008) Double-stranded RNA interferes in a sequence-specific manner with infection of representative members of the two viroid families. *Virology* **371**, 44–53.

Chela-Flores, J. (1994) Are viroids molecular fossils of the RNA world?. *J. Theor. Biol.* **166**, 163–166.

Codoñer, F.M., Daròs, J.A., Solé, R.V. and Elena, S.F. (2006) The fittest versus the flattest: experimental confirmation of the quasispecies effect with subviral pathogens. *PLoS Pathog.* **2**, 1187–1193.

Conejero, V., Bellés, J.M. and García-Breijo, F. (1990) Signal in viroid pathogenesis. In: *Recognition and Response in Plant-Virus Interactions* (R.S.S. Fraser, ed.), pp. 233–261. Berlin: NATO Springer-Verlag.

Daròs, J.A. and Flores, R. (2002) A chloroplast protein binds a viroid RNA *in vivo* and facilitates its hammerhead-mediated self-cleavage. *EMBO J* **21**, 749–759.

Daròs, J.A., Marcos, J.F., Hernández, C. and Flores, R. (1994) Replication of avocado sunblotch viroid: evidence for a symmetric pathway with two rolling circles and hammerhead ribozyme processing. *Proc. Natl Acad. Sci. USA* **91**, 12813–12817.

Daròs, J.A., Elena, S.F. and Flores, R. (2006) Viroids: an Ariadne's thread into the RNA labyrinth. *EMBO Rep.* **7**, 593–598.

De la Peña, M., Navarro, B. and Flores, R. (1999) Mapping the molecular determinant of pathogenicity in a hammerhead viroid: a tetraloop within the *in vivo* branched RNA conformation. *Proc. Natl Acad. Sci. USA* **96**, 9960–9965.

De la Peña, M., Gago, S. and Flores, R. (2003) Peripheral regions of natural hammerhead ribozymes greatly increase their self-cleavage activity. *EMBO J.* **22**, 5561–5570.

De Visser, J.A.G.M., Hermisson, J., Wagner, G.P., Ancel Meyers, L., Bagheri-Chaichian, H., Blanchard, J. *et al.* (2003) Evolution and detection of genetic robustness. *Evolution* **57**, 1959–1972.

Diener, T.O. (1989) Circular RNAs—Relics of precellular evolution. *Proc. Natl Acad. Sci. USA* **86**, 9370–9374.

Diener, T.O. (2003) Discovering viroids—a personal perspective. *Nat. Rev. Microbiol.* **1**, 75–80.

Di Serio, F. (2007) Identification and characterization of *Potato spindle tuber viroid* infecting *Solanum jasminoides* and *S. rantonnetii* in Italy. *J. Plant Pathol.* **89**, 297–300.

Di Serio, F., Aparicio, F., Alioto, D., Ragozzino, A. and Flores, R. (1996) Identification and molecular properties of a 306 nucleotide viroid associated with apple dimple fruit disease. *J. Gen. Virol.* **77**, 2833–2837.

Eigen, M. (1996) On the nature of virus quasispecies. *Trends Microbiol.* **4**, 216–218.

Elena, S.F., Dopazo, J., De la Peña, M., Flores, R., Diener, T.O. and Moya, A. (2001) Phylogenetic analysis of viroid and viroid-like satellite RNAs from plants: A reassessment. *J. Mol. Evol.* **53**, 155–159.

Elena, S.F., Carrasco, P., Daròs, J.A. and Sanjuán, R. (2006) Mechanisms of genetic robustness in RNA viruses. *EMBO Rep.* **7**, 168–173.

Eiras, M., Kitajima, E.W., Flores, R. and Daròs, J.A. (2007) Existence *in vivo* of the loop E motif in potato spindle tuber viroid RNA. *Arch. Virol.* **152**, 1389–1393.

Fadda, Z., Daròs, J.A., Fagoaga, C., Flores, R. and Duran-Vila, N. (2003a) Eggplant latent viroid, the candidate type species for a new genus within the family *Avsunviroidae* (hammerhead viroids). *J. Virol.* **77**, 6528–6532.

Fadda, Z., Daròs, J.A., Flores, R. and Duran-Vila, N. (2003b) Identification in eggplant of a variant of citrus exocortis viroid (CEVd) with a 96 nucleotide duplication in the right terminal region of the rod-like secondary structure. *Virus Res.* **97**, 145–149.

Flores, R., Hernández, C., De la Peña, M., Vera, A. and Daròs, J.A. (2001) Hammerhead ribozyme structure and function in plant RNA replication. *Meth. Enzymol.* **341**, 540–552.

Flores, R., Hernández, C., Martínez de Alba, A.E., Daròs, J.A. and Di Serio, F. (2005) Viroids and viroid-host interactions. *Annu. Rev. Phytopathol.* **43**, 117–139.

Fontana, W. (2002) Modelling "evo-devo" with RNA. *Bioessays* **24**, 1164–1177.

Forster, A.C. and Symons, R.H. (1987) Self-cleavage of plus and minus RNAs of a virusoid and a structural model for the active-sites. *Cell* **49**, 211–220.

Forster, A.C., Davies, C., Sheldon, C.C., Jeffries, A.C. and Symons, R.H. (1988) Self-cleaving viroid and newt RNAs may only be active as dimers. *Nature* **334**, 265–267.

Gadea, J., Mayda, M.E., Conejero, V. and Vera, P. (1996) Characterization of defense-related genes ectopically expressed in viroid-infected tomato plants. *Mol. Plant–Microbe Interact.* **9**, 409–415.

Gandía, M. and Duran-Vila, N. (2004) Variability of the progeny of a sequence variant of citrus bent leaf viroid (CBLVd). *Arch. Virol.* **149**, 407–416.

Gandía, M., Bernad, L., Rubio, L. and Duran-Vila, N. (2007) Host effect on the molecular and biological properties of a CEVd isolate from *Vicia faba*. *Phytopathology* **97**, 1004–1010.

Gómez, G. and Pallás, V. (2007) Mature monomeric forms of Hop stunt viroid resist RNA silencing in transgenic plants. *Plant J.* **51**, 1041–1049.

Góra, A., Candresse, T. and Zagorski, W. (1996) Use of intramolecular chimeras to map molecular determinants of symptom severity of potato spindle tuber viroid (PSTVd) *Arch. Virol.* **141**, 2045–2055.

Góra-Sochacka, A., Kierzek, A., Candresse, T. and Zagórski, W. (1997) The genetic stability of potato spindle tuber viroid (PSTVd) molecular variants. *RNA* **3**, 68–74.

Hammond, R., Smith, D.R. and Diener, T.O. (1989) Nucleotide-sequence and proposed secondary structure of columnea latent viroid: a natural mosaic of viroid sequences. *Nucleic Acids Res.* **17**, 10083–10094.

Haseloff, J. and Symons, R.H. (1981) Chrysanthemum stunt viroid: primary sequence and secondary structure. *Nucleic Acids Res.* **9**, 2741–2752.

Haseloff, J., Mohamed, N.A. and Symons, R.H. (1982) Viroid RNAs of cadang-cadang disease of coconuts. *Nature* **299**, 316–321.

Hernández, C. and Flores, R. (1992) Plus and minus RNAs of peach latent mosaic viroid self-cleave *in vitro* via hammerhead structures. *Proc. Natl Acad. Sci. USA* **89**, 3711–3715.

Herold, T., Hass, B., Singh, R.P., Boucher, A. and Sänger, H.L. (1992) Sequence analysis of new field isolates demonstrates that the chain length of potato spindle tuber viroid (PSTVd) is not strictly conserved as in other viroids. *Plant Mol. Biol.* **19**, 329–333.

Hofacker, I.L. (2003) Vienna RNA secondary structure server. *Nucleic Acids Res.* **31**, 3429–3431.

Hull, R. (2002) *Matthews' Plant Virology*, 4th edn.. London: Academic Press.

Hutchins, C.J., Rathjen, P.D., Forster, A.C. and Symons, R.H. (1986) Self-cleavage of plus and minus RNA transcripts of avocado sunblotch viroid. *Nucleic Acids Res.* **14**, 3627–3640.

Itaya, A., Folimonov, A., Matsuda, Y., Nelson, R.S. and Ding, B. (2001) Potato spindle tuber viroid as inducer of RNA silencing in infected tomato. *Mol. Plant–Microbe Interact.* **14**, 1332–1334.

Itaya, A., Zhong, X.H., Bundschuh, R., Qi, Y.J., Wang, Y., Takeda, R. *et al.* (2007) A structured viroid RNA serves as a substrate for dicer-like cleavage to produce biologically active small RNAs but is resistant to RNA-induced silencing complex-mediated degradation. *J. Virol.* **81**, 2980–2994.

Juhasz, A., Hegyi, H. and Solymosy, F. (1988) A novel aspect of the information-content of viroids. *Biochim. Biophys. Acta* **950**, 455–458.

Keese, P. and Symons, R.H. (1985) Domains in viroids: evidence of intermolecular RNA rearrangements and their contribution to viroid evolution. *Proc. Natl Acad. Sci. USA* **82**, 4582–4586.

Khvorova, A., Lescoute, A., Westhof, E. and Jayasena, S.D. (2003) Sequence elements outside the hammerhead ribozyme catalytic core enable intracellular activity. *Nat. Struct. Biol.* **10**, 708–712.

Kimura, M. and Maruyama, T. (1966) The mutational load with epistatic gene interactions in fitness. *Genetics* **54**, 1337–1351.

Kofalvi, S.A., Marcos, J.F., Cañizares, M.C., Pallás, V. and Candresse, T. (1997) Hop stunt viroid (HSVd) sequence variants from *Prunus* species: evidence for recombination between HSVd isolates. *J. Gen. Virol.* **78**, 3177–3186.

Lakshman, D.K. and Tavantzis, S.M. (1993) Primary and secondary structure of a 360-nucleotide isolate of potato spindle tuber viroid. *Arch. Virol.* **128**, 319–331.

Li, F. and Ding, S.W. (2006) Virus counterdefense: diverse strategies for evading the RNA-silencing immunity. *Annu. Rev. Microbiol.* **60**, 503–531.

Luria, S.E. and Delbrück, M. (1943) Mutations of bacteria from virus sensitivity to virus resistance. *Genetics* **28**, 491–511.

MacDiarmid, R. (2005) RNA silencing in productive virus infections. *Annu. Rev. Phytopathol.* **43**, 523–544.

Malfitano, M., Di Serio, F., Covelli, L., Ragozzino, A., Hernández, C. and Flores, R. (2003) Peach latent mosaic viroid variants inducing peach calico (extreme chlorosis) contain a characteristic insertion that is responsible for this symptomatology. *Virology* **313**, 492–501.

Markarian, N., Li, H.W., Ding, S.W. and Semancik, J.S. (2004) RNA silencing as related to viroid induced symptom expression. *Arch. Virol.* **149**, 397–406.

Martick, M. and Scott, W.G. (2006) Tertiary contacts distant from the active site prime a ribozyme for catalysis. *Cell* **126**, 309–320.

Martín, R., Arenas, C., Daròs, J.A., Covarrubias, A., Reyes, J.L. and Chua, N.H. (2007) Characterization of small RNAs derived from citrus exocortis viroid (CEVd) in infected tomato plants. *Virology* **367**, 135–146.

Martínez de Alba, A.E., Flores, R. and Hernández, C. (2002) Two chloroplastic viroids induce the accumulation of small RNAs associated with posttranscriptional gene silencing. *J. Virol.* **76**, 13094–13096.

Martínez-Soriano, J.P., Galindo-Alonso, J., Maroon, C.J.M., Yucel, I., Smith, D.R. and Diener, T.O. (1996) Mexican papita viroid: putative ancestor of crop viroids. *Proc. Natl Acad. Sci. USA* **93**, 9397–9401.

Matzke, M.A. and Matzke, A.J.M. (2004) Planting the seeds of a new paradigm. *PLoS Biol.* **2**, 582–586.

Mühlbach, H.P. and Sänger, H.L. (1979) Viroid replication is inhibited by α-amanitin. *Nature* **278**, 185–188.

Navarro, B. and Flores, R. (1997) Chrysanthemum chlorotic mottle viroid: unusual structural properties of a subgroup of self-cleaving viroids with hammerhead ribozymes. *Proc. Natl Acad. Sci. USA* **94**, 11262–11267.

Navarro, J.A., Vera, A. and Flores, R. (2000) A chloroplastic RNA polymerase resistant to tagetitoxin is involved in replication of avocado sunblotch viroid. *Virology* **268**, 218–225.

Niblett, C.M., Dickson, E., Fernow, K.H., Horst, R.K. and Zaitlin, M. (1978) Cross protection among four viroids. *Virology* **91**, 198–203.

Nishihara, T., Mills, D.R. and Kramer, F.R. (1983) Localization of the Q-beta replicase recognition site in MDV-1 RNA. *J. Biochem.* **93**, 669–674.

Owens, R.A., Candresse, T. and Diener, T.O. (1990) Construction of novel viroid chimeras containing portions of tomato apical stunt and citrus exocortis viroids. *Virology* **175**, 238–246.

Owens, R.A., Steger, G., Hu, Y., Hammond, R.W. and Riesner, D. (1996) RNA structural features responsible for potato spindle tuber viroid pathogenicity. *Virology* **22**, 144–158.

Palacio-Bielsa, A., Romero-Durban, J. and Duran-Vila, N. (2004) Chracterization of citrus HSVd isolates. *Arch. Virol.* **149**, 537–552.

Papaefthimiou, I., Hamilton, A.J., Denti, M.A., Baulcombe, D.C., Tsagris, M. and Tabler, M. (2001) Replicating potato spindle tuber viroid RNA is accompanied by short RNA fragments that have characteristic of post-transcriptional gene silencing. *Nucleic Acids Res.* **29**, 2395–2400.

Prody, G.A., Bakos, J.T., Buzayan, J.M., Schneider, I.R. and Bruening, G. (1986) Autolytic processing of dimeric plant-virus satellite RNA. *Science* **231**, 1577–1580.

Pruss, G., Ge, X., Shi, X.M., Carrington, J.C. and Vance, V.B. (1997) Plant viral synergism: the potyviral genome encodes a broad-range pathogenicity enhancer that transactivates replication of heterologous viruses. *Plant Cell* **9**, 859–868.

Puchta, H., Ramm, K., Luckinger, R., Hadas, R., Bar-Joseph, M. and Sänger, H.L. (1991) Primary and secondary structure of citrus viroid-IV (CVd-IV), a new chimeric viroid present in dwarfed grapefruit in Israel. *Nucleic Acids Res.* **19**, 6640.

Ratcliff, F.G., MacFarlane, S.A. and Baulcombe, D.C. (1999) Gene silencing without DNA: RNA-mediated cross-protection between viruses. *Plant Cell* **11**, 1207–1215.

Reanwarakorn, K. and Semancik, J.S. (1998) Regulation of pathogenicity in hop stunt viroid-related group II citrus viroids. *J. Gen. Virol.* **79**, 3163–3171.

Rodio, M.E., Delgado, S., Flores, R. and Di Serio, F. (2006) Variants of peach latent mosaic viroid inducing peach calico: uneven distribution in infected plants and requirements of the insertion containing the pathogenicity determinant. *J. Gen. Virol.* **87**, 231–240.

Rodio, M.E., Delgado, S., De Stradis, A., Gómez, M.D., Flores, R. and Di Serio, F. (2007) A viroid RNA with a specific structural motif inhibits chloroplast development. *Plant Cell* **19**, 3610–3626.

Rossetti, V., Pompeu, J., and Rodriguez, O. (1980). Reaction of exocortis-infected and healthy trees to experimental *Phytophthora* inoculations. In: *Proceedings of the 8th IOCV Conference*. pp. 209–214. IOCV, Riverside, USA.

Sanjuán, R., Forment, J. and Elena, S.F. (2006a) *In silico* predicted robustness of viroid RNA secondary structures. I. The effect of single mutations. *Mol. Biol. Evol.* **23**, 1427–1436.

Sanjuán, R., Forment, J. and Elena, S.F. (2006b) *In silico* predicted robustness of viroid RNA secondary structures. II. Interaction between mutation pairs. *Mol. Biol. Evol.* **23**, 2123–2130.

Sänger, H.L. and Spieker, R.L. (1997) RNA recombination between viroids. In: *Plant Viroids and Viroid-like Satellite RNAs from Plants, Animals and Fungi*, p. 13. Madrid: Instituto Juan March de Estudios e Investigaciones.

Sano, T. and Ishiguro, A. (1998) Viability and pathogenicity of intersubgroup viroid chimeras suggest possible involvement of the terminal right region in replication. *Virology* **240**, 238–244.

Sano, T., Candresse, T., Hammond, R.W., Diener, T.O. and Owens, R.A. (1992) Identification of multiple structural domains regulating viroid pathogenicity. *Proc. Natl. Acad. Sci. USA* **89**, 10104–10108.

Schnölzer, M., Haas, B., Ramm, K., Hofmann, H. and Sänger, H.L. (1985) Correlation between structure and pathogenicity of potato spindle tuber viroid (PSTV). *EMBO J.* **4**, 2181–2190.

Semancik, J.S. and Szychowski, J.A. (1994) Avocado sunblotch disease—A persistent viroid infection in which variants are associated with differential symptoms. *J. Gen. Virol.* **75**, 1543–1549.

Semancik, J.S., Szychowski, J.A., Rakowski, A.G. and Symons, R.H. (1993) Isolates of citrus-exocortis viroid recovered by host and tissue selection. *J. Gen. Virol.* **74**, 2427–2436.

Semancik, J.S., Szychowski, J.A., Rakowski, A.G. and Symons, R.H. (1994) A stable 463-nucleotide variant of citrus exocortis viroid produced by terminal repeats. *J. Gen. Virol.* **75**, 727–732.

Serra, P., Barbosa, C.J., Daròs, J.A., Flores, R. and Duran-Vila, N. (2008) Citrus viroid V: molecular characterization and synergistic interactions with other members of the genus *Apscaviroid*. *Virology* **370**, 102–112.

Singh, R.P., Ready, K.F.M. and Nie, X. (2003) Biology. In: *Viroids* (A. Hadidi, R. Flores, J.W. Randles and J.S. Semancik, eds), pp. 30–48. Melbourne: CSIRO Publishing.

Solel, Z., Mogilner, N., Gafny, R. and Bar-Joseph, M. (1995) Induced tolerance to mal secco disease in etrog citron and Rangpur lime by infection with the citrus exocortis viroid. *Plant Dis.* **79**, 60–62.

Spiegelman, S. (1971) An approach to the experimental analysis of precellular evolution. *Q. Rev. Biophys.* **4**, 213–253.

Spieker, R.L. (1996) *In vitro*-generated 'inverse' chimeric *Coleus blumei* viroids evolve *in vivo* into infectious RNA replicons. *J. Gen. Virol.* **77**, 2839–2846.

Szychowski, J.A., Vidalakis, G. and Semancik, J.S. (2005) Host-directed processing of *Citrus exocortis viroid*. *J. Gen. Virol.* **86**, 473–477.

Takeda, A., Tsukuda, M., Mizumoto, H., Okamoto, K., Kaido, M., Mise, K. and Okuno, T. (2005) A plant RNA virus suppresses RNA silencing through viral RNA replication. *EMBO J.* **24**, 3147–3157.

Takeuchi, N. and Hogeweg, P. (2007) Error-threshold exists in fitness landscapes with lethal mutants. *BMC Evol. Biol.* **7**, 15.

Van Nimwegen, E., Crutchfield, J.P. and Huynen, M. (1999) Neutral evolution of mutational robustness. *Proc. Natl. Acad. Sci. USA* **96**, 9716–9720.

Vera, P., Tornero, P. and Conejero, V. (1993) Cloning and expression analysis of a viroid-Induced peroxidase from tomato plants. *Mol. Plant–Microbe Interact.* **6**, 790–794.

Verhoeven, J.T.J., Jansen, C.C.C. and Roenhorst, J.W. (2006) First report of potato virus M and chrysanthemum stunt viroid in *Solanum jasminoides*. *Plant Dis.* **90**, 1359.

Verhoeven, J.T.J., Jansen, C.C.C., Werkman, A.W. and Roenhorst, J.W. (2007) First report of tomato chlorotic dwarf viroid in *Petunia hybrida* from the United States of America. *Plant Dis.* **91**, 324.

Visvader, J.E. and Symons, R.H. (1985) Eleven new sequence variants of citrus exocortis viroid and the correlation of sequence with pathogenicity. *Nucleic Acids Res.* **13**, 2907–2920.

Visvader, J.E. and Symons, R.H. (1986) Replication of in vitro constructed viroid mutants: location of the pathogenicity-modulating domain of citrus exocortis viroid. *EMBO J.* **5**, 2051–2055.

Vogt, U., Pelissier, T., Putz, A., Razvi, F., Fischer, R. and Wassenegger, M. (2004) Viroid-induced RNA silencing of GFP-viroid fusion transgenes does not induce extensive spreading of methylation or transitive silencing. *Plant J.* **38**, 107–118.

Wallace, J.M. and Drake, R.J. (1962) High rate of seed transmission of avocado sunblotch virus from symptomless trees and origin of such trees. *Phytopathology* **52**, 237–241.

Wagner, G.P. and Krall, P. (1993) What is the difference between models of error threshold and Muller's ratchet?. *J. Math. Biol.* **32**, 33–44.

Wang, M.B., Bian, X.Y., Wu, L.M., Liu, L.X., Smith, N.A., Isenegger, D. *et al.* (2004) On the role of RNA silencing in the pathogenicity and evolution of viroids and viral satellites. *Proc. Natl Acad. Sci. USA* **101**, 3275–3280.

Wang, Y., Zhong, X.H., Itaya, A. and Ding, B. (2007) Evidence for the existence of the loop E motif of potato spindle tuber viroid in vivo. *J. Virol.* **81**, 2074–2077.

Wilke, C.O. (2005) Quasispecies theory in the context of population genetics. *BMC Evol. Biol.* **5**, 44.

Wilke, C.O., Wang, J.L., Ofria, C., Lenski, R.E. and Adami, C. (2001) Evolution of digital organisms at high mutation rate leads to survival of the flattest. *Nature* **412**, 331–333.

CHAPTER

3

Mutation, Competition, and Selection as Measured with Small RNA Molecules

Christof K. Biebricher

ABSTRACT

Leviviruses code for a replicase that is able to amplify, together with a host factor, the viral RNA autocatalytically. Short-chained RNA species can be isolated that are amplified efficiently by replicase preparations purified to homogeneity. The products of the replication are the template and the complementary replica strand, both in single-stranded form. The evolutionary behavior of replicating RNA can be precisely predicted from rate parameters of the replication process. The replicating RNA produces a broad mutant distribution, as predicted by Eigen's quasispecies theory, where each type is represented according to its rate of formation by mutation from other types and by its fitness. Under conditions of high replicase and substrate concentrations, the replicase can synthesize short-chained replicable RNA species after a long lag time. The isolated species have a structure in common that allows strand separation during replication.

INTRODUCTION

Darwin's theory of natural selection is one of the greatest milestones in science. It provides answers to deep questions that are otherwise unanswerable. As Dobzhansky put it: "Nothing makes sense in Biology except in light of evolution" (Dobzhansky *et al.*, 1977).

Yet even a century after Darwin we still cannot understand organismic evolution in detail, let alone make quantitative predictions about its course. While neodarwinistic theory does provide insights into evolution processes in quantitative terms, it essentially comprises quantitative descriptions of reproduction, in particular in Mendelian populations, not of evolution itself. For the fundamental processes operating in evolution—mutation and selection—its parameters have to be adjusted to fit, more or less, the evolutionary outcome. They cannot be derived from measurable properties of the organisms themselves or of their genes.

Origin and Evolution of Viruses
ISBN 978-0-12-374153-0

In contrast to our lack of quantitative descriptions of evolution, its molecular basis is very well understood: the information needed for morphogenesis and function is encoded into the genotype as the sequence of nucleotides in each organism's genome (a few exceptions notwithstanding). The key step in reproduction is copying that nucleotide sequence. While replication error rates leading to accepted mutations vary over the genome, and also depend to some degree upon environmental conditions, these errors are more or less random. No teleology directing mutations to an advantageous result has ever been observed.

Information informs only if it is understood; thus genetic information has to be decoded by cellular machinery in order to operate a life-sustaining program of biochemical reactions. The program depends on the environment and leads to the properties of the organism which we observe as its phenotype. Evolutionary success depends on two components of the phenotype: Those that determine survival in the prevailing environment and those that establish the rate of producing viable offspring. The combination of these two determines the population trends that we call fitness.

Environments are typically complex and variable. Species interact by competition for resources, predation or symbiosis; individuals of the same species influence one another socially, and the individuals themselves may be composed of large numbers of specialized cells, all containing the same genetic information, that have to cooperate with one another for the organism they compose to survive and reproduce. While the way that gene expression produces the metabolic apparatus is now felt to be well understood, we know much less about how cells acquire information about the environment and how this information triggers appropriate genetic responses. Biologists still debate the identity of the target of selection: Is it an ecosystem, a species, a variant, a subpopulation, an individual, a gene or merely a "replicator unit" (Dawkins, 1982)? There is no ultimate answer to this question: selection takes place at all of these levels. Which of the selection levels dominates depends on the environment.

Evolution is a dynamic self-organization process in which causal correlations between the performance of the process as a whole and its component subprocesses are not identifiable (Biebricher *et al.*, 1995). We can correlate fitness to the function of a single gene only if this gene happens to be absolutely required for survival under the prevailing conditions. Further, translation of a genotype into a phenotype is far too complicated to be evaluated. Fitness values have to be determined a posteriori, i.e. so as to describe the observed changes in the composition of the two populations under study. Prediction of an evolutionary outcome from fitness values obtained from process-independent parameters is generally impossible. Is the Darwinian concept of evolution then merely tautological, describing the survival of the survivors, as some have criticized? No, it is not. The studies with RNA viruses described in this volume witness that in many cases the molecular basis of fitness can be clearly identified.

Nevertheless, even the simplest RNA viruses are too complicated to allow quantitative descriptions of their mutation, competition, and selection, in particular because their complex interactions with host cells are inevitably involved in the evolutionary process. The 1961 discovery of RNA bacteriophages by Loeb and Zinder (1961), more than 80 years after the discovery of plant RNA viruses, was instrumental in accelerating progress in understanding molecular processes in the infection cycles. Ten years after their discovery they were already by far the best understood viruses. In particular, because of their essential nature as parasitic messenger RNA, they became invaluable experimental tools in studying the expression of genetic information.

THE EXPERIMENTAL SYSTEM

Most RNA bacteriophages belong to the plus-strand virus family *Leviviridae*. Except for a few members of the *Reoviridae* family, no other RNA virus families have been found to infect prokaryotes. Members of the family *Leviviridae* are particularly simple, in all respects, and their genome sizes are the smallest among autonomously infecting viruses.

Shortly after their discovery, several research groups succeeded in detecting a novel RNA-dependent RNA polymerase in levivirus-infected cells (August *et al.*, 1963; Weissmann *et al.*, 1963; Haruna *et al.*, 1963). After being found to be highly specific in amplifying viral RNA it became known as a replicase. The replicase of the coliphage Qβ was found to be particularly stable and thus the most suitable one for *in vitro* studies. Purification of the Qβ replicase to homogeneity (Kamen, 1970; Kondo *et al.*, 1970) revealed four subunits, one coded by the R gene of the phage, the others provided by the host. Together with an additional host factor (Franze de Fernandez *et al.*, 1968) they perform all steps necessary to amplify viral RNA.

The experimental procedure for replication experiments is simple. An RNA template is incubated with replicase purified from Qβ-infected cells, appropriate amounts of the four nucleoside triphosphate precursors and an appropriate buffer. When using viral RNA as template, the progeny RNA synthesized *in vitro* was found to be infectious (Spiegelman *et al.*, 1965). However, when Spiegelman and collaborators tried to dilute out parental RNA by serially diluting aliquots of growing RNA into fresh test tubes containing replicase and precursors, further production of infectious RNA stopped already after the fifth serial transfers, while incorporation of nucleoside triphosphates into RNA continued, even at steadily increasing rates. Spiegelman and collaborators recognized that they had performed "An extracellular Darwinian experiment with a self-duplicating nucleic acid molecule," the title of a paper which became seminal for *in vitro* evolution studies (Mills *et al.*, 1967). After 74 serial transfers, the RNA was analyzed and found to have eliminated 83% of its chain length in the course of increasing its replication rate. Experiments under different replication conditions, e.g. reduced levels of one nucleoside triphosphate or in the presence of inhibitors (Levisohn and Spiegelman, 1969; Saffhill *et al.*, 1970) were performed; the RNA was shown to adapt to these conditions, "revealing an unexpected wealth of phenotypic differences which a replicating nucleic acid can exhibit."

The experiment had an immediate impact. While most scientists were enthusiastic about the new possibilities, some scoffed that the experiment represented "a search for the best carcass." Indeed, the evolution experiments did start with an infectious RNA able to perform many different roles—to serve as messenger, to replicate, and to be packed in a protein coat—and end with a "variant RNA" that had lost most of the information and was only able to replicate. Evolution thus produced degeneration, as have many other experimental evolution trials.

Furthermore, the reaction and its products were declared to be "unphysiological." Indeed, they have to be, for otherwise the consequences for phage reproduction would be disastrous. This criticism applies also to the other evolution studies reported below, i.e. the templates and the reactions are all unphysiological. This does not diminish their value, however, in studying evolution. The experiments I describe abstract the essential features from the enormously complex net of physiological interactions at work in physiological phage infection cycles *in vivo*. While replicase and RNA synthesis precursors must be synthesized during infection cycles, they are environmental factors in the *in vitro* experiments. The *in vitro* experiments provide precisely controllable and reproducible conditions that are indispensable for quantitative studies.

Eigen (1971) provided the theoretical background for quantitative analysis of evolving simple replicators, and we used the quantitative data to test the validity of Eigen's predictions. The nomenclature of the parameters (Eigen and Biebricher, 1987) (Table 1.2) uses Eigen's proposals. As this article shows, the observed features of the evolutionary behavior of replicating RNA follows closely what has been predicted by the theory.

THE MECHANISM OF RNA REPLICATION

Qβ replicase has evolved features absolutely necessary for virus infection:

1 It is highly specific in accepting its own RNA as template. It is vital for phage reproduction that only the viral RNA be amplified while a huge excess of host RNA is ignored. Therefore, the RNA itself cannot be considered to be merely a substrate of the replicase: it shares the catalytic role (Biebricher *et al.*, 1981a). Its template activity, i.e. its efficiency of instructing the replicase to replicate it, is a phenotypic expression of the RNA species that is crucial for its evolutionary success. For practical studies, artificial short-chained RNA species that are efficiently amplified by Qβ replicase have been selected, in most cases by the so-called template-free synthesis procedure described later.

2 Only single-stranded RNA is accepted as template. The replica formed is complementary and antiparallel to the template, suggesting Watson–Crick base-pairing at the replication site. A few other templates, e.g. poly C and C-rich nucleotide co-polymers that are accepted by Qβ replicase, result in a perfect double strand (Hori *et al.*, 1967; Mitsunari and Hori, 1973). Double strands are not replicated (Biebricher *et al.*, 1982), and thus synthesis using these templates stops after transcribing the RNA; neither template nor enzyme is recycled. Autocatalytic amplification of a template takes place only when template and replica strands separate during replication and are released individually in single-stranded form (Dobkin *et al.*, 1979; Biebricher, 1983). An RNA species thus always consists of both complementary sequences.

3 The replicase is highly processive, i.e. it usually does not dissociate from the template before a round is completed, because there is no way to complete a released incomplete replica strand in a subsequent reaction. As a consequence, release of the template after a replication round is the rate-limiting step in the whole cycle.

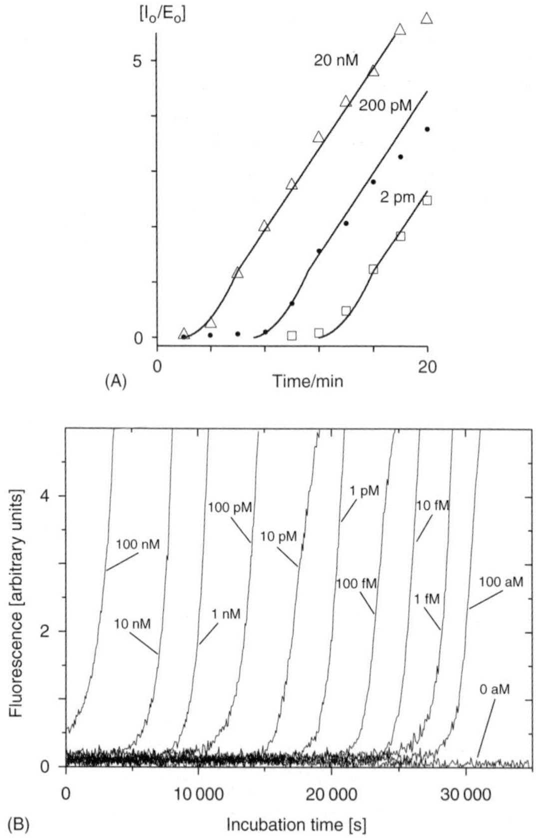

FIGURE 3.1 Incorporation profiles of growing RNA. (A) GMP incorporation of MNV-11 at various dilutions growing with 200 nM Qβ replicase. (B) Ethidium bromide fluorescence of RNA species EcprpG growing with 1 μM RNA polymerase from *E. coli*.

Two growth phases are clearly distinguished (Figure 3.1): an exponential one where enzyme is in excess and a linear one where enzyme is saturated with template. The overall growth rate in the linear growth phase is determined from the slope of the linear part. Since direct measurement of growth rate in the exponential growth rate is inaccurate by the unsatisfactory signal/noise ratio, it is determined by the time displacement Δt of the profile caused by dilution by the factor F_{dil} according to $\kappa = \ln F_{\text{dil}}/\Delta t$.

At the start of an amplification experiment, enzyme is typically present in large excess over RNA template. Newly synthesized replica as well as released template strands quickly bind to enzyme molecules, and exponential growth of the RNA concentration results. Once the enzyme is saturated with template, the RNA concentration increases linearly with time and the main products are free plus and minus strands. The overall replication rate in the linear growth phase is lower than in the exponential growth phase, because recycling of the enzyme becomes ratelimiting. Free complementary strands react to form double strands that are inactive as templates. Eventually, a steady state is reached where the concentration of single strands does not change, because the synthesis of new strands is balanced by loss in double strand formation. At the final steady state, only the concentration of double strands increases.

The essential chemical steps of the replication were identified in a series of experiments (Biebricher *et al.*, 1981b, 1982, 1983, 1984, 1985, 1991), and rate coefficients for some of them were measured; for others, reasonable estimates could be introduced to enable kinetic modeling (Figure 3.2, Box 3.1). A kinetic model set up

Box 3.1 Quantitative measurements of the replication rate

Typical growth profiles are shown in Figure 3.1. The detailed anlaysis of the steps in replication is quite complicated (Biebricher *et al.*, 1981b). However, the simple mechanism shown in Figure 3.2 is an adequate description for the replication time course when the replicase concentration exceeds 150 nM and the concentrations of the triphosphates are higher than 300 μM each.

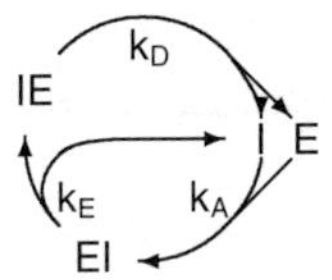

$$\mathrm{d[EI]/dt} = k_A\,[\mathrm{E}][\mathrm{I}] - k_E\,[\mathrm{EI}]$$
$$\mathrm{d[IE]/dt} = k_E\,[\mathrm{EI}] - k_D\,[\mathrm{IE}]$$

$$\mathrm{d[E_c]/dt} = k_A\,[\mathrm{E}][\mathrm{I}] - k_D\,[\mathrm{IE}] = -\mathrm{d[E]/dt}$$
$$\mathrm{d[I]/dt} = k_E\,[\mathrm{EI}] + k_D\,[\mathrm{IE}] - k_A\,[\mathrm{E}][\mathrm{I}]$$

$$\mathrm{d[I_o]/dt} = k_E\,[\mathrm{EI}]$$

FIGURE 3.2 Simplified mechanism of RNA replication. Shown are the steps involving binding and releasing of enzyme. The synthesis steps are combined to a single step. Top: mechanism; bottom: rate equations.

The relative population growth $d[I_o]/dt$ of type i is $A_i = {}^ik_E[{}^i\mathrm{EI}]/[{}^i\mathrm{I_o}]$ which is a well-known law of population genetics: the relative rate of population growth is equal to the proportion of the population in the reproducing age times birth rate. In the exponential growth phase, the intermediates and the total population show after an equilibration period coherent growth (Biebricher *et al.*, 1983):

$$\begin{aligned} d[\mathrm{I}]/[\mathrm{I}]dt = d[\mathrm{EI}]/[\mathrm{EI}]dt &= d[\mathrm{IE}]/[\mathrm{IE}]dt \\ &= d[\mathrm{I_o}]/[\mathrm{I_o}]dt \equiv \kappa \end{aligned}$$

where κ is the overall (exponential) replication rate. In the linear growth phase, virtually all of the enzyme is bound to template, and a steady state is established where the intermediate concentrations do not change and the flux through each step is equal to the total flux v:

$$v = k_A[\mathrm{E}][\mathrm{I}] = k_E[\mathrm{EI}] = k_D[\mathrm{IE}] \equiv \rho[\mathrm{Ec}].$$

In the linear growth phase of a single species, the enzyme is almost totally saturated with template and we obtain $\rho[\mathrm{Ec}] \approx \rho[\mathrm{E_o}]$, where $[\mathrm{E_o}]$ is the total enzyme concentration, free and bound. The relative fecundities A_i are constant in the exponential growth phase ($A_i = {}^i\kappa$), but decrease in the linear growth phase with increasing $[{}^i\mathrm{I_o}]$ ($A_i \approx ?[\mathrm{E_o}]/[\mathrm{I_o}]$). The population change of type i is also dependent on its mortality rate. Under the conditions of the described RNA replication experiments, the mortality is caused by the loss of template molecules by double strand formation; other contributions to mortality like decomposition can be neglected. The loss rate can be described with the equation $-d[\mathrm{I}]/dt = \{1/2\}d[\mathrm{II}]/dt = k_{ds}[\mathrm{I}]^2$. In the exponential growth phase, the concentrations of free strands are very small and the mortality is negligible.

The net population growth is the balance between fecundity and mortality: $E_i = A_i - D_i$. In the exponential growth phase, we obtain $E_i = {}^i\kappa$. In the linear growth phase, A_i decreases and D_i increases until steady state is reached where $A_i = D_i$ and $E_i = 0$. The steady state concentrations can be calculated when the rates values have been determined (Biebricher *et al.*, 1984, 1991).

this way is an oversimplification in the sense that the identified steps are not elementary chemical reactions. Binding of protein to RNA, for example, is not a simple bimolecular event: In reality, a cascade of chemical steps distorting bond angles, establishing short-range van der Waals interactions, hydrogen bonds, and pushing out water molecules is involved for both macromolecular components. Fortunately, incorporating this level of detail in the model was not necessary to rationalize the experimental RNA concentration profiles. On the contrary, the kinetic model was able to describe the complicated experimental RNA growth profiles precisely, and is thus adequate for drawing conclusions about the roles of different parts of the replication process in determining the fitness of mutant RNA species.

SELECTION OF RNA SPECIES

When two or more RNA species are present in the starting template population, they compete with one another. If the sequences and the physical properties of the species differ sufficiently, the outcome of the selection process can be followed. The simplest case is exponential growth of small populations, which prevails when all resources required for amplification are present in excess. Under these conditions, each species grows as it would in the absence of the others. Each species grows independently with its own characteristic growth rate, but as this goes on the composition of the population changes: The population gradually becomes enriched in species with higher growth rates and relatively depleted in species that grow more slowly. The outcome of experiments under strictly exponential conditions can be readily precalculated. The fitness of each species under exponential growth conditions is characterized by its fecundity, i.e. its replication rate, alone.

This strong selection makes working with an RNA replicase rather difficult. Assume that an RNA species with a replication rate 1/10 that of an optimized species has to be amplified by a factor of 10. While that happens, a single strand of the optimized species is amplified by a factor of 10^{10}, i.e. to macroscopic appearance! This illustrates that the techniques of amplification with replicase are technically not as easy as they might seem to be; severe precautions have to be made to avoid contamination of an RNA population with optimized species (Biebricher *et al.*, 1993). Synchronized amplification techniques like the polymerase chain reaction (PCR) are much easier to handle. Purification of an RNA species by physicochemical methods, e.g. by electrophoresis, must always be followed by a cloning procedure, because otherwise only the fastest species will be found. If we know that the separation method has reduced impurities to a level of say 1/1000, then it suffices to start an amplification experiment with fewer than 1000 template strands to get a pure species as product.

The complicated population dynamic can be illustrated by the growth profiles shown in Figure 3.3 obtained by computer simulation. The figure also shows the change of the parameters that are important for selection, A_i, D_i, E_i, and $\bar{E}$. Experimental determinations are in full agreement with the calculated values. The outcomes of selection experiments carried out under linear growth conditions are at first glance surprising: amplification of a mixed population by a factor of ten can result in a change in the population composition by many orders of magnitude, and often a species with a lower replication rate is selected. As observed in organismic evolution, species with low fecundity can be quite successful if they are able to outcompete their competitors for limiting resources.

In the linear growth phase of RNA replication, the limiting resource is the replicase itself, and the species that is fastest in binding to newly liberated replicase molecules will be selected whatever its replication rate may be. The quantitative description is somewhat more complicated. The kinetic model described previously is helpful: to describe competition in the linear growth phase, it is sufficient to set up the rate equations such that two species

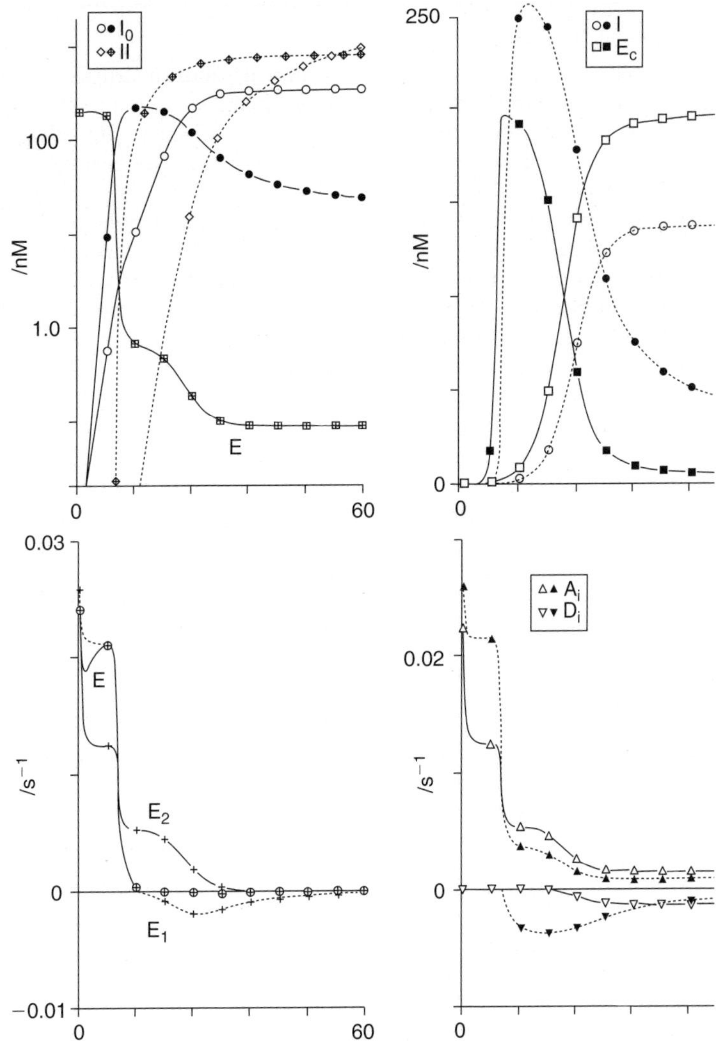

FIGURE 3.3 Competition among RNA species. Calculated growth profiles of two species: 1 (solid symbols) with standard rate values (MNV-11) and 2 (open symbols) having $^{2}k_A = 4 \times {}^{1}k_A$ and $^{2}k_D = 1/4 \times {}^{1}k_D$. Starting conditions were [E] = 200 nM and [^{1}A] = [^{2}A] = 1 pM. In the exponential growth phase (0–8 min), smaller k_D values are detrimental and species 1 grows more rapidly (see the semi-logarithmic plot at upper left). It saturates the enzymes and enters the steady state, where its net growth (lower left) stops. Species 2 continues to grow exponentially at a lower rate (due to the smaller amount of free enzyme); it conquers most of the enzyme because of its higher binding rate (note the diminishing concentration of free enzyme). After 60 min the final steady state is reached where each species occupies a constant part of the enzyme. Calculations were done by numerical integration of the rate equations, using a more detailed mechanism than shown in Figure 3.2. Approximate analytical solutions of the rate equations of the simplified mechanism can be found for certain cases. From Biebricher *et al.* (1985) *Biochemistry* 24, 6550–6560, with permission.

share the resources and the calculated profile again precisely matches the experimental results. Instead of fitness, it is better to work with selection (rate) values, the relative change of the relative population in time. The definitions are listed in Table 3.2. The selection values, which vary with time and concentration of the competitors in the linear growth phase,

Box 3.2 Competition among species

For competition experiments two (or more) different RNA species share the resources enzyme and precursors; their concentrations and rates are distinguished by indices. Since the absolute concentrations vary, especially when serial transfers are used, relative concentrations, i.e. the proportions of the species of the total population, x_i, are used to describe the population composition. Type conversion, i.e. that reproduction results in a different type, is not possible due to the species barrier.

Particularly easy is the calculation of the parameters important for selection in the exponential growth phase. The composition changes according to $([^1I_o]/[^2I_o])t=([^1I_o]/[^2I_o])_{t=0}\ \exp(^1\kappa - {}^2\kappa)t$ (Kramer *et al.*, 1974; Biebricher *et al.*, 1985). Instead of choosing the net growth rates E_i as selection values, it is more instructive to relate the net growth of each species to the total population change: we obtain the selection rate value $\varsigma_i = E_i - \bar{E}$, where $\bar{E}$ is the weighted average of the net synthesis rates. A positive ς_i value means that the species i is enriched in the population, a negative one means the population is depleting in species i.

The calculation of selection values is more difficult in the linear growth rate. It can be shown that in the early linear growth phase the intrinsic selection rate value is proportional to $^ik_A[E]$. At higher concentrations, the mortalities contribute also until finally an ecosystem is formed where each species occupies a constant fraction of the total population. The ratios of the free and bound types can be calculated (Biebricher *et al.*, 1985, 1991).

TABLE 3.1 Symbols and parameters used for kinetic studies of RNA replication

		Units
Concentrations		
$[^iI]$	Concentration of free single-stranded RNA of type *i*	mol/L
$[^iEI]$	Concentration of active replication complex	mol/L
$[^iIE]$	Concentration of inactive replication complex	mol/L
[E]	Concentration of free enzyme	mol/L
$[^iE_c]$	Total concentration of template strands of type *i* complexed to enzyme	mol/L[c]
$[E_o]$	Total concentration of enzyme, bound or free	mol/L[b]
$[^iI_o]$	Total concentration of template strands of type *i*	mol/L[c]
$\Sigma_i[^iI_o]$	Total concentration of RNA	mol/L[a]
$[^{ii}II]$	Concentration of double strands (homoduplex) of type *i*	mol/L[c]
$[^{ij}II]$	Concentration of double strands between plus strand of type *i* and minus strand of type *j* (heteroduplex)	mol/L
Rate constants		
^ik_A	Association rate constant for binding of replicase to RNA of type *i*	L/mol s^{-1}[a]
^ik_E	Rate constant for synthesizing and releasing a replica from a replication complex of type *i*	S^{-1}
^ik_D	Dissociation rate of inactive replication complex $[s^{-1}]$[a]	
$^{ij}k_{ds}$	Rate constant for double strand formation between plus strand of type *i* and minus strand of type *j*	L/mol s^{-1}[a]
$^i\kappa$	Overall replication rate constant of type *i* in the exponential growth phase	s^{-1}[a]
$^i\rho$	Experimentally measured relative rate of RNA synthesis per template strand of type *i*	s^{-1}[a]

[a]Parameter can be readily measured.
[b]Parameter set at the beginning of the experiment.
[c]Parameter can be readily measured when types can be easily distinguished.

TABLE 3.2 Parameters used for the evolution studies

Evolution parameter	Definitions	Units
x_i	Mutant frequency; fraction of type *i* in the total population	b
A_i	Relative fecundity of type *i*	s^{-1}
D_i	Relative mortality of type *i*	s^{-1}
E_i	Relative net excess production rate of type *i*; $E_i = A_i - D_i$	s^{-1b}
$\overline{E}$	Relative net excess production rate for all types	s^{-1a}
W_{ii}	Intrinsic selection rate value	s^{-1}
Q_{ij}	Probability of producing type *i* per reproduction process of type *j*	
Q_{ii}	Probability of producing a correct copy per reproduction process of type *i*	
${}^{i}\mu$	Mutational gain rate (synthesis by miscopying other templates)	s^{-1}
${}^{i}\varsigma$	Selection rate value	s^{-1}
${}^{i}\xi$	Evolution rate value; relative rate for relative increase of type *i*; ${}^{i}\xi = {}^{i}\varsigma + {}^{i}\mu$	s^{-1b}

[a] Parameter can be readily measured.
[b] Parameter can be measured when types can be distinguished.

can be precisely determined from the computed concentration profiles (Figure 3.3), but simple equations can only be found for special conditions.

In the late linear growth phase, the loss terms caused by double-strand formation must also be taken into account in the model. For the case that the nucleotide sequences of two competing species are rather different, formation of heteroduplex strands can be neglected. Species with low concentrations of free single strands are favored by low loss rates through double strand formation and so the population can eventually reach a steady state, where its relative composition no longer changes: a stable ecosystem has been formed (Figure 3.3 and Box 3.2).

Even under the controlled external conditions of *in vitro* evolution experiments, selection patterns can thus be quite complex, basically because the growing RNA species change their own environment. A typical example would be starting with two RNA species (MNV-11 and MDV-1), for which a computer simulation is shown in Figure 3.4. Initially both species are present in small equimolar amounts and exponential growth begins for both. When the enzyme is saturated, MNV-11 has conquered, because of its higher replication rate, most of the enzyme, and shortly afterwards it reaches the steady state of double strand formation and its selection value vanishes. However, MDV-1 continues to grow, and eventually it displaces MNV-11 from the enzyme due to its higher enzyme binding rate. Eventually an ecosystem is formed where both species co-exist; their selection values have both vanished.

MUTATION IN REPLICATING RNA

Any alteration of a genotype is called a mutation. Mutation may occur by chemical modification of a base, such as deamination of a cytidylate to a uridylate, but most mutations are produced by an erroneous replication, i.e. the progeny genotype differs at one or more positions from the parental genotype. Luria and Delbrück (1943) showed in a classic experiment that the mutation event in bacteria is stochastic and that the mutated type may spread by error propagation. Different mutant types compete one with the other and selection occurs. Mutation can be studied quantitatively if selection is excluded by restricting amplification to a single replication round. Mutation rates can be measured, defined as the probability of incorporating a non-cognate base per incorporation event.

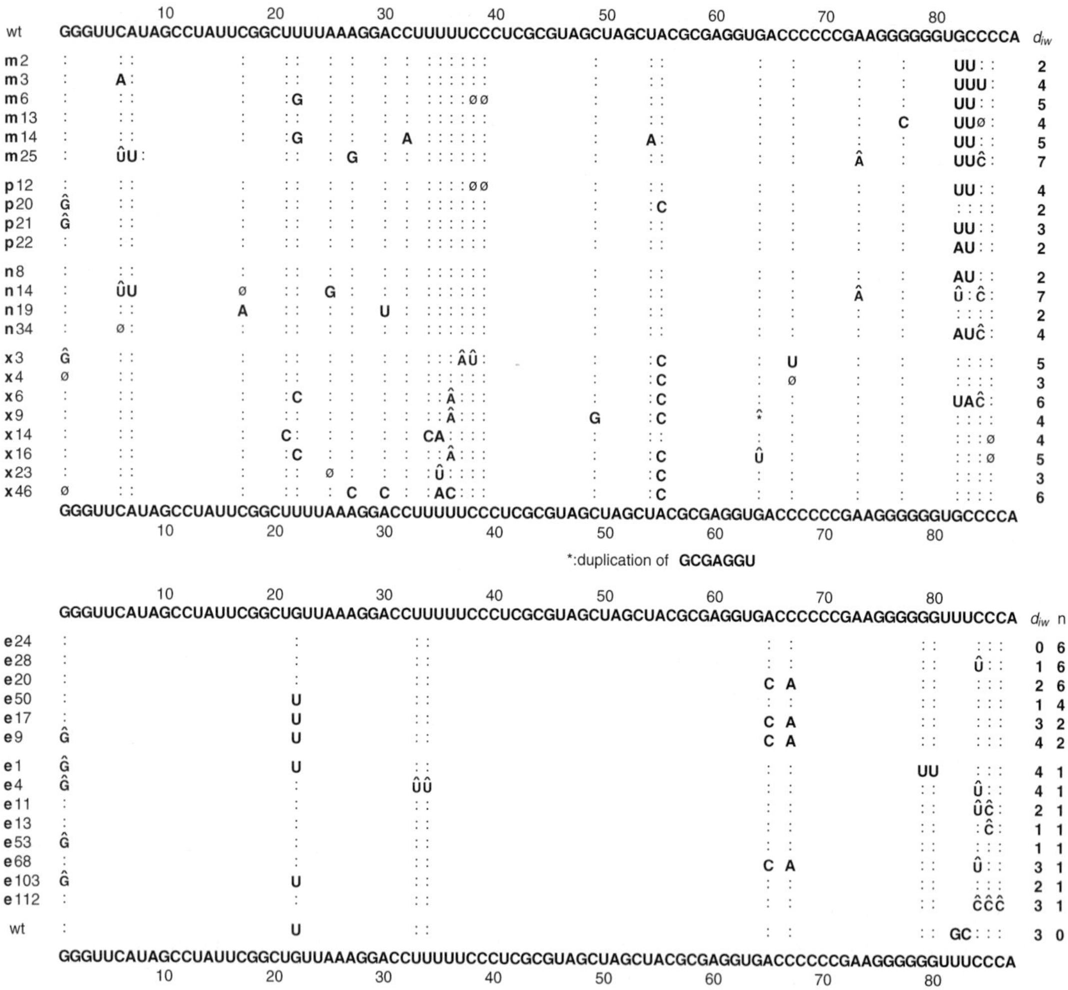

FIGURE 3.4 Mutant spectrum of MNV-11. Top: Linear growth phase. Mutants within a population are indicated by number, different MNV-11 populations by letters: m, linear growth for 5 h; n,p: growth in the linear phase for 2 h followed by separation of plus (p) and minus (m) strands; 1, growth in the exponential growth phase (100 replication rounds); e, exponential growth in the presence of 1 μM ethidium bromide; x, growth in the linear growth phase for 8 h in the presence of 50 mM $(NH_4)_2SO_4$. HD is the Hamming distance, i.e. the number of base exchanges. Bottom: Exponential phase. The last number in each column is the number of clones found in the population. Data from Rohde *et al.* (1995).

It is generally believed that DNA replaced RNA as genetic material during evolution because of its superior replication fidelity and chemical stability. Several energy-consuming error correction systems were invented to accomplish this fidelity. The systematic error caused by the tautomerization reaction of the pyrimidines is mainly corrected directly after phosphodiester formation by proof-reading: a mismatch—particularly the one caused by a base that is returning from the wrong tautomeric structure to its favored one—is removed during the replication process itself. Extensive post-replicative repair systems detect and remove imperfections in newly formed DNA double helix. Neither fidelity-enhancing method is implemented in RNA synthesis: because cellular RNA is normally produced in many copies per cell and is degraded after some time anyway, occasional errors do not cause permanent harm. Remarkably, no repair mechanism has yet been found among viruses with RNA genomes, where mutations can be lethal. Indeed, their mutation rates are so high that only a small fraction of the copies are identical to their parents. The high price for this—that most offspring of RNA viruses are defective—is apparently offset by the higher potential to adapt to changing environments, perhaps caused by host defenses, that is ultimately provided by error-prone replication.

Leviviruses were the first examples where the high mutant diversity of RNA viruses was detected. Watanabe collected a large number of leviviruses from all continents (Yonesaki *et al.*, 1982). The leviviruses could be grouped into four classes. Their organization was nearly identical, yet the RNA fingerprints of species belonging to different classes did not indicate any sequence relationships. Today, several levivirus species have been sequenced, alignment of the sequences of species from different classes is difficult, because the information of the archetype founding the phylus has been almost totally diffused by mutations. Within each virus species, however, RNA fingerprints were remarkably stable and well-defined, indicating a clearly defined wild-type sequence (Billeter *et al.*, 1969; Fiers *et al.*, 1976). The apparent simplicity of this result, however, was shattered when it was shown that clones derived from single phage plaques of a virus population showed differences in their fingerprint patterns (Domingo *et al.*, 1978). It came as a shock to realize that viral populations are predominantly composed of an array of mutants, in which only a small fraction is what one would call a wild-type genome based on the dominant occupation of each nucleotide position in the sequence. Passaging the phage with a series of lysates restored the wild-type sequence. This result left only one explanation: The wild-type sequence is nothing more than the average of all of the sequences present in the viral population.

Eigen and Schuster (1977) predicted this result with straightforward theoretical considerations. Error propagation causes a spread of mutations in the population, leading eventually to a stable population, the "quasispecies," where each mutant type maintains a constant share of the population, its mutant frequency, that depends on its production by mutation and its selective value. A high mutant frequency does not necessarily correlate with a particularly high mutation rate ("hot spot"); nearly neutral multi-error mutants may have substantial mutant frequencies even though their rates of production by mutation are quite small. The theoretical background is covered in detail by Schuster (Chapter 1) in this volume, and Domingo *et al.* (Chapter 4) discuss the evolution of virus populations *in vivo*.

From *in vitro* studies (Batschelet *et al.*, 1976) and *in vivo* data (Drake, 1993) it was possible to estimate average RNA mutation rates per incorporated nucleotide; the two studies give values between 10^{-3} and 10^{-4}. On average, therefore, each phage RNA replica contains about one mutation. On the other hand, for the much shorter RNA sequences used in *in vitro* experiments, the vast majorities of the copies should be correct.

MUTANT SPECTRA

A natural mutant spectrum of the replicating RNA species MNV-11 was investigated by Rohde *et al.* (1995). Each experiment began with a homogeneous RNA population created by cloning. A large number of serial transfers, under constant growth conditions, were then made to allow establishment of an equilibrium population. The same procedure was then repeated, starting with the same RNA clone, for different growth conditions, e.g. higher ionic strength or a different growth phase. In order to determine the sequence and other properties of the mutants in each equilibrated population, representative collections of the mutants had to be cloned. This cloning could not be accomplished by amplifying single RNA strands with Qβ replicase because of the high error rate and intrinsic bias introduced by the replicase. Lethal or seriously disadvantaged mutants, for example, would not show up at all, because they would undergo evolutionary optimization as the clones were amplified to levels where sequencing is possible. Cloning RNA first into DNA, and then amplifying the DNA, does not have these drawbacks, however, and was thus adopted for analyzing the equilibrium populations. The cloning procedure was designed in such a way that the same RNA sequence that provided each clone could be reconstructed from the DNA clone by DNA-directed RNA synthesis (Biebricher and Luce, 1993). It was shown that the RNA populations obtained by transcription were quite homogeneous. To be true, the fidelity of transcription is no better than that of replication by viral replicase, but because transcription uses only the DNA and never the RNA copy as a template, error propagation is avoided. The sequences of some of the mutants are shown in Figure 3.4.

The mutant spectra were found to be quite broad. When the linear growth phase was investigated, for example, "wild-type" RNA comprised less than 40% of the quasispecies population. Mutations were not distributed randomly. At some positions mutations were frequent, while some regions were conserved, indicating parts of the RNA that are required for replication to occur. Single-error mutants were rare, and multi-error mutants appeared with up to 10% of the positions altered. Base transitions, transversions, deletions, and insertions were observed, in one case even duplication of a 7-base segment.

Box 3.3 Mutation and selection

In Darwinian evolution, selection is complemented by mutation. In each replication round of type i, there is a probability Q_{ji} to produce mutant j as copy. The relative fecundities A_i must therefore be corrected by the probabilities Q_{ii} that the progeny is a correct copy of the template. Q_{ji}-values were measured as the reversion rate of lethal mutants (Fersht, 1976; Loeb *et al.*, 1978; see also Chapter 7) after a single round of replication. While they are often generalized as error rate of the enzyme, one has to keep in mind that the error rates vary from position to position. At the usual population sizes and genome chain lengths, multi-error mutations can be neglected.

With the short-chained model RNA templates, this probability is close to unity, e.g. for RNA species MNV-11 the average Q_{ii} value is 0.97. For typical RNA viruses, however, the correction is rather dramatic, because the probability of producing correct offspring is much smaller than unity. The relative mortalities D_i are also influenced by the presence of other species. Under the experimental conditions used, loss is almost totally due to double strand formation. With a few exceptions, double strand formation has been found not to discriminate between mutants, i.e. the loss is approximately proportional to the total RNA strand concentration of the opposite polarity. By combining the synthesis and loss terms, we can define the intrinsic selection value $W_{ii} = Q_{ii}A_i - D_i$. The relative population change $dx_i/(x_i dt)$ due to selection (fecundity and mortality) is the relative selection rate value $\varsigma_i = W_{ii} - \bar{E}$, usually simply called "selection value."

The population of mutants is not only affected by selection; strands of type i are also produced by erroneous replication of other mutants. The relative population change $dx_i/(x_i dt)$ by the sum of all this contribution is called the mutational gain $\mu_i = \Sigma_{j=i}\ Q_{ij}A_j x_j/x_i$. It is always positive and usually dominated by the contributions of one or a few neighboring types that are highly populated.

The evolution rate $dx_i/(x_i dt) \equiv \xi_i = \varsigma_i + \mu_i$ is the total relative population change due to all contributions. Eventually, a steady state is formed where all evolution rates vanish; each mutant frequency has then reached a stable value which does not change any more under constant conditions. The mutant distribution in the steady state is called a quasispecies. For a detailed quantitative description of the evolution of replicating RNA molecules, the reader should consult the literature (Biebricher *et al.* 1991).

It is clear that the observed mutant spectra are not simply correlated to mutation rates. Base transitions were not found more frequently than transversions, and multi-error mutants were strongly overrepresented in comparison to what one would expect on the assumption that mutation rates governed the mutant spectra. Mutations themselves are essentially independent events, and if they would generate the observed mutant spectra one would find a high frequency of one-error mutants, a much smaller frequency of two-error mutants, and multi-error mutants would be extremely rare. One has to conclude that the mutant spectra were governed instead by selection values (Box 3.3). When this is true, frequently found mutants are expected to be neutral or nearly so. This was found to be true: the mutant replication rates were measured and found to be close to that of the wild-type. The rate measurements also showed that the "wild-type" found most frequently in equilibrated populations from the linear growth phase was not the mutant with the highest overall replication rate, but rather the best compromise among the rates of replication, replicase binding, and double strand formation.

The main reason for the high incidence of multi-error mutants must be that structural elements within the RNA are crucial for maintaining replication efficiency (Zamora *et al.*, 1995) and disturbance of such a structural element can be compensated by other mutations to restore replication efficiency. Darwinian evolution of replicating RNA species offers more than an opportunity to do qualitative evolution experiments *in vitro*. It is also possible to predict evolutionary outcomes by deriving quantitative selection values from the physicochemical parameters of the competing RNA species, as outlined above. These parameters can readily be measured for individual RNA species. In addition, interactions among mutants such as formation of heteroduplex strands between single strands of different mutants must be taken into account. Since it has been found that the rate constants for homoduplex and heteroduplex formation are essentially the same (Biebricher, unpublished measurements), these interaction can be quantified. Only when double strand formation with other well-populated mutants is affected does a selective advantage result. Nevertheless, even in this system with its minimal number of biochemical reactions, calculating selection values is a challenging exercise. The reason for this is that the experiments can be modeled using constant selective values only for the conditions of infinite dilution of the competing populations and unlimited resources. Under normal laboratory conditions, however, even in the constant environment of the test tube, the rates of production by mutation ("mutational gain") and the selection values change continuously as the population changes in composition and concentration (Eigen and Biebricher, 1987; Biebricher *et al.*, 1991). Once again, computer simulations by numerical integration of the rate equations are of great help for getting insight, even though it is of course not possible, or reasonable, to try to account in computer simulations for all of the mutation possibilities that exist in the experiments.

Of particular interest is evolution in the pure exponential growth phase, because the selection values are then indeed constant and equal to the overall replication rate coefficient, which can be readily measured. The mutant distribution of such a quasispecies is shown in Figure 3.4 (bottom). Among the 35 clones that were sequenced, the wild-type sequence was not found, because its replication rate is not the maximal one. There are fewer constraints on species evolving in the exponential phase, simply because competition and loss are excluded. A consequence of this is that the master sequence in the exponential growth phase is degenerate, the typical result being that several different mutants are nearly equally populated. These were not found in the mutant distributions in the linear growth phase, because their rate of binding replicase was reduced.

Adaptation to minor changes of growth conditions was found to be quite rapid. This is because the route of adaptation is different than one might naively assume: when growth conditions change, there is no delay until appropriate mutations occur. Selection of the best-adapted mutant already in the quasispecies is much faster than generation of new mutants. Its frequency rapidly increases and with it the (absolute) rate of producing mutants from it. The "center of gravity" within the existing quasispecies floats quickly through sequence space to a new position. Floating continues until a new evolutionary stable mutant spectrum emerges. What is thus observed is what has been described in organismic evolution as "punctuated equilibrium" (Gould and Eldredge, 1977).

The chance that a specific mutant is present depends strongly on the population size. When the population is small, more steps are required to reach a new equilibrium and adaptation takes longer. Furthermore, the route must then traverse a long staircase, on which each intermediate must have a selective advantage *per se* or it can not be a part of the climb. We saw in analyzing the quasispecies, however, that this is seldom the case: multi-error mutants were advantageous because the adverse effects of one mutation were compensated by subsequent ones. In a large population this is no problem, because some downward steps on the fitness landscape can be tolerated. The likelihood of generating a multi-error mutant depends on the number of steps necessary, the number of possible routes to reach it, and on the depth of the valleys that have to be crossed. Very deep canyons (i.e. where one of the intermediates represents a lethal mutation) must be crossed with a single jump, i.e. the two-error mutant be formed in one replication round.

Several adaptation experiments of short-chained RNA species have been reported. The first quantitative one was on replication of the species MDV-1 in the presence of a low concentration of ethidium bromide (Kramer *et al.*, 1974), which resulted in selection of a three-error mutant. Adaptation was achieved slowly, because each transfer began with a population of only 10^6 RNA strands. The first mutant was already present in the quasispecies population and the next mutations occurred in the 7th and 12th transfers, respectively. A disadvantage of the serial transfer technique is that small aliquots are used for inoculation of succeeding transfers. The probability of finding a newly formed mutant in an aliquot may be quite small, depending on its size and the time when the mutant emerged. Furthermore, each step in these experiments involved amplification in both the exponential and the linear growth phase. It was therefore not possible to calculate selection values.

Eigen and collaborators (Strunk and Ederhof, 1997) developed a machine that avoids these disadvantages. It always remains in the exponential growth phase, because the RNA concentration is measured in real time, triggering a serial transfer before the enzyme is saturated. The 1:10 aliquot used for the next transfer insures that mutant populations do not drop to low values. Using this machine a variant MNV-11 resistant to RNase A was selected after a rather large sequence change, including a deletion. The evolution route taken in this process has not yet been reported.

Site-directed mutagenesis experiments with levirus genomes have shown that almost any mutation of the genome affects the fitness of the virus. Studies of the revertants and pseudo-revertants revealed an intricate influence of the RNA structure on replication, translation, and regulation of the virus (Arora *et al.*, 1996; Klovins *et al.*, 1997; Poot *et al.*, 1997). These experiments brought many insights into the subtle control of the biochemical processes involving RNA and illustrated that the fitness of a viral type is a highly complex function that makes quantitative predictions almost hopeless.

RECOMBINATION AMONG RNA MOLECULES

In organisms with DNA genomes, the high replication fidelity makes large mutational jumps impossible as evolutionary routes.

An alternative route is taken, DNA recombination. DNA from other organisms is occasionally inserted, and sections of the native genome are occasionally deleted, duplicated, inversed, or transposed to remote positions. Normal cells contain many enzymes involved in catalyzing DNA recombination, underscoring the importance of this process.

RNA recombination is far less frequent, except in certain families of RNA viruses where it occurs at extremely high frequencies during replication (Lai *et al.*, 1985; Kim and Kao, 2001; Wain-Hobson *et al.*, 2003). Early double infection experiments with leviviruses containing defects in different cistrons showed complementation, but no defect-free recombinant progeny could be isolated. Later experiments have shown that recombination does occur, but only at a very low rate (Palasingam and Shaklee, 1992). RNA recombination has been observed with many different viruses (King *et al.*, 1982; Lai *et al.*, 1985; Lai , 1992), often caused by errors in the replication process itself. Several models have been proposed, the simplest and most plausible being "copy choice," i.e. a jump from one template to another (or to the same template, but on a different position) during replication (Lai, 1992; Kim and Kao, 2001). The replication mechanism of retroviruses includes a recombination between two parental strands (Panganiban and Fiore, 1988; Peliska and Benkovic, 1992).

An RNA species replicated by Qβ replicase has been isolated that is obviously a recombinant between part of the replicase gene of Qβ and host cell tRNA (Munishkin *et al.*, 1988). RNA recombination *in vitro* is a very rare event, but has also been reported (Biebricher and Luce, 1992). Even a very rare event, however, can quickly become evident in an evolution experiment if an advantageous mutant is created. Thus MNV-11 grown to equilibrium builds up a stable mutant spectrum (see above), but under conditions of high ionic strength and growth in the late linear growth phase a new RNA species with a higher chain length (135) than MNV-11 (86) eventually emerges and is rapidly selected. Repetition of the experiment under identical conditions showed that the eventual result is reproducible, indicating an instructed process, while the time lapse to emergence of the new species is not.

RNA recombination events are more frequent at higher ionic strength. In the numerous cases we observed (Biebricher and Luce, 1992; Zamora *et al.*, 1995), usually a short repetitive sequence was found, indicating a copy choice mechanism. Chetverin *et al.* (1997) described examples where this is not the case. Since only the sequence changes that are genetically fixed can be observed, a clear decision between different models is not possible. It is quite possible that several, rare mechanisms occur.

CREATING BIOLOGICAL INFORMATION FROM SCRATCH

So far we have described experiments that showed Darwinian adaptation to the environment, i.e. optimization of a pre-existing biological function. Evolution, however, is able not only to adapt but also to create. Is it possible to generate a self-replicating RNA without offering a template? In the last years, many RNA species with novel functions have indeed been selected starting from completely random RNA sequences. In other words, new biological function has been formed without any ancestry at all. In these experiments human ingenuity (to set up the experiments in the first place), random chance, and Darwinian evolution are the driving forces that create information from nowhere. However, even a century after Darwin, doubts continue to be expressed that such information can be formed without human interference. The main conceptual difficulty that such doubters experience derives from the vanishingly low probability of creating a predefined sequence by chance. To find a specific sequence of chain length 50 would require a population of $4^{50} = 10^{30}$ strands. Fortunately, there is not just a single winner in the sequence lottery: the large number of total blanks is compensated by the large number of minor wins. Sequences with low fitness values, once any are created by chance, are optimized by adaptive evolution on a much quicker time-scale.

Two basic strategies were found to create replicable RNA without any template being present in the starting mixture: in the first it was intentionally extracted from a huge library of randomly assembled sequences (Biebricher and Orgel, 1973; Brown and Gold, 1995). The second was an unexpected finding: incubation of an RNA replicase at high concentration in the presence of high nucleotide triphosphate concentrations produced replicable RNA after long incubation times despite the absence of detectable RNA in the starting material times (Sumper and Luce, 1975). Different RNA species are selected in each experiment of this kind (Figure 3.5; Biebricher *et al.*, 1981a; Biebricher, 1987). Evidence has been presented to show that in the absence of template the replicase condenses nucleotides, at a rate 5 orders of magnitude less than that of template-instructed synthesis, to produce a random mixture of sequences (Biebricher *et al.*, 1986). Once any accepted template is produced, no matter how inefficient it may be, it is amplified and optimized. Indeed, replicability is a particularly sensitive function to select for, because the overwhelming majority of unaccepted RNA is ignored by the replicase.

Impurity RNA would have some genetic origin; however, the emerging species show no base homology to the genomes of the virus or of the host; moreover, they cannot be detected in infected or non-infected cells (Avota *et al.*, 1998). *In vitro*, template-free synthesis was even found to be suppressed by addition of non-replicable RNA or DNA. Aggregation of enzyme molecules increased the efficiency of template-free synthesis. Modification of the non-replicable RNA by instructed terminal elongation has been observed (Biebricher and Luce, 1992). *In vivo*, only weakly replicable RNA species were derived from host RNA, in particular from 16S ribosomal RNA (Avota *et al.*, 1998).

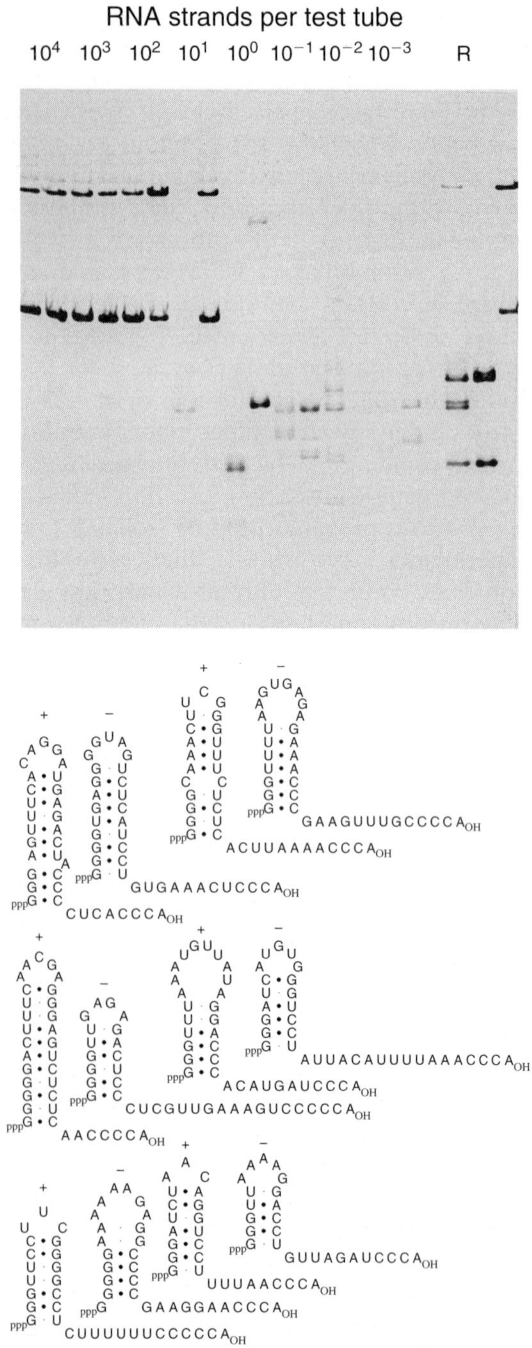

FIGURE 3.5 Template-free synthesis of replicating RNA. Top: Electropherogram of an replication experiment after 16h incorporation at various template concentration. The products of template-instructed and template-free synthesis are clearly different. Bottom: Sequence of some early products of template-free synthesis.

Sequence analysis and quantitative characterization of the properties of the early products of template-free synthesis were quite instructive (Biebricher and Luce, 1993). As mentioned earlier, the first feature noticed was that these oligo RNA strands differ in chain length and sequence in each experiment. The low probability of assembling long replicable RNA strands favors small early products. Experimentally, strands with 25–40 nucleotides dominated. Their replication rates were low compared with those of optimized RNA. During subsequent serial transfers, early RNA products underwent rapid evolutionary optimization. During the optimization the molecular weight increased, in nearly all cases by recombination-like events such as duplications or insertion of sections of the complementary sequence. The optimization rate depended on experimental conditions. At high ionic strength optimization was fast. Otherwise it was so slow that a short inefficient template could be amplified at high amplification factors ($>10^{20}$) during many serial transfers without changes of the average sequence.

STRUCTURAL SIGNALS FOR REPLICATION

The large number of short-chain replicable RNA species found offered a possibility to investigate the minimum sequence requirement for replication. Sequence comparison of the replicable species, however, did not reveal anything like a consensus sequence at all: except for the invariant ends—pppGG[G] at the 5′ termini, CCA at the 3′ termini (a terminal A is attached without template instruction)—no homologies could be found. However, when the secondary structures were calculated, it appeared that the structures of all replicable RNA had a stem at the 5′ termini, while the 3′ termini were unpaired. The alternative folding, with 5′ and 3′ paired with each other, is energetically disfavored (Biebricher and Luce, 1993). The constraints for the more stable structure are more severe than it might seem at first: if base-pairing only involved the canonical base pairs, then a stem at the 5′ strand of one strand would correspond to a 3′ stem for the complementary sequence. Only non-canonical base pairs and outlooped bases at strategic positions makes the conserved replicable structure possible. Site-directed mutation replacing these positions with canonical base pairs destroyed the template activity of the RNA entirely (Zamora *et al.*, 1995).

What is the reason for this structure? It is not known yet, but there are arguments in favor of stems at the 5′ termini (Biebricher, 1994). As a replica is formed, the structure is transiently double stranded. With progressing elongation replica and template separate. Rapid stem formation reduces the danger that replica and template re-form a double strand. There are additional features common to many, but not all, replicable RNA species. A pyrimidine cluster in the interior of the sequence seems to be favorable for enzyme–RNA binding (Brown and Gold, 1995). However, binding strength to replicase is only poorly correlated with template activity. Some RNA sequences, notably 16S rRNA, bind quite well to Qβ replicase but have no template activity, while binding of some early products of template-free synthesis is only weak even though their replication rates are substantial.

The structural features we have described appear to be necessary for RNA replication. To test whether they are sufficient for replication we designed and synthesized RNA strands with sequences predicted to give these structural features (Zamora *et al.*, 1995). Their template activities, however, were found to be barely measurable. Upon incubation with replicase, replicable RNA did grow out. The selected RNA species differed from each other, but were all clearly mutants of their respective initial templates. In some cases two or three base exchanges sufficed to make the species replicable, while in other cases recombination events were also involved. In all cases the above-described structural features were not only conserved, but even enhanced. We conclude that there are unidentified additional requirements for adequate template activity. Clearly it is unlikely to strike a fitness peak when designing templates on the basis of the structural features we have been able to identify. During amplification, drifting of the designed sequences to nearby fitness peaks is thus inevitable. Other research groups have found that simple structural features suffice for single round transcription of RNA (Tretheway *et al.*, 2001; Ugarov *et al.*, 2003).

CONCLUSIONS

Many quantitative insights into the nature of evolution have been gained from studying the model system provided by Qβ replicase. Are the often surprising results obtained from these experiments only a misleading caprice of nature, with no relevance to evolution in general? There are many reasons to think that this is not the case. The principal one is that studies using other enzymes, including some that are not viral, lead to similar results. If, as is generally believed, the origin of viruses is formation of intracellular parasites that eventually develop an apparatus for horizontal gene transfer among different hosts, then viral genes must often derive from cellular ones. RNA replication with Qβ replicase requires catalytic participation of RNA; if template RNA is able to instruct Qβ replicase to replicate it, it seems possible that other RNA polymerases, e.g. transcriptase, can be instructed to carry out RNA replication as well. Indeed, it has been shown that RNA templates exist that are accepted by the DNA-dependent RNA polymerases of the bacteriophage T7 and T3 (Konarska and Sharp, 1989; Biebricher and Luce, 1996) and *E. coli* (Biebricher and Orgel, 1973; Wettich and Biebricher, 2001). Since for these enzymes no physiological RNA templates exist, replicable RNA species were selected by the described methods from random nucleotide libraries (Biebricher and Orgel, 1973; Wettich and Biebricher, 2001) or obtained by template-free synthesis (Sumper and Luce, 1975; Biebricher and Luce, 1996; Wettich and Biebricher, 2001). The templates are specific for their cognate enzyme and not accepted by other RNA polymerases. (One exception that has been found is an RNA species replicated by T7 RNA polymerase as well as by T3 RNA polymerase.) The features such as exponential and linear growth and strand separation during replication were nearly identical to what has been observed previously with Qβ replicase.

Recently, newly developed RNA amplification methods with lower sequence specificity than Qβ replicase have been shown to be superior for artificial selection of functional RNA by evolutive biotechnology (Guatelli *et al.*, 1990; Breaker and Joyce, 1994). For quantitative studies of natural evolution under controlled conditions, however, amplification of RNA by Qβ replicase is still unsurpassed. All predictions of Eigen's theory of the evolution of simple replicators have been verified.

The evolution of viruses, on the other hand, is much more complicated. While it has been possible to devise kinetic models of the infection cycle of simple bacteriophage like that of Qβ, including RNA replication and protein synthesis, and to get good agreement with experimentally measured profiles of RNA and gene products (Eigen *et al.*, 1991), this takes place inside the host cell. Evolution of the phage, however, takes place in another phase, the medium. The fitness of a phage type is dictated by the liberation of the burst size of infective progeny which includes still many additional steps. While the evolution of RNA replicators is now well understood, much remains still to be learned about the evolution of viruses.

REFERENCES

Arora, R., Priano, C., Jacobson, A.B. and Mills, D.R. (1996) Cis-acting elements within an RNA coliphage genome—fold as you please, but fold you must. *J. Mol. Biol.* **258**, 433–446.

August, J.T., Cooper, S., Shapiro, L. and Zinder, N.D. (1963) RNA phage-induced RNA polymerase. *Cold Spring Harb. Symp. Quant. Biol.* **28**, 95–97.

Avota, E., Berzins, V., Grens, E., Vishnevsky, Y., Luce, R. and Biebricher, C.K. (1998) The natural 6S RNA found in Qβ-infected cells is derived from host and phage RNA. *J. Mol. Biol.* **276**, 7–17.

Batschelet, E., Domingo, E. and Weissmann, C. (1976) The proportion of revertant and mutant phage in a growing population, as a function of mutation and growth rate. *Gene* **1**, 27–32.

Biebricher, C.K. (1983) Darwinian selection of self-replicating RNA. *Evol. Biol.* **16**, 1–52.

Biebricher, C.K. (1987) Replication and evolution of short-chained RNA species replicated by Qβ replicase. *Cold Spring Harb. Symp. Quant. Biol.* **52**, 299–306.

Biebricher, C.K. (1994) The role of RNA structure in RNA replication. *Ber. Bunsenges* **98**, 1122–1126.

Biebricher, C.K. and Luce, R. (1992) In vitro recombination and terminal elongation of RNA by Qβ replicase. *EMBO J.* **11**, 5129–5135.

Biebricher, C.K. and Luce, R. (1993) Sequence analysis of RNA species synthesized by Qβ replicase without template. *Biochemistry* **32**, 4848–4854.

Biebricher, C.K. and Luce, R. (1996) Template-free synthesis of RNA species replicating with T7 RNA polymerase. *EMBO J.* **15**, 3458–3465.

Biebricher, C.K. and Orgel, L.E. (1973) An RNA that multiplies indefinitely with DNA-dependent RNA polymerase: Selection from a random copolymer. *Proc. Natl Acad. Sci. USA* **70**, 934–938.

Biebricher, C.K., Eigen, M. and Luce, R. (1981a) Product analysis of RNA generated de novo by Qβ replicase. *J. Mol. Biol.* **148**, 369–390.

Biebricher, C.K., Eigen, M. and Luce, R. (1981b) Kinetic analysis of RNA generated de novo by Qβ replicase. *J. Mol. Biol.* **148**, 391–410.

Biebricher, C.K., Diekmann, S. and Luce, R. (1982) Structural analysis of selfreplicating RNA synthesized by Qβ replicase. *J. Mol. Biol.* **154**, 629–648.

Biebricher, C.K., Eigen, M. and Gardiner, W.C. (1983) Kinetics of RNA replication. *Biochemistry* **22**, 2544–2559.

Biebricher, C.K., Eigen, M. and Gardiner, W.C. (1984) Kinetics of RNA replication: Plus-minus asymmetry and double-strand formation. *Biochemistry* **23**, 3186–3194.

Biebricher, C.K., Eigen, M. and Gardiner, W.C. (1985) Kinetics of RNA replication: Competition and selection among self-replicating RNA species. *Biochemistry* **24**, 6550–6560.

Biebricher, C.K., Eigen, M. and Luce, R. (1986) Template-free RNA synthesis by Qβ replicase. *Nature* **321**, 89–91.

Biebricher, C.K., Eigen, M. and Gardiner, W.C. (1991) Quantitative analysis of selection and mutation in self-replicating RNA. In: *Biologically inspired Physics* (L. Peliti, ed.) NATO ASI Series B, 263, pp. 317–337. New York: Plenum Press.

Biebricher, C.K., Eigen, M. and McCaskill, J.S. (1993) Template-directed and template-free RNA synthesis by Qβ replicase. *J. Mol. Biol.* **231**, 175–179.

Biebricher, C.K., Nicolis, G. and Schuster, P. (1995) *Self-organization in the physico-chemical and life sciences.* Luxemburg: Office for Official Publications of the European Communities.

Billeter, M.A., Dahlberg, J.E., Goodman, H.M., Hindley, J. and Weissmann, C. (1969) Sequence of the first 175 nucleotides from the 5′ terminus of Qβ RNA synthesized *in vitro*. *Nature* **224**, 1083–1086.

Breaker, R.R. and Joyce, G.F. (1994) Emergence of a replicating species from an *in vitro* RNA evolution reaction. *Proc. Natl Acad. Sci. USA* **91**, 6093–6097.

Brown, D. and Gold, L. (1995) Selection and characterization of RNAs replicated by Qβ replicase. *Biochemistry* **34**, 14775–14782.

Chetverin, A.B., Chetverina, H.V., Demidenko, A.A. and Ugarov, V.L. (1997) Nonhomologous RNA recombination in a cell-free system: Evidence for a transesterification mechanism guided by secondary structure. *Cell* **88**, 503–513.

Dawkins, R. (1982) *The Extended Phenotype.* San Francisco: Freeman.

Dobkin, C., Mills, D.R., Kramer, F.R. and Spiegelman, S. (1979) RNA replication: Required intermediates and the dissociation of template, product and Qβ replicase. *Biochemistry* **18**, 2038–2044.

Dobzhansky, T., Ayala, F.J., Stebbins, G.L. and Valentine, J.W. (1977) *Evolution.* San Francisco: Freeman.

Domingo, E., Sabo, D., Taniguchi, T. and Weissmann, C. (1978) Nucleotide sequence heterogeneity of an RNA phage population. *Cell* **13**, 735–744.

Drake, J.W. (1993) Rates of spontaneous mutation among RNA viruses. *Proc. Natl Acad. Sci. USA* **90**, 4171–4175.

Eigen, M. (1971) Self-organisation of matter and the evolution of biological macromolecules. *Naturwissenschaften* **58**, 465–523.

Eigen, M. and Biebricher, C.K. (1988) Sequence space and quasispecies distribution. In: *RNA Genetics, Vol. III: Variability of RNA Genomes* (E. Domingo, P. Ahlquist and J.J. Holland, eds), pp. 211–245. Boca Raton: CRC Press.

Eigen, M. and Schuster, P. (1977) The hypercycle—a principle of natural selforganization. Part A: Emergence of the hypercycle. *Naturwissenschaften* **64**, 541–565.

Eigen, M., Biebricher, C.K., Gebinoga, M. and Gardiner, W.C. (1991) The hypercycle: Coupling of RNA and protein biosynthesis in the infection cycle of an RNA bacteriophage. *Biochemistry* **30**, 11005–11018.

Fersht, A.R. (1976) Fidelity of replication of phage ϕX174 by DNA polymerase III holoenzyme: Spontaneous mutation by misincorporation. *Proc. Natl Acad. Sci. USA* **76**, 4946–4950.

Fiers, W., Contreras, R., Duerinck, F., Haegemann, G., Iserentant, D., Merregaert, J. *et al.* (1976) Complete nucleotide sequence of bacteriophage MS2 RNA: primary and secondary structure of the replicase gene. *Nature* **260**, 500–507.

Franze de Fernandez, M.T., Eoyang, L. and August, J.T. (1968) Factor fraction required for the synthesis of bacteriophage Qβ RNA. *Nature* **219**, 588–590.

Gould, S.J. and Eldredge, N. (1977) Punctuated equilibria: The tempo and mode of evolution reconsidered. *Palaeobiology* **3**, 115–151.

Guatelli, J.C., Whitfield, K.M., Kwoh, D.Y., Barringer, K.E., Richman, D.D. and Gingeras, T.R. (1990) Isothermal in vitro amplification of nucleic acids by a multienzyme reaction modeled after retroviral replication. *Proc. Natl Acad. Sci. USA* **87**, 1874–1878.

Haruna, I., Nozu, K., Ohtaka, Y. and Spiegelman, S. (1963) An RNA "replicase" induced by and selective for a viral RNA: isolation and properties. *Proc. Natl Acad. Sci. USA* **50**, 905–911.

Hori, K.L., Eoyang, L., Banerjee, A.K. and August, J.T. (1967) Template activity of synthetic ribopolymers in the Qβ RNA polymerase reaction. *Proc. Natl Acad. Sci. USA* **57**, 1790–1797.

Kamen, R. (1970) Characterization of the subunits of Qβ replicase. *Nature* **228**, 527–533.

Kim, M.-J. and Kao, C. (2001) Factors regulating template switch in vitro by viral RNA-dependent RNA polymerases: Implications for RNA-RNA recombination. *Proc. Natl Acad. Sci. USA* **98**, 4972–4977.

King, A.M.Q., McCahon, D., Slade, W.R. and Newman, J. W.I. (1982) Recombination in RNA. *Cell* **29**, 921–928.

Klovins, J., Tsareva, N.A., de Smith, M.H., Berzins, V. and van Duin, J. (1997) Rapid evolution of translational control mechanisms in RNA genomes. *J. Mol. Biol.* **265**, 372–384.

Konarska, M.M. and Sharp, P.A. (1989) Replication of RNA by the DNAdependent RNA polymerase of phage T7. *Cell* **57**, 423–431.

Kondo, M., Gallerani, R. and Weissmann, C. (1970) Subunit structure of Qβ replicase. *Nature* **228**, 525–527.

Kramer, F.R., Mills, D.R., Cole, P.E., Nishihara, T. and Spiegelman, S. (1974) Evolution in vitro, Sequence and phenotype of a mutant RNA resistant to ethidium bromide. *J. Mol. Biol.* **89**, 719–736.

Lai, M.M.C. (1992) RNA recombination in animal and plant viruses. *Microbiol. Rev.* **56**, 61–79.

Lai, M.M.C., Baric, R.S., Makino, S., Keck, J.G., Egbert, J., Leibowitz, J.L. and Stohlmann, S.A. (1985) Recombination between nonsegmental RNA genomes of muric coronaviruses. *J. Virol.* **56**, 449–456.

Levisohn, R. and Spiegelman, S. (1969) Further extracellular Darwinian experiments with replicating RNA molecules: diverse variants isolated under different selective conditions. *Proc. Natl Acad. Sci. USA* **63**, 807–811.

Loeb, T. and Zinder, N.D. (1961) A bacteriophage containing RNA. *Proc. Natl Acad. Sci. USA* **47**, 282–289.

Loeb, L.A., Weymouth, L.A., Kunkel, T.A., Gopinathan, K.P., Beckman, R.A. and Dube, D.K. (1978) On the fidelity of DNA replication. *Cold Spring Harbor Symp. Quant. Biol.* **43**, 921–927.

Luria, S.E. and Delbrück, M. (1943) Mutation of bacteria from virus sensitivity to virus resistance. *Genetics* **28**, 486–491.

Mills, D.R., Peterson, R.L. and Spiegelman, S. (1967) An extracellular Darwinian experiment with a self-duplicating nucleic acid molecule. *Proc. Natl Acad. Sci. USA* **58**, 217–224.

Mitsunari, Y. and Hori, K. (1973) Qβ replicase-associated, poly(C)-dependent poly(G) polymerase. *J. Biochem.* **74**, 263–271.

Munishkin, A.V., Voronin, L.A. and Chetverin, A.B. (1988) An in vivo recombinant RNA capable of autocatalytic synthesis by Qβ replicase. *Nature* **333**, 473–475.

Palasingam, K. and Shaklee, P.N. (1992) Reversion of Qβ RNA phage mutants by homologous RNA recombination. *J. Virol.* **66**, 2435–2442.

Panganiban, A.T. and Fiore, D. (1988) Ordered interstrand and intrastrand DNA transfer during reverse transcription. *Science* **241**, 1064–1069.

Peliska, J.A. and Benkovic, S.J. (1992) Mechanism of DNA strand transfer reactions catalyzed by HIV-1 reverse transcriptase. *Science* **258**, 1112–1118.

Poot, R.A., Tsareva, N.V., Boni, I.V. and van Duin, J. (1997) RNA folding kinetics regulates translation of phage MS2 maturation gene. *Proc. Natl Acad. Sci. USA* **94**, 10110–10115.

Rohde, N., Daum, H. and Biebricher, C.K. (1995) The mutant distribution of an RNA species replicated by Qβ replicase. *J. Mol. Biol.* **249**, 754–762.

Saffhill, R., Schneider-Bernloehr, H., Orgel, L.E. and Spiegelman, S. (1970) In vitro selection of bacteriophage Qβ RNA variants resistant to ethidium bromide. *J. Mol. Biol.* **51**, 531–539.

Spiegelman, S., Haruna, I., Holland, I.B., Beaudreau, G. and Mills, D.R. (1965) The synthesis of a self-propagating and infectious nucleic acid with a purified enzyme. *Proc. Natl Acad. Sci. USA* **54**, 919–927.

Strunk, G. and Ederhof, T. (1997) Machines for automated evolution experiments in vitro based on the serial-transfer concept. *Biophys. Chem.* **66**, 193–202.

Sumper, M. and Luce, R. (1975) Evidence for de novo production of selfreplicating and environmentally adapted RNA structures by bacteriophage Qβ replicase. *Proc. Natl Acad. Sci. USA* **72**, 162–166.

Tretheway, D.M., Yoshinari, S. and Dreher, T.W. (2001) Autonomous role of 3′- terminal CCCA in directing transcription of RNA by Qβ replicase. *J. Virol.* **75**, 11373–11383.

Ugarov, V.I., Demidenko, A.A. and Chetverin, A.B. (2003) Qβ replicase discriminates between legitimate and illegitimate templates by having different mechanisms of initiation. *J. Biol. Chem.* **278**, 44139–44146.

Wain-Hobson, S., Renoux-Elbe, C., Vartanian, J.P. and Meyerhans, A. (2003) Network analysis of human and simian immunodeficiency virus sequences sets reveals massive recombination resulting in shorter pathways. *J. Gen. Virol.* **84**, 885–895.

Weissmann, C., Simon, L., Borst, P. and Ochoa, S. (1963) Induction of RNA synthetase in *E. coli* after infection by the RNA phage MS2. *Cold Spring Harbor Symp. Quant. Biol.* **28**, 99–104. 491–511.

Wettich, A. and Biebricher, C.K. (2001) RNA species that replicate with DNAdependent RNA polymerase from E. coli. *Biochemistry* **40**, 3308–3315.

Yonesaki, T., Furuse, K., Haruna, I. and Watanabe, I. (1982) Relationships among four groups of RNA coliphages based on the template specificity of GA replicase. *Virology* **116**, 379–381.

Zamora, H., Luce, R. and Biebricher, C.K. (1995) Design of artificial short-chained RNA species that are replicated by Qβ replicase. *Biochemistry* **34**, 1261–1266.

CHAPTER

4

Viral Quasispecies: Dynamics, Interactions, and Pathogenesis*

Esteban Domingo, Cristina Escarmís, Luis Menéndez-Arias, Celia Perales, Mónica Herrera, Isabel S. Novella, and John J. Holland

ABSTRACT

Quasispecies theory is providing a solid, evolving conceptual framework for insights into virus population dynamics, adaptive potential, and response to lethal mutagenesis. The complexity of mutant spectra can influence disease progression and viral pathogenesis, as demonstrated using virus variants selected for increased replicative fidelity. Complementation and interference exerted among components of a viral quasispecies can either reinforce or limit the replicative capacity and disease potential of the ensemble. In particular, a progressive enrichment of a replicating mutant spectrum with interfering mutant genomes prompted by enhanced mutagenesis may be a key event in the sharp transition of virus populations into error catastrophe that leads to virus extinction. Fitness variations are influenced by the passage regimes to which viral populations are subjected, notably average fitness decreases upon repeated bottleneck events and fitness gains upon competitive optimization of large viral populations. Evolving viral quasispecies respond to selective constraints by replication of subpopulations of variant genomes that display higher fitness than the parental population in the presence of the selective constraint. This has been profusely documented with fitness effects of mutations associated with resistance of pathogenic viruses to antiviral agents. In particular, selection of HIV-1 mutants resistant to one or multiple antiretroviral inhibitors, and the compensatory effect of mutations in the same genome, offers a compendium of the molecular intricacies that a virus can exploit for its survival. This chapter reviews the basic principles of quasispecies dynamics as they can serve to explain the behavior of viruses.

*Dedicated to Manfred Eigen on the occasion of his 80th birthday, for the insights that his pioneer studies have represented for virology.

Origin and Evolution of Viruses
ISBN 978-0-12-374153-0

FROM EARLY REPLICONS TO PRESENT-DAY RNA VIRUSES

The quasispecies theory of molecular evolution was first proposed to describe the error-prone replication, self-organization, and adaptability of primitive replicons such as those thought to have populated the earth some 4000 million years before the present (Eigen, 1971, 1992; Eigen and Schuster, 1979; see Chapter 1). Quasispecies was formulated initially as a deterministic theory involving mutant distributions of infinite population size in equilibrium. Extensions and generalizations to ensembles of genomes of finite population size replicating in changing environments have been developed (Eigen, 2000; Wilke *et al.*, 2001a, 2001b; Saakian and Hu, 2006). Virologists use the term "viral quasispecies" to mean complex distributions of non-identical but closely related viral genomes subjected to genetic variation, competition and selection, and which act as a unit of selection (reviewed in different chapters of Domingo, 2006). More simple and general, a quasispecies has been defined as a population of similar genomes (Nowak, 2006). Quasispecies dynamics is most clearly manifested in systems such as RNA viruses that display short duplication times, generally high fecundity, and error-prone replication, traits that have been maintained despite a probable ancient origin of most extant RNA viruses in coevolution with a cellular world. Increasing numbers of careful analyses of viral populations have supported quasispecies dynamics for animal and plant RNA viruses (for recent examples see Ge *et al.*, 2007; Zhang *et al.*, 2007 and references included in these articles; see also other chapters of this book).

As discussed by Villarreal in Chapter 21, there are two main hypotheses regarding the origin of RNA viruses and other RNA genetic elements: that they are remnants of an ancient RNA world, or that they are modern derivatives of cells, originated in cellular RNAs that acquired autonomous replication. Viroids and other subviral RNA replicons may be direct descendants of early RNA (or RNA-like) replicons that preceded an organized cellular world (Robertson *et al.*, 1992) (Chapter 2). Cells and viruses share a considerable number of essential functional domains or modules: polymerases, proteases, enzymes involved in nucleotide and nucleic acid metabolism, etc. However, on the basis of key proteins involved in viral replication, that are absent in cells, and also based on the evidence of extensive genetic exchange between diverse viruses, the concept of an ancient virus world has been proposed (Koonin *et al.*, 2006). A primordial pool of genetic elements could have been the ancestor of viral and cellular genes. Cells and viruses share a ubiquitous ability to modify, lose, or acquire new genes or gene segments through genomic rearrangements, insertions, deletions, and other recombination events. Shuffling of functional modules among cells, viruses, and other replicons (plasmids, episomes, transposons, retrotransposons) is probably a frequent occurrence through fusion, transfection, conjugation, and other types of horizontal gene transfers (Botstein, 1980; Hickey and Rose, 1988; Zimmern, 1988; Davis, 1997; Holland and Domingo, 1998; Bushman, 2002). Sequence comparisons strongly suggest that all extant viruses have deep, ancient evolutionary roots (Gorbalenya, 1995; Villarreal, 2005) (Chapter 21).

ERROR-PRONE REPLICATION NECESSITATES LIMITED GENETIC COMPLEXITY TO PROTECT AGAINST ERROR CATASTROPHE

One of the critical features that distinguishes cells from viruses is the difference in the complexity of their genetic material, even after accounting for repeated DNA in animal and plant cells. Complexity in this case means the amount of genetic information encoded in their genetic material. A typical mammalian cell includes a number of chromosomes amounting to a total of about 3×10^9 base pairs (bp) of DNA. The chromosomal DNA of *Escherichia coli* has a complexity of about 4×10^6bp. In contrast, RNA viruses have genomes in the range of 3.0×10^3 to 3.2×10^4 nucleotides. Point mutation rates for eukaryotic cells have been estimated to be in the range of 10^{-10}–10^{-11} substitutions per nucleotide (s/nt), while for bacterial cells, values may reach up to 10^{-9} s/nt (Friedberg *et al.*, 2006). Mutation rates for a number of genomic sites of RNA viruses, determined using both genetic and biochemical procedures, are in the range of 10^{-3}–10^{-5} s/nt (Drake, 1993; Drake *et al.*, 1998; Drake and Holland, 1999; Domingo, 2007) (Chapter 7). Despite mutation rates varying with a number of environmental parameters, the above values mean that, in the process of RNA replication or retrotranscription, each progeny genomic molecule of about 10 kb will contain on average 0.1 to several mutations. These determinations of mutation rates and frequencies suggest that even the viral progeny of a single infected cell will be genetically heterogeneous (Domingo *et al.*, 1978; Holland *et al.*, 1982; Temin, 1989, 1993; Domingo, 2006, 2007; see also other chapters of this book).

Penetration into the composition of mutant spectra, either by determining the nucleotide sequence of many clones from the same population, or by other "diving" strategies, has quantitated large genotypic and phenotypic diversity within mutant spectra (Duarte *et al.*, 1994a, 1994b; Nájera *et al.*, 1995; Marcus *et al.*, 1998; Pawlotsky *et al.*, 1998; Quiñones-Mateu *et al.*, 1998; Wyatt *et al.*, 1998; Fernandez *et al.*, 2007; Garcia-Arriaza *et al.*, 2007; Ge *et al.*, 2007; Zhang *et al.*, 2007). Diversity can extend to multiple mutant and recombinant genomes within an infected organ, and even within a single infected cell. Diversity of genetic forms is a prerequisite for evolution, including the major transitions undergone by our biosphere (Eigen, 1992; Maynard Smith and Szathmary, 1995). RNA viruses have an exhuberant diversity to offer as a substrate for evolution. A virus population, by virtue of consisting of dynamic mutant spectra rather than a defined genomic sequence, has the potential to adapt readily to a range of environments.

One of the predictions of quasispecies dynamics of RNA viruses is the existence of an error threshold, defined as an average copying fidelity value at which a transition between an organized mutant spectrum and sequences lacking information contents occurs (reviewed in Eigen and Schuster, 1979; Eigen and Biebricher, 1988; Biebricher and Eigen, 2005; Nowak, 2006; Chapters 1 and 9). This transition has been coined "entry into error catastrophe," a term first used by L. Orgel to describe errors during protein synthesis that could contribute to a collapse of cellular regulatory networks in the process of aging (Orgel, 1963). Both, the concept expressed by Orgel and the one applied to genetic information of viruses address deterioration of meaningful information with a biological consequence, due to errors in an informational macromolecule. The error threshold relationship establishes a limitation for the maximum complexity of genetic information that can be stably maintained by a replicon displaying a given copying accuracy (Chapter 1). Theoretical calculations of the range of mutation rates that should be compatible with maintenance of the information carried by the simple RNA bacteriophages were compatible with the mutation rates and frequencies found experimentally (compare Batschelet *et al.*, 1976; Domingo *et al.*, 1976, 1978, with Eigen and Schuster, 1979; Eigen and Biebricher, 1988). In addition to intrinsic copying fidelity levels of viral polymerases, other biochemical features of virus replication may have evolved to preserve a minimal replication accuracy. It has been hypothesized that the "rule of six" (genome of polyhexameric length) in Mononegavirales that edit their phosphoprotein mRNA, may have evolved to prevent the negative effects of illegitimate editing that could result in error catastrophe (Kolakofsky *et al.*, 2005). Some biological systems exploit enhanced mutagenesis as a defense mechanism against invading molecular parasites. A mechanism known as "repeat-induced point mutations (RIP)" operates in some filamentous fungi such as *Neurospora crassa* resulting in the production of mutations in repeat DNA copies that penetrate into the cells (Bushman, 2002; Galagan and Selker, 2004). Also, the APOBEC3 family of cytidine deaminases are innate immunity factors that induce hypermutation in retroviral DNA. Such activities can be regarded as a form of natural "error catastrophe" against retroviral genomes (see Chapter 8). Thus, a mutagenesis-based antiviral approach to drive virus to extinction has a parallel in natural mechanims which have contributed to the survival of organisms in the face of perturbing molecular parasites.

Increased genetic complexity as is embodied in cells required a correspondingly higher copying accuracy of the genetic material. This appears to have been accomplished with a number of pathways for post-replicative repair mechanisms as well as with the acquisition of a 3′–5′ proofreading-repair exonuclease activity by most cellular DNA polymerases (Goodman and Fygenson, 1998). No evidence of a 3′–5′ exonuclease activity in viral RNA polymerases and reverse transcriptases has been obtained from either biochemical or structural studies with viral enzymes (Steinhauer *et al.*, 1992; Ferrer-Orta *et al.*, 2006). A possible exception was presented in an early report by (Ishihama *et al.*, 1986) showing that the influenza virus RNA polymerase was able to remove excess GMP residues added to a capped oligonucleotide primer. A 3′-end repair mechanism has been described in a satellite RNA of the plant virus turnip crinkle carmovirus, involving synthesis of short oligoribonucleotides by the viral replicase using the 3′-end of the viral genome as template, and, probably, template-independent priming at the 3′-end of the damaged RNA to generate wild-type, negative strand, satellite RNA (Nagy *et al.*, 1997). Also, some coronaviruses encode a polymerase which includes a 3′–5′ exonucleolytic activity (i.e. nsp14 of SARS) (Minskaia *et al.*, 2006). In the coronavirus murine hepatitis virus, mutations in the MSP14 exoribonuclease decreased replication fidelity (Eckerle *et al.*, 2007).

Virus Entry into Error Catastrophe and its Application to Lethal Mutagenesis

The limitations imposed on average mutation rates to maintain the genetic information transmitted by simple RNA replicons (Swetina and Schuster, 1982; Eigen and Biebricher, 1988; Nowak and Schuster, 1989) (Chapter 1) encouraged the first experiments to investigate whether chemical mutagenesis was detrimental to RNA virus replication. The first studies indicated that chemical mutagenesis could increase the mutation frequency by at most three-fold at defined genomic sites of poliovirus (PV) and vesicular stomatitis virus (VSV) (Holland *et al.*, 1990), and 13-fold in the case of a retroviral vector (Pathak and Temin, 1992). Also, increased mutagenesis had an adverse effect on fitness recovery of VSV clones (Lee *et al.*, 1997). These early results suggested that RNA viruses replicate near the error catastrophe threshold, with a copying fidelity that allows a generous production of error copies.

Additional studies in cell culture and *in vivo* have established that enhanced mutagenesis can result in virus extinction (reviewed in Anderson *et al.*, 2004; Domingo, 2005). Loeb and colleagues coined the term "lethal mutagenesis" to refer to the loss of virus infectivity associated with the action of mutagenic agents (Loeb *et al.*, 1999). Mutagenic nucleoside analogues, some used in antimicrobial and anticancer therapy, are currently actively studied as promoters of lethal mutagenesis of viruses, including an ongoing clinical trial with AIDS patients (Harris *et al.*, 2005).

Lethal mutagenesis is attracting increasing interest, and several theoretical models have addressed the mechanisms underlying lethal mutagenesis and the relationship between the observations on viral extinction and the original concept of error catastrophe (several models are reviewed in Chapter 1, and one model is described in Chapter 9). Key to the validation of these models as applied to RNA viruses is the experimental finding that a low viral load and low replicative fitness (relative replication capacity) favor extinction (Sierra *et al.*, 2000; Pariente *et al.*, 2001), and that a mutagenic activity (not merely an inhibitory activity) is necessary to achieve extinction (Pariente *et al.*, 2003). This was shown by absence of extinction when the virus was subjected to equivalent inhibitory activities with cocktails of non-mutagenic inhibitors (Pariente *et al.*, 2003). However, since low viral loads favor extinction, the inhibitory activity that is associated with the action of some mutagenic agents may contribute to lethal mutagenesis. In this respect, a combination of a mutagenic nucleoside analogue and the antiretroviral inhibitor AZT was required to extinguish high fitness HIV-1 during infections in cell culture (Tapia *et al.*, 2005). Even strong reductions in population size of highly debilitated foot-and-mouth disease virus (FMDV) and lymphocytic choriomeningitis virus (LCMV) populations did not result in virus extinction unless a mutagenic activity intervened (Sierra *et al.*, 2000; Pariente *et al.*, 2001; Pariente *et al.*, 2003). A second finding to be considered in the development of theoretical models is the negative interference exerted by mutants that either coinfect the cells along with standard virus, or are generated inside the cell by mutagenesis. The interfering activity of such "defector" genomes as contributing to viral extinction has been documented both experimentally with FMDV and LCMV, and by *in silico* simulations (González-López *et al.*, 2004; Grande-Pérez *et al.*, 2005b; Perales *et al.*, 2007). Production of a fraction of non-infectious hepatitis C virus (HCV) in infected patients as a result of ribavirin (1-β-D-ribofuranosyl-1,2,3-triazole-3-carboxamide) therapy is a key parameter in the models of HCV clearance following treatment with ribavirin and interferon alpha (IFN-α) (Dixit *et al.*, 2004; Dahari *et al.*, 2007) (see Chapter 15).

An argument that has been used to deny a connection between lethal mutagenesis and the transition into error catastrophe has been the absence of hypermutated molecules in mutagenized populations of RNA viruses. However, any hypermutated genome transiently generated during mutagenesis is unlikely to be replication-competent and to be included in any sampling of viral genomes. This has been recognized by us (Grande-Pérez *et al.*, 2005a) and others (Perelson and Layden, 2007). Despite this, a genome with a mutation frequency lying in the lower range of typically hypermutated genomes was identified in a population of 5-fluorouracil (5-FU)-treated LCMV (Grande-Pérez *et al.*, 2005a). The absence or very low frequency of hypermutated genomes in standard genome samplings of pre-extinction viral populations cannot constitute an argument against a mutagenesis-driven transition into error catastrophe.

Concerning the relationship between the concept of error catastrophe and extinction of viruses by lethal mutagenesis, M. Eigen pointed out the following: (i) dependence of copying fidelity on sequence context and the type of mutagen; (ii) fitness landscape of the quasispecies distribution, including the perturbing effects of specific types of mutants that may arise during mutagenesis (as discussed above); (iii) participation of multiple viral functions (not only RNA replication) in determining the replicative collapse of the system. As pointed out by Eigen, "Theory cannot remove complexity, but it shows what kind of 'regular' behavior can be expected and what experiments have to be done to get a grasp on the irregularities" (Eigen, 2002).

In line with the application of the error threshold relationship to real viruses (Eigen, 2002), it is obvious that virus extinction will not occur through "evaporation" into the entire sequence space theoretically available to a viral genome. This is physically impossible. As mutagenesis progresses during viral replication myriads of end-point genomes harboring lethal or highly deleterious mutations will impede further expansions into sequence space by such genomes. This is a consequence of the multiple viral functions (not only RNA replication) that affect replicative competence (Eigen, 2002). These differences between the mechanisms that mediate extinction of real viruses and the original concept of error catastrophe can be expressed by distinguishing "phenotypic" and "extinction" thresholds from an "error theshold," as has been done in some theoretical treatments (for example, Huynen *et al.*, 1996; Manrubia *et al.*, 2005). Apart from these rather obvious adaptations of error catastrophe to a real biological system, the experimental studies carried out in the laboratory of one of us (E.D.) do not provide any basis to dissociate lethal mutagenesis from error catastrophe, as initially developed by

Eigen, Schuster, and colleagues, and even less to consider that the approach to error catastrophe will impede viral extinction. In the section on "Intra-mutant spectrum suppression can contribute to lethal mutagenesis" in this chapter, we summarize our current view on the mechanisms that underlie virus extinction through lethal mutagenesis based on experimental results, and the main challenges facing, in our view, this new antiviral strategy.

INTRA-POPULATION COMPLEMENTATION AND INTERFERENCE IN VIRAL QUASISPECIES: MUTANT DISTRIBUTIONS AS THE UNITS OF SELECTION

A viral quasispecies can have a biological behavior that is not predictable from the behavior of its components considered individually. Several observations with viruses as they replicate in cell culture or *in vivo* suggest that intra-population interactions can modulate the replicative capacity of the ensemble of mutants or of individual mutants introduced in a spectrum of mutants. Fitness of biological clones of bacteriophase Qβ (Domingo *et al.*, 1978) and of VSV (Duarte *et al.*, 1994a) was lower than the fitness of the average populations from which the clones were derived. These quantifications of clonal fitness suggest that an ensemble of related mutants may collectively acquire a selective replicative advantage, perhaps because competent gene products may complement suboptimal or defective products expressed by subsets of components of the mutant spectrum. Specific mutants, including deleterious and lethal mutants, can be maintained in viral populations *in vivo,* and can be transmitted to susceptible hosts (Moreno *et al.*, 1997; Yamada *et al.*, 1998; Aaskov *et al.*, 2006; Vignuzzi *et al.*, 2006).

A seemingly opposite manifestation of the internal interactions within viral quasispecies is the suppression of the replication of specific mutants by the surrounding mutant spectrum. This possibility was suggested by theoretical models according to which a simple replicon of inferior fitness to another could nevertheless dominate the population by virtue of being surrounded by a more favorable mutant spectrum (Swetina and Schuster, 1982) (reviewed in Eigen and Biebricher, 1988; Nowak, 2006) (Chapter 1). The first experimental documentation of this prediction with real viruses was by de la Torre and Holland who showed that a standard VSV population interfered with the replication of a VSV clone of superior fitness, unless the latter was present above a certain frequency in the population (de la Torre and Holland, 1990). Suppressive effects of this type have been subsequently documented in several virus-host systems (reviewed in Domingo, 2006). Remarkable examples include suppression by attenuated PV of neuropathology in monkeys associated with virulent PV present in the vaccine preparation (Chumakov *et al.*, 1991), suppression of pathogenic LCMV by non-pathogenic variants (Teng *et al.*, 1996), the lowered replication rates of drug-resistant viruses (Crowder and Kirkegaard, 2005), and complementing-interfering effects of specific FMDV mutants (Perales *et al.*, 2007).

The Mutant Spectrum as a Determinant of Viral Pathogenesis. Picornaviral Polymerase Mutants

The complexity of the mutant spectrum of a virus (that is, the average number of mutations that distinguish the individual components of the mutant distribution) can affect the course of viral disease and the response to treatment. Most notably, prolonged persistence of HCV infection correlated with high mutant spectrum complexity (Farci *et al.*, 2000); other aspects of quasispecies behavior of HCV were reviewed in Domingo and Gomez, 2007) (see also Chapter 15).

Studies with a PV mutant with an amino acid substitution in the viral polymerase which increases about five-fold its template-copying fidelity have been particularly revealing. The mutant PV produces a narrower mutant spectrum (with a lower average number of mutations per genome) than wild-type PV. In infections of susceptible mice (transgenic for the human PV receptor) the mutant replicated in the animals but failed to reach the brain and to produce the neuropathology that was associated with the infection with wild-type PV (Pfeiffer and Kirkegaard, 2005; Vignuzzi *et al.*, 2006). Remarkably, restoration of the standard mutant spectrum complexity by subjecting the mutant PV to 5-FU-induced mutagenesis led to a neuropathogenic mutant spectrum (Vignuzzi *et al.*, 2006). Moreover, Sabin's attenuated PV vaccine shows relatively low mutant frequency compared with wild-type strains, and this observation could be due to differences in polymerase fidelity (Vignuzzi, personal communication; see also Chapter 6). These observations are highly relevant (Biebricher and Domingo, 2007). Foremost, the results show the biological relevance of high mutation rates, in that they may affect pathology by allowing the virus to reach specific target organs, thereby increasing viral loads and chances of transmission. The observed phenotypic transitions of PV demand consideration of the virus as a quasispecies, since PV behavior could not be explained by taking into account consensus genomic nucleotide sequences alone. We come to the conclusion that virus evolution can affect viral pathogenesis in at least two ways (Domingo, 2007): (i) The information for increased pathology or for adaptation to multiple environments can be contained in the genetic material of the virus (in most of its individual clones) irrespective of the mutant spectrum to which they belong (Kimata *et al.*, 1999; Greene *et al.*, 2005, among other examples). (ii) The information for increased pathology can be contained in a distribution of mutants as such, as documented above for HCV and PV. Again, these observations reinforce the biological advantage of high mutation rates for the long-term survival of RNA viruses, and the consideration of entire quasispecies as the units of selection (see also Domingo, 2006, 2007, and other chapters of this volume).

The PV polymerase mutant displaying higher fidelity than the wild type was obtained by passaging the virus in the presence of increasing concentrations of the nucleoside analogue ribavirin (Pfeiffer and Kirkegaard, 2003; Vignuzzi *et al.*, 2006). The amino acid replacement in the polymerase (G64S) is located away from the catalytic domain of the enzyme, and an action at a distance was invoked to explain the general effect of this substitution on the copying fidelity (Arnold *et al.*, 2005) (see Chapter 6). A mutant of FMDV, selected also by passaging the virus in the presence of increasing concentrations of ribavirin, displayed higher fitness than the wild-type virus when virus replication took place in the presence of ribavirin but not in its absence (Sierra *et al.*, 2007). This phenotypic change was mapped to amino acid substitution M296I in the viral polymerase, and the mutant enzyme displayed decreased capacity to use ribavirin triphosphate as substrate (instead of GTP or ATP), but did not show an apparent alteration of general template-copying fidelity (Sierra *et al.*, 2007). Substitution M296I is located at a loop whose flexibility seems to be required to adapt its conformation and interactions to the size and shape of template residues and incoming nucleotide substrates. Ile at this position may restrict the loop flexibility and affect nucleotide recognition (Ferrer-Orta *et al.*, 2007). M296 is quite distant from the site (G62) where the equivalent, ribavirin-selected substitution in PV lays. These results suggest that in the picornaviral polymerase multiple sites (perhaps domains) might be involved either in specific interactions with nucleotide analogues or in recognition of nucleotide substrates.

Comparison of the structure of the FMDV polymerase complexed with RNA (Ferrer-Orta *et al.*, 2004), and with RNA and a number of nucleotides and nucleotide analogues (Ferrer-Orta *et al.*, 2007) has documented the involvement of multiple amino acids of the FMDV polymerase in the recognition of nucleotides. Several interactions are key to catalysis, as shown by modification of the polymerase activity of the corresponding mutants produced by site-directed mutagenesis. Interestingly, some interactions are common to standard nucleotides and nucleotide analogues, while other interactions are specific for a given nucleotide analogue (Ferrer-Orta *et al.*, 2007). These results suggest that multiple sites in the polymerase can modulate substrate recognition, thereby affecting the fidelity properties of picornaviral (and probably other) polymerases (Arnold *et al.*, 2005; Ferrer-Orta *et al.*, 2007). These and other recent studies on the mechanism of substrate discrimination by viral RNA polymerases and reverse transcriptases are providing important information that may help in the design of drugs able to lower the copying fidelity of viral polymerases to facilitate lethal mutagenesis (see also Chapter 6).

Intra-Mutant Spectrum Suppression can Contribute to Lethal Mutagenesis

Populations of RNA viruses subjected to increased mutagenesis by nucleoside analogues display decreases in specific infectivity due to accumulation of viral genomes harboring deleterious or lethal mutations (Crotty *et al.*, 2001; Grande-Pérez *et al.*, 2002, 2005b; Airaksinen *et al.*, 2003; González-López *et al.*, 2004, 2005; Arias *et al.*, 2005). Mutagenized, pre-extinction FMDV RNA interfered with the replication of standard FMDV RNA, resulting in a delay and in a decrease in the production of progeny virus (González-López *et al.*, 2004). Since the interfering FMDV displayed at least a 0.1-fold fitness relative to the standard FMDV (González-López *et al.*, 2005), the suppression observed could not be due to mechanisms invoking competition between genomes of comparable replication capacity (such as positive clonal interference). It was suggested that the expression (normal or aberrant) of altered viral proteins could contribute to the suppression of replication of standard FMDV, and also to the extinction of FMDV RNA. To test this hypothesis, a number of capsid and polymerase mutants of FMDV were examined regarding their capacity to interfere with standard FMDV, in experiments involving coelectroporation of cells with the relevant RNAs (Perales *et al.*, 2007). The results showed that an excess of several replication-competent mutants caused a strong and specific interference on FMDV replication. Furthermore, mixtures of some capsid and polymerase mutants evoked a very strong, synergistic interference (Perales *et al.*, 2007). Notably, some of the mutants tested had been isolated from mutagenized FMDV populations in their way towards extinction. These results with FMDV are in agreement with observations on enhanced mutagenesis of LCMV which resulted in populations in which the loss of infectious progeny production preceded the loss of replicating viral RNA (Grande-Pérez *et al.*, 2005b). A deleterious effect on infectivity exerted by defective LCMV genomes was also supported by numerical simulations using realistic parameters of LCMV replication (Grande-Pérez *et al.*, 2005b).

The picture emerging from the studies with FMDV and LCMV is that the transition towards viral extinction associated with lethal mutagenesis can have at least two phases: an initial one, with a limited input of mutations in the viral genomes, in which a subset of defective genomes that have been termed "defectors" interfere with replication of standard genomes, and can contribute to viral extinction. This is termed the "lethal defection model" of virus extinction, proposed on the basis of experiments with LCMV (Grande-Pérez *et al.*, 2005b), and supported by the strong interference on FMDV replication exerted by combinations of specific capsid and polymerase mutants of FMDV (Perales *et al.*, 2007). In a second phase, as the number of mutations per genome increases due to continuing mutagenesis, the proportion of lethal mutations increases, resulting in further decreases in specific infectivity (González-López *et al.*, 2005; Grande-Pérez *et al.*, 2005b). In Chapter 6, Cameron and colleagues describe elegant experiments that show that low-fidelity mutants of poliovirus manifest an acceleration of the onset of lethal mutagenesis. Genomes with either deleterious or lethal mutations have been isolated from mutagenized FMDV and LCMV populations on their way towards extinction (Sierra *et al.*, 2000; Pariente *et al.*, 2001; Arias *et al.*, 2005). Some detrimental mutations may be maintained in the viral populations by complementation and whenever the genomes harboring them increase in frequency they may exert an interfering activity provided that the type of genetic lesion belongs to the interfering class (Perales *et al.*, 2007). Interestingly, some genomes harboring multiple mutations (for example a triple mutant in the polymerase of FMDV) that render the genome replication-incompetent may differ in a single nucleotide position from a replication-competent, strongly interfering mutant (Arias *et al.*, 2005; Perales *et al.*, 2007). Viral genomes with interfering or lethal mutations may occupy proximal or distant positions in sequence space, relative to the standard, non-mutated genome. Thus, there might be a gradual but overlapping transition between a phase of dominance of interfering mutants and a phase of increasing presence of lethal mutants, until a replicative collapse and virus extinction occur, in agreement with the theory of error catastrophe (see Chapter 1 for a discussion of the contribution of lethal mutants to error catastrophe). Recent biochemical data have documented that viral proteins are frequently multifunctional and that they often form oligomeric complexes. Thus, mutated forms of a given protein may affect multiple viral functions and result in inactive protein complexes (several examples can be found in Mesters *et al.*, 2006; Sobrino and Mettenleiter, 2008). Abnormal behavior of altered viral proteins may be one of the molecular mechanisms underlying virus transition into error catastrophe, very much in line with the cascade of events initially proposed as a model for aging (Orgel, 1963).

The transition of FMDV and LCMV towards extinction by lethal mutagenesis occurred with a 10^2- to 10^3-fold decrease in specific infectivity (PFU/total viral RNA), and without a modification of the consensus sequence of the population (González-López *et al.*, 2004; Grande-Pérez *et al.*, 2005a) in agreement with results with poliovirus (Crotty *et al.*, 2001). Loss of infectivity was very sharp, and extinction occurred generally after 1–20 passages, depending on viral fitness and the mutagen-inhibitor combination treatment (compare the extinction kinetics in Sierra *et al.*, 2000; Pariente *et al.*, 2001, 2003; Grande-Pérez *et al.*, 2005a). Extinction can be preceded by minimal increase in the average mutation frequency of the mutant spectra (Crotty *et al.*, 2001; Grande-Pérez *et al.*, 2005b; Tapia *et al.*, 2005). These experiments have not provided evidence that as the mutational load in the viral genome increases, the virus acquires resistance to extinction. It remains to be seen whether the presence of M296I in the FMDV RdRp, which was selected by ribavirin, confers any significant resistance to lethal mutagenesis. The vulnerability of FMDV to extinction by lethal mutagenesis offers a significant contrast with the resistance of FMDV to extinction despite accumulation of mutations as a result of plaque-to-plaque transfers (Escarmis *et al.*, 2002, 2008; Lazaro *et al.*, 2003). The key difference between the two scenarios is that resistance to extinction (despite accumulation of mutations accompanying serial bottleneck events) results from the selection for a next transfer of a virus able to replicate thanks to the presence of compensatory mutations. This is in contrast to mutagenesis of a complex population whose suppressive effects do not allow the rescuing of replication-competent individuals (Manrubia *et al.*, 2005).

The course of events preceding viral extinction that we have outlined here has a number of experimentally testable predictions, currently under study. Clarification of the mechanisms underlying virus extinction may help in the design of improved protocols of administration of mutagenic agents and antiviral inhibitors for lethal mutagenesis. In our view, the main challenges facing progress in lethal mutagenesis are: (i) finding and design of new mutagenic base or nucleoside analogues that target viral (but not cellular) polymerases, that can be used in combination with antiviral inhibitors; (ii) evaluation of how widespread is the occurrence of mutagen-resistant virus mutants, and whether lethal mutagenesis may fail either because of the presence of mutagen-resistant mutations (Pfeiffer and Kirkegaard, 2003; Sierra *et al.*, 2007) or other mechanisms (Sanjuan *et al.*, 2007); (iii) understanding of the molecular basis of template-copying fidelity of nucleic acid polymerases, and the design of drugs that can lower specifically the copying fidelity of viral polymerases; (iv) the application of lethal mutagenesis to model systems *in vivo* (Ruiz-Jarabo *et al.*, 2003a; Harris *et al.*, 2005). Concerning possible applications of lethal mutagenesis *in vivo,* measurements of the "critical drug efficacy"—as developed for treatments of infections by HIV-1 and HCV (Callaway and Perelson, 2002; Y. Huang *et al.*, 2003; Dahari *et al.*, 2007)—for mutagen-inhibitor combinations, should guide in establishing protocols adequate for viral clearance, to avoid stabilization of viral levels at a therapy-induced set point.

FITNESS AND ITS MODULATION BY VIRAL POPULATION SIZE

One of the consequences of the quasispecies dynamics of RNA viruses is fitness variations in a constant environment triggered by changes in viral population size. Fitness is a complex parameter that measures the degree of adaptation of a living organism or simple replicons to a specific environment (as general reviews see Williams, 1992, and Reznick and Travis, 1996). For viruses, fitness values have been measured as the relative ability of two competing viruses to produce infectious progeny (Holland *et al.*, 1991; reviewed in Domingo and Holland, 1997; Quiñones-Mateu and Arts, 2006). In the standard protocol, competitions are started by infecting cells or organisms with a mixture of a reference wild-type virus (given arbitrarily a fitness value of 1) and the virus to be tested, in known proportions. The progeny viruses are used to initiate a second round of infection, and the process is repeated a number of times (serial infections). Then, the logarithm of the proportion of the two competing viruses at each passage defines a fitness vector, the slope of which is the logarithm of the fitness of the test virus relative to the reference virus (Figure 4.1). The two competing viruses must be distinguishable by some phenotypic trait (e.g. a clear difference in the ability to replicate in the presence of an antibody or a drug)

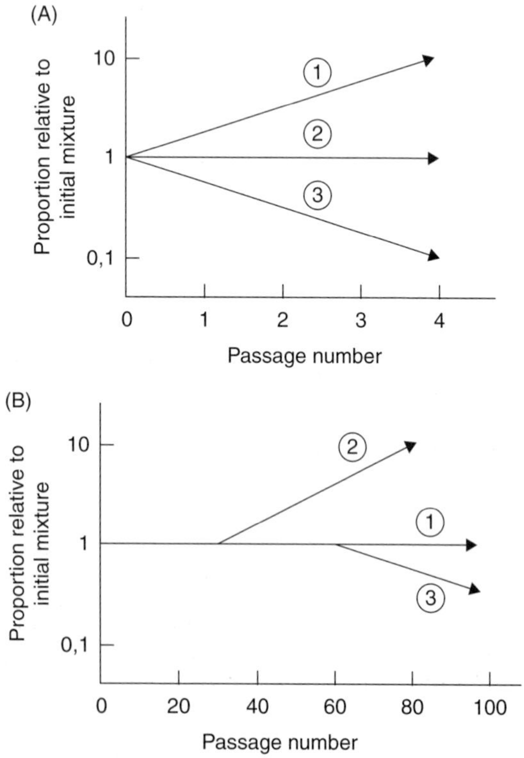

FIGURE 4.1 Schematic representation of fitness vectors and some patterns of fitness variation. (A) Plot of the proportion of the test virus and the reference virus, relative to the initial mixture, as a function of passage number. The plot gives a fitness vector. The test virus can show higher relative fitness than the reference virus (line 1), equal fitness (neutrality, line 2), or lower fitness than the reference virus (line 3). See text for comments and literature references. (B) Possible outcomes of a competition between two neutral variants. The two variants may co-exist for many generations (line 1). Occasionally one variant may displace the other in a rather unpredictable manner (lines 2 and 3), in agreement with the competitive exclusion principle of population genetics. Further information and references are given in the text.

or by some genetic change, such as nucleotide substitutions that allow the proportion of the two viruses to be determined by densitometry of a sequencing gel or by their specific amplification by real time reverse transcription-polymerase chain reaction (RT-PCR) using discriminatory oligonucleotide primers.

Fitness determinations of viruses subjected to different passage regimes have established an important effect of population size of the virus involved in the infections, on fitness evolution.

Fitness Decrease Upon Bottleneck Passages. Viral Virulence May Not Correlate with Fitness

Animal viruses are likely to undergo genetic bottlenecks during transmission; most of the evidence suggesting bottleneck effects comes from sequence analysis of infected hosts (for instance, Frost *et al.*, 2001), but Pfeiffer and Kirkegaard demonstrated bottlenecks during PV transmission from inoculated sites to the brain in transgenic mice expressing the human PV receptor (Pfeiffer and Kirkegaard, 2006). In addition, there is direct evidence demonstrating that plant viruses experience significant bottlenecks during movement from the site of infection (Ali *et al.*, 2006; Jridi *et al.*, 2006) (see Chapter 12).

RNA virus populations subjected to severe serial bottleneck events in cell culture—such as those occurring upon serial plaque-to-plaque transfers—undergo, on average, a decrease in fitness (Chao, 1990; Duarte *et al.*, 1992; Escarmís *et al.*, 1996; Yuste *et al.*, 1999; de la Iglesia and Elena, 2007). This is due to the stochastic accumulation of deleterious mutations (Figure 4.2), predicted by Müller (1964) to occur for small populations of asexual organisms lacking in mechanisms, such as sex or recombination, that could eliminate or compensate for such debilitating mutations (Maynard-Smith, 1976). Subjecting RNA viruses to repeated plaque-to-plaque transfers has all the ingredients to accentuate the effects of Müller's ratchet: a

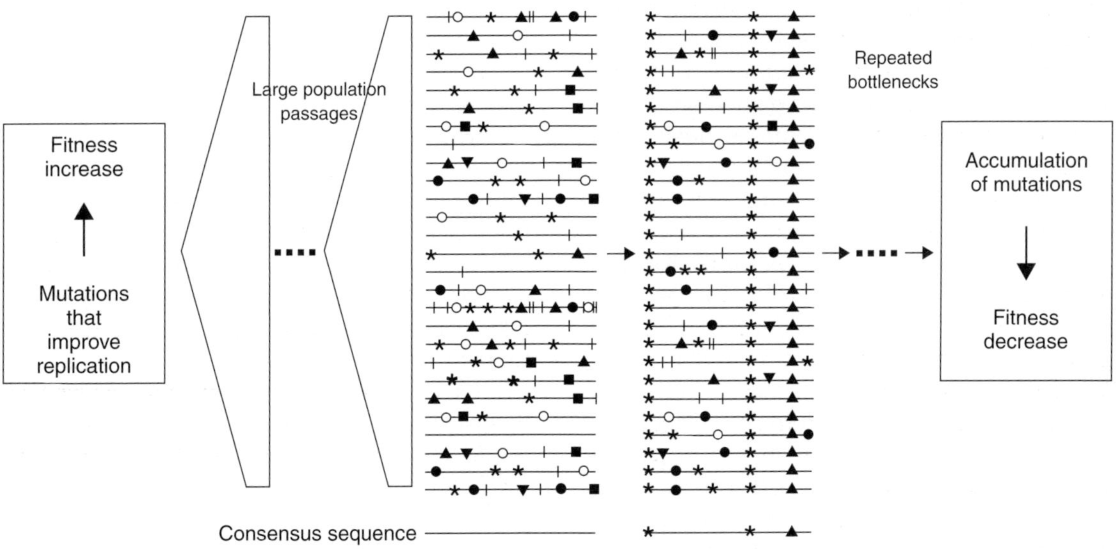

FIGURE 4.2 Schematic representation of viral quasispecies and the effect of viral population size on replicative fitness. Horizontal lines represent genomes and symbols on the lines represent mutations. Random sampling of genomes (bottleneck events, small arrows) lead to accumulation of mutations and fitness decrease. Large population passages (large arrows) lead to increases in replicative fitness. Fitness losses or gains depend on the initial fitness of the viral population and the size of the bottleneck. See text for details and references.

viral population reduced to a single genome at the onset of plaque formation (extreme genetic drift), and high mutation rates.

A study by Novella *et al.* (1995c) using VSV established that the extent of fitness loss for any given bottleneck size depends on the initial fitness of the viral clone under study. The higher the initial fitness, the less severe must the bottleneck be to avoid fitness losses. Debilitated viral clones often gain fitness even when subjected to considerable bottlenecking (Novella *et al.*, 1995a, 1995c). Rather constant, stable fitness values could be attained by choosing the appropriate bottleneck size, although occasional fitness jumps were observed (Novella *et al.*, 1996).

Escarmís *et al.* (1996, 2008) examined the genetic lesions associated with Müller's ratchet by determining genomic nucleotide sequences of FMDV clones prior to and after undergoing repeated (up to 409) plaque-to-plaque transfers. The result was that fitness loss was associated with unusual mutations that had never been seen in natural FMDV isolates or laboratory populations subjected to passages involving large viral populations. Particularly striking were an internal polyadenylate extension preceding the second functional AUG initiation codon of the FMDV genome, and amino acid substitutions at internal capsid residues. Additions or deletions of nucleotides have been frequently observed at homopolymeric tracts, particularly on pyrimidine runs in templates copied by proofreading-repair-deficient polymerases (Kunkel, 1990; Bebenek and Kunkel, 1993). The experimental results suggest that only when the repeated bottlenecks limit the action of negative selection (elimination or decrease in proportion of low fitness genomes) can such internal polyadenylate extensions (and other deleterious mutations) be maintained in the FMDV genome (Escarmís *et al.*, 1996, 2006). In contrast, sequence analysis of VSV genomes subjected to plaque-to-plaque passages did not show unusual mutations, with the possible exception of mutations in the RNA termini, which are uncommon in viruses evolving in regimes of acute replication (Novella and Ebendick-Corpus, 2004).

Fitness decrease upon subjecting FMDV to plaque-to-plaque transfers was biphasic: an initial decrease was followed by a highly fluctuating pattern with a constant average fitness value. The fluctuating pattern followed a Weibull statistical distribution (Weibull, 1951; Lazaro *et al.*, 2003). A Weibull distribution describes disparate physical and biological processes. In the case of plaque-to-plaque transfers of a virus this type of distribution probably results from the multiple host–virus interactions that occur as the virus life cycle is completed, and alterations of such interactions as mutations accumulate in multifunctional viral proteins (Lazaro *et al.*, 2003). The studies of evolution of FMDV when subjected to many repeated serial bottleneck transfers revealed a remarkable resistance of the virus to extinction despite a linear accumulation of mutations in its genome (Escarmís *et al.*, 2002), as well as the existence of multiple evolutionary pathways for fitness recovery (Escarmís *et al.*, 1999) (see also "Intra-mutant spectrum suppression can contribute to lethal mutagenesis," above).

Fitness has often been considered a component of parasite virulence, defined as the capacity of parasites to inflict damage upon their hosts. Indeed, very frequently an increase in viral fitness parallels an increase of virulence. However, a comparative quantitative analysis of fitness and virulence (cell-killing capacity) of an FMDV clone subjected to plaque-to-plaque transfers, and of its parental clone, revealed that fitness and virulence can be two unrelated traits (Herrera *et al.*, 2007). The molecular basis for the different trajectories followed by fitness and virulence resided in the fact that fitness was affected by mutations anywhere in the viral genome while determinants of cell-killing capacity were multigenic but restricted to some specific genomic regions of the viral genome. As a consequence, the random accumulation of mutations associated with bottleneck transfers had a more negative impact on fitness than on virulence of this FMDV clone (Herrera *et al.*, 2007). That viral fitness and virulence can follow different trajectories is supported by several observations with animal and plant viruses. VSV populations that were subjected to a regime of persistent infection in sandfly cells showed overall decrease in both fitness and virulence in mammalian cells, but the decrease in virulence continued throughout the experiment, while the decrease in fitness peaked at intermediate passages and was followed by some degree of recovery (Zárate and Novella, 2004). Simian immunodeficiency virus SIVmac239 attains similar high viral loads in the sooty mangabey and the rhesus macaque, yet it is only virulent for the rhesus macaque (Kaur *et al.*, 1998). At an epidemiological level, greater fitness of historical versus current HIV-1 isolates was taken as evidence of HIV-1 attenuation over time, assuming a direct correlation between fitness and virulence (Arien *et al.*, 2005). However, no trend towards HIV-1 attenuation since the time of introduction of the virus into Switzerland was observed (Muller *et al.*, 2006). These and other studies with viral and non-viral parasites (reviewed in Herrera *et al.*, 2007) suggest that evolution in nature can drive parasites to attain virulence levels that are not necessarily coupled to fitness. This distinction between fitness and virulence should be taken into consideration in the formulation of models for parasite virulence.

Fitness Gain Upon Large Population Passages: Limitations, Exclusions, Memory and Molecular Transitions

In contrast to bottleneck passages, large population infections generally result in fitness gains of RNA viruses (Martinez *et al.*, 1991; Clarke *et al.*, 1993; Novella *et al.*, 1995b; Escarmís *et al.*, 1999). Fitness increase in this case is expected from a gradual optimization of mutant spectra when their different components, arising by mutation and in some cases also by recombination, are allowed unrestricted competition in a constant environment (Figure 4.2). High replicative fitness may help a virus to overcome selective constraints—including antiviral agents or immune responses (Quiñones-Mateu *et al.*, 2006; Grimm *et al.*, 2007)—and to delay extinction by lethal mutagenesis (Sierra *et al.*, 2000; Pariente *et al.*, 2001).

When the relative fitness of the evolving quasispecies reaches a high value, even quite large population sizes can constitute an effective bottleneck and prevent continuing fitness increase (Novella *et al.*, 1999a, 1999b). This limiting high fitness level was manifested by stochastic fluctuations in fitness values expected from random generation of mutations in a continuously evolving mutant swarm. These perturbations illustrate how difficult it is to attain a true population equilibrium even when viruses replicate in a constant environment. A rare combination of mutations—one that may occur only once over many rounds of viral replication—may transfer one genome and its descendants to a distant region of sequence space, and trigger the dominance of one viral subpopulation over another, thereby disrupting a period of population equilibrium. In competitions between two VSV clones of similar fitness coexisting at or near equilibrium, a rapid and unpredictable displacement of one VSV population by the other (Clarke *et al.*, 1994) provided support for a classical concept of population biology: the competitive exclusion principle (Gause, 1971). Furthermore, in the competition passages preceding mutual exclusion, both the winners and the losers gained fitness at comparable rates, in support of yet another concept of population genetics: the Red Queen hypothesis (Van Valen, 1973; Clarke *et al.*, 1994; reviewed in Domingo, 2006) (see Figure 4.1).

Parallel fitness gains were also observed for minority memory genomes and their majority counterparts in evolving FMDV quasispecies (Arias *et al.*, 2004). Memory genomes are subpopulations of genomes that remain in a replicating viral quasispecies at a frequency about 10^2- to 10^3-fold higher than the frequency that can be attributed to mutational pressure alone, and reflect those genomes that were dominant at a previous stage of the evolution of the same viral lineage (Ruiz-Jarabo *et al.*, 2000; review in Domingo, 2000). Memory has been documented with a number of genetic markers of FMDV (Ruiz-Jarabo *et al.*, 2000, 2002, 2003b) and HIV-1 (Briones *et al.*, 2003, 2006), and similar results have been described for VSV (Novella *et al.*, 2007). Memory is a consequence of fitness variations inherent to quasispecies dynamics, likely to exert its main influence on the composition of mutant spectra that have been subjected to various alternating selective pressures (Domingo, 2000).

Relative viral fitness may depend on the multiplicity of infection (m.o.i.) used during selection or competition. High m.o.i. promotes coinfection, and the higher the level of coinfection the more likely that complementation will take place. Complementation effectively hides beneficial (and deleterious) variation from the effects of selection (Sevilla *et al.*, 1998; Wilke and Novella, 2003; Wilke *et al.*, 2004). In addition, high m.o.i. effects may relate to the use of alternative receptors or to interfering interactions occurring within the mutant spectra of viral quasispecies (Sevilla *et al.*, 1998; Perales *et al.*, 2007) (see section on "Intra-population complementation and interference in viral quasispecies: mutant distributions as the units of selection").

Defective viruses can be maintained in the course of high m.o.i. passages by complementation. An extensively documented case is the generation and maintenance of helper-dependent defective-interfering (DI) RNA and particles, which follow the process of mutation, competition and selection typical of quasispecies dynamics (Holland *et al.*, 1982; Roux *et al.*, 1991). Other types of defective genomes can also be maintained in viral populations by complementation (Charpentier *et al.*, 1996; Moreno *et al.*, 1997; Yamada *et al.*, 1998). Some defective genomes can be transmitted from infected into susceptible hosts, rendering the maintenance of defective genomes by complementation an event of potential epidemiological significance (Aaskov *et al.*, 2006).

A striking, extreme case of complementation between defective genomes was provided by evolution of standard FMDV towards two defective forms that were infectious and killed cells by complementation in the absence of standard FMDV (García-Arriaza *et al.*, 2004, 2005, 2006). These studies have provided evidence of a continuous dynamics of generation of defective FMDV genomes harboring in-frame internal deletions within genomic regions encoding *trans*-acting proteins, giving rise to swarms of genomes with non-identical, related deletions (García-Arriaza *et al.*, 2006). Each virion encapsidates only one type of defective genome and, therefore, the same cell must be infected by at least two different particles to permit complementation and formation of progeny defective genomes (Manrubia *et al.*, 2006). The high m.o.i.-dependent evolution of FMDV towards two defective forms that can complement each other has been regarded as experimental support of a first step in a process towards viral genome segmentation. Interestingly, multipartite segmented genomes are rare among the animal and bacterial viruses but are frequent among plant viruses, and the latter are characterized by high m.o.i. as they spread in their host plants (Lazarowitz, 2007).

The main conclusion we derive from the results summarized in the preceding paragraphs is that even in a relatively constant biological and physical environment, as is usually provided by *in vitro* cell culture systems, the degree of adaptation of viral quasispecies may undergo remarkable quantitative variations, prompted by the stochastic generation of mutant genomes, and different opportunities for competitive optimization of mutant spectra.

FITNESS VARIATIONS IN CHANGING ENVIRONMENTS

The experiments of fitness variation of viruses in cell culture summarized in the previous section have been instrumental in defining some basic influences that guide fitness evolution of viral quasispecies. However, in their replication in a natural setting, viruses encounter multiple and changing environments, and they often have to cope with conflicting selective constraints. Because of polymorphisms in key host proteins involved in cellular and humoral immune responses, and in many other cell surface antigens, viruses do not face the same selective constraints in different individuals of the same host species. Biological environments are heterogeneous and vary with time within each infected individual. Furthermore, a considerable number of viruses are capable of infecting different host species, extending even further the range of environments they face.

Arboviruses that replicate in mammalian and insect hosts constitute a classical example of obligate environmental alternacy *in vivo* (Scott *et al.*, 1994; Weaver, 1998) (Chapter 16). Early work documented that extensive replication of viruses in insect cells led to attenuation of infectivity for mammalian cells (Peleg, 1971; Mudd *et al.*, 1973). Prolonged persistence of VSV in sandfly cells cultured at low temperatures resulted in several orders of magnitude greater fitness in insect cells than in mammalian cells (Novella *et al.*, 1995a; Zárate and Novella, 2004). In contrast, acute VSV replication in sandfly cells led to fitness increase in mammalian cells (Novella *et al.*, 1999a), and replication of West Nile virus in mosquito cells resulted in populations that, while not improved, showed no fitness losses in vertebrate cells (Ciota *et al.*, 2007). Thus, we cannot assume selective differences between insect and mammalian cells types, and when we observe tradeoffs, these may be due to different strategies of replication (persistent versus acute), not to difference in cell type *per se*.

A single passage of sandfly cell-adapted VSV in mammalian cells led to an increase in fitness in mammalian cells to near original values. It would be interesting to test whether this capacity for fitness shift would be similar for non-arboviral RNA viruses able to grow in insect cells in culture. VSV adapted to sandfly cells was highly attenuated for mice. Again, a single passage in mammalian cells restored the virulence phenotype *in vivo* (Novella *et al.*, 1995a).

Several groups have studied the evolutionary consequences of alternating environments during arbovirus replication (reviewed in Wilke *et al.*, 2006; Ciota *et al.*, 2007). The overall results showed that extensive alternating replication between mammalian and insect cells led to fitness improvement in both environments; the only exception was VSV adapted to alternation between persistent insect replication and acute mammalian replication: adaptation during alternation is dominated by the persistent environment and there is fitness loss in the mammalian environment (Zárate and Novella, 2004) (for details, see Chapter 16).

Studies of fitness variations *in vivo* have been approached in at least three ways. Some studies have involved growth-competition experiments between two viruses replicating in host organisms. In other studies, the outcome of competitions between viruses that were isolated *in vivo* has been analyzed in primary or established cell cultures. In yet another line of research, the effect of fitness variations in cell culture on the replicative potential of viruses *in vivo* has been examined.

Carrillo *et al.* (1998) isolated two variant FMDVs present at low frequency in the course of replication of a clonal virus preparation in swine. One of the variants was a MAb-resistant mutant (MARM), while the other was isolated from blood during the early viremic phase of the acute infection. The ability of the two variants to compete *in vivo* with the parental clonal population was examined by coinfection of swine with mixtures of the parental clone and each of the two variants individually. None of the two variants became completely dominant in a single coinfection *in vivo*, but fitness differences were clearly documented. The parental FMDV clone manifested a selective advantage over the MARM in that the parental clone was dominant in most lesions (vesicles) in the diseased swine. In contrast, the parental clone and the variant from the early viremic phase were about equally represented in the lesions of the animals infected with equal amounts of the two viruses (Carrillo *et al.*, 1998).

The lentivirus equine infectious anemia virus (EIAV) experiences continuous quasispecies fluctuations during persistent infections in horses (Clements *et al.*, 1988). EIAV quasispecies were characterized in a pony experimentally infected with a biological clone of the virus. New quasispecies were associated with recurrent episodes of disease. A large deletion in the principal neutralizing domain of the virus was identified during the third febrile episode and became dominant during the fourth febrile episode. This drastic genetic change did not appear to diminish significantly the fitness of EIAV *in vivo* and in cell culture (Leroux *et al.*, 1997). The complexity of sequential EIAV populations *in vivo*, was characterized with a non-hierarchical clustering method to analyze quasispecies, termed PAQ (partition analysis of quasispecies) (Baccam *et al.*, 2001). This procedure to dissect the composition of mutant spectra should allow the recognition of subpopulations within viral quasispecies as they evolve towards fitness gain or loss.

Fitness Variations in Viral Disease Emergence and Reemergence. The Case of Human Influenza Virus

The multiple environments in which viruses have to replicate *in vivo* may promote the selective expansion of subpopulations from viral quasispecies thereby leading to variant viruses that display altered relative fitness in different host organs, as compared with their parental populations. Such variations in the potential replicative capacity constitute one of the ingredients that may affect the emergence and reemergence of viral disease (reviews in Smolinski *et al.*, 2003; Peters, 2007). The genetic lottery of blind variation through mutation, recombination, and genome segment reassortment is played in the face of a background of multiple ecological, sociological, and demographic factors. In recent decades viral disease emergences that have affected humans have occurred at a rate of about one per year. Salient examples are acquired immune deficiency syndrome (AIDS), severe acute respiratory syndrome (SARS), encephalitis associated with West Nile virus, the expansion of dengue fever, or periodic influenza pandemics (Smolinski *et al.*, 2003; Peters, 2007).

Multiple genetic changes may favor the adaptation of a virus to a new host. Once adaptation has taken place, the adapted virus may lose or maintain the pathogenic potential for the former (donor) host (as an example of maintenance of virulence for a donor and recipient host in FMDV see Nuñez *et al.*, 2007). A core (or basal) genetic composition of a viral pathogen may be in itself a predictor of pathogenic potential, as profusely documented with natural or laboratory-generated, attenuated variants of many viral pathogens. To take influenza A virus and the threat of a human influenza pandemics as examples, out of the 16 hemagglutinin (H) and 9 neuraminidase (N) subtypes circulating among animal reservoirs, some potentially threatening forms being more carefully kept under surveillance include H_5N_1, H_7N_7, H_9N_2, and H_2N_2 viruses. The expansion of the H_5N_1 subtype among wild and domestic avian species and human contacts since 2005 has resulted in over 300 human cases in nearly 50 countries, with more than 50% deaths, as well as the killing of millions of poultry (Wright *et al.*, 2007). Key parameters for an avian influenza virus to give rise to a human influenza pandemic include acquisition of receptor-recognition specificity for human cells, and the capacity for efficient human-to-human transmission (Parrish and Kawaoka, 2005; Suzuki, 2005). This capacity can be expressed as the basic reproductive ratio (Ro) which is the average number of infected contacts from each infected host (review in Nowak and May, 2000). "Epidemiologic fitness" has been used to describe (through samplings of definitory genomic sequences, diagnostic surveys, etc.) the capacity of a virus to become dominant (relative to related viruses or variants) during epidemic outbreaks (Domingo, 2007). In the case of human influenza virus, the acquisition of high epidemiological fitness depends on multiple gene products. Critical substitutions in H may modify the receptor-binding specificity of influenza viruses, and such substitutions have been found in minority subpopulations of influenza virus in several surveys. In one study, two substitutions in H identified in a human influenza virus from a fatal human case, were shown to modify the receptor-binding preference of H of a H_5N_1 virus from sialic acid-α2,3 galactose (associated with replication in avian hosts) to both sialic acid-α2,3 galactose and sialic acid-α2,6 galactose, both associated with binding to human-type receptors, each expressed preferentially in different sites of the human respiratory tract (Auewarakul *et al.*, 2007). Thus, in influenza virus, and probably many other pathogenic viruses, both epidemiologic fitness and replicative fitness are multigenic traits (Grimm *et al.*, 2007).

Several studies have compared the amino acid sequence of multiple influenza virus proteins to search for markers (amino acid substitutions) of human isolates and human pandemic strains (from 1918, 1957, 1968 and recent human H_5N_1 isolates). In one such proteomics survey, several amino acid changes located in PB2, PA, NP, M1, and NS1 distinguished avian influenza viruses from their human counterparts (Finkelstein *et al.*, 2007). Some markers were conserved in the influenza viruses that caused the 1918, 1957, and 1968 pandemics. Other studies have identified HA and PB2 as critical for adaptation of avian virus to humans, that may occur by a step-wise process reflected in acquisition of diagnostic amino acid markers. Evidence of human-to-human transmission of avian influenza virus H_5N_1 has been obtained in some family case clusters but not in others (Yang *et al.*, 2007). Influenza constitutes the paradigm of a viral disease which, favored by a continuum of genetic variation, reemerges periodically to cause pandemics, and for which extensive epidemiological surveillance is currently in operation.

Fitness and Drug Resistance in HIV-1

An increasing number of measurements of viral fitness involve human immunodeficiency virus 1 (HIV-1) variants isolated from quasispecies replicating *in vivo*. Particularly relevant are fitness comparisons among multiple mutants harboring amino acid substitutions related to resistance to reverse transcriptase and protease inhibitors (see also Chapter 14).

HIV-1 Reverse Transcriptase (RT) Inhibitors

Since the discovery of AZT (3′-azido-3′-deoxythymidine, zidovudine) as an effective inhibitor of HIV replication (Mitsuya *et al.*, 1985), drug therapy has been widely used in the treatment of AIDS. The loss of therapeutic effect due to the acquisition of resistance was recognized for AZT in 1989, when Larder and colleagues showed that HIV isolates from patients with advanced HIV disease became less sensitive to the drug during the course of treatment (Larder *et al.*, 1989). High-level resistance to AZT is achieved through the accumulation of several mutations including M41L, D67N, K70R, L210W, T215Y/F, and K219Q/E (for a review, see Larder, 1994). The first substitution arising during AZT treatment is usually K70R, followed by T215Y. The K70R mutation appears frequently, since it requires only one nucleotide change, and does not have a major impact on viral fitness (Harrigan *et al.*, 1998). The simultaneous presence of Leu41 and Tyr215 in the viral RT-coding region confers high-level resistance to AZT, without having a major effect on viral fitness. In contrast, other combinations of AZT resistance mutations (e.g. M41L/K70R) confer reduced replication capacity (Jeeninga *et al.*, 2001). Interestingly, transmitted HIV-1 carrying D67N or K219Q evolve rapidly to AZT resistance *in vitro* (selecting for K70R) and show a high replicative fitness in the presence of zidovudine (García-Lerma *et al.*, 2004). On the other hand, L210W improved infectivity and relative fitness of an M41L/T215Y mutant in the presence of AZT, but decreased infectivity and relative fitness when introduced into a D67N/K70R/K219Q background (Hu *et al.*, 2006).

Drug-resistant mutations occur in the mutant spectra of HIV-1 quasispecies from untreated patients (Nájera *et al.*, 1995). The replacement of Tyr215 by Cys, Asp, or Ser has been observed *in vivo* in the absence of zidovudine treatment (Goudsmit *et al.*, 1997;

Yerly *et al.*, 1998). In the absence of inhibitor, T215S and T215D confer a small but significant advantage over the wild-type virus, as determined *in vitro* in growth competition experiments. However, the replicative advantage conferred by T215S was lost in the presence of zidovudine-resistance mutations such as M41L and L210W (García-Lerma *et al.*, 2001).

Other nucleoside inhibitors of HIV-1 RT are listed in Table 4.1. High-level resistance to the nucleoside analogue 3TC (2′, 3′-dideoxy-3′-thiacytidine, lamivudine) is rapidly achieved by the substitution M184V, located at the YMDD motif, which is part of the catalytic core of the enzyme. During 3TC treatment, the substitution M184I appears first, but then

TABLE 4.1 Amino Acid Substitutions Associated with HIV-1 Resistance to Antiretroviral Drugs

Inhibitors	Amino acid substitutions associated with drug resistance[a]
Nucleoside analogue RT inhibitors	
Zidovudine (AZT)	M41L, D67N, K70R, L210W, T215Y/F, K219Q/E
Didanosine (ddI)	K65R, L74V
Lamivudine (3TC)	(E44D/V118I), K65R, M184V/I
Stavudine (d4T)	M41L, D67N, K70R, V75T, V118I, L210W, T215Y/F, K219Q/E
Zalcitabine (ddC)	K65R, T69D, L74V, M184V
Abacavir	K65R, L74V, Y115F, M184V
Emtricitabine	(K65R/Q151M), M184V/I
Tenofovir	K65R, K70E
Multiple nucleoside analogues	(i) M41L, D67N, K70R, L210W, T215F/Y, K219Q/E; (ii) A62V, V75I, F77L, F116Y, Q151M; (iii) Insertions between codons 69–70 (e.g. T69SSS, T69SSG, T69SSA, etc.), M41L, A62V, K70R, L210W, T215Y/F
Non-nucleoside analogue RT inhibitors	
Nevirapine	L100I, K101P, K103N/S, V106A/M, V108I, Y181C/I, Y188C/L/H, G190A/C/E/Q/S/T
Delavirdine	K103H/N/T, V106M, Y181C, Y188L, G190E, P236L
Efavirenz	L100I, K103H/N, V106M, V108I, Y188L, G190A/S/T, P225H
PR inhibitors[b]	L10I/R/V, **G48V**, I54L/V, A71T/V, G73S, V77I, V82A, I84V, **L90M**
Saquinavir	
Ritonavir	L10I/R/V, K20M/R, V32I, L33F, M36I, M46I/L, I54V/L, A71V/T, V77I, **V82A/F/S/T, I84V**, L90M
Indinavir	L10I/R/V, K20M/R, L24I, V32I, M36I, **M46I/L**, I54V, A71T/V, G73A/S, V77I, **V82A/F/S/T, I84V**, L90M
Nelfinavir	L10F/I, **D30N**, M36I, M46I/L, A71T/V, V77I, V82A/F/S/T, I84V, N88D/S, **L90M**
Amprenavir	L10F/I/R/V, V32I, M46I/L, I47V, **I50V**, I54M/V, **I84V**, L90M
Lopinavir	L10F/I/R/V, G16E, K20I/M/R, L24I, V32I, L33F, E34Q, M36I/L, K43T, **M46I/L**, **I47A/V**, G48M/V, I50V, **I54L/V**/A/M/S/T, Q58E, L63T, A71T, G73T, T74S, **V82A**/F/S/T, I84V, L89I/M
Atazanavir	L10F/I/V, K20I/M/R, L24I, L33F/I/V, M36I/L/V, M46I/L, G48V, **I50L**, I54L/V, L63P, A71I/T/V, G73A/C/S/T, V82A/F/S/T, **I84V, N88S**, L90M
Tipranavir	L10I/S/V, I13V, K20M/R, **L33F/I/V**, E35G, M36I/L/V, K43T, M46L, I47V, I54A/M/V, Q58E, H69K, T74P, **V82L/T**, N83D, **I84V**, L90M
Darunavir	V11I, V32I, L33F, I47V, **I50V**, I54L/M, G73S, L76V, I84V, L89V
Fusion inhibitors	G36D/E/S, I37T/V, V38A/M/E, Q40H, N42T, N43D/K/S
Enfuvirtide	

[a] For additional information, see (Clark *et al.*, 2007; Clotet *et al.*, 2007; Johnson *et al.*, 2007).

[b] Primary resistance mutations are shown in bold. Most PR inhibitors (saquinavir, indinavir, amprenavir, lopinavir, atazanavir, tipranavir, and darunavir) are usually prescribed in combination with a low dose of ritonavir, that has a boosting effect on the PR inhibitor concentration in plasma.

it is lost due to the outgrowth of the M184V-containing viruses (Keulen *et al.*, 1997). Growth competition experiments showed a selective advantage of viruses with Val184 over those with Ile184. The low efficiency of 3TC-resistant HIV-1, carrying RT mutations M184V or M184I, has been attributed to the low processivity of the mutant RT (Back *et al.*, 1996), which was accentuated in peripheral blood mononuclear cells (PBMCs) (Keulen *et al.*, 1997).

Other nucleoside analogue resistance mutations (e.g. K65R, K70E, or L74V) also have a significant impact on viral fitness, which correlates with a defect in RT processivity (Sharma and Crumpacker, 1997; Miller *et al.*, 1998; Sharma and Crumpacker, 1999; White *et al.*, 2002). The presence of K65R together with L74V or M184V has a strong deleterious effect on viral replication, due to the poor ability of K65R/L74V to use natural nucleotides relative to the wild type (Deval *et al.*, 2004), or to the negative impact of the simultaneous presence of K65R and M184V on the RT's processivity, as well as in the initiation of reverse transcription (White *et al.*, 2002; Frankel *et al.*, 2007). These observations are consistent with the low prevalence of the K65R mutation among isolates from antiretroviral-drug experienced patients, and give rational support to the benefit in combining mutations that impair virus replication.

Drug combinations are very effective in blocking HIV replication, leading to a more than 10000-fold reduction of viral load. Early studies showed that multiple drug resistance to AZT and other inhibitors can be achieved through the accumulation of mutations appearing in monotherapy (Schmit *et al.*, 1996; Shafer *et al.*, 1998). However, the response of a viral quasispecies to multiple constraints (e.g. different antiviral drugs) is often difficult to predict. Simultaneous treatment with AZT and ddI led to viruses with reduced sensitivity to AZT, ddC, ddI, ddG, and d4T (Shirasaka *et al.*, 1995; Iversen *et al.*, 1996). The resistant viruses contained substitutions A62V, V75I, F77L, F116Y, and Q151M. Substitution Q151M, which results from two nucleotide changes, is the first to appear and confers partial resistance to AZT, ddI, ddC, and d4T. Fitness assays involving the determination of replication kinetics or growth competition experiments showed that mutations at codons 62, 75, 77, and 116 improved the replication capacity of the resistant virus (Maeda *et al.*, 1998; Kosalaraksa *et al.*, 1999).

With the increasing complexity of the antiretroviral regimens, novel mutational patterns conferring resistance to multiple antiretroviral drugs have been identified. Thus, HIV-1 variants with insertions or deletions in the "fingers" subdomain of the RT have been found in patients failing therapy with multiple RT inhibitors (Mas *et al.*, 2000; for a recent review, see Menéndez-Arias *et al.*, 2006). The presence of the amino acid changes T69S and T215Y in the RT, together with a dipeptide insertion between positions 69 and 70 (usually Ser-Ser), and the subsequent accumulation of additional mutations (e.g. M41L, A62V, T69S, and K70R) leads to the emergence of virus displaying high-level resistance to thymidine analogues (Matamoros *et al.*, 2004; Cases-González *et al.*, 2007). Dual infection/competition experiments revealed that in the presence of low concentrations of AZT, removal of the two serine residues forming the dipeptide insertion in a multidrug-resistant isolate does not cause a detrimental effect on the replication capacity of the virus (Quiñones-Mateu *et al.*, 2002). However, in the absence of drug, the insertions improved the fitness of virus-carrying thymidine analogue mutations (e.g. M41L, L210W, T215Y, etc.). Although, multidrug-resistant mutants are able to maintain high viral loads in the presence of antiretroviral therapy, it should be noted that *in vivo* wild-type HIV variants outcompete those bearing the insertion, as demonstrated when therapy is interrupted (Briones *et al.*, 2000; Lukashov *et al.*, 2001).

Non-nucleoside RT inhibitors bind to a hydrophobic cavity which is 8–10Å away from the polymerase active site, and lined by the side-chains of Tyr181, Tyr188, Phe227, and Trp229 (Kohlstaedt *et al.*, 1992). High-level resistance appears quickly after treatment and involves amino acid changes in residues located at the inhibitor binding site (Table 4.1). Again, resistance mutations often lead to reduced *in vitro* replication capacity. Examples are the nevirapine-resistance mutation V106A and the delavirdine-resistance mutation P236L that impair RNase H activity (Gerondelis *et al.*, 1999; Archer *et al.*, 2000; Dykes *et al.*, 2001; Iglesias-Ussel *et al.*, 2002; Collins *et al.*, 2004), as well as several mutations at codons 138 and 190, whose effects appear to be related to impaired DNA synthesis and RNase H degradation (Pelemans *et al.*, 2001; Huang *et al.*, 2003; Collins *et al.*, 2004; Wang *et al.*, 2006).

HIV-1 Protease (PR) Inhibitors

The HIV-1 PR is a homodimeric enzyme composed of two polypeptide chains of 99 residues. The substrate binding site is located at the interface between both subunits. The side-chains of Arg8, Leu23, Asp25, Gly27, Ala28, Asp29, Asp30, Val32, Ile47, Gly48, Gly49, Ile50, Phe53, Leu76, Thr80, Pro81, Val82, and Ile84 form the substrate-binding pocket and can interact with specific inhibitors (Wlodawer and Vondrasek, 1998), such as those used in the clinical treatment of AIDS. Approved PR inhibitors share relatively similar chemical structures and cross-resistance is commonly observed in the clinical setting (Menendez-Arias, 2002). It is not unexpected that many resistance mutations affect residues of the inhibitor-binding pocket of the PR (Table 4.1). Studies carried out *in vivo* and *in vitro* have shown that several amino acid substitutions involved in drug resistance may have a deleterious effect on viral fitness. Examples are D30N, I47A, I50V, G48V, and V82A (Eastman *et al.*, 1998; Martinez-Picado *et al.*, 1999; Kantor *et al.*, 2002; Prado *et al.*, 2002; Yusa *et al.*, 2002; Colonno *et al.*, 2004).

The deleterious effects caused by drug resistance mutations can be rescued by other amino acid replacements. For example, multidrug-resistant virus arising during prolonged therapy with indinavir contained PR with the substitutions M46I, L63P, V82T, and I84V (Condra *et al.*, 1995; Martinez-Picado *et al.*, 1999). Crystallographic studies of the mutant enzyme revealed that substitutions at codons 82 and 84 were critical for the acquisition of resistance, while the amino acid changes at codons 46 and 63, which are away from the inhibitor-binding site appear as compensatory mutations (Chen *et al.*, 1995; Schock *et al.*, 1996). Although compensatory mutations within the PR-coding region increase the catalytic efficiency of the enzyme, there are other molecular mechanisms that lead to fitness recovery during PR inhibitor treatments. Examples are: (i) mutations at Gag cleavage sites that increase polyprotein processing (Doyon *et al.*, 1996; Zhang *et al.*, 1997; Pettit *et al.*, 2002), (ii) mutations that affect the frameshift signal between the *gag* and *pol* genes that lead to an increased expression of *pol* products (Doyon *et al.*, 1998), or (iii) mutations outside of the cleavage sites that could affect the conformation of the Gag polyprotein and make the cleavage sites more accessible to the viral PR (Gatanaga *et al.*, 2002; Myint *et al.*, 2004).

Novel Antiretroviral Drugs

For many years, the RT and the PR were the only targets of approved antiretroviral drugs. In 2003, enfuvirtide, a synthetic peptide that impairs virus–host cell membrane fusion, was licensed for clinical use. Resistance to enfuvirtide is mediated by amino acid substitutions at codons 36–38 of the envelope glycoprotein gp41. The amino acid sequences found at those positions in drug-sensitive viruses (DIV, SIV, GIV, or GIM) are replaced by SIM, DIM, or DTV in the drug-resistant clones (Rimsky *et al.*, 1998). As observed with PR and RT inhibitors, resistance mutations cause a fitness loss, which was estimated to be approximately 10% in replication kinetics and growth competition experiments (Lu *et al.*, 2004; Reeves *et al.*, 2005). However, it should be noted that mutations in the V3 loop of the envelope glycoprotein gp120 can also affect the viral susceptibility to enfuvirtide (Reeves *et al.*, 2002), and further studies will be necessary to evaluate its impact on viral fitness *in vivo*.

In August 2007, a CCR5 coreceptor antagonist known as maraviroc was approved for clinical use. Maraviroc has potent antiviral activity against CCR5-tropic HIV-1 variants, including primary isolates from various clades (Dorr *et al.*, 2005). Maraviroc-resistant HIV variants contained unique amino acid changes in the V3 loop (e.g. A316T and I323V) and other positions within the envelope glycoprotein, gp120, but continued to be phenotypically CCR5-tropic and sensitive to CCR5 antagonists in preclinical development, such as vicriviroc (Westby *et al.*, 2007).

Other antiretroviral drugs showing promising results in clinical trials are the integrase inhibitors raltegravir (licensed in October 2007) and elvitegravir. However, the information available on specific drug resistance mutations and their effects on the viral replication capacity is still preliminary.

This review of fitness effects of drug-resistance mutations in HIV-1 provides a dramatic illustration of the adaptive potential of a viral quasispecies. Acquisition of critical amino acid replacements for drug resistance, fitness effects that favor selection of compensatory mutations either in a viral enzyme or in its target substrate, occurrence of clusters of mutations for multidrug resistance are but some of the mechanisms displayed by HIV-1 to persist in the human population.

OVERVIEW

The virological significance of quasispecies theory becomes more apparent each year. Initial reports of extremely high error rates and great population diversity of RNA viruses were hotly disputed as being incorrect and inconsistent with often-observed stability in virus markers such as antigenicity, disease characteristics, host range, etc. High error rates and intra-population heterogeneity for RNA viruses are now widely accepted. Fortunately, early quasispecies theory presented a timely, remarkably prescient theoretical framework within which the behavior of replicating and evolving RNA virus populations could begin to be understood. Following elaboration by Eigen, Biebricher, Schuster, and colleagues, quasispecies-derived theory has been rapidly progressing and evolving. Its ground-breaking initial theoretical structure for exploring consequences of extreme biological error rates was informed by elegant molecular replication/mutation kinetic studies with small RNA replicons *in vitro.* Original quasispecies theory was formally applicable to these *in vitro* experiments, and was necessarily generalized, idealized, and, in many specifics, openly unrealistic for real viruses. Some simplifying assumptions not applicable to viruses in the real world include: infinite virus populations; global optima in the selective landscape; one most-fit master sequence in a single, unvarying selective landscape; fitness restricted to competition solely between one master sequence and diverse variants of equally low fitness; and, finally, omission of complexities such as replicative interference, lethal mutations, complementation, recombination, etc.

Early modeling could not reasonably encompass all real-life complexities. To attempt inclusion of all would render any model (or alternative collection of models) hopelessly unwieldy, uninformative, and poorly predictive due to requisite alternate weightings of factors. Simplified assumptions not conforming to complex realities need not detract from the ability of models to serve as starting points and guideposts toward new directions for experiment and theory. Quasispecies theory has indisputably led virology to powerful new insights, deductions, and directions. A few critics have suggested that the non-real world parameters in early quasispecies models, and the non-realistic (and foregone) conclusions that can be contrived from them, are reason to reject the general validity and broad significance of quasispecies. Such circular arguments are specious and trivial relative to the experimental and conceptual advances already-made, and yet-to-be-made, via quasispecies theory with its straightforward conclusions and more subtle implications.

Increasingly sensitive analyses of viral quasispecies in recent decades have produced many remarkable insights. The most basic, far-reaching, awesomely predictive tenet of quasispecies theory will never be overshadowed; numerous variant genomes are bound together through extreme mutation rates, forming obligatorily co-selected partnerships in a vast, error-prone mutant spectrum from which they cannot escape, and from which they inevitably and coordinately may exert myriad, changing, ultimately unforseeable effects on all life forms. This tenet has been unquestionably and elegantly confirmed recently by the U. C. San Francisco, Stanford and Penn State groups (as reviewed above and elsewhere in this volume).

A significant postulate of early quasispecies models was that of "error catastrophe." This posits that replicase-generated quasispecies mutation rates are, through evolutionary selection, poised at, or near, an error threshold. Prolonged violation of this threshold (through replicase dysfunction, mutagens, elevated temperatures, nucleotide pool alterations, etc.) leads to virus extinction via a fast and irreversible transition, that has sometimes been equated with a phase transition in physics, as discussed in the opening chapter of this volume. Because the simplified model employed nonrealistic parameters and envisioned indefinite mutational drift, critics deny the existence of error thresholds and sharp transitions to error catastrophe. No real-world virus could conform to the simplifying assumptions employed in that model, but recent data from lethal mutagenesis experiments do demonstrate devolution to error catastrophe. Historical precedent for the term "error catastrophe " lies with Orgel's suggestion in 1963 of cascading coordinate collapse of cellular information within (and between) various interdependent cellular nucleic acid and protein *trans*-networks. We also employ the term in a broad manner for lethal mutagenesis. This is especially appropriate with mutagens such as 5-FU (which modify both viral and cellular nucleic acids and their encoded functions and structures). We cannot presently rule out some roles for mutagenized cellular, as well as viral, macromolecules, during lethal mutagenesis. Regardless, complex interactions of altered viral macromolecular networks are definitely involved. Extinction is mediated by "*trans*-acting networks" among abundant lethal defector genomes. The senior author's group in Madrid demonstrated (reviewed above) that strongly mutagenized RNA virus populations do collapse to extinction via a sharp transition, but without the non-lethal, continuous mutational drift exemplified in the original quasispecies simulations. Extremely rapid extinctions are observed for low-fitness input virus strains, which transition into error catastrophe during a single round of infection/mutagenesis. Lethal mutagenesis of FMDV and LCMV is mediated by full-length, replicating, interfering, lethal defector genomes. Total (defective and viable) genomic RNA mutation frequencies are elevated to varying extents, whereas specific infectivity of total genomic RNA is decreased by several orders of magnitude without any change of RNA consensus sequence. In light of quasispecies "variant-ensemble" behavior, it is not surprising that defective genomes can predominate within *trans*-acting networks during lethal mutagenesis, and continue to replicate even after extinction of LCMV infectivity. Defector *trans*-effects can provide positive complementation in concert with, or alternation with, (orthogonal) interference. Standard concepts of virus fitness are only tangentially applicable within such collapsing *trans*-networks. Catastrophic decay of viral digital information proceeds on (at least) two levels: (1) genetic quenching due to egregious fixation of genomic mutations in a *trans*-network environment that does not always select for optimal function of self-encoded proteins, and (2) phenotypic *trans*-quenching of potentially viable genomes via altered, defector-encoded (interfering) proteins. Possible roles of RNA recombination remain to be explored.

Defector-driven transitions will be challenging to dissect, and no theoretical model can possibly capture even their main intricacies. During lethal mutagenesis at high multiplicities of infection, each infected cell is a single compartment in which a separate, discrete error catastrophe event may devolve. Each discrete *trans*-network is disrupted and obscured during virus passages or RNA transfections following initial infection/mutagenesis. Multiplicity of infection (for both viable and defector virions) is clearly a crucial variable during passages. Virus strains with low replicative fitness (and mutagen-debilitated genomes) theoretically should be (and are) more vulnerable than highly fit strains to defector-mediated error catastrophe. Low-fitness strains cannot quickly produce high yields during temporary escape from defector networks. Future investigations with controlled compartmentalization (e.g. characterization of isolated infectious centers, microinjection of single cells, etc.), together with molecular genetic construction/reconstruction of defined *trans*-networks will illuminate pre-extinction events.

The Madrid group has already verified that ordinary, viable FMDV variant RNAs, and mixtures of variant RNAs bearing defined mutations in the capsid and polymerase genes can exert *trans*-complementation and interference effects on standard FMDV RNA following co-electroporation of cells. Clearly, at high multiplicities of co-electroporation, such mixtures of defined variant and control RNAs generate unique, mutually supportive or suppressive (complementing/interfering) *trans*-acting networks within each individual, coinfected cell. This provides strong analogies to events during the transitions of lethal mutagenesis. The compelling differences, of course, are that the latter devolve to extinction due to: (1) mutagen-elevated mutation, AND elevated mutation-fixation rates in a poorly-selective *trans*-milieu; (2) potent *trans*-quenching of surviving-and-collapsing infectious virus via interfering (lethal defector-encoded) proteins.

Thus, the quasispecies postulate of a rapid transition to extinction has been experimentally verified, albeit the complex defector mechanisms for real viruses differ significantly from those originally modeled, as indeed recognized and anticipated by Eigen (see above). It is evident that the details of lethal mutagenesis will likely differ among families of RNA viruses (e.g. those having mono-, bi-, or multipartitite genomes, strong or weak complementation, homologous or only non-homologous recombination, naked or enveloped capsids, etc.). However, it seems probable that error catastrophe will be observed in all. Although no theoretical model can possibly capture all the ingredients involved in the replicative collapse of a mutagenized viral population, it was the original error threshold which inspired the experiments currently being performed in several laboratories.

The tenets of Eigen, Biebricher, Schuster, and colleagues, derived from first principles and tractable models, have had enormous influence in virology. This pervasive influence is in no manner weakened nor negated by original simplifying assumptions.

ACKNOWLEDGMENTS

Work in Madrid was supported by grants BFU 2005-00863 from MEC, Proyecto Intramural de Frontera 2005–20F-0221 from CSIC, 36558/06, 36460/05 and 36523/05 from FIPSE, and Fundación R. Areces. CIBERehd is funded by the Instituto de Salud Carlos III. Work in Toledo was supported by grants AI45686 and AI065960 from NIH. CP is the recipient of a I3P contract from CSIC, financed by Fondo Social Europeo.

REFERENCES

Aaskov, J., Buzacott, K., Thu, H.M., Lowry, K. and Holmes, E.C. (2006) Long-term transmission of defective RNA viruses in humans and Aedes mosquitoes. *Science* **311**, 236–238.

Airaksinen, A., Pariente, N., Menendez-Arias, L. and Domingo, E. (2003) Curing of foot-and-mouth disease virus from persistently infected cells by ribavirin involves enhanced mutagenesis. *Virology* **311**, 339–349.

Ali, A., Li, H., Schneider, W.L., Sherman, D.J., Gray, S., Smith, D. and Roossinck, M.J. (2006) Analysis of genetic bottlenecks during horizontal transmission of Cucumber mosaic virus. *J. Virol.* **80**, 8345–8350.

Anderson, J.P., Daifuku, R. and Loeb, L.A. (2004) Viral error catastrophe by mutagenic nucleosides. *Annu. Rev. Microbiol.* **58**, 183–205.

Archer, R.H., Dykes, C., Gerondelis, P., Lloyd, A., Fay, P., Reichman, R.C., Bambara, R.A. and

Demeter, L.M. (2000) Mutants of human immunodeficiency virus type 1 (HIV-1) reverse transcriptase resistant to nonnucleoside reverse transcriptase inhibitors demonstrate altered rates of RNase H cleavage that correlate with HIV-1 replication fitness in cell culture. *J. Virol.* **74**, 8390–8401.

Arias, A., Ruiz-Jarabo, C.M., Escarmis, C. and Domingo, E. (2004) Fitness increase of memory genomes in a viral quasispecies. *J. Mol. Biol.* **339**, 405–412.

Arias, A., Agudo, R., Ferrer-Orta, C., Perez-Luque, R., Airaksinen, A., Brocchi, E. *et al.* (2005) Mutant viral polymerase in the transition of virus to error catastrophe identifies a critical site for RNA binding. *J. Mol. Biol.* **353**, 1021–1032.

Arien, K.K., Troyer, R.M., Gali, Y., Colebunders, R.L., Arts, E.J. and Vanham, G. (2005) Replicative fitness of historical and recent HIV-1 isolates suggests HIV-1 attenuation over time. *AIDS* **19**, 1555–1564.

Arnold, J.J., Vignuzzi, M., Stone, J.K., Andino, R. and Cameron, C.E. (2005) Remote site control of an active site fidelity checkpoint in a viral RNA-dependent RNA polymerase. *J. Biol. Chem.* **280**, 25706–25716.

Auewarakul, P., Suptawiwat, O., Kongchanagul, A., Sangma, C., Suzuki, Y., Ungchusak, K. *et al.* (2007) An avian influenza H5N1 virus that binds to a human-type receptor. *J. Virol.* **81**, 9950–9955.

Baccam, P., Thompson, R.J., Fedrigo, O., Carpenter, S. and Cornette, J.L. (2001) PAQ: partition analysis of quasispecies. *Bioinformatics* **17**, 16–22.

Back, N.K., Nijhuis, M., Keulen, W., Boucher, C.A., Oude Essink, B.O., van Kuilenburg, A.B. *et al.* (1996) Reduced replication of 3TC-resistant HIV-1 variants in primary cells due to a processivity defect of the reverse transcriptase enzyme. *EMBO J.* **15**, 4040–4049.

Batschelet, E., Domingo, E. and Weissmann, C. (1976) The proportion of revertant and mutant phage in a growing population, as a function of mutation and growth rate. *Gene* **1**, 27–32.

Bebenek, K. and Kunkel, T.A. (1993) The fidelity of retroviral reverse transcriptases. In: *Reverse Transcriptase* (A.M. Skalka and S.P. Goff, eds), pp. 85–102. New York: Cold Spring Harbor Laboratory Press.

Biebricher, C.K. and Domingo, E. (2007) The advantage of the high genetic diversity in RNA viruses. *Future Virol.* **2**, 35–38.

Biebricher, C.K. and Eigen, M. (2005) The error threshold. *Virus Res.* **107**, 117–127.

Botstein, D. (1980) A theory of modular evolution for bacteriophages. *Ann. NY Acad. Sci.* **354**, 484–491.

Briones, C., Mas, A., Gomez-Mariano, G., Altisent, C., Menendez-Arias, L., Soriano, V. and Domingo, E. (2000) Dynamics of dominance of a dipeptide insertion in reverse transcriptase of HIV-1 from patients subjected to prolonged therapy. *Virus Res.* **66**, 13–26.

Briones, C., Domingo, E. and Molina-París, C. (2003) Memory in retroviral quasispecies: experimental evidence and theoretical model for human immunodeficiency virus. *J. Mol. Biol.* **331**, 213–229.

Briones, C., de Vicente, A., Molina-Paris, C. and Domingo, E. (2006) Minority memory genomes can influence the evolution of HIV-1 quasispecies in vivo. *Gene* **384**, 129–138.

Bushman, F. (2002) *Lateral DNA Transfer. Mechanisms and Consequences.* New York: Cold Spring Harbor Laboratory Press.

Callaway, D.S. and Perelson, A.S. (2002) HIV-1 infection and low steady state viral loads. *Bull. Math. Biol.* **64**, 29–64.

Carrillo, C., Borca, M., Moore, D.M., Morgan, D.O. and Sobrino, F. (1998) *In vivo* analysis of the stability and fitness of variants recovered from foot-and-mouth disease virus quasispecies. *J. Gen. Virol.* **79**, 1699–1706.

Cases-Gonzalez, C.E., Franco, S., Martinez, M.A. and Menendez-Arias, L. (2007) Mutational patterns associated with the 69 insertion complex in multi-drug-resistant HIV-1 reverse transcriptase that confer increased excision activity and high-level resistance to zidovudine. *J. Mol. Biol.* **365**, 298–309.

Ciota, A.T., Lovelace, A.O., Ngo, K.A., Le, A.N., Maffei, J.G., Franke, M.A. *et al.* (2007) Cell-specific adaptation of two flaviviruses following serial passage in mosquito cell culture. *Virology* **357**, 165–174.

Clark, S.A., Calef, C. and Mellors, J.W. (2007) Mutations in retroviral genes associated with drug resistance. In: *HIV Sequence Compendium 2006–2007* (ed. by T. Leitner, B. Foley, B. Hahn, P. Marx, F. McCutchan, J. Mellors, S. Wolinsky and B. Korber), pp. 58–158. Theoretical Biology and Biophysics Group, Los Alamos National Laboratory. Los Alamos, New Mexico, USA.

Clarke, D.K., Duarte, E.A., Moya, A., Elena, S.F., Domingo, E. and Holland, J. (1993) Genetic bottlenecks and population passages cause profound fitness differences in RNA viruses. *J. Virol.* **67**, 222–228.

Clarke, D.K., Duarte, E.A., Elena, S.F., Moya, A., Domingo, E. and Holland, J. (1994) The red queen reigns in the kingdom of RNA viruses. *Proc. Natl Acad. Sci. USA* **91**, 4821–4824.

Clements, J.E., Gdovin, S.L., Montelaro, R.C. and Narayan, O. (1988) Antigenic variation in lentiviral diseases. *Annu. Rev. Immunol.* **6**, 139–159.

Clotet, B., Menéndez-Arias, L., Schapiro, J.M., Kuritzkes, D., Burger, D., Telenti, A., Brun-Vezinet, F., Geretti, A.M., Boucher, C.A., and Richman, D.D. (eds.) (2007) *Guide to management of HIV drug resistance, antiretrovirals pharmacokinetics and viral hepatitis in HIV infected subjects*, 7th edn. Fundació de Lluita contra la SIDA. Barcelona, Spain.

Colonno, R., Rose, R., McLaren, C., Thiry, A., Parkin, N. and Friborg, J. (2004) Identification of I50L as the signature atazanavir (ATV)-resistance mutation in treatment-naive HIV-1-infected patients receiving ATV-containing regimens. *J. Infect. Dis.* **189**, 1802–1810.

Collins, J.A., Thompson, M.G., Paintsil, E., Ricketts, M., Gedzior, J. and Alexander, L. (2004) Competitive fitness of nevirapine-resistant human immunodeficiency virus type 1 mutants. *J. Virol.* **78**, 603–611.

Condra, J.H., Schleif, W.A., Blahy, O.M., Gabryelski, L.J., Graham, D.J., Quintero, J.C. *et al.* (1995) In vivo emergence of HIV-1 variants resistant to multiple protease inhibitors. *Nature* **374**, 569–571.

Crotty, S., Cameron, C.E. and Andino, R. (2001) RNA virus error catastrophe: direct molecular test by using ribavirin. *Proc. Natl Acad. Sci. USA* **98**, 6895–6900.

Crowder, S. and Kirkegaard, K. (2005) Trans-dominant inhibition of RNA viral replication can slow growth of drug-resistant viruses. *Nat. Genet.* **37**, 701–709.

Chao, L. (1990) Fitness of RNA virus decreased by Muller's ratchet. *Nature* **348**, 454–455.

Charpentier, N., Davila, M., Domingo, E. and Escarmis, C. (1996) Long-term, large-population passage of aphthovirus can generate and amplify defective noninterfering particles deleted in the leader protease gene. *Virology* **223**, 10–18.

Chen, I.S.Y., Koprowski, H., Srinivasan, A. and Vogt, P.K. (eds) (1995). *Transacting Functions of Human Retroviruses.* Berlin: Springer.

Chumakov, K.M., Powers, L.B., Noonan, K.E., Roninson, I.B. and Levenbook, I.S. (1991) Correlation between amount of virus with altered nucleotide sequence and the monkey test for acceptability of oral poliovirus vaccine. *Proc. Natl Acad. Sci. USA* **88**, 199–203.

Dahari, H., Ribeiro, R.M. and Perelson, A.S. (2007) Triphasic decline of hepatitis C virus RNA during antiviral therapy. *Hepatology* **46**, 16–21.

Davis, J.J. (1997) Origins, acquisition and dissemination of antibiotic resistance determinants. In: *Antibiotic Resitance: Origins, Evolution, Selection and Spread* (D.J. Chadweick and J. Goode, eds), pp. 15–35. New York: John Wiley.

de la Iglesia, F. and Elena, S.F. (2007) Fitness declines in tobacco etch virus upon serial bottleneck transfers. *J. Virol.* **81**, 4941–4947.

de la Torre, J.C. and Holland, J.J. (1990) RNA virus quasispecies populations can suppress vastly superior mutant progeny. *J. Virol.* **64**, 6278–6281.

Deval, J., White, K.L., Miller, M.D., Parkin, N.T., Courcambeck, J., Halfon, P. *et al.* (2004) Mechanistic basis for reduced viral and enzymatic fitness of HIV-1 reverse transcriptase containing both K65R and M184V mutations. *J. Biol. Chem.* **279**, 509–516.

Dixit, N.M., Layden-Almer, J.E., Layden, T.J. and Perelson, A.S. (2004) Modelling how ribavirin improves interferon response rates in hepatitis C virus infection. *Nature* **432**, 922–924.

Domingo, E. (2000) Viruses at the edge of adaptation. *Virology* **270**, 251–253.

Domingo, E., ed. (2005) Virus entry into error catastrophe as a new antiviral strategy. *Virus Res.* **107**, 115–228.

Domingo, E., ed. (2006) Quasispecies: concepts and implications for virology. *Curr. Top. Microbiol. Immunol.* **299**.

Domingo, E. (2007) Virus evolution. In: *Fields Virology* (D.M. Knipe, P.M. Howley *et al.*, eds 5th edn., pp. 389–421. Philadelphia: Lippincott Williams & Wilkins.

Domingo, E. and Gomez, J. (2007) Quasispecies and its impact on viral hepatitis. *Virus Res.* **127**, 131–150.

Domingo, E. and Holland, J.J. (1997) RNA virus mutations and fitness for survival. *Annu. Rev. Microbiol.* **51**, 151–178.

Domingo, E., Flavell, R.A. and Weissmann, C. (1976) In vitro site-directed mutagenesis: generation and properties of an infectious extracistronic mutant of bacteriophage Qb. *Gene* **1**, 3–25.

Domingo, E., Sabo, D., Taniguchi, T. and Weissmann, C. (1978) Nucleotide sequence heterogeneity of an RNA phage population. *Cell* **13**, 735–744.

Dorr, P., Westby, M., Dobbs, S., Griffin, P., Irvine, B., Macartney, M. *et al.* (2005) Maraviroc (UK-427,857), a potent, orally bioavailable and selective small-molecule inhibitor of chemokine receptor CCR5 with broad-spectrum anti-human immunodeficiency virus type 1 activity. *Antimicrob. Agents Chemother.* **49**, 4721–4732.

Doyon, L., Croteau, G., Thibeault, D., Poulin, F., Pilote, L. and Lamarre, D. (1996) Second locus involved in human immunodeficiency virus type 1 resistance to protease inhibitors. *J. Virol.* **70**, 3763–3769.

Doyon, L., Payant, C., Brakier-Gingras, L. and Lamarre, D. (1998) Novel Gag-Pol frameshift site in human immunodeficiency virus type 1 variants resistant to protease inhibitors. *J. Virol.* **72**, 6146–6150.

Drake, J.W. (1993) Rates of spontaneous mutation among RNA viruses. *Proc. Natl Acad. Sci. USA* **90**, 4171–4175.

Drake, J.W. and Holland, J.J. (1999) Mutation rates among RNA viruses. *Proc. Natl Acad. Sci. USA* **96**, 13910–13913.

Drake, J.W., Charlesworth, B., Charlesworth, D. and Crow, J.F. (1998) Rates of spontaneous mutation. *Genetics* **148**, 1667–1686.

Duarte, E., Clarke, D., Moya, A., Domingo, E. and Holland, J. (1992) Rapid fitness losses in mammalian RNA virus clones due to Muller's ratchet. *Proc. Natl Acad. Sci. USA* **89**, 6015–6019.

Duarte, E.A., Novella, I.S., Ledesma, S., Clarke, D.K., Moya, A., Elena, S.F. *et al.* (1994a) Subclonal components of consensus fitness in an RNA virus clone. *J. Virol.* **68**, 4295–4301.

Duarte, E.A., Novella, I.S., Weaver, S.C., Domingo, E., Wain-Hobson, S., Clarke, D.K. *et al.* (1994b) RNA virus quasispecies: significance for viral disease and epidemiology. *Infect. Agents Dis.* **3**, 201–214.

Dykes, C., Fox, K., Lloyd, A., Chiulli, M., Morse, E. and Demeter, L.M. (2001) Impact of clinical reverse transcriptase sequences on the replication capacity of HIV-1 drug-resistant mutants. *Virology* **285**, 193–203.

Eckerle, L.D., Lu, X., Sperry, S.M., Choi, L. and Denison, M.R. (2007) High fidelity of murine hepatitis virus replication is decreased in msp14 exoribonuclease mutants. *J. Virol.* **81**, 12135–12144.

Eastman, P.S., Mittler, J., Kelso, R., Gee, C., Boyer, E., Kolberg, J. *et al.* (1998) Genotypic changes in human immunodeficiency virus type 1 associated with loss of suppression of plasma viral RNA levels in subjects treated with ritonavir (Norvir) monotherapy. *J. Virol.* **72**, 5154–5164.

Eigen, M. (1971) Self-organization of matter and the evolution of biological macromolecules. *Naturwissenschaften* **58**, 465–523.

Eigen, M. (1992) *Steps towards Life*. Oxford: Oxford University Press.

Eigen, M. (2000) Natural selection: a phase transition?. *Biophys. Chem.* **85**, 101–123.

Eigen, M. (2002) Error catastrophe and antiviral strategy. *Proc. Natl Acad. Sci. USA* **99**, 13374–13376.

Eigen, M. and Biebricher, C.K. (1988) Sequence space and quasispecies distribution. In: *RNA Genetics* (E. Domingo, P. Ahlquist and J.J. Holland, eds), Vol. 3, pp. 211–245. Boca Raton, FL: CRC Press.

Eigen, M. and Schuster, P. (1979) *The Hypercycle. A Principle of Natural Self-organization*. Berlin: Springer.

Escarmís, C., Dávila, M., Charpentier, N., Bracho, A., Moya, A. and Domingo, E. (1996) Genetic lesions associated with Muller's ratchet in an RNA virus. *J. Mol. Biol.* **264**, 255–267.

Escarmís, C., Dávila, M. and Domingo, E. (1999) Multiple molecular pathways for fitness recovery of an RNA virus debilitated by operation of Muller's ratchet. *J. Mol. Biol.* **285**, 495–505.

Escarmís, C., Gómez-Mariano, G., Dávila, M., Lázaro, E. and Domingo, E. (2002) Resistance to extinction of low fitness virus subjected to plaque-to-plaque transfers: diversification by mutation clustering. *J. Mol. Biol.* **315**, 647–661.

Escarmís, C., Lázaro, E. and Manrubia, S.C. (2006) Population bottlenecks in quasispecies dynamics. *Curr. Top. Microbiol. Immunol.* **299**, 141–170.

Escarmís, C., Lázaro, E., Arias, A. and Domingo, E. (2008) Repeated bottleneck transfers can lead to non-cytocidal forms of a cytopathic virus. Implications for viral extinction. *J. Mol. Biol.* **376**, 367–379.

Farci, P., Shimoda, A., Coiana, A., Diaz, G., Peddis, G., Melpolder, J.C. *et al.* (2000) The outcome of acute hepatitis C predicted by the evolution of the viral quasispecies. *Science* **288**, 339–344.

Fernandez, G., Clotet, B. and Martinez, M.A. (2007) Fitness landscape of human immunodeficiency virus type 1 protease quasispecies. *J. Virol.* **81**, 2485–2496.

Ferrer-Orta, C., Arias, A., Perez-Luque, R., Escarmis, C., Domingo, E. and Verdaguer, N. (2004) Structure of foot-and-mouth disease virus RNA-dependent RNA polymerase and its complex with a template-primer RNA. *J. Biol. Chem.* **279**, 47212–47221.

Ferrer-Orta, C., Arias, A., Escarmis, C. and Verdaguer, N. (2006) A comparison of viral RNA-dependent RNA polymerases. *Curr. Opin. Struct. Biol.* **16**, 27–34.

Ferrer-Orta, C., Arias, A., Perez-Luque, R., Escarmis, C., Domingo, E. and Verdaguer, N. (2007) Sequential structures provide insights into the fidelity of RNA replication. *Proc. Natl Acad. Sci. USA* **104**, 9463–9468.

Finkelstein, D.B., Mukatira, S., Mehta, P.K., Obenauer, J.C., Su, X., Webster, R.G. and Naeve, C.W. (2007) Persistent host markers in pandemic and H5N1 influenza viruses. *J. Virol.* **81**, 10292–10299.

Frankel, F.A., Invernizzi, C.F., Oliveira, M. and Wainberg, M.A. (2007) Diminished efficiency of HIV-1 reverse transcriptase containing the K65R and M184V drug resistance mutations. *Aids* **21**, 665–675.

Friedberg, E.C., Walker, G.C., Siede, W., Wood, R.D., Schultz, R.A. and Ellenberger, T. (2006) *DNA Repair and Mutagenesis*. Washington, DC: American Society for Microbiology.

Frost, S.D., Dumaurier, M.J., Wain-Hobson, S. and Brown, A.J. (2001) Genetic drift and within-host metapopulation dynamics of HIV-1 infection. *Proc. Natl Acad. Sci. USA* **98**, 6975–6980.

Galagan, J.E. and Selker, E.U. (2004) RIP: the evolutionary cost of genome defense. *Trends Genet.* **20**, 417–423.

Garcia-Lerma, J.G., Nidtha, S., Blumoff, K., Weinstock, H. and Heneine, W. (2001) Increased ability for selection of zidovudine resistance in a distinct class of wild-type HIV-1 from drug-naive persons. *Proc. Natl Acad. Sci. USA* **98**, 13907–13912.

García-Arriaza, J., Manrubia, S.C., Toja, M., Domingo, E. and Escarmís, C. (2004) Evolutionary transition toward defective RNAs that are infectious by complementation. *J. Virol.* **78**, 11678–11685.

García-Arriaza, J., Domingo, E. and Escarmís, C. (2005) A segmented form of foot-and-mouth disease virus interferes with standard virus: a link between interference and competitive fitness. *Virology* **335**, 155–164.

García-Arriaza, J., Ojosnegros, S., Dávila, M., Domingo, E. and Escarmis, C. (2006) Dynamics of mutation and recombination in a replicating population of complementing, defective viral genomes. *J. Mol. Biol.* **360**, 558–572.

Garcia-Arriaza, J., Domingo, E. and Briones, C. (2007) Characterization of minority subpopulations in the mutant spectrum of HIV-1 quasispecies by successive specific amplifications. *Virus Res.* **129**, 123–134.

Garcia-Lerma, J.G., MacInnes, H., Bennett, D., Weinstock, H. and Heneine, W. (2004) Transmitted human immunodeficiency virus type 1 carrying the D67N or K219Q/E mutation evolves rapidly to zidovudine resistance in vitro and shows a high replicative fitness in the presence of zidovudine. *J. Virol.* **78**, 7545–7552.

Gatanaga, H., Suzuki, Y., Tsang, H., Yoshimura, K., Kavlick, M.F., Nagashima, K. *et al.* (2002) Amino acid substitutions in Gag protein at non-cleavage sites are indispensable for the development of a high multitude of HIV-1 resistance against protease inhibitors. *J. Biol. Chem.* **277**, 5952–5961.

Gause, G.F. (1971) *The Struggle for Existence*. New York: Dover.

Ge, L., Zhang, J., Zhou, X. and Li, H. (2007) Genetic structure and population variability of tomato yellow leaf curl china virus. *J. Virol.* **81**, 5902–5907.

Gerondelis, P., Archer, R.H., Palaniappan, C., Reichman, R.C., Fay, P.J., Bambara, R.A. and Demeter, L.M. (1999) The P236L delavirdine-resistant human immunodeficiency virus type 1 mutant is replication defective and demonstrates alterations in both RNA 5′-end- and DNA 3′-end-directed RNase H activities. *J. Virol.* **73**, 5803–5813.

González-López, C., Arias, A., Pariente, N., Gómez-Mariano, G. and Domingo, E. (2004) Preextinction viral RNA can interfere with infectivity. *J. Virol.* **78**, 3319–3324.

González-López, C., Gómez-Mariano, G., Escarmís, C. and Domingo, E. (2005) Invariant aphthovirus consensus nucleotide sequence in the transition to error catastrophe. *Infect. Genet. Evol.* **5**, 366–374.

Goodman, M.F. and Fygenson, K.D. (1998) DNA polymerase fidelity: from genetics toward a biochemical understanding. *Genetics* **148**, 1475–1482.

Gorbalenya, A.E. (1995) Origin of RNA viral genomes; approaching the problem by comparative sequence analysis. In: *Molecular Basis of Virus Evolution* (A. Gibbs, C.H. Calisher and F. Garcia-Arenal, eds), pp. 49–66. Cambridge: Cambridge University Press.

Goudsmit, J., de Ronde, A., de Rooij, E. and de Boer, R. (1997) Broad spectrum of in vivo fitness of human immunodeficiency virus type 1 subpopulations differing at reverse transcriptase codons 41 and 215. *J. Virol.* **71**, 4479–4484.

Grande-Pérez, A., Sierra, S., Castro, M.G., Domingo, E. and Lowenstein, P.R. (2002) Molecular indetermination in the transition to error catastrophe: systematic elimination of lymphocytic choriomeningitis virus through mutagenesis does not correlate linearly with large increases in mutant spectrum complexity. *Proc. Natl Acad. Sci. USA* **99**, 12938–12943.

Grande-Pérez, A., Gómez-Mariano, G., Lowenstein, P.R. and Domingo, E. (2005a) Mutagenesis-induced, large fitness variations with an invariant arenavirus consensus genomic nucleotide sequence. *J. Virol.* **79**, 10451–10459.

Grande-Pérez, A., Lazaro, E., Lowenstein, P., Domingo, E. and Manrubia, S.C. (2005b) Suppression of viral infectivity through lethal defection. *Proc. Natl Acad. Sci. USA* **102**, 4448–4452.

Greene, I.P., Wang, E., Deardorff, E.R., Milleron, R., Domingo, E. and Weaver, S.C. (2005) Effect of alternating passage on adaptation of sindbis virus to vertebrate and invertebrate cells. *J. Virol.* **79**, 14253–14260.

Grimm, D., Staeheli, P., Hufbauer, M., Koerner, I., Martinez-Sobrido, L., Solorzano, A. *et al.* (2007) Replication fitness determines high virulence of influenza A virus in mice carrying functional Mx1 resistance gene. *Proc. Natl Acad. Sci. USA* **104**, 6806–6811.

Harrigan, P.R., Bloor, S. and Larder, B.A. (1998) Relative replicative fitness of zidovudine-resistant human immunodeficiency virus type 1 isolates in vitro. *J. Virol.* **72**, 3773–3778.

Harris, K.S., Brabant, W., Styrchak, S., Gall, A. and Daifuku, R. (2005) KP-1212/1461, a nucleoside designed for the treatment of HIV by viral mutagenesis. *Antiviral Res.* **67**, 1–9.

Herrera, M., Garcia-Arriaza, J., Pariente, N., Escarmis, C. and Domingo, E. (2007) Molecular basis for a lack of correlation between viral fitness and cell killing capacity. *PLoS Pathog.* **3**, e53.

Hickey, D.A. and Rose, M.R. (1988) The role of gene transfer in the evolution of eukaryotic sex. In: *The Evolution of Sex* (R.E. Michod and B.R. Levin, eds), pp. 161–175. Sunderland, MA: Sinauer.

Holland, J. and Domingo, E. (1998) Origin and evolution of viruses. *Virus Genes* **16**, 13–21.

Holland, J.J., Spindler, K., Horodyski, F., Grabau, E., Nichol, S. and VandePol, S. (1982) Rapid evolution of RNA genomes. *Science* **215**, 1577–1585.

Holland, J.J., Domingo, E., de la Torre, J.C. and Steinhauer, D.A. (1990) Mutation frequencies at defined single codon sites in vesicular stomatitis virus and poliovirus can be increased only slightly by chemical mutagenesis. *J. Virol.* **64**, 3960–3962.

Holland, J.J., de la Torre, J.C., Clarke, D.K. and Duarte, E. (1991) Quantitation of relative fitness and great adaptability of clonal populations of RNA viruses. *J. Virol.* **65**, 2960–2967.

Hu, Z., Giguel, F., Hatano, H., Reid, P., Lu, J. and Kuritzkes, D.R. (2006) Fitness comparison of thymidine analog resistance pathways in human immunodeficiency virus type 1. *J. Virol.* **80**, 7020–7027.

Huang, W., Gamarnik, A., Limoli, K., Petropoulos, C.J. and Whitcomb, J.M. (2003) Amino acid substitutions at position 190 of human immunodeficiency virus type 1 reverse transcriptase increase susceptibility to delavirdine and impair virus replication. *J. Virol.* **77**, 1512–1523.

Huang, Y., Rosenkranz, S.L. and Wu, H. (2003) Modeling HIV dynamics and antiviral response with consideration of time-varying drug exposures, adherence and phenotypic sensitivity. *Math. Biosci.* **184**, 165–186.

Huynen, M.A., Stadler, P.F. and Fontana, W. (1996) Smoothness within ruggedness: the role of neutrality in adaptation. *Proc. Natl Acad. Sci. USA* **93**, 397–401.

Iglesias-Ussel, M.D., Casado, C., Yuste, E., Olivares, I. and Lopez-Galindez, C. (2002) In vitro analysis of human immunodeficiency virus type 1 resistance to nevirapine and fitness determination of resistant variants. *J. Gen. Virol.* **83**, 93–101.

Ishihama, A., Mizumoto, K., Kawakami, K., Kato, A. and Honda, A. (1986) Proofreading function associated with the RNA-dependent RNA polymerase from influenza virus. *J. Biol. Chem.* **261**, 10417–10421.

Iversen, A.K., Shafer, R.W., Wehrly, K., Winters, M.A., Mullins, J.I., Chesebro, B. and Merigan, T.C. (1996) Multidrug-resistant human immunodeficiency virus type 1 strains resulting from combination antiretroviral therapy. *J. Virol.* **70**, 1086–1090.

Jeeninga, R.E., Keulen, W., Boucher, C., Sanders, R.W. and Berkhout, B. (2001) Evolution of AZT resistance in HIV-1: the 41–70 intermediate that is not observed in vivo has a replication defect. *Virology* **283**, 294–305.

Johnson, V.A., Brun-Vézinet, F., Clotet, B., Günthard, H.F., Kuritzkes, D.R., Pillay, D., Schapiro, J.M. and Richman, D.D. (2007) Update of the drug resistance mutations in HIV-1: 2007. *Top. HIV Med.* **15**, 119–125.

Jridi, C., Martin, J.F., Marie-Jeanne, V., Labonne, G. and Blanc, S. (2006) Distinct viral populations differentiate and evolve independently in a single perennial host plant. *J. Virol.* **80**, 2349–2357.

Kantor, R., Fessel, W.J., Zolopa, A.R., Israelski, D., Shulman, N., Montoya, J.G. *et al.* (2002) Evolution of primary protease inhibitor resistance mutations during protease inhibitor salvage therapy. *Antimicrob. Agents Chemother.* **46**, 1086–1092.

Kaur, A., Grant, R.M., Means, R.E., McClure, H., Feinberg, M. and Johnson, R.P. (1998) Diverse host responses and outcomes following simian immunodeficiency virus SIVmac239 infection in sooty mangabeys and rhesus macaques. *J. Virol.* **72**, 9597–9611.

Keulen, W., Back, N.K., van Wijk, A., Boucher, C.A. and Berkhout, B. (1997) Initial appearance of the 184Ile variant in lamivudine-treated patients is caused by the mutational bias of human immunodeficiency virus type 1 reverse transcriptase. *J. Virol.* **71**, 3346–3350.

Kimata, J.T., Kuller, L., Anderson, D.B., Dailey, P. and Overbaugh, J. (1999) Emerging cytopathic and antigenic simian immunodeficiency virus variants influence AIDS progression. *Nat. Med.* **5**, 535–541.

Kohlstaedt, L.A., Wang, J., Friedman, J.M., Rice, P.A. and Steitz, T.A. (1992) Crystal structure at 3.5 A resolution of HIV-1 reverse transcriptase complexed with an inhibitor. *Science* **256**, 1783–1790.

Kolakofsky, D., Roux, L., Garcin, D. and Ruigrok, R.W. (2005) Paramyxovirus mRNA editing, the "rule of six" and error catastrophe: a hypothesis. *J. Gen. Virol.* **86**, 1869–1877.

Koonin, E.V., Senkevich, T.G. and Dolja, V.V. (2006) The ancient Virus World and evolution of cells. *Biol. Direct* **1**, 29.

Kosalaraksa, P., Kavlick, M.F., Maroun, V., Le, R. and Mitsuya, H. (1999) Comparative fitness of multi-dideoxynucleoside-resistant human immunodeficiency virus type 1 (HIV-1) in an In vitro competitive HIV-1 replication assay. *J. Virol.* **73**, 5356–5363.

Kunkel, T.A. (1990) Misalignment-mediated DNA synthesis errors. *Biochemistry* **29**, 8003–8011.

Larder, B.A. (1994) Interactions between drug resistance mutations in human immunodeficiency virus type 1 reverse transcriptase. *J. Gen. Virol.* **75**(Pt 5), 951–957.

Larder, B.A., Darby, G. and Richman, D.D. (1989) HIV with reduced sensitivity to zidovudine (AZT) isolated during prolonged therapy. *Science* **243**, 1731–1734.

Lazaro, E., Escarmis, C., Perez-Mercader, J., Manrubia, S.C. and Domingo, E. (2003) Resistance of virus to extinction on bottleneck passages: Study of a decaying and fluctuating pattern of fitness loss. *Proc. Natl Acad. Sci. USA* **100**, 10830–10835.

Lazarowitz, S.D. (2007) Plant viruses. In: *Fields Virology* (D.M. Dnipe and P.M. Howley, eds), pp. 641–705. Philadelphia: Lippincot Williams and Wilkins.

Lee, C.H., Gilbertson, D.L., Novella, I.S., Huerta, R., Domingo, E. and Holland, J.J. (1997) Negative effects of chemical mutagenesis on the adaptive behavior of vesicular stomatitis virus. *J. Virol.* **71**, 3636–3640.

Leroux, C., Issel, C.J. and Montelaro, R.C. (1997) Novel and dynamic evolution of equine infectious anemia virus genomic quasispecies associated with sequential disease cycles in an experimentally infected pony. *J. Virol.* **71**, 9627–9639.

Loeb, L.A., Essigmann, J.M., Kazazi, F., Zhang, J., Rose, K. D. and Mullins, J.I. (1999) Lethal mutagenesis of HIV with mutagenic nucleoside analogs. *Proc. Natl Acad. Sci. USA* **96**, 1492–1497.

Lu, J., Sista, P., Giguel, F., Greenberg, M. and Kuritzkes, D.R. (2004) Relative replicative fitness of human immunodeficiency virus type 1 mutants resistant to enfuvirtide (T-20). *J. Virol.* **78**, 4628–4637.

Lukashov, V.V., Huismans, R., Jebbink, M.F., Danner, S.A., de Boer, R.J. and Goudsmit, J. (2001) Selection by AZT and rapid replacement in the absence of drugs of HIV type 1 resistant to multiple nucleoside analogs. *AIDS Res. Hum. Retroviruses* **17**, 807–818.

Maeda, Y., Venzon, D.J. and Mitsuya, H. (1998) Altered drug sensitivity, fitness and evolution of human immunodeficiency virus type 1 with pol gene mutations conferring multi-dideoxynucleoside resistance. *J. Infect. Dis.* **177**, 1207–1213.

Manrubia, S.C., Escarmis, C., Domingo, E. and Lazaro, E. (2005) High mutation rates, bottlenecks and robustness of RNA viral quasispecies. *Gene* **347**, 273–282.

Manrubia, S.C., Garcia-Arriaza, J., Domingo, E. and Escarmís, C. (2006) Long-range transport and universality classes in *in vitro* viral infection spread. *Europhys. Lett.* **74**, 547–553.

Marcus, P.I., Rodriguez, L.L. and Sekellick, M.J. (1998) Interferon induction as a quasispecies marker of vesicular stomatitis virus populations. *J. Virol.* **72**, 542–549.

Martinez-Picado, J., Savara, A.V., Sutton, L. and D'Aquila, R.T. (1999) Replicative fitness of protease inhibitor-resistant mutants of human immunodeficiency virus type 1. *J. Virol.* **73**, 3744–3752.

Martinez, M.A., Carrillo, C., Gonzalez-Candelas, F., Moya, A., Domingo, E. and Sobrino, F. (1991) Fitness alteration of foot-and-mouth disease virus mutants: measurement of adaptability of viral quasispecies. *J. Virol.* **65**, 3954–3957.

Mas, A., Parera, M., Briones, C., Soriano, V., Martínez, M.A., Domingo, E. and Menéndez-Arias, L. (2000) Role of a dipeptide insertion between codons 69–70 of HIV-1 reverse transcriptase in the mechanism of AZT resistance. *EMBO J.* **19**, 5752–5761.

Matamoros, T., Franco, S., Vazquez-Alvarez, B.M., Mas, A., Martinez, M.A. and Menendez-Arias, L. (2004) Molecular determinants of multi-nucleoside analogue resistance in HIV-1 reverse transcriptases containing a dipeptide insertion in the fingers subdomain: effect of mutations D67N and T215Y on removal of thymidine nucleotide analogues from blocked DNA primers. *J. Biol. Chem.* **279**, 24569–24577.

Maynard-Smith, J. (1976) *The Evolution of Sex*. Cambridge: Cambridge University Press.

Maynard Smith, J. and Szathmary, E. (1995) *The Major Transitions in Evolution*. Oxford: W.H. Freeman.

Menendez-Arias, L. (2002) Targeting HIV: antiretroviral therapy and development of drug resistance. *Trends Pharmacol. Sci.* **23**, 381–388.

Menendez-Arias, L., Matamoros, T. and Cases-Gonzalez, C.E. (2006) Insertions and deletions in HIV-1 reverse transcriptase: consequences for drug resistance and viral fitness. *Curr. Pharm. Des.* **12**, 1811–1825.

Mesters, J.R., Tan, J. and Hilgenfeld, R. (2006) Viral enzymes. *Curr. Opin. Struct. Biol.* **16**, 776–786.

Miller, M.D., Lamy, P.D., Fuller, M.D., Mulato, A.S., Margot, N.A., Cihlar, T. and Cherrington, J.M. (1998) Human immunodeficiency virus type 1 reverse transcriptase expressing the K70E mutation exhibits a decrease in specific activity and processivity. *Mol. Pharmacol.* **54**, 291–297.

Minskaia, E., Hertzig, T., Gorbalenya, A.E., Campanacci, V., Cambillau, C., Canard, B. and Ziebuhr, J. (2006) Discovery of an RNA virus 3′− > 5′ exoribonuclease that is critically involved in coronavirus RNA synthesis. *Proc. Natl Acad. Sci. USA* **103**, 5108–5113.

Mitsuya, H., Weinhold, K.J., Furman, P.A., St Clair, M.H., Lehrman, S.N., Gallo, R.C. *et al.* (1985) 3′-Azido-3′-deoxythymidine (BW A509U): an antiviral agent that inhibits the infectivity and cytopathic effect of human T-lymphotropic virus type III/lymphadenopathy-associated virus in vitro. *Proc. Natl Acad. Sci. USA* **82**, 7096–7100.

Moreno, I.M., Malpica, J.M., Rodriguez-Cerezo, E. and Garcia-Arenal, F. (1997) A mutation in tomato aspermy cucumovirus that abolishes cell-to-cell movement is maintained to high levels in the viral RNA population by complementation. *J. Virol.* **71**, 9157–9162.

Mudd, J.A., Leavitt, R.W., Kingsbury, D.T. and Holland, J.J. (1973) Natural selection of mutants of vesicular stomatitis virus by cultured cells of *Drosophila melanogaster*. *J. Gen. Virol.* **20**, 341–351.

Muller, M.J. (1964) The relation of recombination to mutational advance. *Mut. Res.* **1**, 2–9.

Muller, V., Ledergerber, B., Perrin, L., Klimkait, T., Furrer, H., Telenti, A. *et al.* (2006) Stable virulence levels in the HIV epidemic of Switzerland over two decades. *Aids* **20**, 889–894.

Myint, L., Matsuda, M., Matsuda, Z., Yokomaku, Y., Chiba, T., Okano, A. *et al.* (2004) Gag non-cleavage site mutations contribute to full recovery of viral fitness in protease inhibitor-resistant human immunodeficiency virus type 1. *Antimicrob. Agents Chemother,* **48**, 444–452.

Nagy, P.D., Carpenter, C.D. and Simon, A.E. (1997) A novel 3′-end repair mechanism in an RNA virus. *Proc. Natl Acad. Sci. USA* **94**, 1113–1118.

Nájera, I., Holguín, A., Quiñones-Mateu, M.E., Muñoz-Fernández, M.A., Nájera, R., López-Galíndez, C. and Domingo, E. (1995) Pol gene quasispecies of human immunodeficiency virus: mutations associated with drug resistance in virus from patients undergoing no drug therapy. *J. Virol.* **69**, 23–31.

Novella, I.S. and Ebendick-Corpus, B.E. (2004) Molecular basis of fitness loss and fitness recovery in vesicular stomatitis virus. *J. Mol. Biol.* **342**, 1423–1430.

Novella, I.S., Clarke, D.K., Quer, J., Duarte, E.A., Lee, C.H., Weaver, S.C. *et al.* (1995a) Extreme fitness differences in mammalian and insect hosts after continuous replication of vesicular stomatitis virus in sandfly cells. *J. Virol.* **69**, 6805–6809.

Novella, I.S., Duarte, E.A., Elena, S.F., Moya, A., Domingo, E. and Holland, J.J. (1995b) Exponential increases of RNA virus fitness during large population transmissions. *Proc. Natl Acad. Sci. USA* **92**, 5841–5844.

Novella, I.S., Elena, S.F., Moya, A., Domingo, E. and Holland, J.J. (1995c) Size of genetic bottlenecks leading to virus fitness loss is determined by mean initial population fitness. *J. Virol.* **69**, 2869–2872.

Novella, I.S., Cilnis, M., Elena, S.F., Kohn, J., Moya, A., Domingo, E. and Holland, J.J. (1996) Large-population passages of vesicular stomatitis virus in interferon-treated cells select variants of only limited resistance. *J. Virol.* **70**, 6414–6417.

Novella, I.S., Hershey, C.L., Escarmis, C., Domingo, E. and Holland, J.J. (1999a) Lack of evolutionary stasis during alternating replication of an arbovirus in insect and mammalian cells. *J. Mol. Biol.* **287**, 459–465.

Novella, I.S., Quer, J., Domingo, E. and Holland, J.J. (1999b) Exponential fitness gains of RNA virus populations are limited by bottleneck effects. *J. Virol.* **73**, 1668–1671.

Novella, I.S., Ebendick-Corpus, B.E., Zarate, S. and Miller, E.L. (2007) Emergence of mammalian cell-adapted vesicular stomatitis virus from persistent infections of insect vector cells. *J. Virol.* **81**, 6664–6668.

Nowak, M.A. (2006) *Evolutionary Dynamics*. Cambridge, MA and London: The Belknap Press of Harvard University Press.

Nowak, M.A. and May, R.M. (2000) *Virus dynamics. Mathematical Principles of Immunology and Virology*. New York: Oxford University Press.

Nowak, M. and Schuster, P. (1989) Error thresholds of replication in finite populations mutation frequencies and the onset of Muller's ratchet. *J. Theor. Biol.* **137**, 375–395.

Nuñez, J.I., Molina, N., Baranowski, E., Domingo, E., Clark, S., Burman, A. *et al.* (2007) Guinea pig-adapted foot-and-mouth disease virus with altered receptor recognition can productively infect a natural host. *J. Virol.* **81**, 8497–8506.

Orgel, L.E. (1963) The maintenance of the accuracy of protein synthesis and its relevance to ageing. *Proc. Natl Acad. Sci. USA* **49**, 517–521.

Pariente, N., Sierra, S., Lowenstein, P.R. and Domingo, E. (2001) Efficient virus extinction by combinations of a mutagen and antiviral inhibitors. *J. Virol.* **75**, 9723–9730.

Pariente, N., Airaksinen, A. and Domingo, E. (2003) Mutagenesis versus inhibition in the efficiency of extinction of foot-and-mouth disease virus. *J. Virol.* **77**, 7131–7138.

Parrish, C.R. and Kawaoka, Y. (2005) The origins of new pandemic viruses: the acquisition of new host ranges by canine parvovirus and influenza A viruses. *Annu. Rev. Microbiol.* **59**, 553–586.

Pathak, V.K. and Temin, H.M. (1992) 5-Azacytidine and RNA secondary structure increase the retrovirus mutation rate. *J. Virol.* **66**, 3093–3100.

Pawlotsky, J.M., Germanidis, G., Neumann, A.U., Pellerin, M., Frainais, P.O. and Dhumeaux, D. (1998) Interferon resistance of hepatitis C virus genotype 1b: relationship to nonstructural 5A gene quasispecies mutations. *J. Virol.* **72**, 2795–2805.

Peleg, J. (1971) Growth of viruses in arthropod cell cultures: applications. I. Attenuation of Semliki Forest (SF) virus in continuously cultured *Aedes aegypti* mosquito cells (Peleg) as a step in production of vaccines. *Curr. Top. Microbiol. Immunol.* **55**, 155–161.

Pelemans, H., Aertsen, A., Van Laethem, K., Vandamme, A.M., De Clercq, E., Perez-Perez, M.J. *et al.* (2001) Site-directed mutagenesis of human immunodeficiency virus type 1 reverse transcriptase at amino acid position 138. *Virology* **280**, 97–106.

Perales, C., Mateo, R., Mateu, M.G. and Domingo, E. (2007) Insights into RNA virus mutant spectrum and lethal mutagenesis events: replicative interference and complementation by multiple point mutants. *J. Mol. Biol.* **369**, 985–1000.

Perelson, A.S. and Layden, T.J. (2007) Ribavirin: is it a mutagen for hepatitis C virus?. *Gastroenterology* **132**, 2050–2052.

Peters, C.J. (2007) In: *Fields Virology* (D.M. Knipe, P.M. Howley *et al.*, eds), 5th edn. pp. 605–625. Philadelphia: Lippincott Williams and Wilkins.

Pettit, S.C., Henderson, G.J., Schiffer, C.A. and Swanstrom, R. (2002) Replacement of the P1 amino acid of human immunodeficiency virus type 1 Gag processing sites can inhibit or enhance the rate of cleavage by the viral protease. *J. Virol.* **76**, 10226–10233.

Pfeiffer, J.K. and Kirkegaard, K. (2003) A single mutation in poliovirus RNA-dependent RNA polymerase confers resistance to mutagenic nucleotide analogs via increased fidelity. *Proc. Natl Acad. Sci. USA* **100**, 7289–7294.

Pfeiffer, J.K. and Kirkegaard, K. (2005) Increased fidelity reduces poliovirus fitness under selective pressure in mice. *PLoS Pathog.* **1**, 102–110.

Pfeiffer, J.K. and Kirkegaard, K. (2006) Bottleneck-mediated quasispecies restriction during spread of an RNA virus from inoculation site to brain. *Proc. Natl Acad. Sci. USA* **103**, 5520–5525.

Prado, J.G., Wrin, T., Beauchaine, J., Ruiz, L., Petropoulos, C.J., Frost, S.D. *et al.* (2002) Amprenavir-resistant HIV-1 exhibits lopinavir cross-resistance and reduced replication capacity. *Aids* **16**, 1009–1017.

Quiñones-Mateu, M.E., Mas, A., Lain de Lera, T., Soriano, V., Alcami, J., Lederman, M.M. and Domingo, E. (1998) LTR and tat variability of HIV-1 isolates from patients with divergent rates of disease progression. *Virus Res.* **57**, 11–20.

Quiñones-Mateu, M.E., Tadele, M., Parera, M., Mas, A., Weber, J., Rangel, H.R. *et al.* (2002) Insertions in the reverse transcriptase increase both drug resistance and viral fitness in a human immunodeficiency virus type 1 isolate harboring the multi-nucleoside reverse transcriptase inhibitor resistance 69 insertion complex mutation. *J. Virol.* **76**, 10546–10552.

Quiñones-Mateu, M.E. and Arts, E. (2006) Virus fitness: concept, qunatification and application to HIV population dynamics. *Curr. Top. Microbiol. Immunol.* **299**, 83–140.

Reeves, J.D., Gallo, S.A., Ahmad, N., Miamidian, J.L., Harvey, P.E., Sharron, M. *et al.* (2002) Sensitivity of HIV-1 to entry inhibitors correlates with envelope/coreceptor affinity, receptor density and fusion kinetics. *Proc. Natl Acad. Sci. USA* **99**, 16249–16254.

Reeves, J.D., Lee, F.H., Miamidian, J.L., Jabara, C.B., Juntilla, M.M. and Doms, R.W. (2005) Enfuvirtide resistance mutations: impact on human immunodeficiency virus envelope function, entry inhibitor sensitivity and virus neutralization. *J. Virol.* **79**, 4991–4999.

Reznick, D. and Travis, J. (1996) The empirical study of adaptation in natural populations. In: *Adaptation* (M.R. Rose and G.V. Lander, eds), pp. 243–289. San Diego: Academic Press.

Rimsky, L.T., Shugars, D.C. and Matthews, T.J. (1998) Determinants of human immunodeficiency virus type 1 resistance to gp41-derived inhibitory peptides. *J. Virol.* **72**, 986–993.

Robertson, B.H., Jansen, R.W., Khanna, B., Totsuka, A., Nainan, O.V., Siegl, G. *et al.* (1992) Genetic relatedness of hepatitis A virus strains recovered from different geographical regions. *J. Gen. Virol.* **73**, 1365–1377.

Roux, L., Simon, A.E. and Holland, J.J. (1991) Effects of defective interfering viruses on virus replication and pathogenesis in vitro and in vivo. *Adv. Virus Res.* **40**, 181–211.

Ruiz-Jarabo, C.M., Arias, A., Baranowski, E., Escarmís, C. and Domingo, E. (2000) Memory in viral quasispecies. *J. Virol.* **74**, 3543–3547.

Ruiz-Jarabo, C.M., Arias, A., Molina-París, C., Briones, C., Baranowski, E., Escarmís, C. and Domingo, E. (2002) Duration and fitness dependence of quasispecies memory. *J. Mol. Biol.* **315**, 285–296.

Ruiz-Jarabo, C.M., Ly, C., Domingo, E. and de la Torre, J. C. (2003a) Lethal mutagenesis of the prototypic arenavirus lymphocytic choriomeningitis virus (LCMV). *Virology* **308**, 37–47.

Ruiz-Jarabo, C.M., Miller, E., Gómez-Mariano, G. and Domingo, E. (2003b) Synchronous loss of quasispecies memory in parallel viral lineages: a deterministic feature of viral quasispecies. *J. Mol. Biol.* **333**, 553–563.

Saakian, D.B. and Hu, C.K. (2006) Exact solution of the Eigen model with general fitness functions and degradation rates. *Proc. Natl Acad. Sci. USA*, **103**, 4935–4939.

Sanjuan, R., Cuevas, J.M., Furio, V., Holmes, E.C. and Moya, A. (2007) Selection for robustness in mutagenized RNA viruses. *PLoS Genet.* **3**, e93.

Scott, T.W., Weaver, S.C. and Mallampali, V.L. (1994) Evolution of mosquito-borne viruses. In: *Evolutionary Biology of Viruses* (S.S. Morse, ed.), pp. 293–324. New York: Raven Press.

Schmit, J.C., Cogniaux, J., Hermans, P., Van Vaeck, C., Sprecher, S., Van Remoortel, B. *et al.* (1996) Multiple drug resistance to nucleoside analogues and nonnucleoside reverse transcriptase inhibitors in an efficiently replicating human immunodeficiency virus type 1 patient strain. *J. Infect. Dis.* **174**, 962–968.

Schock, H.B., Garsky, V.M. and Kuo, L.C. (1996) Mutational anatomy of an HIV-1 protease variant conferring cross-resistance to protease inhibitors in clinical trials. Compensatory modulations of binding and activity. *J. Biol. Chem.* **271**, 31957–31963.

Sevilla, N., Ruiz-Jarabo, C.M., Gómez-Mariano, G., Baranowski, E. and Domingo, E. (1998) An RNA virus can adapt to the multiplicity of infection. *J. Gen. Virol,* **79**, 2971–2980.

Shafer, R.W., Winters, M.A., Palmer, S. and Merigan, T.C. (1998) Multiple concurrent reverse transcriptase and protease mutations and multidrug resistance of HIV-1 isolates from heavily treated patients. *Ann. Intern. Med.* **128**, 906–911.

Sharma, P.L. and Crumpacker, C.S. (1997) Attenuated replication of human immunodeficiency virus type 1 with a didanosine-selected reverse transcriptase mutation. *J. Virol.* **71**, 8846–88451.

Sharma, P.L. and Crumpacker, C.S. (1999) Decreased processivity of human immunodeficiency virus type 1 reverse transcriptase (RT) containing didanosine-selected mutation Leu74Val: a comparative analysis of RT variants Leu74Val and lamivudine-selected Met184Val. *J. Virol.* **73**, 8448–8456.

Shirasaka, T., Kavlick, M.F., Ueno, T., Gao, W.Y., Kojima, E., Alcaide, M.L. *et al.* (1995) Emergence of human immunodeficiency virus type 1 variants with resistance to multiple dideoxynucleosides in patients receiving therapy with dideoxynucleosides. *Proc. Natl Acad. Sci. USA,* **92**, 2398–2402.

Sierra, M., Airaksinen, A., González-López, C., Agudo, R., Arias, A. and Domingo, E. (2007) Foot-and-mouth disease virus mutant with decreased sensitivity to ribavirin: implications for error catastrophe. *J. Virol.* **81**, 2012–2024.

Sierra, S., Dávila, M., Lowenstein, P.R. and Domingo, E. (2000) Response of foot-and-mouth disease virus to increased mutagenesis. *Influence of viral load and fitness in loss of infectivity. J. Virol.* **74**, 8316–8323.

Smolinski, M.S., Hamburg, M.A. and Lederberg, J. (2003) *Microbial Threats to Health. Emergence, Detection and Response.* Washington DC: The National Academies Press.

Sobrino, F. and Mettenleiter, T. (2008) *Animal Viruses: Molecular Biology*. UK: Horizon Scientific Press.

Steinhauer, D.A., Domingo, E. and Holland, J.J. (1992) Lack of evidence for proofreading mechanisms associated with an RNA virus polymerase. *Gene* **122**, 281–288.

Suzuki, Y. (2005) Sialobiology of influenza: molecular mechanism of host range variation of influenza viruses. *Biol. Pharm. Bull.* **28**, 399–408.

Swetina, J. and Schuster, P. (1982) Self-replication with errors. A model for polynucleotide replication. *Biophys. Chem.* **16**, 329–345.

Tapia, N., Fernandez, G., Parera, M., Gomez-Mariano, G., Clotet, B., Quiñones-Mateu, M. *et al.* (2005) Combination of a mutagenic agent with a reverse transcriptase inhibitor results in systematic inhibition of HIV-1 infection. *Virology* **338**, 1–8.

Temin, H.M. (1989) Is HIV unique or merely different?. *J. AIDS* **2**, 1–9.

Temin, H.M. (1993) The high rate of retrovirus variation results in rapid evolution. In: *Emerging Viruses* (S.S. Morse, ed.), pp. 219–225. Oxford: Oxford University Press.

Teng, M.N., Oldstone, M.B. and de la Torre, J.C. (1996) Suppression of lymphocytic choriomeningitis virus-induced growth hormone deficiency syndrome by disease-negative virus variants. *Virology* **223**, 113–119.

Van Valen, L. (1973) A new evolutionary law. *Evol. Theory* **1**, 1–30.

Vignuzzi, M., Stone, J.K., Arnold, J.J., Cameron, C.E. and Andino, R. (2006) Quasispecies diversity determines pathogenesis through cooperative interactions in a viral population. *Nature* **439**, 344–348.

Villarreal, L.P. (2005) *Viruses and the Evolution of Life.* Washington DC: ASM Press.

Wang, J., Dykes, C., Domaoal, R.A., Koval, C.E., Bambara, R.A. and Demeter, L.M. (2006) The HIV-1 reverse transcriptase mutants G190S and G190A, which confer resistance to non-nucleoside reverse transcriptase inhibitors, demonstrate reductions in RNase H activity and DNA synthesis from tRNA(Lys, 3) that correlate with reductions in replication efficiency. *Virology* **348**, 462–474.

Weaver, S.C. (1998) Recurrent emergence of Venezuelan equine encephalomyelitis. In: *Emerging Infections* (W.M. Sheld and J. Hughes, eds), Vol. 1, pp. 27–42. Washington DC: ASM Press.

Weibull, W.J. (1951) A statistical distribution function of wide applicability. *Appl. Mech.* **18**, 293–297.

Westby, M., Smith-Burchnell, C., Mori, J., Lewis, M., Mosley, M., Stockdale, M. *et al.* (2007) Reduced maximal inhibition in phenotypic susceptibility assays indicates that viral strains resistant to the CCR5 antagonist maraviroc utilize inhibitor-bound receptor for entry. *J. Virol.* **81**, 2359–2371.

White, K.L., Margot, N.A., Wrin, T., Petropoulos, C.J., Miller, M.D. and Naeger, L.K. (2002) Molecular mechanisms of resistance to human immunodeficiency virus type 1 with reverse transcriptase mutations K65R and K65R + M184V and their effects on enzyme function and viral replication capacity. *Antimicrob. Agents Chemother* **46**, 3437–3446.

Wilke, C.O. and Novella, I..S. (2003) Phenotypic mixing and hiding may contribute to memory in viral quasispecies. *BMC Microbiol.* **3**, 11.

Wilke, C.O., Ronnewinkel, C. and Martinetz, T. (2001a) Dynamic fitness landscapes in molecular evolution. *Phys. Rep.* **349**, 395–446.

Wilke, C.O., Wang, J.L., Ofria, C., Lenski, R.E. and Adami, C. (2001b) Evolution of digital organisms at high mutation rates leads to survival of the flattest. *Nature* **412**, 331–333.

Wilke, C.O., Reissig, D.D. and Novella, I..S. (2004) Replication at periodically changing multiplicity of infection promotes stable coexistence of competing viral populations. *Evolution Int. J. Org. Evolution*, **58**, 900–905.

Wilke, C.O., Foster, R. and Novella, I..S. (2006) Quasispecies in time-dependent environments. *Curr. Top. Microbiol. Immunol.* **299**, 33–50.

Williams, G.C. (1992) *Natural Selection. Domains, Levels and Challenges.* New York, Oxford: Oxford University Press.

Wlodawer, A. and Vondrasek, J. (1998) Inhibitors of HIV-1 protease: a major success of structure-assisted drug design. *Annu. Rev. Biophys. Biomol. Struct.* **27**, 249–284.

Wright, P.F., Neumann, G. and Kawaoka, Y. (2007) Orthomyxoviruses. In: *Fields Virology* (D.M. Dnipe, P.M. Howley, *et al.*, eds) 5th edn, pp. 1691–1740. Philadelphia: Lippincott Williams & Wilkins.

Wyatt, C.A., Andrus, L., Brotman, B., Huang, F., Lee, D.H. and Prince, A.M. (1998) Immunity in chimpanzees chronically infected with hepatitis C virus: role of minor quasispecies in reinfection. *J. Virol.* **72**, 1725–1730.

Yamada, K., Mori, A., Seki, M., Kimura, J., Yuasa, S., Matsuura, Y. and Miyamura, T. (1998) Critical point mutations for hepatitis C virus NS3 proteinase. *Virology* **246**, 104–112.

Yang, Y., Halloran, M.E., Sugimoto, J.D. and Longini, I.M. (2007) Detecting human-to-human transmission of avian influenza A (H5N1). *Emerging Infect. Dis.* **13**, 1348–1353.

Yerly, S., Rakik, A., De Loes, S.K., Hirschel, B., Descamps, D., Brun-Vezinet, F. and Perrin, L. (1998) Switch to unusual amino acids at codon 215 of the human immunodeficiency virus type 1 reverse transcriptase gene in seroconvertors infected with zidovudine-resistant variants. *J. Virol.* **72**, 3520–3523.

Yusa, K., Song, W., Bartelmann, M. and Harada, S. (2002) Construction of a human immunodeficiency virus type 1 (HIV-1) library containing random combinations of amino acid substitutions in the HIV-1 protease due to resistance by protease inhibitors. *J. Virol.* **76**, 3031–3037.

Yuste, E., Sánchez-Palomino, S., Casado, C., Domingo, E. and López-Galíndez, C. (1999) Drastic fitness loss in human immunodeficiency virus type 1 upon serial bottleneck events. *J. Virol.* **73**, 2745–2751.

Zárate, S. and Novella, I..S. (2004) Vesicular stomatitis virus evolution during alternation between persistent infection in insect cells and acute infection in mammalian cells is dominated by the persistence phase. *J. Virol.* **78**, 12236–12242.

Zhang, L., Huang, Y., Yuan, H., Chen, B.K., Ip, J. and Ho, D.D. (1997) Genotypic and phenotypic characterization of long terminal repeat sequences from long-term survivors of human immunodeficiency virus type 1 infection. *J. Virol.* **71**, 5608–5613.

Zhang, X., Hasoksuz, M., Spiro, D., Halpin, R., Wang, S., Vlasova, A. *et al.* (2007) Quasispecies of bovine enteric and respiratory coronaviruses based on complete genome sequences and genetic changes after tissue culture adaptation. *Virology* **363**, 1–10.

Zimmern, D. (1988) Evolution of RNA viruses. In: *RNA Genetics* (E. Domingo, J.J. Holland and P. Ahlquist, eds), Vol. 2, pp. 211–240. Florida: CRC Press Inc.

CHAPTER

5

Comparative Studies of RNA Virus Evolution

Edward C. Holmes

ABSTRACT

The comparative analysis of genes and genomes is frequently used to reveal the patterns and processes of RNA virus evolution. Herein, I review some of the various computational (*in silico*) methods that comprise this approach and outline their multi-faceted contributions to understanding evolutionary change in RNA viruses. I focus on five areas where the most important developments, and controversies, have taken place: phylogenetic analysis, the estimation of recombination, measuring the rates and dates of viral evolution, inferring the selection pressures acting on RNA viruses, and reconstructing population (epidemiological) dynamics. Finally, I also highlight those areas where future research is most urgently required, particularly given the rapidly growing number of viral genomes sequences.

INTRODUCTION

The last 30 years, correspondent with the development of gene sequencing and then polymerase chain reaction (PCR), has witnessed the maturation of three pathways for the study of RNA virus evolution: the theoretical, relying on mathematical models, the experimental, largely based on *in vitro* studies, and the comparative, utilizing the computational (and largely phylogenetic) analysis of gene and/or genome sequence data. Although all three approaches have their advantages and limitations, and each has contributed greatly to the study of viral evolution, this chapter will comprise a broad discussion of the various computational tools currently available for the comparative *in silico* analysis of viral sequence data, and the evolutionary inferences that have been made from them. Rather than giving detailed mechanistic descriptions of the wide variety of methods used to analyze sequence data—whose mathematical and computational details are beyond the scope of this chapter—I will concentrate on their general properties (and limitations) and what they have told us about viral evolution as a whole. A non-exhaustive list of some of the most popular computer software is provided in Table 5.1.

Origin and Evolution of Viruses
ISBN 978-0-12-374153-0

TABLE 5.1 Computer Software Packages Available for the Comparative Analysis of RNA Virus Genomes

Analysis/program	Publication	URL
Phylogenetic analysis		
MODELTEST	Posada and Crandall (1998)	darwin.uvigo.es/software/modeltest.html
PHYLIP	Felsenstein (2005)	evolution.genetics.washington.edu/phylip.html
PAUP*	Swofford (2003)	paup.csit.fsu.edu/about.html
RAXML	Stamatakis (2006)	icwww.epfl.ch/~stamatak
GARLI	Zwickl (2006)	www.bio.utexas.edu/faculty/antisense/garli/Garli.html
PHYLML	Guindon and Gascuel (2003)	atgc.lirmm.fr/phyml
MRBAYES	Ronquist and Huelsenbeck (2003)	mrbayes.csit.fsu.edu
MEGA	Tamura *et al.* (2007)	www.megasoftware.net
SPLITSTREE	Huson and Bryant (2004)	www.splitstree.org
Recombination analysis		
RDP2	Martin *et al.* (2005)	darwin.uvigo.es
TOPALi	Milne *et al.* (2004)	www.bioss.sari.ac.uk/~frank/Genetics/topal.html
LARD	Holmes *et al.* (1999)	evolve.zoo.ox.ac.uk/software.html?id=lard
LDHAT	McVean *et al.* (2002)	www.stats.ox.ac.uk/~mcvean/LDhat/index.html
LIKEWIND	Archibald and Roger (2002)	hades.biochem.dal.ca/Rogerlab/Software/software.html
GARD	Kosakovsky Pond *et al.* (2006)	www.datamonkey.org/GARD/
VISRD	Forslund *et al.* (2004)	www.cmp.uea.ac.uk/Research/cbg/visrd.html
Analyzing selection pressures		
HYPHY (DATAMONKEY)	Kosakovsky Pond *et al.* (2005)	www.hyphy.org
PAML	Yang (1997)	abacus.gene.ucl.ac.uk/software/paml.html
ADAPTSITE	Suzuki *et al.* (2001)	www.cib.nig.ac.jp/dda/yossuzuk/welcome.html
Substitution Rates and Population Dynamics		
BEAST (and TRACER)	Drummond and Rambaut (2003)	beast.bio.ed.ac.uk
TIPDATE	Rambaut (2000)	evolve.zoo.ox.ac.uk/software.html?id=tipdate

This list is not, by any means, exhaustive. More analytical software is available at: evolve.zoo.ox.ac.uk and evolution.genetics.washington.edu/phylip.html. A comprehensive list of recombination detection programs is available at www.bioinf.manchester.ac.uk/recombination/programs.shtml.

Given the remarkable increase in the availability of gene and genome sequence data from RNA viruses, and what will surely occur following the development of new methods for genome sequencing (Margulies *et al.*, 2005), it is also clear that the comparative techniques described here will play an increasingly important role in studies of viral evolution. Perhaps for the first time since their inception, comparative studies of RNA virus evolution are now more limited by the availability of computing power than sequences to analyze.

Lastly, although I will base this chapter around the theme of what comparative methodologies can tell us about viral evolution, it is equally the case that because of their remarkably rapid rates of evolutionary change, RNA viruses have often comprised valuable test data sets for a wide variety of computational tools (see, for example, Huelsenbeck *et al.*, 2001), and that this pattern is likely to continue.

PHYLOGENETIC ANALYSES OF RNA VIRUSES

By far the most common, and undoubtedly one of the most valuable, forms of computational analysis of viral sequence data involves the inference of phylogenetic trees, itself the heart of the comparative method (Harvey and Pagel, 1993; Pagel, 1999). Indeed, phylogenetic analysis is perhaps the area in which evolutionary ideas have most effectively entered the virological arena, with phylogenetic trees now a common feature in virology journals. As well as their increasing use, there have also been major developments in the methods of phylogenetic analysis currently available and their tractability for large data sets. Although these methods are perhaps daunting to the inexperienced, their development brings both greater analytical power and more statistical rigor.

Although phylogeny is often used strictly for taxonomic purposes, the phylogenetic analyses of RNA viruses provides powerful insights into both the patterns and processes of evolutionary change (Sharp, 2002). For example, by examining the distribution of mutations on viral phylogenies, and specifically whether pairs of mutations co-occur more often than expected by chance, it has been possible to investigate the nature of epistasis in viral evolution (Shapiro *et al.*, 2006). Further, and more commonly, phylogenetic analysis is now the leading way to investigate the origin and spread of RNA viruses, as amply demonstrated by the multitude of papers using phylogenetic tools to investigate the origins of the primate lentiviruses, most notably HIV (see, for example Keele *et al.*, 2006; Santiago *et al.*, 2005).

A more recent illustration of the power of phylogenetic inference is provided by work on yellow fever virus (YFV), a single-strand, positive-sense RNA virus of the family *Flaviviridae*. Yellow fever has been the scourge of human populations in the tropics for centuries. Indeed, since the 1600s yellow fever epidemics have periodically swept through tropical cities, often with very high mortality rates, such that the discovery that the disease was transmitted by mosquitoes, the isolation of the virus, and finally the development of a vaccine were considered milestones in the history of medicine. Although there have been suggestions that YFV may have originated in the New World, most medical historians speculate that this disease was first transported to the Americas at the time of the slave trade from an origin somewhere in Africa. This hypothesis was confirmed in a large-scale phylogenetic analysis of YFV gene sequences (Bryant *et al.*, 2007). Not only did this analysis support the African origin of American YFV, but it showed that the virus spread westwards across both Africa and the Americas, and that the YFV lineages imported into the Americas at the time of the slave trade still circulate there to this day (Figure 5.1). More dramatically, the same study utilized dating techniques based on a relaxed molecular clock (see below) to show that the time-scale of YFV evolution is in perfect accord with the history of slavery.

In broad terms, there are two key aspects to obtaining an accurate phylogenetic history; using an appropriate (and realistic) model of nucleotide substitution (that is, simple statistical models that specify the probabilities of each type of base change and the rates at which different sites in the sequence alignment evolve), and employing an efficient search algorithm to find the globally optimal tree. Although there is still a vigorous debate about which method of phylogenetic inference is best (see Felsenstein, 2004, for an elegant history of ideas), and the circumstances where each performs optimally, phylogenies inferred using maximum likelihood (ML) have many desirable properties, particularly when they are based on an appropriate model of nucleotide substitution (which itself can be very easily determined with the MODELTEST package; Posada and Crandall, 1998; Table 5.1). Moreover, important developments have been made with ML methods. Until recently, perhaps the most popular software package to infer ML trees was PAUP* (Swofford, 2003). However, a number of ML packages are now available, including GARLI (Zwickl, 2006), PHYML (Guindon and Gascuel, 2003), and RAXML (Stamatakis, 2006), and have made important advances with respect to computational speed so that ML trees can now be

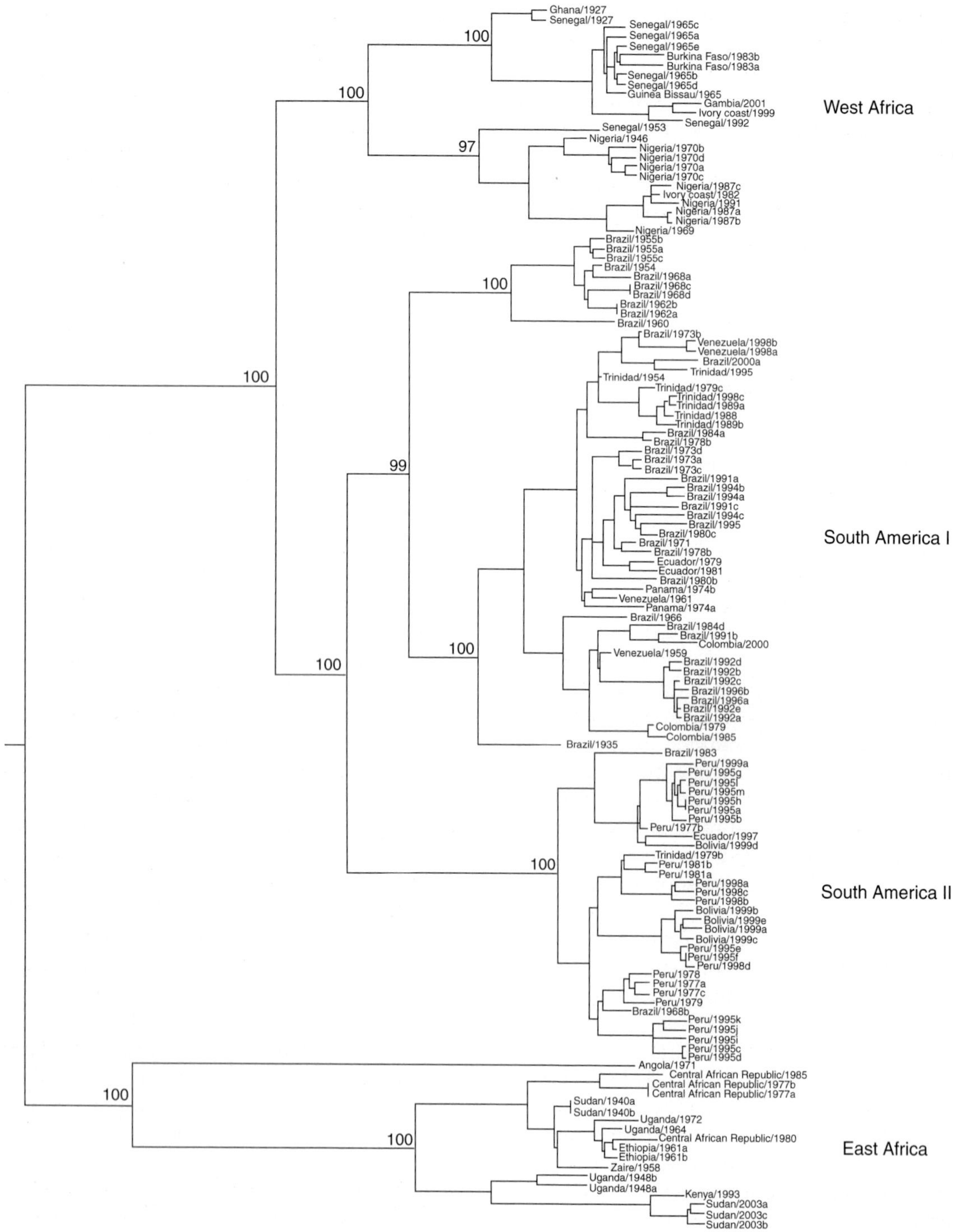

FIGURE 5.1 Maximum *a posteriori* (MAP) phylogenetic tree (estimated using the BEAST package) of 133 prM/E gene sequences of yellow fever virus (YFV) sampled from Africa and Latin America. The major geographic groupings of YFV are indicated and posterior probability values are shown for key nodes on the tree. Tip times correspond to year of virus sampling. Taken from Bryant *et al.* (2007) with permission.

computed on many hundreds, if not thousands, of sequences.

As well as maximum likelihood, there is a burgeoning interest in Bayesian methods of phylogenetic inference (Huelsenbeck *et al.*, 2001), most obviously manifest in the MRBAYES package (Ronquist and Huelsenbeck, 2003). Although some have expressed concerns over the philosophical basis of Bayesian inference, as well the level of statistical support that can be drawn from posterior probabilities (Suzuki *et al.*, 2002), there is little doubt that Bayesian methods have advantages over traditional ML in terms of speed and that they posses a built-in measure of statistical uncertainly (the posterior probability). However, rather than simply relying on default parameters, it is crucial that Bayesian analyses are run for sufficient time to reach statistical convergence otherwise incorrect inferences are possible (and computer programs such as TRACER are extremely helpful in assessing when this condition is met; Table 5.1).

It is also clear that when recombination is frequent in RNA viruses, which is undoubtedly the case in retroviruses such as HIV, the simple branching phylogenetic tree may not always be the optimal method of inferring and depicting evolutionary history because there is not a single evolutionary pathway linking sequences. The alternative in these cases is to use network methods that, in theory, are able to depict all of the variable ancestries of a set of sequences (Bandelt and Dress, 1992; Bryant and Moulton, 2004; Huson and Bryant, 2004). However, although these methods, such as Split Decomposition, are undoubtedly extremely useful for data visualization, they have yet to be subjected to the strong testing that characterizes most "standard" phylogenetic methods. The development of statistically rigorous networking methods will clearly be a major research goal in the coming years.

Similarly, standard methods of phylogenetic inference are not ideally suited for investigating the "deep" evolutionary relationships among RNA viruses, such as inter-family phylogenies. The problem here is that the sequences in question are usually so divergent as to contain no usable phylogenetic signal, even if they can be successfully aligned (Zanotto *et al.*, 1996). This, in turn, represents a major barrier to studies of the origin of RNA viruses, one of the most interesting, yet least explored, questions in evolutionary biology (Koonin *et al.*, 2006). Although there have been major developments in the inference of phylogenetic trees using gene content and/or gene order (Sankov, 2001; Bourque and Pevzner, 2002), these are unlikely to be of value for RNA viruses because their small genomes. As such, perhaps the most profitable research avenue will be the development of phylogenetic methods that utilize similarities and differences in protein structure (Thorne *et al.*, 1996). Although progress has been made in this area, we are still some way from a method that can accurately and efficiently infer phylogenetic history from the structure of proteins.

These issues notwithstanding, by far the biggest challenge facing those who develop phylogenetic software is the sheer scale of the genome data that are now available. Further, given the rapidly developing field of genome sequencing, the challenge posed by large data sets will only ever increase, despite increases in computational speed. For example, the initiative to sequence complete genomes of influenza A virus begun in 2005 (Ghedin *et al.*, 2005) has, at the time of writing, resulted in a data base of over 2500 sequences. Although such a huge data set is information-rich, it equally poses significant challenges to any phylogenetic study, although important advances in clustering algorithms are being made (Frey and Dueck, 2007).

MEASURING RECOMBINATION

One of the most important discoveries stemming from the use of comparative techniques is that RNA viruses recombine far more frequently than previously anticipated (Awadalla, 2003; Posada *et al.*, 2002). This new perspective is due to combination of increasing amounts of gene and genome sequence data and improved computational tools. Although there are still active debates about the frequency of recombination in some systems, perhaps the most notable being Dengue virus where it has been proposed as a potential barrier to effective vaccination (Monath *et al.*, 2005; Seligman and Gould, 2004), it is clear that the process occurs commonly in positive-sense RNA viruses and extremely so in retroviruses such as HIV (Hu and Temin, 1990). Indeed, the potential for recombination to rapidly generate resistance to multiple antiviral agents in HIV is now a major area of study (Nora *et al.*, 2007). More controversial is the role of recombination in monopartite negative-sense RNA viruses. In this case, the majority of both comparative and experimental studies suggest that recombination only occurs very rarely, if at all, in these viruses (Chare *et al.*, 2003; Pringle and Parry, 1982). Such a low rate is in accord with the notion that the RNA molecule in these viruses is never disassociated from protein, which in turn will prevent the template switching necessary for RNA recombination (Lai, 1992). However, despite this sound biological reasoning, claims of more frequent recombination in negative-sense RNA viruses are made on occasion (Gibbs *et al.*, 2001; Schierup *et al.*, 2005).

There are two types of recombination analysis that can be conducted on viral sequence data: (i) identifying specific recombinants, their parentage and their break-points, and (ii) estimating the frequency (or rate) of recombination, particularly compared to that of mutation, without identifying individual recombinants. There is little doubt that the former approach is both simpler and more widely used. Indeed, some software packages, most notably RDP2, allow users to use multiple programs within a single computer interface (Martin *et al.*, 2005; Table 5.1). The statistical properties of the many methods available to perform such analyses—particularly the rate at which they produce false-positive and false-negative results—are also well known (Brown *et al.*, 2001; Posada, 2002). However, it is equally clear that these methods are entirely dependent on the number and type of sequences analyzed and only estimate the rate of successful recombination (i.e. those recombinants with sufficient fitness to propagate), rather than the rate at which recombination occurs intrinsically in viral genomes. In this respect, this class of methods are limited in their power and must underestimate the true rate at which recombination occurs. Consequently, although methods that estimate recombination rate, for example by examining the extent of linkage disequilibrium (LD) (McVean *et al.*, 2002) are often more complex, they also possess more analytical power. The future of computational studies of recombination in RNA viruses clearly lies with the development of methods that can more accurately estimate recombination rates rather than simply finding break-points.

There have also been conflicts between the different approaches used to study recombination in RNA viruses. This is perhaps most apparent with measles virus. Phylogenetic approaches for detecting recombination have shown that recombination is at best rare in measles virus (Chare *et al.*, 2003), as expected given its status as a monopartite negative-sense RNA virus. In contrast, those methods that estimate recombination rate using measures of LD have uncovered relatively frequent recombination in measles virus (Schierup *et al.*, 2005). Although the history of recombination studies in RNA viruses makes it foolish to rule out a role for recombination in measles virus evolution, the high rate of recombination suggested by studies of LD should also be manifest as at least some clear-cut break-points, but few exist. It therefore seems likely that patterns of LD in measles virus are more to do with population structure than abundant recombination.

At present, computational studies of reassortment in multipartite RNA viruses proceed in an analogous manner to those of RNA recombination. Although the use of longer (whole segment) sequences makes reassortment rather easier to detect than RNA recombination, so that there is little active debate over its occurrence, it is also the case that methods available to measure the rate of reassortment in viruses like influenza are still in their infancy (Macken *et al.*, 2006) and that further work is required in this area.

THE ANALYSIS OF EVOLUTIONARY RATES AND TIMES TO COMMON ANCESTRY

While inferring the phylogenetic history of populations of RNA viruses is the first, and often most useful, step in the comparative analysis of gene sequence data, it cannot usually provide information on the rate and time-scale of viral evolution. Consequently, additional methods are required to explore the temporal dynamics of viral evolution. Although these methods require additional assumptions, and so are more error-prone, there is no doubt that the analysis of evolutionary rates and times to common ancestry has matured into one of the most successful aspects of contemporary evolutionary biology.

In many respects, RNA viruses represent the ideal organisms to reconstruct the time-scale of evolutionary change. Although they lack a fossil record, their evolution can be recorded over the time-scale of human observation—so that they represent so-called "measurably evolving populations" (Drummond *et al.*, 2003b). This makes the estimation of evolutionary rates a relatively easy exercise. In contrast, studies of the time-scale of bacterial evolution are far more complex because, as well as a lack of a fossil record, rates of change are not sufficiently rapid to be measurable in the short-term.

The signal of evolutionary rate in RNA viruses is encoded in the distribution of branch lengths of viruses sampled at different times, be it years, months or even days. Given this information, there are a variety of methods that can be used to estimate substitution rates. While simple linear regression is perhaps the most commonly used method in this context, and does provide a useful overview (see Lukashov and Goudsmit, 2002, for a specific example), it also suffers from two major limitations. First, it does not fully take account of the phylogenetic relationships of the sequences in question. Specifically, because all sequences are compared in a pairwise fashion to the oldest sequence, there is extensive pseudo-replication, such that certain (deep) branches in the tree are compared multiple times. Second, linear regression implicitly assumes a constant molecular clock, an assumption that only appears to fit a subset of RNA viruses (Jenkins *et al.*, 2002).

Resolving the problem of phylogenetic non-independence was one of the principle motivations behind the development of likelihood-based methods such as TIPDATE (Rambaut, 2000). Here, rather than undertaking multiple pairwise comparisons, a count is made of the number of substitutions on each branch of a phylogenetic tree with dated nodes (although frequent recombination clearly compromises any analysis based on a single phylogeny). Similarly, some of these likelihood methods also allow the assumption of a molecular clock to be relaxed, by enabling evolutionary rates to vary across lineages (see below). In doing so, these methods greatly improve both analytical power and accuracy (Drummond *et al.*, 2003a).

The most recent class of methods developed to estimate rates are set within a Bayesian Markov Chain Monte Carlo (MCMC) framework, as manifest in the BEAST package (Drummond and Rambaut, 2003). The beauty of the Bayesian MCMC approach is that as well as incorporating phylogenetic information and allowing for variable substitution rates (and a variety of models of nucleotide substitution), it accounts for differences in the demographic history of RNA viruses (that is, rates of population growth and decline—see below) and allows rate estimates to be based on many millions of sampled trees (rather than a single phylogeny), therein providing a more rigorous statistical framework. The major limitation of these methods is that they are computationally intensive.

Those rate estimates undertaken in RNA viruses to date have revealed, with relatively few exceptions, that rates of molecular evolutionary change are normally in the proximity of 10^{-3} to 10^{-4} nucleotide substitutions per

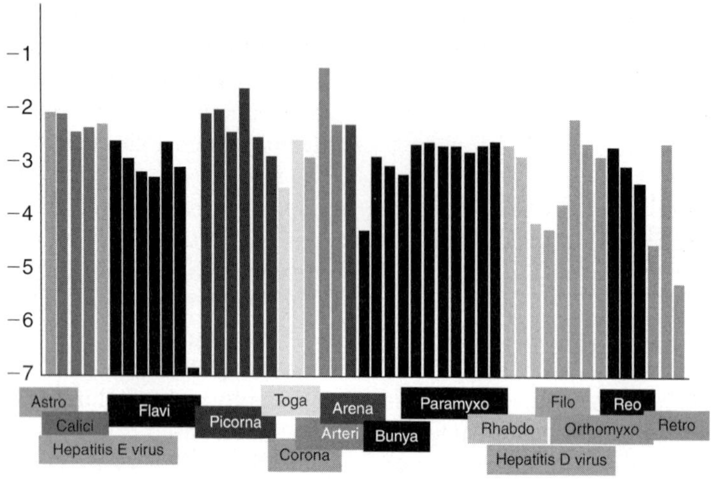

FIGURE 5.2 Rates of nucleotide substitution at synonymous sites among RNA viruses. For simplicity, viruses are color-coded by family. The y-axis records the numbers of synonymous substitutions per site, per year. (See Plate 3 for the color version of this figure.) Taken from Hanada *et al.* (2004) with permission.

site, per year with a range of little more than one logarithm (Jenkins *et al.*, 2002; Hanada *et al.*, 2004; Figure 5.2). Notably these rates are also generally robust to whatever analytical method is used. Although the calculation is only approximate, such rates are compatible with a background mutation rate (that is, prior to the imposition of natural selection) of approximately one error per genome, per replication (Drake and Holland, 1999).

That the inferred substitution rates of RNA viruses are so high seemingly confirms a general evolutionary "rule," that RNA viruses evolve rapidly and DNA viruses evolve slowly. This, in turn, reflects underlying differences in polymerase fidelity, in which error correction is absent from RNA polymerase yet present in DNA polymerases. In recent years, however, a number of studies have shown that this fundamental evolutionary division between RNA and DNA replicating molecules is overly simplistic. Perhaps the most dramatic finding in this respect is that single-stranded DNA viruses, all of which have genome sizes of less than approximately 6 kb, evolve at rates broadly similar to those seen in RNA viruses, even though their small size dictates that they replicate using host DNA polymerases (Shackelton *et al.*, 2005; Shackelton and Holmes, 2006; Ge *et al.*, 2007). There are two possible, but not mutually exclusive hypotheses, to explain the high rates in ssDNA viruses; that single-stranded DNA replication does not allow the full range of DNA repair processes so that error rates can approach those in RNA systems, or that competition with host genes for replication materials necessitates rapid viral replication and this, in turn, results in a higher error-rate because there is a trade-off between the speed and fidelity of replication (so that small viruses always evolve quickly) (Elena and Sanjuan, 2005). Although there is currently no clear explanation for the high rates in ssDNA viruses, there is as yet no biochemical demonstration of a trade-off between replication speed and fidelity, and far lower substitution rates are observed in the dsDNA papillomaviruses, which possess genomes of only ~8 kb in length.

There are also reports of RNA viruses that evolve anomalously slowly. The best documented of these is simian foamy virus (SFV), a retrovirus that infects a wide range of primate species, and which has substitution rates equivalent to those seen in mammalian mitochondrial DNA (Switzer *et al.*, 2005). Although the explanation for this low rate is also unclear, it is more likely to represent a greatly reduced rate of viral replication per unit time than an improved copying fidelity of reverse transcriptase.

Hand in hand with the estimation of substitution rates comes an ability to infer the time-scale of viral evolution. Indeed, some methods, most notably the Bayesian coalescent methods available in the BEAST package, make it possible to co-estimate substitution rates and divergence times. Further, as with the estimation of substitution rates, perhaps the biggest advance that has been made in recent years has been the development of methods that allow rate variation among lineages to be incorporated through the use of a "relaxed" molecular clock (Drummond *et al.*, 2006). Consequently, it is no longer necessary to assume absolute rate constancy (the "strict" molecular clock) when investigating the time-scale of viral evolution.

Before continuing, it is important to clarify what is actually being measured in these studies. Although it is tempting to talk in this manner, studies which consider a sample of sequences from a specific virus are *not* usually estimating the *age* of that virus. Rather, they are estimating the age of the most recent common ancestor (MRCA) of the sample of sequences available for study. As a case in point, although the MRCA of sampled isolates of YFV was found to be approximately 750 years (Bryant *et al.*, 2007), this does not mean that YFV is only 750 years old; rather, no viral lineages older than this have survived to be sampled, and the virus itself could have originated many millennia before. The high birth and death rate of lineages that is likely to characterize RNA viruses as a whole (Holmes, 2003), makes it likely that lineages of YFV would have been produced frequently in the past and then died out because of a lack of susceptible hosts.

This distinction notwithstanding, perhaps the most intriguing observation from studies of the time-scale of viral evolution is that these are often remarkably recent, with the MRCAs of many RNA viruses dating back a few centuries at most. Although the rapidity of RNA virus evolution means that MRCAs of many thousands of years are unlikely to materialize, because any sequences that originated over this time-scale would be too divergent to analyze (Holmes, 2003), the clustering of MRCAs to within a few centuries of the present is still notable. There are three possible explanations for such a shallow evolutionary time-scale: (i) that these viruses first appeared at this time, perhaps as this reflects the point of cross-species transmission (such as the species jump of SIVcpz in chimpanzees to HIV-1 in humans), (ii) that the "real age" of these viruses is a good deal older, but that pre-existing genetic diversity has been purged from the population by a selective sweep (a genetic bottleneck), or (iii) that shallow MRCA values simply reflect the process of neutral genetic drift played out in a population with rapid generation times. While in some cases, such as HIV, it is clear that recent common ancestry does indeed reflect the recently of emergence, this is not a viable explanation for most cases where shallow MRCAs are observed. Similarly, while it is undoubtedly the case that periodic selective sweeps leave important signals in the patterns of genetic diversity, it has been difficult at best to associate recent common ancestry with the occurrence of large-scale selection events, particularly for viruses that have near global distributions (although this is clearly an area that needs to be explained in greater detail). It therefore seems likely that neutral evolutionary dynamics are the most likely explanation for recent common ancestry. To be more specific, the mean time (with a large variance) to common ancestry for a haploid population under genetic drift is $2N_e$ generations (where N_e specifies the effective population size); for acute viral infections with short generations that experience recurrent population bottlenecks, a common occurrence in RNA viruses, this may mean that all lineages sampled will have ancestries of no more than a few centuries.

NATURAL SELECTION ON RNA VIRUSES

One of the most important aspects of studies of viral evolution is to estimate selection pressures at both the lineage and site (individual amino acid) specific level. Indeed, estimating the fitness of mutations, either individually or in combination, is perhaps the most important (and difficult) task in evolutionary genetics. Again, the rapidity with which RNA viruses evolve mean that this task can often be performed more successfully in these than many other organisms, as it provides a unique insight into the process of allele fixation as it occurs.

Although a wide variety of population genetic measures of selection pressure are available, by far the most common method is to estimate of the relative numbers of non-synonymous (d_N) and synonymous (d_S) nucleotide substitutions *per site* (with the key ratio d_N/d_S sometimes denoted w) (Yang and Bielawski, 2000). A wide variety of methods are now available to undertake such analyses, and their statistical properties have been thoroughly investigated (Anisimova *et al.*, 2001; Kosakovsky Pond and Frost, 2005; Table 5.1). These estimates can be made in two ways. First, if sufficient mutant *fixations* (nucleotide substitutions) have occurred, for example when comparing sequences from viruses assigned to different species, or which infect different hosts, or those that are separated by relatively long branches, then a simple computation of the d_N/d_S ratio is sufficient to obtain a broad-scale picture of the selection pressures acting on gene sequences. In particular, the lower the d_N/d_S ratio, the greater the strength of purifying (negative) selection acting on gene sequences; hence, a d_N/d_S of 0.1 indicates that 90% of the non-synonymous mutations were deleterious and removed by purifying selection (if non-synonymous mutations were neutral, then d_N/d_S would of course attain a value of ~1.0.

More interesting, and invariably more controversial, are instances when $d_N/d_S > 1.0$, a fairly regular occurrence in RNA virus evolution. This is usually taken as evidence for the action of positive selection (adaptive evolution) as it means that non-synonymous mutations were fixed faster than synonymous ones, which can only occur if the former are advantageous (although false-positive results can occur when nucleotide compositions are very skewed). The central debate in this area is whether many of the cases of positive selection described in RNA viruses are merely false-positives that have arisen through the use of inferior methods, or whether such estimates are inherently conservative such that positive selection acts far more frequently than can be determined using these computational methods. The answer appears to comprise an element of both viewpoints. There is now no doubt that some of the lineage and site-specific methods used to infer the occurrence of positive selection can be liable to false-positive results under certain conditions, although this is still an area of active debate (Suzuki and Nei, 2004; Wong *et al.*, 2004). In particular, putative selected sites that occur sporadically on the tips of evolutionary trees are unlikely to represent *bona fide* occurrences of positive selection (see below). On the other hand, it is equally apparent that all currently available analytical methods are conservative when it comes to identifying adaptive evolution, particularly those that rely on a simple pairwise comparisons of d_N/d_S (and which also suffer badly from pseudo-replication). The most obvious limitation is that natural selection that has resulted in the fixation of one amino acid change on a single lineage—which may be the most common form of selection in RNA viruses—will not be identified as positively selected using a simple computation of d_N/d_S, which require recurrent non-synonymous changes to make this inference. Similarly, any positive selection on synonymous sites, which is expected given large-scale RNA secondary structures (Simmonds *et al.*, 2004), cannot be detected using these methods.

It is therefore possible to make a more "liberal" estimate of the number of positively selected sites in a sequence by considering the *rate* at which they are fixed in a population, or their "transition times" (Zanotto *et al.*, 1999; Shih *et al.*, 2007). Such an inference is based on firm population genetic theory; the faster a mutation goes to fixation, the more likely that this will have been achieved by natural selection than genetic drift. Even for viruses that undergo periodic epidemic troughs, therein reducing N_e, mutations that are fixed over the time-scale of weeks or months are likely to have done so through natural selection rather than genetic drift (although the statistical basis to this approach needs to be formalized). The complicating factor is hitch-hiking. Hence, although a group of mutations may appear to achieve rapid fixation together, at face value indicative of natural selection, it is possible that only a single of these mutations is selectively advantageous, with the remainder fixed because they are in physical linkage with the advantageous mutation. For viruses with low rates of recombination, hitch-hiking is a major consideration and will inevitably lead to more false-positive results.

The second way in which measures of d_N/d_S can be used to infer the types of selection pressures acting on gene sequences, and one directly related to the analysis of transition times, is to consider whether they fall on internal or external (tip) branches of phylogenetic trees. In this protocol, it is the distribution of (fixed) substitutions compared to (transient) polymorphisms that is the key to understanding selection pressures (and it is important to remember that studies of intraspecies genetic variation in RNA viruses will largely consider polymorphisms). The first method to perform such a test was that of McDonald and Kreitman (1991), initially applied to *Drosophila*. Although other methods have been developed since this time, the underlying principles have not changed: the higher the fitness of a non-synonymous mutation then, on average, the deeper it will fall on a phylogenetic tree because it is likely to have been driven to fixation by natural selection, so that d_N/d_S will be elevated on internal compared to external branches. In contrast, if evolution is dominated by purifying selection, then most non-synonymous mutations will be deleterious and hence young (because they are likely to be removed by purifying selection) and therefore tend to fall on the tips of trees, so that d_N/d_S will be elevated on external branches. Work in recent years has shown that, despite the fairly regular occurrence of adaptive evolution, RNA virus evolution at the broad scale is dominated by purifying selection, such that most non-synonymous mutations sampled are likely to be transient deleterious ones (Pybus *et al.*, 2007). Such an abundance of deleterious mutations is to be expected given the high mutation rates of RNA viruses coupled with their small, efficiently organized genomes. Strikingly, this is also true of the intra-host evolution of HIV, a textbook example of adaptive evolution (Edwards *et al.*, 2006); although positive selection is a major force in shaping the intra-host evolution of HIV, purifying selection occurs more frequently.

A more subtle way in which the strength of natural selection (relative to that of genetic drift) can be measured at the gene sequence level is through studies of patterns of codon usage bias and their determinants. Although generalities are dangerous, it seems that, on average, codon usage biases in RNA viruses are more determined by neutral mutation pressure than natural selection (Jenkins and Holmes, 2003). In broad terms, the relative strength of genetic drift versus natural selection is determined by the compound parameter N_es, where s represents the selection coefficient, a measure of fitness. Hence, where $N_es, < 1$ genetic drift will dominate evolutionary dynamics; hence, genetic drift works most efficiently when effective population sizes or selection coefficients are small. It is this relationship that explains why genetic drift is largely thought to control codon usage bias in mammals (small N_e), while natural selection controls this process in many bacterial species (large N_e). The absence of clear-cut evidence for selection for codon bias in RNA viruses (although see below) is therefore likely to be a reflection of relatively low N_e values in the long term, perhaps because of the regular population bottlenecks that accompany inter-host transmission.

Finally, recent years have also witnessed a major debate concerning whether positive selection can be detected through the analysis of single genome sequences (Plotkin and Dushoff, 2003; Plotkin *et al.*, 2004). The heart of this method is a measurement of "codon volatility" such that the footprint of adaptive evolution at non-synonymous sites is a preference for "volatile" codons that facilitate amino acid change. While it is evident natural selection, at least on occasion, can be manifest in codon volatility, it is less certain that codon volatility is an unambiguous measure of positive selection (Hahn *et al.*, 2004; Sharp, 2005). Statistical tests based the comparison of multiple genome sequences are therefore likely to remain the industry standard.

THE POPULATION DYNAMICS OF RNA VIRUSES

Thus far I have considered the evolutionary processes acting at the scale of the viral gene sequence. However, it is also clear that viral genes and genomes contain an exquisite record of their past history of population growth (or decline) at the epidemiological scale. The formation of coalescent theory in the early 1980s (Kingman, 1982; Tajima, 1983) heralded the development of a suite of methods to infer such demographic histories from viral gene sequence data (Nee *et al.*, 1995; Pybus *et al.*, 1999, 2001; Pybus and Rambaut, 2002).

The most recent manifestation of these methods are those based on Bayesian MCMC, again available in the BEAST package (Drummond and Rambaut, 2003), and which have been used to infer the epidemiological dynamics of a number of viral infections (see, for example, Biek *et al.*, 2006). These methods allow the user to determine whether the demographic history of a viral population best fits a number of specific epidemiological models. If so, parameters of interest can be estimated, such as the rate of population growth (often measured in terms of the number of new infections per individual, per year), the epidemic doubling time of the infection, and the effective number of infections ($N_e t$, a compound parameter reflecting effective population size and the generation time, t).

Although powerful, the extra assumptions required compared to simple phylogenetic analysis again means that great caution must be exercised the use of these methods. In particular, a major limiting assumption is that they require the viral population from which the sequences are drawn to exhibit panmixia (random mating). If such an assumption is broken, perhaps because a cluster of very closely related sequences from a single outbreak have been included in a broad-scale population analysis, then an incorrect inference of evolutionary dynamics can be made (in the case of analyzing a cluster of closely related sequences a model of population decline would be erroneously supported).

If generalities can be made from these analyses performed to date, it is that RNA viruses exhibit a rather limited number of modes of population growth, namely: exponential population growth, with epidemic doubling times ranging from weeks to years; logistic population growth, in which growth rates exhibit an initially rapid phase, followed by a lower growth rate secondary phase; and more complex epidemiological dynamics, involving large-scale fluctuations in population size through time. Which of these dynamical patterns a virus occupies reflects its intrinsic epidemiology as well as its duration of infection (Grenfell *et al.*, 2004). For example, those viruses that cause long-term, chronic infections (such as HIV and hepatitis C virus) are often characterized by slow population dynamics, manifest as either logistic population growth or relatively slow rates of exponential growth (Walker *et al.*, 2005; Nakano *et al.*, 2006). In contrast, acute infections, such as measles, often exhibit more complex fluctuating dynamics, with phases of population growth occurring with a distinct periodicity.

At present, it is not possible to precisely estimate rates of population growth in viruses with complex dynamics. Rather, a graphical view of changes in population size through time can be achieved through the use of a Bayesian skyline plot, which depicts changing patterns of $N_e t$ across different time segments (Drummond *et al.*, 2005; Figure 5.3). When using the Bayesian skyline plot to make inferences of population dynamics through time it

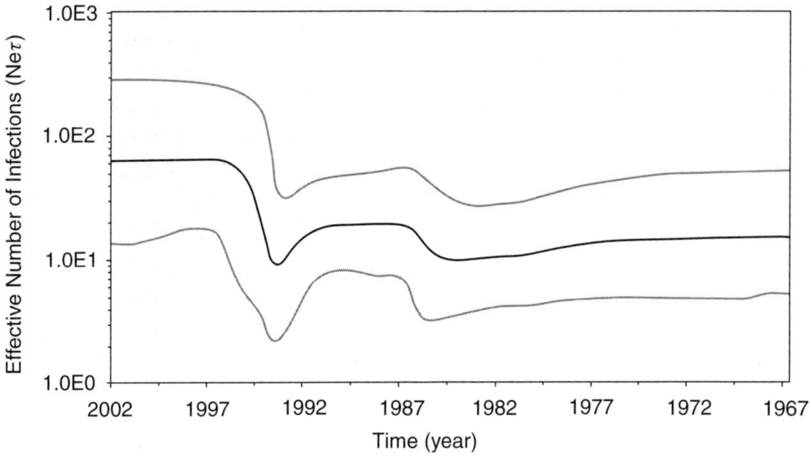

FIGURE 5.3 Bayesian skyline plot of DENV-1 in Bangkok, Thailand, inferred using E (envelope) gene sequences. The plot depicts changes in the effective number of infections—$N_e t$—through time, indicative of changing epidemiological dynamics (population growth rates). The black line depicts the mean estimate of $N_e t$ while the 95% HPD (highest probability density values) are shown in gray.

is also important to recall that the quality of the inference depends strongly on the graininess of the temporal sampling. Specifically, for those viruses with slow dynamics, the key epidemiological signals can perhaps be recovered with samples collected on a yearly basis. However, for those viral epidemics that have a more distinct periodicity, such as the biannual epidemics of measles or the annual (winter) epidemics of influenza, it is clear that a far more fine-grained temporal sampling is required to fully extract all epidemiological information from these sequences. Given the increase in gene and genome sequence data sets from specific viral epidemics, including those with more fine-grained sampling, it is evident that methods to estimate the population dynamics of RNA viruses will continue to develop rapidly over the next few years and that they will be used in mainstream epidemiological research.

I have briefly outlined the computational methods that are currently at the forefront of the comparative approach to viral evolution. While these methods dominate the current literature, it is possible that the production of very large numbers of complete genome sequences will radically change the scope of what can be achieved through *in silico* analysis and therein stimulate a whole new era of method (and theoretical) development. The age of comparative genomics promises to be an exciting one for students of viral evolution.

ACKNOWLEDGMENTS

This work was funded by NIH grant GM080533-01.

REFERENCES

Anisimova, M., Bielawski, J.P. and Yang, Z. (2001) Accuracy and power of the likelihood ratio test in detecting adaptive molecular evolution. *Mol. Biol. Evol.* **18**, 1585–1592.

Archibald, J.M. and Roger, A.J. (2002) Gene conversion and the evolution of euryarchaeal chaperonins: a maximum likelihood-based method for detecting conflicting phylogenetic signals. *J. Mol. Evol.* **55**, 232–245.

Awadalla, P. (2003) The evolutionary genomics of pathogen recombination. *Nat. Rev. Genet.* **4**, 50–60.

Bandelt, H.-J. and Dress, A.W.M. (1992) Split decomposition: a new and useful approach to phylogenetic analysis of distance data. *Mol. Phylogenet. Evol.* **1**, 242–252.

Biek, R., Drummond, A.J. and Poss, M. (2006) A virus reveals population structure and recent demographic history of its carnivore host. *Science* **311**, 538–541.

Brown, C.J., Garner, E.C., Dunker, K.A. and Joyce, P. (2001) The power to detect recombination using the coalescent. *Mol. Biol. Evol.* **18**, 1421–1424.

Bryant, D. and Moulton, V. (2004) Neighbor-net: an agglomerative method for the construction of phylogenetic networks. *Mol. Biol. Evol.* **21**, 255–265.

Bryant, J.E., Holmes, E.C. and Barrett, A.D.T. (2007) Out of Africa: A molecular perspective on the introduction of Yellow Fever Virus into the Americas. *PLoS Pathog.* **3**, e75.

Bourque, G. and Pevzner, P.A. (2002) Genome-scale evolution: reconstructing gene orders in the ancestral species. *Genome Res.* **12**, 26–36.

Chare, E.R., Gould, E.A. and Holmes, E.C. (2003) Phylogenetic analysis reveals a low rate of homologous recombination in negative-sense RNA viruses. *J. Gen. Virol.* **84**, 2691–2703.

Drake, J.W. and Holland, J.J. (1999) Mutation rates among RNA viruses. *Proc. Natl Acad. Sci. USA* **96**, 13910–13913.

Drummond, A.J. and Rambaut, A. (2003) BEAST version 1.3. Available from http://evolve.zoo.ox.ac.uk/beast/.

Drummond, A., Pybus, O.G. and Rambaut, A. (2003a) Inference of viral evolutionary rates from molecular sequences. *Adv. Parasitol.* **54**, 331–358.

Drummond, A.J., Pybus, O.G., Rambaut, A., Forsberg, R. and Rodrigo, A.G. (2003b) Measurably evolving populations. *Trends Ecol. Evol.* **18**, 481–488.

Drummond, A.J., Rambaut, A., Shapiro, B. and Pybus, O.G. (2005) Bayesian coalescent inference of past population dynamics from molecular sequences. *Mol. Biol. Evol.* **22**, 1185–1192.

Drummond, A.J., Ho, S.Y.W., Phillips, M.J. and Rambaut, A. (2006) Relaxed phylogenetics and dating with confidence. *PLoS Biol.* **4**, e88.

Edwards, C.T.T., Holmes, E.C., Pybus, O.G., Wilson, D.J., Viscidi, R.P., Abrams, E.J. *et al.* (2006) Evolution of the HIV-1 envelope is dominated by purifying selection. *Genetics* **174**, 1441–1453.

Elena, S.F. and Sanjuan, R. (2005) Adaptive value of high mutation rates of RNA viruses: separating causes from consequences. *J. Virol.* **79**, 11555–11558.

Felsenstein, J. (2004) *Inferring Phylogenies*. Sunderland, MA: Sinauer Associates.

Felsenstein, J. (2005) *PHYLIP (Phylogeny Inference Package) version 3.6. Distributed by the author*. Department of Genome Sciences, University of Washington Seattle .

Forslund, K., Huson, D.H. and Moulton, V. (2004) VisRD—visual recombination detection. *Bioinformatics* **20**, 3654–3655.

Frey, B.J. and Dueck, D. (2007) Clustering by passing messages between data points. *Science* **315**, 972–976.

Ghedin, E., Sengamalay, N.A., Shumway, M., Zaborsky, J., Feldblyum, T., Subbu, V., Spiro, D.J., Sitz, J., Koo, H., Bolotov, P., Dernovoy, D., Tatusova, T., Bao, Y., St. George, K., Taylor, J., Lipman, D.J., Fraser, C.M., Taubenberger, J.K. and Salzberg, S.L. (2005) Large-scale sequencing of human influenza reveals the dynamic nature of viral genome evolution. *Nature* **437**, 1162–1166.

Gibbs, M.J., Armstrong, J.S. and Gibbs, A.J. (2001) Recombination in the hemagglutinin gene of the 1918 "Spanish flu". *Science* **293**, 1842–1845.

Ge, L., Zhang, J., Zhou, X. and Li, H. (2007) Genetic structure and population variability of tomato yellow leaf curl China virus. *J. Virol.* **81**, 5902–5907.

Grenfell, B.T., Pybus, O.G., Gog, J.R., Wood, J.L.N., Daly, J.M., Mumford, J.A. and Holmes, E.C. (2004) Unifying the epidemiological and evolutionary dynamics of pathogens. *Science* **303**, 327–332.

Guindon, S. and Gascuel, O. (2003) A simple, fast, and accurate algorithm to estimate large phylogenies by maximum likelihood. *Mol. Biol. Evol.* **52**, 696–704.

Hahn, M.W., Mezey, J.G., Begun, D.J., Gillespie, J.H., Kern, A.D., Langley, C.H. and Moyle, L.C. (2004) Evolutionary genomics: codon bias and selection on single genomes. *Nature* **433**, E5–6.

Hanada, K., Suzuki, Y. and Gojobori, T. (2004) A large variation in the rates of synonymous substitution for RNA viruses and its relationship to a diversity of viral infection and transmission modes. *Mol. Biol. Evol.* **21**, 1074–1080.

Harvey, P.H. and Pagel, M.D. (1993) *The Comparative Method in Evolutionary Biology*. Oxford: Oxford University Press.

Holmes, E.C. (2003) Molecular clocks and the puzzle of RNA virus origins. *J. Virol.* **77**, 3893–3897.

Holmes, E.C., Worobey, M. and Rambaut, A. (1999) Phylogenetic evidence for recombination in dengue virus. *Mol. Biol. Evol.* **16**, 405–409.

Huelsenbeck, J.P., Ronquist, F., Nielsen, R. and Bollback, J.P. (2001) Bayesian inference of phylogeny and its impact on evolutionary biology. *Science* **294**, 2310–2314.

Hu, W.S. and Temin, H.M. (1990) Retroviral recombination and reverse transcription. *Science* **250**, 1227–1233.

Huson, D.H. and Bryant, D. (2004) Application of phylogenetic networks in evolutionary studies. *Mol. Biol. Evol.* **23**, 254–267.

Jenkins, G.M. and Holmes, E.C. (2003) The extent of codon usage bias in human RNA viruses and its evolutionary origin. *Virus Res.* **92**, 1–7.

Jenkins, G.M., Rambaut, A., Pybus, O.G. and Holmes, E.C. (2002) Rates of molecular evolution in RNA viruses: a quantitative phylogenetic analysis. *J. Mol. Evol.* **54**, 152–161.

Keele, B.F., Van Heuverswyn, F., Li, Y., Bailes, E., Takehisa, J., Santiago, M.L., Bibollet-Ruche, F., Chen, Y., Wain, L.V., Liegeois, F., Loul, S., Ngole, E.M., Bienvenue, Y., Delaporte, E., Brookfield, J.F., Sharp, P.M., Shaw, G.M., Peeters, M. and Hahn, B.H. (2006) Chimpanzee reservoirs of pandemic and nonpandemic HIV-1. *Science* **313**, 523–526.

Kingman, J.F.C. (1982) On the genealogy of large populations. *J. Appl. Probab.* **19A**, 27–43.

Koonin, E.V., Senkevich, T.G. and Dolja, V.V. (2006) The ancient virus world and evolution of cells. *Biol. Direct*, **1**, 29.

Kosakovsky Pond, S.L. and Frost, S.D.W. (2005) Not so different after all: A comparison of methods for detecting amino-acid sites under selection. *Mol. Biol. Evol.* **22**, 1208–1222.

Kosakovsky Pond, S.L., Frost, S.D. and Muse, S.V. (2005) HyPhy: hypothesis testing using phylogenies. *Bioinformatics* **21**, 676–679.

Kosakovsky Pond, S.L., Posada, D., Gravenor, M.B., Woelk, C.H. and Frost, S.D. (2006) GARD: a genetic algorithm for recombination detection. *Bioinformatics* **22**, 3096–3098.

Lai, M.M.C. (1992) RNA recombination in animal and plant viruses. *Microbiol. Rev.* **56**, 61–79.

Lukashov, V.V. and Goudsmit, J. (2002) Recent evolutionary history of human immunodeficiency virus type 1 subtype B: reconstruction of epidemic onset based on sequence distances to the common ancestor. *J. Mol. Evol.* **54**, 680–691.

Macken, C.A., Webby, R.J. and Bruno, W.J. (2006) Genotype turnover by reassortment of replication complex genes from avian Influenza A virus. *J. Gen. Virol.* **87**, 2803–2815.

Margulies, M., Egholm, M., Altman, W.E., Attiya, S., Bader, J.S., Bemben, L.A. *et al.* (2005) Genome sequencing in microfabricated high-density picolitre reactors. *Nature* **437**, 376–380.

Martin, D.P., Williamson, C. and Posada, D. (2005) RDP2: recombination detection and analysis from sequence alignments. *Bioinformatics* **21**, 260–262.

McDonald, J.H. and Kreitman, M. (1991) Adaptive evolution at the *Adh* locus in Drosophila. *Nature* **351**, 652–654.

McVean, G., Awadalla, P. and Fearnhead, P. (2002) A coalescent-based method for detecting and estimating recombination from gene sequences. *Genetics* **160**, 1231–1241.

Milne, I., Wright, F., Rowe, G., Marshal, D.F., Husmeier, D. and McGuire, G. (2004) TOPALi: Software for automatic identification of recombinant sequences within DNA multiple alignments. *Bioinformatics* **20**, 1806–1807.

Monath, T.P., Kanesa-Thasan, N., Guirakhoo, F., Pugachev, K., Almond, J., Lang, J. *et al.* (2005) Recombination and flavivirus vaccines: a commentary. *Vaccine* **23**, 2956–2958.

Nakano, T., Lu, L., He, Y., Fu, Y., Robertson, B.H. and Pybus, O.G. (2006) Population genetic history of hepatitis C virus 1b infection in China. *J. Gen. Virol.* **87**, 73–82.

Nee, S., Holmes, E.C., Rambaut, A. and Harvey, P.H. (1995) Inferring population history from molecular phylogenies. *Philos. Trans. R. Soc. B.* **349**, 25–31.

Nora, T., Charpentier, C., Tenaillon, O., Hoede, C., Clavel, F. and Hance, A.J. (2007) Contribution of recombination to the evolution of human immunodeficiency viruses expressing resistance to antiretroviral treatment. *J. Virol.* **May 9**. [Epub ahead of print].

Pagel, M. (1999) Inferring the historical patterns of biological evolution. *Nature* **401**, 877–884.

Plotkin, J.B. and Dushoff, J. (2003) Codon bias and frequency-dependent selection on the hemagglutinin epitopes of influenza A virus. *Proc. Natl Acad. Sci. USA*, **100**, 7152–7157.

Plotkin, J.B., Dushoff, J. and Fraser, H.B. (2004) Detecting selection using a single genome sequence of *M. tuberculosis* and *P. falciparum*. *Nature* **428**, 942–945.

Posada, D. (2002) Evaluation of methods for detecting recombination from DNA sequences: empirical data. *Mol. Biol. Evol.* **19**, 708–717.

Posada, D. and Crandall, K.A. (1998) Modeltest: testing the model of DNA substitution. *Bioinformatics* **14**, 817–818.

Posada, D., Crandall, K.A. and Holmes, E.C. (2002) Recombination in evolutionary genomics. *Annu. Rev. Genet.* **36**, 75–97.

Pringle, C.R. and Parry, J.E. (1982) Measurement of surface antigen by specific bacterial adherence and scanning electron microscopy (SABA/SEM) in cells infected by vesiculovirus ts mutants. *J. Gen. Virol.* **59**, 207–211.

Pybus, O.G. and Rambaut, A. (2002) GENIE: estimating demographic history from molecular phylogenies. *Bioinformatics* **18**, 1404–1405.

Pybus, O.G., Holmes, E.C. and Harvey, P.H. (1999) The mid-depth method and HIV-1: a practical approach to testing hypotheses of viral epidemic history. *Mol. Biol. Evol.* **16**, 953–959.

Pybus, O.G., Charleston, M.A., Gupta, S., Rambaut, A., Holmes, E.C. and Harvey, P.H. (2001) The epidemic behaviour of the hepatitis C virus. *Science* **292**, 2323–2325.

Pybus, O.G., Rambaut, A., Freckleton, R.P., Belshaw, R., Drummond, A.J. and Holmes, E.C. (2007) Phylogenetic evidence for deleterious mutation load in RNA viruses and its contribution to viral evolution. *Mol. Biol. Evol.* **24**, 845–852.

Rambaut, A. (2000) Estimating the rate of molecular evolution: incorporating non-contemporaneous sequences into maximum likelihood phylogenies. *Bioinformatics* **16**, 395–399.

Ronquist, F. and Huelsenbeck, J.P. (2003) MRBAYES 3: Bayesian phylogenetic inference under mixed models. *Bioinformatics* **19**, 1572–1574.

Sankoff, D. (2001) Gene and genome duplication. *Curr. Opin. Genet. Dev.* **11**, 681–684.

Santiago, M.L., Range, F., Keele, B.F., Li, Y., Bailes, E., Bibollet-Ruche, F., Fruteau, C. *et al.* (2005) Simian immunodeficiency virus infection in free-ranging sooty mangabeys (*Cercocebus atys atys*) from the Tai Forest, Cote d'Ivoire: implications for the origin of epidemic human immunodeficiency virus type 2. *J. Virol.* **79**, 12515–12527.

Schierup, M.H., Mordhorst, C.H., Muller, C.P. and Christensen, L.S. (2005) Evidence of recombination among early-vaccination era measles virus strains. *BMC Evol. Biol.* **5**, 52.

Seligman, S.J. and Gould, E.A. (2004) Live flavivirus vaccines: reasons for caution. *Lancet* **363**, 2073–2075.

Shackelton, L.A. and Holmes, E.C. (2006) Phylogenetic evidence for the rapid evolution of human B19 erythrovirus. *J. Virol.* **80**, 3666–3669.

Shackelton, L.A., Parrish, C.R., Truyen, U. and Holmes, E.C. (2005) High rate of viral evolution associated with the emergence of canine parvoviruses. *Proc. Natl Acad. Sci. USA*, **102**, 379–384.

Shapiro, B., Rambaut, A., Pybus, O.G., Drummond, A. and Holmes, E.C. (2006) A phylogenetic method for detecting positive epistasis in gene sequences and its application to RNA virus evolution. *Mol. Biol. Evol.* **23**, 1724–1730.

Sharp, P.M. (2002) Origins of human virus diversity. *Cell* **108**, 305–312.

Sharp, P.M. (2005) Gene "volatility" is most unlikely to reveal adaptation. *Mol. Biol. Evol.* **22**, 807–809.

Shih, A.C., Hsiao, T.C., Ho, M.S. and Li, W.-H. (2007) Simultaneous amino acid substitutions at antigenic sites drive influenza A hemagglutinin evolution. *Proc. Natl Acad. Sci. USA*, **104**, 6283–6288.

Simmonds, P., Tuplin, A. and Evans, D.J. (2004) Detection of genome-scale ordered RNA structure (GORS) in genomes of positive-stranded RNA viruses: implications for virus evolution and host persistence. *RNA* **10**, 1337–1351.

Stamatakis, A. (2006) RAxML-VI-HPC: maximum likelihood-based phylogenetic analyses with thousands of taxa and mixed models. *Bioinformatics* **22**, 2688–2690.

Suzuki, Y. and Nei, M. (2004) False-positive selection identified by ML-based methods: Examples from the *Sig1* gene of the diatom *Thalassiosira weissflogii* and the *tax* gene of the human T-cell lymphotropic virus. *Mol. Biol. Evol.* **21**, 914–921.

Suzuki, Y., Gojobori, T. and Nei, M. (2001) ADAPTSITE: detecting natural selection at single amino acid sites. *Bioinformatics* **17**, 660–661.

Suzuki, Y., Glazko, G.V. and Nei, M. (2002) Overcredibility of molecular phylogenies obtained by Bayesian phylogenetics. *Proc. Natl Acad. Sci. USA* **99**, 16138–16143.

Switzer, W.M., Salemi, M., Shanmugam, V., Gao, F., Cong, M.-E., Kuiken, C. *et al.* (2005) Ancient co-speciation of simian foamy viruses and primates. *Nature* **434**, 376–380.

Swofford, D.L. (2003) *PAUP*. Phylogenetic Analysis Using Parsimony (*and other methods) Version 4*. Sunderland, MA: Sinauer Associates.

Tajima, F. (1983) Evolutionary relationship of DNA sequences in finite populations. *Genetics* **105**, 437–460.

Tamura, K., Dudley, J., Nei, M. and Kumar, S. (2007) MEGA4: Molecular Evolutionary Genetics Analysis (MEGA) Software Version 4.0. *Mol. Biol. Evol.* May, 7. [Epub ahead of print].

Thorne, J.L., Goldman, N. and Jones, D.T. (1996) Combining protein evolution and secondary structure. *Mol. Biol. Evol.* **13**, 666–673.

Walker, P.R., Pybus, O.G., Rambaut, A. and Holmes, E. C. (2005) Comparative population dynamics of HIV-1 subtypes B and C: Subtype-specific differences in patterns of epidemic growth. *Infect. Genet. Evol.* **5**, 199–208.

Wong, W.S., Yang, Z., Goldman, N. and Nielsen, R. (2004) Accuracy and power of statistical methods for detecting adaptive evolution in protein coding sequences and for identifying positively selected sites. *Genetics* **168**, 1041–1051.

Yang, Z. (1997) PAML: a program package for phylogenetic analysis by maximum likelihood. *CABIOS* **13**, 555–556.

Yang, Z. and Bielawski, J.P. (2000) Statistical methods for detecting molecular adaptation. *Trends Ecol. Evol.* **15**, 496–502.

Zanotto, P.M.de.A., Gibbs, M.J., Gould, E.A. and Holmes, E.C. (1996) A reassessment of the higher taxonomy of viruses based on RNA polymerases. *J. Virol.* **70**, 6083–6096.

Zanotto, P.M.de.A., Kallas, E.G., de Souza, R.F. and Holmes, E.C. (1999) Genealogical evidence for positive selection in the *nef* gene of HIV-1. *Genetics* **153**, 1077–1089.

Zwickl, D.J. (2006) *Genetic algorithm approaches for the phylogenetic analysis of large biological sequence datasets under the maximum likelihood criterion. Ph.D. dissertation*. The University of Texas at Austin.

CHAPTER

6

Nucleic Acid Polymerase Fidelity and Viral Population Fitness

Eric D. Smidansky, Jamie J. Arnold and Craig E. Cameron

ABSTRACT

Viral polymerases are essential for the maintenance and expression of the genomes of all viruses. The fidelity of polymerase-catalyzed nucleotide addition varies between classes of nucleic acid polymerases. Here we present a kinetic, thermodynamic, and structural description of the process employed by polymerases to modulate the accuracy of nucleotide addition. Direct connections between polymerase fidelity and virus biology are discussed that lead to the general conclusion that polymerase fidelity is tuned by natural selection to optimize viral population genotypic diversity and, consequently, viral competitiveness in the dynamic and hostile environment of the cell. Finally, we discuss the potential of exploiting the optimized nature of viral polymerase fidelity for development of strategies to treat and prevent viral infections.

INTRODUCTION

Chapter Goals and Perspectives

This chapter examines nucleic acid polymerase fidelity, or accuracy of nucleotide incorporation. While a vast body of primary literature and excellent recent reviews (Showalter and Tsai, 2002; Sousa and Mukherjee, 2003; Sweasy, 2003; Kunkel, 2004; Rothwell and Waksman, 2005; Beard and Wilson, 2006; Radhakrishnan *et al.*, 2006; Showalter *et al.*, 2006) address nucleic acid polymerase fidelity broadly (including eukaryotic, bacterial, and viral polymerases), experimental data summarized here will be largely limited to viral polymerases. In particular, RNA virus polymerases employed in genome replication (Ferrer-Orta *et al.*, 2006) will be emphasized because these are simple biological systems that permit the influence of polymerase fidelity on genotypic diversity and viral fitness to be examined. *Poliovirus* (PV) and its polymerase, termed 3Dpol, (Cameron *et al.*, 2002) will serve as the primary model to illustrate important concepts about polymerase fidelity and consequences for fitness, adaptation and evolution.

Several important themes will surface repeatedly in this chapter. One is that fundamental features of polymerase mechanism and fidelity are conserved evolutionarily and, therefore, general, unifying principles can be identified (Steitz, 1999; Rothwell and Waksman,

Origin and Evolution of Viruses
ISBN 978-0-12-374153-0

2005). Another is that polymerase fidelity has its most profound meaning in its influence on viral population fitness (Biebricher and Eigen, 2006; Vignuzzi *et al.*, 2006; Bull *et al.*, 2007). Polymerase fidelity is a major determinant of, and, in fact, is dictated by, viral population fitness needs and so fidelity is directly linked to virus adaptation and evolution (Vignuzzi *et al.*, 2006). Therefore, the product of viral polymerase activity is not simply a genome but genotypic diversity. Finally, it will be emphasized that because polymerase fidelity is tightly linked to viral population fitness, therapeutic modulation of fidelity offers a potent means of viral inhibition.

Classes and Functions of Polymerases

Nucleic acid polymerases are classified according to whether they use DNA or RNA as template and whether ribo- or 2′-deoxyribonucleotides (rNTPs or dNTPs) are chosen as substrate for addition to the primer 3′-OH. Therefore, four biochemical classes of polymerases exist. Those requiring DNA as template and adding dNTPs to the primer terminus are termed DNA-dependent DNA polymerases, or DdDps. Those using RNA as template but adding dNTPs are RdDps, and so on, producing the additional classes RdRp and DdRp (Figure 6.1) (Beese *et al.*, 1993; McAllister and Raskin, 1993; Sousa, 1996; Hansen *et al.*, 1997; Brautigam and Steitz, 1998; Doublie *et al.*, 1998; Huang *et al.*, 1998; Franklin *et al.*, 2001; Thompson and Peersen, 2004; Yin and Steitz, 2004).

Nucleic acid polymerases accomplish a range of nucleic acid tasks and fidelity differs substantially between different polymerases, with nucleotide incorporation error frequencies varying by an astounding ten orders of magnitude (Kunkel, 2004). Functionally, polymerases can be categorized broadly as being replicative or reparative (Sousa and Mukherjee, 2003; Rothwell and Waksman, 2005). Replicative polymerases assemble genomes and are processive (i.e. complete multiple, sequential nucleotide incorporation cycles before dissociating from primer-template (PT)). Reparative polymerases identify nucleic acid defects and correct them and vary from being modestly processive to distributive (i.e. complete a single-nucleotide incorporation and then dissociate from PT), depending upon the size of the nucleic acid lesion being repaired (Rothwell and Waksman, 2005). Replicative polymerases function *in vivo* as part of complex, macromolecular assemblages that include other proteins supplying accessory functions, for example, enforcement of processivity (Yang *et al.*, 2004; Bebenek *et al.*, 2005). Many replicative DdDps have, in addition to polymerase activity, exonuclease activity (proofreading) that permits removal and correction of 90–99% of misincorporations as they occur, leading to very low error frequencies of up to ~10^{-10} (Kunkel, 2004). In contrast, RNA virus replicative polymerases lack exonuclease activity and, consequently, misincorporate at far higher frequencies, a condition thought to permit rapid evolution of RNA viruses (Crotty *et al.*, 2001). For example, PV 3Dpol, a small (52 kDa), replicative RdRp, produces transition mutations at frequencies of 10^{-4} and transversions at 10^{-7} (Arnold and Cameron, 2004a; Freistadt *et al.*, 2007). It should be noted that although viral RdRps are considered "error prone," their error frequencies are comparable to those of highly accurate DdDps prior to exonuclease correction (Arnold and Cameron, 2004a).

Conserved Polymerase Active Site Features

Steitz and co-workers (Kohlstaed *et al.*, 1992) first observed in x-ray crystal structures that the basic three-dimensional architecture of nucleic acid polymerases resembles a cupped right hand consisting of palm, thumb, and fingers subdomains (Figure 6.1). The orientation of the cupped right hand has the palm subdomain as floor, the thumb extending up to the right, and the fingers curling up to the left. The fingers and thumb are relatively open in many polymerases, such as in HIV 1

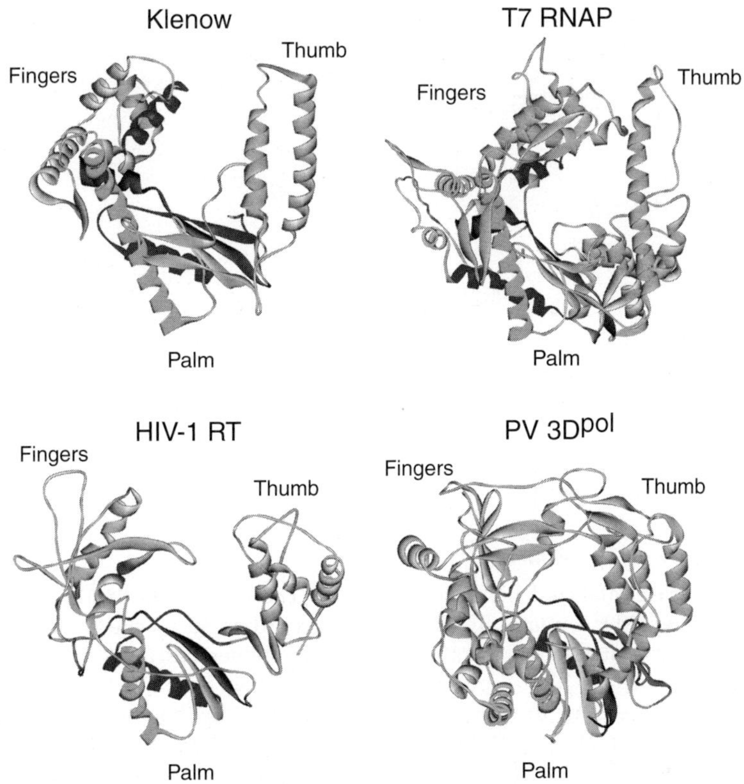

FIGURE 6.1 Structures of the four classes of nucleic acid polymerases. Structures of the polymerase domains of the DdDp large (Klenow) fragment of *E. coli* DNA polymerase I (1KFD) (Beese *et al.*, 1993), RdDp HIV 1 reverse transcriptase (HIV-1 RT) (1RTD) (Huang *et al.*, 1998), DpRp T7 RNAP (1S76) (Yin and Steitz, 2004), and RdRp PV 3Dpol (1RA6) (Thompson and Peersen, 2004). The conserved structural motifs in the palm subdomain are colored as follows: motif A, red; motif B, green; motif C, yellow; motif D, blue; and motif E, purple (HIV-1 RT and PV 3Dpol only). Helix O in Klenow and T7 RNAP is colored in blue. The analogous helix in B-family polymerases is helix P. The images were rendered using the program WebLab Viewer Pro (Molecular Simulations Inc., San Diego, CA). (See Plate 4 for the color version of this figure.)

reverse transcriptase (HIV-1 RT) (Figure 6.1) (Kohlstaedt *et al.*, 1992), but overarch the top, forming essentially a closed channel in others, as in PV 3Dpol (Figure 6.1) (Thompson and Peersen, 2004). Across the range of polymerases, the palm subdomain nearly always includes conserved components of the catalytic site. The thumb is important for interactions between polymerase and primer-template, and the fingers subdomain is prominent in recognition of the incoming nucleotide (Steitz, 1999).

Polymerase active site architecture is highly conserved (Steitz, 1998; Patel and Loeb, 2001) (Figure 6.2). The mechanism of nucleotide incorporation has been described by Steitz as a two-metal ion mechanism (Steitz and Steitz, 1993) because two magnesium cations (Mg^{2+}) are an invariant active site structural feature, helping to organize active site alignments and aid in catalysis (Steitz, 1998). The most prominent and conserved active site amino acid residues are two acidic residues, which can be Asp or Glu (Steitz, 1999), and a basic residue, most frequently Lys (Castro *et al.*, 2007) but sometimes Arg (Kraynov *et al.*, 2000) or His (Wang *et al.*, 2006). In PV 3Dpol the acidic residues are D233 from conserved palm structural motif A and D328 from motif C, and the basic residue is K359 from motif D (Castro

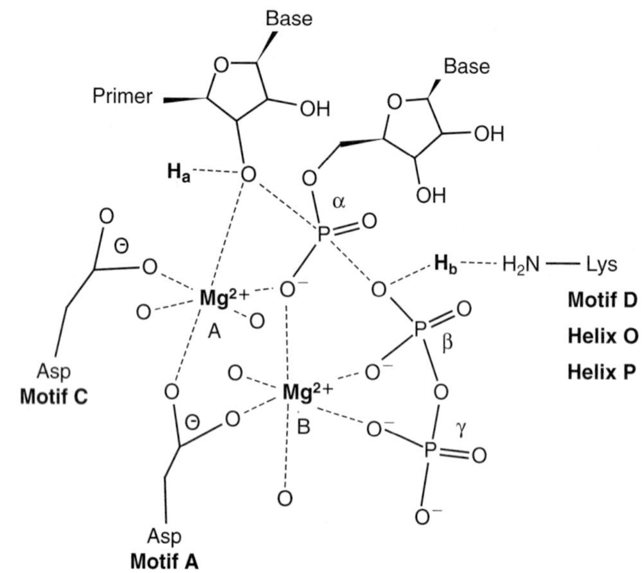

FIGURE 6.2 Polymerase-catalyzed nucleotidyl transfer. The nucleoside triphosphate enters the active site with a divalent cation (Mg^{2+}, metal B). This metal is coordinated by the β- and γ-phosphates of the nucleotide, an Asp residue located in structural motif A of all polymerases, and likely water molecules (indicated as oxygen ligands to metal without specific designation). Metal B orients the triphosphate in the active site and may contribute to charge neutralization during catalysis. Once the nucleotide is in place, the second divalent cation binds (Mg^{2+}, metal A). Metal A is coordinated by the 3′-OH, the α-phosphate, as well as Asp residues of structural motifs A and C. Metal A lowers the pK_a of the 3′-OH (denoted as H_a) facilitating catalysis at physiological pH. A conserved basic residue, usually a Lys located in structural motif D of RdRps and RdDps or helix O of DdDps and DdRps, serves as a general acid and donates a proton (denoted as H_b) to the pyrophosphate leaving group, assisting in the efficiency of nucleotidyl transfer. Adapted from Liu and Tsai (2001) and Steitz (1993).

et al., 2007). Metal ion B enters the active site bound to the triphosphate moiety of the incoming nucleotide. The coordination properties of metal ion B help orient the nucleotide relative to the conserved amino acid residues and also aid in dissipation of charge build up during phosphoryl transfer (Steitz, 1998). Metal ion A binds to an active site location near the primer terminus 3′-OH, after occupancy by the incoming nucleotide, and helps guide active site alignments and decreases the acidity of the 3′-OH, permitting deprotonation and subsequent phosphoryl transfer (Steitz, 1998).

Chemistry of Phosphoryl Transfer

The goal of the nucleotide incorporation reaction is covalent capture of the information content of the base moiety of the incoming nucleotide. Metal ion A reduces the affinity of the primer terminus 3′-OH for its proton, allowing removal of that proton by a nearby proton acceptor (Steitz, 1998) (Figure 6.2). This deprotonation event produces a highly reactive and unstable 3′-O^- nucleophile which seeks stability by attacking the closest available electrophilic center, the α-phosphorus atom (Steitz, 1998). This attack sets up a competition for bonding to the α-phosphorus between the strong attacking 3′-O^- nucleophile and the weaker nucleophile already covalently bonded, the pyrophosphate. The 3′-O^- wins the competition, covalently bonds to the α-phosphorus, and the pyrophosphate leaving group departs. The net result is that a phosphate group (along with covalently attached sugar and base) is transferred from the incoming nucleotide to the primer terminus, and so the chemistry is termed phosphoryl transfer.

An open question has been whether polymerases rely on more than two-metal ion catalysis for phosphoryl transfer. In many model polymerases, phosphoryl transfer is not rate-limiting (Joyce and Benkovic, 2004) and thus not accessible to biochemical analysis. This limitation was overcome by the discovery in PV 3Dpol that phosphoryl transfer in the presence of Mg^{2+} is partially rate-limiting (Arnold and Cameron, 2004a) and therefore available for examination by *in vitro* nucleotide incorporation assays. The chemical nature and spatial location of conserved active site amino acid residues (Figure 6.2) suggested the participation of protonic catalysis. PV 3Dpol phosphoryl transfer was therefore assayed for proton transfers during catalysis (Castro *et al.*, 2007). Solvent deuterium isotope effect and proton inventory experiments revealed that two proton transfers indeed occur. Conserved active site lysine 359 was found to serve as a general acid, donating its proton (Castro *et al.*, unpublished observations) to the pyrophosphate leaving group, expediting departure and accelerating chemistry. Similar experiments with polymerases from the other three classes, RT HIV-1 RdDp, T7 DdRp, and RB69 DdDp revealed that protonic catalysis during phosphoryl transfer is a general feature of the polymerase single-nucleotide incorporation reaction mechanism (Castro *et al.*, 2007). Therefore, phosphoryl transfer by nucleic acid polymerases involves more than two-metal ion catalysis: an active site amino acid residue functions as a general acid, playing a direct role in covalent chemistry (Figure 6.2). The second proton transfer presumably activates the primer terminus 3′-OH for nucleophilic attack. The identity of the acceptor is not known and could be a conserved active site Asp or Glu, a structural water molecule or a non-bridging oxygen of the triphosphate moiety of the incoming nucleotide (Kiefer *et al.*, 1998; Florian *et al.*, 2003).

POLYMERASE FIVE-STEP KINETIC MECHANISM AND FIDELITY

Single-Nucleotide Incorporation Five-Step Kinetic Mechanism

The polymerase binds to a PT duplex, forming a binary complex, and a single-nucleotide incorporation cycle, consisting of five kinetically observable steps, ensues (Kuchta *et al.*, 1987; Patel *et al.*, 1991; Kati *et al.*, 1992) (Scheme 6.1). Binding to PT is weak for many polymerases, causing the rate of polymerase movement on and off the PT to be on a similar time-scale to subsequent nucleotide incorporation events, which complicates kinetic analysis (Capson *et al.*, 1992). In contrast, PV 3Dpol (Arnold and Cameron, 2004a) and certain other polymerases (Kati *et al.*, 1992), undergo a conformational change upon initial binding to PT, committing the enzyme to tight, long-term binding. For PV 3Dpol, the half-life of polymerase-PT binding is ~2h (Arnold and Cameron, 2000), causing this binding event to be irrelevant to subsequent nucleotide incorporation steps and greatly simplifying kinetic analysis (Arnold and Cameron, 2004a).

After formation of the polymerase–PT binary complex, the incoming nucleotide substrate, guided by the templating base, binds (step 1 in Scheme 6.1), forming a polymerase–PT–nucleotide ternary complex. This complex changes in conformation, via a set of physical movements of enzyme and substrates, to produce a catalytically competent ternary complex (step 2). Step 2 is often termed the "prechemistry conformational change." The complicated physical adjustments of step 2 provide the reactive group alignments necessary

$$ER_n + NTP \overset{\text{Step 1}}{\rightleftharpoons} ER_nNTP \overset{\text{Step 2}}{\rightleftharpoons} {}^*ER_nNTP \overset{\text{Step 3}}{\rightleftharpoons} {}^*ER_{n+1}PP_i \overset{\text{Step 4}}{\rightleftharpoons} ER_{n+1}PP_i \overset{\text{Step 5}}{\rightleftharpoons} ER_{n+1} + PP_i$$

SCHEME 6.1

for phosphoryl transfer to occur in step 3, in which covalent capture of a portion of the incoming nucleotide is completed. Step 3 is commonly termed "chemistry." Covalent chemistry is followed by a second complicated set of physical movements (step 4) that essentially undo the step 2 physical movements and also accomplish translocation of the polymerase to the next templating position. Finally, the removed portion of the just-added nucleotide, the pyrophosphate group, is released (step 5). This completes the catalytic cycle, resetting the polymerase–PT binary complex, with primer now lengthened by one nucleotide, in a conformation accessible to the next incoming nucleotide, and located at the next templating site (Kuchta *et al.*, 1987; Patel *et al.*, 1991; Kati *et al.*, 1992). The PV 3Dpol five-step kinetic mechanism is canonical; it uses the same mechanism for single-nucleotide incorporation as polymerases in other classes (Arnold and Cameron, 2004a). Therefore, insights gained from studying the 3Dpol nucleotide incorporation mechanism should be generalizable.

Polymerase Fidelity is Dictated by Biology

Fidelity (nucleotide substrate specificity) describes frequency of errors in nucleotide incorporation by a polymerase (Kunkel, 2004). The correct nucleotide substrate changes with each catalytic cycle, as the templating base changes (Johnson, 1993). Functionally tolerated mistakes are cumulative in the genome because the product from one replication cycle serves as the template for the next. Importantly, unlike other enzyme activities, in which there is presumably no fitness benefit to errors in specificity, a defined, consistently achieved error frequency is crucial to viral population fitness and evolution (Vignuzzi *et al.*, 2006). In spite of the many structural, thermodynamic, and kinetic hindrances to incorrect nucleotide incorporation, errors must, and do, occur at a defined, fitness-driven frequency. This is made possible because thermodynamics dictates that correct vs. incorrect nucleotide incorporations are not all-or-none but, rather, probability distributed. Therefore, polymerase fidelity is intimately and directly tied to biology, to the current and long-term fitness needs of a viral population (Crotty *et al.*, 2001; Vignuzzi *et al.*, 2006). Therefore, the product of a viral polymerase is not a population of genomes with a single sequence but a population of genomes with a precisely defined level of sequence diversity.

Estimating and Measuring Polymerase Fidelity

Polymerase fidelity is often estimated by frequency of appearance of phenotypically identifiable mutations, such as point mutation-induced emergence of guanidine resistance in picornaviruses in cell culture (Anderson-Sillman *et al.*, 1984; Baltera and Tershak, 1989), or by sequencing viral genomes. However, fidelity is an attribute of a polymerase, not of viral population phenotype change, nor of viral genome diversity. Therefore, indirect estimates of fidelity suffer much information loss. For example, error counts from sequencing reflect not only polymerase misincorporation events but also subsequent purging, for a variety of reasons, of genomes too defective to provide adequate function and so underestimate polymerase fidelity. However, while indirect estimates from phenotype change or sequencing fail to provide reliable absolute quantitative information about polymerase fidelity, such measures may provide valuable comparative information about the relative fidelity of different polymerase alleles (Pfeiffer and Kirkegaard, 2003; Arnold *et al.*, 2005).

The most direct approach to define the upper limit of intrinsic polymerase nucleotide incorporation error rate is by *in vitro* biochemical analysis (Johnson, 1993). The five-step kinetic mechanism shown in Scheme 6.1 can be collapsed to a simpler, minimal kinetic mechanism comprising only two steps, labeled $K_{d,app}$ and k_{pol} (Scheme 6.2), and these kinetic constants are combined to quantify polymerase fidelity:

$$\text{Fidelity} = [(k_{pol}/K_{d,app})_{correct} + (k_{pol}/K_{d,app})_{incorrect}] / [(k_{pol}/K_{d,app})_{incorrect}] \quad (1)$$

$$ER_n + NTP \underset{}{\overset{K_{d,app}}{\rightleftarrows}} ER_nNTP \xrightarrow{k_{pol}} ER_{n+1}PP_i$$

SCHEME 6.2

$K_{d,app}$ is the apparent dissociation constant for the nucleotide from the polymerase–PT–nucleotide ternary complex (Johnson, 1992). $K_{d,app}$ is therefore a measure of binding affinity of polymerase–PT binary complex for an incoming nucleotide. It is called "apparent" because binding affinity is measured indirectly, by kinetic means. The lower the $K_{d,app}$ value, the greater the affinity and the tighter the binding. Intuitively, the greater the affinity a binary complex has for a nucleotide, the "better" a substrate that nucleotide is. Binding affinity should therefore be stronger for a correct nucleotide than for an incorrect nucleotide. k_{pol}, the maximum observed rate constant for a single-nucleotide incorporation, is a measure of the maximum speed of a reaction (Johnson, 1992). Intuitively, a reaction involving a correct nucleotide should be faster than for an incorrect nucleotide. Combining the two kinetic constants to give an overall measure of substrate specificity, placing k_{pol} in the numerator and $K_{d,app}$ in the denominator of a quotient accomplishes this; a "better" (i.e. correct) nucleotide substrate should exhibit a relatively larger k_{pol} (numerator) and a relatively smaller $K_{d,app}$ (denominator), yielding a relatively larger quotient than for a "poorer" (i.e. incorrect) nucleotide substrate (Patel *et al.*, 1991; Johnson, 1993; Showalter *et al.*, 2006). To summarize, comparison of $k_{pol}/K_{d,app}$ for correct incorporation to $k_{pol}/K_{d,app}$ for incorrect incorporation provides a quantitative measure of fidelity in polymerase nucleotide incorporation assays.

BIOCHEMICAL ANALYSIS OF POLYMERASE FUNCTION

In Vitro Single-Turnover Polymerase Assays Reveal the Mechanistic Basis of Fidelity

Polymerases discriminate against misincorporation errors one nucleotide incorporation cycle at a time during genome assembly (Johnson, 1993). Consequently, the keys to understanding the mechanistic basis of polymerase fidelity are embedded within the five-step kinetic mechanism for single-nucleotide incorporation (Patel *et al.*, 1991). Therefore, to dissect out the mechanisms by which a polymerase discriminates between a correct and an incorrect nucleotide, the ability to assay a single polymerase turnover is needed. Steady-state polymerase assays do not provide information at the single-nucleotide incorporation level of resolution and do not reveal the primary mechanisms underlying fidelity (Johnson, 1992; Werneburg *et al.*, 1996).

Single-turnover polymerase assays provide rich details about the mechanistic basis of fidelity. Individual steps in the single-nucleotide incorporation cycle (Scheme 6.1) can be interrogated and the complete kinetic mechanism can be solved (Johnson, 1992). Measurements of individual steps for wild-type polymerase incorporation with a correct or incorrect nucleotide present and for fidelity mutant polymerases allow the mapping of nucleotide discrimination to conserved amino acid residues acting on specific events in the nucleotide incorporation cycle. Each individual nucleotide, differing in base or sugar configuration, can be examined one at a time, templated by a defined base, allowing identification of which functional groups on nucleotide base and sugar moieties are utilized by the polymerase to discern correct from incorrect. In addition, ambiguous nucleotide analogues, such as ribavirin, can be examined in a single-nucleotide incorporation cycle to gain further insights into the functional basis for nucleotide discrimination (Arnold *et al.*, 2005).

Requirements for *In Vitro* Single-Turnover Polymerase Assays

A prerequisite for examining a polymerase single-nucleotide incorporation cycle is the ability to assemble the polymerase onto a simple PT so that nucleotide addition to the primer terminus mimics that which occurs during *in vivo* primer elongation. A favorable experimental attribute of PV 3Dpol is that it exhibits simple requirements for productive binding to a PT. A 10-base self-annealing RNA oligonucleotide, termed "sym-sub" for "symmetrical-substrate," which produces two identical four-base single-stranded overhangs and a six-base-pair duplex region, serves as an efficient PT *in vitro* (Arnold and Cameron, 2000) (Figure 6.3A). 3Dpol can bind in proximity to either 3′-primer terminus and efficiently proceed with elongation when supplied with a nucleotide (Arnold and Cameron, 2000).

The most important external influence on ternary complex organization is use, in appropriate concentrations, of the biologically relevant metal ion co-factor which, in nearly all nucleic acid polymerases, is Mg^{2+} (Brautigam and Steitz, 1998; Patel and Loeb, 2001). Polymerase active sites have evolved to make use of the precise ionic radius, coordinating ability, and electrophilicity of Mg^{2+} to achieve the necessary alignments between enzyme and substrate functional groups for efficient, and accurate, phosphoryl transfer. In contrast, the presence of certain other metal cations, such as Mn^{2+}, distorts active site organization, leading to highly defective nucleotide discrimination (Tabor and Richardson, 1989; Brautigam and Steitz, 1998, Arnold *et al.*, 2004b).

Because single-nucleotide incorporation events occur on a millisecond time-scale, instrumentation designed to initiate and stop or monitor reactions very rapidly are required to examine single turnovers. Chemical quench flow and stopped flow devices provide access to these time-scales (Anderson, 2003; Patel *et al.*, 2003). A favorable attribute of PV 3Dpol is that it is only moderately fast in its nucleotide incorporation reactions (Arnold and Cameron, 2004a). Its rates of incorporation are therefore readily accessible to study using single-turnover kinetic instruments.

In principle, an *in vitro* polymerase single-turnover assay consists of (1) allowing enzyme to bind to PT, with 5′ terminus of primer radiolabeled, (2) initiating incorporation by adding nucleotide, (3) quenching the reaction at various time points until the end-point is approached, (4) fractionating radiolabeled,

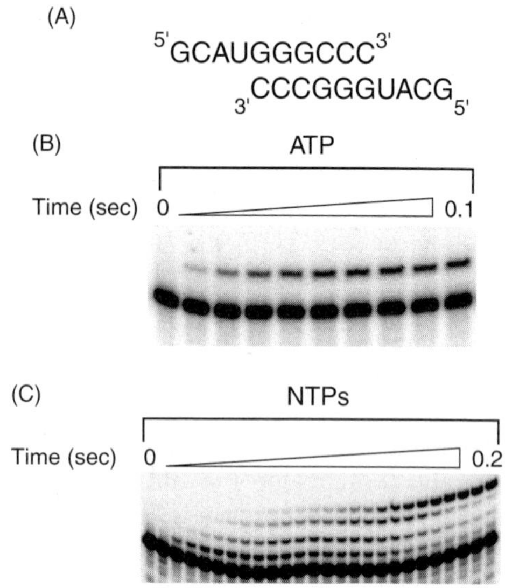

FIGURE 6.3 Use of a symmetrical primer/template (sym/sub) to study 3Dpol-catalyzed nucleotide incorporation. (A) sym/sub-U. (B) AMP incorporation into sym/sub-U. Reactions contained 500 μM ATP. (C) Multiple nucleotide incorporation into sym/sub-U. Reactions contained 500 μM NTPs. 2 μM 3Dpol was incubated with 2 μM sym/sub (1 μM duplex) and rapidly mixed with 500 μM nucleotide. Reactions were quenched by addition of EDTA to a final concentration of 0.3 M. Products were resolved by electrophoresis on a denaturing, highly crosslinked 23% polyacrylamide gel.

elongated primer on a polyacrylamide gel (Figure 6.3B,C) quantitating the amount of nucleotide added as a function of time so that a time-course for the reaction can be plotted and (6) analyzing the reaction time-course to estimate the kinetic parameters that reveal how efficiently the nucleotide was incorporated by the polymerase (Arnold and Cameron, 2004a, 2004b).

KINETICS AND THERMODYNAMICS OF POLYMERASE FIDELITY

How is Polymerase Fidelity Enforced?

A large body of experimental data points to steps 2 and 3 (Scheme 6.1) as the primary fidelity-governing steps (Joyce and Benkovic, 2004 and refs therein). Step 1, ground-state binding of the incoming nucleotide to the polymerase–PT binary complex, contributes little to fidelity (Johnson, 1993; Kuchta *et al.*, 1987; Wolfenden, 2003). Initial ground-state binding is driven by interactions between polymerase amino acid residues and the metal-complexed triphosphate moiety of the incoming nucleotide, which are the same for all nucleotides, correct or incorrect (Castro *et al.*, 2005). Steps 4 and 5 occur after covalent chemistry. In polymerases that lack intrinsic exonuclease activity, there is presumably little opportunity to influence the outcome of nucleotide incorporation in these steps. The largest free-energy change in the single correct nucleotide incorporation cycle of wild-type PV 3Dpol results from the conformational change after phosphoryl transfer (step 4) (Arnold and Cameron, 2004a). This renders correct incorporation essentially irreversible. However, intriguingly, step 4 may offer the possibility of proofreading via reversal of chemistry in the case of sugar misincorporation. For PV 3Dpol, the reverse reaction of chemistry, pyrophosphorolysis, is far more efficient when a nucleotide with an incorrect (deoxy) sugar has been incorporated and, therefore, postchemistry proofreading by pyrophosphorolysis may occur (Korneeva, 2007).

Whereas steps 4 and 5 appear to contribute little to single-nucleotide incorporation fidelity, they may influence processive, multiple-cycle nucleotide incorporation fidelity. Even if an incorrect nucleotide gets past the many barriers that decrease the likelihood of successful completion of chemistry for a single incorporation event, the need for processive incorporation by most polymerases presents additional opportunities to influence error frequency (Kuchta *et al.*, 1987; Zinnen *et al.*, 1994; Joyce and Benkovic, 2004).

While steps 2 and 3 appear to be the primary fidelity-controlling steps, a fundamental controversy that is not yet resolved is whether it is primarily step 2, the prechemistry conformational change, or step 3, phosphoryl transfer, that primarily governs polymerase nucleotide incorporation accuracy. Aspects of this controversy illuminate important basic principles underlying fidelity enforcement and so are instructive to review.

Joyce and Benkovic (2004) have carefully examined a wide range of studies involving DNA polymerase kinetics and fidelity and concluded that existing data do not permit a unified description of the kinetic and structural bases of fidelity. They find that part of the difficulty in arriving at a consensus view is that experimental data bearing upon whether or not the prechemistry conformational change (step 2) is rate-limiting (and, therefore, fidelity-governing) are difficult to interpret rigorously. Specifically, use of the magnitude of the phosphorothioate effect (Herschlag *et al.*, 1991), which compares the maximum rate constant for incorporation of a natural nucleotide to that for a nucleotide analogue in which the non-bridging oxygen of the α-phosphorus is replaced by sulfur, causing decreased reactivity and, hence, slower chemistry, is an ambiguous indicator of which steps are rate-limiting. While pointing out that polymerase fidelity is most fundamentally determined by the relative heights of activation energy barriers for correct vs. incorrect nucleotide incorporation, Joyce and Benkovic suggest that differential stability of intermediate species along the pathway for correct vs. incorrect incorporation (fidelity checkpoints) will define specific mechanistic stages at which nucleotide discrimination is accomplished, and that this may vary in different types of polymerases. They conclude that the notion of prechemistry fidelity checkpoints is thermodynamically sound and leave open the possibility that either step 2 or step 3 can serve as fidelity governor.

Prechemistry Conformational Change (Step 2) as Fidelity Regulator

Tsai and Johnson (2007) have described experiments using the replicative T7 DdDp, suggesting that step 2, the prechemistry conformational change step, is fundamentally where fidelity is physically controlled. Analyzing signals from a fluorophore attached to the fingers subdomain, which serves, in part, for recognition of correctness of an incoming nucleotide, these authors describe data indicating distinctly different physical events after initial nucleotide binding (step 1) but before phosphoryl transfer (step 3) for correct vs. incorrect nucleotide presence. They suggest the existence of two physically different conformational states stimulated by nucleotide binding, one for correct nucleotide occupancy and another for incorrect, and that the physical differences in these conformational states then determine the efficiency of successful covalent chemistry in step 3. They speculate that the efficiency of catalysis for an incorrect nucleotide is depressed by step 2 physical events that actively misalign catalytic site reactive groups, which then promotes rapid rejection of an incorrect nucleotide prior to phosphoryl transfer and/or slow, inefficient phosphoryl transfer.

A rate-limiting, and thus fidelity-controlling, prechemistry conformational change step has been reported in Klenow fragment DdDp (Kuchta *et al.*, 1987; Eger *et al.*, 1991) and HIV-1 RT RdDp (Zinnen *et al.*, 1994). Recent work with the DNA repair DdDp Pol β suggests the existence of conformational energy barriers after nucleotide binding but before phosphoryl transfer (Arora *et al.*, 2005). Rothwell *et al.* (2005) used fluorescence resonance energy transfer (FRET) to monitor motions of the family A polymerase Klentaq1 fingers subdomain after nucleotide binding. The most important observation from their work was that the open-to-closed conformational change affecting the position of the fingers subdomain occurred upon correct nucleotide binding but could not be detected upon incorrect nucleotide binding, consistent with the findings of Tsai and Johnson (2007).

Phosphoryl Transfer (Step 3) as Fidelity Regulator

Tsai and co-workers have reported data on Pol β (Dunlap and Tsai, 2002), Klenow fragment, and African swine fever virus DNA polymerase X (Bakhtina *et al.*, 2007) suggesting that step 3, phosphoryl transfer, dictates frequency of misincorporation. They describe experiments in which fluorescent signals from a template-located reporter reveal the prechemistry conformational change, step 2, to be much faster than step 3, phosphoryl transfer, and, therefore, incapable of influencing fidelity.

Both Prechemistry Conformational Change (Step 2) and Chemistry (Step 3) Regulate Fidelity

As alluded to above, an important, and useful, finding arising from solving the complete kinetic mechanism for single-nucleotide incorporation by PV 3Dpol was that the prechemistry conformational change (step 2) and phosphoryl transfer (step 3) are both partially rate-limiting (Arnold and Cameron, 2004a, 2004b). This valuable kinetic property renders both steps accessible to kinetic investigation for this RdRp.

As with other polymerases, in PV 3Dpol there is no difference in initial, ground-state binding (step 1) for correct or incorrect nucleotides, indicating that step 1 is not capable of contributing to fidelity (Arnold and Cameron, 2004a). This is thought to reflect the use of the triphosphate, which is the same for all nucleotides, correct or incorrect, for ground-state binding, rather than base or sugar configuration (Gohara *et al.*, 2004). However, fidelity enforcement does take place in both steps 2 and 3. The events that accomplish this are summarized in Figure 6.4. Initial nucleotide binding in step 1 stimulates reorientation of polymerase structural components (subdomain movements and amino acid residue backbone and side-chain adjustments) and triphosphate repositioning in step 2 to align reactive groups in preparation for phosphoryl transfer in step 3. Studies with PV 3Dpol mutants reveal that during step 2, amino acid residues in the nucleotide-binding pocket form non-covalent interactions with specific functional groups on base and sugar moieties of the bound nucleotide (Gohara *et al.*, 2000, 2004; Arnold *et al.*,

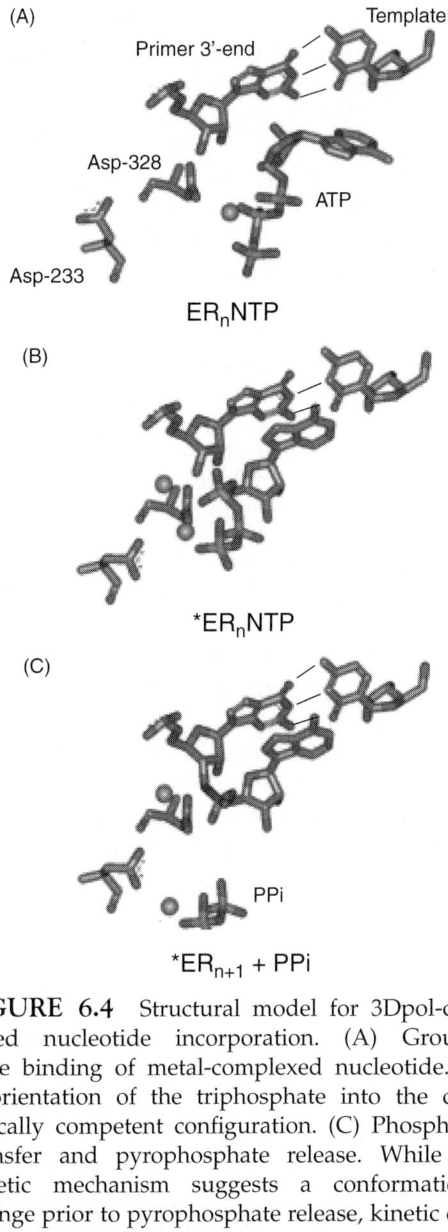

FIGURE 6.4 Structural model for 3Dpol-catalyzed nucleotide incorporation. (A) Ground-state binding of metal-complexed nucleotide. (B) Reorientation of the triphosphate into the catalytically competent configuration. (C) Phosphoryl transfer and pyrophosphate release. While the kinetic mechanism suggests a conformational change prior to pyrophosphate release, kinetic data does not provide any information to permit a molecular description of this step. Images were generated from the model previously described (Gohara *et al.*, 2000). Nucleotide and side-chain motions were derived from (Johnson *et al.*, 2003) by approximate rotation and translation movements. Atom colors correspond to the following: red, oxygen; blue; nitrogen; gray, carbon; magenta, Mg^{2+} or Mn^{2+}. The images were rendered with WebLab Viewer Pro (Accelrys Inc., San Diego, CA). (See Plate 5 for the color version of this figure.) Reproduced with permission from *Biochemistry* (Arnold *et al.*, 2004b).

2005; Korneeva and Cameron, 2007). Optimal interactions occur when a correct nucleotide is bound whereas defective interactions occur in the presence of an incorrect nucleotide, allowing discrimination to take place. Defective interactions with an incorrect nucleotide result in suboptimal alignments in the active site and thus destabilization of the activated ternary complex (Gohara *et al.*, 2000, 2004). The differential responses of step 2 to correct vs. incorrect nucleotide presence provide fidelity enforcement (Arnold and Cameron, 2004a). As a consequence of active site misalignments and inability to maintain the triphosphate in a catalytically competent conformation, efficiency of phosphoryl transfer in step 3 suffers substantially when a bound nucleotide is incorrect, providing further fidelity enforcement (Arnold and Cameron, 2004a). The net result of defects in the physical events of step 2 and extremely inefficient phosphoryl transfer in step 3 is a very low frequency of successful completion of the incorporation cycle when a bound nucleotide is incorrect.

Polymerase fidelity, then, is a direct result of higher activation energy barriers on the pathway to incorrect nucleotide incorporation than to correct incorporation, causing the former to be far less frequent in occurrence (Joyce and Benkovic, 2004; Castro *et al.*, 2005) (Figure 6.5). The more difficult thermodynamics for incorrect incorporation is, in turn, a direct result of defective interactions between polymerase amino acid residues and nucleotide base and sugar functional groups during step 2, which result in active site disorganization and instability of triphosphate positioning, followed by a low frequency of successful phosphoryl transfer in step 3. Therefore, fidelity enforcement is manifested in both step 2 and step 3 in PV 3Dpol (Arnold and Cameron, 2004a).

In summary, available evidence suggests that Step 2 "reads" correct or incorrect nucleotide presence and adjusts enzyme and substrate reactive group alignments for upcoming phosphoryl transfer. A "read" of correct, based on formation of optimal amino acid residue–nucleotide interactions permits step 2 to align all necessary ternary complex

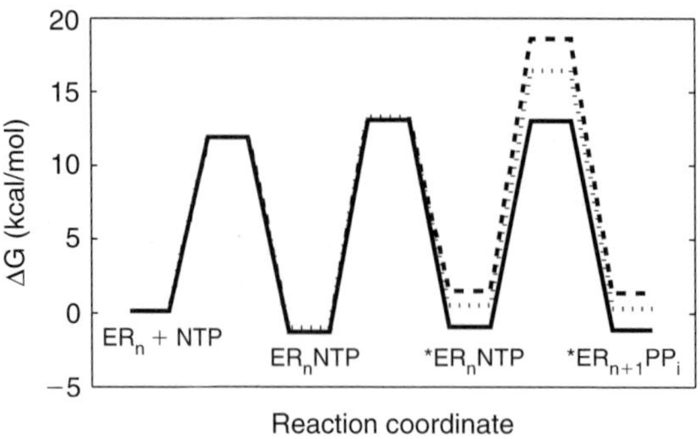

FIGURE 6.5 Comparison of the free energy profile for correct and incorrect 3Dpol-catalyzed nucleotide incorporation in the presence of Mg^{2+}. The free energy profile for correct and incorrect nucleotide incorporation are shown as follows: solid line for AMP incorporation, small dotted line for 2′-dAMP incorporation, and large dotted line for GMP incorporation. The concentrations of the substrates and products used were 2000 μM NTP and 20 μM PP_i. The free energy for each reaction step was calculated from $\Delta G = RT[\ln(kT/h) - \ln(k_{obs})]$, where R is 1.99 cal K^{-1} mol^{-1}, T is 303 K, k is 3.30×10^{-24} cal K^{-1}, h is 1.58×10^{-34} cal s and k_{obs} is the first-order rate constant. The free energy for each species was calculated from $\Delta G = RT[\ln(kT/h) - \ln(k_{obs,for})] - RT[\ln(kT/h) - \ln(k_{obs,rev})]$. Reproduced with permission from *Biochemistry* (Arnold *et al.*, 2004b).

components optimally. A "read" of incorrect, based on defective interactions between key fidelity-governing amino acid residues and bound nucleotide, results in suboptimal active site alignments. The alignments provided by step 2 result in the transition state stabilization achieved during phosphoryl transfer in step 3. Therefore, both step 2 and step 3 influence misincorporation frequency.

STRUCTURAL PERSPECTIVES ON THE SINGLE-NUCLEOTIDE ADDITION CYCLE AND FIDELITY

Identification of Nucleotide-Sensing Amino Acid Residues

Sequence alignments of animal virus RdRps reveal the presence of several absolutely conserved amino acid residues that can be mapped to the nucleotide-binding pocket (Koonin, 1991; Hansen *et al.*, 1997; Gohara *et al.*, 2000, 2004). Six of these interact with the nucleotide substrate (Figure 6.6). In order to determine the importance of these residues for nucleotide selection, PV 3Dpol derivatives were created in which some of these residues were changed to Ala (Gohara *et al.*, 2000, 2004). Analysis of these permitted the identification of essential amino acid residues and the interactions that are important for correct nucleotide selection.

As described above, step 1 is binding of the nucleotide in the ground state. In this ground-state configuration, the ribose cannot bind in a productive orientation because the interaction between Asp238 and Asn297 observed in the unliganded enzyme occludes the ribose-binding pocket (Gohara *et al.*, 2000, 2004) (Figure 6.6). A conformational change occurs that orients the triphosphate for phosphoryl transfer (step 2). This transition is partially rate-limiting for correct nucleotide incorporation (Arnold and Cameron, 2004a, 2004b). In addition, the stability of the complex in this conformation will dictate the efficiency of phosphoryl transfer as any misalignment of the triphosphate will produce either a suboptimal orientation or a suboptimal distance for catalysis (Arnold and Cameron, 2004a, 2004b; Gohara *et al.*, 2004). In order to maintain the triphosphate in the appropriate orientation, an extensive hydrogen-bonding network is involved that can be traced to residues in the ribose-binding pocket (Figure 6.7) (Gohara *et al.*, 2004). Formation of this network requires reorientation of Asp238 and Asn297 as well as interaction of the oxygen of the β-phosphate with the 3′-OH of the nucleotide

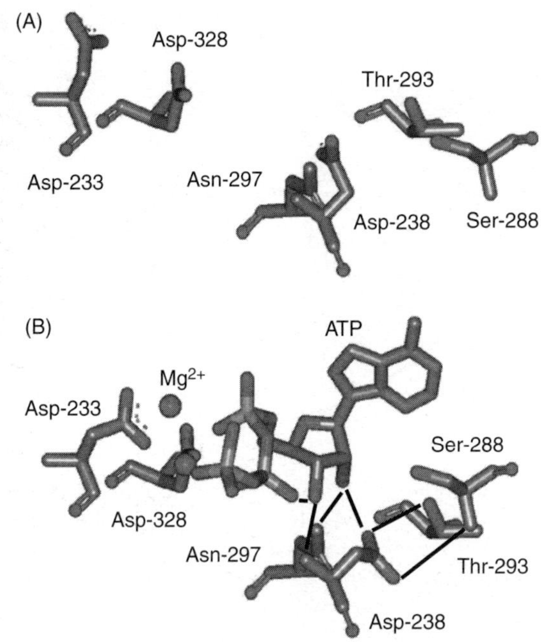

FIGURE 6.6 Nucleotide-binding pocket of 3Dpol. (A) Residues located in the NTP-binding pocket as observed in the unliganded structure of 3Dpol (Hansen *et al.*, 1997; Thompson and Peersen, 2004). Asp233 and Asp238 are from structural motif A; Ser288, Thr293, and Asn297 are from motif B; and Asp328 is from motif C. (B) Model for interaction of 3Dpol with bound nucleotide (Gohara *et al.*, 2000) ATP and metal ions required for catalysis are labeled. In this model, the side-chains for Asp233 and Asp238 have been rotated to permit interactions with ATP. Asp238, Ser288, and Thr293 have been positioned to interact. The image was created by using the program WebLab Viewer (Molecular Simulations Inc., San Diego, CA). (See Plate 6 for the color version of this figure.) Reproduced with permission from *Biochemistry* (Gohara *et al.*, 2004).

substrate (Gohara *et al.*, 2004). The position of the ribose is held firmly by interactions between the 3′-OH and the backbone of Asp238 and by interactions between the 2′-OH and Asn297 (Gohara *et al.*, 2004). Indeed, this set of interactions has been observed in sequential structures of FMDV polymerase that has undergone successive replication events in the crystal (Ferrer-Orta *et al.*, 2007). Thus, Asp238 is a key component in the line of communication between the ribose-binding pocket and the catalytic center that functions by modulating the conformation of the triphosphate moiety of the nucleotide substrate (Gohara *et al.*, 2004). The appropriate organization of this complex will permit binding and/or alignment of the second divalent cation co-factor, permitting phosphoryl transfer, translocation, and pyrophosphate release. Information on the nature of interactions in the ribose-binding pocket can be disseminated to the catalytic center by using the conformation of the Asp238 and the corresponding orientation of the triphosphate.

The orientation of the triphosphate moiety of the nucleotide substrate is fundamental for nucleotide incorporation not only for the RdRp but also for other polymerases (Beese *et al.*, 1993; Doublie *et al.*, 1998; Li *et al.*, 1998; Cheetham and Steitz, 1999; Yin and Steitz, 2002; Johnson *et al.*, 2003). Stabilization of the triphosphate conformation requires conserved structural motif A (Gohara *et al.*, 2004). Stabilization of the triphosphate–metal complex in the active conformation requires a network of hydrogen bonds provided mostly by the backbone of the residues in motif A (Gohara *et al.*, 2004). Therefore, any movement of the motif A side-chains located in the sugar-binding pocket will be transmitted through the rest of motif A, consequently perturbing the position of both the sugar and triphosphate and reducing the

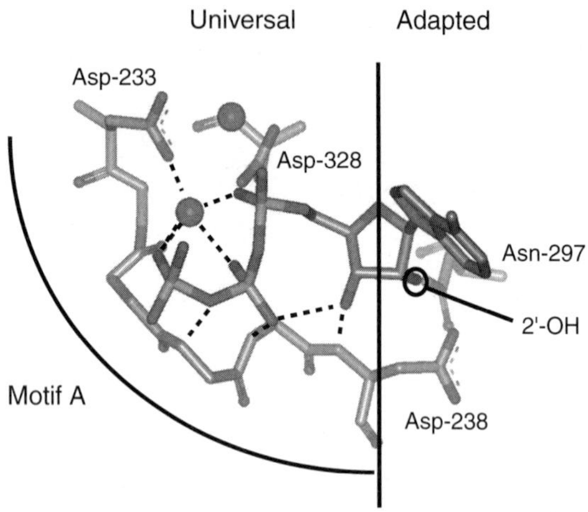

FIGURE 6.7 Structural basis for fidelity. The nucleotide-binding pocket of all nucleic acid polymerases with a canonical "palm"-based active site is highly conserved. The site can be divided into two parts: a region that has "universal" interactions mediated by conserved structural motif A that organize the metals and triphosphate for catalysis, and a region that has "adapted" interactions mediated by conserved structural motif B that dictate whether ribo- or 2′-deoxribonucleotides will be utilized. In the classical polymerase, there is a motif A residue located in the sugar-binding pocket capable of interacting with motif B residue(s) involved in sugar selection. This motif A residue in other polymerases could represent the link between the nature of the bound nucleotide (correct vs. incorrect) to the efficiency of nucleotidyl transfer as described herein for Asp238 of 3Dpol (Gohara *et al.*, 2000). (See Plate 7 for the color version of this figure.) Reproduced with permission from *Biochemistry* (Gohara *et al.*, 2004).

efficiency of phosphoryl transfer (Gohara *et al.*, 2004). The position of residues in motif A can be altered by either the base or the sugar of the nucleotide (Gohara *et al.*, 2004). Similar to Asn297 (motif B) for PV 3Dpol (Gohara *et al.*, 2004), T7 DdRp uses His784 (motif B) for hydrogen bonding to the 2′-OH of the NTP substrate (Brieba and Sousa, 2000). In HIV-RT RdDp, Phe160 (motif B) (Gutierrez-Rivas *et al.*, 1999) interacts with the 2′-OH of the 2′-dNTP substrate. Similarly, the presence of a motif B residue in DdDps will cause movement of motif A via the motif A residue located in the sugar-binding pocket (Beese *et al.*, 1993; Doublie *et al.*, 1998; Li *et al.*, 1998; Cheetham and Steitz, 1999; Yin and Steitz, 2002; Johnson *et al.*, 2003). Thus, in all polymerases, at least one residue in the conserved structural motif B has evolved to sense the presence of a 2′-OH as appropriate for the nucleotide substrate specificity of the enzyme (Gohara *et al.*, 2004).

Structural Basis for a Conserved Mechanism Linking Binding of the Correct Nucleotide to the Efficiency of Phosphoryl Transfer

The nucleotide-binding pocket of all nucleic acid polymerases with a canonical "palm"-based active site is highly conserved. The site can be divided into two parts: a region that has "universal" interactions mediated by conserved structural motif A that organizes the metals and triphosphate for catalysis and a region that has "adapted" interactions mediated by conserved structural motif B that dictate whether ribo- or 2′-deoxyribonucleotides will be utilized (Figure 6.7) (Gohara *et al.*, 2004). These two motifs intersect in the sugar-binding pocket, providing a mechanism for inappropriate base pairing and/or sugar configuration to be identified and cause the appropriate reduction in phosphoryl transfer efficiency by moving the triphosphate moiety of the nucleotide substrate into a suboptimal orientation (Gohara *et al.*, 2004).

POLYMERASE FIDELITY INFLUENCES VIRAL POPULATION FITNESS

Polymerase Fidelity Mutants

Although a fundamental intrinsic property of polymerases, fidelity acquires meaning only in the context of viral population fitness and, consequently, infectivity and pathogenesis. Polymerase fidelity mutants are therefore needed that permit the connections to be identified between polymerase structure and mechanism, fidelity modulation, viral genotypic diversity production and, finally, influences on viral population fitness. Many important fidelity-governing amino acid positions have been identified and studied in a range of model nucleic acid polymerases in different biochemical classes (Harris *et al.*, 1998; Gutierrez-Rivas *et al.*, 1999; Kim *et al.*, 1999; Minnick *et al.*, 1999; Brieba and Sousa, 2000; Yang *et al.*, 2005; Kim *et al.*, 2006; Zhang *et al.*, 2006; Loh *et al.*, 2007; Pursell *et al.*, 2007). However, because of the biological complexity of the system from which most of these polymerases originate, and/or because of lack of robust *in vivo* experimental systems, it has been difficult to assess the impact of polymerase fidelity on population biology.

The fact that the importance of fidelity occurs at the level of population fitness gives distinct advantages to studying viral systems because in both tissue culture and animal models rates of viral evolution are rapid and quantifiable. One of the fundamental features of virus evolution is that low polymerase fidelity permits rapid and thorough exploration of genotypic sequence space and, as a consequence, efficient exploitation of phenotypic opportunities (Domingo *et al.*, 1999). It is far more tractable to study fidelity in a population biology context for viruses than for bacteria or eukaryotes.

PV 3Dpol fidelity mutants have been identified that exhibit robust activity and defined, small (higher or lower) changes in fidelity and that have revealed the participation of amino acid residues both in direct contact with, and remote from, single functional groups on the bound nucleotide that communicate correct or incorrect, and therefore underlie fidelity (Figure 6.8). Knowledge of the complete kinetic mechanism for wild-type PV 3Dpol (Arnold and Cameron, 2004a) serves as an invaluable baseline to evaluate mechanistic changes observed in 3Dpol fidelity mutants. It is the contrasts between PV wild-type and fidelity mutant polymerases that provide insight into the mechanistic bases of fidelity enforcement. Importantly, the PV fidelity mutant polymerases are testable for influences on viral population fitness in well-developed cell culture and animal model systems (Arnold *et al.*, 2005; Vignuzzi *et al.*, 2006).

G64S 3Dpol Exhibits Enhanced Fidelity and Confers an Antimutator Phenotype

G64S 3Dpol is a high-fidelity polymerase with an antimutator phenotype. This polymerase was obtained by serial passage of poliovirus in the presence of the nucleoside analogue ribavirin (Pfeiffer and Kirkegaard, 2003; Arnold *et al.*, 2005) which ambiguously templates both A and G upon incorporation into a viral genome. A ribavirin-resistant virus emerged that harbored a 3Dpol variant with a single Gly-to-Ser substitution at position 64. It was subsequently shown that G64S 3Dpol discriminated more stringently against incorrect nucleotide incorporation than wild-type 3Dpol and that the phenotype of ribavirin resistance resulted from a lower frequency of incorporation of ribavirin and, hence, reduced production of functionally-defective viral genomes and, consequently, retention of viral population vigor (Pfeiffer and Kirkegaard, 2003; Arnold *et al.*, 2005). The fact that two labs independently obtained the G64S high fidelity 3Dpol as a result of imposing ribavirin selection on PV-infected cell cultures suggests that there may be few amino acid changes readily available for natural selection to increase polymerase fidelity.

(A)

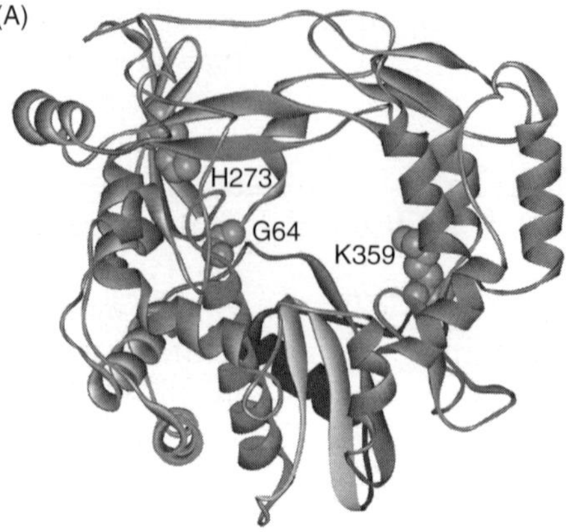

(B)

Allele	Mutation Frequency		Replication Rate[3]
	Sequencing[1]	Kinetics[2]	
WT	1.9	1/6,000	$90 \pm 5\ s^{-1}$
G64S	0.5	1/8,600	$30 \pm 5\ s^{-1}$
H273R	3.0	1/4,000	$160 \pm 10\ s^{-1}$
K359L	nd[4]	1/450,000	$0.50 \pm 0.05\ s^{-1}$
K359H	nd	1/13,500	$5.0 \pm 0.5\ s^{-1}$
K359R	nd	1/9,000	$5.0 \pm 0.5\ s^{-1}$

[1]The calculated average number of mutations per genome based upon sequencing 36,000 nucleotides of capsid coding sequence from 18 viral isolates.
[2]The calculated transition mutation frequency based upon the ratio of the kinetic parameters for correct and incorrect nucleotide incorporation by the PV RdRp allele.
[3]The maximal observed rate constant, kpol, for correct nucleotide incorporation.
[4]Not determined.

FIGURE 6.8 Sites in PV RdRp controlling rates of mutation and replication. (A) The structure of PV polymerase showing the sites of all mutations G64, H273, K359. (B) Mutation frequency and replication rates for PV polymerase alleles. (See Plate 8 for the color version of this figure.)

Position 64 is located in the fingers subdomain, not in direct proximity to the catalytic site (Figure 6.8A). However, it is physically and functionally linked by a hydrogen bonding network to fidelity-influencing residues at the active site (Arnold *et al.*, 2005) (Figure 6.9). Substitution at position 64 away from Gly is believed to cause defects in orientation of the triphosphate moiety of the incoming nucleotide and, consequently, to efficiency of phosphoryl transfer owing to misalignment of conserved structural motif A of the palm subdomain. G64S 3Dpol therefore demonstrates that sites remote from the catalytic center can alter fidelity and illustrates the principle of functional connectivity between spatially distinct, fidelity-governing amino acid residues.

Step 2, (Scheme 6.1) the prechemistry conformational change, is affected by the G64S mutation (Arnold *et al.*, 2005). The Gly-to-Ser change decreases the equilibrium constant across step 2 by three-fold, leading to destabilization of the catalytically competent ternary complex. The result of greater difficulty in successfully traversing step 2 is increased fidelity relative to wild-type. The G64S mutation thus reveals the use of triphosphate reorientation and stability of the isomerized triphosphate in step 2 as a fidelity checkpoint.

The increased G64S 3Dpol fidelity leads to PV populations in cell culture having four-fold fewer mutations per genome than wild-type PV populations (Figure 6.8) (Arnold *et al.*, 2005). In other words, the G64S 3Dpol

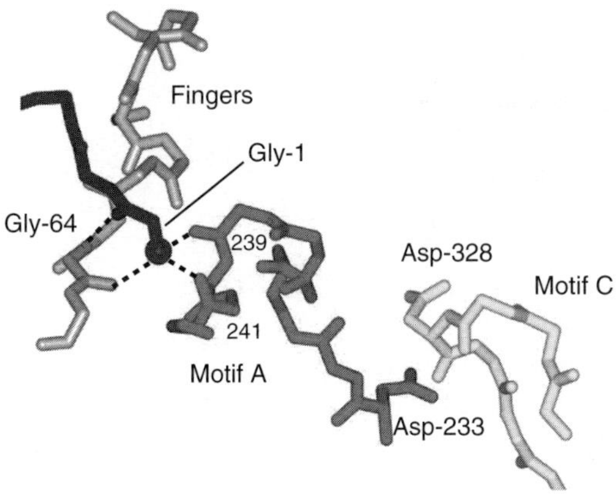

FIGURE 6.9 A link between the fingers subdomain and conserved structural motif A of the palm subdomain of PV 3Dpol. Gly64 backbone orients the N-terminus, and the N-terminus interacts with motif A. The N-terminus of 3Dpol is in blue, Gly64 is in orange, motif A is in red, motif C is in yellow and hydrogen bonds are shown as dashed lines. Misalignment of position 64 will cause defects to the orientation of the triphosphate as well as the efficiency of phosphoryl transfer owing to misalignment of motif A. (See Plate 9 for the color version of this figure.) Reproduced with permission from the *Journal of Biological Chemistry* (Arnold *et al.*, 2005).

produces a less diverse quasispecies than wild-type. The decreased genotypic diversity of the population manifests as an antimutator phenotype in cell culture (Pfeiffer and Kirkegaard, 2003; Arnold *et al.*, 2005), which is revealed in several ways. As mentioned, the virus exhibits decreased sensitivity to ribavirin-induced mutagenesis. Conversely, it suffers from enhanced sensitivity to other classes of antiviral compounds (such as the WIN compounds that bind to mature virus capsids and inhibit uncoating) due to inadequate genotypic diversity to produce resistance. In addition, there is a decrease in frequency of guanidine resistance emergence in the G64S PV populations relative to wild-type.

In contrast to the obvious loss of fitness by G64S PV relative to wild-type PV in the presence of inhibitors, G64S PV exhibits the same fitness as wild-type in cell culture in the absence of inhibitors (Arnold *et al.*, 2005) (Figure 6.10). However, competition experiments with wild-type PV reveal decreased G64S PV population fitness in the absence of inhibitors (Arnold *et al.*, 2005). When co-infected with wild-type PV in cell culture, G64S PV is rapidly out-competed and replaced by wild-type virus. The reduced viral population fitness of G64S PV is manifested in mice as restricted tissue tropism and failure to establish infection and replicate effectively in the primary pathogenic sites of wild-type poliovirus replication, spinal cord and brain (Figure 6.10) (Pfeiffer and Kirkegaard, 2005; Vignuzzi *et al.*, 2006). Artificial expansion of genomic diversity by serial passage in the presence of ribavirin completely restores pathogenicity and tissue tropism to wild-type levels in mice (Vignuzzi *et al.*, 2006).

Related to observations of the increased fidelity PV G64S variant are findings in wild-type coxsackievirus B3. This virus exhibits more sensitivity to inhibition by ribavirin in tissue culture than poliovirus (Figure 6.11A) (Graci, 2007; Graci *et al.*, 2007). However this finding suggests that coxsackievirus B3 populations require, and tolerate, less genotypic diversity than poliovirus populations which, in turn, implies that the coxsackievirus B3 polymerase should demonstrate higher fidelity than poliovirus polymerase. Significantly, this was found to be the case (Figure 6.11B) (Graci, 2007). Coxsackievirus B3 therefore appears to restrict its genome sequence space by demanding higher fidelity from its polymerase.

H273R 3Dpol Exhibits Decreased Fidelity and Confers a Mutator Phenotype*

H273R is a low-fidelity remote-site 3Dpol variant. Position 273 is located in the hinge region of the fingers subdomain, approximately 20 Å from the active site (Figure 6.8). Position 273 is linked by hydrogen bonding to active site residues and thus affects the hydrogen-bond

*This section is entirely Korneeva, 2007.

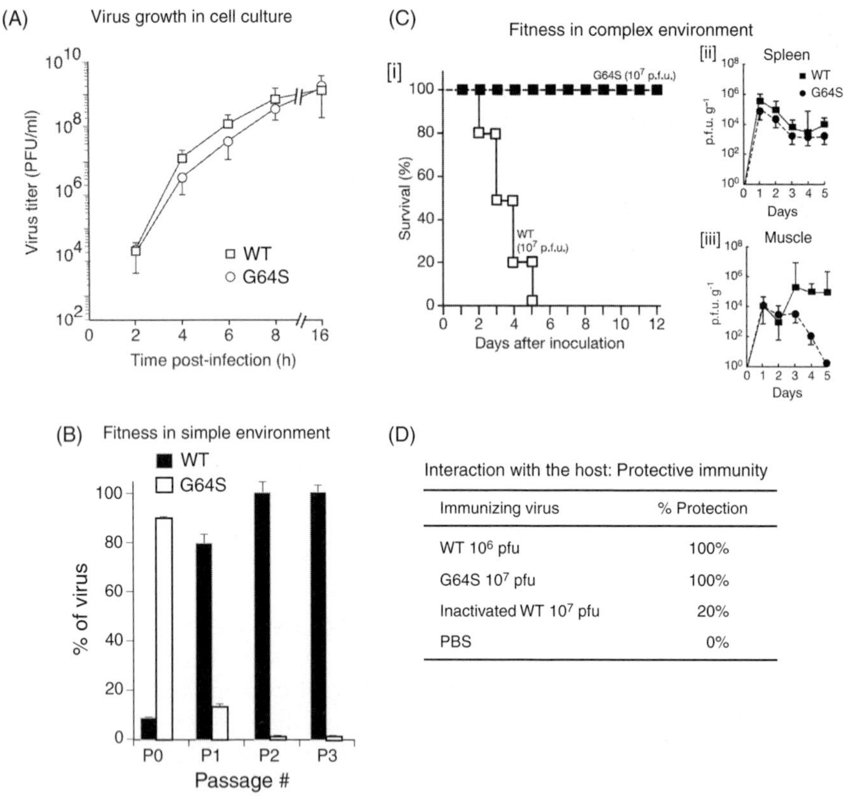

FIGURE 6.10 The fitness of wild-type and G64S PV varies in different environments. (A) Wild-type and G64S replicate equally in cell culture. One-step growth curves for wild-type and G64S PV. (B) wild-type PV out-competes G64S PV. Percentage of wild-type PV (black bars) and G64S PV (white bars) remaining after 0–3 serial passages. The initial virus mixture contained a ratio of wild-type PV:G64S PV of 1:10. (C) Genomic diversity in viral population is critical for pathogenesis and viral tissue tropism. (i) Percentage of mice surviving intramuscular injection with G64S and wild-type PV (10^7 pfu); $n = 20$ mice per group. Virus titers in pfu per gram from either (ii) spleen or (iii) muscle of mice infected intravenously with the wild-type or G64S PV. (D) Adaptive immunity can be elicited by different PV alleles. Mice ($n = 5$) were immunized with the indicated doses (pfu) 4 weeks prior to challenge with $5LD_{50}$ of wild-type PV by intraperitoneal injections. Wild-type PV was inactivated by 2 h UV treatment. Reproduced with permission from *Nature* (panels A–C) (Vignuzzi *et al.*, 2006).

network connecting the fingers subdomain with the active site and the ribose-binding pocket. Position 273 is in the vicinity of position 64 (see above). Thus, changing His273 to Arg may affect the H-bond network that stabilizes the fingers subdomain, functionally linking it to the polymerase active site. An Arg273 3Dpol crystal structure was almost identical to the His273 wild-type crystal structure, suggesting the possibility that the H273R mutation affects mostly dynamics.

The H273R substitution increases the equilibrium constant across step 2, the prechemistry conformational change step, leading to decreased fidelity relative to wild-type. Thus, increased ease in traversing step 2 of the single-nucleotide incorporation cycle results in reduced stringency in correct vs. incorrect nucleotide discrimination and, as a consequence, decreased fidelity in H273R 3Dpol.

The decreased fidelity of H273R 3Dpol gives rise to a mutator phenotype in cell culture, which manifests as excessive population genotypic diversity and reduced fitness. H273R 3Dpol is more susceptible to inhibition by ribavirin, due to acceleration of onset of lethal mutagenesis, showing up to a 100-fold decrease in H273R PV titer relative to wild-type at high

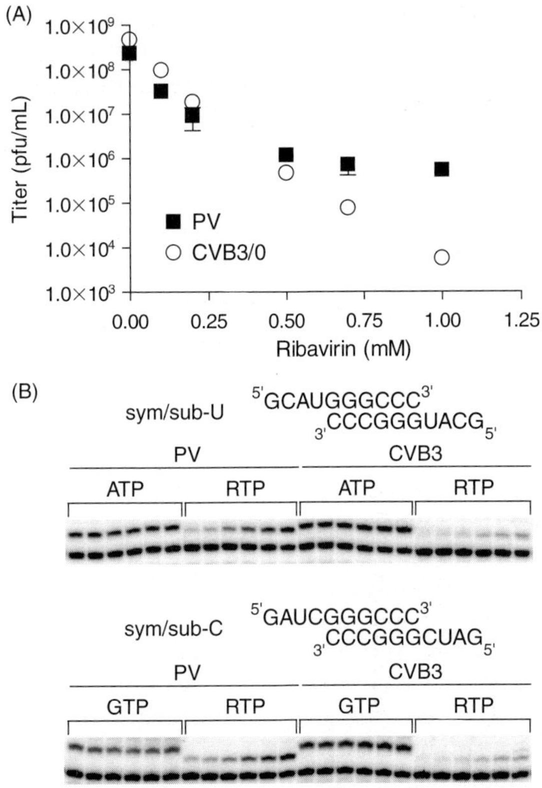

FIGURE 6.11 Coxsackievirus B3 polymerase is more faithful due to sequence limitations in the viral genome. (A) CVB3 is more susceptible to ribavirin than PV. Titer of surviving virus (pfu/mL) at 20 h post-infection for PV (black squares) and CVB3 (open circles) in the presence of increasing concentrations of ribavirin. (B) CVB3 3Dpol incorporates ribavirin less efficiently than PV 3Dpol. AMP and RMP incorporation into sym/sub-U and GMP and RMP incorporation into sym/sub-C. While the kinetics of correct nucleotide incorporation (AMP and GMP) are similar between CVB3 and PV 3Dpol, ribavirin incorporation by CVB3 3Dpol is less efficient, indicative of a more faithful polymerase for CVB3 3Dpol. CVB3 RNA is more susceptible to increased mutation. Relative specific infectivity (pfu/μg RNA) of PV (black squares) and CVB3 (open circles) RNA as a function of the number of PMP incorporations per genome. RNA transcripts were generated by *in vitro* transcription in the presence of nucleoside analogue P to generate RNA transcripts with varying amounts of PMP incorporations per genome. The number of incorporations per genome was determined by digestion of transcript RNA and quantified by HPLC. RNA transcripts were transfected into HeLa cells and the titer of virus determined at 20 h post-transfection.

ribavirin concentrations. In addition, H273R exhibits a three-fold increase in appearance of guanidine resistance. In direct sequencing of viral genomes, H273R PV had 1.6-fold more mutations per genome than wild-type (Figure 6.8). H273R 3Dpol thus produces a more diverse quasispecies than wild-type.

In the absence of selection, H273R PV functions indistinguishably from wild-type in cell culture. No differences were noted in infectious center or one-step growth curve assays. Furthermore, no defects in viral RNA synthesis were detected in subgenomic replicon assays. However, when placed in competition with wild-type in the absence of selection, decreased fitness of H273R PV was revealed, being rapidly displaced by wild-type in early serial passages.

Even though indistinguishable from wild-type PV in tissue culture, H273R PV shows a highly attenuated phenotype in mice. H273R PV is much less neuropathogenic than wild-type. All mice survive infection and there is restricted tissue tropism; no virus is found in spinal cord or brain. H273R PV was even more attenuated in mice than G64S PV (see above).

An intriguing additional fitness defect observed with H273R PV is production of more empty viral particles than wild-type, suggesting a possible link between excessive (lethal) genome diversity and aborted RNA packaging.

Fidelity, Genotypic Diversity and Viral Population Fitness

Wild-type, high-fidelity G64S and low-fidelity H273R PV 3Dpols represent an exceptional experimental system for understanding the connections between polymerase fidelity, viral genomic mutation rates, and virus population fitness. It is interesting that the amino acid substitutions leading to both higher (G64S) and lower (H273R) fidelity 3Dpols caused changes in step 2, prechemistry conformational change, events, underscoring the importance of these polymerase structural adjustments to control of misincorporation frequency.

Important insights emerge from studying the poliovirus fidelity variants. Optimal population fitness requires that genotypic diversity remains within a narrow range. Small decreases or increases in mutation rate result in reduced viral population fitness. Both fidelity mutant viruses were as robust as wild-type in cell culture in the absence of selection when infected individually but each was readily outcompeted by wild-type under conditions of co-infection. Both fidelity mutant viruses exhibited severely attenuated infectivity and pathogenesis in mice, a more diverse and challenging host environment than cell culture. These observations combine to indicate that viral genotypic diversity must remain within a narrow range to retain phenotypic vigor and point toward the important conclusions that mutation rate, and therefore polymerase fidelity, are tuned by natural selection to optimize population fitness and that population genotypic diversity is the mediator between polymerase fidelity and population fitness. If virus evolution has produced optimal polymerase fidelity and mutation rates, then even small decreases in mutation rate, as in G64S PV (Pfeiffer and Kirkegaard, 2003, 2005; Arnold *et al.*, 2005; Vignuzzi *et al.*, 2006), or increases in mutation rate, as in H273R PV, should result in fitness losses (Korneeva, 2007).

A recurring theme in this chapter is that polymerase fidelity is most fundamentally a viral population fitness-controlling parameter. Viral population fitness needs drive, via the level of population genotypic diversity that is optimal, a tuned, optimized level of polymerase fidelity (Vignuzzi *et al.*, 2006).

CONCLUSIONS AND FUTURE DIRECTIONS

Polymerase Fidelity is Tuned by Natural Selection to Achieve Optimal Population Genotypic Diversity

A critical attribute of polymerase fidelity is that it is modifiable by natural selection according to population fitness needs. The fidelity of a polymerase reflects its specific nucleic acid task and is tuned to serve the adaptive needs of the virus (Kunkel, 2004). Fidelity does not tend to be maximal but, rather, optimal (Crotty *et al.*, 2001; Joyce and Benkovic, 2004; Vignuzzi *et al.*, 2006). In the course of virus infection, the need for virus population phenotypic diversity, and therefore population genotypic diversity, is dynamic. The level of diversity and type of diversity needed may change in different tissues as different host challenges are encountered. Natural selection tunes fidelity to best serve the fitness needs of a viral population. A viral population with a lower tolerance for genotypic diversity will require a higher fidelity polymerase to achieve this. Conversely, a viral population with a higher requirement for genome change will require a lower fidelity polymerase.

Natural selection identifies an optimal fidelity level for viral population fitness and alters activation energy barriers in the catalytic cycle of a polymerase to achieve that level of fidelity (Arnold and Cameron, 2004a; Arnold *et al.*, 2005). The raw material natural selection has to work with to modify polymerase fidelity is amino acid substitution; the effector of amino acid substitution is change in activation energy barrier height; and the arbiter of appropriate differential activation energy barrier heights for correct vs. incorrect nucleotide incorporation is viral population fitness.

The process of virus infection is a competition between generation of adequate (but not excessive) viral population genotypic diversity and host cell responses on a specific time scale. Therefore, the frequency of nucleotide misincorporation by viral polymerases is highly evolved, finely tuned and completely dictated by the biological needs of the virus population as it goes through an infection cycle in its host (Vignuzzi *et al.*, 2006). Too little genotypic variability generated on the time scale of critical virus–host cell interactions (fidelity too high) means the virus will fail to acquire adequate phenotypes to cope with the dynamic challenges presented by the host. Too much genotypic variability generated on the virus-host interaction time scale (fidelity too low) will produce excessive numbers of defective viral genomes that contribute nothing to viral population fitness and function only to retard the progress of virus infection, tipping balances in favor of host responses.

Paradigm Shift: Targeting Viral Population Fitness Vulnerabilities for Treatment and Prevention of Viral Infections

A conceptual shift that is emerging from the sum of many recent studies is that an important target for antiviral therapy is viral population fitness. A virus is most accurately viewed as a population of genotypically, and thus phenotypically, diverse members (Domingo *et al.*, 1999; Biebricher and Eigen, 2006; Bull *et al.*, 2007). Furthermore, it is clear that adequate viral population robustness to succeed in infection and pathogenesis in a challenging host environment utterly depends upon genotypic diversity remaining within a narrow range (Pfeiffer and Kirkegaard, 2005; Vignuzzi *et al.*, 2006; Korneeva, 2007). The findings of Crotty *et al.* (2001) demonstrate, for example, that poliovirus populations exist at the edge of genotypic diversity tolerance. Small reductions in genotypic diversity result in defective adaptability whereas increases in genotypic diversity result in genome dysfunction. This chapter has emphasized that polymerase fidelity is the primary controller of optimal viral population genotypic diversity. Viral population vigor is extremely sensitive to amount of genotypic variation generated and thus to polymerase fidelity. In essence, the stringent requirements of viral populations for defined levels of genotypic diversity create a dangerous vulnerability to therapeutic modulation of polymerase fidelity (Castro *et al.*, 2005). As described, wild-type 3Dpol fidelity appears exquisitely tuned to maximize population fitness through amount of genotypic diversity produced, and even small, several-fold increases or decreases in fidelity lead to dire consequences for viral infectivity and pathogenesis (Crotty *et al.*, 2001; Pfeiffer and Kirkegaard, 2003; Arnold *et al.*, 2005; Korneeva, 2007).

Data from PV G64S 3Dpol studies indicate that increasing fidelity decreases viral fitness and imply that it may be possible to target small molecules to remote polymerase sites, causing interference with nucleotide incorporation at the active site, suggesting a new class of viral inhibitors (Arnold *et al.*, 2005). PV H273R 3Dpol data reveal that decreasing polymerase fidelity undermines viral infectivity and pathogenesis and also indicates the extreme sensitivity of population fitness to altered genotypic diversity. Experiments with this fidelity mutant suggest that therapeutic reduction of polymerase fidelity holds promise as an antiviral approach. Interestingly, PV mutants with altered fidelity, while attenuated in mice, provide a means of defense against lethal challenge by wild-type virus through protective immunity (Vignuzzi, Cameron and Andino, unpublished observation) (Figure 6.10D). An additional intriguing approach for development of protective immunity involves use as live vaccine of virus variants having the polymerase active site general acid (Figure 6.2) changed to a less efficient proton donor (Lys changed to Arg, His or Leu in the case of PV 3Dpol) (Figure 6.8) (Cameron, unpublished observation). This active site residue, because of its essential function as a proton donor during phosphoryl transfer, is highly conserved and, therefore, available for mutation and vaccine development in any virus. In addition, loss of efficient protonic catalysis during phosphoryl transfer produces a virus that replicates too slowly to mount a successful infection, yet is authentic in every feature presented to the host immune system and so elicits a robust immune response for protection against subsequent wild-type virus infection (Cameron, unpublished observation).

In total, these data indicate that an understanding of viral polymerase mechanism and fidelity is fundamental to development of antiviral therapies and that there is a direct link between polymerase fidelity, viral population fitness, and antiviral therapy opportunities. Polymerase fidelity determines the amount of genotypic diversity that develops in a virus population within a host and, therefore, the fate of that population.

Polymerase Dynamics: The Key to a Complete Understanding of Polymerase Fidelity

The major current barrier to more fully understanding nucleic acid polymerase mechanism and, consequently, fidelity is the almost complete lack of data revealing the nature of polymerase conformational movements and dynamics. One of the most fundamental functional attributes of protein enzymes is that they are completely reliant on defined motions on different time scales, chosen by natural selection, to accomplish catalysis (Hammes, 2002; Benkovic and Hammes-Schiffer, 2003; Hammes-Schiffer and Benkovic, 2006). For example, essentially all of the kinetic mechanism step 2 prechemistry conformational change events are motional in nature. In stark contrast to this, nearly all current polymerase structural data are static. While x-ray crystal structures have provided, and continue to provide, many critical insights into polymerase function, their severe limitation is that they provide motionless snapshots of events that are, by definition, totally reliant on continuous movement. Therefore, experimental and computational approaches providing insight into the influences of the motions and dynamics of nucleic acid polymerase molecules on the nucleotide incorporation cycle will be extremely fruitful in advancing our understanding of mechanism and fidelity.

Benkovic and co-workers have demonstrated in NMR experiments that enzyme dynamical motions are important in catalysis (Epstein *et al.*, 1995; Cameron and Benkovic, 1997). In light of these findings, it is likely that fidelity will ultimately be best understood as a phenomenon governed heavily by polymerase dynamical motions and that the continuous dynamical motions of a polymerase evolve to sample correct conformations (Eisenmesser *et al.*, 2005) and, simultaneously, to disfavor incorrect conformations. A prediction, then, is that the evolved dynamical motions of polymerases more efficiently sample conformational changes for correct nucleotide incorporation than for incorrect and therefore that dynamical motions contribute to fidelity.

The size of relatively small, simple polymerases (for example, 52 kDa for poliovirus polymerase) is now appropriate for high-resolution NMR solution structure determination. (Boehr *et al.*, 2006; Mittermaier and Kay, 2006; Foster *et al.*, 2007). In parallel, computational methods such as molecular dynamics which are capable of allowing accurate simulation of protein dynamical motions are advancing powerfully (Florian *et al.*, 2005). Additionally, technologies permitting examination of single-molecule motions (rather than ensemble behavior) are already beginning to provide new insights (Smiley and Hammes, 2006). The combination of these and other technical advances promises to provide views of real-time, functional motions of nucleic acid polymerases which will lead to improvements in understanding fidelity far surpassing the current level.

ACKNOWLEDGMENTS

Our studies of polymerase structure, function and mechanism are funded by grant AI45818 from NIAID, NIH to CEC.

REFERENCES

Anderson, K.A. (2003) Detection and characterization of enzyme intermediates: utility of rapid chemical quench methodology and single enzyme turnover experiments. In: *Kinetic Analysis of Macromolecules* (K.A. Johnson, ed.), pp. 19–47. New York: Oxford University Press.

Anderson-Sillman, K., Bartal, S. and Tershak, D.R. (1984) Guanidine-resistant poliovirus mutants produce modified 37-kilodalton proteins. *J. Virol.* **50**, 922–928.

Arnold, J.J. and Cameron, C.E. (2000) Poliovirus RNA-dependent RNA Polymerase (3Dpol.) Assembly of stable, elongation-competent complexes by using a symmetrical primer-template substrate (sym/sub). *J. Biol. Chem.* **275**, 5329–5339.

Arnold, J.J. and Cameron, C.E. (2004a) Poliovirus RNA-dependent RNA polymerase (3Dpol): pre-steady-state kinetic analysis of ribonucleotide incorporation in the presence of Mg^{2+}. *Biochemistry* **42**, 5126–5137.

Arnold, J.J., Gohara, D.W. and Cameron, C.E. (2004b) Poliovirus RNA-dependent RNA polymerase (3Dpol); pre-steady-state kinetic analysis of ribonucleotide incorporation in the presence of Mn^{2+}. *Biochemistry* **42**, 5138–5748.

Arnold, J.J., Vignuzzi, M., Stone, J.K., Andino, R. and Cameron, C.E. (2005) Remote site control of an active site fidelity checkpoint in a viral RNA-dependent RNA polymerase. *J. Biol. Chem.* **280**, 25706–25716.

Arora, K., Beard, W.A., Wilson, S.H. and Schlick, T. (2005) Mismatch-induced conformational distortions in polymerase β support an induced-fit mechanism for fidelity. *Biochemistry* **44**, 13328–13341.

Bakhtina, M., Roettger, M.P., Kumar, S. and Tsai, M-D. (2007) A unified kinetic mechanism applicable to multiple DNA polymerases. *Biochemistry* **46**, 5463–5472.

Baltera, R.F., Jr. and Tershak, D.R. (1989) Guanidine-resistant mutants of poliovirus have distinct mutations in peptide 2C. *J. Virol.* **63**, 4441–4444.

Beard, W.A. and Wilson, S.H. (2006) Structure and mechanism of DNA pol β. *Chem. Rev.* **106**, 361–382.

Bebenek, A., Carver, G.T., Kadyrov, F.A., Kissling, G.E. and Drake, J.W. (2005) Processivity clamp gp45 and ssDNA-binding-protein gp32 modulate the fidelity of bacteriophage RB69 DNA polymerase in a sequence-specific manner, sometimes enhancing and sometimes compromising accuracy. *Genetics* **169**, 1815–1824.

Beese, L.S., Friedman, J.M. and Steitz, T.A. (1993) Crystal structures of the Klenow fragment of DNA polymerase I complexed with deoxynucleotide triphosphate and pyrophosphate. *Biochemistry* **32**, 14095–14101.

Benkovic, S.J. and Hammes-Schiffer, S. (2003) A perspective on enzyme catalysis. *Science* **301**, 1196–1202.

Biebricher, C.K. and Eigen, M. (2006) What is a quasispecies? In: *Quasispecies: Concepts and Implications for Virology* (E. Domingo, ed.), pp. 1–31. Berlin: Springer.

Boehr, D.D., Dyson, H.J. and Wright, P.E. (2006) An NMR perspective on enzyme dynamics. *Chem Rev.* **106**, 3055–3079.

Brautigam, C.A. and Steitz, T.A. (1998) Structural and functional insights provided by crystal structures of DNA polymerases and their substrate complexes. *Curr. Opin. Struct. Biol.* **8**, 54–63.

Brieba, L.G. and Sousa, R. (2000) Roles of histidine 784 and tyrosine 639 in ribose discrimination by T7 RNA polymerase. *Biochemistry* **39**, 919–923.

Bull, J.J., Sanjuan, R. and Wilke, C.O. (2007) Theory of lethal mutagenesis for viruses. *J. Virol.* **81**, 2930–2939.

Cameron, C.E. and Benkovic, S.J. (1997) Evidence for a functional role of the dynamics of glycine-121 of Escherichia coli dihydrofolate reductase obtained from kinetic analysis of a site-directed mutant. *Biochemistry* **36**, 15792–15800.

Cameron, C.E., Gohara, D.W. and Arnold, J.J. (2002) Poliovirus RNA-dependent RNA polymerase (3Dpol): structure, function and mechanism. In: *Molecular Biology of Picornaviruses* (B.L. Semler and E. Wimmer, eds), pp. 255–267. Washington D.C.: ASM Press.

Capson, T.L., Peliska, J.A., Kaboord, B.F., Frey, M.W., Lively, D., Dahlberg, M. and Benkovic, S.J. (1992) Kinetic characterization of the polymerase and exonuclease activities of the gene 43 protein of bacteriophage T4. *Biochemistry* **31**, 10984–10994.

Castro, C., Arnold, J.J. and Cameron, C.E. (2005) Incorporation fidelity of the viral RNA-dependent RNA polymerase: a kinetic, thermodynamic and structural perspective. *Virus Res.* **107**, 141–149.

Castro, C., Smidansky, E., Maksimchuk, K.R., Arnold, J.J., Korneeva, V.S., Gotte, M., Konigsberg, W. and Cameron, C.E. (2007) Two proton transfers in the transition state for nucleotidyl transfer catalyzed by RNA- and DNA-dependent RNA and DNA polymerases. *Proc. Natl Acad. Sci. USA* **104**, 4267–4272.

Cheetham, G.M. and Steitz, T.A. (1999) Structure of a transcribing T7 RNA polymerase initiation complex. *Science* **286**, 2305–2309.

Crotty, S., Cameron, C.E. and Andino, R. (2001) RNA virus error catastrophe: direct molecular test by using ribavirin. *Proc. Natl Acad. Sci. USA* **98**, 6895–6900.

Domingo, E., Escarmis, C., Menendez-Arias, L. and Holland, J.J. (1999) Viral quasispecies and fitness variations. In: *Origin and Evolution of Viruses* (E. Domingo, R. Webster and J. Holland, eds), pp. 141–161. London: Academic Press.

Doublie, S., Tabor, S., Long, A.M., Richardson, C.C. and Ellenberger, T. (1998) Crystal structure of a bacteriophage T7 DNA replication complex at 2, 2, Å resolution. *Nature* **391**, 251–258.

Dunlap, C.A. and Tsai, M-D. (2002) Use of 2-aminopurine and tryptophan fluorescence as probes in kinetic analyses of DNA polymerase β. *Biochemistry* **41**, 11226–11235.

Eger, B.T., Kuchta, R.D., Carroll, S.S., Benkovic, P.A., Dahlberg, M.E., Joyce, C.M. and Benkovic, S.J. (1991) Mechanism of DNA replication fidelity for three mutants of DNA polymerase I: Klenow fragment KF (exo^+), KF (polA5), and KF (exo^-). *Biochemistry* **30**, 1441–1448.

Eisenmesser, E.Z., Millet, O., Labeikovsky, W., Korzhnev, D.M., Wolf-Watz, M., Bosco, D.A. *et al.* (2005) Intrinsic dynamics of an enzyme underlies catalysis. *Nature* **438**, 117–121.

Epstein, D.M., Benkovic, S.J. and Wright, P.E. (1995) Dynamics of the dihydrofolate reductase-folate complex: catalytic sites and regions known to undergo conformational change and exhibit diverse dynamical features. *Biochemistry* **34**, 11037–11048.

Ferrer-Orta, C., Arias, A., Escarmis, C. and Verdaguer, N. (2006) A comparison of viral RNA-dependent RNA polymerases. *Curr. Opin. Struct. Biol.* **16**, 1–8.

Ferrer-Orta, C., Arias, A., Perez-Luque, R., Escarmis, C., Domingo, E. and Verdaguer, N. (2007) Sequential structures provide insights into the fidelity of RNA replication. *Proc. Natl Acad. Sci. USA* **104**, 9463–9468.

Florian, J., Goodman, M.F. and Warshel, A. (2003) Computer simulation of the chemical catalysis of DNA polymerases: discriminating between alternative nucleotide insertion mechanism for T7 DNA polymerase. *J. Am. Chem. Soc.* **125**, 8163–8177.

Florian, J., Goodman, M.F. and Warshel, A. (2005) Computer simulations of protein functions: searching for the molecular origin of the replication fidelity of DNA polymerases. *Proc. Natl Acad. Sci. USA* **102**, 6819–6824.

Foster, M.P., McElroy, C.A. and Amero, C.D. (2007) Solution NMR of large molecules and assemblies. *Biochemistry* **46**, 331–340.

Franklin, M.C., Wang, J. and Steitz, T.A. (2001) Structure of the replicating complex of a Pol α family DNA polymerase. *Cell* **105**, 657–667.

Freistadt, M.S., Vaccaro, J.A. and Eberle, K.E. (2007) Biochemical characterization of the fidelity of poliovirus RNA-dependent RNA polymerase. *Virol. J.* (Epub ahead of print, May 24, 2007)

Gohara, D.W., Crotty, S., Arnold, J.J., Yoder, J.D., Andino, R. and Cameron, C.E. (2000) Poliovirus RNA-dependent RNA polymerase (3Dpol). Structural, biochemical, and biological analysis of conserved structural motifs A and B. *J. Biol. Chem.* **275**, 25523–25532.

Gohara, D.W., Arnold, J.J. and Cameron, C.E. (2004) Poliovirus RNA-dependent RNA polymerase (3Dpol): kinetic, thermodynamic, and structural analysis of ribonucleotide selection. *Biochemistry* **43**, 5149–5158.

Graci, J.D. (2007). Coxsackievirus B3 is more susceptible to lethal mutagenesis than poliovirus due to lower error threshold. In: *Evaluation of Nucleoside Analogs with Ambiguous Hydrogen-bonding Capacity as Antiviral Lethal Mutagens.* PhD thesis, pp. 156–179. The Pennsylvania State University.

Graci, J.D., Harki, D.A., Korneeva, V.S., Edathil, J.P., Too, K., Franco, D. *et al.* (2007) Lethal mutagenesis of poliovirus mediated by a mutagenic pyrimidine analogue. *J. Virol.* Aug 8; (Epub ahead of print).

Gutierrez-Rivas, M., Ibanez, A., Martinez, M.A., Domingo, E. and Menendez-Arias, L. (1999) Mutational analysis of Phe160 within the "palm" subdomain of human immunodeficiency virus type 1 reverse transcriptase. *J. Mol. Biol.* **290**, 615–625.

Hammes, G.G. (2002) Multiple conformational changes in enzyme catalysis. *Biochemistry* **41**, 8221–8228.

Hammes-Schiffer, S. and Benkovic, S. (2006) Relating protein motion to catalysis. *Annu. Rev. Biochem.* **75**, 519–541.

Hansen, J.L., Long, A.L. and Schultz, S.C. (1997) Structure of the RNA-dependent RNA polymerase of poliovirus. *Structure* **5**, 1109–1122.

Harris, D., Kaushik, N., Pandey, P.K., Yadav, P.N.S. and Pandey, V.N. (1998) Functional analysis of amino acid residues constituting the dNTP binding pocket of HIV-1 reverse transcriptase. *J. Biol. Chem.* **273**, 33624–33634.

Herschlag, D., Piccirilli, J.A. and Cech, T.R. (1991) Ribozyme-catalyzed and nonenzymatic reactions of phosphate diesters: rate effects upon substitution of sulfur for a nonbridging oxygen atom. *Biochemistry* **30**, 4844–4854.

Huang, H., Chopra, R., Verdine, G.L. and Harrison, S.C. (1998) Structure of a covalently trapped catalytic complex of HIV-1 reverse transcriptase: implications for drug resistance. *Science* **282**, 1669–1675.

Johnson, K.A. (1992). Transient-state kinetic analysis of enzyme reaction pathways. In: *The Enzymes*, Vol. XX, 3rd edn. (D.S. Sigman, ed.), pp. 1–61. Academic Press, San Diego, CA.

Johnson, K.A. (1993) Conformational coupling in DNA polymerase fidelity. *Annu. Rev. Biochem.* **62**, 685–713.

Johnson, S.J., Taylor, J.S. and Beese, L.S. (2003) Processive DNA synthesis observed in a polymerase crystal suggests a mechanism for the prevention of frameshift mutations. *Proc. Natl Acad. Sci. USA* **100**, 3900–3985.

Joyce, C.M. and Benkovic, S.J. (2004) DNA polymerase fidelity: kinetics, structure and checkpoints. *Biochemistry* **43**, 14317–14324.

Kati, W.M., Johnson, K.A., Jerva, L.F. and Anderson, K. S. (1992) Mechanism and fidelity of HIV reverse transcriptase. *J. Biol. Chem.* **267**, 25988–25997.

Kiefer, J.R., Mao, C., Braman, J.C. and Beese, L.S. (1998) Visualizing DNA replication in a catalytically active *Bacillus* DNA polymerase crystal. *Nature* **391**, 304–307.

Kim, B., Ayran, J.C., Sagar, S.G., Adman, E.T., Fuller, S.M., Tran, N.H. and Horrigan, J. (1999) New human immunodeficiency virus, type 1 reverse transcriptase (HIV-1 RT) mutants with increased fidelity of DNA synthesis. *J. Biol. Chem.* **274**, 27666–27673.

Kim, T.W., Brieba, L.G., Ellenberger, T. and Kool, E.T. (2006) Functional evidence for a small and rigid active site in a high fidelity DNA polymerase. *J. Biol. Chem.* **281**, 2289–2295.

Kohlstaedt, L.A., Wang, J., Friedman, J.M., Rice, P.A. and Steitz, T.A. (1992) Crystal Structure at 3.5 of HIV-1 reverse transcriptase complexed with an inhibitor. *Science* **256**, 1781–1790.

Korneeva, V.S. (2007) Residue Arg-273 as a modulator of the polymerase fidelity. In: Poliovirus RNA-dependent RNA polymerase (in)fidelity: mechanisms consequences and applications. Ph.D. thesis, pp. 155–211. The Pennsylvania State University.

Korneeva, V.S. and Cameron, C.E. (2007) Structure-function relationships of the viral RNA-dependent RNA polymerase: fidelity, replication speed, and initiation mechanism determined by a residue in the ribose-binding pocket. *J Biol. Chem* **282**, 16135–16145.

Koonin, E.V. (1991) The phylogeny of RNA-dependent RNA polymerases of positive-strand RNA viruses. *J. Gen. Virol.* **72**, 2197–2206.

Kraynov, V.S., Showalter, A.K., Liu, J., Zhong, X. and Tsai, M-D. (2000) DNA polymerase β: contributions of template-positioning and dNTP triphosphate-binding residues in catalysis and fidelity. *Biochemistry* **39**, 16008–16015.

Kuchta, R.D., Mizrahi, V., Benkovic, P. and Benkovic, S.J. (1987) Kinetic mechanism of DNA polymerase I (Klenow). *Biochemistry* **26**, 8410–8417.

Kunkel, T.A. (2004) DNA replication fidelity. *J. Biol. Chem.* **279**, 16895–16898.

Li, Y., Korolev, S. and Waksman, G. (1998) Crystal structures of open and closed forms of binary and ternary complexes of the large fragment of *Thermus aquaticus* DNA polymerase I: structural basis for nucleotide incorporation. *EMBO J.* **17**, 7514–7525.

Liu, J. and Tsai, M.D. (2001) DNA polymerase beta: pre-steady-state kinetic analyses of dATP alpha S stereoselectivity and alteration of the stereoselectivity by various metal ions and by site-directed mutagenesis. *Biochemistry* **40**, 9014–9022.

Loh, E., Choe, J. and Loeb, L.A. (2007) Highly tolerated amino acid substitutions increase the fidelity of *Escherichia coli* DNA Polymerase I. *J. Biol. Chem.* **282**, 12201–12209.

McAllister, W.T. and Raskin, C.A. (1993) The phage RNA polymerases are related to DNA polymerases and reverse transcriptases. *Mol. Microbiol.* **10**, 1–6.

Minnick, D.T., Bebenek, K., Osheroff, W.P., Turner, R. M., Jr., Astatke, M., Liu, L. *et al.* (1999) Side chains that influence fidelity at the polymerase active site of *Escherichia coli* DNA polymerase I (Klenow fragment). *J. Biol. Chem.* **274**, 3067–3075.

Mittermaier, A. and Kay, L.E. (2006) New tools provide new insights in NMR studies of protein dynamics. *Science* **312**, 224–227.

Patel, P.H. and Loeb, L.A. (2001) Getting a grip on how DNA polymerases function. *Nat. Struct. Biol.* **8**, 656–659.

Patel, S.S., Wong, I. and Johnson, K.A. (1991) Pre-steady-state kinetic analysis of processive DNA replication including complete characterization of an exonuclease-deficient mutant. *Biochemistry* **30**, 511–525.

Patel, S.S., Bandwar, R.P. and Levin, M.K. (2003) Transient-state kinetics and computational analysis of transcription initiation. In: *Kinetic Analysis of Macromolecules* (K.A. Johnson, ed.), pp. 87–129. New York: Oxford University Press.

Pfeiffer, J.K. and Kirkegaard, K. (2003) A single mutation in poliovirus RNA-dependent RNA polymerase confers resistance to mutagenic nucleotide analogs via increased fidelity. *Proc. Natl Acad. Sci. USA* **100**, 7289–7294.

Pfeiffer, J.K. and Kirkegaard, K. (2005) Increased fidelity reduces poliovirus fitness and virulence under selective pressure in mice. *PloS Pathog.* **1**, 102–110.

Pursell, Z.F., Isoz, I., Lundstrom., E-B., Johansson, E. and Kunkel, T.A. (2007) Regulation of B family DNA polymerase fidelity by a conserved active site residue: characterization of M644W, M644L and M644F mutants of yeast polymerase ε. *Nucleic Acids Res.* **35**, 3076–3086.

Radhakrishnan, R., Arora, K., Wang, Y., Beard, W.A., Wilson, S.H. and Schlick, T. (2006) Regulation of DNA repair fidelity by molecular checkpoints: "gates" in DNA polymerase βs substrate selection. *Biochemistry* **45**, 15142–15156.

Rothwell, P.J. and Waksman, G. (2005) Structure and mechanism of DNA polymerases. *Adv. Protein Chem.* **71**, 401–440.

Rothwell, P.J., Mitaksov, V. and Waksman, G. (2005) Motions of the fingers subdomain of Klentaq1 are fast and not rate limiting: implications for the molecular basis of fidelity in DNA polymerases. *Mol. Cell* **19**, 345–355.

Showalter, A.K. and Tsai, M-D. (2002) A reexamination of the nucleotide incorporation fidelity of DNA polymerases. *Biochemistry* **41**, 10571–10576.

Showalter, A.K., Lamarche, B.J., Bakhtina, M., Su, M-I., Tang, K-H. and Tsai, M-D. (2006) Mechanistic comparison of high-fidelity and error-prone DNA polymerases and ligases involved in DNA repair. *Chem. Rev.* **106**, 340–360.

Smiley, R.D. and Hammes, G.G. (2006) Single molecule studies of enzyme mechanisms. *Chem. Rev.* **106**, 3080–3094.

Sousa, R. (1996) Structural and mechanistic relationships between nucleic acid pols. *Trends Biochem. Sci.* **21**, 186–190.

Sousa, R. and Mukherjee, S. (2003) T7 RNA polymerase. *Prog. Nucl. Acid Res.* **73**, 1–41.

Steitz, T.A. (1993) DNA- and RNA-dependent DNA polymerases. *Curr. Opin. Struct. Biol.* **3**, 31–38.

Steitz, T.A. (1998) A mechanism for all polymerases. *Nature* **391**, 231–232.

Steitz, T.A. (1999) DNA polymerases: structural diversity and common mechanisms. *J. Biol. Chem.* **274**, 17395–17398.

Steitz, T.A. and Steitz, J.A. (1993) A general two-metal-ion mechanism for catalytic RNA. *Proc. Natl Acad. Sci. USA* **90**, 6498–6502.

Sweasy, J.B. (2003) Fidelity mechanisms of DNA polymerase β. *Prog. Nucl. Acid Res. Mol. Biol.* **73**, 137–169.

Tabor, S. and Richardson, C.C. (1989) Effect of manganese ions on the incorporation of dideoxynucleotides by bacteriophage T7 DNA polymerase and *Escherichia coli* DNA Polymerase I. *Proc. Natl Acad. Sci. USA* **86**, 4076–4080.

Thompson, A.A. and Peersen, O.B. (2004) Structural basis for proteolysis-dependent activation of the poliovirus RNA-dependent RNA polymerase. *EMBO J.* **23**, 3462–3471.

Tsai, Y-C. and Johnson, K.A. (2007) A new paradigm for DNA polymerase specificity. *Biochemistry* **45**, 9675–9687.

Vignuzzi, M., Stone, J.K., Arnold, J.J., Cameron, C.E. and Andino, R. (2006) Quasispecies diversity determines pathogenesis through cooperative interactions in a viral population. *Nature* **439**, 344–348.

Wang, D., Bushnell, D.A., Westover, K.D., Kaplan, C.D. and Kornberg, R.D. (2006) Structural basis of transcription: role of the trigger loop in substrate specificity and catalysis. *Cell* **127**, 941–954.

Werneburg, B.G., Ahn, J., Zhong, X., Hondal, R.J., Kraynov, V.S. and Tsai, M-D. (1996) DNA polymerase β: pre-steady-state kinetic analysis and roles of arginine-283 in catalysis and fidelity. *Biochemistry* **35**, 7041–7050.

Wolfenden, R. (2003) Thermodynamic and extrathermodynamic requirements of enzyme catalysis. *Biophys. Chem.* **105**, 559–572.

Yang, G., Wang, J. and Konigsberg, W. (2005) Base selectivity is impaired by mutants that perturb hydrogen bonding networks in the RB69 polymerase active site. *Biochemistry* **44**, 3338–3346.

Yang, J., Zhuang, Z., Roccasecca, R.M., Trakselis, M.A. and Benkovic, S.J. (2004) The dynamic processivity of the T4 DNA polymerase during replication. *Proc. Natl Acad. Sci. USA* **101**, 8289–8294.

Yin, Y.W. and Steitz, T.A. (2002) Structural basis for the transition from initiation to elongation transcription in T7 RNA polymerase. *Science* **298**, 1387–1395.

Yin, Y.W. and Steitz, T.A. (2004) The structural mechanism of translocation and helicase activity in T7 RNA polymerase. *Cell* **116**, 393–404.

Zhang, H., Rhee, C., Bebenek, A., Drake, J.W., Wang, J. and Konigsberg, W. (2006) The L561A substitution in the nascent base-pair binding pocket of RB69 DNA polymerase reduces base discrimination. *Biochemistry* **45**, 2211–2220.

Zinnen, S., Hsieh, J-C. and Modrich, P. (1994) Misincorporation and mispaired primer extension by human immunodeficiency virus reverse transcriptase. *J. Biol. Chem.* **269**, 24195–24202.

CHAPTER

7

The Complex Interactions of Viruses and the RNAi Machinery: A Driving Force in Viral Evolution

Ronald P. van Rij and Raul Andino

ABSTRACT

RNA interference (RNAi) is a mechanism for sequence-specific gene silencing triggered by double-stranded (ds) RNA. RNAi constitutes an effective antiviral defense mechanism in many organisms. Accordingly, viruses interact with the RNAi pathway at different levels. As a counter-defense, viruses have evolved suppressors of the RNAi pathway. Many DNA viruses, on the other hand, exploit the related microRNA pathway by producing microRNAs to regulate viral and host gene expression. In this chapter we summarize recent findings on these interactions and discuss how they may have shaped viral evolution.

INTRODUCTION

It has been about ten years since the initial report that double-stranded (ds) RNA triggers a mechanism for posttranscriptional gene silencing, RNA interference (RNAi) (Fire *et al.*, 1998). Ever since, novel insights in RNA-silencing pathways continue to revolutionize our way of thinking about gene regulation. RNAi has had a major impact on our understanding of virus–host interactions as well. RNAi functions as a major antiviral defense mechanism in several organisms, including plants and insects (Voinnet, 2001; Sanchez-Vargas *et al.*, 2004; Voinnet, 2005a). Conversely, viruses have found ways to exploit the RNA-silencing machinery for their own advantage (Sullivan and Ganem, 2005b; Cullen, 2006a). Experimentally, RNAi provides avenues to study host factors that are required for viral replication (Garrus *et al.*, 2001; Cherry *et al.*, 2005). Furthermore, RNAi may be used to target viral sequences, or essential host factors for degradation, and thus represents a novel class of therapeutic or prophylactic drugs (reviewed in Gitlin and Andino, 2003; Hannon and Rossi, 2004; van Rij and Andino, 2006). In this chapter we will discuss the interactions between viruses and RNA-silencing machinery and we will speculate how this antiviral mechanism may drive viral evolution.

Origin and Evolution of Viruses
ISBN 978-0-12-374153-0

RNA-SILENCING MECHANISMS

A number of small RNA-mediated gene-silencing mechanisms have recently been identified, of which RNA interference is the best characterized. RNAi is typically initiated by exogenous encoded, cytoplasmic dsRNA leading to degradation of the corresponding target RNA (Meister and Tuschl, 2004). The related, but distinct, microRNA (miRNA) pathway is initiated by endogenously encoded miRNAs, leading to translational regulation of mRNAs (Ambros, 2004). Current evidence suggests that viruses interact at different levels with the RNAi and the miRNA pathway. We will therefore discuss these pathways in more detail below, with an emphasis on the mechanism in *Drosophila melanogaster* and mammals. Even though the basic mechanism of RNA-silencing pathways appears to be highly conserved, there are major differences among different species (Table 7.1). For more detailed reviews on RNA-silencing pathways in plants and nematodes, see Baulcombe (2004), Brodersen and Voinnet (2006), Steiner and Plasterk (2006), and Miska and Ahringer (2007).

In addition to these posttranscriptional mechanisms of gene silencing, small RNAs have been implicated in transcriptional gene silencing, de novo methylation of DNA, and changes in chromatin structure in several organisms (Lippman and Martienssen, 2004). More recently, a novel, distinct class of small RNAs of 26–30 nucleotides were identified based on their association with the piwi subclass of Argonaute proteins (piRNAs). Expression of piwi proteins and piRNAs is restricted to the germline, where they control activation of transposons, including retrotransposons, and endogenous retroviruses (Sarot *et al.*, 2004; O'Donnell and Boeke, 2007; Pelisson *et al.*, 2007; Zamore, 2007). Even though piRNAs may, therefore, be important for viral pathogenesis, we will not further discuss this pathway in this chapter, since little is known about the biogenesis and mechanism of piRNAs.

RNA Interference

RNAi is a mechanism for gene silencing triggered by the presence of dsRNA in the cytoplasm (Meister and Tuschl, 2004) (Figure 7.1A). dsRNA is cleaved by the RNase III-like enzyme Dicer (Dicer-2 in *Drosophila*) into short interfering (si) RNAs, 21 base pair double-stranded RNAs, with two-base 3′ overhangs (Bernstein

TABLE 7.1 Functional diversification and specialization among Argonaute and Dicer family members in different model organisms

Pathway	Gene family	Plants (*A. thaliana*)	*C. elegans*	*Drosophila melanogaster*	Mammals
	Dicer[a]	4	1	2	1
	Argonaute[a,b]	10	27 b	2	4
miRNA	Dicer	DCL1	Dicer	Dicer-1	Dicer
	Argonaute	Ago-1	Alg-1, Alg-2	Ago-1	Ago-1, 2, 3, 4[c]
RNAi	Dicer	DCL3, 4	Dicer	Dicer-2	Dicer
	Argonaute	Ago-1	Rde-1	Ago-2	Ago-2
Antiviral RNAi	Dicer	DCL1, 2, 3, 4	Dicer	Dicer-2	–
	Argonaute	Ago-1	Rde-1	Ago-2	–

[a] The number of Dicer and Argonaute family members in each organism is given.

[b] For plants, insects, and mammals, the number of members in the Argonaute subclass of Argonaute proteins, thus excluding piwi proteins, is reported. The number of *C. elegans* genes is the composite of the Ago, Piwi and *C. elegans* specific subclass of Ago genes (*Yigit et al.*, 2006).

[c] Undefined. The microRNA pathway is fully functional in $Ago2^{-/-}$ mice; Ago-1, 2, 3, and 4 are likely redundant in miRNA function (Liu *et al.*, 2004).

et al., 2001). Dicer progressively cleaves dsRNA from the ends of the dsRNA molecule (Zhang *et al.*, 2002), acting like a molecular ruler; the 3′ two-base overhang binds its PAZ domain and the typical length of an siRNA is measured by the distance of this domain to the catalytically active RNase III domains (Zhang *et al.*, 2004; Macrae *et al.*, 2006). siRNAs are incorporated into the multiprotein complex RISC (RNA-induced silencing complex), one of the strands of the siRNA (the passenger strand) is degraded and the other strand (the guide strand) is stabilized in RISC. Loading of RISC occurs by Dicer in association with R2D2 (in *Drosophila*; Liu *et al.*, 2003) or TRBP (in mammals; Chendrimada *et al.*, 2005; Haase

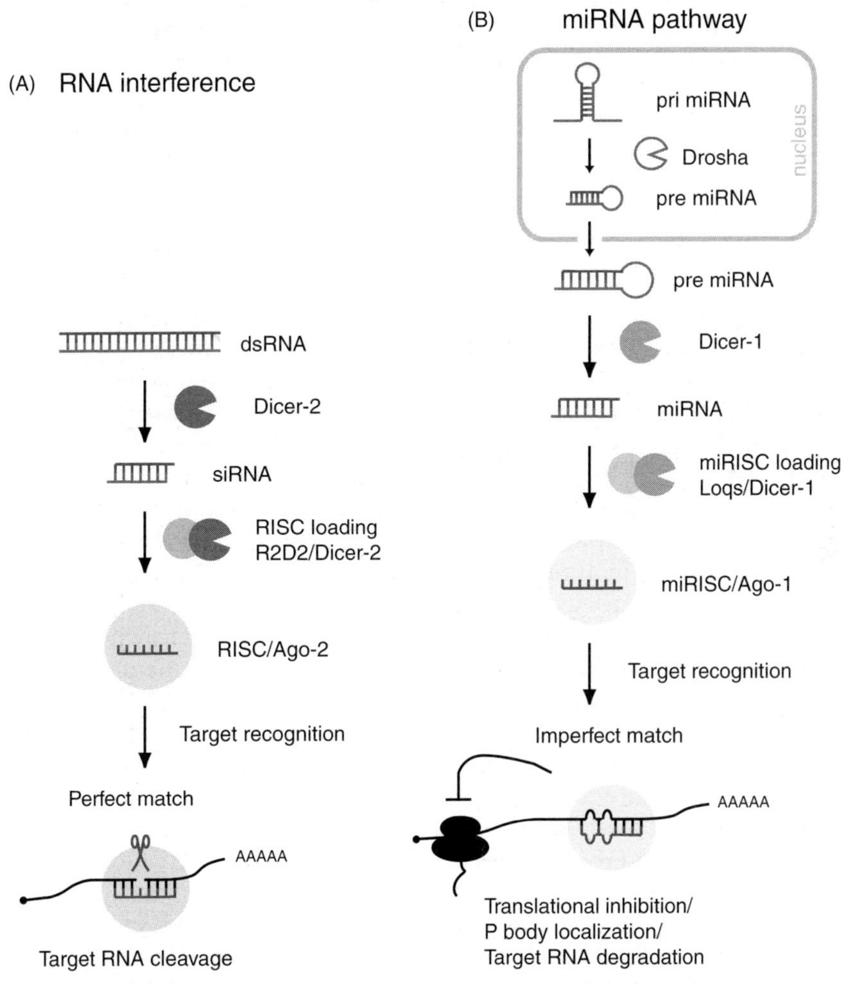

FIGURE 7.1 Schematic representation of RNA-silencing pathways in *Drosophila melanogaster*. (A) RNA interference is initiated by the presence of dsRNA in the cytoplasm of the cell, eventually leading to degradation of perfect complementary RNA in the cell. (B) The miRNA pathway is instructed by endogenously encoded miRNAs that guide recognition of imperfect complementary target sites, which are located in the 3′ untranslated regions (3′ UTRs) of endogenous genes. Target recognition may lead to the translational inhibition, mRNA degradation, and/or localization of the repressed mRNA to processing bodies (P bodies). See text for details. Note that although this figure represents the RNA-silencing pathway in *Drosophila*, the basic mechanism of RNAi is highly conserved mechanism among different organisms. However, functional diversification and specialization among RNAi genes has occurred throughout evolution (see Table 7.1). (See Plate 10 for the color version of this figure.)

et al., 2005). Once incorporated in RISC the guide strand directs the RISC activity onto specific mRNAs. Recognition of a fully complementary sequence triggers the endonucleic cleavage of the target RNA (Slicer activity) at position 10 (counting from the 5′ end of the guide strand). Cleavage is mediated by the RNase H-like piwi domain of Argonaute (Ago-2 in *Drosophila* and mammals) (Liu *et al.*, 2004). Although biochemical analyses indicate that RISC is a large multiprotein complex, purified, bacterially expressed Argonaute 2 is capable of cleaving target RNA *in vitro* (Rivas *et al.*, 2005). The role(s) of associated proteins in RISC have not been elucidated.

miRNA Pathway

Whereas the RNAi pathway seems to be mainly dependent on exogenous dsRNA, the miRNA pathway is triggered by endogenously encoded miRNAs (Ambros, 2004) (Figure 7.1B). miRNAs are transcribed as primary miRNA transcripts (pri-miRNA) of several hundred to several thousand nucleotides that usually encode multiple miRNAs. Maturation of pri-miRNA into mature miRNA occurs via two steps. In the nucleus, the initial processing of pri-miRNA into pre-miRNA is mediated by the RNase III-like enzyme Drosha, component of a multiprotein complex termed "microprocessor complex" (Lee *et al.*, 2003; Denli *et al.*, 2004). Pre-miRNAs are stem-loop structures of ~70 nt with a two-base 3′ overhang. After export from the nucleus into the cytoplasm via Exportin 5 (Yi *et al.*, 2003; Lund *et al.*, 2004), pre-miRNAs are cleaved by Dicer (Dicer-1 in *Drosophila*) into an siRNA-like structure (21 nt duplex RNA, with two-base 3′ overhangs at both ends) (Bernstein *et al.*, 2001; Lee *et al.*, 2004). The combined action of Dicer-1 and Loquacious/R3D1 (Forstemann *et al.*, 2005; Jiang *et al.*, 2005; Saito *et al.*, 2005) (or Dicer/TRBP in mammals; Chendrimada *et al.*, 2005; Haase *et al.*, 2005) facilitates the incorporation of the mature miRNA into miRISC, where the miRNA functions as a specificity determinant. The other strand of the miRNA (termed miRNA*) is usually degraded, but for some miRNAs both the mature miRNA and the miRNA* are loaded in a functional miRISC.

Recognition of target RNAs by miRISC is usually mediated via imperfect base pairing to a target site in 3′ untranslated regions of endogenous genes. The specificity of the interaction is determined by the 5′ half of the miRNA, nucleotides 2 through 7 or 8, called the "seed" region (Doench and Sharp, 2004; Brennecke *et al.*, 2005; Lewis *et al.*, 2005). Upon recognition of a target RNA, inhibition of translation occurs via a mechanism in which Argonaute competes with eIF4E for binding to the m7G cap (Valencia-Sanchez *et al.*, 2006; Kiriakidou *et al.*, 2007; Mathonnet *et al.*, 2007; Pillai *et al.*, 2007). In addition to their effect on translation, target recognition may result in a Slicer-independent degradation of the mRNA. Translationally repressed mRNA appear to localize to the processing bodies (P bodies) in a miRNA-dependent fashion, and under some conditions return to the translating pool of mRNAs (Liu *et al.*, 2005; Bhattacharyya *et al.*, 2006). However, it is currently unclear whether P body localization is the cause or consequence of translational inhibition. In *Drosophila*, Ago-1 is dedicated to the miRNA pathway (Okamura *et al.*, 2004), whereas Ago-1 through 4 may be functionally redundant for miRNA activity in mammals (Liu *et al.*, 2004).

Hundreds of miRNAs have been identified, each with an average of ~200 predicted targets that together may potentially regulate expression of an estimated 30% of genes in the genome (John *et al.*, 2004; Bentwich *et al.*, 2005; Berezikov *et al.*, 2005; Rajewsky, 2006). Many miRNAs are regulated during development and miRNA expression is often highly organ and cell type specific (Wienholds *et al.*, 2005), thus providing a widespread and dynamic platform for the regulation of gene expression.

ANTIVIRAL FUNCTION OF RNA SILENCING

Double-stranded RNA is an explicit danger signal: it is produced during replication of many viruses, and is not normally present in healthy cells. RNA silencing may therefore act as a nucleic acid-based antiviral immune system, especially in plants and non-vertebrate animals that lack the classical adaptive immune response and the dsRNA-activated interferon response typically found in vertebrates. In theory, antiviral RNA silencing can inhibit virus replication by at least two mechanisms. First, Dicer could directly cleave and deplete viral dsRNA intermediate or dsRNA virus structures which are essential for virus replication. In addition, antiviral RNA silencing could depend on RISC-like activity, whereby viral siRNAs, incorporated in RISC, guide the recognition and degradation of newly produced viral RNAs. In both cases it is predicted that viral-derived siRNAs should be detectable during viral infection, and that functional inactivation of RNAi genes, such as Dicer and Argonaute, should result in an increase in viral replication and potentially in an increase in pathogenicity. We will focus this review on animal viruses. However, no review of antiviral RNA silencing is complete without a brief description of antiviral RNA silencing in plants (for more details see reviews by Voinnet, 2001; Baulcombe, 2004; Voinnet, 2005a).

Antiviral RNAi in Plants

The Argonaute (Ago) and Dicer-like (DCL) gene families in *Arabidopsis* have greatly expanded during evolution to ten and four members, respectively (Table 7.1). Different family members play a role in specific RNA-silencing pathways, including RNA silencing as a defense against viruses. Indeed, virus-derived siRNA have been identified during replication of a number of RNA and DNA viruses (Hamilton and Baulcombe, 1999; Chellappan *et al.*, 2004; Molnar *et al.*, 2005). The substrates for DCLs for the production of viral siRNAs are currently largely unknown. It has been hypothesized that the dsRNA intermediate in replication or local base paired structures in ssRNA viruses are cleaved by Dicer, or, during replication of DNA viruses, overlapping transcripts derived from bidirectional transcription may the source of dsRNA (Hamilton and Baulcombe, 1999; Chellappan *et al.*, 2004; Molnar *et al.*, 2005). In addition, a translational leader sequence with an extensive fold-back structure was identified as the major source of viral siRNAs during infection with the pararetrovirus cauliflower mosaic virus (Moissiard and Voinnet, 2006). Of note, all four DCLs involved in distinct endogenous RNA-silencing pathways, which appear to localize in differ subcellular compartments, have been implicated in antiviral immunity in plants. However, the relative contribution of each DCL to viral siRNA production differs among viruses. Specifically, all four DCLs are involved in production of viral siRNA during replication of nuclear DNA viruses (geminiviruses and the pararetrovirus cauliflower mosaic virus). RNA viruses are mainly targeted by DCL4, and when its activity is reduced, by the subordinate redundant activity of DCL2 (Blevins *et al.*, 2006; Bouche *et al.*, 2006; Deleris *et al.*, 2006; Moissiard and Voinnet, 2006). Thus, multiple viral dsRNA structures may act as substrates for the production of viral siRNAs, and functional specialization of DCLs in antiviral RNAi may therefore facilitate targeting different virus families in plants. Of note, during replication of RNA viruses, virus-derived siRNAs are not evenly distributed over the genome, nor evenly over (+) and (−) strand (Molnar *et al.*, 2005). This observation suggests that the dsRNA replication intermediate is not necessarily the major trigger of antiviral RNAi activity. Further in-depth analysis, e.g. by deep sequencing the viral siRNA population, will be needed to further define the substrates for Dicer in viral infections.

Even though it has been known for quite some time that RNA silencing is involved in antiviral defense in plants, only recently has evidence accumulated implying Argonaute 1 as the effector responsible for the antiviral function. Argonaute 1, the only plant Argonaute with a demonstrated Slicer activity (Baumberger and Baulcombe, 2005), associates with viral siRNAs (Zhang *et al.*, 2006). In agreement, discrete viral cleavage products, likely Slicer products, were identified in infected plants (Pantaleo *et al.*, 2007). Furthermore, Argonaute 1 hypomorphic plants are hypersensitive to viral infection (Morel *et al.*, 2002), even though the pleiotropic phenotype of the Ago-1 mutant plants complicates the interpretation of this result. Supporting the concept that RNAi is an essential antiviral mechanism in plants is the finding that most plant viruses have evolved suppressors of RNA silencing (Li *et al.*, 2002; Roth *et al.*, 2004; Voinnet, 2005a) (discussed below). The observation that cucumber mosaic virus-silencing suppressor (2b) directly interacts with Argonaute 1, and thereby inhibits its cleavage activity (Zhang *et al.*, 2006), lends further support to the importance of Slicer activity for antiviral RNAi.

One important feature of the plant antiviral response is the ability to mediate a systemic sequence specific antiviral response. This systemic response seems to be a composite of a short-range silencing signal that can spread over 10–15 cells, and a long-range silencing signal that is dependent on the RNA-dependent RNA polymerase RDR6 and DCL2 and DCL4 (Himber *et al.*, 2003; Schwach *et al.*, 2005; Voinnet, 2005b; Moissiard *et al.*, 2007). The importance of this mechanism for amplification and spread of the silencing signal during antiviral response is underlined by the observation that RDR6 mutant plants are hypersensitive to infection by some viruses (Mourrain *et al.*, 2000).

Antiviral RNAi in Insects

The existence of an RNA-based immune system in insects was initially suggested by the observation that infection of mosquitoes or mosquito cell lines with a Sindbis virus vector expressing dengue virus sequences protects from subsequent challenge with dengue virus (Gaines *et al.*, 1996; Olson *et al.*, 1996; Sanchez-Vargas *et al.*, 2004). This protection seemed RNA-based, as translation of the dengue virus sequence was not required, and an antisense insert also conferred protection. Later, it was shown that viral siRNAs could be detected in cells infected with the Sindbis vector expressing dengue virus sequences, and in related alphavirus infections in other insect species (Uhlirova *et al.*, 2003; Sanchez-Vargas *et al.*, 2004; Garcia *et al.*, 2005). An increase in O'nyong nyong virus replication after knock-down of an Argonaute family member in the mosquito *Anopheles gambiae* further confirms that RNAi can confer antiviral immunity in insects (Keene *et al.*, 2004).

RNA silencing has been dissected biochemically and genetically in great detail in *Drosophila melanogaster.* Similar to the situation in plants, the Argonaute and Dicer gene families have expanded during *Drosophila* evolution, albeit to a much lesser extent. *Drosophila* encodes two Dicer proteins and two Argonaute proteins (Table 7.1). RNAi initiated with exogenous dsRNA depends on Dicer-2 and Ago-2, whereas the miRNA pathway is mediated by Dicer-1 and Ago-1 (Lee *et al.*, 2004; Okamura *et al.*, 2004). Dicer-2, R2D2, and Argonaute-2 mutant flies are hypersensitive to viral infection to several (+)-stranded RNA viruses, Flock House virus (FHV), cricket paralysis virus (CrPV) and *Drosophila* C virus (DCV) (Galiana-Arnoux *et al.*, 2006; van Rij *et al.*, 2006; Wang *et al.*, 2006) and the dsRNA *Drosophila* X virus (Zambon *et al.*, 2006). Thus, as was seen in plants, antiviral RNA silencing in *Drosophila* depends on Dicer and Slicer activities for its effector mechanism. Further supporting a role for RNAi in antiviral defense are the observations that virus-derived siRNAs could be detected in FHV and CrPV virus infection of *Drosophila* (Galiana-Arnoux *et al.*, 2006; Wang *et al.*, 2006), and that suppressors of RNA silencing were identified in three viruses that infect different species of insects, FHV, CrPV, and DCV (Li *et al.*, 2002; van Rij *et al.*, 2006; Wang *et al.*, 2006) (discussed below). Interestingly, it has been proposed that this evolutionary arms race between virus and host may explain the rapid evolution of the Dcr-2, R2D2, and Ago-2 genes, but not of the miRNA genes Dcr-1, Loqs, and Ago-1, in *Drosophila* species (Obbard *et al.*, 2006).

Antiviral RNAi in Nematodes

While *Caenorhabditis elegans* is an excellent model for the genetic dissection of RNA-silencing pathways, it has been of limited use for the study of antiviral immunity, due to the lack of viruses that naturally infects nematodes. Nevertheless, the role of RNAi in antiviral immunity has been analyzed using a transgenic model for FHV replication, or by vesicular stomatitis virus infection of *C. elegans* cell culture (Lu *et al.*, 2005; Schott *et al.*, 2005; Wilkins *et al.*, 2005). Virus replication correlated with RNAi activity. Increased viral replication was observed in several RNA-silencing mutants, including a mutant for Rde-1, the Argonaute responsible for RNAi. Conversely, mutants with enhanced RNAi activity were less sensitive to virus replication. Unlike plants and insects, *C. elegans* encodes a single Dicer gene. An important implication of these studies, therefore, is the notion that a functional diversification of the Dicer family *per se*, is not required for antiviral RNA silencing. It would be of great interest to extrapolate these findings to a *bona fide* infection model in the entire organism. This would allow the evaluation as to how some intriguing aspects of the *C. elegans* RNAi mechanism, such as the generation of secondary siRNAs, systemic spread of the silencing signal, and the inheritance of RNAi to the offspring contributes to antiviral immunity in the entire organism.

Antiviral RNAi in Mammals

While RNA silencing is crucial for antiviral immunity in plants and insects, there is no conclusive evidence supporting an antiviral function for RNAi in mammals. The detection of virus-derived siRNAs would provide direct evidence that viral sequences are cleaved by Dicer. Thus far, however, identification of virus-derived siRNAs have been unsuccessful in tissue culture models for infection with hepatitis C virus (HCV), human immunodeficiency virus 1 (HIV-1) and yellow fever virus (Pfeffer *et al.*, 2005). A report of the detection of a specific virus-derived siRNA in HIV-1 infection (Bennasser *et al.*, 2005), has been subject of debate. This siRNA corresponds to a well-defined extensive structure (Rev responsive element), in which the two strands of the putative viral siRNA do not base pair *in vivo* and is therefore unlikely to be cleaved by Dicer (Cullen, 2006b).

The RNA viruses encephalomyocarditis virus, lymphocytic choriomeningitis virus, coxsackie virus B3, influenza A virus, and the DNA virus vaccinia virus (also known to produce dsRNA during replication) replicated to similar, or even lower, viral titers in macrophages from Dicer-deficient mice as in wild-type macrophages (Otsuka *et al.*, 2007). Similarly, we did not observe a difference in Sindbis virus replication between Dicer knockout and wild-type murine fibroblasts (R.P. van Rij and R. Andino, unpublished observations).

An alternative approach to answer the question whether RNAi has antiviral activity in mammals has been through the search for RNAi suppressive factors in mammalian viruses. The acquisition of RNAi suppressor activities by viruses implies that the RNAi machinery exerts strong selection pressure on the virus. The interferon antagonists NS1 from influenza A virus and the E3L protein from vaccinia virus could suppress RNAi in *Drosophila* S2 cells and in plants (Lichner *et al.*, 2003; Bucher *et al.*, 2004; Li *et al.*, 2004). The functional and physiological significance of these observations, however, is far from clear. For example, overexpression of the dsRNA-binding domain from RNase III from *Escherichia coli* was sufficient to suppress RNAi in plants (Lichner *et al.*, 2003), even though *E. coli* is incapable of performing RNAi. Furthermore, a defect of replication of influenza lacking NS1 is typically observed in wild-type cells; however this defect is absent in type I interferon (IFN)-deficient cells (Bergmann *et al.*, 2000). Thus, while the dsRNA binding activity of NS1 may explain its RNAi suppressor activity *in vitro*, the *in vivo* function of NS1 appears to be the evasion of the interferon response rather than suppressing of RNAi. In addition, direct attempts to detect RNAi suppressive activity of NS1 in mammalian cells failed (Kok and Jin, 2006). Similarly, La Crosse virus NSs protein was implied as a suppressor of RNAi in mammalian cells, but recombinant viruses lacking NSs were attenuated in IFN competent systems, but not in fibroblasts or mice lacking the type I interferon receptor (Soldan *et al.*, 2005; Blakqori *et al.*, 2007).

The retroviral RNA-binding proteins HIV-1 tat and primate foamy virus-1 tas were suggested to be inhibitors of RNA silencing (Bennasser *et al.*, 2005; Lecellier *et al.*, 2005), based on experiments under conditions of overexpression. Recently it was reported that the tat and tas proteins do not suppress RNAi when expressed at lower, more physiological levels (Cullen, 2006b).

Thus, identification of viral RNAi suppressors under non-physiological conditions, i.e. overexpression, should be interpreted cautiously, especially when studying RNA-binding proteins. The detection of RNAi suppressive activity *in vitro* should not be taken as definitive evidence for an antiviral function of mammalian RNAi. Thus, while it is clearly established that RNAi controls virus infection in plants and insects, it is still unresolved whether RNAi plays a role in antiviral defense in the mammalian system. Perhaps the evolution of the complex mammalian immune system, consisting of adaptive and an innate response, including dsRNA-activated responses, such as interferon responses, have functionally substituted the antiviral activity of RNAi.

VIRUS-ENCODED miRNAs

With the benefit of hindsight, it is not surprising that viruses have found ways to exploit a mechanism for gene regulation as versatile as the miRNA pathway; a hairpin structure as little as ~70 nt is already sufficient to instruct miRISC to target host genes, and it does not require protein expression, and will therefore not be immunogenic. Furthermore, minor alterations of the miRNA sequence, especially in the seed region, may alter the set of host genes that are targeted by the viral miRNA.

Given that the biogenesis of miRNAs starts with Drosha processing in the nucleus, and that (human) Dicer is unlikely to cleave a hairpin structure from within a long viral RNA molecule with extensive tertiary structure (Zhang *et al.*, 2002), it is unlikely that cytoplasmic RNA viruses encode miRNAs. Indeed viral miRNAs have thus far only been identified in nuclear DNA viruses, the polyomavirus SV40 and in many members of the herpesvirus family (Pfeffer *et al.*, 2004, 2005; Sullivan and Ganem, 2005b; Sullivan *et al.*, 2005; Cullen, 2006a). Especially herpesviruses, which establish long-term persistent or latent infection, encode a large number of miRNAs, up to 23 in Epstein-Bar virus (EBV). While it will be a formidable task to determine which host genes are regulated by these viral miRNA, and how these interactions contribute to viral pathogenesis, the function of the SV40, herpes simplex virus 1 (HSV-1) miRNA, and cytomegalovirus (CMV) miR-UL122 have been defined.

The SV40 miRNA is expressed from the late transcript and is complementary to the early viral transcript. The viral miRNA targets the early transcript for degradation, reducing the expression of small and large T antigens, without a concomitant reduction in viral titer. Infected cells were less susceptible to lysis by T antigen-specific cytotoxic T cells, suggesting that SV40 exploits the RNAi machinery to regulate viral gene expression, and thereby reduce its susceptibility to the adaptive immune response (Sullivan *et al.*, 2005).

In contrast, the targets for the HSV-1 and CMV miRNAs are cellular mRNAs. The non-coding, latency-associated transcript (LAT) gene, is the only gene that is expressed in latent infection, and encodes a miRNA. miR-LAT targets transforming growth factor beta (TGF-β) and SMAD-3 which are functionally linked to the TGF-β pathway, and thereby prevents apoptosis, and ensures maintenance of latent infection (Gupta *et al.*, 2006). CMV miR-UL122 downregulates the major histocompatibility complex class I-related chain B (MICB), the stress-induced ligand of the natural killer cell activating receptor NKG2D. The reduced expression of MICB results in a decreased binding of NKG2D and, consequently, reduced killing by NK cells. CMV miR-UL122 thus provides a miRNA-based mechanism for immune evasion (Stern-Ginossar *et al.*, 2007). Similarly, viral miRNAs appear to be encoded by plant viruses. A viral miRNA, derived from a translational leader sequence with an extensive fold-back structure, has recently been implied as a regulator of host gene expression in pararetrovirus cauliflower mosaic virus (Moissiard and Voinnet, 2006).

Adenovirus expresses high levels of ~160 nt virus-associated (VA) RNA I and II that block activation of protein kinase R. In addition, VA RNAs are substrates for Dicer, and small RNAs derived from the terminal stem of VA I and II accumulate in infected cells and are incorporated into a functional RISC (Andersson *et al.*, 2005; Aparicio *et al.*, 2006). Additionally, due to their high expression and some structural features shared with pre-miRNAs, VA RNAs inhibit RNAi initiated by short hairpin RNAs and miRNAs, by competitive binding to the nuclear export factor Exportin 5 and to Dicer (Lu and Cullen, 2004; Andersson *et al.*, 2005). The main function of VA RNAs seems to be inhibition of the PKR response, thus enhancing viral mRNA translation. The competitive inhibition of Exportin 5 and Dicer was initially suggested function as a suppressor of the RNAi pathway. However, blocking adenoviral small RNAs results in a modest decrease in viral titer (Aparicio *et al.*, 2006). It is therefore possible that these small RNAs indeed act as miRNAs, that participate in the virus replication cycle in the infected host. Further studies are needed to address these issues.

CELLULAR miRNAs AND VIRUSES

While viruses may encode their own set of miRNAs, cellular miRNA may interact with viruses via several, quite distinct, mechanisms. miRNAs are able to control replication of viruses in which miRNA complementary sequences were artificially engineered into viral genome (Gitlin *et al.*, 2005; Brown *et al.*, 2006). In addition, miR32 mediates translational inhibition of the retrovirus primate foamy virus type 1 (PFV-1) via a classical miRNA mechanism, translational inhibition without concomitant RNA degradation in a human cell line (Lecellier *et al.*, 2005). It is unclear, however, whether the interaction between PFV-1 and miR32 occurs in a natural infection, or whether this interaction is merely fortuitous due to the use of a human cell line. For example, is miR32 expressed in the natural target cells, oropharyngeal tissues, in the chimpanzee (Murray *et al.*, 2006)? If so, why did the virus fail to mutate the miRNA target site, as an engineered escape variant was viable *in vitro* (Lecellier *et al.*, 2005)?

Similarly, the ubiquitously expressed, conserved miRNAs miR24 and miR93 inhibit replication of the vesicular stomatitis virus (VSV) by targeting the L gene, encoding the viral polymerase, and the P gene, a cofactor for the polymerase. Accordingly, Dicer-deficient mice, thus lacking these miRNAs, are more susceptible to viral infection, with an increased mortality and approximately 100-fold higher viral titers in the brain. Finally, mutant virus in which the miRNA target sites were mutated replicated to higher titers in wild-type, but not in Dicer-deficient macrophages, and were more pathogenic in wild-type mice (Otsuka *et al.*, 2007). Strikingly, the miRNA target sites are not conserved among different serotypes of VSV; the more pathogenic New Jersey serotype serendipitously contains the same miRNA seed mutation that Otsuka *et al.* introduced into the Indiana serotype to abolish seed pairing to miR-24 (Muller and Imler, 2007).

These results may be interpreted as proof for an innate antiviral activity of host miRNAs, and that VSV New Jersey may represent an immune escape variant. Why then, does the VSV Indiana serotype and PFV-1 retain miRNA target site in their genomes, especially since replication and reverse transcription, respectively, are mediated by error-prone viral polymerases? Alternatively, the interaction with miRNAs may be beneficial for the virus. For example, the virus may "allow" the miRNA machinery to target its genome to fine-tune its own gene expression.

Host miRNAs may also indirectly affect viral replication by inhibiting expression of an essential host factor. The histone acetyltransferase PCAF, a co-factor for the HIV-1 transactivator tat, is a target for miR17-5p and miR20a. While the miR17-5p/miR20a–PCAF interaction may be part of a basic cellular gene regulatory network, it inhibits HIV-1 replication, as an increase in viral replication is observed upon Dicer and Drosha knockdown. The observation that the expression of the polycistronic miR-17/92 cluster (which includes miR17-5p and miR20a) is decreased in HIV-1-infected cells, may lend support for the physiological role of this interaction (Triboulet *et al.*, 2007). However, whether and how HIV-1 actively inhibits expression of this miRNA cluster, merits further investigation.

Hepatitis C virus (HCV) provides yet another, unanticipated, interaction between a cellular miRNA and an RNA virus. HCV depends on miR122 (which is highly expressed in the liver, where HCV replicates) for replication (Jopling *et al.*, 2005). This interaction is atypical in two aspects: it stimulates replication rather than represses it, and the miRNA target site is located in the 5′ UTR. While the mechanistic details of this interaction are still unclear, these results demonstrate that our understanding of the role of miRNA in viral pathogenesis is far from complete.

EVOLUTIONARY IMPLICATIONS OF RNAi

In the previous sections we illustrated the multiple mechanisms by which viruses and the RNA-silencing machinery interact during viral replication. In this section we discuss how these interactions could shape and drive viral evolution.

Antiviral RNAi and Viral Suppressors

Antiviral RNA silencing suppresses virus replication via a Slicer-dependent mechanism in plants and insects. At any point during viral replication a portion of the RISC complexes in the cell will, therefore, be loaded with virus-specific siRNAs. On the other hand, the RNAi machinery is exquisitely sensitive to mutations in the target sequences; diverse point mutations within the target region may abolish the inhibitory activity of antiviral siRNAs (Gitlin *et al.*, 2002; Boden *et al.*, 2003; Das *et al.*, 2004; Gitlin *et al.*, 2005; Westerhout *et al.*, 2005; Wilson and Richardson, 2005). However, assuming that viral siRNA are derived from relatively large portions of the virus genome it is unlikely that viruses can accumulate sufficient mutations to escape RNAi by simply mutating the target sequences. It is not known, however, whether there is any bias in the selection of functional virus siRNAs and whether this can contribute to the diversity of virus population, in which the quasispecies structure is shaped by the effect of the RNAi response. It is possible that the evolutionary pressure of escaping the build-up of specific antiviral siRNAs may partly underlie the high mutation rates observed in viral polymerases.

The most dramatic effect of selection pressure exerted by antiviral RNAi on the genetic make-up of viruses is the evolution, or acquisition, of diverse silencing suppressors as a counter-defense mechanism (Figure 7.2). Most, if not all, plant viruses encode suppressors of RNA silencing, underlining the widespread and dramatic effect of RNA silencing on viral pathogenesis in plants (Li *et al.*, 2002; Roth *et al.*, 2004; Voinnet, 2005a). More recently, RNA-silencing suppressors have been identified in insect viruses as well. Viruses have gone through great lengths to generate suppressors of RNAi. For example, nodaviruses, such as FHV, have a minimal genome, encoding only three proteins, an RNA-dependent RNA polymerase, capsid, and B2, a potent suppressor of RNA silencing (Li *et al.*, 2002). Furthermore, single plant viruses may encode multiple silencing suppressors that may target distinct steps in RNA-silencing processes. Among these are citrus tristeza virus, a large RNA virus, and geminiviruses, viruses with a single-stranded circular DNA genome (Voinnet *et al.*, 1999; Lu *et al.*, 2004; Vanitharani *et al.*, 2004).

RNA binding seems to be a ubiquitous feature of plant virus RNA-silencing suppressors. Indeed, many viruses have either a size-independent dsRNA-binding activity or

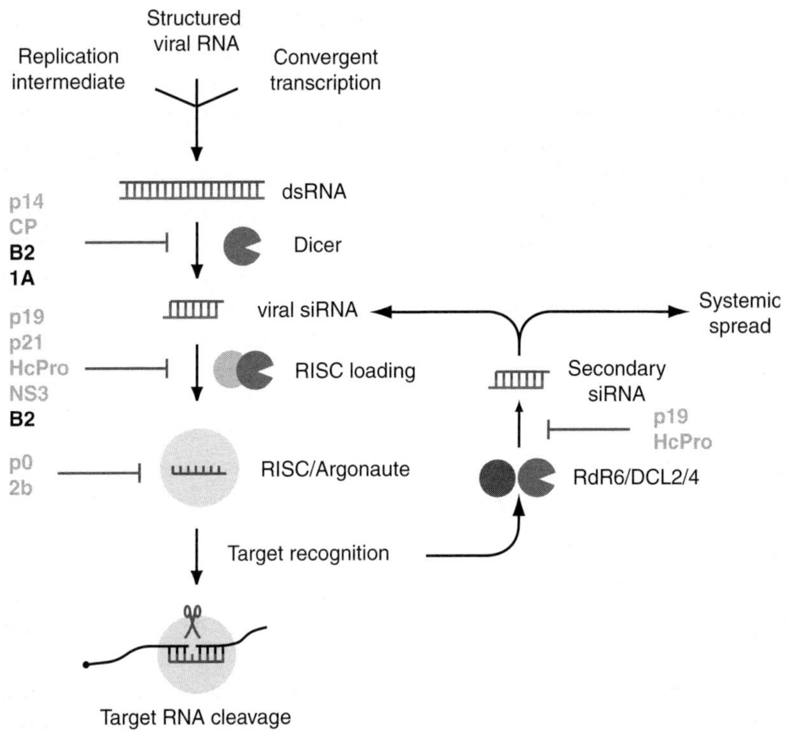

FIGURE 7.2 Viral suppressors inhibit distinct steps of the RNA-silencing pathway. Schematic representation of the mechanism of action of selected insect and plant virus RNA-silencing suppressors. Green and black font indicate silencing suppressors of plant and insect origin, respectively. Note that many more silencing suppressors have been identified in plant viruses, although their mechanism of action remains to be established. See Li and Ding (2006) for a more comprehensive overview. Silencing suppressors with specificity for siRNA, such as p19 and HcPro, sequester siRNAs and inhibit their incorporation into a functional RISC complex *in vitro* (Lakatos *et al.*, 2006). However, they may additionally inhibit the RDR6-DCL2/DCL4-dependent production of secondary siRNAs, and prevent secondary siRNAs from moving to neighboring uninfected cells or into the vasculature (Moissiard *et al.*, 2007). Amplification and spread of the RNA-silencing signal has not been observed in insects. (See Plate 11 for the color version of this figure.)

a 21 bp siRNA-binding activity (Merai *et al.*, 2006). The best characterized of these RNAi suppressors is tombusvirus p19, for which a crystal structure is available. p19 binds as a dimer specifically to siRNA (Vargason *et al.*, 2003; Ye *et al.*, 2003; Zamore, 2004). Specificity for the 21 bp length of siRNAs is provided by an alpha-helical "reading head." A tryptophan stacks on top of the first base of each siRNA strand, which allows p19 to act as a molecular caliper to measure the length of an siRNA (Figure 7.3A). Three other structurally and evolutionary unrelated plant virus RNAi suppressors, closterovirus p21, potyvirus HcPro, and tenuivirus NS3, also seem to inhibit RNAi by virtue of an siRNA-binding activity (Lakatos *et al.*, 2006; Hemmes *et al.*, 2007). Although the structural determinants that provide siRNA specificity are unknown for p21, HcPro, and NS3, an important difference resides in the recognition of the hallmarks of an siRNA: the 5′ phosphate and the 3′ two-base overhang. The binding affinity of p19 for an siRNA is enhanced by the 5′ phosphate, whereas the two-base overhang seems to be dispensable for RNAi suppression (Vargason *et al.*, 2003; Ye *et al.*, 2003). The two-base overhang is also not essential for NS3 binding (Hemmes *et al.*, 2007). In contrast, for p21 and HcPro the reverse seems to hold true; these proteins

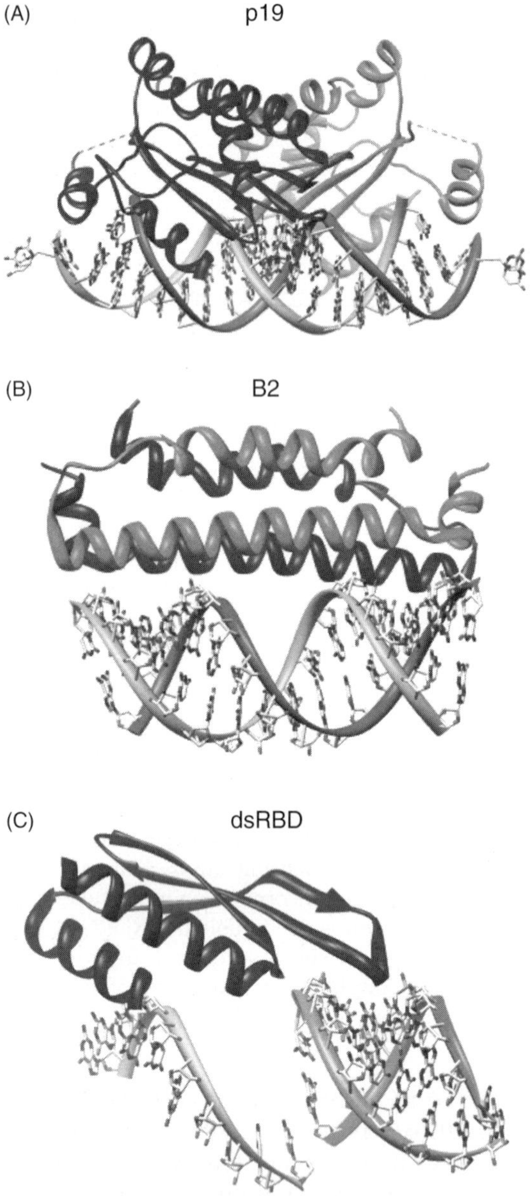

FIGURE 7.3 Structures of viral RNA-silencing suppressor with dsRNA-binding activity in complex with dsRNA.Structures of (A) Carnation Italian ringspot virus (tombusvirus) p19 (protein data bank ID 1RPU) (Vargason *et al.*, 2003). (B) A canonical dsRBD from *Xenopus laevis* RNA protein A (PDB 1DI2) (Ryter and Schultz, 1998). Note that dsRBDs are highly conserved structure, and the DCV suppressor 1A will likely have a similar structure. (C) Flock House virus (FHV) B2 (PDB 2AZ2) (Chao *et al.*, 2005). The images were produced using UCSF Chimera package (www.cgl.ucsf.edu/chimera) (Pettersen *et al.*, 2004). (See Plate 12 for the color version of this figure.)

recognize the 3′ overhangs, which thus provides specificity for a typical siRNA (Lakatos *et al.*, 2006). Despite the differences in the mechanism of siRNA binding, all three suppressors seem to sequester siRNA, thereby inhibiting the formation of the RNA-silencing initiator complex, and preventing incorporation of an siRNA into an active RISC complex (Lakatos *et al.*, 2006). Silencing suppressors with a size-independent dsRNA-binding activity, such as turnip crinkle virus coat protein (CP) and aureusvirus p14, likely inhibit processing of viral dsRNA into siRNA (Merai *et al.*, 2005, 2006).

Binding of dsRNA or siRNA is by no means the only mechanism by which plant viruses suppress RNAi. Polerovirus P0 is an inhibitor of RNAi that interacts with E3 ubiquitin ligase complex via an F-box motif (Pazhouhandeh *et al.*, 2006), and targets Argonaute family members for degradation (Baumberger *et al.*, 2007). The degradation mechanism, however, is insensitive to proteasome inhibitors, and therefore does not seem to involve the standard ubiquitination and proteasome pathway. Furthermore, the RNA-silencing suppressor from cucumber mosaic virus, 2b, physically interacts with Ago-1 and directly inhibits its Slicer activity (Zhang *et al.*, 2006).

Insect viruses have not been tested as extensively as plant viruses for RNAi suppressive activity. Thus far, RNAi suppressors have been identified in viruses that infect different insect species, FHV, CrPV, and DCV (Li *et al.*, 2002; van Rij *et al.*, 2006; Wang *et al.*, 2006). Furthermore, suppressors of RNA silencing were identified in betanodaviruses (from the same virus family as the alphanodavirus FHV) that infect fish. Of note, fish do have dsRNA-activated interferon response which indicates that the evolution of such immune responses does not preclude an antiviral function of the more "ancient" RNA-silencing system.

The two best-defined RNAi suppressors from insect viruses, DCV 1A and FHV B2, rely on dsRNA binding for their suppressive activity (Li *et al.*, 2002; van Rij *et al.*, 2006), however via distinct mechanisms. DCV 1A contains a canonical dsRNA-binding domain (dsRBD), which provides high-affinity, sequence non-specific binding to dsRNA (Figure 7.3B) (van Rij *et al.*, 2006).

DCV 1A efficiently binds to long dsRNA, but not to siRNA, and consequently inhibits cleavage of dsRNA by Dicer. Due to the low affinity for siRNAs, DCV 1A does not efficiently inhibit siRNA-initiated RNAi, or the miRNA pathway. Within the dicistrovirus family, CrPV is the closest relative to DCV, and also encodes an RNAi suppressor (Wang *et al.*, 2006). Strikingly, the suppressor maps to the same location in the viral genome, but there is no sequence similarity between the two viruses in this region, and they suppress RNAi via distinct mechanisms; CrPV does not inhibit processing of dsRNA into siRNA (van Rij *et al.*, 2006). Thus it seems that within a single virus family an RNAi suppressor activity has independently evolved in two closely related viruses (Figure 7.4).

FHV B2 binds dsRNA via a different structural solution (Figure 7.3C) (Chao *et al.*, 2005; Lingel *et al.*, 2005). B2 can bind as a dimer to dsRNA between 17 and 25 bp with comparable affinity, independent of the two-base 3′ overhangs (Chao *et al.*, 2005; Lingel *et al.*, 2005), and binds longer dsRNA with an even higher affinity (Lu *et al.*, 2005). FHV B2 thus has two different mechanism of action, it can bind siRNAs and inhibit RNAi via a similar mechanism as p19, p21, and HcPro. In addition, FHV B2 may bind to long dsRNA and prevent processing of dsRNA into siRNAs (Chao *et al.*, 2005; Lingel *et al.*, 2005; Lu *et al.*, 2005). FHV B2 dsRNA binding shares some characteristics with canonical dsRBDs: the protein binds sequence non-specific to two minor grooves and the intervening major groove, on one face of a dsRNA. However, the overall protein architecture is completely different from a dsRBD. B2 from the distantly related betanodaviruses, which infect fish, and from nodamuravirus, an alphanodavirus that infects rodents and insects, suppress RNAi via a similar mechanism as FHV B2 (Sullivan and Ganem, 2005a; Fenner *et al.*, 2006a, 2006b).

B2 proteins from alphanodaviruses and the betanodaviruses show little sequence similarity. However, since the mechanism of action and genomic location are similar, it is likely that the B2-silencing suppressors from alpha- and betanodaviruses have a common evolutionary origin, even though the amino acid sequences have diverged beyond detectable similarity. Similarly, the silencing suppressors from the aureusvirus and tombusvirus genera of the plant tombusvirus family, p14 and p19 respectively, seem to derive from a common ancestor, despite limited sequence similarity between the genera. Both p14 and p19 bind dsRNA, and regions of similarity correspond with regions that are important for dsRNA binding. However, in contrast to p19, p14 does not display a size specificity, a feature that corresponds with the lack of the alpha-helical reading head that interacts with the end of an siRNA (Merai *et al.*, 2005). Thus, the silencing suppressors in these (and other (Li and Ding, 2006)) examples were already present in an early ancestral virus for these virus families. In contrast, the acquisition of silencing suppressor in the dicistrovirus family seems to be a relatively recent evolutionary event that likely occurred after DCV and CrPV diverged from their most recent common ancestor.

An interesting aspect of many viral RNAi suppressors is that, based on the available structures, there appears to be no similarity with known eukaryotic proteins. This raises the question of how these proteins arose, and whether there are homologous unknown host proteins with similar structures regulating RNAi activities.

In conclusion, insect and plant viruses have evolved mechanisms to suppress the antiviral RNA silencing response. The molecular mechanisms may differ between different virus families, and even within virus families. This resembles the variety of mechanisms by which mammalian viruses suppress the different branches of the antiviral immune response, and underscores the importance of RNAi as an antiviral response.

Cellular miRNA

Genomic 3′ UTRs are under miRNA selection pressure during evolution; miRNA target sites are under-represented in 3′ UTRs of genes that are co-expressed with that particular miRNA in the same cell type (Farh *et al.*, 2005; Stark

(A)

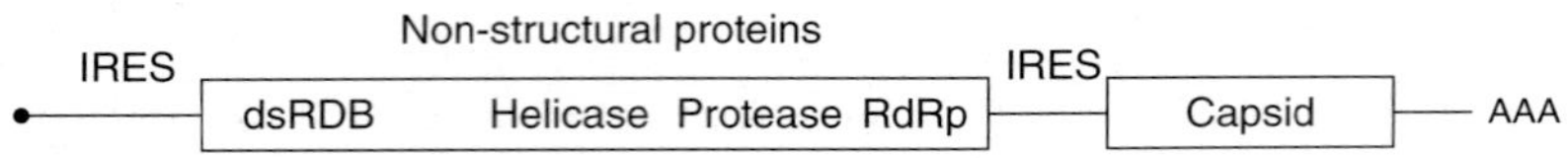

(B)

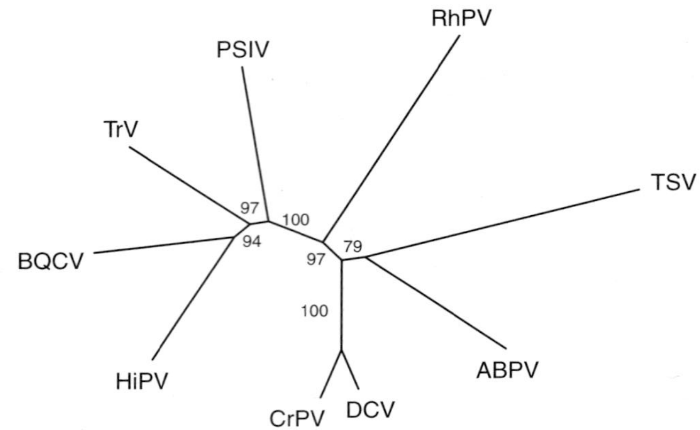

(C)

```
CrPV  1 MSFQQTNNNATNNINSLEELAAQELIAAQFEGNLDGFFCTFYVQSKPQLLDLESECYCMDDFDCGCDRIKREEELRKLIFLTSDVYGYNFEEWKGLVWKF 100
DCV                                                                              1 M-SDKSMACLNRILMNKMMFV-D-ISTL-M  30

                    CrPV RNA silencing suppressor <
CrPV    V=QNYCPEHRYGSTFGNGLLIVSPRFFMDHLDWFQQWKLVSSNDECRAFLRKRTQLLMSGDVESNPGPVQSRPVYACDNDPRAIRLEKALQRRDEKISTL 199
DCV     -ADY-QK-VK-DFDAVESPREAPVFRCTCRFLGYTIMTQGIGKKNPKQEAARQML--L-----T-----------YRY----YT-----IE---D--K-- 130
                                               DCV RNA silencing suppressor <

CrPV    IKKLRQEIKNNRIYTQGFF=======DDLKGAKGEVGQLNGNLTRICDFLENSLPTLTAQIQTTVLTTTDKYVNLKEDLLKVAILLVLVRLLMVWKKYRA 292
DCV     --E--RQ---RK--S--M-DKLTKQIS-GIKDGVGSE-M------------T--G-Q-N--A--ID------S----IM-IVLVIL------------ 230
```

FIGURE 7.4 Independent convergent evolution of an RNAi suppressor function in closely related viruses. (A) Schematic representation of the *Drosophila* C virus (DCV) genome. The (+)-stranded RNA genome of approximately 9200 nt is modified by a 5′ covalent protein attachment and a 3′ poly A tail. Translation is driven by two internal ribosome entry sites. A canonical dsRNA-binding domain (dsRBD) is present in the DCV genome, but not in cricket paralysis virus (CrPV) or other members of the dicistrovirus family. (B) Phylogenetic tree of viruses from the dicistrovirus family based on the amino acid sequence of open reading frame (ORF) 1. Alignments were generated using Clustal W (Thompson *et al.*, 1994). Neighbor-joining phylogenetic trees were generated using Protdist and Neighbor as implemented in the Phylip package (Felsenstein, 1989). Statistical support for the tree was evaluated by bootstrapping using Phylip's Seqboot and Consense. DCV, *Drosophila* C virus (accession number NP44945); CrPV, cricket paralysis virus (NP647481); HiPV, Himetobi P virus (AB017037); BQCV, black queen cell virus (AF183905); TrV, triatoma virus (AF178440); PSIV, *Plautia stali* intestine virus (AB006531); RhPV, *Rhopalosiphum padi* virus (AF022937); TSV, taura syndrome virus (AF277675); ABPV, acute bee paralysis virus (AF150629). Numbers indicate bootstrap values, the number of times the depicted clustering occurred in 100 replicate trees. (C) Protein alignment of the first N-terminal 292 amino acids (aa) of CrPV and 230 aa of DCV. —indicates identity; = indicates gaps inserted for proper alignment. RNA-silencing suppressors were mapped to the first 99 aa of DCV, which includes the dsRBD, in bold, and the first 140 aa of CrPV (van Rij *et al.*, 2006; Wang *et al.*, 2006). The RNA-silencing suppressors are highly dissimilar in sequence (and therefore do not readily align) and mechanism of action (van Rij *et al.*, 2006). The remainder of ORF-1 and ORF-2 of DCV and CrPV are closely related, with 54.9% identity at the amino acid level in ORF-1 and 64.1% identity in ORF-2.

et al., 2005), and genes that are involved in basic cellular processes tend to have smaller 3′ UTRs that are depleted of miRNA target sites (Stark *et al.*, 2005). As illustrated earlier, miRNAs may interact with viral genomes (Gitlin *et al.*, 2005; Lecellier *et al.*, 2005; Brown *et al.*, 2006), and therefore miRNAs may affect viral evolution. Given that specificity is mainly provided by a short 7-nt seed match (Doench and Sharp, 2004; Brennecke *et al.*, 2005; Lewis *et al.*, 2005), and that the error-prone nature of viral polymerases generates a swarm of viral mutants during replication, target sites for miRNAs may be generated by chance during replication. Newly emerging miRNA target sites may be selected against by the presence of that miRNA in the target cell. If this is indeed the case, it can be predicted that miRNA target sites for such miRNAs will be depleted from viral sequences, whereas there will be little selection against target sites for miRNAs that are not expressed in the cell type in which the virus replicates. Analyses using publicly available miRNA expression data, indeed suggest that HIV-1 sequences are depleted of target sites for miRNAs that are highly expressed in their target cells, CD4+ T cells. Conversely, targets sites do occur for miRNAs that are not expressed in CD4+ T cells (R.P. van Rij, unpublished observations). These results provide support to the idea that miRNA may indeed exert selection pressure on the possible genomic sequences that a viral population may sample. If this holds true in a larger analysis, these results imply that the host miRNA profile will directly affect the sequence space that is available to a viral quasispecies.

In a more extreme case, miRNA target sites may define viral tropism. Tissue-specific expression of a miRNA may prevent infection by a viral genome that contains a conserved miRNA target site for that miRNA. An analogous regulation for cell type-specific expression was recently suggested for the *Drosophila* transcription factor Nerfin-1, which may be expressed specifically in the nervous system due to miRNA-mediated repression in non-nervous system tissues (Stark *et al.*, 2005).

Virally Encoded miRNA

A short 7-nt seed region is the specificity determinant of a miRNA, and may be solely sufficient for target regulation (Doench and Sharp, 2004; Brennecke *et al.*, 2005; Lewis *et al.*, 2005). Therefore, a single miRNA may have an enormous potential to regulate gene expression; indeed, roughly 200 targets have been predicted per cellular miRNA (Rajewsky, 2006). Viral miRNAs often derive from non-coding regions of the genome and are therefore not under the same evolutionary constraints as protein coding sequences. Single point mutations in miRNA genes may dramatically alter the set of genes that are regulated by the miRNA. This is especially true for the seed region, but regions outside the seed region may also affect target site recognition, or miRNA processing, and, consequently, mature miRNA expression (Gottwein *et al.*, 2006). In a more extreme case, *de novo* evolution or loss of viral miRNAs may further alter the repertoire of host genes that are regulated by the virus through RNA silencing. Adaptive evolution of viral miRNAs, therefore, may present viruses with a versatile tool to regulate both host (and viral) gene expression, and may contribute to successful virus speciation, co-evolution of virus and host, and cross-species infections. Indeed, miRNAs that were cloned from CMV, EBV, and Kaposi's sarcoma-associated herpesvirus (KSHV) members of the herpesvirus family, which shows characteristics of co-evolution of virus and host over prolonged evolutionary timespans, show no sequence similarity to each other. In that respect, it might be more surprising that nearly half of the viral miRNAs are conserved between rhesus lymphocryptic virus and EBV, two members of the lymphocryptovirus genus, that diverged an estimated >13 million years ago (Cai *et al.*, 2006). Conservation was most pronounced in the seed region, and three miRNAs (including a miRNA and miRNA* pair) were completely identical. Conservation of miRNAs over such extremely long periods of independent evolution indicates positive selection for miRNA function, and conservation of their targets. Similarly, conserved miRNAs have been predicted in other pairs of closely related viruses that infect different primate species, human and chimpanzee CMV, and polyomaviruses SV40 and SA12 (Cullen, 2006a; Sullivan, 2007).

On a shorter evolutionary timeframe, adaptive evolution of viral miRNAs may even contribute to pathogenesis within an infected host. Polymorphisms in KSHV miRNAs from different viral isolates have been identified (Gottwein *et al.*, 2006; Marshall *et al.*, 2007), of which at least one was shown to affect Drosha processing, and, consequently, mature miRNA levels in the cell. If, and how these polymorphisms contribute to viral pathogenesis remains to be further established.

CONCLUDING REMARKS

We are only beginning to understand the mechanistic details and full regulatory potential of RNA-silencing pathways, and how viruses interact with these pathways. On the other hand, viruses have "studied" the RNAi machinery over millions of years of virus–host co-evolution. Selection pressure exerted by RNAi as an adaptive nucleic acid-based antiviral system resulted in the evolution of viral suppressors of RNAi in plants and insect viruses. Viruses suppress RNAi via a multitude of mechanisms, interacting with different players of the RNAi machinery, and even interacting via different molecular mechanisms with the same molecule (e.g. dsRNA, Figure 7.3). The mechanism of suppression of RNAi can even differ within single virus families, suggestive of a relatively recent acquisition of RNAi suppressive activity, perhaps as a part of the evolutionary adaptation to a new host. Nuclear DNA viruses have evolved a strategy to exploit the miRNA pathway by encoding miRNA in their genomes. While viral miRNAs provide viruses with a versatile platform to regulate host gene expression, some viral miRNAs have remained remarkably conserved during millions years of evolution. Finally, the finding that hepatitis C virus seems to require a physical interaction with a cellular miRNA for replication suggests that there is much to be learned about viruses, RNA silencing, and their interactions.

ACKNOWLEDGMENTS

Research on RNA silencing in the laboratory of Raul Andino is supported by NIH grants AI40085 and AI64738. We thank Laurens Kraal for help with miRNA target predictions, Drs Adam Lauring and Carla Saleh and members of the Ketting lab for useful discussions.

REFERENCES

Ambros, V. (2004) The functions of animal microRNAs. *Nature* **431**, 350–355.

Andersson, M.G., Haasnoot, P.C., Xu, N., Berenjian, S., Berkhout, B. and Akusjarvi, G. (2005) Suppression of RNA interference by adenovirus virus-associated RNA. *J. Virol.* **79**, 9556–9565.

Aparicio, O., Razquin, N., Zaratiegui, M., Narvaiza, I. and Fortes, P. (2006) Adenovirus virus-associated RNA is processed to functional interfering RNAs involved in virus production. *J. Virol.* **80**, 1376–1384.

Baulcombe, D. (2004) RNA silencing in plants. *Nature* **431**, 356–363.

Baumberger, N. and Baulcombe, D.C. (2005) Arabidopsis ARGONAUTE1 is an RNA Slicer that selectively recruits microRNAs and short interfering RNAs. *Proc. Natl Acad. Sci. USA*, **102**, 11928–11933.

Baumberger, N., Tsai, C.H., Lie, M., Havecker, E. and Baulcombe, D.C. (2007) The polerovirus silencing suppressor P0 targets Argonaute proteins for degradation. *Curr. Biol.* in press.

Bennasser, Y., Le, S.Y., Benkirane, M. and Jeang, K.T. (2005) Evidence that HIV-1 encodes an siRNA and a suppressor of RNA silencing. *Immunity* **22**, 607–619.

Bentwich, I., Avniel, A., Karov, Y., Aharonov, R., Gilad, S., Barad, O. *et al.* (2005) Identification of hundreds of conserved and nonconserved human microRNAs. *Nat. Genet.* **37**, 766–770.

Berezikov, E., Guryev, V., van de Belt, J., Wienholds, E., Plasterk, R.H. and Cuppen, E. (2005) Phylogenetic shadowing and computational identification of human microRNA genes. *Cell* **120**, 21–24.

Bergmann, M., Garcia-Sastre, A., Carnero, E., Pehamberger, H., Wolff, K., Palese, P. and Muster, T. (2000) Influenza virus NS1 protein counteracts PKR-mediated inhibition of replication. *J. Virol.* **74**, 6203–6206.

Bernstein, E., Caudy, A.A., Hammond, S.M. and Hannon, G.J. (2001) Role for a bidentate ribonuclease in the initiation step of RNA interference. *Nature* **409**, 363–366.

Bhattacharyya, S.N., Habermacher, R., Martine, U., Closs, E.I. and Filipowicz, W. (2006) Relief of microRNA-mediated translational repression in human cells subjected to stress. *Cell* **125**, 1111–1124.

Blakqori, G., Delhaye, S., Habjan, M., Blair, C.D., Sanchez-Vargas, I., Olson, K.E. *et al.* (2007) La Crosse bunyavirus nonstructural protein NSs serves to suppress the type I interferon system of mammalian hosts. *J. Virol.* **81**, 4991–4999.

Blevins, T., Rajeswaran, R., Shivaprasad, P.V., Beknazariants, D., Si-Ammour, A., Park, H.S. *et al.* (2006) Four plant Dicers mediate viral small RNA biogenesis and DNA virus induced silencing. *Nucleic Acids Res.* **34**, 6233–6246.

Boden, D., Pusch, O., Lee, F., Tucker, L. and Ramratnam, B. (2003) Human immunodeficiency virus type 1 escape from RNA interference. *J. Virol.* **77**, 11531–11535.

Bouche, N., Lauressergues, D., Gasciolli, V. and Vaucheret, H. (2006) An antagonistic function for Arabidopsis DCL2 in development and a new function for DCL4 in generating viral siRNAs. *EMBO J.* **25**, 3347–3356.

Brennecke, J., Stark, A., Russell, R.B. and Cohen, S.M. (2005) Principles of microRNA-target recognition. *PLoS Biol.* **3**, e85.

Brodersen, P. and Voinnet, O. (2006) The diversity of RNA silencing pathways in plants. *Trends Genet.* **22**, 268–280.

Brown, B.D., Venneri, M.A., Zingale, A., Sergi Sergi, L. and Naldini, L. (2006) Endogenous microRNA regulation suppresses transgene expression in hematopoietic lineages and enables stable gene transfer. *Nat. Med.* **12**, 585–591.

Bucher, E., Hemmes, H., de Haan, P., Goldbach, R. and Prins, M. (2004) The influenza A virus NS1 protein binds small interfering RNAs and suppresses RNA silencing in plants. *J. Gen. Virol.* **85**, 983–991.

Cai, X., Schafer, A., Lu, S., Bilello, J.P., Desrosiers, R.C., Edwards, R. *et al.* (2006) Epstein-Barr virus microRNAs are evolutionarily conserved and differentially expressed. *PLoS Pathog.* **2**, e23.

Chao, J.A., Lee, J.H., Chapados, B.R., Debler, E.W., Schneemann, A. and Williamson, J.R. (2005) Dual modes of RNA-silencing suppression by Flock House virus protein B2. *Nat. Struct. Mol. Biol.* **12**, 952–957.

Chellappan, P., Vanitharani, R. and Fauquet, C.M. (2004) Short interfering RNA accumulation correlates with host recovery in DNA virus-infected hosts, and gene silencing targets specific viral sequences. *J. Virol.* **78**, 7465–7477.

Chendrimada, T.P., Gregory, R.I., Kumaraswamy, E., Norman, J., Cooch, N., Nishikura, K. and Shiekhattar, R. (2005) TRBP recruits the Dicer complex to Ag02 for microRNA processing and gene silencing. *Nature* **436**, 740–744.

Cherry, S., Doukas, T., Armknecht, S., Whelan, S., Wang, H., Sarnow, P. and Perrimon, N. (2005) Genome-wide RNAi screen reveals a specific sensitivity of IRES-containing RNA viruses to host translation inhibition. *Genes Dev.* **19**, 445–452.

Cullen, B.R. (2006a) Viruses and microRNAs. *Nat. Genet.* **38**(Suppl), S25–S30.

Cullen, B.R. (2006b) Is RNA interference involved in intrinsic antiviral immunity in mammals?. *Nat. Immunol.* **7**, 563–567.

Das, A.T., Brummelkamp, T.R., Westerhout, E.M., Vink, M., Madiredjo, M., Bernards, R. and Berkhout, B. (2004) Human immunodeficiency virus type 1 escapes from RNA interference-mediated inhibition. *J. Virol.* **78**, 2601–2605.

Deleris, A., Gallego-Bartolome, J., Bao, J., Kasschau, K.D., Carrington, J.C. and Voinnet, O. (2006) Hierarchical action and inhibition of plant dicer-like proteins in antiviral defense. *Science* **313**, 68–71.

Denli, A.M., Tops, B.B., Plasterk, R.H., Ketting, R.F. and Hannon, G.J. (2004) Processing of primary microRNAs by the Microprocessor complex. *Nature* **432**, 231–235.

Doench, J.G. and Sharp, P.A. (2004) Specificity of microRNA target selection in translational repression. *Genes Dev.* **18**, 504–511.

Farh, K.K., Grimson, A., Jan, C., Lewis, B.P., Johnston, W.K., Lim, L.P. *et al.* (2005) The widespread impact of mammalian MicroRNAs on mRNA repression and evolution. *Science* **310**, 1817–1821.

Felsenstein, J. (1989) PHYLIP-Phylogeny inference package. *Cladistics* **5**, 164–166.

Fenner, B.J., Goh, W. and Kwang, J. (2006a) Sequestration and protection of double-stranded RNA by the betanodavirus b2 protein. *J. Virol.* **80**, 6822–6833.

Fenner, B.J., Thiagarajan, R., Chua, H.K. and Kwang, J. (2006b) Betanodavirus B2 is an RNA interference antagonist that facilitates intracellular viral RNA accumulation. *J. Virol.* **80**, 85–94.

Fire, A., Xu, S., Montgomery, M.K., Kostas, S.A., Driver, S. E. and Mello, C.C. (1998) Potent and specific genetic interference by double-stranded RNA in *Caenorhabditis elegans*. *Nature* **391**, 806–811.

Forstemann, K., Tomari, Y., Du, T., Vagin, V.V., Denli, A. M. *et al.* (2005) Normal microRNA maturation and germ-line stem cell maintenance requires Loquacious, a double-stranded RNA-binding domain protein. *PLoS Biol.* **3**, e236.

Gaines, P.J., Olson, K.E., Higgs, S., Powers, A.M., Beaty, B. J. and Blair, C.D. (1996) Pathogen-derived resistance to dengue type 2 virus in mosquito cells by expression of the premembrane coding region of the viral genome. *J. Virol.* **70**, 2132–2137.

Galiana-Arnoux, D., Dostert, C., Schneemann, A., Hoffmann, J.A. and Imler, J.L. (2006) Essential function in vivo for Dicer-2 in host defense against RNA viruses in drosophila. *Nat. Immunol.* **7**, 590–597.

Garcia, S., Billecocq, A., Crance, J.M., Munderloh, U., Garin, D. and Bouloy, M. (2005) Nairovirus RNA sequences expressed by a Semliki Forest virus replicon induce RNA interference in tick cells. *J. Virol.* **79**, 8942–8947.

Garrus, J.E., von Schwedler, U.K., Pornillos, O.W., Morham, S.G., Zavitz, K.H., Wang, H.E. *et al.* (2001) Tsg101 and the vacuolar protein sorting pathway are essential for HIV-1 budding. *Cell* **107**, 55–65.

Gitlin, L. and Andino, R. (2003) Nucleic Acid-based immune system: the antiviral potential of Mammalian RNA silencing. *J. Virol.* **77**, 7159–7165.

Gitlin, L., Karelsky, S. and Andino, R. (2002) Short interfering RNA confers intracellular antiviral immunity in human cells. *Nature* **418**, 430–434.

Gitlin, L., Stone, J.K. and Andino, R. (2005) Poliovirus escape from RNAi: siRNA-target recognition and implications for therapeutic approaches. *J. Virol.* **79**, 1027–1035.

Gottwein, E., Cai, X. and Cullen, B.R. (2006) A novel assay for viral microRNA function identifies a single nucleotide polymorphism that affects Drosha processing. *J. Virol.* **80**, 5321–5326.

Gupta, A., Gartner, J.J., Sethupathy, P., Hatzigeorgiou, A. G. and Fraser, N.W. (2006) Anti-apoptotic function of a microRNA encoded by the HSV-1 latency-associated transcript. *Nature* **442**, 82–85.

Haase, A.D., Jaskiewicz, L., Zhang, H., Laine, S., Sack, R., Gatignol, A. and Filipowicz, W. (2005) TRBP, a regulator of cellular PKR and HIV-1 virus expression, interacts with Dicer and functions in RNA silencing. *EMBO Rep.* **6**, 961–967.

Hamilton, A.J. and Baulcombe, D.C. (1999) A species of small antisense RNA in posttranscriptional gene silencing in plants. *Science* **286**, 950–952.

Hannon, G.J. and Rossi, J.J. (2004) Unlocking the potential of the human genome with RNA interference. *Nature* **431**, 371–378.

Hemmes, H., Lakatos, L., Goldbach, R., Burgyan, J. and Prins, M. (2007) The NS3 protein of Rice hoja blanca tenuivirus suppresses RNA silencing in plant and insect hosts by efficiently binding both siRNAs and miRNAs. *RNA* **13**, 1079–1089.

Himber, C., Dunoyer, P., Moissiard, G., Ritzenthaler, C. and Voinnet, O. (2003) Transitivity-dependent and—independent cell-to-cell movement of RNA silencing. *EMBO J.* **22**, 4523–4533.

Jiang, F., Ye, X., Liu, X., Fincher, L., McKearin, D. and Liu, Q. (2005) Dicer-1 and R3D1-L catalyze microRNA maturation in Drosophila. *Genes Dev.* **19**, 1674–1679.

John, B., Enright, A.J., Aravin, A., Tuschl, T., Sander, C. and Marks, D.S. (2004) Human MicroRNA targets. *PLoS Biol.* **2**, e363.

Jopling, C., MinKyung, Y., Lancaster, A.M., Lemon, S.M. and Sarnow, P. (2005) Modulation of Hepatitis C virus RNA abundance by a liver-specific microRNA. *Science* **309**, 1577–1581.

Keene, K.M., Foy, B.D., Sanchez-Vargas, I., Beaty, B.J., Blair, C.D. and Olson, K.E. (2004) From the Cover: RNA interference acts as a natural antiviral response to O'nyong-nyong virus (Alphavirus; Togaviridae) infection of *Anopheles gambiae*. *Proc. Natl Acad. Sci. USA* **101**, 17240–17245.

Kiriakidou, M., Tan, G.S., Lamprinaki, S., De Planell-Saguer, M., Nelson, P.T. and Mourelatos, Z. (2007) An mRNA m7G cap binding-like motif within human Ag02 represses translation. *Cell* **129**, 1141–1151.

Kok, K.H. and Jin, D.Y. (2006) Influenza A virus NS1 protein does not suppress RNA interference in mammalian cells. *J. Gen. Virol.* **87**, 2639–2644.

Lakatos, L., Csorba, T., Pantaleo, V., Chapman, E.J., Carrington, J.C., Liu, Y.P. *et al.* (2006) Small RNA binding is a common strategy to suppress RNA silencing by several viral suppressors. *EMBO J.* **25**, 2768–2780.

Lecellier, C.H., Dunoyer, P., Arar, K., Lehmann-Che, J., Eyquem, S., Himber, C. *et al.* (2005) A cellular microRNA mediates antiviral defense in human cells. *Science* **308**, 557–560.

Lee, Y., Ahn, C., Han, J., Choi, H., Kim, J., Yim, J. *et al.* (2003) The nuclear RNase III Drosha initiates microRNA processing. *Nature* **425**, 415–419.

Lee, Y.S., Nakahara, K., Pham, J.W., Kim, K., He, Z., Sontheimer, E.J. and Carthew, R.W. (2004) Distinct roles for Drosophila Dicer-1 and Dicer-2 in the siRNA/miRNA silencing pathways. *Cell* **117**, 69–81.

Lewis, B.P., Burge, C.B. and Bartel, D.P. (2005) Conserved seed pairing, often flanked by adenosines, indicates that thousands of human genes are microRNA targets. *Cell* **120**, 15–20.

Li, F. and Ding, S.W. (2006) Virus counterdefense: diverse strategies for evading the RNA-silencing immunity. *Annu. Rev. Microbiol.* **60**, 503–531.

Li, H.W., Li, W.X. and Ding, S.W. (2002) Induction and suppression of RNA silencing by an animal virus. *Science* **296**, 1319–1321.

Li, W.X., Li, H., Lu, R., Li, F., Dus, M., Atkinson, P. *et al.* (2004) Interferon antagonist proteins of influenza and vaccinia viruses are suppressors of RNA silencing. *Proc. Natl Acad. Sci. USA* **101**, 1350–1355.

Lichner, Z., Silhavy, D. and Burgyan, J. (2003) Double-stranded RNA-binding proteins could suppress RNA interference-mediated antiviral defences. *J. Gen. Virol.* **84**, 975–980.

Lingel, A., Simon, B., Izaurralde, E. and Sattler, M. (2005) The structure of the flock house virus B2 protein, a viral suppressor of RNA interference, shows a novel mode of double-stranded RNA recognition. *EMBO Rep.* **6**, 1149–1155.

Lippman, Z.. and Martienssen, R. (2004) The role of RNA interference in heterochromatic silencing. *Nature* **431**, 364–370.

Liu, J., Carmell, M.A., Rivas, F.V., Marsden, C.G., Thomson, J.M., Song, J.J. *et al.* (2004) Argonaute2 is the catalytic engine of mammalian RNAi. *Science* **305**, 1437–1441.

Liu, J., Valencia-Sanchez, M.A., Hannon, G.J. and Parker, R. (2005) MicroRNA-dependent localization of targeted mRNAs to mammalian P-bodies. *Nat. Cell Biol.* **7**, 719–723.

Liu, Q., Rand, T.A., Kalidas, S., Du, F., Kim, H.E., Smith, D.P. and Wang, X. (2003) R2D2, a bridge between the initiation and effector steps of the Drosophila RNAi pathway. *Science* **301**, 1921–1925.

Lu, R., Folimonov, A., Shintaku, M., Li, W.X., Falk, B. W., Dawson, W.O. and Ding, S.W. (2004) Three distinct suppressors of RNA silencing encoded by a 20-kb viral RNA genome. *Proc. Natl Acad. Sci. USA* **101**, 15742–15747.

Lu, R., Maduro, M., Li, F., Li, H.W., Broitman-Maduro, G., Li, W.X. and Ding, S.W. (2005) Animal virus replication and RNAi-mediated antiviral silencing in *Caenorhabditis elegans*. *Nature* **436**, 1040–1043.

Lu, S. and Cullen, B.R. (2004) Adenovirus VA1 noncoding RNA can inhibit small interfering RNA and MicroRNA biogenesis. *J. Virol.* **78**, 12868–12876.

Lund, E., Guttinger, S., Calado, A., Dahlberg, J.E. and Kutay, U. (2004) Nuclear export of microRNA precursors. *Science* **303**, 95–98.

Macrae, I.J., Zhou, K., Li, F., Repic, A., Brooks, A. N., Cande, W.Z. *et al.* (2006) Structural basis for double-stranded RNA processing by Dicer. *Science* **311**, 195–198.

Marshall, V., Parks, T., Bagni, R., Wang, C.D., Samols, M.A., Hu, J. *et al.* (2007) Conservation of virally encoded microRNAs in Kaposi sarcoma-associated herpesvirus in primary effusion lymphoma cell lines and in patients with Kaposi sarcoma or multicentric Castleman disease. *J. Infect. Dis.* **195**, 645–659.

Mathonnet, G., Fabian, M.R., Svitkin, Y.V., Parsyan, A., Huck, L., Murata, T. *et al.* (2007) MicroRNA inhibition of translation initiation in vitro by targeting the cap-binding complex eIF4F. *Science* **317**, 1764–1767.

Meister, G. and Tuschl, T. (2004) Mechanisms of gene silencing by double-stranded RNA. *Nature* **431**, 343–349.

Merai, Z., Kerenyi, Z., Molnar, A., Barta, E., Valoczi, A., Bisztray, G. *et al.* (2005) Aureusvirus P14 is an efficient RNA silencing suppressor that binds double-stranded RNAs without size specificity. *J. Virol.* **79**, 7217–7226.

Merai, Z., Kerenyi, Z., Kertesz, S., Magna, M., Lakatos, L. and Silhavy, D. (2006) Double-stranded RNA binding may be a general plant RNA viral strategy to suppress RNA silencing. *J. Virol.* **80**, 5747–5756.

Miska, E.A. and Ahringer, J. (2007) RNA interference has second helpings. *Nat. Biotechnol.* **25**, 302–303.

Moissiard, G. and Voinnet, O. (2006) RNA silencing of host transcripts by cauliflower mosaic virus requires coordinated action of the four *Arabidopsis* Dicer-like proteins. *Proc. Natl Acad. Sci. USA,* **103**, 19593–19598.

Moissiard, G., Parizotto, E.A., Himber, C. and Voinnet, O. (2007) Transitivity in *Arabidopsis* can be primed, requires the redundant action of the antiviral Dicer-like 4 and Dicer-like 2 and is compromised by viral-encoded suppressor proteins. *RNA* **13**, 1268–1278.

Molnar, A., Csorba, T., Lakatos, L., Varallyay, E., Lacomme, C. and Burgyan, J. (2005) Plant virus-derived small interfering RNAs originate predominantly from highly structured single-stranded viral RNAs. *J. Virol.* **79**, 7812–7818.

Morel, J.B., Godon, C., Mourrain, P., Beclin, C., Boutet, S., Feuerbach, F. *et al.* (2002) Fertile hypomorphic ARGONAUTE (ag01) mutants impaired in post-transcriptional gene silencing and virus resistance. *Plant Cell* **14**, 629–639.

Mourrain, P., Beclin, C., Elmayan, T., Feuerbach, F., Godon, C., Morel, J.B. *et al.* (2000) Arabidopsis SGS2 and SGS3 genes are required for posttranscriptional gene silencing and natural virus resistance. *Cell* **101**, 533–542.

Muller, S. and Imler, J.L. (2007) Dicing with viruses: microRNAs as antiviral factors. *Immunity* **27**, 1–3.

Murray, S.M., Picker, L.J., Axthelm, M.K. and Linial, M. L. (2006) Expanded tissue targets for foamy virus replication with simian immunodeficiency virus-induced immunosuppression. *J. Virol.* **80**, 663–670.

O'Donnell, K.A. and Boeke, J.D. (2007) Mighty Piwis defend the germline against genome intruders. *Cell* **129**, 37–44.

Obbard, D.J., Jiggins, F.M., Halligan, D.L. and Little, T.J. (2006) Natural selection drives extremely rapid evolution in antiviral RNAi genes. *Curr. Biol.* **16**, 580–585.

Okamura, K., Ishizuka, A., Siomi, H. and Siomi, M.C. (2004) Distinct roles for Argonaute proteins in small RNA-directed RNA cleavage pathways. *Genes Dev.* **18**, 1655–1666.

Olson, K.E., Higgs, S., Gaines, P.J., Powers, A.M., Davis, B.S., Kamrud, K.I. *et al.* (1996) Genetically engineered resistance to dengue-2 virus transmission in mosquitoes. *Science* **272**, 884–886.

Otsuka, M., Jing, Q., Georgel, P., New, L., Chen, J., Mols, J. *et al.* (2007) Hypersusceptibility to vesicular stomatitis virus infection in Dicer1-deficient mice is due to impaired miR24 and miR93 expression. *Immunity* **27**, 123–134.

Pantaleo, V., Szittya, G. and Burgyan, J. (2007) Molecular bases of viral RNA targeting by viral small interfering RNA-programmed RISC. *J. Virol.* **81**, 3797–3806.

Pazhouhandeh, M., Dieterle, M., Marrocco, K., Lechner, E., Berry, B., Brault, V. *et al.* (2006) F-box-like domain in the polerovirus protein P0 is required for silencing suppressor function. *Proc. Natl Acad. Sci. USA* **103**, 1994–1999.

Pelisson, A., Sarot, E., Payen-Groschene, G. and Bucheton, A. (2007) A novel repeat-associated small interfering RNA-mediated silencing pathway downregulates complementary sense gypsy transcripts in somatic cells of the *Drosophila* ovary. *J. Virol.* **81**, 1951–1960.

Pettersen, E.F., Goddard, T.D., Huang, C.C., Cough, G.S., Greenblatt, D.M., Meng, E.C. and Ferrin, T.E. (2004) UCSF chimera—a visualization system for exploratory research and analysis. *J. Comput. Chem.* **25**, 1605–1612.

Pfeffer, S., Zavolan, M., Grasser, F.A., Chien, M., Russo, J.J., Ju, J. *et al.* (2004) Identification of virus-encoded microRNAs. *Science* **304**, 734–736.

Pfeffer, S., Sewer, A., Lagos-Quintana, M., Sheridan, R., Sander, C., Grasser, F.A. *et al.* (2005) Identification of microRNAs of the herpesvirus family. *Nat. Methods,* **2**, 269–276.

Pillai, R.S., Bhattacharyya, S.N. and Filipowicz, W. (2007) Repression of protein synthesis by miRNAs: how many mechanisms?. *Trends Cell Biol.* **17**, 118–126.

Rajewsky, N. (2006) microRNA target predictions in animals. *Nat. Genet.* **38**(Suppl), S8–S13.

Rivas, F.V., Tolia, N.H., Song, J.J., Aragon, J.P., Liu, J., Hannon, G.J. and Joshua-Tor, L. (2005) Purified Argonaute2 and an siRNA form recombinant human RISC. *Nat. Struct. Mol. Biol.* **12**, 340–349.

Roth, B.M., Pruss, G.J. and Vance, V.B. (2004) Plant viral suppressors of RNA silencing. *Virus Res.* **102**, 97–108.

Ryter, J.M. and Schultz, S.C. (1998) Molecular basis of double-stranded RNA–protein interactions: structure of a dsRNA-binding domain complexed with dsRNA. *EMBO J.* **17**, 7505–7513.

Saito, K., Ishizuka, A., Siomi, H. and Siomi, M.C. (2005) Processing of pre-microRNAs by the Dicer-1-Loquacious complex in *Drosophila* cells. *PLoS Biol.* **3**, e235.

Sanchez-Vargas, I., Travanty, E.A., Keene, K.M., Franz, A. W., Beaty, B.J. *et al.* (2004) RNA interference, arthropod-borne viruses and mosquitoes. *Virus Res.* **102**, 65–74.

Sarot, E., Payen-Groschene, G., Bucheton, A. and Pelisson, A. (2004) Evidence for a piwi-dependent RNA silencing of the gypsy endogenous retrovirus by the Drosophila melanogaster flamenco gene. *Genetics* **166**, 1313–1321.

Schott, D.H., Cureton, D.K., Whelan, S.P. and Hunter, C.P. (2005) An antiviral role for the RNA interference machinery in *Caenorhabditis elegans*. *Proc. Natl Acad. Sci. USA* **102**, 18420–18424.

Schwach, F., Vaistij, F.E., Jones, L. and Baulcombe, D.C. (2005) An RNA-dependent RNA polymerase prevents meristem invasion by potato virus X and is required for the activity but not the production of a systemic silencing signal. *Plant Physiol.* **138**, 1842–1852.

Soldan, S.S., Plassmeyer, M.L., Matukonis, M.K. and Gonzalez-Scarano, F. (2005) La Crosse virus nonstructural protein NSs counteracts the effects of short interfering RNA. *J. Virol.* **79**, 234–244.

Stark, A., Brennecke, J., Bushati, N., Russell, R.B. and Cohen, S.M. (2005) Animal MicroRNAs confer robustness to gene expression and have a significant impact on 3′UTR evolution. *Cell* **123**, 1133–1146.

Steiner, F.A. and Plasterk, R.H. (2006) Knocking out the Argonautes. *Cell* **127**, 667–668.

Stern-Ginossar, N., Elefant, N., Zimmermann, A., Wolf, D. G., Saleh, N., Biton, M. *et al.* (2007) Host immune system gene targeting by a viral miRNA. *Science* **317**, 376–381.

Sullivan, C.S. (2007) High conservation of Kaposi sarcoma-associated herpesvirus microRNAs implies important function. *J. Infect. Dis.* **195**, 618–620.

Sullivan, C. and Ganem, D. (2005a) A virus encoded inhibitor that blocks RNA interference in mammalian cells. *J. Virol.* **79**, 7371–7379.

Sullivan, C.S. and Ganem, D. (2005b) MicroRNAs and viral infection. *Mol. Cell* **20**, 3–7.

Sullivan, C.S., Grundhoff, A.T., Tevethia, S., Pipas, J.M. and Ganem, D. (2005) SV40-encoded microRNAs regulate viral gene expression and reduce susceptibility to cytotoxic T cells. *Nature* **435**, 682–686.

Thompson, J.D., Higgins, D.G. and Gibson, T.J. (1994) CLUSTAL W: improving the sensitivity of progressive multiple sequence alignment through sequence weighting, position-specific gap penalties and weight matrix choice. *Nucleic Acids Res.* **22**, 4673–4680.

Triboulet, R., Mari, B., Lin, Y.L., Chable-Bessia, C., Bennasser, Y., Lebrigand, K. *et al.* (2007) Suppression of microRNA-silencing pathway by HIV-1 during virus replication. *Science* **315**, 1579–1582.

Uhlirova, M., Foy, B.D., Beaty, B.J., Olson, K.E., Riddiford, L.M. and Jindra, M. (2003) Use of Sindbis virus-mediated RNA interference to demonstrate a conserved role of Broad-Complex in insect metamorphosis. *Proc. Natl Acad. Sci. USA* **100**, 15607–15612.

Valencia-Sanchez, M.A., Liu, J., Hannon, G.J. and Parker, R. (2006) Control of translation and mRNA degradation by miRNAs and siRNAs. *Genes Dev.* **20**, 515–524.

van Rij, R.P. and Andino, R. (2006) The silent treatment: RNAi as a defense against virus infection in mammals. *Trends Biotechnol.* **24**, 186–193.

van Rij, R.P., Saleh, M.C., Berry, B., Foo, C., Houk, A., Antoniewski, C. and Andino, R. (2006) The RNA silencing endonuclease Argonaute 2 mediates specific antiviral immunity in Drosophila melanogaster. *Genes Dev.* **20**, 2985–2995.

Vanitharani, R., Chellappan, P., Pita, J.S. and Fauquet, C.M. (2004) Differential roles of AC2 and AC4 of cassava geminiviruses in mediating synergism and suppression of posttranscriptional gene silencing. *J. Virol.* **78**, 9487–9498.

Vargason, J.M., Szittya, G., Burgyan, J. and Tanaka Hall, T.M. (2003) Size selective recognition of siRNA by an RNA silencing suppressor. *Cell* **115**, 799–811.

Voinnet, O. (2001) RNA silencing as a plant immune system against viruses. *Trends Genet.* **17**, 449–459.

Voinnet, O. (2005a) Induction and suppression of RNA silencing: insights from viral infections. *Nat. Rev. Genet.* **6**, 206–220.

Voinnet, O. (2005b) Non-cell autonomous RNA silencing. *FEBS Lett.* **579**, 5858–5871.

Voinnet, O., Pinto, Y.M. and Baulcombe, D.C. (1999) Suppression of gene silencing: a general strategy used by diverse DNA and RNA viruses of plants. *Proc. Natl Acad. Sci. USA* **96**, 14147–14152.

Wang, X.H., Aliyari, R., Li, W.X., Li, H.W., Kim, K., Carthew, R. *et al.* (2006) RNA interference directs innate immunity against viruses in adult Drosophila. *Science* **312**, 452–454.

Westerhout, E.M., Ooms, M., Vink, M., Das, A.T. and Berkhout, B. (2005) HIV-1 can escape from RNA interference by evolving an alternative structure in its RNA genome. *Nucleic Acids Res.* **33**, 796–804.

Wienholds, E., Kloosterman, W.P., Miska, E., Alvarez-Saavedra, E., Berezikov, E., de Bruijn, E., Horvitz, H.R.,

Kauppinen, S. and Plasterk, R.H. (2005) MicroRNA expression in zebrafish embryonic development. *Science* **309**, 310–311.

Wilkins, C., Dishongh, R., Moore, S.C., Whitt, M.A., Chow, M. and Machaca, K. (2005) RNA interference is an antiviral defence mechanism in *Caenorhabditis elegans*. *Nature* **436**, 1044–1047.

Wilson, J.A. and Richardson, C.D. (2005) Hepatitis C virus replicons escape RNA interference induced by a short interfering RNA directed against the NS5b coding region. *J. Virol.* **79**, 7050–7058.

Ye, K., Malinina, L. and Patel, D.J. (2003) Recognition of small interfering RNA by a viral suppressor of RNA silencing. *Nature* **426**, 874–878.

Yi, R., Qin, Y., Macara, I.G. and Cullen, B.R. (2003) Exportin-5 mediates the nuclear export of pre-microRNAs and short hairpin RNAs. *Genes Dev.* **17**, 3011–3016.

Yigit, E., Batista, P.J., Bei, Y., Pang, K.M., Chen, C.C., Tolia, N.H. *et al.* (2006) Analysis of the C. elegans Argonaute family reveals that distinct Argonautes act sequentially during RNAi. *Cell* **127**, 747–757.

Zambon, R.A., Vakharia, V.N. and Wu, L.P. (2006) RNAi is an antiviral immune response against a dsRNA virus in *Drosophila melanogaster*. *Cell Microbiol.* **8**, 880–889.

Zamore, P.D. (2004) Plant RNAi: How a viral silencing suppressor inactivates siRNA. *Curr. Biol.* **14**, R198–R200.

Zamore, P.D. (2007) RNA silencing: genomic defence with a slice of pi. *Nature* **446**, 864–865.

Zhang, H., Kolb, F.A., Brondani, V., Billy, E. and Filipowicz, W. (2002) Human Dicer preferentially cleaves dsRNAs at their termini without a requirement for ATP. *EMBO J.* **21**, 5875–5885.

Zhang, H., Kolb, F.A., Jaskiewicz, L., Westhof, E. and Filipowicz, W. (2004) Single processing center models for human Dicer and bacterial RNase III. *Cell* **118**, 57–68.

Zhang, X., Yuan, Y.R., Pei, Y., Lin, S.S., Tuschl, T., Patel, D.J. and Chua, N.H. (2006) Cucumber mosaic virus-encoded 2b suppressor inhibits *Arabidopsis* Argonaute1 cleavage activity to counter plant defense. *Genes Dev.* **20**, 3255–3268.

CHAPTER

8

The Role of the APOBEC3 Family of Cytidine Deaminases in Innate Immunity, G-to-A Hypermutation, and Evolution of Retroviruses

Mario L. Santiago and Warner C. Greene

ABSTRACT

The APOBEC3 family of cytidine deaminases comprises a novel group of innate immunity factors that counteract both exogenous and endogenous retroviruses. Initially identified as critical targets of the HIV-1 protein Vif, APOBEC3G (A3G) and APOBEC3F (A3F) are encapsidated into budding virions, where these proteins attack sequential steps of the reverse transcription process in the target cell by virtue of its ability to: (1) bind RNA and (2) catalyze G-to-A hypermutation in nascent single-stranded DNA. In resting CD4T cells and monocytes, A3G in a low-molecular-mass (LMM) form also acts as a potent post-entry barrier against incoming HIV-1. However, T cell activation and differentiation of monocytes into macrophages recruits A3G into enzymatically inactive high-molecular-mass (HMM) complexes, thereby forfeiting its post-entry antiviral activity against HIV-1. Augmenting the antiviral properties of A3G, either by blocking the Vif–A3G interaction or carefully disrupting HMM complex formation, serve as potentially novel therapeutic approaches against HIV-1. Intriguingly, these same antiviral properties of A3G, as well as other APOBEC3 family members, may significantly contribute to the G-to-A mutational bias in retroviral genomes, raising the intriguing possibility that these innate factors may help directly shape the evolutionary diversity of retroviruses.

INTRODUCTION

Biological interactions with retroviruses or related retroelements form a significant component of human evolution. Indeed, approximately 42% of the human genome is composed of retroviral-like DNA (Lander *et al.*, 2001). This interaction is dynamic and reciprocal, as humans and retroviruses have

Origin and Evolution of Viruses
ISBN 978-0-12-374153-0

evolved mechanisms to counteract each other's biological activities. Unraveling the extent and molecular basis of these co-evolutionary mechanisms has become increasingly urgent. One compelling reason is the overwhelming burden of the global acquired immune deficiency syndrome (AIDS) pandemic, which in 2006 alone caused 2.9 million deaths, with 39.5 million people remaining infected (UNAIDS, 2006).

Pandemic AIDS is caused by human immunodeficiency virus (HIV-1), a member of the retroviral family known as primate lentiviruses. Commonly referred to as simian immunodeficiency viruses (SIVs), primate lentiviruses originated from Old World non-human primates in sub-Saharan Africa and radiated into at least six phylogenetic groups that roughly mirror the primate phylogeny (Bailes *et al.*, 2002). One lineage from chimpanzees, SIVcpz, is the precursor of pandemic HIV-1 (Santiago *et al.*, 2002; Keele *et al.*, 2006); another, SIVsm from sooty mangabeys, gave rise to HIV-2, which remains mainly restricted to West Africa (Chen *et al.*, 1996; Santiago *et al.*, 2005). Other lineages, such as SIVagm from African green monkeys, are not known to have "jumped" the species barrier to humans. Species-specific barriers for infection serve to limit the zoonotic spread of these retroviruses. Elucidating the nature of these barriers provides important insights into the interplay of these viruses with both natural and unnatural hosts (Mariani *et al.*, 2003; Stremlau *et al.*, 2004; Schindler *et al.*, 2006).

Several therapeutic modalities currently exist for HIV-1 infection. However, the cost of these drugs has previously limited their deployment in the developing world where HIV infection is hitting the hardest. This fact, combined with the inability of these drugs to completely eradicate HIV-1 infection, the wide spectrum of side-effects associated with antiretroviral therapy, and the inevitable emergence of drug resistance, has led to major efforts aimed at developing new and more effective combinations of antiviral drugs as well efficacious HIV-1 vaccines. Because the development of vaccines that target cellular and humoral immune responses has been slow, interest in harnessing elements of the innate immune response is garnering increasing attention.

RETROVIRUSES AND CYTIDINE DEAMINASES

Since the initial identification of HIV (Barre-Sinoussi *et al.*, 1983), this pathogenic retrovirus has become a prototype for studies on many aspects of retroviral biology, including virus–host interactions. In contrast to simple retroviruses that encode only Gag, Pol, and Env, HIV-1 encodes six additional gene products, Vif, Vpr, Vpu, Tat, Rev, and Nef (Figure 8.1A). While the major functions of Vpr, Vpu, Tat, Rev, and Nef were largely unraveled within a few years after the HIV-1 genome was decoded, the mechanism underlying the function of Vif remained shrouded in mystery for almost two decades, despite knowledge that its deletion results in virions with significantly reduced infectivity (Strebel *et al.*, 1987).

Investigating Vif action Reveals a Novel Antiviral Restriction Factor

Approximately 15 years ago, Gabuzda and colleagues discovered that the function of Vif is tightly linked to the nature of the virus-producing cell (Gabuzda *et al.*, 1992). Several cell lines, termed permissive, produce infectious HIV-1 virions in the absence of Vif (ΔVif HIV-1). Conversely, ΔVif HIV-1 virions derived from non-permissive cells—including CD4T cells and macrophages, the natural targets of HIV-1—are non-infectious. The reason for this difference remained unknown until 2001, when permissive–non-permissive heterokaryons were prepared and tested by the Kabat and Malim laboratories. Both studies revealed that non-permissive cells express a dominant antiviral factor that renders ΔVif HIV-1 virion progeny non-infectious (Madani and Kabat, 1998; Simon *et al.*, 1998). Using subtractive hybridization techniques with isogenic CEM-S (permissive) and CEM-SS (non-permissive)

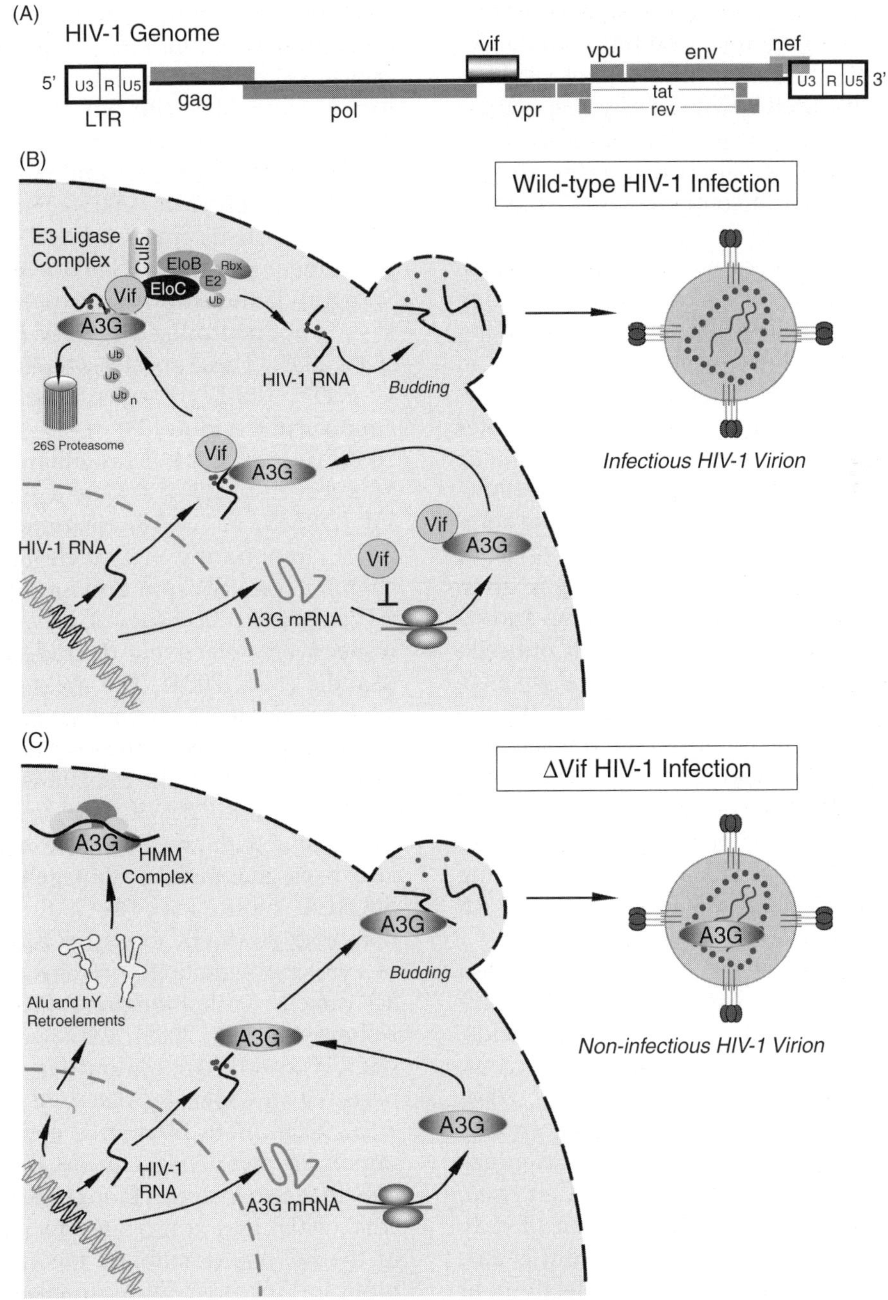

FIGURE 8.1 A3G impairs HIV-1 virion infectivity, but this activity is counteracted by the Vif protein. (A) The HIV-1 genome encodes nine open reading frames. The Vif protein, which counteracts A3G and A3F, is highlighted. Vif is translated from a multi-cistronic spliced transcript that is synthesized during the later stages of the viral life cycle. (B) Infection with wild-type HIV-1. Vif targets A3G for proteosome-mediated degradation by binding to Elongin C and Cullin 5 in the E3 ligase complex, while at the same time impairing A3G mRNA translation. This leads to the packaging of HIV-1 RNA without bound A3G, leading to the budding of infectious virus. (C) Infection with ΔVif HIV-1. In the absence of Vif, newly synthesized A3G, but not A3G that already exists in a cellular HMM complex, is incorporated into budding virions by associating with HIV-1 RNA and possibly, the nucleocapsid segment of Gag. The presence of A3G in virions renders them non-infectious.

cell lines, Ann Sheehy in the Malim laboratory identified apolipoprotein B editing complex 3G (APOBEC3G, or A3G) as the host antiviral factor that is thwarted by Vif (Sheehy *et al.*, 2002).

A3G is a 384-amino-acid (aa), 46-kDa protein belonging to a large family of cytidine deaminases that contain a conserved Zn^{2+}-binding motif C/HxE–PCxC (Jarmuz *et al.*, 2002). This family of enzymes includes APOBEC1, which regulates cholesterol metabolism by editing the apolipoprotein B100 mRNA transcript; APOBEC2, which is expressed primarily in cardiac and skeletal muscle but whose function is unknown; and AID, which regulates immunoglobulin class-switching and somatic hypermutation. Vif interacts directly with A3G and links A3G to the E3 ubiquitin ligase complex (Yu *et al.*, 2003). This complex consists of A3G, Vif, and the E3 ligase, and its formation leads to the polyubiquitylation of A3G and its degradation by the 26S proteasome (Conticello *et al.*, 2003; Marin *et al.*, 2003; Sheehy *et al.*, 2003; Stopak *et al.*, 2003; Mehle *et al.*, 2004b) (Figure 8.1B). In addition, Vif impairs the translational stability of A3G (Stopak *et al.*, 2003). These phenomena result in the nearly complete depletion of intracellular levels of A3G in the producer cell during wild-type HIV-1 infection, reducing the half-life of cellular A3G from >7h to <2h (Stopak *et al.*, 2003).

In the absence of Vif, A3G is incorporated into HIV-1 virions, likely reflecting its propensity to bind single-stranded nucleic acids, particularly viral RNA (Schafer *et al.*, 2004; Svarovskaia *et al.*, 2004; Zennou *et al.*, 2004; Khan *et al.*, 2005; Soros *et al.*, 2007), and its specific interaction with the HIV-1 nucleocapsid protein (Alce and Popik, 2004; Cen *et al.*, 2004; Luo *et al.*, 2004). Incorporation of only seven molecules of A3G in the virion is sufficient to significantly reduce in infectivity at the next round of infection (Xu *et al.*, 2007) (Figure 8.1C). Thus, the functional Vif:A3G interface forms a compelling new target for the development of a novel class of antivirals. Efforts are currently underway to determine the atomic structure of this region, which promises to further propel rational drug discovery efforts.

Deletional mutagenesis coupled with co-immunoprecipitation suggests that the critical interaction domain in A3G required for its assembly with Vif (Wichroski *et al.*, 2005; Simon *et al.*, 2005) resides at the N-terminal region between amino acids 54 and 124 (Conticello *et al.*, 2003). Further insights emerged when cross-species studies were performed. Human A3G is not neutralized by SIVagm Vif and A3G from African green monkeys is not neutralized by HIV-1 Vif (Mariani *et al.*, 2003). These species-specific effects map to a D^{HUM}-to-K^{AGM} substitution at a single amino acid (residue 128) in A3G (Bogerd *et al.*, 2004; Xu *et al.*, 2004; Schrofelbauer *et al.*, 2004; Mangeat *et al.*, 2004) and to a short region of Vif (DRMR; aa 14–18) (Schrofelbauer *et al.*, 2006). Importantly, Vif alleles of SIV from chimpanzees (SIVcpz) and sooty mangabeys (SIVsm), the precursors of HIV-1 and HIV-2, respectively, effectively degrade human A3G (Gaddis *et al.*, 2004). The ability of these Vif alleles to successfully counteract human A3G likely played a role in the successful establishment of these zoonotic infections.

The regions in Vif that are critical for interacting with the E3 ligase complex have also been mapped by mutagenesis. The Vif SLQxLA motif (aa 144–150) shares close homology to the BC-box of SOCS (suppressor of cytokine signaling) proteins, which interact directly with Elongin C (Yu *et al.*, 2003; Kobayashi *et al.*, 2005), while a cryptic Zn^{2+} (HCCH) coordination motif (aa 108–139) connects Vif to Cullin 5 (Xiao *et al.*, 2006; Mehle *et al.*, 2006). Both Elongin C and Cullin 5 are important components of an E3 ligase complex that also includes Elongin B and Rbx1 (Yu *et al.*, 2003) that is recruited by Vif. Mutation of the BC box domain or the cryptic HCCH motif in Vif render Vif incapable of degrading A3G and lead to greater A3G antiviral activity, even though Vif remains competent for binding to A3G (Yu *et al.*, 2003; Mehle *et al.*, 2004a, 2006). These findings suggest that Vif binding to A3G alone is insufficient to overcome its antiviral activity. Thus, the Vif:Cullin 5 and Vif:Elongin C interaction sites form additional targets for antiviral drug design. This notion is further supported by a recent study showing that certain human Cullin 5 polymorphisms are linked to accelerated CD4 T-cell loss and HIV-1 disease progression (An *et al.*, 2007).

APOBEC3G Functions as a Potent Cellular Restriction Factor in HIV-1-Resistant Target Cells

Monocytes and resting CD4T cells are refractory to HIV-1 infection in contrast to macrophages and activated CD4T cells (Zack *et al.*, 1990; Rich *et al.*, 1992). This difference in susceptibility is due at least in part to a failure of the virus to effectively reverse transcribe its RNA genome. This effect is commonly attributed to the limited deoxynucleotide pools present in these cells (Kootstra *et al.*, 2000). Other factors, like MURR1 have also been proposed to play a role (Ganesh *et al.*, 2003). However, this model was sharply revised in 2005 when it was discovered that A3G actively restricts at the level of viral reverse transcription (Chiu *et al.*, 2005) (Figure 8.2A). Intriguingly, A3G exists in an LMM form in both monocytes and resting CD4T cells. Upon the activation of these T cells or the differentiation of monocytes into macrophages, LMM A3G is recruited into an enzymatically inactive >6-MDa HMM complex (Chiu *et al.*, 2005). The HMM A3G complex contains both RNA and protein components. RNase A treatment of lysates containing HMM A3G complexes shifts A3G into an enzymatically active LMM form. RNA interference-mediated depletion of LMM A3G in resting CD4T cells is sufficient to render these cells permissive for HIV-1 infection. These results indicate that LMM A3G can function as a potent post-entry restriction factor for HIV in resting CD4T cells and likely in monocytes (Chiu *et al.*, 2005). However, this restricting activity is forfeited when A3G is recruited into the HMM A3G complex.

The A3G HMM complex contains at least 95 different proteins, as determined by tandem mass spectrometry (Kozak *et al.*, 2006; Chiu *et al.*, 2006). Careful analysis suggests that these components fall into at least three previously defined ribonucleoprotein complexes: (1) Staufen RNA granules, (2) Ro ribonucleoproteins (RNPs), and (3) reservoirs for prespliceosomes and translational regulators (Chiu *et al.*, 2006). Localization of A3G in P bodies and stress granules (Wichroski *et al.*, 2006; Gallois-Montbrun *et al.*, 2007) has been reported but different groups are obtaining varying results. The fact that Staufen RNA granules, stress bodies, and P bodies represent a continuum of granular-like structures where cargoes may be transferred back and forth could contribute to these differing results. Strikingly, in the presence of A3G, at least two RNA species, *Alu* and hY RNAs, are selectively recruited and enriched in Staufen and Ro/La RNPs, respectively (Figure 8.2B).

Alu elements are recognized as the most successful group of retroelements in humans, accounting for approximately 10% of the human genome (Lander *et al.*, 2001). *Alu* elements belong to a group of retroelements termed non-autonomous short interspersed nuclear elements (SINEs). These elements encode no protein but are able to retrotranspose from one genetic site to another by stealing the reverse transcriptase and endonuclease (integrase) function of a set of autonomous retroelements termed LINE (long interspersed nuclear elements). An *in vitro* assay has been developed to measure *Alu* retrotransposition (Dewannieux *et al.*, 2003). In this system, A3G mediates strong inhibition of *Alu* activity (Chiu *et al.*, 2006; Hulme *et al.*, 2007), but not LINE-1 (Turelli *et al.*, 2004b; Muckenfuss *et al.*, 2006; Stenglein and Harris, 2006). Since A3G is primarily cytoplasmic and *Alu* RNA is recruited to Staufen RNA granules in an A3G-dependent manner, it seems likely that A3G interrupts retrotransposition by sequestering transcribed *Alu* RNAs in the cytoplasm, denying *Alu* RNAs access to the nuclear reverse transcription machinery of the LINE-1 (Chiu *et al.*, 2006). Unfortunately, the assembly of HMM A3G complexes to combat *Alu* retrotransposition opens the door for HIV-1 infection as the post-entry restricting activity of LMM A3G is forfeited (Figure 8.2B).

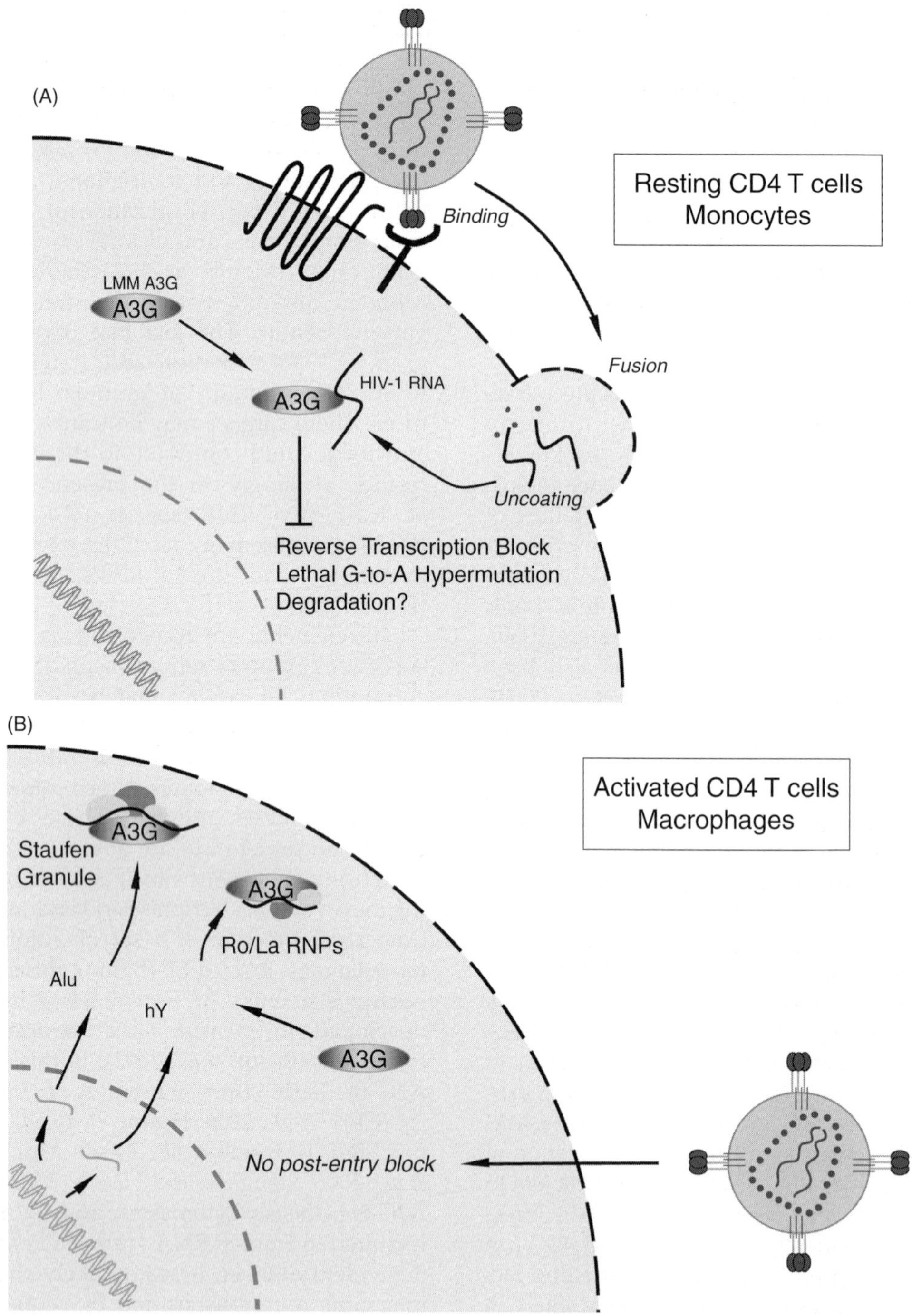

FIGURE 8.2 Distinct functions of cellular A3G. (A) Low-molecular-mass (LMM) A3G functions as a potent post-entry restriction factor against incoming HIV-1. Infection of resting CD4 T cells and monocytes with HIV-1 appears to result in the binding of LMM A3G to incoming HIV-1 RNA, which results in a block in reverse transcription. "Rogue" reverse transcription may then be subject to lethal editing by hypermutation, resulting in possible degradation. (B) High-molecular-mass (HMM) A3G restricts retroelements. Increased expression of mobile elements such as *Alu* and hY results in the sequestration of A3G into distinct complexes such as Staufen-containing granules or Ro/La ribonucleoprotein complexes. While this sequestration mechanism inhibits the activity of these retroelements, the post-entry restriction block posed by LMM A3G is forfeited, allowing HIV-1 to infect these cells.

The Human APOBEC3 Family as Innate Antiretroviral Restriction Factors

A3G, APOBEC1/AID, and APOBEC2 reside on chromosomes 22, 12, and 6, respectively, but six other APOBEC3 members—A3A, A3B, A3C, A3DE, A3F, and A3H—have also been found on human chromosome 22 (Jarmuz *et al.*, 2002; Conticello *et al.*, 2005; OhAinle *et al.*, 2006). Interestingly, mice contain only a single APOBEC3 gene, murine *Apobec3* (or mA3), present in a syntenic region (chromosome 15). These findings indicate a relatively recent tandem duplication of this region in primates (Sheehy *et al.*, 2002). This expansion likely reflects an evolutionary response aimed at achieving greater control over retroelement retrotransposition: transposable element activity is 100-fold higher in mouse cells than in human cells (Maksakova *et al.*, 2006). Moreover, phylogenetic analysis of primate A3G proteins reveals a high rate of positive selection (non-synonymous mutations) that predate the diversification of primate lentiviruses, suggesting that the APOBEC3 family evolved to counteract endogenous retroelements (Sawyer *et al.*, 2004; Zhang and Webb, 2004). In fact, multiple human APOBEC3 members inhibit either *Alu* or LINE-1 retrotransposition *in vitro* (Bogerd *et al.*, 2006; Chiu *et al.*, 2006; Muckenfuss *et al.*, 2006; Stenglein and Harris, 2006; Hulme *et al.*, 2007; Kinomoto *et al.*, 2007).

The recent expansion of the APOBEC3 family cannot be explained by simple tandem duplication (Conticello *et al.*, 2005; OhAinle *et al.*, 2006). A3G, A3B, A3F, and mA3 have two Zn^{2+} coordination (deaminase) motifs, while A3A, A3C, A3H, AID, and APOBEC2 have only a single deaminase motif (Figure 8.3A). Phylogenetic analyses suggest that the primordial APOBEC3 likely contains only a single deaminase domain, which might have evolved from AID or APOBEC2 during the onset of vertebrate speciation (Conticello *et al.*, 2005). This founding APOBEC3 member then diversified into three distinct classes: Z1a, Z1b, and Z2. Single-domained Z1a, Z1b, and Z2 APOBEC3 classes are still found in dogs and horses, while fused Z1a–Z2 double-domain versions developed in rodents, cows, and pigs. Interestingly, primate APOBEC3 members are mostly Z1a–Z1b combinations, except for A3H, which belongs to the Z2 class (OhAinle *et al.*, 2006) (Figure 8.3A). These analyses suggested that recombination with unequal crossover played a major role in the evolution of these family members. The advantages or disadvantages of having a single- versus a double-domain structure are unknown. APOBEC2, a single-domain deaminase, has been crystallized as a rod-shaped tetramer, held together by hydrophobic beta-sheet interactions (Prochnow *et al.*, 2007). The APOBEC2 tetramer appears to be analogous to dimers of the double-domained A3G (Wedekind *et al.*, 2006; Zhang *et al.*, 2007; Santiago *et al.*, 2007), suggesting that the topologically functional APOBEC3 unit might be shared among this larger family of mammalian cytidine deaminases. The evolutionary shuffling of these domains may have created novel specificities to combat a variety of endogenous and exogenous retroviral pathogens (Hache *et al.*, 2005; Langlois *et al.*, 2005).

Much effort has been expended defining the function of the various APOBEC3 family members. One important finding is that A3F functions in concert with A3G to inhibit HIV-1 (Wiegand *et al.*, 2004; Bishop *et al.*, 2004; Zheng *et al.*, 2004; Liddament *et al.*, 2004). A3F appears to be coordinately expressed with A3G in the cellular targets of HIV-1, as it is also packaged into virions and inhibited by Vif, albeit to a lesser extent, possibly because it has a different binding interface (Tian *et al.*, 2006; Simon *et al.*, 2005). A3G is about 10- to 50-fold more active than A3F, although this may not have been the case for A3F homologues in other non-human primates (Zennou and Bieniasz, 2006). A similar scenario exists for A3H: the human counterpart appears to lack antiviral activity relative to the homologue present in rhesus macaques (OhAinle *et al.*, 2006). A3A, A3B, A3C, and A3DE also counteract HIV-1 to varying extents, as will be discussed later in this review.

(A)

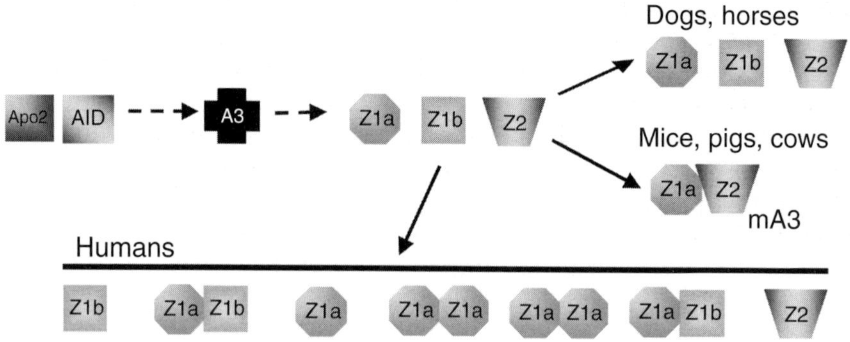

(B)

	A3A	A3B	A3C	A3DE	A3F	A3G	A3H
ΔVif HIV-1	−	+	+	+	++	+++	−
ΔVif SIVmac	NT	+++	+++*	+	+++	+++	−
HBV	NT	++	−	NT	++	+++	NT
HTLV-I	NT	−	−	NT	−	(+)*	NT
SFV	NT	+	−	NT	+++	+++	NT
Alu	+++	+++	+	NT	−	+++	NT
LINE-1	+++	+++	+	−	+/−*	−	NT
AAV	+++	−	−	NT	−	−	NT
MLV	−	++	−	−	+/−	+++	−
MPMV	NT	NT	NT	NT	+++	+++	NT

FIGURE 8.3 Molecular evolution of the APOBEC3 family. (A) AID and APOBEC2, being present since gnastosomes, are the likely progenitors of the single-domained primordial APOBEC3 precursor (A3) which diversified into Z1a, Z1b, and Z2 APOBEC3 versions, the descendants of which can still be found in dogs and horses. Gene duplication, coupled with recombination with unequal crossover, generated a Z1a-Z2 double-domained cytidine deaminase in pigs, cows, and rodents (the mouse APOBEC3 is referred to as mA3). An expansion of the Z1a and Z1b members in primates gave rise to six single- or double-domained members named A3A to G, while A3H appears to be a direct descendant of Z2. (B) The biological activities of various APOBEC3 family members against a variety of viruses. SIV, simian immunodeficiency virus; HBV, hepatitis B virus; HTLV-I, human T-lymphotropic virus-I; SFV, spumavirus; LINE-1, long interspersed element-1; AAV, adenovirus; MLV, murine leukemia virus; MPMV, Mason-Pfizer monkey virus;. NT, not tested; +++, highly active; ++, moderately active; + slightly active; −, not active. *Varying results have been reported by other groups (see text for references).

The disproportionate activities, expression profiles, and even the intracellular localizations of the various APOBEC3 members suggest that each may have evolved to counteract specific exogenous or endogenous retroviral agents (Figure 8.3B). This notion finds support when the activities of the different members are compared in viral model systems other than HIV-1, like SIV (Yu *et al.*, 2004a), human T lymphotropic virus-I (HTLV-I) (Mahieux *et al.*, 2005), murine leukemia virus (MLV) (Doehle *et al.*, 2005b), spumavirus (SFV) (Delebecque *et al.*, 2006), hepatitis B virus (HBV) (Bonvin *et al.*, 2006), retroelements (Bogerd *et al.*, 2006; Stenglein and Harris, 2006; Kinomoto *et al.*, 2007) and even adenoviruses (Chen *et al.*, 2006). These studies describe a fascinating spectrum of antiviral activities for the APOBEC3 family of enzymes and provide a rationale for its expansion and rapid evolution. Intriguingly, some of these viruses appear to have evolved alternative, "non-Vif" strategies to counteract the antiviral effects of these APOBEC3 members. For example, spumaviruses encode a gene known as *Bet* that appears to function similarly to Vif but does not target A3G for proteasomal degradation (Russell *et al.*, 2005; Lochelt *et al.*, 2005), although this conclusion has recently been challenged (Delebecque *et al.*, 2006). The viral protease of MLV appears to actively cleave the non-spliced form of mA3 (Abudu *et al.*, 2006). At the same time, MLV, as well as complex retroviruses such as HTLV, may have also evolved strategies to exclude packaging of deaminases in budding virions (see below for more detailed discussion).

MECHANISMS OF A3G ANTIVIRAL ACTION

The effective incorporation of A3G into budding ΔVif HIV-1 virions potently arrests viral replication in the next target cells. The antiviral mechanism of A3G has been attributed to two fundamental properties: (1) its ability to bind single-stranded RNA and (2) its inherent deaminase activity on single-stranded DNA (ssDNA) templates. Mutagenesis studies on critical residues on the N- and C-terminal catalytic domains suggest that these two properties of A3G distinctly map to these two domains, respectively: the N-terminal region appears to dictate RNA binding, while the C-terminal domain is responsible for deaminating ssDNA (Hache *et al.*, 2005; Navarro *et al.*, 2005; Iwatani *et al.*, 2006). The relative contribution of these properties to the overall antiviral activity of A3G has been the subject of considerable investigation, with some groups suggesting that deaminase-independent mechanisms are sufficient, while others report an essential contribution of deaminase-dependent mechanisms to achieve the antiviral effect. However, it seems more likely that *both* RNA binding and deamination are important to achieve A3G's full antiviral activity, and that these properties enable A3G to attack sequential steps of the reverse transcription process (Figure 8.4).

The Deaminase-independent Antiviral Properties of A3G may be Linked to RNA Binding

RNA binding appears to play a major role in rendering ΔVif HIV-1 virions non-infectious. This conclusion follows from studies showing that A3G mutants with an inactivated C-terminal catalytic domain can substantially reduce the infectivity of ΔVif HIV-1 virions (Newman *et al.*, 2005; Bishop *et al.*, 2006). A similar scenario was reported for A3F (Bishop *et al.*, 2006; Holmes *et al.*, 2007). The precise mechanism of a deaminase-independent antiviral action of A3G and A3F in HIV-1 virions remains to be completely resolved. Newly synthesized A3G is incorporated into HIV-1 virions, presumably by binding to HIV-1 RNA and the nucleocapsid component of Gag at the plasma membrane site of budding (Burnett and Spearman, 2007). It is then held in an enzymatically inactive intravirion complex that includes HIV-1 RNA and virion proteins, but not components of cellular HMM complexes (Soros *et al.*, 2007). A3G activation in the target cell requires the action of the viral RNase H, an enzyme that is much less efficient than RNase A (Soros *et al.*, 2007). This raises the possibility that a fraction of viral genomic RNA may still be bound to A3G, thereby physically impeding reverse transcription (Figure 8.4A). Alternatively, A3G and A3F inhibit tRNALys[1,3] priming, again providing a mechanism for a block at the level of reverse transcription (Guo *et al.*, 2006; Yang *et al.*, 2007b) (Figure 8.4A).

The role of non-enzymatic mechanisms of APOBEC3 antiviral function has been evaluated in other systems. For example, A3G-mediated inhibition of *Alu* retrotransposition (Chiu *et al.*, 2006; Hulme *et al.*, 2007), HTLV-I (Strebel, 2005; Sasada *et al.*, 2005), and HBV (Turelli *et al.*, 2004a; Nguyen *et al.*, 2007) does not appear to involve the enzymatic activity of A3G, since catalytically inactive mutants of A3G retain an ability to counteract these viruses. Intriguingly, some retroviruses, such as HTLV-I, MLV, and Mason-Pfizer monkey virus (MPMV), evolved mechanisms to escape A3G, mA3, or rhA3G (rhesus A3G) by virion exclusion, respectively. HTLV-I encodes a segment of nucleocapsid that excludes A3G from its genomic RNA (Derse *et al.*, 2007). On the other hand, MLV RNA appears to preferentially exclude mA3 but not A3G by an unknown mechanism (Doehle *et al.*, 2005b;

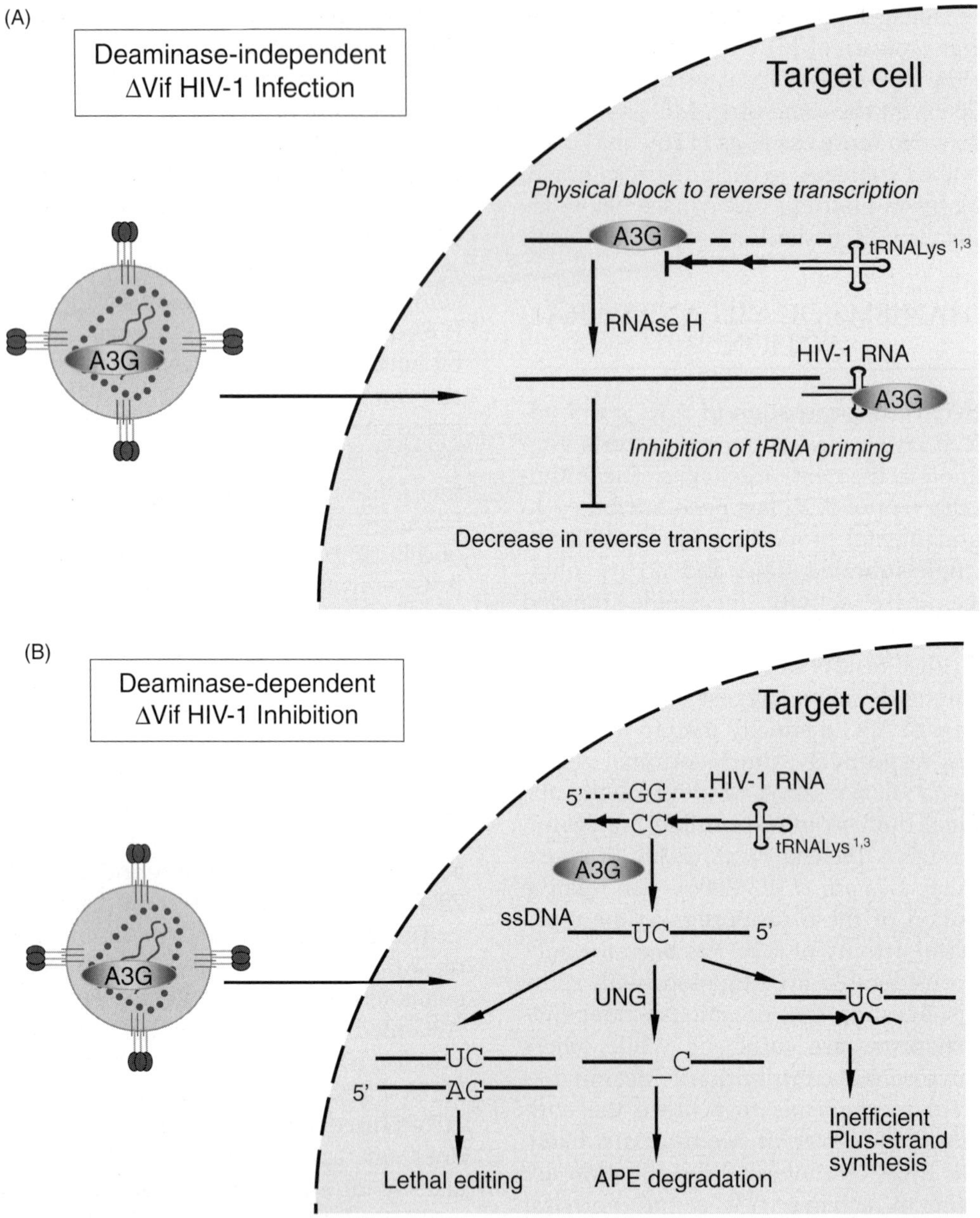

FIGURE 8.4 Proposed mechanisms for A3G antiviral action. (A) Deaminase-independent mechanisms. A3G gets packaged by binding HIV-1 RNA, which locks it into an enzymatically inactive intravirion complex that requires viral RNase H for activation. It is possible that a fraction of viral genomic RNA remains bound to A3G, thereby impeding reverse transcription. If A3G is liberated, its binding to $tRNALys^{1,3}$ could impair subsequent priming reactions during strand transfer. (B) Deaminase-dependent mechanisms. If reverse transcription is successful, A3G can deaminate deoxycytidines in single-stranded DNA into uracils, which can lead to lethal editing, degradation by apurinic/apyrimidinic endonucleases (APE) after base excision by uracil-*N*-glycosylase (UNG), inefficient plus-strand synthesis by RT due the presence of uracils, or the generation of aberrant viral cDNA ends before proviral integration (not shown).

Abudu *et al.*, 2006). A similar phenomenon has been noted for MPMV and rhesus A3G (Doehle *et al.*, 2006). These escape mechanisms highlight the importance of A3G's RNA binding properties for its antiviral action.

Host-induced Error Catastrophe: The Contribution of A3G's Deaminase Activity to its Overall Antiviral Properties

If RNA binding solely accounts for the antiviral activity of A3G, why would the deaminase function remain evolutionary conserved? Theoretically, the DNA-mutating properties of A3G may have detrimental effects in the genome, and millions of years of evolution would predictably select against this property. Thus, the enzymatic properties of A3G are likely required for some as-yet-unknown physiological function. Intriguingly, mice with a functional deletion of mA3 appear normal with respect to reproductive viability and immunological repertoire (Mikl *et al.*, 2005) but are more susceptible to exogenous retroviral infection (Okeoma *et al.*, 2007). Thus, although we cannot exclude the existence of other physiological functions, the conserved deaminase property of A3G might provide it with a potent "second-punch" antiviral mechanism, as reported by many groups (Harris *et al.*, 2003; Lecossier *et al.*, 2003; Mangeat *et al.*, 2003; Zhang *et al.*, 2003; Shindo *et al.*, 2003; Esnault *et al.*, 2006) (Figure 8.4B).

Mechanistically, the C-terminal Zn^{2+} motif of A3G (and A3F) catalyzes the deamination of deoxycytidines, creating deoxyuridines in newly reverse-transcribed single-stranded DNA. During double-stranded DNA synthesis, dU is paired with dA, leading to G-to-A mutation in the resulting plus strand of the provirus (Figure 8.4B). Extensive G-to-A hypermutation, likely resulting in lethal inactivation of coding sequences, has been detected in ΔVif HIV-1 nascent reverse transcripts produced in the presence of A3G (Harris *et al.*, 2003; Lecossier *et al.*, 2003; Mangeat *et al.*, 2003; Zhang *et al.*, 2003). A3G preferentially deaminates ssDNA in the 5′-CC-3′ (the targeted deoxycytidine is underlined) dinucleotide context, while A3F preferentially deaminates TC (Beale *et al.*, 2004; Bishop *et al.*, 2004; Liddament *et al.*, 2004) and, to a small but significant extent, GC (Liddament *et al.*, 2004; Suspene *et al.*, 2005). This translates to preferred 5′-GG→AG hypermutations for A3G and GA→AA (and to a lesser extent, GC→AC) for A3F. Interestingly, G-to-A hypermutated HIV-1 transcripts also show an increasing 5′→ 3′ gradient relative to the central and 3′ LTR polypurine tract, correlating with the earliest regions of the genome that are reverse-transcribed, and may therefore be more available for APOBEC3-mediated deamination (Yu *et al.*, 2004b; Suspene *et al.*, 2006). Host-induced hypermutation appears to be a form of "error catastrophe," whereby the maximum mutational load compatible for production of infectious progeny is exceeded (Eigen, 2002).

While lethal inactivation through hypermutation is an attractive mechanism for conferring antiviral activity, this might not fully explain the substantial reduction in the copy number of ΔVif HIV-1 reverse transcripts in the target cell (Mangeat *et al.*, 2003; Mariani *et al.*, 2003; Bishop *et al.*, 2006). Some of the uracils generated by A3G/A3F may be excised by uracil *N*-glycosylase (UNG), leading to abasic sites that could cause template degradation after the action of apurinic/apyrimidinic (APE) endonucleases (Dianov *et al.*, 2003). However, two groups reported that depletion of target cells of UNG2 or inactivating UNG did not increase the levels of reverse transcription in A3G-positive ΔVif HIV-1 (Kaiser and Emerman, 2006; Mbisa *et al.*, 2007). These results raise the question as to whether UNG is indeed involved in the inhibitory effects of A3G, although it should be noted that multiple UNGs are present in mammalian cells (Priet *et al.*, 2005; Yang *et al.*, 2007a). However, deamination may also influence subsequent steps in proviral DNA synthesis. Accumulation of uracils in minus-strand DNA has been reported to lead to decreased plus-strand synthesis by

HIV-1 RT, potentially due to aberrant initiation in polypurine tract sites (Klarmann *et al.*, 2003) (Figure 8.4B). Moreover, the C-terminal deaminase domain might cause defects in tRNA cleavage during plus-strand DNA transfer, leading to the formation of aberrant viral DNA ends that may be impaired for integration (Mbisa *et al.*, 2007).

There is also evidence that an enzymatic mechanism is important for APOBEC3-mediated antiviral responses against retroviruses other than HIV-1. Archival MLV sequences in the genome exhibit detectable levels of G-to-A hypermutation that is consistent with the deamination preferences of mA3 (Esnault *et al.*, 2005). Similarly, G-to-A hypermutation, albeit at lower levels, has also been detected in HBV (Noguchi *et al.*, 2005; Suspene *et al.*, 2005), HTLV (Mahieux *et al.*, 2005), and SFV (Russell *et al.*, 2005) genomes produced in cells expressing APOBEC3 family members. Thus, to various degrees, the APOBEC3 family has a nearly universal capacity to deaminate retroviral transcripts. Whether these mechanisms contribute to antiviral protection is an experimental topic that remains unresolved.

Cellular Restriction of HIV-1 by LMM A3G: Enzymatic Versus Non-enzymatic Mechanisms

Preliminary mutagenesis studies in our laboratory suggest that mutations in the N-terminal domain disrupt HMM complex formation and disable *Alu* retrotransposition inhibition, but do not reconstitute the HIV-1 post-entry block. However, this latter finding is likely explained by defects in the RNA-binding properties of these mutants, and RNA binding by A3G is likely centrally involved in the ability of LMM A3G to restrict incoming HIV-1 (Santiago *et al.*, 2007). This mechanism is reminiscent of a recently reported factor known as ZAP, a zinc-finger RNA-binding protein, that restricts MLV and Sindbis virus by preventing the accumulation of viral mRNA in the cytoplasm (Guo *et al.*, 2004). The rules governing the binding of A3G to RNA are unknown. Since the known substrates of A3G include *Alu*, hY, and HIV-1 RNA, RNA secondary structure may play a major role in determining its binding specificity.

While RNA binding allows A3G to access its substrate and possibly block cDNA synthesis by the viral reverse transcriptase, its enzymatic activity may operate if "rogue" reverse transcription successfully occurs. G-to-A hypermutation has been noted in reverse transcripts from unstimulated peripheral blood mononuclear cells (PBMCs) infected with HIV-1 *in vitro* (Janini *et al.*, 2001; Vartanian *et al.*, 1997). Sequence analysis of reverse transcripts slowly formed in resting CD4T cells revealed that about 8% had the hallmarks of G-to-A hypermutation mediated by A3G (and A3F) (Chiu *et al.*, 2005). This suggests that LMM A3G possesses detectable enzymatic activity *in vivo*. It seems quite likely that A3G employs a dual strategy involving sequential non-enzymatic and enzymatic actions to achieve its antiviral effects.

HIV-1 EVOLUTIONARY DYNAMICS

HIV-1 displays an overall genetic diversity that is unparalleled with respect to other known pathogens and this property poses a major impediment for developing therapeutics and vaccines. Within the same individual, swarms of related HIV-1 virions, often termed a "quasispecies," can generate sequence diversities that range up to 5%, five-fold higher than the annual antigenic drift of influenza viruses that necessitates annual vaccine revisions (Korber *et al.*, 2001). On the other hand, circulating pandemic (Group M) strains can vary by up to 25% from each other, in part due to constant mutation since the 1930s, when HIV-1 started to diversify in the human population (Korber *et al.*, 2000). HIV-1 is estimated to mutate at a rate of 1.1 mutations per viral genome per infection cycle (Gao *et al.*, 2004), a rate that is at least a million-fold higher than that of the human genome itself.

Error-prone Reverse Transcription, Recombination, and G-to-A Substitutions

The extraordinary diversity of HIV-1 has been attributed to a variety of factors. First, the HIV-1 reverse transcriptase is highly error prone, at least 20-fold higher than MLV RT, mostly due to the lack of $3' \rightarrow 5'$ exonucleolytic proofreading activity (Roberts *et al.*, 1988). Second, HIV-1 is highly recombinogenic, in part due to the packaging of two copies of viral genomic RNA per virion, the ability of RT to "jump" between these templates, and the high frequency of proviral coinfections *in vivo* (Temin, 1991; Jung *et al.*, 2002). Recent estimates reveal that at least 10–30 crossovers occur in T cells and macrophages per replication cycle (Levy *et al.*, 2004). Intriguingly, the HIV-1 genome appears to accommodate strong positive selection forces induced by the cellular and humoral arms of the immune system. Intuitively, one may suspect that such a high genetic plasticity is detrimental for a retrovirus because the genomic and structural integrity of the various viral components needed to complete the virus life cycle may have been compromised. However, HIV-1 appears to utilize this for its own advantage: while only <1% of the circulating virions are likely infectious, the non-infectious particles may function as "decoys" for the immune system (Tsai *et al.*, 1996). Thus, HIV-1 appears to propagate just at the limits of an error catastrophe (Smith *et al.*, 2005).

While error-prone reverse transcription and recombination are key mechanisms for generating HIV-1 diversity, *in vitro* analyses of mutational patterns in HIV-1 reveal a biased occurrence of G-to-A changes (Mansky and Temin, 1995). In a population-level study of 127 HIV-infected patients, G→A transitions are the most common nucleotide substitution observed (Pace *et al.*, 2006). Importantly, G→A mutations account for 21% of known mutations that lead to resistance against currently available protease and reverse transcriptase inhibitors, higher than other permutations (Berkhout and de Ronde, 2004; Hache *et al.*, 2006). G→A mutations are also linked to non-synonymous changes in epitope targets of the adaptive immune response, particularly the V3 loop (Leitner *et al.*, 1997), and even coreceptor switch from CCR5 to CXCR4 utilization (Pastore *et al.*, 2004). Thus, understanding the mechanisms involved in generating this diversity is of considerable importance (Figure 8.5).

Initially, G-to-A hypermutation in the HIV-1 genome is proposed to occur by a mechanism that involves dislocation mutagenesis (Vartanian *et al.*, 1991). In the context of GA→AA substitutions, hypermutation is explained by a −1 slippage that leads to dG:dT mismatches that are highly prone to HIV-1 RT bypass. However, this does not explain the propensity of GG→AG mutations that is subsequently reported. In this case, "A" transitions were found to occur preferentially at the ends in a run of G's, and imbalances in intracellular dCTP pools increases the likelihood of hypermutation (Vartanian *et al.*, 1994). However, this scenario has been difficult to reproduce in physiological systems: a ratio of at least 10^4:1 of dTTP/dCTP is required to recapitulate G-to-A hypermutation *in vitro* (Martinez *et al.*, 1994). This difficulty has been cited as a potential pitfall of this mechanism in initiating G-to-A hypermutation (Bourara *et al.*, 2000). Nonetheless, these mechanisms might still induce non-lethal G-to-A transitions under certain circumstances *in vivo*, particularly in settings that skew the dNTP pool, such as highly active antiretroviral therapy.

Addiction to "A": HIV-1 Codon Usage Bias and its Potential Advantages

Lentiviruses are intrinsically A-rich: up to 40% of the HIV-1 RNA genome is composed of adenosines (Berkhout and van Hemert, 1994). This property is shared with several retroviral genera, including spumaviruses and caulimoviruses, but not with HTLV. The reason for this A-bias remains unknown. However, since >90% of the HIV-1 genome codes for an open reading frame, it was proposed that A-bias is a direct product of tRNA/codon usage.

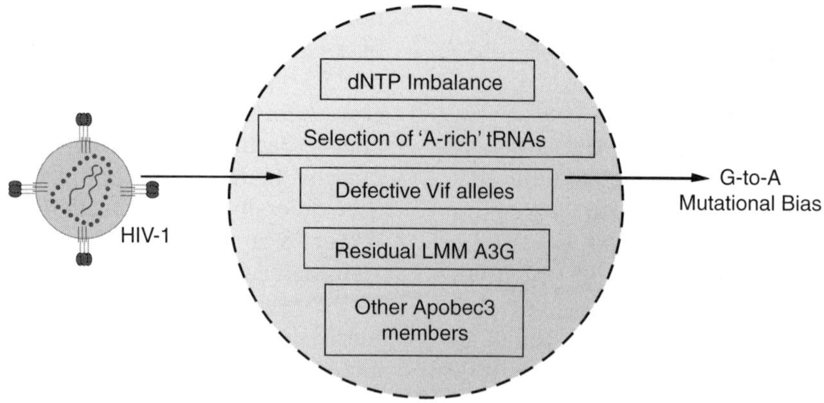

FIGURE 8.5 Mechanisms for generating G-to-A substitutions to drive HIV-1 evolution. Skewed deoxynucleotide pools, particularly an imbalance in dCTP/dTTP ratio, results in increased levels of G-to-A hypermutation in *in vitro* systems. Biased selection of A-rich aminoacyl tRNA pools is maintained by an unknown selection pressure among lentiviruses (see text for details). Defective Vif alleles, suboptimal target cell activation that results in residual LMM A3G in producer cells, and other Apobec3 family members are also potential mechanisms that may allow for non-lethal G-to-A substitutional bias in lentiviruses.

Compared with cellular genes, HIV-1 genes have an intrinsically different codon usage profile (Berkhout and van Hemert, 1994). In fact, tRNAs that are used by the most frequently expressed cellular genes are GC-rich, while those utilized by HIV-1 env, for example, are more biased to A. Such a difference has led to the codon optimization of a variety of HIV-1 genes for enhanced expression in vaccine development studies (Haas *et al.*, 1996). Thus, the A-richness of retroviral genomes might reflect a concerted effort by lentiviruses to deviate from the normal aminoacyl tRNAs used by the cellular translational machinery (van Hemert and Berkhout, 1995).

Let us speculate about mechanisms that might explain A-bias. First, such A-bias could allow maximal viral replication without sacrificing the efficiency of the cellular translational machinery that drives this process. A-richness might minimize the inadvertent creation of alternative open reading frames that may be detrimental to the viral life cycle (F. Bibollet-Ruche, personal communication). Normal mRNA clearance pathways in the cell might also be delayed by a distinct viral nucleotide composition (Kofman *et al.*, 2003), which could facilitate tight regulation, as converting a well-expressed Thy-1 antigen with the most prevalent HIV-1 codons is reported to impair its expression (Haas *et al.*, 1996). AT-rich RNA genomes are also more amenable to recombine and form hairpin structures that may be critical for participating in highly regulated RNA export pathways. On the other hand, GC-rich DNA fragments may activate Toll-like receptor pathways that enhance protective immunity and curb viral replication (Schlaepfer *et al.*, 2004). These and other possibilities support the notion that A-richness is an intrinsic property of lentiviruses and is a by-product of some level of purifying selection (Muller and Bonhoeffer, 2005). Curiously, this phenomenon has routinely been attributed to RT misincorporation, despite the discovery of the APOBEC3 family (Beale *et al.*, 2004; Berkhout and de Ronde, 2004; Muller and Bonhoeffer, 2005).

APOBEC3-induced G-to-A Substitutions: A Driver of HIV-1 Evolution?

RNA editing is initially proposed as a mechanism for G-to-A changes in the HIV-1 genome even before A3G has been identified (Bourara *et al.*, 2000). Experiments show that G-to-A changes in spliced transcripts do not have a proviral DNA counterpart in short-term infections of H9 cells with an HIV-1 molecular clone. This proposal was eventually supported when the site preference for A3G and A3F deamination (CC and TC, respectively) matched the most frequent dinucleotide context (up to 94% GG→AG and GA→AA) of hypermutated retroviral transcripts observed by different groups. It is therefore proposed that G-to-A mutations induced by the APOBEC3 family, perhaps by A3G and A3F, form a significant component of the evolutionary history of HIV-1 (for example, see Hache *et al.*, 2006).

Several arguments have been put forward against this hypothesis. First, lentiviral genomes are intrinsically A-rich (see above), and this is complemented by a poor representation (<20%) of C, but not G, ribonucleotides (Berkhout and van Hemert, 1994). Second, while most lethally inactivated hypermutated sequences exhibit *GG*→*AG* and *GA*→*AA* dinucleotide preferences consistent with the actions of A3G and A3F, an analysis of "background" G-to-A transitions in non-hypermutated *env* sequences ($n = 178$) reveal no such specific dinucleotide preferences (Beale *et al.*, 2004). Third, the 3′→5′ "slide-and-jump" mechanism of A3G action enables multiple deaminations to occur on ssDNA (Pham *et al.*, 2007), making it more likely for deaminated transcripts to be lethally inactivated, thereby producing dead-end viruses that will not propel HIV-1 diversity. Finally, how can this mechanism occur in the presence of Vif?

A recent analysis of circulating Vif alleles shows that a significant fraction of HIV-1 genomes have defective, and sometimes even non-functional, Vif proteins. This mechanism may explain how A3G and A3F can mediate low levels of hypermutation (Simon *et al.*, 2005; Pace *et al.*, 2006) (Figure 8.1C). Interestingly, these defective Vif alleles may have also arisen from G-to-A transitions (Simon *et al.*, 2005; Pace *et al.*, 2006). *In vitro*, high levels of A3G can override the levels of produced Vif (Mariani *et al.*, 2003). In fact, G-to-A hypermutated HIV-1 proviral sequences are detectable in primary isolates, suggesting that the deaminase activity of A3G functionally operates *in vivo* (Janini *et al.*, 2001; Kieffer *et al.*, 2005). If so, one would predict that high A3G levels would reduce plasma viral loads, a well-accepted surrogate marker for a favorable clinical outcome (Mellors *et al.*, 1996).

Hypermutated HIV-1 transcripts have been found in long-term non-progressors (LTNPs), a subset of HIV-1 infected patients that have low to undetectable baseline viral loads in the absence of combination drug therapy (Huang *et al.*, 1998; Wei *et al.*, 2004). This might be due to sampling error, as the viral loads are lower, facilitating the detection of defective transcripts by PCR. However, another study shows that detection of GG→AG hypermutated transcripts correlates with at least a 0.7-log decrease in viral load (Pace *et al.*, 2006). Coincident with this, another study found a strong correlation between plasma viral loads and A3G mRNA levels from activated PBMCs in a cohort of 25 HIV-1-infected patients (Jin *et al.*, 2005). However, these data have not been reproduced in another study, which investigated mRNA levels in resting CD4T cells (Cho *et al.*, 2006). Since HIV-1 primarily targets activated T cells, A3G expression in this context might be a more relevant parameter. These findings reveal that G-to-A hypermutations occur *in vivo* and can be directly linked to A3G and A3F. Thus, the lower end of the activity of these deaminases may confer nonlethal G-to-A substitutions (Hache *et al.*, 2006).

Could the complex status of A3G in suboptimally activated cellular targets also play a role in diversifying HIV-1? LMM A3G in peripheral blood resting T cells appears to have some enzymatic activity (Chiu *et al.*, 2005). However, LMM A3G likely does not contribute to HIV-1 molecular evolution in this setting, owing to the near-complete postentry block in these cells. Interestingly, hypermutated transcripts are more easily detected if PBMCs are mitogen-activated and infected with HIV-1 simultaneously; infection after 12 hours of mitogen stimulation yields mostly wild-type sequences (Janini *et al.*, 2001). Thus, residual LMM A3G in cells undergoing activation may have the ability to deaminate nascent reverse transcripts.

In contast to resting CD4T cells circulating in the peripheral blood, resting CD4T cells residing in lymphoid tissues are susceptible to HIV-1 infection, a property that likely reflects the action of locally produced cytokines including IL-2, IL-15, and other endogenous factors (Kreisberg *et al.*, 2006). However, these factors could, at most, only result in 5–22% activation, leading to very low levels of HIV-1 production (Kreisberg *et al.*, 2006). While the addition of these components sweeps A3G into HMM complexes, a substantial fraction of A3G remains in the LMM form. Although infected cells from these suboptimally activated mixed cultures appear to have a predominantly HMM form of A3G (thereby conferring permissivity), low levels of LMM A3G remain detectable (Stopak *et al.*, 2007). Similarly, maturing dendritic cells exhibit A3G in an intermediate-molecular mass (IMM) form, with some A3G moieties fractionating at the LMM range (Stopak *et al.*, 2007). G-to-A hypermutated transcripts (17% of clones sequenced) has been detected in relatively non-permissive immature dendritic cells (Pion *et al.*, 2006). Recently, a subset of monocytes (CD16+) is also found to be more permissive to HIV-1 infection (Ellery *et al.*, 2007). These cells also harbor IMM A3G with a small fraction in the LMM range. Therefore, low levels of LMM A3G in suboptimally activated tissue-resident CD4T cells, dendritic cells, and CD16 monocytes might mediate low levels of non-lethal G-to-A mutations (Figure 8.5), thereby driving HIV-1 evolution. This hypothesis remains untested.

The remaining APOBEC3 family members, particularly A3A, A3B, A3C, and A3DE, may theoretically also impact HIV-1 evolution, since they are expressed in the cellular targets of HIV-1, albeit to varying degrees. These APOBEC3 family members exhibit deaminase activity *in vitro*. A3A confers a post-entry restriction block in primary monocytes (Peng *et al.*, 2007), and this enzyme does not appear to assemble into HMM complexes, implying functional activity in macrophages (our unpublished results). Interestingly, while A3A can be packaged into virions and appears to deaminate with a TC preference, it does not apparently deaminate viral ssDNA or decrease HIV-1 infectivity (Chen *et al.*, 2006). In contrast, A3B is active against HIV-1 *in vitro* even in the presence of Vif (Bishop *et al.*, 2004, Doehle *et al.*, 2005a). Although A3B is expressed at very low levels in T cells and macrophages, even low levels of A3B virion incorporation in wild-type HIV-1 may have a cumulative effect on G-to-A mutation rates in the TC context (Doehle *et al.*, 2005a). Intriguingly, A3B appears to be absent in some human populations (27% East Asians, 58% Amerindians, and 93% Oceanic populations) due to a 29.5-kb deletion within the gene (Kidd *et al.*, 2007). Thus, it would be intriguing to compare the levels of G-to-A changes in HIV-1 isolates from A3B+ and A3B− haplogroups.

A3C is another single-domain APOBEC3 member that can be incorporated into HIV-1 and SIV virions. However, in contrast to A3A, it appears to confer some antiviral activity, and it may also be partially resistant to Vif (Yu *et al.*, 2004a; Langlois *et al.*, 2005). These studies show that A3C can deaminate HIV-1, with a roughly equal preference for CC and TC (Langlois *et al.*, 2005). The complex status of A3C in the cellular targets of HIV-1 remains unknown, but recently it has been shown to bind 7SK RNA, where it is sequestered in the nucleolus in an enzymatically inactive form. Thus, the cellular activity of A3C may be limited and subject to regulation (He *et al.*, 2006). Finally, A3DE encodes a functional deaminase that is somewhat sensitive to Vif-mediated degradation, is incorporated into virions, and could inhibit ΔVif HIV-1 infectivity, albeit to a lesser extent than A3G or A3F. It preferentially deaminates AC, a dinucleotide preference that is distinct from that of other APOBEC3 members but that has been found in clinical HIV-1 isolates (Dang *et al.*, 2006). Thus, the site preference for A3DE completes the four possible dinucleotide preferences for G-to-A transitions for the APOBEC3 family: GG→AA can occur with A3G and A3C; GA→AA with A3F, A3B, A3C, and A3A; GC→AC with A3F; and GT→AT with A3DE. This and the arguments presented above justify further studies of the role of the entire APOBEC3 family as major drivers of HIV-1 evolution.

CONCLUSIONS AND PERSPECTIVES

The importance of the innate immune system in controlling retroviral infections has been highlighted by the discovery of the APOBEC3 family of cytidine deaminases, of which A3G is the best characterized. The ability of LMM A3G to counteract HIV-1 at the post-entry level in an LMM form and to inhibit retroelements in an HMM form provides an elegant snapshot of the host's economizing effort to minimize the threat posed by both exogenous and endogenous retroviral elements. Similarly, the ability of a variety of retroviruses to counteract A3G (e.g., by encoding HIV-1 Vif to target it for degradation or by encoding Gag proteins that exclude its encapsidation in virions) illustrates the aggressive steps taken by viruses to combat these innate restriction factors. The proposed antiviral mechanisms of the APOBEC3 family, which appear to involve both its RNA binding and deaminase activities, remain to be fully deciphered and harnessed for therapeutic benefit. Finally, confirming the possibility that a host protein family can directly drive the evolutionary trajectory of HIV-1 may provide a surprising and fascinating twist in our general understanding of retroviral evolution.

ACKNOWLEDGMENTS

We thank Y.L. Chiu, V. Soros, S. Wissing, R. Bransteitter, K. Stopak, W. Yonemoto, and other members of the Greene laboratory (GIVI) for helpful discussions. Special thanks go to F. Bibollet-Ruche (UAB) for helpful suggestions on the A-bias of lentiviral genomes. This work was supported by the University of California San Francisco–Gladstone Institute of Virology and Immunology Center for AIDS Research Grant AI0277635P30, the National Institutes of Health Grants R01 AI065329-01, RR18928-01 and P01 HD40543 to W.C.G. and a University AIDS-Wide Program fellowship F05-GI-225 to M.L.S.

REFERENCES

Abudu, A., Takaori-Kondo, A., Izumi, T., Shirakawa, K., Kobayashi, M., Sasada, A. *et al.* (2006) Murine retrovirus escapes from murine APOBEC3 via two distinct novel mechanisms. *Curr. Biol.* **16**, 1565–1570.

Alce, T.M. and Popik, W. (2004) APOBEC3G is incorporated into virus-like particles by a direct interaction with HIV-1 Gag nucleocapsid protein. *J. Biol. Chem.* **279**, 34083–34086.

An, P., Duggal, P., Wang, L.H., O'Brien, S.J., Donfield, S., Goedert, J.J. *et al.* (2007) Polymorphisms of CUL5 are associated with CD4+ T cell loss in HIV-1 infected individuals. *PLoS Genet.* **3**, e19.

Bailes, E., Chaudhuri, R.R., Santiago, M.L., Bibollet-Ruche, F., Hahn, B.H. and Sharp, P.M. (2002) The evolution of primate lentiviruses and the origin of AIDS. In: *The Molecular Epidemiology of Human Viruses* (T. Leitner, ed.), pp. 65–95. Boston, MA: Kluwer Academic Publishers.

Barre-Sinoussi, F., Chermann, J.C., Rey, F., Nugeyre, M.T., Chamaret, S., Gruest, J. *et al.* (1983) Isolation of a T-lymphotropic retrovirus from a patient at risk for acquired immune deficiency syndrome (AIDS). *Science* **220**, 868–871.

Beale, R.C., Petersen-Mahrt, S.K., Watt, I.N., Harris, R.S., Rada, C. and Neuberger, M.S. (2004) Comparison of the differential context-dependence of DNA deamination by APOBEC enzymes: correlation with mutation spectra in vivo. *J. Mol. Biol.* **337**, 585–596.

Berkhout, B. and de Ronde, A. (2004) APOBEC3G versus reverse transcriptase in the generation of HIV-1 drug-resistance mutations. *AIDS* **18**, 1861–1863.

Berkhout, B. and van Hemert, F.J. (1994) The unusual nucleotide content of the HIV RNA genome results in a biased amino acid composition of HIV proteins. *Nucleic Acids Res.* **22**, 1705–1711.

Bishop, K.N., Holmes, R.K., Sheehy, A.M., Davidson, N. O., Cho, S.J. and Malim, M.H. (2004) Cytidine deamination of retroviral DNA by diverse APOBEC proteins. *Curr. Biol.* **14**, 1392–1396.

Bishop, K.N., Holmes, R.K. and Malim, M.H. (2006) Antiviral potency of APOBEC proteins does not correlate with cytidine deamination. *J. Virol.* **80**, 8450–8458.

Bogerd, H.P., Doehle, B.P., Wiegand, H.L. and Cullen, B.R. (2004) A single amino acid difference in the host APOBEC3G protein controls the primate species specificity of HIV type 1 virion infectivity factor. *Proc. Natl Acad. Sci. USA* **101**, 3770–3774.

Bogerd, H.P., Wiegand, H.L., Doehle, B.P., Lueders, K.K. and Cullen, B.R. (2006) APOBEC3A and APOBEC3B are potent inhibitors of LTR-retrotransposon function in human cells. *Nucleic Acids Res.* **34**, 89–95.

Bonvin, M., Achermann, F., Greeve, I., Stroka, D., Keogh, A., Inderbitzin, D. *et al.* (2006) Interferon-inducible expression of APOBEC3 editing enzymes in human hepatocytes and inhibition of hepatitis B virus replication. *Hepatology* **43**, 1364–1374.

Bourara, K., Litvak, S. and Araya, A. (2000) Generation of G-to-A and C-to-U changes in HIV-1 transcripts by RNA editing. *Science* **289**, 1564–1566.

Burnett, A. and Spearman, P. (2007) APOBEC3G multimers are recruited to the plasma membrane for packaging into human immunodeficiency virus type 1 virus-like particles in an RNA-dependent process requiring the NC basic linker. *J. Virol.* **81**, 5000–5013.

Cen, S., Guo, F., Niu, M., Saadatmand, J., Deflassieux, J. and Kleiman, L. (2004) The interaction between HIV-1 Gag and APOBEC3G. *J. Biol. Chem.* **279**, 33177–33184.

Chen, H., Lilley, C.E., Yu, Q., Lee, D.V., Chou, J., Narvaiza, I. *et al.* (2006) APOBEC3A is a potent inhibitor of adeno-associated virus and retrotransposons. *Curr. Biol.* **16**, 480–485.

Chen, Z., Telfier, P., Gettie, A., Reed, P., Zhang, L., Ho, D.D. and Marx, P.A. (1996) Genetic characterization of new West African simian immunodeficiency virus SIVsm: geographic clustering of household-derived SIV strains with human immunodeficiency virus type 2 subtypes and genetically diverse viruses from a single feral sooty mangabey troop. *J. Virol.* **70**, 3617–3627.

Chiu, Y.L., Soros, V.B., Kreisberg, J.F., Stopak, K., Yonemoto, W. and Greene, W.C. (2005) Cellular APOBEC3G restricts HIV-1 infection in resting CD4+ T cells. *Nature* **435**, 108–114.

Chiu, Y.L., Witkowska, H.E., Hall, S.C., Santiago, M., Soros, V.B., Esnault, C. *et al.* (2006) High-molecular-mass APOBEC3G complexes restrict Alu retrotransposition. *Proc. Natl Acad. Sci. USA* **103**, 15588–15593.

Cho, S.J., Drechsler, H., Burke, R.C., Arens, M.Q., Powderly, W. and Davidson, N.O. (2006) APOBEC3F and APOBEC3G mRNA levels do not correlate with human immunodeficiency virus type 1 plasma viremia or CD4+ T-cell count. *J. Virol.* **80**, 2069–2072.

Conticello, S.G., Harris, R.S. and Neuberger, M.S. (2003) The Vif protein of HIV triggers degradation of the human antiretroviral DNA deaminase APOBEC3G. *Curr. Biol.* **13**, 2009–2013.

Conticello, S.G., Thomas, C.J., Petersen-Mahrt, S.K. and Neuberger, M.S. (2005) Evolution of the AID/APOBEC family of polynucleotide (deoxy)cytidine deaminases. *Mol. Biol. Evol.* **22**, 367–377.

Dang, Y., Wang, X., Esselman, W.J. and Zheng, Y.H. (2006) Identification of APOBEC3DE as another antiretroviral factor from the human APOBEC family. *J. Virol.* **80**, 10522–10533.

Delebecque, F., Suspene, R., Calattini, S., Casartelli, N., Saib, A., Froment, A. *et al.* (2006) Restriction of foamy viruses by APOBEC cytidine deaminases. *J. Virol.* **80**, 605–614.

Derse, D., Hill, S.A., Princler, G., Lloyd, P. and Heidecker, G. (2007) Resistance of human T cell leukemia virus type 1 to APOBEC3G restriction is mediated by elements in nucleocapsid. *Proc. Natl Acad. Sci. USA* **104**, 2915–2920.

Dewannieux, M., Esnault, C. and Heidmann, T. (2003) LINE-mediated retrotransposition of marked Alu sequences. *Nat. Genet.* **35**, 41–48.

Dianov, G.L., Sleeth, K.M., Dianova, I.I. and Allinson, S.L. (2003) Repair of abasic sites in DNA. *Mutat. Res.* **531**, 157–163.

Doehle, B.P., Schafer, A. and Cullen, B.R. (2005a) Human APOBEC3B is a potent inhibitor of HIV-1 infectivity and is resistant to HIV-1 Vif. *Virology* **339**, 281–288.

Doehle, B.P., Schafer, A., Wiegand, H.L., Bogerd, H.P. and Cullen, B.R. (2005b) Differential sensitivity of murine leukemia virus to APOBEC3-mediated inhibition is governed by virion exclusion. *J. Virol.* **79**, 8201–8207.

Doehle, B.P., Bogerd, H.P., Wiegand, H.L., Jouvenet, N., Bieniasz, P.D., Hunter, E. and Cullen, B.R. (2006) The betaretrovirus Mason-Pfizer monkey virus selectively excludes simian APOBEC3G from virion particles. *J. Virol.* **80**, 12102–12108.

Eigen, M. (2002) Error catastrophe and antiviral strategy. *Proc. Natl Acad. Sci. USA* **99**, 13374–13376.

Ellery, P.J., Tippett, E., Chiu, Y.L., Paukovics, G., Cameron, P.U., Solomon, A. *et al.* (2007) The CD16+ monocyte subset is more permissive to infection and preferentially harbors HIV-1 in vivo. *J. Immunol.* **178**, 6581–6589.

Esnault, C., Heidmann, O., Delebecque, F., Dewannieux, M., Ribet, D., Hance, A.J. *et al.* (2005) APOBEC3G cytidine deaminase inhibits retrotransposition of endogenous retroviruses. *Nature* **433**, 430–433.

Esnault, C., Millet, J., Schwartz, O. and Heidmann, T. (2006) Dual inhibitory effects of APOBEC family proteins on retrotransposition of mammalian endogenous retroviruses. *Nucleic Acids Res.* **34**, 1522–1531.

Gabuzda, D.H., Lawrence, K., Langhoff, E., Terwilliger, E., Dorfman, T., Haseltine, W.A. and Sodroski, J. (1992) Role of vif in replication of human immunodeficiency virus type 1 in CD4+ T lymphocytes. *J. Virol.* **66**, 6489–6495.

Gaddis, N.C., Sheehy, A.M., Ahmad, K.M., Swanson, C. M., Bishop, K.N., Beer, B.E. *et al.* (2004) Further investigation of simian immunodeficiency virus Vif function in human cells. *J. Virol.* **78**, 12041–12046.

Gallois-Montbrun, S., Kramer, B., Swanson, C.M., Byers, H., Lynham, S., Ward, M. and Malim, M.H. (2007) Antiviral protein APOBEC3G localizes to ribonucleoprotein complexes found in P bodies and stress granules. *J. Virol.* **81**, 2165–2178.

Ganesh, L., Burstein, E., Guha-Niyogi, A., Louder, M. K., Mascola, J.R., Klomp, L.W. *et al.* (2003) The gene product Murr1 restricts HIV-1 replication in resting CD4+ lymphocytes. *Nature* **426**, 853–857.

Gao, F., Chen, Y., Levy, D.N., Conway, J.A., Kepler, T. B. and Hui, H. (2004) Unselected mutations in the human immunodeficiency virus type 1 genome are mostly non-synonymous and often deleterious. *J. Virol.* **78**, 2426–2433.

Guo, F., Cen, S., Niu, M., Saadatmand, J. and Kleiman, L. (2006) Inhibition of formula-primed reverse transcription by human APOBEC3G during human immunodeficiency virus type 1 replication. *J. Virol.* **80**, 11710–11722.

Guo, X., Carroll, J.W., Macdonald, M.R., Goff, S.P. and Gao, G. (2004) The zinc finger antiviral protein directly binds to specific viral mRNAs through the CCCH zinc finger motifs. *J. Virol.* **78**, 12781–12787.

Haas, J., Park, E.C. and Seed, B. (1996) Codon usage limitation in the expression of HIV-1 envelope glycoprotein. *Curr. Biol.* **6**, 315–324.

Hache, G., Liddament, M.T. and Harris, R.S. (2005) The retroviral hypermutation specificity of APOBEC3F and APOBEC3G is governed by the C-terminal DNA cytosine deaminase domain. *J. Biol. Chem.* **280**, 10920–10924.

Hache, G., Mansky, L.M. and Harris, R.S. (2006) Human APOBEC3 proteins, retrovirus restriction and HIV drug resistance. *AIDS Rev.* **8**, 148–157.

Harris, R.S., Bishop, K.N., Sheehy, A.M., Craig, H.M., Petersen-Mahrt, S.K., Watt, I.N. *et al.* (2003) DNA deamination mediates innate immunity to retroviral infection. *Cell.* **113**, 803–809.

He, W.J., Chen, R., Yang, Z. and Zhou, Q. (2006) Regulation of two key nuclear enzymatic activities by the 7SK small nuclear RNA. *Cold Spring Harb. Symp. Quant. Biol.* **71**, 301–311.

Holmes, R.K., Koning, F.A., Bishop, K.N. and Malim, M.H. (2007) APOBEC3F can inhibit the accumulation of HIV-1 reverse transcription products in the absence of hypermutation. Comparisons with APOBEC3G. *J. Biol. Chem.* **282**, 2587–2595.

Huang, Y., Zhang, L. and Ho, D.D. (1998) Characterization of gag and pol sequences from long-term survivors of human immunodeficiency virus type 1 infection. *Virology* **240**, 36–49.

Hulme, A.E., Bogerd, H.P., Cullen, B.R. and Moran, J.V. (2007) Selective inhibition of Alu retrotransposition by APOBEC3G. *Gene* **390**, 199–205.

Iwatani, Y., Takeuchi, H., Strebel, K. and Levin, J.G. (2006) Biochemical activities of highly purified, catalytically active human APOBEC3G: correlation with antiviral effect. *J. Virol.* **80**, 5992–6002.

Janini, M., Rogers, M., Birx, D.R. and McCutchan, F.E. (2001) Human immunodeficiency virus type 1 DNA sequences genetically damaged by hypermutation are often abundant in patient peripheral blood mononuclear cells and may be generated during near-simultaneous infection and activation of CD4(+) T cells. *J. Virol.* **75**, 7973–7986.

Jarmuz, A., Chester, A., Bayliss, J., Gisbourne, J., Dunham, I., Scott, J. and Navaratnam, N. (2002) An anthropoid-specific locus of orphan C to U RNA-editing enzymes on chromosome 22. *Genomics* **79**, 285–296.

Jin, X., Brooks, A., Chen, H., Bennett, R., Reichman, R. and Smith, H. (2005) APOBEC3G/CEM15 (hA3G) mRNA levels associate inversely with human immunodeficiency virus viremia. *J. Virol.* **79**, 11513–11516.

Jung, A., Maier, R., Vartanian, J.P., Bocharov, G., Jung, V., Fischer, U. *et al.* (2002) Multiply infected spleen cells in HIV patients. *Nature* **418**, 144.

Kaiser, S.M. and Emerman, M. (2006) Uracil DNA glycosylase is dispensable for human immunodeficiency virus type 1 replication and does not contribute to the antiviral effects of the cytidine deaminase Apobec3G. *J. Virol.* **80**, 875–882.

Keele, B.F., Van Heuverswyn, F., Li, Y., Bailes, E., Takehisa, J., Santiago, M.L. *et al.* (2006) Chimpanzee reservoirs of pandemic and nonpandemic HIV-1. *Science* **313**, 523–526.

Khan, M.A., Kao, S., Miyagi, E., Takeuchi, H., Goila-Gaur, R., Opi, S. *et al.* (2005) Viral RNA is required for the association of APOBEC3G with human immunodeficiency virus type 1 nucleoprotein complexes. *J. Virol.* **79**, 5870–5874.

Kidd, J.M., Newman, T.L., Tuzun, E., Kaul, R. and Eichler, E.E. (2007) Population stratification of a common APOBEC gene deletion polymorphism. *PLoS Genet.* **3**, e63.

Kieffer, T.L., Kwon, P., Nettles, R.E., Han, Y., Ray, S.C. and Siliciano, R.F. (2005) G→A hypermutation in protease and reverse transcriptase regions of human immunodeficiency virus type 1 residing in resting CD4+ T cells in vivo. *J. Virol.* **79**, 1975–1980.

Kinomoto, M., Kanno, T., Shimura, M., Ishizaka, Y., Kojima, A., Kurata, T. *et al.* (2007) All APOBEC3 family proteins differentially inhibit LINE-1 retrotransposition. *Nucleic Acids Res.* **35**, 2955–2964.

Klarmann, G.J., Chen, X., North, T.W. and Preston, B.D. (2003) Incorporation of uracil into minus strand DNA affects the specificity of plus strand synthesis initiation during lentiviral reverse transcription. *J. Biol. Chem.* **278**, 7902–7909.

Kobayashi, M., Takaori-Kondo, A., Miyauchi, Y., Iwai, K. and Uchiyama, T. (2005) Ubiquitination of APOBEC3G by an HIV-1 Vif-Cullin5-Elongin B-Elongin C complex is essential for Vif function. *J. Biol. Chem.* **280**, 18573–18578.

Kofman, A., Graf, M., Deml, L., Wolf, H. and Wagner, R. (2003) Codon usage-mediated inhibition of HIV-1 gag expression in mammalian cells occurs independently of translation. *Tsitologiia* **45**, 94–100.

Kootstra, N.A., Zwart, B.M. and Schuitemaker, H. (2000) Diminished human immunodeficiency virus type 1 reverse transcription and nuclear transport in primary macrophages arrested in early G_1 phase of the cell cycle. *J. Virol.* **74**, 1712–1717.

Korber, B., Muldoon, M., Theiler, J., Gao, F., Gupta, R., Lapedes, A. *et al.* (2000) Timing the ancestor of the HIV-1 pandemic strains. *Science* **288**, 1789–1796.

Korber, B., Gaschen, B., Yusim, K., Thakallapally, R., Kesmir, C. and Detours, V. (2001) Evolutionary and immunological implications of contemporary HIV-1 variation. *Br. Med. Bull.* **58**, 19–42.

Kozak, S.L., Marin, M., Rose, K.M., Bystrom, C. and Kabat, D. (2006) The anti-HIV-1 editing enzyme APOBEC3G binds HIV-1 RNA and messenger RNAs that shuttle between polysomes and stress granules. *J. Biol. Chem.* **281**, 29105–29119.

Kreisberg, J.F., Yonemoto, W. and Greene, W.C. (2006) Endogenous factors enhance HIV infection of tissue naive CD4T cells by stimulating high molecular mass APOBEC3G complex formation. *J. Exp. Med.* **203**, 865–870.

Lander, E.S., Linton, L.M., Birren, B., Nusbaum, C., Zody, M.C., Baldwin, J. *et al.* (2001) Initial sequencing and analysis of the human genome. *Nature* **409**, 860–921.

Langlois, M.A., Beale, R.C., Conticello, S.G. and Neuberger, M.S. (2005) Mutational comparison of the single-domained APOBEC3C and double-domained APOBEC3F/G anti-retroviral cytidine deaminases provides insight into their DNA target site specificities. *Nucleic Acids Res.* **33**, 1913–1923.

Lecossier, D., Bouchonnet, F., Clavel, F. and Hance, A.J. (2003) Hypermutation of HIV-1 DNA in the absence of the Vif protein. *Science* **300**, 1112.

Leitner, T., Kumar, S. and Albert, J. (1997) Tempo and mode of nucleotide substitutions in gag and env gene fragments in human immunodeficiency virus type 1 populations with a known transmission history. *J. Virol.* **71**, 4761–4770.

Levy, D.N., Aldrovandi, G.M., Kutsch, O. and Shaw, G.M. (2004) Dynamics of HIV-1 recombination in its natural target cells. *Proc. Natl Acad. Sci. USA* **101**, 4204–4209.

Liddament, M.T., Brown, W.L., Schumacher, A.J. and Harris, R.S. (2004) APOBEC3F properties and hypermutation preferences indicate activity against HIV-1 in vivo. *Curr. Biol.* **14**, 1385–1391.

Lochelt, M., Romen, F., Bastone, P., Muckenfuss, H., Kirchner, N., Kim, Y.B. *et al.* (2005) The antiretroviral activity of APOBEC3 is inhibited by the foamy virus accessory Bet protein. *Proc. Natl Acad. Sci. USA* **102**, 7982–7987.

Luo, K., Liu, B., Xiao, Z., Yu, Y., Yu, X., Gorelick, R. and Yu, X.F. (2004) Amino-terminal region of the human immunodeficiency virus type 1 nucleocapsid is required for human APOBEC3G packaging. *J. Virol.* **78**, 11841–11852.

Madani, N. and Kabat, D. (1998) An endogenous inhibitor of human immunodeficiency virus in human lymphocytes is overcome by the viral Vif protein. *J. Virol.* **72**, 10251–10255.

Mahieux, R., Suspene, R., Delebecque, F., Henry, M., Schwartz, O., Wain-Hobson, S. and Vartanian, J.P. (2005) Extensive editing of a small fraction of human T-cell leukemia virus type 1 genomes by four APOBEC3 cytidine deaminases. *J. Gen. Virol.* **86**, 2489–2494.

Maksakova, I.A., Romanish, M.T., Gagnier, L., Dunn, C. A., van de Lagemaat, L.N. and Mager, D.L. (2006) Retroviral elements and their hosts: insertional mutagenesis in the mouse germ line. *PLoS Genet.* **2**, e2.

Mangeat, B., Turelli, P., Caron, G., Friedli, M., Perrin, L. and Trono, D. (2003) Broad antiretroviral defence by human APOBEC3G through lethal editing of nascent reverse transcripts. *Nature* **424**, 99–103.

Mangeat, B., Turelli, P., Liao, S. and Trono, D. (2004) A single amino acid determinant governs the species-specific sensitivity of APOBEC3G to Vif action. *J. Biol. Chem.* **279**, 14481–14483.

Mansky, L.M. and Temin, H.M. (1995) Lower in vivo mutation rate of human immunodeficiency virus type 1 than that predicted from the fidelity of purified reverse transcriptase. *J. Virol.* **69**, 5087–5094.

Mariani, R., Chen, D., Schrofelbauer, B., Navarro, F., Konig, R., Bollman, B. *et al.* (2003) Species-specific exclusion of APOBEC3G from HIV-1 virions by Vif. *Cell* **114**, 21–31.

Marin, M., Rose, K.M., Kozak, S.L. and Kabat, D. (2003) HIV-1 Vif protein binds the editing enzyme APOBEC3G and induces its degradation. *Nat. Med.* **9**, 1398–1403.

Martinez, M.A., Vartanian, J.P. and Wain-Hobson, S. (1994) Hypermutagenesis of RNA using human immunodeficiency virus type 1 reverse transcriptase and biased dNTP concentrations. *Proc. Natl Acad. Sci. USA* **91**, 11787–11791.

Mbisa, J.L., Barr, R., Thomas, J.A., Vandegraaff, N., Dorweiler, I.J., Svarovskaia, E.S. *et al.* (2007) HIV-1 cDNAs produced in the presence of APOBEC3G exhibit defects in plus-strand DNA transfer and integration. *J. Virol.* **81**, 7099–7110.

Mehle, A., Goncalves, J., Santa-Marta, M., McPike, M. and Gabuzda, D. (2004a) Phosphorylation of a novel SOCS-box regulates assembly of the HIV-1 Vif-Cu15 complex that promotes APOBEC3G degradation. *Genes Dev.* **18**, 2861–2866.

Mehle, A., Strack, B., Ancuta, P., Zhang, C., McPike, M. and Gabuzda, D. (2004b) Vif overcomes the innate antiviral activity of APOBEC3G by promoting its degradation in the ubiquitin-proteasome pathway. *J. Biol. Chem.* **279**, 7792–7798.

Mehle, A., Thomas, E.R., Rajendran, K.S. and Gabuzda, D. (2006) A zinc-binding region in Vif binds Cu15 and determines cullin selection. *J. Biol. Chem.* **281**, 17259–17265.

Mellors, J.W., Rinaldo, C.R., Jr., Gupta, P., White, R.M., Todd, J.A. and Kingsley, L.A. (1996) Prognosis in HIV-1 infection predicted by the quantity of virus in plasma. *Science* **272**, 1167–1170.

Mikl, M.C., Watt, I.N., Lu, M., Reik, W., Davies, S.L., Neuberger, M.S. and Rada, C. (2005) Mice deficient in APOBEC2 and APOBEC3. *Mol. Cell Biol.* **25**, 7270–7277.

Muckenfuss, H., Hamdorf, M., Held, U., Perkovic, M., Lower, J., Cichutek, K. *et al.* (2006) APOBEC3 proteins inhibit human LINE-1 retrotransposition. *J. Biol. Chem.* **281**, 22161–22172.

Muller, V. and Bonhoeffer, S. (2005) Guanine-adenine bias: a general property of retroid viruses that is unrelated to host-induced hypermutation. *Trends Genet.* **21**, 264–268.

Navarro, F., Bollman, B., Chen, H., Konig, R., Yu, Q., Chiles, K. and Landau, N.R. (2005) Complementary function of the two catalytic domains of APOBEC3G. *Virology* **333**, 374–386.

Newman, E.N., Holmes, R.K., Craig, H.M., Klein, K.C., Lingappa, J.R., Malim, M.H. and Sheehy, A.M. (2005) Antiviral function of APOBEC3G can be dissociated from cytidine deaminase activity. *Curr. Biol.* **15**, 166–170.

Nguyen, D.H., Gummuluru, S. and Hu, J. (2007) Deamination-independent inhibition of hepatitis B virus reverse transcription by APOBEC3G. *J. Virol.* **81**, 4465–4472.

Noguchi, C., Ishino, H., Tsuge, M., Fujimoto, Y., Imamura, M., Takahashi, S. and Chayama, K. (2005) G to A hypermutation of hepatitis B virus. *Hepatology* **41**, 626–633.

OhAinle, M., Kerns, J.A., Malik, H.S. and Emerman, M. (2006) Adaptive evolution and antiviral activity of the conserved mammalian cytidine deaminase APOBEC3H. *J. Virol.* **80**, 3853–3862.

Okeoma, C.M., Lovsin, N., Peterlin, B.M. and Ross, S.R. (2007) APOBEC3 inhibits mouse mammary tumour virus replication in vivo. *Nature* **445**, 927–930.

Pace, C., Keller, J., Nolan, D., James, I., Gaudieri, S., Moore, C. and Mallal, S. (2006) Population level analysis of human immunodeficiency virus type 1 hypermutation and its relationship with APOBEC3G and vif genetic variation. *J. Virol.* **80**, 9259–9269.

Pastore, C., Ramos, A. and Mosier, D.E. (2004) Intrinsic obstacles to human immunodeficiency virus type 1 coreceptor switching. *J. Virol.* **78**, 7565–7574.

Peng, G., Greenwell-Wild, T., Nares, S., Jin, W., Lei, K.J., Rangel, Z.G. *et al.* (2007) Myeloid differentiation and susceptibility to HIV-1 are linked to APOBEC3 expression. *Blood* **110**, 393–400.

Pham, P., Chelico, L. and Goodman, M.F. (2007) DNA deaminases AID and APOBEC3G act processively on single-stranded DNA. *DNA Repair (Amst.)* **6**, 689–692.

Pion, M., Granelli-Piperno, A., Mangeat, B., Stalder, R., Correa, R., Steinman, R.M. and Piguet, V. (2006) APOBEC3G/3F mediates intrinsic resistance of monocyte-derived dendritic cells to HIV-1 infection. *J. Exp. Med.* **203**, 2887–2893.

Priet, S., Gros, N., Navarro, J.M., Boretto, J., Canard, B., Querat, G. and Sire, J. (2005) HIV-1-associated uracil DNA glycosylase activity controls dUTP misincorporation in viral DNA and is essential to the HIV-1 life cycle. *Mol. Cell* **17**, 479–490.

Prochnow, C., Bransteitter, R., Klein, M.G., Goodman, M.F. and Chen, X.S. (2007) The APOBEC-2 crystal structure and functional implications for the deaminase AID. *Nature* **445**, 447–451.

Rich, E.A., Chen, I.S., Zack, J.A., Leonard, M.L. and O'Brien, W.A. (1992) Increased susceptibility of differentiated mononuclear phagocytes to productive infection with human immunodeficiency virus-1 (HIV-1). *J. Clin. Invest.* **89**, 176–183.

Roberts, J.D., Bebenek, K. and Kunkel, T.A. (1988) The accuracy of reverse transcriptase from HIV-1. *Science* **242**, 1171–1173.

Russell, R.A., Wiegand, H.L., Moore, M.D., Schafer, A., McClure, M.O. and Cullen, B.R. (2005) Foamy virus Bet proteins function as novel inhibitors of the APOBEC3 family of innate antiretroviral defense factors. *J. Virol.* **79**, 8724–8731.

Santiago, M.L., Rodenburg, C.M., Kamenya, S., Bibollet-Ruche, F., Gao, F., Bailes, E. *et al.* (2002) SIVcpz in wild chimpanzees. *Science* **295**, 465.

Santiago, M.L., Range, F., Keele, B.F., Li, Y., Bailes, E., Bibollet-Ruche, F. *et al.* (2005) Simian immunodeficiency virus infection in free-ranging sooty mangabeys (*Cercocebus atys atys*) from the Tai Forest, Cote d'Ivoire: implications for the origin of epidemic human immunodeficiency virus type 2. *J. Virol.* **79**, 12515–12527.

Santiago, M.L., Soros, V.B., Chiu, Y.L., Prochnow, C., Bransteitter, R., Neidleman, J. *et al.* (2007) *APOBEC3G determinants underlying high molecular mass complex formation.* New York: Cold Spring Harbor. Paper presented at: Retroviruses

Sasada, A., Takaori-Kondo, A., Shirakawa, K., Kobayashi, M., Abudu, A., Hishizawa, M. *et al.* (2005) APOBEC3G targets human T-cell leukemia virus type 1. *Retrovirology* **2**, 32.

Sawyer, S.L., Emerman, M. and Malik, H.S. (2004) Ancient adaptive evolution of the primate antiviral DNA-editing enzyme APOBEC3G. *PLoS Biol.* **2**, E275.

Schafer, A., Bogerd, H.P. and Cullen, B.R. (2004) Specific packaging of APOBEC3G into HIV-1 virions is mediated by the nucleocapsid domain of the gag polyprotein precursor. *Virology* **328**, 163–168.

Schindler, M., Munch, J., Kutsch, O., Li, H., Santiago, M.L., Bibollet-Ruche, F., Muller-Trutwin, M.C. *et al.* (2006) Nef-mediated suppression of T cell activation was lost in a lentiviral lineage that gave rise to HIV-1. *Cell* **125**, 1055–1067.

Schlaepfer, E., Audige, A., von Beust, B., Manolova, V., Weber, M., Joller, H. *et al.* (2004) CpG oligodeoxynucleotides block human immunodeficiency virus type 1 replication in human lymphoid tissue infected ex vivo. *J. Virol.* **78**, 12344–12354.

Schrofelbauer, B., Chen, D. and Landau, N.R. (2004) A single amino acid of APOBEC3G controls its species-specific interaction with virion infectivity factor (Vif). *Proc. Natl Acad. Sci. USA* **101**, 3927–3932.

Schrofelbauer, B., Senger, T., Manning, G. and Landau, N.R. (2006) Mutational alteration of human immunodeficiency virus type 1 Vif allows for functional interaction with nonhuman primate APOBEC3G. *J. Virol.* **80**, 5984–5991.

Sheehy, A.M., Gaddis, N.C., Choi, J.D. and Malim, M.H. (2002) Isolation of a human gene that inhibits HIV-1 infection and is suppressed by the viral Vif protein. *Nature* **418**, 646–650.

Sheehy, A.M., Gaddis, N.C. and Malim, M.H. (2003) The antiretroviral enzyme APOBEC3G is degraded by the proteasome in response to HIV-1 Vif. *Nat. Med.* **9**, 1404–1407.

Shindo, K., Takaori-Kondo, A., Kobayashi, M., Abudu, A., Fukunaga, K. and Uchiyama, T. (2003) The enzymatic activity of CEM15/Apobec-3G is essential for the regulation of the infectivity of HIV-1 virion but not a sole determinant of its antiviral activity. *J. Biol. Chem.* **278**, 44412–44416.

Simon, J.H., Gaddis, N.C., Fouchier, R.A. and Malim, M.H. (1998) Evidence for a newly discovered cellular anti-HIV-1 phenotype. *Nat. Med.* **4**, 1397–1400.

Simon, V., Zennou, V., Murray, D., Huang, Y., Ho, D.D. and Bieniasz, P.D. (2005) Natural variation in Vif: differential impact on APOBEC3G/3F and a potential role in HIV-1 diversification. *PLoS Pathog.* **1**, e6.

Smith, R.A., Loeb, L.A. and Preston, B.D. (2005) Lethal mutagenesis of HIV. *Virus Res.* **107**, 215–228.

Soros, V.B., Yonemoto, W. and Greene, W.C. (2007) Newly synthesized APOBEC3G is incorporated into HIV virions, inhibited by HIV RNA and subsequently activated by RNase H. *PLoS Pathog.* **3**, e15.

Stenglein, M.D. and Harris, R.S. (2006) APOBEC3B and APOBEC3F inhibit L1 retrotransposition by a DNA deamination-independent mechanism. *J. Biol. Chem.* **281**, 16837–16841.

Stopak, K., de Noronha, C., Yonemoto, W. and Greene, W.C. (2003) HIV-1 Vif blocks the antiviral activity of APOBEC3G by impairing both its translation and intracellular stability. *Mol. Cell* **12**, 591–601.

Stopak, K.S., Chiu, Y.L., Kropp, J., Grant, R.M. and Greene, W.C. (2007) Distinct patterns of cytokine regulation of APOBEC3G expression and activity in primary lymphocytes, macrophages and dendritic cells. *J. Biol. Chem.* **282**, 3539–3546.

Strebel, K. (2005) APOBEC3G & HTLV-1: inhibition without deamination. *Retrovirology* **2**, 37.

Strebel, K., Daugherty, D., Clouse, K., Cohen, D., Folks, T. and Martin, M.A. (1987) The HIV 'A' (sor) gene product is essential for virus infectivity. *Nature* **328**, 728–730.

Stremlau, M., Owens, C.M., Perron, M.J., Kiessling, M., Autissier, P. and Sodroski, J. (2004) The cytoplasmic body component TRIM5alpha restricts HIV-1 infection in Old World monkeys. *Nature* **427**, 848–853.

Suspene, R., Guetard, D., Henry, M., Sommer, P., Wain-Hobson, S. and Vartanian, J.P. (2005) Extensive editing of both hepatitis B virus DNA strands by APOBEC3 cytidine deaminases in vitro and in vivo. *Proc. Natl Acad. Sci. USA* **102**, 8321–8326.

Suspene, R., Rusniok, C., Vartanian, J.P. and Wain-Hobson, S. (2006) Twin gradients in APOBEC3 edited HIV-1 DNA reflect the dynamics of lentiviral replication. *Nucleic Acids Res.* **34**, 4677–4684.

Svarovskaia, E.S., Xu, H., Mbisa, J.L., Barr, R., Gorelick, R.J., Ono, A. *et al.* (2004) Human apolipoprotein B mRNA-editing enzyme-catalytic polypeptide-like 3G (APOBEC3G) is incorporated into HIV-1 virions through interactions with viral and nonviral RNAs. *J. Biol. Chem.* **279**, 35822–35828.

Temin, H.M. (1991) Sex and recombination in retroviruses. *Trends Genet.* **7**, 71–74.

Tian, C., Yu, X., Zhang, W., Wang, T., Xu, R. and Yu, X. F. (2006) Differential requirement for conserved tryptophans in human immunodeficiency virus type 1 Vif for the selective suppression of APOBEC3G and APOBEC3F. *J. Virol.* **80**, 3112–3115.

Tsai, W.P., Conley, S.R., Kung, H.F., Garrity, R.R. and Nara, P.L. (1996) Preliminary in vitro growth cycle and transmission studies of HIV-1 in an autologous primary cell assay of blood-derived macrophages and peripheral blood mononuclear cells. *Virology* **226**, 205–216.

Turelli, P., Mangeat, B., Jost, S., Vianin, S. and Trono, D. (2004a) Inhibition of hepatitis B virus replication by APOBEC3G. *Science* **303**, 1829.

Turelli, P., Vianin, S. and Trono, D. (2004b) The innate antiretroviral factor APOBEC3G does not affect human LINE-1 retrotransposition in a cell culture assay. *J. Biol. Chem.* **279**, 43371–43373.

UNAIDS (2006) Report on the global AIDS epidemic 2006. http://www.unaids.org/en/.

van Hemert, F.J. and Berkhout, B. (1995) The tendency of lentiviral open reading frames to become A-rich: constraints imposed by viral genome organization and cellular tRNA availability. *J. Mol. Evol.* **41**, 132–140.

Vartanian, J.P., Meyerhans, A., Asjo, B. and Wain-Hobson, S. (1991) Selection, recombination and G—A hypermutation of human immunodeficiency virus type 1 genomes. *J. Virol.* **65**, 1779–1788.

Vartanian, J.P., Meyerhans, A., Sala, M. and Wain-Hobson, S. (1994) G→A hypermutation of the human immunodeficiency virus type 1 genome: evidence for dCTP pool imbalance during reverse transcription. *Proc. Natl Acad. Sci. USA* **91**, 3092–3096.

Vartanian, J.P., Plikat, U., Henry, M., Mahieux, R., Guillemot, L., Meyerhans, A. and Wain-Hobson, S. (1997) HIV genetic variation is directed and restricted by DNA precursor availability. *J. Mol. Biol.* **270**, 139–151.

Wedekind, J.E., Gillilan, R., Janda, A., Krucinska, J., Salter, J.D., Bennett, R.P. *et al.* (2006) Nanostructures of APOBEC3G support a hierarchical assembly model of high molecular mass ribonucleoprotein particles from dimeric subunits. *J. Biol. Chem.* **281**, 38122–38126.

Wei, M., Xing, H., Hong, K., Huang, H., Tang, H., Qin, G. and Shao, Y. (2004) Biased G-to-A hypermutation in HIV-1 proviral DNA from a long-term non-progressor. *Aids* **18**, 1863–1865.

Wichroski, M.J., Ichiyama, K. and Rana, T.M. (2005) Analysis of HIV-1 viral infectivity factor-mediated proteasome-dependent depletion of APOBEC3G: correlating function and subcellular localization. *J. Biol. Chem.* **280**, 8387–8396.

Wichroski, M.J., Robb, G.B. and Rana, T.M. (2006) Human retroviral host restriction factors APOBEC3G and APOBEC3F localize to mRNA processing bodies. *PLoS Pathog.* **2**, e41.

Wiegand, H.L., Doehle, B.P., Bogerd, H.P. and Cullen, B.R. (2004) A second human antiretroviral factor,

APOBEC3F, is suppressed by the HIV-1 and HIV-2 Vif proteins. *EMBO J.* **23**, 2451–2458.

Xiao, Z., Ehrlich, E., Yu, Y., Luo, K., Wang, T., Tian, C. and Yu, X.F. (2006) Assembly of HIV-1 Vif-Cu15 E3 ubiquitin ligase through a novel zinc-binding domain-stabilized hydrophobic interface in Vif. *Virology* **349**, 290–299.

Xu, H., Svarovskaia, E.S., Barr, R., Zhang, Y., Khan, M. A., Strebel, K. and Pathak, V.K. (2004) A single amino acid substitution in human APOBEC3G antiretroviral enzyme confers resistance to HIV-1 virion infectivity factor-induced depletion. *Proc. Natl Acad. Sci. USA* **101**, 5652–5657.

Xu, H., Chertova, E., Chen, J., Ott, D.E., Roser, J.D., Hu, W.S. and Pathak, V.K. (2007) Stoichiometry of the antiviral protein APOBEC3G in HIV-1 virions. *Virology* **360**, 247–256.

Yang, B., Chen, K., Zhang, C., Huang, S. and Zhang, H. (2007a) Virion-associated uracil DNA glycosylase-2 and apurinic/apyrimidinic endonuclease are involved in the degradation of APOBEC3G-edited nascent HIV-1 DNA. *J. Biol. Chem.* **282**, 11667–11675.

Yang, Y., Guo, F., Cen, S. and Kleiman, L. (2007b) Inhibition of initiation of reverse transcription in HIV-1 by human APOBEC3F. *Virology* **365**, 92–100.

Yu, Q., Chen, D., Konig, R., Mariani, R., Unutmaz, D. and Landau, N.R. (2004a) APOBEC3B and APOBEC3C are potent inhibitors of simian immunodeficiency virus replication. *J. Biol. Chem.* **279**, 53379–53386.

Yu, Q., Konig, R., Pillai, S., Chiles, K., Kearney, M., Palmer, S. *et al.* (2004b) Single-strand specificity of APOBEC3G accounts for minus-strand deamination of the HIV genome. *Nat. Struct. Mol. Biol.* **11**, 435–442.

Yu, X., Yu, Y., Liu, B., Luo, K., Kong, W., Mao, P. and Yu, X.F. (2003) Induction of APOBEC3G ubiquitination and degradation by an HIV-1 Vif-Cu15-SCF complex. *Science* **302**, 1056–1060.

Zack, J.A., Arrigo, S.J., Weitsman, S.R., Go, A.S., Haislip, A. and Chen, I.S. (1990) HIV-1 entry into quiescent primary lymphocytes: molecular analysis reveals a labile, latent viral structure. *Cell* **61**, 213–222.

Zennou, V. and Bieniasz, P.D. (2006) Comparative analysis of the antiretroviral activity of APOBEC3G and APOBEC3F from primates. *Virology* **349**, 31–40.

Zennou, V., Perez-Caballero, D., Gottlinger, H. and Bieniasz, P.D. (2004) APOBEC3G incorporation into human immunodeficiency virus type 1 particles. *J. Virol.* **78**, 12058–12061.

Zhang, H., Yang, B., Pomerantz, R.J., Zhang, C., Arunachalam, S.C. and Gao, L. (2003) The cytidine deaminase CEM15 induces hypermutation in newly synthesized HIV-1 DNA. *Nature* **424**, 94–98.

Zhang, J. and Webb, D.M. (2004) Rapid evolution of primate antiviral enzyme APOBEC3G. *Hum. Mol. Genet.* **13**, 1785–1791.

Zhang, K.L., Mangeat, B., Ortiz, M., Zoete, V., Trono, D., Telenti, A. and Michielin, O. (2007) Model structure of human APOBEC3G. *PLoS ONE* **2**, e378.

Zheng, Y.H., Irwin, D., Kurosu, T., Tokunaga, K., Sata, T. and Peterlin, B.M. (2004) Human APOBEC3F is another host factor that blocks human immunodeficiency virus type 1 replication. *J. Virol.* **78**, 6073–6076.

CHAPTER

9

Lethal Mutagenesis

James J. Bull, Rafael Sanjuán, and Claus O. Wilke

ABSTRACT

Most mutations with phenotypic effect are harmful, and an increase in mutation rate generally leads to a decrease in population fitness. This relationship lies at the heart of the concept of lethal mutagenesis, whereby a viral population is driven to extinction through the administration of a mutagen. Here we explain the theoretical basis for lethal mutagenesis, describe how lethal mutagenesis differs from Eigen's error threshold, and show that experimental results on lethal mutagenesis are consistent with theoretical expectations. The most important insight derived from the theory of lethal mutagenesis is that extinction arises from a combination of genetic and demographic processes, so that there is no hard mutation threshold at which extinction is guaranteed, and no genetic signature that distinguishes lethal from non-lethal mutagenesis. Another important property is that the decline in population fitness following an increase in mutation rate may happen slowly. Thus, elevation of the mutation rate to a level sufficient to cause extinction may be followed by many generations before population size begins to decline.

INTRODUCTION

Lethal mutagenesis is the concept and practise of elevating the mutation rate of a virus to the point that the viral population dies out. This is potentially a workable mechanism for an infection within one host, where drugs can be applied in sufficient concentrations to elevate mutation rates above background. Nearly a century of work in genetics has shown that most mutations with phenotypic effect are harmful, many even being lethal, so there is little question that lethal mutagenesis can work in principle. The question is then the quantitative one of just how much elevation is required. Our chapter addresses the quantitative basis of lethal mutagenesis.

A high mutation rate is obviously necessary to achieve lethal mutagenesis, because at low mutation rate, selection can purge deleterious mutations as fast as they arise, allowing the virus to persist. Yet the virology literature is somewhat contradictory about the fitness effects of a high mutation rate. When considering lethal mutagenesis, it is universally acknowledged that mutation is bad overall for the virus, hence that "more is better" to

Origin and Evolution of Viruses
ISBN 978-0-12-374153-0

extinguish it. In another context, however, most discussions of virus adaptive evolution argue that the high mutation rates of RNA viruses bestow them with higher evolvability than possessed by DNA viruses—that the high mutation rate is beneficial. Since a random sample of mutations will contain some good and some bad, neither of these extremes will be accurate: a high mutation rate has good and bad components. No model is complete without both, therefore, but we will find that the derivation of thresholds for lethal mutagenesis is most tractable by assuming an "all bad" model, and then adding in beneficial mutations.

Progress in understanding the foundations of lethal mutagenesis has been hampered by confusing it with Eigen's error catastrophe. The error catastrophe is a genetic transition in certain types of fitness landscapes in which high-fitness, mutation-sensitive genotypes are lost in favor of low-fitness, mutation-robust genotype networks (Eigen, 1971, 1987, 2002). Both lethal mutagenesis and error catastrophes are associated with elevated mutation rates, but they are different. Indeed, all models of error catastrophe have assumed that the population size remains constant, and if extinction was a possibility in those models, an error catastrophe would actually reduce the chance of extinction by lethal mutagenesis.

BASIC MODEL

Lethal mutagenesis is a process of extinction. As with any population dynamic process, if there are fewer surviving offspring than parents, the population size will decline. If this continues long enough, the population will die out. We use the term "extinction threshold" to indicate the point at which the population can merely maintain itself. Lethal mutagenesis can effect extinction by introducing so many bad mutations that most offspring die and the population passes below the extinction threshold. It is worth commenting that extinction is the subject of much interest in ecology and has also been considered at least indirectly in the literature on the evolutionary benefits of sex and recombination (that asexual organisms tend to go extinct) (Maynard Smith, 1978). In much of that literature, extinction is considered to be a consequence of small (or finite) population sizes. For lethal mutagenesis, however, extinction can be achieved deterministically, hence is not due to any chance events in finite populations.

The extinction threshold is most easily derived when all mutations are lethal. At this extreme, surviving offspring and thus all parents are mutation-free. In the next generation of offspring, some get no mutations, some get one, others get two, and so on. Of the individuals that grow up to be adults and reproduce, we only need to count those that got zero mutations. If mutations are distributed among genomes according to a Poisson process with an average of U mutations per genome per generation, the mutation-free fraction is simply e^{-U}.

To know whether this fraction is enough to permit the population to persist in the long run, we also need to know if at least one offspring per infected cell will survive, on average, which depends on the total number of offspring. If each infected cell produces b viral progeny that would go on to infect new cells (in the absence of mutation), then

$$be^{-U} = 1 \tag{1a}$$

is the extinction threshold, and

$$be^{-U} < 1 \tag{1b}$$

ensures extinction (Bull *et al.*, 2007). In a closed culture system, b would be approximately the burst size per infected cell. However, in a body, immune processes and the reticulocyte system no doubt remove a large fraction of free virus, so b would be correspondingly reduced below the burst size.

Result (1b), which will be generalized below, reveals several properties of lethal mutagenesis that apply more generally.

1. Extinction may be gradual. Starting with N parents, there will be Nbe^{-U} surviving

offspring, and the population size will be $N(be^{-Ut})$ after t generations (for the simplest model of population growth). Increasing U will accelerate the process, but extinction need not happen quickly.

2. The extinction threshold is not determined just by mutation rate. The ecology of the infection is important too, as reflected in b. Most antiviral drugs work by reducing b in some fashion.
3. The threshold is not affected by neutral mutations, however U applies only to the deleterious ones.

DELETERIOUS MUTATIONS

A far more realistic assumption than strict lethality of all mutations is that mutations have various harmful effects, not all lethal. Somewhat surprisingly, the extinction threshold for this case is the same as when all mutations are lethal, no matter what the distribution of harmful fitness effects is (some unrealistic cases provide exceptions to this, but we will not concern ourselves with those). An explanation for the generality of this result is as follows. Inequality (1b) gives the conditions for loss of the mutation-free genotype, by the same argument that applies when all mutations are lethal: it guarantees that the number of mutation-free, surviving offspring is less than the number of mutation-free parents. Once the mutation-free class is lost, condition (1b) gives conditions for the loss of the one-mutation genotypes. And so on. At some point, all individuals are carrying enough mutations that they can no longer reproduce enough to replace themselves.

Underlying this result is a generality about the average population fitness at equilibrium that was realized decades ago. When the evolutionary process has reached an equilibrium, such that the number of new mutations is balanced by the loss of old ones from selection against them, population mean fitness is e^{-U}, where U is the deleterious mutation rate (Kimura and Maruyama, 1966; Wilke and Adami, 2003). This surprising result is due to the fact that mutations of smaller effect have weaker selection against them than those of large effect. Consequently, they can accumulate to higher levels than the more severe mutations, and the two effects balance exactly. The extinction threshold is the point at which average fitness (at equilibrium) exactly balances fecundity. To understand why fecundity enters the threshold, consider that, if the loss to mutation is 80% of the offspring, then this loss dooms a species that only produces three offspring per parent but may be trivial for one that produces 100.

Although the extinction threshold is the same for non-lethal, harmful mutations as for purely lethal mutations, the dynamics of extinction is very dependent on the fitness effects of the mutations. When all mutations are lethal, the adult population consists of only mutation-free individuals, and population mean fitness is at equilibrium immediately. When deleterious mutations are not lethal, an elevation of mutation rate leads to a gradual accumulation of mutations in the population. Equilibrium fitness and mutation load may take many generations to be approached closely. Consequently, even once the mutation rate is increased beyond the extinction threshold, it may take many generations before the population size even begins declining, much less goes extinct (Figure 9.1). The pattern of the decline and of mutation accumulation is highly sensitive to interactions (epistasis) among the mutations, and there are no obvious generalities (Bull *et al.*, 2007).

The preceding argument that $be^{-U} < 1$ ensures a progressive loss of the 0, 1, 2, . . . mutation classes might seem to imply that lethal mutagenesis involves a progressive accumulation of mutations—that a genetic "signature" of lethal mutagenesis exists in the form of a cascading or perhaps accelerating mutational build-up that progressively debilitates the genomes. For example, the process known as Muller's ratchet that is commonly involved to explain the extinction of finite, asexual populations involves a progressive loss of mutation-free genotypes, then of one-mutation genotypes, and so on, indefinitely

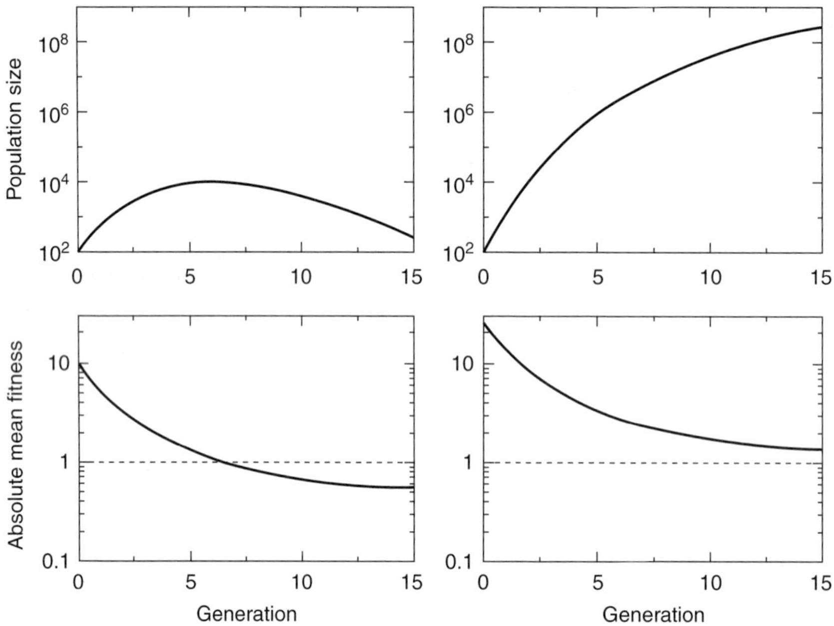

FIGURE 9.1 Mutagenesis, fitness change, and population growth. All simulations used a genome-wide mutation rate per generation of $U = 3.0$, with a fitness cost of 0.2 per mutation. Thus, a genome with no mutations had a relative fitness of 1.0, a genome with one mutation had a fitness of 0.8, a genome with two mutations had a fitness of $0.8 \times 0.8 = 0.64$, and so on. Top graphs represent population sizes of surviving adults over time, starting with 100 individuals in generation 0. Lower graphs represent (absolute) fitness changes over time. On the left side, a fecundity of $b = 10$ was assumed; on the right, $b = 25$. In generation 0, the 100 parents each produced b offspring. All parents in generation 0 were free of mutations, but their progeny carried on average three deleterious mutations (according to a Poisson distribution). Those progeny survived according to the number of mutations they carried to become the parents in generation, whereupon they also produced b progeny. Their progeny accumulated additional mutations and survived accordingly to become the parents of generation 2. And so on. As mutations accumulated over time, the average fitness (survival of progeny) declined. On the left, the fecundity ($b = 10$) was not high enough to offset the ultimate decline in fitness ($be^{-U} = 10 \times e^{-3} = 10 \times 0.05 = 0.5 < 1$), and the population increased at first but then ultimately declined. On the right, the fecundity of 25 was enough to offset the fitness effect of mutagenesis ($be^{-U} = 25 \times 0.05 = 1.25 > 1$), so the population never declined. Population size in generation $t + 1$ was taken simply as the product of population size in generation t and absolute mean fitness in generation $t + 1$; it is merely intended to show the difference between population growth and decline.

(Maynard Smith, 1978). No genetic signature exists for lethal mutagenesis, however. At least while the population size remains large (and recall that lethal mutagenesis can operate in the largest of populations), there is no genetic difference between lethal mutagenesis and the same level of mutagenesis with population persistence (Figure 9.1). The underlying mathematical reason for this observation is that the equation governing changes in *relative* mutant frequencies decouples from the equation describing the *absolute* size of the virus population (Wilke *et al.*, 2001). Thus, one cannot merely look at mutation frequencies to tell if lethal mutagenesis is operating. The equilibrium population fitness is e^{-U}, and this equilibrium obtains independent of population size. (The population may go extinct before equilibrium is attained.)

The absence of a genetic signature of lethal (as opposed to non-lethal) mutagenesis is borne out by experimental results. For example, Grande-Pérez *et al.* (2002) propagated

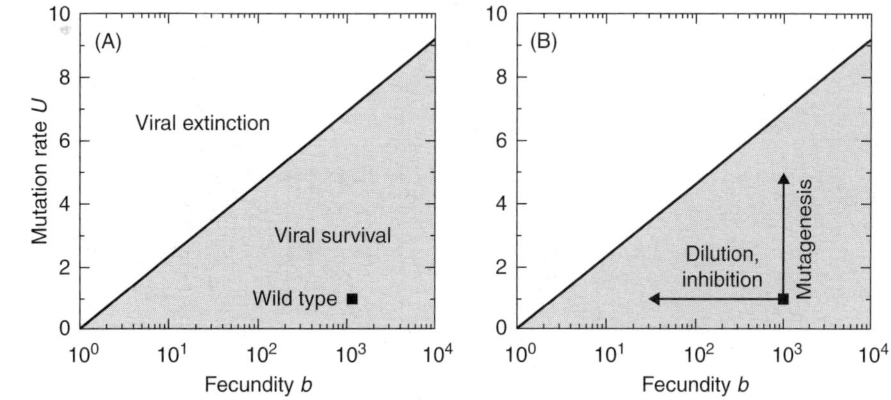

FIGURE 9.2 Phase diagram of lethal mutagenesis. Under all experimental conditions, a virus strain has a fecundity b and a mutation rate U. Each combination of b and U corresponds to a point in the diagram. For example, assume that the wild-type virus has $b = 10^3$ and $U = 1.0$. Then, the location of the wild-type virus in the diagram is as indicated in (A). The lethal mutagenesis condition $be^{-U} = 1$ is indicated as solid diagonal line. All virus strains that fall above this line will go extinct eventually, while all virus strains that fall below this line (i.e. into the shaded area in the plot) will survive. (B) shows how antiviral treatment can be visualized in this diagram. Mutagenesis corresponds to a vertical movement, whereas dilution or other viral inhibition corresponds to a horizontal movement. Both mutagenesis and dilution/inhibition may or may not lead to extinction, depending on whether the arrow extends into the area above the solid line.

lymphocytic choriomeningitis virus (LCMV) in the presence of the mutagenic base analogues 5-fluorouracil (5-FU) or 5-azacytidine (AZC) in a number of different concentrations, and determined the critical concentration at which extinction occurred. (Extinction set in around 25 μg/mL for 5-FU, whereas extinction was not achieved at any concentration of AZC.) They then carried out extensive sequencing of clones to find the genetic signature of this transition to extinction but could find none, concluding that "direct evidence of melting genomic information is still lacking." As they acknowledged, their sampling scheme was biased in favor of viable genomes, but this fact alone would not explain the lack of a genetic signature.

GRAPHICAL VISUALIZATION OF THE EXTINCTION THRESHOLD

A convenient way to visualize the extinction threshold $be^{-U} < 1$ is to draw a phase diagram (Figure 9.2). We plot the fecundity b along the horizontal axis and U along the vertical axis, and shade all points that correspond to $be^{-U} > 1$; the shaded points correspond to combinations of b and U that admit viral survival. All the unshaded areas of the plot correspond to viral extinction. Any virus strain falls somewhere in this diagram. For example, assume that the wild-type strain of some virus has a fecundity $b = 10^3$ and a deleterious mutation rate $U = 1.0$. Then, this strain falls into the regime that admits viral survival, at the location indicated in Figure 9.2A. Now, assume we subject this strain to mutagenesis, such that its mutation rate is elevated to $U = 5.0$. We can indicate this experimental treatment by a vertical arrow in the diagram, as shown in Figure 9.2B. Likewise, if we subject the strain to additional dilution or inhibition, such that its fecundity is reduced to $b = 30$, we can indicate this experimental treatment by a horizontal arrow in the diagram, again as shown in Figure 9.2B. In this particular example, neither mutagenesis to $U = 5.0$ nor dilution/inhibition to $b = 30$ are sufficient to cause extinction; the virus will survive both treatments. In general, mutagenesis is always represented by a vertical movement in the diagram, whereas dilution or inhibition are represented by a horizontal movement in

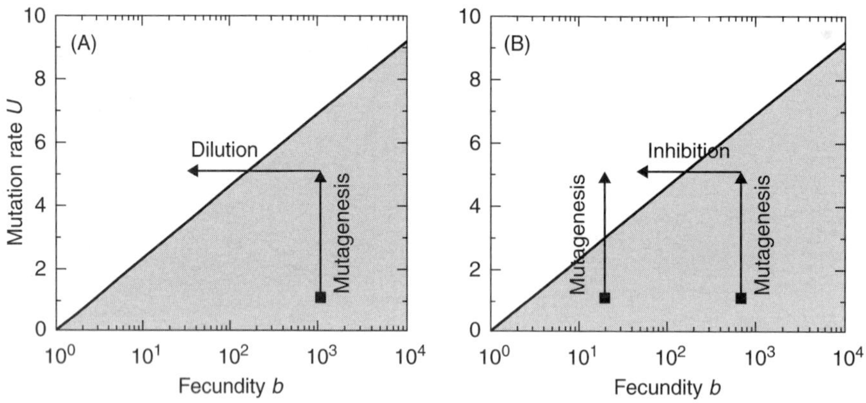

FIGURE 9.3 Application of the lethal mutagenesis phase diagram to specific lethal mutagenesis experiments. (A) Mutagenesis alone may not lead to viral extinction, but mutagenesis combined with dilution does. This diagram corresponds to the experimental situation presented for example in figure 2 of Sierra *et al.* (2000). (B) Mutagenesis alone leads to extinction for a low-fecundity (i.e. low-fitness) strain, but has to be combined with viral inhibition to result in extinction for a high-fecundity strain. This diagram corresponds to the experimental situation presented for example in figure 4 of Pariente *et al.* (2001).

the diagram. Finally, a drug that can act both as mutagen and as viral inhibitor at the same time would lead to a diagonal movement in the diagram.

We can use the lethal mutagenesis phase diagram to interpret the outcome of lethal mutagenesis experiments. For example, Sierra *et al.* (2000) found that propagation of foot-and-mouth disease virus (FMDV) C-S8c1 in the presence of the mutagenic base analogues 5-FU or AZC (mutagenic drugs) did not lead to extinction, whereas propagation in the presence of these mutagens combined with 10-fold dilution led to extinction (figure 2 in Sierra *et al.*, 2000). This experimental finding is described in Figure 9.3A. (Note that we do not claim to know the true values of b or U in the presence or absence of treatment; the diagram is only meant as a qualitative representation of the experimental results.) For FMDV C-S8c1, mutagenesis alone by either 5-FU or AZC is not sufficient to move the virus into the extinction regime. However, mutagenesis moves the virus closer to the extinction threshold, and combined with 10-fold dilution, is sufficient to cause extinction. Interestingly, dilution alone, even up to 1000-fold, was also not sufficient to cause extinction. In the same paper, the authors repeated their experiment with two other FMDV strains, the high-fitness strain MARLS and the low-fitness strain C^9_{22} (Sierra *et al.*, 2000; their figure). Briefly, the results are again consistent with lethal mutagenesis theory: for the high-fitness strain, mutagenesis combined with ten-fold dilution was not obviously sufficient to cause extinction, whereas for the low-fitness strain, extinction can be achieved even in the absence of dilution.

It is not clear how one would combine mutagenesis with dilution to treat a viral infection *in vivo*. However, one can administer one or more viral inhibitors with the mutagen in a form of multidrug therapy. With respect to viral extinction or survival, the inhibitor has the same effect as dilution. Several studies have tried this approach (e.g. Pariente *et al.*, 2001, 2003). Here, we focus on one experiment of Pariente *et al.* (2001), summarized in their figure 4. They grew FMDV strain MARLS in two different host cell types, baby hamster kidney 21 (BHK-21) cells and Chinese hamster ovary (CHO) cells. MARLS had higher fitness in BHK-21 cells than in CHO cells (Pariente *et al.*, 2001). In both cell types, Pariente *et al.* administered various combinations of the

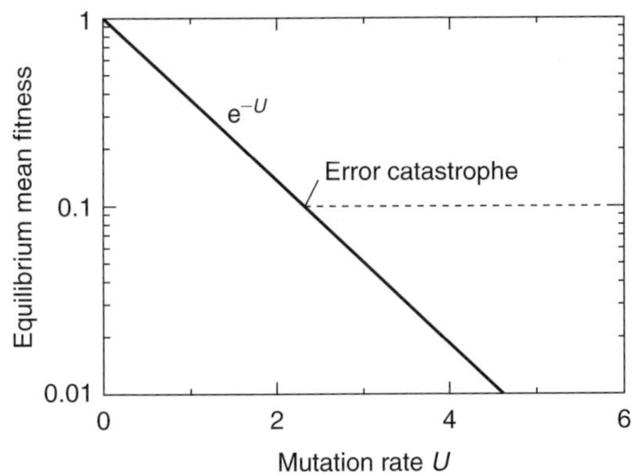

FIGURE 9.4 Equilibrium mean fitness as a function of the mutation rate U (after Bull *et al.*, 2007). For most fitness landscapes, the equilibrium mean fitness decays as e^{-U}, where U is the deleterious mutation rate (i.e. neutral mutations do not contribute to U). In the Eigen single-peak fitness landscape (Eigen, 1971; Swetina and Schuster, 1982), at a specific mutation rate (whose value depends on the height of the single peak) the population undergoes a transition called the error catastrophe, and the population disperses over the sequence space. Note that (i) the error catastrophe does not coincide with a sudden loss of fitness for the population as a whole, and (ii) the mean fitness does not decline further after the error catastrophe. Observation (i) is true for all error catastrophes, observation (ii) is an artifact of the Eigen fitness landscape. In general, mean fitness will still decline after the error catastrophe, but the severity of the decline will be reduced (Bull *et al.*, 2005).

mutagen 5-FU and the viral inhibitors guanidine hydrochloride (G) and heparin (H), and found that in BHK-21 cells, where MARLS had high fitness, joint administration of 5-FU and both inhibitors G and H was necessary to achieve extinction. Administration of either one or two of the three drugs 5-FU, G, and H led to a reduction of viral titer but not extinction. On the other hand, in CHO cells, where MARLS has lower fitness, administration of just 5-FU was sufficient to cause extinction. We visualize the interpretation of these results in the framework of the theory of lethal mutagenesis in Figure 9.3B. The low-fitness virus (i.e. MARLS in CHO cells) is already close to the extinction threshold, and mutagenesis alone is sufficient to cause extinction. The high-fitness virus, on the other hand, is fairly distant from the extinction threshold, and mutagenesis combined with substantial inhibition is necessary to cause extinction.

EXTINCTION IS THWARTED BY AN ERROR CATASTROPHE

In the 1970s and beyond, Manfred Eigen introduced a novel evolutionary result in which a "master" sequence, free of deleterious mutation dominated the population at low mutation rates but was lost at high mutation rates (Eigen, 1971, 1987, 2002). His model specifically assumed that all genotypes were viable, but genotypes with one or more mutations were of the same, lower fitness than the master sequence. In the simplest models, the master sequence could not be regenerated by back mutation. Frequency of the master sequence declined as the mutation rate was increased, and at some threshold (the "error" threshold), it was completely absent from the population. This loss was called an error catastrophe. For comparison to lethal mutagenesis, the error threshold is a description of equilibrium behavior of a population. That is, if a mutation rate increase is imposed on a population, the population will experience a gradual increase in the number of segregating mutations, but over time this increase slows and approaches an equilibrium state. (In models of infinite population size, it strictly takes an infinite number of generations to reach the new mutation-selection balance equilibrium.) The error threshold is the lowest mutation rate at which the master sequence is absent from the equilibrium population. The error threshold and error catastrophe are strictly genetic results; they depend on the relative fitnesses of the different genotypes, but there is no "ecological" component similar to the "b" of the extinction threshold.

As noted in the opening of this chapter, the error catastrophe has often been equated with extinction by lethal mutagenesis. We can now show how they are different. Plotting the equilibrium mean fitness as a function of U in this system, the impact of an error catastrophe on population mean fitness is easily visualized (Figure 9.4). Recalling the Kimura–Maruyana result (Kimura and Maruyama, 1966), equilibrium mean fitness follows e^{-U} down to the error threshold. At the error threshold, increases in U have no further effect on mean fitness because the mutation-free genotype is lost and all mutated genotypes have the same fitness (in the simplest of Eigen's models). Note that, in particular, it is *not true* that the error threshold is accompanied by a sudden and substantial loss of fitness, as is frequently stated in the literature. To the contrary, fitness is continuous at the error threshold, not just in the simple Eigen fitness landscape depicted in Figure 9.4, but in all fitness landscapes. In more complicated landscapes, multiple error thresholds may occur, and each catastrophe increases robustness against mutations somewhat (Tannenbaum and Shakhnovich, 2004; Bull *et al.*, 2005).

An extinction threshold may be superimposed on this graph. Where the extinction threshold lies will depend on the fecundity per cell as well as U, and in principle, the extinction threshold can lie anywhere. It will generally *not* coincide with an error threshold. Furthermore, if the extinction threshold was otherwise below the error threshold in the Eigen model, then the error catastrophe would avoid extinction by halting the decay of mean fitness with further increases in U, simply because all mutant genotypes in that model have the same fitness.

In light of this comparison, we return to the study of Grande-Pérez *et al.* (2002) described above. The authors cast their results in a somewhat different context than we explained it above. The study actually looked for the genetic signature of an error catastrophe coinciding with lethal mutagenesis, because the paper considered the two processes equivalent. The explanation given for the failure to observe a genetic signature of lethal mutagenesis was sampling bias: only replication-competent sequences could be sampled, so any signal in non-viable sequences would be lost. While this bias against observing non-replicating genomes is valid, recent theoretical work on the error catastrophe in the presence of lethal mutants found a clear genetic signature of the error catastrophe that should be observable even when lethal mutants cannot be sampled (Takeuchi and Hogeweg, 2007). To us, the simpler and more obvious explanation of the results by Grande-Pérez *et al.* (2002) is that extinction by lethal mutagenesis is unrelated to the error catastrophe, and that therefore the absence of a genetic signature of lethal mutagenesis is the expected experimental outcome.

The virus used in the study of Grande-Perez *et al.* (2002), lymphocytic choriomeningitis virus (LCMV), forms persistent infections in BHK-21 cells. Our theory of lethal mutagenesis does likely not apply in unaltered form to this mode of replication. In particular, our theory neglects any effects mediated by viral co-infection at high multiplicity of infection. However, a persistent infection in the presence of mutagen leads to the accumulation of many mutant viral genomes within the same cell, and we cannot neglect co-infection in this case. Grande-Perez *et al.* (2005) have proposed that in this situation and for moderate amounts of mutagen, lethal mutagenesis is mediated by the accumulation of defective particles. Nevertheless, the basic concept of lethal mutagenesis, that is, the rule that for viral survival the total number of viable offspring that go on to infect a new cell has to be larger than one, will be valid even in more complex situations involving co-infection and defective particles.

BENEFICIAL MUTATIONS

The assumption that all mutations are deleterious is unrealistic. We can suggest how a finite number of beneficial mutations will affect the threshold. If the combined fitness effect on a wild-type genome carrying all beneficial mutations is K (that is, the fitness of the wild-type genome is 1, the fitness of a wild-type genome carrying the complete set of beneficial mutations is K), then the modified extinction threshold is

$$bKe^{-U} = 1 \tag{2}$$

This assumes that the beneficial mutations do not interact with the deleterious ones, hence merely push the baseline fitness from 1 ultimately up to K. This threshold also assumes the population size is large enough that a large number of individuals experience the full set of beneficial mutations (i.e. we can ignore stochastic effects). In more complicated cases, it will generally be true that beneficial mutations impede lethal mutagensis, but the effect will not necessarily be this simple.

ESTIMATING THE PARAMETERS OF LETHAL MUTAGENESIS

PubMed currently lists nearly 3800 references with the key phrase "lethal mutagenesis." Yet it remains true that there has been no empirical determination in any system of the parameters required to satisfy the extinction threshold ($be^{-U} < 1$) and thus that underlie lethal mutagenesis. Although only two parameters need be measured, U and b, each one presents its own difficulties.

It should first be understood how the model partitions different biological effects into these two parameters. U is the deleterious mutation rate, and b is the number of virions per infected cell that go on to infect other cells. As mutations accumulate and fitness declines during mutagenesis, infected cells will produce fewer and fewer virions, on average. The model attributes this effect to mutation, however, and does not alter b. Thus, the value of b in the threshold accrues to the wild-type (mutation-free) genotype. Suppose, however, that the mutagenic drug has the additional effect of harming the virus in a way separate from mutation (as attributed to the drug ribavirin, for example) or inhibits the cell's ability to make virus; the model partitions those effects into b. Thus, all immediate and cumulative effects of mutation are subsumed into e^{-U}; other effects enter b. A further complication is that b may not be constant during the course of the infection—it may vary due to changes in the immune response and may change as the infection starts to wane, if treatment is successful. For extinction, the relevant value of b is the maximum from the onset of treatment until the virus is extinguished, for the mutation-free genome.

The most straightforward quantity to measure is U, the genome-wide deleterious mutation rate per generation. U includes all forms of deleterious effects but excludes neutral and beneficial mutations. No single approach that has been used previously to measure mutation rates is without difficulties in applications to U, but some are better than the others.

An upper limit on U is the total genomic mutation rate per generation, U_T. The deleterious mutation rate would then be the product of U_T times the fraction of mutations that are deleterious. Although both quantities should be obtained *in vivo* for an accurate determination of the extinction threshold, *in vitro* estimates may be sufficiently accurate to justify their use. In particular, 61% of random mutations in vesicular stomatitis virus (VSV) were found to be deleterious *in vitro* (Sanjuan *et al.*, 2004). If one accepts a conservative value of 50% as the fraction of mutations that are deleterious, then it remains to know U_T as a function of drug dose.

Estimating U might seem to be easy in this era of easy genome sequencing. The main complication in measuring U_T is that it applies to a single round of infection. The basic idea is to measure the total number of mutations generated after a single round of infection, but most attempts to measure mutation numbers have measured numbers of mutations after several rounds of infection. The number of mutations after multiple infection cycles does not generally equal the mutation number per single round. Once mutated viruses have been allowed to re-infect, the numbers of observed mutations are changed due to (i) selection against those with the more harmful mutations, and (ii) new mutations being added to the genomes that carry mutations from previous generations. The two effects cannot be separated easily. When only a single round of infection is allowed, neither of these complications occurs. A single round of infection can be enforced, for example, by using genetic constructs that assemble non-infectious particles unless complemented by the host (Mansky, 1998). When virions produced by a complementing host are used to infect non-complementing hosts, the resulting progeny will be unable to undergo further rounds of infection, so their genomic mutations will be from a single round of infection. Two other commonly used measures of mutation rate, the Luria–Delbruck method and the mutation accumulation method, do not rely on sequence data but pose other problems in estimating U for only and all deleterious mutations (Bull *et al.*, 2007).

An alternative method of estimating the deleterious effects of mutagenesis merely estimates the equilibrium relative fitness (e^{-U}), and bypasses the mutation rate estimate altogether. If a population is grown to equilibrium between mutation and selection, then the mean fitness should be estimable by transfecting a standard number of genomes into cells and determining the number of progeny that result. If this value is divided by the same measure for the wild-type virus, the resulting ratio should be e^{-U}. This method is essentially that of Crotty *et al.* (2001), except that they did not grow the virus to equilibrium prior to making the determinations. However, their study clearly demonstrated the technical feasibility of this approach.

The estimation of b is challenging. The quantity b is the number of viral progeny per infected cell that go on to establish new infections (for the wild-type genome), but it is perhaps more easily thought of as the product of burst size and survival to infection. Thus, burst size will be a strict upper limit to b. Of course, burst size may vary from tissue to tissue, and it is not even clear if burst size measured *in vitro* will apply *in vivo*. Although the average burst size would be the relevant quantity if virus–cell infection followed strict mass-action process, spatial structure of the host may require more severe mutagenesis to achieve extinction.

An indirect approach to estimating b would be to monitor the density of the infection soon after it is established. By knowing the approximate lifespan of an infected cell and how rapidly the density of infected cells increase, it becomes possible to calculate an approximate b. However, in none of these cases does it appear that an accurate determination of b is possible. This is not necessarily a serious problem, because b enters the extinction threshold as a linear term, but U as an exponent (Figure 9.2). Thus, relatively small increases in U can overwhelm any uncertainty in b.

SUBLETHAL MUTAGENESIS

The models of lethal mutagenesis consider that mutations are deleterious, except as noted above. They thus lead to the result that viral fitness will be depressed by mutagenesis even if extinction does not result. If those models could be trusted, the unavoidable conclusion is that sublethal mutagenesis would reduce the infection even in the absence of extinction. We caution that any such conclusion is premature. First, short-term mutagenesis followed by drug withdrawal would be expected to enhance viral fitness much as short-term mutagenesis has been used by geneticists to enhance the recovery of beneficial mutations. Second, it is at least feasible that elevation of the mutation rate in the long term could result in irreversible viral adaptation to the host and may even "addict" the infected host to drug treatment (whereby withdrawing drug treatment leads to a far worse infection than would occur in the complete absence of treatment). These speculations offer some exciting possibilities for future theory and empirical work.

AN ALTERNATIVE MODEL OF MUTATION

In our derivation of the extinction threshold $be^{-U} < 1$, we have implicitly assumed that mutations arise in offspring sequences while the parent (i.e. template) sequence is being copied, and that the template sequence remains free of new mutations. These assumptions do not capture all conceivable models of mutation. For example, ionizing radiation from an external source can cause mutations to accumulate in a fashion that is decoupled from the replication process. Therefore, mutagenesis induced by radiation rather than by base analogues will not be governed by the extinction threshold $be^{-U} < 1$. However, the general principles of lethal mutagenesis as we have described them in Figure 9.2 remain valid. The exact contour separating the parameter region of viral survival from the region of viral extinction may take on a shape different from what is shown in Figure 9.2, but it is always possible to draw a phase diagram analogous to Figure 9.2 in which mutagenesis corresponds to vertical movement, dilution or inhibition to horizontal movement, and extinction is achieved when the combination of mutagenesis and dilution/inhibition move the virus out of the shaded area of viral survival. In fact, Zeldovich *et al.* (2007) recently calculated the shape of the curve separating extinction from survival in a model of protein evolution and mutagenesis in which mutations are generated by ionizing radiation, and found that the fecundity and the mutation rate both enter the extinction threshold linearly (whereas in $be^{-U} < 1$, the mutation rate enters exponentially). The main qualitative conclusion from this alternative extinction threshold is that mutagenesis by radiation will rapidly cause extinction if the time-scale on which mutations happen is short compared with the time-scale at which the viruses replicate, that is, if every genome receives multiple mutations between replication events, whereas extinction is impossible if mutations occur slowly compared to replication.

FUTURE DIRECTIONS

Despite the difficulty of documenting *in vivo* that lethal mutagenesis is operating in any system, and thus the possibility that we will never be able to show conclusively that it works for any treatment, there are some obvious steps to be taken toward the goal of further understanding lethal mutagenesis. Most obviously, it is feasible to evaluate lethal mutagenesis *in vitro*, and such studies would go a long way toward solidifying our understanding of the process *in vivo*. At the most comprehensive level, these studies could include estimating the total mutation rate (U_T), the mean fitness at equilibrium (e^{-U}) and the fecundity term b. Another area that needs addressing is the nature of evolution when mutation rates have been elevated to just below the extinction threshold. Is a high mutation rate usually deleterious, even with viral persistence, or does it allow the evolution of genotypes that are increasingly difficult to control and perhaps enhance within-host adaptation (e.g. an expansion of tissue tropisms, rapid escape from immune recognition). None of these questions can be answered with models, and the answers may very well differ from system to system. For example, the elevation of mutation rates may affect adaptation differently for DNA viruses than for RNA viruses.

To the extent that prior views of lethal mutagenesis were dominated by the error threshold model, we may ask what is gained with the new understanding of lethal mutagenesis—that there is no genetic signature, and that it involves an ecological as well as a genetic component. For the most part, the new understanding is not revolutionary. It merely suggests that we should not be looking for any magic genetic process, but that the key to viral extinction lies in the boring details of just how fast mutations are being thrown at the genome and how many "excess" progeny are normally produced. Ironically, this is probably the view that most virologists would have been led to in the first place, had it not been for the seductive concept of a phase transition from physics and the profoundly misleading term "error catastrophe." Nonetheless, however we have arrived here, lethal mutagenesis is a viable therapy provided only that the mutation rate of the virus can be (selectively) boosted to high enough levels. If it becomes easier to make mutagenic drugs, it is a therapy that is potentially applicable to a wide variety of viruses.

REFERENCES

Bull, J.J., Meyers, L.A. and Lachmann, M. (2005) Quasispecies made simple. *PLoS Comput. Biol.* **1**, e61.

Bull, J.J., Sanjuan, R. and Wilke, C.O. (2007) Theory of lethal mutagenesis for viruses. *J. Virol.* **81**, 2930–2939.

Crotty, S., Cameron, C.E. and Andino, R. (2001) RNA virus error catastrophe: direct molecular test by using ribavirin. *Proc. Natl Acad. Sci. USA* **98**, 6895–6900.

Eigen, M. (1971) Selforganization of matter and the evolution of biological macromolecules. *Naturwissenschaften* **58**, 465–523.

Eigen, M. (1987) New concepts for dealing with the evolution of nucleic acids. *Cold Spring Harb. Symp. Quant. Biol.* **52**, 307–320.

Eigen, M. (2002) Error catastrophe and antiviral strategy. *Proc. Natl Acad. Sci. USA* **99**, 13374–13376.

Grande-Perez, A., Sierra, S., Castro, M.G., Domingo, E. and Lowenstein, P.R. (2002) Molecular indetermination in the transition to error catastrophe: systematic elimination of lymphocytic choriomeningitis virus through mutagenesis does not correlate linearly with large increases in mutant spectrum complexity. *Proc. Natl Acad. Sci. USA* **99**, 12938–12943.

Grande-Perez, A., Lazaro, E., Lowenstein, P., Domingo, E. and Manrubia, S.C. (2005) Suppression of viral infectivity through lethal defection. *Proc. Natl Acad. Sci. USA* **102**, 4448–4452.

Kimura, M. and Maruyama, T. (1966) The mutational load with epistatic gene interactions in fitness. *Genetics* **54**, 1337–1351.

Mansky, L.M. (1998) Retrovirus mutation rates and their role in genetic variation. *J. Gen. Virol.* **79**, 1337–1345.

Maynard Smith, J. (1978) *The Evolution of Sex*. Cambridge and New York: Cambridge University Press.

Pariente, N., Sierra, S., Lowenstein, P.R. and Domingo, E. (2001) Efficient virus extinction by combinations of a mutagen and antiviral inhibitors. *J. Virol.* **75**, 9723–9730.

Pariente, N., Airaksinen, A. and Domingo, E. (2003) Mutagenesis versus inhibition in the efficiency of extinction of foot-and-mouth disease virus. *J. Virol.* **77**, 7131–7138.

Sanjuan, R., Moya, A. and Elena, S.F. (2004) The distribution of fitness effects caused by single-nucleotide substitutions in an RNA virus. *Proc. Natl Acad. Sci. USA* **101**, 8396–8401.

Sierra, S., Davila, M., Lowenstein, P.R. and Domingo, E. (2000) Response of foot-and-mouth disease virus to increased mutagenesis: influence of viral load and fitness in loss of infectivity. *J. Virol.* **74**, 8316–8323.

Swetima, J. and Schuster, P. (1982) Model studies on RNA replication. 2. Self-replication with errors – A model for polynucleotide replication. *Biophys. Chem.* **16**, 329–345.

Takeuchi, N. and Hogeweg, P. (2007) Error-threshold exists in fitness landscapes with lethal mutants. *BMC Evol. Biol.* **7**, 15. author reply 15.

Tannenbaum, E. and Shakhnovich, E.I. (2004) Solution of the quasispecies model for an arbitrary gene network. *Phys. Rev. E Stat. Nonlin. Soft Matter Phys.* **70**, 021903.

Wilke, C.O. and Adami, C. (2003) Evolution of mutational robustness. *Mut. Res.* **522**, 3–11.

Wilke, C.O., Ronnewinkel, C. and Martinetz, T. (2001) Dynamic fitness landscapes in molecular evolution. *Phys. Rep.* **349**, 395–446.

Zeldovich, K., Chen, P. and Shakhnovich, E.I. (2007) The hypercube of life: How protein stability imposes limits on organism complexity and speed of molecular evolution. arXiv:0705.4062v1.

CHAPTER

10

Evolution of dsDNA Tailed Phages

Roger W. Hendrix

ABSTRACT

Understanding of the evolution of the double-stranded DNA (dsDNA) tailed bacteriophages has been advanced by the availability of multiple genome sequences which reveal, through comparative analysis, evidence for genetic events in the evolutionary past of the phages. The data show that the phages are genetic mosaics with respect to each other, exchanging modules of sequence by horizontal exchange mediated by non-homologous recombination. The non-homologous recombination events apparently occur essentially randomly across the genomes, producing primarily genetic monsters that are rapidly removed from the population by natural selection. The minority of such recombinants that survive are those that have recombined at sites such as gene boundaries that do not compromise phage fitness. Most phage genes are under strong purifying selection, but there is evidence that phages may tolerate small amounts of DNA that provide no selective benefit due to constraints of capsid size and DNA packaging mechanism, opening the possibility of genetic drift toward novel functions without counterselection of non-functional intermediates. These views of phage evolution require very large numbers of intrinsically improbable events in the phages' history, and they are made more plausible by recent evidence on the size of the global phage population and the length of the phages' evolutionary history.

NATURAL POPULATIONS, ABUNDANCE AND TURNOVER

The evolution of the double-stranded DNA (dsDNA) tailed bacteriophages, as it has come to be understood over the past few years, only makes sense when it is considered in the context of the remarkable size of the global population of phages. Knowledge of the population size has been slow in coming due to the lack of an effective way to count phages in the natural environment. The plaque assay, the well-established and reliable method for quantifying phages in a laboratory setting, underestimates the number of phages in an environmental sample by many orders of magnitude because it sees only the tiny fraction of the population able to infect the bacterial strain chosen for the assay. It was the realization that tailed phages could be counted directly by electron

Origin and Evolution of Viruses
ISBN 978-0-12-374153-0

microscopy (Bergh *et al.*, 1989) that led to the first accurate estimates of environmental phage populations. Tailed phage particles in coastal seawater from a Norwegian fjord were estimated at 10^7 particles per milliliter. Numerous additional measurements have been made at other marine, freshwater, and terrestrial locations, giving particle concentrations within one to two orders of magnitude of the original measurement depending on the location (Wommack *et al.*, 1999; Suttle, 2005). Regardless of the concentration of phages, the number of prokaryotic cells counted in environmental samples is 5- to 10-fold lower than the number of phage particles, leading to the conclusion that the tailed bacteriophages constitute the majority of biological organisms on the planet.

The total global population of tailed phages can be estimated from the environmental concentrations measured by electron microscopy or now more frequently by fluorescent dye-based methods (Noble and Fuhrman, 2000). The number obtained—10^{31} individual viral particles—is necessarily approximate but is in reasonable agreement with an independent estimation of the global number of bacterial and archaeal cells of $4–6 \times 10^{30}$ (Whitman *et al.*, 1998). The enormous size of the nearly incomprehensible number 10^{31} begins to become apparent when we consider that if 10^{31} phages were laid end-to-end they would extend into space for 200 million light years. The significance of this number becomes clear in light of measurements of the turnover time of the population in surface waters of the ocean. The number obtained, presumably due to a combination of loss by infection of new hosts, by predation, and by physical damage to the phages, is 4–10 days (Suttle and Chen, 1992; Noble and Fuhrman, 1997). In other words, assuming that that turnover time applies to the entire global population, then 10^{31} new phages must be produced every week. This would require roughly 10^{24} productive infections per second on a global scale. Each of those infections is an opportunity for evolutionary events to occur, either by generation of mutations during growth or by recombination with DNA of any co-infecting phages, with host DNA, or with the DNA of resident prophages. We argue below that these processes may have been going on for 3.5 billion years or more, which suggests that there are essentially no practical constraints on the number of diversity-generating events we can imagine in the ancestry of these phages, regardless of how intrinsically improbable the individual events may be.

GENOME COMPARISONS AND EVOLUTIONARY MECHANISMS

The first genomic comparisons of phages were carried out in the late 1960s on *Escherichia coli* phage λ and some of its close relatives. In these experiments virion DNA from two phages was mixed, melted to single strands, re-annealed, and examined by electron microscopy (Simon *et al.*, 1971). These DNA heteroduplexes showed graphically where the sequences matched and had annealed into dsDNA and where the DNA remained single-stranded. The result was a dramatic one—the dsDNA regions were distributed in a patchwork fashion between the single-stranded denaturation bubbles, implying that these phages are genetic mosaics with respect to each other. It thus appeared as if sequences from different phages must be reassorting with each other in the history of the phages by some sort of recombination mechanism. Within the limits of resolution it appeared that the boundaries between homologous and heterologous sequence often occurred at the same location in different pairwise comparisons, suggesting that there were hotspots for recombination. These observations led to the "modular theory of phage evolution" (Susskind and Botstein, 1975), which proposes that phages generate new genome compositions by exchanging modules of sequence, possibly through the agency of "linker sequences" at the proposed recombination hotspots. With the advent of large-scale DNA sequencing it became possible to carry out such genome comparisons at single nucleotide resolution and with large numbers of genomes simultaneously. Genome sequences have now been determined and analyzed for nearly 300 tailed phages, and additional sequences are appearing at an increasing rate.

The phages examined now include many from other groups of hosts beyond the lambdoid phages of *E. coli* examined in the early experiments, and the results outlined below apply to all of these groups.

Sequence level comparisons of genomes show evidence of the expected sorts of mutational differences—point mutations, small insertions, and deletions. However, the most prominent feature revealed by the comparisons is the genetic mosaicism first seen in the heteroduplex experiments (Figure 10.1). This can be seen in the substitution of a new allele of a particular gene at the same position in the gene order as its homologue in the comparison phage, but it can often mean that a completely novel gene can be found inserted into a flanking context that is homologous but uninterrupted in the comparison phage. In addition to single-gene events there are examples in which a whole set of tail genes or head genes has been swapped into a new context, as with phage SfV which has the head genes of HK97 and the tail genes of Mu (Lawrence *et al.*, 2002), or phage N15, which has the head and tail genes of λ and the other half of the genome with an unknown origin (Ravin *et al.*, 2000). The overall picture is that there has been pervasive and promiscuous mixing of sequences across the phage genomes. In an alignment of five or six similar genomes, such as the genomes of lambdoid phages of *E. coli* (Juhala *et al.*, 2000) or of a closely related group of phages that infect *Mycobacterium smegmatis* (Hatfull *et al.*, 2006), 50 or more mosaic joints can be identified—that is, locations in one of the pairwise alignments where there has been a recombination event between two non-identical sequences occurring in the ancestry of one of the phages being compared (Figure 10.1). This a minimum estimate of how many such

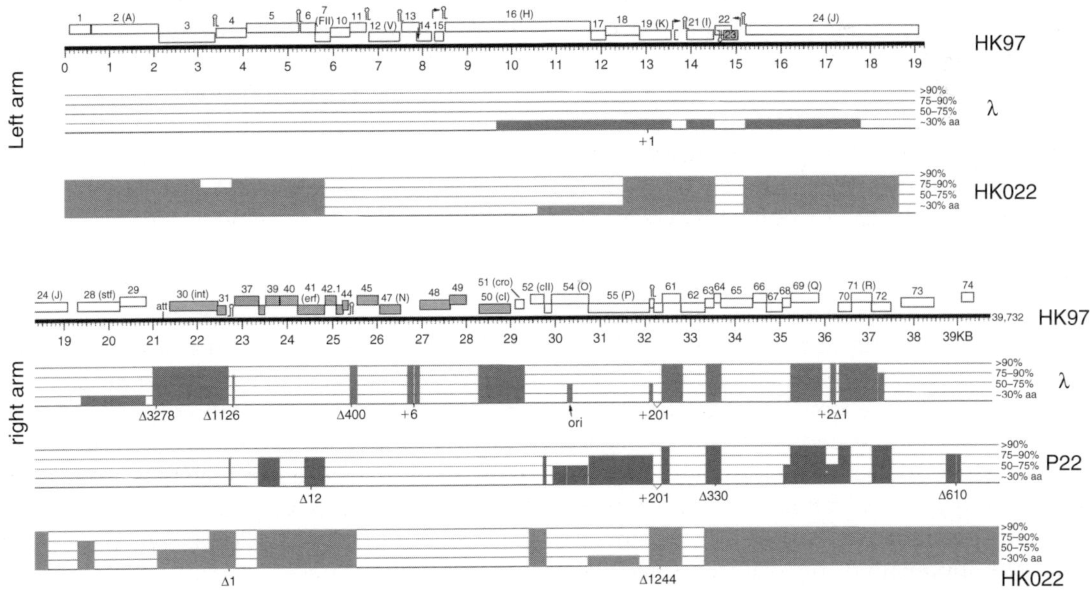

FIGURE 10.1 Mosaic relationships among lambdoid phages. The genome sequence of phage HK97 is represented as a scale bar, broken near the middle, with genes represented as rectangles above the bar. Open rectangles are genes transcribed rightwards; shaded rectangles are genes transcribed leftwards. The histograms below the HK97 genome show the locations and degree of sequence identity of regions that match between HK97 and the three comparison phages, λ, P22 and HK022. The numbers below the histograms preceded by "Δ" or "+" indicate the size of deletion or insertion, respectively, in the HK97 sequence relative to the sequence of the comparison phage. (See Plate 13 for the color version of this figure.) Modified with permission from Juhala *et al.* (2000).

recombination events there have been, since this approach only detects events that have happened in the time since the last common ancestor of the two sequences in the comparison. It of course also can only detect recombination joints that have allowed the phage carrying them to survive natural section. We argue below that the vast majority of products of non-homologous recombination fail to survive natural selection.

Examination of the locations of recombination joints shows that they are found in most cases at gene boundaries. However, the idea that there is something special about gene boundaries that directs recombination to that location is not supported. In almost all cases there is no sequence between genes that could provide homology for homologous recombination or a target for a site-specific recombination system. There are frequent examples in which the recombination appears to have been "sloppy," producing a quasi-duplication by out-of-register recombination, for example, or a recombination joint near but not exactly at the gene end (Juhala *et al.*, 2000). Perhaps the most telling examples are recombination joints located in the interior of a coding region. In these cases the recombination joint is typically found at a point corresponding to a domain boundary in the encoded protein (Juhala *et al.*, 2000; Dobbins *et al.*, 2004). The model that fits all the data available is that non-homologous recombination occurs essentially randomly across the genomes, at gene boundaries and at interior positions in the genes, in register with respect to the gene organization of the recombining genomes, and probably much more frequently, out of register. The expectation is that nearly all of these non-homologous recombinants are functionally compromised and are rapidly removed from the population by natural selection. The minority that survive to be observed by us are ones that are at least as functionally competent as their precursors. This process of generating diversity by non-homologous recombination, followed by a severe winnowing by natural selection, produces a product with a high level of order and organization that belies the essentially random processes that produced it. In other words, it is a classical Darwinian process.

In many cases the stretch of sequence that constitutes an exchanging module in the evolution of the phages is a single gene. It can also be less than a gene, as with the protein domains mentioned above that exchange independently. In addition, there are often groups of sequences that appear to "travel together" through phage evolution. The most obvious example is the head genes, which encode proteins that must interact intimately with each other in constructing the head. They are presumed to have co-evolved to facilitate those interactions and would be unlikely to tolerate reassortment with other head genes that have co-evolved down a different path. The tail genes of moderate-sized phages like λ also tend to stay together, though not as stringently as the head genes. We also see co-segregation of genes with the DNA sites that their encoded proteins bind, as with the phage λ *cI* and *cro* repressor genes and their operators (Juhala *et al.*, 2000). In all these cases we believe that recombination is not prevented from disrupting these genetic groupings; in fact there is evidence that recombination can occur in the head gene region of the lambdoid phages (Juhala *et al.*, 2000). Rather we believe that recombinants with compromised function are lost from the population.

In contrast to phages with moderate-sized genomes, a group of larger phages that includes *E. coli* phage T4 has a much larger group of genes, constituting more than half the genome, that segregates together through phage evolution. These core genes, as they are called, include both the head and the tail genes as well as the genes of DNA replication and nucleotide metabolism (Filee *et al.*, 2006). Why these genes should all co-segregate is perhaps not immediately clear, but all these groups of encoded proteins are thought to form complexes, and those complexes either certainly or possibly interact with each other (Mosig and Eiserling, 2006). In contrast to the core genes, the "T4-type" phages have a large number of genes, somewhat interspersed among the core genes, that are swapping in and out of the genomes at a rapid evolutionary rate, as seen from the fact that their inventory is very different even between phages for which the similarity of the core genes is very high, implying recent common ancestry. The non-core genes are typically rather small, and in many cases make no matches to the sequence databases. There are functions known for a few of these genes, and many of them have homologues in one or more of the other phages in the T4-type group, suggesting they may provide a useful function to these phages. For most of these genes, however, we do not know either what their functions are or even if they provide a useful function to the phage. It will be an interesting challenge to understand the functional meaning of these genes' evanescent appearance in the genomes. The overall picture of the evolution of the T4-type phages, however, is understandable in terms of the same mechanisms outlined above for other phages: promiscuous non-homologous recombination followed by stringent selection for function, which in this case means primarily selection for a full and self-compatible set of core genes.

SELECTIVE PRESSURES ON PHAGE GENES

About 90–95% of a typical phage genome is occupied by protein coding genes and much of the remaining sequence contains regulatory sequences such as promoters. Phages appear to be models of genetic economy and efficiency. We might predict that every gene in the genome is there because it provides a selective benefit to the phage. Part of this prediction can be tested in the case of genes with slightly diverged orthologues in two phages. Weigele *et al.* (2007) measured the ratio of synonymous mutations to non-synonymous mutations for all the orthologues in several pairs of mycobacteriophages. The ratio was about 4:1 in favor of synonymous mutations. For marine cyanophages this analysis gave an even stronger bias (10:1) in favor of synonymous mutations, possibly reflecting a larger effective population size. The conclusion for both cases is that the genes that are found in more than one of these phages are under strong purifying selection, in accord with the prediction. However, we have seen that phage genomes can contain, at least transiently, DNA sequence that is unlikely to benefit the phage directly—for example, quasi-duplications produced by out-of-register recombination. Most phages also contain regions that have no obvious genes, or genes that appear to be pastiches of parts of genes found in other phages. With the caveat that it is difficult to prove absence of function, it does appear that phages tolerate small amounts of DNA that is not providing a selective benefit to the phage. An alternative (and untested) possibility is that the phage tolerates apparently useless DNA because it makes the genome big enough to be packaged efficiently. For a phage like λ, packaging efficiency is known to be sensitive to the amount of DNA in the genome (Feiss and Siegele, 1979). It is an interesting speculation that phages may tolerate DNA without regard to its specific sequence. Such DNA might provide an opportunity for stepwise development of genes with novel functions, by random mutation and reassortment of gene parts, in the absence of counterselection of the many functionless intermediates.

POPULATION STRUCTURE, METAGENOMICS

Well over 5000 tailed phages have been isolated and reported in the literature, and perhaps 300 have been characterized either by detailed genetic and biochemical experiments or by genome sequencing. This is clearly a minute sample of the 10^{31} individuals that we believe are present at any one time, and it means there are severe limitations on what we can say about the genetic structure of the population. There are nevertheless some features of the population emerging, albeit in a rather ill-focused and non-quantitative way.

It is important to note first that there are substantial biases in our sampling of the phage population. The most obvious of these is that almost all of the well-characterized phages grow on a group of host cells that constitute a small slice of bacterial (and archaeal) diversity. These phages are heavily biased toward ones that grow on *Escherichia, Salmonella, Bacillus, Mycobacterium, Streptococcus,* the "dairy" bacteria, and a few others, reflecting the particular interests of the scientists who isolated them. For obvious reasons no phage has been sequenced whose host is among the >99% of bacterial types that have not been successfully cultivated in the lab. Many phages have been amplified from environmental samples by the technique of enrichment culture, which selects for the one phage type in a sample that competes most effectively in lytic growth under the laboratory culture conditions used. There are certainly other biases in effect. One example discovered recently is that *Bacillus* phage G, which is the largest phage known, with a longest dimension of ~450 nm, makes invisibly small plaques with standard laboratory concentrations of top agar; plaques are visible only when very low concentrations of top agarose are used, probably because the unusually large particles are only able to diffuse through the more open gel (Serwer *et al.*, 2007). This result implies that biologists have been isolating phages for the 90 years since their discovery under conditions that are biased against finding very large phages.

Given those constraints, what can we say about the genetic structure of the tailed phage population? One extreme view would be that there is a limited number of successful ways to make a phage and so all phages will fit into one of those types. For the classically studied phages of *E. coli,* those standard types would include phages resembling λ, T4, T7, T5, P2, N4, and a few more. At the other extreme, it might be that the phage population is a smooth continuum of types, with no preferred gene inventory, genome organization, or regulatory arrangements, and the appearance of discrete types is an artefact of sparse sampling. When we examine actual phages and phage genomes it becomes clear that the truth lies between these extremes. There are, for example, many large lytic phages with structure and DNA metabolism genes related to those of T4; there are numerous temperate phages that have the genome size and organization of phage λ, as well as some genes with similar sequences; and there is a handful of phages that strongly resemble T7. Even though these phages for the most part grow in the same host, they do not exchange genes equally; that is, there is evidence for much more gene exchange among members of one of these types than for exchange between types. This may constitute an isolation mechanism that tends to keep the different types separated. However, as the number of genomes in each of these phage types increases, the diversity within the group continues to expand rather than cohering around some hypothetical "optimized" type. For example, there is a prophage of *Pseudomoas putida* which over most of its genome is very similar to the quintessentially lytic phage T7 but has acquired an integrase gene and the ability to repress expression of its genes. Also, *E. coli* phage N15 has head and tail genes—constituting ~50% of the genome—that are very similar to those of λ but the other half of the genome is very different from λ and those genes have clearly been swimming in a different gene pool (Ravin *et al.*, 2000; Casjens *et al.*, 2004). At this point it is not yet clear whether the phage types that we perceive will become better defined or instead will overlap more extensively as the phage genome database continues to fill up. There are now more than 50 genome sequences for phages that infect *Mycobacterium smegmatis,* and these can be grouped into several different types, as outlined for *E. coli* phages above. It is striking that there is no clear correspondence between any of the apparent phage types of the mycobacteriophages and any of the specific types discussed above for *E. coli.* Thus the number and diversity of phage genome types, however one chooses to define them, is likely very large and at this point, barely glimpsed.

There have been several recent metagenomic studies of phages, in which ostensibly all the phages in an environmental sample such as sea water are pooled and the total DNA in the sample is sequenced. Although there is still certainly some sampling bias in these procedures, they should be much more representative of the total population than individual genome sequencing, for the reasons outlined above. Information about gene content and gene arrangement for individual genomes is lost in metagenomic analysis, but it is the best tool available for assessing sequence diversity in the phage population as a whole. The most striking result from these studies is that >60% of the sequences recovered have no match to the sequence databases. (Similar results have been seen for the genes of the genomes of some individual phages, namely ones for which no close relatives had been sequenced.) For comparison, <10% of the genes of the bacterial metagenome from a comparable sample have novel sequences. These results testify to the remarkable diversity of the phage metagenome. The diversity has been estimated in various ways with somewhat variable results, but all methods agree that viruses, of which most are phages, have the greatest genetic diversity of any biological group on Earth.

DEEP EVOLUTIONARY CONNECTIONS

There has long been a suspicion that the tailed phages are a monophyletic group, based on their similar virion morphologies and on a number of common features of their life cycles. In fact, the International Committee on the Taxonomy of Viruses has grouped all of the tailed phages together in the order *Caudovirales* because of those shared features (Fauquet *et al.*, 2005). However, there is no gene or predicted protein sequence that is recognizably shared across all phages for which we have genome sequences. Despite the great diversity in gene inventory across the known tailed phages, they all do encode a few functionally analogous virion structural proteins, and we can ask whether there is detectable sequence similarity in these proteins that could provide evidence for common ancestry for the tailed phages as a whole, or at least for common ancestry for these capsid genes across the whole tailed phage group.

The two structural proteins that are best conserved with regard to sequence are the portal and the large subunit of the terminase. These proteins are central components of the DNA packaging pump and might be expected to be constrained from rapid change in their sequences by their intricate functional role. When we probe the sequence databases with the amino acid sequence of a terminase or a portal, we find matches to the sequences of the corresponding proteins from a majority, but not all, of the phages in the database. The degree of similarity in different pairwise matches varies from near identity to only barely detectable. It is a reasonable though still unproven assumption for these proteins that they all share common ancestry across all the tailed phages but have diverged to such an extent that common sequence with the group can no longer be detected in some members of the group. A similar picture is seen for the major capsid protein sequences except that here the sequences are somewhat more diverged and separate into a handful of sequence-related groups that are not detectably related between groups. These separate groups ostensibly represent separate lineages of capsid proteins, and the question becomes whether these separate lineages share common ancestry farther back in the history of the tailed phages.

Structural studies of phage capsids are giving the beginnings of an affirmative answer to this question. When the capsid of phage HK97 was solved by x-ray crystallography (Wikoff *et al.*, 2000) the subunit had a novel fold. That same fold was subsequently seen in the phage T4 capsid protein (Fokine *et al.*, 2005), even though the T4 and HK97 proteins have no detectable sequence similarity. The capsid protein similarity is taken as a strong indication of common ancestry for these two capsid protein lineages. Additional information comes from cryo-EM structures of capsid subunits of four additional phages representing four additional lineages. Within the more limited resolution of cryo-EM these proteins also appear to have the HK97 fold, bolstering the argument for common ancestry across all tailed phages. Remarkably, a cryo-EM image of the capsid of herpes simplex virus also shows evidence for the HK97 fold in this capsid protein (Baker *et al.*, 2005). This last result was not entirely unexpected because there are several striking similarities between tailed phages and herpesviruses regarding their capsid structures and their mechanisms of capsid assembly (Newcomb *et al.*, 1999, 2001). In addition, there is weak but probably significant sequence similarity between the terminase proteins of herpesviruses and tailed phages. Taken together, these results make a somewhat speculative case that the capsid proteins of all the tailed phages as well as the herpesviruses share common ancestry. The high degree of divergence among the sequences of these proteins argues that they are ancient, and their presence in viruses infecting all three domains of life suggests that they may have been diverging since around the time of the divergence of cellular life into the three current domains, roughly 3.5 billion years ago.

A similar set of connections among capsid proteins with a completely different fold has been seen for a group of viruses with members infecting all three domains of life. This has also been interpreted to suggest that there were already viruses resembling contemporary viruses infecting cells 3.5 billion years ago and more (Hendrix, 1999). The abundance of horizontal exchange of genes described earlier in this chapter, however, cautions us that it is probably an oversimplification to think of the capsid protein lineages inferred from these capsid protein structures as lineages for the viruses as a whole. Clearly more work remains to clarify this aspect of the evolution of viruses.

REFERENCES

Baker, M.L., Jiang, W., Rixon, F.J. and Chiu, W. (2005) Common ancestry of herpesviruses and tailed DNA bacteriophages. *J. Virol.* **79**, 14967–14970.

Bergh, O., Borsheim, K.Y., Bratbak, G. and Heldal, M. (1989) High abundance of viruses found in aquatic environments. *Nature* **340**, 467–468.

Casjens, S.R., Gilcrease, E.B., Huang, W.M., Bunny, K.L., Pedulla, M.L., Ford, M.E. *et al.* (2004) The pK02 linear plasmid prophage of Klebsiella oxytoca. *J. Bacteriol.* **186**, 1818–1832.

Dobbins, A.T., George, M., Jr., Basham, D.A., Ford, M.E., Houtz, J.M., Pedulla, M.L. *et al.* (2004) Complete genomic sequence of the virulent Salmonella bacteriophage SP6. *J. Bacteriol.* **186**, 1933–1944.

Fauquet, C., Mayo, M.A., Maniloff, J., Desselberger, U. and Ball, L.A. (eds) (2005). *Virus Taxonomy: Classification and Nomenclature of Viruses.* New York: Elsevier Academic Press.

Feiss, M. and Siegele, D.A. (1979) Packaging of the bacteriophage lambda chromosome: dependence of cos cleavage on chromosome length. *Virology* **92**, 190–200.

Filee, J., Bapteste, E., Susko, E. and Krisch, H.M. (2006) A selective barrier to horizontal gene transfer in the T4-type bacteriophages that has preserved a core genome with the viral replication and structural genes. *Mol. Biol. Evol.* **23**, 1688–1696.

Fokine, A., Leiman, P.G., Shneider, M.M., Ahvazi, B., Boeshans, K.M., Steven, A.C. *et al.* (2005) Structural and functional similarities between the capsid proteins of bacteriophages T4 and HK97 point to a common ancestry. *Proc. Natl Acad. Sci. USA* **102**, 7163–7168.

Hatfull, G.F., Pedulla, M.L., Jacobs-Sera, D., Cichon, P.M., Foley, A., Ford, M.E. *et al.* (2006) Exploring the mycobacteriophage metaproteome: phage genomics as an educational platform. *PLoS Genet.* **2**, e92.

Hendrix, R.W. (1999) Evolution: the long evolutionary reach of viruses. *Curr. Biol.* **9**, R914–917.

Juhala, R.J., Ford, M.E., Duda, R.L., Youlton, A., Hatfull, G.F. and Hendrix, R.W. (2000) Genomic sequences of bacteriophages HK97 and HK022: pervasive genetic mosaicism in the lambdoid bacteriophages. *J. Mol. Biol.* **299**, 27–51.

Lawrence, J.G., Hatfull, G.F. and Hendrix, R.W. (2002) Imbroglios of viral taxonomy: genetic exchange and failings of phenetic approaches. *J. Bacteriol.* **184**, 4891–4905.

Mosig, G. and Eiserling, F. (2006) T4 and related phages: structure and development. In: *The Bacteriophages* (R. Calendar, ed.), pp. 225–267. Oxford: Oxford University Press.

Newcomb, W.W., Homa, F.L., Thomsen, D.R., Trus, B.L., Cheng, N., Steven, A. *et al.* (1999) Assembly of the herpes simplex virus procapsid from purified components and identification of small complexes containing the major capsid and scaffolding proteins. *J. Virol.* **73**, 4239–4250.

Newcomb, W.W., Juhas, R.M., Thomsen, D.R., Homa, F.L., Burch, A.D., Weller, S.K. and Brown, J.C. (2001) The UL6 gene product forms the portal for entry of DNA into the herpes simplex virus capsid. *J. Virol.* **75**, 10923–10932.

Noble, R.T. and Fuhrman, J.A. (1997) Virus decay and its causes in coastal waters. *Appl. Environ. Microbiol.* **63**, 77–83.

Noble, R.T. and Fuhrman, J.A. (2000) Rapid virus production and removal as measured with fluorescently labeled viruses as tracers. *Appl. Environ. Microbiol.* **66**, 3790–3797.

Ravin, V., Ravin, N., Casjens, S., Ford, M.E., Hatfull, G.F. and Hendrix, R.W. (2000) Genomic sequence and analysis of the atypical temperate bacteriophage N15. *J. Mol. Biol.* **299**, 53–73.

Serwer, P., Hayes, S.J., Thomas, J.A. and Hardies, S.C. (2007) Propagating the missing bacteriophages: a large bacteriophage in a new class. *Virol. J.* **4**, 21.

Simon, M.N., Davis, R.W. and Davidson, N. (1971) Heteroduplexes of DNA molecules of lambdoid phages: physical mapping of their base sequence relationships by electron microscopy. In: *The Bacteriophage Lambda* (A.D. Hershey, ed.), pp. 313–328. Cold Spring Harbor: Cold Spring Harbor Laboratory.

Susskind, M.M. and Botstein, D. (1975) Mechanism of action of Salmonella phage P22 antirepressor. *J. Mol. Biol.* **98**, 413–424.

Suttle, C.A. (2005) Viruses in the sea. *Nature* **437**, 356–361.

Suttle, C.A. and Chen, F. (1992) Mechanisms and rates of decay of marine viruses in seawater. *Appl. Environ. Microbiol.* **58**, 3721–3729.

Weigele, P.R., Pope, W.H., Pedulla, M.L., Houtz, J.M., Smith, A.L., Conway, J.F. *et al.* (2007) Genomic and structural analysis of Syn9, a cyanophage infecting marine Prochlorococcus and Synechococcus. *Environ. Microbiol.* **9**, 1675–1695.

Whitman, W.B., Coleman, D.C. and Wiebe, W.J. (1998) Prokaryotes: the unseen majority. *Proc. Natl Acad. Sci. USA* **95**, 6578–6583.

Wikoff, W.R., Liljas, L., Duda, R.L., Tsuruta, H., Hendrix, R.W. and Johnson, J.E. (2000) Topologically linked protein rings in the bacteriophage HK97 capsid. *Science* **289**, 2129–2133.

Wommack, K.E., Ravel, J., Hill, R.T., Chun, J. and Colwell, R.R. (1999) Population dynamics of chesapeake bay virioplankton: total-community analysis by pulsed-field gel electrophoresis. *Appl. Environ. Microbiol.* **65**, 231–240.

CHAPTER

11

More About Plant Virus Evolution: Past, Present, and Future

Adrian Gibbs, Mark Gibbs, Kazusato Ohshima, and Fernando García-Arenal

ABSTRACT

Gene sequencing was invented in the 1980s, enabling the evolutionary relationships of organisms to be studied in detail. The ways in which these studies provide the intellectual framework for research into the life of viruses continues to expand.

Plants, animals and other organisms present viruses with very different environments, both structurally and biochemically, and this may be the reason why so few virus groups span host kingdoms, but a few do, and studies of these reveal the shared and unique constraints and opportunities provided by different types of host, and also the diverse ways that viruses overcome the constraints. The RNA-silencing system seems to provide the primary plant defense against viruses, and although RNA-silencing mechanisms are present in all eukaryotes, they are most developed in plants where they also modulate the expression of plant genes. Plants have a rigid cellular structure with the cells connected by plasmodesmata too narrow for virions to pass through. This has required viruses to adopt specific mechanisms to aid their systemic spread within plants. The diverse measures adopted by viruses to suppress RNA silencing and to aid their spread through plants indicate that such mechanisms have evolved independently on several occasions.

Likewise a great range of symbiotic, commensal, and satellite relationships are found among plant viruses, and again the diversity of the relationships, of the virus groups involved, and of the resulting phenotypes, emphasizes that viruses of plants are polyphyletic.

Studies of mutations in model experimental systems, and of gene sequence variation in natural viral populations, are clarifying the mechanisms that produce "quasispecies," even though the concept seems to be still largely misunderstood. The relative contribution of different evolutionary processes, including mutation, drift, recombination, and selection, to viral population change is becoming better understood.

The taxonomies of tobamoviruses and of their principle hosts seem to be congruent, indicating that they have probably co-evolved, and hence may be of the same age, around

Origin and Evolution of Viruses
ISBN 978-0-12-374153-0

100 million years. However potyviruses and their hosts show no such relationships, indeed gene sequence differences in viral populations, of which the history is known, indicate that the genus *Potyvirus* may be only a few thousand years old. Our understanding of more distant relationships remains very speculative as it depends on comparisons of "molecular phenotypic" characters (e.g. structure and function) rather than of gene sequences.

Viruses have been studied for more than a century, their molecules are well known, but our understanding of the molecular basis of plant virus biology is still in its infancy, and we have little idea of how viruses will respond to "climate change" and "transgene pollution."

INTRODUCTION

A study of the transmission of tobacco mosaic virus (TMV) inspired Martinus Beijerinck in 1898 to propose the existence of a virus, a "*contagium vivum fluidum*" (Beijerinck, 1898). Since then plant virologists have isolated and described many species; 129 were known by 1939 (Holmes, 1939), and now there are data from more than 500 species (Brunt *et al.*, 1996; Fauquet *et al.*, 2005). In the early years, much was learned of the transmission, biology and symptoms of different viruses and about the morphology and antigenic relationships of their virions. Before work on their evolution was possible, these features were used to cluster isolates into species and "groups," and these groupings were mostly confirmed by the earliest molecular data on the composition and sequences of viral proteins (Gibbs, 1968). Experimental evidence showing that plant viruses could adapt to new hosts was first obtained in the 1920s, but the role of mutations, the structure of their populations and the relationships, if any, between the genera were unknown, and plant virus origins were merely a subject of speculation until the 1980s, when methods for determining the nucleotide sequences of genomes became routine, and the study of their relationships, and hence their evolution became an attainable goal.

Most plant viruses have small, single-stranded RNA genomes which are replicated in the cytoplasm of host cells via RNA intermediates. There are only two groups of plant viruses with DNA genomes, neither of them as large as those of some viruses of bacteria and animals. Many viruses of fungi and animals have large double-stranded RNA genomes, but these are uncommon in plants. The plant viruses with DNA genomes include the very diverse geminiviruses, which have small single-stranded DNA genomes, and the caulimovirids, sometimes called the pararetroviruses, which include the caulimoviruses and badnaviruses, both of which replicate via RNA intermediates, like the hepadnaviruses of animals (Beck and Nassal, 2007), rather than exploiting the cellular DNA replication machinery as do most animal viruses with DNA genomes. The badnaviruses are both exogenous and endogenous (Mette *et al.*, 2002; Geering *et al.*, 2005; Staginnus and Richert-Poggeler, 2006); namely infectious and transmissible as virions, rather than intercalated in the host's genome. Although retroviruses are common in animals, none have been found in plants but gene sequencing has shown that many plant species contain large numbers of retrovirus-like elements, most of them incomplete (Wright and Voytas, 2002; Marco and Marín, 2005; Yano *et al.*, 2005).

This chapter is not an exhaustive review of what is known of plant virus evolution, instead we focus on what we find most interesting among current studies of plant virus evolution, gained from comparative genomics, population studies and gene sequence analyses that probe various evolutionary time-scales.

PLANTS, VIRUS HOSTS FROM A DIFFERENT UNIVERSE

For obvious reasons, research on virus evolution has focused on species that infect vertebrates and especially those that infect people. Vertebrate immune systems shape virus populations as viruses are eliminated by the adaptive response and immune memory mediated by mobile lymphocytes and antibodies. These defenses place viruses under selection and in some instances mutants are favored, leading to the successive replacement of particular genotypes in populations (Grenfell *et al.*, 2004). Immune selection is focussed onto the genes of exposed proteins such as the hemaglutinins and envelope proteins, and some viruses have acquired genes that interfere with the host's immune system (Hughes and Friedman, 2005). By contrast, plant defenses against pathogens are significantly different. Plants have no mobile defense cells nor an adaptive protein recognition system, instead they rely on a sophisticated nucleic acid targeting defense system called RNA silencing, and a generic protein system (Soosaar *et al.*, 2005). Pathogen resistance (R) proteins are a key component of the latter, which when induced by fungi, bacteria, or viruses lead to hypersensitive cell death that limits infection. Plant protein-based responses are apparently much less specific than vertebrate antibody responses (Dangl and Jones, 2001). Even so, some interactions between virus and host proteins are believed to be evolutionarily significant as there is evidence of positive selection on some plant virus proteins. Linkages between selected viral codons and host defenses are starting to be understood, for example particular amino acids in the genome-linked protein (VPg) of potato virus Y allow it to overcome the resistance conferred by pathogen resistance genes (pvr2) of *Capsicum* species (Moury, 2004; Moury *et al.*, 2002; Tsompana *et al.*, 2005).

RNA silencing appears to be an antiviral mechanism in all eukaryotes, and is also a mechanism for modulating gene expression (Lecellier *et al.*, 2005; Wang *et al.*, 2006). In plants it has become highly developed and is probably the main bulwark against viruses, when the hypersensitive response is not effective (Hamilton and Baulcombe, 1999; Deleris *et al.*, 2006). Dicer enzymes that are similar to ribonuclease III initiate RNA silencing by recognizing and cleaving double-stranded RNA that is either a replicative form of the viral genomic RNA or a folded partially self-complementary single-stranded RNA, of either viral or host source. Pieces that are up to 26 bases long are produced and used in an enzyme–RNA complex to recognize other copies of the same RNA sequence for destruction. RNA silencing is highly specific and so the recognition mechanism is believed to involve hybridization with the RNA pieces. Vertebrates, invertebrates, and fungi have one or two Dicer genes, whereas plants have at least four Dicer genes (Margis *et al.*, 2006). When the short RNAs are derived from virus RNA, the Dicers target the viral RNA and prevent infection by the same virus and in some instances closely related viruses (Lindbo *et al.*, 1993; Ratcliff, Harrison, and Baulcombe, 1997; Voinnet, 2005). Some plants have long been known to recover from infection with certain viruses, a phenomenon shown most clearly by nepoviruses (Gibbs and Harrison, 1976); new leaves show few or no symptoms and there are decreasing concentrations of virions in successive leaves. It has now been shown that this is due to RNA silencing, which spreads throughout the plant as RNA fragments are transported in the symplast from cell to cell (Himber *et al.*, 2003). RNA silencing has clearly had a profound effect on plant virus evolution, as almost all plant viruses have acquired the ability to suppress RNA silencing. This appears to be the sole function of some plant virus proteins, but many virus proteins with other primary functions also act as suppressors (Voinnet, 2005); such suppressors have evolved independently in several different virus lineages. Some of the proteins interfere with the Dicers, others with the small RNAs or double-stranded RNAs. Some of the suppressor genes, such as the potyvirus-encoded helper component protease can function to enhance the replication of other viruses, and so synergistic interaction between viruses and temporary commensal relationships may arise through RNA-silencing suppression.

As yet no other evolutionary effect on plant viruses has been attributed to RNA silencing. However we speculate that selection due to RNA silencing may go further. Positive selection due to RNA silencing, if it occurs, will probably not focus on specific codons and it might be detected from the distribution and rates of synonymous substitution. Negative selection due to RNA silencing may preclude the presence of particular sequences in virus genomes. Indeed preliminary tests (by M.J.G.) have shown that some sequences are missing from virus genomes. It is also possible that it was strong selection for a subtle and innovative RNA defense mechanism that epistatically produced mechanisms allowing plants to generate developmental novelty, and may have been one of the factors leading to the ecological domination of much of the land surface of the planet by the flowering plants.

Plant, animal, and bacterial viruses move between host individuals in specific ways that usually require viral functions. One important difference between animal and plant cells is that plant cells have rigid cell walls composed mostly of cellulose. No virus has acquired the ability to penetrate and infect intact plant cells unaided, instead most rely on wounding or on vector organisms that penetrate the cell walls. No plant virus protein has been shown to interact with a plant cell surface molecule or to be exposed on a cell surface and most plant viruses also appear to lack the capacity to exit cells independently. By contrast, phages have evolved to penetrate bacterial cell walls and lyse bacteria, and binding cell surface molecules is essential to vertebrate virus cell entry (Smith and Helenius, 2004). There are still many unanswered questions about some modes of transmission. For example most carmoviruses and tombusviruses are transmitted to roots by aquatic fungi, chytrids, but some seem to infect their hosts abiotically by an unknown mechanism.

Once within a plant, viruses move in the symplasm from cell to cell through the plasmodesmata. These are pores in plant cell walls that are traversed by thin extensions of the plasmalemma, cytoplasm and endoplasmic reticulum so that these components form the continuous symplasm within the plant (Overall and Blackman, 1996; Cilia and Jackson, 2004). The symplasm enables metabolites to move between cells and, with the assistance of specific proteins and pathways, larger molecules are trafficked too. However the virions of plant viruses are too large to pass through most plasmodesmata, and although viruses may spread through plants as genomes or virions, and they require the plasmodesmata to be modified and enlarged by movement proteins, which interact with host proteins to achieve this outcome. There are several unrelated families of movement proteins and at least two modes of movement through plasmodesmata, so this function has probably evolved more than once. For example the movement protein of TMV binds viral single-stranded RNA to form a stable complex, increases the "size exclusion limit" of the plasmodesmata, brings the TMV RNA into contact with the plasmodesmata, and goes with the viral RNA into an adjacent cell (Lucas, 2006). By contrast cowpea mosaic comovirus induces membranous tubules to replace or penetrate plasmodesmata, allowing whole virions to pass between cells (Waigmann *et al.*, 2004). The genetic basis of host adaptation is seen most clearly in the *Bunyaviridae*, a group of RNA genome viruses with vertebrate or plant hosts and insect vectors. There are four genera of bunyavirids with animal hosts, and one that infects plant, the tospoviruses (Kormelink *et al.*, 1992). All bunyavirids have the same genome plan, but the tospoviruses have an extra gene, which encodes the movement protein.

Only a few viruses of vascular land plants, unlike those of animals, are contagious and rely on direct contact between plants; the tobamoviruses are the exception. Many animal viruses spread by contact, in body fluids and in aerosols to the mucous membrane surfaces of animals, however no plant viruses are spread in this way. Several plant viruses are transmitted to progeny via pollen, and more still to seed from the maternal parent. Although most plant and animal viruses spread throughout individual infected hosts, once a plant has been systemically infected it is likely to remain infected for the rest of its life, whereas those of animals are usually cleared by various defensive immune responses.

Viruses attain very large population sizes but ecological bottlenecks strongly limit their populations (see below). The persistent and systemic nature of infections probably counter the effects of bottlenecks for some species so that their populations exist as diverse clusters of co-existing strains (Gog and Grenfell, 2002). Persistently infected plants also usually have distinct populations of slightly different variants in separate groups of ontogenetically related cells and, although these are connected through the symplast, they remain separate. This phenomenon is probably responsible for the distinctive symptoms produced by different viruses (Hull, 2001). Plants may also be co-infected with two or more virus species (Rochow, 1977; Pruss *et al.*, 1997; Seal *et al.*, 2006). Probably as a consequence, and because plant viruses are connected through the symplast, there are many examples of long-term relationships between plant viruses that are either symbiotic or commensal. There are also examples of synergisms between plant viruses, and, perhaps most importantly, of inter-species recombination. Examples of each of these interactions will be discussed next.

SYMBIOTIC AND COMMENSAL RELATIONSHIPS AND VIRUS SYNERGIES

At least ten kinds of mutualistic or commensal symbioses between species of plant viruses have been recognized, each involving viruses from a different combination of genera (Chin *et al.*, 1993; Murant, 1993; Fauquet *et al.*, 2005; Roossinck, 2005). By contrast we know of only two similar associations between vertebrate-infecting viruses (Berns, 1990; Taylor, 1991). In each plant virus symbiosis, at least one virus gains some capability it otherwise lacked through co-infection or association with the other virus. The most commonly acquired capability is transmission, namely one virus is aided by, or depends upon, the other virus for transmission, however, as such transmission dependency has been relatively easy to detect whereas other interactions have been cryptic, this view may represent a research sampling bias. The immediate relatives of the symbiotic virus are in some cases independent and not mutualistic, for example potato aucuba mosaic potexvirus is transmitted by aphids from co-infections with a potyvirus, but is related to potexviruses which are independently transmitted by plant contact not by aphids. However, in four groups, the *Sequiviridae*, *Luteoviridae* and the nanoviruses and umbraviruses, there are several symbiotic species suggesting the capacity to form a symbiosis is conserved, if not the symbioses themselves (Fauquet *et al.*, 2005). In many instances the symbioses cause a well-known crop disease. Rice tungro disease is caused by a co-infection of rice tungro bacilliform badnavirus and rice tungro spherical waikavirus; the badnavirus depends on the waikavirus for its transmission by leafhoppers, and the two viruses together cause the most severe tungro symptoms (Hibino, 1996).

Umbraviruses and luteoviruses, from the genera *Polerovirus* or *Enamovirus*, form long-term associations that are probably the best known among plant viruses. Each of the seven known umbravirus species associates with one luteovirus species (Taliansky and Robinson, 2003); umbraviruses can replicate independently, but are not transmitted independently and do not have coat protein genes (Gibbs *et al.*, 1996a, 1996b; Taliansky *et al.*, 1996); their luteovirus partners complement this deficiency when umbravirus genomic RNA is encapsidated in the luteovirus coat proteins allowing transmission by aphids. The luteovirus and luteovirus-encapsidated umbravirus particles are absorbed and pass through the hemolymph of the aphids without replicating to be excreted in the aphid saliva. The particles are protected from degradation in the aphid by a protein produced by endosymbiotic bacteria (van den Heuvel *et al.*, 1994). All luteoviruses are capable of independent replication and transmission, except one, pea enation enamovirus, which is unable to establish an infection without its associated umbravirus, showing its association is mutualistic (Demler *et al.*, 1994; Falk *et al.*, 1999). As the other luteoviruses are capable of full independence, we expect competition between strains that exist with and without umbraviruses, as well as competition between the umbraviruses and their luteovirus partners for access to the coat proteins and the vectors. Such competition may influence population genetics and rates of evolution. Evidence that umbravirus movement proteins may assist luteoviruses to invade plant tissues, and that umbravirus-associated satellite RNAs may modulate host symptoms, possibly affecting transmission, suggests that all umbravirus–luteovirus associations may also be mutualistic, and this may offset some of the costs luteoviruses may suffer in supporting umbraviruses (Taliansky and Robinson, 1997; Ryabov *et al.*, 2001). Indeed evidence that the replication genes of umbraviruses and some luteoviruses are related suggests that they may be recombinants.

Many plant viruses are known to interact synergistically in mixed infections; virion concentrations may be increased and symptoms enhanced. Some potyviruses are synergistic with a broad range of unrelated viruses, including pararetroviruses, potexviruses, and comoviruses. Studies of co-infections with a potexvirus show the potyvirus helper component proteinase suppresses RNA silencing of the potexvirus (Pruss *et al.*, 1997; Anandalakshmi *et al.*, 1998). Mixed infections of potyviruses and potexviruses may kill the infected plant and so the fitness of neither virus is improved.

Recombination and modular evolution have clearly been important in the evolution of luteoviruses, and it is possible that some luteovirus lineages have arisen through symbiogenesis, a process where symbionts coalesce to produce new species (Gibbs, 1995). This phenomenon may also explain other evidence of modular evolution among plant viruses (Roossinck, 2005).

Another related phenomenon is the presence of so many and different satellite viruses and satellite nucleic acids. The first satellite virus to be described was that of tobacco necrosis virus (Kassanis and Nixon, 1960); it replicates only in cells infected with tobacco necrosis virus and encodes little more than its own coat protein, but represents a distinct evolving entity. Satellite nucleic acids, by contrast, rely on the helper virus to provide virions. Satellite viruses and nucleic acids have been found associated with a great variety of viruses (Simon *et al.*, 2004), indeed so many are known that they constitute a significant component of the "Subviral RNA Database" (Rocheleau and Pelchat, 2006). They are widespread in populations of viruses with RNA genomes (Pinel *et al.*, 2003), and also those with DNA genomes (Briddon *et al.*, 2004; Amin *et al.*, 2006), and although some seem to have little effect on the interaction between the helper virus and the plant, others have dramatic effects and their presence seems to be responsible for much of the damage associated with the infection.

GENETIC DIVERSITY OF PLANT VIRUS POPULATIONS AND THE EVOLUTIONARY PROCESSES THAT PRODUCE AND CONTROL THAT DIVERSITY

The first evidence that plant virus populations are heterogeneous was the isolation, in the 1920s, of symptom variants from parts of systemically infected plants showing atypical symptoms (Kunkel, 1947). Early transmission experiments also showed that the major components of virion preparations could change when the conditions in which a virus-infected plant was grown were changed. For instance, growing viruses in different hosts changed their properties, a phenomenon described as host adaptation (Yarwood, 1979). The reversibility of such adaptations and the first molecular characterization of the phenomenon (Donis-Keller *et al.*, 1981) showed that adaptation resulted from selection among variants present, or newly generated, in the original virus stock. Population heterogeneity of laboratory stocks of tobamoviruses was estimated from the percentage of symptom mutants (Gierer and Mundry, 1958) or from molecular analyses of the genomic RNA (Rodríguez-Cerezo and García-Arenal, 1989). Early molecular analyses also showed that stocks did not remain genetically homogeneous, mutants were generated continuously, either when derived from single lesions (García-Arena *et al.*, 1984) or when obtained by transcribing infectious RNA from cloned cDNAs of TMV or the satellite RNA of cucumber mosaic virus (CMV-satRNA) (Aldaoud *et al.*, 1989; Kurath and Palukaitis, 1990).

The fast appearance and accumulation of mutants in RNA virus stocks had previously been described for bacterial or animal RNA viruses and had been associated with large error rates of RNA-dependent RNA polymerases (RdRp), which were estimated to be in the range of 10^{-4}–10^{-6} misincorporations per nucleotide position and per replication round; the rate is several orders of magnitude greater than for DNA-dependent DNA polymerases of large DNA phages or of cellular organisms (Domingo and Holland, 1997). These mutation rates would result in genomic mutation rates of about one mutation per genome and per replication round (Drake and Holland, 1999). However, these estimates of RdRp error rates were obtained using mutational targets that were potentially unrepresentative because they were either very small or were a transgene. Also, the selection and phenotypic masking of deleterious mutants was not estimated (Drake *et al.*, 1998). More recently, the spontaneous mutation rate of TMV was determined by detecting mutants lethal for cell-to-cell movement. The mutational target was the movement protein gene, a gene that is 804 nt long and that co-evolves with the viral replication complex. Mutants were detected in conditions of minimal selection against deleterious ones (Malpica *et al.*, 2002). The mutation rates found were large but smaller than those previously reported for lytic RNA viruses (i.e. 0.05–0.1 vs. ~1 mutations per genome and replication round), but this could be a more realistic estimate as suggested by a new estimate of the mutation rate of vesicular stomatitis virus (VSV; see (Furió *et al.*, 2005). More importantly, the mutational spectrum for an RNA genome was reported for the first time: more than one-third of mutants were multiple mutants, and about two-thirds of them were insertions and deletions, so that a large fraction of the mutations are likely to be very deleterious or immediately lethal. An analysis of the mutational spectrum of VSV has also shown that most point mutations are deleterious (Sanjuan *et al.*, 2004). These data show that most mutations in RNA viruses are probably not adaptive, and support the view that the high mutation rate of RNA viruses, rather than being a strategy to promote their evolution, is required to replace their chemically unstable genome (Drake *et al.*, 1998). The rate and character of mutations in TMV is also consistent with two classical observations: the characteristically small specific infectivity of RNA viruses and the vulnerability of RNA virus populations to increased mutation rates, that rapidly lead to their extinction.

Mutation rates are an order of magnitude smaller for retroviruses than for RNA viruses (Drake *et al.*, 1998), and this may also be true for plant viruses, like caulimoviruses, that have a DNA genome that replicates by reverse transcription of an RNA intermediate. For viruses with large double-stranded (ds)DNA genomes mutation rates per base are much smaller than for RNA viruses or for retroviruses (about 10^{-8}), and mutation rates per genome are about 0.003 per replication round (Drake *et al.*, 1998). It is not known if these values also apply to the many small single-stranded (ss)DNA plant viruses, for which no estimate of mutation rate is available.

Another source of genetic variation in plant viruses is genetic exchange by recombination or reassortment of genomic segments. It has been reported to occur in natural populations of plant viruses with either RNA or DNA genomes, and both within and between species (Desbiez and Lecoq, 2004; García-Andrés *et al.*, 2007; Valli *et al.*, 2007). Table 11.1 shows the results of a search for recombinants in the gene sequences of some representative viruses. Genetic exchange may result in

TABLE 11.1 Complete genomic sequences of positive-sense plant viruses analyzed for recombination

Family and genus	Species and number of isolates analyzed	Protein gene and number of recombination sites	Reference
Family Potyviridae			
Genus Potyvirus	Potato virus Y (n = 51)	Protein 1 (5)	Ogawa *et al.*, 2008; also this study[a]
		Helper-component proteanse (2)	
		Genome-linked viral protein (1)	
		Nuclear inclusion proteinase (1)	
		Nuclear inclusion b (3)	
		Coat protein (>2)	
	Turnip mosaic virus (*n* = 92)	Protein 1 (7)	Ohshima *et al.*, 2007
		Helper-component protease (7)	
		Protein 3 (2)	
		Cylindrical inclusion (11)	
		6-kDa 2 (2)	
		Genome-linked viral protein (6)	
		Nuclear inclusion protease (3)	
		Nuclear inclusion b (2)	
		Coat protein (4)	
	Bean yellow mosaic virus (*n* = 7)	Protein 3 (1)	Chare and Holmes, 2006; also this study[a]
		Cylindrical inclusion (2, tentative)	
		Nuclear inclusion b (2)	
		Coat protein (1)	
Family Flexviridae			
Genus *Potexvirus*	Potato virus X (*n* = 14)	166-kDa protein (1)	Chare and Holmes, 2006; also this study[a]
		Coat protein (2)	
Family Geminiviridae			
Genus *Mastrevirus*	Maize streak virus (*n* = 26)	Movement protein (3)	Martin *et al.*, 2005[a]
		Coat protein (2)	
		Short intergenic region (4)	
		Replication-associated protein (9)	
		Long intergenic region (2)	
Genus *Tobamovirus*	Tobacco mosaic virus (*n* = 17[b])	180/130-kDa protein (1)	This study

[a] Recombination sites analyzed by the RDP3 package (Martin *et al.*, 2005b).
[b] Tobacco mosaic virus isolate sequences were analyzed for recombination with those of tomato mosaic virus.

larger phenotypic effects than mutation, and is often associated with phenomena such as host switches, host range expansion, or the emergence of new viral diseases. The only estimate of recombination rates for a plant virus has come from the analysis of the progenies of co-infections with different genotypes of cauliflower mosaic virus in which neutral molecular markers had been introduced (Froissart *et al.*, 2005). Recombination rates were 2–5 $\times$ 10^{-5} per base and replication cycle, thus they were similar to mutation rates in RNA viruses, both animal- and plant-infecting (García-Arenal *et al.*, 2001). Recombinants are common in virus populations (Table 11.1) (Tan *et al.*, 2004; Ohshima *et al.*, 2007), but analyses of the genetic structure of field virus populations often indicates constraints to genetic exchange (Bonnet *et al.*, 2005). Indeed, experiments with both DNA and RNA viruses have shown that heterologous gene combinations are selected against, supporting the notion that gene complexes co-adapt in viral genomes (Martin *et al.*, 2005a; Escriu *et al.*, 2007). This is an important idea in genome evolution that had been proposed for plant viruses a long time ago (Hanada and Harrison, 1977). Thus, epistatic interactions would constrain the plasticity of the small genomes of plant viruses, and further limit their ability to respond to selection.

The development in the 1970s of methods for analyzing nucleic acids has resulted in many studies of the diversity of virus populations. It was found that the frequency distribution of genotypes in virus populations were of the so-called "gamma statistical distribution"; one major genotype plus a set of minor variants newly generated by mutation and/or kept at a low level by selection, as shown initially for tobacco mild green mosaic virus (TMGMV) (Rodríguez-Cerezo and García-Arenal, 1989). The shape of the gamma distribution depended on both the virus and the host plant (Schneider and Roossinck, 2000; Schneider and Roossinck, 2001). This genetic structure had been previously reported for RNA viruses infecting bacteria or animals and has been named a "quasispecies" (Domingo and Holland, 1997), as it corresponded to that predicted by Eigen's quasispecies theory, proposed to describe the evolution of an infinitely large population of asexual replicators that had a large mutation rate (Eigen and Schuster, 1977). Recently the "quasispecies concept" has been used to describe any genetically heterogeneous virus population, with no concern or awareness of its further implications, or for the specific conditions required for the quasispecies concept to apply (Eigen, 1996). In spite of the limited appreciation of its implications, the quasispecies concept was crucial in making virologists aware of the intrinsic heterogeneity of virus populations, an early discovery overlooked in the 1980s when virology focussed on the molecular analyses of viral genomes.

Although plant viruses have a great potential to vary, and their populations include large numbers of mutants, population diversity is usually small. Genetic diversity is most accurately estimated from nucleotide sequence data, but can also be measured, with larger errors, by analyzing restriction fragment lengths, or RNase T1 fragment polymorphisms (Nei, 1987). Nucleotide diversities cannot be estimated from ribonuclease protection assay (RPA) data unless the method has been calibrated directly using known sequences (Aranda *et al.*, 1995), and cannot be estimated by methods such as "single strand conformation polymorphism analysis." These methodological limitations have been often overlooked by researchers, and this has handicapped the quantitative analyses of population structures.

The genetic stability of plant virus populations has been observed since the earliest times, especially by comparing isolates passaged during long periods under laboratory conditions (Goelet *et al.*, 1982; Dawson *et al.*, 1986; Hillman *et al.*, 1991), or isolated for long periods of time (Fraile *et al.*, 1997) or in regions that had been unconnected for thousands of years (Blok *et al.*, 1987; Skotnicki *et al.*, 1993). Data accumulated during the last 20 years show that no natural populations of plant viruses evolve as quickly as some populations of animal viruses, and they have values of nucleotide diversity per site mostly less than 0.07 (see table 2 in García-Arenal *et al.*, 2001). Most genetic resistance of crop plants to viruses is durable, despite the common occurrence of resistance breaking isolates (Harrison, 2002; García-Arenal and McDonald, 2003). No correlation was found between population diversity and any other trait, such as the mode of transmission or the nature of the host plant, or whether the viral genome is DNA or RNA.

Many factors may ensure that the population of a virus with a high potential for variation does not vary. Selection can be one such reason. Evidence for various types of selection pressures on plant viruses have been reviewed (García-Arenal *et al.*, 2001). Here we will stress the importance of negative selection on virus-encoded proteins. The degree of negative selection in genes, or the degree of functional constraints that maintain the function of the encoded protein sequence, is usually estimated from the ratio between the nucleotide diversities at non-synonymous (d_{NS}) versus synonymous (d_S) positions. Analysis of this ratio for structural and non-structural proteins of a number of RNA and DNA viruses (see table 1 in García-Arenal *et al.*, 2001) shows that it is similar for RNA and DNA viruses and that, interestingly, it does not depend on the function of the encoded protein. This is in contrast with proteins of cellular organisms, in which certain classes of proteins (e.g., histones) are always more conserved than others (Nei, 1987). Moreover, d_{NS}/d_S ratios for viral genes all fall within the range reported for genes of cellular organisms (Nei, 1987). Thus, variation of the genes and encoded proteins of viruses is as constrained as those of their eukaryotic hosts and vectors, which suggests that the constraints arise from functional interactions between viral-, host-, and vector-encoded factors. Another important well-documented source of constraint could be the multifunctionality of virus-encoded proteins, which might result in selective constraints for one function having epistatic effects on another, so that the protein would be never optimized for just one of its functions. An extreme case of multiple functional constraints occurs in genomic regions with overlapping reading frames, which are common in the tightly packaged genomes of viruses (Keese and Gibbs, 1993; García-Arenal *et al.*, 2001). Although the non-synonymous (d_{NS}) vs. synonymous (d_S) ratio provides evidence of selection of the nucleotides, via the amino acids they encode, this may not be the only factor affecting this ratio as it depends on both the total and relative rates of accumulation of NS and S changes (Sharp, 1997), and, for example, in ribosomal RNA genes and in viral genomes, these may be selected to maintain crucial secondary structures of nucleic acids (Sharp, 1997).

Another evolutionary process that may limit the diversity of viral populations is genetic drift. Because populations may not be large enough to ensure that all extant variants will be present in the next generation, random extinctions would determine the genetic composition of each new generation and might result in a new balance; this random process is called genetic drift. Populations of plant viruses can reach very large sizes, there may be 10^{11}–10^{12} TMV particles in an infected tobacco leaf, but this may not be the number relevant for viral evolution, as was proposed long ago (Harrison, 1981). Indeed, the relevant evolutionary parameter is not the census size of the population, but the effective population size, which could be grossly defined as the fraction of the population that passes its genes to the new generation. In a virus population the effective population size may be much smaller than the actual population size for several reasons, for instance the small intrinsic infectivity of RNA viruses caused by the large proportion of lethal mutants. Also changes of population size during the viral life cycle resulting in population bottlenecks, would affect the effective population size. It has been shown that virus populations pass through severe bottlenecks during plant colonization, and that the effective size of the population that initiates colonization of a new leaf could be as small as units or tens of individual genomes (French and Stenger, 2003; Sacristán *et al.*, 2003; Ali *et al.*, 2006). It has also been shown that severe population bottlenecks occur during transmission by aphids to new plants (Ali *et al.*, 2006). Hence, a new population in, for example, a newly infected leaf, or plant, or geographical area, etc, may come from a very small number of genomes randomly chosen from the mother population. This so-called "founder effect" results in smaller diversities within populations and larger diversities between populations. Genetic drift can result in the elimination from the population of the fittest genotypes and the accumulation of deleterious mutations, eventually leading to population extinction (i.e. mutational meltdown), as shown experimentally for various RNA viruses, including the plant virus tobacco etch virus (Iglesia and Elena, 2007). Mutation accumulation and population extinction was also shown to occur in nature in a TMV population infecting *Nicotiana glauca* when TMGMV entered the same plant population. It resulted from a reduction in the TMV population size caused by co-infection with TMGMV, to our knowledge the only report of mutational meltdown occurring in viral populations in nature (Fraile *et al.*, 1997). Hence, random genetic drift, as opposed to selection, can be an important evolutionary factor for plant viruses, a possibility not contemplated in early studies of viral evolution or in the quasispecies theory, which is a deterministic model of evolution.

THE COMPARATIVE PHYLOGENETICS OF PLANT VIRUSES

Gene sequencing has become a standard and central technique for studying all aspects of modern biology. Over the last three decades it has provided amazing but sometimes baffling new insights to the evolution of all organisms at many different levels of evolutionary time. The international gene sequence databases now contain a great number of gene sequences of plant viruses, for examples there are sequences of over 4000 isolates of the *Potyviridae*. Most of these sequences were probably obtained during routine attempts to identify viruses, however they can also be studied to provide information of population variation, difference between species within genera and families, and even the origins of viruses. We will first discuss information obtained from intra-familial comparisons of two different plant virus families.

The Tobamoviruses

The tobamoviruses, and especially TMV, the type species and "mother of virology," have been the subject of many taxonomic studies, starting with their host range differences and the serological relationships of their proteins (Bawden, 1956; Smith, 1957; Van Regenmortel, 1986; Van Regenmortel, 1999), which were congruent with the amino acid compositions of their coat proteins (Gibbs, 1968). Gene sequence comparisons have resolved many of the anomalies found in the earlier analyses, for example many TMV "strains" are now designated as distinct tobamovirus species as they have separate evolutionary histories and infect different hosts. There are now over 700 gene sequences of tobamovirus in the international sequence databases, and these can be used to differentiate around 20 tobamovirus species (Gibbs *et al.*, 1999). Consistent relationships are found between all, except odontoglossum ringspot virus (ORSV), whichever method of phylogenetic analysis is used, and whether the complete genomic sequences, part sequences or encoded amino acid sequences are compared. By contrast, ORSV was found to be a recombinant; its replicase genes are closest to those of the brassica-infecting tobamoviruses, whereas its movement and coat protein genes are closest to those from *Solanaceae* (Lartey *et al.*, 1996; Gibbs *et al.*, 1999).

The taxonomy of the tobamoviruses inferred from their gene sequences Figure 11.1 correlates well with groupings based on other criteria. Fukuda and colleagues (1980) identified two major groups of tobamoviruses: TMV and the other group 1 tobamoviruses had their origin of virion assembly region within the movement protein gene, whereas group 2 tobamoviruses had their origin of virion assembly region within the coat protein gene. As a consequence group 1 tobamoviruses produced virions of one length (c. 300 µm long) whereas those of group 2 tobamoviruses also produced shorter particles containing the mRNA of the coat protein. Lartey and colleagues (1996) noted that group 1 tobamoviruses mostly infect species of the *Solanaceae*, namely species from the asterid lineage of plants (Qiu *et al.*, 1999; Soltis *et al.*, 1999) also known as "tenuinucelli" (Young and Watson, 1970), whereas the group 2 tobamoviruses are mostly isolated from species of rosid lineage plants, also known as "crassinucelli." They concluded that the tobamoviruses and their hosts had co-evolved, and the group 1 and 2 tobamoviruses may have diverged when the "core Eudicot" plants radiated 100–115 million years ago (Chaw *et al.*, 2004).

Lartey and colleagues also considered the relationships of another newly identified group of tobamoviruses, mostly isolated from brassicas (i.e. rosids), which they called group 3. By clever analysis of overlapping genes they concluded that group 3 tobamoviruses were

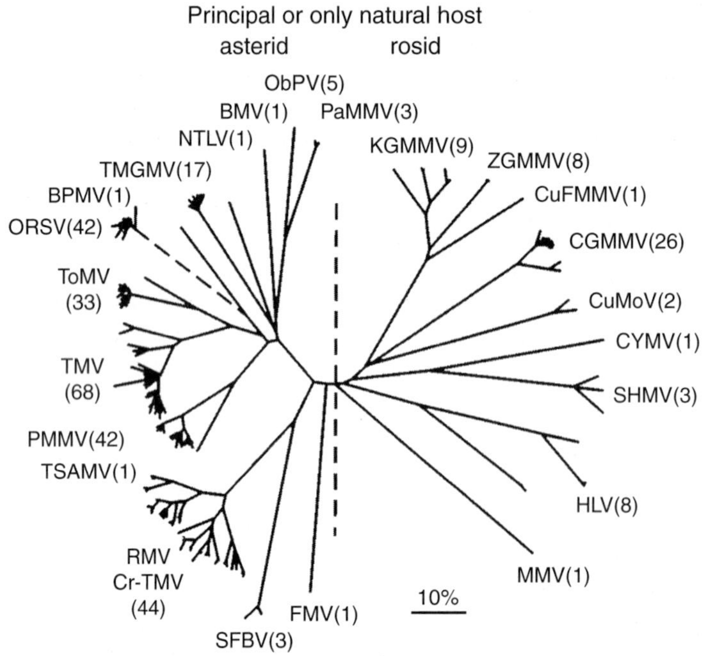

FIGURE 11.1 Neighbor-joining tree (Saitou and Nei, 1987) showing the relationships of the coat protein gene sequences of 320 tobamovirus isolates. The virus acronyms for all species of "asterid-favoring" tobamovirus are BPMV, bell pepper mottle virus; BMV, *Brugmansia* mosaic virus; Cr-TMV, crucifer TMV; FMV, frangipani mosaic virus; NTLV, Nigerian tobacco latent virus; ObPV, Obuda pepper virus; PaMMV, paprika mild mottle virus; PMMV, pepper mild mottle virus; RMV, ribgrass mosaic virus; SFBV, *Streptocarpus* flowerbreak virus; TMGMV, tobacco mild green mosaic virus; TMV, tobacco mosaic virus; ToMV, tomato mosaic virus; TSAMV, tropical soda apple mosaic virus; and of all species of "rosid-favoring" tobamovirus, CGMMV, cucumber green mottle mosaic virus; CuFMMV, cucumber fruit mottle mosaic virus; CuMoV, cucumber mottle virus; CYMV, *Clitoria* yellow mottle virus; HLV, *Hibiscus* latent, chlorotic ringspot and S viruses; KGMMV, kyuri green mottle mosaic virus; MMV, maracuja mosaic virus; SHMV, sunnhemp mosaic virus; ZGMMV, zucchini green mottle mosaic virus. In parentheses is the number of sequence/isolates in each cluster. ORSV (*Odontoglossum* ringspot virus) is a recombinant; its coat and movement protein gene sequences cluster as shown (broken line) with those of the tobamoviruses from solanaceous plants, but its polymerase gene sequences cluster with those from the "ribgrass-crucifer" tobamoviruses. The taxonomic grouping (asterid or rosid) of the principal or only host was from http://www.ncbi.nlm.nih.gov/sites/entrez?db = taxonomy.

"derived" rather than "ancestral" to those of the other two groups, and that they were a sublineage of group 1 (i.e. asterid) tobamoviruses. Some uncertainty remains, however, as the first group 3 tobamovirus to be described was ribgrass mosaic virus (RMV). Ribgrass, *Plantago lanceolata*, is an asterid plant. Thus the fact that most group 3 tobamoviruses have been isolated from brassicas (i.e. rosids) may reflect the fact that brassicas are crop plants and their viruses are being well studied, whereas *Plantago* spp. (i.e. asterids) are "weeds" and therefore their viruses are less well known. The ability of group 3 tobamoviruses to infect plants of both lineages may suggest that group 3 tobamoviruses are less host-specific than group 1 and 2 tobamoviruses. However it could also indicate that the concordance between virus and host lineages merely reflects a predisposition of group 1 and 2 to infect and adapt to particular lineages of plants rather than co-evolution.

If the tobamoviruses are as ancient as their hosts then sequences obtained from isolates collected over, say, the last century will not provide an estimate of the rate of TMGMV evolution, and this is indeed the case. Gene sequences of TMGMV isolates obtained from herbarium specimens of *Nicotiana glauca* collected in eastern Australia over the past century (Fraile *et al.*, 1997) showed no changes that correlated with the year in which they were collected. Samples collected millennia or centuries apart will probably be required to measure the rate of change directly. The idea that tobamoviruses are ancient is also congruent with the observation (Holmes, 1950) that the species of *Nicotiana* that respond to infection by TMV in a hypersensitive manner are all natives of the Americas: *N. glutinosa* a native of Peru, *N. repanda* of Mexico, *N. rustica* of Ecuador and Peru and *N. langsdorfii* of Brazil. Several species of other genera of the *Solanaceae* native to South America behave in the same way, including *Solanum capsicastrum* of Brazil and *S. tuberosum* of Bolivia and Peru. By contrast, *Nicotiana* species that respond to TMV infection with bright chlorosis and mottling, and accumulate the greatest concentrations of virions, are mostly found in North America, southern South America, and Australia. Holmes argued that the hypersensitive response may reflect exposure to, and selection by TMV, and stated that this "would seem to imply that the original habitat of tobacco-mosaic virus was within an area of the New World, centering about some part of Peru, Bolivia, or Brazil." He also noted that in this region there are now three species of *Nicotiana* (*N. glauca*, *N. raimondii* and *N. wigandioides*) that tolerate TMV infection, show few or no symptoms and may be the long-term host of TMV; *N. tabacum* itself is an amphidiploid species found only in crops or as a "crop fugitive," and is unlikely to have been the original long-term host of TMV.

The *Solanaceae* is a mostly tropical family. Its earliest fossils are from the Cretaceous, 65 million years ago (D'Arcy, 1991), and its present distribution is Gondwanan (Symon, 1991); the major center of diversity is Central and South America and there are minor centers in Eurasia, especially around the Himalayas, and also Australia. Hence the *Solanaceae* probably originated before the Indian subcontinent separated from Gondwana around 80 million years ago, and hypersensitivity to TMV may have arisen or been lost more than once, and may have spread by hybridization between species. Whichever is correct, it suggests that the sort of processes that enabled the Americas to become the center of TMV-resistance genes probably required tens of millions of years not thousands.

Although the angiosperms first appeared in the fossils from 120–140 million years ago, they are probably older, but most modern families did not appear before 60–80 million years ago (Raven, 1983), and the major tobamovirus radiations producing the clusters of species now found in the *Solanaceae*, the legumes and cucurbits, may have occurred when these modern plant families radiated. This period includes both the final stages of dismemberment of Gondwana, and also the Cretaceous–Tertiary extinction, and either of these events may have been responsible for the deep branches of some tobamovirus lineages.

The Potyviruses

Unlike the tobamoviruses, the relationships of potyviruses show no correlation with those of their hosts; for example different species of the largest lineage of potyviruses, the bean common mosaic (BCMV) group, are isolated from aroids, cucurbits, legumes, orchids, passionflowers, and others. The *Potyviridae* is the largest family of known plant viruses, and most of its species are from the genus *Potyvirus*. Over 100 out of the 500 or so recognized species of plant virus are potyviruses. They infect angiosperms in all parts of the world and in all climatic zones, and are especially damaging to crops. Potyviruses are transmitted by migrating aphids when they probe plants while searching for their preferred hosts; many aphid species may transmit each potyvirus. Some potyviruses are also transmitted in seeds to the progeny of infected plants and also, of course, to vegetative propagules. They have flexuous filamentous virions, and each contains a single copy of the genome, which is a single-stranded positive sense RNA molecule about 10 kb long. More than 5000 potyvirus sequences have been reported, most of them include the coat protein (CP) gene. This is a large set of data that could be used to date the radiation of the potyviruses, or certain outbreaks or even the emergence of potyviruses during expansion of agriculture in recent centuries.

The CP gene has a variable N-terminal part that is often repetitive and seems to have evolved by replicase slippage in a saltatory way (Ward *et al.*, 1995). The remainder of the CP (i.e. the core and C-terminal regions) has no unusual sequences of this sort, and seems to have evolved coherently and only by point mutations, indels, and occasional homologous recombination. The relationships of the aligned "coherent CP" (cCP) sequences of a representative set of 194 potyvirid isolates (Figure 11.2) show that the potyvirids are of two types in that all the sequences from different potyviruses form a star-burst cluster with a surprisingly uniform radial branch length and a pairwise sequence difference of 36.0% ± 2.15% (Gibbs *et al.*, 2007), whereas all the other potyvirids form longer hierarchically diverging lineages with larger peak pairwise differences. These phylogenies indicate that the potyviruses have evolved in a mode that is different from the other potyvirids. The phylogeny suggests that the potyviruses have radiated most recently, they initially speciated rapidly and then subspeciated to produce lineages or species groups, like the BCMV group, but all potyvirus lineages nonetheless have evolved at similar rates and, as a result, all extant potyvirus species are similar distances from the initial radiation. Nucleotide substitutions in the potyvirus cCP genes are not "saturated," and so it is probably valid to extrapolate from contemporary evolutionary changes to establish when the initial radiation occurred.

A mite-borne potyvirid, wheat streak mosaic tritimovirus (WSMV) (Stenger *et al.*, 2002) was

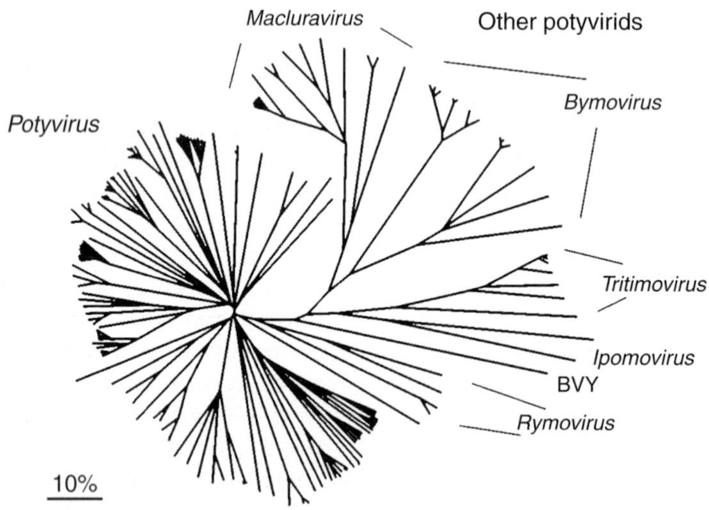

FIGURE 11.2 Neighbor-joining tree (Saitou and Nei, 1987) showing the relationships of the "coherent coat protein" gene sequences of 194 potyvirids. The star-burst sequences are all from potyviruses. The "other potyvirids" are species of *Macluravirus* (maclura mosaic, narcissus latent and cardamom mosaic viruses), of *Bymovirus* (barley mild mosaic, barley yellow mosaic, oat mosaic, wheat spindle streak mosaic and wheat yellow mosaic viruses), of *Tritimovirus* (brome streak mosaic, oat necrotic mottle and wheat streak mosaic viruses), of *Ipomovirus* (cucumber vein yellowing and sweet potato mild mottle viruses) and of *Rymovirus* (ryegrass mosaic, agropyron mosaic and hordeum mosaic viruses). The ungrouped virus is blackberry virus Y (BVY). The gene sequences were aligned via their encoded amino acid sequences using the Transalign program (kindly supplied by Georg Weiller) and CLUSTALX (Jeanmougin *et al.*, 1998) with default parameters, and gave sequences with 753 nucleotides and gaps (251 codons).

first recorded in American wheat crops in the 1920s (McKinney, 1937). It probably entered North America from Europe (Stenger *et al.*, 2002; Dwyer *et al.*, 2007) in seed (Jones *et al.*, 2005). The mean pairwise sequence difference of the cCP sequences of 68 isolates collected in North America and Australia is 2.9%. If a similar rate of divergence applied to all potyviruses, then it is possible that the main potyvirus radiation occurred no more than 10000 years ago, and potyviruses are diverging more than 10000 times more quickly than tobamoviruses.

ORIGINS OF PLANT VIRUS FAMILIES

The discovery of the -GDD- sequence motif in many viral polymerases (Kamer and Argos, 1984; Argos, 1988) showed that viruses, previously thought to be quite unrelated, had related genes. Haseloff and colleagues (1984) showed that TMV, alfalfa mosaic alfamovirus, cucumber mosaic cucumovirus, and Sindbis alphavirus have related replication proteins, even though they were previously not known to be related in any way other than that their genomes were ssRNA. Most surprising was that the first three of these only infected plants whereas Sindbis alphavirus replicates only in vertebrates and its invertebrate vector mosquitoes! They concluded that "Reassortment of functional modules of coding and regulatory sequence from preexisting viral or cellular sources, perhaps via RNA recombination, may be an important mechanism in RNA virus evolution." This phenomenon was originally detected in the bacteriophages of coliform bacteria and called "modular evolution" (Botstein, 1980) and involves genetic recombination (Lai, 1992), which has subsequently been shown to be one of the dominant features of viral evolution.

Despite the unresolved nature of some of the phylogeny, where it was possible to trace gene lineages, a deep history of viruses was revealed. Many plant RNA viruses were found to have "modular" origins with related replication genes, movement protein genes, or particle protein genes combined with unrelated genes (Gibbs, 1987; Koonin and Dolja, 1993; Goldbach and de Haan, 1994). Comparative genomics indicated that recombination between ancestral viruses from different genera was probably responsible for the creation of many of the present day genera. The acquisition of genes from hosts, such as the movement protein genes, was perhaps another mechanism leading to the emergence of new genera or families. Gene creation *de novo* by overprinting was probably a third mechanism behind the evolution of new groups (Keese and Gibbs, 1992, 1993; Gibbs and Keese, 1995; Lartey *et al.*, 1996).

At least six groups of viruses include some members that infect vertebrates and other members that infect plants. These groups are the *Reoviridae* (viruses with dsRNA genomes), the *Rhabdoviridae* and *Bunyaviridae* (viruses with negative sense RNA genomes), the alpha-like (ALVG) and picorna-like supergroups (viruses with positive sense RNA genomes) and the *Geminiviridae*, *Circoviridae*, and nanoviruses (viruses with single-stranded circular DNA genomes) (Anzola *et al.*, 1989; Kormelink *et al.*, 1992; Goldbach and de Haan, 1994; Wetzel *et al.*, 1994; Meehan *et al.*, 1997; Gibbs and Weiller, 1999; Zanotto *et al.*, 1996; Gibbs *et al.*, 2000). There has been much speculation on the origins of these various groups. It is possible that ancestral viruses switched hosts either from plants to vertebrates, or vice versa, that an invertebrate or fungal vector aided one or more of the host switches, or that ancestral viruses originally infected invertebrates or fungi and were subsequently transmitted to the plants or vertebrates by species that became vectors (Hacker *et al.*, 2005). Another possibility is that one or more of the viral groups is truly ancient and predates the divergence of the hosts, and the viruses have co-evolved with the host lineages. Only in two cases has evidence been found to distinguish between the options. Genomic and phylogenetic analyses indicate the tospoviruses, members of the *Bunyaviridae* which infect plants, probably evolved from an ancestral virus from the family that infected vertebrates (Kormelink *et al.*, 1992), and there is phylogenetic evidence that the circoviruses, that infect vertebrates, evolved from an ancestral nanovirus that infected plants (Gibbs and Weiller, 1999). There is of course also the possibility that viral and other genes interact via mechanisms that are at present unknown (Sharma *et al.*, 2006).

The tobamoviruses, discussed above, belong to the ALVG, which includes vertebrate-infecting viruses such as the rubiviruses, hepatitis E virus, and alphaviruses, and plant-infecting viruses such as the furoviruses, potexviruses, tymoviruses, capilloviruses, closteroviruses, hordeiviruses, tobraviruses, alfamoviruses, bromoviruses, cucumoviruses, ilarviruses, idaeoviruses, and the endornaviruses (Goldbach and de Haan, 1994). The relationships are supported by comparisons of the RNA-dependent RNA polymerases, 5′ terminal methyltransferases, and helicases (Koonin and Dolja, 1993; Gibbs *et al.*, 2000), and these replication enzyme genes have formed a module with a long phylogenetic history. Within the ALVG there is a subset that includes the tobraviruses (Goulden *et al.*, 1992), and probably also the hordeiviruses and furoviruses, that have rod-shaped and filamentous virions and coat proteins that are related in sequence and structure (Dolja *et al.*, 1991). Thus all the species of the ALVG share a replication enzyme module, and some of them also share the coat protein gene. In the multi-component ALVG viruses the replication enzyme module is now divided among separate genome segments, and another subset of viruses have acquired a papain-like serine protease gene, that is inserted between the methyltransferase and helicase genes. If it is correct that the tobamoviruses co-evolved with the angiosperms, then their links with other viruses of the ALVG occurred before 120–140 million years ago.

POSTSCRIPT

The relationships discussed in this review rely mostly on comparisons of the sequences of genes or the proteins they encode and, where appropriate, some have been tested statistically (Zanotto *et al.*, 1996). This is possible because each unit of these sequences is a separate quantum of data, and so a sequence is a potentially rich store of discriminatory information. Others have attempted to extend such comparisons into even deeper evolutionary time (Koonin and Dolja, 2006; Koonin *et al.*, 2006) mostly relying on the assumption that proteins with similar structure and function may be related even though they have no significant sequence homology. However such characters are, in essence, phenotypic rather than genetic. Molecular phenotypic characters may be no more phylogenetically informative than other more traditional phenotypic characters, and therefore conclusions based on them are probably more speculative than those based on sequence comparisons. Spanners provide a simple analogy. The shape of the functional "motif" of spanners might suggest that they are all related, yet they require that shape to fit hexagonal nuts, so too RNA polymerases may appear similar but the -GDD- motif they contain may be the only combination of extant amino acids able to fulfill a crucial step in the function of a polymerase; this motif may be the result of convergent evolution rather than a signature of shared ancestry.

There have been great advances in our understanding of viruses over the past century. Nonetheless many questions remain. When and if answers are obtained they will be enriched if they are placed in an evolutionary framework. As Theodosius Dobzhansky stated "Nothing makes sense in biology except in the light of evolution" (Dobzhansky, 1973). Questions worth answering include:

1. From where do viruses "emerge"? Do many viruses of plants, like badnaviruses, and some animal and bacterial viruses, alternate "endogenous" and "exogenous" lifestyles?
2. What combination of viral and host factors determine viral host ranges?
3. How do viruses respond to, and circumvent, the defenses of plants? How quickly do they respond?

4. How will plant viruses and their vectors respond to "climate change"?
5. How will viruses respond to the use of transgenes in crop plants?
6. Will "climate change" and "transgene pollution" increase the pace of virus evolution and increase the frequency with which damaging new viral diseases arise in crops?

The speed with which current evolutionary adventures will impact on agriculture is uncertain, and likely to remain so because it seems that fewer scientists are funded to observe the consequences of such adventures than are funded to generate them (MacLean *et al.*, 1997; Tepfer and Balázs, 1997).

NOTE ADDED IN PROOF

New evidence indicates that the evolutionary rate of wheat streak mosaic virus mentioned above (1.1×10^{-4} substitutions/site/year) is not unique. Duffy and Holmes (*Journal of Virology* **82**, 957–965, 2008) have reported that tomato yellow leaf curl begomovirus is evolving at 4.6×10^{-4} subs/site/year, similarly comparisons of isolates of rice yellow mottle sobemovirus collected over a 40-year period show that it is evolving at a rate of $4–8 \times 10^{-4}$ subs/site/year (Fargette, Pinel, Rakotomalala, Sangu, Traoré, Séréme, Sorho, Issaka, Hébrard, Séré, Kanyeka and Konaté, in press), and we have dated trees of potyvirus coat protein genes using historical isolation and outbreak events and found an evolutionary rate of around $1–2 \times 10^{-4}$ subs/site/year. Thus some viruses of plants are evolving as rapidly as some viruses of animals.

REFERENCES

Aldaoud, R., Dawson, W.O. and Jones, G.E. (1989) Rapid, random evolution of the genetic structure of replicating tobacco mosaic virus populations. *Intervirology* **30**, 227–233.

Ali, A., Li, H., Schneider, W.L., Sherman, D.J., Gray, S., Smith, D. and Roossinck, M.J. (2006) Analysis of genetic bottlenecks during horizontal transmission of Cucumber mosaic virus. *J. Virol.* **80**, 8345–8350.

Amin, I., Mansoor, S., Amrao, L., Hussain, M., Irum, S., Zafar, Y. *et al.* (2006) Mobilisation into cotton and spread of a recombinant cotton leaf curl disease satellite. *Arch. Virol.* **151**, 2055–2065.

Anandalakshmi, R., Pruss, G.J., Ge, X., Marathe, R., Mallory, A.C., Smith, T.H. and Vance, V.B. (1998) A viral suppressor of gene silencing in plants. *Proc. Natl Acad. Sci. USA* **95**, 13079–13084.

Anzola, J.V., Dall., D.J., Xu, Z. and Nuss, D.L. (1989) Complete nucleotide sequence of wound tumor necrosis virus genomic segments encoding nonstructural polypeptides. *Virology* **171**, 222–228.

Aranda, M.A., Fraile, A., García-Arenal, F. and Malpica, J.M. (1995) Experimental evaluation of the ribonuclease protection assay method for the assessment of genetic heterogeneity in populations of RNA viruses. *Arch. Virol.* **140**, 1373–1383.

Argos, P. (1988) A sequence motif in many polymerases. *Nucleic Acids Res.* **16**, 9909–9916.

Bawden, F.C. (1956) *Plant Viruses and Virus Diseases*, 3rd edn. Waltham, MA: Chronica Botanica.

Beck, J. and Nassal, M. (2007) Hepatitis B virus replication. *World J. Gastroenterol.* **13**, 48–64.

Beijerinck, M.W. (1898) Ueber ein contagium vivum fluidum als Ursache der Fleckenkrankheit der Tabaksblätter. *Verhandelingen der Koninklyke akademie van Wettenschapppen te Amsterdam* **65**, 3–21.

Berns, K.I. (1990) Parvovirus replication. *Microbiol. Rev.* **54**, 316–329.

Blok, J., Mackenzie, A., Guy, P. and Gibbs, A.J. (1987) Nucleotide sequence comparisons of turnip yellow mosaic virus isolates from Australia and Europe. *Arch. Virol.* **97**, 283–295.

Bonnet, J., Fraile, A., Sacristan, S., Malpica, J.M. and Garcia-Arenal, F. (2005) Role of recombination in the evolution of natural populations of Cucumber mosaic virus, a tripartite RNA plant virus. *Virology* **332**, 359–368.

Botstein, D. (1980) A theory of modular evolution for bacteriophages. *Ann. N.Y. Acad. Sci.* **354**, 484–491.

Briddon, R.W., Bull, S.E., Amin, I., Mansoor, S., Bedford, I.D., Rishi, N. *et al.* (2004) Diversity of DNA 1: a satellite-like molecule associated with monopartite begomovirus-DNA beta complexes. *Virology* **324**, 462–474.

Brunt, A.A., Crabtree, K., Dallwitz, M., Gibbs, A. and Watson, L. (1996) *Viruses of Plants*. Oxford: C.A.B. International.

Chare, E.R. and Holmes, E.C. (2006) A phylogenetic survey of recombination frequency in plant RNA viruses. *Arch. Virol.* **151**, 933–946.

Chaw, S.-M., Chang, C.-C., Chen, H.-L. and Li, W.H. (2004) Dating the Monocot–Dicot divergence and the origin of core Eudicots using whole chloroplast genomes. *J. Mol. Evol.* **58**, 424–441.

Chin, L.-S., Forster, J.L. and Falk, B.W. (1993) The beet western yellows virus ST9-associated RNA shares structural and nucleotide sequence homology with carmo-like viruses. *Virology* **192**, 473–482.

Cilia, M.L. and Jackson, D. (2004) Plasmodesmata form and function. *Curr. Opin. Cell Biol.* **16**, 500–506.

D'Arcy, W.G. (1991) *"The Solanaceae since 1976, with a review of its biogeography." Solanaceae* (J.G. Hawkes, R.N. Lester, M. Nee and N. Estrada, eds), *III. Taxonomy, Chemistry, Evolution*. London: Royal Botanic Gardens and Linnaeansociety of London.

Dangl, J.L. and Jones, J.D.G. (2001) Plant pathogens and integrated defense responses to infection. *Nature, Lond.* **411**, 826–833.

Dawson, W.O., Beck, D.L., Knorr, D.A. and Grantham, G.L. (1986) cDNA cloning of the complete genome of tobacco mosaic virus and production of infectious trancripts. *Proc. Natl Acad. Sci. USA* **83**, 1832–1836.

Deleris, A., Gallego-Bartolome, J., Bao, J., Kasschau, K.D., Carrington, J.C. and Voinnet, O. (2006) Hierarchical action and inhibition of plant Dicer-like proteins in antiviral defense. *Science* **313**, 68–71.

Demler, S.A., Borkhsenious, O.N., Rucker, D.G. and de Zoeten, G.A. (1994) Assessment of the autonomy of replicative and structural functions encoded by the luteo-phase of pea enation mosaic virus. *J. Gen. Virol.* **75**, 997–1007.

Desbiez, C. and Lecoq, H. (2004) The nucleotide sequence of Watermelon mosaic virus (WMV, Potyvirus) reveals interspecific recombination between two related potyviruses in the 5′ part of the genome. *Arch. Virol.* **149**, 1619–1632.

Dobzhansky, T. (1973) Nothing in biology makes sense except in the light of evolution. *Am. Biol. Teacher* **35**, 125–129.

Dolja, V.V., Boyko, V.P., Agranovsky, A.A. and Koonin, E.V. (1991) Phylogeny of capsid proteins of rod-shaped and filamentous RNA plant viruses: two families with distinct patterns of sequence and probably structure conservation. *Virology* **184**, 79–86.

Domingo, E. and Holland, J.J. (1997) RNA virus mutations and fitness for survival. *Annu. Rev. Microbiol.* **51**, 151–178.

Donis-Keller, H., Browning, K.S. and Clarck, J.M. (1981) Sequence heterogeneity in satellite tobacco necrosis virus RNA. *Virology* **110**, 43–54.

Drake, J.W. and Holland, J.J. (1999) Mutation rates among lytic RNA viruses. *Proc. Natl Acad. Sci. USA* **96**, 13910–13913.

Drake, J.W., Charlesworth, B., Charlesworth, D. and Crow, J.F. (1998) Rates of spontaneous mutation. *Genetics* **148**, 1667–1686.

Dwyer, G.I., Gibbs, M.J., Gibbs, A.J. and Jones, R.A.C. (2007) Wheat streak mosaic virus in Australia: Relationship to isolates from the Pacific Northwest of the USA and its dispersion via seed transmission. *Plant Dis.* **91**, 164–170.

Eigen, M. (1996) On the nature of virus quasispecies. *Trends Microbiol.* **4**, 216–218.

Eigen, M. and Schuster, P. (1977) The hypercycle. A principle of natural self-orgenization. Pt.A: emergence of the hypercycle. *Naturwissenchaften* **64**, 541–565.

Escriu, F., Fraile, A. and García-Arenal, F. (2007) Constraints to genetic exchange support gene coadaptation in a tripartite RNA virus. *PLoS Pathog.* **3**, e8.

Falk, B.W., Tian, T. and Yeh, H.-H. (1999) Luteovirus-associated viruses and subviral RNAs. In: *Satellites and Defective Viral RNAs* (P.K. Vogt and A.O. Jackson, eds), pp. 159–175. Berlin: Springer.

Fauquet, C.M., Mayo, M.A., Maniloff, J., Desselberger, U. and Ball, L.A. (2005). Virus Taxonomy: classification and Nomenclatura of viruses 8th Report of the international committee on the toxonomy of viruses 1 vol. San Diego: Elsevier-Academic Press.

Fraile, A., Escriu, F., Aranda, M.A., Malpica, J.M., Gibbs, A.J. and García-Arenal, F. (1997) A century of tobamovirus evolution in an Australian population of Nicotiana glauca. *J. Virol.* **71**, 8316–8320.

French, R. and Stenger, D.C. (2003) Evolution of wheat streak mosaic virus: dynamics of population growth within plants may explain limited variation. *Annu. Rev. Phytopathol.* **41**, 199–214.

Froissart, R., Roze, D., Uzest, M., Galibert, L., Blanc, S. and Michalakis, Y. (2005) Recombination every day: abundant recombination in a virus during a single multi-cellular host infection. *PLoS Biol.* **3**, e89.

Fukuda, M., Okada, Y., Otsuki, Y. and Takebe, I. (1980) The site of initiation of rod assembly on the RNA of a tomato and a cowpea strain of tobacco mosaic virus. *Virology* **101**, 493–502.

Furió, V., Moya, A. and Sanjuán, R. (2005) The cost of replication fidelity in an RNA virus. *Proc. Natl Acad. Sci. USA* **102**, 10233–10237.

García-Andrés, S., Accotto, P.G., Navas-Castillo, J. and Moriones, E. (2007) Founder effect, plant host, and recombination shape the emergent population of begomoviruses that cause the tomato yellow leaf curl disease in the Mediterranean basin. *Virology* **359**, 302–312.

García-Arena, F., Palukaitis, P. and Zaitlin, M. (1984) Strains and mutants of tobacco mosaic virus are both found in virus derived from single-lesion-passaged inoculum. *Virology* **132**, 131–137.

García-Arenal, F. and McDonald, B.A. (2003) An analysis of the durability of resistance of plant to viruses. *Phytopathology* **93**, 941–952.

García-Arenal, F., Fraile, A. and Malpica, J.M. (2001) Variability and genetic structure of plant virus populations. *Annu. Rev. Phytopathol.* **39**, 157–186.

Geering, A.D.W., Olszewski, N.E., Harper, G., Lockhart, B.E.L., Hull, R. and Thomas, J.E. (2005) Banana contains a diverse array of endogenous badnaviruses. *J. Gen. Virol.* **86**, 511–520.

Gibbs, A. (1968) Plant virus classification. *Adv. Virus Res.* **14**, 263–328.

Gibbs, A. (1987) Molecular evolution of viruses; 'trees,' 'clocks' and 'modules.' *J. Cell Sci.* **7**, 319–337.

Gibbs, A. and Harrison, B. (1976) *Plant Virology: The Principles*. London: Edward Arnold.

Gibbs, A. and Keese, P. (1995) In search of the origins of viral genes. In: *Molecular Basis of Virus Evolution* (A.J. Gibbs, C.H. Calisher and F. García-Arenal, eds), pp. 76–90. Cambridge: Cambridge University Press.

Gibbs, A.J., Keese, P.L., Gibbs, M.J. and Garcia-Arenal, F. (1999) Plant virus evolution; past, present and future. In: *Origin and Evolution of Viruses* (E. Domingo, R. Webster and J. Holland, eds), pp. 263–285. New York: Academic Press.

Gibbs, M.J. (1995) *The genome of carrot mottle mimic umbravirus and the evolution of the carmo and sobemo virus families*. Oxford: D. Phil. thesis, University of Oxford.

Gibbs, M.J. and Weiller, G.F. (1999) Evidence that a plant virus switched hosts to infect a vertebrate and then recombined with a vertebrate-infecting virus. *Proc. Natl Acad. Sci. USA* **96**, 8022–8027.

Gibbs, M.J., Cooper, J.I. and Waterhouse, P.M. (1996a) The genome organization and affinities of an Australian isolate of carrot mottle umbravirus. *Virology* **224**, 310–313.

Gibbs, M.J., Ziegler, A., Robinson, D.J., Waterhouse, P.M. and Cooper, J.I. (1996b) Carrot mottle mimic virus (CMoMV): A second umbravirus associated with carrot motley dwarf disease recognised by nucleic acid hybridisation. *Mol. Plant Pathol.* On-Line http://www.bspp.org.uk/mppol/1996/1111gibbs.

Gibbs, M.J., Koga, R., Moriyama, H., Pfeiffer, P. and Fukuhara, T. (2000) Phylogenetic analysis of some large double-stranded RNA replicons from plants suggests they evolved from a defective single-stranded RNA virus. *J. Gen. Virol.* **81**, 227–233.

Gibbs, M.J., Ohshima, K., and Gibbs, A.J. (2007). Potyvirus: a genus for the Holocene. in preparation.

Gierer, A. and Mundry, K.W. (1958) Production of mutants of tobacco mosaic virus by chemical alteration of its ribonucleic acid in vitro. *Nature, Lond.* **182**, 1457–1458.

Goelet, P., Lomonossoff, G.P., Butler, P.J.G., Akam, M.E., Gait, M.J. and Karn, J.N. (1982) Nucleotide sequence of tobacco mosaic virus RNA. *Proc. Natl Acad. Sci. USA* **79**, 5818–5822.

Gog, J.R. and Grenfell, B.T. (2002) Dynamics and selection of many-strain pathogens. *Proc. Natl Acad. Sci. USA* **99**, 17209–17214.

Goldbach, R. and de Haan, P. (1994) RNA viral supergroups and the evolution of RNA viruses. In: *The Evolutionary Biology of Viruses* (S. Morse, ed.), pp. 161–184. New York: Raven Press.

Goulden, M.G., Davies, J.W., Wood, K.R. and Lomonossoff, G.P. (1992) Structure of tobraviral particles: a model suggested from sequence conservation in tobraviral and tobamoviral coat proteins. *J. Mol. Biol.* **227**, 1–8.

Grenfell, B.T., Pybus, O.G., Gog, J.R., Wood, J.L., Daly, J.M., Mumford, J.A. and EC, H. (2004) Unifying the epidemiological and evolutionary dynamics of pathogens. *Science* **303**, 327–332.

Hacker, C.V., Brasier, C.M. and Buck, K.W. (2005) A double-stranded RNA from a Phytophthora species is related to the plant endornaviruses and contains a putative UDP glycosyltransferase gene. *J. Gen. Virol.* **86**, 1561–1570.

Hamilton, A.J. and Baulcombe, D.C. (1999) A species of small antisense RNA in post-transcriptional gene silencing in plants. *Science* **286**, 950–952.

Hanada, K. and Harrison, B.D. (1977) Effects of virus genotype and temperature on seed transmission of nepoviruses. *Ann. Appl. Biol.* **85**, 79–92.

Harrison, B.D. (1981) Plant virus ecology: ingredients, interactions and environment influences. *Ann. Appl. Biol.* **99**, 195–209.

Harrison, B.D. (2002) Virus variation in relation to resistance-breaking in plants. *Euphytica* **124**, 181–192.

Haseloff, J., Goelet, P., Zimmern, D., Ahlquist, P., Dasgupta, R. and Kaesberg, P. (1984) Striking similarities in amino acid sequence among nonstructural proteins encoded by RNA viruses that have dissimilar genomic organization. *Proc. Natl Acad. Sci. USA* **81**, 4358–4362.

Hibino, H. (1996) Biology and epidemiology of rice viruses. *Annu. Rev. Phytopathol.* **34**, 249–274.

Hillman, B.I., Anzola, J.V., Halpern, B.T., Cavileer, T.D. and Nuss, D.L. (1991) First field isolation of wound tumor virus from a plant host: minimal sequence divergence from the type strain isolated from an insect vector. *Virology* **185**, 896–900.

Himber, C., Dunoyer, P., Moissiard, G., Ritzenthaler, C. and Voinnet, O. (2003) Transitivity-dependent and -independent cell-to-cell movement of RNA silencing. *EMBO J.* **22**, 4523–4533.

Holmes, F.O. (1939) *Handbook of Phytopathogenic Viruses*. Minneapolis, Minnesota: Burgess Publishing.

Holmes, F.O. (1950) Indications of a New-World origin of tobacco-mosaic virus. *Phytopathology* **41**, 341–349.

Hughes, A.L. and Friedman, R. (2005) Poxvirus genome evolution by gene gain and loss. *Mol. Phylogenet. Evol.* **35**, 186–195.

Hull, R. (2001) *Matthews' Plant Virology*, 4th edn. New York: Academic Press.

Iglesia, F., de la, and Elena, S.F. (2007) Fitness declines in Tobacco etch virus upon serial bottlenecks transfers. *J. Virol.* **81**, 4941–4947.

Jeanmougin, F., Thompson, J.D., Gibson, T.J., Gouy, M. and G, H.D. (1998) Multiple sequence alignment with Clustal X. *Trends in Biochemical Sciences* **23**, 403–405.

Jones, R.A.C., Coutts, B.A., Mackie, A.E. and Dwyer, G.I. (2005) Seed transmission of Wheat streak mosaic virus shown unequivocally in wheat. *Plant Dis.* **89**, 1048–1050.

Kamer, G. and Argos, P. (1984) Primary structural comparisons of RNA-dependent polymerases from plant, animal and bacterial viruses. *Nucleic Acids Res.* **12**, 7269–7282.

Kassanis, B. and Nixon, H.L. (1960) Activation of one plant virus by another. *Nature, Lond.* **187**, 713–714.

Keese, P. and Gibbs, A. (1992) Origins of genes: big bang or continuous creation?. *Proc. Natl Acad. Sci. USA* **89**, 9489–9493.

Keese, P. and Gibbs, A. (1993) Plant viruses: master explorers of evolutionary space. *Curr. Opin. Genet. Dev.* **3**, 873–877.

Koonin, E.V. and Dolja, V.V. (1993) Evolution and taxonomy of positive-strand RNA viruses: implications of comparative analysis of amino acid sequences. *Crit. Rev. Biochem. Mol. Biol.* **28**, 375–430.

Koonin, E.V. and Dolja, V.V. (2006) Evolution of complexity in the viral world: The dawn of a new vision. *Virus Res.* **117**, 1–4.

Koonin, E.V., Senkevich, T.G. and Dolja, V.V. (2006) The ancient virus world and evolution of cells *Biol. Direct* **1**, 29.

Kormelink, R., De Haan, P., Meurs, C., Peters, D. and Goldbach, R. (1992) The nucleotide sequence of the M RNA segment of tomato spotted wilt virus, a bunyavirus with two ambisense RNA segments. *J. Gen. Virol.* **73**, 2795–2804.

Kunkel, L.O. (1947) Variation in phytopathogenic viruses. *Annu. Rev. Microbiol.* **1**, 85–100.

Kurath, G. and Palukaitis, P. (1990) Serial passage of infectious transcripts of a cucumber mosaic virus satellite RNA clone results in sequence heterogeneity. *Virology* **176**, 8–15.

Lai, M.M. (1992) RNA recombination in animal and plant viruses. *Microbiol. Rev.* **56**, 61–79.

Lartey, R.T., Voss, T.C. and Melcher, U. (1996) Tobamovirus evolution: gene overlaps, recombination and taxonomic implications. *Mol. Biol. Evol.* **13**, 1327–1338.

Lecellier, C.H., Dunoyer, P., Arar, K., Lehmann-Che, J., Eyquem, S., Himber, C. *et al.* (2005) A cellular microRNA mediates antiviral defense in human cells. *Science* **308**, 557–560.

Lindbo, J.A., Silva-Rosales, L., Proebsting, W.M. and Dougherty, W.G. (1993) Induction of a highly specific antiviral state in transgenic plants: Implications for regulation of gene expression and virus resistance. *Plant Cell* **5**, 1749–1759.

Lucas, W. (2006) Plant viral movement proteins: agents for cell-to-cell trafficking of viral genomes. *Virology* **344**, 169–184.

MacLean, G.D., Waterhouse, P.M., Evans, G. and Gibbs, M.J. (1997) Commercialization of transgenic crops: risk, benefit and trade considerations. Canberra: Australian Government Publishing Service.

Malpica, J.M., Fraile, A., Moreno, I., Obies, C.I., Drake, J.W. and Garcia-Arenal, F. (2002) The rate and character of spontaneous mutation in an RNA virus. *Genetics* **162**, 1011–1505.

Marco, A. and Marín, I. (2005) Retrovirus-like elements in plants. *Recent Res. Dev. Plant Sci.* **3**, 1–10.

Margis, R., Fusaro, A., Smith, N., Curtin, S., Watson, J., Finnegan, E. and Waterhouse, P. (2006) The evolution and diversification of Dicers in plants. *FEBS Lett.* **580**, 2442–2450.

Martin, D.P., van der Walt, E., Posada, D. and Rybicki, E.P. (2005a) The evolutionary value of recombination is constrained by genome modularity. *PLoS Genet.* **1**, e51.

Martin, D.P., Williamson, C. and Posada, D. (2005b) RDP2: recombination detection and analysis from sequence alignments. *Bioinformatics* **21**, 260–262.

McKinney, H.H. (1937) Mosaic diseases of wheat and related cereals. US Department of Agriculture Circular pp. 1–23.

Meehan, B.M., Creelan, J.L., McNulty, M.S. and Todd, D. (1997) Sequence of porcine circovirus DNA: affinities with plant circoviruses. *J. Gen. Virol.* **78**, 221–227.

Mette, M.F., Kanno, T., Aufsatz, W., Jakowitsch, J., van der Winden, J., Matzke, M.A. and Matzke, A.J.M. (2002) Endogenous viral sequences and their potential contribution to heritable virus resistance in plants. *EMBO J.* **21**, 461–469.

Moury, B. (2004) Differential selection of genes of cucumber mosaic virus subgroups. *Mol. Biol. Evol.* **21**, 1602–1611.

Moury, B., Morel, C., Johansen, E. and Jacquemond, M. (2002) Evidence for diversifying selection in Potato virus Y and in the coat protein of other potyviruses. *J. Gen. Virol.* **83**, 2563–2573.

Murant, A.F. (1993) Complexes of transmission-dependent and helper viruses. In: *Diagnosis of Plant Virus Diseases* (R.E.F. Matthews, ed.), pp. 333–357. Boca Raton, FL: CRC Press.

Nei, M. (1987) *Molecular Evolutionary Genetics*. New York: Columbia University Press.

Ogawa, T., Tomitaka, Y., Nakagawa, A., and Ohshima, K. (2008). Genetic structure of a population of Potato virus Y inducing potato tuber necrotic ringspot disease in Japan; comparison with North American and European populations. *Virus Research* **131**, 199–212.

Ohshima, K., Tomitaka, Y., Wood, J.T., Minematsu, Y., Kajiyama, H., Tomimura, K. and Gibbs, A.J. (2007) Patterns of recombination in Turnip mosaic virus genomic sequences indicate hotspots of recombination. *J. Gen. Virol.* **88**, 298–315.

Overall, R. and Blackman, L. (1996) A model of the macromolecular structure of plasmodesmata. *Trends Plant Sci.* **1**, 307–311.

Pinel, A., Abubakar, Z., Traoré, O., Konaté, G. and Fargette, D. (2003) Molecular epidemiology of the RNA satellite of rice yellow mottle virus in Africa. *Arch. Virol.* **148**, 1721–1733.

Pruss, G., Ge, X., Shi, X.M., Carrington, J.C. and Bowman, V.V. (1997) Plant viral synergism: the potyviral genome encodes a broad-range pathogenicity enhancer that transactivates replication of heterologous viruses. *Plant Cell* **9**, 859–868.

Qiu, Y.L., Lee, J., Bernasconi-Quadroni, F., Soltis, D.E., Soltis, P.S., Zanis, M. *et al.* (1999) The earliest angiosperms: Evidence from mitochondrial, plastid and nuclear genomes. *Nature, Lond.* **402**, 404–407.

Ratcliff, F., Harrison, B.D. and Baulcombe, D.C. (1997) A similarity between viral defense and gene silencing in plants. *Science* **276**, 1558–1560.

Raven, P.H. (1983) The migration and evolution of floras in the southern hemisphere. *Bothalia* **14**, 325–328.

Rocheleau, L. and Pelchat, M. (2006) The Subviral RNA Database: a toolbox for viroids, the hepatitis delta virus and satellite RNAs research. *BMC Microbiol.* **6**, 24.

Rochow, W.F. (1977) Dependent virus transmission from mixed infections. In: *Aphids as Virus Vectors* (K.F. Harris and K. Maramorosch, eds), pp. 253–273. New York: Academic Press.

Rodríguez-Cerezo, E. and García-Arenal, F. (1989) Genetic heterogeneity of the RNA genome population of the plant virus U5-TMV. *Virology* **170**, 418–423.

Roossinck, M. (2005) Symbiosis versus competition in plant virus evolution. *Nature Rev. Microbiol.* **3**, 917–924.

Ryabov, E.V., Fraser, G., Mayo, M.A., Barker, H. and Taliansky, M. (2001) Umbravirus gene expression helps potato leafroll virus to invade mesophyll tissues and to be transmitted mechanically between plants. *Virology* **286**, 363–372.

Sacristán, S., Malpica, J.M., Fraile, A. and García-Arenal, F. (2003) Estimation of population bottlenecks during systemic movement of Tobacco mosaic virus in tobacco plants. *J. Virol.* **77**, 9906–9911.

Saitou, N. and Nei, M. (1987) The neighbor-joining method: a new method for reconstructing phylogenetic trees. *Mol. Biol. Evol.* **4**, 406–425.

Sanjuan, R., Moya, A. and Elena, S.F. (2004) The contribution of epistasis to the architecture of fitness in an RNA virus. *Proc. Natl Acad. Sci. USA* **101**, 15376–15379.

Schneider, W.L. and Roossinck, M.J. (2000) Evolutionary related Sindbis-like plant viruses maintain different levels of population diversty in a common host. *J. Virol.* **74**, 3130–3134.

Schneider, W.L. and Roossinck, M.J. (2001) Genetic diversity in RNA virus quasispecies is controlled by host-virus interactions. *J. Virol.* **75**, 6566–6571.

Seal, S.E., van den Bosch, F. and Jeger, M.J. (2006) Factors influencing begomovirus evolution and their increasing global significance: implications for sustainable control. *Crit. Rev. Plant Sci.* **25**, 23–46.

Sharma, R., Damgaard, D., Alexander, T.W., Dugan, M.E., Aalhus, J.L., Stanford, K. and McAllister, T.A. (2006) Detection of transgenic and endogenous plant DNA in digesta and tissues of sheep and pigs fed Roundup Ready canola meal. *J. Agric. Food Chem.* **54**, 1699–1709.

Sharp, P.M. (1997) In search of molecular darwinism. *Nature, Lond.* **385**, 111–112.

Simon, A.E., Roossinck, M.J. and Havelda, Z. (2004) Plant virus satellite and defective interfering RNAs: new paradigms for a new century. *Annu. Rev. Phytopathol.* **42**, 415–437.

Skotnicki, M.L., Mackenzie, A.M., Ding, S.W., Mo, J.Q. and Gibbs, A.J. (1993) RNA hybrid mismatch polymorphisms in Australian populations of turnip yellow mosaic tymovirus. *Arch. Virol.* **132**, 83–99.

Smith, A.E. and Helenius, A. (2004) How viruses enter animal cells. *Science* **304**, 237–242.

Smith, K.M. (1957) *A Textbook of Plant Virus Diseases*, 2nd edn. London: Churchill.

Soltis, P.S., Soltis, D.E. and Chase, M.W. (1999) Angiosperm phylogeny inferred from multiple genes: A research tool for comparative biology. *Nature* **402**, 402–404.

Soosaar, J.L., Burch-Smith, T.M. and Dinesh-Kumar, S. P. (2005) Mechanisms of plant resistance to viruses. *Nat. Rev. Microbiol.* **3**, 789–798.

Staginnus, C. and Richert-Poggeler, K.R. (2006) Endogenous pararetroviruses: two-faced travelers in the plant genome. *Trends Plant Sci.* **11**, 485–491.

Stenger, D.C., Seifers, D.L. and French, R. (2002) Patterns of polymorphism in wheat streak mosaic virus: sequence space explored by a clade of closely related viral genotypes rivals that between the most divergent strains. *Virology* **3–2**, 58–70.

Symon, D.E. (1991) Gondwanan elements of the Solanaceae. In: *Solanaceae III: Taxonomy, Chemistry, Evolution* (J.G. Hawkes, R.N. Lester, M. Nee and N. Estrada, eds). London: Royal Botanic Gardens and Linnaean Society of London.

Taliansky, M.E. and Robinson, D.J. (1997) Trans-acting untranslated elements of groundnut rosette virus satellite RNA are involved in symptom production. *J. Gen. Virol.* **78**, 1277–1285.

Taliansky, M.E. and Robinson, D.J. (2003) Molecular biology of umbraviruses: phantom warriors. *J. Gen. Virol.* **84**, 1951–1960.

Taliansky, M.E., Robinson, D.J. and Murant, A.F. (1996) Complete nucleotide sequence and organization of the RNA genome of groundnut rosette umbravirus. *J. Gen. Virol.* **77**, 2335–2345.

Tan, Z., Wada, Y., Chen, J. and Ohshima, K. (2004) Inter- and intralineage recombinants are common in natural populations of Turnip mosaic virus. *J. Gen. Virol.* **85**, 2683–2696.

Taylor, J.M. (1991) Human hepatitis delta virus. *Curr. Top. Microbiol. Immunol.* **168**, 141–166.

Tepfer, M. and Balázs, E. (1997) *Virus-resistant Plants: Potential Ecological Impact*. Berlin: Springer.

Tsompana, M., Abad, J., Purugganan, M. and Moyer, J.W. (2005) The molecular population genetics of the Tomato spotted wilt virus (TSWV) genome. *Mol. Ecol.* **14**, 53–66.

Valli, A., López-Moya, J.J. and García, J.A. (2007) Recombination and gene duplication in the evolutionary diversification of P1 proteins in the family Potyviridae. *J. Gen. Virol.* **88**, 1016–1028.

van den Heuvel, J.F., Verbeek, M. and van der Wilk, F. (1994) Endosymbiotic bacteria associated with circulative transmission of potato leafroll virus by. *Myzus persicae*. *J. Gen. Virol.* **75**, 2559–2565.

Van Regenmortel, M.H.V. (1986) Tobacco mosaic virus: antigenic structure. In: *The Plant Viruses. 2. The Rod-shaped Plant Viruses* (M.H.V. van Regenmortel and H. Fraenkel-Conrat, eds), pp. 79–104. New York: Plenum Press.

Van Regenmortel, M.H.V. (1999) The antigenicity of tobacco mosaic virus Philos. *Trans. R Soc. Lond. B Biol. Sci.* **354**, 559–568.

Voinnet, O. (2005) Induction and suppression of RNA silencing: insights from viral infections. *Nat. Rev. Genet.* **6**, 206–220.

Waigmann, E., Ueki, S., Trutnyeva, K. and Citovsky, V. (2004) The ins and outs of nondestructive cell-to-cell and systemic movement of plant viruses. *Crit. Rev. Plant Sci.* **23**, 195–250.

Wang, X.H., Aliyari, R., Li, W.X., Li, H.W., Kim, K., Carthew, R. *et al.* (2006) RNA interference directs innate immunity against viruses in adult Drosophila. *Science* **312**, 452–454.

Ward, C.W., Weiller, G., Shukla, D.D. and Gibbs, A.J. (1995) Molecular systematics of the Potyviridae, the largest plant virus family. In: *Molecular Basis of Virus Evolution* (A.J. Gibbs, C.H. Calisher and F. Garcia-Arenal, eds), pp. 477–500. Cambridge: Cambridge University Press.

Wetzel, T., Deitzgen, R.G. and Dale, J.L. (1994) Genomic organization of lettuce necrotic yellows rhabdovirus. *Virology* **200**, 401–412.

Wright, D.A. and Voytas, D.F. (2002) Athila4 of Arabidopsis and Calypso of soybean define a lineage of endogenous plant retroviruses. *Genome Res.* **12**, 122–131.

Yano, S.T., Panbehi, B., Das, A. and Laten, H.M. (2005) Diaspora, a large family of Ty3-gypsy retrotransposons in *Glycine max*, is an envelope-less member of an endogenous plant retrovirus lineage. *BMC Evol. Biol.* **5**, 30.

Yarwood, C.E. (1979) Host passage effects with plant viruses. *Adv. Virus Res.* **25**, 169–190.

Young, D.J. and Watson, L. (1970) The classification of dicotyledons: a study of the upper levels of the hierarchy. *Aust. J. Bot.* **18**, 387–433.

Zanotto, P.M.d.A., Gibbs, M.J., Gould, E.A. and Holmes, E.C. (1996) A reevaluation of the higher taxonomy of viruses based on RNA polymerases. *J. Virol.* **70**, 6083–6096.

CHAPTER

12

Mutant Clouds and Bottleneck Events in Plant Virus Evolution

Marilyn J. Roossinck

ABSTRACT

Plant viruses can develop high levels of variation in their populations, but this is not always the case. The level of diversity in single plant infections varies dramatically with different viruses and different hosts. Recent studies on diversity in the DNA geminivurses indicate that they have variation levels that are comparable to RNA viruses, in spite of their replication by the host DNA polymerase. Genetic bottlenecks occur during systemic infection of plant viruses and transmission events. Bottlenecks can have important effects on plant virus evolution due to genetic drift, and can ultimately result in isolation and evolution of new variants and speciation events.

INTRODUCTION

The majority of viruses found in plants have RNA genomes (Hull, 2002), although the geminiviruses, a group of DNA plant viruses, have posed the most significant disease problems in recent years due to their widespread emergence in crops (Rojas *et al.*, 2005; Seal *et al.*, 2006). The dominance of RNA genomes may be related to a greater adaptability of RNA. Most acute plant viruses must be generalists in order to survive, since plants are found in diverse communities in nature and their vectors often feed on many of the plants in these communities, resulting in horizontal transmission to distantly related host species. Plant viruses, with the exception of the algal viruses, all have very small genomes; most are under 15 kilobases (kb). This is most likely because plant viruses must move through the restricted connections between plant cells called plasmodesmata.

Most of what is known about plant viruses is from the study of viruses that cause disease in the monocultural settings of agricultural plants, and the diversity, incidence, and host spectrums of plant viruses in wild plants is largely unknown (Wren *et al.*, 2006). Monoculture could lead to highly specialized viruses, but the host ranges of characterized viruses of crop plants range from extremely broad, like cucumber mosaic virus (CMV) that infects over 1200 species (Edwardson and Christie, 1991), to very narrow, like barley stripe mosaic virus that naturally infects only barley, and occasionally wheat (Timian, 1974).

Plant virus evolution has a long history of study, beginning with early observations of

Origin and Evolution of Viruses
ISBN 978-0-12-374153-0

phenotypic change in viruses upon passage in plants (Price, 1934; McKinney, 1935). The role of the host in plant virus evolution was first studied in the mid-twentieth century, when Bawden described isolates of a tobamovirus that changed phenotypically after passage in different hosts (Bawden, 1958). The requirements for host adaptation are well documented, and range from replication functions, cell-to-cell and systemic movement functions, and dissemination functions. Host adaptation has been mapped to most of the genes of plant viruses in various host–virus combinations (Roossinck and Schneider, 2005).

MUTATION RATES AND FREQUENCIES

Mutation rate primarily refers to the fidelity of the polymerase, or the rate at which mutations are introduced during replication, although mutations also may be introduced by abiotic mutagens or by RNA editing. Mutation frequency is a very different measurement, and describes the amount of sequence variation seen in a virus population, generally after a given amount of time, but often with no knowledge of the number of generations the virus has undergone, or the forces of selection and drift that have affected the population. Accurate measurement of either is difficult. The most reliable data is sequence determination, but care must be taken to assure that mutations are not introduced *in vitro*, masking the true viral mutations (Schneider and Roossinck, 2000). Single stranded conformational polymorphism (SSCP) (Kong *et al.*, 2000), restriction enzyme length polymorphism (RFLP) (Naraghi-Arani *et al.*, 2001), and T1 RNase fingerprinting (Rodríguez-Cerezo *et al.*, 1989, 1991) also have been used to measure diversity. SSCP, RFLP, and fingerprinting methods reveal an overall picture of diversity, but these methods only detect mutations that are fixed in the population at some level.

No precise measurement of the substitution mutation rate of a plant viral RNA-dependent RNA polymerase (RdRp) has been done, or indeed of any virus in an intact host. Studies done for animal viruses *in vitro* or in cell culture have estimated rates of 10^{-3} to 10^{-5} substitutions per nucleotide per round of replication. An estimation of substitution mutation rates for tobacco mosaic virus (TMV) suggested that plant viruses are probably similar to other RNA viruses (Malpica *et al.*, 2002). The rates of insertions and deletions (indels) of the CMV RdRp have been measured, and were found to vary, depending on the secondary structure of the reporter RNA and on the host. Insertions were below the level of reliable detection, but deletion rates ranged from 1×10^{-4} to 3×10^{-6}. Deletion rates were significantly higher in pepper than in tobacco, and in structured vs. non-structured regions of the RNA (Pita *et al.*, 2007). These have important implications for the evolution of viruses in different hosts, and could account for genomic regions that are "hot-spots" for mutations. Although deletions are most often deleterious, they can be responsible for large changes in coding capacity, by creating alternative open reading frames.

Mutation frequency has been measured in a number of plant virus systems. This reflects both the mutation rate and other forces that act on the population: selection (both positive and negative) and genetic bottlenecks. It is not possible to extrapolate a mutation rate from a mutation frequency because generation times or generation sizes for plant viruses are not known. The mode of replication, whether fully exponential or partially linear, is also unknown. If the incoming viral RNA is only copied once to produce a pregenome, that is then copied multiple times, the replication is essentially linear (French and Stenger, 2003), as has been shown for bacteriophage ϕ6 (Chao *et al.*, 2002). However, given the rapid increase in virus titer that is possible an exponential mode of replication seems more likely, where the infecting viral genome is copied into many pregenomes that in turn are copied into many new genomes, which in turn are copied into pregnomes, etc. (Figure 12.1). The replication of plant viruses may employ either of these modes or some combination of them. It also seems likely that different viruses employ different modes of replication.

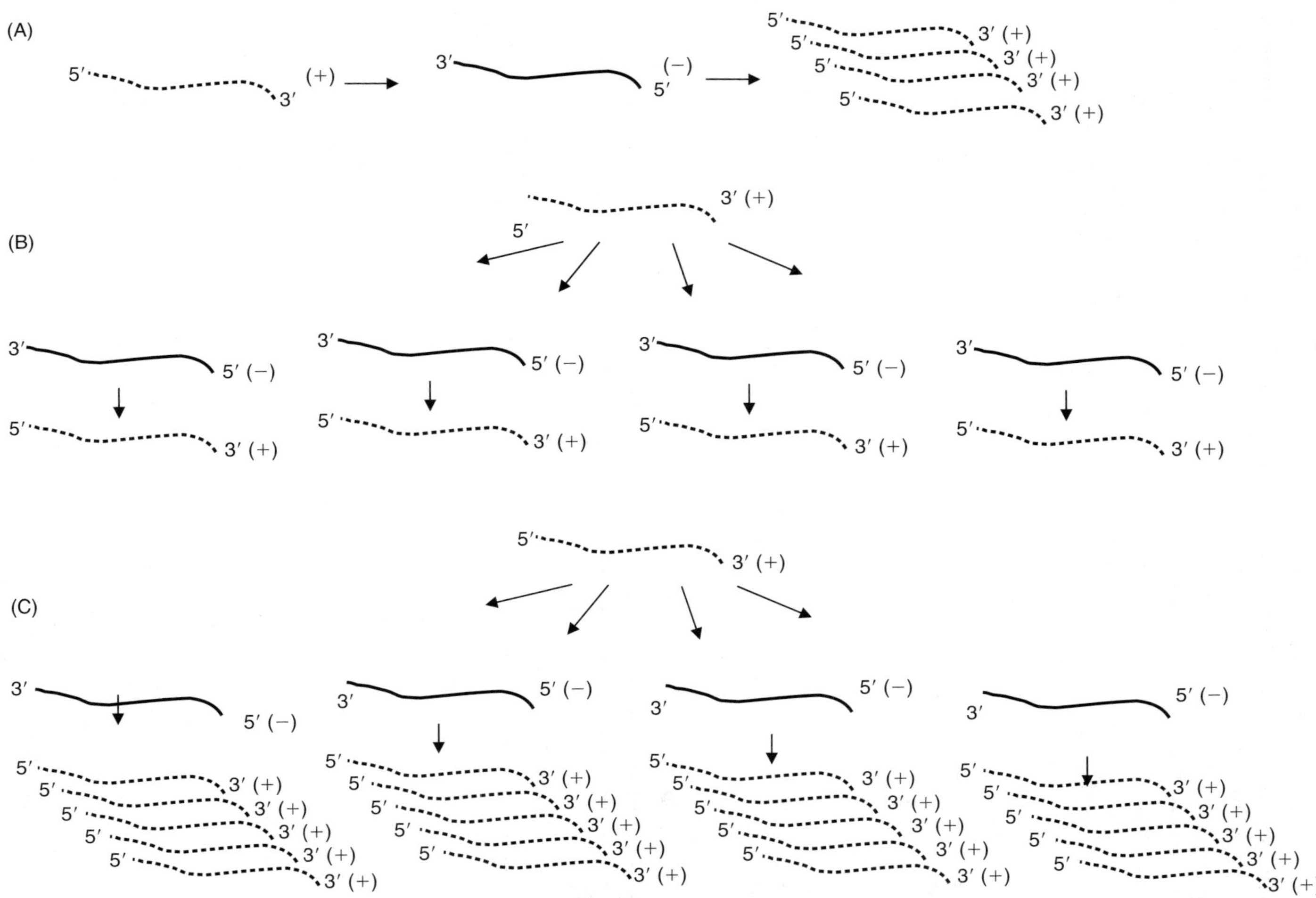

FIGURE 12.1 Linear vs. exponential replication. (A) The incoming viral genome is copied into one (−) strand pregenome that is the template for all of the (+) strands. (B) The incoming genome is copied into many (−) strand pregenomes, that in turn are each copied into one progeny (+) strand. (C) Fully exponential replication.

A few reviews on variation in plant virus populations have been published recently (García-Arenal *et al.*, 2003; Roossinck and Schneider, 2005). In some cases the level of variation in plant viruses was surprisingly low. For example, isolates of the same virus from 50 to 100 years apart showed remarkably few differences in consensus sequences (Gibbs *et al.*, 1999). There is much less information about intra-isolate population variation, especially from natural isolates (Roossinck and Schneider, 2005; Roossinck and Ali, 2007). In experimental evolution studies the level of diversity reported for plant RNA viruses ranges from 0.05 mutations per kb of RNA for cowpea chlorotic mottle virus (Schneider and Roossinck, 2000), to 2.7 for Kyuri green mottle mosaic virus (Kim *et al.*, 2005). Dramatic differences can be seen in the same virus if it evolves in different hosts: for example CMV mutation frequencies ranged from 0.6 to 1.8 per kb in the host plants *Nicotiana benthamiana* and pepper, respectively (Schneider and Roossinck, 2001). Experimental evolution studies were recently published for a plant DNA virus as well. Mutation frequencies ranged from 0.3 to 0.5 per kb in the geminivirus tomato yellow leaf curl China virus (Ge *et al.*, 2007), which is similar to levels found in RNA viruses. Similar or slightly higher levels of variation were seen in a natural isolate of another geminivirus, maize streak virus (Isnard *et al.*, 1998). This suggests that some DNA viruses may also exhibit a quasispecies-like nature, and supports the theory that geminiviruses may have evolved ways to increase the mutation rate of the plant host DNA polymerase that they use to replicate (Roossinck, 1997). The variation seen in diversity levels among plant viruses in summarized in Table 12.1.

Viroids are small, non-coding parasitic RNAs that use host RNA polymerases for replication. They are found only in plants, although the hepatitis delta agent has some similarities to viroids. Several studies indicate that viroids can have highly diverse intra-host populations, much like RNA viruses, and that levels of diversity vary significantly in different hosts (Semancik and Duran-Vila, 1999; Gandía and Duran-Vila, 2004; Vidalakis *et al.*, 2005).

TABLE 12.1 Ranges of diversity for plant virus populations

Genome type	Population type[a]	Diversity[b]
RNA	Field isolate	0.04–3.1[c]
RNA	Experimental evolution	0.05–2.7
DNA	Field isolate	0.4–1.0
DNA	Experimental evolution	0.3–0.5

[a] Population of viruses from a single host.
[b] The range of diversity reported from a number of studies is shown in mutations per kb.
[c] This does not include the diversity of banana mild mosaic virus mentioned in the text, which has up to 200 mutations per kb, because it is not clear if this represents one or more than one virus.

Studies of diversity in naturally occurring virus isolates have been done for a few viruses, mostly from field isolates of crop plants. In some studies, field isolates have been passaged after collection. Unless they are analyzed directly from the field sample the mutant spectrum can be changed significantly. In general, levels of diversity in field isolates are similar to levels seen in experimental evolution studies, but no comparisons have been made for individual viruses. A recent study of banana mild mosaic virus isolated from several different accessions of banana showed strikingly high levels of sequence diversity within single plants, with up to 20% divergence (Teycheney *et al.*, 2005). This virus is thought to be transmitted only vertically through vegatatively propagated tissue. One study has looked at the populations of a plant virus, tobacco mild green mottle virus, in a wild plant, *Nicotiana glauca*, although a precise measurement of mutation frequency was not done (Fraile *et al.*, 1996). Hence it is difficult to draw any conclusions about the diversity of viruses in their natural plant hosts, or how this diversity leads to the evolution and emergence of new crop diseases.

GENETIC BOTTLENECKS

Plant viruses have several opportunities during their life cycle to undergo the stochastic reductions of population diversity known

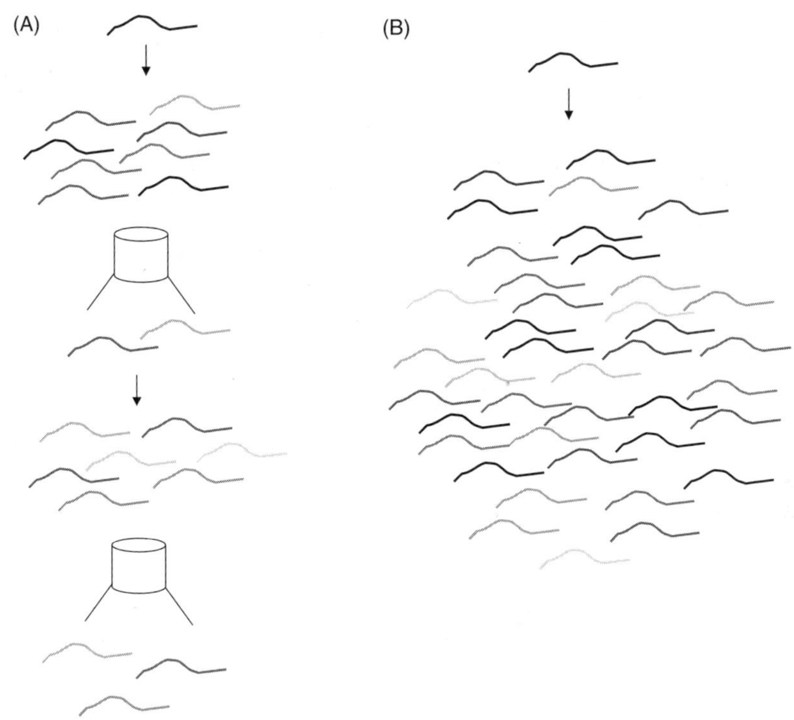

FIGURE 12.2 Schematic representation of the consequences of genetic bottlenecks. (A) The viral genome replicates and generates variants, then passes through a bottleneck. The remaining variants again replicate and generate more variants, that in turn pass through a bottleneck. (B) A population replicating without being subjected to bottleneck events, where diversity is allowed to accumulate, and is only restricted by selection. (See Plate 14 for the color version of this figure.)

as genetic bottlenecks (Figure 12.2). Natural infection is initiated by a vector, most often a plant-feeding insect such as an aphid. For most plant viruses amplification primarily occurs in the mesophyll cells. After some level of accumulation, the virus may move systemically, to other leaves of the plant. In some cases it may also move to the roots. These movements require transport into and out of the plant vascular system, a process that is tightly regulated and requires special proteins encoded by the virus. Some viruses are transmitted sexually through pollen and vertically through seeds as well. Any of these steps can constitute a genetic bottleneck.

Bottlenecks are important in virus evolution because they can result in genetic drift and in loss of fitness. When a diverse population undergoes a severe bottleneck the few variants that survive to form a founding population may not be the most adapted. This process, known as Muller's ratchet, has been demonstrated for a number of bacterial and animal viruses during artificially imposed bottlenecks (Chao, 1990; Duarte *et al.*, 1992; Escarmís *et al.*, 1996). However, few studies have been done on naturally occurring bottlenecks and their effects on virus evolution.

Three recent studies have demonstrated that RNA plant viruses undergo severe bottlenecks during the process of systemic infection. In two studies the effective population sizes were estimated based on the variants recovered from plants after infection (French and Stenger, 2003; Sacristán *et al.*, 2003). In another study a more direct measurement of bottlenecks was undertaken. Twelve marker-bearing mutants were generated in CMV to simulate an artificial quasispecies. Although all 12 mutants were always recovered from inoculated leaves of tobacco, an average of only seven were recovered from the first systemically infected leaves, and only three from the secondarily infected leaves (Li and Roossinck, 2004). This study also indicated that movement from initially infected tissue is completed within two days of inoculation, since detachment of inoculated leaves at this time did not affect the number of mutants recovered from systemically infected leaves. Hence the process of systemic movement could be considered as a single event, rather than a continuous process.

Bottlenecks during transmission events were examined using a similar set of CMV mutants. In this study the host was zucchini squash, and the vectors were two different species of aphids. Insects were allowed to feed on tissue infected with the mutant population, then transferred to fresh plants. Severe bottlenecks also were identified in this study, and the study further demonstrated that the population was not restricted during aphid acquisition, but during transmission (Ali *et al.*, 2006). No studies have been done on potential bottlenecks during sexual or vertical transmission of plant viruses.

EVOLUTIONARY IMPLICATIONS OF MUTANT SWARMS AND BOTTLENECKS

It is clear that highly diverse populations of plant viruses can be generated during virus infections, but this is not always evident in the resulting mutant swarm. In some cases, such as CCMV (Schneider and Roossinck, 2000) a virus may be atop a steep fitness peak that prevents variation through negative selection. With as few as four or five encoded proteins, most plant viruses must make the most of their genetic content, and each protein must serve multiple functions. In addition, the RNA genome itself can have important biological functions, including signals for replication and packaging. This can leave little room for flexibility, or tolerance of mutations.

On the other hand, with so few genes, a diverse mutant swarm can provide extended genetic robustness, because variants in the population may compliment each other and provide extended function. This may be the case for a virus like CMV, where the consensus sequence does not readily change, but the mutant swarm is much greater than for the closely related CCMV (Schneider and Roossinck, 2001). It seems unlikely that CCMV cannot generate mutants, but rather that mutants are rapidly removed by selection. The parasitic satellite RNAs of CMV often have very little variation, yet a change in the helper virus can result in 2% of the nucleotides changing in just ten days (unpublished data). This indicates that in spite of the lack of variation, mutations can be generated rapidly. For survival a virus must strike a balance between mutation frequency and adaptability, and optimal fitness. Perhaps viruses do this in different ways, but as yet we do not know what controls these factors.

The role of large mutant swarms in the emergence of new viruses has been frequently discussed. Mutant swarms can theoretically contribute to speciation events, especially if they are subjected to bottlenecks that result in genetic drift, but in plant viruses there is little evidence that this affects emergence. Only one truly novel virus has been reported recently as an emerging virus in plants (Verbeek *et al.*, 2007). All other cases of emerging plant viruses have been attributed to other factors, including changes in insect vector range, movement of plant material by humans, and recombination or reassortment of virus in mixed infections. Recombination or reassortment seem to be the most common methods of plant virus speciation events (Roossinck, 2005).

An important implication of mutant swarms is the generation of variants that can overcome host resistance. For plant viruses host resis-tance comes in many forms. It may be innate immunity in plants expressing portions of viral sequences, it may be incompatibility between the viral proteins and necessary host factors for replication or movement, or it may be induced after virus infection, by gene silencing. Resistance has been introduced into crop plants by breeding and by engineering. In many cases viruses have evolved to overcome resistance, including gene silencing. As we learn more about viruses of plants in natural habitats however, we may find that while virus infection is common, virus disease is rare.

Preventing virus disease is more important than preventing virus infection.

The role of bottlenecks in the evolution of plant viruses is unknown. Bottlenecks during systemic infections of plants could lead to significant drift within a single plant. In a tree infected with the potyvirus plum pox virus for a period of 13 years, widely diverse populations were found in different parts of the tree (Jridi *et al.*, 2006). Although the diversity of the initial population is not known, it seems likely that bottlenecks during systemic infection resulted in different founding populations that then established individual quasispecies in different branches.

Many plants (perhaps most plants) are infected with viruses that do not induce any visible symptoms, do not move systemically and are only transmitted vertically. These viruses could be considered persistant plant viruses, and they belong to two major groups, the cryptic viruses (Boccardo *et al.*, 1987) and the endornaviruses (Fukuhara *et al.*, 2005). Both of these groups have relatives among the fungal viruses, and they may be derived from fungal viruses that became trapped in plants during an fungal endophytic association. These viruses may exist in the absence of any genetic bottlenecks, and a study of their population diversity could provide insights into the role of bottlenecks in plant virus evolution, but unfortunately no population studies have been done for these to date. Although it is in a different family of viruses from the others, banana mild mosaic virus, mentioned above, may be similar in its lack of bottlenecks. It is transmitted through vegetative propagation, so most cells are infected throughout the life of the plant, and it is not known to be transmitted horizontally. This virus has the highest intra-host population diversity of any known plant virus (Teycheney *et al.*, 2005).

Bottlenecks during vector transmission could also contribute to speciation. Although bottlenecks are severe in single insect vector transmissions (Ali *et al.*, 2006), in a natural environment a plant can be infested with large numbers of aphids, so the deleterious fitness effects of these bottlenecks may be masked by repeated transmission events. However, these transmission events could facilitate speciation through reassortment in plant viruses. The genomic segments of plant viruses with divided genomes are usually packaged in separate particles, so transmission of a mixed infection could lead to the acquisition of two different viruses, followed by the deposition of only a subset of capsids to a new host, leading to speciation by reassortment.

The field of plant virus evolution has made great strides in the past decade. Besides the importance of understanding plant virus evolution for the successful production of crop plants, plants are hosts that are easy and inexpensive to grow, and can be produced as unlimited numbers of genetically identical hosts. Plant viruses provide an excellent experimental model system for the study of virus evolution in general, most of which can be extrapolated to other virus systems.

ACKNOWLEDGMENTS

The author was supported in full by the Samuel Roberts Noble Foundation.

REFERENCES

Ali, A., Li, H., Schneider, W.L. *et al.* (2006) Analysis of genetic bottlenecks during horizontal transmission of cucumber mosaic virus. *J. Virol.* **80**, 8345–8350.

Bawden, F.C. (1958) Reversible changes in strains of tobacco mosaic virus from leguminous plants. *J. Gen. Microbiol.* **18**, 751–766.

Boccardo, G., Lisa, V., Luisoni, E. *et al.* (1987) Cryptic plant viruses. *Adv. Virus Res.* **32**, 171–214.

Chao, L. (1990) Fitness of RNA virus decreased by Muller's ratchet. *Nature* **348**, 454–455.

Chao, L., Rang, C.U. and Wong, L.E. (2002) Distribution of spontaneous mutants and inferences about the replication mode of the RNA bacteriophage ø6. *J. Virol.* **76**, 3276–3281.

Duarte, E., Clarke, D., Moya, A. *et al.* (1992) Rapid fitness losses in mammalian RNA virus clones due to Muller's ratchet. *Proc. Natl Acad. Sci. USA* **89**, 6015–6019.

Edwardson, J.R. and Christie, R.G. (1991) *Cucumoviruses. CRC Handbook of Viruses Infecting Legumes*. Boca Raton: CRC Press. Legumis, pp. 293–319

Escarmís, C., Dávila, M., Charpentier, N. *et al.* (1996) Genetic lesions associated with Muller's Ratchet in an RNA virus. *J. Mol. Biol.* **264**, 255–267.

Fraile, A., Malpica, J.M., Aranda, M.A. *et al.* (1996) Genetic diversity in tobacco mild green mosaic tobamovirus infecting the wild plant *Nicotiana glauca*. *Virology* **223**, 148–155.

French, R. and Stenger, D.C. (2003) Evolution of Wheat streak mosaic virus: Dynamics of population growth within plants may explain limited variation. *Annu. Rev. Phytopathol.* **41**, 199–214.

Fukuhara, T., Koga, R., Aioki, N. *et al.* (2005) The wide distribution of endornaviruses, large double-stranded RNA replicons with plasmid-like properties. *Arch. Virol.*, epub 13 Dec. 05.

Gandía, M. and Duran-Vila, N. (2004) Variability of the progeny of a sequence variant of citrus bent leaf viroid (CBLVd). *Arch. Virol.* **149**, 407–416.

García-Arenal, F., Fraile, A. and Malpica, J. (2003) Variation and evolution of plant virus populations. *Int. Microbiol.* **6**, 225–232.

Ge, L., Zhang, J., Zhou, X. *et al.* (2007) Genetic structure and population variability of tomato yellow leaf curl China virus. *J. Virol.* **81**, 5902–5907.

Gibbs, A.J., Keese, P.L., Givvs, M.J. *et al.* (1999) Plant virus evolution: past, present and future. In: *Origin and Evolution of Viruses* (E. Domingo, R. Webster and J. Holland, eds), pp. 263–285. London: Academic Press.

Hull, R. (2002) *Matthews' Plant Virology*. San Diego: Academic Press.

Isnard, M., Granier, M., Frutos, R. *et al.* (1998) Quasispecies nature of three maize streak virus isolates obtained through different modes of selection from a population used to assess response to infection of maize cultivars. *J. Gen. Virol.* **79**, 3091–3099.

Jridi, C., Martin, J.-F., Marie-Jeanne, V. *et al.* (2006) Distinct viral populations differentiate and evolve independently in a single perennial host plant. *J. Virol.* **80**, 2349–2357.

Kim, T., Youn, M.Y., Min, B.E. *et al.* (2005) Molecular analysis of quasispecies of Kyuri green mottle mosaic virus. *Virus Res.* **110**, 161–167.

Kong, P., Rubio, L., Polek, M. *et al.* (2000) Population structure and genetic diversity within California citrus tristeza virus (CTV) isolates. *Virus Genes* **21**, 139–145.

Li, H. and Roossinck, M.J. (2004) Genetic bottlenecks reduce population variation in an experimental RNA virus population. *J. Virol.* **78**, 10582–10587.

Malpica, J.M., Fraile, A., Moreno, I. *et al.* (2002) The rate and character of spontaneous mutations in an RNA virus. *Genetics* **162**, 1505–1511.

McKinney, H.H. (1935) Evidence of virus mutation in the common mosaic of tobacco. *J. Agric. Res.* **51**, 951–981.

Naraghi-Arani, P., Daubert, S. and Rowhani, A. (2001) Quasispecies nature of the genome of grapevine fanleaf virus. *J. Gen. Virol.* **82**, 1791–1795.

Pita, J.S., deMiranda, J.R., Schneider, W.L. *et al.* (2007) Environment determines fidelity for an RNA virus replicase. *J. Virol.* **81**, 9072–9077.

Price, W.C. (1934) Isolation and study of some yellow strains of cucumber mosaic. *Phytopathology* **24**, 743–761.

Rodríguez-Cerezo, E., Moya, A. and García-Arenal, F. (1989) Variability and evolution of the plant RNA virus pepper mild mottle virus. *J. Virol.* **63**, 2198–2203.

Rodríguez-Cerezo, E., Elena, S.F., Moya, A. *et al.* (1991) High genetic stability in natural populations of the plant RNA virus tobacco mild green mosaic virus. *J. Mol. Evol.* **32**, 328–332.

Rojas, M.R., Hagen, C., Lucas, W.J. *et al.* (2005) Exploiting chinks in the plant's armor: Evolution and emergences of geminiviruses. *Annu. Rev. Phytopathol.* **43**, 361–394.

Roossinck, M.J. (1997) Mechanisms of plant virus evolution. *Ann. Rev. Phytopathol.* **35**, 191–209.

Roossinck, M.J. (2005) Symbiosis versus competition in the evolution of plant RNA viruses. *Nat. Rev. Microbiol.* **3**, 917–924.

Roossinck, M.J. and Ali, A. (2007) Mechanisms of plant virus evolution and identification of genetic bottlenecks: impact on disease management. In: *Biotechnology and Plant Disease Management* (Z.K. Punja, S.H. DeBoer and H. Sanfaçon, eds), pp. 109–124. Wallingford: CABI.

Roossinck, M.J. and Schneider, W.L. (2005) Mutant clouds and occupation of sequence space in plant RNA viruses. In: *Quasispecies: Concepts and Implications for Virology* (E. Domingo, ed.), pp. 299, 337–348. Heidelberg: Springer.

Sacristán, S., Malpica, J., Fraile, A. *et al.* (2003) Estimation of population bottlenecks during systemic movement of tobacco mosaic virus in tobacco plants. *J. Virol.* **77**, 9906–9911.

Schneider, W.L. and Roossinck, M.J. (2000) Evolutionarily related sindbis-like plant viruses maintain different levels of population diversity in a common host. *J. Virol.* **74**, 3130–3134.

Schneider, W.L. and Roossinck, M.J. (2001) Genetic diversity in RNA viral quasispecies is controlled by host-virus interactions. *J. Virol.* **75**, 6566–6571.

Seal, S.E., vandenBosch, F. and Jeger, M.J. (2006) Factors influencing begomovirus evolution and their increasing global significance: Implications for sustainable control. *Crit. Rev. Plant Sci.* **25**, 23–46.

Semancik, J.S. and Duran-Vila, N. (1999) Viroids in plants: Shadows and footprints of a primitive RNA. In: *Origin and Evolution of Viruses* (E. Domingo, R. Webster and J. Holland, eds), pp. 37–64. San Diego: Academic Press.

Teycheney, P.-Y., Laboureau, N., Iskra-Caruana, M.-L. *et al.* (2005) High genetic variability and evidence for plant-to-plant transfer of banana mild mosaic virus. *J. Gen. Virol.* **86**, 3179–3187.

Timian, R.G. (1974) The range of symbiosis of barley and barley stripe mosaic virus. *Phytopathology* **64**, 342–345.

Verbeek, M., Dullemans, A.M., vandenHeuvel, J.F.J.M. *et al.* (2007) Identification and characterisation of tomato torrado virus, a new plant picorna-like virus from tomato. *Arch. Virol.* **152**, 881–890.

Vidalakis, G., Davis, J.Z. and Semancik, J.S. (2005) Intra-population diversity between citrus viroid II variants described as agents of cachexia disease. *Ann. App. Biol.* **146**, 449–458.

Wren, J.D., Roossinck, M.J., Nelson, R.S. *et al.* (2006) Plant virus biodiveristy and ecology. *PLoS Biol.* **4**, 1–2.

CHAPTER

13

Retrovirus Evolution

Simon Wain-Hobson

ABSTRACT

Retroviruses manifest a very rich ensemble of genome structures. The evolution of retroviruses varies enormously, with fixation rates varying by as much as a million fold. The emergence of novel genome structures follows remorselessly with the fixation of point mutations and is most apparent for the lentivirus subgroup that has burst on the scene recently. Accordingly, bio-logic suggests that new genome structures will emerge among the lentiviruses, most notably HIV-1.

INTRODUCTION

There is little more ambiguous than words. Yet it is through words that we communicate. Take retrovirus and evolution that make up the title. Once upon a time "retrovirus" was pretty clear and referred to a group of viruses with two copies of polyA+ genomic RNA in the virion. Their genetic bauplan was three orfs, *gag-pol-env*, bounded by two long terminal repeats (LTRs). The situation was complicated a little with the identification of endogenous retroviruses in the mid-1970s and thoroughly disjointed by the discoveries in 1982 and 1983 that both hepatitis B virus (HBV) and cauliflower mosaic virus (CaMV) replicate by way of an RNA intermediate. The infectious form of both of these viruses is DNA. However, given that the history of "retroviruses" goes back to the early 1900s, that they induced cancer, while the remarkable phenomenon of reverse transcription that broke "The Central Dogma" of the day, i.e. DNA makes RNA makes protein, was established using these oncogenic retroviruses, classical "retrovirology" understandably had the advantage of terrain.

HBV was referred to as a "pararetrovirus" before becoming the prototype of the "hepadnavirus" group. CaMV lent its name to the caulimovirus group. Retrovirology has come to be synonymous with classical retroviruses.

Yet just at this moment, retrovirology was about to be discombobulated once again, and this time by an insider. Prior to HIV, the key retroviruses were oncogenic, otherwise known as the avian and murine leukemia viruses (ALV and MLV) and mouse mammary tumor virus (MMTV). The discovery in 1980 of human T cell leukemia virus (HTLV) associated with

Origin and Evolution of Viruses
ISBN 978-0-12-374153-0

adult T cell leukemia virus fitted the bill, so all was well. However, with the isolation of HIV-1, a lentivirus, and a major new disease, AIDS, indeed the infection of a generation, everything changed. It coincided with an unprecedented firepower in molecular biology that has led us to generate almost more data than we know what to do with. Within 2 years of isolation detection kits became available, with the first designer antivirals appearing by 1995. HIV is the number 1 virus for the *Journal of Virology*. The younger generations of students know little of ALV, MLV, and MMTV. As ever such a magnet can warp vision, and anything and everything new is tried and tested on HIV, a little like the involvement of every new virus in multiple sclerosis (Vandvik and Norrby, 1989; Winkelmann, 1993; Ablashi *et al.*, 2000).

"Evolution" has almost as many meanings as researchers working on it. Outside of a biological setting, evolution actually refers to dy/dx, the change of one parameter with respect to another. Of course in biology we are interested in the temporal component, hence it is dy/dt that is important. Yet for Darwin the "stuff of evolution" was adaptation, in fact repeated adaptation that over time led to something as wondrously complex as the eye probably 40 times or so.

Coming to the point, this chapter will deal with the evolution of the classical retroviruses. Evolution will be considered from the macroscopic point of view which, considering that genomics dominates contemporary microbiology, is tantamount to the generation of novel genomic organizations. And this is where we will start.

GENOME ORGANIZATION AS A STARTING POINT

Retroviruses are enveloped viruses composed of a lipid bilayer surrounding a nucleocapsid within which resides the two copies of genomic plus stranded genomic RNA. Understandably the genome must code for capsid and envelope proteins. Given the provirus, the integrated DNA replication intermediate, a reverse transcriptase and integrase are called for. It turns out that three open reading frames (orfs) are sufficient to generate a fully functional retrovirus. Retroviral genetic maps are invariably given in the proviral form because it is that of the gene (Figure 13.1). With the signal exception of the spumavirus group of retroviruses, the provirus is a single, multiply spliced gene. The words gene and orf are used interchangeably in retrovirology which is, of course, an abuse of language. The convention is that genes/orfs are referred to in italics while the corresponding proteins are given in normal type. The *gag* orf encodes the nucleocapsid proteins in the form of a polyprotein precursor. *gag* is followed by *pol*, encoding a large polyprotein incorporating the mature reverse transcriptase; RNase H and integrase proteins. Maturation of the polyproteins is the accomplishment of the viral protease that is generally part of *pol*, but can exist in a separate orf termed *pro* located between *gag* and *pol*. Uniquely for Rous sarcoma virus (RSV) is the protease attached to the C-terminus of the Gag polyprotein. The third orf is *env*, and encodes a polyprotein envelope precursor that is matured by host cell proteases. For ALV, MLV, and MMTV, the historical benchmark retroviruses, *gag* is always followed by *pol* which is always followed by *env*, and *gag-pol-env* represents the lowest common denominator for retrovirus genome organization.

However, as can be seen from Figure 13.1, this effective yet minimalist view of retroviral genomics is far from common. The most complex retroviruses bear up to nine distinct orfs encoding 15 mature proteins. Some orfs are unique to a virus, others common to a cluster. Some orfs are apparently analogues even though there is no sequence identity between them. While the *gag-pol-env* synteny is never violated, a multitude of orfs can separate *pol* and *env*. Frequently *env* is followed by 1–4 orfs, although an orf preceding *gag* is rare.

Such exuberance in terms of genetic organization is probably without precedent among small viral genomes, although coronavirus genome organization shows considerable

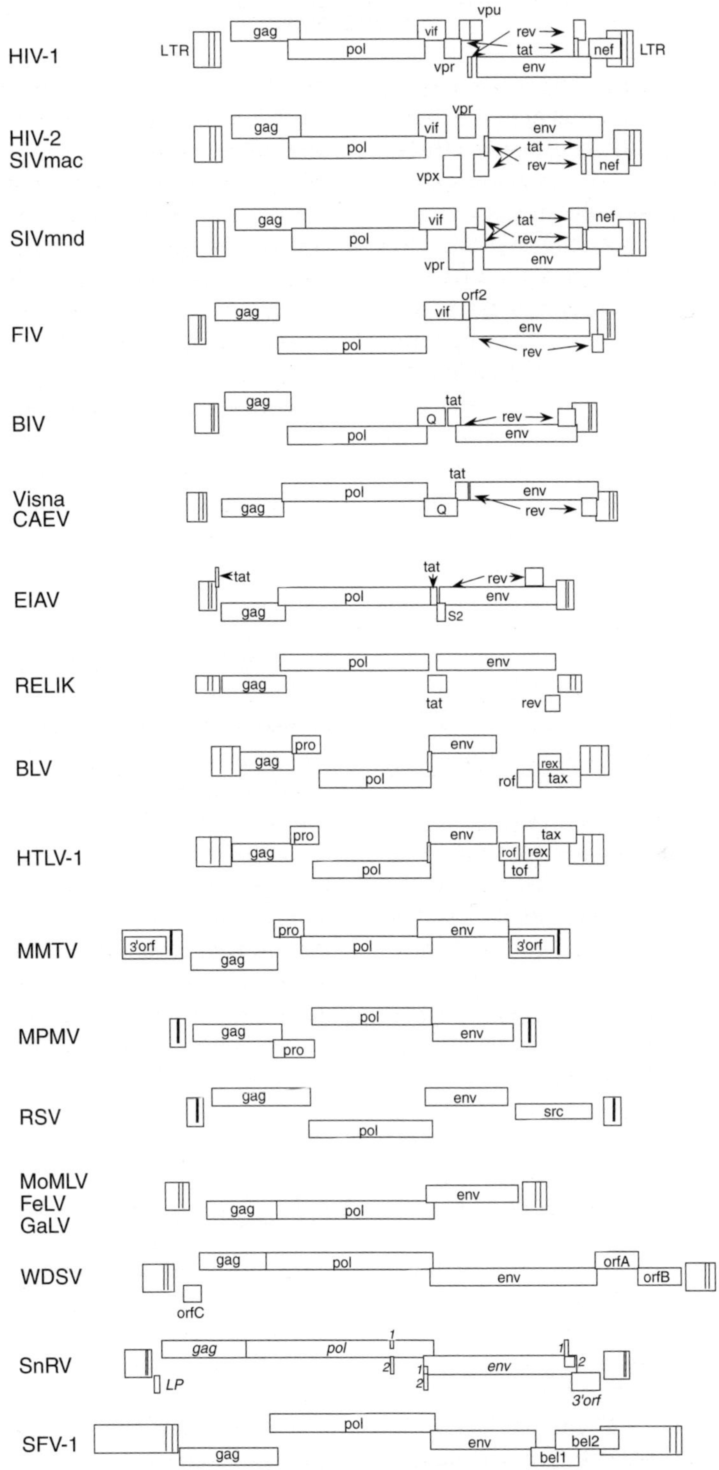

FIGURE 13.1 Retroviral genome configurations. These are given in the DNA, or proviral, form found integrated into host chromosomal DNA. The open reading frames are represented by boxes flanked by the LTR (long terminal repeat), which encodes most of the information necessary for transcription and reverse transcription. In some cases there is considerable amino acid identity, as in the case of *pol*. However, although *env* indicates the envelope orf there is no sequence identity whatsoever between, say, those of HIV, HTLV-1, RSV, and WDSV. The same is true of the retroviral transactivators *tat*, *tax*, and *orf1*. Even for a given virus, strain variation can attain staggering proportions (see Figure 13.5).

plasticity. Of course herpesviruses genomes span 119–236 kb while poxvirus genomes vary even more—134–360 kb. By contrast the threefold range in the number of retroviral orfs is accompanied by a relatively modest ~50% variation in the size of the coding region.

Where do all these "extra" genes come from? The origin of the RSV *src* gene is no secret, it being transduced from the avian genome (Stehelin *et al.*, 1976), while the primate lentiviral *vpx* gene is arguably derived from a duplication of the *vpr* gene (Tristem *et al.*, 1990). For the rest, sequence homology is non-existent. It is possible that many are of cellular origin, however, given the tempo of retroviral mutation (*vide infra*), all traces have been lost.

STABILITY OF GENE ORGANIZATION YET AT LEAST SIX WAYS TO EXPRESS POL

Apart from gene transduction, the genomic organization of a particular retrovirus is invariant. While it is possible to isolate viable strains carrying deletions say in *env* (JDV) or *nef* (HIV, SIV), deletions are common fare among microorganisms. As the provirus represents a single gene, large rearrangements and inversions are presumably deleterious. When making retrovirus-wide comparisons of the common *gag*, *pol*, and *env* orfs, the internal order of the major proteins is invariably conserved. Thus for Gag the order is always matrix antigen, capsid antigen, and nucleic acid binding protein, for Pol the order is protease, reverse transcriptase/RNaseH, integrase while for Env, the order is surface protein, transmembrane protein.

Nonetheless, there are a couple of noteworthy cases of reorganization. The first concerns the protease. As can be seen, it always precedes the reverse transcriptase/RNaseH (Figure 13.2A). However, the coding sequences can be part of *gag*, *pol* or in a separate reading frame, called *pro*. In view of this it is not surprising to learn that there are several distinct mechanisms by which these distinct reading frames are connected. The mechanisms range from −1 ribosomal frameshifting once, twice, splicing, and suppression of a terminator codon. If the larger family of retroelements are considered for once, there are two additional Gag-Pol syntenies involving +1 ribosomal frameshifing and the two being part of one long polyprotein precursor. These examples show that while the gene/orf order may be preserved, there are radical differences underlying their expression.

The second example concerns the retroviral dUTPase (McGeoch, 1990). Interestingly, only a subset of lentiviruses and the MMTV/SRV group encode such an enzyme and then in two different locations (Figure 13.2B). For the lentiviruses, it fits snugly in phase between the RT/RNaseH and integrase coding regions, while for the MMTV/SRV group it is located upstream of the viral protease. While phylogenic analysis shows that they cluster together, they do not constitute a robust monophyletic group, meaning that it is not possible to distinguish between a single introduction or independent introductions into each group (McGeoch, 1990). This is the only example of radical change in the synteny of retroviral coding sequences. Of course for large DNA viruses, such as the mycophages, marine phaeoviruses, and herpesvirus, to name but three, extensively rearranged genomes are commonplace (Weir, 1998; Delaroque *et al.*, 2003; Pedulla *et al.*, 2003).

ERROR AND RECOMBINATION

When it comes to making replication errors, DNA and RNA virus differ hugely. RNA viruses are unable to undertake proofreading or mismatch repair while DNA viruses can, either because DNA replication is performed by the host cell or else the genome is large enough to encode the necessary enzymes to undertake high-fidelity replication. In this latter context, poxviruses replicating in the cytoplasm is perhaps the most extreme example. Where do retroviruses, half DNA, half RNA viruses sit when it comes to error? Errors

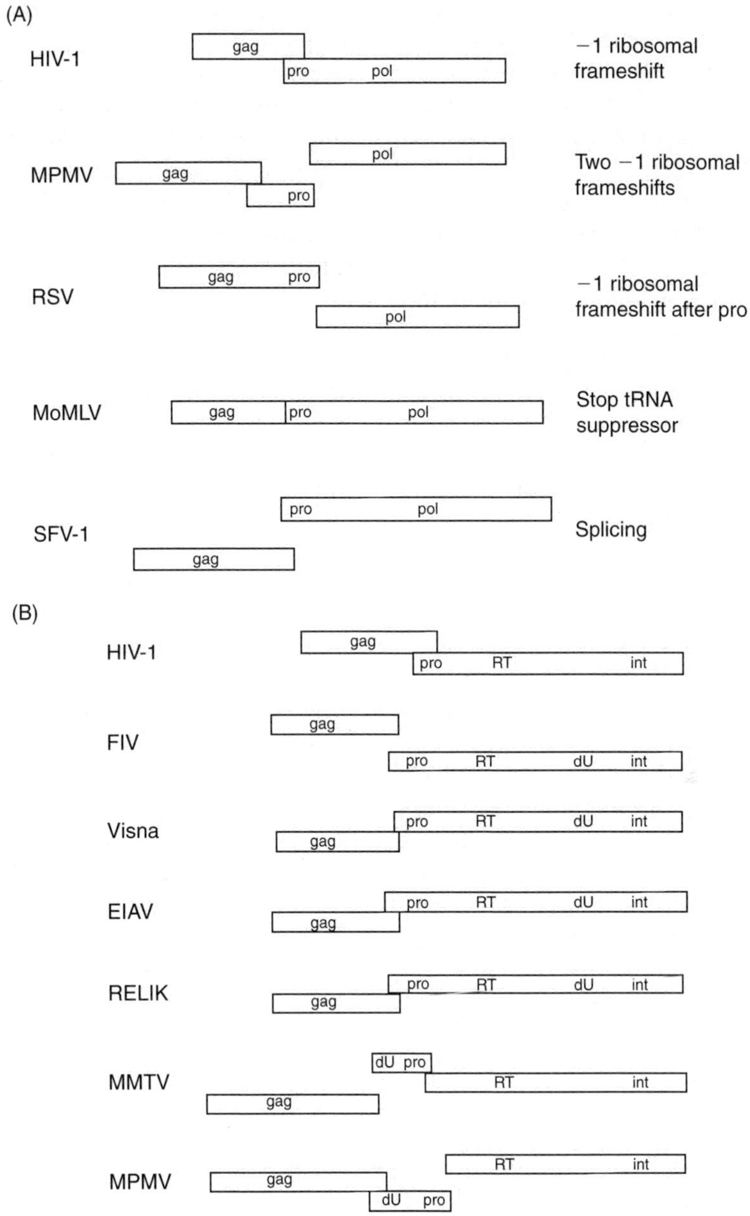

FIGURE 13.2 Variable synteny for the retroviral protease (*pro*) and dUTPase (*dU*) coding sequences. (A) Not only is the *pro* coding region in different reading frames, the mechanism by which it is expressed in the form of a Gag-Pol polyprotein precursor is radically different. (B) The dUTPase coding sequence is found only in a subset of retroviral genomes and then not consistently. For example, all the primate lentiviruses do not carry a dUTPase.

made during reverse transcription per se are not corrected whatsoever, while those made during plus-strand synthesis could be proofread when the DNA is translocated to the nucleus. It turns out that although retroviral mutation rates are in the range of 0.3–0.05 per genome per cycle (Mansky and Temin, 1995; Mansky and Wisniewski, 1998; Mansky, 2000), they are approximately an order of magnitude lower than those for RNA viruses, 1–3 per genome per cycle (Drake and Holland, 1999). While it would be tempting to ascribe this difference to proofreading of mutations of the plus-strand errors, as more errors are made during minus-strand synthesis, it probably reflects some subtle intrinsic difference between the polymerases.

As in almost all polymerases, transitions are produced more readily than transversions, while deletions are more frequent than insertions (essentially duplications). For the HIV reverse transcriptase and presumably the lentiviruses as a whole, deletions arise more frequently than for other RTases. Being diploid viruses, recombination is to be reckoned with. Retroviral recombination goes back to 1971 (Vogt, 1971; Kawai and Hanafusa, 1972; Weiss *et al.*, 1973) just one year following the description of reverse transcription. It occurs essentially by a copy choice mechanism (Coffin, 1979). Precise recombination rates have been worked out for two retroviruses, HIV-1 and MLV, the values being 3–4 per virus per cycle respectively for both viruses, despite an initially lower rate for MLV (Jetzt *et al.*, 2000; Zhuang *et al.*, 2006). Importantly, these values are ~10–100× the point mutation rate for both retroviruses. This means that as soon as a point mutation is made it will be recombined in the following round.

There are a couple of conditions to be met before concluding that everything is recombined for a retrovirus. First, infected cells have to be multiply infected, meaning that they must harbor multiple proviruses. Second, the proviruses must be genetically distinct. If, for example, a multiple infected cell received all its HIV content from a productively infected neighboring cell that harbored a single provirus, then recombination would proceed but effectively produce the parental virus despite recombination. So does multi-infection occur and are the proviruses divergent *in vivo*?

We have only the beginnings of an answer to these questions and then only for HIV-1. Using fluorescent *in situ* hybridization (FISH) a single report showed that for two patients, one with high and one with low plasma viremia, the average proviral copy number in splenic mononuclear cells was 3–4 with a range of 1–8 (Jung *et al.*, 2002). The majority of cells (~72%) were multiply infected. Laser microdissection of the FISH positive nuclei and transfer to PCR tubes allowed amplification and sequencing of the HIV-1 DNA. The sequence complexity of the V1V2 hypervariable regions analyzed, probably the most sensitive indicator of Env variation, was stunning. Up to 28% amino acid variation was found within a single nucleus. This is greater than the average protein variation between humans and birds! To add to a complex situation, numerous distinct recombinants were also found within the same nucleus.

What does this mean? First, it indicates that phenotypic mixing is possible, generating heterogeneous virions (Figure 13.3B). Second, it means that HIV virions are non-clonal objects. Worse, these non-clonal structures are ephemeral as production, infection and reverse transcription involving recombination are a matter of 10–18 hours (Figure 13.3). Such complexity inside a single nucleus (synonymous here with boundary or discrete volume) indeed conjures up the quasispecies. Indeed, is this the true quasispecies where genomes and gene products clearly interact? The population of genomes is of course limited to two or at most eight functional genomes, which can produce a rich variety of heterokaryons. However, they will recombine in the next round of infection producing a new ensemble of recombinant genomes. This leads to the third point. How is fitness maintained? Perhaps this question should be rephrased into a more prosaic form, what fraction of recombinants is neutral and what fraction deleterious? The concise answer is we simply don't know. However, when recombinants of HIV and SIV are made the resulting "SHIV" that is eventually recovered invariably carries "additional" substitutions as though the jump in sequence space made by the molecular virologist landed the SHIV in a relatively unfit region. It seems possible that most recombinants may in fact reduce fitness. Certainly, it seems logical given that functional sequence space can represent only a small fraction of total sequence space. Could the retroviral

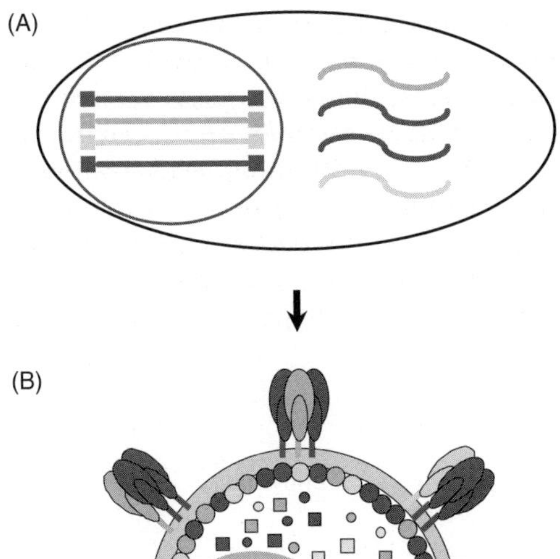

FIGURE 13.3 The non-clonal and ephemeral nature of HIV. (A) A cell bearing four genetically distinct proviruses is represented by different shading. (B) Upon productive infection, only two RNAs can be packaged per virion and there is no reason to consider that proteins from all four proviruses are incorporate into the virion. The resulting structure is much like a harlequin. No two virions will be the same. It is not known if this impacts the biology of the virus. (C) Upon infection of the target cell, recombination occurs leaving a mosaic provirus. All this occurs in as little as 18 hours. (See Plate 15 for the color version of this figure.)

Red Queen be constantly running to make up for the devastating effects of recombination? Could this be the major driving force leading to the relentless accumulation of point mutations for HIV, no matter the multiplicity of infection, the disease stage, or the nature of existing immune responses?

At the more macroscopic level, recombinants are found at all levels of HIV and SIV—intrapatient recombinants, those among different strains of the same clade, the ever-expanding number of circulating recombinant forms (CRFs) worldwide that can be readily shown to be mixes of the original founder clade viruses, recombinants of CRFs, intergroup recombinants, e.g. HIV-1 M-O (Peeters *et al.*, 1999), and a plethora of recombinants between HIV-1 precursor viruses with SIVs of chimpanzees (Bibollet-Ruche *et al.*, 2004). Many of the diverse SIVs from small equatorial monkeys are demonstrably recombinants (Jin *et al.*, 1994). Indeed we are probably at a level where the null hypothesis should be that everything is recombinant unless shown otherwise.

One of the hottest topics in retrovirus variation right now is APOBEC3-mediated hypermutation, which is extensively treated in the excellent chapter by Warner Greene (Chapter 8). Briefly, these are cellular cytidine deaminases that if packaged into the virion can edit hundreds of cytidine residues, leaving behind uracil. This is tantamount to a host restriction system. It is particularly a problem for the lentiviruses to the point that they have evolved the *vif* gene that prevents APOBEC3 incorporation (Sheehy *et al.*, 2002). While other retroviruses including HBV do undergo editing *in vitro* and *in vivo* (Mahieux *et al.*, 2005; Suspene *et al.*, 2005; Delebecque *et al.*, 2006), there is no evidence that this seriously impacts their replication. For the mouse there is a single APOBEC3 gene and knockout mice are both viable and fertile, which suggests that it does not restrict the transposition of endogenous retroviruses (Mikl *et al.*, 2005). This fits in well with the fact that although the bird genome carries a plethora of endogenous retroviruses, there is no APOBEC3 counterpart (Harris and Liddament, 2004). Might a little APOBEC3 editing help the lentiviruses evolve? Could it occasionally act as the equivalent of a bacterial mutator? As is frequently the case, speculation is understandably inversely proportional to the amount of data.

RETROVIRAL FIXATION RATES VARY HUGELY

It comes as no surprise to any virologist that fixation rates vary with the protein under study—the flu paradigm is known to all. To get this quickly out of the way, the retroviral surface envelope protein is generally the most variable, although some of the small accessory proteins fix mutations at a comparable rate. The nucleocapsid Gag proteins are understandably less tolerant of change only in the temporal sense, while the enzymes that make up Pol bring up the rear. The retroviral integrase has the lowest amino acid fixation rate and so it not a surprise that this region is invariably used to perform retrovirus-wide phylogenic analyses. A typical phylogenic tree based on the integrase gene is shown in Figure 13.4A.

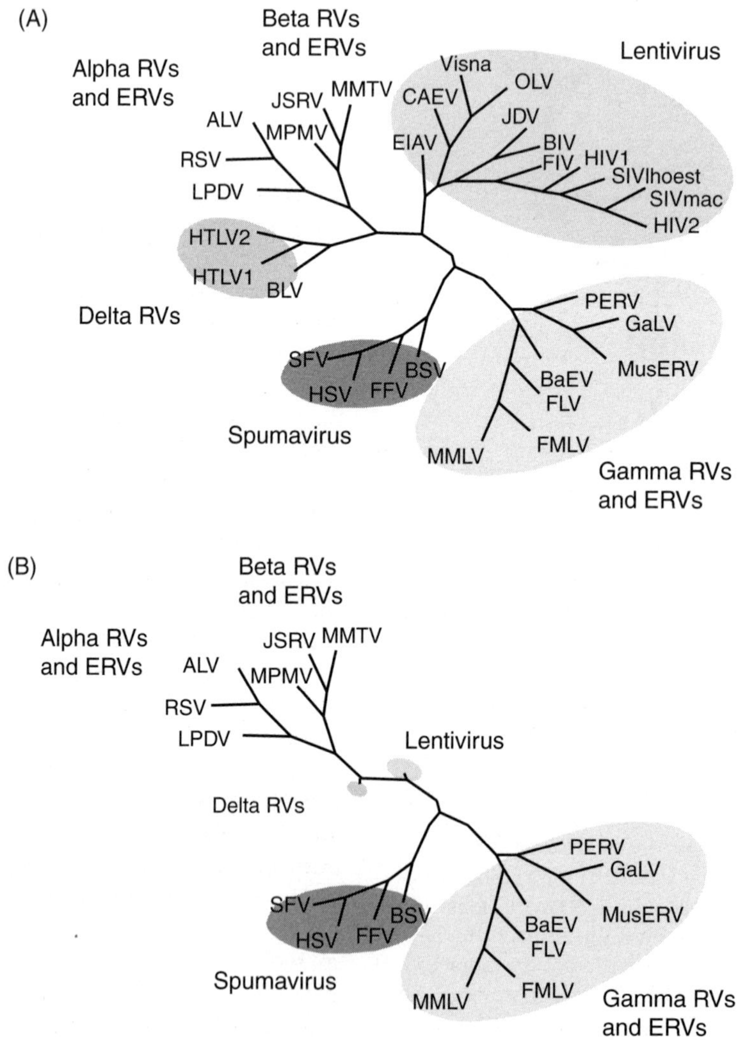

FIGURE 13.4 (A) Standard Protpars tree for the retroviral integrase sequence. Shading helps distinguish the retroviral families. (B) The lentiviral and HTLV lineages only have been shortened to take into account their much faster fixation rates compared to that for the spumavirus lineages. It is tempting to conclude that in a temporal sense the lentiviruses and delta retroviruses have recently emerged from some sort of endogenous retroviral baseline. (See Plate 16 for the color version of this figure.)

However, there is a complication. Between retroviruses, fixation rates vary hugely. For the lentiviruses, read HIV/SIV simply because the databases are so much larger, the amino acid fixation rates are of the order of 0.1–1% per year depending on the protein (Johnson *et al.*, 1991). For HTLV overall fixation rates are more like 1% per century while for the foamy viruses or spumaviruses it estimated that the rates could be as slow as ~2% per million years (Switzer *et al.*, 2005). Obviously this cannot be ascribed to the RTase as the HTLV point mutation rate is $\sim 7 \times 10^{-6}$ per base per cycle (Mansky, 2000) which is only four-fold down on that for HIV ($\sim 3 \times 10^{-5}$). It transpires that viruses of the HTLV family replicate a little by reverse transcription and extensively by Tax-driven clonal expansion of the host cell, i.e. mitosis. Tax is a viral protein that impacts the cell by tripping it into cycling even during the non-malignant carrier state. Quite why the foamy viruses fix mutations so slowly has not been clearly worked out, yet it might well involve clonal expansion, as the consensus opinion is that retrovirus and RNA polymerase complexes are not accompanied by proofreading enzymes.

Nonetheless the consequences for phylogeny are considerable. Understandably a phylogenic tree being built essentially from a distance matrix, is a scalar construction. Yet in our mind's eye it is inevitably read in a temporal sense, which is also understandable, for after all we are interested in evolution, i.e. dx/dt. There is the implicit assumption that lineages are lengthening at roughly equal rates. However, when the rates differ by several orders of magnitude scalar trees hide information. If the lentiviral and HTLV and normalized to the spumavirus lineage as a function of their fixation rates, then both the large and diverse lentivirus group as well as the small HTLV group collapse to next to nothing (Figure 13.4B). In this tree, the other retroviral lineages have been kept constant either for lack of data about fixation rates, or else the possibility of recombination with endogenous retroviruses is such as to confound any evaluation. How are we to interpret the hobbled lentivirus and HTLV viruses? As mere blips in retrovirology? If so they certainly represent a recent blip from some sort of a retroviral base line. Yet given the complexity of lentiviral genomes, considerable changes in synteny have occurred in a very short time compared to the relatively fewer differences in the spumavirus group over much greater periods of time.

HOW FAR CAN VARIATION GO FOR A RETROVIRUS?

Given the phenomenal fixation rates, particularly for the lentiviruses, the question often asked is, are there limits to the change a virus can absorb? This was dealt with extensively in my chapter of the original edition of this book and so there is no need to belabor the point—functional sequence space is so vast that there are effectively no limits. One example will suffice: feline immunodeficiency virus produces profound immunodeficiency and an AIDS-like disease in domestic and big wild cats. A simple alignment of the envelope polyprotein precursor from the PPR and Pallas cat isolates shows a stunning 10% match for the complete Env sequence (Figure 13.5) and approaches the 5% expectation value given 20 amino acids. Despite this both bear all the hallmarks of a lentiviral envelope sequence—they are highly glycosylated, rich in cysteine residues, have a long cytoplasmic tail and contain the requisite three hydrophobic segments. In general for retroviral proteins, the envelope proteins are the most divergent while the enzymes encoded by *pol* are the least variable. Thus the rate of fixation of integrase is perhaps a factor of 10 less than for envelope (Johnson *et al.*, 1991). So long as the protein fold and a few crucial residues are preserved, the rest will, and does, change given time.

To appreciate this it is worth mentioning a contrary situation. The "Paracelsus challenge" defied or encouraged researchers to radically change the fold of a known protein by changing less than 50% of its amino acid

```
       1         10        20        30        40        50        60        70        80        90        100
       •         •         •         •         •         •         •         •         •         •         •
OMA    MAEGGRVDVVERADEELGRQGVEGHEYAFGMNPDWIGPYEGEMLLDFDILQYVTEEGPFRPGHNPFRAPGITEQERQELCVMLQDKLKEIKGTITEGPH-
PPR    MAEG-----------------------FAANRQWIGPEEAEELLDFDKATQMNEEGPLNPGVNPFRVPAVTEADKQEYCKILQPRLQEIRNEIQEVKLE
       ****                       *  *  **** * * *****      ****  ** **** *  **   ** *  **  * **   * *

OMA    KIPPGKYRRLRYLQYSDMQVTQSLALLVFDISHYLRNKLGKEVYDIEGDRQAEYKFE----KRVKGRTYNNCRCRLLLIGAGFFYTCLIIGLGCLIRETS
PPR    EGNAGKFRRARFLRYSD-ESILSLIHLFIGYCTYLVNRR--RLGSLRHDINIEAPQEEQYSSREQGTTENIKYGRRCLIGTASLYLLLFIGVAIYLGTTN
           ** ** * * ***     ** *     **                *   *   *     *  * * *    *  ***        * **      *
       > surface protein
OMA    GVILALDPPWVIPVTKMDEINFQCHGNYEECPVLESVATWKTDFQWNYSRPFNETIGLEQYVDQIQAKALQDLLGSCQK--------------------
PPR    AQIVWRLPPLVVPVEESEIIFWDCWA--PEEPACQDFLGAMIHLKASTNISIQEGPTLGNWAREIWGTLFKKATRHCRRNKIWKRWNETITGPVGCANNT
         *    ** * **     *   *     *                    *   *      *           *

OMA    LSKNKLGVLQWRCFYDRGMKQLLGLQKIRICPIGGYMLVRKIDENNYTLSMCTEEIDIKILNMTLSQE---KYEHYPFND--IVWMGNRYFNMTTANITQ
PPR    CYNISVIIPDYQCYLDRVDTWLQGKVNISLCLTGGKMLY---NRDTKQLSYCTDPLQIPLINYTFGPNQTCMWNTSQIQDPEIPKCGWWNQIAYYNSCRW
                   * **    * *   *  *  ** **        ** **    *  ** *                  *  *   *

OMA    QQVNISIKCDIIVPTVVKVKKEFAGYNNDFLGPWGGLKYRSILIRYKDWANVTDPPLDLNCTGLPGIALNGTEANYTCAQNATITYGDICTQPELYVPC-
PPR    ESTNVKFYCQRTQSQPGTWIRTISSWRQKNRWEWRP-DFESEKVKISLQCNSTHNLTFAMRSSGDYGEVMGAWIEFGCHRNKSR----FHTEARFRIRCR
          *    *                        *      *        * *                          *  *       *      *

OMA    YSPNYSMPVMVQCKLHQEYHPNDTYRNSSNDMQVMRCRIMKEVELRFGDEFISLNFTLLRDPFLAHLRGAINFTCNLTG-QFWAYKFNNATWGYEGNGSA
PPR    WNVGDNTSLIDTCGKNLNVSGANPVDCTMYANKMYNCSLQNGFTMKVDDLIMHFNMTKA----VEMYNIAGNWSCKSDLPQNWGYMNC--------NCTN
                 *                    *          *    * *              *  *     * * *           *
                                                      surface protein <    < transmembrane protein
OMA    WNESLNWLVPYRNYTKEMYVWGAYSAINYNHILLKDYKLVKKPLYTPLKYLPP----RKKR    GLGLTLALVTATTAGLIGTTTGTSALAVSLKLKE
PPR    GTSNDNKMACPE---DKGILRNWYNPVAGLRQALEKYQVVKQPEYIVVPTEVMTYKYKQKR     AAIHIMLALATVLSIAGAGTGATAIGMVTQYQQ
           *                *         *  *  ** * *               **         *  **  * *  ** *

OMA    VMLQQSQINEATLGMLKILQRRLKQAERVILTLHQRVSRIERYLEIQYQLRGMCPFKD--ICEIP----GNGNFTNYNDSWAIG-----RWAEQAEEDWQ
PPR    VLATHQEALDKITEALKINNLRLVTLEHQMLVIGLKVEAIEKFLYTAFAMQELGCNQNQFFCEIPKELWLRYNMTLNQTIWNHGNITLGEWYNQTKYLQQ
       *               ***  **  *   *     *  ** *                ****      * *      *  *        *  *    *

OMA    QFGTIVKQCNRTNENLKNDLEKLS-IDSWLSWNP-----LGNVFQMLITLIIIIGMGVILKGCILNCCK-ILMASMGYKRVAEEMVILPDSELDSE----
PPR    KFYEIIMDIEQNNVQGKQGLQKLQNWQDWMGWIGKIPQYLKGLLGGILGIGLGILLLILCLPTLVDCIRNCISKVLGYTVIAMPEIDDEEETVQMELRKN
        * *     *  *  * **   *  *      *           *              *         **   *           *

OMA    ------SEIELN-VTEKE-----KKPMVN-SGKEESDE-EF-------
PPR    GRQCGMSEKEEECNTENTDISPSITINFRMDATANAMEINFILEDATA
             ** *    **                        *  *
```

FIGURE 13.5 Envelope variation for two strains of feline immunodeficiency virus (FIV). The accession numbers for the PPR and Pallas cat OMA strains are M36968 and U56928 respectively. Amino acid identities are scored as asterisks. The membrane spanning segments are underlined, as are the *N*-linked glycosylation sites. Despite the differences, which include extensive changes in the positions of the disulfide bonds, both were derived from viable strains.

residues (Rose and Creamer, 1994). The challenge was first met in 1997. The β1 domain of the *Streptococcal* IgG-binding protein G which displays an extensive β-sheet structure was redesigned and transformed into the *E. coli* protein Rop, a homodimeric four-helix bundle protein (Dalal *et al.*, 1997). The first solution was perhaps not surprisingly called Janus. This designer protein is a fascinating example of cultural evolution impinging on biology. It represents a statistically improbable saltation as opposed to, the familiar and presumably "incremental" polymerase mutations that allow FIV Env to drift beyond way beyond the 50% level separating Janus from its natural counterpart.

PHYLOGENY

Retrovirology has been a hotbed of methodological improvement in the realm of phylogeny over the past decade. A rich variety of methods have emerged; those which have been particularly useful being those enabling the detection of complex recombinant structures. Despite this problems remain. Many simple assumptions and corrections haven't been developed using real data sets. For example, a complete retrovirus tree includes a branch for the lentiviruses and the δ-retroviruses (HTLV-BLV group) that have profound differences in base composition, codon usage, mutation matrices, and fixation rates as mentioned above. Applying corrections for back mutations is not easy because there are hardly any good data sets on which to train methods.

Recombination is a major problem for sequences jump repeatedly, rather than gradually diffuse or radiate through sequence space. Of course sequence space is of such high dimensionality that words essentially fail us, words like jump, diffuse, and radiate being used vainly to communicate some sort of sense. Recombination introduces homoplasies that are otherwise relatively rare (Pelletier *et al.*, 1995; Wain-Hobson *et al.*, 2003). Ignoring these tends to inflate the number of mutations in a data set, which means that the sequences would appear older than they really were. Unfortunately network analyses that allow a clear description of recombination work only on small numbers of informative sites and small data sets. Nonetheless they show that recombination can sometimes account for up to 50% of all substitutions in a sequence set (Wain-Hobson *et al.*, 2003). For HIV, the simplest conclusion was that the number of true mutations in a set of PCR-derived sequences is equal to the number of variable sites.

An additional problem plagues HIV phylogeny. It is clear that relatively old proviruses can be reactivated from the carrying T cell, presumably by antigenic stimulation (Bello *et al.*, 2005). This means that not all progeny are derived from the most recently infected cells. This would serve to reduce the intra-patient cross-sectional diversity, and perhaps underestimate the time of infection. When sequences are close they are scored as such by whatever phylogenic method. Equally, when sequences are highly divergent they remain so. However, interpreting deep relationships or asking too precise a question is going to be fraught with assumptions.

GENOME DESIGN IMPACTS THE TEMPO OF EVOLUTION

The design of a retroviral genome can strongly impact replication and hence evolution. It has been known for some years that the HTLV-1 protein Tax transactivates a number of host cell genes, among them the IL-2 and IL-2 receptor genes helping set up an autocrine system (Yoshida, 1993). It has been shown that Tax intervenes directly in cell proliferation by inhibiting the cyclin D/Rb/p16INT pathway. A consequence of this, clonal expansion of HTLV-1 provirus-bearing T cells, has recently been shown via PCR-based studies of the HTLV-1 integration sites (Wattel *et al.*, 1995). In asymptomatic carriers and patients with neurological disease devoid of malignancy, clonal expansion was found to be the norm. That HTLV-1 replicates mainly via mitosis reconciles the problem of the simultaneous occurrence of high proviral load with genetic stability as proviral synthesis is accomplished essentially by the host cell replication machinery, which is endowed with proofreading systems.

However, this raises another problem: how does a clonally expanded infected cell produce little virus? Here the answers are less complete. Intrinsically, HTLV-1 production is low compared with that of HIV and is an inevitable consequence of their different genome organization (Figure 13.6). Both HIV and HTLV-1 encode analogous viral transcription transactivators, Tat and Tax, and temporal regulators of splicing, Rev and Rex respectively. The regulator proteins have a negative effect upon Tat and Tax which also curtails expression of the genomic RNA, and the large mRNA encoding the key virion proteins, Gag, Pol, and Env. HIV encodes both a Rev-dependent and -independent *tat* mRNA (Figure 13.6) allowing continual Tat-driven transcription and Rev co-ordinated splicing, and consequently continued expression of virion, *gag* and *pol* mRNAs. By contrast, the HTLV-1 analogues Tax and Rex are encoded by the same mRNA (Figure 13.6). Thus Rex tends to increase Gag, Pol, and Env expression at the expense of both Tax and Rex. The result is an equilibrium with greatly reduced virus production. In short, via genome organization, molecular switches with negative feedback loops can be generated which grossly

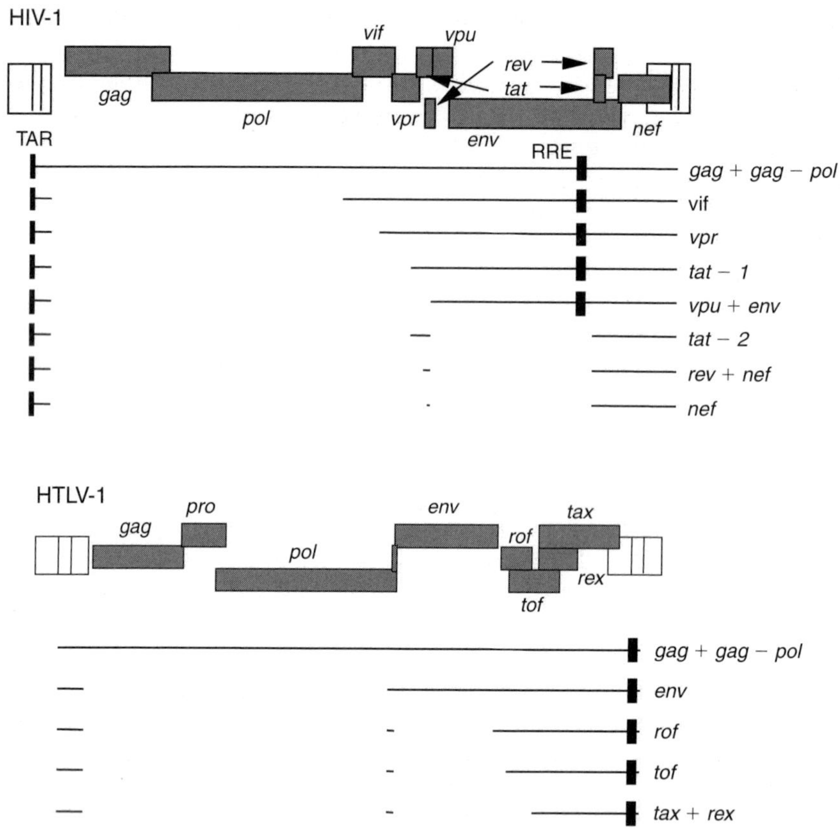

FIGURE 13.6 Transcription maps of HTLV-1 and HIV-1. Below the genome configuration is the transcription map, each line designating a unique mRNA. When broken, the mRNA is made up of ≥2 exons. The HTLV-1 transactivator, Tax, acts on a DNA element in LTR U3 region and cellular genes. That of HIV, Tat, recognizes the RNA TAR element in the R region of the LTR. The analogous HTLV-1 and HIV-1 regulators of splicing, Rex and Rev, act on specific intragenomic RNA structures, RxRE and RRE respectively. Note that Tax and Rex are derived from the same mRNA. The HIV-1 Tat protein comes in two functionally equivalent forms of 72 and 86 residues. The 86 residue form (exons 2+3) comes from a small Rev-independent mRNA while the 76 residue form (exon 2) comes from a large Rev-dependent mRNA.

impact the way a virus replicates and evolves. One final point must be added; the Tax-transactivated HTLV-1 promoter is much less powerful than the Tat-transactivated HIV-1 LTR.

ENDOGENOUS RETROVIRUSES AND MULTIPLE INTRODUCTIONS

Within the phylogenic analysis above (Figure 13.4) are featured a number of complete endogenous retrovirus (ERV) sequences. These range from recently inactivated proviruses such as the HERV-Ks to baboon endogenous retrovirus (BaEV) that circulates in the peripheral blood of baboons. The retrovirus is produced but unable to infect its own cells, a simple example of negative selection against insertional gene inactivation. As can be seen the majority of these mammalian ERVs are clustered into two groups for which there are exogenous retroviral counterparts. Massive PCR-based efforts have shown that the extent of ERVs spreads across a vast spectrum of the living world, from insects, turtles, snakes, and of course mammals (Gifford and Tristem, 2003; Belshaw *et al.*, 2004, 2007; Gifford *et al.*, 2005). There is no doubt that retroviruses have invaded germinal cells many times in the past. Equally there is no doubt that some ERVs can escape their host and infect another species, the case of gibbon ape leukemia virus (GaLV) being exemplary. Its closest relative is a murine ERV (Wolgamot *et al.*, 1998). While there were numerous reports of endogenous counterparts to the lentiviruses, foamy virus, the alignments were borderline (Cordonnier *et al.*, 1995).

A recent and glorious surprise is afforded by the aptly named sequence RELIK, for rabbit endogenous lentivirus, where the K refers to the tRNA lysine (K) primer used to initiate reverse transcription (Katzourakis *et al.*, 2007). Through a bioinformatics analysis a number of extensively degraded copies of an endogenous retrovirus were identified in the rabbit genome. After careful reconstruction of a consensus sequence it transpired that these relics were those of an endogenous lentivirus, the first unambiguous example to date (Figure 13.1). A simple comparison of the genome organization with named orfs is enough to convince. What is so nice about RELIK are the fine details that any lentivirologist would pick up in a moment. For example, the envelope protein is highly glycosylated, full of cysteine residues, has a furin-like cleavage site and encodes a long intracytoplasmic tail. It encodes a dUTP sequence precisely located between RTase/RNaseH and integrase (Figure 13.2). The primer-binding site (tRNAlys3) and polypurine tract are typical of a lentivirus.

Despite all the above traits, in terms of gene organization it is the least complex lentivirus genome. Apart from *gag*, *pol*, and *env*, there are only the *tat* and *rev* genes. There are none of the so-called, yet misnamed, accessory genes typical of the primate lentiviruses. It is tempting to consider RELIK as a precursor to the lentiviruses, although one must be ever careful of the possibility of loss of function. A phylogenic analysis based on the RTase/integrase domains shows it to be the closest relative to the lentiviruses, yet it is the outlier to all extant lentiviruses. Given that RELIK is the outlier, it is plausible that the monophyletic lentiviral group is derived from some ancestor of this lentivirus. This concords with the relative simplicity of the genetic organization of RELIK.

From a detailed analysis of the integration sites, the authors estimate that the invasion of the rabbit germline might have taken place some 7 million years ago. As there is no reason to doubt this finding we are placed in a very interesting evolutionary spot. If the present set of lentiviruses have remained exogenous for so long, then from what is known of the fixation rates of the primate lentiviruses, of the order of 0.1% amino acids replacements per year for the most conserved integrase gene, then it should not be possible to find 40% amino acid sequence identity between the HIV-1 and RELIK integrases. In the light of RELIK it seems more parsimonious to assume that the lentiviruses are very recently derived from an endogenous lentiviral RV and have been mutating furiously ever since. As mammalian genomics studies develop rapidly, the hope is that a much less degraded endogenous retrovirus will turn up in the near future.

BACK TO THE FUTURE—OR RECOVERING RETROVIRAL ANCESTORS

Even though the RELIK sequences were highly degraded, a consensus sequence was derived that is qualitatively satisfying. The ultimate would be to synthesize the provirus in the hope of resurrecting a functional lentivirus. Successful recovery of an active human ERV (HERV-K) was recently achieved following mutagenesis of a minimally defective HERV-K molecular clone (Lee and Bieniasz, 2007). Other workers have been calculating consensus or ancestral sequences for the HIV-1 envelope protein in the hope of making an immunogen that is equidistant to a maximum number of strains circulating worldwide. Chemical synthesis of the entire coding sequence and cloning into an expression vector generated functional envelope proteins that were remarkably "HIV-like" (Nickle *et al.*, 2003; Rolland *et al.*, 2007; Kothe *et al.*, 2007). Of course, the ultimate in this area is the complete chemical syntheses of an infectious molecular clone of poliovirus (Cello *et al.*, 2002) and subsequently ΦX174.

The interesting evolutionary point here, which is a general one, is that total chemical synthesis of a virus, or "therapy," of a partly degraded retrovirus constitutes a hiatus in Darwinian evolution. The mantra has always been "descent with modification," and with reason. Otherwise stated there is an uninterrupted series of DNA replication rounds linking this author to one of the earliest single-celled DNA organisms floating around 3 billion years ago. Total chemical synthesis takes biological evolution into the realm of cultural evolution, something totally new. While the resulting viruses *are* poliovirus and ΦX174, from an evolutionary setting where biology and history are inseparable, they are worlds apart. No other discipline is as advanced. The bacteriologists threaten to make a "minimalist bacterium," but so far they are not there.

In fact ever since 1962 and the demonstration that phenol-extracted papillomavirus DNA could be infectious (Ito, 1962; Chambers and Ito, 1964), virologists have been playing games with the Darwinian mantra.

HOW GOOD IS THE RETROVIRAL TREE?

Given that retroviral replication is endlessly interrupted by recombination, while many of the corrections applied to phylogeny are rather ad hoc and approximate, is it possible that the phylogeny is completely wrong? How can it be checked? It is possible to encode retroviral organization in terms of a series of answers to questions that command discrete responses, for example yes or no, that can be coded as 1 or 0. Does *env* overlap *pol*? Does the genome encode a *bet* orf? Features like the density of glycosylation sites in the Env protein, which in fact varies hugely, represent a continuous variable which is not useful. Questions with discrete answers have the advantage of being independent of the precise nucleotide sequence, polymorphisms, multiple substitutions, and saturation.

When this is done for the retroviruses (Renoux-Elbe *et al.*, 2002) it turns out that the arborescence is reasonably similar to that provided by the retroviral integrase gene, although there are a few differences (Figure 13.7). However, overall it is a comforting result. The answers can be overlaid on the non-classical tree (Figure 13.8). The minimal retroviral organization was always *gag-pol-env* and so there are no surprises that this configuration exemplified by MLV is found branching near the base of the tree. However, the lentiviruses turn out to be the largest and most varied tree. Indeed there is almost as much novelty within this group as in all the others and throws into light the trivial distinctions that were used in the past to lump or split retroviruses. The minimal lentivirus turns out to be *gag-pol-tat-rev-env* with the acquisition of *vif* coming immediately after. However, as all but one lentivirus (EIAV) encodes a *vif* gene, this could be a legitimate case of loss of function. Since the tree in Figure 13.8 was calculated, "RELIK" was discovered (Katzourakis *et al.*, 2007). As mentioned above, RELIK has this minimalist lentivirus organization, i.e. *gag-pol-tat-rev-env* yet does not encode a *vif* gene. This is consistent with the hypothesis that *vif* was a subsequent acquisition compared to *tat* and *rev.*

As always care has to be exercised when interpreting trees. Basically sequence space is so huge and multidimensional that simple two-dimensional constructs cannot do justice to the complexity. While not shown here, many of the bootstrap values are rather small, indicating that parts of the tree, particularly towards the root, are not robust (Renoux-Elbe *et al.*, 2002). It should perhaps be mentioned that the endogenous retrovirus HERV-K encodes a *rev*-like orf. It is of course tempting to conclude that *rev* appeared before *tat.*

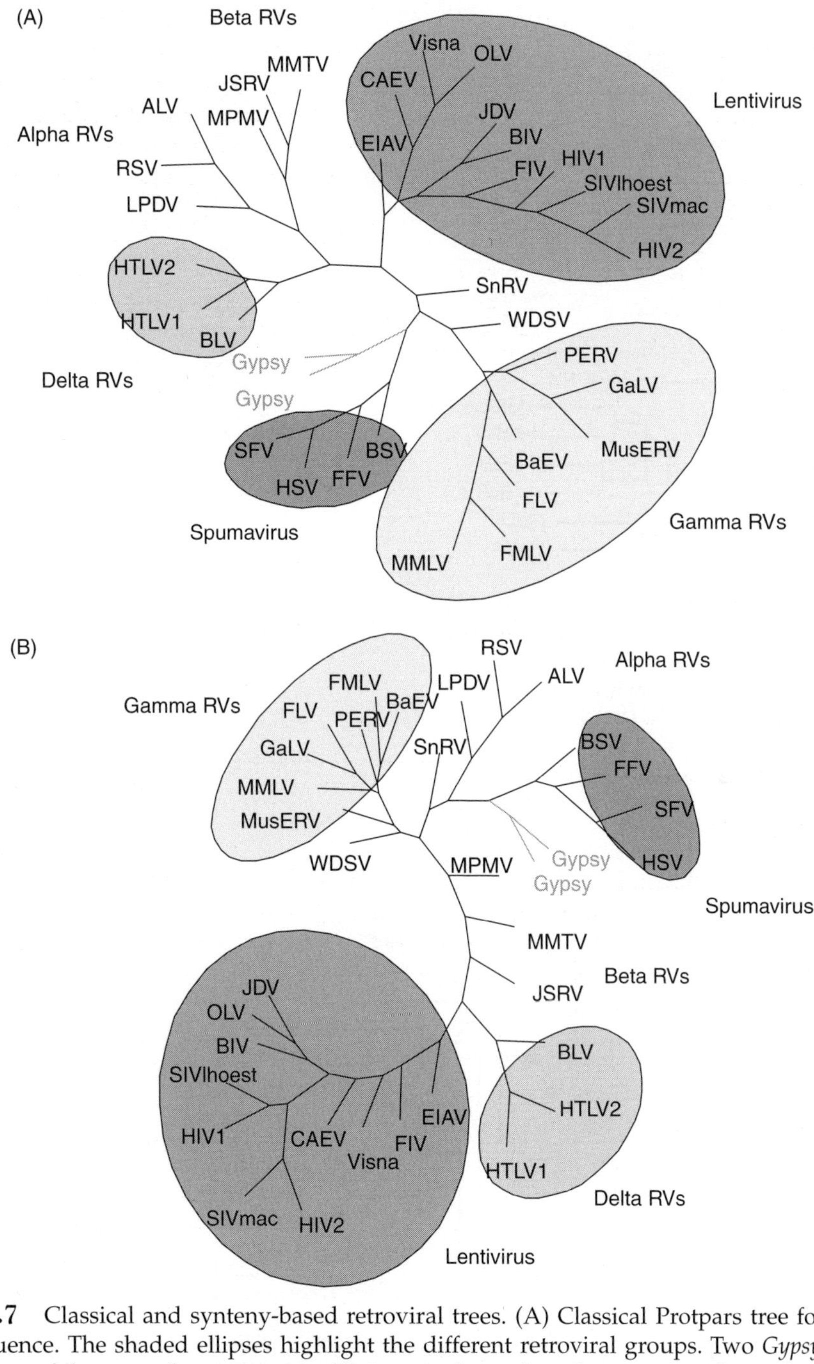

FIGURE 13.7 Classical and synteny-based retroviral trees. (A) Classical Protpars tree for the retroviral integrase sequence. The shaded ellipses highlight the different retroviral groups. Two *Gypsy* retroviral elements from *Drosophila* are used as outgroups. (B) A syntenic tree based on a series of questions giving binary (1, 0) answers. While there is general agreement, the positions of the alpha- and betaretroviruses vary. (See Plate 17 for the color version of this figure.)

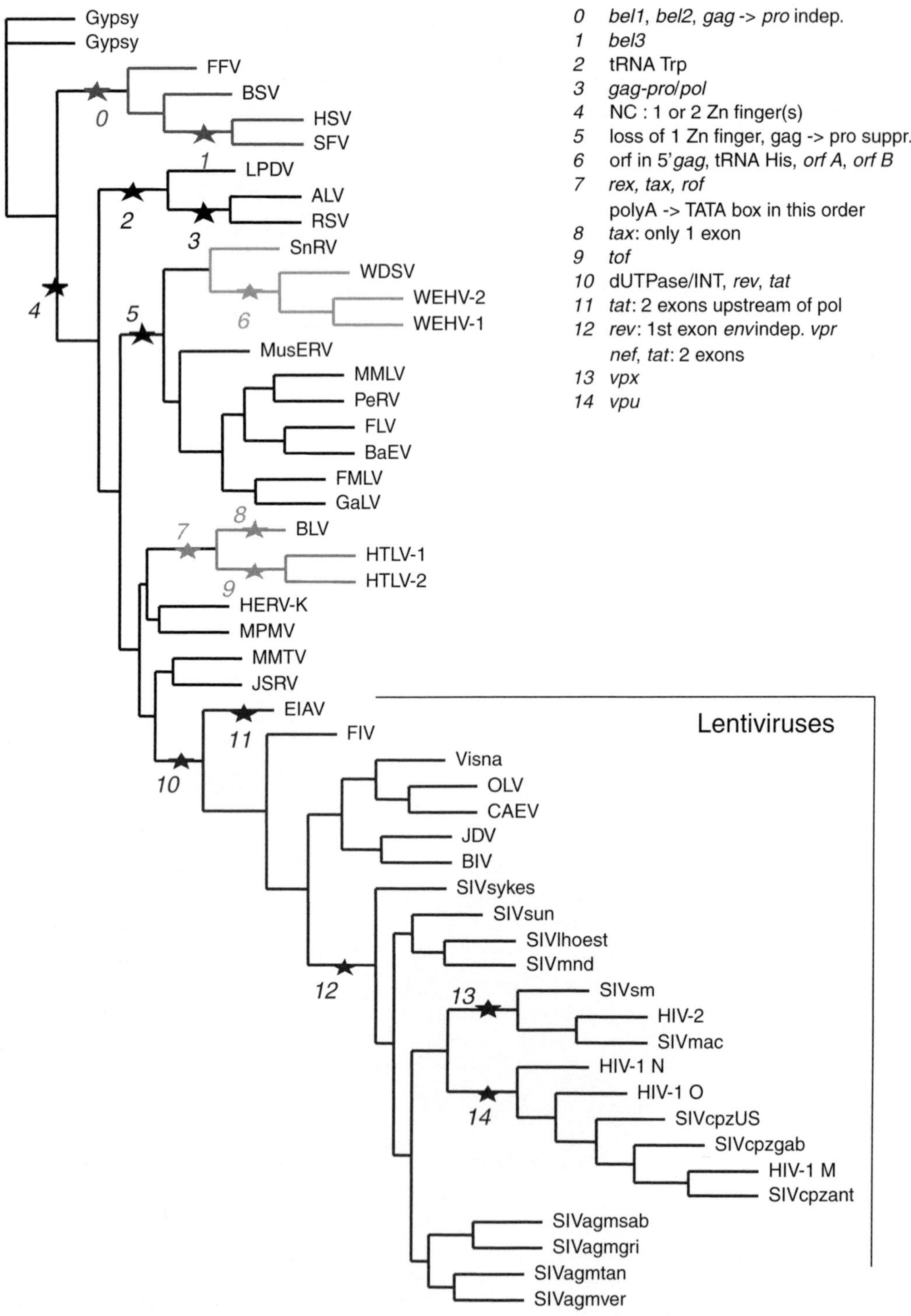

FIGURE 13.8 Annotated phylogenic retroviral tree based on the binary answer mode alignment. Each of the 15 stars corresponds to one or more characters, i.e. organizational traits or evolutionary events at the origin of one group of retroviruses. The lentiviral group represents the largest group of retroviruses in terms of genome complexity. (See Plate 18 for the color version of this figure.)

While it is satisfying that HERV-K is found in the twilight zone not too far from the lentiviruses, given the low bootstrap values it is not possible to be more precise.

THE FUTURE OF RETROVIRAL EVOLUTION?

An additional point can be made concerning Figures 13.7 and 13.8. The standard retroviral tree concerns the accumulation of point mutations while the "unconventional" tree concerns the accumulation of novel features. That the general branching orders coincide suggests that the acquisition of novelty *inevitably* accompanies the remorseless accumulation of point mutations. As any retrovirus radiates through sequence space it will inevitably accumulate novel features along with massive numbers of point mutations. As the lentiviruses sport the highest fixation rates of any retrovirus, organizational novelty is to be expected here, and probably with the accompanying HIV-1 pandemic more than anywhere else.

It transpires that almost every small equatorial African monkey harbors a distinct SIV, albeit pathogenic. To date there are >34 SIVs if not more (Van Heuverswyn *et al.*, 2006). Through hunting and butchering it is eminently reasonable to postulate yet another zoonotic transmission, and perhaps secondary transmission between humans. While a HIV-3 cannot be ruled out, it seems hard to beat HIV-1 that has in the space of a half-century climbed to the top of the virological Richter scale (Hale *et al.*, 2001; Weiss and McMichael, 2004).

If one of these small monkey SIVs ever crosses to humans another pandemic is not axiomatic. There is one unfortunate "experiment" that was unwittingly performed on humans and yet, fortunately, did not take off. With the identification of SV40 in rhesus macaque primary kidney cultures, manufacture of the attenuated Sabin strains of poliovirus was switched to African green monkey (Agm) kidney cultures. It is no secret that the incidence of SIVagm in some troops of Agms can be as high as 70%. As there was no inactivating step in the vaccine preparation, some individuals must have received SIVagm. Yet HIV-1 simply is not derived from any SIVagm of any sort despite the fact that some strains can grow on human peripheral blood mononuclear cells (Gautam *et al.*, 2007). It was, as Robin Weiss remarked, a close shave.

The only real question of any importance for us as potential hosts for retrovirus replication is whether variation will be the Achilles' heel of an eventual HIV vaccine. Is it going to be a problem? Certainly studies show that an attenuated SIV vaccine can contain the challenge virus despite a certain degree of envelope variation (Johnson *et al.*, 1999; Blancou *et al.*, 2004). However, breakthrough does occur when more diverse challenge strains are used. There are simply not enough data to be more precise on the degree of variation. Superinfection by HIV-1 is often given as proof that vaccination will not work. This is not the same as vaccination. It is increasingly clear that during the very early stages of infection there is a loss of HIV/SIV-specific immunity, meaning that a seropositive individual is not the same as a vaccinated volunteer. Whatever the state of experimental findings and subtending riders, HIV variation is not good news.

To conclude, retroviruses are remarkable mutation machines and their genetic complexity and lifestyles vary enormously. Almost anything is possible given time. With the massive effort devoted to sequencing complete genomes, more endogenous retroviruses will be described in less familiar species, and why not clear counterparts to HTLV and spumaviruses? The koala endogenous retrovirus is an interesting case in point (Gifford, 2006), although there must be many more. Although there are a handful of retroviruses in fish, the aquatic world is so diverse, that there must be many more. Concerning extant retroviruses, the lentiviruses represent the most diverse or complex group. As their fixation rates are so high, the probability is higher that new surprises will show up within this group rather than others. The biologist J.D. Bernal remarked almost 50 years ago that "everything that can happen will happen, and nobody will be safe from it." While I would not share the rather dark 3′ end to his remark because we now can, after all, vaccinate against a large number of microbes and successfully treat one human cancer out of two, the 5′ part is a good précis of retrovirus evolution.

REFERENCES

Ablashi, D.V., Eastman, H.B., Owen, C.B., Roman, M.M., Friedman, J., Zabriskie, J.B. *et al.* (2000) Frequent HHV-6 reactivation in multiple sclerosis (MS) and chronic fatigue syndrome (CFS) patients. *J. Clin. Virol.* **16**, 179–191.

Bello, G., Casado, C., Sandonis, V., Alonso-Nieto, M., Vicario, J.L., Garcia, S. *et al.* (2005) A subset of human immunodeficiency virus type 1 long-term non-progressors is characterized by the unique presence of ancestral sequences in the viral population. *J. Gen. Virol.* **86**, 355–364.

Belshaw, R., Pereira, V., Katzourakis, A., Talbot, G., Paces, J., Burt, A. and Tristem, M. (2004) Long-term reinfection of the human genome by endogenous retroviruses. *Proc. Natl Acad. Sci. USA* **101**, 4894–4899.

Belshaw, R., Watson, J., Katzourakis, A., Howe, A., Woolven-Allen, J., Burt, A. and Tristem, M. (2007) Rate of recombinational deletion among human endogenous retroviruses. *J. Virol.* **81**, 9437–9442.

Bibollet-Ruche, F., Gao, F., Bailes, E., Saragosti, S., Delaporte, E., Peeters, M. *et al.* (2004) Complete genome analysis of one of the earliest SIVcpzPtt strains from Gabon (SIVcpzGAB2). *AIDS Res Hum Retroviruses* **20**, 1377–1381.

Blancou, P., Chenciner, N., Ho Tsong Fang, R., Monceaux, V., Cumont, M.C., Guetard, D. *et al.* (2004) Simian immunodeficiency virus promoter exchange results in a highly attenuated strain that protects against uncloned challenge virus. *J. Virol.* **78**, 1080–1092.

Cello, J., Paul, A.V. and Wimmer, E. (2002) Chemical synthesis of poliovirus cDNA: generation of infectious virus in the absence of natural template. *Science* **297**, 1016–1018.

Chambers, V.C. and Ito, Y. (1964) Morphology of Shope papilloma virus associated with nucleic acid-induced tumors of cottontail rabbits. *Virology* **23**, 434–436.

Coffin, J.M. (1979) Structure, replication, and recombination of retrovirus genomes: some unifying hypotheses. *J. Gen. Virol.* **42**, 1–26.

Cordonnier, A., Casella, J.F. and Heidmann, T. (1995) Isolation of novel human endogenous retrovirus-like elements with foamy virus-related pol sequence. *J. Virol.* **69**, 5890–5897.

Dalal, S., Balasubramanian, S. and Regan, L. (1997) Protein alchemy: changing beta-sheet into alpha-helix. *Nat. Struct. Biol.* **4**, 548–552.

Delaroque, N., Boland, W., Muller, D.G. and Knippers, R. (2003) Comparisons of two large phaeoviral genomes and evolutionary implications. *J. Mol. Evol.* **57**, 613–622.

Delebecque, F., Suspene, R., Calattini, S., Casartelli, N., Saib, A., Froment, A. *et al.* (2006) Restriction of foamy viruses by APOBEC cytidine deaminases. *J. Virol.* **80**, 605–614.

Drake, J.W. and Holland, J.J. (1999) Mutation rates among RNA viruses. *Proc. Natl Acad. Sci. USA* **96**, 13910–13913.

Gautam, R., Carter, A.C., Katz, N., Butler, I.F., Barnes, M., Hasegawa, A. *et al.* (2007) In vitro characterization of primary SIVsmm isolates belonging to different lineages. In vitro growth on rhesus macaque cells is not predictive for in vivo replication in rhesus macaques. *Virology* **362**, 257–270.

Gifford, R.J. (2006) Evolution at the host-retrovirus interface. *Bioessays* **28**, 1153–1156.

Gifford, R. and Tristem, M. (2003) The evolution, distribution and diversity of endogenous retroviruses. *Virus Genes* **26**, 291–315.

Gifford, R., Kabat, P., Martin, J., Lynch, C. and Tristem, M. (2005) Evolution and distribution of class II-related endogenous retroviruses. *J. Virol.* **79**, 6478–6486.

Hale, P., Makgoba, M.W., Merson, M.H., Quinn, T.C., Richman, D.D., Vella, S. *et al.* (2001) Mission now possible for AIDS fund. *Nature* **412**, 271–272.

Harris, R.S. and Liddament, M.T. (2004) Retroviral restriction by APOBEC proteins. *Nat. Rev. Immunol.* **4**, 868–877.

Ito, Y. (1962) Relationship of components of papilloma virus to papilloma and carcinoma cells. *Cold Spring Harb. Symp. Quant. Biol.* **27**, 387–394.

Jetzt, A.E., Yu, H., Klarmann, G.J., Ron, Y., Preston, B.D. and Dougherty, J.P. (2000) High rate of recombination throughout the human immunodeficiency virus type 1 genome. *J. Virol.* **74**, 1234–1240.

Jin, M.J., Hui, H., Robertson, D.L., Muller, M.C., Barre-Sinoussi, F., Hirsch, V.M. *et al.* (1994) Mosaic genome structure of simian immunodeficiency virus from west African green monkeys. *EMBO J.* **13**, 2935–2947.

Johnson, P.R., Hamm, T.E., Goldstein, S., Kitov, S. and Hirsch, V.M. (1991) The genetic fate of molecularly cloned simian immunodeficiency virus in experimentally infected macaques. *Virology* **185**, 217–228.

Johnson, R.P., Lifson, J.D., Czajak, S.C., Cole, K.S., Manson, K.H., Glickman, R. *et al.* (1999) Highly attenuated vaccine strains of simian immunodeficiency virus protect against vaginal challenge: inverse relationship of degree of protection with level of attenuation. *J. Virol.* **73**, 4952–4961.

Jung, A., Maier, R., Vartanian, J.P., Bocharov, G., Jung, V., Fischer, U. *et al.* (2002) Multiply infected spleen cells in HIV patients. *Nature* **418**, 144.

Katzourakis, A., Tristem, M., Pybus, O.G. and Gifford, R.J. (2007) Discovery and analysis of the first endogenous lentivirus. *Proc. Natl Acad. Sci. USA* **104**, 6261–6265.

Kawai, S. and Hanafusa, H. (1972) Genetic recombination with avian tumor virus. *Virology* **49**, 37–44.

Kothe, D.L., Decker, J.M., Li, Y., Weng, Z., Bibollet-Ruche, F., Zammit, K.P., Salazar, M.G. *et al.* (2007) Antigenicity and immunogenicity of HIV-1 consensus subtype B envelope glycoproteins. *Virology* **360**, 218–234.

Lee, Y.N. and Bieniasz, P.D. (2007) Reconstitution of an infectious human endogenous retrovirus. *PLoS Pathog.* **3**, e10.

Mahieux, R., Suspene, R., Delebecque, F., Henry, M., Schwartz, O., Wain-Hobson, S. and Vartanian, J.P. (2005) Extensive editing of a small fraction of human T-cell leukemia virus type 1 genomes by four APOBEC3 cytidine deaminases. *J. Gen. Virol.* **86**, 2489–2494.

Mansky, L.M. (2000) In vivo analysis of human T-cell leukemia virus type 1 reverse transcription accuracy. *J. Virol.* **74**, 9525–9531.

Mansky, L.M. and Temin, H.M. (1995) Lower in vivo mutation rate of human immunodeficiency virus type 1 than that predicted from the fidelity of purified reverse transcriptase. *J. Virol.* **69**, 5087–5094.

Mansky, L.M. and Wisniewski, R.M. (1998) The bovine leukemia virus encapsidation signal is composed of RNA secondary structures. *J. Virol.* **72**, 3196–3204.

McGeoch, D.J. (1990) Protein sequence comparisons show that the 'pseudoproteases' encoded by poxviruses and certain retroviruses belong to the deoxyuridine triphosphatase family. *Nucleic Acids Res.* **18**, 4105–4110.

Mikl, M.C., Watt, I.N., Lu, M., Reik, W., Davies, S.L., Neuberger, M.S. and Rada, C. (2005) Mice deficient in APOBEC2 and APOBEC3. *Mol. Cell. Biol.* **25**, 7270–7277.

Nickle, D.C., Jensen, M.A., Gottlieb, G.S., Shriner, D., Learn, G.H., Rodrigo, A.G. and Mullins, J.I. (2003) Consensus and ancestral state HIV vaccines. *Science* **299**, 1515–1518.

Pedulla, M.L., Ford, M.E., Houtz, J.M., Karthikeyan, T., Wadsworth, C. *et al.* (2003) Origins of highly mosaic mycobacteriophage genomes. *Cell* **113**, 171–182.

Peeters, M., Liegeois, F., Torimiro, N., Bourgeois, A., Mpoudi, E., Vergne, L. *et al.* (1999) Characterization of a highly replicative intergroup M/O human immunodeficiency virus type 1 recombinant isolated from a Cameroonian patient. *J. Virol.* **73**, 7368–7375.

Pelletier, E., Saurin, W., Cheynier, R., Letvin, N.L. and Wain-Hobson, S. (1995) The tempo and mode of SIV quasispecies development in vivo calls for massive viral replication and clearance. *Virology* **208**, 644–652.

Renoux-Elbe, C., Cheynier, R. and Wain-Hobson, S. (2002) Phylogeny derived from coding retroviral genome organization. *J. Mol. Evol.* **54**, 376–385.

Rolland, M., Jensen, M.A., Nickle, D.C., Yan, J., Learn, G.H., Heath, L. *et al.* (2007) Reconstruction and function of ancestral center-of-tree human immunodeficiency virus type 1 proteins. *J. Virol.* **81**, 8507–8514.

Rose, G.D. and Creamer, T.P. (1994) Protein folding: predicting. *Proteins* **19**, 1–3.

Sheehy, A.M., Gaddis, N.C., Choi, J.D. and Malim, M.H. (2002) Isolation of a human gene that inhibits HIV-1 infection and is suppressed by the viral Vif protein. *Nature* **418**, 646–650.

Stehelin, D., Varmus, H.E., Bishop, J.M. and Vogt, P.K. (1976) DNA related to the transforming gene(s) of avian sarcoma viruses is present in normal avian DNA. *Nature* **260**, 170–173.

Suspene, R., Guetard, D., Henry, M., Sommer, P., Wain-Hobson, S. and Vartanian, J.P. (2005) Extensive editing of both hepatitis B virus DNA strands by APOBEC3 cytidine deaminases in vitro and in vivo. *Proc. Natl Acad. Sci. USA* **102**, 8321–8326.

Switzer, W.M., Salemi, M., Shanmugam, V., Gao, F., Cong, M.E., Kuiken, C. *et al.* (2005) Ancient cospeciation of simian foamy viruses and primates. *Nature* **434**, 376–380.

Tristem, M., Marshall, C., Karpas, A., Petrik, J. and Hill, F. (1990) Origin of vpx in lentiviruses. *Nature* **347**, 341–342.

Van Heuverswyn, F., Li, Y., Neel, C., Bailes, E., Keele, B.F., Liu, W. *et al.* (2006) Human immunodeficiency viruses: SIV infection in wild gorillas. *Nature* **444**, 164.

Vandvik, B. and Norrby, E. (1989) Paramyxovirus SV5 and multiple sclerosis. *Nature* **338**, 769–771.

Vogt, P.K. (1971) Genetically stable reassortment of markers during mixed infection with avian tumor viruses. *Virology* **46**, 947–952.

Wain-Hobson, S., Renoux-Elbe, C., Vartanian, J.P. and Meyerhans, A. (2003) Network analysis of human and simian immunodeficiency virus sequence sets reveals massive recombination resulting in shorter pathways. *J. Gen. Virol.* **84**, 885–895.

Wattel, E., Vartanian, J.P., Pannetier, C. and Wain-Hobson, S. (1995) Clonal expansion of human T-cell leukemia virus type I-infected cells in asymptomatic and symptomatic carriers without malignancy. *J. Virol.* **69**, 2863–2868.

Weir, J.P. (1998) Genomic organization and evolution of the human herpesviruses. *Virus Genes* **16**, 85–93.

Weiss, R.A. and McMichael, A.J. (2004) Social and environmental risk factors in the emergence of infectious diseases. *Nat. Med.* **10**, S70–76.

Weiss, R.A., Mason, W.S. and Vogt, P.K. (1973) Genetic recombinants and heterozygotes derived from endogenous and exogenous avian RNA tumor viruses. *Virology* **52**, 535–552.

Winkelmann, J.C. (1993) HTLV-1 and multiple sclerosis: the link is missing. *J. Lab. Clin. Med.* **122**, 230–231.

Wolgamot, G., Bonham, L. and Miller, A.D. (1998) Sequence analysis of *Mus dunni* endogenous virus reveals a hybrid VL30/gibbon ape leukemia virus-like structure and a distinct envelope. *J. Virol.* **72**, 7459–7466.

Yoshida, M. (1993) HTLV-1 Tax: regulation of gene expression and disease. *Trends Microbiol.* **1**, 131–135.

Zhuang, J., Mukherjee, S., Ron, Y. and Dougherty, J.P. (2006) High rate of genetic recombination in murine leukemia virus: implications for influencing proviral ploidy. *J. Virol.* **80**, 6706–6711.

CHAPTER

14

Intra-host Dynamics and Evolution of HIV Infection

Viktor Müller and Sebastian Bonhoeffer

ABSTRACT

Evolutionary and population dynamics are rarely linked together, although the two are intimately connected. Here we attempt to outline these connections on the example of the human immunodeficiency virus (HIV), which has been studied intensively in both aspects, and focus on those aspects for which our understanding has been promoted substantially by the use of mathematical techniques. Special emphasis will be given to the effects of drug treatment, both on HIV population dynamics and on the evolution to drug resistance. Although much of this chapter has an explicit mathematical basis, we have attempted to make the content accessible to the non-mathematical reader.

Population dynamics has important effects on the nature and rate of evolution. While in infinite populations, the fixation and accumulation of mutations is governed by the rates of mutation and selection only, in finite populations stochastic effects also play a role. The strength of these effects in turn depends on the size and structure of a population—which belong to the realm of population dynamics. Furthermore, the clock of evolution ticks in generations and its calibration to "real time" requires an estimate of the generation time, which again population dynamics can provide.

We start by an introduction to the within-host population dynamics of HIV infection, describing virus and cell populations and their interactions. This section builds upon an established mathematical framework, which will be fleshed out by the interpretation of the model results. The mathematical framework is essential for tackling a system of such vast complexity: it allows us to extract presumed key processes with an unambiguous definition of assumptions, and thereby to test hypotheses and to interpret experimental observations within a given hypothesis. We describe the steady state that characterises asymptomatic infection, and discuss the factors that set the steady-state virus load, which is an important predictor of disease progression. We present models of drug treatment on short and long time-scales, and show how population dynamics provided important insight

Origin and Evolution of Viruses
ISBN 978-0-12-374153-0

into the population size and structure, and the turnover of the virus. Moreover, we show how this information can be combined with estimations of the mutation rate to characterize the evolutionary potential of HIV, and to model the rise of drug-resistant mutants during therapy. Next we discuss how stochastic processes can confound the estimation of the evolutionary potential and present analyses demonstrating that such confounding factors are probably relevant in HIV infections. Finally, we discuss the implications of recombination for the evolution of HIV. We show that the evolutionary advantage of this "primitive" form of sexual reproduction is far from being clear, and neither is its effect on the evolution of drug resistance fully understood. Recombination adds further complexity to the evolution of HIV, and is itself interrelated with population size and structure. In all, we intend to describe the toolbox of population and evolutionary dynamics required to estimate the evolutionary potential of an organism, and show that the complexities of a real biological system make this a hard task even for a well-studied organism like HIV.

WITHIN-HOST POPULATION DYNAMICS OF HIV INFECTION

HIV belongs to the genus of lentiviruses (*lenti-*, Latin for "slow") named after the long incubation period of infection by its members. Untreated HIV infection typically remains asymptomatic for many years and the level of virus detectable in blood plasma stays relatively stable, increasing only slowly over the years (Sabin *et al.*, 2000; Hubert *et al.*, 2000). However, the apparent tranquillity of the asymptomatic period conceals a highly dynamic balance between massive virus production and clearance. The dynamic nature of HIV replication was exposed by the analysis of clinical trials of the first effective antiretrovirals (Ho *et al.*, 1995; Wei *et al.*, 1995). After the start of therapy, the plasma level of HIV-1 RNA (the "virus load") dropped one to two orders of magnitude in just two weeks. Importantly, the employed drugs do not affect the clearance of virus particles or the removal of infected cells, but act by blocking the infection of further cells. The rapid drop of the virus level during therapy thus reflected the natural clearance of HIV. In the absence of treatment, the same fast clearance must be balanced by equally fast production of new virus particles to achieve the observed stability of the virus load. In addition to demonstrating the efficacy of the drugs, these experiments thus revealed also the dynamical nature of HIV replication during clinical latency. By fitting simple mathematical models to the virus load data, the authors were also able to derive the first estimates for the daily production of virus particles in an infected individual (Ho *et al.*, 1995; Wei *et al.*, 1995). These two key publications were important milestones in "quantitative virology," which involves the use of mathematical models to interpret clinical and experimental observations. This thriving field has been reviewed in several publications in great detail (Nowak and May, 2000; Perelson, 2002; Wodarz and Nowak, 2002; Müller and Bonhoeffer, 2003). Here we will review the basic foundations of the modeling framework, then focus on issues that concern the evolutionary potential of the virus.

The Basic Model of Virus Dynamics

The simplest model of virus dynamics considers uninfected target cells, T, productively infected cells, I, and virus particles, V. Uninfected cells are produced at a constant rate, σ, and die at rate $\delta_T T$. Virus particles infect uninfected cells at a rate proportional to the product of their abundances, βTV, and infected cells die at rate δI. Virus is produced from infected cells at rate pI and is cleared at rate cV. This gives rise to the following system of ordinary differential equations:

$$\mathrm{d}T/\mathrm{d}t = \sigma - \delta_T T - \beta TV \qquad (1)$$

$$dI/dt = \beta TV - \delta I \quad (2)$$

$$dV/dt = pI - cV \quad (3)$$

The basic model (with varying notations) has been used widely as a starting point for describing HIV dynamics (reviewed in Nowak and May, 2000). The scheme of the model is shown in Figure 14.1. The model can describe quantities in the whole body or in a given volume of blood plasma or tissue, depending on the scaling of the parameters. The equations define the rate of change of each quantity (variable) and each term corresponds to a production or decay process. The reciprocal of the decay rate defines the average lifespan, which is thus $1/\delta_T$, $1/\delta$, and $1/c$ for uninfected and infected cells, and virus particles, respectively. The basic model captures the dynamics of a single well-mixed compartment with large homogeneous populations of

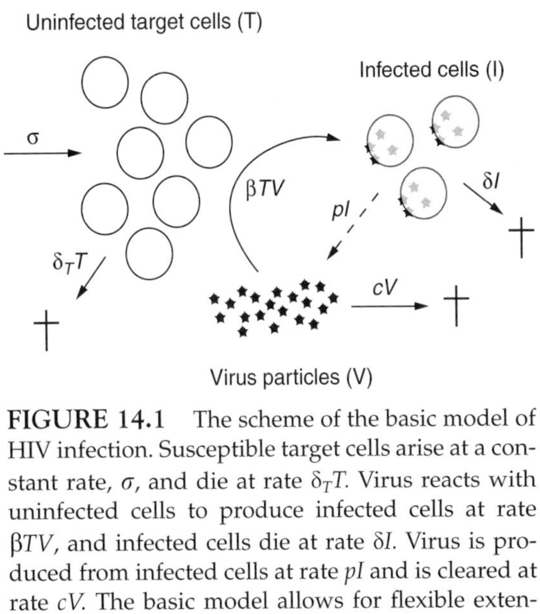

FIGURE 14.1 The scheme of the basic model of HIV infection. Susceptible target cells arise at a constant rate, σ, and die at rate $\delta_T T$. Virus reacts with uninfected cells to produce infected cells at rate βTV, and infected cells die at rate δI. Virus is produced from infected cells at rate pI and is cleared at rate cV. The basic model allows for flexible extensions, e.g. the source of susceptible cells may be activation of quiescent cells or proliferation, the death rate of infected cells may be influenced by an HIV-specific immune response that may itself be activated as a function of virus levels, etc. A pragmatic approach involves expanding those processes that display large variations between patients (Müller *et al.*, 2001b).

cells and viruses that undergo asynchronous infection and cell cycles (i.e. generations are overlapping). Clearly, the full complexity of HIV is not captured by this model. However, the basic model can be used as a consensus starting point and be extended to incorporate more complexity. Building a single "full model" of HIV dynamics that describes all details of the biological system is not feasible; in fact, a complex model can itself quickly become intractable, making it impossible to dissect the roles of individual processes in the system—which would defeat the purpose of modeling. The implementation of complexity therefore needs to be guided by the particular research question and to be kept at the minimum possible level.

The basic model, albeit very simple, is non-linear and a general solution for the time course of its variables cannot be derived. The analysis of the model can nevertheless provide valuable insight. In the absence of virus, the basic model attains an uninfected equilibrium with $\hat{T}_U = \sigma/\delta_T$, $I = 0$, $V = 0$. If virus is added to the system (in the form of infected cells or virus particles), the infection may take hold or it may peter out, depending on the parameters. The conditions for successful infection can be summarized in the form of the basic reproductive ratio, R_0 (analogous to the condition of a disease outbreak at the population level). This quantity describes the number of cells infected by a single infected cell that is added to an uninfected equilibrium. If $R_0 < 1$, infected cells cannot replace themselves during their lifetime and therefore their numbers dwindle steadily towards zero. If $R_0 > 1$, the number of infected cells increases to a transient peak and then the system settles to an infected equilibrium. In the basic model, the basic reproductive ratio can be calculated as $R_0 = \beta\sigma p/\delta\delta_T c$. The infected equilibrium is at $\hat{T} = \delta c/\beta p$, $\hat{I} = \sigma/\delta - \delta_T c/\beta p$, $\hat{V} = p\hat{I}/c$. An established infection unambiguously indicates $R_0 > 1$; the lack of infection may indicate the failure of viral invasion, but also the lack of real exposure. In principle, highly exposed uninfected individuals may have $R_0 < 1$, which would indicate systemic resistance to HIV. Alternatively, an efficient innate immune response or a lack of appropriate target cells at the site of entry may block the access of the virus to the main (susceptible) target cell population in these individuals. In particular, homozygous carriers of the CCR5Δ32 deletion mutant lack susceptible target cells at the sites of entry important in sexual transmission, and are therefore apparently resistant to infection by this route (Liu *et al.*, 1996). However, this resistance to infection can be breached if viruses able to infect target cells that reside in the systemic circulation are transmitted directly into blood (Sheppard *et al.*, 2002).

Some Implications of the Basic Model

In the case of an established infection it is *a priori* unlikely that the basic reproductive ratio would be just marginally above one. Indeed, analyses of acute infections with simian immunodeficiency virus (SIV) in macaques (Nowak *et al.*, 1997b) and HIV in humans (Little *et al.*, 1999; Stafford *et al.*, 2000) using the basic model have yielded estimates for R_0 ranging between 3 (a lower bound) and about 70. For $R_0 >> 1$ (which thus seems plausible) the equilibrium of the basic model has some interesting features that provide insight into the infection dynamics. First, a high basic reproductive ratio implies a large reduction in the abundance of uninfected cells, because the ratio of steady-state target cell levels before and after infection can be expressed as $\hat{T}_U/\hat{T} = R_0$. Note that the depletion of uninfected target cells does not necessarily imply the depletion of the total pool of target cells. For a non-cytopathic virus, the steady-state level of infected cells is close to the uninfected equilibrium of target cells, i.e. the overall cell count is hardly affected. In contrast, a cytopathic virus can reduce this level by approximately the factor of reduction that it causes in the lifespan of infected cells. That is, a depletion of the overall target cell count is possible only for a combination of a high reproductive ratio (to reduce uninfected cell counts) and a strong cytopathic effect (to reduce infected cell counts). In this case, one would generally expect a high percentage of infected cells within the total target cell pool (unless the cytopathic effect is extremely strong even relative to a high R_0). HIV infects primarily $CD4^+$ T lymphocytes and depletion of these cells from the blood was the first recognized hallmark of the infection. *In vitro* assays have indicated a marked cytopathic effect (Kwa *et al.*, 2001; Speirs *et al.*, 2005); however, the frequency of productively infected $CD4^+$ T cells was estimated to be very low (0.01–1%) in the blood (Brenchley *et al.*, 2004) and in peripheral lymph nodes (Chun *et al.*, 1997; Haase *et al.*, 1996; Cavert *et al.*, 1997). Within the frame of the basic model and its variants such low frequencies could only be reconciled with substantial target cell depletion by assuming bystander killing of uninfected cells (De Boer and Perelson, 1998). However, recent findings indicate that a major source of virus production may be the pool of memory $CD4^+$ T cells located primarily in the gut (reviewed in Centlivre *et al.*, 2007). In this population, up to 60% of cells may be infected during the peak of primary infection and most of these cells then die within a few days, leaving behind a lasting CD4 depletion in the gut-associated lymphoid tissue. The major target cell population of HIV may thus experience direct cytopathic effects, while the blood CD4 count may be depleted, at least in part, due to indirect effects.

A second implication of $R_0 >> 1$ is that the steady-state level of infected cells and virus particles can be approximated as $\hat{I} \approx \sigma/\delta$, $\hat{V} \approx \sigma p/\delta c$. That is, the level of infected cells is determined largely by the supply of susceptible target cells and the death rate of the infected cells, but is, surprisingly, independent of the "infection rate," β, which describes the efficacy of the infection process (Bonhoeffer *et al.*, 1997a). An intuitive explanation for this behavior is that efficient infection results in the infection of most susceptible target cells soon after their production. The rate of new infections is thus limited by the generation of new target cells and cannot be accelerated by further increases in the infection efficiency. This finding has implications for both the identification of factors that influence the virus load and the understanding of the effect of long-term antiviral therapy. Considering that the death rate of infected cells (δ) and the ratio of virus particles and infected cells (p/c) seems to be relatively invariant between patients, much of the considerable variation in the steady-state virus load may emerge from differences in the production rate of susceptible target cells, σ (Bonhoeffer *et al.*, 2003). Such variability may arise from differences in the proportion of produced $CD4^+$ T cells that are susceptible, e.g. due to differences in the expression of the co-receptors that are required for virus entry (Reynes *et al.*, 2000; Lin *et al.*, 2002), or from differences in the ability of the virus to induce the activation of $CD4^+$ T cells, which increases its target cell supply (Hazenberg *et al.*, 2000). Identifying the factors that influence virus load is highly relevant because virus load is a strong predictor of disease progression (Mellors *et al.*, 1996; Arnaout *et al.*, 1999).

Drug treatments act roughly by decreasing the infection rate, β. There is an interesting criticality in the outcome of this intervention (Bonhoeffer *et al.*, 1997a): as long as R_0 remains considerably larger than 1, the virus load should hardly be affected; conversely, if R_0 is pushed below 1, the virus level should tend to 0 and be eradicated eventually. Only for a very narrow range of drug efficacies could a new low-level steady state be reached. Thus the basic HIV model cannot account for ongoing low-level virus replication during treatment, which has been observed in some patients (Ramratnam *et al.*, 2000, 2004). Remarkably, this problem of a low steady-state virus load cannot be solved easily by modifying the basic model (Callaway and Perelson, 2002). After rejecting a large number of model variants, Callaway and Perelson suggested two possibilities to explain a low steady state. Either a "sanctuary" (cell type or anatomical location) must exist where drugs have diminished effect, allowing for continued virus replication, or the death rate of infected cells must be density-dependent, which may compensate reduced production with reduced death at a small population size. An earlier explanation by Grossman *et al.* belongs essentially to the latter category: in their model the replenishment of infected cells by "infection bursts" becomes more efficient at low population sizes due to density dependence (Grossman *et al.*, 1999). A third possibility involves episodic ("blips") rather than steady-state virus replication, enabled by fluctuations in the availability of target cells due to, e.g. stochastic antigenic activation (Fraser *et al.*, 2001; Jones and Perelson, 2007). In this scenario, virus replication is suppressed to $R_0 < 1$ most of the time during treatment, but some virus persists in long-lived reservoirs, e.g. in latently infected cells. However, R_0 is also affected by the availability of target cells, and fluctuations may temporarily increase it above one, allowing residual virus to start new rounds of infections. These "blips" may in turn replenish the long-lived reservoirs.

HIV-Specific Immune Responses

In the basic model, HIV-specific immune responses can only take effect by influencing the parameters of the model, but cannot react to changes in the other variables of the system. This approximation is valid if either the effect of immune responses is negligible, or the time scale of investigation is too short to allow changes in the level of the immune effectors, or if the responses are "saturated" at high antigen concentrations. During primary infection, it is unclear whether peak virus replication is limited by the emerging immune response (Regoes *et al.*, 2004), or by the exhaustion of the target cell population (Davenport *et al.*, 2007; De Boer, 2007), as had been proposed many years ago (Phillips, 1996). During chronic infection, immune responses seem to play an important role (e.g. blocking $CD8^+$ T cells dramatically increases virus levels in SIV-infected macaques (Jin *et al.*, 1999; Schmitz *et al.*, 1999)), but there is some indication that they may be operating at the maximum saturated level (Müller *et al.*, 2001b). Saturation may reflect the lack of CD4 help, which arises from the preferential depletion of HIV-specific CD4$^+$ T cells (Douek *et al.*, 2002). Preservation of these cells through primary infection (e.g. by drug treatment) may, in principle, result in efficient immune control (Wodarz and Nowak, 1999). In the following section, we will investigate the decline of the virus load during therapy. On the time-scale of first-phase decline (1–2 weeks) the levels of immune responses probably do not change much. On the time-scale of long-term treatment (years) HIV-specific immune responses wane in the absence of antigen. Therefore, the use of target cell-limited models (the basic model and its derivatives) will be justified.

Estimating Rates of Production and Decay

Having considered some fundamental properties and implications of the basic model, we now turn to the estimation of kinetic rates of viral and cellular processes. These quantitative analyses will shed some light on the nature of relevant cell and virus populations, and provide information on the within-host evolutionary potential of HIV.

Quantitative insights have been derived from perturbations of the infected steady state. Importantly, observing just virus and cell counts at the steady state is not informative: any cell and virus level could be maintained by the balance of both equally fast or equally slow production and decay. Drug treatment perturbs the steady state by cutting the supply of infected cells and infectious virus. In particular, inhibitors of the reverse transcriptase (RT) block the infection of new cells, which can be implemented in the model by reducing the rate of new infections to $(1 - \varepsilon_{RT})\beta TV$, where ε_{RT} describes the efficacy of an RT inhibitor and can vary between 0 and 1. Protease (PR) inhibitors render newly produced virus particles non-infectious, which can be modeled by introducing separate equations for infectious and non-infectious virions. The total production of virus particles is split between the two virus pools according to the efficacy of PR inhibitors, ε_{PI}. $(1 - \varepsilon_{PI})pI$ gives the production of infectious virions, while the rest ($\varepsilon_{PI}\, pI$) enters the non-infectious pool. The new drug classes of integrase and fusion inhibitors and co-receptor antagonists act analogously to RT inhibitors in the framework of the basic model.

In the simplest approximation, the efficacy of effective antiretrovirals was assumed to be 100%, corresponding to complete inhibition. This breaks the non-linearity of the basic model. For RT inhibitors, the model yields a simple exponential decline of infected cells (governed by the equation $dI/dt = -\delta I$), while for PR inhibitors the titer of infectious virus should decline exponentially ($dV_I/dt = -cV_I$). The decline of the total virus level, which is easiest to measure, is slightly more complicated, influenced by both decay rates (δ and c) and dominated by the slower of the two. The estimation of parameters works by finding the values of the parameters for which the difference between the observations and the predictions of the model is the smallest. The first two seminal papers on short-term effective drug treatment could only estimate a single compound decay rate, which therefore corresponds to the slower of the death of infected cells and the decay of virus particles (Ho *et al.*, 1995; Wei *et al.*, 1995). This rate was estimated to be around 0.34 per day, setting the lower bound for the turnover of both virus particles and infected cells to about one third of the pool cleared and replenished each day. Estimating the rate of production thus depends critically also on the steady-state pool size. In this example, calculating with a mean of the total blood virus pool around 3×10^8 particles, Wei *et al.* estimated minimum production as 10^8 virions per day (Wei *et al.*, 1995), while the estimate of Ho *et al.* based on a higher pool size including extracellular fluid was closer to 10^9 virions per day (Ho *et al.*, 1995). In addition, the turnover of CD4$^+$ T cells was estimated from the slope of their increase after the start of therapy and from the extrapolation of the baseline CD4 count to total body counts to be around 2×10^9 cells per day.

This gave rise to a situation with "more bodies than bullets," especially considering that most virus particles seem to be non-infectious (Piatak *et al.*, 1993) and therefore count more as blank cartridges than bullets. More frequent measurements of virus load during PR inhibitor treatment later allowed for the separate estimation of the two decay rates (Perelson *et al.*, 1996). As had been hypothesized, the death of virus-producing cells proved to be the slower process, with a decay rate around 0.5 per day. The clearance rate of virus particles was estimated to be around 3 per day both from simultaneous fitting of both decay rates to the total virus load data and from fitting of the virus decay rate to the decline of infectivity titres in the blood plasma. This increased estimates for daily virus production by an order of magnitude to around 10^{10} virions per day. The average length of the virus generation time was estimated to be 2.6 days (as the sum of the average lifespan of a virion and the average lifespan of an infected cell).

Perturbation of the steady-state virus load in the blood by plasma apheresis (removal of virus particles) yielded an even higher estimate around 23 per day for the decay of virus particles (Ramratnam *et al.*, 1999). One possibility to reconcile this finding with earlier estimates is that virus particles located in the lymphoid tissues (which form the majority of the total virus pool (Pantaleo *et al.*, 1993)) have a distinct decay rate. The amount of infected cells and virus particles in lymphoid tissues declines parallel to the blood (Cavert *et al.*, 1997), which indicates active exchange between the two compartments. The basic model can be adjusted to describe this situation by introducing separate variables for virus and cell levels in each compartment. Using such a model it has been demonstrated that plasma apheresis may correctly estimate virus decay in the blood, while the earlier slower estimates around 3 per day may characterise the decay of virus in lymphoid tissues (Müller *et al.*, 2001a). This result increases estimated total production even further. Decay in a blood virus pool of around 3×10^8 particles must be balanced by the production of about 7×10^9 virions per day, while decay in the lymphoid tissue population of about 10^{10} virus particles (Haase *et al.*, 1996; Cavert *et al.*, 1997) needs the production of further 3×10^{10} virions per day. The number of "bullets" thus seems to be increasing as the estimations are refined. Experiments tracking the level of SIV particles infused into uninfected macaques yielded even faster rates of virus clearance from the blood (Zhang *et al.*, 1999, 2002), but this may reflect a higher initial capacity of clearance, which is partly saturated in chronic infection.

The estimation of infected cell death rates has also been refined and re-interpreted. In the first approximation drug treatment was assumed to be 100% effective, but this is likely not true. Partially effective inhibitors result in slower decline of the virus level: compared with full inhibition, the rate of decline is roughly multiplied by the efficacy of the drugs (Bonhoeffer *et al.*, 1997b). This principle has been used to estimate the relative efficacy of different drug regimens (Mittler *et al.*, 2001; Louie *et al.*, 2003), and it implies that the true death rate of infected cells has been underestimated. The higher the efficacy, the closer we get to the real death rate. Our current best estimate, derived from a study using especially potent treatment, is around 1 per day (Markowitz *et al.*, 2003), which implies that the efficacy in the earlier studies (yielding an estimate around 0.5 per day) was at most 50%. This reduces the estimated generation time of HIV to around 2 days, implying about 180 generations per year.

Besides incomplete efficacy, the estimation of the death rate of infected cells may be confounded by additional factors. First, the pool of virus-producing cells is likely to be heterogeneous, and subpopulations decaying at different rates may give rise to multiple time-scales. This effect has been invoked primarily for the interpretation of long-term virus decay during therapy (see below), but one modeling study suggested that it could play a role already in the first few weeks (Ferguson *et al.*, 1999). Ferguson *et al.* fitted a complex model of antigen-driven $CD4^+$ T cell dynamics to data from multiple patients simultaneously, and estimated the turnover of a highly activated infected cell population to be around 4 per day, while another population of low activation was responsible for the slower time scale of decay. Thus even the intermediate time scale of short term virus decay (around 3 per day) may be attributed to infected cells, while the virus compartment model attributed this time-scale to virus particles in lymphoid tissues (Müller *et al.*, 2001a). Finally, the decline slope of virus levels may be completely decoupled from the death rate of infected cells. There is an "eclipse phase" between the infection of a cell (integration of the provirus) and the start of substantial production of viral proteins. If transition through this phase is slower than the turnover of cells that are already producing virus, then the decline of the virus load during short-term treatment is dominated by the time-scale of the eclipse phase (Klenerman *et al.*, 1996; De Boer, 2007). The length of the phase has been estimated around 1 per day (Reddy and Yin, 1999; Dixit *et al.*, 2004), which is consistent with the dominant turnover rate around 1 per day during short-term treatment. If this is indeed true, then the decline of the blood virus level during therapy provides no clue about the death rate of virus-producing cells, except that it must be faster than 1 per day. Such a decoupling between observed virus decay during therapy and the death rate of infected cells is supported by a further observation or, more precisely, by the absence of an observation that would be expected. A major mechanism of host immunity against HIV is the killing of infected cells by cytotoxic T lymphocytes (CTLs) (recently reviewed in Davenport *et al.*, 2007), and the strength of this mechanism is likely to differ between individuals (although it is not easy to measure; Sun *et al.*, 2003). The death rate of virus-producing cells should be affected by the rate of CTL-mediated killing, but the slope of first-phase virus decay during treatment has shown little variation in most studies (reviewed in Bonhoeffer *et al.*, 2003; though one recent study using Bayesian estimation with population priors set by earlier observations found considerable variation; Wu *et al.*, 2005). Furthermore, the slope of the virus load after the peak of primary SIV infection in macaques is not affected by prior vaccination that raises the level of HIV-specific CTLs (Abdel-Motal *et al.*, 2005). There is thus no correlation between the rate of CTL-mediated killing and the slope of virus decay, which indicates that either the death rate is decoupled from the virus slope or it is hardly affected by CTLs.

After the first few weeks of therapy, the decline of the virus level becomes slower. This again can be explained by the presence of an additional population of infected cells (or, less likely, of virus particles) that have a slower decay rate corresponding to the new time-scale. Second-phase virus decay could be explained by a population of persistently infected cells (e.g. macrophages) that have an average lifespan of about two weeks (Perelson *et al.*, 1997). An alternative explanation invokes the local structure of target cell activation, infection, and virus production (Grossman *et al.*, 1998, 1999). While the basic model and its variants inherently assume continuous rounds of virus production and global mixing, HIV may be transmitted primarily locally within the lymphoid tissues and production may occur in "infection bursts" involving several rounds of infection that deplete recently activated target cells in their immediate vicinity. The first phase of virus decay during therapy may correspond to a decrease in the magnitude of such bursts, while the second phase may reflect a decrease in the frequency of bursts under this model. Finally, a third explanation is possible if HIV-specific immune responses do have an impact on the turnover of infected cells, and the magnitude of these responses falls rapidly, resulting in decelerating death rates (Arnaout *et al.*, 2000).

The decline of the virus load continues to decelerate after the "second phase." Rather than being characterized by yet another time-scale and rate of exponential decay, the process seems to be gradual, indicating a range of lifespans within the remaining sources of virus. This observation can be explained by residual virus production from a heterogeneous population of long-lived latently infected cells (Müller *et al.*, 2002; Strain *et al.*, 2003; Kim and Perelson, 2006). Encountering their specific antigen can re-activate these cells into short-lived productive infection and thereby drain them from the persisting pool. Those cells that are specific for common antigens are re-activated quickly, while cells specific for rare antigens have lower probability of re-activation. The remaining pool is thus gradually enriched in cells that have a low rate of activation and therefore a low rate of decay. If latently infected cells are capable of bystander proliferation without being activated into virus production, then a new steady state can be attained, which is independent of *de novo* cycles of infection (Kim and Perelson, 2006).

What have we learned so far? We have chosen to give a "historical account" of the estimation of kinetic parameters to illustrate the dependence of the estimates on the models used or, more generally, on our picture of the infection process. Even a moderately complex model can be fitted reasonably well to any series of data, i.e. parameter fitting will always yield a "quantitative" answer. However, the estimated numbers are only as good as the hypothesis behind the mathematical model that was used to obtain them. It is therefore a dangerous illusion to believe that getting "numbers" in itself proves or improves the credibility of a hypothesis. The true use of quantitative analyses in hypothesis testing is to see whether numbers derived from observing different aspects of a system are consistent with each other. For example, the inconsistency of the estimation of virus decay rates from different measurements indicated that the complex decay process involves at least three time-scales already in its first weeks. However, the mapping of the time-scales to distinct cell or virus populations, or life stages, requires further empirical evidence and hypotheses. The observation that first-phase virus decay is dominated by a single exponential rate until the virus load is reduced by about two logs indicates that a single cell population is responsible for about 99% of virus production. However, this numerical result cannot point out the identity of this cell population.

The Identity and Population Size of Productively Infected Cells

It has quickly become clear that the infected cells found in the blood cannot be responsible for mass virus production, because the resistance mutations that arose after a few weeks of monotherapy spread very quickly in the pool of free plasma virus, but only much slower among infected $CD4^+$ T cells circulating in the blood (Wei *et al.*, 1995). Furthermore, a genetic comparison of free plasma viruses and T-cell-associated viruses from the blood revealed considerable dissimilarity confirming a distinct origin of plasma viruses (Malaspina *et al.*, 2002). A logical next choice is the pool of infected cells in the peripheral lymph nodes, which has been estimated by image analysis to be around 10^7–10^8 cells (Haase *et al.*, 1996; Cavert *et al.*, 1997; Chun *et al.*, 1997). This is consistent with the estimated daily production of virus particles needed to balance clearance in the lymphoid tissues (3×10^{10}), considering that the rate of production per cell has been estimated around to be of order 10^2 (Haase *et al.*, 1996) to 10^3 (Hockett *et al.*, 1999) virions per day. Remarkably, free plasma virus was found to be genetically distinct also from cell-associated virus in the lymphoid tissues (Malaspina *et al.*, 2002). Plasma virus may be produced by gut-associated $CD4^+$ T cells, but this remains to be confirmed. Such a situation would imply a different source for the majority of virus production during primary and chronic infection. The population size of gut-associated virus-producing cells during chronic infection is unknown. If they are indeed responsible for the production of about 7×10^9 virions entering the blood each day, and we assume the same per capita virus production rate, then there must be about one-quarter as many infected cells in the gut lining as in the peripheral lymph nodes. Mattapallil *et al.* found that in acute SIV infection memory $CD4^+$ cells were affected by the virus very similarly in the peripheral lymph nodes and in the gut lining, suggesting the similarity of the two cell populations (Mattapallil *et al.*, 2005). Overall, the order of magnitude of the total population size of short-lived productively infected cells seems to remain around 10^7–10^8 cells. With a minimum turnover rate of one per day, the population size also corresponds to a minimum estimate of daily production and death.

The turnover of uninfected target cells has also been studied extensively. Careful mathematical modeling has ironically made the quantitative interpretation of experimental results more, rather than less, difficult. Rates of cell death and proliferation have typically been estimated intuitively with simple concepts, but detailed modeling has revealed a more complex picture which makes the extraction of rates rather difficult. For a recent review see (Borghans and de Boer, 2007).

Linking Population and Evolutionary Dynamics

The rate and level at which new mutations arise and can be maintained in a population depend on population size and turnover (which we have discussed so far), and also on the mutation rate. During the life cycle of HIV, the genome of the virus is copied from RNA to DNA (during reverse transcription), from DNA to DNA (during the synthesis of the complementary DNA strand) and from DNA to RNA (during the production of progeny virus). From these steps the major source of mutations seems to be reverse transcription (O'Neil *et al.*, 2002), which implies that the number of infected cells constitutes the relevant population size for HIV evolution. The rate of mutations per nucleotide per replication cycle has been estimated to be between 3×10^{-5} (Mansky, 1996; Mansky and Temin, 1995) and 9×10^{-5} (O'Neil *et al.*, 2002). With a daily turnover of 10^7 infected cells, each single mutant is expected to be generated 300 to 900 times per day. With a turnover of 10^8 cells/day, every double mutant may be generated once every 1–11 days. Furthermore, even deleterious mutants are not necessarily lost immediately; in a large population they are expected to be maintained at a mutation-selection equilibrium determined by the mutation rate and their replicative capacity relative to the predominant type. This principle has been used in the study of the emergence of drug resistance under therapy (Bonhoeffer and Nowak, 1997; Ribeiro *et al.*, 1998; Ribeiro and Bonhoeffer, 2000). Before the start of therapy, most resistance mutations are thought to confer reduced replication capacity. The expected steady-state frequency of single mutants can be approximated as μ/s, where μ is the mutation rate and s is the selective disadvantage of a mutant. In the case of resistance mutations, differences in replicative capacity are most likely to involve the efficiency of new infections, i.e. the infection rate of a mutant can be written as $\beta^* = (1 - s)\beta$. Many mutations are likely to confer only a slight disadvantage (Smith *et al.*, 2004). This, in conjunction with the high mutation rate, facilitates the existence of drug resistance mutations prior to therapy (Najera *et al.*, 1994, 1995; de Jong *et al.*, 1996; Kapoor *et al.*, 2004). Assuming that mutations have equal and multiplicative effect on replicative capacity, the mutation-selection frequency of double mutants is expected to be $2(\mu/s)^2$, and the frequency of n-error mutants can be calculated as $n!(\mu/s)^n$. For example, employing the higher estimate for the mutation rate (9×10^{-5}), a combination of two mutations, each conferring 10% loss in replicative capacity ($s = 0.1$) may be maintained at a frequency 1.6×10^{-6}, which would imply its probable pre-existence even in a population of 10^7 cells. However, a combination of three such mutations would have an expected frequency of 4.4×10^{-9}, which makes its pre-existence unlikely even in a population of 10^8 cells. Note that these considerations are valid also for mutations not involved in drug resistance, which explains the immense variation of HIV within a single patient (Korber *et al.*, 2001). We must nevertheless emphasize that quantitative predictions are liable to considerable error. In addition to uncertainties about the size and turnover of the population(s) of infected cells, the mutation rate may also vary widely, e.g. depending on the target nucleotide and its neighbours (Mansky and Temin, 1995;

Mansky, 1996; O'Neil *et al.*, 2002), mutations in viral genes and drug treatment (Mansky *et al.*, 2003) and host factors (Zhang *et al.*, 2003). The estimation of replicative capacity is also notoriously difficult (Bonhoeffer *et al.*, 2002).

The framework of virus dynamics allows us to model the rise of the resistant mutants when therapy is initiated by duplicating the equations for infected cells and virus particles to implement both mutant and wild-type viruses (Nowak *et al.*, 1997a). Using these models, it has been shown that, assuming efficient therapy, the emergence of drug resistance is more likely to be due to the pre-existence of mutations than due to *de novo* generation of mutants during treatment (Bonhoeffer and Nowak, 1997; Ribeiro and Bonhoeffer, 2000). The reason for this is that under efficient therapy (i.e. one that reduces R_0 below 1) the sum of all residual virus production during treatment is less than the population size at the start of therapy. This result is robust with respect to the estimation of mutation rates and selection coefficients.

So far, we have relied on dynamical models to estimate the evolutionary potential of HIV and to describe the emergence of mutants. However, this framework does not allow the prediction of long-term molecular evolution, which is addressed typically by population genetic models. Dynamical models nevertheless provide an estimation for the generation time, which sets the time-scale of evolutionary models. A remarkable study has also attempted to link dynamical and population genetic models by tracking the accumulation of replication cycles within the viral population over time (Kelly *et al.*, 2003). This study has demonstrated that long-lived populations of infected cells may have a disproportional effect on the evolutionary dynamics, even though most of the virus production may originate from short-lived cells. If viral lineages pass through both cell populations, then replication in the long-lived cells can substantially slow down the rate of evolution. Kelly *et al.* have also demonstrated that the main infected cell population that produces the majority of the virus could be below $R_0 = 1$, if a small long-lived population maintains the overall infection. Similarly, evolutionary dynamics may be slowed down (even though viral diversity may be increased) by the presence of archived sequences in latently infected cells (Karlsson *et al.*, 1999; Briones *et al.*, 2003, 2006). Re-activation of these cells can re-introduce ancient virus sequences into the population.

Finally, a general shortcoming of the basic HIV model and its derivatives is that they are inherently deterministic. Even if an event has very low probability (e.g. the generation of triple mutants), it will always occur, albeit at a slow rate. Such an approximation is justified for very large population size, but may not always be applicable to HIV, as we will discuss below.

EFFECTIVE POPULATION SIZE

As explained in the previous sections, in HIV infection, population size influences the probability that a drug resistance mutation is present at the start of therapy or that it appears during treatment. However, changes in the frequency of mutations need not be the result of selection, but can also be the result of genetic drift (i.e. the statistical influences of chance). The relative importance of drift versus selection increases with decreasing population size. Considering the estimate of at least 10^7–10^8 productively infected cells derived above, the role of drift on the evolutionary dynamics of HIV may at first sight seem limited. However, it is important to note that the total (census) population size is not the relevant quantity to estimate the relative role of drift. This is best illustrated by a simple (albeit unrealistic) example: Assume that all 10^7–10^8 infected cells in the current generation were infected by virus released from 1000 (randomly chosen) cells in the previous generation, then, with regard to the effects of drift and selection, this population behaves as if it consisted of 1000 cells only.

The role of drift is frequently assessed on the basis of an "effective" population size, N_e, a concept originally introduced by Sewall

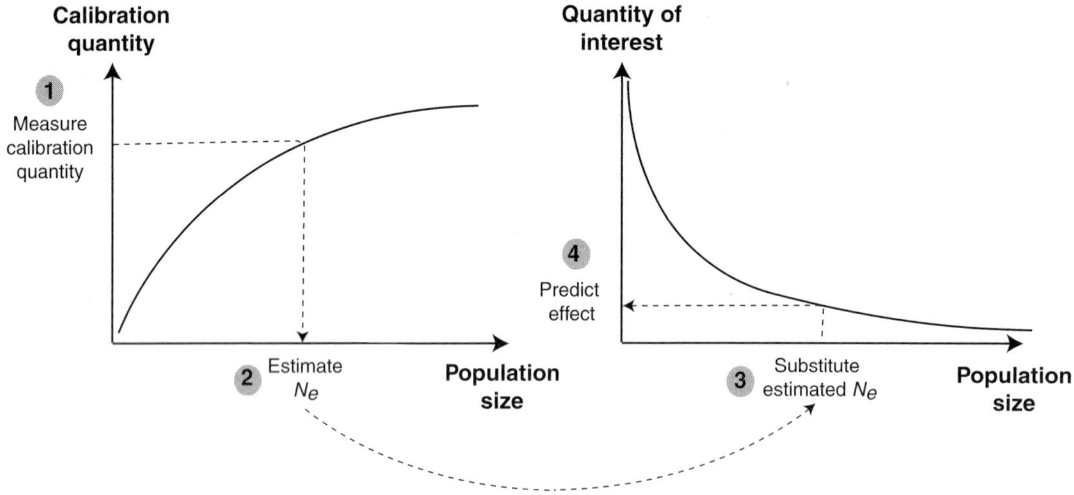

FIGURE 14.2 Schematic illustration of the concept of the effective population size, N_e. The effective population size was introduced in order to build a bridge between complex population genetic processes in natural populations and their representation in idealized mathematical models. In order to obtain an estimate for a quantity of interest that is difficult to measure directly in the natural population the following steps are taken. First one derives on the basis of the idealized model the functional relation between population size and a calibration quantity (such as genetic diversity) that can be measured in the population. Then one determines the effective population size, N_e, as that population size that within the realm of the idealized model yields the measured value of the calibration quantity in the natural population. Next, one determines the relationship between population size and the quantity of interest and substitutes the estimated value of N_e to predict the value of the quantity of interest in the natural population. Importantly, the same model must underlie both the estimation of N_e and the prediction of the effect. In other words, care must be taken when using N_e that both the calibration quantity and the quantity of interest must be subject to the same evolutionary process. Adapted with permission from Kouyos *et al.* (2006).

Wright in the 1930s (Wright, 1931). In essence, the effective population size is a tool to map complex population dynamical processes onto an idealized mathematical model in order to predict quantities of interest in that idealized model that are difficult to measure in the actual population. Figure 14.2 illustrates the concept of N_e.

Estimates of N_e in HIV-1 Infection

Considerable debate currently surrounds the question whether the evolution of HIV-1 within a patient is largely determined by drift or selection. Estimates of the effective population size range from a few hundred to a million (Leigh-Brown, 1997; Nijhuis *et al.*, 1998; Rouzine and Coffin, 1999; Seo *et al.*, 2002; Achaz *et al.*, 2004; Shriner *et al.*, 2004). The discrepancy between these estimates is in part due to the fact that there is not a single value of N_e that is universally applicable, but rather N_e depends on the nature of the process in question. Estimates of N_e that range from a few hundred to thousands are generally based on processes that assume neutral evolution (i.e. absence of selective differences between mutants) (Leigh-Brown, 1997; Nijhuis *et al.*, 1998; Seo *et al.*, 2002; Achaz *et al.*, 2004; Shriner *et al.*, 2004). Two considerations must be kept in mind when applying these estimates. First, any measured quantity used to estimate N_e (such as genetic diversity) must have evolved under a process of neutral evolution, which is questionable for some of the cited studies (Kouyos *et al.*, 2006). Second, the estimate of N_e can only be applied to processes that are also under neutral evolution. For example, an N_e estimated based on genetic diversity at neutral sites cannot be applied to estimate genetic diversity at selected sites (Kouyos *et al.*, 2006). This is of particular relevance in the context of the discussion whether the evolution of drug resistance mutations is governed by drift or selection. Applying estimates of N_e that are based on neutral processes is not appropriate, because drug resistance mutations are under selection (both before and during therapy).

The estimate of N_e of up to a million (Rouzine and Coffin, 1999) does not suffer from the above problem, as it was derived based on an approach that explicitly allows for selection. For technical reasons this estimate of N_e is a lower bound. The difficulty with this estimate, however, is that its derivation is based on the assumption that recombination can be neglected. More recent estimates of the rate of recombination (Levy *et al.*, 2004) and the frequency of superinfected cells (Jung *et al.*, 2002) (see also section below), suggest that this assumption is questionable. A higher effective recombination rate effectively lowers the value of the lower bound of the estimate of N_e, and therefore unfortunately renders this estimate less informative.

Factors that Reduce the Effective Population Size

Several factors that may be relevant to HIV are known to make N_e smaller than the census population size (Kouyos *et al.*, 2006). First, transient decreases in population size greatly increase the relative importance of drift versus selection. Fitness loss due to population bottlenecks has been demonstrated for the segmented bacteriophage Φ6 (Chao, 1990), vesicular stomatitis virus (Duarte *et al.*, 1992), foot-and-mouth disease virus (Escarmis *et al.*, 1996), and HIV (Yuste *et al.*, 1999). In HIV strong bottlenecks may for example occur during transmission (Edwards *et al.*, 2006; Frater *et al.*, 2006; Delwart *et al.*, 2002). Selective sweeps can also act as bottlenecks by purging genetic diversity at loci linked to the locus under selection.

Bottlenecks have an over-proportional effect of reducing N_e below the census population size. Wright showed that N_e of a population undergoing periodical fluctuations in population size is given by the harmonic time average of the census population size (Wright, 1938), which tends strongly towards the smallest population size. Importantly the effect of bottlenecks is much stronger for genetic diversity at neutral than at selected sites. The intuitive reason is that the expected genetic diversity at neutral sites is much larger than that at selected sites, and therefore after a bottleneck it takes much longer to restore genetic diversity at neutral than at selected sites.

As outlined by an example in the beginning of this section, a second factor reducing N_e relative to the census population size is variance in the production of offspring (Wright, 1938). This is relevant for HIV-1 infection since for example latent or defectively infected cells produce little or no virus. Moreover, the virus production rate may depend on the activation status of the target cell. Both processes contribute to the variance in reproductive output, and thus lower N_e.

A third factor reducing N_e is population structure. In particular, Frost *et al.* (2001) argued that a more realistic description of HIV dynamics in terms of localized, interconnected subpopulations with frequent extinction and recolonization events leads to a reduction of N_e.

Is Intra-Host HIV-1 Evolution Stochastic or Deterministic?

Whatever the correct estimate of N_e may be, the debate in the literature has highlighted the important point that despite the vast number of virions and infected cells the evolutionary dynamics of HIV within an infected host may in certain cases be governed more by drift than by selection. The low estimates of N_e based on neutral genetic diversity, for example, suggest, that the dynamics of building up neutral diversity would be appropriately described by random drift in small populations. However, other evolutionary processes, such as the emergence of drug resistance mutations immediately after the start of therapy, may be more accurately described by an N_e closer to the census population size (Kouyos *et al.*, 2006). Great care must therefore be taken in evaluating the specific biological details when assessing whether selection or drift dominates in a particular evolutionary process of HIV-1 infection.

RECOMBINATION

Retroviruses are unique in that they package two full-length genomic RNA molecules into each virion. Provided a virion contains two distinct genomic RNA molecules recombinant proviral DNA can be produced. This "primitive" form of sexual reproduction (Temin, 1991) has important implications for the evolutionary potential of HIV.

Mechanism of Recombination

The predominant mechanism of recombination in HIV-1, referred to as copy-choice (Vogt, 1973), is based on template switching between the two genomic RNA molecules during reverse transcription. A second mechanism, referred to as the "strand displacement assimilation model" (Junghans *et al.*, 1982), has also been described. Although there is some experimental support also for this mechanism, it is considered to be of lesser importance in HIV-1 (Galetto and Negroni, 2005) and will therefore not be discussed in detail here.

In order to result in the production of recombinant proviruses, both mechanisms require infection by virions carrying genetically distinct RNA molecules (see Figure 14.3). This in turn requires that individual cells become superinfected by two or more genetically distinct viral strains. While the infection by distinct viral strains need not occur simultaneously, simultaneous expression of the corresponding proviruses is necessary for the production of virions harbouring distinct genomic RNA molecules. The simultaneous expression of the proviruses may also result in phenotypic mixing (Novick and Szilard, 1951; Vogt, 1973), i.e. the chimeric assembly of viral proteins derived from distinct proviruses during the production of new virions. Moreover, the viral genomic RNAs must be able to form a functional dimer that can be packaged into the budding virion.

Rate of Recombination

The effective rate of recombination depends both on the frequency of superinfection of cells and the frequency of recombination during synthesis of proviral DNA. Jung *et al.* estimated the number of proviruses per cell using fluorescence *in situ* hybridization (Jung *et al.*, 2002). Analyzing infected splenocytes from two patients they observed a range of 1–8 proviruses per cell with a mean around 3. How far these estimates are representative also for other tissues remains to be determined. However, these results imply that downregulation of the entry receptor CD4 by HIV following infection appears not to prevent superinfection of cells by multiple viruses. Because the effective rate of recombination depends on the frequency of superinfected cells, it may in turn depend on virus load and may thus be reduced during drug therapy (Boerlijst *et al.*, 1996; Fraser, 2005).

The frequency of template switching during copy-choice recombination has been estimated to be around 3–4 in HeLa-CD4 fibroblastic cells (Yu *et al.*, 1998; Jetzt *et al.*, 2000), around 9 in CD4 T lymphocytes and around 30 in macrophages (Levy *et al.*, 2004) per genome per replication cycle. Several factors may influence the frequency of template switching such as single-strand RNA breaks, strong pause sites, and regions of RNA secondary structure (Galetto and Negroni, 2005). Recombination appears to be widespread along the genome (Jetzt *et al.*, 2000; Zhuang *et al.*, 2002). A study investigating recombination in a narrow region of the *env* gene showed a recombination frequency up to ten times higher than the surrounding regions (Galetto *et al.*, 2004), supporting the notion of localized hotspots of recombination in the HIV genome.

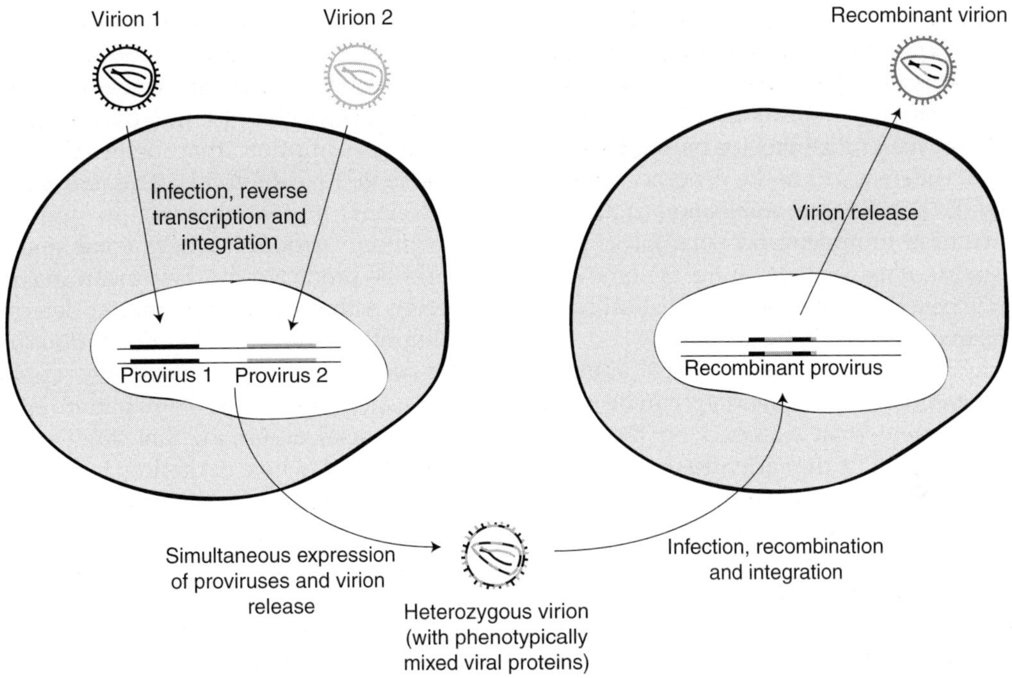

FIGURE 14.3 Schematic illustration of recombination in retroviruses. A prerequisite for recombination is that a single cell carries two genetically distinct proviruses. This happens if the cell is infected by virions carrying genetically distinct genomic RNA molecules (or, alternatively, if the cell is infected by several genetically identical viruses, but mutation during reverse transcription leads to the production of genetically distinct proviruses). When both proviruses are simultaneously expressed "heterozygous" virions may be produced that carry two distinct genomic RNA molecules. Moreover, the proteins produced from the proviruses may get mixed during virion assembly, a process that is referred to as phenotypic mixing (Novick and Szilard, 1951; Vogt, 1973). Once such a heterozygous virion infects another cell, recombination (either through a copy choice (Vogt, 1973) or strand displacement assimilation (Junghans *et al.*, 1982)) can occur and result in the production of a recombinant provirus.

Cause and Consequences of Recombination

It is frequently stated (Gu *et al.*, 1995; Kellam and Larder, 1995; Moutouh *et al.*, 1996; Burke, 1997; Wain-Hobson *et al.*, 2003) that recombination allows retroviruses such as HIV to rapidly adapt to novel selection pressures (such as those constituted by the immune responses or drug therapy). However, whether recombination in retroviruses evolved for that reason is unclear. A simpler explanation is that recombination is the consequence of the evolution of copy-choice replication, because copy choice offers the selective advantage of enabling transcription beyond single strand breaks in the genomic RNA strands (Coffin, 1979). This hypothesis proposes an evolutionary benefit of diploidy in retroviruses, which is independent of genetic shuffling. Consequently, it is conceivable that recombination is the consequence of copy-choice replication, but need not represent an actual evolutionary benefit by itself.

Under what conditions recombination in HIV-1 may actually offer a selective advantage in terms of increasing adaptability (and may therefore itself be selected for) is much more difficult to assess. This question is intimately linked with one of the major open questions of evolutionary biology, namely, what is the evolutionary benefit of reproductive strategies that involve the shuffling of parental genetic material (Barton and Charlesworth, 1998; Otto and Lenormand, 2002). The only genetic effect of recombination is to break up statistical associations between mutations (or more generally alleles) at different loci in the genome. Thus to identify the benefit of recombination (if there is any), we need to understand what forces generate statistical associations in the genome and why it is beneficial to break up statistical associations that these forces have built up.

There are two forms of statistical associations. Combinations of mutations can be more or less frequent than expected on the basis of the frequency of the individual mutations in the population. An overrepresentation is termed positive linkage disequilibrium and an underrepresentation is termed negative linkage disequilibrium. (As a technical note, we emphasize that the definition of the sign of linkage disequilibrium requires the definition of a reference type (the "wild-type"), because whenever certain combinations are overrepresented other combinations of alleles at the relevant loci are necessarily underrepresented.)

Many hypotheses have been proposed for the evolution of sexual reproduction and recombination (Kondrashov, 1993; Barton and Charlesworth, 1998; Otto and Lenormand, 2002). Here we discuss three hypotheses and their relevance to evolution of HIV within the host: (i) the mutational deterministic hypothesis (Feldman *et al.*, 1980; Kondrashov, 1982; Kondrashov, 1988), (ii) the Hill–Robertson effect (Hill and Robertson, 1966; Barton and Otto, 2005), (iii) and the Red Queen hypothesis (Jaenike, 1978; Hamilton, 1980; Hamilton *et al.*, 1990).

Evolutionary Hypotheses for the Benefit of Recombination

(i) Mutational Deterministic Hypothesis and Related Epistasis-Based Hypotheses

The mutational deterministic hypothesis postulates that the evolutionary benefit of recombination is due to the improved purging of deleterious mutations from a population by breaking down statistical associations that arise through fitness interactions between mutations. Specifically the hypothesis rests on the assumption that deleterious mutations tend to exhibit negative epistasis (i.e. deleterious mutations tend to reduce fitness more strongly when combined than would be predicted by their mean individual effects). Although the mutational deterministic hypothesis focusses on deleterious mutations, an analogous argument also applies to beneficial mutations with negative epistasis (i.e. beneficial mutations that tend to have a weaker effect when occurring together than would be predicted by their individual effects (for a review see Kondrashov, 1993).

Epistatic interactions generate statistical associations between alleles at different loci. In particular negative epistasis leads to negative linkage disequilibrium (Eshel and Feldman, 1970), i.e. an underrepresentation of mutants carrying higher number of mutations, as their fitness is lower than would be predicted on the basis of the effect of each mutation alone. Breaking up negative linkage disequilibria (generated by epistasis) in a population leads to (i) an increase in the frequency of mutants with high number of mutations and (ii) an increase in variance of fitness, which in turn increases the response to selection according to Fisher's fundamental theorem of natural selection (Fisher, 1930).

The effect of epistasis on the evolution of drug resistance in HIV has been investigated by means of population genetic and population dynamical models (Bretscher *et al.*, 2004; Fraser, 2005). In line with the population genetic theory concerning the mutational deterministic hypothesis, Bretscher *et al.* (2004) found (i) that the preexistence frequency of combinations of drug resistant mutations in the absence of drugs and (ii) their rate of fixation in the presence of drugs is increased only if the mutations exhibit negative epistasis (i.e. exhibit a weaker effect of fitness than would be predicted by the multiplication of their individual effects). Fraser (2005) came to a seemingly contradictory conclusion that recombination always had a decelerating effect on the emergence of resistance independent of epistasis. However, this discrepancy results from differences between models that describe the population dynamics in discrete versus continuous time. In discrete time statistical associations between mutations at different loci vanish if the effect of the mutations is multiplicative, while in continuous-time models this is the case when the effect of mutations is additive. All examples used in the continuous time model in fact represent positive epistasis on an additive scale, and therefore are not in contradiction with population genetic theory. A large-scale analysis of the fitness effects of drug resistance mutations in HIV-1 reported that at least in the absence of drugs resistance mutations tend to interact predominantly with positive epistasis (on a multiplicative scale and therefore also on an additive scale) (Bonhoeffer *et al.*, 2004). Hence, provided epistasis is the dominating force generating linkage disequilibria among drug resistance mutations, recombination may in fact lower the level of drug resistance mutations in drug-naïve patients, and thereby act against the emergence of resistance.

Hill–Robertson Effect

The Hill–Robertson effect describes how drift (in finite populations) in concert with selection can generate negative linkage disequilibria (Hill and Robertson, 1966), which in turn favors the evolution of recombination. In a nutshell the Hill–Robertson effect works in the following way. Drift (i.e. random sampling in a finite population) generates both positive and negative deviations from the linkage disequilibria generated by selection. Selection, however, acts more efficiently at eliminating the positive than the negative deviations, because positive deviations correspond to a larger variance in fitness and consequently to more efficient selection according to Fisher's law (Fisher, 1930). Thus selection operating on randomly generated linkage disequilibrium deviations results in an overall force driving the linkage disequilibrium towards more negative values.

The Hill–Robertson effect can override epistasis as a force in generating linkage disequilibria (Otto and Barton, 2001). The effect of effective population size on the evolution of drug resistance in HIV has recently been addressed by two studies (Althaus and Bonhoeffer, 2005; Carvajal-Rodriguez *et al.*, 2007). These studies show that for very small population sizes recombination has no effect. This is because small populations lack the required polymorphisms for recombination to act upon as typically an advantageous mutation will go to fixation before the next one occurs. As population size increases, recombination reduces the time to fixation of resistance mutations (in the presence of drugs) independent of epistasis. This is the case because in this parameter region the Hill–Robertson effect can override the effect of positive epistasis. As a result recombination efficiently reduces the negative linkage disequilibria that the Hill–Robertson effect creates, and thus accelerates the rate of selection. For very large population sizes, epistasis overrides the Hill–Robertson effect. In this parameter region, the effect of recombination thus depends on whether combinations of mutations exhibit positive or negative epistasis.

Whether recombination within HIV-1-infected patients does in fact correspond to the parameter regime of small, intermediate or large population size remains to be determined. In two-locus models the Hill–Robertson effect causes substantial selection for high recombination rates only when the population size is $<10^4$–10^5 (Otto and Barton, 2001). As discussed in greater detail in the preceding section, it is unclear whether the effective population size of HIV-1 (at least for sites under selection) is indeed as small. However, Iles *et al.* (2003) showed that in multilocus models the Hill–Robertson effect may generate favorable conditions for recombination for considerably larger population sizes. Therefore it is plausible that the Hill–Robertson effect also plays a role in intra-host HIV evolution.

Red Queen Hypothesis

Although the mutational deterministic hypothesis and the Hill–Robertson effect are the theories that have received most attention, they are not the only ones that may be relevant in the context of within host evolution of HIV. In particular, the so-called Red Queen hypothesis (Jaenike, 1978; Hamilton, 1980; Hamilton *et al.*, 1990) may be applicable to HIV. This theory attempts to explain the benefit of sexual reproduction or recombination by focussing on its effect on the dynamics between hosts and their co-evolving parasites. Host–parasite co-evolution is characterized by continual rounds of adaptation and counteradaptation. These dynamics have been termed Red Queen dynamics in reference to Lewis Carroll's children's book *Through the Looking Glass* in which the Red Queen states that "it takes all the running you can do, to stay in the same place." In particular, the Red Queen hypothesis posits that recombination in hosts has evolved as a means to allow better escape from co-evolving parasites. In the context of HIV, the adaptive immune response plays the role of the co-evolving parasites that continuously adapt to novel antigenic variants of HIV. The application of the Red Queen hypothesis to retroviral recombination will be an interesting area of future research.

Conclusion: The Effect of Recombination on the Evolution of Drug Resistance

Having been neglected for a long time by mathematical modelers, the effect of recombination in HIV has recently witnessed a rapid surge of interest. Importantly, the models show that recombination does not generally facilitate the evolution of drug resistance as has frequently been assumed. Recombination has both an effect on the expected frequency (or probability of existence) of combinations of drug resistance mutations in the absence of therapy, and on the rate of fixation of these combinations in the presence of drugs. Whether recombination facilitates or impedes the evolution of drug resistance depends on whether stochastic effects (i.e. the Hill–Robertson effect) are the primary force generating statistical associations, or if not, on the sign of epistatic interactions.

A deeper understanding of the effect of recombination in HIV will not only shed light onto the evolution of drug resistance, but may also provide valuable insights into one of the remaining great questions of evolutionary biology, namely, what is the benefit of sexual reproduction.

CONCLUDING REMARKS

We have outlined our current picture of the population and evolutionary dynamics of HIV within an infected individual. A great amount of progress has been made in this research area over the last decade: clearly there is no other virus infection for which a comparable research effort has been undertaken to quantify its dynamical behavior. This quantitative approach has led to a more profound understanding of HIV infection. Nevertheless considerable uncertainty still surrounds many central aspects of the viral population dynamics and its evolutionary response to changing selection pressures. Quantitative virology of HIV was born 13 years ago: it is still in its childhood.

REFERENCES

Abdel-Motal, U.M., Gillis, J., Manson, K., Wyand, M., Montefiori, D., Stefano-Cole, K. *et al.* (2005) Kinetics of expansion of SIV Gag-specific CD8+ T lymphocytes following challenge of vaccinated macaques. *Virology* **333**, 226–238.

Achaz, G., Palmer, S., Kearney, M., Maldarelli, F., Mellors, J.W., Coffin, J.M. and Wakeley, J. (2004) A robust measure of HIV-1 population turnover within chronically infected individuals. *Mol. Biol. Evol.* **21**, 1902–1912.

Althaus, C.L. and Bonhoeffer, S. (2005) Stochastic interplay between mutation and recombination during the acquisition of drug resistance mutations in human immunodeficiency virus type 1. *J. Virol.* **79**, 13572–13578.

Arnaout, R.A., Lloyd, A.L., O'Brien, T.R., Goedert, J.J., Leonard, J.M. and Nowak, M.A. (1999) A simple relationship between viral load and survival time in HIV-1 infection. *Proc. Natl Acad. Sci. USA* **96**, 11549–11553.

Arnaout, R.A., Nowak, M.A. and Wodarz, D. (2000) HIV-1 dynamics revisited: biphasic decay by cytotoxic T lymphocyte killing? *Proc. R Soc. Lond. B Biol. Sci.* **267**, 1347–1354.

Barton, N.H. and Charlesworth, B. (1998) Why sex and recombination? *Science* **281**, 1986–1990.

Barton, N.H. and Otto, S.P. (2005) Evolution of recombination due to random dirift. *Genetics* **169**, 2353–2370.

Boerlijst, M.C., Bonhoeffer, S. and Nowak, M.A. (1996) Viral quasi-species and recombination. *Proc. R Soc. Lond. B Biol. Sci.* **263**, 1577–1584.

Bonhoeffer, S. and Nowak, M.A. (1997) Pre-existence and emergence of drug resistance in HIV-1 infection. *Proc. R Soc. Lond. B Biol. Sci.* **264**, 631–637.

Bonhoeffer, S., Coffin, J.M. and Nowak, M.A. (1997a) Human immunodeficiency virus drug therapy and virus load. *J. Virol.* **71**, 3275–3278.

Bonhoeffer, S., May, R.M., Shaw, G.M. and Nowak, M.A. (1997b) Virus dynamics and drug therapy. *Proc. Natl Acad. Sci. USA* **94**, 6971–6976.

Bonhoeffer, S., Barbour, A.D. and De Boer, R.J. (2002) Procedures for reliable estimation of viral fitness from time-series data. *Proc. R Soc. Lond. B Biol. Sci.* **269**, 1887–1893.

Bonhoeffer, S., Funk, G.A., Gunthard, H.F., Fischer, M. and Müller, V. (2003) Glancing behind virus load variation in HIV-1 infection. *Trends Microbiol.* **11**, 499–504.

Bonhoeffer, S., Chappey, C., Parkin, N.T., Whitcomb, J. M. and Petropoulos, C.J. (2004) Evidence for positive epistasis in HIV-1. *Science* **306**, 1547–1550.

Borghans, J.A. and de Boer, R.J. (2007) Quantification of T-cell dynamics: from telomeres to DNA labeling. *Immunol. Rev.* **216**, 35–47.

Brenchley, J.M., Hill, B.J., Ambrozak, D.R., Price, D.A., Guenaga, F.J., Casazza, J.P. *et al.* (2004) T-cell subsets that harbor human immunodeficiency virus (HIV) in vivo: implications for HIV pathogenesis. *J. Virol.* **78**, 1160–1168.

Bretscher, M.T., Althaus, C.L., Muller, V. and Bonhoeffer, S. (2004) Recombination in HIV and the evolution of drug resistance: for better or for worse? *Bioessays* **26**, 180–188.

Briones, C., Domingo, E. and Molina-Paris, C. (2003) Memory in retroviral quasispecies: experimental evidence and theoretical model for human immunodeficiency virus. *J. Mol. Biol.* **331**, 213–229.

Briones, C., de Vicente, A., Molina-Paris, C. and Domingo, E. (2006) Minority memory genomes can influence the evolution of HIV-1 quasispecies in vivo. *Gene* **384**, 129–138.

Burke, D.S. (1997) Recombination in HIV: an important viral evolutionary strategy. *Emerg. Infect. Dis.* **3**, 253–259.

Callaway, D.S. and Perelson, A.S. (2002) HIV-1 infection and low steady state viral loads. *Bull. Math. Biol.* **64**, 29–64.

Carvajal-Rodriguez, A., Crandall, K.A. and Posada, D. (2007) Recombination favors the evolution of drug resistance in HIV-1 during antiretroviral therapy. *Infect. Genet. Evol.* **7**, 476–483.

Cavert, W., Notermans, D.W., Staskus, K., Wietgrefe, S. W., Zupancic, M., Gebhard, K. *et al.* (1997) Kinetics of response in lymphoid tissues to antiretroviral therapy of HIV-1 infection. *Science* **276**, 960–964.

Centlivre, M., Sala, M., Wain-Hobson, S. and Berkhout, B. (2007) In HIV-1 pathogenesis the die is cast during primary infection. *Aids* **21**, 1–11.

Chao, L. (1990) Fitness of RNA virus decreased by Muller ratchet. *Nature* **348**, 454–455.

Chun, T.W., Carruth, L., Finzi, D., Shen, X., DiGiuseppe, J.A., Taylor, H. *et al.* (1997) Quantification of latent tissue reservoirs and total body viral load in HIV-1 infection. *Nature* **387**, 183–188.

Coffin, J.M. (1979) Structure, replication, and recombination of retrovirus genomes—some unifying hypotheses. *J. Gen. Virol.* **42**, 1–26.

Davenport, M.P., Ribeiro, R.M., Zhang, L., Wilson, D.P. and Perelson, A.S. (2007) Understanding the mechanisms and limitations of immune control of HIV. *Immunol. Rev.* **216**, 164–175.

De Boer, R.J. (2007) Understanding the failure of CD8+ T-cell vaccination against simian/human immunodeficiency virus. *J. Virol.* **81**, 2838–2848.

De Boer, R.J. and Perelson, A.S. (1998) Target cell limited and immune control models of HIV infection: a comparison. *J. Theor. Biol.* **190**, 201–214.

De Jong, M.D., Schuurman, R., Lange, J.M. and Boucher, C.A. (1996) Replication of a pre-existing resistant HIV-1 subpopulation in vivo after introduction of a strong selective drug pressure. *Antivir. Ther.* **1**, 33–41.

Delwart, E., Magierowska, M., Royz, M., Foley, B., Peddada, L., Smith, R. *et al.* (2002) Homogeneous quasispecies in 16 out of 17 individuals during very early HIV-1 primary infection. *Aids* **16**, 189–195.

Dixit, N.M., Markowitz, M., Ho, D.D. and Perelson, A.S. (2004) Estimates of intracellular delay and average drug efficacy from viral load data of HIV-infected individuals under antiretroviral therapy. *Antivir. Ther.* **9**, 237–246.

Douek, D.C., Brenchley, J.M., Betts, M.R., Ambrozak, D.R., Hill, B.J., Okamoto, Y. *et al.* (2002) HIV preferentially infects HIV-specific CD4+ T cells. *Nature* **417**, 95–98.

Duarte, E., Clarke, D., Moya, A., Domingo, E. and Holland, J. (1992) Rapid fitness losses in mammalian RNA virus clones due to Muller ratchet. *Proc. Natl Acad. Sci. USA* **89**, 6015–6019.

Edwards, C.T., Holmes, E.C., Wilson, D.J., Viscidi, R.P., Abrams, E.J., Phillips, R.E. and Drummond, A.J. (2006) Population genetic estimation of the loss of genetic diversity during horizontal transmission of HIV-1. *BMC Evol. Biol.* **6**, 28.

Escarmis, C., Davila, M., Charpentier, N., Bracho, A., Moya, A. and Domingo, E. (1996) Genetic lesions associated with Muller's ratchet in an RNA virus. *J. Mol. Biol.* **264**, 255–267.

Eshel, I. and Feldman, M.W. (1970) On the evolutionary effect of recombination. *Theor. Popul. Biol.* **1**, 88–100.

Feldman, M.W., Christiansen, F.B. and Brooks, L.D. (1980) Evolution of recombination in a constant environment. *Proc. Natl Acad. Sci. USA* **77**, 4838–4841.

Ferguson, N.M., deWolf, F., Ghani, A.C., Fraser, C., Donnelly, C.A., Reiss, P. *et al.* (1999) Antigendriven CD4+ T cell and HIV-1 dynamics: residual viral replication under highly active antiretroviral therapy. *Proc. Natl Acad. Sci. USA* **96**, 15167–15172.

Fisher, R.A. (1930) *The Genetical Thery of Natural Selection*. Oxford: Clarendon Press.

Fraser, C. (2005) HIV recombination: what is the impact on antiretroviral therapy? *J R Soc Interface* **2**, 489–503.

Fraser, C., Ferguson, N.M. de Wolf, F. and Anderson, R. M. (2001) The role of antigenic stimulation and cytotoxic T cell activity in regulating the long-term immunopathogenesis of HIV: mechanisms and clinical implications. *Proc. R Soc. Lond. B Biol. Sci.* **268**, 2085–2095.

Frater, A.J., Edwards, C.T., McCarthy, N., Fox, J., Brown, H., Milicic, A. *et al.* (2006) Passive sexual transmission of human immunodeficiency virus type 1 variants and adaptation in new hosts. *J. Virol.* **80**, 7226–72234.

Frost, S.D., Dumaurier, M.J., Wain-Hobson, S. and Brown, A.J. (2001) Genetic drift and within-host metapopulation dynamics of HIV-1 infection. *Proc. Natl Acad. Sci. USA* **98**, 6975–6980.

Galetto, R. and Negroni, M. (2005) Mechanistic features of recombination in HIV. *Aids Rev.* **7**, 92–102.

Galetto, R., Moumen, A., Giacomoni, V., Veron, M., Charneau, P. and Negroni, M. (2004) The structure of HIV-1 genomic RNA in the gp120 gene determines a recombination hot spot in vivo. *J. Biol. Chem.* **279**, 36625–36632.

Grossman, Z., Feinberg, M.B. and Paul, W.E. (1998) Multiple modes of cellular activation and virus transmission in HIV infection: a role for chronically and latently infected cells in sustaining viral replication. *Proc. Natl Acad. Sci. USA* **95**, 6314–6319.

Grossman, Z., Polis, M., Feinberg, M.B., Levi, I., Jankelevich, S., Yarchoan, R. *et al.* (1999) Ongoing HIV dissemination during HAART. *Nat. Med.* **5**, 1099–1104.

Gu, Z., Gao, Q., Faust, E.A. and Wainberg, M.A. (1995) Possible involvement of cell fusion and viral recombination in generation of human immunodeficiency virus variants that display dual resistance to AZT and 3TC. *J. Gen. Virol.* **76**(Pt 10), 2601–2605.

Haase, A.T., Henry, K., Zupancic, M., Sedgewick, G., Faust, R.A., Melroe, H. *et al.* (1996) Quantitative image analysis of HIV-1 infection in lymphoid tissue. *Science* **274**, 985–989.

Hamilton, W.D. (1980) Sex versus non-sex versus parasite. *Oikos* **35**, 282–290.

Hamilton, W.D., Axelrod, R. and Tanese, R. (1990) Sexual reproduction as an adaptation to resist parasites (a review). *Proc. Natl Acad. Sci. USA* **87**, 3566–3573.

Hazenberg, M.D., Stuart, J.W., Otto, S.A., Borleffs, J.C., Boucher, C.A., de Boer, R.J. *et al.* (2000) T-cell division in human immunodeficiency virus (HIV)-1 infection is mainly due to immune activation: a longitudinal analysis in patients before and during highly active antiretroviral therapy (HAART). *Blood* **95**, 249–255.

Hill, W.G. and Robertson, A. (1966) The effect of linkage on the limits to artificial selection. *Genet. Res.* **8**, 269–294.

Ho, D.D., Neumann, A.U., Perelson, A.S., Chen, W., Leonard, J.M. and Markowitz, M. (1995) Rapid turnover of plasma virions and CD4 lymphocytes in HIV-1 infection. *Nature* **373**, 123–126.

Hockett, R.D., Kilby, J.M., Derdeyn, C.A., Saag, M.S., Sillers, M., Squires, K. *et al.* (1999) Constant mean viral copy number per infected cell in tissues regardless of high, low, or undetectable plasma HIV RNA. *J. Exp. Med.* **189**, 1545–1554.

Hubert, J.B., Burgard, M., Dussaix, E., Tamalet, C., Deveau, C., Le Chenadec, J. *et al.* (2000) Natural history of serum HIV-1 RNA levels in 330 patients with a known date of infection. The SEROCO Study Group. *Aids* **14**, 123–131.

Iles, M.M., Walters, K. and Cannings, C. (2003) Recombination can evolve in large finite populations given selection on sufficient loci. *Genetics* **165**, 2249–2258.

Jaenike, J. (1978) An hypothesis to account for the maintenance of sex within populations. *Evol. Theory* **3**, 191.

Jetzt, A.E., Yu, H., Klarmann, G.J., Ron, Y., Preston, B.D. and Dougherty, J.P. (2000) High rate of recombination throughout the human immunodeficiency virus type 1 genome. *J. Virol.* **74**, 1234–1240.

Jin, X., Bauer, D.E., Tuttleton, S.E., Lewin, S., Gettie, A., Blanchard, J. *et al.* (1999) Dramatic rise in plasma viremia after CD8(+) T cell depletion in simian immunodeficiency virus-infected macaques. *J. Exp. Med.* **189**, 991–998.

Jones, L.E. and Perelson, A.S. (2007) Transient viremia, plasma viral load, and reservoir replenishment in HIV-infected patients on antiretroviral therapy. *J. Acquir. Immune Defic. Syndr.* **45**, 483–493.

Jung, A., Maier, R., Vartanian, J.P., Bocharov, G., Jung, V., Fischer, U. *et al.* (2002) Recombination—Multiply infected spleen cells in HIV patients. *Nature* **418**, 144.

Junghans, R.P., Boone, L.R. and Skalka, A.M. (1982) Retroviral DNA H-structures—displacement-assimilation model of recombination. *Cell* **30**, 53–62.

Kapoor, A., Jones, M., Shafer, R.W., Rhee, S.Y., Kazanjian, P. and Delwart, E.L. (2004) Sequencing-based detection of low-frequency human immunodeficiency virus type 1 drugresistant mutants by an RNA/DNA heteroduplex generator-tracking assay. *J. Virol.* **78**, 7112–7123.

Karlsson, A.C., Gaines, H., Sallberg, M., Lindback, S. and Sonnerborg, A. (1999) Reappearance of founder virus sequence in human immunodeficiency virus type 1-infected patients. *J. Virol.* **73**, 6191–6196.

Kellam, P. and Larder, B.A. (1995) Retroviral recombination can lead to linkage of reverse transcriptase mutations that confer increased zidovudine resistance. *J. Virol.* **69**, 669–674.

Kelly, J.K., Williamson, S., Orive, M.E., Smith, M.S. and Holt, R.D. (2003) Linking dynamical and population genetic models of persistent viral infection. *Am. Nat.* **162**, 14–28.

Kim, H. and Perelson, A.S. (2006) Viral and latent reservoir persistence in HIV-1-infected patients on therapy. *PLoS Comput. Biol.* **2**, e135.

Klenerman, P., Phillips, R.E., Rinaldo, C.R., Wahl, L.M., Ogg, G., May, R.M. *et al.* (1996) Cytotoxic T lymphocytes and viral turnover in HIV type 1 infection. *Proc. Natl Acad. Sci. USA* **93**, 15323–15328.

Kondrashov, A.S. (1982) Selection against harmful mutations in large sexual and asexual populations. *Genet. Res.* **40**, 325–332.

Kondrashov, A.S. (1988) Deleterious mutations and the evolution of sexual reproduction. *Nature* **336**, 435–440.

Kondrashov, A.S. (1993) Classification of hypotheses on the advantage of amphimixis. *J. Hered.* **84**, 372–387.

Korber, B., Gaschen, B., Yusim, K., Thakallapally, R., Kesmir, C. and Detours, V. (2001) Evolutionary and immunological implications of contemporary HIV-1 variation. *Br. Med. Bull.* **58**, 19–42.

Kouyos, R.D., Althaus, C.L. and Bonhoeffer, S. (2006) Stochastic or deterministic: what is the effective population size of HIV-1? *Trends Microbiol.* **14**, 507–511.

Kwa, D., Vingerhoed, J., Boeser-Nunnink, B., Broersen, S. and Schuitemaker, H. (2001) Cytopathic effects of non-syncytium-inducing and syncytium-inducing human immunodeficiency virus type 1 variants on different CD4(+)-T-cell subsets are determined only by coreceptor expression. *J. Virol.* **75**, 10455–10459.

Leigh-Brown, A.J. (1997) Analysis of HIV-1 env gene sequences reveals evidence for a low effective number in the viral population. *Proc. Natl Acad. Sci. USA* **94**, 1862–1865.

Levy, D.N., Aldrovandi, G.M., Kutsch, O. and Shaw, G.M. (2004) Dynamics of HIV-1 recombination in its natural target cells. *Proc. Natl Acad. Sci. USA* **101**, 4204–4209.

Lin, Y.L., Mettling, C., Portales, P., Reynes, J., Clot, J. and Corbeau, P. (2002) Cell surface CCR5 density determines the postentry efficiency of R5 HIV-1 infection. *Proc. Natl Acad. Sci. USA* **99**, 15590–15595.

Little, S.J., McLean, A.R., Spina, C.A., Richman, D.D. and Havlir, D.V. (1999) Viral dynamics of acute HIV-1 infection. *J. Exp. Med.* **190**, 841–850.

Liu, R., Paxton, W.A., Choe, S., Ceradini, D., Martin, S.R., Horuk, R. *et al.* (1996) Homozygous defect in HIV-1 coreceptor accounts for resistance of some multiply-exposed individuals to HIV-1 infection. *Cell* **86**, 367–377.

Louie, M., Hogan, C., Di Mascio, M., Hurley, A., Simon, V., Rooney, J. *et al.* (2003) Determining the relative efficacy of highly active antiretroviral therapy. *J. Infect Dis.* **187**, 896–900.

Malaspina, A., Moir, S., Nickle, D.C., Donoghue, E.T., Ogwaro, K.M., Ehler, L.A. *et al.* (2002) Human immunodeficiency virus type 1 bound to B cells: relationship to virus replicating in CD4+ T cells and circulating in plasma. *J. Virol.* **76**, 8855–8863.

Mansky, L.M. (1996) Forward mutation rate of human immunodeficiency virus type 1 in a T lymphoid cell line. *AIDS Res. Hum. Retroviruses* **12**, 307–314.

Mansky, L.M. and Temin, H.M. (1995) Lower in vivo mutation rate of human immunodeficiency virus type 1 than that predicted from the fidelity of purified reverse transcriptase. *J. Virol.* **69**, 5087–5094.

Mansky, L.M., Le Rouzic, E., Benichou, S. and Gajary, L.C. (2003) Influence of reverse transcriptase variants, drugs, and Vpr on human immunodeficiency virus type 1 mutant frequencies. *J. Virol.* **77**, 2071–2080.

Markowitz, M., Louie, M., Hurley, A., Sun, E., Di Mascio, M., Perelson, A.S. and Ho, D.D. (2003) A novel antiviral intervention results in more accurate assessment of human immunodeficiency virus type 1 replication dynamics and T-cell decay in vivo. *J. Virol.* **77**, 5037–5038.

Mattapallil, J.J., Douek, D.C., Hill, B., Nishimura, Y., Martin, M. and Roederer, M. (2005) Massive infection and loss of memory CD4+ T cells in multiple tissues during acute SIV infection. *Nature* **434**, 1093–1097.

Mellors, J.W., Rinaldo, C.R., Jr., Gupta, P., White, R.M., Todd, J.A. and Kingsley, L.A. (1996) Prognosis in HIV-1 infection predicted by the quantity of virus in plasma. *Science* **272**, 1167–1170.

Mittler, J., Essunger, P., Yuen, G.J., Clendeninn, N., Markowitz, M. and Perelson, A.S. (2001) Short-term measures of relative efficacy predict longer-term reductions in human immunodeficiency virus type 1 RNA levels following nelfinavir monotherapy. *Antimicrob. Agents Chemother.* **45**, 1438–1443.

Moutouh, L., Corbeil, J. and Richman, D.D. (1996) Recombination leads to the rapid emergence of HIV-1 dually resistant mutants under selective drug pressure. *Proc. Natl Acad. Sci. USA* **93**, 6106–6111.

Müller, V. and Bonhoeffer, S. (2003) Mathematical approaches in the study of viral kinetics and drug resistance in HIV-1 infection. *Curr. Drug Targets Infect. Disord.* **3**, 329–344.

Müller, V., Marée, A.F. and De Boer, R.J. (2001a) Release of virus from lymphoid tissue affects human immunodeficiency virus type 1 and hepatitis C virus kinetics in the blood. *J. Virol.* **75**, 2597–2603.

Müller, V., Maree, A.F. and De Boer, R.J. (2001b) Small variations in multiple parameters account for wide variations in HIV-1 set-points: a novel modelling approach. *Proc. R Soc. Lond. B Biol. Sci.* **268**, 235–242.

Müller, V., Vigueras-Gomez, J.F. and Bonhoeffer, S. (2002) Decelerating decay of latently infected cells during prolonged therapy for human immunodeficiency virus type 1 infection. *J. Virol.* **76**, 8963–8965.

Najera, I., Richman, D.D., Olivares, I., Rojas, J.M., Peinado, M.A., Perucho, M. *et al.* (1994) Natural occurrence of drug resistance mutations in the reverse transcriptase of human immunodeficiency virus type 1 isolates. *AIDS Res. Hum. Retroviruses* **10**, 1479–1488.

Najera, I., Holguin, A., Quinones-Mateu, M.E., Munoz-Fernandez, M.A., Najera, R., Lopez-Galindez, C. and Domingo, E. (1995) Pol gene quasispecies of human immunodeficiency virus: mutations associated with drug resistance in virus from patients undergoing no drug therapy. *J. Virol.* **69**, 23–31.

Nijhuis, M., Boucher, C.A., Schipper, P., Leitner, T., Schuurman, R. and Albert, J. (1998) Stochastic processes strongly influence HIV-1 evolution during suboptimal proteaseinhibitor therapy. *Proc. Natl Acad. Sci. USA* **95**, 14441–14446.

Novick, A. and Szilard, L. (1951) Virus strains of identical phenotype but different genotype. *Science* **113**, 34–35.

Nowak, M.A. and May, R.M. (2000) *Virus Dynamics: Mathematical Principles of Immunology and Virology*. Oxford: Oxford University Press.

Nowak, M.A., Bonhoeffer, S., Shaw, G.M. and May, R.M. (1997a) Anti-viral drug treatment: dynamics of resistance in free virus and infected cell populations. *J. Theor. Biol.* **184**, 203–217.

Nowak, M.A., Lloyd, A.L., Vasquez, G.M., Wiltrout, T.A., Wahl, L.M., Bischofberger, N. *et al.* (1997b) Viral dynamics of primary viremia and antiretroviral therapy in simian immunodeficiency virus infection. *J. Virol.* **71**, 7518–7525.

O'Neil, P.K., Sun, G., Yu, H., Ron, Y., Dougherty, J.P. and Preston, B.D. (2002) Mutational analysis of HIV-1 long terminal repeats to explore the relative contribution of reverse transcriptase and RNA polymerase II to viral mutagenesis. *J. Biol. Chem.* **277**, 38053–38061.

Otto, S.P. and Barton, N.H. (2001) Selection for recombination in small populations. *Evolution* **55**, 1921–1931.

Otto, S.P. and Lenormand, T. (2002) Resolving the paradox of sex and recombination. *Nat. Rev. Genet.* **3**, 252–261.

Pantaleo, G., Graziosi, C., Demarest, J.F., Butini, L., Montroni, M., Fox, C.H. *et al.* (1993) HIV infection is active and progressive in lymphoid tissue during the clinically latent stage of disease. *Nature* **362**, 355–358.

Perelson, A.S. (2002) Modelling viral and immune system dynamics. *Nat. Rev. Immunol.* **2**, 28–36.

Perelson, A.S., Neumann, A.U., Markowitz, M., Leonard, J.M. and Ho, D.D. (1996) HIV-1 dynamics in vivo: virion clearance rate, infected cell life-span, and viral generation time. *Science* **271**, 1582–1586.

Perelson, A.S., Essunger, P., Cao, Y., Vesanen, M., Hurley, A., Saksela, K., Markowitz, M. and Ho, D.D. (1997) Decay characteristics of HIV-1-infected compartments during combination therapy. *Nature* **387**, 188–191.

Phillips, A.N. (1996) Reduction of HIV concentration during acute infection: independence from a specific immune response. *Science* **271**, 497–499.

Piatak, M., Jr., Saag, M.S., Yang, L.C., Clark, S.J., Kappes, J.C., Luk, K.C. *et al.* (1993) High levels of HIV-1 in plasma during all stages of infection determined by competitive PCR. *Science* **259**, 1749–1754.

Ramratnam, B., Bonhoeffer, S., Binley, J., Hurley, A., Zhang, L., Mittler, J.E. *et al.* (1999) Rapid production and clearance of HIV-1 and hepatitis C virus assessed by large volume plasma apheresis. *Lancet* **354**, 1782–1785.

Ramratnam, B., Mittler, J.E., Zhang, L., Boden, D., Hurley, A., Fang, F. *et al.* (2000) The decay of the latent reservoir of replication-competent HIV-1 is inversely correlated with the extent of residual viral replication during prolonged anti-retroviral therapy. *Nat. Med.* **6**, 82–85.

Ramratnam, B., Ribeiro, R., He, T., Chung, C., Simon, V., Vanderhoeven, J. *et al.* (2004) Intensification of antiretroviral therapy accelerates the decay of the HIV-1 latent reservoir and decreases, but does not eliminate, ongoing virus replication. *J. Acquir. Immune Defic. Syndr.* **35**, 33–37.

Reddy, B. and Yin, J. (1999) Quantitative intracellular kinetics of HIV type 1. *AIDS Res. Hum. Retroviruses* **15**, 273–283.

Regoes, R.R., Antia, R., Garber, D.A., Silvestri, G., Feinberg, M.B. and Staprans, S.I. (2004) Roles of target cells and virus-specific cellular immunity in primary simian immunodeficiency virus infection. *J. Virol.* **78**, 4866–4875.

Reynes, J., Portales, P., Segondy, M., Baillat, V., Andre, P., Reant, B. *et al.* (2000) CD4+ T cell surface CCR5 density as a determining factor of virus load in persons infected with human immunodeficiency virus type 1. *J. Infect. Dis.* **181**, 927–932.

Ribeiro, R.M. and Bonhoeffer, S. (2000) Production of resistant HIV mutants during antiretroviral therapy. *Proc. Natl Acad. Sci. USA* **97**, 7681–7686.

Ribeiro, R.M., Bonhoeffer, S. and Nowak, M.A. (1998) The frequency of resistant mutant virus before antiviral therapy. *Aids* **12**, 461–465.

Rouzine, I.M. and Coffin, J.M. (1999) Linkage disequilibrium test implies a large effective population number for HIV in vivo. *Proc. Natl Acad. Sci. USA* **96**, 10758–10763.

Sabin, C.A., Devereux, H., Phillips, A.N., Hill, A., Janossy, G., Lee, C.A. and Loveday, C. (2000) Course of viral load throughout HIV-1 infection. *J. Acquir. Immune Defic. Syndr.* **23**, 172–177.

Schmitz, J.E., Kuroda, M.J., Santra, S., Sasseville, V.G., Simon, M.A., Lifton, M.A. *et al.* (1999) Control of viremia in simian immunodeficiency virus infection by CD8+ lymphocytes. *Science* **283**, 857–860.

Seo, T.K., Thorne, J.L., Hasegawa, M. and Kishino, H. (2002) Estimation of effective population size of HIV-1 within a host: a pseudomaximum-likelihood approach. *Genetics* **160**, 1283–1293.

Sheppard, H.W., Celum, C., Michael, N.L., O'Brien, S., Dean, M., Carrington, M. *et al.* (2002) HIV-1 infection in individuals with the CCR5-Delta32/Delta32 genotype: acquisition of syncytium-inducing virus at seroconversion. *J. Acquir. Immune Defic. Syndr.* **29**, 307–313.

Shriner, D., Shankarappa, R., Jensen, M.A., Nickle, D.C., Mittler, J.E., Margolick, J.B. and Mullins, J.I. (2004) Influence of random genetic drift on human immunodeficiency virus type 1 env evolution during chronic infection. *Genetics* **166**, 1155–1164.

Smith, R.A., Anderson, D.J. and Preston, B.D. (2004) Purifying selection masks the mutational flexibility of HIV-1 reverse transcriptase. *J. Biol. Chem.* **279**, 26726–26734.

Speirs, C., van Nimwegen, E., Bolton, D., Zavolan, M., Duvall, M., Angleman, S. *et al.* (2005) Analysis of human immunodeficiency virus cytopathicity by using a new method for quantitating viral dynamics in cell culture. *J. Virol.* **79**, 4025–4032.

Stafford, M.A., Corey, L., Cao, Y., Daar, E.S., Ho, D.D. and Perelson, A.S. (2000) Modeling plasma virus concentration during primary HIV infection. *J. Theor. Biol.* **203**, 285–301.

Strain, M.C., Günthard, H.F., Havlir, D.V., Ignacio, C.C., Smith, D.M., Leigh-Brown, A.J. *et al.* (2003) Heterogeneous clearance rates of long-lived lymphocytes infected with HIV: intrinsic stability predicts lifelong persistence. *Proc. Natl Acad. Sci. USA* **100**, 4819–4824.

Sun, Y., Iglesias, E., Samri, A., Kamkamidze, G., Decoville, T., Carcelain, G. and Autran, B. (2003) A systematic comparison of methods to measure HIV-1 specific CD8 T cells. *J. Immunol. Methods* **272**, 23–34.

Temin, H.M. (1991) Sex and recombination in retroviruses. *Trends Genet.* **7**, 71–74.

Vogt, P. (1973) The genome of avian RNA tumour viruses: a discussion of four models. *Possible Episomes in Eukaryotes*. Amsterdam: North Holland Publishing Co.

Wain-Hobson, S., Renoux-Elbe, C., Vartanian, J.P. and Meyerhans, A. (2003) Network analysis of human and simian immunodeficiency virus sequence sets reveals massive recombination resulting in shorter pathways. *J. Gen. Virol.* **84**, 885–895.

Wei, X., Ghosh, S.K., Taylor, M.E., Johnson, V.A., Emini, E.A., Deutsch, P. *et al.* (1995) Viral dynamics in human immunodeficiency virus type 1 infection. *Nature* **373**, 117–122.

Wodarz, D. and Nowak, M.A. (1999) Specific therapy regimes could lead to long-term immunological control of HIV. *Proc. Natl Acad. Sci. USA* **96**, 14464–14469.

Wodarz, D. and Nowak, M.A. (2002) Mathematical models of HIV pathogenesis and treatment. *Bioessays* **24**, 1178–1187.

Wright, S. (1931) Evolution in Mendelian populations. *Genetics* **16**, 97–159.

Wright, S. (1938) Size of population and breeding structure in relation to evolution. *Science* **87**, 430–431.

Wu, H., Huang, Y., Acosta, E.P., Rosenkranz, S.L., Kuritzkes, D.R., Eron, J.J. *et al.* (2005) Modeling long-term HIV dynamics and antiretroviral response: effects of drug potency, pharmacokinetics, adherence, and drug resistance. *J. Acquir. Immune Defic. Syndr.* **39**, 272–283.

Yu, H., Jetzt, A.E., Ron, Y., Preston, B.D. and Dougherty, J.P. (1998) The nature of human immunodeficiency virus type 1 strand transfers. *J. Biol. Chem.* **273**, 28384–28391.

Yuste, E., Sanchez-Palomino, S., Casado, C., Domingo, E. and Lopez-Galindez, C. (1999) Drastic fitness loss in human immunodeficiency virus type 1 upon serial bottleneck events. *J. Virol.* **73**, 2745–2751.

Zhang, L., Dailey, P.J., He, T., Gettie, A., Bonhoeffer, S., Perelson, A.S. and Ho, D.D. (1999) Rapid clearance of simian immunodeficiency virus particles from plasma of rhesus macaques. *J. Virol.* **73**, 855–860.

Zhang, L., Dailey, P.J., Gettie, A., Blanchard, J. and Ho, D.D. (2002) The liver is a major organ for clearing simian immunodeficiency virus in rhesus monkeys. *J. Virol.* **76**, 5271–5273.

Zhang, H., Yang, B., Pomerantz, R.J., Zhang, C., Arunachalam, S.C. and Gao, L. (2003) The cytidine deaminase CEM15 induces hypermutation in newly synthesized HIV-1 DNA. *Nature* **424**, 94–98.

Zhuang, J.L., Jetzt, A.E., Sun, G.L., Yu, H., Klarmann, G., Ron, Y., Preston, B.D. and Dougherty, J.P. (2002) Human immunodeficiency virus type 1 recombination: Rate, fidelity, and putative hot spots. *J. Virol.* **76**, 11273–11282.

CHAPTER

15

The Impact of Rapid Evolution of Hepatitis Viruses

J. Quer, M. Martell, F. Rodriguez, A. Bosch, R. Jardi, M. Buti, and J.I. Esteban

ABSTRACT

Hepatitis viruses comprise a group of very diverse pathogens that primarily infect the liver, but belong to very different virus families with very different replication strategies (hepatitis A virus (HAV) *Picornaviridae*, hepatitis B virus (HBV) *Hepadnaviridae*, hepatitis C virus (HCV) *Flaviviridae*, hepatitis delta virus (HDV) genus *Deltavirus*, not assigned to a family yet, and hepatitis E virus (HEV) *Hepeviridae*). All of them have in common a high genome plasticity, and have received special attention because of their worldwide distribution in human population, infecting hundreds of million people, causing either acute and/or chronic infections that in many cases lead to liver cirrhosis and hepatocellular carcinoma, which is one of the leading causes of death worldwide. The huge number of infected people all over the world is the best proof of how different replication and transmission strategies, with the common factor of variability, may succeed in terms of viral persistence.

HEPATITIS A VIRUS

Features of HAV

Hepatitis A virus (HAV), the etiological agent of hepatitis A, belongs to the genus *Hepatovirus* within the *Picornaviridae* family, and is a non-enveloped agent with an icosahedral capsid of ~30 nm in diameter containing a positive single-stranded (ss)RNA genome of 7.5 kb (Feinstone *et al.*, 1973; Fauquet *et al.*, 2005). The genome contains a single open reading frame (ORF) encoding a polyprotein of 2225 amino acids preceded by a 5′ non-coding-region (5′ NCR), which represents ~10% of the genome, and followed by a much shorter 3′ NCR that contains a poly A tract (Baroudy *et al.*, 1985; Cohen *et al.*, 1987). This genome is uncapped but covalently linked to a small viral protein (VPg) (Weitz *et al.*, 1986). The singly translated polyprotein is subsequently cleaved into 11 proteins through a cascade of proteolytic events brought about mainly by the viral 3C protease (Figure 15.1) (Schultheiss *et al.*, 1994, 1995). The genetic distance between the

Origin and Evolution of Viruses
ISBN 978-0-12-374153-0

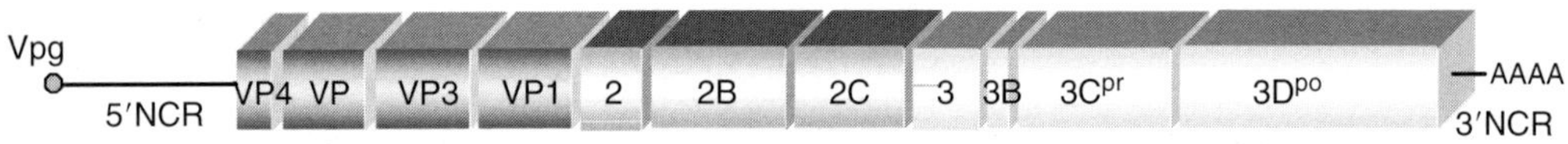

FIGURE 15.1 Hepatitis A virus genome organization.

genus *Hepatovirus* and the other genera of the family reflects not merely a difference in the nucleotide and amino acid composition but a difference in the molecular and biological characteristics of HAV.

The main characteristics that differentiate HAV from other picornaviruses are, first, the usage of an internal ribosome entry site (IRES) that shows a very low efficiency in directing translation (Brown *et al.*, 1994; Whetter *et al.*, 1994; Ehrenfeld and Teterina, 2007); second, the codification of a single protease, 3C, which limits the HAV capacity to compete for the cellular translational machinery (Leong *et al.*, 2002); and finally, its high adaptation to use rare codons (Borman and Kean, 1997; Jackson, 2002) to avoid, as much as possible, competition for the cellular tRNAs (Sanchez *et al.*, 2003a) and likely contributing to its slow replication and to its low yields (see reviews by Cristina and Costa-Mattioli, 2007; and Pinto *et al.*, 2007).

HAV Generation of Variation

Mutation

As other RNA viruses, HAV exists *in vivo* as a distribution of closely related variants referred to as quasispecies (Costa-Mattioli *et al.*, 2003a; Sanchez *et al.*, 2003b). The HAV rate of evolution has been estimated to be 1×10^{-3} to 1×10^{-4} substitutions per site per year (Sanchez *et al.*, 2003b) similar to that of other RNA viruses, which implies a continuous generation of variant viral genomes. However, very low capsid variability has been reported, a feature that correlates with very low antigenic variation (a single serotype exists) suggesting that negative selection strongly constrains potentially newly arising protein variants. It has been proposed that such constraints may be related to cohesive stability needed to overcome the challenges posed by the acidic pH of the stomach during the entry phase, to escape from erythrocyte attachment (Sanchez *et al.*, 2004), and the detergent action of biliary salts during the exit phase. Quasispecies analysis has revealed a dynamics of mutation-selection occurring at and around the rare codons, which encode 15% of its surface amino acid residues, confirming a seminal role of the codon usage on HAV evolution subject to negative selection (Sanchez *et al.*, 2003b).

Genotypes and Subtypes

HAV has been classified into a single serotype (Lemon and Binn, 1983) and six genotypes (Robertson *et al.* 1992; Costa-Mattioli *et al.*, 2003a), three (I, II, and III) of human origin and the remainder (IV, V, and VI) of simian origin. Genotypes I and II contain two subtypes each (Ia, Ib, IIa, and IIb) defined by a nucleotide divergence of 7–7.5%. Nucleotide divergence among genotypes is not homogeneous along the genome, being minimal (<5%) at the 5′ NCR (the most conserved genomic region), and having similar degrees of nucleotide variability to those of other picornavirus in the capsid coding region (Sanchez *et al.*, 2003a).

Recombination

The first case of recombination in HAV *in vivo* was reported by Costa-Mattioli *et al.* in 2003. Since then it was considered that recombination was cell culture-related (Lemon *et al.*, 1991; Beard *et al.*, 2001; Gauss-Muller and Kusov, 2002). The recombinant isolate (9F94) took place at the VP1 capsid protein involving genotype II and IB variants (Costa-Mattioli *et al.*, 2003b). This finding indicates that recombination may occur in nature, thus acting as an important factor to generate HAV diversity that can have important implications for its evolution, biology, and control (Cristina and Costa-Mattioli, 2007).

HAV Transmission and Prevention

Hepatitis A is present in the feces of infected patients at very high concentration (up to 10^{11} genome copies/g) for 2 weeks after the onset of symptoms and lasting at least 4 more weeks (Costafreda *et al.*, 2006). Although exceptionally transmitted by the parenteral route (Noble *et al.*, 1984; Sherertz *et al.*, 1984), HAV is primarily transmitted by the fecal–oral route during person-to-person contact (Mast and Alter, 1993), although contaminated water, food, and fomites may be vehicles for its transmission, especially when poor sanitary conditions occur (Abad *et al.*, 1994; Sanchez *et al.*, 2002; Bosch *et al.*, 2007). The current seroprevalence of anti-HAV antibodies in Spain is low in infants, children, and unvaccinated adolescents or young adults but increases with age, reaching 68% among people older than 50 (Bruguera *et al.*, 1999).

Inactivated HAV vaccines have been available since the early 1990s and provide long-lasting immunity against hepatitis A infection (Bell and Feinstone, 2004) with the induction of protective high-titer specific antibodies. The efficacy of these vaccines is related to the existence of a single serotype of HAV, and the only drawback to the implementation of universal massive vaccination campaigns is the economic cost of vaccine production (Salleras, 1999; Shouval, 1999; Franco and Vitiello, 2003; Dagan *et al.*, 2005; Wasley *et al.*, 2005). Vaccination against hepatitis A should be recommended to high-risk groups, including healthcare workers, travelers to high endemic areas, men having sex with men, drug users, and patients receiving blood products. In addition, the inclusion of hepatitis A vaccines in mass vaccination programs in developed countries receiving high numbers of immigrants from endemic countries is particularly advisable (as is the case in Spain). However, given the quasispecies nature of HAV (Sanchez *et al.*, 2003b), which could lead, in populations with continued exposure to the virus, to the selection of new antigenic variants escaping immune protection, the introduction of mass vaccination programs in highly endemic areas remains controversial.

HEPATITIS B VIRUS

Features of HBV

HBV infects more than 2000 million people worldwide. HBV may cause acute and chronic liver disease that vary greatly among individuals (Ganem and Prince, 2004). In some subjects, HBV infection is self-limited (95% of infected adults, 10–70% early childhood) with virus clearance without clinical sequelae, while in others (more than 350 million people), viral clearance fails (this is specially common in newborns to HBV-infected mothers), and chronic infection (CHB) is established. Up to 30% of CHB infections are active, putting the subject at risk of progressing to liver cirrhosis (LC) and primary hepatocellular carcinoma (HCC) (WHO, 2000a). HBV is transmitted by sexual contact and by parenteral exposure to infected blood, and is more prevalent in the Eastern than in the Western Hemisphere. In high-prevalence regions (>8% of the population infected) (WHO, 2006) such as South and East Asia, mother-to-child perinatal transmission with establishment of a life-long highly infectious carrier state, has long been responsible for the observed high rates of endemicity. In these areas, the majority of HCC (50–80%) cases arise in HBV carriers (Ryu, 2003).

HBV is an enveloped DNA virus, the smallest of all human DNA viruses (diameter of 42 nm), with an icosahedral nucleocapsid of approximately 30 nm diameter enclosing the viral DNA genome and DNA polymerase. It belongs to genus *Hepadnavirus* which within the family *Hepadnaviridae*.

HBV has a partially double-stranded (ds)-DNA genome of only 3.2 kb, with a complete negative strand ((−)-DNA) and an incomplete positive strand ((+)-DNA) (Okamoto *et al.*, 1988). This genome is held in a circular non-covalently closed conformation by a short cohesive overlap between the 5′ ends of the two strands (Figure 15.2A), a structure referred as relaxed circular DNA (rcDNA). The (−)-DNA is 5′ covalently linked to a molecule of the viral polymerase, P protein. The (+)-DNA is 5′ covalently linked to a small RNA oligonucleotide (derived from the RNA precursor). The complete genome encodes for at least seven viral proteins through four highly overlapping (67%) ORFs S, C, P, and X.

The ORF S (Surface) and Envelope Proteins

This ORF has three in-frame ATG start codons, defining three regions: preS1, preS2, and S (Figure 15.2A). Three groups of co-C-terminal S-derived peptides with different degrees of glycosylation are encoded: small or SHBs (only S region, the most abundant), middle or MHBs (preS2 and S), and large or LHBs (preS1, preS2, and S) envelope proteins. All together compose the hepatitis B surface antigen (HBsAg), which is produced in excess and may self-assemble into "empty" non-infectious structures as ~22 nm rods or spheres or become part of true virions (also known as Dane particles) which constitute a minor population (Coleman, 2006). LHBs proteins seem to specify the virus–host range by recognizing cell surface receptors, while SHBs (shared by all envelope proteins) is the immunodominant component of the envelope inducing specific antibodies (anti-HBs response), especially the determinant "a" from the major hydrophilic region (MHR) (positions 100–170) that protrude from the viral surface (Coleman, 2006) (Figure 15.2A,B).

The ORF P and the Polymerase Protein

This ORF covers 80% of hepadnaviral genome. It encodes a single multifunctional protein P which has four distinct domains: an N-terminal domain called terminal protein which primes the synthesis of (−)-DNA strand, a reverse transcriptase (RT)/DNA polymerase and the RNase H (RH) (Figure 15.2A,B).

The ORF C and Core and Pre-core Proteins

This ORF has two in-frame ATG start codons. The core protein (HBcAg) is a product of 185 amino acids, encoded from the second start codon (position 1901). The pre-core protein has 214 amino acids and is translated from the first start codon (position 1814). The post-translational cleavage of the pre-core protein results in a protein, the hepatitis B "e" antigen (HBeAg) that is not used to generate the mature virion, but is secreted from infected cells as a soluble protein acting as an immunomodulator.

The ORF X

This ORF is well conserved among all mammalian hepadnaviruses. HBx protein is a multifunctional viral regulator that modulates transcription, signaling pathways, protein degradation, and cell responses to genotoxic stress, acting on cell cycle checkpoints, cell death, and carcinogenesis (Tang *et al.*, 2006). It has been shown that HBx stimulates HBV replication (Bouchard *et al.*, 2001) probably by RNA transcription enhancement (Tang *et al.*, 2005).

Other Genomic Elements

The HBV genome contains four promoters (preS1, preS2/S, C, and X) which control the transcription by the cell RNA polymerase II of six functionally co-terminal distinct viral mRNAs (Figure 15.2A) copied from the viral (−)-DNA, which is itself the product of reverse transcription of one of the largest viral mRNAs (the pregenome or pgRNA) under the C promoter (Seeger and Mason, 2000). Interestingly, the promoter C (nt 1620–1814) is composed of a minimal or basic core promoter (BCP) sequence that contains the main binding site for a variety of transcription factors such as HNF3, HNF4, SP1, and the chicken ovalbumin upstream promoter transcription factor 1 (COUP-TF1). This binding site is essential for the HBV liver cell specificity (Li *et al.*, 1995; Raney *et al.*, 1997; Gunther *et al.*, 1999).

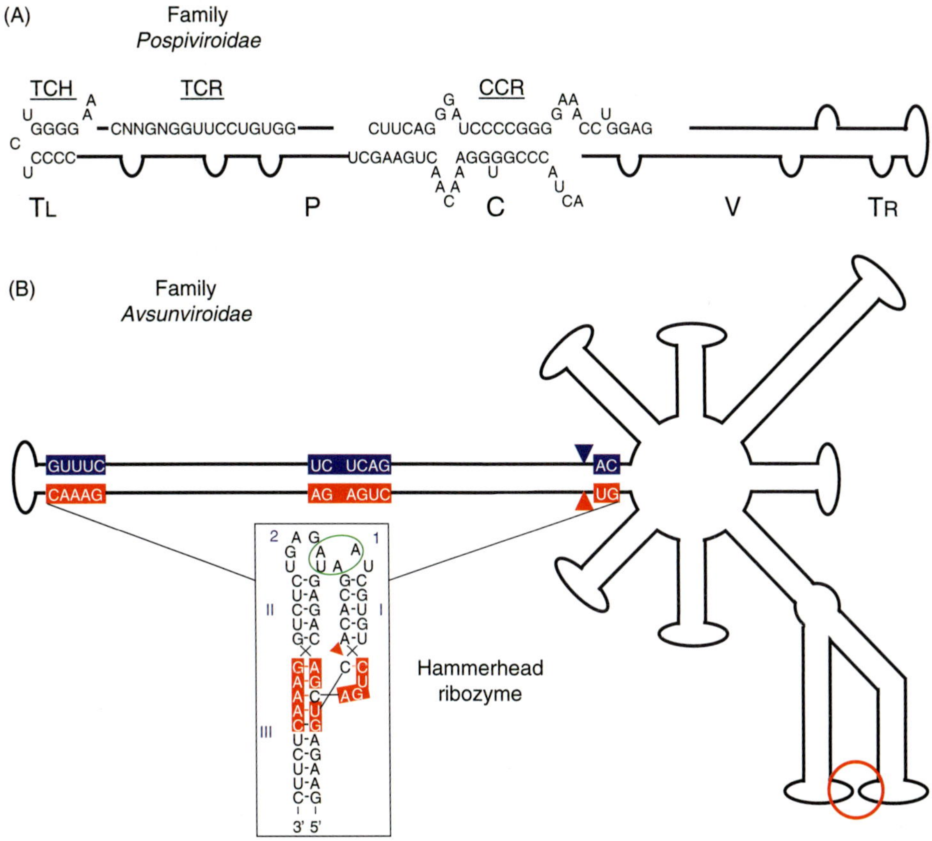

PLATE 1 Viroid structure. (A) Scheme of the rod-like genomic RNA characteristic of the family *Pospiviroidae* with the central (C), pathogenic (P), variable (V) and terminal left and right (T_L and T_R, respectively) domains. The central conserved region (CCR) (genus *Pospiviroid*), the terminal conserved region (TCR) (genera *Pospi-*, *Apsca-*, and part of *Coleviroid*) and the terminal conserved hairpin (TCH) (genera *Hostu-* and *Cocadviroid*), are displayed. (B) Scheme of the branched genomic RNA of PLMVd (family *Avsunviroidae*) in which the sequences conserved in most natural hammerhead ribozymes are displayed on a red and blue background for (+) and (−) polarities, respectively, and the self-cleavage sites by arrowheads. Red circle denotes a kissing-loop interaction. The structure of the (+) hammerhead ribozyme is displayed in the boxed inset, with Roman and Arabic numerals depicting helices I, II, and III, and loops 1 and 2, respectively, and the arrowhead the self-cleavage site. Short black and red lines indicate canonical and non-canonical base pairs, respectively, and the green oval a tertiary interaction between loops 1 and 2 that enhances the catalytic activity.

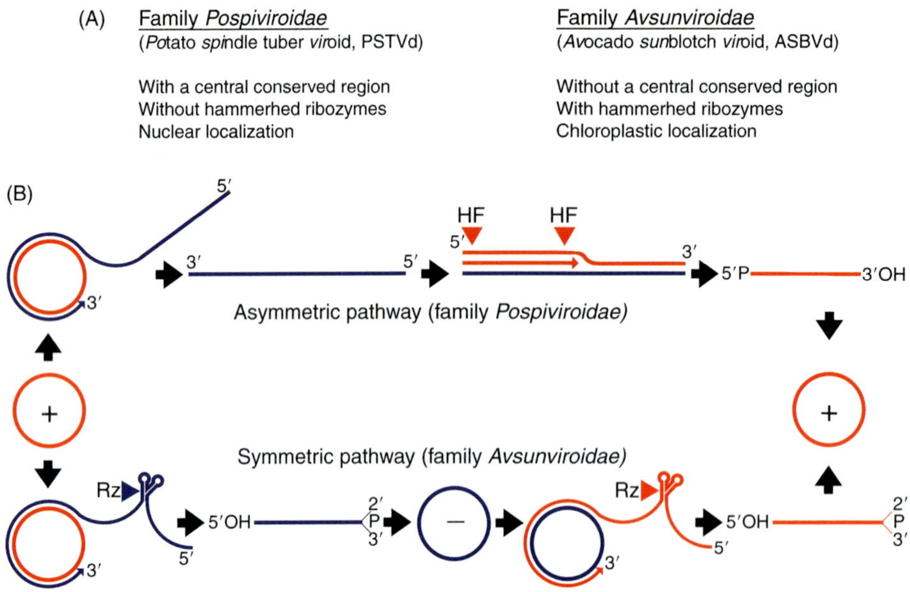

PLATE 2 (A) Distinctive properties of the *Pospiviroidae* and *Avsunviroidae* families. (B) Asymmetric and symmetric pathways of the rolling-circle replication mechanism followed by members of the families *Pospiviroidae* and *Avsunviroidae*, respectively. Red and blue lines refer to (+) and (−) strands, respectively. Arrowheads point to cleavage sites of a host factor (HF) or ribozymes (Rz), and the resulting 5′ and 3′ groups are indicated.

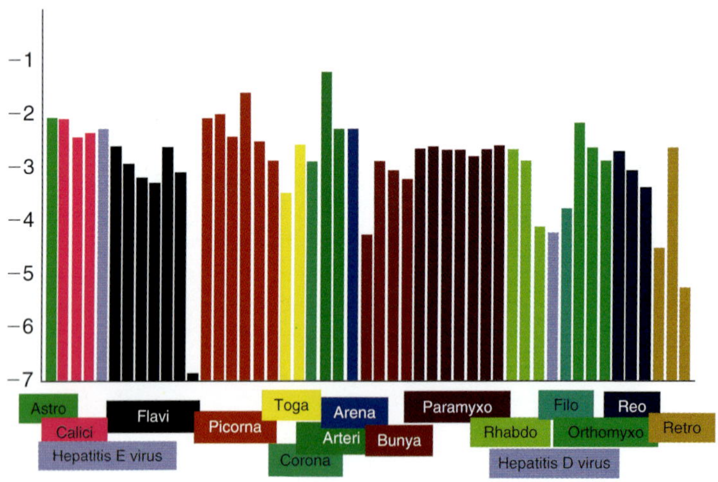

PLATE 3 Rates of nucleotide substitution at synonymous sites among RNA viruses. For simplicity, viruses are color-coded by family. The y-axis records the numbers of synonymous substitutions per site, per year. Taken from Hanada *et al.* (2004) with permission.

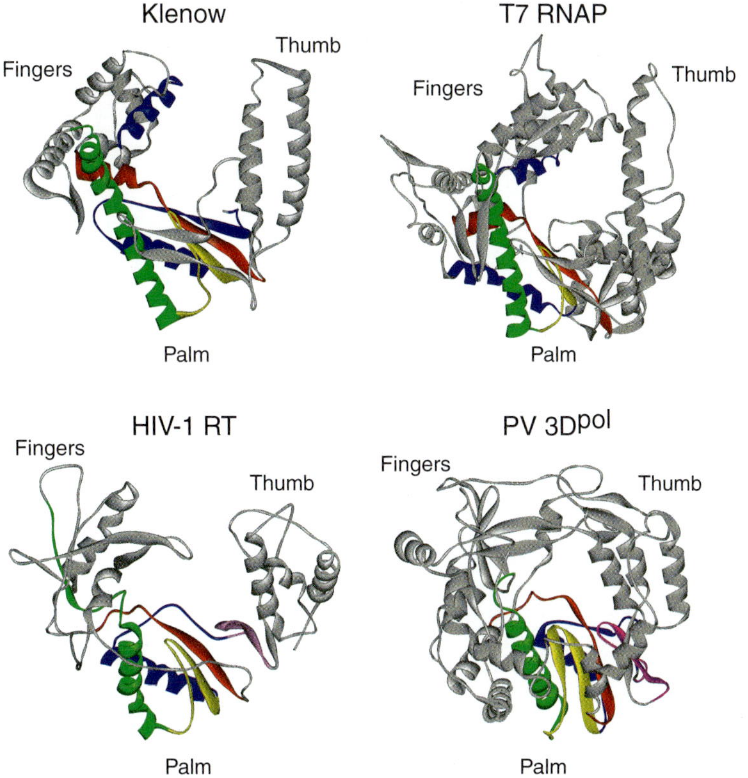

PLATE 4 Structures of the four classes of nucleic acid polymerases. Structures of the polymerase domains of the DdDp large (Klenow) fragment of *E. coli* DNA polymerase I (1KFD) (Beese *et al.*, 1993), RdDp HIV 1 reverse transcriptase (HIV-1 RT) (1RTD) (Huang *et al.*, 1998), DpRp T7 RNAP (1S76) (Yin and Steitz, 2004), and RdRp PV 3Dpol (1RA6) (Thompson and Peersen, 2004). The conserved structural motifs in the palm subdomain are colored as follows: motif A, red; motif B, green; motif C, yellow; motif D, blue; and motif E, purple (HIV-1 RT and PV 3Dpol only). Helix O in Klenow and T7 RNAP is colored in blue. The analogous helix in B-family polymerases is helix P. The images were rendered using the program WebLab Viewer Pro (Molecular Simulations Inc., San Diego, CA).

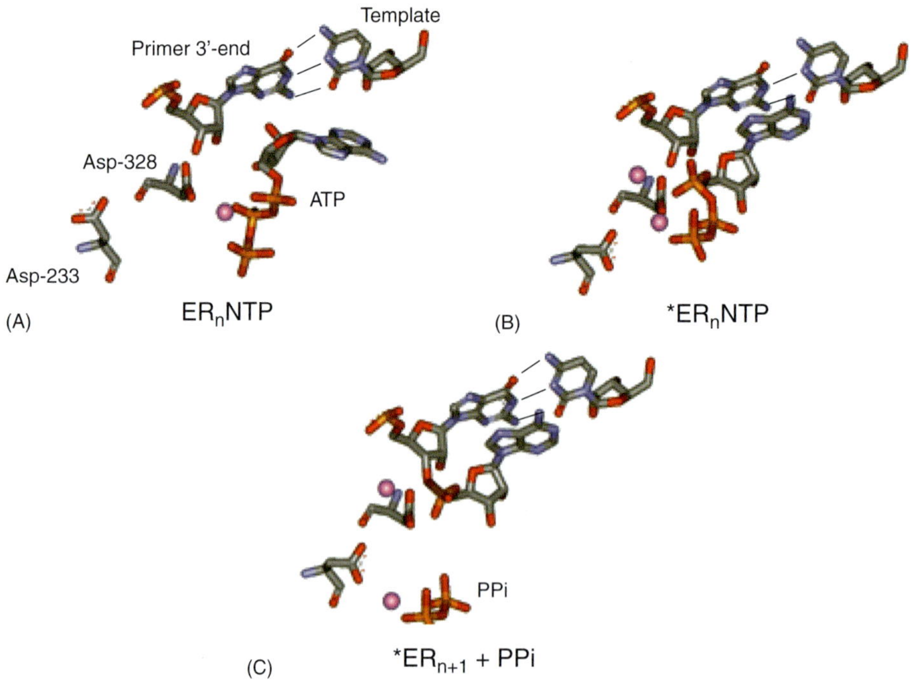

PLATE 5 Structural model for 3Dpol-catalyzed nucleotide incorporation. (A) Ground-state binding of metal-complexed nucleotide. (B) Reorientation of the triphosphate into the catalytically competent configuration. (C) Phosphoryl transfer and pyrophosphate release. While the kinetic mechanism suggests a conformational change prior to pyrophosphate release, kinetic data does not provide any information to permit a molecular description of this step. Images were generated from the model previously described (Gohara *et al.*, 2000). Nucleotide and side-chain motions were derived from (Johnson *et al.*, 2003) by approximate rotation and translation movements. Atom colors correspond to the following: red, oxygen; blue; nitrogen; gray, carbon; magenta, Mg^{2+} or Mn^{2+}. The images were rendered with WebLab Viewer Pro (Accelrys Inc., San Diego, CA). Reproduced with permission from *Biochemistry* (Arnold *et al.*, 2004).

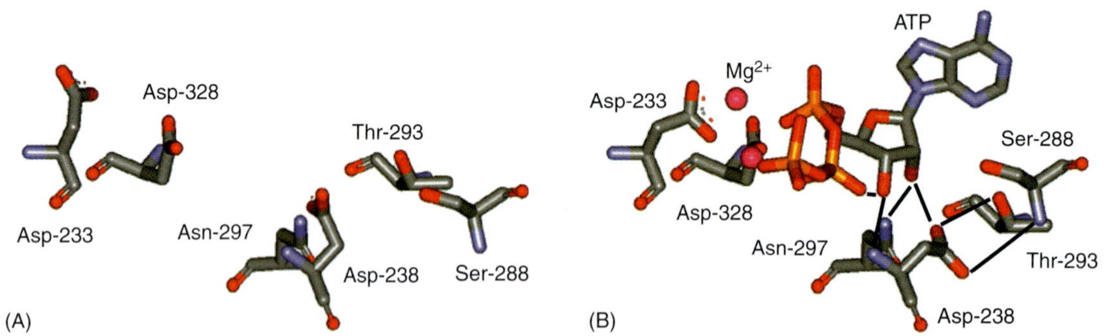

PLATE 6 Nucleotide-binding pocket of 3Dpol. (A) Residues located in the NTP-binding pocket as observed in the unliganded structure of 3Dpol (Hansen *et al.*, 1997; Thompson and Peersen, 2004). Asp233 and Asp238 are from structural motif A; Ser288, Thr293, and Asn297 are from motif B; and Asp328 is from motif C. (B) Model for interaction of 3Dpol with bound nucleotide (Gohara *et al.*, 2000) ATP and metal ions required for catalysis are labeled. In this model, the side-chains for Asp233 and Asp238 have been rotated to permit interactions with ATP. Asp238, Ser288, and Thr293 have been positioned to interact. The image was created by using the program WebLab Viewer (Molecular Simulations Inc., San Diego, CA). Reproduced with permission from *Biochemistry* (Gohara *et al.*, 2004).

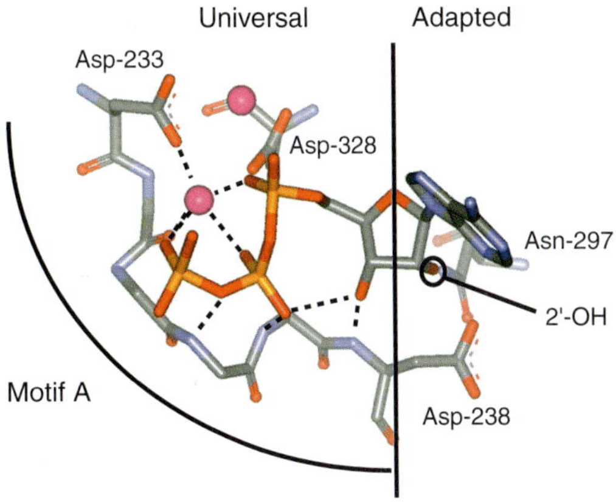

PLATE 7 Structural basis for fidelity. The nucleotide-binding pocket of all nucleic acid polymerases with a canonical "palm"-based active site is highly conserved. The site can be divided into two parts: a region that has "universal" interactions mediated by conserved structural motif A that organize the metals and triphosphate for catalysis, and a region that has "adapted" interactions mediated by conserved structural motif B that dictate whether ribo- or 2′-deoxribonucleotides will be utilized. In the classical polymerase, there is a motif A residue located in the sugar-binding pocket capable of interacting with motif B residue(s) involved in sugar selection. This motif A residue in other polymerases could represent the link between the nature of the bound nucleotide (correct vs. incorrect) to the efficiency of nucleotidyl transfer as described herein for Asp238 of 3Dpol (Gohara *et al.*, 2000). Reproduced with permission from *Biochemistry* (Gohara *et al.*, 2004).

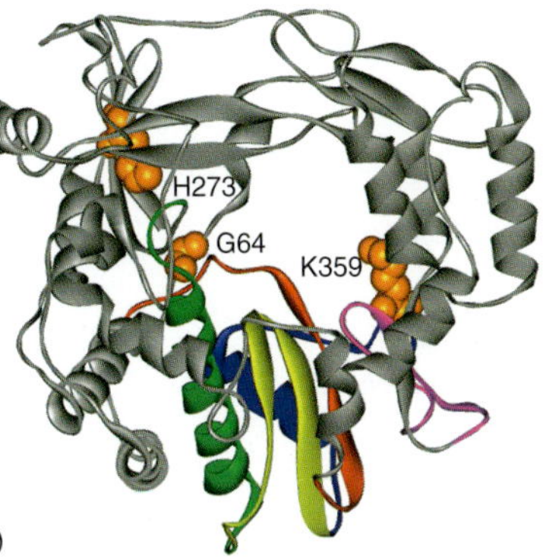

(A)

(B)

Allele	Mutation Frequency		Replication Rate[3]
	Sequencing[1]	Kinetics[2]	
WT	1.9	1/6,000	$90 \pm 5\ s^{-1}$
G64S	0.5	1/8,600	$30 \pm 5\ s^{-1}$
H273R	3.0	1/4,000	$160 \pm 10\ s^{-1}$
K359L	nd[4]	1/450,000	$0.50 \pm 0.05\ s^{-1}$
K359H	nd	1/13,500	$5.0 \pm 0.5\ s^{-1}$
K359R	nd	1/9,000	$5.0 \pm 0.5\ s^{-1}$

[1] The calculated average number of mutations per genome based upon sequencing 36,000 nucleotides of capsid coding sequence from 18 viral isolates.
[2] The calculated transition mutation frequency based upon the ratio of the kinetic parameters for correct and incorrect nucleotide incorporation by the PV RdRp allele.
[3] The maximal observed rate constant, kpol, for correct nucleotide incorporation.
[4] Not determined.

PLATE 8 Sites in PV RdRp controlling rates of mutation and replication. (A) The structure of PV polymerase showing the sites of all mutations G64, H273, K359. (B) Mutation frequency and replication rates for PV polymerase alleles.

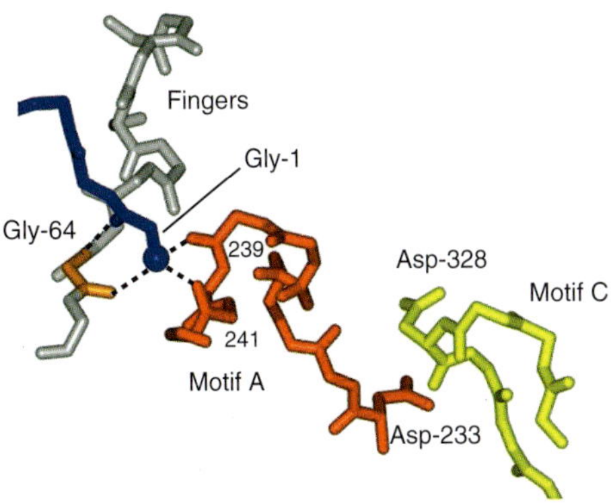

PLATE 9 A link between the fingers subdomain and conserved structural motif A of the palm subdomain of PV 3Dpol. Gly64 backbone orients the N-terminus, and the N-terminus interacts with motif A. The N-terminus of 3Dpol is in blue, Gly64 is in orange, motif A is in red, motif C is in yellow and hydrogen bonds are shown as dashed lines. Misalignment of position 64 will cause defects to the orientation of the triphosphate as well as the efficiency of phosphoryl transfer owing to misalignment of motif A. Reproduced with permission from the *Journal of Biological Chemistry* (Arnold *et al.*, 2005).

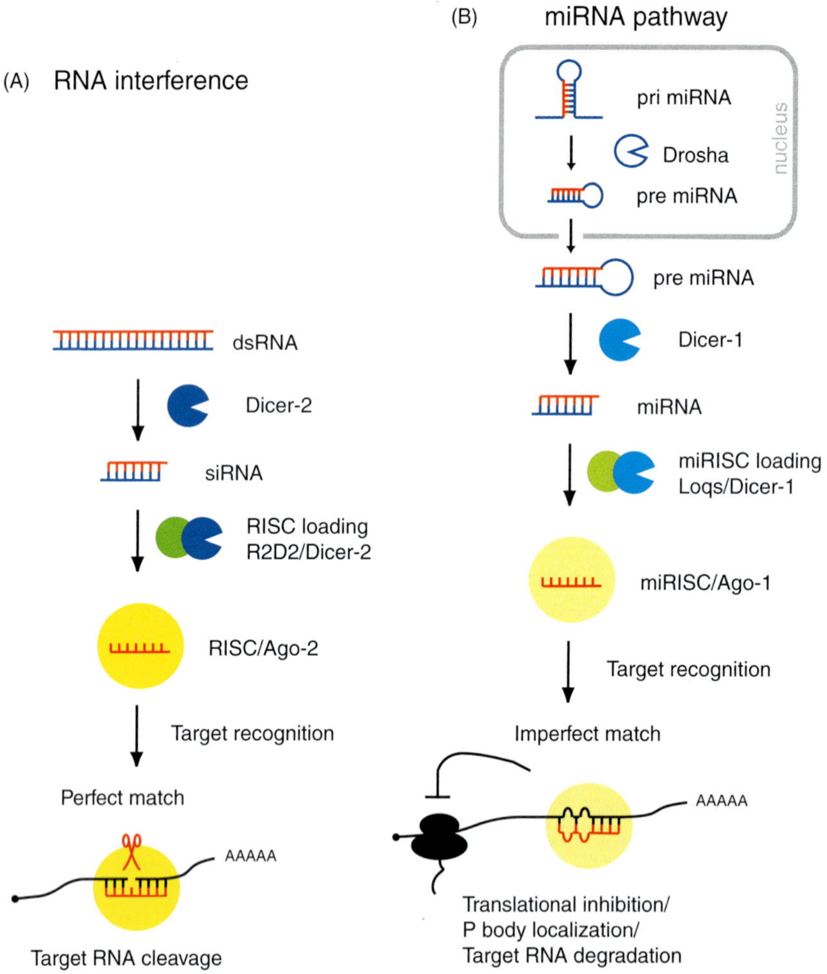

PLATE 10 Schematic representation of RNA-silencing pathways in *Drosophila melanogaster*. (A) RNA interference is initiated by the presence of dsRNA in the cytoplasm of the cell, eventually leading to degradation of perfect complementary RNA in the cell. (B) The miRNA pathway is instructed by endogenously encoded miRNAs that guide recognition of imperfect complementary target sites, which are located in the 3′ untranslated regions (3′ UTRs) of endogenous genes. Target recognition may lead to the translational inhibition, mRNA degradation, and/or localization of the repressed mRNA to processing bodies (P bodies). See text for details. Note that although this figure represents the RNA-silencing pathway in *Drosophila*, the basic mechanism of RNAi is highly conserved mechanism among different organisms. However, functional diversification and specialization among RNAi genes has occurred throughout evolution (see Table 7.1).

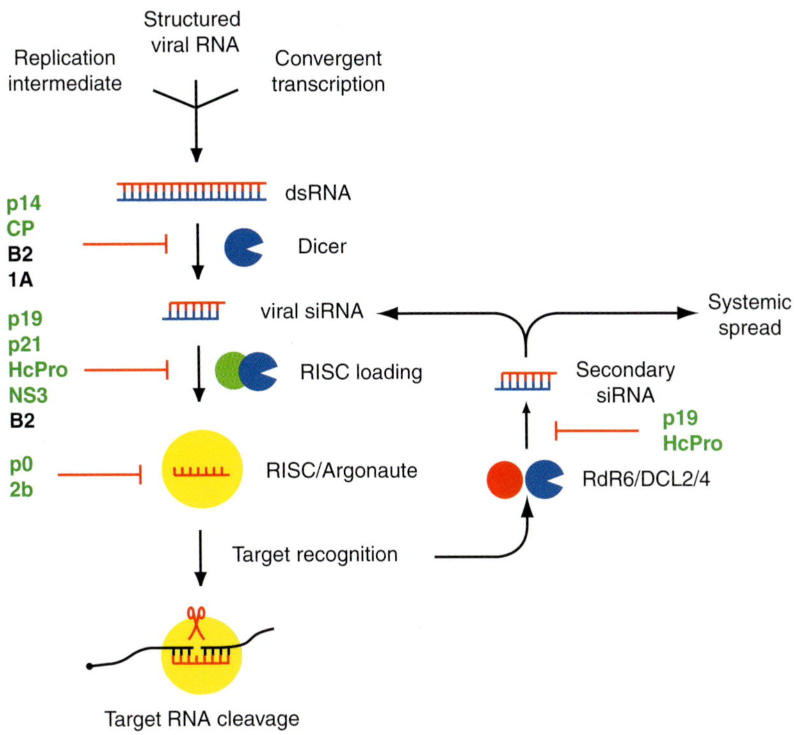

PLATE 11 Viral suppressors inhibit distinct steps of the RNA-silencing pathway. Schematic representation of the mechanism of action of selected insect and plant virus RNA-silencing suppressors. Green and black font indicate silencing suppressors of plant and insect origin, respectively. Note that many more silencing suppressors have been identified in plant viruses, although their mechanism of action remains to be established. See Li and Ding (2006) for a more comprehensive overview. Silencing suppressors with specificity for siRNA, such as p19 and HcPro, sequester siRNAs and inhibit their incorporation into a functional RISC complex *in vitro* (Lakatos *et al.*, 2006). However, they may additionally inhibit the RDR6-DCL2/DCL4-dependent production of secondary siRNAs, and prevent secondary siRNAs from moving to neighboring uninfected cells or into the vasculature (Moissiard *et al.*, 2007). Amplification and spread of the RNA-silencing signal has not been observed in insects.

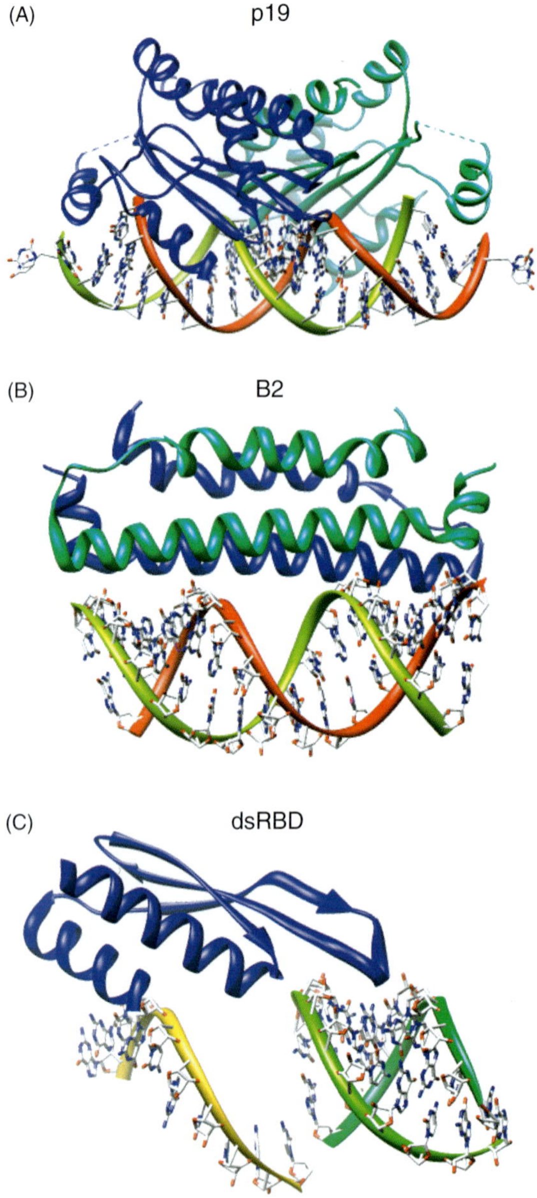

PLATE 12 Structures of viral RNA-silencing suppressor with dsRNA-binding activity in complex with dsRNA. Structures of (A) Carnation Italian ringspot virus (tombusvirus) p19 (protein data bank ID 1RPU) (Vargason *et al.*, 2003). (B) A canonical dsRBD from *Xenopus laevis* RNA protein A (PDB 1DI2) (Ryter and Schultz, 1998). Note that dsRBDs are highly conserved structure, and the DCV suppressor 1A will likely have a similar structure. (C) Flock House virus (FHV) B2 (PDB 2AZ2) (Chao *et al.*, 2005). The images were produced using UCSF Chimera package (www.cgl.ucsf.edu/chimera) (Pettersen *et al.*, 2004).

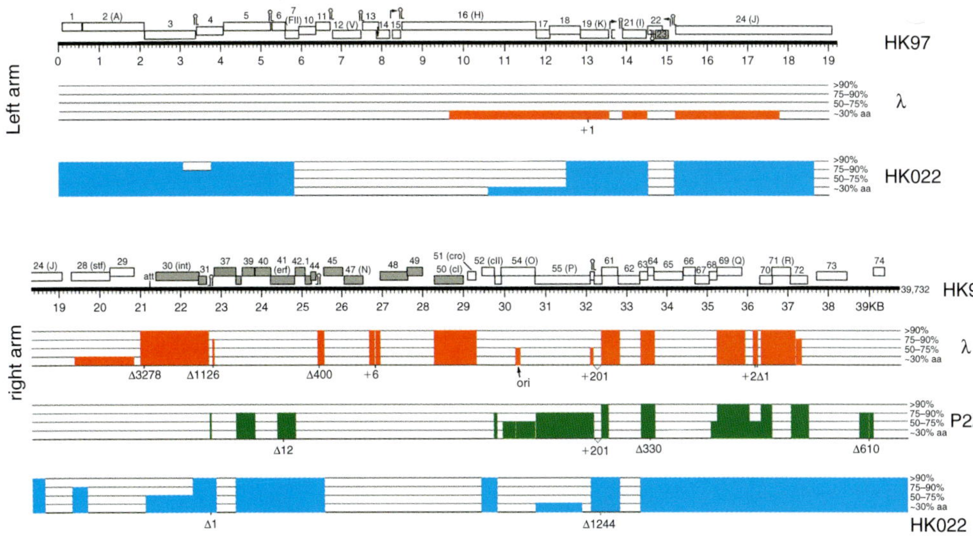

PLATE 13 Mosaic relationships among lambdoid phages. The genome sequence of phage HK97 is represented as a scale bar, broken near the middle, with genes represented as rectangles above the bar. Open rectangles are genes transcribed rightwards; shaded rectangles are genes transcribed leftwards. The histograms below the HK97 genome show the locations and degree of sequence identity of regions that match between HK97 and the three comparison phages, λ, P22 and HK022. The numbers below the histograms preceded by "Δ" or "+" indicate the size of deletion or insertion, respectively, in the HK97 sequence relative to the sequence of the comparison phage. Modified with permission from Juhala *et al.* (2000).

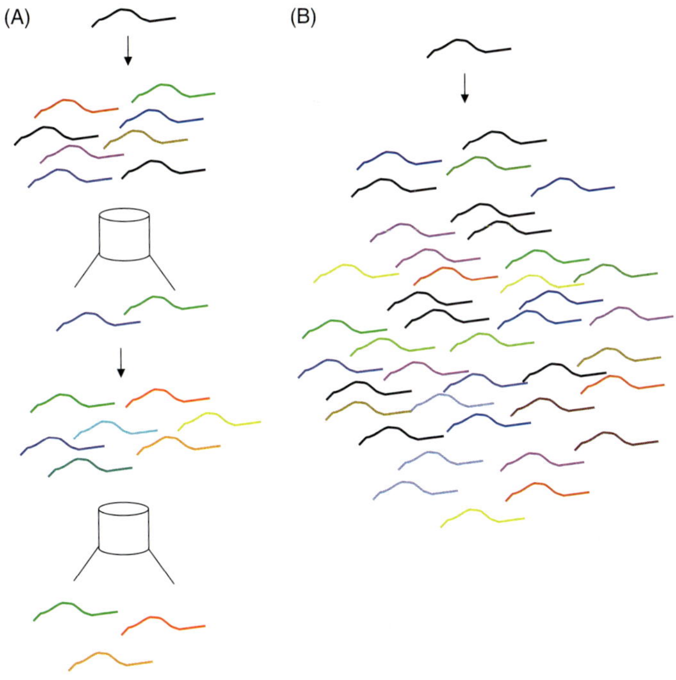

PLATE 14 Schematic representation of the consequences of genetic bottlenecks. (A) The viral genome replicates and generates variants, then passes through a bottleneck. The remaining variants again replicate and generate more variants, that in turn pass through a bottleneck. (B) A population replicating without being subjected to bottleneck events, where diversity is allowed to accumulate, and is only restricted by selection.

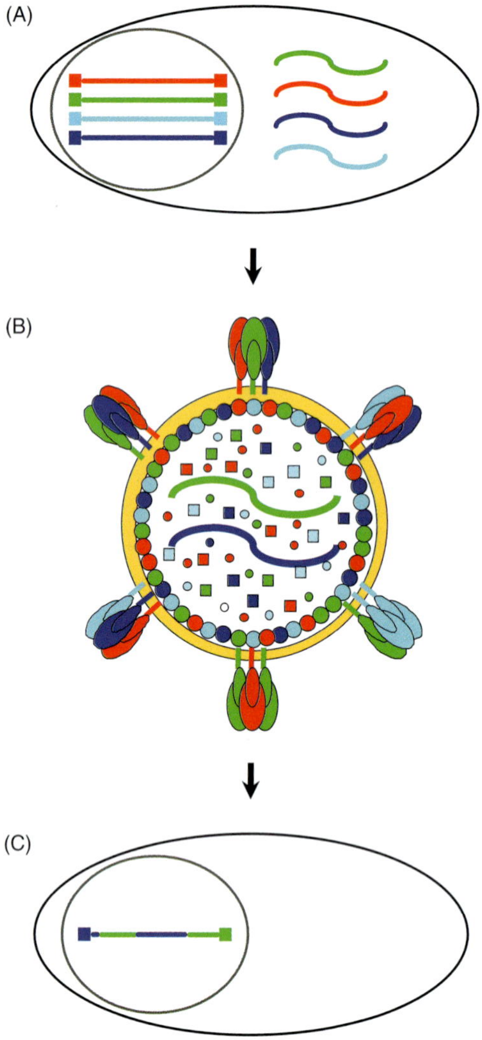

PLATE 15 The non-clonal and ephemeral nature of HIV. (A) A cell bearing four genetically distinct proviruses is represented by different shading. (B) Upon productive infection, only two RNAs can be packaged per virion and there is no reason to consider that proteins from all four proviruses are incorporated into the virion. The resulting structure is much like a harlequin. No two virions will be the same. It is not known if this impacts the biology of the virus. (C) Upon infection of the target cell, recombination occurs leaving a mosaic provirus. All this occurs in as little as 18 hours.

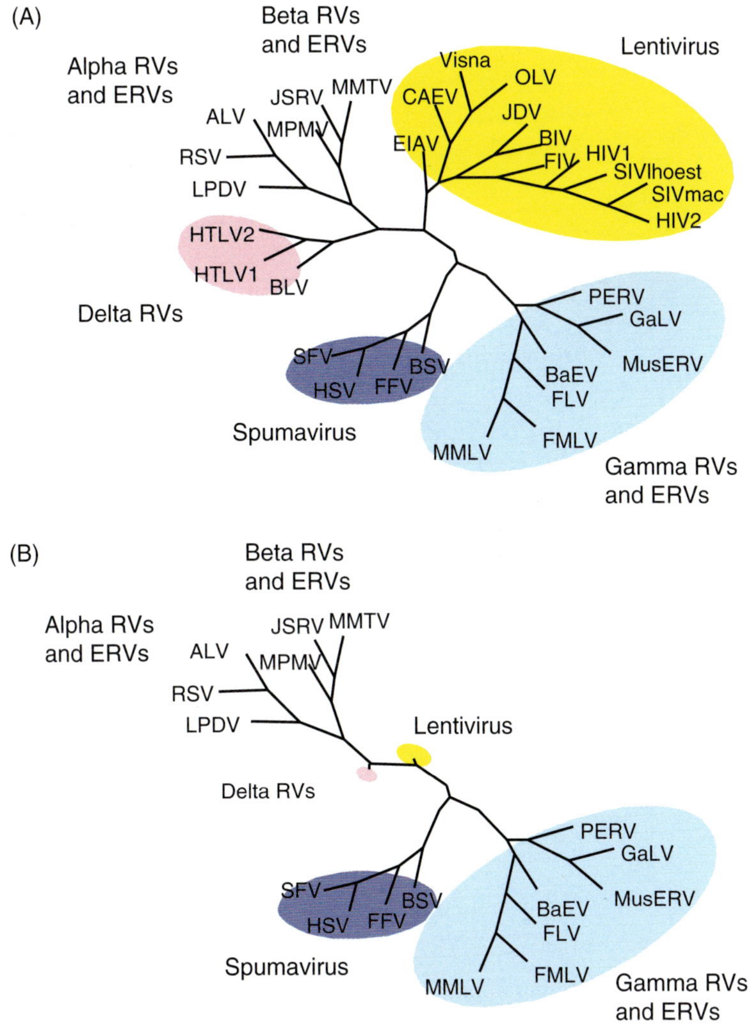

PLATE 16 (A) Standard Protpars tree for the retroviral integrase sequence. Shading helps distinguish the retroviral families. (B) The lentiviral and HTLV lineages only have been shortened to take into account their much faster fixation rates compared to that for the spumavirus lineages. It is tempting to conclude that in a temporal sense the lentiviruses and delta retroviruses have recently emerged from some sort of endogenous retroviral baseline.

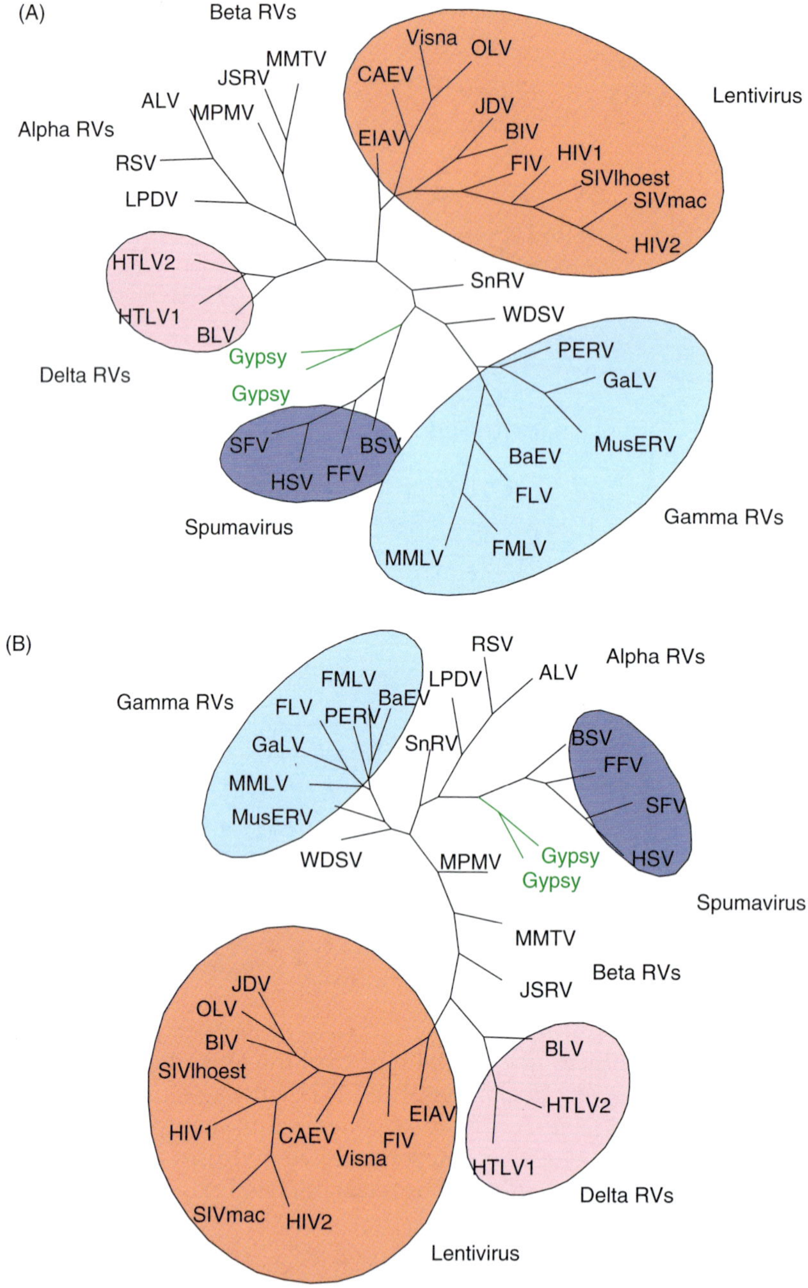

PLATE 17 Classical and synteny-based retroviral trees. (A) Classical Protpars tree for the retroviral integrase sequence. The shaded ellipses highlight the different retroviral groups. Two *Gypsy* retroviral elements from *Drosophila* are used as outgroups. (B) A syntenic tree based on a series of questions giving binary (1, 0) answers. While there is general agreement, the positions of the alpha- and betaretroviruses vary.

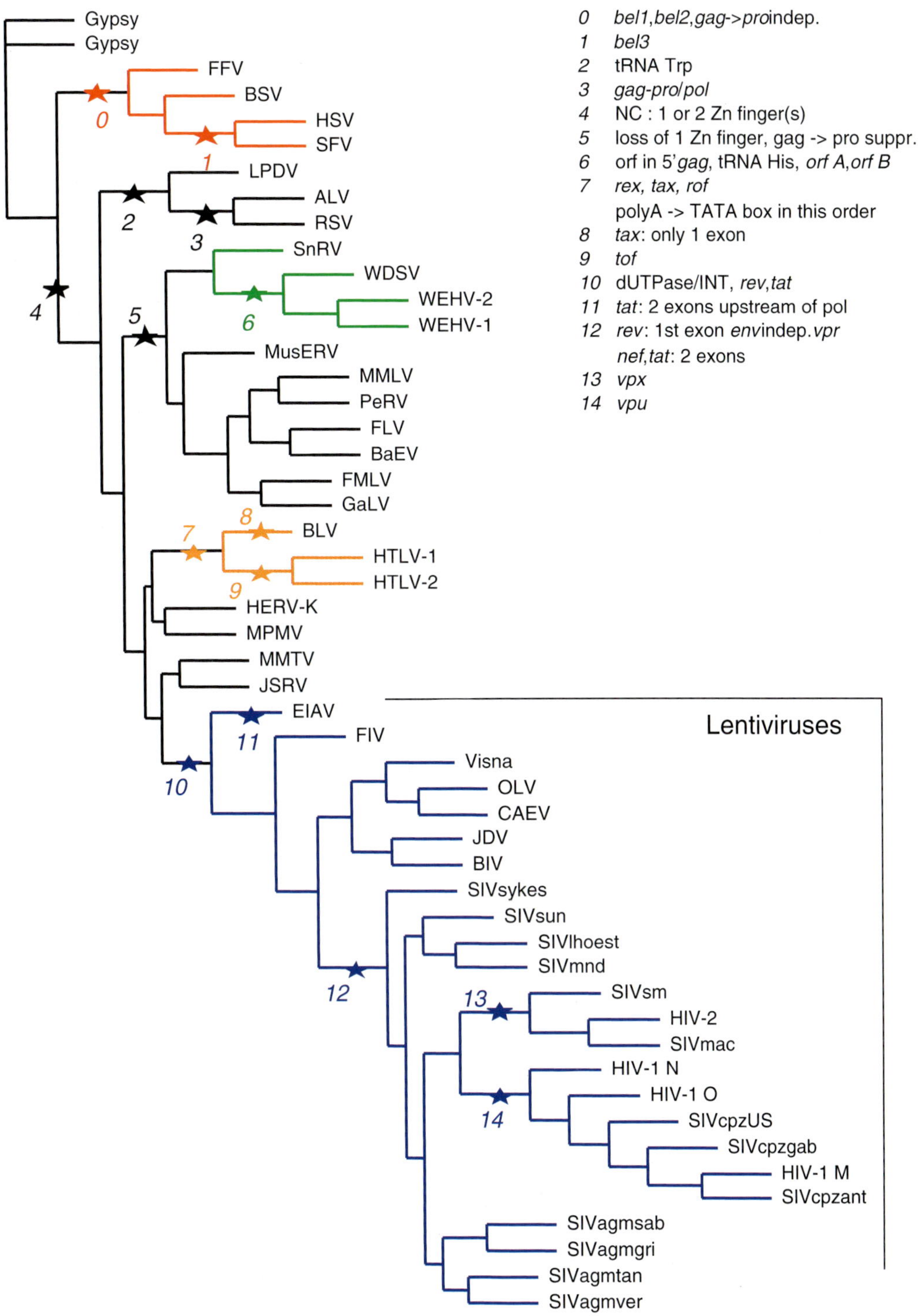

PLATE 18 Annotated phylogenic retroviral tree based on the binary answer mode alignment. Each of the 15 stars corresponds to one or more characters, i.e. organizational traits or evolutionary events at the origin of one group of retroviruses. The lentiviral group represents the largest group of retroviruses in terms of genome complexity.

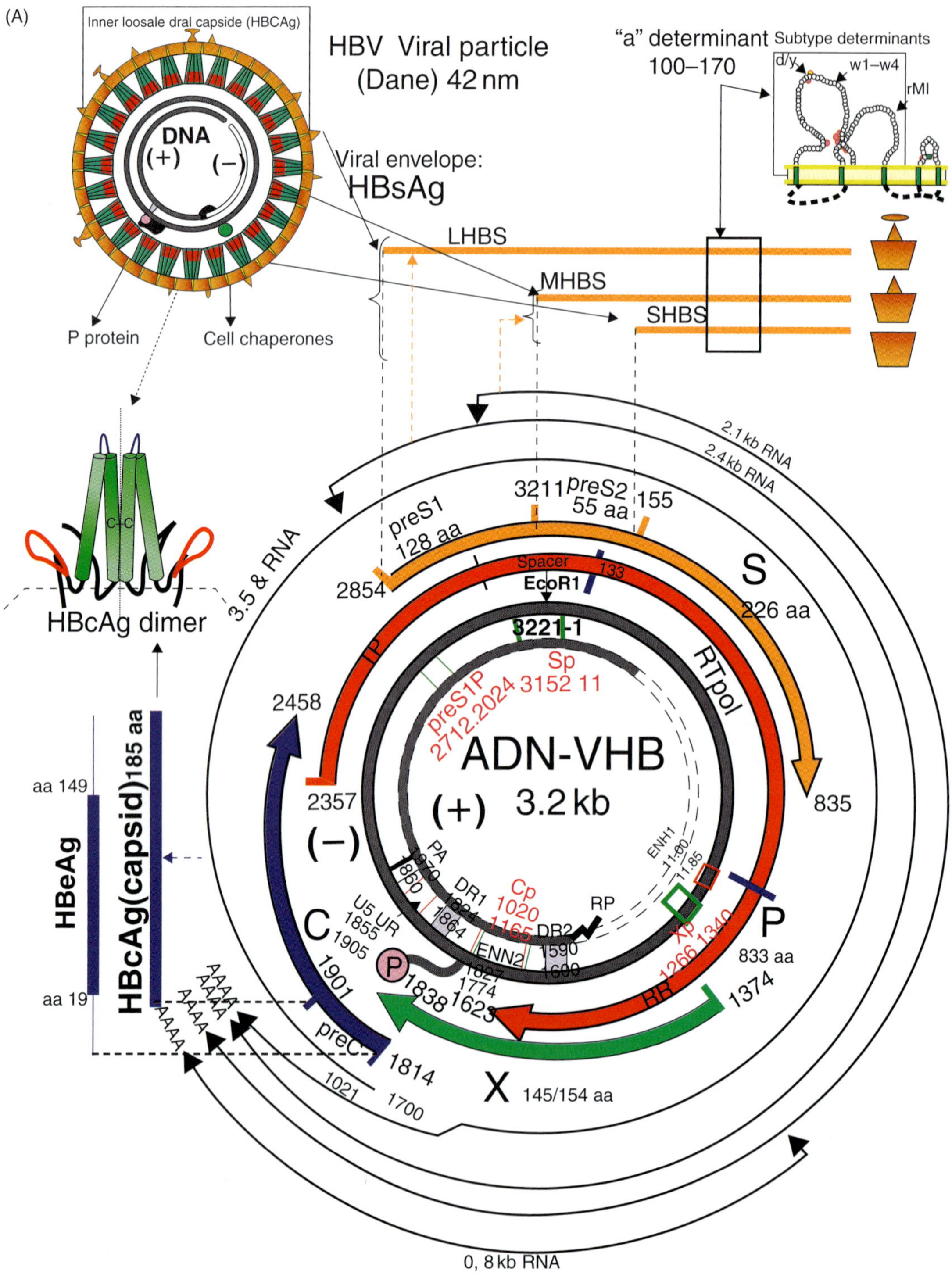

PLATE 19A Hepatitis B virus structure and viral particle representation. Main HBV antigens (HBsAg and HBeAg) have been represented together with their corresponding encoding regions, ORF S and ORF C respectively. HBV polymerase domains are indicated in the ORF P. Location and draft representation of determinant "a" more detailed in Figure 15.2B, and HBcAg are also included.

(B)

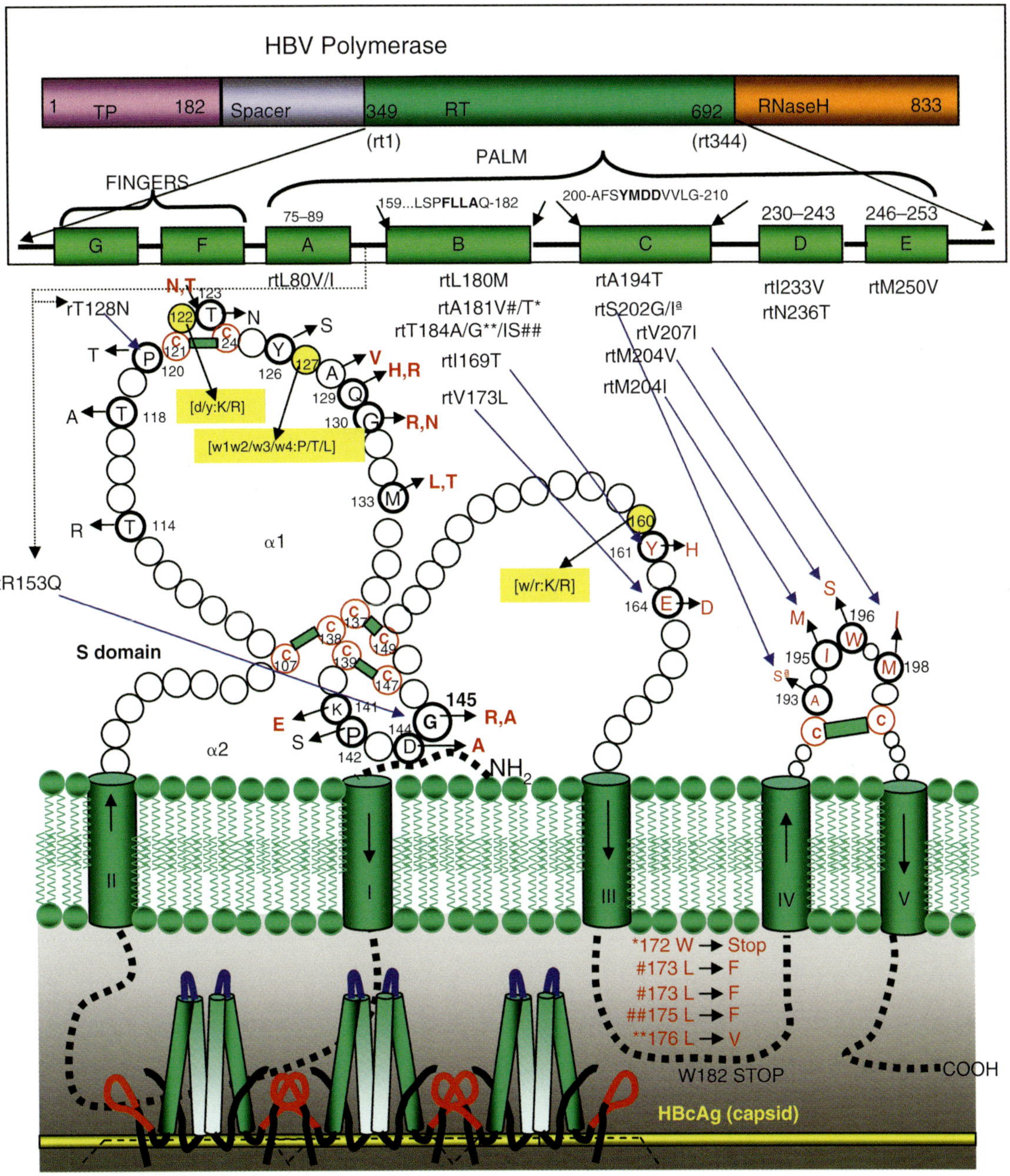

PLATE 19B Schematic representation of HBV polymerase (top), showing the catalytic domains A to G of HBV reverse transcriptase (RT). The panel below shows the main variants associated with resistance to antiviral therapies and their relative positions along the major hydrophilic region of determinant "a," loops α1 and α2, of HBV surface antigen (S domain) and some clinically important variants (such as the ones induced by vaccination and HBIG). HBsAg is represented inserted in the lipid HBV envelope (dominions I–V). Amino acids are represented with single letter codes. Overlapped RT and surface variants are linked by arrows and marked with specials symbols that represent specific amino acid changes: * W172Stop, # L173F, ** L175F, and ## L176V. The amino acid positions related to subtype specificities are indicated with yellow highlight.

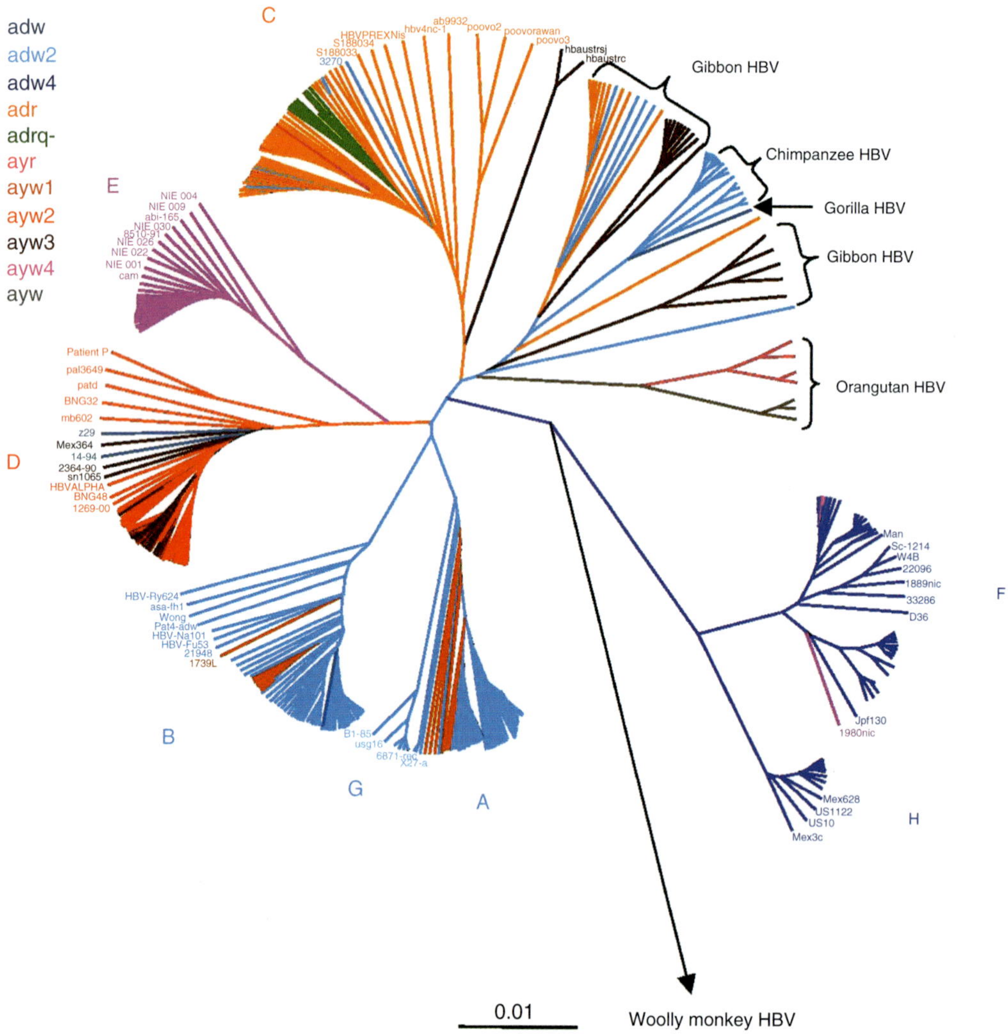

PLATE 20 Hepatitis B virus phylogenetic tree. UPGMA dendrogram based on 639 human and other primates HBV S gene sequences. The subtype is indicated by the color of the branches. Note that genotypes A and G are extremely related. Reproduced from Norder *et al.* (2004), with permission.

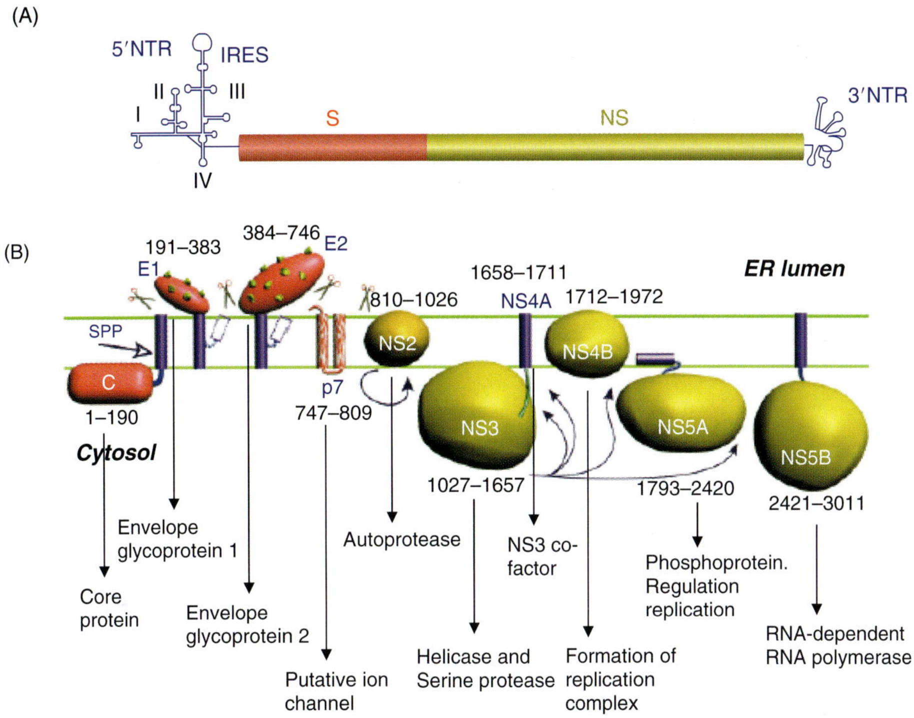

PLATE 21 Hepatitis C virus structure. (A) HCV genome organization with 5′ and 3′ untranslated regions limiting the single ORFs that codify for a polyprotein of 3011 amino acids with C-terminal (one-third) having the structural proteins (S) and the rest that contains the nonstructural proteins (NS). (B) Polyprotein processing with scissors indicating endoplasmic reticulum signal peptidase cleavage sites; cyclic arrow, autocatalytic cleavage of the NS2 NS3 junction; and black arrows indicating the cleavage of NS3/NS4A proteinase complex. Reproduced from Penin *et al.* (2004) with author′s permission.

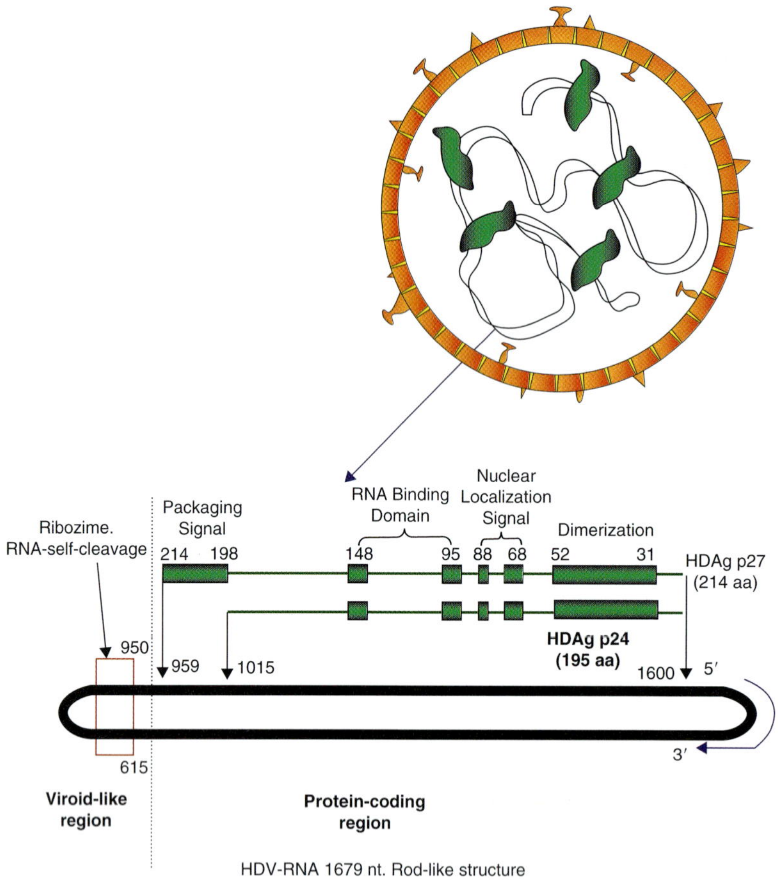

PLATE 22 Hepatitis δ virus. Schematic representation of hepatitis delta particle and genomic organization. The direction of the genomic sense is clockwise. The dotted line represents the boundary between the viroid-like region and the protein-coding region. Antigenomic strand RNA (residues 1600–959) codes for two forms of the delta antigen (HDAg). These two species of HDAg differ in having a termination codon (AUG) in one of the RNAs (residue 1015). Residue 615, site of ribozyme self-cleavage in the genomic strand; residue 950, site of ribozyme self-cleavage in the antigenomic strand; box, region of local tertiary structure and interaction. The main features of the two forms of HDAg are indicated.

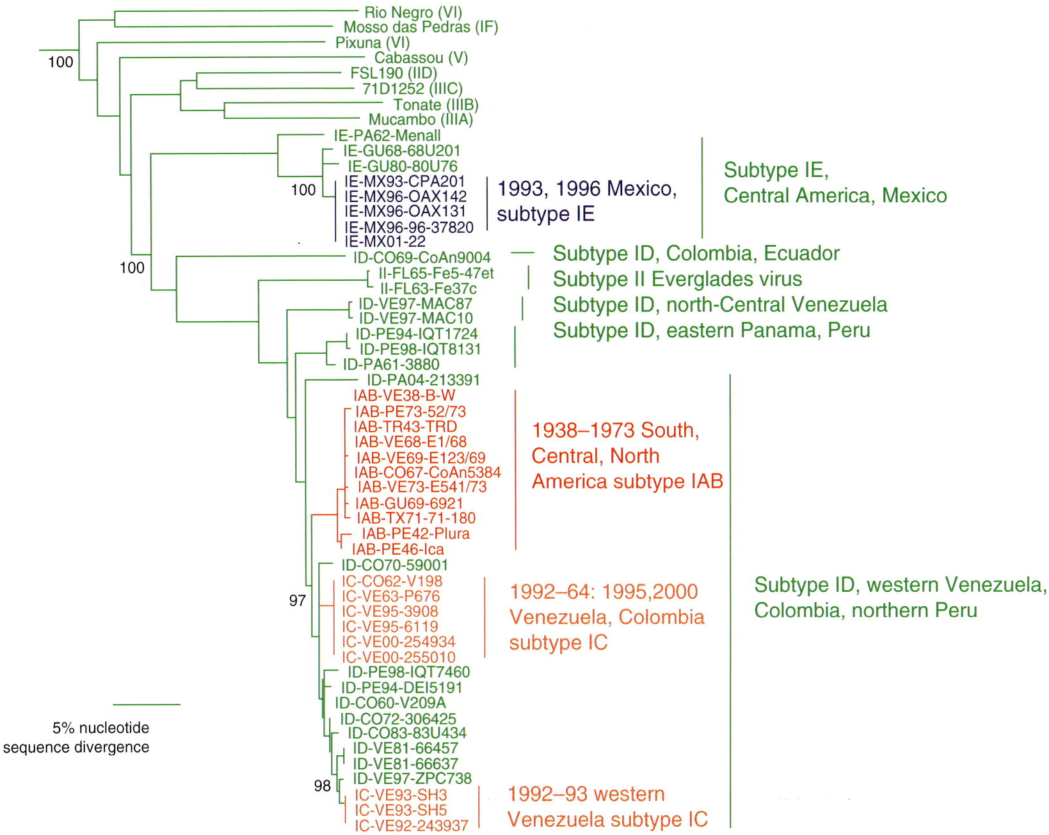

PLATE 23 Phylogenetic trees of VEE complex alphaviruses derived from structural polyprotein amino acid sequences using the neighbor-joining program. Green branches and viruses represent enzootic strains that circulate in forest or swamp habitats and are generally avirulent for equids and incapable of using them as amplification hosts. Green labels on right show six major lineages of VEEV, only two of which have given rise to epidemic strains. Red and orange strains of VEEV are epidemic strains that produce high titer viremia in equids to increase amplification and spillover to humans. The blue Mexican strains are enzootic in the sense that they do not produce adequate viremia in equids for efficient amplification, but epizootic in the sense that they are neurovirulent and cause equine encephalitis. Similar topologies were produced using maximum parsimony and Bayesian methods. Bootstrap values indicate support for clades to the right. The tree was rooted using homologous Eastern equine encephalitis virus sequences.

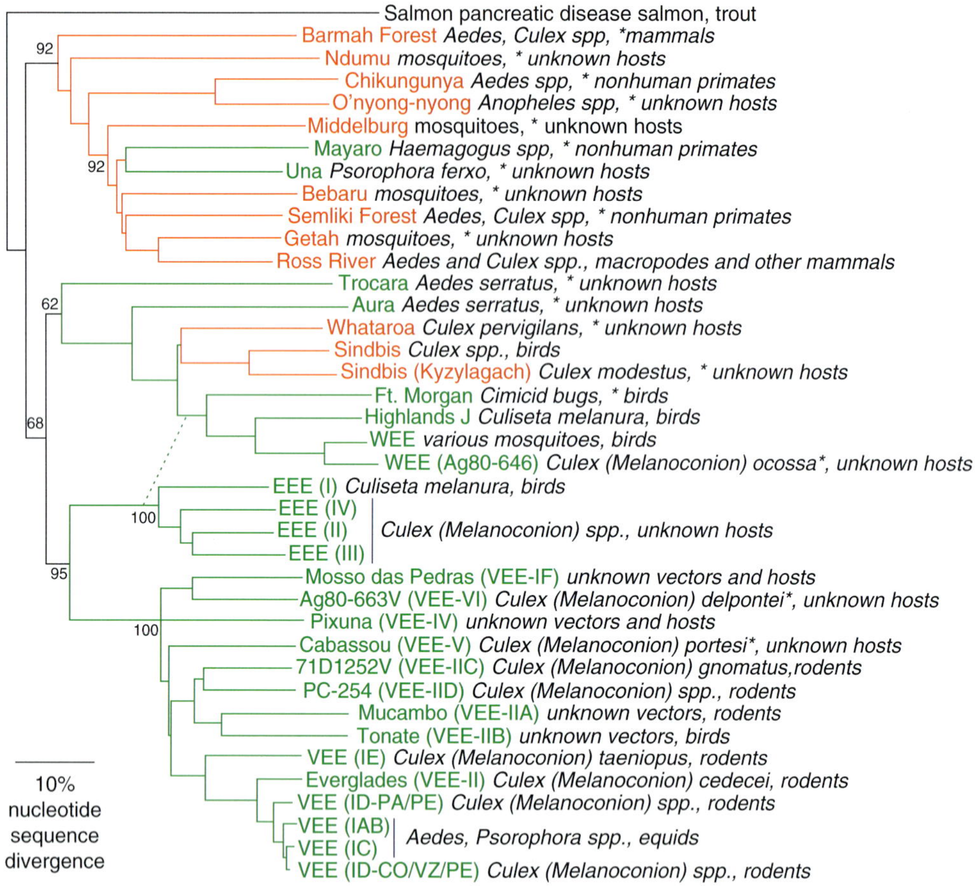

PLATE 24 Phylogenetic tree of all species and major lineages of alphaviruses derived from E1 envelope glycoprotein sequences. Subtypes are written in parentheses after virus names. Reservoir hosts and vectors are listed after viruses. New World viruses are printed in green, Old World viruses in red. Dashed line represents the recombination event that led to the ancestor of WEE, Highlands J and Ft. Morgan viruses. The tree was drawn using the neighbor-joining program with the HKY distance formula using PAUP 4.0. Similar topologies were produced using maximum parsimony and Bayesian methods. Bootstrap values indicate support for clades to the right.

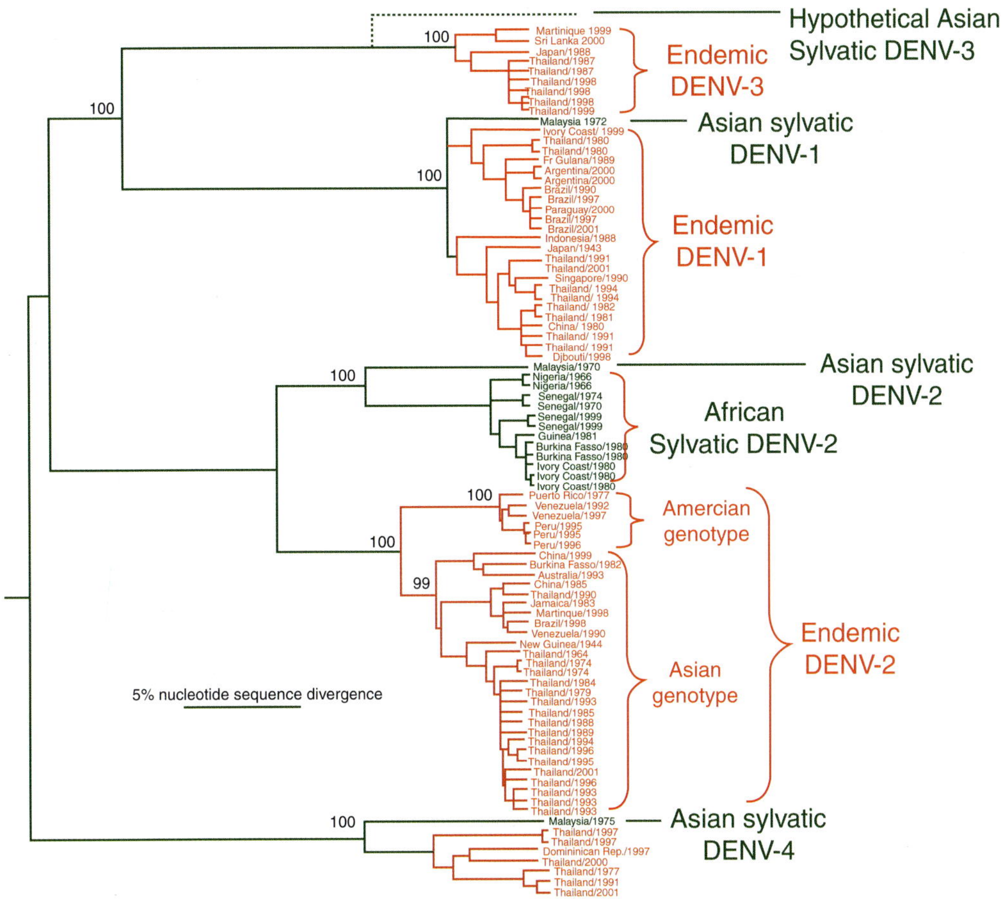

PLATE 25 Phylogenetic tree derived from complete open reading frame sequences of dengue virus strains from the GenBank library. The tree was drawn using the neighbor-joining method. Numbers indicate bootstrap values for major clades to the right. Virus strains and internal branches representing hypothetical ancestors are colored green for sylvatic and red for endemic strains. The hypothetical sylvatic dengue virus-3 is believed to exist based on seroconversions in sentinel monkeys exposed in Malaysia (Rudnick, 1984).

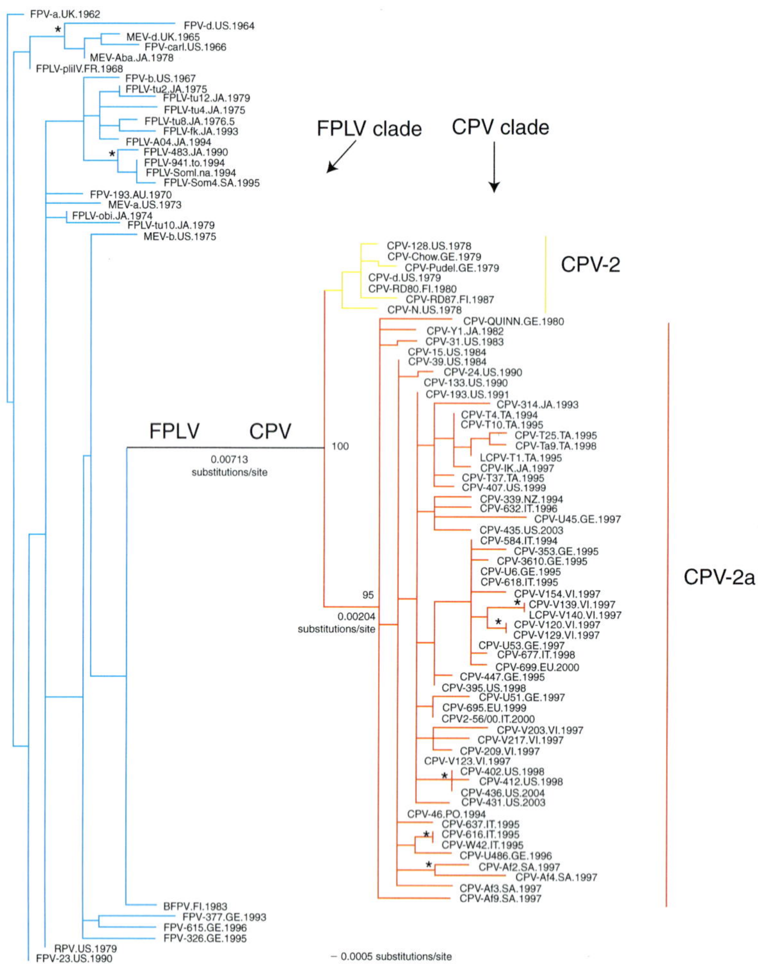

PLATE 26 Phylogenetic tree of 91 VP2 gene sequences from carnivore parvoviruses, rooted with the oldest sampled sequence. Bootstrap values are shown for relevant nodes, and nodes with >70% support are marked with an asterisk. Horizontal branch lengths are drawn to scale. The name of each isolate is followed by the location and year of isolation. Locations are coded as follows: UK, United Kingdom; US, United States; JA, Japan; FR, France; AU, Australia; FI, Finland; GE, Germany; TA, Taiwan; NZ, New Zealand; VI, Vietnam; EU, Europe (no further information available); PO, Poland; IT, Italy; SA, South Africa. The FPLV clade is shown in blue, the CPV-2 subclade is shown in yellow, and the CPV-2a subclade is shown in red. BFPV, blue fox parvovirus. From Shackelton *et al.* (2005) with permission.

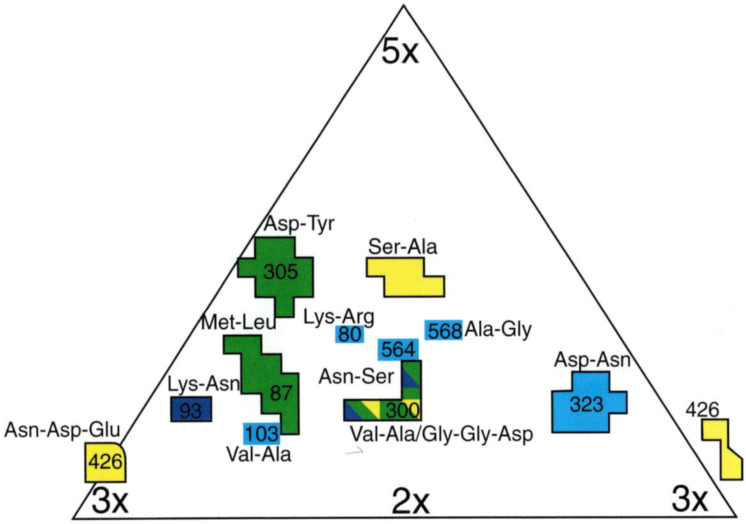

PLATE 27 A model of one asymmetric unit of the CPV capsid, showing the locations of substituted amino acids. Sites that changed along the FPV→CPV branch are in blue, with the darker shade indicating surface-exposed residues and the lighter shade showing the relative positions of subsurface sites. Sites that underwent substitutions along the CPV→CPV-2a branch are in green, and sites responsible for variants within the CPV-2a subclade are in yellow; all are surface-exposed. From Shackelton *et al.* (2005) with permission.

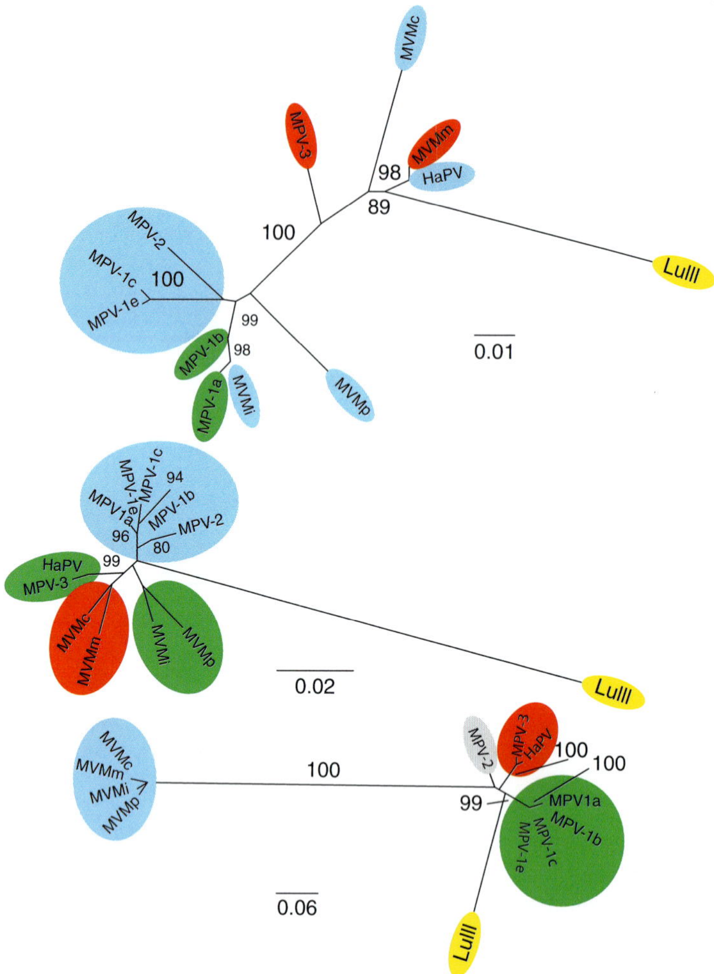

PLATE 28 Phylogenetic analyses showing recombination among the rodent parvoviruses. Nucleotide sequences spanning the complete coding region (nucleotides 114–4728 of mouse parvovirus 1 reference sequence, GeneBank accession number NC_001630) were aligned. Phylogenetic trees were constructed from nucleotides (a) 1–854, (b) 855–2226, and (c) 2227–4654. Colors indicate different rodent virus types. Nodes with >70% bootstrap support are shown and branch lengths are scaled according to nucleotide substitutions per site. Abbreviations: LuIII virus (LuIII), mice minute virus (MMVx), minute virus of mice strain m (MMVm), minute virus of mice (MMVy), mouse parvovirus minute virus immunosuppressive variant (MMViv), mice minute virus (MMVz), minute virus of mice, a lymphotropic variant of MVM (MMVlyv), mouse parvovirus 1 (Mouse1), mouse parvovirus 1e (Mouse1e), mouse parvovirus 1c (Mouse1c), mouse parvovirus 1b (Mouse1b), mouse parvovirus 2 (Mouse2), mouse parvovirus 3 (Mouse3), hamster parvovirus (Hamster). From Shackelton *et al.* (in press) with permission.

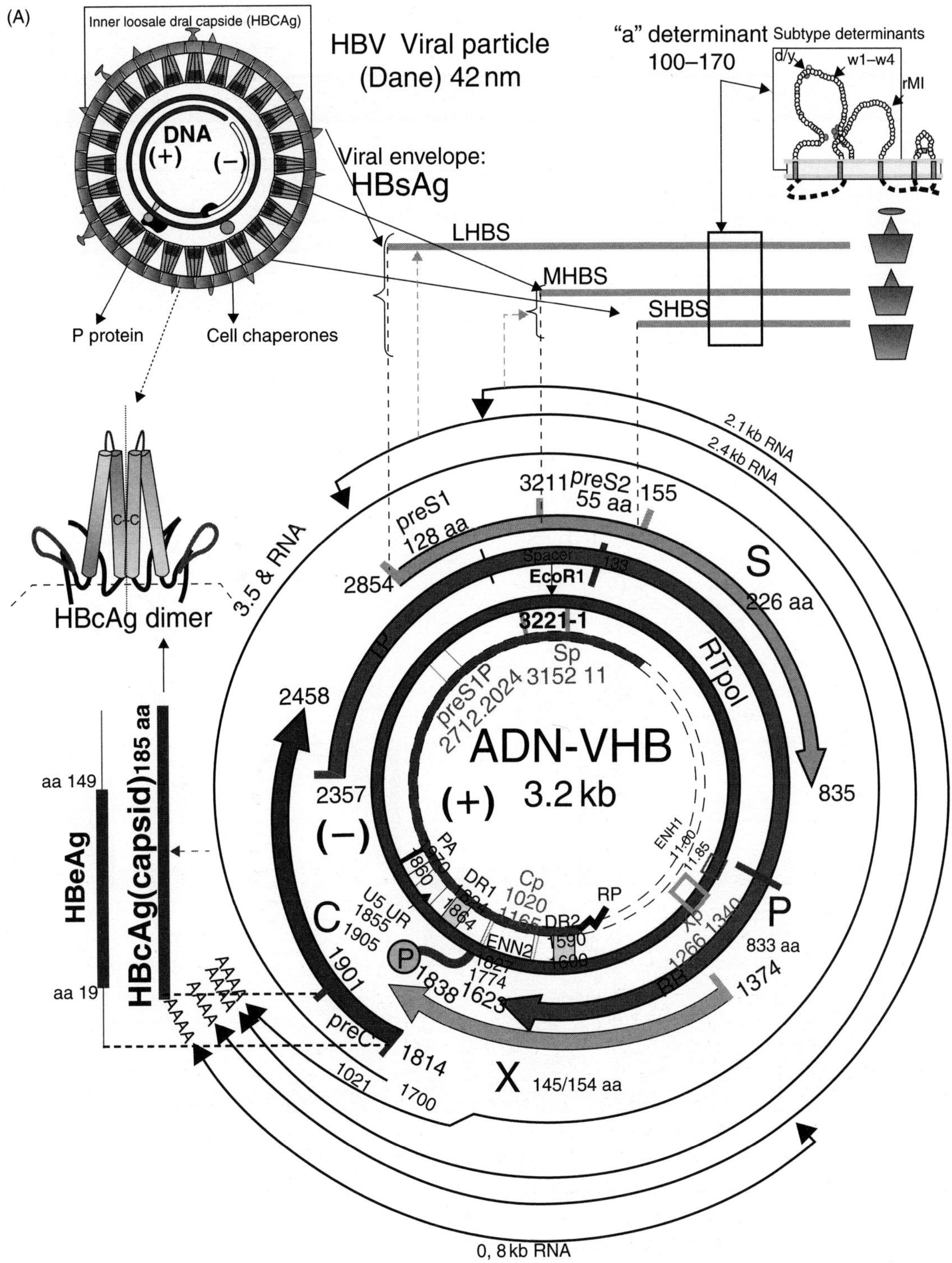

FIGURE 15.2A Hepatitis B virus structure and viral particle representation. Main HBV antigens (HBsAg and HBeAg) have been represented together with their corresponding encoding regions, ORF S and ORF C respectively. HBV polymerase domains are indicated in the ORF P. Location and draft representation of determinant "a" more detailed in Figure 15.2B, and HBcAg are also included. (See Plate 19A for the color version of this figure.)

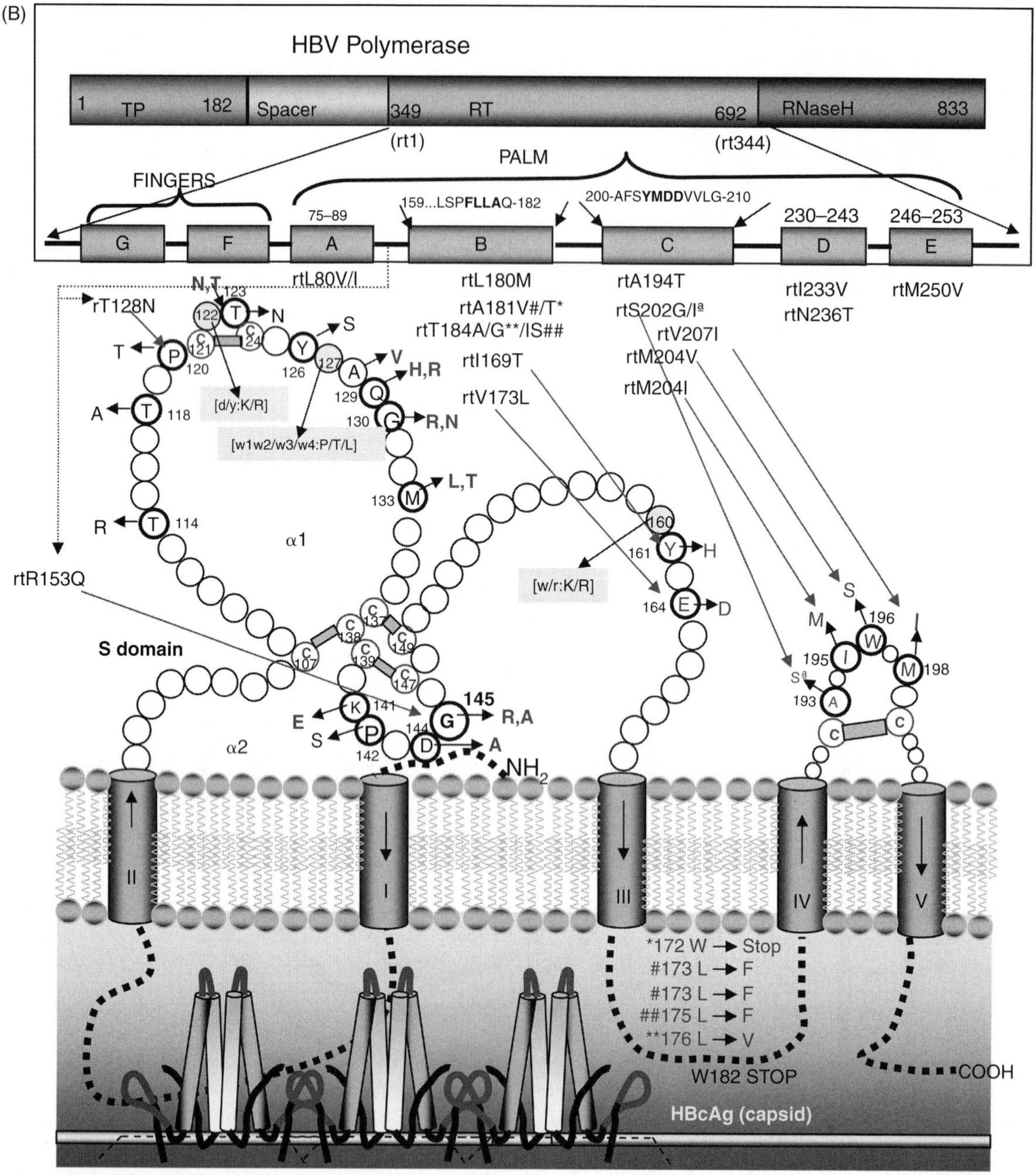

FIGURE 15.2B Schematic representation of HBV polymerase (top), showing the catalytic domains A to G of HBV reverse transcriptase (RT). The panel below shows the main variants associated with resistance to antiviral therapies and their relative positions along the major hydrophilic region of determinant "a," loops α1 and α2, of HBV surface antigen (S domain) and some clinically important variants (such as the ones induced by vaccination and HBIG). HBsAg is represented inserted in the lipid HBV envelope (dominions I–V). Amino acids are represented with single letter codes. Overlapped RT and surface variants are linked by arrows and marked with special symbols that represent specific amino acid changes: *W172Stop, # L173F, **L175F, and ## L176V. The amino acid positions related to subtype specificities are indicated with yellow highlight. (See Plate 19B for the color version of this figure.)

HBV Entry and Replication Cycle

The HBV cell membrane receptor is still unknown. Attachment to hepatocytes seems to be dependent on a short sequence (aa 9–18) of the preS1 domain in the LHBs (Glebe *et al.*, 2005), a process followed by receptor-mediated endocytosis and proteolytic cleavage in the endosomal domain to release non-enveloped icosahedral core capsids into the cytosol which are then transported to the cell nucleus via microtubules (MT) where they interact with nuclear pores, dissociate and release the viral genome into the nucleus (Stoeckl *et al.*, 2006).

Once inside nuclear karyoplasm, the partially double-stranded rcDNA is converted to a plasmid-like fully double-stranded and covalently closed circular DNA (cccDNA). Naturally infected hepatocytes contain a non-uniform (cccDNA) and up to 50 or more cccDNA copies "per" cell, organized into minichromosomes by histone and non-histone proteins (Bock *et al.*, 2001; Zhu *et al.*, 2001) The cccDNA seems to be under a complex network of epigenetic regulation (Pollicino *et al.*, 2006) and may play an important role in viral persistence. Cellular RNA polymerase II uses cccDNA as the template for transcription of several genomic and subgenomic viral mRNAs. Particle formation requires reverse transcription within the nucleocapsid (Nassal and Schaller, 1996; Beck and Nassal, 2007) with several instances of template switching in a quite complex mechanism (Beck and Nassal, 2007; Bruss, 2007).

HBV DNA is able to integrate in the host genome at random locations, being a parallel process of the replication cycle of any hepadnavirus. It has been proposed that viral DNA could be linearized by cellular topoisomerase I followed by non-homologous recombination with cellular DNA (Pourquier *et al.*, 1999). Nearly 90% of hepatocellular carcinomas (HCC) detected in HBV endemic areas contain integrated viral sequences (Feitelson *et al.*, 2002). Integration seems to be an early phenomenon in HBV infection (Brechot *et al.*, 2000). Integrated sequences are complex and very highly rearranged structures.

HBV generation of variation

Mutation

The main characteristic of the HBV-RT is the lack of a proofreading/editing 3′–5′ exonuclease activity, which results in an estimated nucleotide misincorporation rate of around 10^{-4} (Park *et al.*, 2003). This high error rate together with a replication rate which produces 10^{13} viral particles per day, implies that, during active infection, 10^9 mutations take place every day over the 3200-bp genome (Locarnini, 2003; Whalley *et al.*, 2001). However, the rate of accumulation of mutations (rate of evolution) (Domingo and Gomez, 2007) calculated for the HBeAg region is lower, estimated at $1.4–7.9 \times 10^{-5}$ substitutions/site per year (Okamoto *et al.*, 1987; Osiowy *et al.*, 2006). This discordance may be related to the highly compact genomic structure, with overlapping ORFs including regulatory sequences ("ε" signal, promoters, enhancers, etc.) which tend to constrain HBV genetic variability (Mizokami *et al.*, 1997), and also by the strict dependence of P protein activity on cellular factors, such as heat shock protein 90 (Hsp90), chaperones (Hu *et al.*, 1997; Hu and Boyer, 2006) and toother polymerases (Beck and Nassal, 2007).

Cellular RNA polymerase II, for which error rate has not been calculated (Nesser *et al.*, 2006), uses the cccDNA molecule as a template for transcription of several genomic and subgenomic viral mRNAs. Since this enzyme is responsible for pgRNA synthesis it might also contribute to HBV genome variability. Other errors such as deletions and insertions may be generated by multiple template switching (primer translocations) during viral replication. Occasional pgRNA splicing or topoisomerase I cleavage/ligation of HBV DNA might give an additional source of HBV variation (Gunther *et al.*, 1999).

The final result of these sources of variation is the presence in a single patient of a pool of closely related variants or viral quasispecies (Ngui *et al.*, 1999) which predicts the daily production of all possible single and double mutants in a single patient. This highly variable population of genomes is the source for the selection of escape mutants to the host immune response and to antiviral agents (Weinberger *et al.*, 2000).

One of the most important features of HBV variability is that, because of the presence of overlapping ORFs along the HBV genome, a mutation in one gene can lead to an amino acid change in another. This is the case of some RT polymerase mutants, that also produce mutants in the overlapping HBsAg region and vice versa.

Genotypes

HBV has been classified into eight genotypes (A–H) (Norder *et al.*, 1994; Stuyver *et al.*, 2000; Arauz-Ruiz *et al.*, 2002) with 10–13% of nucleotide divergence (Norder *et al.*, 2004; Simmonds and Midgley, 2005). Subtypes have been described for all genotypes, except for E, G, and H, (24 subgenotypes as shown in Figure 15.3 and Table 15.1) with at least 4% of intra-genotype differences (Kramvis and Kew, 2005). The lack of subtypes of HBV genotype E has been attributed to its recent genesis since it is detected in Africa but absent in African Americans from Venezuela and Brazil (Schaefer, 2007). No subgenotypes of genotype G have been reported yet, probably because of the low number of complete genotype G sequences reported.

Origin and Spread of HBV in the World

HBV genotype distribution is closely related to ethnicity, with genotype A being predominant in white populations and genotypes B and C in Asian populations (Westland *et al.*, 2003). Genotypes A and D have global distribution. Genotype D is mainly found in the Mediterranean basin, genotypes B and C are predominantly found in East and South-East Asia, genotype E is predominant in West Africa, genotype F, which is the most divergent, is mainly found in Central and South America, genotype G in North America and Europe and genotype H in North America (Norder *et al.*, 2004; Wai and Fontana, 2004; Schaefer, 2007) (Figure 15.3).

Several hypotheses about the origin of HBV have been formulated. One suggests that HBV originated in the Americas and spread into the Old World over the last 400 years after contact from Europeans during colonization (Bollyky and Holmes, 1999). Another suggests that HBV co-evolved with modern humans as they migrated from Africa, approximately 100 000 years ago (Magnius and Norder, 1995; Norder *et al.*, 1994). The main criticism against both hypothesis is that rates of evolution of HBV should be many orders of magnitude lower than the ones calculated from HBV carriers. Recently, it has been reported some evidences of cross-species transmissions, between apes and between human and apes (Hu *et al.*, 2000) suggesting that diversification of HBV could be a recent event, with frequent cross-species transmission between humans and primates. This possibility is supported by the fact that the highest human HBV prevalent areas (South America, Sub-Saharan Africa, and South-East Asia) are those in which human–primates contacts and cross-species transmissions are most likely. This theory could explain the specificity of certain HBV genotypes found in these areas (F, E, and B/C respectively) (Simmonds, 2001). In Europe and North America a mixture of HBV genotypes is found, probably as a result of more recent contacts (immigration, intravenous drug addiction, etc.).

Comparing accumulation of mutations in HBeAg+ region ($1.4–5 \times 10^{-5}$ mutations/nt per year from patients) and genetic distances, the origin of HBV in hominoid primates has been dated around 6000 years ago. The fact that human and non-human primate HBV lineages have similar divergence suggest that species groups were infected at approximately the same time (Fares and Holmes, 2002). HBV serotypes and genotypes diverged from a common ancestor approximately 2300–3100 years ago (Simmonds, 2001). Based in the prevalence, geographic distribution, and characteristics of recombinant genotypes, it has been suggested that genotypes A and D have

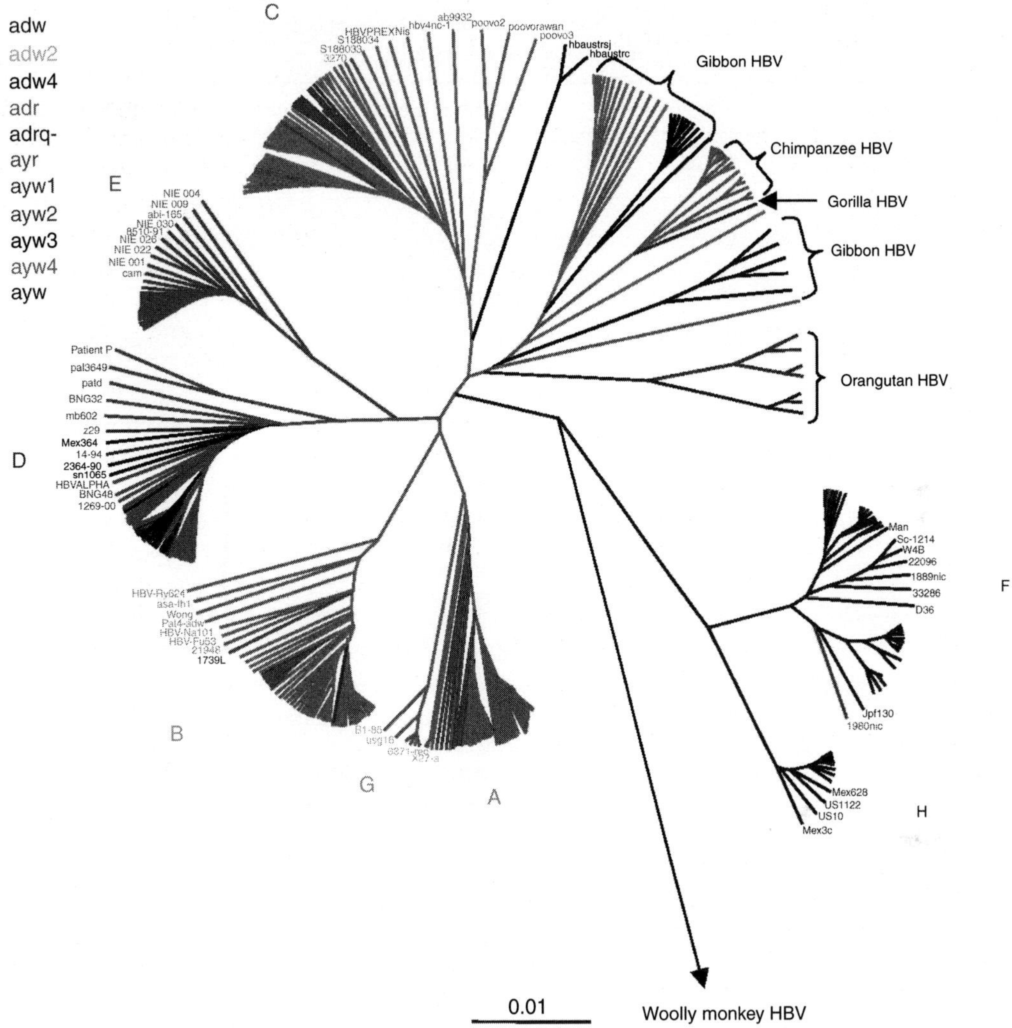

FIGURE 15.3 Hepatitis B virus phylogenetic tree. UPGMA dendrogram based on 639 human and other primates HBV S gene sequences. The subtype is indicated by the color of the branches. Note that genotypes A and G are extremely related. (See Plate 20 for the color version of this figure.) Reproduced from Norder *et al.* (2004), with permission.

been in contact over a relatively long period of time, while genotypes B and C have had a relatively recent epidemiological contact, clearly in Asia (Fares and Holmes, 2002).

Recombination

Recombination between HBV genotypes is not exceptional and is an additional source of HBV variability. Recombination events seem to be clustered to specific regions of HBV genome (Simmonds and Midgley, 2005). Infection with two different HBV genotypes is frequent and detected in 4.4–17.5% of HBV-infected patients (Schaefer, 2007). Super infection with HBV isolates of the same or different genotype has been described (Kao *et al.*, 2001), sometimes accompanied by acute exacerbation of CHB disease (Hannoun *et al.*, 2002; Janssen *et al.*, 2005). Infection with HBV

TABLE 15.1 HBV Genotypes, Subgenotypes and Subtype Specificities with DNA Size, and Geographic Distribution

Genotype	DNA size	Subgenotype	Geographic area	Subtype specificity
A	**3221 nt**: 6 nt Core insertion	A1 (Aa)	Africa, Asia, South America	ayw1
		A2 (Ae)	Europe, USA	adw2
		A3	Africa (Gabon, Cameroon)	
		A4	Africa (Mali, Nigeria)	
		A5	Africa (Nigeria)	
B	**3215 nt**	B1 (Bj)	Japan	adw2
		B2 (Ba)	Asia without Japan	adw2
		B3	Indonesia, Philippines	adw2
		B4	Vietnam	ayw1
		B5	Indonesia, Philippines	adw2
C	**3215 nt**	C1 (Cs)	South-East Asia (Vietnam, Thailand, Southern China)	adrq− , adrq+, ayr (all "r")
		C2 (Ce)	Far East (Korea, Japan, Northern China)	
		C3	Micronesia	
		C4	Australia	
		C5	Philippines, Vietnam	
D	**3182 nt**: 33 nt preS1 deletion	D1	Mongolia, Belarus, Europe?	ayw2, ayw3, ayw4, cases of adw3
		D2	India, Europe (Mediterranean area), Russia, USA	
		D3	South Africa, East India, Serbia, Alaska	
		D4	Australia, Somalia	
		D5	East India	
E	**3212 nt**: 3 nt preS1 deletion	No genotypes reported	West Africa	ayw4
F	**3215 nt**	F1	South and Central America	adw4q− , adw2, ayw4
		F2	South America	
		F3	Bolivia	
		F4	Argentina	
		F not classified	Europe (Spain), Polynesia	
G	**3248 nt**: 3 nt preS1 deletion, 36 nt Core insertion (from codon2), pc2 Stop, pc28 Stop	No genotypes reported	USA, Mexico, Europe (France)	adw2
H	**3215 nt**	No genotypes reported	Nicaragua, Mexico, California	adw4

Prior to consensus definition of genotypes, HBV strains were serologically classified into nine HBsAg subtypes (Le Bouvier, 1971) based in specificities from surface proteins amino acid polymorphisms specificities such as "d" or "y" (Lys/Arg122), "w" or "r" (Lys/Arg160), w1/w2, w3 or w4 (Pro/Thr/Leu127) and the specificity "q" (amino acids 177 and 178). Although there is a good correlation between HBV subtyes and genotypes, several subtypes are encoded by more than one genotype, and one genotype by several subtypes (Norder *et al.*, 2004).

genotype G seems to be often associated with genotype A, thus suggesting that the strong homology between genotypes A and G may be due to co-infection and recombination, genotype G being selected by seroconversion to anti-HBe (Kato *et al.*, 2002). Recombinant genotypes have been detected, even being dominant in certain geographic regions.

Recombination may induce a fast evolutionary change, to help adaptation to a new environment. Alternatively, a hypothetical mosaicism of the HBV genome has been proposed, describing HBV genotypes as a modular genome made of a mixture of small segments, like functional modules with different domains (e.g. binding sites for transcription factors or antigenic epitopes) which come from many different HBV genotypes. Thus, a certain combination of these modules would make up an HBV genotype (Bowyer and Sim, 2000).

Biological and Clinical Implications of HBV Variability

Naturally occurring mutations in the structural and non-structural genes as well as regulatory elements have been associated with unique clinical manifestations that may modify the natural course of infection or induce resistance to antivirals (Baumert *et al.*, 2007).

In general, mutations in the preS/S regions are characterized by misassembly and alterations of B- and T-cell epitopes inducing immune escape, diagnostic failure, vaccine escape, and clinical implications such as fibrosing cholestatic hepatitis. Mutations occur mainly in the preS region, including deletions that differentiate genotypes but have no obvious clinical implications (Pumpens *et al.*, 2002) probably because the preS region overlaps with the dispensable spacer domain of P protein, while the S region overlaps with the essential RT domain. The fact that the majority of preS deletions affect B-cell epitopes recognized by antibodies points to a B-cell escape involved in the selection of these variants (Tai *et al.*, 2002; Wang *et al.*, 2006).

All the envelope proteins share SHBs, the small protein that includes the major antigenic site called determinant "a." Several specific mutations in determinant "a" have been involved in escape from humoral immune response, vaccination, and/or administration of HBV-specific immunoglobulin (HBIG) (Gunther *et al.*, 1999; Ngui *et al.*, 1999; Pumpens *et al.*, 2002; Torresi, 2002; Zuckerman and Zuckerman, 2003; Weber, 2005). These mutations may induce HBsAg false-negative results in diagnostics tests.

The most important sR145 variant seems to alter the projecting loop (aa 139–147) of determinant "a" impeding the recognition of neutralizing antibodies induced after vaccination or from HBV immunoprophylaxis treatment (HBIG) (Carman, 1996; Ghany *et al.*, 1998; Protzer-Knolle *et al.*, 1998; Terrault *et al.*, 1998; Coleman, 2006). After the introduction of a universal vaccination program (1984–1994) the proportion of HBsAg mutants increased, from 8% to 25% over 10 years (Hsu *et al.*, 1999) showing a selective role of HBV vaccine over HBsAg mutations (Hsu *et al.*, 2004).

Mutations in the pre-core region are the best-characterized mutants and are related to anti-HBe seroconversion. The selection of these types of mutants in the host may be due in part to the immunomodulatory properties of HBeAg, resulting in a survival advantage for the virus once seroconversion to anti-HBe has taken place.

These mutations are divided into two groups. The first group (translational variants) include mutations in the pre-core region, referred to as HBV "pre-C defective" variants, which abrogate the HBeAg synthesis (Gunther *et al.*, 1999). The best characterized of these mutations are stop codons leading to loss of HBeAg, and associated with disease severity and progression. They have been detected mainly in patients with chronic and fulminant hepatitis B (FHB) (Liang *et al.*, 1991; Fang *et al.*, 1998; Kramvis and Kew, 1999), although they have also been detected in asymptomatic HBV carriers or self-limiting hepatitis (Jardi *et al.*, 2004; Chauhan *et al.*, 2006). The second group includes clusters of mutations in the basal core promoter (BCP) that result in enhanced viral replication (Baumert *et al.*, 2007), transcriptional reduction in core expression, and decrease in HBeAg synthesis. They have been related to more severe hepatitis including liver cirrhosis, HCC and fulminant hepatitis B, modulation of drug resistance, and HBeAg seronegativity (Chan *et al.*, 1999; Jardi *et al.*, 2004).

Single amino acid mutations in HBV core region are mainly (75%) found in immunodominant B, $CD4^+$ and $CD8^+$ T-cell epitopes (Carman *et al.*, 1995; Rodriguez-Frias *et al.*, 1995; Radecke *et al.*, 2000; Tanaka *et al.*, 2005). Internal deletions involve the loss of B- and T-cell epitopes. All together these allow production of immune escape epitopes that may facilitate persistence (Tsubota *et al.*, 1998). In general, HBcAg sequence is very stable during the immune tolerance phase (absence of immune selection pressure) of HBV infection despite intense viral replication (Uchida *et al.* 1994; Bozkaya *et al.* 1997). Amino acid changes are frequently detected following activation of hepatitis during loss of HBeAg, emerging during seroconversion (active immune response in chronic hepatitis). This seems to be a clear consequence of the quasispecies nature of HBV, with a well-adapted population which shows a stable consensus sequence during the immune tolerance phase, but changes dynamically as a result of immune selective pressure inducing a fluctuation in the proportion of viral subpopulations present in the quasispecies (Gunther *et al.*, 1999). In fact longitudinal studies reveal this dynamic process, with the emergence and disappearance or even re-emergence of HBcAg mutations (Uchida *et al.*, 1994; Alexopoulou *et al.*, 1997; Gunther *et al.*, 1998).

Several mutations have been correlated with outcome of infection. For instance, cooccurrence of the two main BCP mutations, A1762T and G1764A, enhance viral replication *in vitro* an *in vivo* (Tacke *et al.*, 2004; Baumert *et al.*, 2005; Chauhan *et al.*, 2006; Dal Molin *et al.*, 2006). Some deletions and insertions in BCP introduce new binding sites for the liver-enriched transcription factors HNF1 and HNF3 (Gunther *et al.* 1996. 1999; Kurosaki *et al.*, 1996) that result in increase of pgRNA transcription thus enhancing HBV replication (Gunther *et al.*, 1996). The high viral replication phenotype of these HBV-BCP mutations has been associated with the most severe forms of HBV infection (LC and/or HCC) (Kao *et al.*, 2003; Chen *et al.*, 2006; Iloeje *et al.*, 2006; Liu *et al.* 2006; Jang *et al.*, 2007). Besides, intense cellular immune response as a result of this high level replication, together with induction of hepatocyte apoptosis by some BCP mutations have been suggested as the mechanisms involved in FHB (Baumert *et al.*, 2005). Regarding the pathogenic role of HBcAg mutations, their accumulation does not seems to correlate with the outcome of liver disease in adults (Gunther *et al.*, 1999).

Amino acid substitutions in the RT/polymerase gene are involved in resistances to antiviral treatments with RT inhibitors, and these mutants arise from preexisting variants within the quasispecies (Thibault *et al.*, 2002; Jardi *et al.*, 2007). Selective pressure during antiviral therapy would decrease the relative wild-type cccDNA fitness in the liver and favor their replacement drug-resistant cccDNA (Locarnini, 2003). In general, all these "resistant" variants have lower fitness than wild-type and are usually replaced again after drug withdrawal. However, long-term antiviral therapy with inhibitors such as lamivudine (LMV) increases the probability for additional compensatory mutations, restoring replication fitness of the mutant and worsening disease outcome (Papatheodoridis *et al.*, 2002; Lai *et al.*, 2003; Yuen *et al.*, 2003; Sheldon *et al.*, 2005). Indeed, mutation rtL180M (domain B) restores the replication capacity of the main resistant mutations to LMV rtM204V and rtM204I, in the YMDD RT catalytic motif (C domain) (Das *et al.*, 2001; Ono *et al.*, 2001; Delaney *et al.*, 2003; Sheldon *et al.*, 2006).

Catalytic domains B and C (where the main polymerase variants have been detected) overlaps with the main antigenic domain "a" of S gene (MHR). Therefore, mutation in one of these regions can affect the other (Figure 15.2). Interestingly, mutation rtR153Q that gives partial resistance to LMV (Locarnini, 2003) and compensates viral replication in the presence of the main LMV variants (Torresi *et al.*, 2002), is associated with sG145R mutation which is the main vaccine escape variant. This is especially dangerous in liver transplant recipients treated with HBIG and LMV (used to prevent re-infection of the graft) that may select sG145R (equivalent to rtR153) in the case of HBIG and rtM204V/I in the case of LMV with the double resistance capacity (to HBIG and LMV) (Bock *et al.*, 2002).

Clinical Significance of Genotypes

Several studies have suggested a close association between HBV genotype or subtype with natural history, clinical features of infection and response to antiviral therapy (Sanchez-Tapias *et al.* 2002; Janssen *et al.*, 2005; Ozasa *et al.*, 2006; Rodriguez-Frias *et al.*, 2006; Sakamoto *et al.*, 2006). In general, it has been found that genotype C induces a more severe liver disease than B, C, D, and A, in that order (then $C > B > D > A$), and F more severe disease than A and D ($F > D > A$). On the other hand, progression to chronicity after acute infection seems more likely and faster with genotypes B and C than A, and A faster than D (Ozasa *et al.*, 2006; Rodriguez-Frias *et al.*, 2006). Genotype B induces a greater Th1 and lesser Th2 response than genotype C, thus providing an immunological evidence for the higher chance of HBeAg seroconversion in patients with genotype B (Yuen *et al.*, 2007). Genotype D patients had more severe disease and more HCC than genotype A patients (Thakur *et al.*, 2002). In liver transplantation (LT), genotype A patients have the lowest risk for HBV recurrence while patients with genotype D have the highest risk of recurrence and mortality after LT (Devarbhavi *et al.*, 2002). Sustained remission after anti-HBe seroconversion was higher in genotype A than in D and death related to liver disease was more frequent among patients infected with genotype F (Sanchez-Tapias *et al.*, 2002).

Subgenotype prevalences are also important to help explain discordances in several studies, such as the case of HCC which develops later in genotype C-infected Taiwanese patients than in those infected with genotype B (Ba or B2), in contrast to Japanese patients infected with other B subtypes (subtypes Bj or B1) (Orito *et al.*, 2001; Sugauchi *et al.*, 2004). A fulminant outcome of acute HBV infection has been associated with subtype Bj in the presence of HBeAg-negative HBV variants (Ozasa *et al.*, 2006). Subtype C1 has been shown to carry a higher proportion of basic core promoter (BCP) mutants (associated with a high risk of developing HCC) and low pre-core defective variants than subtype C2 (Chan *et al.*, 2005; Wang *et al.*, 2007).

Mechanisms of Chronicity

HBV is preferentially hepatotropic but not directly cytopathic. A cellular immune response seems to be responsible for disease pathogenesis of HBV infection. After infection, HBV viral antigens are not detectable in serum or liver until 4–7 weeks post infection. Cytokine activation involved in early antiviral response, such IFN-γ, IL-2, or TNF-α and the recruitment of inflammatory cells is delayed until the logarithmic phase of replication (Wieland *et al.*, 2004). A putative explanation is that HBV replication occurs within nucleocapsid particles that protect single-stranded RNA or DNA viral intermediates, generally strong activators of innate and intracellular antiviral responses (Bertoletti and Gehring, 2006; Wieland *et al.*, 2004). Patients and animals who develop CHB, often without acute hepatitis symptoms, lack the large IFN-γ and TNF-α production, observed in acute resolving infections, failing to develop an efficient Th1 antiviral immune response (Menne *et al.*, 2002; Wieland *et al.*, 2004). Thus, activation of innate immune elements during the initial phase of HBV replication seems to be a key factor for the induction of an effective T-cell response that determines the outcome of HBV infection (Bertoletti and Gehring, 2006).

Resolution of acute hepatitis B infection is associated with an intense, polyclonal, multispecific CD4+ T helper and CD8+ CTL response (revised in Bertoletti and Gehring, 2006) while in CHB these responses are often much more difficult to detect, suggesting that the chronic state is associated with the establishment of some form of immunological tolerance (Kakimi *et al.*, 2002).

HEPATITIS C VIRUS

Features of HCV

Hepatitis C virus (HCV) infects some 210 million people worldwide or 3% of the human population and is the leading cause of chronic hepatitis, end-stage liver disease, and hepatocellular carcinoma (Quer and Esteban, 2005). Currently responsible for over 50% of liver transplantations (Willems *et al.*, 2002), the number of patients with HCV-associated cirrhosis and its complications is expected to increase exponentially during the next decades (Mizokami *et al.*, 2006).

Efficiently transmitted by the parenteral route, blood transfusion, unsafe medical, or surgical procedures and intravenous drug use (IDU) have been the predominant modes of transmission throughout the world (Alter *et al.*, 1992; Alter, 1995; Mansell and Locarnini, 1995; NIH, 2002). With the implementation of blood donor screening for anti-HCV and improvement in health-related standards, IDU has become the predominant mode of HCV transmission in the developed world, although transfusion of unscreened blood and unsafe parenteral procedures with non-disposable equipment, continue to spread infection in the rest of the world (WHO, 2000b; Quer and Esteban, 2005).

Current treatment of chronic hepatitis C with the combination of pegylated interferon-α and ribavirin leads to viral eradication in about half of patients (40% of those infected with HCV genotype 1 and about 80% of those with genotypes 2 and 3 (Manns *et al.*, 2001; Fried *et al.*, 2002). Because of the limited efficacy of this treatment, selective inhibitors of the enzymatic activity of non-structural proteins of the virus (NS3 protease and NS5 polymerase) have been developed and are currently being evaluated in clinical trials.

The hepatitis C virus, a *Hepacivirus* within the family *Flaviviridae*, is a small, enveloped virus, with a 9.6 kb positive ssRNA genome. Its genome contains a single ORF, encoding a polyprotein precursor, flanked by 5′ and 3′ NCRs (Penin *et al.*, 2004) (Figure 15.4). The 5′ NCR is highly conserved among different HCV isolates and has a highly structured stem–loop structure which with the first 30–40 nucleotides of the core-coding acts as an IRES for the binding of the ribosomal 40S subunit and cap-independent translation of the coding region (Tsukiyama-Kohara *et al.*, 1992; Wang *et al.*, 1993). The 3′ NCR is a tripartite structure with an upstream variable sequence, followed by a poly U stretch sequence of variable length and a downstream highly conserved 98-nt-long sequence which is essential for viral replication (Kolykhalov *et al.*, 1996, 2000; Tanaka *et al.*, 1996; Yamada *et al.*, 1996; Yanagi *et al.*, 1999) The polyprotein precursor encoded by the ORF is co- and posttranslationally processed by cellular and viral proteases to produce ten mature structural and non-structural (NS) proteins (Grakoui *et al.*, 1993c). The structural proteins core (the mature form of which binds the genomic RNA to form the viral nucleocapsid) (Yasui *et al.*, 1998; Tellinghuisen and Rice, 2002) and the type 1 transmembrane envelope glycoproteins E1 and E2 (Deleersnyder *et al.*, 1997) are released by the action of a host signal peptidase. They are separated from NS proteins by a small membrane polypeptide, p7, with ion channel activity and with potential role in viral particle release and maturation (Griffin *et al.*, 2003; Pavlovic *et al.*, 2003). The non-structural proteins NS2 through NS5B are processed by viral proteases (Bartenschlager, 1999), and are essential for replication of viral RNA (Lohmann *et al.*, 1999). The NS2–NS3 junction is cleaved by the zinc-dependent autocatalytic protease comprising NS2 and the N-terminal third of NS3 (Grakoui *et al.*, 1993a; Hijikata *et al.*, 1993; Yasui *et al.*, 1998). NS3 is a multifunctional molecule, with serine protease activity at the

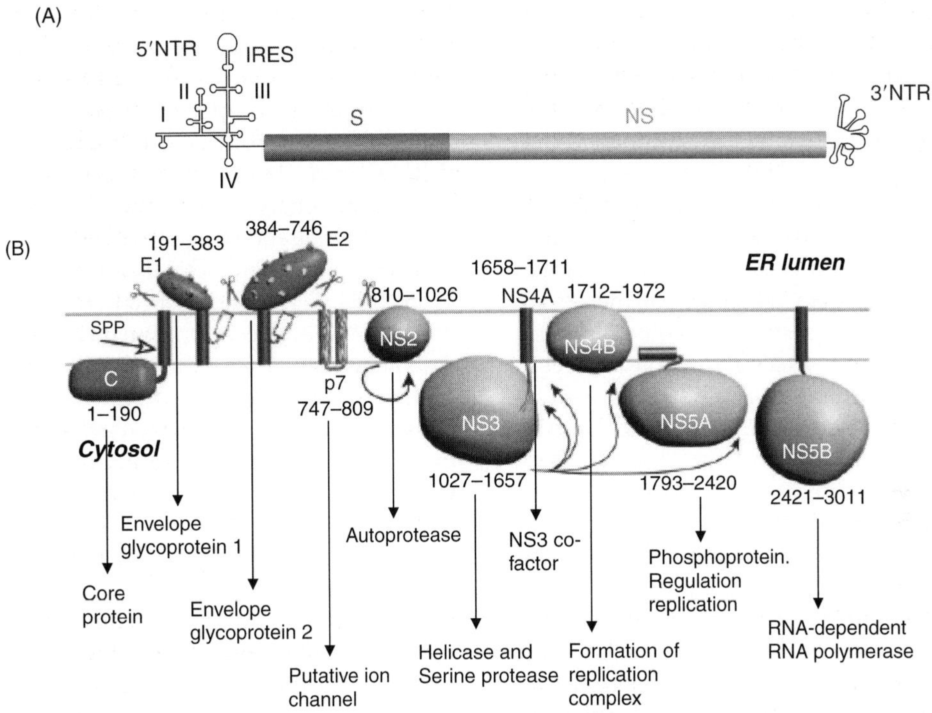

FIGURE 15.4 Hepatitis C virus structure. (A) HCV genome organization with 5′ and 3′ untranslated regions limiting the single ORFs that codify for a polyprotein of 3011 amino acids with C-terminal (one-third) having the structural proteins (S) and the rest that contains the non-structural proteins (NS). (B) Polyprotein processing with scissors indicating endoplasmic reticulum signal peptidase cleavage sites; cyclic arrow, autocatalytic cleavage of the NS2 NS3 junction; and black arrows indicating the cleavage of NS3/NS4A proteinase complex. (See Plate 21 for the color version of this figure.) Reproduced from Penin *et al.* (2004) with author's permission.

N-terminal one-third, which, in association with the co-factor NS4A (Bartenschlager *et al.*, 1994; Failla *et al.*, 1994; Lin *et al.*, 1994; Tanji *et al.*, 1995a), is responsible for downstream cleavage of the NS proteins, and NTPase/RNA helicase activities at the C-terminal two-thirds essential for translation and replication of the HCV genome (Bartenschlager *et al.*, 1993; Eckart *et al.*, 1993; Grakoui *et al.*, 1993b; Kim *et al.*, 1995; Kolykhalov *et al.*, 2000). NS4B is an integral membrane protein involved in the formation of the replication complex (Egger *et al.*, 2002; Gosert *et al.*, 2003). NS5A is a highly phosphorylated membrane-associated protein, of unknown function supposedly playing a role in regulation of replication, by analogy with other RNA viruses (Tanji *et al.*, 1995b; Reed and Rice, 1999). Finally, NS5B has been identified as the RNA-dependent RNA polymerase (RdRp) the key enzyme for the synthesis of new RNA genomes (Behrens *et al.*, 1996; Lohmann *et al.*, 1997; Yamashita *et al.*, 1998).

The existence of previously unknown core protein products encoded by alternative reading frames have recently been reported (Walewski *et al.*, 2001; Roussel *et al.*, 2003; Xu *et al.*, 2003; Branch *et al.*, 2005), but their production in the liver during natural infection remains to be clarified.

The mechanism of HCV entry into the host cell remains poorly defined, although a variety of model systems have been used using viral particles from clinical isolates, HCV pseudotype particles (HCVpp) lacking lipoprotein association, and cell culture-derived HCV particles (HCVcc). From these studies, several receptors have been proposed to be involved in attachment and entry of HCV (Diedrich, 2006), including the tetraspanin CD81 receptor (Pileri *et al.*, 1998; Flint *et al.*, 1999), the type I scavenger receptor class B (SR-BI) receptor (Scarselli *et al.*, 2002), the low-density lipoprotein receptor (LDLr) (Agnello *et al.*, 1999; Monazahian *et al.*, 1999), and also L-SIGN (Gardner *et al.*, 2003), DC-SIGN (Lozach *et al.*, 2003; Pohlmann *et al.*, 2003), the asialoglycoprotein receptor (ASGPR) (Saunier *et al.*, 2003), as well as heparan sulfate (Barth *et al.*, 2003) and Claudin-1 (Evans *et al.*, 2007).

Studies with clinical HCV isolates suggest that the LDL receptor is the main HCV receptor, and that lipoproteins, acquired during viral egress (Gastaminza *et al.*, 2006), play a crucial role in HCV entry (Andre *et al.*, 2002, 2005; Favre and Muellhaupt, 2005), while CD81 and SR-BI may act as co-factors (Cocquerel *et al.*, 2006; Diedrich, 2006; Favre and Muellhaupt, 2005), but exact entry of HCV has not yet been clearly defined.

HCV post-receptor attachment stages imply the internalization by receptor-mediated endocytosis (Dimmock, 1982) into vesicles. These mature to endosomes (Hoekstra and Kok, 1989; Marsh and Helenius, 1989) triggering fusion of the viral envelope to the endosome membrane allowing release of the viral genome into the cytoplasm, and the process appears to be pH-dependent (Quer *et al.*, 2005a; Tscherne *et al.*, 2006; Meertens *et al.*, 2006).

HCV Generation of Variation

Like most RNA viruses, HCV evolves rapidly due to high mutation rates and high level viral replication through an error-prone RNA polymerase without proofreading capacity. Consequently, in the infected individual, the viral population replicates and circulates as a quasispecies composed of a complex mixture of different but closely related genomes (Martell *et al.*, 1992) which shape is subject to continuous changes due both to competitive selection (Holland *et al.*, 1982, 1992; Steinhauer and Holland, 1987; Domingo and Holland, 1988, 1994; Eigen and Biebricher, 1988; Duarte *et al.*, 1994) and cooperation (Vignuzzi *et al.*, 2006) between arising mutants. Both mutation and, to a much lesser extent, recombination, constitute the known variation mechanisms by which HCV persists and adapts to changing environmental conditions.

Mutation

High mutation rates and high level viral replication (10^{11} new HCV particles/day), are likely to be the factors responsible for viral persistence in the majority of exposed individuals (50 to 80%). Mutation rate (misincorporation/nucleotide site/replication cycle) has not yet been established for HCV because of the lack of a highly efficient cell culture system for wild-type isolates, but the average rate of fixation of mutations has consistently been found to range between 1.1 and 1.5×10^{-3} mutations per site and per year (Okamoto *et al.*, 1992; Quer *et al.*, 2005b). The rate of fixation of mutation is not evenly distributed throughout the genome with highly variable regions within the envelope coding genes and well-conserved regions like the 5′ NCR (Figure 15.5). The different degree of variability likely reflects different positive selective pressures from the host's immune response as well as different constrains to maintain critical enzymatic or structural domains required for efficient generation of progeny viruses.

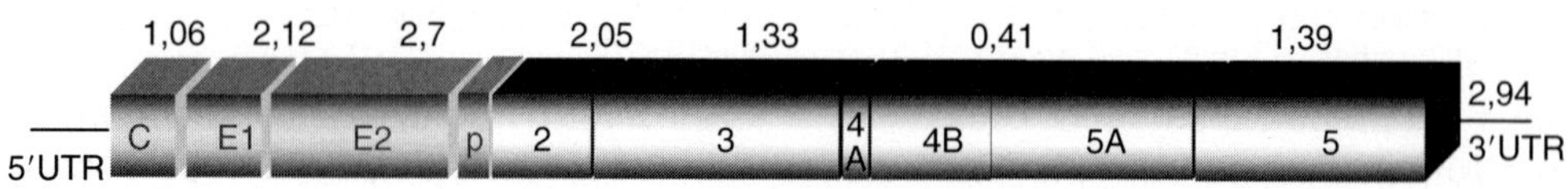

FIGURE 15.5 Hepatitis C virus rate of accumulation (or fixation) of mutations (mutations/site/year) ($\times 10^{-3}$) along the genome. From Okamoto *et al.* (1992).

In this regard, an important constraint to sequence variation in certain genomic regions is the requirement for RNA secondary structures essential to maintain enzymatic activity and/or RNA–protein interactions. The RNA of several genomic regions form secondary and tertiary structures which are part of the viral phenotype (Gomez *et al.*, 2004), since they often correlate with functions such as replication control, mRNA processing, and translation (Rajagopalan and Malter 1997; Bashirullah *et al.*, 1998). The highly ordered genome of HCV has complex RNA secondary structures scattered throughout the genome, including the complete 5′ NCR (Wang *et al.*, 1994, 1995; Smith *et al.*, 1995), several regions in core the coding sequence (Simmonds, 2004; Simmonds and Smith, 1999; Tuplin *et al.*, 2002; Walewski *et al.*, 2002) and part of the 3′ NCR (Yamada *et al.*, 1996; Blight and Rice, 1997; Ito and Lai, 1997). The requirement for base-pairing in such structures severely limits the number of "neutral" sites, since even synonymous changes might disrupt RNA folding. Therefore, truly "neutral" sites, where sequence changes have no significant effect on virus phenotype, may be not be very common and virus diversification over time must be limited or require compensatory mutations that maintain structured regions (Quer *et al.*, 2001).

HCV Genotypes

Since the first publications of complete HCV genomes (Kato *et al.*, 1990; Choo *et al.*, 1991), intensive sequencings of HCV isolates throughout the world have been done, with over 30 000 published sequences including 181 full-length genomes (GenBank, EMBL, and DDBJ). More recently, three specific HCV sequence databases, from Japan (http://s2as02.genes.nig.ac.jp/), the European Union (http://euhcvdb.ibcp.fr/), and the US (http://hcv.lanl.gov/ and http://hcv-db.org), have become available, providing a resource to study genetic variability of HCV and to establish platforms for a consensus on a unified nomenclature system for new HCV variants (Simmonds *et al.*, 2005).

Analysis of extensive sets of sequences from HCV isolates throughout the world has revealed the existence of six major genetic groups or genotypes (Figure 15.6), and a large number of subtypes c within the six main genotypes (Chamberlain *et al.*, 1997; Simmonds *et al.*, 1994a). Genotypes are numbered from 1 to 6 and subtypes designated as a, b, c, etc., in both cases in order of discovery (Simmonds *et al.*, 2005). Overall sequence divergence between genotypes ranges from 31 to 34% and from 20 to 23% between subtypes (Pawlotsky, 2004). Quasispecies studies have shown that cloned E1/E2 sequences from single patients may differ by up to 10%, depending on the duration of infection.

Origin and Spread of HCV in the World

Epidemiological and phylogenetic studies have shown that the initial epidemic spread of HCV in Japan occurred in the 1920s–1930s through mass campaigns of parenteral antischistosomal therapy (PAT), followed during World War II by intravenous drug use (IDU), transfusion, and unsafe invasive medical and surgical procedures. Similarly, in Europe, the initial spread of the virus started during the last century through the use of unsafe parenteral injections, invasive medical and surgical procedures and transfusion of blood products. An epidemic explosion of IDU shortly followed the iatrogenic spread in the early 1960s both in the United States and Europe (Alter, 1997; Nakano *et al.*, 2004; Mizokami *et al.*, 2006).

HCV genotypes have different geographical distribution. HCV genotypes 1a, 1b, 2, and 3a are highly prevalent "epidemic" strains found globally. Again, epidemiological and molecular evolutionary methods have shown that the spread of genotype 1b in Spain and France, and that of genotype 3a in the former Soviet Union, coincided with local outbreaks of unsafe parenteral treatments. These strains spread rapidly around the world during the twentieth century most likely through infected blood products, unsafe parenteral procedures, and IDU, and have relatively low levels of

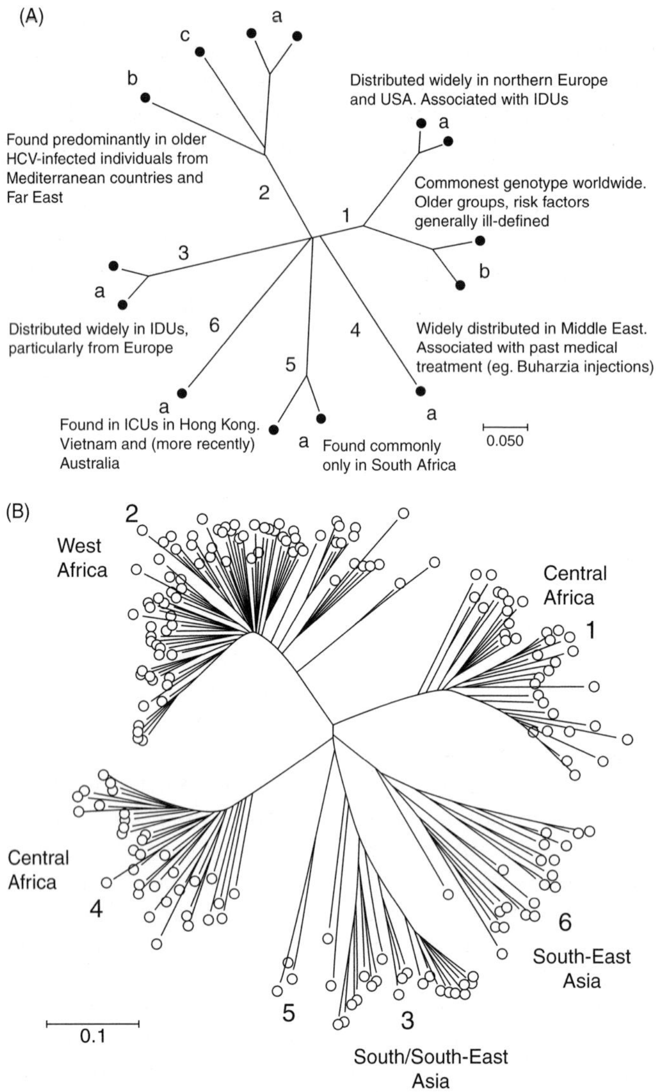

FIGURE 15.6 Hepatitis C virus evolutionary trees. (A) Principal genotypes found in industrialized countries and their main association with specific risk groups. (B) All known genotypes and subtypes found in endemic areas. Reproduced from Simmonds (2004) with permission.

genetic variation (Pybus *et al.* 2001, 2003; Cochrane *et al.*, 2002; Ndjomou *et al.*, 2003). In the early 1990s, HCV genotypes 1, 2, and 3 accounted for most infections in blood donors and patients in Europe (McOmish *et al.*, 1994). After genotype 1, the next most prevalent was genotype 3a, except in southern Italy, where genotype 2c accounted for 25–30% of infections among older adults (Ansaldi *et al.*, 2005). Genotype 4, found at low frequencies (4–6%) in Southern Europe, was spread on a large scale in Egypt through mass campaigns of PAT. Recent studies have reported that HCV genotype 5a, once believed to be restricted to South Africa, has been endemic for a long time in isolated areas of Central France and West Flanders (Verbeeck *et al.*, 2006). In contrast, HCV strains found in restricted geographic areas are highly divergent. These "endemic" strains reflect long-term transmission at low levels in particular populations (Simmonds and Smith, 1997; Pybus *et al.*, 2001) and represent the source populations for the epidemic strains. Genotypes 1, 2, and 4 appear to be endemic to regions of West and Central Africa and the Middle East, genotype 5 is endemic in South Africa, whereas divergent endemic strains of genotypes 3 and 6 are found in South-East Asia (Figure 15.6) (Pawlotsky 2003a).

The divergence of subtype variants is estimated to have occurred less than 100 years ago, whereas the numerous subtypes of HCV are proposed to have diverged some 300 years ago while the major HCV genotypes would have originated at least 500–2000 years ago.

Recombination in Natural Isolates

Recombination as a mechanism of generation of variation has important implications in viral evolution such as rescuing viable genomes from debilitated parental genomes, as exemplified by the neurotropism recovery of poliovirus vaccines, and the generation of viruses with new infectious capabilities (Domingo and Holland, 1997). Homologous recombination requires that a cell be infected simultaneously with two or more distinct viral strains, as may occur during co-infection or superinfection (Blackard and Sherman, 2007), and it has been demonstrated in most members of the *Flaviviridae* family (Holmes *et al.*, 1999; Becher *et al.*, 2001; Worobey and Holmes, 2001; Cristina and Colina, 2006).

Successful recombination events depend on the properties of the viral replicase, co-infection of a single cell with both variants at the same time, and on the viability and fitness of the recombinant as compared to that of non-recombinant variants in the context of a viral quasispecies. Until the spontaneous HCV recombinant RF1_2k/1b was identified in St Petersburg (Kalinina *et al.*, 2002), inter- and intragenotypic recombination in HCV was considered an exceptional event *in vivo* and of little, if any, relevance because any resultant recombinant was presumed not viable (Simmonds *et al.*, 1994b; Smith and Simmonds, 1997). This assumption was reinforced by the observation that chimeric 1a-1b HCV replicons failed to replicate in cell culture (Gu *et al.*, 2003; Gates *et al.*, 2004) and by data on superinfection exclusion on replicon-containing cells (Tscherne *et al.*, 2007). However, shortly after the naturally arising 2k/1b homologous recombinant was first reported (Kalinina *et al.*, 2002, 2004) it was shown to be rapidly spreading in Europe, being reported among IDUs from Estonia (Tallo *et al.*, 2007) to Ireland (Moreau *et al.*, 2006). Other natural intergenotypic recombinant virus have been found in Vietnam (a 2i/6p recombinant) (Noppornpanth *et al.*, 2006), the Philippines (2b/1b) (Kageyama *et al.*, 2006), and in France (2/5) (Legrand-Abravanel *et al.*, 2007). Intragenotyping recombinant viruses have also been identified in Peru (1b/1a) (Colina *et al.*, 2004) and a 1a/1c intragenotypic recombinant was identified by examining just 89 full length sequences, deposited at the LANL database (Kuiken *et al.*, 2005; Cristina and Colina, 2006).

Biological Implications of HCV Variability

The numerous implications of the quasispecies structure on the biology of RNA viruses include: correlations between inoculum size and infection outcome, establishment of persistent infection, selection of resistant mutants to antivirals or of vaccine-escape mutants, changes in cell tropism, changes in pathogenic potential modifying the natural course of disease progression and selection of mutants undetectable with standard molecular tests, among others.

HCV Transmission. From Massive Infection to Bottlenecking

Transmission of HCV into a new host provides a unique opportunity to identify virus-related features associated with infection outcome. The mode of HCV transmission may be one of the key factors determining transition from acute to persistent infection.

In cases of massive transmission of HCV particles, as in transfusion-associated infection or during HCV recurrence after liver transplantation for HCV-associated end-stage liver disease, persistent infection ensues in 80 and 100% of the cases, respectively.

In contrast, situations in which a small inoculum size is transmitted, as in sexual intercourse, accidental needlestick exposure, tattooing, acupuncture, or mother-to-infant HCV transmission (Quer and Esteban, 2005), persistence occurs in a much smaller proportion of cases. Only 0.013–10% of healthcare workers undergoing accidental needlestick injury (Puro *et al.*, 1995a, 1995b) and 2–7% of children born to HCV-infected mothers (Della *et al.*, 2005), develop persistent infection. Moreover, at least 40% of female spouses of HCV carriers have detectable NS3-specific T-cell responses in the absence of any clinical or serological evidence of exposure, suggesting frequent subclinical infection with complete viral clearance (Bronowicki *et al.*, 1997). Similarly, specific T-cell responses to nonstructural HCV proteins have been detected in 30% of sexual contacts of patients with acute hepatitis C, again in the absence of detectable viremia or antibody seroconversion (Kamal *et al.*, 2004). The same phenomenon has been documented in up to 70% of children born to HCV-infected women (Della *et al.*, 2005). Finally, chimpanzees inoculated with very small HCV inocula (1–10 copies) frequently develop cellular immune responses without corresponding viremia (Shata *et al.*, 2003).

The lower risk of viral persistence after exposure to small size inocula, might be related to the genetic bottlenecking phenomenon as recently proposed (Quer *et al.*, 2005b). Genetic bottlenecking, in which a limited number of very closely related viral particles is transmitted, would be associated with a drastic reduction in viral genomic sequence diversity and potentially carrying restrictive mutations (Muller, 1964; Chao, 1990; Domingo *et al.*, 1996; Quer *et al.*, 2001), compromising the capacity of the viral population to adapt and generate sufficient mutants to overcome immune-mediated clearance (Novella, 2004). Randomly transmitted particles usually carry a constellation of mutations which may impair their movement through the fitness landscape, limiting the extent of their variability and thus their potential to generate a certain threshold of mutant diversity required to establish persistence (Escarmis *et al.*, 1999; Quer *et al.*, 2001).

Resistant Mutations to Antiviral Inhibitors

New specifically targeted antiviral therapies for hepatitis C (STAT-C), with small peptides inhibiting key enzymes necessary for HCV replication, specially the serine protease domain in NS3 and the RNA-dependent RNA polymerase domain in NS5B, have been developed to improve efficacy of current antiviral treatment. As with HIV, a major cause of failure of antiviral therapy is the selection of resistant mutants. The quasispecies structure of RNA viruses implies the presence in a single host of all single mutants and double or triple mutants depending on the viral load (with 4×10^4 particles all simple mutants would be represented, and all double mutants with viral loads around 1.6×10^9 particles). With high-level HCV replication in the untreated patient (10^{12} new viral particles per day), point mutations at each position in the genome arise at least once every day, so potentially resistant mutants are constantly being generated. During treatment, pre-existing minor mutants resistant to a given antiviral would gain a selective growth advantage over the existing wild-type viral population and rapidly become dominant in the viral quasispecies.

Resistance Mechanisms to IFN-α and Ribavirin

The striking difference in sustained response rates to pegylated interferon-α (Peg-IFN-α) and ribavirin (RBV) combination therapy (38–52% for genotypes 1 and 4, and close to 90% for genotypes 2 or 3) (McHutchison *et al.*, 1998; Manns *et al.*, 2001; Fried *et al.*, 2002;

Hadziyannis *et al.*, 2004), strongly suggests a determinant role of HCV genotype in IFN-α sensitivity. IFN-α does not directly inhibit HCV replication but induces expression of a large number of IFN-stimulated genes (ISGs) whose products (which include protein kinase R (PKR), an inhibitor of the eukaryotic translation initiation factor eIF2; RNA-specific adenosine deaminase 1 (ADAR 1), which destabilizes secondary structures of the viral genome; 2′5′ oligoadenylate synthetases (2′5′ OAS), which activate the endoribonuclease RNaseL, degrading viral RNA; and P56, which inhibits translation through binding eIF3), limit HCV replication (Pawlotsky, 2003b, 2005). Different HCV proteins that have been proposed to be involved in IFN resistance (Moradpour *et al.*, 1998; Heim *et al.*, 1999; Blindenbacher *et al.*, 2003; Gale and Foy, 2005) including a portion of the envelope 2 (E2) (which has a PKR eukaryotic initiation factor 2 alfa phosphorylation homology domain (PePHD) that would interfere with PKR activity), the serine protease NS3/4A (which has been shown to cleave the two adaptor proteins required for downstream signaling from RIG-1 and TLR-3), core (which activates the JAK-STAT signaling inhibitor SOCS-3), and the NS5A (which upregulates IL-8 expression and secretion, attenuating ISG expression and binds both 2′5′ OAS and PKR inhibiting their activity). However, no prediction of treatment response can be made by sequencing any of the genomic sequences encoding these proteins (Quer *et al.*, 2004; Hofmann *et al.*, 2005; Brillet *et al.*, 2007).

The mode of action of RBV is not well known since, although it has little antiviral activity *in vivo*, it increases 50% sustained response rates when combined with IFN-α. Its antiviral activity *in vitro* has been proven in the subgenomic replicon system in a dose-dependent manner where it seems to act as a mutagen, and effect that appears more evident when used in combination with IFN-α (Wohnsland *et al.*, 2007). In addition to its modest antiviral effect, enhanced by IFN-α, RBV seems to have a major effect restoring and stabilizing the intracellular mechanisms of IFN-α action (it decreases IL-8 expression and SOCS3 activation and increases cell membrane expression of type I IFN receptors).

Resistant Mutants to NS3/4A Protease Inhibitors

Selective pressure in the replicon system with anti-protease inhibitors such as BILN2061 (already withdrawn from further trials because of cardiotoxicity) resulted in selection of genomes carrying drug-resistant single mutations such as A156T, R155Q, D168V associated with 357-, 24- and 144-fold reduction in drug susceptibility, respectively (Lu *et al.*, 2004). Similar results have been reported with VX-950 (Telaprevir) for which substitutions A156V/T/S, V36A/M, T54A, and R155K/T induced different degrees of resistance (Wohnsland *et al.*, 2007), and with SCH503034 to which mutation A156T resulted in a > 100-fold resistance, although resistant variants were less fit than their wild-type counterparts at least *in vitro* (Tong *et al.*, 2006a, 2006b; Yi *et al.*, 2006). In the case of SCH6, mutation A156T in the replicon system resulted in high-level resistance (Yi *et al.*, 2006). No information is yet available regarding resistance to new NS3 inhibitors currently being evaluated in phase I trials, such as ITMN 191 and ACH-806/GS-9132.

Interestingly, in replicon system studies, the A156T mutation which induces high-level resistance to SCH6 and a related ketoamide SCH503034, as well as to BILN2061 and VX-950 (Courcambeck *et al.*, 2006), by reducing NS3/4A catalytic efficiency, polyprotein processing, and replicon fitness in culture, is partially compensated by three separate second-site mutations P86L, Q86R, and G162R capable of restoring fitness and polyprotein processing without significantly reducing resistance to the inhibitor (Yi *et al.*, 2006).

Resistant Mutants to NS5B (RdRp) Inhibitors

A number of distinct NS5B RdRp inhibitors, including nucleoside analogues (NM283 Valopicitabine Idenix/Novartis, R1626 Roche), which act as chain terminators, pyrophosphate mimics (JTK109, JTK003 Japan Tobacco; and

HCV796 ViroPharma/Wyeth) which interact with the catalytic metal ions, and non-nucleoside analogues such as benzothiadiazines (GSK6 Glaxo SmithKline, A-782759 Abbott, and NNI-3 Roche) which bind the palm domain near the active site (Tomei *et al.*, 2004; Tedesco *et al.*, 2006), thiophene carboxylic acids (NNI-1 Roche) (Le Pogam *et al.*, 2006) which bind at the outer surface of the thumb domain (thumb I site) and benzimidazoles and indoles which bind to the thumb domain near the fingertips (thumb II site) (Tomei *et al.*, 2003; Di Marco *et al.*, 2005; Kukolj *et al.*, 2005), are continuously being developed.

Specific mutations in the NS5B gene in replicons and in isolates from clinical samples resistant to the compounds have systematically been identified (Neyts, 2006; Wohnsland *et al.*, 2007), including mutation S282T for NM283; C316Y for HCV796, P495 for JTKs, mutations at residues 419 and 423 for thiophene derivatives, M414L/T for benzothiadiazines, and many others. It has been recently reported that mutation M414T conferring resistance to A-782759 pre-existed in a minor population of sequences in replicon cells (Lu *et al.*, 2007).

Future STAT-C Therapies. Combination Therapy

In summary, studies on inhibitor-resistant mutations support the prediction that monotherapy with protease or polymerase inhibitors will inescapably fail, as in HIV infection. Hence, the goals of the future STAT-C therapies will be the combination of multiple agents with different mechanism of action, including protease and polymerase inhibitors along with boosting agents, together with other potential agents (inhibitors of host factors involved in viral replication such as cyclophilin B inhibitors (Ma *et al.*, 2006), inhibitors of virion entry, release and replicase complex formation), leading to therapies that are effective across a range of genotypes. In the mean time, new inhibitors are likely to be used in combination with current standard antiviral treatment with interferon and ribavirin.

Adaptation to Infect other Cell Types. Extrahepatic Replication

Hepatocytes are the major, if not the only cells capable of supporting efficient HCV replication. Numerous studies using rTth enzyme (Myers and Gelfand, 1991) or strand-specific *in situ* hybridization (Pal *et al.*, 2006) have presumably identified negative-stranded RNA in several non-hepatic tissues such as pancreas, thyroid, bone marrow, intestine, adrenal gland, spleen, lymph nodes, cervicovaginal fluid, skin, and brain (Laskus *et al.* 1998, 2000; Deforges *et al.*, 2004; Diaz *et al.*, 2006; Nowicki *et al.* 2005), as well as in white blood cells (dendritic cells, granulocytes, B lymphocytes, monocytes/macrophages and T lymphocytes) (Shimizu *et al.* 1992, 1997; Lerat *et al.*, 1998; Goutagny *et al.*, 2003; Radkowski *et al.*, 2005; MacParland *et al.*, 2006; Kondo *et al.*, 2007). Given the quasispecies nature of HCV, adaptive mutations allowing the virus to replicate in extrahepatic cell types or tissues is theoretically possible, but because of the extensive base-pairing in secondary and tertiary structures along the genome, technical difficulties to prove unequivocally the presence of negative-stranded intermediates cast doubt on the true significance of these findings and clear-cut demonstration of independent viral replication in extrahepatic sites remains to be established.

Adaptation to in vitro Cell Culture

The development of self-replicating HCV subgenomic replicons has revolutionized the study of HCV (Lohmann *et al.*, 1999). The construction of genomic replicons, also referred to as selectable full-length genomes, became possible with the identification of replication enhancing mutations. However, such adaptive mutations that allowed RNA replication, also reduced virus production. The problem was solved with the JFH1 isolate (Kato *et al.*, 2003), obtained from a patient with fulminant hepatitis C caused by HCV genotype 2, which replicates to high levels in Huh-7 cells independently from cell culture adaptations (Wakita *et al.*, 2005). Based on this observation, a full-length genome chimeric genome constructed with core-to-NS2 from a genotype 1 sequence and NS3 to NS5B from JFH1 was shown to release infectious particles into the culture supernatant when transfected into the Huh-7 hepatoma cell line (Lindenbach *et al.*, 2005; Zhong *et al.*, 2005). More recently it has been reported that viruses *in vitro* generated from a chimeric construct of a genotype 2 isolate (C to NS2 from genotype 2a strain HC-J6) with JFH1, were infectious when inoculated to chimpanzees as well as in chimeric mice carrying human liver grafts and that viral particles obtained from the infected animals efficiently infected naive Huh-7 cells (Lindenbach *et al.*, 2006). These subgenomic and genomic replicon systems are invaluable to study many processes of the viral life cycle, the mechanisms of action and resistance to IFN-α and RBV and for development of new antiviral compounds.

Since natural HCV recombination appears more common than initially thought, it is possible that future studies will allow adaptation of natural HCV isolates to grow in cell culture.

Quasispecies and Disease Progression

Quasispecies evolution has been associated with progression from acute to chronic hepatitis C infection (Farci *et al.*, 2000) and chronicity facilitates accumulation of mutations thus increasing the complexity of the population. However, correlation between circulating quasispecies complexity and degree of liver damage has not been clearly established.

Recovery appears to be more common among patients with clinically evident acute hepatitis (Villano *et al.*, 1999) and in those exposed to small and likely less complex inocula (Zuckerman *et al.*, 1994; Grande Gimenez *et al.*, 2001; Ross *et al.*, 2002; Puro *et al.*, 1995a, 1995b; Quer and Esteban, 2005). Although the precise determinants of viral clearance or persistence are not well understood, a significant association has been found between a broad and sustained HCV-specific T-cell response and viral clearance in acute hepatitis C. At the CD4 T-cell level, such response is especially directed against NS proteins, and highly conserved immunodominant epitopes within the NS3 protein have been identified, suggesting a key role of NS3-specific $CD4^+$ responses and spontaneous recovery (Diepolder *et al.*, 1997; Cerny and Chisari, 1999; Rosen *et al.*, 2002; Grakoui *et al.*, 2003).

Recent reports have provided evidence of positive Darwinian pressure leading to evolutionary changes within T-cell epitopes recognized by the host as well as reversion of prior escape mutants unrecognized by the new host (Weiner *et al.*, 1995; Karlsson *et al.*, 1999; Grakoui *et al.*, 2003; Tester *et al.*, 2005). Both selection and reversion of T-cell epitope escape variants are influenced by their relative fitness cost to the replicating quasispecies (Altman and Feinberg, 2004).

Exposure to less complex inocula may be an important determinant of spontaneous viral clearance and treatment of acute hepatitis C with IFN-α leads to sustained viral response in close to 90% of cases irrespective of viral load and genotype. This suggests that, in persistent infection, viral quasispecies diversify with time and poorly understood phenomena of virus–host co-evolution maintain persistence and increase resistance to endogenous and exogenous antiviral treatment.

Molecular Epidemiology

HCV detection is based on serological assays which detect HCV-specific antibodies (anti-HCV) and on molecular assays which detect HCV RNA. Serological tests are less sensitive in immunocompromised individuals, such as transplanted patients or patients with immunodeficiencies, and in hemodyalisis patients (Moradpour *et al.*, 2001). Furthermore, there is a window period of 3–8 weeks between infection and anti-HCV seroconversion (Forns and Costa, 2006). Besides, anti-HCV antibodies decrease and tend to disappear in patients who spontaneously clear HCV infection

(Takaki *et al.*, 2000). Therefore, definition of infection on the basis of seroconversion at a single time-point may underestimate the true rate of infection.

RNA detection by nucleic acid technology (NAT) either by transcription-mediated amplification (TMA) or PCR techniques have shortened the window period to 1–2 weeks after infection, thus improving the safety of blood supply by reducing the risk of transmission; this has practically eradicated transfusion-associated transmission.

HCV RNA detection, *de novo* seroconversion or detection of an acute infected patient provide an opportunity to study recent infections and to identify the potential source of transmission. Phylogenetic analysis of E2 region sequences (specially the PePHD region) is the most informative way to demonstrate genetic linkage between infected patients, if samples are available shortly after the transmission event (up to six months). Evidence for more distant events can still be obtained from analysis of genes such as NS5B and E1, and NS5B sequences have been used to trace the evolutionary history of HCV (Simmonds, 2004).

Among the practical implications of HCV variability and its classification into genotypes, is that only the 5′ UTR region has enough nucleotide conservation among all HCV genotypes to be used for universal detection of HCV RNA (Bukh *et al.*, 1992; Smith *et al.*, 1995), but typing errors occur (between genotype 1 and 4) and subtyping errors might happen in around 15–20% of cases (Bukh *et al.*, 1995; Forns and Bukh, 1999; Pawlotsky, 2002). In recent years, increasing fluxes of immigration from endemic areas to less prevalent regions has introduced genotypes and subtypes previously uncommon in developed countries. This implies that routine serological and molecular methods should be evaluated for their ability to identify new variants.

Mechanisms of Chronicity

Acute HCV infection is associated with symptoms in only 20% of the cases (of whom 50% spontaneously recover and clear infection) while the majority remain asymptomatic despite laboratory evidence of hepatocellular damage and 80% develop persistent infection. The host immune response plays a leading role in both viral control and liver damage and its interplay with the viral quasispecies determines the outcome of infection.

HCV does not integrate into the host genome yet persistence is very common. Hence other mechanisms might explain persistence. In addition, assuming some cytopathic potential, in order to persist, HCV must regulate its lytic potential and avoid elimination by the host immune system. Due to the quasispecies structure of its genome, HCV may use a variety of strategies to fulfill both requirements. HCV might downregulate its cytophatic potential by the generation of defective genomes (Martell *et al.*, 1992), decreasing the overall viral yield and explaining why only 10% of hepatocytes are infected, thus leading to a decrease in liver cell destruction, as has been shown for other RNA viruses (Roux *et al.*, 1991).

Host immune response to HCV includes both innate (mainly type I interferons (IFN-α and IFN-β) and natural killer (NK) cells) and adaptive immunity (antibodies and, most importantly, cellular immune responses ($CD4^+$ and $CD8^+$ T-cells)). These different components of the host immune response are involved in the outcome of infection but their timing and relative contribution differs markedly.

The innate immune response is the first line of defense against infecting pathogens. HCV is efficient at disturbing the intracellular type I IFN response and subvert its antiviral activity, at different levels, including the induction of IFN-β production, the IFN-α/β signaling pathway and the action of several ISG products, as previously stated (see above). It has been suggested that HCV may also evade the NK/NKT-cell response. *In vitro* E2 crosslinks the HCV receptor CD81, thereby inhibiting NK cell cytotoxicity and IFN-γ production. Similarly, overexpression of the CD94-NK2A receptor with concomitant production of IL-10 and TGF-β by NK cells has been proposed to impair their capacity to activate dendritic cells

(Jinushi *et al.*, 2004). Nonetheless, because of the considerable functional and phenotypic heterogeneity of NK/NKT-cell populations, most studies suggesting reduction in the number of these cells in persistent HCV infection remain controversial, and might not compromise their cytolytic function (Morishima *et al.*, 2006).

The mechanisms leading to evasion of HCV from the adaptive immune response remain incompletely understood. Several mechanisms have been proposed, including primary failure to induce a T-cell response or exhaustion of and initially functional response. These might be related to impairment of adequate antigen presentation by DCs or macrophages, associated with HCV-induced dysfunction of APCs or priming by immature APCs in the absence of adequate co-stimulation. Although some studies have suggested DC dysfunction (Bain *et al.*, 2001; Lee *et al.*, 2001; Sarobe *et al.*, 2002), others have not (Longman *et al.*, 2004). Although primary failure of the Th1 response could be due to inadequate presentation of antigen by non-professional cells, a potential cause of tolerance induction in naive $CD4^+$ T-cells, a CD4 response associated with transient control of viremia during acute infection may be detected in most patients eventually developing persistence. Hence, the high chronicity rate of HCV may not be attributed to a primary failure to mount such response in most cases.

HCV infection induces antibody production to all HCV protein in nearly all immunocompetent individuals, 7–8 weeks after infection. However, although neutralizing antibodies (directed to the envelope glycoproteins) have been identified using infectious pseudotype lentiviral particles bearing native HCV envelope proteins, they are common in chronically infected patients, rare in patients with resolved infection and do not protect from HCV rechallenge in chimpanzees, nor do they correlate with infection outcome in humans.

There is strong evidence that viral clearance and liver disease are mediated by a vigorous T-cell response in the liver with cytotoxic and non-cytotoxic effector functions (Neumann-Haefelin *et al.*, 2005). A strong, multispecific, and sustained HCV-specific $CD4^+$ and $CD8^+$ T-cell response directed mainly against non-structural proteins of HCV is associated with a self-limited course of infection (Diepolder *et al.*, 1995, 1997; Missale *et al.*, 1996; Gerlach *et al.*, 1999; Gruner *et al.*, 2000; Lechner *et al.*, 2000; Urbani *et al.*, 2006). However, viral clearance occurs only in a minority of cases when massive infection occurs. In those patients, viral escape mutations or T-cell dysfunction may contribute to viral persistence (Neumann-Haefelin *et al.*, 2005).

Viral amino acid substitutions that inhibit $CD4^+$ and $CD8^+$ T-cell recognition have been observed in patients (Chang *et al.*, 1997; Kaneko *et al.*, 1997; Tsai *et al.*, 1998; Frasca *et al.*, 1999) and chimpanzees (Weiner *et al.*, 1995; Erickson *et al.*, 2001). Besides, it has been reported that several mutations that impaired MHC class I binding of CTL recognition remained fixed for years without diversification, as a consequence of lack of $CD4^+$ T-cells help (Klenerman and Zinkernagel, 1998) and that CTL escape variants mainly occur as an early event during HCV infection (Chang *et al.*, 1997; Tsai *et al.*, 1998; Wang and Eckels, 1999). Besides, certain features of the cellular immune response in chronic hepatitis C such as persistence of functionally impaired CD8 T-cells (Penna *et al.*, 2007), inability to mount *de novo* CTL responses and the presence of unresponsive CD4 T-cells (Rehermann and Nascimbeni, 2005) suggest that adaptive tolerance of Th cells to HCV antigens plays the leading role in maintaining viral persistence. CD4 response associated with transient control of viremia during acute infection may be detected in at least 50% of patients eventually developing persistence. Hence the reason for the high chronicity rate of HCV is related to early functional abrogation of the T helper response, rather than to a primary failure to mount such response.

On the other hand, the weak or absent *in vitro* CD4 response to most HCV antigens is not due to deletion or ignorance. Adequate control of viral replication during IFN-α treatment or removal of the productively infected liver at the time of liver transplantation seem to be sufficient for the rapid reappearance of previously undetectable responses in a substantial proportion of patients (Cramp *et al.*, 2000; Schirren *et al.*, 2001; Barnes *et al.*, 2002), suggesting antigen-driven peripheral tolerance of the CD4 response. Adaptive tolerance of the CD4 T-cell response appears to be an almost inescapable consequence of the extreme genetic variability of HCV and its quasispecies structure (Bukh *et al.*, 1995; Martell *et al.*, 1992), a common feature of most RNA viruses, with tremendous biological relevance (Domingo and Gomez, 2007).

HEPATITIS DELTA VIRUS

Features of HDV

Hepatitis delta virus (HDV) is a single-stranded RNA subviral agent whose packaging, release, and infectivity depend on HBV as its natural helper virus (Ryu *et al.*, 1992). To date, HDV is the only member of the *Deltavirus* genus (for which no viral family has been described). HDV circulates as a spherical particle of approximately 36 nm in diameter, containing a 19-nm nucleocapside, consisting of the delta RNA genome and a structural protein, the hepatitis delta antigen (HDAg), which is encapsidated by the envelope protein of the HBV surface antigen (HBsAg) (Figure 15.7). The HDV genome is a highly structured, approximately 1.68-kb, single-stranded circular RNA that replicates though a circular RNA intermediate, the antigenome. Both the genome and the antigenome molecules possess extensive intramolecular complementarity with nearly 70% of the nucleotides over the complete length of the molecule being base-paired (Kos *et al.*, 1986; Wang *et al.*, 1986; Makino *et al.*, 1987; Gerin *et al.*, 2001; Farci, 2003; Taylor, 2006).

HDV RNA has two structural domains, one (nt 615–950) has similarities with plant pathogens, namely viroids and virusoids, and is called the viroid-like domain (Chang *et al.*, 2005). This region has ribozyme activity which is required for HDV replication (Sharmeen *et al.*, 1988, 1989; Wu *et al.*, 1989; Wu and Lai, 1989; Lai, 1995; Taylor, 2003). The other structural domain contains a single functional ORF, which is located on the antigenomic strand and encodes the structural HDAg (Bonino *et al.*, 1984; Hoofnagle, 1989).

HDV Heterogeneity

Mutation

Hepatitis D virus differs from all other animal RNA viruses in that its genome replication does not utilize a viral polymerase. Transcription of HDV mRNA and synthesis of genomic RNA are mediated by host RNA polymerase II (pol II) (MacNaughton *et al.*, 1991; Fu and Taylor, 1993; Lai, 2005). However, there is still controversy as to which is the host enzyme that catalyzes the synthesis of the HDV antigenome (Chao, 2007). Interestingly, HDV replication is totally independent of any HBV sequence or function.

Hepatitis delta virus RNA exhibits significant genome heterogeneity in individual isolates (Wang *et al.*, 1986; Y.C. Chao *et al.*, 1990, 1991), a feature consistent with the quasispecies distribution of most RNA viruses. Studies of complete RNA sequences from sequential isolates have shown overall evolutionary rates of $3–5 \times 10^{-3}$ substitutions per nucleotide per year (Weiner *et al.*, 1988; Lee *et al.*, 1992), similar to that of other RNA viruses. The evolutionary rates are different along the genome with highly conserved regions corresponding to functional domains of HDV RNA or its encoded HDAg (Lee *et al.*, 1992).

A phenotypic consequence of the genomic variability of HDV is the existence of two different forms of the delta antigen, the small (HDAg-p24) and the large (HDAg-p27) form (195 and 214 amino acids respectively), that are strictly dependent upon genome replication because of specific site mutations (Wang *et al.*, 1986; Weiner *et al.*, 1988; Luo *et al.*, 1990). At the nucleotide level, the difference consists of a single base substitution at

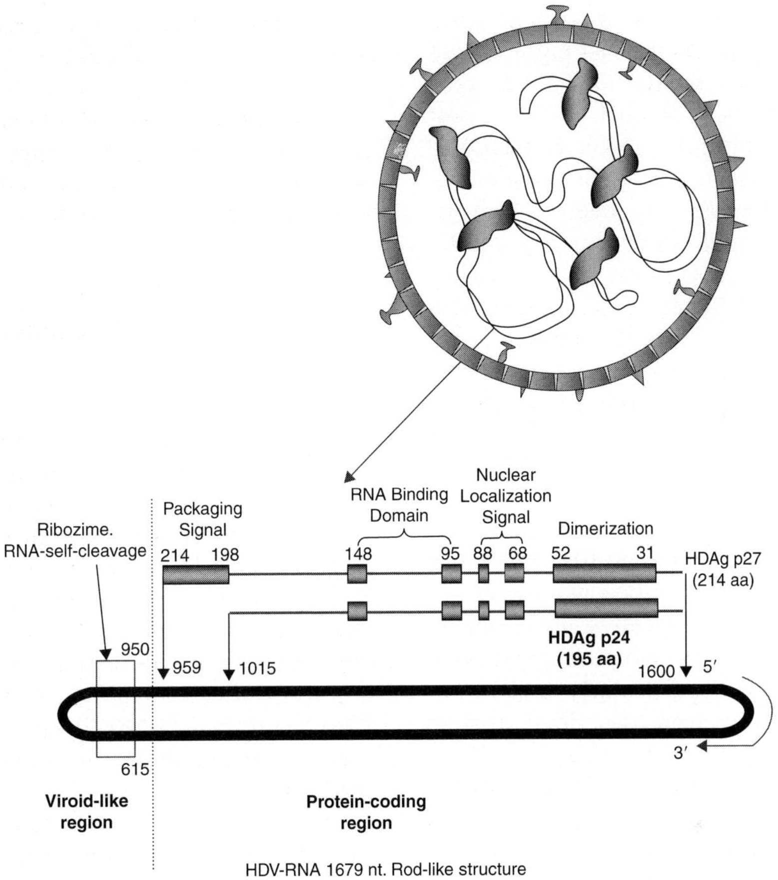

FIGURE 15.7 Hepatitis δ virus. Schematic representation of hepatitis delta particle and genomic organization. The direction of the genomic sense is clockwise. The dotted line represents the boundary between the viroid-like region and the protein-coding region. Antigenomic strand RNA (residues 1600–959) codes for two forms of the delta antigen (HDAg). These two species of HDAg differ in having a termination codon (AUG) in one of the RNAs (residue 1015). Residue 615, site of ribozyme self-cleavage in the genomic strand; residue 950, site of ribozyme self-cleavage in the antigenomic strand; box, region of local tertiary structure and interaction. The main features of the two forms of HDAg are indicated. (See Plate 22 for the color version of this figure.)

position 1015, changing the termination codon UAG present in the short form to UGG, thus allowing continued synthesis to the long form (Bergmann and Gerin, 1986; Wang *et al.*, 1986; Weiner *et al.*, 1988; Luo *et al.*, 1990). These two forms also differ functionally: the short form is required for RNA replication (Kuo *et al.*, 1989), whereas the long form suppresses viral RNA replication and is required for packaging the HDV genome with HBsAg (M. Chao *et al.*, 1990; Chang *et al.*, 1991; Glenn and White 1991; Taylor, 2005).

Genotypes

Three HDV genotypes were initially defined based on genetic analysis of HDV isolates from different areas of the world, with 27–34% nucleotide divergence and 30% divergence at the amino acid level. However recent extensive phylogenetic analyses have concluded that HDV genotypes fall into at least eight clades (HDV-1 to HDV-8) having a varied geographical distribution (Radjef *et al.*, 2004; Deny, 2006; Le Gal *et al.*, 2006). The HDV-1 clade (previously genotype I), the most widespread geographically, has been found in Europe (Saldanha *et al.*, 1990; Niro *et al.*, 1997; Shakil *et al.*, 1997; Cotrina *et al.*, 1998), Northern Africa (Zhang *et al.*, 1996), the Middle East (Lee *et al.*, 1992), South Pacific (M. Chao *et al.*, 1990), and East Asia (Izban and Luse, 1992). HDV-2 (previously genotype IIa) has been found only in East Asia (Taiwan and Japan) (Imazeki *et al.*, 1991; Wu *et al.*, 1995; Lee *et al.*, 1996) and in the Russian region of Yacutia (Ivaniushina *et al.*, 2001). HDV-3 (previously genotype III), the most distantly related to the other genotypes, is exclusively associated with HDV infections in northern South America, where outbreaks of severe hepatitis with particularly high morbidity and mortality were described during the 1990s (Imazeki *et al.*, 1991; Wu *et al.*, 1995; Lee *et al.*, 1996; Nakano *et al.*, 2001). HDV-4 (previously genotype IIb) has been isolated from Taiwan and Japan (Shakil *et al.*, 1997; Wu *et al.*, 1998; Sakugawa *et al.*, 1999; Wang and Chao, 2003; Watanabe *et al.*, 2003). Finally, the HDV-5, HDV-6, HDV-7, and HDV-8 clades are found in Western and Central Africa. This new classification makes the genetic diversity of HDV more closely related to that of HBV.

Recombination

HDV recombination has been recently reported in patients with mixed-genotype HDV infections and in cells co-transfected with genotypes I and IV RNAs (Wang and Chao, 2005) thus increasing its ability to adapt to a changing environment. A recombination junction has not been identified, but crossovers occurred at homologous regions between the two parental RNAs, with no insertions, deletions, or mismatches, suggesting that HDV genome rearrangement occurs through faithful homologous recombination (Wang and Chao, 2005; Chao, 2007).

Origin of HDV and Spread in the World

Interestingly, the finding of a cellular mRNA in infected liver cells that encodes for a protein strongly related to HDAg has suggested that delta RNA may have arisen when a free-living, self-replicating RNA (viroid-like pathogen) "captured" a cellular mRNA encoding this protein (Branch *et al.*, 1989; Brazas and Ganem, 1996; Robertson 1996; Chao, 2007).

The epidemiology of HDV resembles that of HBV with a universal distribution but being endemic in some geographic areas of the world (countries in the Mediterranean basin, Middle East, West Africa, the Amazon basin of South America and Asia, specially Taiwan) (Jacobson *et al.*, 1985). It has been estimated that 5% of HBV carriers are HDV infected, which would represent a reservoir of 15 million people. In the Western World, where the HBV prevalence is lower, HDV is mainly transmitted among IDUs (Buti *et al.*, 1988). Because of the widespread use of vaccination against HBV, the epidemiology of HDV infection has changed in the last 15 years, showing a decline in areas such as Italy and Greece where it had been endemic (Huo *et al.*, 2004; Stroffolini *et al.*, 2004). During the last three years, however, high immigration rates of people from Eastern Europe and Turkey, where there has been an explosive increase of IDU and the prevalence of chronic HDV infection is rising, is causing increases in the number of new HDV infections in countries like Germany. The phenomenon is likely to spread all over Europe through IDU and modify the epidemiology of HDV.

HEPATITIS E VIRUS

Features of HEV

Hepatitis E virus (HEV) is classified into a separate family, *Hepeviridae*, as the prototypic genus *Hepevirus* (Mayo and Ball, 2006). HEV is a spherical, non-enveloped viral particle of around 32–34 nm in diameter. The genome is a ssRNA molecule of positive polarity of approximately 7.2 kb containing a short 5′ NCR followed by three overlapping ORFs and a 3′ poly A tail (Worm *et al.*, 2002; Emerson and Purcell, 2003; Schlauder, 2004).

In vitro analysis suggested that HEV RNA is capped at the 5′ end (Kabrane-Lazizi *et al.*, 1999). After the 5′ NCR of 27–35 nt, ORF-1 encodes about 1693 amino acids encompassing non-structural proteins with enzymatic activity that are involved in viral replication, transcription, and protein processing (Emerson and Purcell 2003). ORF-2 extends 1980 nt, terminating 65 nt upstream of the poly A tail, and encoding a 660-amino-acid protein likely representing the structural capsid protein(s) (Tam *et al.*, 1991). *In vitro* experiments suggest that ORF-2 protein is synthesized as a large glycoprotein precursor of around 88 kDa, which is posttranslationally cleaved into the mature protein (Jameel *et al.*, 1996; Zafrullah *et al.*, 1999). ORF-2 protein contains epitopes inducing neutralizing antibodies which are mainly located near the carboxy end (Tam *et al.*, 1991). ORF-3 overlaps the 3′ end of ORF-1 by only 1 nt and the 5′ end ORF-2 by 328 nt. It encodes a 123-amino-acid protein which is posttranslationally modified by phosphorylation giving a mature protein of around 13.5 kDa of unknown function (Emerson and Purcell, 2003), believed to be involved in the assembly of the viral particle (Jameel *et al.*, 1996), although it is not required for infection of Huh-7 cells or production of infectious virus *in vitro* (Emerson *et al.*, 2006). In addition, ORF-3 protein also bears neutralizing epitopes near its 3′ end (Tam *et al.*, 1991).

HEV Generation of Variation

Mutation

Some studies showed that the genome sequence of HEV is quite stable (Arankalle *et al.*, 1999) with high genomic homology among isolates from the same outbreak, and serial passages in animal models did not result in genetic drift (Schlauder, 2004; Worm *et al.*, 2002). However, studies of inter- and intra-patient diversity revealed the presence of major and minor variants consistent with a quasispecies structure of HEV (Grandadam *et al.*, 2004). Indeed, intra-patient sequence diversity between 0,11 and 3,4% has been reported among acute infected patients during a single source epidemic outbreak (1986–1987) in Tanefdour, Algeria (Grandadam *et al.*, 2004). The rate of accumulation of mutations (rate of evolution) has been estimated to be $1.40–1.72 \times 10^{-3}$ base substitution per site per year (Takahashi *et al.*, 2004). Sequence variability is not uniformly distributed throughout the HEV genome and some regions within ORF-1 (which contains a hypervariable region) are highly polymorphic, whereas both ends of the capsid gene are well conserved. A major neutralization epitope is located in ORF-2 located between amino acid 458 and 607 (Khudyakov *et al.*, 1999; Schofield *et al.*, 2000; Meng *et al.*, 2001).

The biological significance of HEV variability is not well known because studies have been done with partial sequencing of selected genomic regions, and no studies on HEV recombination have yet been performed.

Genotypes

Isolates from different geographical regions are relatively diverse. Based on this genomic heterogeneity, HEV has been classified into four different genotypes (Worm *et al.*, 2002; Meng, 2003; Schlauder, 2004). Inter-genotype difference at the amino acid level of ORF-2 (capsid protein) is only 6.5–11.7%. Genotype I is mainly isolated in endemic areas from Asia and Africa. Genotype II includes the Mexican isolates and some Nigerian strains. Isolates from regions considered as non-endemics (USA, Spain, Italy, Greece, etc.) represent a more diverse cluster of sequences (Pina *et al.*, 2000; Buti *et al.*, 2004) and are grouped into genotype III (Banks *et al.*, 2004). Finally, genotype IV contains strains from human and/or domestic pigs isolated exclusively in Asian countries (Liu *et al.*, 2003; Okamoto 2007). Despite this genotypic diversity, no evidence of serological heterogeneity has been reported and, therefore, there is only one HEV serotype so far described.

HEV Transmision and Prevention

HEV epidemics are mainly caused by a common source of contamination, usually fecally contaminated drinking water resources (Aggarwal and Naik, 1994; Smith, 2001). In most instances, people affected by HEV outbreaks live near rivers with inadequate sanitary conditions (Bile *et al.*, 1994) and a high incidence of HEV seropositivity has been correlated with the use of unboiled river water for drinking, cooking, and washing (Guthmann *et al.*, 2006). Washing, irrigating, and processing of food with HEV-contaminated water may cause HEV outbreaks if the food is eaten uncooked (Cacopardo *et al.*, 1997), and a case of hepatitis E after ingestion of Chinese medicinal herbs has also been reported (Ishikawa *et al.*, 1995). Some evidence indicates that there are animal reservoirs of HEV and that hepatitis E may in fact be a zoonotic disease (Okamoto, 2007) and specifically enzootic in pigs worldwide. Sporadic acute or fulminant hepatitis E has been linked to uncooked pig liver and wild boar meat consumption in Japan (Yazaki *et al.*, 2003; Li *et al.*, 2005). A clear demonstration of acute HEV infection after consumption of HEV-infected uncooked deer meat has been reported (Tei *et al.*, 2003). It has also been reported that shellfish may act as a vehicles for HEV transmission (Cacopardo *et al.*, 1997; Mechnik *et al.*, 2001; Koizumi *et al.*, 2004).

Sporadic cases of HEV infection in non-endemic, industrialized countries have been associated with travel to endemic countries (Hino *et al.*, 1991; Dawson *et al.*, 1992). Recently, HEV-associated hepatitis cases among individuals in industrialized countries who have no history of travel to areas where HEV is endemic have been reported (Ijaz *et al.*, 2005; Amon *et al.*, 2006) and secondary person-to-person HEV transmission cases have been documented in 1–2% of household contacts of patients with confirmed HEV infection (Aggarwal and Naik, 1994). Epidemiological studies have revealed that HEV may be more prevalent than previously thought in industrialized countries (Clemente-Casares *et al.*, 2003) with a prevalence of anti-HEV antibodies of 7.3% among the adult Catalonian population (Buti *et al.*, 2006).

The disease is usually self-limited but it causes high mortality rates in pregnant women (approximately 20%) (Balayan, 1997). Since, only one HEV serotype has been described, production of a broadly crossreactive vaccine should be possible. Such a vaccine would be useful in protection against HEV infection, mainly in pregnant women and in people from endemic regions and travelers to these areas. Nevertheless, the lack of a susceptible cell culture system has hampered the development of live attenuated or killed vaccines (Wang and Zhuang, 2004) and to date no commercially vaccines against HEV are available. Currently, most research on HEV vaccines is focused on ORF-2-derived proteins or peptides that contain neutralizing epitopes common to all different genotypes (Emerson and Purcell 2001; Meng *et al.*, 2001; Worm *et al.*, 2002; Wang and Zhuang, 2004; Deshmukh *et al.*, 2007). A recombinant HEV baculovirus vaccine candidate that protects against intravenous administration of heterologous HEV strains has entered clinical trials and appears very promising (Stevenson 2000; Zhang *et al.*, 2002; Purcell *et al.*, 2003).

REFERENCES

Abad, F.X., Pinto, R.M. and Bosch, A. (1994) Survival of enteric viruses on environmental fomites. *Appl. Environ. Microbiol.* **60**, 3704–3710.

Aggarwal, R. and Naik, S.R. (1994) Hepatitis E: intrafamilial transmission versus waterborne spread. *J. Hepatol.* **21**, 718–723.

Agnello, V., Abel, G., Elfahal, M., Knight, G.B. and Zhang, Q.X. (1999) Hepatitis C virus and other flaviviridae viruses enter cells via low density lipoprotein receptor. *Proc. Natl Acad. Sci. USA* **96**, 12766–12771.

Alexopoulou, A., Karayiannis, P., Hadziyannis, S.J., Aiba, N. and Thomas, H.C. (1997) Emergence and selection of HBV variants in an anti-HBe positive patient persistently infected with quasi-species. *J. Hepatol.* **26**, 748–753.

Alter, M.J. (1995) Epidemiology of hepatitis C in the West. *Semin. Liver Dis.* **15**, 5–14.

Alter, M.J. (1997) Epidemiology of hepatitis C. *Hepatology* **26**, 62S–65S.

Alter, M.J., Margolis, H.S., Krawczynski, K., Judson, F.N., Mares, A., Alexander, W.J. *et al.* (1992) The natural history of community-acquired hepatitis C in the United States. The Sentinel Counties Chronic non-A, non-B Hepatitis Study Team. *N. Engl. J. Med.* **327**, 1899–1905.

Altman, J.D. and Feinberg, M.B. (2004) HIV escape: there and back again. *Nat. Med.* **10**, 229–230.

Amon, J.J., Drobeniuc, J., Bower, W.A., Magana, J.C., Escobedo, M.A., Williams, I.T. *et al.* (2006) Locally acquired hepatitis E virus infection, El Paso, Texas. *J. Med. Virol.* **78**, 741–746.

Andre, P., Komurian-Pradel, F., Deforges, S., Perret, M., Berland, J.L., Sodoyer, M. *et al.* (2002) Characterization of low- and very-low-density hepatitis C virus RNA-containing particles. *J. Virol.* **76**, 6919–6928.

Andre, P., Perlemuter, G., Budkowska, A., Brechot, C. and Lotteau, V. (2005) Hepatitis C virus particles and lipoprotein metabolism. *Semin. Liver Dis.* **25**, 93–104.

Ansaldi, F., Bruzzone, B., Salmaso, S., Rota, M.C., Durando, P., Gasparini, R. and Icardi, G. (2005) Different seroprevalence and molecular epidemiology patterns of hepatitis C virus infection in Italy. *J. Med. Virol.* **76**, 327–332.

Arankalle, V.A., Paranjape, S., Emerson, S.U., Purcell, R.H. and Walimbe, A.M. (1999) Phylogenetic analysis of hepatitis E virus isolates from India (1976–1993). *J. Gen. Virol.* **80**(pt 7), 1691–1700.

Arauz-Ruiz, P., Norder, H., Robertson, B.H. and Magnius, L.O. (2002) Genotype H: a new Amerindian genotype of hepatitis B virus revealed in Central America. *J. Gen. Virol.* **83**(pt 8), 2059–2073.

Bain, C., Fatmi, A., Zoulim, F., Zarski, J.P., Trepo, C. and Inchauspe, G. (2001) Impaired allostimulatory function of dendritic cells in chronic hepatitis C infection. *Gastroenterology* **120**, 512–524.

Balayan, M.S. (1997) Epidemiology of hepatitis E virus infection. *J. Viral Hepat.* **4**, 155–165.

Banks, M., Heath, G.S., Grierson, S.S., King, D.P., Gresham, A., Girones, R. *et al.* (2004) Evidence for the presence of hepatitis E virus in pigs in the United Kingdom. *Vet. Rec.* **154**, 223–227.

Barnes, E., Harcourt, G., Brown, D., Lucas, M., Phillips, R., Dusheiko, G. and Klenerman, P. (2002) The dynamics of T-lymphocyte responses during combination therapy for chronic hepatitis C virus infection. *Hepatology* **36**, 743–754.

Baroudy, B.M., Ticehurst, J.R., Miele, T.A., Maizel, J.V., Jr., Purcell, R.H. and Feinstone, S.M. (1985) Sequence analysis of hepatitis A virus cDNA coding for capsid proteins and RNA polymerase. *Proc. Natl Acad. Sci. USA* **82**, 2143–2147.

Bartenschlager, R. (1999) The NS3/4A proteinase of the hepatitis C virus: unravelling structure and function of an unusual enzyme and a prime target for antiviral therapy. *J. Viral Hepat.* **6**, 165–181.

Bartenschlager, R., Ahlborn-Laake, L., Mous, J. and Jacobsen, H. (1993) Nonstructural protein 3 of the hepatitis C virus encodes a serine-type proteinase required for cleavage at the NS3/4 and NS4/5 junctions. *J. Virol.* **67**, 3835–3844.

Bartenschlager, R., Ahlborn-Laake, L., Mous, J. and Jacobsen, H. (1994) Kinetic and structural analyses of hepatitis C virus polyprotein processing. *J. Virol.* **68**, 5045–5055.

Barth, H., Schafer, C., Adah, M.I., Zhang, F., Linhardt, R.J., Toyoda, H. *et al.* (2003) Cellular binding of hepatitis C virus envelope glycoprotein E2 requires cell surface heparan sulfate. *J. Biol. Chem.* **278**, 41003–41012.

Bashirullah, A., Cooperstock, R.L. and Lipshitz, H.D. (1998) RNA localization in development. *Annu. Rev. Biochem.* **67**, 335–394.

Baumert, T.F., Yang, C., Schurmann, P., Kock, J., Ziegler, C., Grullich, C. *et al.* (2005) Hepatitis B virus mutations associated with fulminant hepatitis induce apoptosis in primary Tupaia hepatocytes. *Hepatology* **41**, 247–256.

Baumert, T.F., Thimme, R. and Von Weizsacker, F. (2007) Pathogenesis of hepatitis B virus infection. *World J Gastroenterol.* **13**, 82–90.

Beard, M.R., Cohen, L., Lemon, S.M. and Martin, A. (2001) Characterization of recombinant hepatitis A virus genomes containing exogenous sequences at the 2A/2B junction. *J. Virol.* **75**, 1414–1426.

Becher, P., Orlich, M. and Thiel, H.J. (2001) RNA recombination between persisting pestivirus and a vaccine strain: generation of cytopathogenic virus and induction of lethal disease. *J. Virol.* **75**, 6256–6264.

Beck, J. and Nassal, M. (2007) Hepatitis B virus replication. *World J. Gastroenterol.* **13**, 48–64.

Behrens, S.E., Tomei, L. and De Francesco, R. (1996) Identification and properties of the RNA-dependent RNA polymerase of hepatitis C virus. *EMBO J.* **15**, 12–22.

Bell, B.P. and Feinstone, S.M. (2004) Hepatitis A vaccine. In: *Vaccine* (S.A. Plotkin, W.A. Orenstein and P.A. Offit, eds), pp. 269–297. Philadelphia, PA: Saunders.

Bergmann, K.F. and Gerin, J.L. (1986) Antigens of hepatitis delta virus in the liver and serum of humans and animals. *J. Infect. Dis.* **154**, 702–706.

Bertoletti, A. and Gehring, A.J. (2006) The immune response during hepatitis B virus infection. *J. Gen. Virol.* **87**, 1439–1449.

Bile, K., Isse, A., Mohamud, O., Allebeck, P., Nilsson, L., Norder, H. *et al.* (1994) Contrasting roles of rivers and wells as sources of drinking water on attack and fatality rates in a hepatitis E epidemic in Somalia. *Am. J. Trop. Med. Hyg.* **51**, 466–474.

Blackard, J.T. and Sherman, K.E. (2007) Hepatitis C virus coinfection and superinfection. *J. Infect. Dis.* **195**, 519–524.

Blight, K.J. and Rice, C.M. (1997) Secondary structure determination of the conserved 98-base sequence at the 3′ terminus of hepatitis C virus genome RNA. *J. Virol.* **71**, 7345–7352.

Blindenbacher, A., Duong, F.H., Hunziker, L., Stutvoet, S.T., Wang, X., Terracciano, L. *et al.* (2003) Expression of hepatitis c virus proteins inhibits interferon alpha signaling in the liver of transgenic mice. *Gastroenterology* **124**, 1465–1475.

Bock, C.T., Schwinn, S., Locarnini, S., Fyfe, J., Manns, M.P., Trautwein, C. and Zentgraf, H. (2001) Structural organization of the hepatitis B virus minichromosome. *J. Mol. Biol.* **307**, 183–196.

Bock, C.T., Tillmann, H.L., Torresi, J., Klempnauer, J., Locarnini, S., Manns, M.P. and Trautwein, C. (2002) Selection of hepatitis B virus polymerase mutants with enhanced replication by lamivudine treatment after liver transplantation. *Gastroenterology* **122**, 264–273.

Bollyky, P.L. and Holmes, E.C. (1999) Reconstructing the complex evolutionary history of hepatitis B virus. *J. Mol. Evol.* **49**, 130–141.

Bonino, F., Hoyer, B., Shih, J.W., Rizzetto, M., Purcell, R.H. and Gerin, J.L. (1984) Delta hepatitis agent: structural and antigenic properties of the delta-associated particle. *Infect. Immun.* **43**, 1000–1005.

Borman, A.M. and Kean, K.M. (1997) Intact eukaryotic initiation factor 4G is required for hepatitis A virus internal initiation of translation. *Virology* **237**, 129–136.

Bosch, A., Lucena, F., Díez, J.M., Gajardo, R., Blasi, M. and Jofre, J. (2007) Human enteric viruses and indicator microorganisms in a water supply associated with an outbreak of infectious hepatitis. *J. Am. Water Works Assoc.* **83**, 80–83.

Bouchard, M.J., Wang, L.H. and Schneider, R.J. (2001) Calcium signaling by HBx protein in hepatitis B virus DNA replication. *Science* **294**, 2376–2378.

Bowyer, S.M. and Sim, J.G. (2000) Relationships within and between genotypes of hepatitis B virus at points across the genome: footprints of recombination in certain isolates. *J. Gen. Virol.* **81**, 379–392.

Bozkaya, H., Akarca, U.S., Ayola, B. and Lok, A.S. (1997) High degree of conservation in the hepatitis B virus core gene during the immune tolerant phase in perinatally acquired chronic hepatitis B virus infection. *J. Hepatol.* **26**, 508–516.

Branch, A.D., Benenfeld, B.J., Baroudy, B.M., Wells, F.V., Gerin, J.L. and Robertson, H.D. (1989) An ultraviolet-sensitive RNA structural element in a viroid-like domain of the hepatitis delta virus. *Science* **243**, 649–652.

Branch, A.D., Stump, D.D., Gutierrez, J.A., Eng, F. and Walewski, J.L. (2005) The hepatitis C virus alternate reading frame (ARF) and its family of novel products: the alternate reading frame protein/F-protein, the double-frameshift protein and others. *Semin. Liver Dis.* **25**, 105–117.

Brazas, R. and Ganem, D. (1996) A cellular homolog of hepatitis delta antigen: implications for viral replication and evolution. *Science* **274**, 90–94.

Brechot, C., Gozuacik, D., Murakami, Y. and Paterlini-Brechot, P. (2000) Molecular bases for the development of hepatitis B virus (HBV)-related hepatocellular carcinoma (HCC). *Semin. Cancer Biol.* **10**, 211–231.

Brillet, R., Penin, F., Hezode, C., Chouteau, P., Dhumeaux, D. and Pawlotsky, J.M. (2007) The nonstructural 5A protein of hepatitis C virus genotype 1b does not contain an interferon sensitivity-determining region. *J. Infect. Dis.* **195**, 432–441.

Bronowicki, J.P., Vetter, D., Uhl, G., Hudziak, H., Uhrlacher, A., Vetter, J.M. and Doffoel, M. (1997) Lymphocyte reactivity to hepatitis C virus (HCV) antigens shows evidence for exposure to HCV in HCV-seronegative spouses of HCV-infected patients. *J. Infect. Dis.* **176**, 518–522.

Brown, E.A., Zajac, A.J. and Lemon, S.M. (1994) *In vitro* characterization of an internal ribosomal entry site (IRES) present within the 5′ nontranslated region of hepatitis A virus RNA: comparison with the IRES of encephalomyocarditis virus. *J. Virol.* **68**, 1066–1074.

Bruguera, M., Salleras, L., Plans, P., Vidal, J., Navas, E., Dominguez, A. *et al.* (1999) [Changes in seroepidemiology of hepatitis A virus infection in Catalonia in the period 1989–1996. Implications for new vaccination strategy]. *Med. Clin. (Barc.)* **112**, 406–408.

Bruss, V. (2007) Hepatitis B virus morphogenesis. *World J Gastroenterol.* **13**, 65–73.

Bukh, J., Purcell, R.H. and Miller, R.H. (1992) Sequence analysis of the 5′ noncoding region of hepatitis C virus. *Proc. Natl Acad. Sci. USA* **89**, 4942–4946.

Bukh, J., Miller, R.H. and Purcell, R.H. (1995) Genetic heterogeneity of hepatitis C virus: quasispecies and genotypes. *Semin. Liver Dis.* **15**, 41–63.

Buti, M., Esteban, R., Jardi, R., Allende, H., Baselga, J.M. and Guardia, J. (1988) Epidemiology of delta infection in Spain. *J. Med. Virol.* **26**, 327–332.

Buti, M., Clemente-Casares, P., Jardi, R., Formiga-Cruz, M., Schaper, M., Valdes, A. *et al.* (2004) Sporadic cases of acute autochthonous hepatitis E in Spain. *J. Hepatol.* **41**, 126–131.

Buti, M., Dominguez, A., Plans, P., Jardi, R., Schaper, M., Espunes, J. *et al.* (2006) Community-based seroepidemiological survey of hepatitis E virus infection in Catalonia, Spain. *Clin. Vaccine Immunol.* **13**, 1328–1332.

Cacopardo, B., Russo, R., Preiser, W., Benanti, F., Brancati, G. and Nunnari, A. (1997) Acute hepatitis E in Catania (eastern Sicily) 1980–1994. The role of hepatitis E virus. *Infection* **25**, 313–316.

Carman, W.F. (1996) Molecular variants of hepatitis B virus. *Clin. Lab. Med.* **16**, 407–428.

Carman, W.F., Thursz, M., Hadziyannis, S., McIntyre, G., Colman, K., Gioustoz, A. *et al.* (1995) Hepatitis B e antigen negative chronic active hepatitis: hepatitis B virus core mutations occur predominantly in known antigenic determinants. *J. Viral Hepat.* **2**, 77–84.

Cerny, A. and Chisari, F.V. (1999) Pathogenesis of chronic hepatitis C: immunological features of hepatic injury and viral persistence. *Hepatology* **30**, 595–601.

Chamberlain, R.W., Adams, N.J., Taylor, L.A., Simmonds, P. and Elliott, R.M. (1997) The complete coding sequence of hepatitis C virus genotype 5a, the predominant genotype in South Africa. *Biochem. Biophys. Res. Commun.* **236**, 44–49.

Chan, H.L., Hussain, M. and Lok, A.S. (1999) Different hepatitis B virus genotypes are associated with different mutations in the core promoter and precore regions during hepatitis B e antigen seroconversion. *Hepatology* **29**, 976–984.

Chan, H.L., Tsui, S.K., Tse, C.H., Ng, E.Y., Au, T.C., Yuen, L. *et al.* (2005) Epidemiological and virological characteristics of 2 subgroups of hepatitis B virus genotype C. *J. Infect. Dis.* **191**, 2022–2032.

Chang, F.L., Chen, P.J., Tu, S.J., Wang, C.J. and Chen, D.S. (1991) The large form of hepatitis delta antigen is crucial for assembly of hepatitis delta virus. *Proc. Natl Acad. Sci. USA* **88**, 8490–8494.

Chang, J., Gudima, S.O. and Taylor, J.M. (2005) Evolution of hepatitis delta virus RNA genome following long-term replication in cell culture. *J. Virol.* **79**, 13310–13316.

Chang, K.M., Rehermann, B., McHutchison, J.G., Pasquinelli, C., Southwood, S., Sette, A. and Chisari, F.V. (1997) Immunological significance of cytotoxic T lymphocyte epitope variants in patients chronically infected by the hepatitis C virus. *J. Clin. Invest.* **100**, 2376–2385.

Chao, L. (1990) Fitness of RNA virus decreased by Muller's ratchet. *Nature* **348**, 454–455.

Chao, M. (2007) RNA recombination in hepatitis delta virus: Implications regarding the abilities of mammalian RNA polymerases. *Virus Res.* **127**, 208–215.

Chao, M., Hsieh, S.Y. and Taylor, J. (1990) Role of two forms of hepatitis delta virus antigen: evidence for a mechanism of self-limiting genome replication. *J. Virol.* **64**, 5066–5069.

Chao, Y.C., Chang, M.F., Gust, I. and Lai, M.M. (1990) Sequence conservation and divergence of hepatitis delta virus RNA. *Virology* **178**, 384–392.

Chao, Y.C., Lee, C.M., Tang, H.S., Govindarajan, S. and Lai, M.M. (1991) Molecular cloning and characterization of an isolate of hepatitis delta virus from Taiwan. *Hepatology* **13**, 345–352.

Chauhan, R., Kazim, S.N., Bhattacharjee, J., Sakhuja, P. and Sarin, S.K. (2006) Basal core promoter, precore region mutations of HBV and their association with e antigen, genotype and severity of liver disease in patients with chronic hepatitis B in India. *J. Med. Virol.* **78**, 1047–1054.

Chen, B.F., Liu, C.J., Jow, G.M., Chen, P.J., Kao, J.H. and Chen, D.S. (2006) High prevalence and mapping of pre-S deletion in hepatitis B virus carriers with progressive liver diseases. *Gastroenterology* **130**, 1153–1168.

Choo, Q.L., Richman, K.H., Han, J.H., Berger, K., Lee, C., Dong, C. *et al.* (1991) Genetic organization and diversity of the hepatitis C virus. *Proc. Natl Acad. Sci. USA* **88**, 2451–2455.

Clemente-Casares, P., Pina, S., Buti, M., Jardi, R., Martin, M., Bofill-Mas, S. and Girones, R. (2003) Hepatitis E virus epidemiology in industrialized countries. *Emerg. Infect. Dis.* **9**, 448–454.

Cochrane, A., Searle, B., Hardie, A., Robertson, R., Delahooke, T., Cameron, S. *et al.* (2002) A genetic analysis of hepatitis C virus transmission between injection drug users. *J. Infect. Dis.* **186**, 1212–1221.

Cocquerel, L., Voisset, C. and Dubuisson, J. (2006) Hepatitis C virus entry: potential receptors and their biological functions. *J. Gen. Virol.* **87**, 1075–1084.

Cohen, J.I., Rosenblum, B., Ticehurst, J.R., Daemer, R.J., Feinstone, S.M. and Purcell, R.H. (1987) Complete nucleotide sequence of an attenuated hepatitis A virus: comparison with wild-type virus. *Proc. Natl Acad. Sci. USA* **84**, 2497–2501.

Coleman, P.F. (2006) Detecting hepatitis B surface antigen mutants. *Emerg. Infect. Dis.* **12**, 198–203.

Colina, R., Casane, D., Vasquez, S., Garcia-Aguirre, L., Chunga, A., Romero, H. *et al.* (2004) Evidence of intratypic recombination in natural populations of hepatitis C virus. *J. Gen. Virol.* **85**, 31–37.

Costa-Mattioli, M., Di Napoli, A., Ferre, V., Billaudel, S., Perez-Bercoff, R. and Cristina, J. (2003a) Genetic variability of hepatitis A virus. *J. Gen. Virol.* **84**, 3191–3201.

Costa-Mattioli, M., Ferre, V., Casane, D., Perez-Bercoff, R., Coste-Burel, M., Imbert-Marcille, B.M. *et al.* (2003b) Evidence of recombination in natural populations of hepatitis A virus. *Virology* **311**, 51–59.

Costafreda, M.I., Bosch, A. and Pinto, R.M. (2006) Development, evaluation, and standardization of a real-time TaqMan reverse transcription-PCR assay for quantification of hepatitis A virus in clinical and shellfish samples. *Appl. Environ. Microbiol.* **72**, 3846–3855.

Cotrina, M., Buti, M., Jardi, R., Quer, J., Rodriguez, F., Pascual, C. *et al.* (1998) Hepatitis delta genotypes in chronic delta infection in the northeast of Spain (Catalonia). *J. Hepatol.* **28**, 971–977.

Courcambeck, J., Bouzidi, M., Perbost, R., Jouirou, B., Amrani, N., Cacoub, P. *et al.* (2006) Resistance of hepatitis C virus to NS3-4A protease inhibitors: mechanisms of drug resistance induced by R155Q, A156T, D168A and D168V mutations. *Antivir. Ther.* **11**, 847–855.

Cramp, M.E., Rossol, S., Chokshi, S., Carucci, P., Williams, R. and Naoumov, N.V. (2000) Hepatitis C virus-specific T-cell reactivity during interferon and ribavirin treatment in chronic hepatitis C. *Gastroenterology* **118**, 346–355.

Cristina, J. and Colina, R. (2006) Evidence of structural genomic region recombination in Hepatitis C virus. *Virol. J.* **3**, 53.

Cristina, J. and Costa-Mattioli, M. (2007) Genetic variability and molecular evolution of Hepatitis A virus. *Virus Res.* **127**, 151–157.

Dagan, R., Leventhal, A., Anis, E., Slater, P., Ashur, Y. and Shouval, D. (2005) Incidence of hepatitis A in Israel following universal immunization of toddlers. *JAMA* **294**, 202–210.

Dal Molin, G., Poli, A., Croce, L.S., D'Agaro, P., Biagi, C., Comar, M. *et al.* (2006) Hepatitis B virus genotypes, core promoter variants. and precore stop codon variants in patients infected chronically in North-Eastern Italy. *J. Med. Virol.* **78**, 734–740.

Das, K., Xiong, X., Yang, H., Westland, C.E., Gibbs, C.S., Sarafianos, S.G. and Arnold, E. (2001) Molecular modeling and biochemical characterization reveal the mechanism of hepatitis B virus polymerase resistance to lamivudine (3TC) and emtricitabine (FTC). *J. Virol.* **75**, 4771–4779.

Dawson, G.J., Mushahwar, I.K., Chau, K.H. and Gitnick, G.L. (1992) Detection of long-lasting antibody to hepatitis E virus in a US traveller to Pakistan. *Lancet* **340**, 426–427.

Deforges, S., Evlashev, A., Perret, M., Sodoyer, M., Pouzol, S., Scoazec, J.Y. *et al.* (2004) Expression of hepatitis C virus proteins in epithelial intestinal cells *in vivo*. *J. Gen. Virol.* **85**, 2515–2523.

Delaney, W.E., Yang, H., Westland, C.E., Das, K., Arnold, E., Gibbs, C.S. *et al.* (2003) The hepatitis B virus polymerase mutation rtV173L is selected during lamivudine therapy and enhances viral replication *in vitro*. *J. Virol.* **77**, 11833–11841.

Deleersnyder, V., Pillez, A., Wychowski, C., Blight, K., Xu, J., Hahn, Y.S. *et al.* (1997) Formation of native hepatitis C virus glycoprotein complexes. *J Virol.* **71**, 697–704.

Della, B.S., Riva, A., Tanzi, E., Nicola, S., Amendola, A., Vecchi, L. *et al.* (2005) Hepatitis C virus-specific reactivity of CD4+ -lymphocytes in children born from HCV-infected women. *J. Hepatol.* **43**, 394–402.

Deny, P. (2006) Hepatitis delta virus genetic variability: from genotypes I, II, III to eight major clades? *Curr. Top. Microbiol. Immunol.* **307**, 151–171.

Deshmukh, T.M., Lole, K.S., Tripathy, A.S. and Arankalle, V.A. (2007) Immunogenicity of candidate hepatitis E virus DNA vaccine expressing complete and truncated ORF2 in mice. *Vaccine* **25**, 4350–4360.

Devarbhavi, H.C., Cohen, A.J., Patel, R., Wiesner, R.H., Dickson, R.C. and Ishitani, M.B. (2002) Preliminary results: outcome of liver transplantation for hepatitis B virus varies by hepatitis B virus genotype. *Liver Transpl.* **8**, 550–555.

Di Marco, S., Volpari, C., Tomei, L., Altamura, S., Harper, S., Narjes, F. *et al.* (2005) Interdomain communication in hepatitis C virus polymerase abolished by small molecule inhibitors bound to a novel allosteric site. *J. Biol. Chem.* **280**, 29765–29770.

Diaz, O., Delers, F., Maynard, M., Demignot, S., Zoulim, F., Chambaz, J. *et al.* (2006) Preferential association of Hepatitis C virus with apolipoprotein B48-containing lipoproteins. *J. Gen. Virol.* **87**, 2983–2991.

Diedrich, G. (2006) How does hepatitis C virus enter cells?. *FEBS J.* **273**, 3871–3885.

Diepolder, H.M., Zachoval, R., Hoffmann, R.M., Wierenga, E.A., Santantonio, T., Jung, M.C. *et al.* (1995) Possible mechanism involving T-lymphocyte response to non-structural protein 3 in viral clearance in acute hepatitis C virus infection. *Lancet* **346**, 1006–1007.

Diepolder, H.M., Gerlach, J.T., Zachoval, R., Hoffmann, R.M., Jung, M.C., Wierenga, E.A. *et al.* (1997) Immunodominant CD4+ T-cell epitope within nonstructural protein 3 in acute hepatitis C virus infection. *J. Virol.* **71**, 6011–6019.

Dimmock, N.J. (1982) Initial stages in infection with animal viruses. *J. Gen. Virol.* **59**, 1–22.

Domingo, E. and Gomez, J. (2007) Quasispecies and its impact on viral hepatitis. *Virus Res.* **127**, 131–150.

Domingo, E. and Holland, J.J. (1988) High error rates, population equilibrium and evolution of RNA replication systems. In: *RNA Genetics* (E. Domingo and J.J. Holland, eds), pp. 3–36. Boca Raton, FL: CRC Press.

Domingo, E. and Holland, J.J. (1994) Mutation rates and rapid evolution of RNA viruses. In: *Evolutionary Biology of Viruses* (S.S. Morse, ed.), pp. 161–184. New York: Raven Press.

Domingo, E. and Holland, J.J. (1997) RNA virus mutations and fitness for survival. *Annu. Rev. Microbiol.* **51**, 151–178.

Domingo, E., Escarmis, C., Sevilla, N., Moya, A., Elena, S.F., Quer, J. *et al.* (1996) Basic concepts in RNA virus evolution. *FASEB J.* **10**, 859–864.

Duarte, E.A., Novella, I.S., Weaver, S.C., Domingo, E., Wain-Hobson, S., Clarke, D.K. *et al.* (1994) RNA virus quasispecies: significance for viral disease and epidemiology. *Infect. Agents Dis.* **3**, 201–214.

Eckart, M.R., Selby, M., Masiarz, F., Lee, C., Berger, K., Crawford, K. *et al.* (1993) The hepatitis C virus encodes a serine protease involved in processing of the putative nonstructural proteins from the viral polyprotein precursor. *Biochem. Biophys. Res. Commun.* **192**, 399–406.

Egger, D., Wolk, B., Gosert, R., Bianchi, L., Blum, H.E., Moradpour, D. and Bienz, K. (2002) Expression of hepatitis C virus proteins induces distinct membrane alterations including a candidate viral replication complex. *J. Virol.* **76**, 5974–5984.

Ehrenfeld, E. and Teterina, N.L. (2007) Initiation of translation of picornavirus RNAs: structure and function of the internal ribosome entry site. In: *Molecular Biology of Picornaviruses* (B.L. Semler and E. Wimmer, eds), pp. 159–170. Washington DC: ASM Press.

Eigen, M. and Biebricher, C.K. (1988) Variability of RNA genomes. Sequence space and quasispecies distribution. In: *RNA Genetics* (E. Domingo, J.J. Holland and P. Ahlquist, eds), vol. 3, pp. 211–245. Boca Ratón., FL: CRC Press, Inc.

Emerson, S.U. and Purcell, R.H. (2001) Recombinant vaccines for hepatitis E. *Trends Mol. Med.* **7**, 462–466.

Emerson, S.U. and Purcell, R.H. (2003) Hepatitis E virus. *Rev. Med. Virol.* **13**, 145–154.

Emerson, S.U., Nguyen, H., Torian, U. and Purcell, R.H. (2006) ORF3 protein of hepatitis E virus is not required for replication, virion assembly, or infection of hepatoma cells *in vitro*. *J. Virol.* **80**, 10457–10464.

Erickson, A.L., Kimura, Y., Igarashi, S., Eichelberger, J., Houghton, M., Sidney, J. *et al.* (2001) The outcome of hepatitis C virus infection is predicted by escape mutations in epitopes targeted by cytotoxic T lymphocytes. *Immunity* **15**, 883–895.

Escarmis, C., Davila, M. and Domingo, E. (1999) Multiple molecular pathways for fitness recovery of an RNA virus debilitated by operation of Muller's ratchet. *J. Mol. Biol.* **285**, 495–505.

Evans, M.J., von Hahn, T., Tscherne, D.M., Syder, A.J., Panis, M., Wolk, B. *et al.* (2007) Claudin-1 is a hepatitis C virus co-receptor required for a late step in entry. *Nature* **446**, 801–805.

Failla, C., Tomei, L. and De Francesco, R. (1994) Both NS3 and NS4A are required for proteolytic processing of hepatitis C virus nonstructural proteins. *J. Virol.* **68**, 3753–3760.

Fang, Z.L., Ling, R., Wang, S.S., Nong, J., Huang, C.S. and Harrison, T.J. (1998) HBV core promoter mutations prevail in patients with hepatocellular carcinoma from Guangxi, China. *J. Med. Virol.* **56**, 18–24.

Farci, P. (2003) Delta hepatitis: an update. *J. Hepatol.* **39**(Suppl 1), S212–S219.

Farci, P., Shimoda, A., Coiana, A., Diaz, G., Peddis, G., Melpolder, J.C. *et al.* (2000) The outcome of acute hepatitis C predicted by the evolution of the viral quasispecies. *Science* **288**, 339–344.

Fares, M.A. and Holmes, E.C. (2002) A revised evolutionary history of hepatitis B virus (HBV). *J. Mol. Evol.* **54**, 807–814.

Fauquet, C., Mayo, M.A., Desselberger U, and Ball, L. A. (2005) Virus Taxonomy: Eighth Report of the International Committee on Taxonomy of Viruses. *Virus Taxonomy*. Amsterdam: Elsevier/Academic Press.

Favre, D. and Muellhaupt, B. (2005) Potential cellular receptors involved in hepatitis C virus entry into cells. *Lipids Health Dis.* **4**, 9.

Feinstone, S.M., Kapikian, A.Z. and Purceli, R.H. (1973) Hepatitis A: detection by immune electron microscopy of a viruslike antigen associated with acute illness. *Science* **182**, 1026–1028.

Feitelson, M.A., Sun, B., Satiroglu Tufan, N.L., Liu, J., Pan, J. and Lian, Z. (2002) Genetic mechanisms of hepatocarcinogenesis. *Oncogene* **21**, 2593–2604.

Flint, M., Maidens, C., Loomis-Price, L.D., Shotton, C., Dubuisson, J., Monk, P. *et al.* (1999) Characterization of hepatitis C virus E2 glycoprotein interaction with a putative cellular receptor, CD81. *J. Virol.* **73**, 6235–6244.

Forns, X. and Bukh, J. (1999) The molecular biology of hepatitis C virus. Genotypes and quasispecies. *Clin. Liver Dis.* **3**, 693–716. vii.

Forns, X. and Costa, J. (2006) HCV virological assessment. *J. Hepatol.* **44**(suppl.), S35–S39.

Franco, E. and Vitiello, G. (2003) Vaccination strategies against hepatitis A in southern Europe. *Vaccine* **21**, 696–697.

Frasca, L., Del Porto, P., Tuosto, L., Marinari, B., Scotta, C., Carbonari, M. *et al.* (1999) Hypervariable region 1 variants act as TCR antagonists for hepatitis C virus-specific CD4+ T cells. *J. Immunol.* **163**, 650–658.

Fried, M.W., Shiffman, M.L., Reddy, K.R., Smith, C., Marinos, G., Goncales, F.L. *et al.* (2002) Peginterferon alfa-2a plus ribavirin for chronic hepatitis C virus infection. *N. Engl. J. Med.* **347**, 975–982.

Fu, T.B. and Taylor, J. (1993) The RNAs of hepatitis delta virus are copied by RNA polymerase II in nuclear homogenates. *J. Virol.* **67**, 6965–6972.

Gale, M., Jr. and Foy, E.M. (2005) Evasion of intracellular host defence by hepatitis C virus. *Nature* **436**, 939–945.

Ganem, D. and Prince, A.M. (2004) Hepatitis B virus infection—natural history and clinical consequences. *N. Engl. J. Med.* **350**, 1118–1129.

Gardner, J.P., Durso, R.J., Arrigale, R.R., Donovan, G.P., Maddon, P.J., Dragic, T. and Olson, W.C. (2003) L-SIGN (CD 209L) is a liver-specific capture receptor for hepatitis C virus. *Proc. Natl Acad. Sci. USA* **100**, 4498–4503.

Gastaminza, P., Kapadia, S.B. and Chisari, F.V. (2006) Differential biophysical properties of infectious intracellular and secreted hepatitis C virus particles. *J Virol.* **80**, 11074–11081.

Gates, A.T., Sarisky, R.T. and Gu, B. (2004) Sequence requirements for the development of a chimeric HCV replicon system. *Virus Res.* **100**, 213–222.

Gauss-Muller, V. and Kusov, Y.Y. (2002) Replication of a hepatitis A virus replicon detected by genetic recombination *in vivo*. *J. Gen. Virol.* **83**, 2183–2192.

Gerin, J.L., Casey, J.L. and Purcell, R.H. (2001) Hepatitis delta virus. In: *Fields' Virology* (D.M. Knipe and P.M. Howley, eds), 4th edn., pp. 3037–3050. Philadelphia, PA: Lippincott, Williams and Wilkins.

Gerlach, J.T., Diepolder, H.M., Jung, M.C., Gruener, N.H., Schraut, W.W., Zachoval, R. *et al.* (1999) Recurrence of hepatitis C virus after loss of virus-specific CD4(+) T-cell response in acute hepatitis C. *Gastroenterology* **117**, 933–941.

Ghany, M.G., Ayola, B., Villamil, F.G., Gish, R.G., Rojter, S., Vierling, J.M. and Lok, A.S. (1998) Hepatitis B virus S mutants in liver transplant recipients who were reinfected despite hepatitis B immune globulin prophylaxis. *Hepatology* **27**, 213–222.

Glebe, D., Urban, S., Knoop, E.V., Cag, N., Krass, P., Grun, S. *et al.* (2005) Mapping of the hepatitis B virus attachment site by use of infection-inhibiting preS1 lipopeptides and tupaia hepatocytes. *Gastroenterology* **129**, 234–245.

Glenn, J.S. and White, J.M. (1991) Trans-dominant inhibition of human hepatitis delta virus genome replication. *J. Virol.* **65**, 2357–2361.

Gomez, J., Nadal, A., Sabariegos, R., Beguiristain, N., Martell, M. and Piron, M. (2004) Three properties of the hepatitis C virus RNA genome related to antiviral strategies based on RNA-therapeutics: variability, structural conformation and tRNA mimicry. *Curr. Pharm. Des.* **10**, 3741–3756.

Gosert, R., Egger, D., Lohmann, V., Bartenschlager, R., Blum, H.E., Bienz, K. and Moradpour, D. (2003) Identification of the hepatitis C virus RNA replication complex in Huh-7 cells harboring subgenomic replicons. *J. Virol.* **77**, 5487–5492.

Goutagny, N., Fatmi, A., de Ledinghen, V., Penin, F., Couzigou, P., Inchauspe, G. and Bain, C. (2003) Evidence of viral replication in circulating dendritic cells during hepatitis C virus infection. *J. Infect. Dis.* **187**, 1951–1958.

Grakoui, A., McCourt, D.W., Wychowski, C., Feinstone, S.M. and Rice, C.M. (1993a) A second hepatitis C virus-encoded proteinase. *Proc. Natl Acad. Sci. USA* **90**, 10583–10587.

Grakoui, A., McCourt, D.W., Wychowski, C., Feinstone, S.M. and Rice, C.M. (1993b) Characterization of the hepatitis C virus-encoded serine proteinase: determination of proteinase-dependent polyprotein cleavage sites. *J. Virol.* **67**, 2832–2843.

Grakoui, A., Wychowski, C., Lin, C., Feinstone, S.M. and Rice, C.M. (1993c) Expression and identification of hepatitis C virus polyprotein cleavage products. *J. Virol.* **67**, 1385–1395.

Grakoui, A., Shoukry, N.H., Woollard, D.J., Han, J.H., Hanson, H.L., Ghrayeb, J. *et al.* (2003) HCV persistence and immune evasion in the absence of memory T cell help. *Science* **302**, 659–662.

Grandadam, M., Tebbal, S., Caron, M., Siriwardana, M., Larouze, B., Koeck, J.L. *et al.* (2004) Evidence for hepatitis E virus quasispecies. *J. Gen. Virol.* **85**, 3189–3194.

Grande Gimenez, M.C., El Far, F., Barsanti, W.S. and Servolo Medeiros, E.A. (2001) Cut and puncture accidents involving health care workers exposed to biological materials. *Brazil J. Infect. Dis.* **5**, 235–242.

Griffin, S.D., Beales, L.P., Clarke, D.S., Worsfold, O., Evans, S.D., Jaeger, J. *et al.* (2003) The p7 protein of hepatitis C virus forms an ion channel that is blocked by the antiviral drug, Amantadine. *FEBS Lett.* **535**, 34–38.

Gruner, N.H., Gerlach, J.T., Jung, M.C., Diepolder, H.M., Schirren, C.A. and Schraut, W.W. (2000) Association of hepatitis C virus-specific CD8+ T cells with viral clearance in acute hepatitis C. *J. Infect. Dis.* **181**, 1528–1536.

Gu, B., Gates, A.T., Isken, O., Behrens, S.E. and Sarisky, R.T. (2003) Replication studies using genotype 1a subgenomic hepatitis C virus replicons. *J. Virol.* **77**, 5352–5359.

Gunther, S., Fischer, L., Pult, I., Sterneck, M. and Will, H. (1999) Naturally occurring variants of hepatitis B virus. *Adv. Virus Res.* **52**, 25–137.

Gunther, S., Piwon, N., Iwanska, A., Schilling, R., Meisel, H. and Will, H. (1996) Type, prevalence and significance of core promoter/enhancer II mutations in hepatitis B viruses from immunosuppressed patients with severe liver disease. *J. Virol.* **70**, 8318–8331.

Gunther, S., Paulij, W., Meisel, H. and Will, H. (1998) Analysis of hepatitis B virus populations in an interferon-alpha-treated patient reveals predominant mutations in the C-gene and changing e-antigenicity. *Virology* **244**, 146–160.

Guthmann, J.P., Klovstad, H., Boccia, D., Hamid, N., Pinoges, L., Nizou, J.Y. *et al.* (2006) A large outbreak of hepatitis E among a displaced population in Darfur, Sudan, (2004): the role of water treatment methods. *Clin. Infect. Dis.* **42**, 1685–1691.

Hadziyannis, S.J., Sette, H., Jr., Morgan, T.R., Balan, V., Diago, M., Marcellin, P. *et al.* (2004) Peginterferon-alpha2a and ribavirin combination therapy in chronic hepatitis C: a randomized study of treatment duration and ribavirin dose. *Ann. Intern. Med.* **140**, 346–355.

Hannoun, C., Krogsgaard, K., Horal, P. and Lindh, M. (2002) Genotype mixtures of hepatitis B virus in patients treated with interferon. *J. Infect. Dis.* **186**, 752–759.

Heim, M.H., Moradpour, D. and Blum, H.E. (1999) Expression of hepatitis C virus proteins inhibits signal transduction through the Jak-STAT pathway. *J. Virol.* **73**, 8469–8475.

Hijikata, M., Mizushima, H., Akagi, T., Mori, S., Kakiuchi, N., Kato, N. *et al.* (1993) Two distinct proteinase activities required for the processing of a putative nonstructural precursor protein of hepatitis C virus. *J. Virol.* **67**, 4665–4675.

Hino, K., Kondo, T., Niwa, H., Uchida, T., Shikata, T., Rikahisa, T. and Mizuno, K. (1991) A small epidemic of enterically transmitted non-A, non-B acute hepatitis. *Gastroenterol. Jpn* **26**(suppl 3), 139–141.

Hoekstra, D. and Kok, J.W. (1989) Entry mechanisms of enveloped viruses. Implications for fusion of intracellular membranes. *Biotechniques* **9**, 273–305.

Hofmann, W.P., Zeuzem, S. and Sarrazin, C. (2005) Hepatitis C virus-related resistance mechanisms to interferon alpha-based antiviral therapy. *J. Clin. Virol.* **32**, 86–91.

Holland, J.J., Spindler, K., Horodyski, F., Grabau, E., Nichol, S. and VandePol, S. (1982) Rapid evolution of RNA genomes. *Science* **215**, 1577–1585.

Holland, J.J., De La Torre, J.C. and Steinhauer, D.A. (1992) RNA virus populations as quasispecies. *Curr. Top. Microbiol. Immunol.* **176**, 1–20.

Holmes, E.C., Worobey, M. and Rambaut, A. (1999) Phylogenetic evidence for recombination in dengue virus. *Mol. Biol. Evol.* **16**, 405–409.

Hoofnagle, J.H. (1989) Type D (delta) hepatitis. *JAMA* **261**, 1321–1325.

Hsu, H.Y., Chang, M.H., Liaw, S.H., Ni, Y.H. and Chen, H.L. (1999) Changes of hepatitis B surface antigen variants in carrier children before and after universal vaccination in Taiwan. *Hepatology* **30**, 1312–1317.

Hsu, H.Y., Chang, M.H., Ni, Y.H. and Chen, H.L. (2004) Survey of hepatitis B surface variant infection in children 15 years after a nationwide vaccination programme in Taiwan. *Gut* **53**, 1499–1503.

Hu, J. and Boyer, M. (2006) Hepatitis B virus reverse transcriptase and epsilon RNA sequences required for specific interaction *in vitro*. *J. Virol.* **80**, 2141–2150.

Hu, J., Toft, D.O. and Seeger, C. (1997) Hepadnavirus assembly and reverse transcription require a multi-component chaperone complex which is incorporated into nucleocapsids. *EMBO J.* **16**, 59–68.

Hu, X., Margolis, H.S., Purcell, R.H., Ebert, J. and Robertson, B.H. (2000) Identification of hepatitis B virus indigenous to chimpanzees. *Proc. Natl Acad. Sci. USA* **97**, 1661–1664.

Huo, T.I., Wu, J.C., Wu, S.I., Chang, A.L., Lin, S.K., Pan, C.H. *et al.* (2004) Changing seroepidemiology of hepatitis B, C and D virus infections in high-risk populations. *J. Med. Virol.* **72**, 41–45.

Ijaz, S., Arnold, E., Banks, M., Bendall, R.P., Cramp, M.E., Cunningham, R. *et al.* (2005) Non-travel-associated hepatitis E in England and Wales: demographic, clinical and molecular epidemiological characteristics. *J. Infect. Dis.* **192**, 1166–1172.

Iloeje, U.H., Yang, H.I., Su, J., Jen, C.L., You, S.L. and Chen, C.J. (2006) Predicting cirrhosis risk based on the level of circulating hepatitis B viral load. *Gastroenterology* **130**, 678–686.

Imazeki, F., Omata, M. and Ohto, M. (1991) Complete nucleotide sequence of hepatitis delta virus RNA in Japan. *Nucleic Acids Res.* **19**, 5439.

Ishikawa, K., Matsui, K., Madarame, T., Sato, S., Oikawa, K. and Uchida, T. (1995) Hepatitis E probably contracted via a Chinese herbal medicine, demonstrated by nucleotide sequencing. *J. Gastroenterol.* **30**, 534–538.

Ito, T. and Lai, M.M. (1997) Determination of the secondary structure of and cellular protein binding to the 3′-untranslated region of the hepatitis C virus RNA genome. *J. Virol.* **71**, 8698–8706.

Ivaniushina, V., Radjef, N., Alexeeva, M., Gault, E., Semenov, S., Salhi, M. *et al.* (2001) Hepatitis delta virus genotypes I and II cocirculate in an endemic area of Yakutia, Russia. *J. Gen. Virol.* **82**, 2709–2718.

Izban, M.G. and Luse, D.S. (1992) The RNA polymerase II ternary complex cleaves the nascent transcript in a 3′–5′ direction in the presence of elongation factor SII. *Genes Dev.* **6**, 1342–1356.

Jackson, R.J. (2002) Proteins involved in the function of picornavirus internal ribosome entry sites. In: *Molecular Biology of Picornaviruses* (B.L. Semler and E. Wimmer, eds), pp. 171–186. Washington DC: ASM Press.

Jacobson, I.M., Dienstag, J.L., Werner, B.G., Brettler, D.B., Levine, P.H. and Mushahwar, I.K. (1985) Epidemiology and clinical impact of hepatitis D virus (delta) infection. *Hepatology* **5**, 188–191.

Jameel, S., Zafrullah, M., Ozdener, M.H. and Panda, S.K. (1996) Expression in animal cells and characterization of the hepatitis E virus structural proteins. *J. Virol.* **70**, 207–216.

Jang, J.W., Lee, Y.C., Kim, M.S., Lee, S.Y., Bae, S.H., Choi, J.Y. and Yoon, S.K. (2007) A 13-year longitudinal study of the impact of double mutations in the core promoter region of hepatitis B virus on HBeAg seroconversion and disease progression in patients with genotype C chronic active hepatitis. *J. Viral Hepat.* **14**, 169–175.

Janssen, H.L., van Zonneveld, M., Senturk, H., Zeuzem, S., Akarca, U.S., Cakaloglu, Y. *et al.* (2005) Pegylated interferon alfa-2b alone or in combination with lamivudine for HBeAg-positive chronic hepatitis B: a randomised trial. *Lancet* **365**, 123–129.

Jardi, R., Rodriguez, F., Buti, M., Costa, X., Valdes, A., Allende, H. *et al.* (2004) Mutations in the basic core promoter region of hepatitis B virus. Relationship with precore variants and HBV genotypes in a Spanish population of HBV carriers. *J. Hepatol.* **40**, 507–514.

Jardi, R., Rodriguez-Frias, F., Schaper, M., Ruiz, G., Elefsiniotis, I., Esteban, R. and Buti, M. (2007) HBV polymerase gene mutations associated with entecavir drug resistance in treatment-naive patients. *J. Viral Hepat.* **14**, 835–840.

Jinushi, M., Takehara, T., Tatsumi, T., Kanto, T., Miyagi, T., Suzuki, T. *et al.* (2004) Negative regulation of NK cell activities by inhibitory receptor CD94/NKG2A leads to altered NK cell-induced modulation of dendritic cell functions in chronic hepatitis C virus infection. *J. Immunol.* **173**, 6072–6081.

Kabrane-Lazizi, Y., Fine, J.B., Elm, J., Glass, G.E., Higa, H., Diwan, A. *et al.* (1999) Evidence for widespread infection of wild rats with hepatitis E virus in the United States. *Am. J. Trop. Med. Hyg.* **61**, 331–335.

Kageyama, S., Agdamag, D.M., Alesna, E.T., Leano, P.S., Heredia, A.M., Abellanosa-Tac-An, I.P. *et al.* (2006) A natural inter-genotypic (2b/1b) recombinant of hepatitis C virus in the Philippines. *J. Med. Virol.* **78**, 1423–1428.

Kakimi, K., Isogawa, M., Chung, J., Sette, A. and Chisari, F.V. (2002) Immunogenicity and tolerogenicity of hepatitis B virus structural and nonstructural proteins: implications for immunotherapy of persistent viral infections. *J. Virol.* **76**, 8609–8620.

Kalinina, O., Norder, H., Mukomolov, S. and Magnius, L.O. (2002) A natural intergenotypic recombinant of hepatitis C virus identified in St. Petersburg. *J. Virol.* **76**(8), 4034–4043.

Kalinina, O., Norder, H. and Magnius, L.O. (2004) Full-length open reading frame of a recombinant hepatitis C virus strain from St Petersburg: proposed mechanism for its formation. *J. Gen. Virol.* **85**, 1853–1857.

Kamal, S.M., Amin, A., Madwar, M., Graham, C.S., He, Q., Al Tawil, A., Rasenack, J., Nakano, T., Robertson, B., Ismail, A. and Koziel, M.J. (2004) Cellular immune responses in seronegative sexual contacts of acute hepatitis C patients. *J. Virol.* **78**(22), 12252–12258.

Kaneko, T., Moriyama, T., Udaka, K., Hiroishi, K., Kita, H., Okamoto, H., Yagita, H., Okumura, K. and Imawari, M. (1997) Impaired induction of cytotoxic T lymphocytes by antagonism of a weak agonist borne by a variant hepatitis C virus epitope. *Eur.J Immunol.* **27**(7), 1782–1787.

Kao, J.H., Chen, P.J., Lai, M.Y. and Chen, D.S. (2001) Acute exacerbations of chronic hepatitis B are rarely associated with superinfection of hepatitis B virus. *Hepatology* **34**(4 Pt 1), 817–823.

Kao, J.H., Chen, P.J., Lai, M.Y. and Chen, D.S. (2003) Basal core promoter mutations of hepatitis B virus increase the risk of hepatocellular carcinoma in hepatitis B carriers. *Gastroenterology* **124**(2), 327–334.

Karlsson, A.C., Gaines, H., Sallberg, M., Lindback, S. and Sonnerborg, A. (1999) Reappearance of founder virus sequence in human immunodeficiency virus type 1-infected patients. *J.Virol.* **73**(7), 6191–6196.

Kato, N., Hijikata, M., Ootsuyama, Y., Nakagawa, M., Ohkoshi, S., Sugimura, T. and Shimotohno, K. (1990) Molecular cloning of the human hepatitis C virus genome from Japanese patients with non-A, non-B hepatitis. *Proc.Natl.Acad.Sci.U.S.A.* **87**, 9524–9528.

Kato, H., Orito, E., Gish, R.G., Bzowej, N., Newsom, M., Sugauchi, F., Suzuki, S., Ueda, R., Miyakawa, Y. and Mizokami, M. (2002) Hepatitis B e antigen in sera from individuals infected with hepatitis B virus of genotype G. *Hepatology* **35**(4), 922–929.

Kato, T., Date, T., Miyamoto, M., Furusaka, A., Tokushige, K., Mizokami, M. and Wakita, T. (2003) Efficient replication of the genotype 2a hepatitis C virus subgenomic replicon. *Gastroenterology* **125**(6), 1808–1817.

Khudyakov, Y.E., Lopareva, E.N., Jue, D.L., Crews, T.K., Thyagarajan, S.P. and Fields, H.A. (1999) Antigenic domains of the open reading frame 2-encoded protein of hepatitis E virus. *J Clin Microbiol.* **37**(9), 2863–2871.

Kim, D.W., Gwack, Y., Han, J.H. and Choe, J. (1995) C-terminal domain of the hepatitis C virus NS3 protein contains an RNA helicase activity. *Biochem.Biophys.Res. Commun.* **215**(1), 160–166.

Klenerman, P. and Zinkernagel, R.M. (1998) Original antigenic sin impairs cytotoxic T lymphocyte responses to viruses bearing variant epitopes. *Nature* **394**(6692), 482–485.

Koizumi, Y., Isoda, N., Sato, Y., Iwaki, T., Ono, K., Ido, K., Sugano, K., Takahashi, M., Nishizawa, T. and Okamoto, H. (2004) Infection of a Japanese patient by genotype 4 hepatitis e virus while traveling in Vietnam. *J Clin Microbiol.* **42**(8), 3883–3885.

Kolykhalov, A.A., Feinstone, S.M. and Rice, C.M. (1996) Identification of a highly conserved sequence element at the 3′ terminus of hepatitis C virus genome RNA. *J Virol* **70**, 3363–3371.

Kolykhalov, A.A., Mihalik, K., Feinstone, S.M. and Rice, C.M. (2000) Hepatitis C virus-encoded enzymatic activities and conserved ARN elements en the 3′ non-translated-region are essential for virus replication *in vivo*. *J.Virol.* **74**, 2046–2051.

Kondo, Y., Sung, V.M., Machida, K., Liu, M. and Lai, M.M. (2007) Hepatitis C virus infects T cells and affects interferon-gamma signaling in T cell lines. *Virology* **361**(1), 161–173.

Kos, A., Dijkema, R., Arnberg, A.C., van der Meide, P.H. and Schellekens, H. (1986) The hepatitis delta (delta) virus possesses a circular RNA. *Nature* **323**(6088), 558–560.

Kramvis, A. and Kew, M.C. (1999) The core promoter of hepatitis B virus. *J Viral Hepat.* **6**(6), 415–427.

Kramvis, A. and Kew, M.C. (2005) Relationship of genotypes of hepatitis B virus to mutations, disease progression and response to antiviral therapy. *J Viral Hepat.* **12**(5), 456–464.

Kuiken, C., Yusim, K., Boykin, L. and Richardson, R. (2005) The Los Alamos hepatitis C sequence database. *Bioinformatics.* **21**(3), 379–384.

Kukolj, G., McGibbon, G.A., McKercher, G., Marquis, M., Lefebvre, S., Thauvette, L., Gauthier, J., Goulet, S., Poupart, M.A. and Beaulieu, P.L. (2005) Binding site characterization and resistance to a class of non-nucleoside inhibitors of the hepatitis C virus NS5B polymerase. *J Biol Chem.* **280**(47), 39260–39267.

Kuo, M.Y., Chao, M. and Taylor, J. (1989) Initiation of replication of the human hepatitis delta virus genome from cloned DNA: role of delta antigen. *J. Virol.* **63**(5), 1945–1950.

Kurosaki, M., Enomoto, N., Asahina, Y., Sakuma, I., Ikeda, T., Tozuka, S., Izumi, N., Marumo, F. and Sato, C. (1996) Mutations in the core promoter region of hepatitis B virus in patients with chronic hepatitis B. *J Med Virol.* **49**(2), 115–123.

Lai, C.L., Dienstag, J., Schiff, E., Leung, N.W., Atkins, M., Hunt, C., Brown, N., Woessner, M., Boehme, R. and Condreay, L. (2003) Prevalence and clinical correlates of YMDD variants during lamivudine therapy for patients with chronic hepatitis B. *Clin Infect.Dis.* **36**(6), 687–696.

Lai, M.M. (1995) The molecular biology of hepatitis delta virus. *Annu.Rev Biochem.* **64**, 259–286.

Lai, M.M. (2005) RNA replication without RNA-dependent RNA polymerase: surprises from hepatitis delta virus. *J. Virol.* **79**(13), 7951–7958.

Laskus, T., Radkowski, M., Wang, L.F., Vargas, H. and Rakela, J. (1998) Search for hepatitis C virus extra-hepatic replication sites in patients with acquired immunodeficiency syndrome: specific detection of negative-strand viral RNA in various tissues. *Hepatology* **28**, 1398–1401.

Laskus, T., Radkowski, M., Wang, L.F., Nowicki, M. and Rakela, J. (2000) Uneven distribution of hepatitis C virus quasispecies in tissues from subjects with end-stage liver disease: confounding effect of viral adsorption and mounting evidence for the presence of low-level extra-hepatic replication. *J. Virol.* **74**, 1014–1017.

Le Bouvier, G.L. (1971) The heterogeneity of Australia antigen. *J. Infect. Dis.* **123**, 671–675.

Le Gal, F., Gault, E., Ripault, M.P., Serpaggi, J., Trinchet, J.C., Gordien, E. and Deny, P. (2006) Eighth major clade for hepatitis delta virus. *Emerg. Infect. Dis.* **12**, 1447–1450.

Le Pogam, S., Kang, H., Harris, S.F., Leveque, V., Giannetti, A.M., Ali, S. *et al.* (2006) Selection and characterization of replicon variants dually resistant to thumb- and palm-binding nonnucleoside polymerase inhibitors of the hepatitis C virus. *J. Virol.* **80**, 6146–6154.

Lechner, F., Wong, D.K., Dunbar, P.R., Chapman, R., Chung, R.T., Dohrenwend, P. *et al.* (2000) Analysis of successful immune responses in persons infected with hepatitis C virus. *J. Exp. Med.* **191**, 1499–1512.

Lee, C.H., Choi, Y.H., Yang, S.H., Lee, C.W., Ha, S.J. and Sung, Y.C. (2001) Hepatitis C virus core protein inhibits interleukin 12 and nitric oxide production from activated macrophages. *Virology* **279**, 271–279.

Lee, C.M., Bih, F.Y., Chao, Y.C., Govindarajan, S. and Lai, M.M. (1992) Evolution of hepatitis delta virus RNA during chronic infection. *Virology* **188**, 265–273.

Lee, C.M., Changchien, C.S., Chung, J.C. and Liaw, Y.F. (1996) Characterization of a new genotype II hepatitis delta virus from Taiwan. *J. Med. Virol.* **49**, 145–154.

Legrand-Abravanel, F., Claudinon, J., Nicot, F., Dubois, M., Chapuy-Regaud, S., Sandres-Saune, K. *et al.* (2007) New natural intergenotypic (2/5) recombinant of hepatitis C virus. *J. Virol.* **81**, 4357–4362.

Lemon, S.M. and Binn, L.N. (1983) Antigenic relatedness of two strains of hepatitis A virus determined by cross-neutralization. *Infect. Immun.* **42**, 418–420.

Lemon, S.M., Murphy, P.C., Shields, P.A., Ping, L.H., Feinstone, S.M., Cromeans, T. and Jansen, R.W. (1991) Antigenic and genetic variation in cytopathic hepatitis A virus variants arising during persistent infection: evidence for genetic recombination. *J. Virol.* **65**, 2056–2065.

Leong, L.E.C., Cornell, C.T. and Semler, B.L. (2002) Processing determinants and functions of cleavage products of picornaviruses. In: *Molecular Biology of Picornaviruses* (B.L. Semler and E. Wimmer, eds), 187–198. Washington DC: ASM Press.

Lerat, H., Rumin, S., Habersetzer, F., Berby, F., Trabaud, M.A., Trepo, C. and Inchauspe, G. (1998) *In vivo* tropism of hepatitis C virus genomic sequences in hematopoietic cells: influence of viral load, viral genotype and cell phenotype. *Blood* **91**, 3841–3849.

Li, M., Xie, Y., Wu, X., Kong, Y. and Wang, Y. (1995) HNF3 binds and activates the second enhancer, ENII, of hepatitis B virus. *Virology* **214**, 371–378.

Li, T.C., Chijiwa, K., Sera, N., Ishibashi, T., Etoh, Y., Shinohara, Y. *et al.* (2005) Hepatitis E virus transmission from wild boar meat. *Emerg. Infect. Dis.* **11**, 1958–1960.

Liang, T.J., Hasegawa, K., Rimon, N., Wands, J.R. and Ben Porath, E. (1991) A hepatitis B virus mutant associated with an epidemic of fulminant hepatitis. *N. Engl. J. Med.* **324**, 1705–1709.

Lin, C., Pragai, B.M., Grakoui, A., Xu, J. and Rice, C.M. (1994) Hepatitis C virus NS3 serine proteinase: trans-cleavage requirements and processing kinetics. *J. Virol.* **68**, 8147–8157.

Lindenbach, B.D., Evans, M.J., Syder, A.J., Wolk, B., Tellinghuisen, T.L., Liu, C.C. *et al.* (2005) Complete replication of hepatitis C virus in cell culture. *Science* **309**, 623–626.

Lindenbach, B.D., Meuleman, P., Ploss, A., Vanwolleghem, T., Syder, A.J., McKeating, J.A. *et al.* (2006) Cell culture-grown hepatitis C virus is infectious *in vivo* and can be recultured *in vitro*. *Proc. Natl Acad. Sci. USA* **103**, 3805–3809.

Liu, C.J., Chen, B.F., Chen, P.J., Lai, M.Y., Huang, W.L., Kao, J.H. and Chen, D.S. (2006) Role of hepatitis B virus precore/core promoter mutations and serum viral load on noncirrhotic hepatocellular carcinoma: a case-control study. *J. Infect. Dis.* **194**, 594–599.

Liu, Z., Chi, B., Takahashi, K. and Mishiro, S. (2003) A genotype IV hepatitis E virus strain that may be indigenous to Changchun, China. *Intervirology* **46**, 252–256.

Locarnini, S. (2003) Hepatitis B viral resistance: mechanisms and diagnosis. *J. Hepatol.* **39**(suppl), S124–S132.

Lohmann, V., Korner, F., Herian, U. and Bartenschlager, R. (1997) Biochemical properties of hepatitis C virus NS5B RNA-dependent RNA polymerase and identification of amino acid sequence motifs essential for enzymatic activity. *J. Virol.* **71**, 8416–8428.

Lohmann, V., Korner, F., Koch, J., Herian, U., Theilmann, L. and Bartenschlager, R. (1999) Replication of subgenomic hepatitis C virus RNAs in a hepatoma cell line. *Science* **285**, 110–113.

Longman, R.S., Talal, A.H., Jacobson, I.M., Albert, M.L. and Rice, C.M. (2004) Presence of functional dendritic cells in patients chronically infected with hepatitis C virus. *Blood* **103**, 1026–1029.

Lozach, P.Y., Lortat-Jacob, H., de Lacroix, D.L., Staropoli, I., Foung, S., Amara, A. *et al.* (2003) DC-SIGN and L-SIGN are high affinity binding receptors for hepatitis C virus glycoprotein E2. *J. Biol. Chem.* **278**, 20358–20366.

Lu, L., Pilot-Matias, T.J., Stewart, K.D., Randolph, J.T., Pithawalla, R., He, W. *et al.* (2004) Mutations conferring resistance to a potent hepatitis C virus serine protease inhibitor *in vitro*. *Antimicrob. Agents Chemother.* **48**, 2260–2266.

Lu, L., Mo, H., Pilot-Matias, T.J. and Molla, A. (2007) Evolution of resistant M414T mutants among hepatitis C virus replicon cells treated with polymerase inhibitor A-782759. *Antimicrob. Agents Chemother.* **51**, 1889–1896.

Luo, G.X., Chao, M., Hsieh, S.Y., Sureau, C., Nishikura, K. and Taylor, J. (1990) A specific base transition occurs on replicating hepatitis delta virus RNA. *J. Virol.* **64**, 1021–1027.

Ma, S., Boerner, J.E., TiongYip, C., Weidmann, B., Ryder, N.S., Cooreman, M.P. and Lin, K. (2006) NIM811, a cyclophilin inhibitor, exhibits potent *in vitro* activity against hepatitis C virus alone or in combination with alpha interferon. *Antimicrob. Agents Chemother.* **50**, 2976–2982.

MacNaughton, T.B., Gowans, E.J., McNamara, S.P. and Burrell, C.J. (1991) Hepatitis delta antigen is necessary for access of hepatitis delta virus RNA to the cell transcriptional machinery but is not part of the transcriptional complex. *Virology* **184**, 387–390.

MacParland, S.A., Pham, T.N., Gujar, S.A. and Michalak, T.I. (2006) De novo infection and propagation of wild-type Hepatitis C virus in human T lymphocytes *in vitro*. *J. Gen. Virol.* **87**, 3577–3586.

Magnius, L.O. and Norder, H. (1995) Subtypes, genotypes and molecular epidemiology of the hepatitis B virus as reflected by sequence variability of the S-gene. *Intervirology* **38**, 24–34.

Makino, S., Chang, M.F., Shieh, C.K., Kamahora, T., Vannier, D.M., Govindarajan, S. and Lai, M.M. (1987) Molecular cloning and sequencing of a human hepatitis delta (delta) virus RNA. *Nature* **329**, 343–346.

Manns, M.P., McHutchison, J.G., Gordon, S.C., Rustgi, V.K., Shiffman, M.L., Reindollar, R. *et al.* (2001)

Peginterferon alfa-2b plus ribavirin compared with interferon alfa-2b plus ribavirin for initial treatment of chronic hepatitis C: a randomised trial. *Lancet* **358**, 958–965.

Mansell, C.J. and Locarnini, S.A. (1995) Epidemiology of hepatitis C in the East. *Semin. Liver Dis.* **15**, 15–32.

Marsh, M. and Helenius, A. (1989) Virus entry into animal cells. *Adv. Virus Res.* **36**, 107–151.

Martell, M., Esteban, J.I., Quer, J., Genesca, J., Weiner, A., Esteban, R. *et al.* (1992) Hepatitis C virus (HCV) circulates as a population of different but closely related genomes: quasispecies nature of HCV genome distribution. *J. Virol.* **66**, 3225–3229.

Mast, E.E. and Alter, M.J. (1993) Epidemiology of viral hepatitis: an overview. *Semin. Virol.* **4**, 273–283.

Mayo, M.A. and Ball, L.A. (2006) ICTV in San Francisco: a report from the Plenary Session. *Arch. Virol.* **151**, 413–422.

McHutchison, J.G., Gordon, S.C., Schiff, E.R., Shiffman, M.L., Lee, W.M., Rustgi, V.K. *et al.* (1998) Interferon alfa-2b alone or in combination with ribavirin as initial treatment for chronic hepatitis C. Hepatitis Interventional Therapy Group. *N. Engl. J. Med.* **339**, 1485–1492.

McOmish, F., Yap, P.L., Dow, B.C., Follett, E.A., Seed, C., Keller, A.J. *et al.* (1994) Geographical distribution of hepatitis C virus genotypes in blood donors: an international collaborative survey. *J. Clin. Microbiol.* **32**, 884–892.

Mechnik, L., Bergman, N., Attali, M., Beergabel, M., Mosenkis, B., Sokolowski, N. and Malnick, S. (2001) Acute hepatitis E virus infection presenting as a prolonged cholestatic jaundice. *J. Clin. Gastroenterol.* **33**, 421–422.

Meertens, L., Bertaux, C. and Dragic, T. (2006) Hepatitis C virus entry requires a critical postinternalization step and delivery to early endosomes via clathrin-coated vesicles. *J. Virol.* **80**, 11571–11578.

Meng, J., Dai, X., Chang, J.C., Lopareva, E., Pillot, J., Fields, H.A. and Khudyakov, Y.E. (2001) Identification and characterization of the neutralization epitope(s) of the hepatitis E virus. *Virology* **288**, 203–211.

Meng, X.J. (2003) Swine hepatitis E virus: cross-species infection and risk in xenotransplantation. *Curr. Top. Microbiol. Immunol.* **278**, 185–216.

Menne, S., Roneker, C.A., Roggendorf, M., Gerin, J.L., Cote, P.J. and Tennant, B.C. (2002) Deficiencies in the acute-phase cell-mediated immune response to viral antigens are associated with development of chronic woodchuck hepatitis virus infection following neonatal inoculation. *J. Virol.* **76**, 1769–1780.

Missale, G., Bertoni, R., Lamonaca, V., Valli, A., Massari, M., Mori, C. *et al.* (1996) Different clinical behaviors of acute hepatitis C virus infection are associated with different vigor of the anti-viral cell-mediated immune response. *J. Clin. Invest.* **98**, 706–714.

Mizokami, M., Orito, E., Ohba, K., Ikeo, K., Lau, J.Y. and Gojobori, T. (1997) Constrained evolution with respect to gene overlap of hepatitis B virus. *J. Mol. Evol.* **44**(suppl 1), S83–S90.

Mizokami, M., Tanaka, Y. and Miyakawa, Y. (2006) Spread times of hepatitis C virus estimated by the molecular clock differ among Japan, the United States and Egypt in reflection of their distinct socioeconomic backgrounds. *Intervirology* **49**, 28–36.

Monazahian, M., Bohme, I., Bonk, S., Koch, A., Scholz, C., Grethe, S. and Thomssen, R. (1999) Low density lipoprotein receptor as a candidate receptor for hepatitis C virus. *J. Med. Virol.* **57**, 223–229.

Moradpour, D., Wakita, T., Wands, J.R. and Blum, H.E. (1998) Tightly regulated expression of the entire hepatitis C virus structural region in continuous human cell lines. *Biochem. Biophys. Res. Commun.* **246**, 920–924.

Moradpour, D., Cerny, A., Heim, M.H. and Blum, H.E. (2001) Hepatitis C: an update. *Swiss Med. Wkly* **131**, 291–298.

Moreau, I., Hegarty, S., Levis, J., Sheehy, P., Crosbie, O., Kenny-Walsh, E. and Fanning, L.J. (2006) Serendipitous identification of natural intergenotypic recombinants of hepatitis C in Ireland. *Virol. J.* **3**, 95.

Morishima, C., Paschal, D.M., Wang, C.C., Yoshihara, C.S., Wood, B.L., Yeo, A.E. *et al.* (2006) Decreased NK cell frequency in chronic hepatitis C does not affect ex vivo cytolytic killing. *Hepatology* **43**, 573–580.

Muller, H.J. (1964) The relation of recombination to mutational advance. *Mutat. Res.* **106**, 2–9.

Myers, T.W. and Gelfand, D.H. (1991) Reverse transcription and DNA amplification by a *Thermus thermophilus* DNA polymerase. *Biochemistry* **30**, 7661–7666.

Nakano, T., Shapiro, C.N., Hadler, S.C., Casey, J.L., Mizokami, M., Orito, E. and Robertson, B.H. (2001) Characterization of hepatitis D virus genotype III among Yucpa Indians in Venezuela. *J. Gen. Virol.* **82**, 2183–2189.

Nakano, T., Lu, L., Liu, P. and Pybus, O.G. (2004) Viral gene sequences reveal the variable history of hepatitis C virus infection among countries. *J. Infect. Dis.* **190**, 1098–1108.

Nassal, M. and Schaller, H. (1996) Hepatitis B virus replication—an update. *J. Viral Hepat.* **3**, 217–226.

Ndjomou, J., Pybus, O.G. and Matz, B. (2003) Phylogenetic analysis of hepatitis C virus isolates indicates a unique pattern of endemic infection in Cameroon. *J. Gen. Virol.* **84**, 2333–2341.

Nesser, N.K., Peterson, D.O. and Hawley, D.K. (2006) RNA polymerase II subunit Rpb9 is important for transcriptional fidelity *in vivo*. *Proc. Natl Acad. Sci. USA* **103**, 3268–3273.

Neumann-Haefelin, C., Blum, H.E., Chisari, F.V. and Thimme, R. (2005) T cell response in hepatitis C virus infection. *J. Clin. Virol.* **32**, 75–85.

Neyts, J. (2006) Selective inhibitors of hepatitis C virus replication. *Antiviral Res.* **71**, 363–371.

Ngui, S.L., Hallet, R. and Teo, C.G. (1999) Natural and iatrogenic variation in hepatitis B virus. *Rev. Med. Virol.* **9**, 183–209.

NIH (2002) National Institutes of Health Consensus Development Conference Statement: Management of hepatitis C 2002. June 10–12, 2002. *Gastroenterology* **123**, 2082–2099.

Niro, G.A., Smedile, A., Andriulli, A., Rizzetto, M., Gerin, J.L. and Casey, J.L. (1997) The predominance of hepatitis delta virus genotype I among chronically infected Italian patients. *Hepatology* **25**, 728–734.

Noble, R.C., Kane, M.A., Reeves, S.A. and Roeckel, I. (1984) Posttransfusion hepatitis A in a neonatal intensive care unit. *JAMA* **252**, 2711–2715.

Noppornpanth, S., Lien, T.X., Poovorawan, Y., Smits, S.L., Osterhaus, A.D. and Haagmans, B.L. (2006) Identification of a naturally occurring recombinant genotype 2/6 hepatitis C virus. *J. Virol.* **80**, 7569–7577.

Norder, H., Courouce, A.M. and Magnius, L.O. (1994) Complete genomes, phylogenetic relatedness, and structural proteins of six strains of the hepatitis B virus, four of which represent two new genotypes. *Virology* **198**, 489–503.

Norder, H., Courouce, A.M., Coursaget, P., Echevarria, J.M., Lee, S.D., Mushahwar, I.K. *et al.* (2004) Genetic diversity of hepatitis B virus strains derived worldwide: genotypes, subgenotypes, and HBsAg subtypes. *Intervirology* **47**, 289–309.

Novella, I.S. (2004) Negative effect of genetic bottlenecks on the adaptability of vesicular stomatitis virus. *J. Mol. Biol.* **336**, 61–67.

Nowicki, M.J., Laskus, T., Nikolopoulou, G., Radkowski, M., Wilkinson, J., Du, W.B. *et al.* (2005) Presence of hepatitis C virus (HCV) RNA in the genital tracts of HCV/HIV-1-coinfected women. *J. Infect. Dis.* **192**, 1557–1565.

Okamoto, H. (2007) Genetic variability and evolution of hepatitis E virus. *Virus Res.* **127**, 216–228.

Okamoto, H., Imai, M., Kametani, M., Nakamura, T. and Mayumi, M. (1987) Genomic heterogeneity of hepatitis B virus in a 54-year-old woman who contracted the infection through materno-fetal transmission. *Jpn J. Exp. Med.* **57**, 231–236.

Okamoto, H., Tsuda, F., Sakugawa, H., Sastrosoewignjo, R.I., Imai, M., Miyakawa, Y. and Mayumi, M. (1988) Typing hepatitis B virus by homology in nucleotide sequence: comparison of surface antigen subtypes. *J. Gen. Virol.* **69**, 2575–2583.

Okamoto, H., Kojima, M., Okada, S., Yoshizawa, H., Iizuka, H., Tanaka, T. *et al.* (1992) Genetic drift of hepatitis C virus during an 8.2-year infection in a chimpanzee: variability and stability. *Virology* **190**, 894–899.

Ono, S.K., Kato, N., Shiratori, Y., Kato, J., Goto, T., Schinazi, R.F. *et al.* (2001) The polymerase L528M mutation cooperates with nucleotide binding-site mutations, increasing hepatitis B virus replication and drug resistance. *J. Clin. Invest.* **107**, 449–455.

Orito, E., Ichida, T., Sakugawa, H., Sata, M., Horiike, N., Hino, K. *et al.* (2001) Geographic distribution of hepatitis B virus (HBV) genotype in patients with chronic HBV infection in Japan. *Hepatology* **34**, 590–594.

Osiowy, C., Giles, E., Tanaka, Y., Mizokami, M. and Minuk, G.Y. (2006) Molecular evolution of hepatitis B virus over 25 years. *J. Virol.* **80**, 10307–10314.

Ozasa, A., Tanaka, Y., Orito, E., Sugiyama, M., Kang, J.H., Hige, S. *et al.* (2006) Influence of genotypes and precore mutations on fulminant or chronic outcome of acute hepatitis B virus infection. *Hepatology* **44**, 326–334.

Pal, S., Sullivan, D.G., Kim, S., Lai, K.K., Kae, J., Cotler, S.J. *et al.* (2006) Productive replication of hepatitis C virus in perihepatic lymph nodes *in vivo*: implications of HCV lymphotropism. *Gastroenterology* **130**, 1107–1116.

Papatheodoridis, G.V., Dimou, E., Laras, A., Papadimitropoulos, V. and Hadziyannis, S.J. (2002) Course of virologic breakthroughs under long-term lamivudine in HBeAg-negative precore mutant HBV liver disease. *Hepatology* **36**, 219–226.

Park, S.G., Kim, Y., Park, E., Ryu, H.M. and Jung, G. (2003) Fidelity of hepatitis B virus polymerase. *Eur. J. Biochem.* **270**, 2929–2936.

Pavlovic, D., Neville, D.C., Argaud, O., Blumberg, B., Dwek, R.A., Fischer, W.B. and Zitzmann, N. (2003) The hepatitis C virus p7 protein forms an ion channel that is inhibited by long-alkyl-chain iminosugar derivatives. *Proc. Natl Acad. Sci. USA* **100**, 6104–6108.

Pawlotsky, J.M. (2002) Use and interpretation of virological tests for hepatitis C. *Hepatology* **36**(suppl 1), S65–S73.

Pawlotsky, J.M. (2003a) Hepatitis C virus genetic variability: pathogenic and clinical implications. *Clin. Liver Dis.* **7**, 45–66.

Pawlotsky, J.M. (2003b) Mechanisms of antiviral treatment efficacy and failure in chronic hepatitis C. *Antiviral Res.* **59**, 1–11.

Pawlotsky, J.M. (2004) Pathophysiology of hepatitis C virus infection and related liver disease. *Trends Microbiol.* **12**, 96–102.

Pawlotsky, J.M. (2005) Current and future concepts in hepatitis C therapy. *Semin. Liver Dis.* **25**, 72–83.

Penin, F., Dubuisson, J., Rey, F.A., Moradpour, D. and Pawlotsky, J.M. (2004) Structural biology of hepatitis C virus. *Hepatology* **39**, 5–19.

Penna, A., Pilli, M., Zerbini, A., Orlandini, A., Mezzadri, S., Sacchelli, L. *et al.* (2007) Dysfunction and functional restoration of HCV-specific CD8 responses in chronic hepatitis C virus infection. *Hepatology* **45**, 588–601.

Pileri, P., Uematsu, Y., Campagnoli, S., Galli, G., Falugi, F., Petracca, R. *et al.* (1998) Binding of hepatitis C virus to CD81. *Science* **282**, 938–941.

Pina, S., Buti, M., Cotrina, M., Piella, J. and Girones, R. (2000) HEV identified in serum from humans with acute hepatitis and in sewage of animal origin in Spain. *J. Hepatol.* **33**, 826–833.

Pinto, R.M., Aragones, L., Costafreda, M.I., Ribes, E. and Bosch, A. (2007) Codon usage and replicative strategies of hepatitis A virus. *Virus Res.* **127**, 158–163.

Pohlmann, S., Zhang, J., Baribaud, F., Chen, Z., Leslie, G.J., Lin, G., Granelli-Piperno, A., Doms, R.W., Rice, C.M. and McKeating, J.A. (2003) Hepatitis C virus glycoproteins interact with DC-SIGN and DC-SIGNR. *J. Virol.* **77**, 4070–4080.

Pollicino, T., Belloni, L., Raffa, G., Pediconi, N., Squadrito, G., Raimondo, G. and Levrero, M. (2006) Hepatitis B virus replication is regulated by the acetylation status of hepatitis B virus cccDNA-bound H3 and H4 histones. *Gastroenterology* **130**, 823–837.

Pourquier, P., Jensen, A.D., Gong, S.S., Pommier, Y. and Rogler, C.E. (1999) Human DNA topoisomerase I-mediated cleavage and recombination of duck hepatitis B virus DNA *in vitro*. *Nucleic Acids Res.* **27**, 1919–1925.

Protzer-Knolle, U., Naumann, U., Bartenschlager, R., Berg, T., Hopf, U., Meyer zum Buschenfelde, K.H. *et al.* (1998) Hepatitis B virus with antigenically altered hepatitis B surface antigen is selected by high-dose hepatitis B immune globulin after liver transplantation. *Hepatology* **27**, 254–263.

Pumpens, P., Grens, E. and Nassal, M. (2002) Molecular epidemiology and immunology of hepatitis B virus infection—an update. *Intervirology* **45**, 218–232.

Purcell, R.H., Nguyen, H., Shapiro, M., Engle, R.E., Govindarajan, S., Blackwelder, W.C. *et al.* (2003) Pre-clinical immunogenicity and efficacy trial of a recombinant hepatitis E vaccine. *Vaccine* **21**, 2607–2615.

Puro, V., Petrosillo, N. and Ippolito, G. (1995a) Risk of hepatitis C seroconversion after occupational exposures in health care workers. Italian Study Group on Occupational Risk of HIV and Other Bloodborne Infections. *Am. J. Infect. Control* **23**, 273–277.

Puro, V., Petrosillo, N., Ippolito, G., Aloisi, M.S., Boumis, E. and Rava, L. (1995b) Occupational hepatitis C virus infection in Italian health care workers. Italian Study Group on Occupational Risk of Bloodborne Infections. *Am. J. Public Health* **85**, 1272–1275.

Pybus, O.G., Charleston, M.A., Gupta, S., Rambaut, A., Holmes, E.C. and Harvey, P.H. (2001) The epidemic behavior of the hepatitis C virus. *Science* **292**, 2323–2325.

Pybus, O.G., Drummond, A.J., Nakano, T., Robertson, B.H. and Rambaut, A. (2003) The epidemiology and iatrogenic transmission of hepatitis C virus in Egypt: a Bayesian coalescent approach. *Mol. Biol. Evol.* **20**, 381–387.

Quer, J. and Esteban, J.I. (2005) Epidemiology. In: *Viral Hepatitis* (H.C. Thomas, S. Lemon and A.J. Zuckerman, eds), 3rd edn, pp. 407–425. Oxford: Blackwell Publishing.

Quer, J., Hershey, C.L., Domingo, E., Holland, J.J. and Novella, I.S. (2001) Contingent neutrality in competing viral populations. *J. Virol.* **75**, 7315–7320.

Quer, J., Murillo, P., Martell, M., Gomez, J., Esteban, J.I., Esteban, R. and Guardia, J. (2004) Subtype mutations in the envelope 2 region including phosphorylation homology domain of Hepatitis C virus do not predict effectiveness of antiviral therapy. *J. Viral Hepat.* **11**, 45–54.

Quer, J., Cos, J., Murillo, P., Esteban, J.I., Esteban, R. and Guardia, J. (2005a) Improved attachment of natural HCV isolate to Daudi cells upon elimination of immune complexes and close pH control. *Intervirology* **48**, 285–291.

Quer, J., Esteban, J.I., Cos, J., Sauleda, S., Ocana, L., Martell, M. *et al.* (2005b) Effect of bottlenecking on evolution of the nonstructural protein 3 gene of hepatitis C virus during sexually transmitted acute resolving infection. *J. Virol.* **79**, 15131–15141.

Radecke, K., Protzer, U., Trippler, M., Meyer zum Buschenfelde, K.H. and Gerken, G. (2000) Selection of hepatitis B virus variants with aminoacid substitutions inside the core antigen during interferon-alpha therapy. *J. Med. Virol.* **62**, 479–486.

Radjef, N., Gordien, E., Ivaniushina, V., Gault, E., Anais, P., Drugan, T. *et al.* (2004) Molecular phylogenetic analyses indicate a wide and ancient radiation of African hepatitis delta virus, suggesting a deltavirus genus of at least seven major clades. *J. Virol.* **78**, 2537–2544.

Radkowski, M., Gallegos-Orozco, J.F., Jablonska, J., Colby, T.V., Walewska-Zielecka, B., Kubicka, J. *et al.* (2005) Persistence of hepatitis C virus in patients successfully treated for chronic hepatitis C. *Hepatology* **41**, 106–114.

Rajagopalan, L.E. and Malter, J.S. (1997) Regulation of eukaryotic messenger RNA turnover. *Prog. Nucleic Acid Res. Mol. Biol.* **56**, 257–286.

Raney, A.K., Johnson, J.L., Palmer, C.N. and McLachlan, A. (1997) Members of the nuclear receptor superfamily regulate transcription from the hepatitis B virus nucleocapsid promoter. *J. Virol.* **71**, 1058–1071.

Reed, K.E. and Rice, C.M. (1999) Identification of the major phosphorylation site of the hepatitis C virus H strain NS5A protein as serine 2321. *J. Biol. Chem.* **274**, 28011–28018.

Rehermann, B. and Nascimbeni, M. (2005) Immunology of hepatitis B virus and hepatitis C virus infection. *Nat. Rev. Immunol.* **5**, 215–229.

Robertson, B.H., Jansen, R.W., Khanna, B., Totsuka, A., Nainan, O.V., Siegl, G. *et al.* (1992) Genetic relatedness of hepatitis A virus strains recovered from different geographical regions. *J. Gen. Virol.* **73**, 1365–1377.

Robertson, H.D. (1996) How did replicating and coding RNAs first get together? *Science* **274**, 66–67.

Rodriguez-Frias, F., Buti, M., Jardi, R., Cotrina, M., Viladomiu, L., Esteban, R. and Guardia, J. (1995) Hepatitis B virus infection: precore mutants and its relation to viral genotypes and core mutations. *Hepatology* **22**, 1641–1647.

Rodriguez-Frias, F., Jardi, R., Buti, M., Schaper, M., Hermosilla, E., Valdes, A. *et al.* (2006) Hepatitis B virus genotypes and G1896A precore mutation in 486 Spanish patients with acute and chronic HBV infection. *J. Viral Hepat.* **13**, 343–350.

Rosen, H.R., Miner, C., Sasaki, A.W., Lewinsohn, D. M., Conrad, A.J., Bakke, A. *et al.* (2002) Frequencies of HCV-specific effector CD4+ T cells by flow cytometry: correlation with clinical disease stages. *Hepatology* **35**, 190–198.

Ross, R.S., Viazov, S., Thormahlen, M., Bartz, L., Tamm, J., Rautenberg, P. *et al.* (2002) Risk of hepatitis C virus transmission from an infected gynecologist to patients: results of a 7-year retrospective investigation. *Arch. Intern. Med.* **162**, 805–810.

Roussel, J., Pillez, A., Montpellier, C., Duverlie, G., Cahour, A., Dubuisson, J. and Wychowski, C. (2003) Characterization of the expression of the hepatitis C virus F protein. *J. Gen. Virol.* **84**, 1751–1759.

Roux, L., Simon, A.E. and Holland, J.J. (1991) Effects of defective interfering viruses on virus replication and pathogenesis *in vitro* and *in vivo*. *Adv. Virus Res.* **40**, 181–211.

Ryu, W.S. (2003) Molecular aspects of hepatitis B viral infection and the viral carcinogenesis. *J. Biochem. Mol. Biol.* **36**, 138–143.

Ryu, W.S., Bayer, M. and Taylor, J. (1992) Assembly of hepatitis delta virus particles. *J. Virol.* **66**, 2310–2315.

Sakamoto, T., Tanaka, Y., Orito, E., Co, J., Clavio, J., Sugauchi, F. *et al.* (2006) Novel subtypes (subgenotypes) of hepatitis B virus genotypes B and C among chronic liver disease patients in the Philippines. *J. Gen. Virol.* **87**, 1873–1882.

Sakugawa, H., Nakasone, H., Nakayoshi, T., Kawakami, Y., Miyazato, S., Kinjo, F. *et al.* (1999) Hepatitis delta virus genotype IIb predominates in an endemic area, Okinawa, Japan. *J. Med. Virol.* **58**, 366–372.

Saldanha, J.A., Thomas, H.C. and Monjardino, J.P. (1990) Cloning and sequencing of RNA of hepatitis delta virus isolated from human serum. *J. Gen. Virol.* **71**, 1603–1606.

Salleras, L.L. (1999) Catalonia, Sapin introduces mass hepatitis A vaccination programme. *Viral Hepatitis* **8**, 3–4.

Sanchez, G., Pinto, R.M., Vanaclocha, H. and Bosch, A. (2002) Molecular characterization of hepatitis a virus isolates from a transcontinental shellfish-borne outbreak. *J. Clin. Microbiol.* **40**, 4148–4155.

Sanchez, G., Bosch, A. and Pinto, R.M. (2003a) Genome variability and capsid structural constraints of hepatitis A virus. *J. Virol.* **77**, 452–459.

Sanchez, G., Bosch, A., Gomez-Mariano, G., Domingo, E. and Pinto, R.M. (2003b) Evidence for quasispecies distributions in the human hepatitis A virus genome. *Virology* **315**, 34–42.

Sanchez, G., Aragones, L., Costafreda, M.I., Ribes, E., Bosch, A. and Pinto, R.M. (2004) Capsid region involved in hepatitis A virus binding to glycophorin A of the erythrocyte membrane. *J. Virol.* **78**, 9807–9813.

Sanchez-Tapias, J.M., Costa, J., Mas, A., Bruguera, M. and Rodes, J. (2002) Influence of hepatitis B virus genotype on the long-term outcome of chronic hepatitis B in western patients. *Gastroenterology* **123**, 1848–1856.

Sarobe, P., Lasarte, J.J., Casares, N., Lopez-Diaz, D.C., Baixeras, E., Labarga, P. *et al.* (2002) Abnormal priming of CD4(+) T cells by dendritic cells expressing hepatitis C virus core and E1 proteins. *J. Virol.* **76**, 5062–5070.

Saunier, B., Triyatni, M., Ulianich, L., Maruvada, P., Yen, P. and Kohn, L.D. (2003) Role of the asialoglycoprotein receptor in binding and entry of hepatitis C virus structural proteins in cultured human hepatocytes. *J. Virol.* **77**, 546–559.

Scarselli, E., Ansuini, H., Cerino, R., Roccasecca, R.M., Acali, S., Filocamo, G. *et al.* (2002) The human scavenger receptor class B type I is a novel candidate receptor for the hepatitis C virus. *EMBO J.* **21**, 5017–5025.

Schaefer, S. (2007) Hepatitis B virus taxonomy and hepatitis B virus genotypes. *World J. Gastroenterol.* **13**, 14–21.

Schirren, C.A., Jung, M.C., Worzfeld, T., Mamin, M., Baretton, G., Gerlach, J.T. *et al.* (2001) Hepatitis C virus-specific CD4+ T cell response after liver transplantation occurs early, is multispecific, compartmentalizes to the liver and does not correlate with recurrent disease. *J. Infect. Dis.* **183**, 1187–1194.

Schlauder, G.G. (2004) Viral hepatitis: molecular biology, diagnosis, epidemiology and control. In: *Perspectives in Medical Virology* (A.J. Zuckerman and I.K. Mushahwar, eds), pp. 199–222. Amsterdam: Elsevier.

Schofield, D.J., Glamann, J., Emerson, S.U. and Purcell, R.H. (2000) Identification by phage display and characterization of two neutralizing chimpanzee monoclonal antibodies to the hepatitis E virus capsid protein. *J. Virol.* **74**, 5548–5555.

Schultheiss, T., Kusov, Y.Y. and Gauss-Muller, V. (1994) Proteinase 3C of hepatitis A virus (HAV) cleaves the HAV polyprotein P2-P3 at all sites including VP1/2A and 2A/2B. *Virology* **198**, 275–281.

Schultheiss, T., Sommergruber, W., Kusov, Y. and Gauss-Muller, V. (1995) Cleavage specificity of purified recombinant hepatitis A virus 3C proteinase on natural substrates. *J. Virol.* **69**, 1727–1733.

Seeger, C. and Mason, W.S. (2000) Hepatitis B virus biology. *Microbiol. Mol. Biol. Rev.* **64**, 51–68.

Shakil, A.O., Hadziyannis, S., Hoofnagle, J.H., Di Bisceglie, A.M., Gerin, J.L. and Casey, J.L. (1997) Geographic distribution and genetic variability of hepatitis delta virus genotype I. *Virology* **234**, 160–167.

Sharmeen, L., Kuo, M.Y., Dinter-Gottlieb, G. and Taylor, J. (1988) Antigenomic RNA of human hepatitis delta virus can undergo self-cleavage. *J. Virol.* **62**, 2674–2679.

Sharmeen, L., Kuo, M.Y. and Taylor, J. (1989) Self-ligating RNA sequences on the antigenome of human hepatitis delta virus. *J. Virol.* **63**, 1428–1430.

Shata, M.T., Tricoche, N., Perkus, M., Tom, D., Brotman, B., McCormack, P. *et al.* (2003) Exposure to low infective doses of HCV induces cellular immune responses without consistently detectable viremia or seroconversion in chimpanzees. *Virology* **314**, 601–616.

Sheldon, J., Camino, N., Rodes, B., Bartholomeusz, A., Kuiper, M., Tacke, F. *et al.* (2005) Selection of hepatitis B virus polymerase mutations in HIV-coinfected patients treated with tenofovir. *Antivir. Ther.* **10**, 727–734.

Sheldon, J., Rodes, B., Zoulim, F., Bartholomeusz, A. and Soriano, V. (2006) Mutations affecting the replication capacity of the hepatitis B virus. *J. Viral Hepat.* **13**, 427–434.

Sherertz, R.J., Russell, B.A. and Reuman, P.D. (1984) Transmission of hepatitis A by transfusion of blood products. *Arch. Intern. Med.* **144**, 1579–1580.

Shimizu, Y.K., Iwamoto, A., Hijikata, M., Purcell, R.H. and Yoshikura, H. (1992) Evidence for *in vitro* replication of hepatitis C virus genome in a human T-cell line, *Proc. Natl Acad. Sci. USA* **89**, 5477–5481.

Shimizu, Y.K., Igarashi, H., Kanematu, T., Fujiwara, K., Wong, D.C., Purcell, R.H. and Yoshikura, H. (1997) Sequence analysis of the hepatitis C virus genome recovered from serum, liver and peripheral blood mononuclear cells of infected chimpanzees. *J. Virol.* **71**, 5769–5773.

Shouval, D. (1999) Israel implements universal hepatitis A immunization. *Viral Hepatitis* **8**, 2.

Simmonds, P. (2001) Reconstructing the origins of human hepatitis viruses. *Philos. Trans. R Soc. Lond. B Biol. Sci.* **356**, 1013–1026.

Simmonds, P. (2004) Genetic diversity and evolution of hepatitis C virus—15 years on. *J. Gen. Virol.* **85**, 3173–3188.

Simmonds, P. and Midgley, S. (2005) Recombination in the genesis and evolution of hepatitis B virus genotypes. *J. Virol.* **79**, 15467–15476.

Simmonds, P. and Smith, D.B. (1997) Investigation of the pattern of diversity of hepatitis C virus in relation to times of transmission. *J. Viral Hepat.* **4**(suppl 1), 69–74.

Simmonds, P. and Smith, D.B. (1999) Structural constraints on RNA virus evolution. *J. Virol.* **73**, 5787–5794.

Simmonds, P., Alberti, A., Alter, H.J., Bonino, F., Bradley, D.W., Brechot, C. *et al.* (1994a) A proposed system for the nomenclature of hepatitis C viral genotypes. *Hepatology* **19**, 1321–1324.

Simmonds, P., Smith, D.B., McOmish, F., Yap, P.L., Kolberg, J., Urdea, M.S. and Holmes, E.C. (1994b) Identification of genotypes of hepatitis C virus by sequence comparisons in the core, E1 and NS-5 regions. *J. Gen. Virol.* **75**, 1053–1061.

Simmonds, P., Bukh, J., Combet, C., Deleage, G., Enomoto, N., Feinstone, S. *et al.* (2005) Consensus proposals for a unified system of nomenclature of hepatitis C virus genotypes. *Hepatology* **42**, 962–973.

Smith, D.B. and Simmonds, P. (1997) Review: molecular epidemiology of hepatitis C virus. *J. Gastroenterol. Hepatol.* **12**, 522–527.

Smith, D.B., Mellor, J., Jarvis, L.M., Davidson, F., Kolberg, J., Urdea, M. *et al.* (1995) Variation of the hepatitis C virus 5′ non-coding region: implications for secondary structure, virus detection and typing. The International HCV Collaborative Study Group. *J. Gen. Virol.* **76**, 1749–1761.

Smith, J.L. (2001) A review of hepatitis E virus. *J. Food Prot.* **64**, 572–586.

Steinhauer, D.A. and Holland, J.J. (1987) Rapid evolution of RNA viruses. *Annu. Rev. Microbiol.* **41**, 409–433.

Stevenson, P. (2000) Nepal calls the shots in hepatitis E virus vaccine trial. *Lancet* **355**, 1623.

Stoeckl, L., Funk, A., Kopitzki, A., Brandenburg, B., Oess, S., Will, H. *et al.* (2006) Identification of a structural motif crucial for infectivity of hepatitis B viruses. *Proc. Natl Acad. Sci. USA* **103**, 6730–6734.

Stroffolini, T., Sagnelli, E., Mele, A., Craxi, A. and Almasio, P. (2004) The aetiology of chronic hepatitis in Italy: results from a multicentre national study. *Dig. Liver Dis.* **36**, 829–833.

Stuyver, L., De Gendt, S., Van Geyt, C., Zoulim, F., Fried, M., Schinazi, R.F. and Rossau, R. (2000) A new genotype of hepatitis B virus: complete genome and phylogenetic relatedness. *J. Gen. Virol.* **81**, 67–74.

Sugauchi, F., Kumada, H., Sakugawa, H., Komatsu, M., Niitsuma, H., Watanabe, H. *et al.* (2004) Two subtypes of genotype B (Ba and Bj) of hepatitis B virus in Japan. *Clin. Infect. Dis.* **38**, 1222–1228.

Tacke, F., Gehrke, C., Luedde, T., Heim, A., Manns, M.P. and Trautwein, C. (2004) Basal core promoter and precore mutations in the hepatitis B virus genome enhance replication efficacy of Lamivudine-resistant mutants. *J. Virol.* **78**, 8524–8535.

Tai, P.C., Suk, F.M., Gerlich, W.H., Neurath, A.R. and Shih, C. (2002) Hypermodification and immune escape of an internally deleted middle-envelope (M) protein of frequent and predominant hepatitis B virus variants. *Virology* **292**, 44–58.

Takahashi, K., Toyota, J., Karino, Y., Kang, J.H., Maekubo, H., Abe, N. and Mishiro, S. (2004) Estimation of the mutation rate of hepatitis E virus based on a set of closely related 7.5-year-apart isolates from Sapporo, Japan. *Hepatol Res.* **29**, 212–215.

Takaki, A., Wiese, M., Maertens, G., Depla, E., Seifert, U., Liebetrau, A. *et al.* (2000) Cellular immune responses persist and humoral responses decrease two decades after recovery from a single-source outbreak of hepatitis C. *Nat. Med.* **6**, 578–582.

Tallo, T., Norder, H., Tefanova, V., Krispin, T., Schmidt, J., Ilmoja, M. *et al.* (2007) Genetic characterization of hepatitis C virus strains in Estonia: fluctuations in the predominating subtype with time. *J. Med. Virol.* **79**, 374–382.

Tam, A.W., Smith, M.M., Guerra, M.E., Huang, C.C., Bradley, D.W., Fry, K.E. and Reyes, G.R. (1991) Hepatitis E virus (HEV): molecular cloning and sequencing of the full-length viral genome. *Virology* **185**, 120–131.

Tanaka, H., Ueda, H., Hamagami, H., Yukawa, S., Ichinose, M., Miyano, M. *et al.* (2005) Mutations in hepatitis B virus core regions correlate with hepatocellular injury in Chinese patients with chronic hepatitis B. *World J. Gastroenterol.* **11**, 4693–4696.

Tanaka, T., Kato, N., Cho, M.J., Sugiyama, K. and Shimotohno, K. (1996) Structure of the 3′ terminus of the hepatitis C virus genome. *J. Virol.* **70**, 3307–3312.

Tang, H., Delgermaa, L., Huang, F., Oishi, N., Liu, L., He, F. *et al.* (2005) The transcriptional transactivation function of HBx protein is important for its augmentation role in hepatitis B virus replication. *J. Virol.* **79**, 5548–5556.

Tang, H., Oishi, N., Kaneko, S. and Murakami, S. (2006) Molecular functions and biological roles of hepatitis B virus x protein. *Cancer Sci.* **97**, 977–983.

Tanji, Y., Hijikata, M., Satoh, S., Kaneko, T. and Shimotohno, K. (1995a) Hepatitis C virus-encoded nonstructural protein NS4A has versatile functions in viral protein processing. *J. Virol.* **69**, 1575–1581.

Tanji, Y., Kaneko, T., Satoh, S. and Shimotohno, K. (1995b) Phosphorylation of hepatitis C virus-encoded nonstructural protein NS5A. *J. Virol.* **69**, 3980–3986.

Taylor, J.M. (2003) Replication of human hepatitis delta virus: recent developments. *Trends Microbiol.* **11**, 185–190.

Taylor, J.M. (2005) Hepatitis delta virus. In: *Topley and Wilson's Microbiology and Microbial Infections* (B.W.J. Mahy and V. ter Muelen, eds), 10th edn, pp. 1269–1275. London: Hodder Arnold.

Taylor, J.M. (2006) Hepatitis delta virus. *Virology* **344**, 71–76.

Tedesco, R., Shaw, A.N., Bambal, R., Chai, D., Concha, N.O., Darcy, M.G. *et al.* (2006) 3-(1,1-dioxo-2H-(1,2,4)-benzothiadiazin-3-yl)-4-hydroxy-2(^{1}H)-quinolinones, potent inhibitors of hepatitis C virus RNA-dependent RNA polymerase. *J. Med. Chem.* **49**, 971–983.

Tei, S., Kitajima, N., Takahashi, K. and Mishiro, S. (2003) Zoonotic transmission of hepatitis E virus from deer to human beings. *Lancet* **362**, 371–373.

Tellinghuisen, T.L. and Rice, C.M. (2002) Interaction between hepatitis C virus proteins and host cell factors. *Curr. Opin. Microbiol.* **5**, 419–427.

Terrault, N.A., Zhou, S., McCory, R.W., Pruett, T.L., Lake, J.R., Roberts, J.P. *et al.* (1998) Incidence and clinical consequences of surface and polymerase gene mutations in liver transplant recipients on hepatitis B immunoglobulin. *Hepatology* **28**, 555–561.

Tester, I., Smyk-Pearson, S., Wang, P., Wertheimer, A., Yao, E., Lewinsohn, D.M. *et al.* (2005) Immune evasion versus recovery after acute hepatitis C virus infection from a shared source. *J. Exp. Med.* **201**, 1725–1731.

Thakur, V., Guptan, R.C., Kazim, S.N., Malhotra, V. and Sarin, S.K. (2002) Profile, spectrum and significance of HBV genotypes in chronic liver disease patients in the Indian subcontinent. *J. Gastroenterol. Hepatol.* **17**, 165–170.

Thibault, V., Aubron-Olivier, C., Agut, H. and Katlama, C. (2002) Primary infection with a lamivudine-resistant hepatitis B virus. *AIDS* **16**, 131–133.

Tomei, L., Altamura, S., Bartholomew, L., Biroccio, A., Ceccacci, A., Pacini, L. *et al.* (2003) Mechanism of action and antiviral activity of benzimidazole-based allosteric inhibitors of the hepatitis C virus RNA-dependent RNA polymerase. *J. Virol.* **77**, 13225–13231.

Tomei, L., Altamura, S., Bartholomew, L., Bisbocci, M., Bailey, C., Bosserman, M. *et al.* (2004) Characterization of the inhibition of hepatitis C virus RNA replication by nonnucleosides. *J. Virol.* **78**, 938–946.

Tong, X., Chase, R., Skelton, A., Chen, T., Wright-Minogue, J. and Malcolm, B.A. (2006a) Identification and analysis of fitness of resistance mutations against the HCV protease inhibitor SCH 503034. *Antiviral Res.* **70**, 28–38.

Tong, X., Guo, Z., Wright-Minogue, J., Xia, E., Prongay, A., Madison, V. *et al.* (2006b) Impact of naturally occurring variants of HCV protease on the binding of different classes of protease inhibitors. *Biochemistry* **45**, 1353–1361.

Torresi, J. (2002) The virological and clinical significance of mutations in the overlapping envelope and polymerase genes of hepatitis B virus. *J. Clin. Virol.* **25**, 97–106.

Torresi, J., Earnest-Silveira, L., Civitico, G., Walters, T.E., Lewin, S.R., Fyfe, J. *et al.* (2002) Restoration of replication phenotype of lamivudine-resistant hepatitis B virus mutants by compensatory changes in the fingers subdomain of the viral polymerase selected as a consequence of mutations in the overlapping S gene. *Virology* **299**, 88–99.

Tsai, S.L., Chen, Y.M., Chen, M.H., Huang, C.Y., Sheen, I.S., Yeh, C.T. *et al.* (1998) Hepatitis C virus variants circumventing cytotoxic T lymphocyte activity as a mechanism of chronicity. *Gastroenterology* **115**, 954–965.

Tscherne, D.M., Jones, C.T., Evans, M.J., Lindenbach, B.D., McKeating, J.A. and Rice, C.M. (2006) Time- and temperature-dependent activation of hepatitis C virus for low-pH-triggered entry. *J. Virol.* **80**, 1734–1741.

Tscherne, D.M., Evans, M.J., von Hahn, T., Jones, C.T., Stamataki, Z., McKeating, J.A. *et al.* (2007) Superinfection exclusion in cells infected with hepatitis C virus. *J. Virol.* **81**, 3693–3703.

Tsubota, A., Kumada, H., Takaki, K., Chayama, K., Kobayashi, M., Kobayashi, M. *et al.* (1998) Deletions in the hepatitis B virus core gene may influence the clinical outcome in hepatitis B e antigen-positive asymptomatic healthy carriers. *J. Med. Virol.* **56**, 287–293.

Tsukiyama-Kohara, K., Iizuka, N., Kohara, M. and Nomoto, A. (1992) Internal ribosome entry site within hepatitis C virus RNA. *J. Virol.* **66**, 1476–1483.

Tuplin, A., Wood, J., Evans, D.J., Patel, A.H. and Simmonds, P. (2002) Thermodynamic and phylogenetic prediction of RNA secondary structures in the coding region of hepatitis C virus. *RNA* **8**, 824–841.

Uchida, T., Aye, T.T., Shikata, T., Yano, M., Yatsuhashi, H., Koga, M. and Mima, S. (1994) Evolution of the hepatitis B virus gene during chronic infection in seven patients. *J. Med. Virol.* **43**, 148–154.

Urbani, S., Amadei, B., Fisicaro, P., Tola, D., Orlandini, A., Sacchelli, L. *et al.* (2006) Outcome of acute hepatitis C is related to virus-specific CD4 function and maturation of antiviral memory CD8 responses. *Hepatology* **44**, 126–139.

Verbeeck, J., Maes, P., Lemey, P., Pybus, O.G., Wollants, E., Song, E. *et al.* (2006) Investigating the origin and spread of hepatitis C virus genotype 5a. *J. Virol.* **80**, 4220–4226.

Vignuzzi, M., Stone, J.K., Arnold, J.J., Cameron, C.E. and Andino, R. (2006) Quasispecies diversity determines pathogenesis through cooperative interactions in a viral population. *Nature* **439**, 344–348.

Villano, S.A., Vlahov, D., Nelson, K.E., Cohn, S. and Thomas, D.L. (1999) Persistence of viremia and the importance of long-term follow-up after acute hepatitis C infection. *Hepatology* **29**, 908–914.

Wai, C.T. and Fontana, R.J. (2004) Clinical significance of hepatitis B virus genotypes, variants, and mutants. *Clin. Liver Dis.* **8**, 321–352, vi.

Wakita, T., Pietschmann, T., Kato, T., Date, T., Miyamoto, M., Zhao, Z. *et al.* (2005) Production of infectious hepatitis C virus in tissue culture from a cloned viral genome. *Nat. Med.* **11**, 791–796.

Walewski, J.L., Keller, T.R., Stump, D.D. and Branch, A.D. (2001) Evidence for a new hepatitis C virus antigen encoded in an overlapping reading frame. *RNA* **7**, 710–721.

Walewski, J.L., Gutierrez, J.A., Branch-Elliman, W., Stump, D.D., Keller, T.R., Rodriguez, A. *et al.* (2002) Mutation Master: profiles of substitutions in hepatitis C virus RNA of the core, alternate reading frame and NS2 coding regions. *RNA* **8**, 557–571.

Wang, C., Sarnow, P. and Siddiqui, A. (1993) Translation of human hepatitis C virus RNA in cultured cells is mediated by an internal ribosome-binding mechanism. *J. Virol.* **67**, 3338–3344.

Wang, C., Sarnow, P. and Siddiqui, A. (1994) A conserved helical element is essential for internal initiation of translation of hepatitis C virus RNA. *J. Virol.* **68**, 7301–7307.

Wang, C., Le, S.Y., Ali, N. and Siddiqui, A. (1995) An RNA pseudoknot is an essential structural element of the internal ribosome entry site located within the hepatitis C virus 5′ noncoding region. *RNA* **1**, 526–537.

Wang, H. and Eckels, D.D. (1999) Mutations in immunodominant T cell epitopes derived from the nonstructural 3 protein of hepatitis C virus have the potential for generating escape variants that may have important consequences for T cell recognition. *J. Immunol.* **162**, 4177–4183.

Wang, H.C., Huang, W., Lai, M.D. and Su, I.J. (2006) Hepatitis B virus pre-S mutants, endoplasmic reticulum stress and hepatocarcinogenesis. *Cancer Sci.* **97**, 683–688.

Wang, K.S., Choo, Q.L., Weiner, A.J., Ou, J.H., Najarian, R.C., Thayer, R.M. *et al.* (1986) Structure, sequence and expression of the hepatitis delta (delta) viral genome. *Nature* **323**, 508–514.

Wang, L. and Zhuang, H. (2004) Hepatitis E: an overview and recent advances in vaccine research. *World J Gastroenterol.* **10**, 2157–2162.

Wang, T.C. and Chao, M. (2003) Molecular cloning and expression of the hepatitis delta virus genotype IIb genome. *Biochem. Biophys. Res. Commun.* **303**, 357–363.

Wang, T.C. and Chao, M. (2005) RNA recombination of hepatitis delta virus in natural mixed-genotype infection and transfected cultured cells. *J. Virol.* **79**, 2221–2229.

Wang, Z., Tanaka, Y., Huang, Y., Kurbanov, F., Chen, J., Zeng, G. *et al.* (2007) Clinical and virological characteristics of hepatitis B virus subgenotypes Ba, C1, and C2 in China. *J. Clin. Microbiol.* **45**, 1491–1496.

Wasley, A., Samandari, T. and Bell, B.P. (2005) Incidence of hepatitis A in the United States in the era of vaccination. *JAMA* **294**, 194–201.

Watanabe, H., Nagayama, K., Enomoto, N., Chinzei, R., Yamashiro, T., Izumi, N. *et al.* (2003) Chronic hepatitis delta virus infection with genotype IIb variant is correlated with progressive liver disease. *J. Gen. Virol.* **84**, 3275–3289.

Weber, B. (2005) Genetic variability of the S gene of hepatitis B virus: clinical and diagnostic impact. *J. Clin. Virol.* **32**, 102–112.

Weinberger, K.M., Bauer, T., Bohm, S. and Jilg, W. (2000) High genetic variability of the group-specific a-determinant of hepatitis B virus surface antigen (HBsAg) and the corresponding fragment of the viral polymerase in chronic virus carriers lacking detectable HBsAg in serum. *J. Gen. Virol.* **81**, 1165–1174.

Weiner, A., Erickson, A.L., Kansopon, J., Crawford, K., Muchmore, E., Hughes, A.L. *et al.* (1995) Persistent hepatitis C virus infection in a chimpanzee is associated with emergence of a cytotoxic T lymphocyte escape variant. *Proc. Natl Acad. Sci. USA,* **92**, 2755–2759.

Weiner, A.J., Choo, Q.L., Wang, K.S., Govindarajan, S., Redeker, A.G., Gerin, J.L. and Houghton, M. (1988) A single antigenomic open reading frame of the hepatitis delta virus encodes the epitope(s) of both hepatitis delta antigen polypeptides p24 delta and p27 delta. *J. Virol.* **62**, 594–599.

Weitz, M., Baroudy, B.M., Maloy, W.L., Ticehurst, J.R. and Purcell, R.H. (1986) Detection of a genome-linked protein (VPg) of hepatitis A virus and its comparison with other picornaviral VPgs. *J. Virol.* **60**, 124–130.

Westland, C., Delaney, W., Yang, H., Chen, S.S., Marcellin, P., Hadziyannis, S. *et al.* (2003) Hepatitis B virus genotypes and virologic response in 694 patients in phase III studies of adefovir dipivoxill. *Gastroenterology* **125**, 107–116.

Whalley, S.A., Murray, J.M., Brown, D., Webster, G.J., Emery, V.C., Dusheiko, G.M. and Perelson, A.S. (2001) Kinetics of acute hepatitis B virus infection in humans. *J. Exp. Med.* **193**, 847–854.

Whetter, L.E., Day, S.P., Elroy-Stein, O., Brown, E.A. and Lemon, S.M. (1994) Low efficiency of the 5′ nontranslated region of hepatitis A virus RNA in directing cap-independent translation in permissive monkey kidney cells. *J. Virol.* **68**, 5253–5263.

WHO (2000a). *Hepatitis B.* Fact sheet number 204. http://www.who.int/mediacentre/factsheets/fs204/en/

WHO (2000b). Hepatitis C. Fact sheet No.164. http://www.who.inf/inf-fs/en/fact164.html, fact 164.

WHO (2006). *Cancer.* Fact sheet number 297. http://www.who.int/mediacentre/factsheets/fs297/en/

Wieland, S., Thimme, R., Purcell, R.H. and Chisari, F.V. (2004) Genomic analysis of the host response to hepatitis B virus infection. *Proc. Natl Acad. Sci. USA* **101**, 6669–6674.

Willems, M., Metselaar, H.J., Tilanus, H.W., Schalm, S.W. and de Man, R.A. (2002) Liver transplantation and hepatitis C. *Transplant. Int.* **15**, 61–72.

Wohnsland, A., Hofmann, W.P. and Sarrazin, C. (2007) Viral determinants of resistance to treatment in patients with hepatitis C. *Clin. Microbiol. Rev.* **20**, 23–38.

Worm, H.C., Schlauder, G.G. and Brandstatter, G. (2002) Hepatitis E and its emergence in non-endemic areas. *Wien. Klin. Wochenschr.* **114**, 663–670.

Worm, H.C., van der Poel, W.H. and Brandstatter, G. (2002) Hepatitis E: an overview. *Microb. Infect.* **4**, 657–666.

Worobey, M. and Holmes, E.C. (2001) Homologous recombination in GB virus C/hepatitis G virus. *Mol. Biol. Evol.* **18**, 254–261.

Wu, H.N. and Lai, M.M. (1989) Reversible cleavage and ligation of hepatitis delta virus RNA. *Science* **243**, 652–654.

Wu, H.N., Lin, Y.J., Lin, F.P., Makino, S., Chang, M.F. and Lai, M.M. (1989) Human hepatitis delta virus RNA subfragments contain an autocleavage activity. *Proc. Natl Acad. Sci. USA* **86**, 1831–1835.

Wu, J.C., Choo, K.B., Chen, C.M., Chen, T.Z., Huo, T.I. and Lee, S.D. (1995) Genotyping of hepatitis D virus by restriction-fragment length polymorphism and relation to outcome of hepatitis D. *Lancet* **346**, 939–941.

Wu, J.C., Chiang, T.Y. and Sheen, I.J. (1998) Characterization and phylogenetic analysis of a novel hepatitis D virus strain discovered by restriction fragment length polymorphism analysis. *J. Gen. Virol.* **79**, 1105–1113.

Xu, Z., Choi, J., Lu, W. and Ou, J.H. (2003) Hepatitis C virus f protein is a short-lived protein associated with the endoplasmic reticulum. *J. Virol.* **77**, 1578–1583.

Yamada, N., Tanihara, K., Takada, A., Yorihuzi, T., Tsutsumi, M., Shimomura, H. *et al.* (1996) Genetic organization and diversity of the 3′ noncoding region of the hepatitis C virus genome. *Virology* **223**, 255–261.

Yamashita, T., Kaneko, S., Shirota, Y., Qin, W., Nomura, T., Kobayashi, K. and Murakami, S. (1998) RNA-dependent RNA polymerase activity of the soluble recombinant hepatitis C virus NS5B protein truncated at the C-terminal region. *J. Biol. Chem.* **273**, 15479–15486.

Yanagi, M., St Claire, M., Emerson, S.U., Purcell, R.H. and Bukh, J. (1999) *In vivo* analysis of the 3′ untranslated region of the hepatitis C virus after *in vitro* mutagenesis of an infectious cDNA clone. *Proc. Natl Acad. Sci. USA* **96**, 2291–2295.

Yasui, K., Wakita, T., Tsukiyama-Kohara, K., Funahashi, S.I., Ichikawa, M., Kajita, T. *et al.* (1998) The native form and maturation process of hepatitis C virus core protein. *J. Virol.* **72**, 6048–6055.

Yazaki, Y., Mizuo, H., Takahashi, M., Nishizawa, T., Sasaki, N., Gotanda, Y. and Okamoto, H. (2003) Sporadic acute or fulminant hepatitis E in Hokkaido, Japan, may be food-borne, as suggested by the presence of hepatitis E virus in pig liver as food. *J. Gen. Virol.* **84**, 2351–2357.

Yi, M., Tong, X., Skelton, A., Chase, R., Chen, T., Prongay, A. *et al.* (2006) Mutations conferring resistance to SCH6, a novel hepatitis C virus NS3/4A protease inhibitor. Reduced RNA replication fitness and partial rescue by second-site mutations. *J. Biol. Chem.* **281**, 8205–8215.

Yuen, M.F., Kato, T., Mizokami, M., Chan, A.O., Yuen, J.C., Yuan, H.J. *et al.* (2003) Clinical outcome and virologic profiles of severe hepatitis B exacerbation due to YMDD mutations. *J. Hepatol.* **39**, 850–855.

Yuen, M.F., Wong, D.K., Zheng, B.J., Chan, C.C., Yuen, J.C., Wong, B.C. and Lai, C.L. (2007) Difference in T helper responses during hepatitis flares in hepatitis B e antigen (HBeAg)-positive patients with genotypes B and C: implication for early HBeAg seroconversion. *J. Viral. Hepat.* **14**, 269–275.

Zafrullah, M., Ozdener, M.H., Kumar, R., Panda, S.K. and Jameel, S. (1999) Mutational analysis of glycosylation, membrane translocation, and cell surface expression of the hepatitis E virus ORF2 protein. *J. Virol.* **73**, 4074–4082.

Zhang, M., Emerson, S.U., Nguyen, H., Engle, R., Govindarajan, S., Blackwelder, W.C. *et al.* (2002) Recombinant vaccine against hepatitis E: duration of protective immunity in rhesus macaques. *Vaccine* **20**, 3285–3291.

Zhang, Y.Y., Tsega, E. and Hansson, B.G. (1996) Phylogenetic analysis of hepatitis D viruses indicating a new genotype I subgroup among African isolates. *J. Clin. Microbiol.* **34**, 3023–3030.

Zhong, J., Gastaminza, P., Cheng, G., Kapadia, S., Kato, T., Burton, D.R. *et al.* (2005) Robust hepatitis C virus infection *in vitro*. *Proc. Natl Acad. Sci. USA* **102**, 9294–9299.

Zhu, Y., Yamamoto, T., Cullen, J., Saputelli, J., Aldrich, C.E., Miller, D.S. *et al.* (2001) Kinetics of hepadnavirus loss from the liver during inhibition of viral DNA synthesis. *J. Virol.* **75**, 311–322.

Zuckerman, J.N. and Zuckerman, A.J. (2003) Mutations of the surface protein of hepatitis B virus. *Antiviral Res.* **60**, 75–78.

Zuckerman, J., Clewley, G., Griffiths, P. and Cockcroft, A. (1994) Prevalence of hepatitis C antibodies in clinical health-care workers. *Lancet* **343**, 1618–1620.

CHAPTER

16

Arbovirus Evolution

Kathryn A. Hanley and Scott C. Weaver

ABSTRACT

Arboviruses (*ar*thropod-*bo*rne *viruses*), though taxonomically diverse, share a cycle of transmission between vertebrate hosts and arthropod vectors. All but one arbovirus species belong to one of five families of RNA viruses, suggesting that the high mutation frequencies of RNA genomes may be a prerequisite for entry into a cycle of alternating replication in the very different environments represented by vertebrate and invertebrate animals. Arboviruses such as dengue, yellow fever, and Venezuelan equine encephalitis viruses pose significant, recognized threats to human and animal health. Moreover, the potential for new arboviruses to emerge from enzootic cycles to cause disease in humans and domestic animals, or for recognized arboviruses to spread into new geographic areas, is already high and may be exacerbated by ongoing changes in human demography and global climate. Thus it is critical to understand the forces that shape the evolution of arboviruses in order to better predict their emergence and to control their spread. In this chapter we review existing scholarship on the origin, diversification and evolution of host and vector use of arboviruses, with a focus on the two best-studied groups: the vector-borne alphaviviruses and flaviviruses. Specifically, we review the contribution of phylogenetic analysis to current understanding of arbovirus evolution and epidemiology, we evaluate the role of recombination and reassortment in the origin of new arboviruses, and we examine the mechanisms of arbovirus emergence and geographic spread. In addition, we discuss the impact of cycles of alternating replication in vertebrate hosts and vectors on the rates and patterns of arbovirus evolution and the selection imposed by alternating transmission cycles on specific arbovirus phenotypes. Finally we consider the challenges that arbovirus evolution may pose to the deployment of existing and novel measures for disease control.

TRANSMISSION CYCLES, EVOLUTIONARY HISTORY AND SYSTEMATICS OF THE ARBOVIRUSES

Transmission Cycles

Arthropod-borne viruses (arboviruses) comprise a taxonomically diverse group united

Origin and Evolution of Viruses
ISBN 978-0-12-374153-0

by their transmission cycles and maintenance mechanisms (Karabatsos, 1985; Calisher and Karabatsos, 1988). Arboviruses are usually defined as viruses that are transmitted biologically among vertebrate hosts by hematophagous arthropod vectors. Biological transmission requires virus replication in the arthropod vector prior to transmission, in contrast to mechanical transmission that occurs through contamination of vector mouthparts with a virus, followed by inoculation of the virus during a subsequent host contact (i.e. a "flying pin"). Biological transmission can include both vertical and horizontal modes (Figure 16.1). Vertical transmission involves the passing of the virus from a female vector to her offspring, both males and females. Horizontal transmission includes both venereal transmission, from a vertically infected male directly to a female vector, as well as oral transmission between an arthropod vector and a vertebrate host during blood feeding. During this latter process, the virus replicates in the salivary glands of the vector and enters the saliva. Salivation is required to modulate hemostasis and allow for vector engorgement, thus the virus is delivered to the vertebrate host during blood feeding. Subsequent replication within the vertebrate host generates viremia (virus in the bloodstream), which can lead to vector infection if a susceptible vector takes up an infectious bloodmeal. While

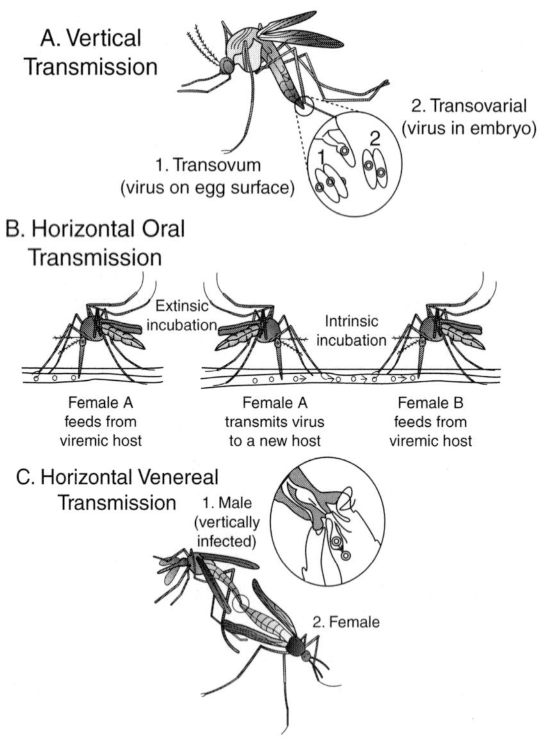

FIGURE 16.1 Transmission of arboviruses by mosquitoes. (A) Vertical transmission occurs when the female mosquito passes the virus to its progeny. Both male and female mosquitoes can be infected via vertical transmission. (B) Horizontal transmission occurs following the infection of a vertebrate host by the bite of an infectious mosquito. After intrinsic incubation during which time the host becomes viremic, oral infection of susceptible mosquitoes occurs following ingestion of viremic blood. (C) Venereal transmission is a form of horizontal transmission that occurs when a vertically infected male copulates with a female and transmits the virus directly to her with no involvement of a vertebrate. From Weaver *et al.* (2004a), with permission.

some arboviruses can be maintained in their arthropod hosts via transovarial transmission for substantial periods, and a few generate persistent infection of vertebrates, the vast majority, if not all, require some degree of horizontal transmission between vertebrate hosts and arthropod vectors in order to be maintained in nature. This fundamental requirement for alternating replication in taxonomically disparate hosts distinguishes arboviruses from other animal viruses, most of which specialize on groups of closely related species.

The need to replicate in the very different environments found in invertebrate and vertebrate hosts undoubtedly poses an evolutionary challenge for arboviruses, and may, as detailed below, account for the limitation of their taxonomic distribution to a few families of RNA viruses. However vector transmission also offers substantial benefits, including the high mobility of flying arthropods, protection from the external environment and freedom from the requirement of being shed into various host secretions. The wide geographic distribution of arboviruses, which are found on all seven continents as well as marine and freshwater environments, and their great epidemiological significance for the health of humans, livestock, and wild animals are a testament to the success of the arbovirus lifestyle.

Origins and Systematics of Arboviruses

Arthropod-borne transmission cycles probably arose several times, as suggested by the inclusion of arboviruses in several different families of RNA viruses, all of which also include members that do not rely on arthropod transmission. These include the alphaviruses (genus *Alphavirus*, one of two genera in the family *Togaviridae*); the flaviviruses (genus *Flavivirus*, one of three genera in the family *Flaviviridae*); the bunyaviruses, nairoviruses, and phleboviruses (three of five genera in the family *Bunyaviridae*); the orbiviruses (one of nine genera in the family *Reoviridae*); the vesiculoviruses (one of six genera in the family *Rhabdoviridae*), and the thogotoviruses (one of four genera in the family *Orthomyxoviridae*). These groups of RNA viruses have a variety of types of RNA genomes and replication strategies. The alphaviruses and flaviviruses have positive- or messenger-sense, single-stranded RNA genomes of about 11–12 kb, but their replication strategies differ. Alphaviruses encode two open reading frames (ORFs), the smaller of which is located at the 3′ end of the genome and is translated from a subgenomic message that results in the expression of large amounts of structural proteins (Strauss and Strauss, 1994; Schlesinger and Schlesinger, 2001; Kuhn, 2007). In contrast, the flavivirus genomes encode single ORFs with the structural protein genes at the 5′ end. Structural similarity between the E1 envelope glycoprotein of alphaviruses and the envelope protein of flaviviruses suggests that these two families are distantly related (Lescar *et al.*, 2001). The bunyaviruses have three-segmented, negative-sense, single-stranded RNA genomes. The orbiviruses (*Reoviridae: Orbivirus*) have ten segments of double-stranded RNA (Mertens *et al.*, 2005), while the thogotoviruses (*Orthomyxoviridae: Thogotovirus*) contain six segments of linear, negative-sense RNA (Kawaoka *et al.*, 2005). The vesiculoviruses (*Rhabdoviridae: Vesiculovirus*) have single-stranded, negative-sense genomes (Tordo *et al.*, 2005). The only DNA arboviruses known is *African swine fever virus* (*Asfarviridae: Asfarvirus*) (Karabatsos, 1985; Calisher and Karabatsos, 1988; van Regenmortel *et al.*, 2000); this lack of DNA arboviruses suggests that the greater genetic plasticity and higher mutation rates exhibited by RNA viruses (Holland and Domingo, 1998) (see also Chapter 6) facilitate their ability to enter a cycle of alternating replication in disparate vertebrate and invertebrate hosts.

Arboviruses have evolved to utilize species in four orders of blood-feeding insects, *Diptera*, *Anoplura*, *Siphonaptera*, and *Hemiptera*, as well as the two families of ticks, *Ixodidae* and *Argasidae*, as vectors. (Kuno and Chang, 2005). As discussed in various examples below, these vector species differ substantially in many features that impact arbovirus evolution, including host range, period of attachment, lifespan, and competence for horizontal and vertical transmission. Moreover, arboviruses infect all four of the terrestrial or semiterrestrial classes of vetebrates: Amphibia, Reptilia, Mammalia, and Aves. The host range of arboviruses varies tremendously; at the extremes, one flavivirus, West Nile virus (WNV), infects 43 species of potential vectors in North America alone, as well as members of all four classes of vertebrate hosts (Granwehr *et al.*, 2004). However, another arbovirus, endemic dengue virus, uses only primates as its reservoir hosts and a small group of mosquitoes in the genus *Aedes* as vectors. Variation in the host utilization and host breadth of arboviruses, coupled with variation in the host breadth of their vectors, shapes both their evolutionary patterns and the tendency for emergence.

INFERENCES ON ARBOVIRUS EVOLUTION FROM PHYLOGENETIC STUDIES

Phylogenetic studies have formed the cornerstone for much of our understanding of arbovirus evolution. Here, we review inferences from phylogenetic analyses of the most extensively studied of the major arbovirus groups, the alphaviruses and the flaviviruses.

Alphaviruses

The *Togaviridae* is the only virus family comprised almost exclusively of arboviruses. Aside from rubella virus (the sole member of the genus *Rubivirus*) and two alphaviruses with no vector transmission yet established (southern elephant seal virus (La Linn *et al.*, 2001) and salmon pancreas disease virus (Weston *et al.*, 1999)), all togaviruses in the genus *Alphavirus* are mosquito-borne (Weaver *et al.*, 2005). In humans and domestic animals, alphaviruses cause a spectrum of disease ranging from inapparent to highly pathogenic syndromes including arthralgia accompanied by rash (most Old World alphaviruses) that can persist for months, and severe, often fatal encephalitis (several New World alphaviruses) (Tsai *et al.*, 2002; Griffin, 2007). The most important causes of severe morbidity and mortality include the New World members Venezuelan (VEEV), eastern (EEEV), and western equine encephalitis virus (WEEV), etiologic agents of encephalitis in humans and equids, and Old World alphaviruses that cause a severe but self-limiting arthralgia and rash syndrome, including Ross River, Chikungunya (CHIKV), O'nyong-nyong, and certain variants of Sindbis virus. Epidemiological studies suggest that the former three viruses can use humans as amplification hosts (those that generate enough viremia and have appropriate vector contacts to increase levels of circulation) during some outbreaks; otherwise, the alphaviruses generally use birds or small mammals as reservoir and amplification hosts, with humans and domestic animals representing dead-end infections (inadequate viremia to infect mosquito vectors and propagate the transmission cycle). A notable exception is VEEV, which exploits equids as highly efficient amplification hosts, resulting in explosive and widespread epidemics (Weaver *et al.*, 2004b). Outbreaks of Venezuelan equine encephalitis appear to involve adaptation of equine-avirulent, sylvatic enzootic strains for equine replication, mediated by small numbers of envelope glycoprotein gene mutations (Anishchenko *et al.*, 2006) (Figure 16.2). Adaptation to new mosquito vectors, also involving envelope glycoprotein amino acid changes, also appears to augment some but not all outbreaks (Brault *et al.*, 2002b, 2004b; Ortiz and Weaver, 2004).

Comprehensive phylogenetic analyses of the genus *Alphavirus* have been used to elucidate patterns of evolution and epidemiology (Powers *et al.*, 2001). Although rubella virus is closely related to the alphaviruses based on genome organization and functions of the major proteins, sequence divergence is

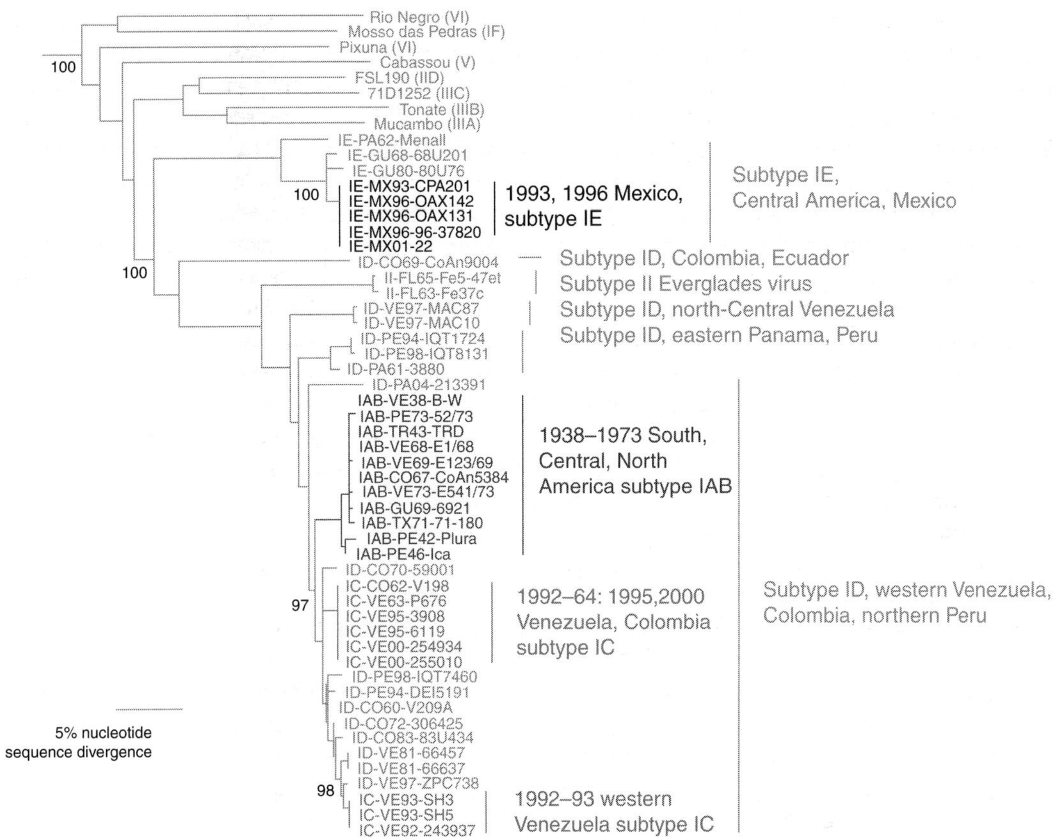

FIGURE 16.2 Phylogenetic trees of VEE complex alphaviruses derived from structural polyprotein amino acid sequences using the neighbor-joining program. Green branches and viruses represent enzootic strains that circulate in forest or swamp habitats and are generally avirulent for equids and incapable of using them as amplification hosts. Green labels on right show six major lineages of VEEV, only two of which have given rise to epidemic strains. Red and orange strains of VEEV are epidemic strains that produce high titer viremia in equids to increase amplification and spillover to humans. The blue Mexican strains are enzootic in the sense that they do not produce adequate viremia in equids for efficient amplification, but epizootic in the sense that they are neurovirulent and cause equine encephalitis. Similar topologies were produced using maximum parsimony and Bayesian methods. Bootstrap values indicate support for clades to the right. The tree was rooted using homologous Eastern equine encephalitis virus sequences. (See Plate 23 for the color version of this figure.)

extensive and homology cannot be demonstrated statistically, aside from conserved motifs in the non-structural proteins. Sequence analyses also have demonstrated homology among the non-structural proteins of alphaviruses and those of several plant virus groups with dissimilar genome organizations, indicating a process of modular evolution leading to these groups (Strauss and Strauss, 1994). Homology was identified between the envelope glycoprotein of flaviviruses and the E1 glycoprotein of alphaviruses, based on the major folds revealed by crystal structures at atomic resolution (Lescar *et al.*, 2001).

The alphaviruses with no known vectors are the most divergent members of the genus and, although rooted trees are inappropriate due to the lack of a closely related outgroup for the alphaviruses, probably represent a basal clade (Figure 16.3). The distribution of these fish and seal viruses in both hemispheres provides no clues on ancestral distributions to estimate the geographic origin of the mosquito-borne members of the genus.

The systematics of the alphaviruses were first defined based on serocomplexes identified by antigenic crossreactivity (Calisher and Karabatsos, 1988); these antigenic complexes still generally correspond to clades defined by phylogenetic studies (Figure 16.3) including the Old World Semliki Forest and New World VEE and EEE complexes. The WEE complex represents a geographically and pathologically diverse group including the new World WEEV, Ft. Morgan (FMV) and highlands J viruses (HJV), some of which cause equine and/or human encephalitis, and the Sindbis-like viruses including Whataroa and Sindbis (SINV) from the Old World, and Aura from the New World. Some of these viruses can cause a human arthralgia syndrome. The dichotomy in disease syndromes and distribution of the WEE complex viruses is reconciled by an ancient recombination event between a

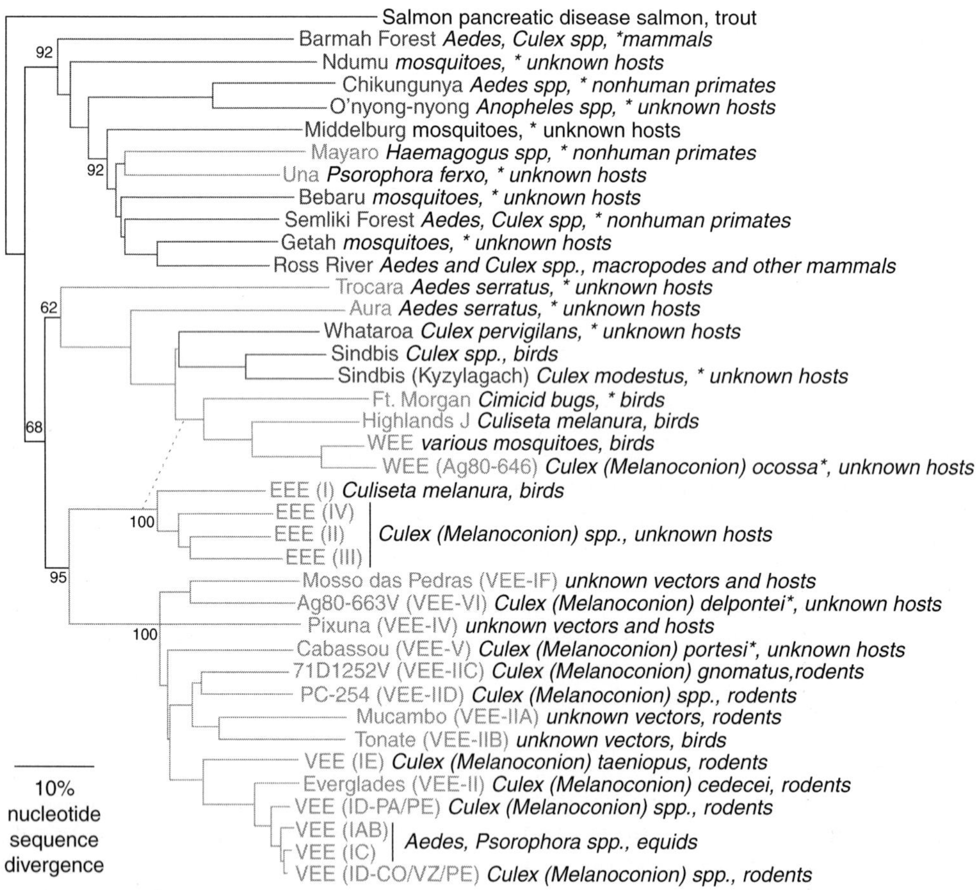

FIGURE 16.3 Phylogenetic tree of all species and major lineages of alphaviruses derived from E1 envelope glycoprotein sequences. Subtypes are written in parentheses after virus names. Reservoir hosts and vectors are listed after viruses. New World viruses are printed in green, Old World viruses in red. Dashed line represents the recombination event that led to the ancestor of WEE, Highlands J and Ft. Morgan viruses. The tree was drawn using the neighbor-joining program with the HKY distance formula using PAUP 4.0. Similar topologies were produced using maximum parsimony and Bayesian methods. Bootstrap values indicate support for clades to the right. (See Plate 24 for the color version of this figure.)

SINV-like virus and an EEEV-like ancestor, leading to the progenitor of the WEE–HJV–FMV group. This recombination event was followed by introduction of a descendant of the SINV recombination ancestor into the Old World (Figure 16.3; see below) (Hahn *et al.*, 1988; Weaver *et al.*, 1997).

Alphavirus Host Utilization Patterns

Examination of host relationships in the alphavirus tree (Figure 16.3) also suggests patterns of host switching (cross-species transfers) and a lack of co-speciation of the viruses with their hosts and vectors (Weaver, 2006). Vector species and genera vary widely within alphavirus clades, serocomplexes and even species, with only a few exceptions: (1) the VEE complex alphaviruses appear to use exclusively members of the Spissipes section (a group of only 23 species), within the subgenus *Culex* (*Melanoconion*), as enzootic vectors. However, some if not all relationships among the vector species (Navarro and Weaver, 2004) are discordant with virus relationships, indicating a lack of co-speciation or co-evolution; (2) with the exception of the North American strains of EEEV, all of the EEEV and VEE complex lineages also appear to use these *Culex* (*Melanoconion*) vectors, suggesting that they were the ancestral vectors in the New World. Either genetic or ecological constraints may limit changes in the vector host range to closely related, consubspecific mosquitoes. The almost complete restriction of alphavirus vectors to the mosquito family (Culicidae), including lack of evidence for an important role of ticks or other non-insect arthropods that are vectors of several other arbovirus taxa, suggests similar constraints for the alphaviruses as a whole.

Alphaviruses use a wide variety of mammalian and avian vertebrate hosts for their maintenance and amplification (Figure 16.3). In contrast to their relationships with vectors, where a given alphavirus typically uses one or a few mosquito species as primary vectors, individual alphavirus species and lineages appear to be more flexible or opportunistic in their vertebrate usage and may exploit several different species or even higher taxa simultaneously. For example, in North America, EEEV infects a variety of passeriform birds in enzootic swamp foci, many of which generate viremia sufficient for horizontal transmission by the highly susceptible and ornithophilic enzootic vector, *Culiseta melanura* (Scott and Weaver, 1989; Weaver, 2001). Although an important role in maintenance has not been established for many other vertebrate groups, alphaviruses like EEEV infect an extremely diverse group of hosts including birds, mammals, amphibians, and reptiles as indicated by seroprevalence. The wider vertebrate host range of the alphaviruses compared to the range of their arthropod vectors suggests greater potential for reservoir than vector host switching during the course of evolution and disease emergence. Studies described below have begun to test experimentally this and related hypotheses (see below).

The uniformity in vector taxa (mosquitoes) used by the alphaviruses is also observed in other arbovirus taxa, and contrasts with the wide range of vertebrate hosts used as reservoirs and amplification hosts, typically including both birds and mammals. This contrasting pattern of vector vs. vertebrate reservoir host usage suggests that adaptation to different vectors, such as other biting flies or ticks, is genetically difficult, and/or that arboviruses have evolved as generalists for their vertebrate hosts but specialists with respect to their vectors. However, the specificity for vectors is often manifested only at the level of initial midgut infection; most alphaviruses replicate in most mosquitoes following intrathoracic inoculation, which bypasses the more selective midgut. Better understanding of the interactions between host factors and arboviruses during infection and replication is needed to understand these differences in vertebrate and vector host specificity.

The taxa used as reservoir hosts appear to influence strongly the genetic structure of alphavirus populations (Mackenzie *et al.*, 1995; Weaver, 1995). Those viruses that use avian hosts, such as EEEV (Brault *et al.*, 1999), HJV (Cilnis *et al.*, 1996), WEEV (Weaver *et al.*, 1997), Barmah Forest virus (Poidinger *et al.*, 1997), and SINV (Norder *et al.*, 1996; Sammels *et al.*, 1999), appear to evolve within a small number of broadly distributed lineages, presumably reflecting efficient dispersal by birds (which generally have much longer flight ranges than mosquitoes). In the case of SINV (Mackenzie *et al.*, 1995) and EEEV (Weaver *et al.*, 1993, 1994), selective sweeps may occur periodically in Australia and North America, respectively, resulting in maintenance over time of these viruses as small numbers of broadly distributed lineages. In contrast, the alphaviruses that use mammals with limited dispersal, such as Ross River (Sammels *et al.*, 1995), CHIKV (Powers *et al.*, 2000), and most of the VEE complex viruses (Powers *et al.*, 2001), evolve within a greater number of geographically limited lineages reflecting very limited dispersal ability. Presumably, efficient dispersal acts to constrain lineage diversity by resulting in frequent mixing of populations and elimination of less fit populations via competition. However, the relatively low infection rates in mosquito vector populations in nature (typically ≤1:100) and identifications of mixed populations of two or more arbovirus lineages with hosts at the same time and place (Weaver *et al.*, 1993; Yanoviak *et al.*, 2005; Charrel *et al.*, 2007b; Kondig *et al.*, 2007) indicate that host resources may not be limiting enough to enforce competitive exclusion in many situations.

Alphavirus Sequence Evolution

Studies on the nature and rates of nucleotide sequence evolution in alphaviruses have yielded estimates that fall below many estimated for single host taxon, non-arthropod-borne RNA viruses (Weaver *et al.*, 1992b). Analyses of sequence change obtained from phylogenies emphasize the preponderance of synonymous substitutions ($d_N/d_S \ll 1$), suggest that strong purifying dominates alphavirus evolution. This presumably represents long term evolutionary relationships of many alphaviruses with their vectors and hosts such that near-optimal fitness peaks have been reached and most amino acid changes in alphavirus proteins have deleterious effects. Even the 26S subgenomic promoter sequence that is conserved but includes a few differences among alphaviruses shows no evidence of adaptive substitutions during its evolution, and most of the promoter sequences are interchangeable between SINV and other species (Hertz and Huang, 1992). Studies of genetic diversity within alphavirus populations indicate quasispecies distributions of genetic variants similar to that exhibited by single host RNA viruses (Weaver *et al.*, 1993), suggesting that mutation frequencies are not the explanation for the genetic stability observed in nature. The requirement for alternate replication in disparate hosts is a possible explanation for this phenomenon and for the slow rate of sequence change (see below).

Flaviviruses

The genus *Flavivirus* comprises a highly diverse group of both vector-borne and non-vector-borne viruses distributed nearly worldwide (Gould *et al.*, 2003). Included in this taxon are important causes of human encephalitis such as Japanese (JEV) and tick-borne encephalitis viruses (TBEV), as well as agents of hemorrhagic fever including yellow fever virus (YFV), which is among the most virulent human pathogens and remains an important cause of mortality in Africa and South America, and dengue virus (DENV), the leading arboviral cause of morbidity and mortality in most tropical and subtropical locations. In addition to their overall greater diversity compared to the alphaviruses, the flaviviruses exhibit a wider range of transmission cycles and vectors; some flaviviruses that infect vertebrates have no known vector and are probably transmitted directly, and large monophyletic groups use either mosquitoes or ticks as vectors (Gaunt *et al.*, 2001) (Figure 16.4). Clades of both the non-vectored and tick-vectored flaviviruses tend to specialize on rodents, bats, or seabirds with little evidence of frequent cross-taxa transfers among vertebrate hosts or to or from the use of vectors. Another group includes viruses that have been isolated only from mosquitoes and that are not known to infect vertebrates in nature (Cammisa-Parks *et al.*, 1992; Crabtree *et al.*, 2003). These viruses appear to specialize on the use of *Culex* or *Aedes* vectors, again with little evidence of cross-genus transfers in vector usage (Figure 16.4). The flaviviruses also appear to have been transferred between the eastern and western hemispheres much more frequently during their evolution than the alphaviruses (Figures 16.3 and 16.4). DNA sequences comprising about two-thirds of a flavivirus-like genome have been identified in an *Ae. aegypti* cell line, and a 492-amino-acid ORF related to the flavivirus polymerase is integrated into the genomic DNA of laboratory-bred and wild *Ae. albopictus* and *Ae. aegypti* mosquitoes (Crochu *et al.*, 2004). These findings suggest the possible interchange between viral RNA and vector DNA genomes, although it is unclear whether the vector sequences are anything more than irrelevant relics of a past integration event.

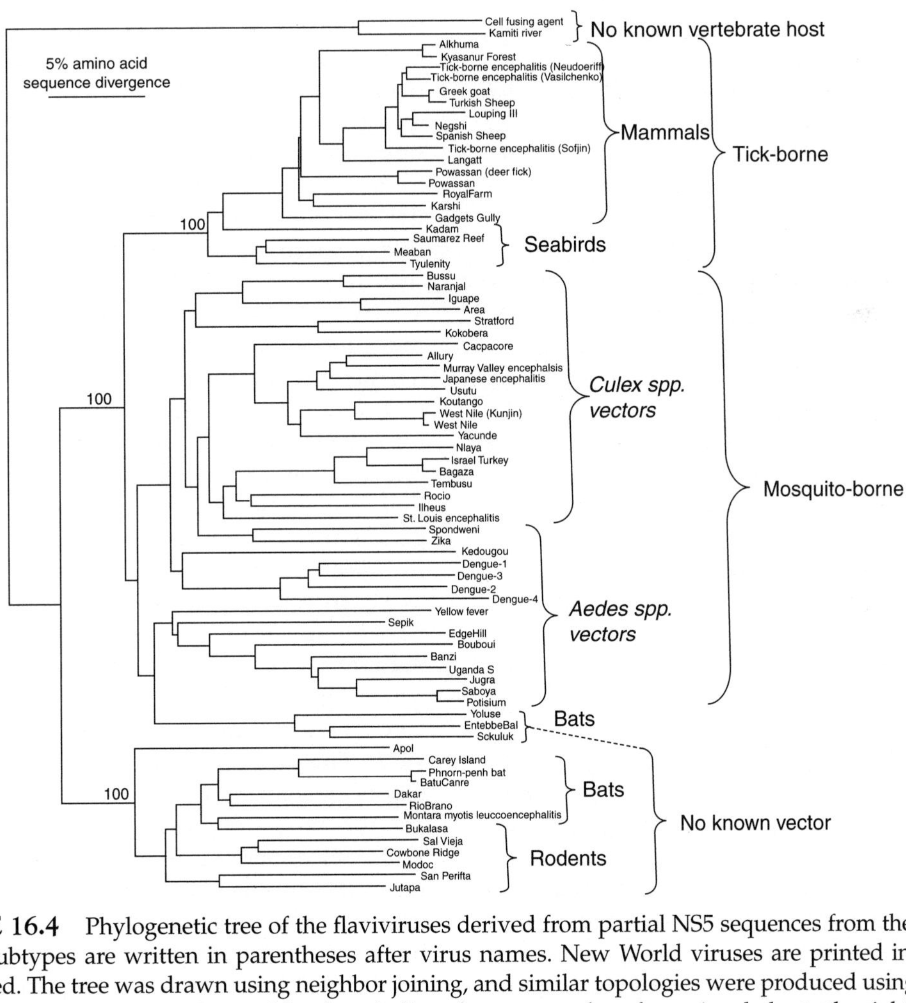

FIGURE 16.4 Phylogenetic tree of the flaviviruses derived from partial NS5 sequences from the GenBank library. Subtypes are written in parentheses after virus names. New World viruses are printed in bold and underlined. The tree was drawn using neighbor joining, and similar topologies were produced using Bayesian methods and maximum parsimony. Numbers indicate bootstrap values for major clades to the right.

Relationships within the *Flavivirus* Genus

The flaviviruses comprise one of three genera in the family *Flaviviridae*. The other genera, *Pestivirus* and *Hepacivirus*, are non-vector-borne animal viruses. Within the genus *Flavivirus*, four major clades include non-vector-borne, tick-borne, and two mosquito-borne groups. Like the alphaviruses, the flaviviruses show a nearly worldwide distribution, excepting Antarctica. They also infect a wide range of vertebrates and arthropod vectors, including ticks. Phylogenetic studies have identified interesting differences in evolutionary patterns among the four flavivirus groups mentioned above. Greater genetic conservation in the tick-borne than mosquito-borne groups has suggested that different selective constraints operate during the evolution of these two groups (Shiu *et al.*, 1991). The tick-borne viruses appear to have evolved in a progressive, clinal pattern from east to west across Asia and Europe (Zanotto *et al.*, 1995), while the mosquito-borne flaviviruses have evolved in a more discontinuous manner, probably in several different regions of the world. Tick-borne flaviviruses comprise three major groups (mammalian, seabird, and Kadam tick-borne flavivirus groups), suggesting a complex relationship between flaviviruses infecting birds and those infecting mammals. Ticks that feed on both birds and mammals may constitute the evolutionary bridge between the three distinct tick-borne flavivirus lineages (Grard *et al.*, 2007).

The mosquito-borne flaviviruses tend to exhibit relatively long time periods between lineage divergence, suggesting a "boom and bust" pattern of intense diversification followed by a "pruning" of the tree via extinction of many lineages (Zanotto *et al.*, 1996). The best examples of this pattern are the four serotypes of DENV, which appear to be undergoing a period of rapid radiation (Holmes and Twiddy, 2003). Detailed maximum likelihood analyses of DENV isolates to analyze rates of synonymous vs. non-synonymous substitution suggest that different genotypes or lineages experience different selective pressures. While most codons show the signature of purifying selection, a few amino acids implicated in virulence and transmissibility show evidence of positive selection (Twiddy *et al.*, 2002a). Amino acid positions subject to weak, positive selection were also identified in the envelope glycoprotein of some but not all DENV serotypes. The majority of these amino acid sites are located in, or near to, putative T- or B-cell epitopes in the envelope glycoprotein, suggesting immune selection, as well as in the NS2B and NS5 genes of DENV-2 (Twiddy *et al.*, 2002b). Cross-sectional phylogenetic analysis of DENV isolates indicate that clade replacement events are linked to the epidemic cycle and suggest an effect of interserotypic immunity. Modeling has shown that moderate cross-protective immunity can lead to persistent out-of-phase oscillations in DEN incidence similar to those observed in nature. Epidemic patterns observed in holoendemic locations such as Bangkok may therefore be the result of cross-immunity (Adams *et al.*, 2006). These kinds of studies implying positive selection should be corroborated with reverse genetic validation of fitness effects in mosquito vectors or surrogate model systems for human infection.

The origins of DENV have been discussed for decades. Gubler (1997) hypothesized that endemic DENV evolved from sylvatic strains that utilize non-human primate hosts and gallery forest *Aedes* vectors (not *Ae. aegypti* or *Ae. albopictus*). The sylvatic cycle is believed to be ancestral because efficient interhuman transmission is thought to require a minimum human population size of 10000–1 million, which did not exist until about 4000 years ago when urban civilizations arose (Gubler, 1997). This led to the hypothesis that endemic DENV evolved from sylvatic DENV in Africa and/or Asia. Wang *et al.* (2000) provided the first phylogenetic support for this hypothesis, when they sequenced DENV-1, -2, and -4 isolated in tropical forest habitats in Malaysia by Rudnick *et al.* during the 1960s and 1970s (Rudnick, 1965; Rudnick *et al.*, 1967), as well as DENV-2 from sylvatic locations in West

Africa analyzed previously (Rico-Hesse, 1990). As in previous studies using other genome regions (Rico-Hesse, 1990), the African sylvatic DENV-2 were quite distinct from all endemic/epidemic isolates, differing by 19% in nucleotide sequences (Figure 16.5). The Malaysian sylvatic isolates of DENV-1, -2, and -4 were also distinct from both endemic isolates and African sylvatic strains; the Malaysian DENV-2 isolates differ from epidemic strains by 17%, while the DENV-1 and -4 Malaysian sylvatic strains differ from their endemic counterparts by about 7% and 14%, respectively.

Although sylvatic DENV-3 has not been isolated in Malaysia, the presence of DENV-3 antibodies in non-human, canopy-dwelling primates indicated that a sylvatic DENV-3 cycle also exists there. Using various methods and estimates of nucleotide substitution rates, it has been estimated that the endemic DENV-2 genotypes diverged from the sylvatic forms about 40–600 years ago (Vasilakis *et al.*, 2007a). The African and Malaysian sylvatic lineages diverged slightly later, as did the sylvatic and endemic DENV-1 and -4 lineages. The greater diversity of sylvatic DENV in Malaysia compared to Africa suggests that an ancestor of all DENV arose in the Asian or Oceanic region (although the exact location cannot be determined) and diverged in the relatively distant past into the four serotypes recognized today. The four independent endemic DENV evolution events, presumably involving switching from canopy-dwelling *Aedes* mosquitoes to *Ae. albopictus* and later to *Ae. aegypti*, suggest that adaptation to new vectors and vertebrate

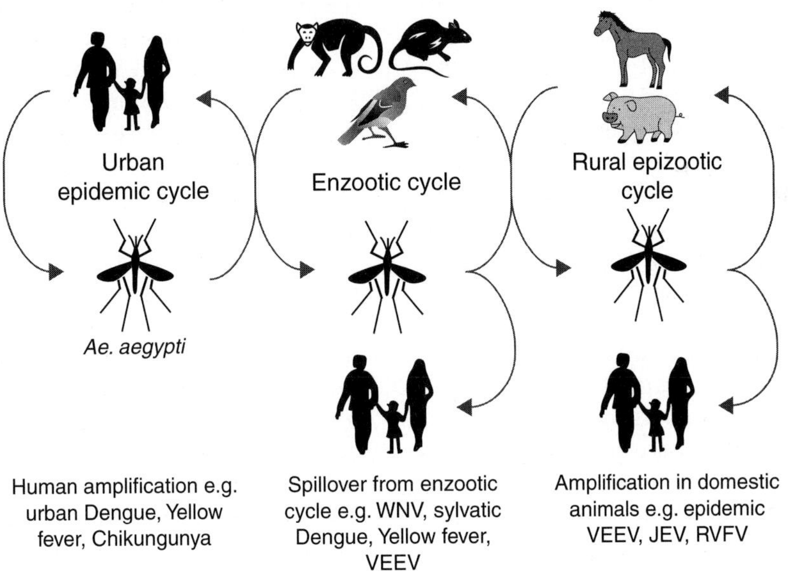

FIGURE 16.5 Cartoon showing mechanisms of human infection by zoonotic arboviruses. At the center is a typical enzootic cycle involving avian, rodent, or non-human primates as reservoir and/or amplification hosts and mosquito vectors. Humans become infected via direct spillover when they enter enzootic habitats and/or when amplification results in high levels of circulation in their proximity. Transmission to humans may involve the enzootic vector or bridge vectors with broader host preferences including humans. At right, secondary amplification involving domestic animals can increase circulation around humans, increasing their chance of infection via spillover. Examples include Rift Valley fever, Japanese and Venezuelan equine encephalitis virus (VEEV). In the case of VEEV, mutations that enhance equine viremia mediate secondary equine amplification. At left, dengue, yellow fever and Chikungunya viruses can use humans directly for amplification, resulting in urban epidemic cycles and massive outbreaks. In the case of dengue viruses, humans also serve as reservoir hosts. From Weaver (2005), with permission.

hosts occurred repeatedly. Because *Ae. aegypti* did not occur in Asia and Oceania at that time, *Ae. albopictus* was probably the original human vector. Acquisition of *Ae. aegypti* as a vector may have occurred only during the past few centuries as commerce distributed this mosquito throughout the tropics (Gubler, 1997). Phylogenies suggest that the rate of evolutionary change and patterns of natural selection do not differ appreciably between the endemic and sylvatic DENV lineages (Vasilakis *et al.*, 2007a).

As for the alphaviruses, the mobility of the reservoir hosts appears to have a strong influence on the population structure and evolution of flaviviruses. Those that use birds as reservoir hosts, like Japanese (Solomon *et al.*, 2003), St. Louis encephalitis (Kramer and Chandler, 2001), and Murray Valley encephalitis viruses (Lobigs *et al.*, 1988), as well as West Nile virus (WNV) (Davis *et al.*, 2005) evolved within broadly distributed lineages that exhibit genetic stability (Mackenzie *et al.*, 1995). After its introduction into North America, WNV diverged into two (NY99 and WN02) genotypes; the rate of epidemiological growth of the WN02 population was estimated at three times that of the NY99 population, leading to its dominance (Snapinn *et al.*, 2007). Recently, there has also been a decrease in the growth rate of the WN02 lineage, suggesting that WNV has reached its peak prevalence in North America.

Kilpatrick *et al.* suggested that heterogeneity in the viremia responses and attraction of mosquitoes among avian host populations affects WNV evolution when a few hosts (super spreaders) give rise to the majority of secondary infections. Models indicate that a single relatively uncommon bird, the American robin (*Turdus migratorius*), may be responsible for the majority of infectious mosquitoes (Kilpatrick *et al.*, 2006a). Although they are preferred by mosquitoes at some sites and attract attention due to high rates of mortality, crows may be relatively unimportant in WNV amplification and spread. This suggests that the virulence for crows of the North American WNV strains may not be under much negative selection. Shifts in the feeding preferences of important mosquito vectors from avian WNV hosts to humans and other dead-end mammalian hosts after mosquito abundances climb during the summer may explain the late season peak in WNV encephalitis (Kilpatrick *et al.*, 2006b).

Flaviviruses with mammalian hosts, such as YFV that uses non-human primate reservoir hosts, tend to be partitioned into smaller, geographically delineated populations (Bryant *et al.*, 2003). However, the endemic dengue viruses, which are perhaps the most mobile arboviruses due to the extensive and rapid travel behavior of human reservoir hosts, exhibit complex patterns of evolution within multiple lineages that are frequently introduced into new locations and also appear to undergo local extinctions (Holmes and Twiddy, 2003; Thu *et al.*, 2004). Genetic studies suggest that population shifts and replacements may be selected by adaptive mutations in the DENV non-structural proteins (Bennett *et al.*, 2003) and in cytotoxic T-cell epitopes (Hughes, 2001). Fitness for transmission may be responsible for some of these population changes; evidence from *Ae. aegypti* susceptibility studies suggests that an Asian DENV-2 genotype that recently colonized the New World is more infectious than the American genotype it is replacing in some locations (Armstrong and Rico-Hesse, 2003; Cologna *et al.*, 2005; Anderson and Rico-Hesse, 2006). This change in the distribution of DENV genotypes has critical public health implications because the Asian genotype is more likely to cause hemorrhagic disease (Watts *et al.*, 1999).

Analyses of flavivirus phylogenies also indicate considerable plasticity in their relationships with vertebrate hosts, and less plasticity in vector usage. Although tick- and mosquito-borne flaviviruses are occasionally isolated from mosquitoes and ticks, respectively, it appears that their principal vectors are very stable taxonomically within these groups (Gould *et al.*, 2003). Even within the mosquito-borne clades, generic vector relations are relatively stable, with the hemorrhagic viruses mainly using *Aedes* spp. and the encephalitic members relying principally on *Culex* spp. (Figure 16.4).

Of particular interest in flavivirus evolution is the presence of a large group of animal viruses with no known arthropod vectors (Figure 16.4). This group appears to have diverged early during the evolution of the flavivirus genus and may represent the ancestral phenotype. Another smaller group of bat viruses comprised of Yokose, Entebbe bat and Sokoluk viruses appears to have lost the need for vector transmission secondarily (Gould *et al.*, 2003). These non-vector-borne flaviviruses represent an ideal system to study the effect of vector transmission on arbovirus evolution, as well as the genetic determinants of arthropod transmission, because they share basic replication strategies and genetics with the vector-borne members of the genus.

RECOMBINATION AND REASSORTMENT IN ARBOVIRUS EVOLUTION

Recombination (the exchange of sequences within a genomic segment of a virus) and/or reassortment (the exchange of RNA segments between viruses with segmented genomes) have been detected in most major groups of arboviruses. In general, RNA viruses with positive sense genomes undergo more frequent recombination than those with negative-sense genomes, probably due to RNA interactions with viral proteins during replication of the latter. The advantages of recombination and reassortment in viral evolution are discussed in greater detail elsewhere (see Chapters 4 and 13), but probably include the ability to sample a greater number of viral genotypes than would be permitted by point mutations alone, as well as the ability to recover from the accumulation of deleterious point mutations that can accompany population bottlenecks of RNA viruses, including arboviruses (Duarte *et al.*, 1992; Weaver *et al.*, 1999).

Recombination

Retrospective evidence of recombination in the deep evolutionary history of arboviruses came from early sequencing studies that demonstrated homology among protein sequences involved in the replication of RNA viruses within different families. These included, for example, relationships of alphavirus nonstructural proteins to those of several plant viruses. These ancient relationships among sequences from RNA virus genomes with completely different genetic organizations, including some with segmented genomes, presumably reflect ancient recombination events that shaped a process termed modular evolution (Strauss and Strauss, 1994).

More recent recombination events have also been detected within several arbovirus groups. One of the first recombinant RNA viruses discovered is western equine encephalitis virus, which is derived from a recombination event between ancestors of SINV and EEEV (Hahn *et al.*, 1988). Further studies revealed that the recombinant ancestor gave rise to a group of closely related alphaviruses that includes Highlands J and Fort Morgan viruses (Figure 16.3). Within the alphavirus genus, recombination is limited to the WEE complex, but the possibility of recombinants between more closely related viruses or strains has received little attention.

The most likely venue for an alphavirus recombination event is difficult to predict; both mosquitoes and vertebrate hosts exhibit superinfection exclusion of sequential infection by closely related alphaviruses (Karpf *et al.*, 1997). However, exclusion is not immediate, so sequential infection of a vertebrate within a few hours by multiple mosquito bites, or sequential infection of a mosquito via multiple, partial blood meals from two different viremic hosts could result in a dual infection.

As in the alphaviruses, the biology of DENV transmission creates opportunities for recombination among serotypes and genotypes. The principal DENV vector, *Aedes aegypti*, shows a propensity take multiple, partial blood meals from several different human hosts and to rely on blood as a carbohydrate nutritional source, rather than on plant nectars like most other mosquitoes (Harrington *et al.*, 2001). Multiple feeding may increase the chances of dual infections in both mosquitoes (from biting more than one viremic human during a short time period) and in humans (from receiving multiple *Ae. aegypti* bites during a short time period due to this vector's endophilic resting and feeding behavior, and its peri-domestic larval habitats. In humans DENV reach titers as high as 10^8 infectious units/mL of blood, suggesting that when individuals are infected with more than one virus, the double infection of cells required for recombination may be common (Murphy *et al.*, 2004). Infection of individual vectors and hosts with multiple DENV serotypes and genotypes has been documented (Gubler *et al.*, 1965; Lorono-Pino *et al.*, 1999; Craig *et al.*, 2003; Wang *et al.*, 2003) and may be common in some outbreaks (Thavara *et al.*, 2006). It is striking, then, that despite apparent opportunities for recombination, no natural inter-serotypic recombinants of DENV have ever been detected (Murphy *et al.*, 2004).

The low fitness of engineered inter-serotypic chimeric viruses may help to account for the absence of natural recombinants (Murphy *et al.*, 2004). Certainly there is no evidence of recombination between different flaviviruses comparable to the origins of the alphavirus WEEV as described above. Similarly Barrett and Monath have proposed that recombination in YFV is, at most, a rare event (Barrett and Monath, 2003). Evidence of intra-serotypic recombination from sequence and phylogenetic studies of DENV has been reported (Holmes and Twiddy, 2003; Craig *et al.*, 2003). However at least one of these putative recombinants was later shown to be an artifact of accidental mixing of sequence files (de Silva and Messer, 2004). Given the current debate over the potential threat that recombination poses to the safe deployment of live-attenuated flavivirus vaccines (discussed below), studies are urgently needed to evaluate the potential role of recombination in DENV evolution. Retrospective evidence of recombination in nature should include analyses of plaque-purified virus populations (to avoid the potential of analyzing a mixed virus population), and PCR amplification and sequencing in single reactions across putative crossover sites.

Reassortment

Reassortment of gene segments has been shown to occur extensively within the family *Bunyaviridae*, and occurs efficiently in dually infected mosquitoes when the two different viruses are ingested within 2 days (Borucki *et al.*, 1999). Reassortant bluetongue viruses can be detected in *Culicoides variipennis* that ingest two different strains within 5 days of each other, while superinfection exclusion prevents reassortment by day 7 (El Hussein *et al.*, 1989). A recombinant *Orthobunyavirus* (family *Bunyaviridae*) was recently characterized from hemorrhagic fever cases during an East African epidemic. This virus, Ngari virus, a reassortant with S and L segments derived from bunyamwera virus and an M segment from an unidentified member of the genus, demonstrates the public health importance of arbovirus reassortment (Gerrard *et al.*, 2004). Several examples of reassortment within the group C complex have also been described (Nunes *et al.*, 2005).

EMERGENCE MECHANISMS OF ZOONOTIC ARBOVIRAL DISEASES

The vast majority of arboviral diseases are zoonotic, with primary, enzootic transmission cycles involving wild animals and with humans and domestic animals representing tangential or dead-end infections that do not influence the long-term evolution of the pathogen. The simplest mechanism of infection is direct "spillover," whereby enzootic transmission in the vicinity of humans or domestic animals, or the epizootic amplification of a virus due to favorable ecological conditions such as large vector populations following rainfall, lead to direct, tangential transmission (Figure 16.4). This can result from a wide host range of the enzootic vector that includes both reservoir hosts and humans or domestic animals, such as transmission of WNV from birds to humans by the principal enzootic vectors in north America, *Culex pipiens* and *Cx. quinque fasciatus* (Turell *et al.*, 2002). However, among arboviruses such as EEEV that utilize vectors with narrow host ranges, such as *Culiseta melanura* that feeds almost exclusively on birds, spillover may be mediated by bridge vectors that bite both birds and humans. The range of vector species infected by an arbovirus and the host preferences of those vectors can therefore have a strong influence on arboviral disease.

Secondary Amplification

The development of domestic animals has provided some arboviruses with the opportunity to undergo secondary amplification and thereby increase levels of circulation and the probability of spillover to humans or domestic animals. A good example of this process is the transmission of JEV, which infects pigs and chickens living in close proximity to humans in many parts of Asia, resulting in local amplification and transmission to humans by mosquitoes that do not necessarily include the avian enzootic reservoir hosts among their preferred blood sources (Endy and Nisalak, 2002). Another is Rift Valley fever virus (RVFV), which is amplified in secondary cycles involving cattle, sheep, and other ungulates (Bouloy, 2001). Indeed the importation of susceptible European cattle into Africa in the early 1900s has been implicated in the initial emergence of RVFV (Bird *et al.*, 2007). The capacity of these viruses to cause human disease is driven by their wide vertebrate host ranges in combination with the broad host ranges of their vectors, as well as farming practices that bring domestic animals and humans into close proximity. There is no evidence that adaptation is required for most of these secondary amplification cycles, i.e. most or all wild-type strains can readily infect these animals. However in the case of JEV, simple demographic changes such as increasing the distance between pig farms and residential areas have been shown to decrease the number of human cases substantially (Chevalier *et al.*, 2004).

A more complex form of secondary amplification mediated by genetic change in the virus is epitomized by VEEV, the most important alphaviral pathogen of the New World. The VEEV strains that undergo sustained, continuous transmission, and long-term evolution are the enzootic variants that circulate primarily in sylvatic or swamp habits, where they utilize rodents as reservoir hosts and specialize almost exclusively on vectors in the *Culex* (*Melanoconion*) subgenus (Weaver *et al.*, 2004b). The reservoir hosts from enzootic regions generate viremia sufficient to infect the mosquito vectors yet generally develop no detectable disease. The enzootic VEEV strains infect people, horses, bovines, and a wide range of other hosts via spillover, with humans suffering severe febrile disease that can be fatal. Horses and bovines living near enzootic habitats become infected but develop little or no disease. The limited dispersal of the *Culex* (*Melanoconion*) vectors generally limits disease resulting from direct spillover to locations close to forest or swamp habitats (Mendez *et al.*, 2001).

The "silent" sylvatic VEEV cycle is occasionally expanded into new habitats when mutations allow the virus to expand its host range and undergo secondary amplification, resulting in explosive equine epizootics and epidemics (Weaver *et al.*, 2004b). The evolution of the equine amplification-competent, epidemic subtype IAB and IC strains leads to large numbers of human cases due to huge amounts of VEEV circulation in agricultural habitats, leading to spillover to humans. These epidemic strains have evolved at least three times, with the IC serotype evolving convergently in the early 1960s and again in 1992 (Figure 16.2). For unknown reasons, only one of the six major enzootic VEEV lineages has generated all of these epidemic strains (Weaver *et al.*, 2004b). The recent emergence of equine-virulent VEEV strains in southern Mexico (Oberste *et al.*, 1998) involved the evolution of a different kind of epizootic phenotype; these subtype IE strains, traditionally viewed as enzootic, apparently acquired an equine-virulent phenotype that is not accompanied by high viremia (Gonzalez-Salazar *et al.*, 2003). This has resulted in small epizootics of disease in equids without the large geographic spread that typically accompanies subtype IAB and IC outbreaks, including the 1969–1971 epidemic that swept through Mexico into Texas (Lord, 1974).

Mutations in the E2 envelope glycoprotein mediate two critical adaptation events involved in VEE emergence: (1) enzootic VEEV strains are selected for the generation of high titer equine viremia, which inadvertently (with respect to selection) results in equine virulence. Recent studies (Anishchenko *et al.*, 2006) indicate that a single point mutation can mediate adaptation for equine viremia; (2) in some cases selection for enhanced infection of mosquito vectors that populate agricultural settings results in enhanced transmission among equine amplification hosts and humans. Adaptation to epizootic mosquito vectors can also involve as little as one mutation in the E2 protein (Brault *et al.*, 2002a, 2004b). The efficiency of VEEV in achieving dramatic host range changes with minor genetic changes epitomizes the natural threats imposed by RNA viruses as emerging pathogens. The dramatic effect of importation of equids to the New World on VEE and of domesticated ungulates into Africa on RVFV emergence (Meegan and Bailey, 1988; Bouloy, 2001) also underscores the ability of arboviruses to exploit anthropogenic changes in unpredictable ways.

Humans as Arboviral Amplification Hosts

Several arboviruses such as Ross River virus, YFV, and CHIKV probably use humans as temporary amplification hosts during epidemic conditions, but there is no evidence that adaptation is involved. In the case of YFV, transfers from non-human primates–sylvatic mosquito cycles into the urban human–*Ae. aegypti* appear to be temporary (Figure 16.5). This form of epidemic emergence once occurred during the summer in many temperate port cities, but now appears to be confined to tropical regions of Africa. Of particular interest is the history of YFV in the New World, where epidemics are described dating back to the seventeenth century (Monath, 1988). Phylogenetic studies have suggested that YFV was introduced into the New World about 300–400 years ago during the slave trade (Wang *et al.*, 1996; Bryant *et al.*, 2007). In addition to causing self-limited epidemics in port cities, YFV established enzootic, sylvatic transmission cycles in many tropical areas of the Americas by utilizing local vectors in the genus *Hemagogus* spp. and neotropical non-human primates. Whether or not this colonization of the New World by YFV required adaptation to these new vectors and reservoir hosts is a question that could provide important predictive insights on the potential for future introductions of arboviruses into new continents or regions.

Chikungunya, YFV, and DENV are also particularly successful at exploiting human amplification because they utilize a highly anthropophilic vector, *Ae. aegypti* (Figure 16.5) (Woodall, 2001). This mosquito itself underwent an evolutionary recent adaptation from the ancestral, sylvatic form found in West Africa, *Ae. aegypti formosus* (Tabachnick and Powell, 1979). The derived form *Ae. aegypti aegypti* now lives in close contact with people un urban settings by relying on artificial water containers for its larval habitats, becoming endophilic to increase contact with people, and relying on blood (instead of plant carbohydrates) for its energetic needs (Harrington *et al.*, 2001). Recently, CHIKV emerged in islands off the eastern coast of Africa to cause major epidemics accompanied by some fatalities (Higgs, 2006; Josseran *et al.*, 2006), and then re-emerged in the Indian subcontinent to cause over one million cases (Kalantri *et al.*, 2006; Charrel *et al.*, 2007a). Transmission on some islands was reported to principally involve *Ae. albopictus* rather then the typical *Ae. aegypti*-borne human cycle, and an amino acid substitution in the E1 envelope glycoprotein that affects cholesterol-dependent fusion of virions has been hypothesized to have mediated this vector-switching event (Schuffenecker *et al.*, 2006).

The four serotypes of DENV are the most important human arboviral pathogens because they are the only viruses to have adapted to use humans as both reservoir and amplification hosts. In many respects DENV are the ultimate human arboviral pathogens. The ancestral forms are sylvatic strains that continue to circulate in sylvatic habitats of West Africa and Asia. These strains utilize sylvatic treehole mosquitoes as vectors and non-human primates as reservoir hosts (Rudnick, 1984; Diallo *et al.*, 2003). Phylogenetic studies indicate that hundreds to thousands of years ago, the four DENV serotypes each underwent ecological and host range changes to establish peri-domestic and later urban transmission cycles (Figures 16.5 and 16.6) (Wang *et al.*, 2000; Holmes and Twiddy, 2003). These endemic and epidemic DENV strains use humans as their sole reservoir hosts, and peri-domestic mosquitoes as vectors, resulting in a huge burden of human dengue disease in the tropics and subtropics (Gubler, 2006). The possible role in endemic emergence of adaptation of sylvatic progenitors to human reservoir hosts has been evaluated using two models for human infection. Neither SCID mice transplanted with human hepatoma cells nor infection of human dendritic cells infected *ex vivo* revealed any consistent differences in replication between endemic and sylvatic strains of DENV-2, arguing against any role for adaptation and suggesting the sylvatic strains can readily re-emerge (Vasilakis *et al.*, 2007b). However, the American genotype of endemic DENV-2 and the sylvatic lineages showed indistinguishable levels of replication that were lower than those of the Asian genotype, suggesting that the lower virulence of the American genotype reflects an ancestral human replication phenotype. Analyses of sequence evolution in the sylvatic lineages also revealed no fundamental differences compared to endemic strains (Vasilakis *et al.*, 2007a). In addition, the partial cross-protectivity exhibited among the four DENV serotypes may have allowed for the co-circulation of closely related DENV strains leading to immune enhancement, which can result in severe hemorrhagic forms of disease (Ferguson *et al.*, 1999).

A more complete understanding of the molecular determinants of host range changes responsible for the emergence of arboviruses like VEEV, DENV, and YFV in the New World, are critical to anticipating future disease trends and designing public health interventions. For example, several candidate DENV vaccines offer the hope of DEN eradication because humans are the only reservoir hosts for the strains circulating in most locations. However, rapid re-emergence of sylvatic DENV would undermine the long-term success of such eradication campaigns. Predicting the ability of sylvatic DENV strains to re-emerge will depend on a more thorough understanding of human pathogenesis and viremia following sylvatic strain infection, characterization of any genetic changes required to adapt to humans and peri-domestic vectors, and the degree and nature of contact between people and the sylvatic cycles of DENV in Asia and West Africa. If contact is frequent and few or no mutations are required for emergence, sylvatic DENV strains in Asia and Africa will represent a readily available source of new urban DENV strains for the foreseeable future.

Prospective studies are also needed to evaluate the ability of other enzootic arboviruses to cross species barriers and initiate endemic transmission cycles involving peri-domestic vectors and human amplification hosts. One example is the flavivirus Zika virus, which circulates enzootically in Africa and perhaps Asia, and has recently caused an epidemic in Micronesia, apparently mediated by *Ae. aegypti* transmission http://www.guampdn.com/apps/pbcs.dll/article?AID = /20070624/NEWS01/706240308/1002). Zika virus has been isolated from humans, non-human primates, and mosquitoes (*Ae. africanus* (Haddow, 1964), *Ae. aegypti* (Marchette *et al.*, 1969)), and epidemics have been documented in Uganda,

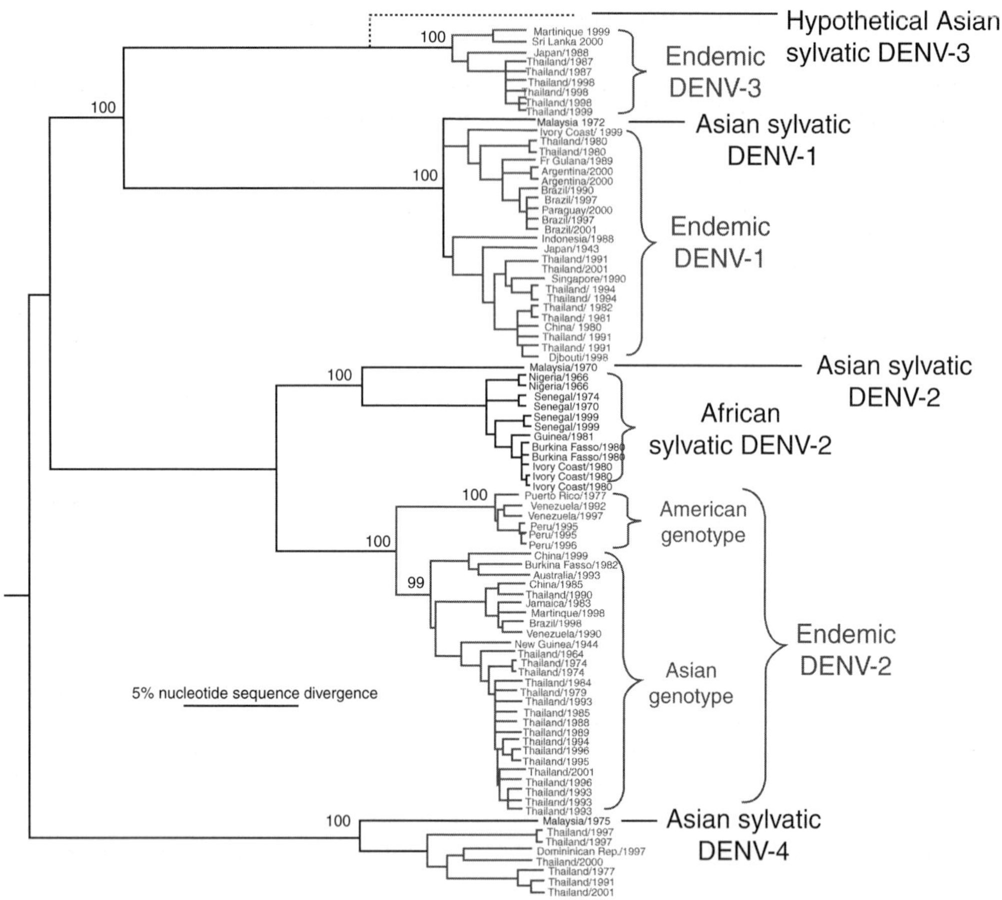

FIGURE 16.6 Phylogenetic tree derived from complete open reading frame sequences of dengue virus strains from the GenBank library. The tree was drawn using the neighbor-joining method. Numbers indicate bootstrap values for major clades to the right. Virus strains and internal branches representing hypothetical ancestors are colored green for sylvatic and red for endemic strains. The hypothetical sylvatic dengue virus-3 is believed to exist based on seroconversions in sentinel monkeys exposed in Malaysia (Rudnick, 1984). (See Plate 25 for the color version of this figure.)

Nigeria, Malaysia, and Indonesia. Like DENV infection, signs and symptoms of infection typically include high fever, headache, rash, malaise, stomach ache, dizziness, and anorexia (Olson *et al.*, 1981). Seroprevalence in Uganda was estimated at 31% (Fagbami, 1979), suggesting that this virus, like many other febrile arboviral infections, hides under the umbrellas of DENV and malaria in locations where laboratory diagnostics are not available.

Another example of a sylvatic arbovirus that may have epidemic emergence potential like DENV is Mayaro virus, a close relative of CHIKV in the genus *Alphavirus* that frequently infects humans in the neotropics (Pinheiro and LeDuc, 1988; Tesh *et al.*, 1999). Because *Ae. aegypti* is susceptible to Mayaro virus (P.V. Aguilar, S.C. Weaver, unpublished), adequate human viremia could result in epidemic emergence in growing Latin American cities within the enzootic regions of South and Central America. A potentially catastrophic scenario is the epidemic emergence of VEEV carried by *Ae. aegypti*; humans are known to develop viremias comparable those of highly efficient equine amplification hosts (Bowen *et al.*, 1976; Weaver *et al.*, 1996), and *Ae. aegypti* is a competent vector in the laboratory of many VEEV strains (D. Ortiz, S.C. Weaver, unpublished). There is no reason to believe that the behaviors that facilitate its transmission of DENV would not allow this mosquito to transmit VEEV among humans in urban settings, and recent epizootics have provided such opportunities by encroaching on large cities such as Maracaibo in Venezuela (Weaver *et al.*, 1996) where DENV is endemic.

GEOGRAPHIC RANGE EXPANSION OF ARBOVIRUSES

Dispersal

As with emergence into new hosts, the expansion of arboviruses into new geographic distributions or new habitats is often driven by anthropogenic change. Perhaps the most dramatic example of this is the introduction of YFV from Africa into the Americas through the enforced movement of people and accidental transport of mosquitoes in the course of the slave trade (Bryant *et al.*, 2007). Global shipping by sea and air are known to move disease vectors, and by inference the viruses they carry, over large distances in patterns that closely match the intensity of shipping traffic (Tatem *et al.*, 2006a, 2006b). Movement of animals and animal products can also drive range expansions, for example the transportation of animals for use in Islamic religious festivals may have contributed to the outbreaks of RVFV and CCHFV in the Middle East (Chevalier *et al.*, 2004; Deyde *et al.*, 2006). However arboviruses may also be introduced into new areas via natural mechanisms. Transport of viruses or infected vectors by migratory birds offers can result in long-range movement in the absence of human activities (Hubalek *et al.*, 1982: Hubalek, 2000, 2004; Ergonul, 2006). Similarly wind dispersal of infected midges is responsible for the ongoing spread of bluetongue virus (BTV) (Purse *et al.*, 2005).

Establishment

Establishment in a new geographic range may involve the movement of a small virus propagule, for example one infected vector, into a region containing a different suite of vectors and hosts and possibly a different climate than that available in the ancestral range. Thus both founder effects and adaptation to novel environments might be expected to shape arbovirus evolution during range expansion. Yet many geographic shifts are accomplished without detectable adaptation by the virus. For example the strain of WNV that invaded the Americas in 1999 was closely related to lineage 1 WNV strains circulating in Israel in 1997 and 1998 and this virus showed little evolution in its initial spread through the US (Lanciotti *et al.*, 1999; Solomon *et al.*, 2003; Beasley *et al.*, 2003). Subsequently, a new, dominant WNV genotype has evolved in the Americas which shows a lower extrinsic incubation period in a local vector population, and therefore more efficient transmission, than the initial invading strain (Ebel *et al.*, 2004; Davis *et al.*, 2005). While this adaptation to local vectors may enhance the competitive superiority of the dominant strain, clearly it was not necessary for the initial establishment of WNV. Such ecological flexibility might have been expected of WNV, since earlier studies of the diversity of this virus in its ancestral Old World range had noted that the virus appeared to be able to shift between hosts and locales without detectable adaptation (Berthet *et al.*, 1997). What could not have been predicted was the extremely high virulence of WNV in both avian and human hosts following the invasion of North America (Briese and Bernard, 2005) which appears to result not from virus evolution but from differences in vector genetics (Fonseca *et al.*, 2004) and behavior (Kilpatrick *et al.*, 2006b), and perhaps also host genetics (Diamond and Klein, 2006; Glass *et al.*, 2006), between the Old and New World.

Another puzzle posed by WNV is the lack of severe disease detected during the movement of the virus through Mexico and Central

America (Briese and Bernard, 2005). Similar to WNV, individual lineages of BTV (Purse *et al.*, 2005), CCHFV (Deyde *et al.*, 2006), and RVFV (Deyde *et al.*, 2006; Bird *et al.*, 2007) have been shown to cross large distances without substantial genetic change and to cause unexpectedly high levels of disease in their new range.

Other arboviruses, however, have shown genetic changes during range expansion that may represent adaptations to their new environments. As discussed above, during its invasion of Reunion CHIKV acquired a mutation in its envelope glycoprotein that enabled a dramatic increase in virus transmission and may represent an adaptation of CHIKV to local mosquito populations (Schuffenecker *et al.*, 2006). YFV in its ancestral range in Africa possesses two or three conserved repeat sequence elements in its 3′ untranslated region (UTR), whereas American YFV possesses only one (Barrett and Monath, 2003). Since the expansion of YFV into the Americas involved a vector switch from *Aedes* into *Haemagogus* mosquitoes, it is tempting to speculate that this change in the 3′ UTR represents an adaptation to American vectors. In both cases, however, founder effects could also explain these genetic changes, so experimental studies that mutate these sequences in the ancestral virus and evaluate the impact of such mutations on vector infectivity are needed.

Even movements over relatively small distances may require adaptation if an arbovirus encounters a new habitat. In a study of the phylogeny of VSV in Costa Rica, Rodriguez *et al.* (1996) found that VSV isolates collected across a large geographic and temporal range clustered by habitat rather than by year of isolation or distance between sites, suggesting that VSV adapts to particular ecological zones. Efforts are underway to develop techniques that link satellite-derived environmental data with molecular phylogenies to better test the impact of climate and other environmental factors on arbovirus evolution (Randolph and Rogers, 2002; Rogers and Randolph, 2006).

Certain arboviruses have shown a notable failure to invade regions where apparently susceptible hosts and vectors occur. In some cases, for example CHIKV's absence from the Americas, this may simply represent inadequate opportunities for dispersal. Air traffic from Africa is relatively light and primarily directed to European ports (Tatem *et al.*, 2006a) and CHIKV outbreaks, though explosive, are infrequent. In this case, intensification of air travel, coupled with diversification of flight patterns, seems likely to produce a range expansion in the near future. Lack of opportunity seems a much less likely explanation for the continued absence of YFV from Asia, where susceptible *Ae. aegypti* are highly abundant, and explaining the global distribution of YFV remains a challenge (Barrett and Monath, 2003).

Global Climate Change

The potential impacts of global climate change on the arbovirus ecology have been reviewed extensively (Shope, 1991; Jetten and Focks, 1997; Patz *et al.*, 1998; Randolph and Rogers, 2000; Gubler *et al.*, 2001; Purse *et al.*, 2005) and include changes in vector distribution, abundance, behavior, and susceptibility as well as host behavior. The range of most arboviruses is predicted to increase under conditions of global warming; for example mathematical models indicate that the distribution, incidence, and duration of transmission of dengue will all increase (Jetten and Focks, 1997; Patz *et al.*, 1998). Climate change has already been implicated in the current increase in of tick-borne infections, although existing evidence suggests that this increase is driven by socio-political changes rather than global warming (Randolph, 2004; Ergonul, 2006; Sumilo *et al.*, 2007). Indeed, in an exception to the rule of increasing ranges, the range of TBEV is predicted to contract under global warming (Randolph and Rogers, 2000).

In contrast to tick-borne viruses, there is strong evidence that global warming is responsible for the unprecedented spread of midge-borne BTV in Europe (Purse *et al.*, 2005), where increasing temperatures have rendered previously resistant midge species susceptible to BTV, thereby expanding the array of competent vectors for this virus and enabling a geographic expansion.

However, both quantitative and qualitative models of the impact of climate change generally fail to account for the ability of arboviruses to rapidly adapt to novel conditions (Chevalier *et al.*, 2004). One intriguing possibility is that global warming could promote the evolution of novel vector–virus associations. This could occur if abnormally high temperatures render vectors susceptible to viruses to which they were previously resistant, as has been documented for BTV (Purse *et al.*, 2005). If arboviruses adapt to these new vectors during periods of higher temperatures, these transmission cycles could persist even in regions or at times when temperatures subside to pre-warming levels. Empirical and theoretical studies of the effect of increased temperature on arbovirus evolution are clearly needed to address current concerns about the impact of global warming on global disease burdens.

EFFECTS OF ARBOVIRUS TRANSMISSION CYCLES ON THEIR EVOLUTION

As described above, the unique requirements for alternating replication in taxonomically disparate hosts may affect the evolution of arboviruses. Specifically, it has been proposed that arboviruses may be constrained in their ability to reach a fitness peak in both host and vector concurrently, a hypothesis sometimes termed the "double-filter concept." This topic has been addressed in various ways for several decades. The concept that arboviruses may be constrained in their ability to replicate optimally in both vertebrates and vectors was supported by experimental studies that identified many mutant arboviruses with replication phenotypes restricted to vertebrate or arthropod cells (Strauss and Strauss, 1994; Schlesinger and Schlesinger, 2001; Lindenbach and Rice, 2001; Kuhn, 2007). A second line of evidence came from studies demonstrating the phenotypic and genetic stability of arboviruses in nature or under experimental transmission conditions. Taylor and Marshall demonstrated that Ross River virus (family *Togaviridae*) undergoes rapid changes in virulence during passage in cell cultures (attenuation) or mice (increased virulence) (Taylor and Marshall, 1975a). Plaque-derived biological clones from both parent populations and the 10th passage in mice were heterogeneous in virulence, indicative of a quasispecies distribution of phenotypes. However, alternate passages between *Ae. aegypti* mosquitoes and mice stabilized the virulence of two different Ross River virus strains (Taylor and Marshall, 1975b). Taylor and Marshall speculated that the conservation of virulence that resulted from alternating mosquito–mouse passages was selected by the *Ae. aegypti* mosquitoes used; this mosquito strain was susceptible to oral infection only at the time of peak mouse viremia, when the authors speculated that a subpopulation of viruses of higher virulence was not abundant enough to be represented in the mosquito's blood meal (Taylor and Marshall, 1975b).

Much of the fundamental understanding of arbovirus adaptation to host cells has come from studies of the arbovirus vesicular stomatitis virus (VSV; family *Rhabdoviridae*). Holland and colleagues developed highly sensitive methods to detect fitness changes in VSV following serial passages in cell cultures (Holland *et al.*, 1991). Subsequent studies with VSV indicated that adaptation required the passage of large virus populations, was generally cell-specific, and was continuous with serial passages (Clarke *et al.*, 1994; Duarte *et al.*, 1994). Similar findings have since been derived using *in vitro* studies other arboviruses such as eastern equine encephalitis and Sindbis viruses (family *Togaviridae*) (Weaver *et al.*, 1999; Greene *et al.*, 2005).

With the advent of molecular viral genetic techniques to detect mutations in RNA viral genomes, phenotypic stability was complemented by studies demonstrating the genetic stability of arboviruses. Baldridge *et al.*

showed that La Crosse virus exhibited a high degree of genetic stability during both horizontal (oral infection of *Ae. triseriatus* mosquitoes) and vertical (transovarial transmission in mosquitoes) transmission (Baldridge *et al.*, 1989). No RNA sequence change was detected by RNA oligonucleotide T1 fingerprinting in any of these passages, corroborating genetic stability in nature. Similar studies examining transovarial transmission of Toscana virus (family *Bunyaviridae*) also indicated genetic stability during over 12 sandfly generations during a 2-year time period (Bilsel *et al.*, 1988).

More direct evidence for the effect of vector transmission on arboviral adaptation and evolution has come from experimental studies designed specifically to test the effect of the alternating host transmission cycle. When VSV was passaged alternately in sand fly and hamster cells, or allowed to specialize on one cell type through serial *in vitro* passages, fitness increases in the selecting cell were observed (Novella *et al.*, 1999). Surprisingly, VSV replicating exclusively in hamster cells also increased its fitness in sandfly cells, indicating that specialization did not result in cell-specific adaptation. These findings suggested that arboviruses do not necessarily compromise their fitness for a given host by adapting to both vertebrate and invertebrate hosts. The number of mutations in the consensus VSV sequence that accumulated during alternated cell culture passages was similar or larger than that observed in populations allowed to specialize on one cell type, arguing against the hypothesis that the alternating cycle constrains rates of sequence change. Similar results demonstrating host range expansion following selection for replication in a single cell type have also been reported for other, non-arthropod-borne RNA viruses (Ruiz-Jarabo *et al.*, 2004).

Studies with alphaviruses yielded different results and conclusions regarding the effect of host cell alteration on adaptation and fitness. Specialization of eastern equine encephalitis (EEEV) (Weaver *et al.*, 1999) and Sindbis (Greene *et al.*, 2005) viruses on vertebrate cells resulted in fitness losses for mosquito cells, and vice versa. Similar results with EEEV were also obtained using avian and mosquito cells (Cooper and Scott, 2001). However, viruses forced to alternate achieved comparable or only slightly lesser fitness increases in both cell types to the specialists, contradictory to the hypothesis that alternation constrains adaptation by arboviruses. For Sindbis virus, specialized viruses passaged in a single cell exhibited more mutations and amino acid changes per passage than those passaged alternately supporting the hypothesis that host alteration constrains evolutionary rates.

Ciota *et al.* (2007a) examined the evolutionary constraints of adaptation of WNV and St. Louis encephalitis viruses (SLEV) passaged in mosquito or avian cell cultures. Fitness improved in both viruses following passage in mosquito, but not avian, cells, despite minimal genetic changes. These results are consistent with cell-specific adaptation and suggest that virus fitness under natural conditions may be optimized for the vertebrate host but not the arthropod vector.

In contrast, detailed studies of the subgenomic promoter of SINV and its response to selective pressure (Hertz and Huang, 1995b) have suggested that the wild-type sequence may be optimal for both mosquito and vertebrate cells. Selection from a pool of mutated promoter sequences appeared to be stronger in mosquito cells, but passage of the SINV populations in both mosquito and hamster cells led to the generation of a promoter consensus similar to the wild-type. These results indicate that the wild-type and similar SINV subgenomic promoter sequences are optimal for function in hamster (Hertz and Huang, 1995a) and mosquito cells (Hertz and Huang, 1995b), suggesting that this sequence element makes little or no evolutionary compromise in maintaining the ability to replicate alternately in the two disparate host organisms.

Studies of the bunyaviruses RVFV and Crimean-Congo hemorrhagic fever virus (CCHFV) have challenged the evolutionary compromise hypothesis, though from two different perspectives. Bird *et al.* (2007) detected the high degree of sequence conservation typical of arboviruses in the complete genomes of 33 ecologically RVFV isolates collected over more than 50 years. However they interpret this low level of diversity as a reflection of a relatively recent common ancestor rather than a constraint imposed by the transmission cycle of the virus. In contrast Deyde *et al.* (2006) detected surprisingly high levels of variation in the complete genome sequences of 13 geographically and temporally diverse CCHFV isolates. Such diversity may contradict the requirement for an evolutionary compromise between host and vector. Alternatively, these high levels of diversity may result from high efficiency vertical transmission in the tick that obviates the need for maintaining high fitness in vertebrates (Deyde *et al.*, 2006).

OTHER IMPACTS OF VECTOR AND HOST BIOLOGY ON ARBOVIRUS EVOLUTION

In addition to the impact on overall rates of genetic evolution, the ecology of arthropod transmission as well biological differences between and among hosts and vectors may impose, or relax, selection pressure on arbovirus characteristics as well. For example, relative to viruses that are transmitted through contact with a substrate or inhalation of aerosols, arboviruses are rarely exposed to the external environment. Hardestam *et al.* (2007) hypothesized that selection for *ex vivo* stability may therefore be relaxed for arboviruses. They compared the stability of rodent-transmitted Hantaan virus (HTNV) held under various environmental conditions with two other members of the family *Bunyaviridae* that are transmitted by arthropods: sandfly-transmitted sandfly fever Sicilian virus (SFSV) and tick-transmitted CCHFV. Contrary to their hypothesis, SFSV showed the highest stability under all conditions whereas CCHFV showed the lowest. Although both SFSV and CCHFV are arboviruses, SFSV is maintained primarily through transovarial transmission in the sandfly while CCHFV maintains a more regular alternation between vertebrates and ticks. Thus these data suggest the intriguing possibility that adaptation for continuous replication in an ectothermic host may generate correlated selection for enhanced *ex vivo* stability.

As illustrated by the examples above, the nature and frequency of transmission by routes other than strict vector–host alternation varies among arboviruses. Under experimental conditions, various arboviruses can be transmitted among vertebrates without the involvement of arthropods via the conjunctival, nasal, oral, rectal, venereal, and congenital routes (Kuno, 2001). This type of direct transmission has also been detected under natural conditions. In western Siberia, for example, water voles have become infected with Omsk hemorrhagic fever virus (OHMV) by contact with water contaminated by diseased muskrats (Kuno and Chang, 2005). Contaminated water could also serve a route of direct transmission between vertebrates and arthropods. For example, it has been suggested that during Rift Valley fever virus (RVFV) outbreaks, infected vertebrates may contaminate the seasonal pools of water, known as dambos and thereby directly infect the mosquito larvae that develop in these pools. Additionally, many arboviruses can be maintained in their arthropod vectors in the absence of vertebrates via venereal and transovarial transmission for varying periods of time (Kuno and Chang, 2005). Moreover, several arboviruses (Kuno and Chang, 2005), most recently WNV (Higgs *et al.*, 2005; McGee *et al.*, 2007), have now been shown to have capacity for transmission during co-feeding by an infected and an uninfected vector before the host becomes viremic, a phenomenon known as non-viremic transmission. The impact of alternative routes of transmission on arbovirus evolution remain to be fully explored.

Selection pressures imposed by hosts vs. vectors, and also by different classes of vector, may vary considerably. Arbovirus infection of vector versus host differs in several important features. Most prominently, arbovirus infection of the vector is generally lifelong whereas infection of the host is usually transient and eliminated by the host immune response. Vector lifespan is usually substantially shorter than that of the host, though ticks may be an exception to this rule (Randolph, 1998). Vectors are always ectothermic whereas hosts are often endothermic. Vectors are often, but not always (Elliot *et al.*, 2003) more mobile than hosts. Finally the titers reached in and transmitted by vectors may differ from that of hosts, although the size of the transmission propagule has proven difficult to investigate in biologically realistic systems. The most widely accepted difference between arbovirus infection of host and vector is that arboviruses are selected to be less virulent, here defined as the fitness cost to the host (Galvani *et al.*, 2003), to their vectors than their hosts because of their dependence on the former for mobility (Ewald, 1994); though see Elliot *et al.* (2003) for further refinement of these predictions. In DENV, the prediction that arboviruses will have low or no fitness costs for their vectors has been borne out. For example, Platt *et al.* (1997) reported that DENV infection resulted in a decrease in mosquito feeding efficiency, whereas Putnam and Scott (1995) reported that it did not. Evidence for the impact of DENV on mosquito survival is similarly mixed: Joshi *et al.* (2002) detected a decrease in survival of *Ae. aegypti* intrathoracically inoculated with DENV relative to controls, but in our experience mosquitoes that acquire DENV through the more natural route of oral infection show similar rates of survival to mosquitoes fed upon an uninfected bloodmeal (Hanley, unpublished data). However alphaviruses appear to have a more substantial impact on vector fitness. EEEV (Weaver *et al.*, 1988) and WEEV (Weaver *et al.*, 1992a) infection cause detectable pathologic changes, as well as decreased survival and fecundity (Scott and Lorenz, 1998; Mahmood *et al.*, 2004). As molecular strategies to interrupt arbovirus transmission in the vector come to fruition (see below), further study of the impact of arboviruses on vector fitness will be needed to predict the impact of such interventions on mosquito population biology.

Evolution of Arbovirus Virulence

Arboviruses cause a wide range of diseases in humans and domestic animals. However, there is relatively little evidence of severe disease in reservoir hosts; most of the apparent disease caused by arboviruses involves humans, equines and other ungulates, and other domestic animals that represent dead-end infections that do not exert long-term evolutionary pressures. The lack of apparent disease in many reservoir hosts within endemic regions may reflect selection for resistance in populations exposed for long time periods to infection and/or selection for attenuation of arboviruses in these species. These competing hypotheses are difficult to evaluate experimentally aside from the use of model cell culture systems (see below). Evidence for selection of host populations resistant to arboviral disease comes from experimental infections of cotton rats both within and outside the range of VEE complex alphaviruses. A sympatric population from Florida is resistant to VEEV-induced disease, while two allopatric populations are highly susceptible yet produce comparable viremia levels (Carrara *et al.*, 2007).

The recent introduction of WNV into North America has provided a unique opportunity to observe prospectively these hypothetical evolutionary pressures on an arbovirus *in vivo*, in nature (Weaver and Barrett, 2004). As it has spread through most of North America since its introduction in 1999, WNV has developed very limited genetic diversity. Phylogenetic analyses have identified temporally and geographically defined variants, and suggest that a dominant genotype has emerged (Davis *et al.*, 2005). Although natural variants with differing mouse virulence have been identified (Estrada-Franco *et al.*, 2003; Beasley *et al.*, 2004), it is unknown whether selection or genetic drift has driven lineage replacement. Brault and colleagues have investigated in detail the virulence of WNV for American crows, which suffer extremely high rates of mortality following natural and experimental infection. North

American WNV strains exhibit higher virulence and viremia levels in crows compared to most Old World lineages (Brault *et al.*, 2004a). Reverse genetic studies indicate that ability of North American isolates to replicate at the high temperatures measured in infected crows could be an important factor leading to the increased avian virulence and emergence of this strain of WNV (Kinney *et al.*, 2006). A single mutation in the NS3 protein (Thr249 to Pro) appears to be largely responsible for the virulence of the North American strains in crows, and comparative sequence analyses of many WNV genomes indicates that this NS3–249 site is subject to adaptive evolution in several Old World locations (Brault *et al.*, 2007). A complete understanding of this positive selection of a mutation encoding increased viremia potential and virulence in a natural arbovirus host will require its evaluation in additional avian hosts from both the New and Old Worlds.

QUASISPECIES DIVERSITY IN ARBOVIRUSES

As discussed earlier, arboviruses are capable of extraordinarily rapid rates of evolution; nonetheless, relative to directly-transmitted RNA viruses, most arbovirus species show an exceptionally low degree of variation in the consensus genome sequence among different hosts. Their potential for rapid diversification is more fully realized within individual hosts, where virus populations exist as a swarm of genetically variable genomes, or quasispecies (Eigen, 1996). Quasispecies are composed of one or more numerically common master sequences an array of rare variants, sometimes termed minority populations (Holland, 2006). Traditional sequencing methods generally detect only the master sequence(s) as a consensus sequence; minority populations can be identified only if a sample of individual genomes from a single host are cloned and sequenced individually or otherwise separated with techniques such as mass spectrometry (Yea *et al.*, 2007) or laser capture microdissection and *in situ* PCR. The diversity of the mutation spectrum within a given quasispecies may enhance rates of adaptation (Domingo *et al.*, 2006), shape pathogenesis (Vignuzzi *et al.*, 2006), and serve as a repository of evolutionary "memory" (Domingo *et al.*, 2006). Quasispecies dynamics have long been of particular interest in the context of vector-transmitted viruses. Initially, quasispecies were thought to offer a mechanism by which arboviruses could shift between the two fitness peaks created by their vertebrate host and arthropod vector. This hypothesis relied on alternating amplification of different subpopulations within the mutant swarm that were better adapted to either host or vector over the course of transmission (Borucki *et al.*, 2001). However subsequent research has demonstrated that under natural circumstances master sequences do not generally differ between host and vector. Bluetongue virus (family *Reoviridae*) in ungulates and midges (Bonneau *et al.*, 2001, Aaskov *et al.*, 2006), dengue virus (family *Flaviviridae*) in humans and mosquitoes (Aaskov *et al.*, 2006; Lin *et al.*, 2004), and WNV (family *Flaviviridae*) in birds and mosquitoes (Jerzak *et al.*, 2005) all show conservation of the consensus sequence in both vertebrates and arthropods.

Recent studies have shifted in focus from the impact of host alternation on the identity of the master sequence in a virus population to the impact of host alternation on the diversity of that population. Table 16.1 summarizes the intra-host diversity resulting from natural infections, e.g. virus populations that are not under selection for replication in a novel host or novel pattern of transmission, for various arboviruses in either host or vector or both. One study of dengue virus reported higher diversity in the host (Lin *et al.*, 2004) while one study of WNV reported higher diversity in the vector (Jerzak *et al.*, 2005). To compare the four studies that measured diversity in both groups, the median trend (host relative to vector) of all measures of diversity (excluding d_N/d_S, see below) was calculated, irrespective of the magnitude of the difference; under this scheme two (Bonneau *et al.*, 2001;

TABLE 16.1 Arbovirus Quasispecies Diversity in Hosts and Vectors

					Mean value in		
Virus	Host	Vector	Viral genes analyzed	Diversity measure	Host	Vector	Reference
Bluetongue virus	Sheep and calf	*Culicoides* midge	Glycoprotein VP2	Mutation frequency[a]	0.00018	0.00017	Bonneau *et al.*, 2001
			Glycoprotein VP2	% Mutant clones	13.5	43.0	
			NS3/NS3A	Mutation frequency	0.00019	0.00018	
			NS3/NS3A	% Mutant clones	13.0	8.3	
Dengue type 1	Human	*Aedes* mosquitoes	Envelope	Mean pairwise genetic diversity	0.008	0.0143	Aaskov *et al.*, 2006
			Envelope	d_N/d_S[b]	0.572	0.547	
Dengue type 3	Human	*Aedes aegypti* mosquitoes	Envelope	Nucleotide sequence diversity[c]	0.38	0.21	Lin *et al.*, 2004
			Capsid	Nucleotide sequence diversity	0.23	0.09	
Dengue type 3	Human	Not done	Envelope	Nucleotide sequence diversity	0.33	–	Wang *et al.*, 2002a
			Envelope	Amino acid diversity[d]	0.75	–	
Dengue type 3	Human	not done	Capsid/pre-membrane	Nucleotide mutation frequency[e]	0.0013	–	Chao *et al.*, 2005
			Capsid/preMembrane	Amino acid mutation frequency[f]	0.0023	–	
			Envelope/NS1	Nucleotide mutation frequency	0.0018	–	
			Envelope/NS1	Amino acid mutation frequency	0.0031	–	
			NS3	Nucleotide mutation frequency	0.0006	–	

			NS3	Amino acid mutation frequency	0.0013	–	
			NS5	Nucleotide mutation frequency	0.0015	–	
			NS5	Amino acid mutation frequency	0.0033	–	
La Crosse virus	Not done	*Aedes triseriatus* mosquitoes	Glycoprotein G1	% Mutant clones	–	43.3	Borucki *et al.*, 2001
West Nile virus	Not done	*Culex quinque-fasciatus*	preMembrane/Envelope	Nucleotide mutation frequency	–	0.0011	Davis *et al.*, 2003
				% Mutant clones	–	50.0	
West Nile virus	Birds (various species)	Mosquitoes (various species)	Envelope/NS1	Mean pairwise distance	0.00021	0.00034	Jerzak *et al.*, 2005
			Envelope/NS1	d_N/d_S	0.159	0.334	
			Envelope/NS1	Nucleotide sequence diversity	0.011	0.020	
			Envelope/NS1	Amino acid diversity	0.012	0.017	
			Envelope/NS1	% Mutant clones	15.5	23.5	

[a] Number of mutations/total number of nucleotides sequenced.
[b] Ratio of non-synonymous to synonymous substitutions.
[c] Number of mutations/total number of nucleotides sequenced (same as mutation frequency).
[d] Number of substitutions/total number of amino acids sequenced (same as mutation frequency).
[e] Proportion of nucleotide mutations relative to the consensus nucleotide sequence.
[f] Proportion of amino acid mutations relative to the consensus nucleotide sequence.

Lin *et al.*, 2004) studies found higher diversity in the host and two (Aaskov *et al.*, 2006; Jerzak *et al.*, 2007) found higher diversity in the vector. While this is still a small number of studies, the emerging pattern is one of balanced diversity. Nor does host alternation appear to result in elevated intra-host diversity; average diversity values in arboviruses were broadly similar to other single-host RNA viruses such as hepatitis C virus (HCV; family *Flaviviridae*) or human immunodeficiency virus type 1 (HIV-1; family *Retroviridae*) (Wang *et al.*, 2002a; Chao *et al.*, 2005).

A surprising recent finding is the presence in many mosquitoes and humans infected with DENV-1 of a subpopulation of genomes containing a stop-codon mutation in the envelope protein gene (Aaskov *et al.*, 2006). This mutation resulted in genome producing a truncated form of the envelope protein that apparently spread among the host population of Myanmar. Because a defective genome can only persist if it is transmitted in concert with full-length viral genomes, these data suggest that the bottleneck during arbovirus transmission is quite wide and allows a large and diverse population to pass from host to vector and from vector to host. In support of this hypothesis, Jerzak *et al.* (2005) found related minority variants of WNV in several different avian hosts, a pattern best explained by the transmission of a large selection of variants over the virus life cycle. However, recent studies of other flaviviruses suggest that genetic bottlenecks accompany the oral infection of mosquito vectors (Scholle *et al.*, 2004) and studies of alphavirus transmission suggest bottlenecks during infection of vertebrates via mosquito saliva (Smith *et al.*, 2006). Experimental transmission studies with these DENV-1 defective genomes are needed to determine how infection multiplicities are maintained at levels needed to support replication.

Patterns of selection on quasispecies diversity have been analyzed using the ratio of non-synonymous to synonymous mutations (d_N/d_S); most studies found d_N/d_S values to be substantially less than 1 (Wang *et al.*, 2002a; Lin *et al.*, 2004; Jerzak *et al.*, 2005; Aaskov *et al.*, 2006), a hallmark of purifying selection. However Chao *et al.* (2005) reported d_N/d_S values for the dengue virus envelope gene that were higher than 1, possibly indicating positive selection in this case. Despite exhibiting the signature of purifying selection, arbovirus quasispecies usually contain a small percentage of defective genomes containing premature stop codons in the coding sequence (Wang *et al.*, 2002a; Jerzak *et al.*, 2005; Aaskov *et al.*, 2006). Defective genomes are of interest not only for the insight they provide into patterns of virus transmission but also because they may interfere with replication of intact genomes and thereby modulate pathogenesis and population dynamics (Bielefeldt-Ohmann and Barclay, 1998; Perales *et al.*, 2007).

Another area of interest has been the impact of quasispecies diversity on disease severity. Among directly transmitted viruses, this relationship has been studied in detail in both HCV and HIV-1, which show opposite correlations between diversity and disease severity. In HCV, severe disease, manifesting as progressing hepatitis, is associated with relatively high quasispecies diversity when compared to less severe, resolving hepatitis (Farci and Purcell, 2000; Farci *et al.*, 2000). In HIV, in contrast, slow virus evolution and homogeneous intra-host populations are associated with rapid progression to AIDS, while virus populations are more diverse in non-progressors (Delwart *et al.*, 1997). Both HIV and HCV result in chronic infections, and both patterns of diversity are thought to result from immune selection over an extended period of time. Such selection seems less likely in acute, resolving infections such as those produced by most arboviruses. Indeed, comparisons of the mutant spectrum of dengue virus in dengue fever cases versus the more severe dengue hemorrhagic fever cases have detected no difference in diversity between the two (Wang *et al.*, 2002a; Chao *et al.*, 2005; Aaskov *et al.*, 2006). Bielefeldt-Ohmann and Barclay (1998) proposed a number of mechanisms by which quasispecies dynamics of Ross River virus (family *Togaviridae*) may exacerbate pathogenesis in humans, including

(i) selection of more virulent variants during epidemic spread in humans, (ii) over-production of cytotoxic or inflammogenic epitopes by defective genomes, (iii) evolution of minority variants carrying inflammogenic epitopes. These mechanisms have yet to be tested in Ross River virus or any other arbovirus.

An exciting development in the study of arbovirus quasispecies is the use of experimental evolution to define the roles of consensus change versus genetic diversity in arbovirus adaptation. For example, Ciota *et al.* (2007a, 2007b) showed that WNV passage in mosquito or chick embryo fibroblast cells in culture showed a steady increase in both quasispecies diversity and fitness in the former while gaining neither diversity nor fitness in the latter. Similarly, Jerzak *et al.* (2007) demonstrated that WNV passaged 20 times in either mosquitoes or chicks *in vivo* gained greater diversity in mosquitoes. Moreover, population diversity was negatively correlated with lethality in mice. These results suggest that purifying selection may be relaxed in mosquito cells relative to cells from a vertebrate host.

Two common methodological practices may potentially bias the measurement of quasispecies diversity *in vivo* in arthropod vectors relative to vertebrate hosts. First, as noted by Jerzak *et al.* (2005), the tissues used for isolation of viral sequences from arthropods are often substantially more diverse than those from vertebrates. In most studies of arbovirus quasispecies, viral sequences have been isolated from whole single (Bonneau *et al.*, 2001; Jerzak *et al.*, 2005, 2007; Aaskov *et al.*, 2006) or even multiple (Davis *et al.*, 2003; Lin *et al.*, 2004; Jerzak *et al.*, 2005) arthropods, while in vertebrates they have been isolated from a single tissue type, such as whole blood (Bonneau *et al.*, 2001), serum (Aaskov *et al.*, 2006; Jerzak *et al.*, 2007), plasma (Wang *et al.*, 2002a, 2002b; Lin *et al.*, 2004; Chao *et al.*, 2005), or kidneys (Jerzak *et al.*, 2005). This disparity has the potential to bias the detection of diversity, because quasispecies may be partitioned at different levels of tissue organization. Such partitioning has long been recognized in HIV, where the identity of the mutant swarm and its overall diversity may differ among different tissue types (Chang *et al.*, 1998), among different compartments of the same organ (Salemi *et al.*, 2005), or even among spatially distinct patches of the same cell type (Sala *et al.*, 1994). Feuer *et al.* (1999) demonstrated that in natural infections, Sin Nombre virus (family *Bunyaviridae*) showed an approximately threefold higher mutation frequency in the spleen of deer mice than in the liver, kidney, or lungs (Feuer *et al.*, 1999), even though no organ-specific mutations were detected.

If partitioning occurs in arboviruses, then samples derived from whole animals will give an accurate picture of host-wide diversity but will fail to distinguish the diversity of virus populations infecting tissues from which transmission may occur, such as blood. Samples derived from a single tissue may target such populations but will underestimate the level of diversity in the whole host. To date only one study has examined partitioning in arbovirus quasispecies diversity: Borucki *et al.* (2001) used single-strand conformational polymorphism (SSCP) to assess quasispecies diversity from the midgut, ovary, and salivary glands of *Aedes triseriatus* mosquitoes orally infected with La Crosse virus (family *Bunyaviridae*). They found that diversity in the midgut tissue was substantially greater than diversity in either the ovary or the salivary gland, suggesting that the midgut may act to filter out a subset of viral variants that are obtained in the bloodmeal. Because SSCP data cannot be used to identify the nucleotide sequence of each variant, these data cannot address whether tissue-specific mutations occurred. Thus studies are needed to explicitly test the impact of partitioning on arbovirus quasispecies diversity and identity; it would be particularly informative to contrast diversity in tissues from which transmission can occur and those that are evolutionary dead-ends for the virus. In addition, new algorithms have recently been developed to identify positive selection in quasispecies (Stewart *et al.*, 2001), and these might be applied to existing and newly developed datasets.

Second, in experimental studies, it is common to infect arthropods by injecting virus directly into the thorax (Lin *et al.*, 2004; Jerzak *et al.*, 2007), a technique that bypasses the midgut and thereby results in high and repeatable rates of infection. If in fact the midgut is an important filter of quasispecies variation (Borucki *et al.*, 2001), then intrathoracic inoculation may artificially boost diversity in other tissues, resulting in higher overall levels of quasispecies diversity than would be achieved through oral infection. The same argument might also be made for direct inoculation of virus into vertebrate hosts. While numerous studies have compared the overall replication of virus in hosts infected via injection versus vector bite (e.g. Smith *et al.*, 2005 and references therein); no studies to date have assessed the impact of the two modes of infection on quasispecies diversity. A notable exception to the tendency to use intrathoracic inoculation is a study of bluetongue virus (Bonneau *et al.*, 2001) in which biting midges, *Culicoides sonorensis*, were used to transmit the virus to a sheep and then between that sheep and a calf. Intriguingly, this study found only slight differences in quasispecies diversity between the ruminant hosts and the midge vectors.

CHALLENGES TO THE CONTROL OF ARBOVIRAL DISEASES

Strategies to control arboviral disease can be organized into three general categories: (i) vaccination to prevent host infection, (ii) antiviral compounds to ameliorate host disease and/or limit transmission to vectors, and (iii) interruption of vector transmission. Here we consider the special challenges that arbovirus evolution poses for each of these strategies.

Vaccines

Vaccines are thought to offer the best hope for control of arboviral disease, and development of vaccines against arboviruses has met with some success. The live attenuated YFV 17D vaccine developed in the 1930s is considered to be the safest and most efficacious vaccine ever produced (Seligman and Gould, 2004). Vaccines approved for human or animal use exist for other flaviviruses, bunyaviruses, and reoviruses, including JEV, TBEV, WNV, Louping ill virus, RVFV, and BTV, but in many cases their efficacy is limited by high reactogenicity, poor immunogenicity, and high cost (Nalca *et al.*, 2003; Sidwell and Smee, 2003; Chang *et al.*, 2004; Stephenson, 2006). Among the alphaviruses, human vaccination is currently available only for EEEV under Investigational New Drug (IND) status from the US Army, but this vaccine is poorly immunogenic (three doses are required) and immunogenicity is short-lived (Nalca *et al.*, 2003). There remain major gaps in protection against some of the most significant emerging arboviruses, including DENV (Whitehead *et al.*, 2007) and CCHFV (Sidwell and Smee, 2003). Moreover, second-generation vaccines for all arbovirus vaccines except YFV are being actively pursued (Hart *et al.*, 2000; Schoepp *et al.*, 2002; Paessler *et al.*, 2003, 2006; Nalca *et al.*, 2003; Sidwell and Smee, 2003; Chang *et al.*, 2004; Seligman and Gould, 2004).

Efforts to develop DENV vaccines illustrate well some of the challenges that arboviruses pose for vaccine development, as well as the particular difficulties inherent to the DENV. Like many other arboviruses, DENV circulates most extensively in the developing world. Thus a DENV vaccine that protects the populations at risk must be inexpensive to manufacture. This requirement is most easily satisfied by a live attenuated vaccine, and the front-runners include live viruses in which attenuation has been generated through classical passage techniques, through incorporation of a 30-nucleotide deletion in the 3′ UTR, through replacement of the YFV-17D membrane and envelope glycoprotein genes with those of DENV, and through chimerization between different DENV serotypes (Whitehead *et al.*, 2007).

However live attenuated RNA virus vaccines retain the capacity to evolve, and experience with the live polio vaccine has shown that reversion to virulence can occur both within an individual vaccinee as well as during transmission of vaccine viruses among an unvaccinated population (Nathanson and Fine, 2002). Thus, an acceptable DENV vaccine must satisfy two criteria for stability. First, the mutations that confer attenuation must be clearly characterized and genetically stable. This has been achieved through the use of multiple mutations, large deletions, or chimerization (Whitehead *et al.*, 2007). Second, a DENV vaccine must not be transmissible by mosquitoes, and all of the live attenuated DENV vaccine candidates currently in human trials have met this requirement (Whitehead *et al.*, 2007). The former two criteria should be applied to all live attenuated arbovirus vaccines. However because sequential infections with different DENV serotypes are associated with enhanced disease, an acceptable DENV vaccine must also be tetravalent, e.g. target all four DENV serotypes.

Authors mindful of the tendency of the components of the trivalent live attenuated poliovirus vaccine to recombine (Nathanson and Fine, 2002) have raised concerns about the potential for recombination by a tetravalent DENV vaccine, and have suggested that recombination among vaccine strains, between vaccine strains and wild-type DENV, or between vaccine strains and other flaviviruses could produce new virus strains of unknown virulence (Seligman and Gould, 2004). However, as discussed above, current evidence suggests that recombination among dengue serotypes either does not occur or results in chimeric viruses that are so attenuated that they do not persist in nature (Murphy *et al.*, 2004). Another concern is that if a DENV vaccination campaign is successful and DENV is eradicated locally or globally, vaccination efforts will relax and new strains of DENV will emerge from sylvatic reservoirs. Studies of the biology of sylvatic DENV suggest that these viruses could readily enter into a human transmission cycle (Vasilakis *et al.*, 2007b), and thus careful surveillance of the sylvatic DENV circulation is needed. In the end, however, the major obstacle to the successful deployment of a DENV vaccine or any other arbovirus vaccine may not be vaccine evolution but rather societal barriers such as political instability, false alarms about vaccine safety, and risk-averse vaccine policies (Stephenson, 2006).

Antiviral Therapies

Passive immunotherapy has been used successfully to treat some arbovirus infections, but may enhance infection by others, for example TBEV (Nalca *et al.*, 2003). Few effective antiviral drugs are available for treatment of arbovirus infections. Ribavirin appears to have limited efficacy in humans against some, but not all, flaviviruses (Lewis and Amsden, 2007) and bunyaviruses (Sidwell and Smee, 2003); no antiviral drugs have proven effective in treating togaviruses (Sidwell and Smee, 2003). The urgent need for treatment options has prompted a search for novel antiviral molecules, including morpholino oligomers (Kinney *et al.*, 2005; Harris *et al.*, 2006; Holden *et al.*, 2006; Deas *et al.*, 2007) and short interfering RNAs (siRNAs) (Barik, 2004; Bai *et al.*, 2005; Murakami *et al.*, 2005; Kumar *et al.*, 2006; Ong *et al.*, 2006; O'Brien, 2007). Notably, siRNAs that trigger RNA interference (RNAi) have been used successfully to prevent lethal encephalitis in mice infected with WNV and JEV (Kumar *et al.*, 2006). Unfortunately, both morpholino oligomers and siRNAs require high, sometimes perfect, sequence identity to their genetic targets for efficacy. Thus, both genetic strategies may rapidly select for resistant virus variants (Deas *et al.*, 2007; O'Brien, 2007), and such resistance may require only a single nucleotide change (Dykxhoorn and Lieberman, 2006). Moreover VEEV strains have evolved resistance to RNAi inhibition without mutations in the siRNA target site, suggesting that undiscovered mechanisms for RNAi resistance exist (O'Brien, 2007). The problem of viral escape may be overcome by using multiple siRNAs (O'Brien, 2007) targeting highly conserved regions of the viral genome (Dykxhoorn and Lieberman, 2006), however it will first be necessary to address the function of such conserved regions and their tolerance for mutation.

Interrupting Transmission

Traditional, pesticide-based methods of vector control have become increasingly ineffective due to the evolution of pesticide resistance among vectors, reduction in vector control programs, societal resistance to the use of toxic and sometimes persistent chemicals, and increases in global travel, among other factors (Blair *et al.*, 2000; Gubler, 2001, 2002; Beaty, 2005). The decline in the efficacy of pesticides has spurred a search for new approaches for interrupting arbovirus transmission. To this end, creation of genetically modified, arbovirus-resistant mosquitoes has been investigated as an alternative approach to arboviral disease control (Beaty, 2005; Blair *et al.*, 2000; Olson *et al.*, 2002). RNAi appears to a major mechanism of arbovirus replication modulation by mosquitoes (Sanchez-Vargas *et al.*, 2004); for example, Keene and colleagues (2004) demonstrated that silencing the RNAi pathway in *Anopheles gambiae* makes this mosquito significantly more susceptible to O'nyong nyong virus (ONNV). Thus considerable effort has been devoted to the generation of siRNAs that can suppress arbovirus replication in vectors. siRNAs have been shown to inhibit the replication of the mosquito-borne DENV (Adelman *et al.*, 2001; Caplen *et al.*, 2002) and SFV (Caplen *et al.*, 2002), as well as the tick-borne hazara virus (Garcia *et al.*, 2005), in cultured mosquito and tick cells, respectively. Moreover, transgenic *Ae. aegypti* expressing DENV-derived siRNAs in the midgut are resistant to DENV infection (Franz *et al.*, 2006). The cultural acceptability of transgenic vector release is questionable, at best (Alphey *et al.*, 2002). In the long term, vector resistance based on RNAi also faces the same potential stumbling block as therapeutic siRNAs, namely rapid evolution of virus resistance via selection for single nucleotide mismatches in the siRNA target site (Dykxhoorn and Lieberman, 2006; Deas *et al.*, 2007; O'Brien, 2007). It is reasonable to suppose that the selection imposed on a targeted arbovirus population by a large population of vectors expressing siRNAs will be substantially stronger than treatment of a single patient with therapeutic siRNAs, and that the evolutionary response will be proportionally faster and more complete. As with therapeutic siRNAs, targeting vector-expressed siRNAs for resistance to multiple, conserved regions of the viral genome may enhance the chances for success. This assumes of course that the larger challenge to the use of transgenic mosquitoes, namely the development of a system to drive effector genes into the vector population (Scott *et al.*, 2002; James, 2005), will be overcome.

REFERENCES

Aaskov, J., Buzacott, K., Thu, H.M., Lowry, K. and Holmes, E.C. (2006) Long-term transmission of defective RNA viruses in humans and Aedes mosquitoes. *Science* **311**, 236–238.

Adams, B., Holmes, E.C., Zhang, C., Mammen, M.P., Jr., Nimmannitya, S., Kalayanarooj, S. and Boots, M. (2006) Cross-protective immunity can account for the alternating epidemic pattern of dengue virus serotypes circulating in Bangkok. *Proc. Natl Acad. Sci. USA* **103**, 14234–14239.

Adelman, Z.N., Blair, C.D., Carlson, J.O., Beaty, B.J. and Olson, K.E. (2001) Sindbis virus-induced silencing of dengue viruses in mosquitoes. *Insect Mol. Biol.* **10**, 265–273.

Alphey, L., Beard, C.B., Billingsley, P., Coetzee, M., Crisanti, A., Curtis, C. *et al.* (2002) Malaria control with genetically manipulated insect vectors. *Science* **298**, 119–121.

Anderson, J.R. and Rico-Hesse, R. (2006) Aedes aegypti vectorial capacity is determined by the infecting genotype of dengue virus. *Am. J. Trop. Med. Hyg.* **75**, 886–892.

Anishchenko, M., Bowen, R.A., Paessler, S., Austgen, L., Greene, I.P. and Weaver, S.C. (2006) Venezuelan encephalitis emergence mediated by a phylogenetically predicted viral mutation. *Proc. Natl Acad. Sci. USA* **103**, 4994–4999.

Armstrong, P.M. and Rico-Hesse, R. (2003) Efficiency of dengue serotype 2 virus strains to infect and disseminate in Aedes aegypti. *Am. J. Trop. Med. Hyg.* **68**, 539–544.

Bai, F., Wang, T., Pal, U., Bao, F., Gould, L.H. and Fikrig, E. (2005) Use of RNA interference to prevent lethal murine west nile virus infection. *J. Infect. Dis.* **191**, 1148–1154.

Baldridge, G.D., Beaty, B.J. and Hewlett, M.J. (1989) Genomic stability of La Crosse virus during vertical and horizontal transmission. *Arch. Virol.* **108**, 89–99.

Barik, S. (2004) Control of nonsegmented negative-strand RNA virus replication by siRNA. *Virus Res.* **102**, 27–35.

Barrett, A.D. and Monath, T.P. (2003) Epidemiology and ecology of yellow fever virus. *Adv. Virus Res.* **61**, 291–315.

Beasley, D.W., Davis, C.T., Guzman, H., Vanlandingham, D.L., Travassos Da Rosa, A.P., Parsons, R.E. *et al.* (2003) Limited evolution of West Nile virus has occurred during its southwesterly spread in the United States. *Virology* **309**, 190–195.

Beasley, D.W., Davis, C.T., Estrada-Franco, J., Navarro-Lopez, R., Campomanes-Cortes, A., Tesh, R.B. *et al.* (2004) Genome sequence and attenuating mutations in West Nile virus isolate from Mexico. *Emerg. Infect. Dis.* **10**, 2221–2224.

Beaty, B.J. (2005) Control of arbovirus diseases: is the vector the weak link?. *Arch. Virol. Suppl.*, 73–88.

Bennett, S.N., Holmes, E.C., Chirivella, M., Rodriguez, D.M., Beltran, M., Vorndam, V. *et al.* (2003) Selection-driven evolution of emergent dengue virus. *Mol. Biol. Evol.* **20**, 1650–1658.

Berthet, F.X., Zeller, H.G., Drouet, M.T., Rauzier, J., Digoutte, J.P. and Deubel, V. (1997) Extensive nucleotide changes and deletions within the envelope glycoprotein gene of Euro-African West Nile viruses. *J. Gen. Virol.* **78**(Pt 9), 2293–2297.

Bielefeldt-Ohmann, H. and Barclay, J. (1998) Pathogenesis of Ross River virus-induced diseases: a role for viral quasispecies and persistence. *Microb. Pathog.* **24**, 373–383.

Bilsel, P.A., Tesh, R.B. and Nichol, S.T. (1988) RNA genome stability of Toscana virus during serial transovarial transmission in the sandfly Phlebotomus perniciosus. *Virus Res.* **11**, 87–94.

Bird, B.H., Khristova, M.L., Rollin, P.E., Ksiazek, T.G. and Nichol, S.T. (2007) Complete genome analysis of 33 ecologically and biologically diverse Rift Valley fever virus strains reveals widespread virus movement and low genetic diversity due to recent common ancestry. *J. Virol.* **81**, 2805–2816.

Blair, C.D., Adelman, Z.N. and Olson, K.E. (2000) Molecular strategies for interrupting arthropod-borne virus transmission by mosquitoes. *Clin. Microbiol. Rev.* **13**, 651–661.

Bonneau, K.R., Mullens, B.A. and Maclachlan, N.J. (2001) Occurrence of genetic drift and founder effect during quasispecies evolution of the VP2 and NS3/NS3A genes of bluetongue virus upon passage between sheep, cattle, and *Culicoides sonorensis*. *J. Virol.* **75**, 8298–8305.

Borucki, M.K., Chandler, L.J., Parker, B.M., Blair, C.D. and Beaty, B.J. (1999) Bunyavirus superinfection and segment reassortment in transovarially infected mosquitoes. *J. Gen. Virol.* **80**(Pt 12), 3173–3179.

Borucki, M.K., Kempf, B.J., Blair, C.D. and Beaty, B.J. (2001) The effect of mosquito passage on the La Crosse Virus genotype. *J. Gen. Virol.* **82**, 2919–2926.

Bouloy, M. (2001) Rift Valley fever virus. In: *The Encyclopedia of Arthropod-transmitted Infections* (M.W. Service, ed.). Wallingford, UK: CAB International.

Bowen, G.S., Fashinell, T.R., Dean, P.B. and Gregg, M.B. (1976) Clinical aspects of human Venezuelan equine encephalitis in Texas. *Bull. PAHO* **10**, 46–57.

Brault, A.C., Powers, A.M., Chavez, C.L., Lopez, R.N., Cachon, M.F., Gutierrez, L.F. *et al.* (1999) Genetic and antigenic diversity among eastern equine encephalitis viruses from North, Central, and South America. *Am. J. Trop. Med. Hyg.* **61**, 579–586.

Brault, A.C., Powers, A.M., Holmes, E.C., Woelk, C.H. and Weaver, S.C. (2002a) Positively charged amino acid substitutions in the E2 envelope glycoprotein are associated with the emergence of Venezuelan equine encephalitis virus. *J. Virol.* **76**, 1718–1730.

Brault, A.C., Powers, A.M. and Weaver, S.C. (2002b) Vector infection determinants of Venezuelan equine encephalitis virus reside within the E2 envelope glycoprotein. *J. Virol.* **76**, 6387–6392.

Brault, A.C., Langevin, S.A., Bowen, R.A., Panella, N.A., Biggerstaff, B.J., Miller, B.R. and Komar, N. (2004a) Differential virulence of West Nile strains for American crows. *Emerg. Infect. Dis.* **10**, 2161–2168.

Brault, A.C., Powers, A.M., Ortiz, D., Estrada-Franco, J.G., Navarro-Lopez, R. and Weaver, S.C. (2004b) Venezuelan equine encephalitis emergence: Enhanced vector infection from a single amino acid substitution in the envelope glycoprotein. *Proc. Natl. Acad. Sci. USA* **101**, 11344–11349.

Brault, A.C., Huang, C.Y.-H., Langevin, S.A., Kinney, R. M., Bowen, R.A., Ramey, W.N. *et al.* (2007) A single positively selected West Nile viral helicase mutation confers increased avian virogenesis in American crows. *Nat. Genet.* **39**, 1162–1166.

Briese, T. and Bernard, K.A. (2005) West Nile virus—an old virus learning new tricks? *J. Neurovirol.* **11**, 469–475.

Bryant, J., Wang, H., Cabezas, C., Ramirez, G., Watts, D., Russell, K. and Barrett, A. (2003) Enzootic transmission of yellow fever virus in Peru. *Emerg. Infect. Dis.* **9**, 926–933.

Bryant, J.E., Holmes, E.C. and Barrett, A.D. (2007) Out of Africa: a molecular perspective on the introduction of yellow fever virus into the Americas. *PLoS Pathog.* **3**, e75.

Calisher, C.H. and Karabatsos, N. (1988) Arbovirus serogroups: Definition and geographic distribution. In: *The Arboviruses: Epidemiology and Ecology* (T.P. Monath, ed.), Vol. I. Boca Raton, Florida: CRC Press.

Cammisa-Parks, H., Cisar, L.A., Kane, A. and Stollar, V. (1992) The complete nucleotide sequence of cell fusing agent (CFA): homology between the nonstructural proteins encoded by CFA and the nonstructural proteins encoded by arthropod-borne flaviviruses. *Virology* **189**, 511–524.

Caplen, N.J., Zheng, Z., Falgout, B. and Morgan, R.A. (2002) Inhibition of viral gene expression and replication in mosquito cells by dsRNA-triggered RNA interference. *Mol. Ther.* **6**, 243–251.

Carrara, A., Coffey, L.L., Aguilar, P.A., Moncayo, A.C., Travassos Da Rosa, A.P.A., Nunes, M.R.T. *et al.* (2007) Venezuelan equine encephalitis virus infection of sympatric and allopatric populations of cotton rats. *Emerg. Infect. Dis.* **13**, 1158–1165.

Chang, G.J., Kuno, G., Purdy, D.E. and Davis, B.S. (2004) Recent advancement in flavivirus vaccine development. *Expert Rev. Vaccines* **3**, 199–220.

Chang, J., Jozwiak, R., Wang, B., Ng, T., Ge, Y.C., Bolton, W. *et al.* (1998) Unique HIV type 1 V3 region sequences derived from six different regions of brain: region-specific evolution within host-determined quasispecies. *AIDS Res. Hum. Retroviruses* **14**, 25–30.

Chao, D.Y., King, C.C., Wang, W.K., Chen, W.J., Wu, H.L. and Chang, G.J. (2005) Strategically examining the full-genome of dengue virus type 3 in clinical isolates reveals its mutation spectra. *Virol. J.* **2**, 72.

Charrel, R.N., De Lamballerie, X. and Raoult, D. (2007a) Chikungunya outbreaks—the globalization of vector-borne diseases. *N. Engl. J. Med.* **356**, 769–771.

Charrel, R.N., Izri, A., Temmam, S., Delaunay, P., Toga, I., Dumon, H. *et al.* (2007b) Cocirculation of 2 genotypes of Toscana virus, southeastern France. *Emerg. Infect. Dis.* **13**, 465–468.

Chevalier, V., De La Rocque, S., Baldet, T., Vial, L. and Roger, F. (2004) Epidemiological processes involved in the emergence of vector-borne diseases: West Nile fever, Rift Valley fever, Japanese encephalitis and Crimean-Congo haemorrhagic fever. *Rev. Sci. Technol.* **23**, 535–555.

Cilnis, M.J., Kang, W. and Weaver, S.C. (1996) Genetic conservation of Highlands J viruses. *Virology* **218**, 343–351.

Ciota, A.T., Lovelace, A.O., Ngo, K.A., Le, A.N., Maffei, J.G., Franke, M.A. *et al.* (2007a) Cell-specific adaptation of two flaviviruses following serial passage in mosquito cell culture. *Virology* **357**, 165–174.

Ciota, A.T., Ngo, K.A., Lovelace, A.O., Payne, A.F., Zhou, Y., Shi, P.Y. and Kramer, L.D. (2007b) Role of the mutant spectrum in adaptation and replication of West Nile virus. *J. Gen. Virol.* **88**, 865–874.

Clarke, D.K., Duarte, E.A., Elena, S.F., Moya, A., Domingo, E. and Holland, J. (1994) The red queen reigns in the kingdom of RNA viruses. *Proc. Natl Acad. Sci. USA* **91**, 4821–4824.

Cologna, R., Armstrong, P.M. and Rico-Hesse, R. (2005) Selection for virulent dengue viruses occurs in humans and mosquitoes. *J. Virol.* **79**, 853–859.

Cooper, L.A. and Scott, T.W. (2001) Differential evolution of eastern equine encephalitis virus populations in response to host cell type. *Genetics* **157**, 1403–1412.

Crabtree, M.B., Sang, R.C., Stollar, V., Dunster, L.M. and Miller, B.R. (2003) Genetic and phenotypic characterization of the newly described insect flavivirus, Kamiti River virus. *Arch Virol.* **148**, 1095–1118.

Craig, S., Thu, H.M., Lowry, K., Wang, X.F., Holmes, E.C. and Aaskov, J. (2003) Diverse dengue type 2 virus populations contain recombinant and both parental viruses in a single mosquito host. *J. Virol.* **77**, 4463–4467.

Crochu, S., Cook, S., Attoui, H., Charrel, R.N., De Chesse, R., Belhouchet, M. *et al.* (2004) Sequences of flavivirus-related RNA viruses persist in DNA form integrated in the genome of Aedes spp. mosquitoes. *J. Gen. Virol.* **85**, 1971–1980.

Davis, C.T., Beasley, D.W., Guzman, H., Raj, R., D'anton, M., Novak, R.J. *et al.* (2003) Genetic variation among temporally and geographically distinct West Nile virus isolates, United States, 2001, 2002. *Emerg. Infect. Dis.* **9**, 1423–1429.

Davis, C.T., Ebel, G.D., Lanciotti, R.S., Brault, A.C., Guzman, H., Siirin, M. *et al.* (2005) Phylogenetic analysis of North American West Nile virus isolates, 2001–2004: evidence for the emergence of a dominant genotype. *Virology* **342**, 252–265.

De Silva, A. and Messer, W. (2004) Arguments for live flavivirus vaccines. *Lancet* **364**, 500.

Deas, T.S., Bennett, C.J., Jones, S.A., Tilgner, M., Ren, P., Behr, M.J. *et al.* (2007) In vitro resistance selection and in vivo efficacy of morpholino oligomers against West Nile virus. *Antimicrob. Agents Chemother.* **51**, 2470–2482.

Delwart, E.L., Pan, H., Sheppard, H.W., Wolpert, D., Neumann, A.U., Korber, B. and Mullins, J.I. (1997) Slower evolution of human immunodeficiency virus type 1 quasispecies during progression to AIDS. *J. Virol.* **71**, 7498–7508.

Deyde, V.M., Khristova, M.L., Rollin, P.E., Ksiazek, T.G. and Nichol, S.T. (2006) Crimean-Congo hemorrhagic fever virus genomics and global diversity. *J. Virol.* **80**, 8834–8842.

Diallo, M., Ba, Y., Sall, A.A., Diop, O.M., Ndione, J.A., Mondo, M. *et al.* (2003) Amplification of the sylvatic cycle of dengue virus type 2, Senegal, 1999–2000: entomologic findings and epidemiologic considerations. *Emerg. Infect. Dis.* **9**, 362–367.

Diamond, M.S. and Klein, R.S. (2006) A genetic basis for human susceptibility to West Nile virus. *Trends Microbiol.* **14**, 287–289.

Domingo, E., Martin, V., Perales, C., Grande-Perez, A., Garcia-Arriaza, J. and Arias, A. (2006) Viruses as quasispecies: biological implications. *Curr. Top. Microbiol. Immunol.* **299**, 51–82.

Duarte, E., Clarke, D., Moya, A., Domingo, E. and Holland, J. (1992) Rapid fitness losses in mammalian RNA virus clones due to Muller's ratchet. *Proc. Natl. Acad. Sci. USA* **89**, 6015–6019.

Duarte, E.A., Novella, I.S., Ledesma, S., Clarke, D.K., Moya, A., Elena, S.F. *et al.* (1994) Subclonal components of consensus fitness in an RNA virus clone. *J. Virol.* **68**, 4295–4301.

Dykxhoorn, D.M. and Lieberman, J. (2006) Silencing viral infection. *PLoS Med.* **3**, e242.

Ebel, G.D., Carricaburu, J., Young, D., Bernard, K.A. and Kramer, L.D. (2004) Genetic and phenotypic variation of West Nile virus in New York, 2000–2003. *Am. J. Trop. Med. Hyg.* **71**, 493–500.

Eigen, M. (1996) On the nature of virus quasispecies. *Trends Microbiol.* **4**, 216–218.

El Hussein, A., Ramig, R.F., Holbrook, F.R. and Beaty, B.J. (1989) Asynchronous mixed infection of Culicoides variipennis with bluetongue virus serotypes 10 and 17. *J. Gen. Virol.* **70**(Pt 12), 3355–3362.

Elliot, S.L., Adler, F.R. and Sabelis, M.W. (2003) How virulent should a parasite be to its vector?. *Ecology* **84**, 2568–2574.

Endy, T.P. and Nisalak, A. (2002) Japanese encephalitis virus: ecology and epidemiology. *Curr. Top. Microbiol. Immunol.* **267**, 11–48.

Ergonul, O. (2006) Crimean-Congo haemorrhagic fever. *Lancet Infect Dis.* **6**, 203–214.

Estrada-Franco, J.G., Navarro-Lopez, R., Beasley, D.W., Coffey, L., Carrara, A.S., Travassos Da Rosa, A. *et al.* (2003) West Nile virus in Mexico: evidence of widespread circulation since July 2002. *Emerg. Infect. Dis.* **9**, 1604–1607.

Ewald, P.W. (1994) *Evolution of Infectious Disease.* New York: Oxford University Press.

Fagbami, A.H. (1979) Zika virus infections in Nigeria: virological and seroepidemiological investigations in Oyo State. *J. Hyg. (Lond.)* **83**, 213–219.

Farci, P. and Purcell, R.H. (2000) Clinical significance of hepatitis C virus genotypes and quasispecies. *Semin. Liver Dis.* **20**, 103–126.

Farci, P., Shimoda, A., Coiana, A., Diaz, G., Peddis, G., Melpolder, J.C. *et al.* (2000) The outcome of acute hepatitis C predicted by the evolution of the viral quasispecies. *Science* **288**, 339–344.

Ferguson, N., Anderson, R. and Gupta, S. (1999) The effect of antibody-dependent enhancement on the transmission dynamics and persistence of multiple-strain pathogens. *Proc. Natl Acad. Sci. USA* **96**, 790–794.

Feuer, R., Boone, J.D., Netski, D., Morzunov, S.P. and St Jeor, S.C. (1999) Temporal and spatial analysis of Sin Nombre virus quasispecies in naturally infected rodents. *J. Virol.* **73**, 9544–9554.

Fonseca, D.M., Keyghobadi, N., Malcolm, C.A., Mehmet, C., Schaffner, F., Mogi, M. *et al.* (2004) Emerging vectors in the Culex pipiens complex. *Science* **303**, 1535–1538.

Franz, A.W., Sanchez-Vargas, I., Adelman, Z.N., Blair, C.D., Beaty, B.J., James, A.A. and Olson, K.E. (2006) Engineering RNA interference-based resistance to dengue virus type 2 in genetically modified *Aedes aegypti*. *Proc. Natl Acad. Sci. USA* **103**, 4198–4203.

Galvani, A.P., Coleman, R.M. and Ferguson, N.M. (2003) The maintenance of sex in parasites. *Proc. Biol. Sci.* **270**, 19–28.

Garcia, S., Billecocq, A., Crance, J.M., Munderloh, U., Garin, D. and Bouloy, M. (2005) Nairovirus RNA sequences expressed by a Semliki Forest virus replicon induce RNA interference in tick cells. *J. Virol.* **79**, 8942–8947.

Gaunt, M.W., Sall, A.A., De Lamballerie, X., Falconar, A.K., Dzhivanian, T.I. and Gould, E.A. (2001) Phylogenetic relationships of flaviviruses correlate with their epidemiology, disease association and biogeography. *J. Gen. Virol.* **82**, 1867–1876.

Gerrard, S.R., Li, L., Barrett, A.D. and Nichol, S.T. (2004) Ngari virus is a Bunyamwera virus reassortant that can be associated with large outbreaks of hemorrhagic fever in Africa. *J. Virol.* **78**, 8922–8926.

Glass, W.G., McDermott, D.H., Lim, J.K., Lekhong, S., Yu, S.F., Frank, W.A. *et al.* (2006) CCR5 deficiency increases risk of symptomatic West Nile virus infection. *J. Exp. Med.* **203**, 35–40.

Gonzalez-Salazar, D., Estrada-Franco, J.G., Carrara, A.S., Aronson, J.F. and Weaver, S.C. (2003) Equine amplification and virulence of subtype IE Venezuelan equine encephalitis viruses isolated during the 1993 and 1996 Mexican epizootics. *Emerg. Infect. Dis.* **9**, 161–168.

Gould, E.A., De Lamballerie, X., Zanotto, P.M. and Holmes, E.C. (2003) Origins, evolution, and vector/host coadaptations within the genus Flavivirus. *Adv. Virus Res.* **59**, 277–2314.

Granwehr, B.P., Lillibridge, K.M., Higgs, S., Mason, P.W., Aronson, J.F., Campbell, G.A. and Barrett, A.D. (2004) West Nile virus: where are we now?. *Lancet Infect. Dis.* **4**, 547–556.

Grard, G., Moureau, G., Charrel, R.N., Lemasson, J.J., Gonzalez, J.P., Gallian, P. *et al.* (2007) Genetic characterization of tick-borne flaviviruses: new insights into evolution, pathogenetic determinants and taxonomy. *Virology* **361**, 80–92.

Greene, I.P., Wang, E., Deardorff, E.R., Milleron, R., Domingo, E. and Weaver, S.C. (2005) Effect of alternating passage on adaptation of sindbis virus to vertebrate and invertebrate cells. *J. Virol.* **79**, 14253–14260.

Griffin, D.E. (2007) Alphaviruses. In: *Fields' Virology* 5th edn. (D.M. Knipe and P.M. Howley, eds), New York: Lippincott, Williams and Wilkins.

Gubler, D.J. (1997) Dengue and dengue hemorrhagic fever: its history and resurgence as a global public health problem. In: *Dengue and Dengue Hemorrhagic Fever* (D.J. Gubler and G. Kuno, eds). New York: CAB International.

Gubler, D.J. (2001) Human arbovirus infections worldwide. *Ann. NY Acad. Sci.* **951**, 13–24.

Gubler, D.J. (2002) The global emergence/resurgence of arboviral diseases as public health problems. *Arch. Med. Res.* **33**, 330–342.

Gubler, D.J. (2006) Dengue/dengue haemorrhagic fever: history and current status. *Novartis Found. Symp.* **277**, 3–16; discussion 16–22, 71–73, 251–253.

Gubler, D.J., Kuno, G., Sather, G.E. and Waterman, S.H. (1965) A case of natural concurrent human infection with two dengue viruses. *Am. J. Trop. Med. Hyg.* **34**, 170–173.

Gubler, D.J., Reiter, P., Ebi, K.L., Yap, W., Nasci, R. and Patz, J.A. (2001) Climate variability and change in the United States: potential impacts on vector- and rodent-borne diseases. *Environ. Health Perspect.* **109**(suppl 2), 223–233.

Haddow, J. (1964) Twelve isolations of Zika virus from *Aedes (Stegomyia) africanus* (Theobald) taken in and above a Uganda forest. *Bull. World Health Organ.* **31**, 57–69.

Hahn, C.S., Lustig, S., Strauss, E.G. and Strauss, J.H. (1988) Western equine encephalitis virus is a recombinant virus. *Proc. Natl Acad. Sci. USA* **85**, 5997–6001.

Hardestam, J., Simon, M., Hedlund, K.O., Vaheri, A., Klingstrom, J. and Lundkvist, A. (2007) Ex vivo stability of the rodent-borne hantaan virus in comparison to that of arthropod-borne members of the bunyaviridae family. *Appl. Environ. Microbiol.* **73**, 2547–2551.

Harrington, L.C., Edman, J.D. and Scott, T.W. (2001) Why do female *Aedes aegypti* (Diptera: Culicidae) feed preferentially and frequently on human blood?. *J. Med. Entomol.* **38**, 411–422.

Harris, E., Holden, K.L., Edgil, D., Polacek, C. and Clyde, K. (2006) Molecular biology of flaviviruses. *Novartis Found. Symp.* **277**, 23–39; discussion 40, 71–73, 251–253.

Hart, M.K., Caswell-Stephan, K., Bakken, R., Tammariello, R., Pratt, W., Davis, N. *et al.* (2000) Improved mucosal protection against Venezuelan equine encephalitis virus is induced by the molecularly defined, live-attenuated V3526 vaccine candidate. *Vaccine* **18**, 3067–3075.

Hertz, J.M. and Huang, H.V. (1992) Utilization of heterologous alphavirus junction sequences as promoters by Sindbis virus. *J. Virol.* **66**, 857–864.

Hertz, J.M. and Huang, H.V. (1995a) Evolution of the Sindbis virus subgenomic mRNA promoter in cultured cells. *J. Virol.* **69**, 7768–7774.

Hertz, J.M. and Huang, H.V. (1995b) Host-dependent evolution of the Sindbis virus promoter for subgenomic mRNA synthesis. *J. Virol.* **69**, 7775–7781.

Higgs, S. (2006) The 2005–2006 Chikungunya epidemic in the Indian Ocean. *Vector Borne Zoonotic Dis.* **6**, 115–116.

Higgs, S., Schneider, B.S., Vanlandingham, D.L., Klingler, K.A. and Gould, E.A. (2005) Nonviremic transmission of West Nile virus. *Proc. Natl Acad. Sci. USA* **102**, 8871–8874.

Holden, K.L., Stein, D.A., Pierson, T.C., Ahmed, A.A., Clyde, K., Iversen, P.L. and Harris, E. (2006) Inhibition of dengue virus translation and RNA synthesis by a morpholino oligomer targeted to the top of the terminal 3′ stem-loop structure. *Virology* **344**, 439–452.

Holland, J.J. (2006) Transitions in understanding of RNA viruses: a historical perspective. *Curr. Top. Microbiol. Immunol.* **299**, 371–401.

Holland, J. and Domingo, E. (1998) Origin and evolution of viruses. *Virus Genes* **16**, 13–21.

Holland, J.J., De La Torre, J.C., Clarke, D.K. and Duarte, E. (1991) Quantitation of relative fitness and great adaptability of clonal populations of RNA viruses. *J. Virol.* **65**, 2960–2967.

Holmes, E.C. and Twiddy, S.S. (2003) The origin, emergence and evolutionary genetics of dengue virus. *Infect. Genet. Evol.* **3**, 19–28.

Hubalek, Z. (2000) European experience with the West Nile virus ecology and epidemiology: could it be relevant for the New World? *Viral Immunol.* **13**, 415–426.

Hubalek, Z. (2004) An annotated checklist of pathogenic microorganisms associated with migratory birds. *J. Wildl. Dis.* **40**, 639–659.

Hubalek, Z., Cerny, V. and Rodl, P. (1982) Possible role of birds and ticks in the dissemination of Bhanja virus. *Folia Parasitol. (Praha)* **29**, 85–95.

Hughes, A.L. (2001) Evolutionary change of predicted cytotoxic T cell epitopes of dengue virus. *Infect. Genet. Evol.* **1**, 123–130.

James, A.A. (2005) Gene drive systems in mosquitoes: rules of the road. *Trends Parasitol.* **21**, 64–67.

Jerzak, G., Bernard, K.A., Kramer, L.D. and Ebel, G.D. (2005) Genetic variation in West Nile virus from naturally infected mosquitoes and birds suggests quasispecies structure and strong purifying selection. *J. Gen. Virol.* **86**, 2175–2183.

Jerzak, G.V., Bernard, K., Kramer, L.D., Shi, P.Y. and Ebel, G.D. (2007) The West Nile virus mutant spectrum is host-dependant and a determinant of mortality in mice. *Virology* **360**, 469–476.

Jetten, T.H. and Focks, D.A. (1997) Potential changes in the distribution of dengue transmission under climate warming. *Am. J. Trop. Med. Hyg.* **57**, 285–297.

Joshi, V., Mourya, D.T. and Sharma, R.C. (2002) Persistence of dengue-3 virus through transovarial transmission passage in successive generations of *Aedes aegypti* mosquitoes. *Am. J. Trop. Med. Hyg.* **67**, 158–161.

Josseran, L., Paquet, C., Zehgnoun, A., Caillere, N., Le Tertre, A., Solet, J.L. and Ledrans, M. (2006) Chikungunya disease outbreak, Reunion Island. *Emerg. Infect. Dis.* **12**, 1994–1995.

Kalantri, S.P., Joshi, R. and Riley, L.W. (2006) Chikungunya epidemic: an Indian perspective. *Natl Med. J. India* **19**, 315–322.

Karabatsos, N. (1985) *International Catalog of Arboviruses Including Certain Other Viruses of Vertebrates.* San Antonio: American Society of Tropical Medicine and Hygiene.

Karpf, A.R., Lenches, E., Strauss, E.G., Strauss, J.H. and Brown, D.T. (1997) Superinfection exclusion of alphaviruses in three mosquito cell lines persistently infected with Sindbis virus. *J. Virol.* **71**, 7119–7123.

Kawaoka, Y., Cox, N.J., Haller, O., Hongo, S., Kaverin, N., Klenk, H.-D. *et al.* (2005) Orthomyxoviridae. In: *Virus Taxonomy, VIIIth Report of the ICTV* (C.M. Fauquet, M.A. Mayo, J. Maniloff, U. Desselberger and L.A. Ball, eds). London: Elsevier/Academic Press.

Keene, K.M., Foy, B.D., Sanchez-Vargas, I., Beaty, B.J., Blair, C.D. and Olson, K.E. (2004) RNA interference acts as a natural antiviral response to O'nyong-nyong virus (Alphavirus; Togaviridae) infection of Anopheles gambiae. *Proc. Natl. Acad. Sci. USA* **101**, 17240–17245.

Kilpatrick, A.M., Daszak, P., Jones, M.J., Marra, P.P. and Kramer, L.D. (2006a) Host heterogeneity dominates West Nile virus transmission. *Proc. Biol. Sci.* **273**, 2327–2333.

Kilpatrick, A.M., Kramer, L.D., Jones, M.J., Marra, P.P. and Daszak, P. (2006b) West Nile virus epidemics in North America are driven by shifts in mosquito feeding behavior. *PLoS Biol.* **4**, e82.

Kinney, R.M., Huang, C.Y., Rose, B.C., Kroeker, A.D., Dreher, T.W., Iversen, P.L. and Stein, D.A. (2005) Inhibition of dengue virus serotypes 1 to 4 in vero cell cultures with morpholino oligomers. *J. Virol.* **79**, 5116–5128.

Kinney, R.M., Huang, C.Y., Whiteman, M.C., Bowen, R.A., Langevin, S.A., Miller, B.R. and Brault, A.C. (2006) Avian virulence and thermostable replication of the North American strain of West Nile virus. *J. Gen. Virol.* **87**, 3611–3622.

Kondig, J.P., Turell, M.J., Lee, J.S., O'guinn, M.L. and Wasieloski, L.P., Jr. (2007) Genetic analysis of South American eastern equine encephalomyelitis viruses isolated from mosquitoes collected in the Amazon Basin region of Peru. *Am. J. Trop. Med. Hyg.* **76**, 408–416.

Kramer, L.D. and Chandler, L.J. (2001) Phylogenetic analysis of the envelope gene of St. Louis encephalitis virus. *Arch. Virol.* **146**, 2341–2355.

Kuhn, R.J. (2007) Togaviridae: The viruses and their replication. In: *Fields' Virology* (D.M. Knipe and P.M. Howley, eds). 5th edn. New York: Lippincott, Williams and Wilkins.

Kumar, P., Lee, S.K., Shankar, P. and Manjunath, N. (2006) A single siRNA suppresses fatal encephalitis induced by two different flaviviruses. *PLoS Med.* **3**, e96.

Kuno, G. (2001) Transmission of arboviruses without involvement of arthropod vectors. *Acta Virol.* **45**, 139–150.

Kuno, G. and Chang, G.J. (2005) Biological transmission of arboviruses: reexamination of and new insights into components, mechanisms, and unique traits as well as their evolutionary trends. *Clin. Microbiol. Rev.* **18**, 608–637.

La Linn, M., Gardner, J., Warrilow, D., Darnell, G.A., McMahon, C.R., Field, I. *et al.* (2001) Arbovirus of marine mammals: a new alphavirus isolated from the elephant seal louse. *Lepidophthirus macrorhini.J. Virol* **75**, 4103–4109.

Lanciotti, R.S., Roehrig, J.T., Deubel, V., Smith, J., Parker, M., Steele, K. *et al.* (1999) Origin of the West Nile virus responsible for an outbreak of encephalitis in the northeastern United States. *Science* **286**, 2333–2337.

Lescar, J., Roussel, A., Wien, M.W., Navaza, J., Fuller, S.D., Wengler, G. and Rey, F.A. (2001) The Fusion glycoprotein shell of Semliki Forest virus: an icosahedral assembly primed for fusogenic activation at endosomal pH. *Cell* **105**, 137–148.

Lewis, M. and Amsden, J.R. (2007) Successful treatment of West Nile virus infection after approximately 3 weeks into the disease course. *Pharmacotherapy* **27**, 455–458.

Lin, S.R., Hsieh, S.C., Yueh, Y.Y., Lin, T.H., Chao, D.Y., Chen, W.J. *et al.* (2004) Study of sequence variation of dengue type 3 virus in naturally infected mosquitoes and human hosts: implications for transmission and evolution. *J. Virol.* **78**, 12717–12721.

Lindenbach, B.D. and Rice, C.M. (2001) *Flaviviridae:* The viruses and their replication. In: *Fields' Virology*, 4th edn. (D.M. Knipe and P.M. Howley, eds), New York: Lippincott, illiams and Wilkins.

Lobigs, M., Marshall, I.D., Weir, R.C. and Dalgarno, L. (1988) Murray Valley encephalitis virus field strains from Australia and Papua New Guinea: studies on the sequence of the major envelope protein gene and virulence for mice. *Virology* **165**, 245–255.

Lord, R.D. (1974) History and geographic distribution of Venezuelan equine encephalitis. *Bull. PAHO* **8**, 100–110.

Lorono-Pino, M.A., Cropp, C.B., Farfan, J.A., Vorndam, A.V., Rodriguez-Angulo, E.M. *et al.* (1999) Common occurrence of concurrent infections by multiple dengue virus serotypes. *Am. J. Trop. Med. Hyg.* **61**, 725–730.

Mackenzie, J.S., Poidinger, M., Lindsay, M.D., Hall, R.A. and Sammels, L.M. (1995) Molecular epidemiology and evolution of mosquito-borne flaviviruses and alphaviruses enzootic in Australia. *Virus Genes* **11**, 225–237.

Mahmood, F., Reisen, W.K., Chiles, R.E. and Fang, Y. (2004) Western equine encephalomyelitis virus infection affects the life table characteristics of *Culex tarsalis* (Diptera: Culicidae). *J. Med. Entomol.* **41**, 982–986.

Marchette, N.J., Garcia, R. and Rudnick, A. (1969) Isolation of Zika virus from *Aedes aegypti* mosquitoes in Malaysia. *Am. J. Trop. Med. Hyg.* **18**, 411–415.

McGee, C.E., Schneider, B.S., Girard, Y.A., Vanlandingham, D.L. and Higgs, S. (2007) Nonviremic transmission of West Nile virus: evaluation of the effects of space, time, and mosquito species. *Am. J. Trop. Med. Hyg.* **76**, 424–430.

Meegan, J.M. and Bailey, C.L. (1988) Rift Valley fever virus. In: *The Arboviruses: Epidemiology and Ecology*, (T.P. Monath, ed.) Vol. IV. Boca Raton, Florida: CRC Press.

Mendez, W., Liria, J., Navarro, J.C., Garcia, C.Z., Freier, J.E., Salas, R. *et al.* (2001) Spatial dispersion of adult mosquitoes (Diptera: Culicidae) in a sylvatic focus of Venezuelan equine encephalitis virus. *J. Med. Entomol.* **38**, 813–821.

Mertens, P.P.C., Wei, C. and Hillman, B. (2005) Reoviridae. In: *Virus Taxonomy, VIIIth Report of the ICTV* (C.M. Fauquet, M.A. Mayo, J. Maniloff, U. Desselberger and L.A. Ball, eds). London: Elsevier/Academic Press.

Monath, T.P. (1988) Yellow fever. In: *The Arboviruses: Epidemiology and Ecology*, (T.P. Monath, ed.) Vol. V. Boca Raton, Florida: CRC Press.

Murakami, M., Ota, T., Nukuzuma, S. and Takegami, T. (2005) Inhibitory effect of RNAi on Japanese encephalitis virus replication in vitro and in vivo. *Microbiol. Immunol.* **49**, 1047–1056.

Murphy, B.R., Blaney, J.E., Jr. and Whitehead, S.S. (2004) Arguments for live flavivirus vaccines. *Lancet* **364**, 499–500.

Nalca, A., Fellows, P.F. and Whitehouse, C.A. (2003) Vaccines and animal models for arboviral encephalitides. *Antiviral Res.* **60**, 153–174.

Nathanson, N. and Fine, P. (2002) Poliomyelitis eradication-a dangerous endgame. *Science* **296**, 269–270.

Navarro, J.C. and Weaver, S.C. (2004) Molecular phylogeny of the Vomerifer and Pedroi Groups in the Spissipes Section of the subgenus Culex (Melanoconion). *J. Med. Entomol.* **41**, 575–581.

Norder, H., Lundstrom, J.O., Kozuch, O. and Magnius, L.O. (1996) Genetic relatedness of Sindbis virus strains from Europe, Middle East, and Africa. *Virology* **222**, 440–445.

Novella, I.S., Hershey, C.L., Escarmis, C., Domingo, E. and Holland, J.J. (1999) Lack of evolutionary stasis during alternating replication of an arbovirus in insect and mammalian cells. *J. Mol. Biol.* **287**, 459–465.

Nunes, M.R., Travassos Da Rosa, A.P., Weaver, S.C., Tesh, R.B. and Vasconcelos, P.F. (2005) Molecular epidemiology of group C viruses (*Bunyaviridae, Orthobunyavirus*) isolated in the Americas. *J. Virol.* **79**, 10561–10570.

O'Brien, L. (2007) Inhibition of multiple strains of Venezuelan equine encephalitis virus by a pool of four short interfering RNAs. *Antiviral Res.* **75**, 20–29.

Oberste, M.S., Fraire, M., Navarro, R., Zepeda, C., Zarate, M.L., Ludwig, G.V. *et al.* (1998) Association of Venezuelan equine encephalitis virus subtype IE with two equine epizootics in Mexico. *Am. J. Trop. Med. Hyg.* **59**, 100–107.

Olson, J.G., Ksiazek, T.G., Suhandiman, and Triwibowo, (1981) Zika virus, a cause of fever in Central Java, Indonesia. *Trans. R. Soc. Trop. Med. Hyg.* **75**, 389–393.

Olson, K.E., Adelman, Z.N., Travanty, E.A., Sanchez-Vargas, I., Beaty, B.J. and Blair, C.D. (2002) Developing arbovirus resistance in mosquitoes. *Insect Biochem. Mol. Biol.* **32**, 1333–1343.

Ong, S.P., Choo, B.G., Chu, J.J. and Ng, M.L. (2006) Expression of vector-based small interfering RNA against West Nile virus effectively inhibits virus replication. *Antiviral Res.* **72**, 216–223.

Ortiz, D.I. and Weaver, S.C. (2004) Susceptibility of *Ochlerotatus taeniorhynchus* (Diptera: Culicidae) to infection with epizootic (subtype IC) and enzootic (subtype ID) Venezuelan equine encephalitis viruses: evidence for epizootic strain adaptation. *J. Med. Entomol.* **41**, 987–993.

Paessler, S., Fayzulin, R.Z., Anishchenko, M., Greene, I.P., Weaver, S.C. and Frolov, I. (2003) Recombinant Sindbis/Venezuelan equine encephalitis virus is highly attenuated and immunogenic. *J. Virol.* **77**, 9278–9286.

Paessler, S., Ni, H., Petrakova, O., Fayzulin, R.Z., Yun, N., Anishchenko, M. *et al.* (2006) Replication and clearance of Venezuelan equine encephalitis virus from the brains of animals vaccinated with chimeric SIN/VEE viruses. *J. Virol.* **80**, 2784–2796.

Patz, J.A., Martens, W.J., Focks, D.A. and Jetten, T.H. (1998) Dengue fever epidemic potential as projected by general circulation models of global climate change. *Environ. Health Perspect.* **106**, 147–153.

Perales, C., Mateo, R., Mateu, M.G. and Domingo, E. (2007) Insights into RNA virus mutant spectrum and lethal mutagenesis events: Replicative interference and complementation by multiple point mutants. *J. Mol. Biol.* **369**, 985–1000.

Pinheiro, F.P. and Leduc, J.W. (1988) Mayaro virus disease. In: *The Arboviruses: Epidemiology and Ecology* (T.P. Monath, ed.). Boca Raton, Florida: CRC Press.

Platt, K.B., Linthicum, K.J., Myint, K.S., Innis, B.L., Lerdthusnee, K. and Vaughn, D.W. (1997) Impact of dengue virus infection on feeding behavior of Aedes aegypti. *Am. J. Trop. Med. Hyg.* **57**, 119–125.

Poidinger, M., Roy, S., Hall, R.A., Turley, P.J., Scherret, J.H., Lindsay, M.D. *et al.* (1997) Genetic stability among temporally and geographically diverse isolates of Barmah Forest virus. *Am. J. Trop. Med. Hyg.* **57**, 230–234.

Powers, A.M., Brault, A.C., Tesh, R.B. and Weaver, S.C. (2000) Re-emergence of Chikungunya and O'nyong-nyong viruses: evidence for distinct geographical lineages and distant evolutionary relationships. *J. Gen. Virol.* **81**, 471–479.

Powers, A.M., Brault, A.C., Shirako, Y., Strauss, E. G., Kang, W., Strauss, J.H. and Weaver, S.C. (2001) Evolutionary relationships and systematics of the alphaviruses. *J. Virol.* **75**, 10118–10131.

Purse, B.V., Mellor, P.S., Rogers, D.J., Samuel, A.R., Mertens, P.P. and Baylis, M. (2005) Climate change and the recent emergence of bluetongue in Europe. *Nat. Rev. Microbiol.* **3**, 171–181.

Putnam, J.L. and Scott, T.W. (1995) Blood-feeding behavior of dengue-2 virus-infected Aedes aegypti. *Am. J. Trop. Med. Hyg.* **52**, 225–227.

Randolph, S.E. (1998) Ticks are not insects: consequences of contrasting vector biology for transmission potential. *Parasitol. Today,* **14**, 186–192.

Randolph, S.E. (2004) Tick ecology: processes and patterns behind the epidemiological risk posed by ixodid ticks as vectors. *Parasitology* **129**(suppl), S37–S65.

Randolph, S.E. and Rogers, D.J. (2000) Fragile transmission cycles of tick-borne encephalitis virus may be disrupted by predicted climate change. *Proc. Biol. Sci.* **267**, 1741–1744.

Randolph, S.E. and Rogers, D.J. (2002) Remotely sensed correlates of phylogeny: tick-borne flaviviruses. *Exp. Appl. Acarol.* **28**, 231–237.

Rico-Hesse, R. (1990) Molecular evolution and distribution of dengue viruses type 1 and 2 in nature. *Virology* **174**, 479–493.

Rodriguez, L.L., Fitch, W.M. and Nichol, S.T. (1996) Ecological factors rather than temporal factors dominate the evolution of vesicular stomatitis virus. *Proc. Natl Acad. Sci. USA* **93**, 13030–13035.

Rogers, D.J. and Randolph, S.E. (2006) Climate change and vector-borne diseases. *Adv. Parasitol.* **62**, 345–381.

Rudnick, A. (1965) Studies of the ecology of dengue in Malaysia: a preliminary report. *J. Med. Entomol.* **2**, 203–208.

Rudnick, A. (1984) The ecology of the dengue virus complex in Peninsular Malaysia. In: *Proceedings of the International Conference on Dengue/DHF* (T. Pang and R. Pathmanathan, eds). Kuala Lumpur, University of Malaysia Press.

Rudnick, A., Marchette, N.J. and Garcia, R. (1967) Possible jungle dengue—recent studies and hypotheses. *Jpn J. Med. Sci. Biol.* **20**, 69–74.

Ruiz-Jarabo, C.M., Pariente, N., Baranowski, E., Davila, M., Gomez-Mariano, G. and Domingo, E. (2004) Expansion of host-cell tropism of foot-and-mouth disease virus despite replication in a constant environment. *J. Gen. Virol.* **85**, 2289–2297.

Sala, M., Zambruno, G., Vartanian, J.P., Marconi, A., Bertazzoni, U. and Wain-Hobson, S. (1994) Spatial discontinuities in human immunodeficiency virus type 1 quasispecies derived from epidermal Langerhans cells of a patient with AIDS and evidence for double infection. *J. Virol.* **68**, 5280–5283.

Salemi, M., Lamers, S.L., Yu, S., De Oliveira, T., Fitch, W.M. and Mcgrath, M.S. (2005) Phylodynamic analysis of human immunodeficiency virus type 1 in distinct brain compartments provides a model for the neuropathogenesis of AIDS. *J. Virol.* **79**, 11343–11352.

Sammels, L.M., Coelen, R.J., Lindsay, M.D. and Mackenzie, J.S. (1995) Geographic distribution and evolution of Ross River virus in Australia and the Pacific Islands. *Virology* **212**, 20–29.

Sammels, L.M., Lindsay, M.D., Poidinger, M., Coelen, R.J. and Mackenzie, J.S. (1999) Geographic distribution and evolution of Sindbis virus in Australia. *J. Gen. Virol.* **80**, 739–748.

Sanchez-Vargas, I., Travanty, E.A., Keene, K.M., Franz, A.W., Beaty, B.J., Blair, C.D. and Olson, K.E. (2004) RNA interference, arthropod-borne viruses, and mosquitoes. *Virus Res.* **102**, 65–74.

Schlesinger, S. and Schlesinger, M.J. (2001) *Togaviridae*: the viruses and their replication. In: *Fields' Virology* (D.M. Knipe and P.M. Howley, eds), 4th edn. New York: Lippincott, Williams and Wilkins.

Schoepp, R.J., Smith, J.F. and Parker, M.D. (2002) Recombinant chimeric western and eastern equine encephalitis viruses as potential vaccine candidates. *Virology* **302**, 299–309.

Scholle, F., Girard, Y.A., Zhao, Q., Higgs, S. and Mason, P.W. (2004) trans-Packaged West Nile virus-like particles: infectious properties in vitro and in infected mosquito vectors. *J. Virol.* **78**, 11605–11614.

Schuffenecker, I., Iteman, I., Michault, A., Murri, S., Frangeul, L., Vaney, M.C. *et al.* (2006) Genome microevolution of chikungunya viruses causing the Indian Ocean outbreak. *PLoS Med.* **3**, e263.

Scott, T.W. and Lorenz, L.H. (1998) Reduction of *Culiseta melanura* fitness by eastern equine encephalomyelitis virus. *Am. J. Trop. Med. Hyg.* **59**, 341–346.

Scott, T.W. and Weaver, S.C. (1989) Eastern equine encephalomyelitis virus: epidemiology and evolution of mosquito transmission. *Adv. Virus Res.* **37**, 277–328.

Scott, T.W., Takken, W., Knols, B.G. and Boete, C. (2002) The ecology of genetically modified mosquitoes. *Science* **298**, 117–119.

Seligman, S.J. and Gould, E.A. (2004) Live flavivirus vaccines: reasons for caution. *Lancet* **363**, 2073–2075.

Shiu, S.Y., Ayres, M.D. and Gould, E.A. (1991) Genomic sequence of the structural proteins of louping ill virus: comparative analysis with tick-borne encephalitis virus. *Virology* **180**, 411–415.

Shope, R. (1991) Global climate change and infectious diseases. *Environ. Health Perspect.* **96**, 171–174.

Sidwell, R.W. and Smee, D.F. (2003) Viruses of the Bunya- and Togaviridae families: potential as bioterrorism agents and means of control. *Antiviral Res.* **57**, 101–111.

Smith, D.R., Carrara, A.S., Aguilar, P.V. and Weaver, S.C. (2005) Evaluation of methods to assess transmission potential of Venezuelan equine encephalitis virus by mosquitoes and estimation of mosquito saliva titers. *Am. J. Trop. Med. Hyg.* **73**, 33–39.

Smith, D.R., Aguilar, P.V., Coffey, L.L., Gromowski, G.D., Wang, E. and Weaver, S.C. (2006) Venezuelan equine encephalitis virus transmission and effect on pathogenesis. *Emerg. Infect. Dis.* **12**, 1190–1196.

Snapinn, K.W., Holmes, E.C., Young, D.S., Bernard, K.A., Kramer, L.D. and Ebel, G.D. (2007) Declining growth rate of West Nile virus in North America. *J. Virol.* **81**, 2531–2534.

Solomon, T., NI, H., Beasley, D.W., Ekkelenkamp, M., Cardosa, M.J. and Barrett, A.D. (2003) Origin and evolution of Japanese encephalitis virus in southeast Asia. *J. Virol.* **77**, 3091–3098.

Stephenson, J.R. (2006) Developing vaccines against flavivirus diseases: past success, present hopes and future challenges. *Novartis Found. Symp.* **277**, 193–201.

Stewart, J.J., Watts, P. and Litwin, S. (2001) An algorithm for mapping positively selected members of quasispecies-type viruses. *BMC Bioinformatics* **2**, 1.

Strauss, J.H. and Strauss, E.G. (1994) The alphaviruses: gene expression, replication, and evolution. *Microbiol. Rev.* **58**, 491–562.

Sumilo, D., Asokliene, L., Bormane, A., Vasilenko, V., Golovljova, I. and Randolph, S.E. (2007) Climate change cannot explain the upsurge of tick-borne encephalitis in the baltics. *PLoS ONE* **2**, e500.

Tabachnick, W.J. and Powell, J.R. (1979) A world-wide survey of genetic variation in the yellow fever mosquito. *Aedes aegypti. Genet. Res.* **34**, 215–229.

Tatem, A.J., Hay, S.I. and Rogers, D.J. (2006a) Global traffic and disease vector dispersal. *Proc. Natl Acad. Sci. USA* **103**, 6242–6247.

Tatem, A.J., Rogers, D.J. and Hay, S.I. (2006b) Global transport networks and infectious disease spread. *Adv. Parasitol.* **62**, 293–343.

Taylor, W.P. and Marshall, I.D. (1975a) Adaptation studies with Ross River virus: laboratory mice and cell cultures. *J. Gen. Virol.* **28**, 59–72.

Taylor, W.P. and Marshall, I.D. (1975b) Adaptation studies with Ross River virus: retention of field level virulence. *J. Gen. Virol.* **28**, 73–83.

Tesh, R.B., Watts, D.M., Russell, K.L., Damodaran, C., Calampa, C., Cabezas, C. *et al.* (1999) Mayaro virus disease: an emerging mosquito-borne zoonosis in tropical South America. *Clin. Infect. Dis.* **28**, 67–73.

Thavara, U., Siriyasatien, P., Tawatsin, A., Asavadachnuko, P., Ananatapreecha, S., Wongwanich, R. and Mulla, M.S. (2006) Double infection of heteroserotypes of dengue viruses in field populations of *Aedes aegypti* and *Aedes albopictus* (Diptera: Culicidae) and serological features of dengue viruses found in patients in southern Thailand. *Southeast Asian J. Trop. Med. Public Health*, **37**, 468–476.

Thu, H.M., Lowry, K., Myint, T.T., Shwe, T.N., Han, A. M., Khin, K.K. *et al.* (2004) Myanmar dengue outbreak associated with displacement of serotypes 2, 3, and 4 by dengue 1. *Emerg. Infect. Dis.* **10**, 593–597.

Tordo, N., Benmansour, A., Calisher, C., Dietzgen, R.G., Fang, R.-X., Jackson, A.O. *et al.* (2005) *Rhabdoviridae*. In: *Virus Taxonomy, VIIIth Report of the ICTV* (C.M. Fauquet, M.A. Mayo, J. Maniloff, U. Desselberger and L.A. Ball, eds). London: Elsevier/Academic Press.

Tsai, T.F., Weaver, S.C. and Monath, T.P. (2002) Alphaviruses. In: *Clinical Virology* (D.D. Richman, R.J. Whitley and F.G. Hayden, eds). Washington, DC: ASM Press.

Turell, M.J., Sardelis, M.R., O'Guinn, M.L. and Dohm, D.J. (2002) Potential vectors of West Nile virus in North America. *Curr. Top. Microbiol. Immunol.* **267**, 241–252.

Twiddy, S.S., Farrar, J.J., Vinh Chau, N., Wills, B., Gould, E.A., Gritsun, T., Lloyd, G. and Holmes, E.C. (2002a) Phylogenetic relationships and differential selection pressures among genotypes of dengue-2 virus. *Virology* **298**, 63–72.

Twiddy, S.S., Woelk, C.H. and Holmes, E.C. (2002b) Phylogenetic evidence for adaptive evolution of dengue viruses in nature. *J. Gen. Virol.* **83**, 1679–1689.

Van Regenmortel, M.H.V., Fauquet, C.M., Bishop, D.H.L., Carstens, E.B., Estes, M.K., Lemon, S.M. *et al.* (eds) (2000) *Virus Taxonomy. Classification and Nomenclature of Viruses. Seventh Report of the International Committee on Taxonomy of Viruses*. San Diego: Academic Press.

Vasilakis, N., Holmes, E.C., Fokam, E.B., Faye, O., Diallo, M., Sall, A.A. and Weaver, S.C. (2007a) Evolutionary processes among sylvatic dengue-2 viruses. *J. Virol.* **81**, 9591–9595.

Vasilakis, N., Shell, E.J., Fokam, E.B., Mason, P.W., Hanley, K.A., Estes, D.M. and Weaver, S.C. (2007b) Potential of ancestral sylvatic dengue-2 viruses to re-emerge. *Virology* **358**, 402–412.

Vignuzzi, M., Stone, J.K., Arnold, J.J., Cameron, C.E. and Andino, R. (2006) Quasispecies diversity determines pathogenesis through cooperative interactions in a viral population. *Nature* **439**, 344–348.

Wang, E., Weaver, S.C., Shope, R.E., Tesh, R.B., Watts, D.M. and Barrett, A.D. (1996) Genetic variation in yellow fever virus: duplication in the 3′ noncoding region of strains from Africa. *Virology* **225**, 274–281.

Wang, E., Ni, H., Xu, R., Barrett, A.D., Watowich, S.J., Gubler, D.J. and Weaver, S.C. (2000) Evolutionary relationships of endemic/epidemic and sylvatic dengue viruses. *J. Virol.* **74**, 3227–3234.

Wang, W.K., Lin, S.R., Lee, C.M., King, C.C. and Chang, S.C. (2002a) Dengue type 3 virus in plasma is a population of closely related genomes: quasispecies. *J. Virol.* **76**, 4662–4665.

Wang, W.K., Sung, T.L., Lee, C.N., Lin, T.Y. and King, C.C. (2002b) Sequence diversity of the capsid gene and the nonstructural gene NS2B of dengue-3 virus in vivo. *Virology* **303**, 181–191.

Wang, W.K., Chao, D.Y., Lin, S.R., King, C.C. and Chang, S.C. (2003) Concurrent infections by two dengue virus serotypes among dengue patients in Taiwan. *J. Microbiol. Immunol. Infect.* **36**, 89–95.

Watts, D.M., Porter, K.R., Putvatana, P., Vasquez, B., Calampa, C., Hayes, C.G. and Halstead, S.B. (1999) Failure of secondary infection with American genotype dengue 2 to cause dengue haemorrhagic fever. *Lancet* **354**, 1431–1434.

Weaver, S.C. (1995) Evolution of alphaviruses. In: *Molecular Basis of Virus Evolution* (A.J. Gibbs, C.H. Calisher and F. Garcia-Arenal, eds). Cambridge: Cambridge University Press.

Weaver, S.C. (2001) Eastern equine encephalitis. In: *The Encyclopedia of Arthropod-transmitted Infections* (M.W. Service, ed.). Wallingford, UK: CAB International.

Weaver, S.C. (2005) Host range, amplification and arboviral disease emergence. *Arch. Virol. Suppl.*, 33–44.

Weaver, S.C. (2006) Evolutionary influences in arboviral disease. In: *Quasispecies: Concept and implications for virology* (E. Domingo, ed.). Heidelberg: Springer-Verlag.

Weaver, S.C. and Barrett, A.D. (2004) Transmission cycles, host range, evolution and emergence of arboviral disease. *Nat. Rev. Microbiol.* **2**, 789–801.

Weaver, S.C., Scott, T.W., Lorenz, L.H., Lerdthusnee, K. and Romoser, W.S. (1988) Togavirus-associated pathologic changes in the midgut of a natural mosquito vector. *J. Virol.* **62**, 2083–2090.

Weaver, S.C., Lorenz, L.H. and Scott, T.W. (1992a) Pathologic changes in the midgut of *Culex tarsalis* following infection with Western equine encephalomyelitis virus. *Am. J. Trop. Med. Hyg.* **47**, 691–701.

Weaver, S.C., Rico-Hesse, R. and Scott, T.W. (1992b) Genetic diversity and slow rates of evolution in New World alphaviruses. *Curr. Topics Microbiol. Immunol.* **176**, 99–117.

Weaver, S.C., Bellew, L.A., Gousset, L., Repik, P.M., Scott, T.W. and Holland, J.J. (1993) Diversity within natural populations of eastern equine encephalomyelitis virus. *Virology* **195**, 700–709.

Weaver, S.C., Hagenbaugh, A., Bellew, L.A., Gousset, L., Mallampalli, V., Holland, J.J. and Scott, T.W. (1994) Evolution of alphaviruses in the eastern equine encephalomyelitis complex. *J. Virol.* **68**, 158–169.

Weaver, S.C., Salas, R., Rico-Hesse, R., Ludwig, G. V., Oberste, M.S., Boshell, J. and Tesh, R.B. (1996) Re-emergence of epidemic Venezuelan equine encephalomyelitis in South America. VEE Study Group. *Lancet* **348**, 436–440.

Weaver, S.C., Kang, W., Shirako, Y., Rumenapf, T., Strauss, E.G. and Strauss, J.H. (1997) Recombinational history and molecular evolution of western equine encephalomyelitis complex alphaviruses. *J. Virol.* **71**, 613–623.

Weaver, S.C., Brault, A.C., Kang, W. and Holland, J.J. (1999) Genetic and fitness changes accompanying adaptation of an arbovirus to vertebrate and invertebrate cells. *J. Virol.* **73**, 4316–4326.

Weaver, S.C., Coffey, L.L., Nussenzveig, R., Ortiz, D. and Smith, D. (2004a) Vector Competence. In: *Microbe–vector Interactions in Vector-borne Diseases* (S.H. Gillespie, G.L. Smith and A. Osbourn, eds). Cambridge: Cambridge University Press.

Weaver, S.C., Ferro, C., Barrera, R., Boshell, J. and Navarro, J.C. (2004b) Venezuelan equine encephalitis. *Annu. Rev. Entomol.* **49**, 141–174.

Weaver, S.C., Frey, T.K., Huang, H.V., Kinney, R.M., Rice, C.M., Roehrig, J.T. *et al.* (2005) Togaviridae. In: *Virus Taxonomy, VIIIth Report of the ICTV*. (C.M. Fauquet, M.A. Mayo, J. Maniloff, U. Desselberger and L.A. Ball, eds). London: Elsevier/Academic Press.

Weston, J.H., Welsh, M.D., Mcloughlin, M.F. and Todd, D. (1999) Salmon pancreas disease virus, an alphavirus infecting farmed Atlantic salmon. *Salmo salar L. Virology* **256**, 188–195.

Whitehead, S.S., Blaney, J.E., Durbin, A.P. and Murphy, B.R. (2007) Prospects for a dengue virus vaccine. *Nat. Rev. Microbiol.* **5**, 518–528.

Woodall, J. (2001) Chikungunya virus. In: *The Encyclopedia of Arthropod-transmitted Infections* (M.W. Service, ed.). Wallingford, UK: CAB International.

Yanoviak, S.P., Aguilar, P.V., Lounibos, L.P. and Weaver, S.C. (2005) Transmission of a Venezuelan equine encephalitis complex Alphavirus by *Culex (Melanoconion) gnomatos* (Diptera: Culicidae) in northeastern Peru. *J. Med. Entomol.* **42**, 404–408.

Yea, C., Bukh, J., Ayers, M., Roberts, E., Krajden, M. and Tellier, R. (2007) Monitoring of hepatitis C virus quasispecies in chronic infection by matrix-assisted laser desorption ionization-time of flight mass spectrometry mutation detection. *J. Clin. Microbiol.* **45**, 1053–1057.

Zanotto, P.M., Gao, G.F., Gritsun, T., Marin, M.S., Jiang, W.R., Venugopal, K. *et al.* (1995) An arbovirus cline across the northern hemisphere. *Virology* **210**, 152–159.

Zanotto, P.M., Gould, E.A., Gao, G.F., Harvey, P.H. and Holmes, E.C. (1996) Population dynamics of flaviviruses revealed by molecular phylogenies. *Proc. Natl Acad. Sci. USA* **93**, 548–553.

CHAPTER

17

Evolution and Variation of the Parvoviruses

Karin Hoelzer and Colin R. Parrish

ABSTRACT

The parvoviruses are small, non-enveloped viruses that contain ~5000 bases of linear single-stranded DNA (ssDNA), and they are widespread in nature, infecting many different animals, from crustaceans to primates. The viruses contain two large genes that encode proteins associated with the control of DNA replication and the capsid proteins, as well as variants of those proteins and some small accessory proteins. These viruses replicate using the host cell polymerases, or for the adeno-associated viruses (AAVs) using helper virus polymerase, but the amount of variation seen in nature or after tissue culture passage appears high, and the rates of sequence change of the viruses can be similar that of some RNA viruses. Recombination clearly occurs among the parvoviruses, and may be an under-appreciated evolutionary mechanism. In specific examples the parvoviruses are shown to have different and often complex evolutionary processes. The viruses appear to be remarkably adaptable in some cases; canine parvovirus for example was able to transfer between hosts in a process that involved complex evolutionary mechanisms.

PARVOVIRUSES AND THEIR PROPERTIES

Parvoviruses are a family of small viruses with a non-enveloped capsid that contains a linear ssDNA genome of between 4500 and 5250 nucleotides. The viruses are very widely distributed in nature, and likely infect most vertebrate and invertebrate hosts (Cotmore and Tattersall, 1987; Murphy *et al.*, 1995; Tattersall *et al.*, 2005). The two subfamilies within the family *Parvoviridae* are the subfamily *Parvovirinae*, members of which infect vertebrate hosts, and the subfamily *Densovirinae*, members of which infect invertebrates. Within the *Parvovirinae* the phylogenetic relationships can be shown using a conserved region of the NS1 gene sequences. The subfamily *Parvovirinae* falls into several distinct, well separated clades (Figure 17.1). The five recognized genera are the parvoviruses (autonomous parvoviruses) including various rodent parvoviruses such as minute virus of mice (MVM) and H1, the parvoviruses of carnivores including canine parvovirus (CPV), feline panleukopenia virus (FPV), and mink enteritis virus (MEV), the erythroviruses (human B19 virus and related

Origin and Evolution of Viruses
ISBN 978-0-12-374153-0

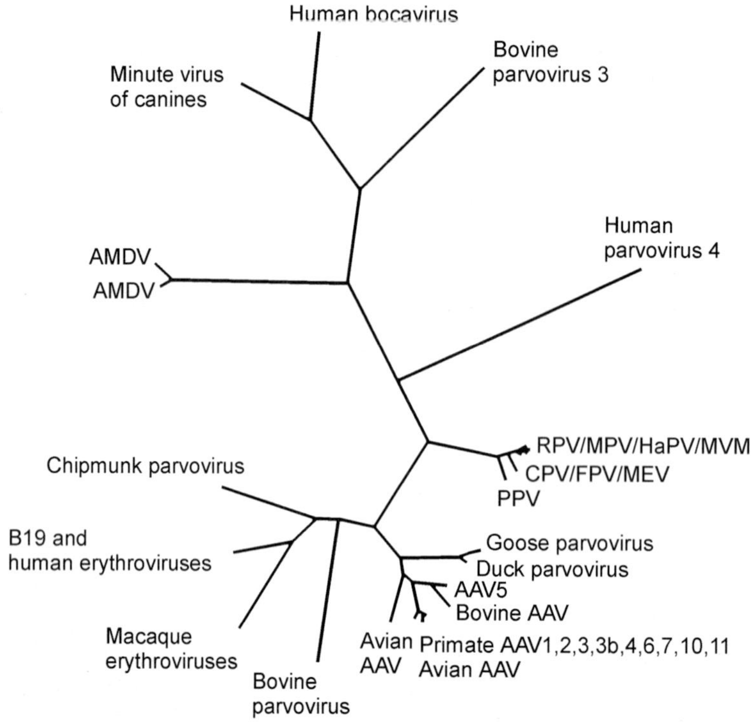

FIGURE 17.1 The evolutionary relationship among the parvoviruses of vertebrates. A conserved region of the NS1 amino acid sequence of representative parvoviruses of vertebrates (residues 277–387 of MVM reference sequence, protein accession number NP_041242) was aligned by eye and using the Clustal W algorithm available in MegAlign. Phylogenetic relationships were inferred using PAUP* version 4.0b10 for Unix (Swofford, 1993). The initial tree was constructed by neighbor-joining and a heuristic tree search algorithm using nearest-neighbor interchange (NNI) was employed to find the optimal tree.

viruses of primates), bocaviruses (bovine parvovirus type 1, minute virus of canines (MVC) and human bocavirus), amdoviruses (Aleutian mink disease virus (AMDV)), and the dependoviruses (adeno-associated viruses (AAV) of humans and other hosts), which primarily replicate in cells that are co-infected with an adenovirus or herpesvirus. In the *Densovirinae* the four genera are the densoviruses, iteraviruses, pefudensoviruses, and brevidensoviruses. Members of the subfamily *Densovirinae* infect many different invertebrate hosts from the classes Insecta and Crustacea. Viruses have been isolated from various insect hosts, including members of the orders Lepidoptera, Diptera, and Orthoptera. Several viruses infect members of the crustacean order Decapoda, but those viruses are still less well characterized. Members of the *Densovirinae* are considerably more diverged than those of the *Parvovirinae*.

GENE STRUCTURE AND REPLICATION

The parvovirus virion contains a single ssDNA molecule. Most members of the parvovirus genus, such as MVM, H1, or CPV preferentially package the negative strand, while others such as the dependoviruses and erythroviruses and the parvovirus LuIII package ssDNA of both polarities with similar frequencies. The bovine parvovirus type 1 preferentially packages negative sense DNA, but approximately 10% of viral capsids contain positive strand DNA (Chen *et al.*, 1988). The detailed processes involved in ssDNA packaging and the molecular determinants responsible for the packaging bias involve the relative efficiency of the left- and right-hand origins of replication in initiating replication (Cotmore and Tattersall, 2005). Parvoviruses use parts of the host cell polymerase complex for viral DNA replication including the α and/or δ polymerases, perhaps among others (Kollek *et al.*, 1982; Berns, 1990; Cossons *et al.*, 1996; Bashir *et al.*, 2000), as well as a variety of cellular proteins including cyclin A and chaperone-associated proteins. Viral DNA replication depends on the host cell passing through S-phase (Cotmore and Tattersall, 1995). Some parvovirus proteins appear to affect the host cell cycle (Morita *et al.*, 2001, 2003; Op De Beeck *et al.*, 2001), generally by leading to cell cycle arrest in G_1 or G_2. The ssDNA genome is filled in through reactions that may involve the cellular DNA-repair responses. In the case of the AAVs the viral genome is converted into circular and concatameric forms in the absence of a helper virus to give active DNA replication. However, parvoviruses do not directly induce mitosis, but must wait until the host cell enters S-phase (Op De Beeck and Caillet-Fauquet, 1997). Dependoviruses require additional factors for productive infection. Genotoxic stimuli such as UV light, cyclohexamine treatment, and heat exposure can lead to permissive replication (reviewed in Berns, 1990), but dependoviruses replicate mostly in the presence of early gene products of their helper viruses.

Replication initiates from the incoming genome which is converted into a double-stranded (ds)DNA intermediate within the nucleus, and then proceeds through a modified rolling-circle mechanism (rolling-hairpin), where the palindromes located in each end of the genome are used for replication of the virus templates (Hong *et al.*, 1994; Berns and Linden, 1995; Cotmore and Tattersall, 1995; Cotmore, 1996). The dependoviruses and erythroviruses have identical palindromes at the 3′ and 5′ end (inverted terminal repeats (ITRs)), while the parvoviruses, bocaviruses, and ambdoviruses have different palindromes, leading to more complex replication mechanisms. The 3′ end of the genome forms a base-paired structure that primes second-strand synthesis, leading to the generation of the first replicative form (RF) DNA, which extends and unfolds the 5′ palindrome. Subsequent replication is NS1-dependent and creates origins by nicking specific sites in the sequence to create a new 5′ end. The NS1 helicase activity unfolds the 5′ end of the DNA, which is then replicated and refolds, a process termed hairpin transfer, giving inverted palindromes. Subsequently replication proceeds through the formation of dimeric or tetrameric RF DNA, where nicking by the viral NS1 protein and strand exchange is involved in the resolution of the DNA replication intermediates (Cotmore and Tattersall, 2003). There are a number of aspects of this replication scheme that would favor higher substitution rates compared to the normal cellular DNA replication. The replication may not require the complete polymerase complex used for the synthesis of the host DNA, the genome is likely not methylated so that the template strand is not identified during replication, the single-stranded form of the genome might be vulnerable to base conversion and the initial fill-in is not error corrected.

VIRAL GENE FUNCTIONS

The parvoviruses are genetically simple, and their genomes contain two large open reading frames (ORFs), where the left-hand ORF encodes the non-structural genes and the right hand ORF encodes the capsid proteins (Figure 17.2). The genomes contain between one and three promoters, depending on the virus. Alternative splicing, alternative transcription start sites, or alternative initiation of translation can give rise to additional gene products, and the two major ORFs give rise to messages for between one and four non-structural proteins, and between two and four capsid proteins. The bocaviruses contain a third ORF, located between the other two ORFs which encodes a phosphorylated NP1 non-structural protein of unknown function (Lederman *et al.*,

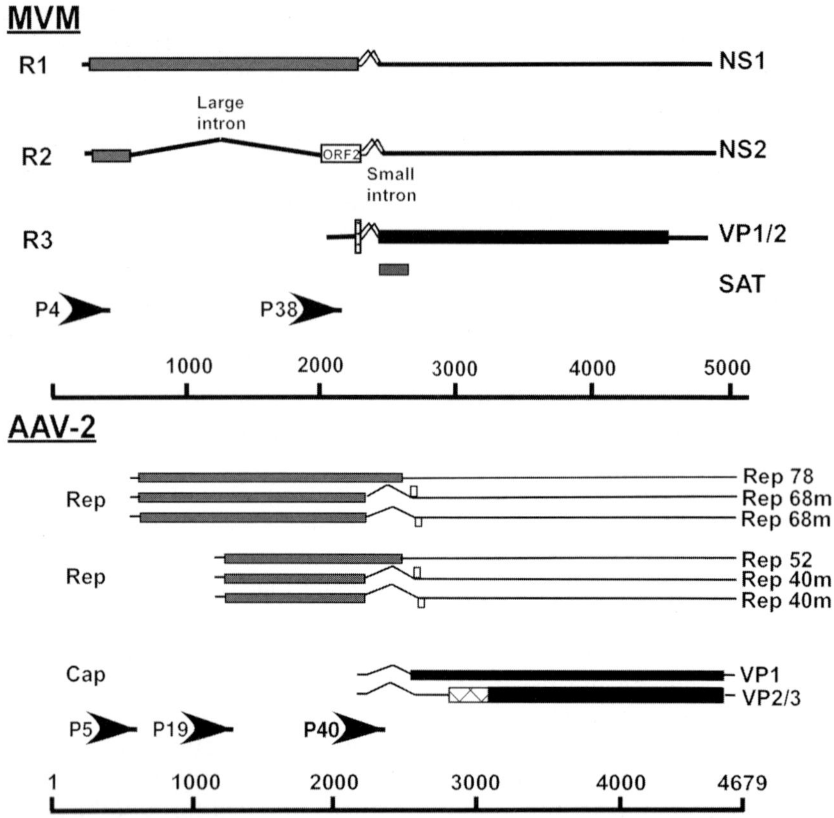

FIGURE 17.2 The genome structures of the parvoviruses minute virus of mice (MVM) and adeno-associated virus, showing the general properties of the parvovirus genome. The members of the *Parvoviridae* vary in having from one to three transcriptional promoters and one or more sites of poly A addition, in the presence of smaller non-structural protein genes, and in whether they have identical or different terminal palindromes. R, transcriptional products; NS or Rep, non-structural proteins involved in replication; VP, viral capsid protein; SAT, small open reading frame of largely unknown function.

1984; Chen *et al.*, 1986; Schwartz *et al.*, 2002; Allander *et al.*, 2005).

NON-STRUCTURAL PROTEINS

The NS1 and Rep proteins in the parvoviruses and dependoviruses, respectively, are multifunctional proteins required for DNA replication, for regulation of viral gene expression, and for integration of some AAV genomes into the host DNA. They have site-specific nickase activity, ATPase, ligase and helicase activities. During replication they are covalently attached to the 5′ end of the viral DNAs. NS1 interacts with various cellular proteins including transcription regulators and members of the replication machinery, and it induces cytopathic effects. The phosphorylation status of the protein is intricately involved in regulating the various NS1 functions (Nuesch *et al.*, 2001, 2003; Lachmann *et al.*, 2003), and its cytopathic potential (Daeffler *et al.*, 2003). The smaller Rep proteins lack the DNA-binding and replication functions of the larger Rep proteins, but do retain the helicase and ATPase function, stimulate the production of ssDNA and are likely to be involved in packaging of the viral DNA (Chiorini *et al.*, 1995; Dubielzig *et al.*, 1999; Smith and Kotin, 2000). The NS2 proteins of MVM or the rat virus LuIII appear dispensable for viral replication in host cells other than mouse or rat cells respectively, but are required for efficient translation or assembly of the virus capsid in the natural host cells (Li and Rhode, 1991; Naeger *et al.*, 1993; Cotmore *et al.*, 1997; Eichwald *et al.*, 2002; Choi *et al.*, 2005; D'Abramo *et al.*, 2005). The functions of NS2 are not completely defined, but the protein interacts with 14-3-3 family member proteins and with the Crm1 proteins involved in nuclear export (Brockhaus *et al.*, 1996; Bodendorf *et al.*, 1999; Miller and Pintel, 2002). NS2 is itself regulated by phosphorylation and may be involved in phosphorylation of the capsid proteins and nuclear export of capsids (D'Abramo *et al.*, 2005). In CPV NS2 does not appear required for virus replication in dog and cat cells in tissue culture or in dogs (Wang *et al.*, 1998).

CAPSID PROTEINS AND GENES

VP1 and VP2 are produced by a variety of strategies, including alternative splicing of the viral mRNA producing ~10% VP1 and 90% VP2 (Cotmore and Tattersall, 1987). The proteins overlap in sequence so that the entire 62–70 kDa VP2 is contained within the VP1 sequence, with VP1 having a unique N-terminal extension (~120 and 150 amino acids depending on the virus). Erythroviruses and ambdoviruses have capsids containing only VP1 and VP2 while the capsids of dependoviruses contain VP1, VP2, and VP3 which are obtained by alternative splicing of the mRNA. The full (DNA containing) capsids of members of the parvovirus genus contain a smaller VP3, which is generated from the VP2 protein by proteolytic cleavages of the N-terminus.

Located in the VP1 unique region of most parvoviruses is a phospholipase 2 (PLA2) enzyme domain required for cell infection (Zadori et al., 2001; Girod *et al.*, 2002; Farr *et al.*, 2005) most likely because it modifies the endosomal membrane and allows the viral particle to penetrate into the cytoplasm (Suikkanen *et al.*, 2003; Farr *et al.*, 2005).

STRUCTURE AND VARIATION OF THE VIRAL CAPSID

The parvovirus capsids are small and appear to contain only a limited number of antigenic sites. However, those have mostly been defined using analysis of naturally or selected antibody escape mutants or by analysis of peptide binding, and it is not clear how accurately those represent the true binding sites of the normal host antibodies. Most antigenic sites appear to be conformation dependent, but linear epitopes defined include the N-terminus of VP2 which is exposed to the exterior of CPV full (DNA-containing) capsids, and the N-terminus of VP1 for the B19 parvoviruses (Yoshimoto *et al.*, 1991; Saikawa *et al.*, 1993; Strassheim *et al.*, 1994; Lopez-Bueno *et al.*, 2003) and AAV2 capsids (Wistuba *et al.*, 1997; Wobus *et al.*, 2000). The role of antigenic variation and immune selection in the evolution of most parvoviruses is not well understood, and in some cases the analysis is complicated as many of the changes which result in antigenic variation also alter host range or other properties of the virus by altering binding to the cellular receptor (Chang *et al.*, 1992; Parker and Parrish, 1997). Genetically distinct viruses can be distinguished by serology with polyclonal sera, but within many virus groups little antigenic variation has been described, even among viruses that have been separated by decades.

Variation of capsid epitopes of natural CPV or MEV isolates has been identified using monoclonal antibody analysis, and some variants have become widely distributed around the world (Parrish *et al.*, 1991; Strassheim *et al.*, 1994; Martella *et al.*, 2004; Nakamura *et al.*, 2004). For MVM antigenic variants were selected in tissue culture after selection with neutralizing monoclonal antibodies, but it is not known whether such variation is common among the viruses in nature (Lopez-Bueno *et al.*, 2003). For many parvoviruses, animals that have recovered from infection resist reinfection by antigenically related viruses, even if variant at one or more epitopes. However, in natural transmission cycles low levels of waning maternal antibody may allow infection and provide an environment allowing selection of antigenic variation.

EPIDEMIOLOGY AND ANTIVIRAL IMMUNITY

The epidemiology and pathogenesis of the virus infections varies widely, and likely influences the variation and evolution of the viruses. Many autonomous parvoviruses, including canine parvovirus (CPV), feline panleukopenia virus (FPV), porcine parvovirus (PPV), and B19 human parvovirus, normally cause acute infections of their hosts lasting <10 days. Virus is usually cleared by the host immune response within this time span, after which infectious virus is not present, and the recovered hosts are no longer infectious for other animals (Musiani *et al.*, 1995; Parrish, 1995; Truyen and Parrish, 2000). However, prolonged replication and persistence may occur in some of those hosts, particularly during fetal infection or when immune suppressed. Persistent infections may affect the development of intra-host variation, permit mixed infections, and may result in prolonged shedding with the re-introduction of older viruses back into circulation. Persistent infection and shedding occurs for some viruses, including rodent parvoviruses where infection in the kidneys results in persistent shedding in the urine. AMDV-infected adult mink are persistently infected and the virus continues to replicate in a number of tissues for long periods (Porter, 1986; Alexandersen *et al.*, 1988; Jacoby *et al.*, 1991, 1996, 2000; Pennick *et al.*, 2005). In neonatal mink AMDV causes an acute disease without persistence developing. Viral replication in neonates occurs in type II pneumocytes, while in older animals replication is limited to macrophages in the lymph nodes and shows reduced transcription and replication. In many genotypes of mink a chronic immune complex-mediated disease develops (Alexandersen, 1986; Alexandersen *et al.*, 1994; Jensen *et al.*, 2000). Chronic and persistent B19 infections can occur in humans who are immune suppressed, or who do not develop effective immunity (Frickhofen and Young, 1989; Brown *et al.*, 1994; Young, 1995). Even in normal infections there is low-level persistence of B19 DNA in the presence of antibodies, although that DNA may not be infectious (Candotti *et al.*, 2004; Lefrere *et al.*, 2005; Manning *et al.*, 2007). Secondary infections by variant erythroviruses have been reported (Hattori *et al.*, 2007; Kaufmann *et al.*, 2007).

The epidemiology of the newly discovered human bocavirus (Mahy, 2006) is not well understood, but clinical disease appears to occur in the presence of other respiratory virus infections (Fry *et al.*, 2007). Whether human bocavirus is persistent is not yet known. Some bovine parvoviruses can be recovered from fetal bovine sera, suggesting that those may also cause longer term or fetal infections (Allander *et al.*, 2001). The recently discovered human parvoviruses PARV4 and PARV5 appear generally apathogenic, although an association with HIV infection is possible (Manning *et al.*, 2007).

MECHANISMS OF TRANSMISSION

Viruses similar to CPV use a fecal–oral transmission route, but replicate systemically in the host tissues before invading the intestine (Macartney *et al.*, 1984a, 1984b; Parrish, 1995). Some rodent parvoviruses also replicate in the intestine, but may be transmitted through urine after replication in the kidney (Ball-Goodrich *et al.*, 1998; Jacoby *et al.*, 1996). Although the human B19 virus replicates primarily in the bone marrow, it is thought to be transmitted by respiratory routes (Brown and Young, 1995). Parenteral transmission is possible as B19 virus is inefficiently inactivated in blood and blood products (Aubin *et al.*, 2000; Parsyan *et al.*, 2006; Bianchi *et al.*, 2007; Hattori *et al.*, 2007). Veneral transmission might be important for porcine parvovirus (PPV) as PPV or viral DNA can be isolated from pig semen (Guerin and Pozzi, 2005). Vertical transmission has been shown for several members of the parvovirus genus including several rodent parvoviruses, FPV, and PPV (Joo *et al.*, 1976; Mengeling *et al.*, 1980). Transplacental transmission can occur for many parvoviruses, although the fetus might be protected by maternal antibodies, depending on the host species and type of placentation (Hayder *et al.*, 1983; Broll and Alexandersen, 1996; Eis-Hubinger *et al.*, 1998; de Haan *et al.*, 2007). Since the non-enveloped parvovirus capsid is very resistant in the environment, viral spread through vectors or formites occurs and might explain the global spread observed for some parvoviruses such as CPV and B19.

IMMUNE RESPONSE AND PROTECTION

There is clearly an important role for cell-mediated immunity in recovery from infection and there may be limited numbers of T cell epitopes in the viral proteins (Kasprowicz *et al.*, 2006). However, for many vertebrate viruses, humoral immunity, including maternal antibody, protects against infection by antigenically related viruses. Antibody treatments can arrest CPV replication in dogs, and can terminate chronic human infections by the B19 parvovirus (Brown *et al.*, 1994; Brown and Young, 1995). In the case of CPV, FPV, PPV, and B19 the virus is functionally inactivated within a few days of the antibody response developing, while in contrast (as described previously) AMDV forms persistent infections despite the strong immune response (Porter, 1986; Alexandersen *et al.*, 1988).

GENETIC RELATIONSHIPS AND VARIATION AMONG THE PARVOVIRUSES

Comparing the sequences of conserved regions of the genome shows that all parvoviruses have sequences in common, and are related through a distant common ancestor (Figure 17.1). The viruses are divided into distinct clades which may show some correlations to the hosts of origin (Lukashov and Goudsmit, 2001). As more viruses are collected and analyzed, a variety of more-or-less related viruses have been seen to infect many species. For example, three distinct bovine parvoviruses (types 1, 2, and 3) infect cattle (Allander *et al.*, 2001), and there are two distantly related parvoviruses infecting dogs (Schwartz *et al.*, 2002). The several erythroviruses from primates (B19 and related human and simian parvoviruses) are most closely related to each other and to the chipmunk parvovirus (Yoo *et al.*, 1999; Lukashov and Goudsmit, 2001). Most viruses from rats, mice, and hamsters were found to be within the same clade as CPV and the related viruses of carnivores and PPV, while AMDV was found to be quite distantly related to all the other vertebrate viruses (Figure 17.1). The LuIII virus proved to be a recombinant between two different rodent viruses, most likely between MPV and hamster PV (Lukashov and Goudsmit, 2001), although whether that occurred in nature or during passage of the virus in tissue culture is not clear (Figure 17.3). True times of divergence of the various viruses are not known as no long-term molecular clock can be estimated. Whether particular vertebrate parvoviruses have co-evolved along with their hosts or have transformed between different hosts in more recent times is also not known, although both origins are likely for different viruses.

HUMAN B19 AND RELATED ERYTHROVIRUSES

There are several human erythrovirus strains that are related to B19. Strain variation of the B19-related viruses so far shows no association with specific clinical signs or tissue tropisms (Erdman *et al.*, 1996; Hemauer *et al.*, 1996; Dorsch *et al.*, 2001; Gallinella *et al.*, 2003; Sanabani *et al.*, 2006). Three genotypes have been identified (Servant *et al.*, 2002), with the prototype B19 viruses classified as genotype 1 (Figure 17.4). Genotype 2 was first identified as the LaLi strain (originally isolated from a Finnish skin sample), and the A6 strain, which are ~11% divergent from the genotype 1 strains at the nucleotide level (Hokynar *et al.*, 2002; Servant *et al.*, 2002; Wong *et al.*, 2003). Genotype 3 was originally from a patient with aplastic crisis in Paris in

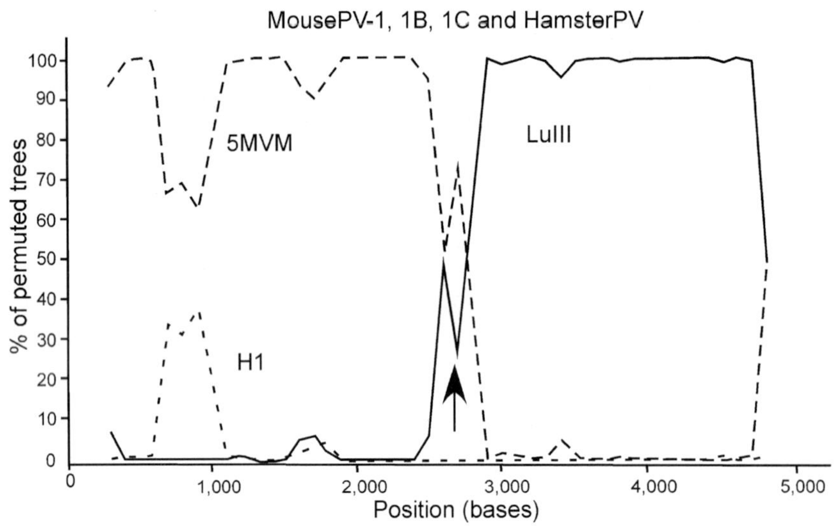

FIGURE 17.3 Bootscan analysis of the phylogenetic relationships among the parvovirus LuIII and parvoviruses from mice and hamsters. The genomes of mouse parvoviruses and the hamster parvovirus (query sequence group) were compared to MVM-related viruses (comparison group 1), LuIII (comparison group 2) and H1 (outgroup). The position of a putative recombination breakpoint is shown (arrow) around the middle of the genome. Modified from Lukashov and Goudsmit (2001) with permission.

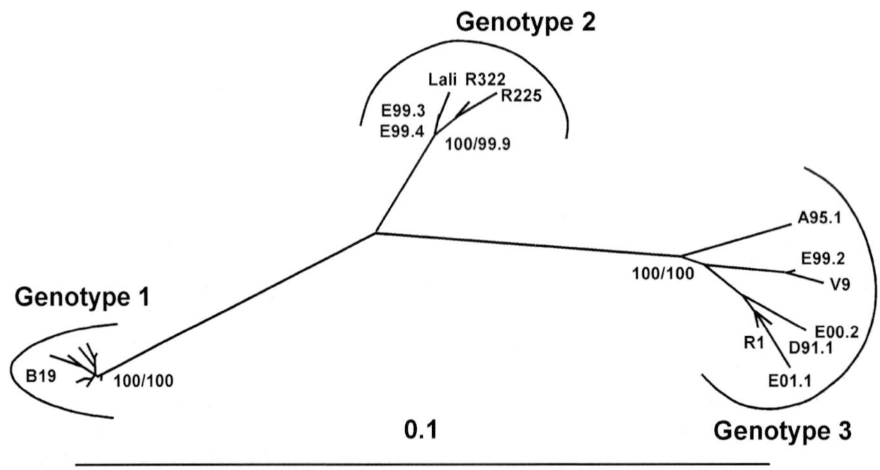

FIGURE 17.4 The diversity in the sequences of human erythroviruses related to the prototype B19 virus, showing the three distinct genotypes that have been characterized. From Servant *et al.* (2002) with permission.

1995 (as the V9 virus) (Nguyen *et al.*, 1999), and is ~12% divergent from B19 at the nucleotide level. The different human erythroviruses are between 5 and 20% different at the DNA sequence level (Hokynar *et al.*, 2002; Servant *et al.*, 2002; Gallinella *et al.*, 2003; Candotti *et al.*, 2004), but most mutations are synonymous and the viruses are >96% homologous in amino acid sequence. Antibodies against genotypes 1 or 2 are cross-reactive (Ekman *et al.*, 2007; Heegaard and Brown, 2002). When examining the B19-related viruses, between 1 and 4% sequence variation was found between isolates within each clade. The simian erythroviruses were 40–50% divergent at the nucleotide level from human erythroviruses.

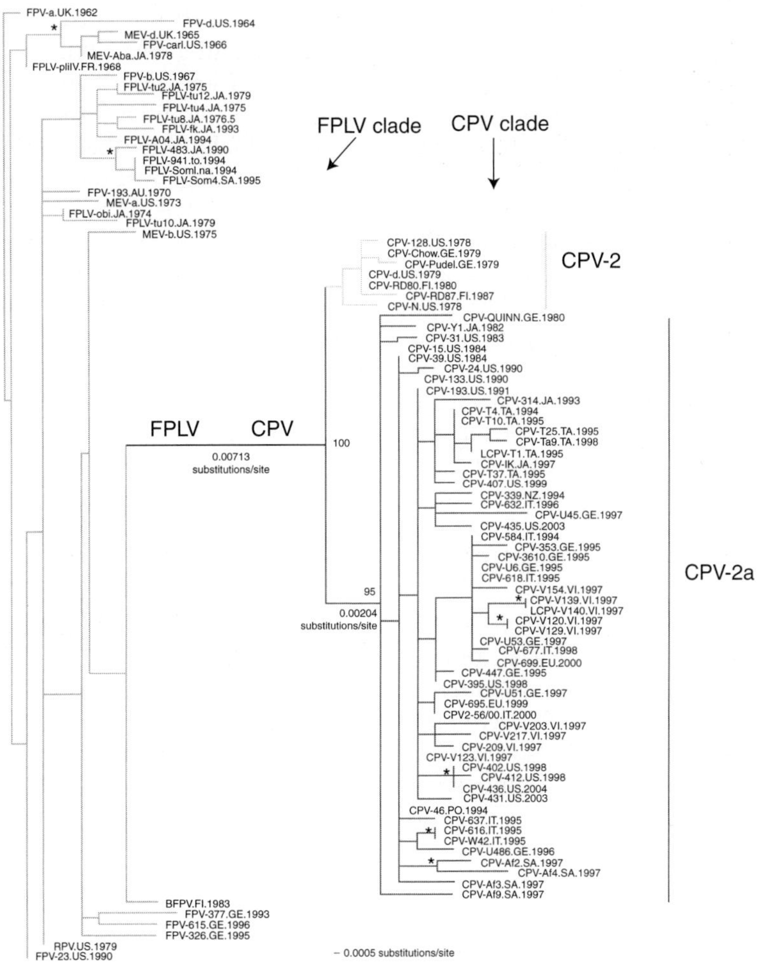

FIGURE 17.5 Phylogenetic tree of 91 VP2 gene sequences from carnivore parvoviruses, rooted with the oldest sampled sequence. Bootstrap values are shown for relevant nodes, and nodes with >70% support are marked with an asterisk. Horizontal branch lengths are drawn to scale. The name of each isolate is followed by the location and year of isolation. Locations are coded as follows: UK, United Kingdom; US, United States; JA, Japan; FR, France; AU, Australia; FI, Finland; GE, Germany; TA, Taiwan; NZ, New Zealand; VI, Vietnam; EU, Europe (no further information available); PO, Poland; IT, Italy; SA, South Africa. The FPLV clade is shown in blue, the CPV-2 subclade is shown in yellow, and the CPV-2a subclade is shown in red. BFPV, blue fox parvovirus. (See Plate 26 for the color version of this figure.) From Shackelton *et al.* (2005) with permission.

CPV AND RELATED VIRUSES

DNA sequences of viruses from dogs, cats, raccoons, mink, and arctic foxes have been examined in a number of studies (Truyen *et al.*, 1995; Horiuchi *et al.*, 1998; Ikeda *et al.*, 2002; Shackelton *et al.*, 2005). CPV was recognized in 1978 as the cause of new diseases of dogs, and all canine isolates are derived from a common ancestor which arose during the late 1960s or early 1970s (Truyen *et al.*, 1995; Shackelton *et al.*, 2005) (Figure 17.5). The sequences of viruses from cats, mink, raccoons, and foxes all fell within a single clade, suggesting that interspecies transmission of those viruses occurs. The specific origin of CPV has not been completely defined but the most closely related virus was an isolate from an arctic fox in Finland (Figure 17.5). It was suggested early on that CPV might be derived from vaccine strains of FPV (Tratschin *et al.*, 1982), but DNA sequence analysis showed the CPV isolates were unrelated to vaccine strains tested (Truyen *et al.*, 1998). Serology and analysis of viral DNA from paraffin-embedded tissues (Truyen *et al.*, 1994) show that CPV was circulating in Europe between 1974 and 1976. Many of the changes between FPV and CPV were identified in the capsid (Figure 17.6). Significant levels of antigenic variation have been demonstrated by monoclonal antibody (MAb) analysis, which showed that CPV isolates differed from viruses of cats, mink, or raccoons in two specific neutralizing epitopes—one present only on CPV, and the other present on FPV isolates (Parrish and Carmichael, 1983; Mochizuki *et al.*, 1989; Parrish *et al.*, 1991; Strassheim *et al.*, 1994). During the later evolution of CPV a number of mutations altered the antigenic structure of capsid. Many of the changes which altered the antigenic structure of the capsid also changed the transferrin receptor (TfR) binding sites, making it difficult to determine which functions are under selection.

The various FPV and CPV isolates differ in their abilities to infect cells in culture and animals, and the subsequent evolution of CPV in dogs further changed their natural host ranges. The CPV type 2 isolates did not replicate in cats, but later viruses (CPV type 2a and its derivatives) replicated efficiently in cats (Truyen *et al.*, 1996a). This was a natural host range for the viruses as CPV type 2a and related viruses were isolated from 10–20% of cats which had natural parvovirus infections in Asia, Germany, Italy, and the USA during the 1990s (Mochizuki *et al.*, 1993; Truyen *et al.*, 1996b; Ikeda *et al.*, 2000; Battilani *et al.*, 2006).

ALEUTIAN MINK DISEASE VIRUS (AMDV)

There is significant variation in the genomes of AMDV isolates, and the sequence variability of the viruses from a single farm or region was seen to be up to 16% at the nucleotide

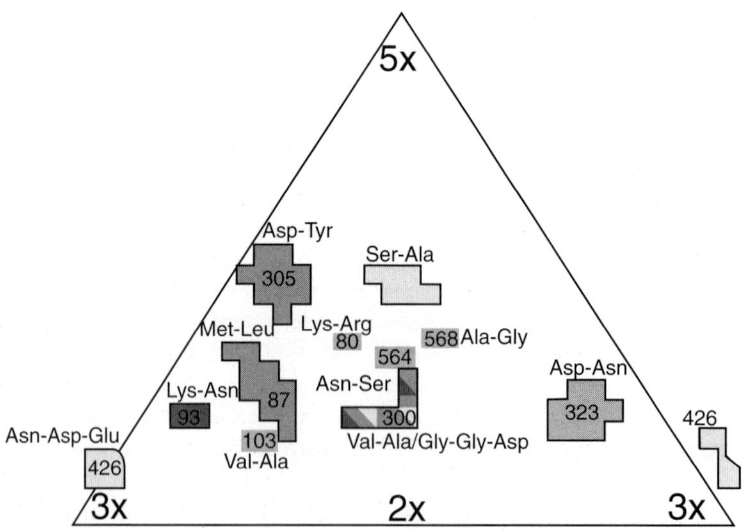

FIGURE 17.6 A model of one asymmetric unit of the CPV capsid, showing the locations of substituted amino acids. Sites that changed along the FPV→CPV branch are in blue, with the darker shade indicating surface-exposed residues and the lighter shade showing the relative positions of subsurface sites. Sites that underwent substitutions along the CPV→CPV-2a branch are in green, and sites responsible for variants within the CPV-2a subclade are in yellow; all are surface-exposed. (See Plate 27 for the color version of this figure.) From Shackelton *et al.* (2005) with permission.

sequence level (Olofsson *et al.*, 1999). AMDV isolates normally grow only in animals, but certain strains have been adapted to grow in feline cells in tissue culture. A 2.5% sequence difference was seen between a wild-type pathogenic strain of AMDV and a tissue culture adapted virus, and a hypervariable sequence within the capsid protein gene was on the surface of the VP2 protein structure (Bloom *et al.*, 1988; Oie *et al.*, 1996; McKenna *et al.*, 1999). AMDV has also been recovered from ferrets, skunks, and raccoons. The raccoon and skunk viruses appear similar to viruses isolated from farmed mink, although it is not known whether there is natural transmission between those hosts (Oie *et al.*, 1996).

RODENT PARVOVIRUSES

The known and suspected rodent parvoviruses are all >70% identical in DNA sequence, while viruses that are considered to be of the same type and given the same name are generally >95% identical (Figure 17.7). The rodent parvoviruses MVM, LuIII, and H1 have been intensively studied for their genetic and biochemical properties, but there is less information available

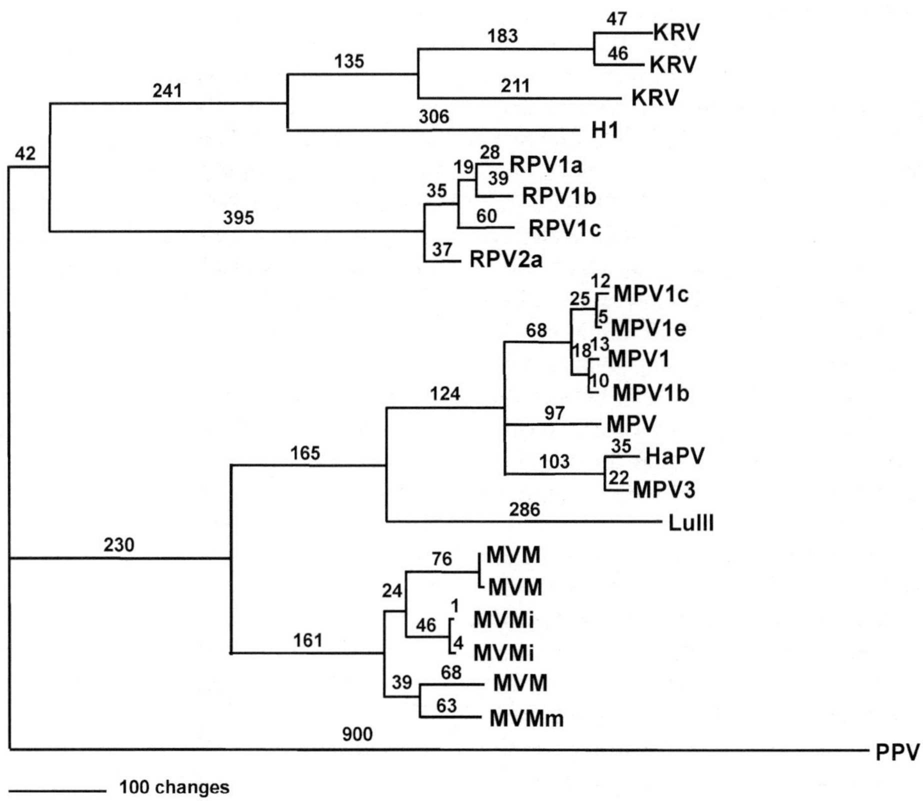

FIGURE 17.7 Phylogenetic relationships among the rodent parvoviruses determined from the almost-complete genome sequences. Nucleotide sequences spanning the complete coding region (nucleotides 300–4251 of the H1 sequence, GeneBank accession number NC_001358) were obtained from GeneBank and aligned using Clustal X. A maximum likelihood tree was constructed using PAUP* version 4.0b10 for Unix (Swofford, 1993), rooted on porcine parvovirus (PPV) which was treated as an outgroup. Horizontal branch lengths are drawn to scale, and the absolute number of substitutions between virus sequences are indicated. Viruses are: Kilham rat virus (KRV), H1 (H1), rat parvovirus (RPV), mouse parvovirus (MPV), and minute virus of mice (MVM), which are divided into distinct clades.

about their variation and evolution in nature. Many early strains were isolated from tissue cultures or transplantable tumors, and their true origin and degree of tissue culture or host adaptation are not fully understood. Those viruses include the MVM prototype strain (MVMp) isolated from a murine adenovirus stock (Crawford, 1966), H1 virus from the HEP-1 human tumor transplanted into rats and likely a rat virus (Toolan, 1960) and LuIII from human cells. A tissue culture isolates of MVM isolated from lymphocytes in 1976 and immune suppressive *in vitro* was named MVMi (Bonnard *et al.*, 1976) while MVM-Cutter (MVM-(c)) was a contaminant of BHK21 cells (Besselsen *et al.*, 1995). A number of viruses have been detected directly in mice, including mouse parvovirus (MPV)) strains 1a, 1b, 1c, 2, and 3, a new MVM strain MVMm which is pathogenic in non-obese diabetic (NOD) mice, as well as viruses of hamsters (hamster parvovirus (HaPV)), and rats (rat parvovirus (RPV)) (Besselsen *et al.*, 1996, 2006; Ball-Goodrich *et al.*, 1998; Wan *et al.*, 2002, 2006). These viruses appear common in wild rodents (Becker *et al.*, 2007), and until these were identified and testing introduced they were also widespread in rodent colonies (Ball-Goodrich and Johnson, 1994; Besselsen *et al.*, 1996; Jacoby *et al.*, 1996; Ball-Goodrich *et al.*, 1998; Wan *et al.*, 2002). Many of the rodent parvoviruses are difficult to grow in tissue culture and the adaptations allowing tissue culture growth are poorly understood.

Apart from the relationships between the viruses at the sequence level, little is known about the details of their evolutionary histories or any host range differences between these viruses, or about their temporal variation. Recombination has been demonstrated among the rodent parvoviruses and porcine parvovirus (Figures 17.3 and 17.8) and may be a common event (Figure 17.9).

ADENO-ASSOCIATED VIRUSES (AAV)

AAVs have been isolated from primates and from many other mammalian and avian hosts. Most depend upon a helper virus for replication, while some related viruses replicate independent of helper viruses. AAV was discovered in the 1960s as contaminant of adenovirus preparations (Atchison *et al.*, 1965; Hoggan *et al.*, 1966). The AAVs of humans and non-human primates are currently divided into five species (AAV1–5) and two tentative species (AAV7 and 8), with the AAV1 species containing two strains (AAV strains 1 and 6) and all other species containing one AAV strain (Tattersall *et al.*, 2005). Screening of tissue samples from humans and other primates by PCR detected large numbers of DNA sequences, and in human and non-human primate tissue the AAV prevalence was ~19% (Gao *et al.*, 2003, 2004). AAV1 and 6 were closely related, and sequence identity between the other strains ranged between 75% and 82 %. The primate AAVs fell into a number of distinct clades (Figure 17.9) (Gao *et al.*, 2004). Clades B and C seemed most prevalent in humans, Clades A and F were only found in humans and clade D only in macaques, while clade E was isolated from both humans and non-human primates. Evolution of AAVs might be at least partly driven by their helper viruses, and some AAVs may also have co-evolved with their hosts but experimental or phylogenetic evidence is missing. Recombination between AAV genomes seems to occur frequently, and in the tissue can give rise to replication-competent vectors following homologous recombination (Figure 17.9) (Allen *et al.*, 1997; Gao *et al.*, 2003).

GENETIC VARIATION AND REPLICATION ERROR RATE

The variation of parvovirus sequences can be high when measured over defined time periods. This has been seen in studies of variation for CPV and B19 parvovirus in nature (Shackelton *et al.*, 2005; Shackelton and Holmes, 2006), in the growth of CPV through serial passage in tissue culture (Badgett *et al.*, 2002), and in the analysis of MVM during replication under selective conditions such as replication in alterative hosts (Lopez-Bueno

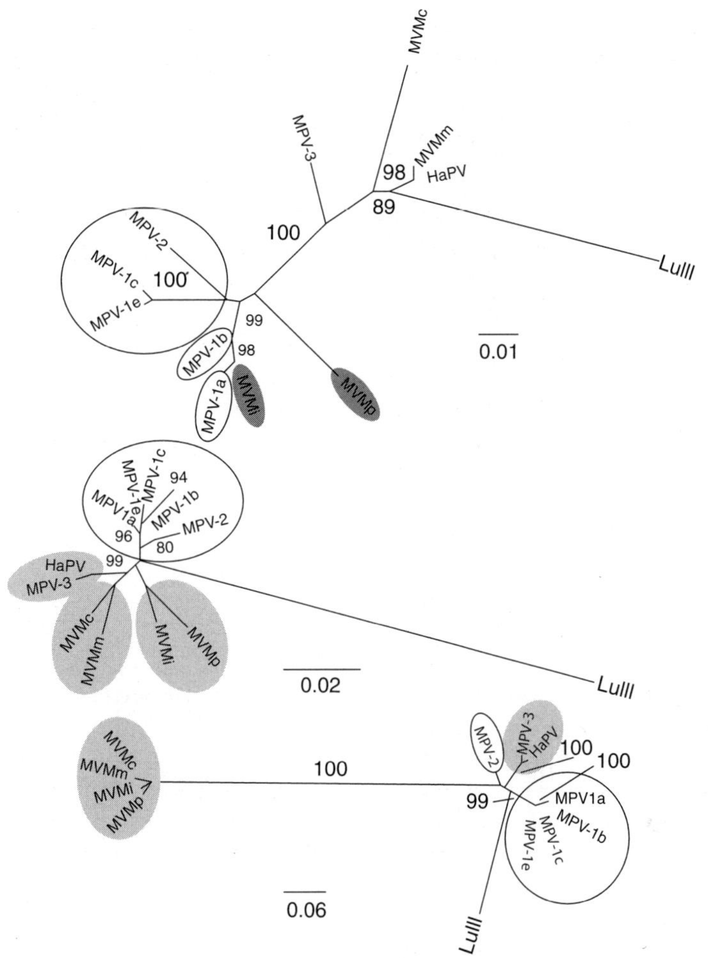

FIGURE 17.8 Phylogenetic analyses showing recombination among the rodent parvoviruses. Nucleotide sequences spanning the complete coding region (nucleotides 114–4728 of mouse parvovirus 1 reference sequence, GeneBank accession number NC_001630) were aligned. Phylogenetic trees were constructed from nucleotides (a) 1–854, (b) 855–2226, and (c) 2227–4654. Colors indicate different rodent virus types. Nodes with >70% bootstrap support are shown and branch lengths are scaled according to nucleotide substitutions per site. Abbreviations: LuIII virus (LuIII), mice minute virus (MMVx), minute virus of mice strain m (MMVm), minute virus of mice (MMVy), mouse parvovirus minute virus immunosuppressive variant (MMViv), mice minute virus (MMVz), minute virus of mice, a lymphotropic variant of MVM (MMVlyv), mouse parvovirus 1 (Mouse1), mouse parvovirus 1e (Mouse1e), mouse parvovirus 1c (Mouse1c), mouse parvovirus 1b (Mouse1b), mouse parvovirus 2 (Mouse2), mouse parvovirus 3 (Mouse3), hamster parvovirus (Hamster). (See Plate 28 for the color version of this figure.) From Shackelton *et al.* (in press) with permission.

et al., 2004) or monoclonal antibody selection (Lopez-Bueno *et al.*, 2003). However, it is not know how much variation occurs during the replication of the parvovirus genome. Although the viral DNA is replicated using host cell DNA polymerases, the fidelity of replication is likely lower than that seen for the host chromosomal DNA replication. If DNA repair mechanisms are activated during parvovirus replication those would also lead to higher substitution rates, perhaps due to the use of alternative DNA polymerases, and to the filling in or replicating of the ssDNA templates (Kunkel, 2004).

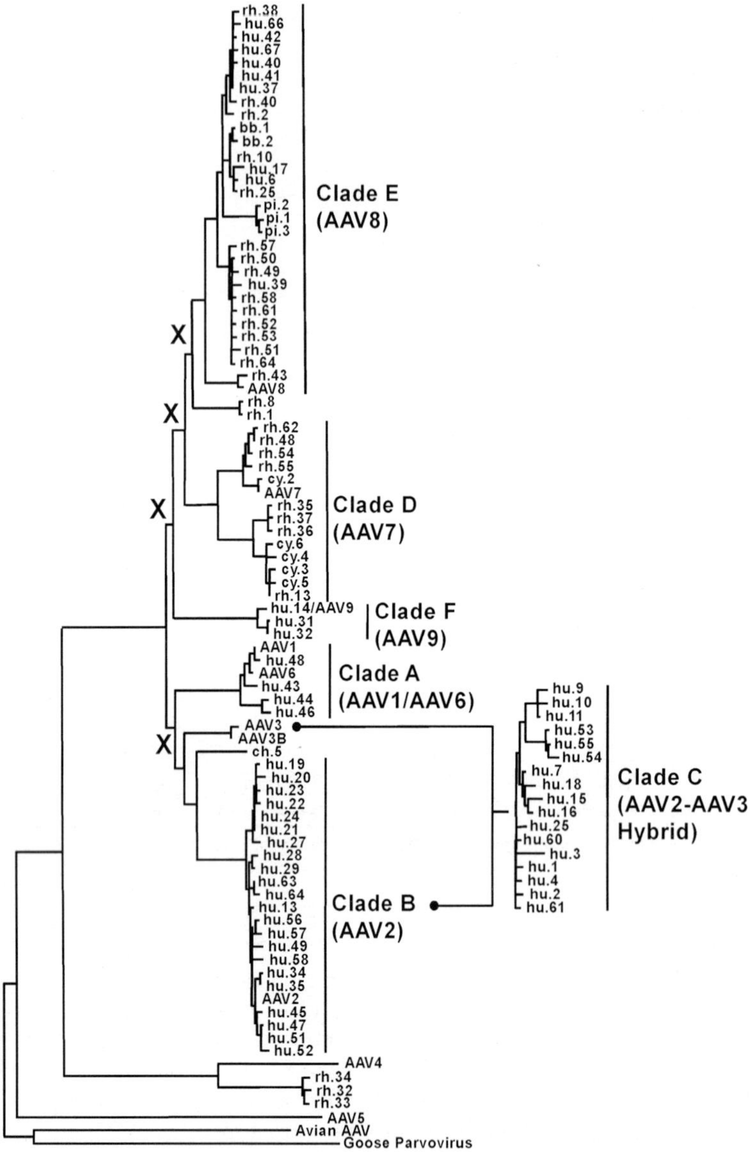

FIGURE 17.9 Phylogenetic relationships between the VP1 protein sequences of primate AAVs. A neighbor-joining phylogeny of VP1 protein sequences was constructed using goose parvovirus and an avian AAV as the outgroup. Clades are indicated by name and by vertical lines. X = major nodes with bootstrap values of <75. Viruses are identified by the serotype name or a reference to the species source (hu, human; rh, rhesus macaque; cy, cynomolgus macaque; bb, baboon; pi, pigtailed macaque; ch, chimpanzee), number = order of sequencing. Clade C originated through the recombination of known clades. The AAV2–AAV3 hybrid clade originated after one recombination event, and its phylogeny is shown. Adapted from Gao *et al.* (2004) with permission.

INTRA-HOST DIVERSITY DURING NATURAL INFECTIONS

The degree of intra-host diversity has been assessed for B19, MVM, and AMDV infections, and most recently also for one CPV-infected cat. While there are probably differences in selective pressures between the chronic infections of ADV and MVM and the acute infection of CPV, all of these studies showed intra-host variation or diversity.

Analyzing individual virus genomes isolated from one CPV-infected cat revealed diversity in a 1745-bp fragment of the VP2 gene (Battilani *et al.*, 2006). Sequences were 99.5–99.9% identical in nucleotide sequence and 99.3–99.8% identical in amino acid sequence, but 10 distinct sequences were observed among 14 analyzed viral clones. Two antigenically distinct CPV variants (CPV type 2a and a variant with the VP2 D426E substitution, named CPV type 2c) were isolated from this animal, indicating that this was likely a co-infection by co-circulating strains. The ratio of non-synonymous to synonymous changes (d_N/d_S ratio), when compared with extant CPV strains circulating in Italy, ranged between 0.08 and 0.4 for individual clones, suggesting purifying selection on the capsid gene.

A spectrum of AMDV sequences was seen in experimentally infected Danish mink, with several different sequences in mink inoculated with a single inoculum. The sequences differed by up to 5% and also differed in the highly variable region (Gottschalck *et al.*, 1991). Since AMDV can establish persistent infections with circulating virus, it is likely that mixed infections occur, and that inocula contained multiple virus strains. Comparing virus non-structural gene sequences (nts 123–2208) from different AMDV isolates showed extensive viral variation in the different preparations (Gottschalck *et al.*, 1994).

Experimental MVM infections in immune-deficient mice using molecularly cloned virus stocks showed emergence of variation over a period of days or weeks (Lopez-Bueno *et al.*, 2003, 2006; Rubio *et al.*, 2005). The original viruses used had been tissue culture adapted, and MVMp variants contained non-synonymous changes in the VP1/2 gene at sites which gave lower affinity binding of sialic acid (Lopez-Bueno *et al.*, 2006; Nam *et al.*, 2006) (Figure 17.10). When treating severe and combined immunodeficient (SCID) mice with a neutralizing monoclonal antibody, escape mutants rapidly arose and quickly dominated the viral population, leading to a delayed onset of clinical disease. Substitutions were

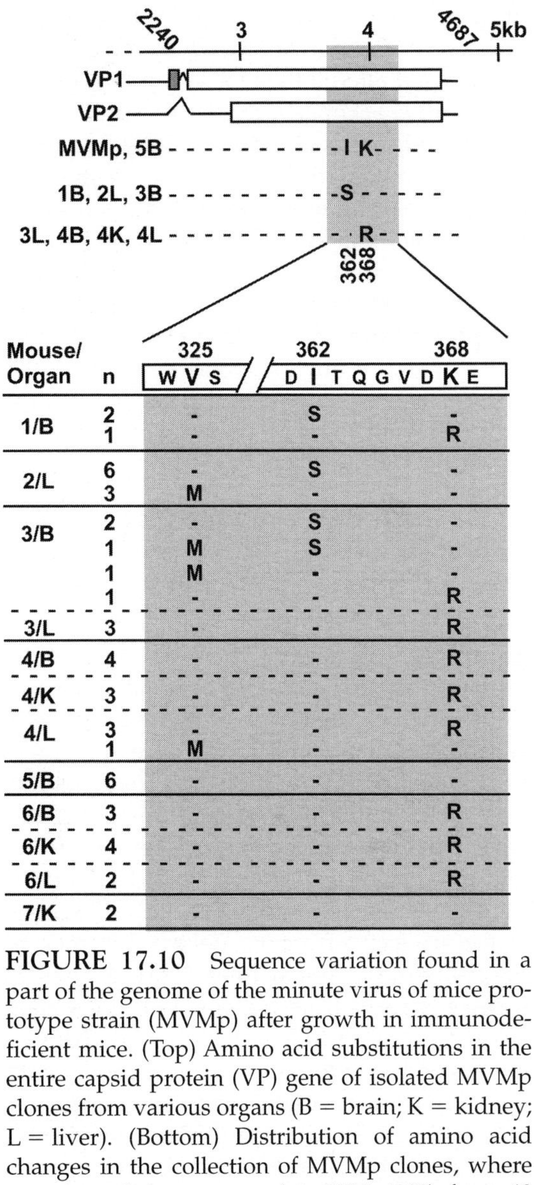

FIGURE 17.10 Sequence variation found in a part of the genome of the minute virus of mice prototype strain (MVMp) after growth in immunodeficient mice. (Top) Amino acid substitutions in the entire capsid protein (VP) gene of isolated MVMp clones from various organs (B = brain; K = kidney; L = liver). (Bottom) Distribution of amino acid changes in the collection of MVMp clones, where a region of the genome (nts 3710–4200) from 48 viral clones from seven mice (numbered 1–7). The amino acid changes at residues 325, 362, and 368 of the VP2 sequence are outlined. *n* = number of clones with identical genotypes in this region. From Lopez-Bueno *et al.* (2006) with permission.

located close to a raised region at the three-fold axis of symmetry (Lopez-Bueno *et al.*, 2003). When polyclonal anti-capsid antibodies were used, those did not allow escape mutations of the MVMi strain to be selected, and non-synonymous changes were restricted to the NS2 protein, including two changes affecting both NS1 and NS2. NS2 changes clustered into two regions and altered the biochemical properties of the protein (Lopez-Bueno *et al.*, 2004).

Adeno-associated viruses show sequence diversity both within hosts and within the population. Mutations are primarily located in variable regions of the capsid proteins which are exposed to the outside of the capsid and some represent antigenic sites or receptor binding sites (Opie *et al.*, 2003; Wu *et al.*, 2006). Immune responses against AAVs *in vivo* suggest a potential role of immune escape (reviewed by Monahan *et al.*, 2002).

VIRUS VARIATION AT A POPULATION LEVEL (SPATIAL HETEROGENEITY)

Sequence variation of the B19 viruses in humans has been analyzed comparing viruses from different regions of the world or from chronic and acute infections. There appears to be a global distribution of the viruses—there was close similarity between some isolates collected from various regions of the world, and also between viruses collected at various times over the past two decades. However, viruses collected from one geographic area are generally more similar to each other. For B19 viruses, viruses from patients with persistent infections appear to have a higher level of variation compared to viruses from patients with acute infections. Comparing seven isolates collected in Italy between 1989 and 1994 from one geographic area showed maximum variation of 0.61% in a 1000-nt sequence within the N-terminal end of the VP genes (Gallinella *et al.*, 1995). Those viruses were 0.7% and 0.77% different on average from the prototype Wi and Au viruses collected in the UK and USA, respectively, and there were nine non-synonymous changes, seven between residues 4 and 114 in the VP1 sequence and two within the VP1/VP2 common region (Gallinella *et al.*, 1995). Viruses from Vietnam showed the presence of two genotypes of the type 1 B19 in that country (Toan *et al.*, 2006).

The sequences of the complete VP1 and VP2 gene region (2343 nt) of 29 isolates from 25 infected patients in various regions of the world were compared to each other, and to the two published Wi and Au sequences. Those viruses included 10 from an outbreak in Ohio, USA, one a mother–child pair, and other isolates from the USA, UK, Brazil, Ireland, Venezuela, Korea, Japan, and China (Erdman *et al.*, 1996). The sequences differed by between 2 and 99 nucleotides (4.2%), and by 0 and 13 (1.7%) in amino acids. Nucleotide variation was found throughout the VP1 and VP2 genes, but non-synonymous changes clustered in three regions, the VP1-unique region, around the junction of the VP1 and VP2-coding regions, and within the VP1 and VP2 overlapping region. Isolates from the outbreak in Ohio were divided into two classes, represented by seven and two samples each (Erdman *et al.*, 1996). Within each group the sequences differed by only a few nucleotides, and where multiple isolates were collected from four individuals only a single sequence difference was found in the sequences from each person. There was some geographic clustering of the strains—viruses from China formed a distinct clade, while those from the USA were generally clustered, as were the isolates from Korea.

Viruses recovered from chronic and acute infections were examined in several studies, and in most cases no specific correlation was seen between any particular disease syndrome and any virus type (Morinet *et al.*, 1986; Mori *et al.*, 1987; Gallinella *et al.*, 1995; Kerr *et al.*, 1995; Takahashi *et al.*, 1999). Higher genomic variability has been seen in the B19 sequences recovered from cases of persistent infection compared to those from acute infections (Hung *et al.*, 2006).

SPATIAL HETEROGENEITY OF CPV AND RELATED VIRUSES

CPV and the closely related viruses have also been examined extensively. CPV-like viruses were first circulating in Europe in the mid to late 1970s, and early CPV type 2 isolates were essentially identical in nucleotide sequence, and were all replaced by the variant CPV type 2a strain which spread worldwide during 1979 and 1980. Other mutations have been seen worldwide in various countries since 1980, including a mutation changing VP2 residue 426 (referred to as CPV-2b) from Asn to Asp first seen around 1984 which spread worldwide. More recently viruses with residue 426 as a Glu (referred to as CPV type 2c) have been identified in Europe (Martella *et al.*, 2004; Decaro *et al.*, 2005), Vietnam (Nakamura *et al.*, 2004), and in North and South America (Perez *et al.*, 2007). The reasons for the global spread of some mutations and the time of their emergence are hard to explain. CPV types 2a and 2b co-circulated throughout the world (although in different proportions) over the 20 years without either becoming fixed. CPV-2a and 2b co-circulated in Brazil as early as 1986 (Pereira *et al.*, 2007) while CPV-2b was the predominant variant circulating there between 1995 and 2001 (Costa *et al.*, 2005). The dominant CPV type circulating in Italy at that time was CPV-2a (Martella *et al.*, 2005), while both CPV-2a and 2b circulated at similar frequencies in 1999/2000. There appear to be Taiwanese-Japanese and Indian CPV populations which are phylogenetically distinct from American and Vietnamese populations (Hirayama *et al.*, 2005; Chinchkar *et al.*, 2006; Doki *et al.*, 2006). Additional mutations that have been observed include Ser297Ala and Glu300Asp, which may affect receptor binding or host range properties of the viruses. However, an analysis of CPV isolates from Brazil and other regions failed to detect clear geographic pattern in extant parvovirus isolates (Pereira *et al.*, 2007).

TEMPORAL VARIATION

Good temporal data analyzing the evolution of parvoviruses are still relatively rare. The best defined are those that examine the evolution of CPV after the host jump into dogs, where all viruses have evolved from a single common ancestor with an accumulation of mutations. Early time points after the colonization of a new host are thought to coincide with accelerated mutation rates due to host adaptation, while other times in the CPV pandemic may be characterized by selective constraints. During the emergence of the CPV lineage the viruses showed a steady rate of change of around 10^{-5} substitutions/site/year (Shackelton *et al.*, 2005). Many of the capsid protein gene differences between CPV and the FPV-like viruses are associated with changes in host range, antigenicity, and sialic acid-binding properties of the viruses, suggesting they are under strong selection (Chang *et al.*, 1992; Horiuchi *et al.*, 1994; Hueffer *et al.*, 2003). The CPV lineage split into two major variants during the mid-1970s, with the CPV type 2 strains emerging worldwide in 1978, and the CPV type 2a variant replacing the CPV type 2 viruses during 1979 and 1980 (Figure 17.5). The rate of variation of FPV was lower than that seen for CPV (Horiuchi *et al.*, 1998; Shackelton *et al.*, 2005). Analysis of Brazilian CPV-2 sequences indicated neutral evolution in the early phase of the pandemic (analyzing isolates collected around 1980), while later time points (isolates collected around and after 1990) were associated with strong purifying selection, with the exception of a change of VP2 residue 297, which was under strong positive selection. A Baysian Skyline plot analysis showed an initial very rapid spread with a subsequent decreasing growth rate spread (Pereira *et al.*, 2007) (Figure 17.11).

CPV isolates collected during 1978 and early 1979 were all antigenically identical worldwide, but the CPV type 2a strain differed in the VP1/VP2 protein by 4–5 amino acids compared to CPV type 2 isolates. CPV type 2a also differed from CPV type 2 in having lost two antigenic epitopes and gained two epitopes on the capsid (Strassheim *et al.*, 1994). Other antigenic variants emerging around 1984, and during the later 1990s and early 2000s, each contained a single amino acid sequence difference within neutralizing epitopes in the capsid (Ikeda *et al.*, 2000, 2002; Buonavoglia *et al.*, 2001; Nakamura *et al.*, 2004). Despite the small numbers of

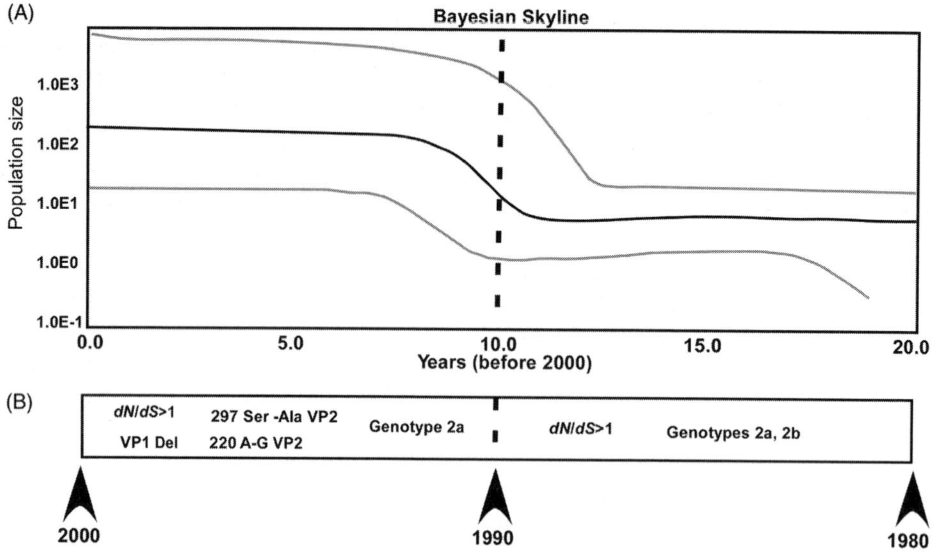

FIGURE 17.11 Evolutionary dynamics of canine parvoviruses in Brazil. (A) Bayesian skyline plot obtained for the VP2 gene sequences of CPV strains from Brazil. The x-axis is in years before 2000, the black line represents the mean and the gray line the 95% high probability density limits. (B) Major genetic events occurring during the period of CPV epidemics. The vertical dashed line delineates the distinct patterns of CPV evolution characterized by two intervals: (1) the period between 1980 and 1990 with stochastic allele fixation and the occurrence of antigenic types CPV-2a and 2b, and (2) the period after 1990 distinguished by deterministic allele fixation ($d_N/d_S > 1$), presence of only the antigenic type CPV-2a strain with positive mutation such as a VP1 intron deletion (VP1Del), and a synonymous mutation in codon 220 of the VP2 gene region. Adapted from Pereira *et al.* (2007) with permission.

differences, those viruses each became globally distributed within a year or two of being detected. Natural antigenic variation of FPV and MEV isolates has also been described (Parrish and Carmichael, 1983; Parrish *et al.*, 1984; Mochizuki *et al.*, 1989). It is unclear what the consequences are of the antigenic variation that is seen in some of the viruses, but in general there appears to be good cross-protection between the antigenically variant strains.

CONCLUSIONS

Our understanding of the evolution of the parvoviruses is still being developed, but there are some interesting conclusions that can be drawn. All of the parvoviruses appear to derive from a common ancestral virus, but the different genera are mostly well separated genetically, suggesting long divergence times. Surveys for non-host DNA sequences often show sequences of previously unknown parvoviruses, suggesting that there are many different viruses still to be discovered. Levels of sequence variation within a virus genus appears to differ significantly between different parvoviruses, and this may be due both to different lifestyles (such as whether they cause acute or persistent infections), and to differences in the levels of sequence variation that are tolerated, particularly in the capsid protein gene. Selection is likely to act on both the gene products and on the DNA sequences themselves. Mixed infections by closely related strains of the same virus or by genetically or serologically different viruses may be common for some viruses, but these are overlooked unless subpopulations in DNA mixtures are specifically tested for. High rates of variation may be seen among many parvoviruses when they are placed under selection. The replication fidelity is not known. Although replicated by host or helper virus DNA polymerases, fidelity is likely considerably lower than those seen during replication of the host genomes. Global spread of genomic variants of some viruses occurs, and has been seen for the distribution of CPV during the early stages of the pandemic, or B19 virus in the human population. Transmission between related host species might be frequent for some parvoviruses (rodent parvoviruses, parvoviruses of the feline parvovirus group, or AAVs of the E clade), while host restrictions appear to be strict for others.

REFERENCES

Alexandersen, S. (1986) Acute interstitial pneumonia in mink kits: experimental reproduction of the disease. *Vet. Pathol.* **23**, 579–588.

Alexandersen, S., Bloom, M.E. and Wolfinbarger, J. (1988) Evidence of restricted viral replication in adult mink infected with Aleutian disease of mink parvovirus. *J. Virol.* **62**, 1495–1507.

Alexandersen, S., Larsen, S., Aasted, B., Uttenthal, A., Bloom, M.E. and Hansen, M. (1994) Acute interstitial pneumonia in mink kits inoculated with defined isolates of Aleutian mink disease parvovirus. *Vet. Pathol.* **31**, 216–228.

Allander, T., Emerson, S.U., Engle, R.E., Purcell, R.H. and Bukh, J. (2001) A virus discovery method incorporating DNase treatment and its application to the identification of two bovine parvovirus species. *Proc. Natl Acad. Sci. USA* **98**, 11609–11614.

Allander, T., Tammi, M.T., Eriksson, M., Bjerkner, A., Tiveljung-Lindell, A. and Andersson, B. (2005) Cloning of a human parvovirus by molecular screening of respiratory tract samples. *Proc. Natl Acad. Sci. USA* **102**, 12891–12896.

Allen, J.M., Debelak, D.J., Reynolds, T.C. and Miller, A.D. (1997) Identification and elimination of replication-competent adeno-associated virus (AAV) that can arise by nonhomologous recombination during AAV vector production. *J. Virol.* **71**, 6816–6822.

Atchison, R.W., Casto, B.C. and Hammon, W.M. (1965) Adenovirus-associated defective virus particles. *Science* **149**, 754–756.

Aubin, J.T., Defer, C., Vidaud, M., Maniez Montreuil, M. and Flan, B. (2000) Large-scale screening for human parvovirus B19 DNA by PCR: application to the quality control of plasma for fractionation. *Vox sanguinis* **78**, 7–12.

Badgett, M.R., Auer, A., Carmichael, L.E., Parrish, C.R. and Bull, J.J. (2002) Evolutionary dynamics of viral attenuation. *J. Virol.* **76**, 10524–10529.

Ball-Goodrich, L.J. and Johnson, E. (1994) Molecular characterization of a newly recognized mouse parvovirus. *J. Virol.* **68**, 6476–6486.

Ball-Goodrich, L.J., Leland, S.E., Johnson, E.A., Paturzo, F.X. and Jacoby, R.O. (1998) Rat parvovirus type 1: the prototype for a new rodent parvovirus serogroup. *J. Virol.* **72**, 3289–3299.

Bashir, T., Horlein, R., Rommelaere, J. and Willwand, K. (2000) Cyclin A activates the DNA polymerase delta -dependent elongation machinery in vitro: A parvovirus DNA replication model. *Proc. Natl Acad. Sci. USA* **97**, 5522–5527.

Battilani, M., Scagliarini, A., Ciulli, S., Morganti, L. and Prosperi, S. (2006) High genetic diversity of the VP2 gene of a canine parvovirus strain detected in a domestic cat. *Virology* **352**, 22–26.

Becker, S.D., Bennett, M., Stewart, J.P. and Hurst, J.L. (2007) Serological survey of virus infection among wild house mice (Mus domesticus) in the UK. *Lab. Anim.* **41**, 229–238.

Berns, K.I. (1990) Parvovirus replication. *Microbiol. Rev.* **54**, 316–329.

Berns, K.I. and Linden, R.M. (1995) The cryptic life style of adeno-associated virus. *Bioessays* **17**, 237–245.

Besselsen, D.G., Besch-Williford, C.L., Pintel, D.J., Franklin, C.L., Hook, R.R., Jr. and Riley, L.K. (1995) Detection of newly recognized rodent parvoviruses by PCR. *J. Clin. Microbiol.* **33**, 2859–2863.

Besselsen, D.G., Pintel, D.J., Purdy, G.A., Besch-Williford, C.L., Franklin, C.L., Hook, R.R., Jr. and Riley, L.K. (1996) Molecular characterization of newly recognized rodent parvoviruses. *J. Gen. Virol.* **77**(Pt 5), 899–911.

Besselsen, D.G., Romero, M.J., Wagner, A.M., Henderson, K.S. and Livingston, R.S. (2006) Identification of novel murine parvovirus strains by epidemiological analysis of naturally infected mice. *J. Gen. Virol.* **87**, 1543–1556.

Bianchi, M., Rago, I., Zini, G., D'Onofrio, G. and Leone, G. (2007) Parvovirus b19 infection after plasma exchange for myasthenia gravis. *Lab. Hematol.* **13**, 34–38.

Bloom, M.E., Kaaden, O.-R., Huggans, E., Cohn, A. and Wolfinbarger, J.B. (1988) Molecular comparisons of in vivo and in vitro derived strains of Aleutian disease of mink parvovirus. *J. Virol.* **62**, 132–138.

Bodendorf, U., Cziepluch, C., Jauniaux, J.C., Rommelaere, J. and Salome, N. (1999) Nuclear export factor CRM1 interacts with nonstructural proteins NS2 from parvovirus minute virus of mice. *J. Virol.* **73**, 7769–7779.

Bonnard, G.D., Manders, E.K., Campbell, D.A., Jr., Herberman, R.B. and Collins, M.J., Jr. (1976) Immunosuppressive activity of a subline of the mouse EL-4 lymphoma. Evidence for minute virus of mice causing the inhibition. *J. Exp. Med.* **143**, 187–205.

Brockhaus, K., Plaza, S., Pintel, D.J., Rommelaere, J. and Salome, N. (1996) Nonstructural proteins NS2 of minute virus of mice associate in vivo with 14-3-3 protein family members. *J. Virol.* **70**, 7527–7534.

Broll, S. and Alexandersen, S. (1996) Investigation of the pathogenesis of transplacental transmission of Aleutian mink disease parvovirus in experimentally infected mink. *J. Virol.* **70**, 1455–1466.

Brown, K.E. and Young, N.S. (1995) Parvovirus B19 infection and hematopoiesis. *Blood Rev.* **9**, 176–182.

Brown, K.E., Young, N.S. and Liu, J.M. (1994) Molecular, cellular and clinical aspects of parvovirus B19 infection. *Crit. Rev. Oncol. Hematol.* **16**, 1–31.

Buonavoglia, C., Martella, V., Pratelli, A., Tempesta, M., Cavalli, A., Buonavoglia, D. *et al.* (2001) Evidence for evolution of canine parvovirus type 2 in Italy. *J. Gen. Virol.* **82**, 3021–3025.

Candotti, D., Etiz, N., Parsyan, A. and Allain, J.P. (2004) Identification and characterization of persistent human erythrovirus infection in blood donor samples. *J. Virol.* **78**, 12169–12178.

Chang, S.F., Sgro, J.Y. and Parrish, C.R. (1992) Multiple amino acids in the capsid structure of canine parvovirus coordinately determine the canine host range and specific antigenic and hemagglutination properties. *J. Virol.* **66**, 6567–6858.

Chen, K.C., Shull, B.C., Moses, E.A., Lederman, M., Stout, E.R. and Bates, R.C. (1986) Complete nucleotide sequence and genome organisation of bovine parvovirus. *J. Virol.* **60**, 1085–1097.

Chen, K.C., Shull, B.C., Lederman, M., Stout, E.R. and Bates, R.C. (1988) Analysis of the termini of the DNA of bovine parvovirus: demonstration of sequence inversion at the left terminus and its implication for the replication model. *J. Virol.* **62**, 3807–3813.

Chinchkar, S.R., Mohana Subramanian, B., Hanumantha Rao, N., Rangarajan, P.N., Thiagarajan, D. and Srinivasan, V.A. (2006) Analysis of VP2 gene sequences of canine parvovirus isolates in India. *Arch. Virol.* **151**, 1881–1887.

Chiorini, J.A., Yang, L., Safer, B. and Kotin, R.M. (1995) Determination of adeno-associated virus Rep68 and Rep78 binding sites by random sequence oligonucleotide selection. *J. Virol.* **69**, 7334–7338.

Choi, E.Y., Newman, A.E., Burger, L. and Pintel, D. (2005) Replication of minute virus of mice DNA is critically dependent on accumulated levels of NS2. *J. Virol.* **79**, 12375–12381.

Cossons, N., Faust, E.A. and Zannis-Hadjopoulos, M. (1996) DNA polymerase delta-dependent formation of a hairpin structure at the 5′ terminal palindrome of the minute virus of mice genome. *Virology* **216**, 258–264.

Costa, A.P., Leite, J.P., Labarthe, N.V. and Garcia, R.C. (2005) Genomic typing of canine parvovirus circulating in the state of Rio de Janeiro, Brazil from 1995 to 2001 using polymerase chain reaction assay. *Vet. Res. Commun.* **29**, 735–743.

Cotmore, S.F. (1996) Parvovirus DNA replication. In: *DNA Replication in Eukaryotic Cells* (M.L. DePamphilis, ed.), p. 799. Cold Spring Harbor, NY: Cold Spring Harbor Laboratory Press.

Cotmore, S.F. and Tattersall, P. (1987) The autonomously replicating parvoviruses of vertebrates. *Adv. Virus Res.* **33**, 91–174.

Cotmore, S.F. and Tattersall, P. (1995) DNA replication in the autonomous parvoviruses. *Semin. Virol.* **6**, 271–281.

Cotmore, S.F. and Tattersall, P. (2003) Resolution of parvovirus dimer junctions proceeds through a novel heterocruciform intermediate. *J. Virol.* **77**, 6245–6254.

Cotmore, S.F. and Tattersall, P. (2005) Genome packaging sense is controlled by the efficiency of the nick site in the right-end replication origin of parvoviruses minute virus of mice and LuIII. *J. Virol.* **79**, 2287–2300.

Cotmore, S.F., D'Abramo, A.M., Jr., Carbonell, L.F., Bratton, J. and Tattersall, P. (1997) The NS2 polypeptide of parvovirus MVM is required for capsid assembly in murine cells. *Virology* **231**, 267–280.

Crawford, L.V. (1966) A minute virus of mice. *Virology* **29**, 605.

D'Abramo, A.M., Jr., Ali, A.A., Wang, F., Cotmore, S.F. and Tattersall, P. (2005) Host range mutants of minute virus of mice with a single VP2 amino acid change require additional silent mutations that regulate NS2 accumulation. *Virology* **340**, 143–154.

Daeffler, L., Horlein, R., Rommelaere, J. and Nuesch, J.P. (2003) Modulation of minute virus of mice cytotoxic activities through site-directed mutagenesis within the NS coding region. *J. Virol.* **77**, 12466–12478.

de Haan, T.R., Beersma, M.F., Claas, E.C., Oepkes, D., Kroes, A.C. and Walther, F.J. (2007) Parvovirus B19 infection in pregnancy studied by maternal viral load and immune responses. *Fetal Diagn. Ther.* **22**, 55–62.

Decaro, N., Elia, G., Martella, V., Desario, C., Campolo, M., Trani, L.D. *et al.* (2005) A real-time PCR assay for rapid detection and quantitation of canine parvovirus type 2 in the feces of dogs. *Vet. Microbiol.* **105**, 19–28.

Doki, M., Fujita, K., Miura, R., Yoneda, M., Ishikawa, Y., Taneno, A. and Kai, C. (2006) Sequence analysis of VP2 gene of canine parvovirus isolated from domestic dogs in Japan in 1999 and 2000. *Comp. Immunol. Microbiol. Infect. Dis.* **29**, 199–206.

Dorsch, S., Kaufmann, B., Schaible, U., Prohaska, E., Wolf, H. and Modrow, S. (2001) The VP1-unique region of parvovirus B19: amino acid variability and antigenic stability. *J. Gen. Virol.* **82**, 191–199.

Dubielzig, R., King, J.A., Weger, S., Kern, A. and Kleinschmidt, J.A. (1999) Adeno-associated virus type 2 protein interactions: formation of pre-encapsidation complexes. *J. Virol.* **73**, 8989–8998.

Eichwald, V., Daeffler, L., Klein, M., Rommelaere, J. and Salome, N. (2002) The NS2 proteins of parvovirus minute virus of mice are required for efficient nuclear egress of progeny virions in mouse cells. *J. Virol.* **76**, 10307–10319.

Eis-Hubinger, A.M., Dieck, D., Schild, R., Hansmann, M. and Schneweis, K.E. (1998) Parvovirus B19 infection in pregnancy. *Intervirology* **41**, 178–184.

Ekman, A., Hokynar, K., Kakkola, L., Kantola, K., Hedman, L., Bonden, H. *et al.* (2007) Biological and immunological relations among human parvovirus B19 genotypes 1–3. *J. Virol.* **81**, 6927–6935.

Erdman, D.D., Durigon, E.L., Wang, Q.-Y. and Anderson, L.J. (1996) Genetic diversity of human parvovirus B19: sequence analysis of the VP1/VP2 gene from multiple isolates. *J. Gen. Virol.* **77**, 2767–2774.

Farr, G.A., Zhang, L.G. and Tattersall, P. (2005) Parvoviral virions deploy a capsid-tethered lipolytic enzyme to breach the endosomal membrane during cell entry. *Proc. Natl Acad. Sci. USA* **102**, 17148–17153.

Frickhofen, N. and Young, N.S. (1989) Persistent parvovirus B19 infections in humans. *Microb. Pathog.* **7**, 319–327.

Fry, A.M., Lu, X., Chittaganpitch, M., Peret, T., Fischer, J., Dowell, S.F. *et al.* (2007) Human bocavirus: a novel parvovirus epidemiologically associated with pneumonia requiring hospitalization in Thailand. *J. Infect. Dis.* **195**, 1038–1045.

Gallinella, G., Venturoli, S., Gentilomi, G., Musiani, M. and Zerbini, M. (1995) Extent of sequence variability in a genomic region coding for capsid proteins of B19 parvovirus. *Arch. Virol.* **140**, 1119–1125.

Gallinella, G., Venturoli, S., Manaresi, E., Musiani, M. and Zerbini, M. (2003) B19 virus genome diversity: epidemiological and clinical correlations. *J. Clin. Virol.* **28**, 1–13.

Gao, G., Alvira, M.R., Somanathan, S., Lu, Y., Vandenberghe, L.H., Rux, J.J. *et al.* (2003) Adeno-associated viruses undergo substantial evolution in primates during natural infections. *Proc. Natl Acad. Sci. USA* **100**, 6081–6086.

Gao, G., Vandenberghe, L.H., Alvira, M.R., Lu, Y., Calcedo, R., Zhou, X. and Wilson, J.M. (2004) Clades of Adeno-associated viruses are widely disseminated in human tissues. *J. Virol.* **78**, 6381–6388.

Girod, A., Wobus, C.E., Zadori, Z., Ried, M., Leike, K., Tijssen, P., Kleinschmidt, J.A. and Hallek, M. (2002) The VP1 capsid protein of adeno-associated virus type 2 is carrying a phospholipase A2 domain required for virus infectivity. *J. Gen. Virol.*.**83**, 973–978.

Gottschalck, E., Alexandersen, S., Cohn, A., Poulsen, L.A., Bloom, M.E. and Aasted, B. (1991) Nucleotide sequence analysis of Aleutian mink disease parvovirus shows that multiple virus types are present in infected mink. *J. Virol.* **65**, 4378–4386.

Gottschalck, E., Alexandersen, S., Storgaard, T., Bloom, M.E. and Aasted, B. (1994) Sequence comparison of the non-structural genes of four different types of Aleutian mink disease parvovirus indicates an unusual degree of variability. *Arch. Virol.* **138**, 213–231.

Guerin, B. and Pozzi, N. (2005) Viruses in boar semen: detection and clinical as well as epidemiological consequences regarding disease transmission by artificial insemination. *Theriogenology* **63**, 556–572.

Hattori, S., Yunoki, M., Tsujikawa, M., Urayama, T., Tachibana, Y., Yamamoto, I. *et al.* (2007) Variability of parvovirus B19 to inactivation by liquid heating in plasma products. *Vox sanguinis* **92**, 121–124.

Hayder, H.A., Storz, J. and Young, S. (1983) Antigenicity of bovine parvovirus in fetal infections. *Am. J. Vet. Res.* **44**, 558–563.

Heegaard, E.D. and Brown, E.B. (2002) Human Parvovirus B19. *Clin. Microbiol. Rev.* **15**, 485–505.

Hemauer, A., von Pobotzki, A., Gigler, A., Cassinotti, P., Siegl, G., Wolf, H. and Modrow, S. (1996) Sequence variability among different parvovirus B19 isolates. *J. Gen. Virol.* **77**, 1781–1785.

Hirayama, K., Kano, R., Hosokawa-Kanai, T., Tuchiya, K., Tsuyama, S., Nakamura, Y. *et al.* (2005) VP2 gene of a canine parvovirus isolate from stool of a puppy. *J. Vet. Med. Sci.* **67**, 139–143.

Hoggan, M.D., Blacklow, N.R. and Rowe, W.P. (1966) Studies of small DNA viruses found in various adenovirus preparations: physical, biological, and immunological characteristics. *Proc. Natl Acad. Sci. USA* **55**, 1467–1474.

Hokynar, K., Soderlund-Venermo, M., Pesonen, M., Ranki, A., Kiviluoto, O., Partio, E.K. and Hedman, K. (2002) A new parvovirus genotype persistent in human skin. *Virology* **302**, 224–228.

Hong, G., Ward, P. and Berns, K.I. (1994) Intermediates of adeno-associated virus DNA replication in vitro. *J. Virol.* **68**, 2011–2015.

Horiuchi, M., Goto, H., Ishiguro, N. and Shinagawa, M. (1994) Mapping of determinants of the host range for canine cells in the genome of canine parvovirus using canine parvovirus/mink enteritis virus chimeric viruses. *J. Gen. Virol.* **75**, 1319–1328.

Horiuchi, M., Yamaguchi, Y., Gojobori, T., Mochizuki, M., Nagasawa, H., Toyoda, Y. *et al.* (1998) Differences in the evolutionary pattern of feline panleukopenia virus and canine parvovirus. *Virology* **249**, 440–452.

Hueffer, K., Govindasamy, L., Agbandje-McKenna, M. and Parrish, C.R. (2003) Combinations of two capsid regions controlling canine host range determine canine transferrin receptor binding by canine and feline parvoviruses. *J. Virol.* **77**, 10099–10105.

Hung, C.C., Sheng, W.H., Lee, K.L., Yang, S.J. and Chen, M.Y. (2006) Genetic drift of parvovirus B19 is found in AIDS patients with persistent B19 infection. *J. Med. Virol.* **78**, 1374–1384.

Ikeda, Y., Mochizuki, M., Naito, R., Nakamura, K., Miyazawa, T., Mikami, T. and Takahashi, E. (2000) Predominance of canine parvovirus (CPV) in unvaccinated cat populations and emergence of new antigenic types of CPVs in cats. *Virology* **278**, 13–19.

Ikeda, Y., Nakamura, K., Miyazawa, T., Takahashi, E. and Mochizuki, M. (2002) Feline host range of canine parvovirus: recent emergence of new antigenic types in cats. *Emerg. Infect. Dis.* **8**, 341–346.

Jacoby, R.O., Johnson, E.A., Paturzo, F.X., Gaertner, D.J., Brandsma, J.L. and Smith, A.L. (1991) Persistent rat parvovirus infection in individually housed rats. *Arch. Virol.* **117**, 193–205.

Jacoby, R.O., Ball-Goodrich, L.J., Besselsen, D.G., McKisic, M.D., Riley, L.K. and Smith, A.L. (1996) Rodent parvovirus infections. *Lab. Anim. Sci.* **46**, 370–380.

Jacoby, R.O., Johnson, E.A., Paturzo, F.X. and Ball-Goodrich, L. (2000) Persistent rat virus infection in smooth muscle of euthymic and athymic rats. *J. Virol.* **74**, 11841–11848.

Jensen, K.T., Wolfinbarger, J.B., Aasted, B. and Bloom, M.E. (2000) Replication of Aleutian mink disease parvovirus in mink lymph node histocultures. *J. Gen. Virol.* **81**, 335–343.

Joo, H.S., Donaldson Wood, C.R. and Johnson, R.H. (1976) Observations on the pathogenesis of porcine parvovirus infection. *Arch. Virol.* **51**, 123–129.

Kasprowicz, V., Isa, A., Tolfvenstam, T., Jeffery, K., Bowness, P. and Klenerman, P. (2006) Tracking of peptide-specific CD4+ T-cell responses after an acute resolving viral infection: a study of parvovirus B19. *J. Virol.* **80**, 11209–11217.

Kaufmann, J., Buccola, J.M., Stead, W., Rowley, C., Wong, M. and Bates, C.K. (2007) Secondary symptomatic parvovirus B19 infection in a healthy adult. *J. Gen. Intern. Med.* **22**, 877–878.

Kerr, J.R., Curran, M.D., Moore, J.E., Erdman, D.D., Coyle, P.V., Nunoue, T. *et al.* (1995) Genetic diversity in the non-structural gene of parvovirus B19 detected by single-stranded conformational polymorphism assay (SSCP) and partial nucleotide sequencing. *J. Virol. Methods* **53**, 213–222.

Kollek, R., Tseng, B.Y. and Goulian, M. (1982) DNA polymerase requirements for parvovirus H-1 DNA replication in vitro. *J. Virol.* **41**, 982–989.

Kunkel, T.A. (2004) DNA replication fidelity. *J. Biol. Chem.* **279**, 16895–16898.

Lachmann, S., Rommeleare, J. and Nuesch, J.P. (2003) Novel PKCeta is required to activate replicative functions of the major nonstructural protein NS1 of minute virus of mice. *J. Virol.* **77**, 8048–8060.

Lederman, M., Patton, J.T., Stout, E.R. and Bates, R.C. (1984) Virally encoded noncapsid protein associated with bovine parvovirus infection. *J. Virol.* **49**, 315–318.

Lefrere, J.J., Servant-Delmas, A., Candotti, D., Mariotti, M., Thomas, I., Brossard, Y. *et al.* (2005) Persistent B19 infection in immunocompetent individuals: implications for transfusion safety. *Blood* **106**, 2890–2895.

Li, X. and Rhode, S.L. (1991) Nonstructural protein NS2 of parvovirus H-1 is required for efficient viral protein synthesis and virus production in rat cells in vivo and in vitro. *Virology* **184**, 117–130.

Lopez-Bueno, A., Mateu, M.G. and Almendral, J.M. (2003) High mutant frequency in populations of a DNA virus allows evasion from antibody therapy in an immunodeficient host. *J. Virol.* **77**, 2701–2708.

Lopez-Bueno, A., Valle, N., Gallego, J.M., Perez, J. and Almendral, J.M. (2004) Enhanced cytoplasmic sequestration of the nuclear export receptor CRM1 by NS2 mutations developed in the host regulates parvovirus fitness. *J. Virol.* **78**, 10674–10684.

Lopez-Bueno, A., Rubio, M.P., Bryant, N., McKenna, R., Agbandje-McKenna, M. and Almendral, J.M. (2006) Host-selected amino acid changes at the sialic acid binding pocket of the parvovirus capsid modulate cell binding affinity and determine virulence. *J. Virol.* **80**, 1563–1573.

Lukashov, V.V. and Goudsmit, J. (2001) Evolutionary relationships among parvoviruses: virus-host coevolution among autonomous primate parvoviruses and links between adeno-associated and avian parvoviruses. *J. Virol.* **75**, 2729–2740.

Macartney, L., McCandlish, I.A., Thompson, H. and Cornwell, H.J. (1984a) Canine parvovirus enteritis 1: Clinical, haematological and pathological features of experimental infection. *Vet. Rec.* **115**, 201–210.

Macartney, L., McCandlish, I.A., Thompson, H. and Cornwell, H.J. (1984b) Canine parvovirus enteritis 2: Pathogenesis. *Vet. Rec.* **115**, 453–460.

Mahy, B. (2006) A new pathogenic human parvovirus. *Rev. Med. Virol.* **16**, 279–280.

Manning, A., Willey, S.J., Bell, J.E. and Simmonds, P. (2007) Comparison of tissue distribution, persistence, and molecular epidemiology of parvovirus B19 and novel human parvoviruses PARV4 and human bocavirus. *J. Infect. Dis.* **195**, 1345–1352.

Martella, V., Cavalli, A., Pratelli, A., Bozzo, G., Camero, M., Buonavoglia, D. *et al.* (2004) A canine parvovirus mutant is spreading in Italy. *J. Clin. Microbiol.* **42**, 1333–1336.

Martella, V., Decaro, N., Elia, G. and Buonavoglia, C. (2005) Surveillance activity for canine parvovirus in Italy. *J. Vet. Med. B Infect. Dis. Vet. Public Health* **52**, 312–315.

McKenna, R., Olson, N.H., Chipman, P.R., Baker, T. S., Booth, T.F., Christensen, J. *et al.* (1999) Three-dimensional structure of Aleutian mink disease parvovirus: implications for disease pathogenicity. *J. Virol.* **73**, 6882–6891.

Mengeling, W.L., Paul, P.S. and Brown, T.T. (1980) Transplacental infection and embryonic death following maternal exposure to porcine parvovirus near the time of conception. *Arch. Virol.* **65**, 55–62.

Miller, C.L. and Pintel, D.J. (2002) Interaction between parvovirus NS2 protein and nuclear export factor Crm1 is important for viral egress from the nucleus of murine cells. *J. Virol.* **76**, 3257–3266.

Mochizuki, M., Konishi, S., Ajiki, M. and Akaboshi, T. (1989) Comparison of feline parvovirus subspecific strains using monoclonal antibodies against a feline panleukopenia virus. *Nippon Juigaku Zasshi* **51**, 264–272.

Mochizuki, M., Harasawa, R. and Nakatani, H. (1993) Antigenic and genomic variabilities among recently prevalent parvoviruses of canine and feline origin in Japan. *Vet. Microbiol.* **38**, 1–10.

Monahan, P.E., Jooss, K. and Sands, M.S. (2002) Safety of adeno-associated virus gene therapy vectors: a current evaluation. *Expert Opin. Drug Saf.* **1**, 79–91.

Mori, J., Beattie, P., Melton, D.W., Cohen, B.J. and Clewley, J.P. (1987) Structure and mapping of the DNA of human parvovirus B19. *J. Gen. Virol.* **68**(Pt 11), 2797–2806.

Morinet, F., Tratschin, J.-D., Perol, Y. and Siegl, G. (1986) Comparison of 17 isolates of the human parvovirus B19 by restriction enzyme analysis. *Arch. Virol.* **90**, 165–172.

Morita, E., Tada, K., Chisaka, H., Asao, H., Sato, H., Yaegashi, N. and Sugamura, K. (2001) Human parvovirus b19 induces cell cycle arrest at G_2 phase with accumulation of mitotic cyclins. *J. Virol.* **75**, 7555–7563.

Morita, E., Nakashima, A., Asao, H., Sato, H. and Sugamura, K. (2003) Human parvovirus B19 non-structural protein (NS1) induces cell cycle arrest at G_1 phase. *J. Virol.* **77**, 2915–2921.

Murphy, F.A., Fauquet, C.M., Mayo, M.A., Jarvis, S.A., Ghabrial, S.A., Summers, M.D. *et al.* (1995) International Committee on Taxonomy of Viruses, International Union of Microbiological Societies. Virology Division Virus taxonomy/G: classification and nomenclature of viruses : Sixth report of the International Committee on Taxonomy of Viruses. *Arch. Virol. (Suppl.) viii* **586**, iii.

Musiani, M., Zerbini, M., Gentilomi, G., Plazzi, M., Gallinella, G. and Venturoli, S. (1995) Parvovirus B19 clearance from peripheral blood after acute infection. *J. Infect. Dis.* **172**, 1360–1363.

Naeger, L.K., Salome, N. and Pintel, D.J. (1993) NS2 is required for efficient translation of viral mRNA in minute virus of mice-infected murine cells. *J. Virol.* **67**, 1034–1043.

Nakamura, M., Tohya, Y., Miyazawa, T., Mochizuki, M., Phung, H.T., Nguyen, N.H. *et al.* (2004) A novel antigenic variant of canine parvovirus from a Vietnamese dog. *Arch. Virol.* **149**, 2261–2269.

Nam, H.J., Gurda-Whitaker, B., Gan, W.Y., Ilaria, S., McKenna, R., Mehta, P. *et al.* (2006) Identification of the sialic acid structures recognized by minute virus of mice and the role of binding affinity in virulence adaptation. *J. Biol. Chem.* **281**, 25670–25677.

Nguyen, Q.T., Sifer, C., Schneider, V., Allaume, X., Servant, A., Bernaudin, F. *et al.* (1999) Novel human erythrovirus associated with transient aplastic anemia. *J. Clin. Microbiol.* **37**, 2483–2487.

Nuesch, J.P., Christensen, J. and Rommelaere, J. (2001) Initiation of minute virus of mice DNA replication is regulated at the level of origin unwinding by atypical protein kinase C phosphorylation of NS1. *J. Virol.* **75**, 5730–5739.

Nuesch, J.P., Lachmann, S., Corbau, R. and Rommelaere, J. (2003) Regulation of minute virus of mice NS1 replicative functions by atypical PKClambda in vivo. *J. Virol.* **77**, 433–442.

Oie, K.L., Durrant, G., Wolfinbarger, J.B., Martin, D., Costello, F., Perryman, S. *et al.* (1996) The relationship between capsid protein (VP2) sequence and pathogenicity of Aleutian mink disease parvovirus (ADV): a possible role for raccoons in the transmission of ADV infections. *J. Virol.* **70**, 852–861.

Olofsson, A., Mittelholzer, C., Treiberg Berndtsson, L., Lind, L., Mejerland, T. and Belak, S. (1999) Unusual, high genetic diversity of Aleutian mink disease virus. *J. Clin. Microbiol.* **37**, 4145–4149.

Op De Beeck, A. and Caillet-Fauquet, P. (1997) Viruses and the cell cycle. *Progr. Cell Cycle Res.* **3**, 1–19.

Op De Beeck, A., Sobczak-Thepot, J., Sirma, H., Bourgain, F., Brechot, C. and Caillet-Fauquet, P. (2001) NS1- and minute virus of mice-induced cell cycle arrest: involvement of p53 and p21(cip1). *J. Virol.* **75**, 11071–11078.

Opie, S.R., Warrington, K.H., Jr., Agbandje-McKenna, M., Zolotukhin, S. and Muzyczka, N. (2003) Identification of amino acid residues in the capsid proteins of adeno-associated virus type 2 that contribute to heparan sulfate proteoglycan binding. *J. Virol.* **77**, 6995–7006.

Parker, J.S.L. and Parrish, C.R. (1997) Canine parvovirus host range is determined by the specific conformation of an additional region of the capsid. *J. Virol.* **71**, 9214–9222.

Parrish, C.R. (1995) Pathogenesis of feline panleukopenia virus and canine parvovirus. *Baillière's Clin. Haematol.* **8**, 57–71.

Parrish, C.R. and Carmichael, L.E. (1983) Antigenic structure and variation of canine parvovirus type-2, feline panleukopenia virus, and mink enteritis virus. *Virology* **129**, 401–414.

Parrish, C.R., Aquadro, C., Strassheim, M.L., Evermann, J.F., Sgro, J.-Y. and Mohammed, H. (1991) Rapid antigenic-type replacement and DNA sequence evolution of canine parvovirus. *J. Virol.* **65**, 6544–6552.

Parrish, C.R., Gorham, J.R., Schwartz, T.M. and Carmichael, L.E. (1984) Characterisation of antigenic variation among mink enteritis virus isolates. *Am. J. Vet. Res.* **45**, 2591–2599.

Parsyan, A., Addo-Yobo, E., Owusu-Ofori, S., Akpene, H., Sarkodie, F. and Allain, J.P. (2006) Effects of transfusion on human erythrovirus B19-susceptible or—infected pediatric recipients in a genotype 3-endemic area. *Transfusion* **46**, 1593–1600.

Pennick, K.E., Stevenson, M.A., Latimer, K.S., Ritchie, B.W. and Gregory, C.R. (2005) Persistent viral shedding during asymptomatic Aleutian mink disease parvoviral infection in a ferret. *J. Vet. Diagn. Invest.* **17**, 594–597.

Pereira, C.A., Leal, E.S. and Durigon, E.L. (2007) Selective regimen shift and demographic growth increase associated with the emergence of high-fitness variants of canine parvovirus. *Infect. Genet. Evol.* **7**, 399–409.

Perez, R., Francia, L., Romero, V., Maya, L., Lopez, I. and Hernandez, M. (2007) First detection of canine parvovirus type 2c in South America. *Vet Microbiol.* **124**, 147–152.

Porter, D.D. (1986) Aleutian disease: a persistent parvovirus infection of mink with a maximal but ineffective host humoral immune response. *Prog. Med. Virol.* **33**, 42–60.

Rubio, M.P., Lopez-Bueno, A. and Almendral, J.M. (2005) Virulent variants emerging in mice infected with the apathogenic prototype strain of the parvovirus minute virus of mice exhibit a capsid with low avidity for a primary receptor. *J. Virol.* **79**, 11280–11290.

Saikawa, T., Anderson, S., Momoeda, M., Kajigaya, S. and Young, N.S. (1993) Neutralizing linear epitopes of B19 parvovirus cluster in the VP1 unique and VP1-VP2 junction regions. *J. Virol.* **67**, 3004–3009.

Sanabani, S., Neto, W.K., Pereira, J. and Sabino, E.C. (2006) Sequence variability of human erythroviruses present in bone marrow of Brazilian patients with various parvovirus B19-related hematological symptoms. *J. Clin. Microbiol.* **44**, 604–606.

Schwartz, D., Green, B., Carmichael, L.E. and Parrish, C.R. (2002) The canine minute virus (minute virus of canines) is a distinct parvovirus that is most similar to bovine parvovirus. *Virology* **302**, 219–223.

Servant, A., Laperche, S., Lallemand, F., Marinho, V., De Saint Maur, G., Meritet, J.F. and Garbarg-Chenon, A. (2002) Genetic diversity within human erythroviruses: identification of three genotypes. *J. Virol.* **76**, 9124–9134.

Shackelton, L.A. and Holmes, E.C. (2006) Phylogenetic evidence for the rapid evolution of human B19 erythrovirus. *J. Virol.* **80**, 3666–3669.

Shackelton, L.A., Parrish, C.R., Truyen, U. and Holmes, E.C. (2005) High rate of viral evolution associated with the emergence of carnivore parvovirus. *Proc. Natl Acad. Sci. USA* **102**, 379–384.

Smith, R.H. and Kotin, R.M. (2000) An adeno-associated virus (AAV) initiator protein, Rep78, catalyzes the cleavage and ligation of single-stranded AAV ori DNA. *J. Virol.* **74**, 3122–3129.

Strassheim, L.S., Gruenberg, A., Veijalainen, P., Sgro, J.-Y. and Parrish, C.R. (1994) Two dominant neutralizing antigenic determinants of canine parvovirus are found on the threefold spike of the virus capsid. *Virology* **198**, 175–184.

Suikkanen, S., Antila, M., Jaatinen, A., Vihinen-Ranta, M. and Vuento, M. (2003) Release of canine parvovirus from endocytic vesicles. *Virology* **316**, 267–280.

Swofford, D.L. (1993) *PAUP—phylogentic analysis with parsimony*. Champaign, IL: Center for Biodiversity, Illinois Natural History Survey.

Takahashi, N., Takada, N., Hashimoto, T. and Okamoto, T. (1999) Genetic heterogeneity of the immunogenic viral capsid protein region of human parvovirus B19 isolates obtained from an outbreak in a pediatric ward. *FEBS Lett.* **450**, 289–293.

Tattersall, P., Bergoin, M., Bloom, M.E. *et al.* (2005) Parvoviridae. In: *Virus Taxonomy* (C.M. Fauquet, M.A. Mayo, L. Maniloff *et al.*, eds), pp. 353–369. Amsterdam: Elsevier.

Toan, N.L., Duechting, A., Kremsner, P.G., Song le, H., Ebinger, M., Aberle, S. *et al.* (2006) Phylogenetic analysis of human parvovirus B19, indicating two subgroups of genotype 1 in Vietnamese patients. *J. Gen. Virol.* **87**, 2941–2949.

Toolan, H.W. (1960) Experimental production of mongoloid hamsters. *Science* **131**, 1446–1448.

Tratschin, J.-D., McMaster, G.K., Kronauer, G. and Siegl, G. (1982) Canine parvovirus: relationship to wild-type and vaccine strains of feline panleukopenia virus and mink enteritis virus. *J. Gen. Virol.* **61**, 33–41.

Truyen, U. and Parrish, C.R. (2000) Epidemiology and pathology of autonomous parvoviruses. *Contrib. Microbiol.* **4**, 149–162.

Truyen, U., Platzer, G., Parrish, C.R., Hanichen, T., Hermanns, W. and Kaaden, O.R. (1994) Detection of canine parvovirus DNA in paraffin-embedded tissues by polymerase chain reaction. *Zentralbl. Veterinarmed. [B]* **41**, 148–152.

Truyen, U., Gruenberg, A., Chang, S.F., Obermaier, B., Veijalainen, P. and Parrish, C.R. (1995) Evolution of the feline-subgroup parvoviruses and the control of canine host range in vivo. *J. Virol.* **69**, 4702–4710.

Truyen, U., Evermann, J.F., Vieler, E. and Parrish, C.R. (1996a) Evolution of canine parvovirus involved loss and gain of feline host range. *Virology* **215**, 186–189.

Truyen, U., Platzer, G. and Parrish, C.R. (1996b) Antigenic type distribution among canine parvoviruses in dogs and cats in Germany. *Vet. Rec.* **138**, 365–366.

Truyen, U., Geissler, K., Parrish, C.R., Hermanns, W. and Siegl, G. (1998) No evidence for a role of modified live virus vaccines in the emergence of canine parvovirus. *J. Gen. Virol.* **79**, 1153–1158.

Wan, C.H., Soderlund-Venermo, M., Pintel, D.J. and Riley, L.K. (2002) Molecular characterization of three newly recognized rat parvoviruses. *J. Gen. Virol.* **83**, 2075–2083.

Wan, C.H., Bauer, B.A., Pintel, D.J. and Riley, L.K. (2006) Detection of rat parvovirus type 1 and rat minute virus type 1 by polymerase chain reaction. *Lab. Anim.* **40**, 63–69.

Wang, D., Yuan, W., Davis, I. and Parrish, C.R. (1998) Nonstructural protein-2 and the replication of canine parvovirus. *Virology* **240**, 273–281.

Wistuba, A., Kern, A., Weger, S., Grimm, D. and Kleinschmidt, J.A. (1997) Subcellular compartmentalization of adeno-associated virus type 2 assembly. *J. Virol.* **71**, 1341–1352.

Wobus, C.E., Hugle-Dorr, B., Girod, A., Petersen, G., Hallek, M. and Kleinschmidt, J.A. (2000) Monoclonal antibodies against the adeno-associated virus type 2 (AAV-2) capsid: epitope mapping and identification of capsid domains involved in AAV-2-cell interaction and neutralization of AAV-2 infection. *J. Virol.* **74**, 9281–9293.

Wong, S., Young, N.S. and Brown, K.E. (2003) Prevalence of parvovirus b19 in liver tissue: no association with fulminant hepatitis or hepatitis-associated aplastic anemia. *J. Infect. Dis.* **187**, 1581–1586.

Wu, Z., Asokan, A., Grieger, J.C., Govindasamy, L., Agbandje-McKenna, M. and Samulski, R.J. (2006) Single amino acid changes can influence titer, heparin binding, and tissue tropism in different adeno-associated virus serotypes. *J. Virol.* **80**, 11393–11397.

Yoo, B.C., Lee, D.H., Park, S.M., Park, J.W., Kim, C.Y., Lee, H.S. *et al.* (1999) A novel parvovirus isolated from Manchurian chipmunks. *Virology* **253**, 250–258.

Yoshimoto, K., Rosenfeld, S., Frickhofen, N., Kennedy, D., Hills, R., Kajigaya, S. and Young, N. (1991) A second neutralizing epitope of B19 parvovirus implicates the spike region in the immune response. *J. Virol.* **65**, 7056–7060.

Young, N.S. (1995) B19 parvovirus. *Baillière's Clin. Haematol.* **8**, 25–56.

Zadori, Z., Szelei, J., Lacoste, M.-C., Raymond, P., Allaire, M., Nabi, I.R. and Tijssen, P. (2001) A viral phospholipase A2 is required for parvovirus infectivity. *Dev. Cell* **1**, 291–302.

CHAPTER

18

Genome Diversity and Evolution of Papillomaviruses

Hans-Ulrich Bernard

ABSTRACT

Papillomaviruses are small DNA viruses that infect cutaneous and mucosal epithelia of humans, other mammals, and of at least some birds. Due to their important role in human neoplastic disease, a huge nucleotide sequence database with hundreds of complete or partial genome sequences has become established. The analysis of these sequences led to the notion that papillomaviruses genomes evolve very slowly, and nucleotide changes accumulate at a rate of about 1% per 100 000 years, only slightly faster than the genomic changes of the hosts. The viruses form clades restricted to specific host species, although ancient diversification also gave rise to remotely related clades within individual host species. As phylogenetic trees of mammalian and bird papillomaviruses have semblance to phylogenetic trees of the infected animal species, host-linked evolution suggests that the papillomavirus genome organization is at least as old as the phylogeny of mammals and birds themselves, i.e. about 150 million years. While papillomaviruses are thought to be unrelated to other DNA viruses, a major homology between replication initiation proteins points to a common ancestor that papillomaviruses share with several other viruses and virus-like elements. The mechanisms of papillomavirus evolution are still largely unclear due to the very similar genetic composition and biology of all papillomaviruses.

INTRODUCTION

The papillomaviruses (*Papillomaviridae*) form one of the seven families of human viruses with DNA genomes. They are thought not to be related to the polyomaviruses (*Polyomaviridae*), with whom they had once been lumped into a single family, papovaviridae, a taxonomic unit that had been based mostly on the similarity of the viral particles and the common double-stranded DNA genomes. As genomes of viruses in the two families do not show much nucleotide sequence similarity, the papovaviridae have been eliminated as a valid taxon. When this book went to print, an exception to this generally hold view was about to be published (Woolford *et al.*, 2007) and will be discussed in the last paragraph of this chapter

Origin and Evolution of Viruses
ISBN 978-0-12-374153-0

dealing with the origin of papillomaviruses and their relationship with other families of viruses.

Several human papillomaviruses are important human carcinogens, and therefore attracted considerable scientific attention for the last three decades. As a result, countless clinical samples have been screened for new papillomavirus isolates, which led to the establishment of a huge database of the diversity of papillomaviruses in humans. Understandably, much less effort has gone into the search of papillomaviruses in animals. All carefully studied mammals and birds appear to be the host of one or several unique papillomaviruses, but these viruses have not yet been found in reptiles, amphibians, fishes, and non-vertebrates. It is not yet clear whether the much larger diversity of papillomaviruses in humans, which vastly exceeds that in any animal, is an idiosyncrasy of the human host or just a reflection of more intense research.

THE BIOLOGY AND PATHOGENESIS OF PAPILLOMAVIRUSES

Papillomaviruses have double-stranded DNA genomes with sizes of 7.5–8 kb. Most of them encode eight genes, as shown in Figure 18.1. All eight genes are transcribed from the same DNA strand, starting at one or few promoters, often located in the long control region (LCR), and the gene products are translated from differentially spliced and polycistronic mRNAs. All papillomaviruses encode the E1 protein, a principal regulator of replication, the E2 protein, a regulator of transcription and genome partitioning, and the L2 and L1 proteins, which are minor and major capsid proteins, respectively. The vast majority of all papillomaviruses contain the transforming genes E5, E6, and E7 in genomic positions as indicated in Figure 18.1, but a few human, mammalian, and bird papillomaviruses have been detected that show major aberrations regarding presence and genomic organization of these three

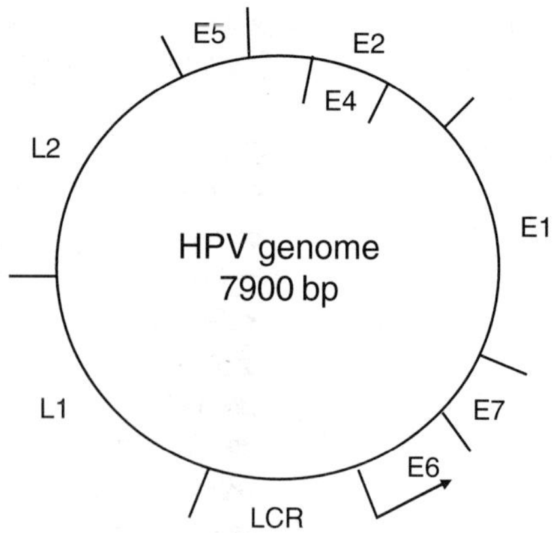

FIGURE 18.1 The genomic organization of a typical papillomavirus.

genes (Scobie *et al.*, 1997; Terai *et al.*, 2002; Chen *et al.*, 2007).

Papillomaviruses reproduce exclusively in epithelial cells of the skin or various mucosas, apparently due to transcriptional requirements that are matched only by these cells (Bernard, 2002). Certain papillomaviruses of hoofed animals can also infect the mesenchyme, but do not produce virions in this tissue.

Since medical research is focussed on those papillomaviruses that are causally associated with clinically detectable lesions, papillomaviruses are normally considered "transforming" or "carcinogenic" viruses. These attributes are appropriate and papillomaviruses are indisputably a necessary prerequisite for many forms of epithelial neoplasia including common warts, genital warts, and anogenital cancer. While these lesions do not occur in the absence of the viruses, it is equally well confirmed that infection with papillomaviruses is not sufficient to induce all etiological changes toward neoplasia. Many papillomaviruses, and even those found in benign and malignant lesions, are widespread in subclinical infections, and some kind of yet poorly defined latent life cycle might be a much more typical representation of the papillomavirus biology than their role in tumorigenesis. An example for this complex and poorly understood fact is the sexual transmission of certain papillomaviruses between the male penis, which is nearly always asymptomatically infected, and the female vagina, which is often also asymptomatically infected with the exception of a very specific tissue of the cervix uteri, the transformation zone of the cervix, which has a high propensity for neoplastic changes under the influence of papillomavirus infection.

The vast majority of the medically relevant research, however, does not deal with asymptomatic infection but with the question of how papillomavirus oncoproteins target the infected epithelial cell and induce hyperplasia. This is achieved by the functions of the oncoproteins E5 (Suprynowicz *et al.*, 2005), E6 (Mantovani and Banks, 2001), and E7 (Munger *et al.*, 2001). These proteins have complex pleiotropic functions, which include their action on membrane-bound receptors and the cell cycle regulators p53 and Rb, respectively. These mechanisms and details of the extensive molecular information about these proteins have been widely reviewed and will not be discussed in this chapter.

THE PHYLOGENY-BASED TAXONOMY OF PAPILLOMAVIRUSES

Taxonomy is founded on phylogenetic considerations rather than being a prerequisite for them, but in order to discuss the phylogeny of papillomaviruses, taxonomic concepts and observations regarding this virus family have to be introduced. Novel papillomavirus isolates have traditionally been described as "types." By definition, a papillomavirus type is a complete isolate of a papillomavirus genome, whose L1 gene nucleotide sequence is at least 10% different from that of any other previously described type (de Villiers, 1997; de Villiers *et al.*, 2004). There is no taxonomic description of papillomaviruses in the context of cell culture or serology. The term "serotype" is not used, as papillomaviruses do not appear to induce a consistent humoral immune response. The "10%-diversity" criterion was arbitrarily chosen, but it turned out to be a useful choice, as it defines natural taxa. The nucleotide sequences of papillomavirus isolates are either much more divergent from previous isolates (often, for instance, as much as 30%, and therefore identifying new types), or much less diverse (0–2%, identifying variants of types, see below). Only a few isolates differ from a related type by close to 10%, and have been termed "subtypes." The few known subtypes, while not conforming to the taxonomic definition of "types," are nevertheless distinct and well separated taxa (Calleja-Macias *et al.*, 2005a, Hazard *et al.*, 2007). No papillomavirus genome has ever been found that forms a link between any two papillomavirus types, and all evolutionary intermediates are apparently extinct.

These discontinuities of sequence similarities make papillomavirus types strong candidates to be considered "species." However, many closely related papillomavirus types are biologically indistinguishable. A rule established by the International Committee on Taxonomy of Viruses (ICTV) requires that any two virus isolates have to be biologically distinct in order to be described as two virus species, a condition that is not fulfilled by many papillomaviruses. This impasse between the views of papillomavirus taxonomists and the ICTV was resolved by leaving the "type concept" unaltered and the description of new types under the control of papillomavirologists, while a papillomavirus species became defined as a cluster of phylogenetically related types with similar or identical biological properties. A complete list of papillomavirus types and their taxonomic grouping into species has been published by de Villiers *et al.* (2004). Most genome sequences have been published in an interpretive format on a website (Farmer *et al.*, 1995), which has been unfortunately (and hopefully only temporarily) discontinued due to funding problems.

The taxonomy of papillomaviruses is based on the nucleotide sequences of their L1 genes and phylogenetic trees calculated from alignments of these sequences (Figure 18.2 In the case of most human papillomaviruses, these trees form a two-level hierarchy of minor and major branches (in many older

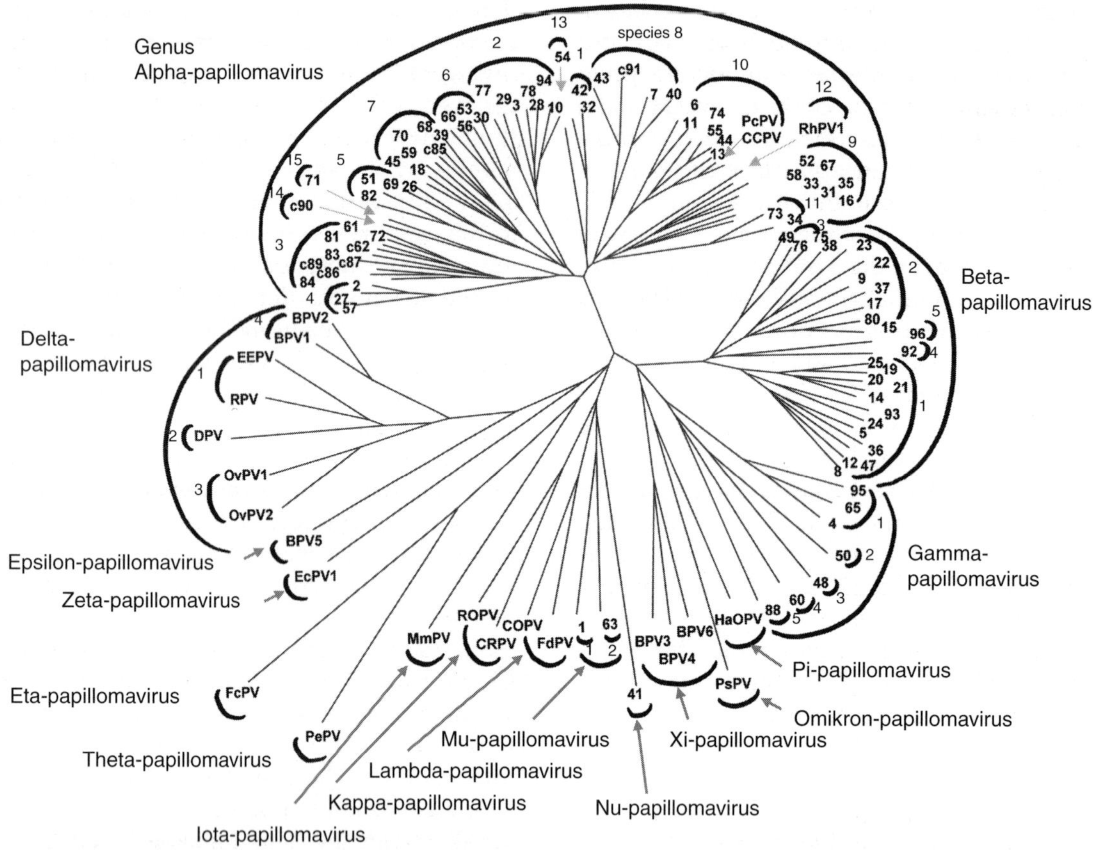

FIGURE 18.2 Phylogenetic tree of 118 papillomavirus types. All numbers refer to HPV types, c-numbers to candidate types, i.e. HPV genomes isolated as PCR amplicons. All abbreviations, including letters, refer to animal papillomaviruses. BPV, bovine papillomavirus; COPV, canine oral papillomavirus; CRPV, cottontail rabbit papillomavirus; DPV, deer papillomavirus; EcPV, *Equus caballus* or horse papillomavirus; EEPV, European elk papillomavirus; FcPV, *Fringilla coelebs* or chaffinch papillomavirus; FdPV, *Felis domesticus* or cat papillomavirus; HaOPV, hamster oral papillomavirus; MnPV, *Mastomys natalensis* or African rat papillomavirus; OvPV ovine or sheep papillomavirus; PePV, *Psittacus erythacus* or African gray parrot papillomavirus; PsPV, *Phocoena spinipinnis* or porpoise papillomavirus; RPV, reindeer papillomavirus; ROPV, rabbit oral papillomavirus. The outermost semi-circular symbols identify papillomavirus genera, for example the genus alpha-papillomavirus. The inner semi-circular symbols identify papillomavirus species. For example, the HPV species 8 within the genus alpha-papillomaviruses lumps the HPV types 7, 40, 43, and cand91. Reprinted from "Classification of papillomaviruses" by de Villiers *et al.* (2004), with permission.

publications also referred to as "groups" and "supergroups"), the minor branches identifying the species, as described above, the major branches defined as "genera." The duality of this branching order is less pronounced among the non-primate mammalian papillomaviruses, but clusters of branches have nevertheless been used to describe species and genera of animal papillomaviruses. Papillomavirus species are named by numbers and genera by Greek letters. For example, human papillomavirus type 16 (the most common virus in anogenital cancers) belongs to the genus of alpha-papillomaviruses and within this genus to species 9. At this point, i.e. 3 years after the publication of the official taxonomy, the new names of genera (e.g. alpha- and beta-papillomaviruses) are becoming widely used. On the other side, the numbering of species might remain a formality, and it is not yet clear whether it will become adopted by most basic researchers and clinicians, or whether they will continue to use the traditional numbering of types as the lingua franca.

RECENT EVOLUTIONARY CHANGES: ALPHA-PAPILLOMAVIRUSES AND THEIR VARIANTS

Slightly more than 100 human papillomavirus (HPV) types have so far been formally described (de Villiers *et al.*, 2004; Chen *et al.*, 2007) and polymerase chain reaction (PCR) amplicons point to the existence of at least another 100 types, whose genomes have not yet been completely isolated. None of these HPV types was ever found in an animal, and no animal papillomavirus in humans, an observation that forms one of the foundations of the concept of evolution of papillomaviruses together with a specific host (variably referred to as host-linked evolution and co-evolution hypothesis). HPV types are members of five genera (alpha-, beta-, gamma-, mu- and nu-papillomaviruses). Alpha-papillomaviruses contain the vast majority of those HPV types that affect human health. Approximately 60 HPV types comprise this genus, and these include those HPV types that are most frequently found in common warts (HPV-2, 27, 57), in genital and laryngeal warts (HPV-6 and 11), and in cervical and other groups of anogenital cancers (HPV-16, 18, 31 and 15 other HPV types) (Matsukura and Sugase, 2001; Munoz *et al.*, 2003). Many of these HPV types have been re-isolated countless times, and several independent research groups have sequenced repeatedly samples of the same HPV type obtained from patients living in virtually all parts of the world and in all three major ethnic groups (unfortunately excluding Australian Aboriginals). These studies were initiated by our group (Ho *et al.*, 1993; Ong *et al.*, 1993; Heinzel *et al.*, 1995; Chan *et al.*, 1997b; Calleja-Macias *et al.*, 2004, 2005b; Prado *et al.*, 2005), but have been confirmed and extended by labs addressing regional cohorts or worldwide comparison numerous times (see for example Stewart *et al.*, 1996; Yamada *et al.*, 1997; Chen *et al.*, 2005; additional studies reviewed in Bernard, 1994; Bernard *et al.*, 2006). All of these investigations unequivocally agree on the following observations and interpretations:

- All alpha-HPV types occur in all human ethnic groups, and were likely infecting all ethnic groups throughout human evolution.
- All HPV types show only very minor sequence diversity, 1–12 nucleotide exchanges in amplicons of 300–400 bp of the non-coding regions, an even lower fraction of maximally 2% in genes. In a full-genomic comparison of all HPV-16 variants, altogether 4% of all nucleotide positions were found to be variable (Chen *et al.*, 2005). These sequence divergent re-isolates of HPV-types became termed "variants," and were defined as having 0–2% sequence diversity from the original isolate (prototype).
- Combinations of variable sites in different parts of the virus genome are linked, no recombination could be observed.
- All research groups found identical or similar variants, and the total diversity ranges from a few ten up to 100 variants for any HPV type.
- Certain variants formed minor clades and were enriched in or restricted to specific ethnic groups, giving rise to terms like African, Asian, or Asian-American variants.
- Variants spread with historic migrations of humans. For example, the spread of people of Asian ancestry into North and South America is reflected in the relationship between Asian and American Indian HPV-16 and 18 variants.
- No mutation patterns were ever observed that suggested an evolution over relatively short periods, e.g. during a few ten or 100 years. For example, migrants to the Americas from Africa and Europe brought during the last five centuries specific variants from

these continents, which are now widespread in American cohorts, still maintaining some correlation between the ethnicity of the infected person and the geographic origin of the viral variant (Xi *et al.*, 2006).

The most plausible interpretation of all these observations is that all HPV types existed already with genomic sequences at least 98% identical to the sequence of today's HPV types in humans at the time of the origin of our species and its spread inside Africa and out of Africa, 50 000 to a few hundred thousand years ago. In other words, all medically relevant HPV types evolved in primates predating the emergence of *Homo sapiens*. Modern people were always infected with all HPV types, and did not acquire any new infections of papillomaviruses from domestic or wild animals. Humans have always been infected with common warts, genital warts, and anogenital cancer. The evolution of some variants of each HPV type occurred apparently during the spread of people out of Africa, supposedly in the last 50 000 years. There is some, but not a precise linkage between specific phylogenetic branches of variants of specific HPV types and human ethnic groups, which is probably a reflection of bottlenecks of the expansion of ancient human populations, geographic spread of humans in prehistoric times, and human migrations in historic times.

GENETIC AND BIOLOGICAL DIVERSITY OF ALPHA-PAPILLOMAVIRUSES FROM HUMANS, APES, AND MONKEYS

Based on the considerations outlined in the last paragraph, close relatives of HPV types must once have existed in pre-human primates and therefore relatives may still be found in apes and monkeys. This is indeed the case, although only few ape and monkey papillomaviruses have so far been described. Intriguingly, a single papillomavirus type has been isolated from each of the two related chimpanzee species, the common chimpanzee and the pigmy chimpanzee. Both papillomavirus types, CCPV and PCPV, are the closest relatives of one another, and they are both most closely related to HPV-13 and belong to the same HPV species that includes HPV-6 and 11 (de Villiers *et al.*, 2004). One may conclude from this that the common precursor of these four papillomavirus types existed in the primate that was the common precursor of humans and the two chimpanzee species, about seven million years ago.

Apes and the ancestors of humans evolved from monkeys at a yet earlier time, roughly 20 million years ago. From the hypothesis that papillomaviruses evolve linked to their host, one may expect that papillomaviruses of monkeys may be more distantly related to human papillomaviruses than the two chimpanzee papillomaviruses. This could indeed be confirmed. Fifteen different papillomaviruses (13 from rhesus macaques, one from a long-tailed macaque, and one from a colobus monkey, CgPV-1; all except two only known from PCR amplicons, not from complete genomes) were found to belong to the alpha-papillomavirus genus, but none of them was a member of any of the species (minor branches) formed by HPV types. Instead, they formed four new minor branches, one of them lumping the single long-tailed macaque virus with several rhesus monkeys (Chan *et al.*, 1997a). In summary, the topology of the tree of 16 monkey and ape papillomaviruses and about 40 HPV types of the alpha-papillomavirus genus mimics the topology of a phylogenetic tree of the six primate host species in strong support of the hypothesis of linked evolution of virus and host.

In spite of major efforts, no alpha-papillomavirus could be detected in New World monkeys or non-primate mammals using PCR amplification with a variety of consensus primers that detect conserved parts of the L1 gene (our unpublished observation; L.L. Villa, personal communication). This makes it likely that alpha-papillomaviruses evolved in primates and separated from other papillomaviruses subsequent to the divergence of Old World and New World monkeys and before the diversification of Old World monkeys, i.e. 25–35 million years ago (Schrago and Russo, 2003).

HUMAN PAPILLOMAVIRUSES OF THE BETA, GAMMA, MU AND NU GENERA

Papillomaviruses isolated from humans form another four genera in addition to the alpha-papillomaviruses, namely the beta-, gamma-, mu-, and nu-papillomaviruses. The latter two taxa contain only three types, namely the little studied skin-infecting virus HPV-41 (nu-papillomavirus) and the two skin types HPV-1 and HPV-63 (mu-papillomaviruses). HPV-1 infects preferentially the foot soles and can give rise to fairly large flat (plantar) warts, which can be a significant health problem. Phylogenetically these three HPV types are not very informative.

The beta-papillomaviruses contain 25 formally described HPV-types including the well-studied types HPV-5 and HPV-8. Beta-papillomaviruses appear to infect asymptomatically the skin of a large fraction of the population (Boxman *et al.*, 1999). Strangely, these viruses can cause epidermodysplasia verruciformis, an aggressive growth of benign and malignant neoplasias of the skin, in genetically predisposed individuals (Ramoz *et al.*, 2002). The phenomenon of widespread asymptomatic infection of the skin is shared with the gamma-papillomaviruses, which contain the little studied type HPV-4, found in plain warts. Due to their asymptomatic and apparently commensalic association with humans, many beta- and gamma-papillomaviruses have not yet been isolated but were only detected as PCR amplicons (Antonsson *et al.*, 2000, 2003). PCR studies also frequently detected known and novel beta- and gamma-papillomaviruses in solar keratoses, specifically in immune suppressed patients. It is not yet known whether they are fortuitously or causally associated with these lesions (de Villiers, 1998; Forslund *et al.*, 2003).

Two papillomaviruses could be isolated from colobus monkeys, and while CgPV-1 was classified as alpha-papillomavirus, as discussed above, CgPV-2 turned out to be a beta-papillomavirus (Chan *et al.*, 1997c). In addition, PCR-based studies identified numerous novel beta- and gamma-papillomaviruses in the skin of long-tailed macaques and gorillas (Antonsson and Hansson, 2002). The same PCR protocol found yet other papillomaviruses in skin samples of various other non-primate mammals, but these did not belong to the beta- or gamma-papillomaviruses. These patterns of the distribution of beta- and gamma-papillomaviruses suggest that these two genera orginated just like alpha-papillomaviruses as defined clades in hosts at the root of primate evolution, and continued to diversify in linkage with each newly evolving primate species into host specific papillomavirus types.

DEEPER BRANCHING OF THE PHYLOGENETIC TREE OF MAMMALIAN AND BIRD PAPILLOMAVIRUSES

Papillomaviruses have been found in all carefully investigated mammals irrespective of whether these were wild-caught, living in zoos, or domesticated as farm animals or pets. This includes hosts as diverse as echidnas, koalas, manatees, dolphins, and bats (Rector *et al.*, 2004, 2006; Antonsson and McMillan, 2006; Rehtanz *et al.*, 2006; van Doorslaer *et al.*, 2006). Only two papillomaviruses have been detected in birds, namely in the European chaffinch and the African gray parrot (Tachezy *et al.*, 2002). Most of these animal papillomaviruses were detected in cutaneous neoplastic lesions, with some also found in mucosal epithelia (Scobie *et al.*, 1997) or asymptomatic skin (Antonsson and Hansson, 2002; Antonsson and McMillan, 2006).

All of these mammalian and bird papillomaviruses provide further support for the concept of host-linked evolution: Only a few of these virus types cluster within one major branch (genus), and relationship among the host species is clearly a prerequisite for such papillomavirus assemblages. Fibropapillomaviruses (genus delta-papillomaviruses) form the best example of such a cluster, which includes not only the well-studied type BPV-1 (bovine papillomavirus type 1), but also several related types from domesticated hoofed animals (other BPV types, sheep papillomavirus) and several species of deer (de Villiers *et al.*, 2004). Yet other papillomaviruses from cattle form the phylogenetically distinct episilon-papillomaviruses (e.g BPV-4, Scobie *et al.*, 1997), which lack the typical E6 gene.

All of these mammalian and bird papillomaviruses form major phylogenetic branches, which are widely separated from the five branches (genera) with primate papillomaviruses and from one another, as long as the host species is remotely related to all other host species. The deepest branching is found between all mammalian and the two bird papillomaviruses, suggestive of an evolutionary split close to the divergence of bird and mammalian lineages, roughly 150 million years ago. It seems unavoidable to propose that major features of the genomic organization of papillomaviruses as well as many domains of the specific proteins encoded by them have not changed since the time of the dinosaurs, which is certainly interesting to contemplate, as specialists and lay people normally consider virus genomes hypervariable due to the interest in certain rapidly mutating RNA viruses, like the human immunodeficiency and influenza viruses.

THE INTERPRETATION OF HOST-LINKED EVOLUTION

Humans had always had exposure to animal papillomaviruses since such viruses are common in pets, farm animals, and venison. Nevertheless, no animal papillomavirus has ever been found in humans or vice versa. Even a recent survey of animal keepers who were infected with numerous novel HPV types could not detect a clue that any of these viruses originated in mammals, although the caged animal population taken care of by these people carried a large number of novel papillomaviruses of their own, each strictly host specific (Antonsson and Hansson, 2002). A previous suspect that was thought to be an animal papillomavirus infecting humans, HPV-7, has been described as "butcher-papillomavirus," as it frequently leads to hand lesions in meat handlers. However, a phylogenetic analysis anchored this virus among the human alphapapillomaviruses, where it forms a species with several related HPV types. This suggests that unknown effects of animal blood or meat on the human skin led to the neoplasia in synergy with the infection by this specific HPV type, and there is no evidence that the slaughtering and meat handling led to an interspecific virus transfer.

All of these observations suggest a barrier of infectivity of papillomaviruses between humans and all other mammals. Similar barriers of transfer may exist between most mammalian species, since wild animals frequently contact one another (e.g. as prey of raptors), but no mammalian papillomavirus type was ever found in two wild mammals. The only known exception to this rule is formed by the bovine papillomavirus-1, which has been found in a variety of domesticated hoofed animals from cows to horses, and the constant physical proximity between these related hosts may have occasionally permitted it to pass host barriers. This may be a rare event as suggested by unique BPV-1 variants in specific horse lesions (Chambers *et al.*, 2003).

Papillomaviruses are able to infect cells from various host species and from various differentiation states quite indiscriminately (Muller *et al.*, 1995), which, however, does not result in productive infections. It has been shown that the transcription apparatus of a host cell puts limits on productive papillomavirus infection, as seen in the case of the phenomenon of epithelial specificity. The enhancers and promoters of papillomaviruses become active only in epithelial cells, and it is thought that an exact match between the papillomavirus transcriptional elements and the host cell's transcription apparatus is a prerequisite for transcription to occur in a cell type-specific way. A need for similar matches of the molecular apparatus of viruses and that of host cells may restrict the complete papillomavirus life cycle to a unique host species. Unfortunately, this hypothesis has never been addressed systematically in a way similar to the establishment of the concept of "permissive" and "non-permissive" hosts in adenovirus and polyomavirus biology.

ALTERNATIVE PHYLOGENETIC ASSEMBLAGES AND MECHANISMS OF PAPILLOMAVIRUS EVOLUTION

The present official taxonomy of papillomaviruses is based on the alignment of the L1 genes, as has been discussed above. It had been noted since the early 1990s that phylogenetic trees of papillomaviruses that were based on alignments of the nucleotide sequence of any of the early genes show similar topologies of the major branches, but important differences in the assemblage of minor branches (Chan *et al.* 1992, 1995; van Ranst *et al.*, 1992). Most notably, an L1 tree clusters the 18 medically interesting HPV types that are associated with cervical cancer, on three different minor branches (establishing thereby three papillomavirus species, HPV species 6, 7, and 9), while trees based on E6, E1, or E2 suggest that these types are united in one or possibly two clades, clustering separately from "low-risk" and "cutaneous" alpha-papillomaviruses (van Ranst *et al.*, 1992; Narechania *et al.*, 2005; Bravo and Alonso, 2007). Such a finding could be medically interesting, as one might extrapolate molecular data obtained from an intensely studied type like HPV-16 to 17 additional types with a minimum of experimental investigations.

These phylogenetic studies of different genes open the discussion of whether different papillomavirus genes are exposed to different evolutionary mechanisms. Analyses of the rates of non-synonymous/synonymous mutations suggested that all papillomavirus genes, particularly E6, E5, and E2, are under positive selection, and it has been proposed that the underlying mechanisms may include immune evasion and interactions between viral and cellular proteins (Chen *et al.*, 2005; Narechania *et al.*, 2005). Some phylogenetic studies also aimed to detect sequence traces of inter-type genomic recombination, which has never been observed among extant genomes, but which may nevertheless occur very rarely (on a time-scale of a few events in all known papillomavirus genomes over a period of millions of years) (Varsani *et al.*, 2006). Databases that use most of the sequence information of papillomavirus genomes likely lead to more robust trees than those that use only a fraction of the available information. This research has already generated useful insight into nodes between papillomavirus genera close to the root of phylogenetic trees (Bravo and Alonso, 2007). While studies of such phylogenetic assemblages and tree topologies are appreciated, it is premature, however, to use this information to propose a completely revised taxonomy originating from one single lab, as such a process would be impossible to follow for most papillomavirus researchers, and could lead to new taxonomies every few years whenever the tree topology changed after addition of some new papillomavirus types.

THE RELEVANCE OF PAPILLOMAVIRUS PHYLOGENY FOR MEDICAL RESEARCH

Epidemiological and etiological studies whose beginning dates back to the 1980s have established the correlation between pathology and HPV taxonomy, notably the characterization of 18 high-risk HPV types (those that were found more frequently in malignant anogenital lesions than in asymptomatic infections, and therefore apparently carcinogenic) and low-risk HPV types (summarized in Matsukura and Sugase, 2001; Munoz *et al.*, 2005). The molecular basis for the pathological association of many of these HPV types has never been studied as it would be cost-inefficient to investigate relatively rare HPV types with the same effort as targeted at the more common types HPV-16 and HPV-18. It is clearly satisfactory and helpful to find that these 18 high-risk HPVs form three closely related phylogenetic clusters based on the L1 sequences, and may be even more closely related based on other gene sequences, as it supports to the assumption that shared pathogenicity is based on shared molecular properties due to shared evolutionary history. Papillomavirus research has created a foundation to develop diagnostic, prophylactic and therapeutic procedures for the common HPV types, and it has to be hoped that sequence relationships will help to expand these medical targets. To give an example, HPV-18, a well-studied HPV type, is closely related to HPV-45, an important carcinogenic HPV type that is rare in Europe and North America but one of the most prevalent HPV types in some other parts of the world. HPV-45 has not been studied at a molecular level, but sequence similarity with HPV-18 suggests a nearly identical molecular biology and pathogenicity of both types.

As discussed above, there is positive evidence for the differing pathogenic properties of different HPV types. Beyond this, there is increasing evidence that different variants of HPV-16 and 18 are medically so different from one another as to be considered epidemiologically and etiologically distinct pathogens. Published data suggest that the high incidence of anogenital cancer in many developing countries such as all of Latin America is not only based on behavioral differences and poor public health systems, but that the cohorts in these countries are exposed to different endemic HPV-16 and 18 variants that have a higher carcinogenicity than those in developed countries (Xi *et al.*, 1998, 2007; Villa *et al.*, 2000; Da Costa *et al.*, 2002; reviewed in Bernard *et al.*, 2006).

THE ORIGIN OF PAPILLOMAVIRUSES AND THE RELATIONSHIP OF PAPILLOMAVIRIDAE WITH OTHER FAMILIES OF VIRUSES

Papillomaviruses do not share homologous genes with the typical members of all other virus families, and many common functions, such as complex formation of proteins from papilloma-, polyoma, and adenoviruses with the cellular proteins p53 and Rb and structural similarities of papilloma- and polyoma capsids may have evolved by convergence rather than by common phylogenetic origin. However, one general exception to this rule appears to exist. Polyoma- and papillomaviruses have long been known to have a conspicuous sequence similarity among nearly 60 amino acid residues within a 200-amino-acid domain between the polyomavirus T-antigen and the papillomavirus E1 protein (total size of the complete E1 protein: 630 amino acids) (Clertant and Seif, 1984). It was recently found that a protein from single-stranded DNA viruses and a planaria extrachromosomal element can be added to this alignment (Rebrikov *et al.*, 2002). The respective proteins from each of these taxa appear to be involved in DNA replication, encoding a helicase function in the case of the papilloma- and polyomaviruses. It is attractive to speculate that these four different viruses and virus-like elements, respectively, had a common ancestor in the form of an efficient regulatory unit for initiation of replication, which gave rise to the extant taxa through alternative pathways of molecular evolution.

A very recent description of a novel virus of the western barred bandicoot, a rare Australian marsupial, invites alternative speculations about the relationship between papilloma- and polyomaviruses (Woolford *et al.*, 2007). The authors found in papillomas and carcinomas of bandicoots circular DNA genomes with a genome size of 7.3 kb (typical of papillomaviruses) encoding homologues to L2 and L1 genes in the typical sequential alignment (also typical of papillomaviruses). However, about half of the genome encoded a transcription unit facing in the opposite direction of the L genes (typical of polyomaviruses), which encoded a protein with homologies to the T-antigen of polyomaviruses. In spite of the significant homologies, none of the viral genes showed very close nucleotide sequence similarity to the complimentary genes of any known papilloma- or polyomavirus. It seems unlikely, particularly in view of the known complexities of the interplay between viral molecular properties, that this virus originated from a recent fortuitous recombination between a yet unknown papillomavirus with an unknown polyomavirus, although this scenario cannot be formally ruled out. It seems more likely that this virus represents a phylogenetic branch of papilloma- and polyomavirus evolution suggesting a complex tree topology whose deeper branching may be much more complex than anticipated. It has to be hoped that papilloma- and polyomavirus isolates from other marsupials or even completely unrelated hosts may eventually reveal the deeper evolutionary origin of these virus families.

REFERENCES

Antonsson, A. and Hansson, B.G. (2002) Healthy skin of many animal species harbours papillomaviruses, which are closely related to their human counterparts. *J. Virol.* **76**, 12537–12542.

Antonsson, A. and McMillan, N.A. (2006) Papillomavirus in healthy skin of Australian animals. *J. Gen. Virol.* **87**, 3195–3200.

Antonsson, A., Forslund, O., Ekberg, H., Sterner, G. and Hansson, B.G. (2000) The ubiquity and impressive genomic diversity of human skin papillomaviruses suggest a commensalic nature of these viruses. *J. Virol.* **74**, 11636–11641.

Antonsson, A., Erfurt, C., Hazard, K., Holmgren, V., Simon, M., Kataoka, A. *et al.* (2003) Prevalence and type spectrum of human papillomaviruses in healthy skin samples collected in three continents. *J. Gen. Virol.* **84**, 1881–1886.

Bernard, H.U. (1994) Coevolution of papillomaviruses and human populations. *Trends Microbiol.* **2**, 140–143.

Bernard, H.U. (2002) Gene expression of genital human papillomaviruses and potential antiviral approaches. *Antiviral Ther.* **7**, 219–237.

Bernard, H.U., Calleja-Macias, I.E. and Dunn, S.T. (2006) Genome variation of human papillomavirus types: Phylogenetic and medical implications. *Int. J. Cancer*, **118**, 1071–1076.

Boxman, I.L., Mulder, L.H., Russell, A., Bouwes-Bavinck, J.N., Green, A. and Ter Schegget, J. (1999) Human papillomavirus type 5 is commonly present in immunosuppressed and immunocompetent individuals. *Br. J. Dermatol.* **141**, 246–249.

Bravo, I.G. and Alonso, A. (2007) Phylogeny and evolution of papillomaviruses based on the E1 and E2 proteins. *Virus Genes* **34**, 249–262.

Calleja-Macias, I.E., Kalantari, M., Huh, J., Ortiz-Lopez, R., Rojas-Martines, A., Gonzales-Guerrero, J.F. *et al.* (2004) High prevalence of specific variants of human papillomavirus-16, 18, 31, and 35 in a Mexican population. *Virology* **319**, 315–323.

Calleja-Macias, I.E., Kalantari, M., Allan, B., Williamson, A.L., Chung, L.P., Collins, R.J. *et al.* (2005a) Papillomavirus subtypes are natural and old taxa: Phylogeny of the human papillomavirus (HPV) types 44/55 and 68a/b. *J. Virol.* **79**, 6565–6569.

Calleja-Macias, I.E., Kalantari, M., Villa, L.L., Prado, J.C., Allan, B., Williamson, A.L. *et al.* (2005b) Worldwide genomic diversity of the high-risk human papillomaviruses-31, 35, 52, and 58, which are closely related to HPV-16. *J. Virol.* **79**, 13630–13640.

Chambers, G., Ellsmore, V.A., O'Brien, P.M., Reid, S.W., Love, S., Campo, M.S. and Nasir, L. (2003) Sequence variants of bovine papillomavirus E5 detected in equine sarcoids. *Virus Res.* **96**, 141–145.

Chan, S.Y., Bernard, H.U., Ong, C.K., Chan, S.P., Hofmann, B. and Delius, H. (1992) Phylogenetic analysis of 48 papillomavirus types and 28 subtypes and variants: a showcase for the molecular evolution of DNA viruses. *J. Virol.* **66**, 5714–5725.

Chan, S.Y., Delius, H., Halpern, A.L. and Bernard, H.U. (1995) Analysis of genomic sequences of 95 papillomavirus types: Uniting typing, phylogeny, and taxonomy. *J. Virol.* **69**, 3074–3083.

Chan, S.Y., Bernard, H.U., Ratterree, M., Birkebak, T.A., Faras, A.J. and Ostrow, R.S. (1997a) Genomic diversity and evolution of papillomaviruses in Rhesus monkeys. *J. Virol.* **71**, 4938–4943.

Chan, S.Y., Chew, S.H., Egawa, K., Grussendorf-Conen, E.I., Honda, Y., Ruebben, A., Tan, K.C. and Bernard, H.U. (1997b) Phylogenetic analysis of the Human papillomavirus type 2 (HPV-2), HPV-27, and HPV-57 group, which is associated with common warts. *Virology* **239**, 296–302.

Chan, S.Y., Ostrow, R.S., Faras, A.J. and Bernard, H.U. (1997c) Genital papillomaviruses (PVs) and epidermodysplasia verruciformis PVs occur in the same monkey species: Implications for PV evolution. *Virology* **228**, 213–217.

Chen, Z., Terai, M., Fu, L., Herrero, R., DeSalle, R. and Burk, R.D. (2005) Diversifying selection in human papillomavirus type 16 lineages based on complete genome analyses. *J. Virol.* **79**, 7014–7023.

Chen, Z., Schiffman, M., Herrero, R., Desalle, R. and Burk, R.D. (2007) Human papillomavirus (HPV) types 101 and 103 isolated from cervicovaginal cells lack an E6 open reading frame (ORF) and are related to gamma-papillomaviruses. *Virology* **360**, 447–453.

Clertant, P. and Seif, I. (1984) A common function for polyoma virus large-T and papillomavirus E1 proteins?. *Nature* **311**, 276–279.

Da Costa, M.M., Hogeboom, C.J., Holly, E.A. and Palefsky, J.M. (2002) Increased risk of high-grade anal neoplasia associated with a human papillomavirus type 16 E6 sequence variant. *J. Infect. Dis.* **185**, 1229–1937.

de Villiers, E-M. (1997) Papillomavirus and HPV typing. *Clin. Dermatol.* **15**, 199–206.

de Villiers, E-M. (1998) Human papillomavirus infections in skin cancer. *Biomed. Pharmacother.* **52**, 26–33.

de Villiers, E.M., Fauquet, C., Broker, T.R., Bernard, H.U. and zur Hausen, H. (2004) Classification of papillomaviruses. *Virology* **324**, 17–27.

Farmer, A.D., Calef, C.E., Millman, K. and Myers, G.L. (1995) The human papillomavirus database. *J. Biomed. Sci.* **2**, 90–104.

Forslund, O., Ly, H., Reid, C. and Higgins, G. (2003) A broad spectrum of human papillomavirus types is present in the skin of Australian patients with non-melanoma skin cancers and solar keratosis. *Br. J. Dermatol.* **149**, 64–73.

Hazard, K., Andersson, K., Dillner, J. and Forslund, O. (2007) Human papillomavirus subtypes are not uncommon. *Virology* .

Heinzel, P.A., Chan, S.Y., Ho, L., O'Connor, M., Balaram, P., Campo, M.S. *et al.* (1995) Variation of human papillomavirus type 6 (HPV-6) and HPV-11 genomes sampled throughout the world. *J. Clin. Microb.* **33**, 1746–1754.

Ho, L., Chan, S.Y., Burk, R.D., Das, B.C., Fujinaga, K., Icenogle, J.P. *et al.* (1993) The genetic drift of human papillomavirus type 16 is a means of reconstructing prehistoric viral spread and movement of ancient human populations. *J. Virol.* **67**, 6413–6414.

Mantovani, F. and Banks, L. (2001) The human papillomavirus E6 protein and its contribution to malignant progression. *Oncogene* **20**, 7874–7887.

Matsukura, T. and Sugase, M. (2001) Relationships between 80 human papillomavirus genotypes and different grades of cervical intraepithelial neoplasia: association and causality. *Virology* **283**, 139–147.

Muller, M., Gissmann, L., Cristiano, R.J., Sun, X.Y., Frazer, I.H., Jenson, A.B. *et al.* (1995) Papillomavirus capsid binding and uptake by cells from different tissues and species. *J. Virol.* **69**, 948–954.

Munger, K., Basile, J.R., Duensing, S., Eichten, A., Gonzalez, S.L., Grace, M. and Zacny, V.L. (2001) Biological activities and molecular targets of the human papillomavirus E7 oncoprotein. *Oncogene* **20**, 7888–7898.

Munoz, N., Bosch, F.X., de Sanjosé, S., Herrero, R., Castellsagué, X., Shah, K.V., Snijders, P.J.F. and Meijer, C.J.L.M. (2003) Epidemiological classification of human papillomavirus types associated with cervical cancer. *N. Engl. J. Med.* **348**, 518–527.

Narechania, A., Chen, Z., DeSalle, R. and Burk, R.D. (2005) Phylogenetic incongruence among oncogenic genital alpha human papillomaviruses. *J. Virol.* **79**, 15503–15510.

Ong, C.K., Chan, S.Y., Campo, M.S., Fujinaga, K., Mavromara, P., Labropoulou, V. *et al.* (1993) Evolution of human papillomavirus type 18: An ancient phylogenetic root in Africa and intratype diversity reflect coevolution with human ethnic groups. *J. Virol.* **67**, 6424–6431.

Prado, J.C., Calleja-Macias, I.E., Bernard, H.U., Kalantari, M., Macay, S.A., Allan, B. *et al.* (2005) Worldwide genomic diversity of the human papillomaviruses-53, 56, and 66, a group of high-risk HPVs unrelated to HPV-16 and HPV-18. *Virology* **340**, 95–104.

Ramoz, N., Rueda, L.A., Bouadjar, B., Montoya, L.S., Orth, G. and Favre, M. (2000) Mutations in two adjacent novel genes are associated with epidermodysplasia verruciformis. *Nat. Genet.* **32**, 579–581.

Rebrikov, D.V., Bogdanova, E.A., Bulina, M.E. and Lukyanov, S.A. (2002) A new planarian extrachromosomal virus-like element revealed by subtractive hybridization. *Mol. Biol.* **36**, 813–820.

Rector, A., Bossart, G.D., Ghim, S.J., Sundberg, J.P., Jenson, A.B. and Van Ranst, M. (2004) Characterization of a novel close-to-root papillomavirus from a Florida manatee by using multiply primed rolling-circle amplification: *Trichechus manatus latirostris* papillomavirus type 1. *J. Virol.* **78**, 12698–12702.

Rector, A., Mostmans, S., Van Doorslaer, K., McKnight, C.A., Maes, R.K., Wise, *et al.* (2006) Genetic characterization of the first chiropteran papillomavirus, isolated from a basosquamous carcinoma in an Egyptian fruit bat: the *Rousettus aegyptiacus* papillomavirus type 1. *Vet. Microbiol.* **117**, 267–275.

Rehtanz, M., Ghim, S.J., Rector, A., Van Ranst, M., Fair, P.A., Bossart, G.D. and Jenson, A.B. (2006) Isolation and characterization of the first American bottlenose dolphin papillomavirus: *Tursiops truncatus* papillomavirus type 2. *J. Gen. Virol.* **87**, 3559–3565.

Schrago, C.G. and Russo, C.A. (2003) Timing the origin of New World monkeys. *Mol. Biol. Evol.* **20**, 1620–1625.

Scobie, L., Jackson, M.E. and Campo, M.S. (1997) The role of exogenous p53 and E6 oncoproteins in *in vitro* transformation by bovine papillomavirus type 4 (BPV-4): significance of the absence of an E6 ORF in the BPV-4 genome. *J. Gen. Virol.* **78**, 3001–3008.

Stewart, A.C., Eriksson, A.M., Manos, M.M., Munoz, N., Bosch, F.X., Peto, J. and Wheeler, C.M. (1996) Intratype variation in 12 human papillomavirus types: a worldwide perspective. *J. Virol.* **70**, 3127–3136.

Suprynowicz, F.A., Disbrow, G.L., Simic, V. and Schlegel, R. (2005) Are transforming properties of the bovine papillomavirus E5 protein shared by E5 from high-risk human papillomavirus type 16?. *Virology* **332**, 102–113.

Tachezy, R., Rector, A., Havelkova, M., Wollants, E., Fiten, P., Opdenakker, G. *et al.* (2002) Avian papillomaviruses: the parrot *Psittacus erithacus* papillomavirus (PePV) genome has a unique organization of the early protein region and is phylogenetically related to the chaffinch papillomavirus. *BMC Microbiol.* **2**, 19–25.

Terai, M., DeSalle, R. and Burk, R.D. (2002) Lack of canonical E6 and E7 open reading frames in bird papillomaviruses: *Fringilla coelebs* papillomavirus and *Psittacus erithacus timneh* papillomavirus. *J. Virol.* **76**, 10020–10023.

Trenfield, K., Spradbrow, P.B. and Vanselow, B.A. (1990) Detection of papillomavirus DNA in precancerous lesions of the ears of sheep. *Vet. Microbiol.* **25**, 103–116.

Van Doorslaer, K., Rector, A., Vos, P. and Van Ranst, M. (2006) Genetic characterization of the *Capra hircus* papillomavirus: a novel close-to-root artiodactyl papillomavirus. *Virus Res.* **118**, 164–169.

Van Ranst, M., Kaplan, J.B. and Burk, R.D. (1992) Phylogenetic classification of human papillomaviruses: correlation with clinical manifestations. *J. Gen. Virol.* **73**, 2653–2660.

Varsani, A., van der Walt, E., Heath, L., Rybicki, E.P., Williamson, A.L. and Martin, D.P. (2006) Evidence of ancient papillomavirus recombination. *J. Gen. Virol.* **87**, 2527–2531.

Villa, L.L., Sichero, L., Rahal, P., Caballero, O., Ferenczy, A., Rohan, T. and Franco, E.L. (2000) Molecular variants of human papillomavirus types 16 and 18 preferentially associated with cervical neoplasia. *J. Gen. Virol.* **81**, 2959–2968.

Woolford, L., Rector, A., van Ranst, M., Ducki, A., Bennett, M.D., Nicholls, P.K. *et al.* (2007) A novel virus detected in papillomas and carcinomas of the endangered western barred bandicoot (*Perameles bougainville*) exhibits genomic features of both the Papillomaviridae and Polyomaviridae. *J. Virol.* **81**, 13280–13290.

Xi, L.F., Critchlow, C.W., Wheeler, C.M., Koutsky, L.A., Galloway, D.A., Kuypers, J. *et al.* (1998) Risk of anal *carcinoma in situ* in relation to human papillomavirus type 16 variants. *Cancer Res.* **58**, 3839–3844.

Xi, L.F., Kiviat, N.B., Hildesheim, A., Galloway, D.A., Wheeler, C.M., Ho, J. and Koutsky, L.A. (2006) Human papillomavirus type 16 and 18 variants: race-related distribution and persistence. *J. Natl Cancer Inst.* **98**, 1045–1052.

Xi, L.F., Koutsky, L.A., Hildesheim, A., Galloway, D.A., Wheeler, C.M., Winer, R.L. *et al.* (2007) Risk for high-grade cervical intraepithelial neoplasia associated with variants of human papillomavirus types 16 and 18. *Cancer Epidemiol. Biomarkers Prev.* **16**, 4–10.

Yamada, T., Manos, M.M., Peto, J., Greer, C.E., Munoz, N., Bosch, F.X. and Wheeler, C.M. (1997) Human papillomavirus type 16 sequence variation in cervical cancers: a worldwide perspective. *J. Virol.* **71**, 2463–2472.

CHAPTER

19

Origin and Evolution of Poxviruses

John W. Barrett and Grant McFadden

ABSTRACT

The family *Poxviridae* is a large and diverse family of double-stranded DNA viruses with ubiquitous distribution. Poxviruses parasitize invertebrates, birds, reptiles and mammals suggesting that the family *Poxviridae* is an ancient virus family. The genomes are linear and encode a large complement of genes. Essential viral functions are clustered in the central region of the genome and a core group of critical genes are conserved among all poxviruses. In contrast, species-specific genes, necessary for host infection and virus virulence, are maintained near the termini. The success of the poxvirus family members is based on the acquisition of cellular genes that have evolved under viral/host selection to permit the virus to modulate the host immune response.

INTRODUCTION

The family *Poxviridae* represents one of the most numerous and geographically widespread virus families that infects animals (vertebrates and invertebrates). Until the recent discovery of mimiviruses, the poxviruses represented the largest, most complex viruses known to virology. The poxviridae include a number of historically significant viruses, such as variola virus, the causative agent of smallpox, which was the most devastating human infectious disease in history. Cowpox virus is reported to have been the original vaccine strain to protect against variola, but the exact origins of the original smallpox vaccine virus are still shrouded in mystery. Vaccinia virus, which is closely related to variola, is the poxvirus that came to be used to vaccinate human populations that led to the global eradication of smallpox; and the development of the field of vaccinology. Shope fibroma virus, a rabbit-specific poxvirus, was the first DNA virus to be demonstrated to produce transmissible tumors in animals. Unique members of the family *Poxviridae* have evolved on all continents that support suitable hosts, and poxviruses have co-evolved with a variety of hosts to utilize their unique opportunities. The large poxvirus genome size and their ability to "capture" host immune regulators and manipulate them to subvert the immune system help explain the evolutionary success of the family *Poxviridae*. These gene captures have also contributed to the wide

Origin and Evolution of Viruses
ISBN 978-0-12-374153-0

genetic variation between poxvirus members that is encountered today.

The poxviruses share a conserved genetic structure including possessing linear, double-stranded DNA genomes in which the two strands are joined by palindromic hairpin termini. Poxvirus genomes range widely in size, varying from 135 kb to greater than 300 kb. The virion structure is large (150 nm × 300 nm) and can be ovoid or brick-shaped, but tends to be morphologically quite similar from one poxvirus to another. Because poxviruses are found in insects, reptiles, birds, and mammals and they reside in the cytoplasm of infected cells where they replicate and undergo morphogenesis independent of the cell nucleus, they are considered to be an ancient virus family that must have existed in a recognizable "pox" form before the divergence of invertebrates and invertebrates.

POXVIRUS GENOME ORGANIZATION

The genome organization of poxviruses are remarkably well conserved. The linear double-stranded DNA molecule is covalently linked at its ends by palidromic hairpins. Immediately adjacent to the hairpins are highly conserved concatemeric resolution sequences involved in replication of concatemeric viral genomes that are later resolved into the correct unit-length segments of the genome. The termini of poxvirus genomes are duplicated and inverted into structures termed terminal inverted repeats (TIRs). The length and organization of the TIRs vary widely among the various poxviruses; they range in length from several hundred nucleotides in orthopoxviruses like variola virus that do not encode any TIR open reading frames (ORFs) to more than 10 kb in the leporipoxviruses, which encode duplicated copies of a dozen predicted ORFs. As more complete sequences of poxviruses have become available, it has become clear that the canonical poxvirus genome is organized in a manner such that genes that encode essential conserved viral functions, such as transcription, structural components, replication factors, and nucleotide biosynthesis, are found clustered with the central portion of the genome (Figure 19.1). It is within this approximately 100 kb central domain of the genome that bioinformatics studies have identified a core of about 90 genes that are highly conserved in all members of the subfamily *Chordopoxvirinae* (Gubser *et al.*, 2004). When the entomopoxviruses are included in the analysis, this core number of poxvirus-specific genes is reduced to 45 (Upton *et al.*, 2003; McLysaght *et al.*, 2003; Gubser *et al.*, 2004). In contrast, no individual poxvirus gene that maps outside the central core is found in all members.

The family *Poxviridae* is divided into two subfamilies: the subfamily *Entomopoxvirinae*, members of which infect invertebrates, and the subfamily *Chordopoxvirinae*, members of which infect vertebrates.

ENTOMOPOXVIRINAE

The entomopoxviruses are restricted to invertebrates. There are three recognized genera (Table 19.1). The virions of members of each genus are morphologically distinctive. Members of genus A (alphaentomopoxviruses) have large genomes (~260–370 kb) and infect members of the Insecta order Coleoptera (beetles). Their virions tend to be ovoid and about 450 × 250 nm. Genus B (betaentomopoxviruses) genomes are ~230 kb and infect species of the orders Lepidoptera (butterflies and moths) and Orthoptera (straight wings [locusts, grasshoppers and crickets]). Viruses of this genus have been studied for their use in biocontrol programs of forestry and agricultural pests. The genome of the type species, *Amsacta moorei* entomopoxvirus, has been sequenced completely (Bawden *et al.*, 2000). It is the only complete sequence available for members that have been classified into recognized genera. A complete genome sequence is also available for a poxvirus that infects the grasshopper, *Melanoplus sanguinipes* (Afonso *et al.*, 1999), however this member currently belongs to an unassigned entomopoxvirus

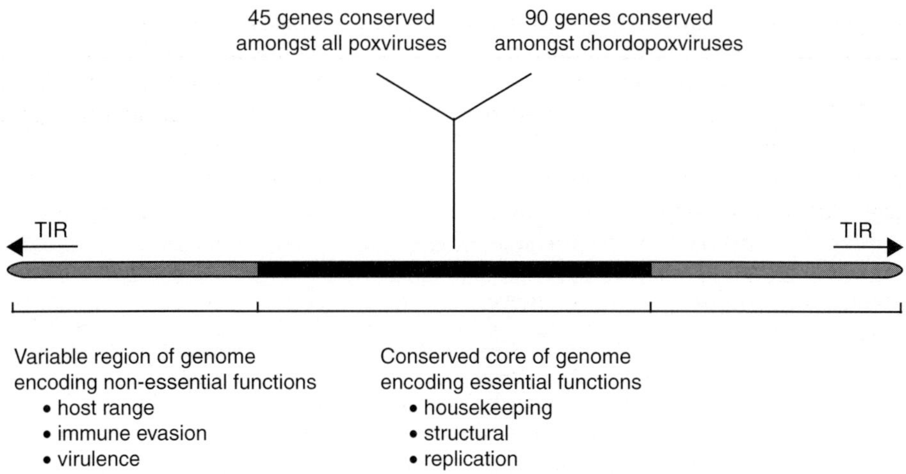

FIGURE 19.1 Schematic of generalized poxvirus genome organization. The gray bars indicate the variable ends regions and the black bar represents the conserved central core. The terminal inverted repeats (TIR) are represented by black arrows.

group. The virions of the betaentomopoxviruses are ovoid and about 250 × 230 nm. Genus C (gammaentomopoxviruses) members have genomes that range in size from 250 to 380 kb, virions that are brick-shaped and are about 320 × 230 nm. These members infect the Insecta order Diptera (true flies).

The only complete genomic sequence information available for the entomopoxviruses come from the sequencing of two members that infect a caterpiller, *Amsacta moorei*, and a locust, *Melanoplus sanguinipes* (Table 19.1). This genetic information has provided insight into the differences in the genetic organization and gene complement between the entomopoxviruses and the chordopoxviruses. Based on these two complete sequences, the entomopoxviruses are 81% AT rich. They do not share any significant gene order conservation even between themselves, suggesting that the entomopoxviruses are more divergent than the chordopoxviruses and thus implies an older evolutionary history (Gubser *et al.*, 2004).

Generally, their genomes are larger than the mammalian poxviruses and they encode a host of proteins specific to their life cycle in an invertebrate host, including DNA repair enzymes and molecules to protect against UV damage (Afonso *et al.*, 1999; Bawden *et al.*, 2000). Other unique features of interest include the sequestration of virions within a proteinaceous body called the spheroid. The appearance of the mature virion within the spheroid is analogous to the compartmentalization of virions of some orthopoxviruses (cowpox, ectromelia, raccoonpox) in A type inclusions (ATI), but there are significant differences. The spheroid is a paracrystalline structure while the ATI is not. In addition, the ATIs are uniquely acid sensitive. In contrast, the entomopoxvirus life cycle and mode of transmission requires that the spheroid be ingested by the insect host and the virions are released upon dissolution of the spheroid in the alkaline environment of the insect gut. Although there are significant differences between the entomopoxviruses and the chordopoxviruses, the entomopoxviruses still reside in the cytoplasm of infected cells, their genes are transcribed in a temporal manner and the genomes and virions are structurally similar. The entomopoxviruses are more diverse than the chordopoxviruses of birds and mammals, and likely diverged from the pox-like ancestor before the separation of the chordopox genera (Gubser *et al.*, 2004). The subfamily *Entomopoxvirinae* has recently been reviewed elsewhere (Becker and Moyer, 2007).

TABLE 19.1 Geographic distribution and host range of members of the family *Poxviridae*

Genus and species	Reservoir host	Geographic distribution	Other infected hosts	Sequence accession number
Chordopoxvirinae				
Orthopoxvirus[a]				
Camelpox virus	Camels	Africa, Asia	None	AY009089
Cowpox virus	Rodents	Europe, western Asia	Cats, cattle, humans, zoo animals	NC_003663
Ectromelia virus	Rodents	Europe	None	NC_004105
Horsepox virus		Central Asia	Horses	DQ792504
Monkeypox virus	Squirrels	Western and central Africa	Monkeys, humans, zoo animals, prairie dogs	NC_003310
Raccoonpox virus	Raccoons	Eastern United States	cats	
Skunk poxvirus	Skunks	United States	None	
Tatera poxvirus	Gerbils	Western Africa	None	NC_008291
Uasin Gishu virus	Unknown	Eastern Africa	Horses	
Vaccinia virus[b]	Unknown	Worldwide	Humans, rabbits, cows. River buffalo	NC_006998
Variola virus	Humans	Previously worldwide, now eradicated	None	DQ437586
Volepox virus	Voles	Western United States	None	
Parapoxvirus[a]				
Ausdyck virus	Camels	Africa, Asia	None	
Bovine papular stomatitis virus	Cattle	Worldwide	Humans	NC_005337
Orf virus[b]	Sheep, goats	worldwide	Humans	DQ184476
Pseudocowpox virus	Cattle	worldwide	Humans	
Red deer poxvirus	Red deer	New Zealand	None	
Seal parapoxvirus	Seals	worldwide	Humans	
Capripoxvirus[a]				
Sheeppoxvirus[b]	Sheep	Asia, Africa	None	AY077833
Goatpoxvirus	Goats	Asia, Africa	None	NC_004003
Lumpy skin disease virus	Cape buffalo	Africa	Cattle	NC_003027
Suipoxvirus[a]				
Swinepox virus[b]	Swine	worldwide	None	NC_003389
Leporipoxvirus[a]				
Myxoma virus[b]	Tapeti	South American	European rabbits	NC_001132
	Brush rabbit	California, Mexico	European rabbits	
Rabbit (shope) fibroma virus	Cottontail rabbit	Eastern United States	European rabbits	NC_001266
Hare fibroma virus		Europe	None	
Squirrel fibroma virus		Eastern United States	Woodchuck	
Western squirrel fibroma virus		California	None	
Yatapoxvirus[a]				
Tanapox virus		Equatorial Africa	Humans	EF420156 TPV-Kenya
Yaba monkey tumor virus[b]	Primates	Western Africa	Humans	NC_005179
Molluscipoxvirus[a]				
Molluscum contagiosum[b]	Humans	Worldwide	None	NC_001731
Avipoxvirus[a]				
Numerous species	Birds	Worldwide	None	NC_005309, canarypox virus NC_002188, fowlpox virus

(Continued)

TABLE 19.1 (Continued)

Genus and species	Reservoir host	Geographic distribution	Other infected hosts	Sequence accession number
Unclassified				
Crocodilepox	Crocodiles Caimans	Australia, Africa	none	NC_008030
Macropod poxvirus	Kangaroos Quokkas	Australia	none none	
Cotia poxvirus	Mice	Brazil	none	
Squirrelpox virus	Grey squirrel	Europe	Red squirrel	
Deerpox virus	Mule deer	North America		AY689437
Entomopoxvirinae				
Alphaentomopoxvirus[a]				
Anomala cuprea entomopoxvirus	Cupreous cockchafer			
Melolontha melolonotha entomopoxvirus[b]	Common cockchafer			
Othnonius bateis entomopoxvirus	Scarab beetle			
Ips typographus entomopoxvirus	European spruce bark beetle			
Betaentomopoxvirus[a]				
Amsacta moorei entomopoxvirus[b]	Red hairy caterpiller			NC_002520
Choristoneura biennis entomopoxvirus	Two year cycle spruce budworm			
Choristoneura fumiferana entomopoxvirus	Spruce budworm			
Heliothis armigeria entomopoxvirus	Cotton bollworm			
Ocnogyna baetica entomopoxvirus				
Elasmopalphus lignosellus entomopoxvirus	Lesser cornstalk borer			
Adoxophyes honmai entomopoxvirus	Tea tortrix			
Pseudaletia unipuncta entomopoxvirus	Army worm			
Euxoa auxiliaris entomopoxvirus	Army cutworm			
Gammaentomopoxvirus[a]				
Chironomus luridus entomopoxvirus[b]	Midge			
Goeldichironomus haloprasimus entomopoxvirus	Midge			
Unassigned				
Melanoplus sanguinipes entomopoxvirus	Locust			NC_001993

[a] Genus.
[b] Type species.

CHORDOPOXVIRINAE

Poxviruses that infect vertebrates belong to the subfamily *Chordopoxvirinae*. This subfamily is divided into eight genera and includes several poxvirus members that are unique and are unassigned to any specific genus (Table 19.1). Viral membership within a given genus is based on shared defining features, including antigenic similarity and the ability of an infection by one viral member to induce resistance to subsequent infection by another member of that genus. For example, it was the ability of vaccinia virus to function as a vaccine to block subsequent infection by the related genus *Orthopoxvirus* member, variola virus, that allowed for the eradication of smallpox. Today there is a complete genomic sequence data for almost every chordopox member (Table 19.1). In some cases, multiple viral strains of individual species have been sequenced, including over 50 variola virus strains (Esposito *et al.*, 2006). This huge amount of genetic information offers clues to the evolutionary history of the chordopoxviruses. The members of the family *Chordopoxvirinae* range in size from 135 kb (orf virus; Delhon *et al.*, 2004) to 365 kb (canarypox virus; Tulman *et al.*, 2004). These genomes encode between 130 putative genes (orf virus) to 328 genes (canarypox virus), respectively. However, in contrast with the invertebrate poxviruses, the genetic organization and gene order of the chordopoxviruses is much more conserved (Upton *et al.*, 2003).

Although the features of the poxvirus genome organization is outlined below, it is worth pointing out that poxvirus genomes are packed with coding ORFs that are driven by short viral-specific promoters that are generally found within short distances of the ATG start site. The genomic sequences have revealed that poxviruses encode a large number of putative viral genes with little or no overlap and very short intergenic non-coding spaces between the ORFs. As more sequence data is acquired and the function of individual gene products is confirmed, it is becoming clear that another strategy that the poxviruses have evolved to increase their coding potential is the evolution of genes with multiple functions (Table 19.2). For example, a single poxvirus gene product called M-T2 demonstrates one function intracellularly (apoptosis blockade) and another function (TNF inhbition) extracellularly (Schreiber *et al.*, 1997; Sedger *et al.*, 2006). As another example, a variola virus protein has been shown to bind both TNF and chemokines, using distinct protein domains (Alejo *et al.*, 2006). Although the list of multiple function poxvirus proteins is still short, more examples continue to be discovered and it seems clear that as more gene products are teased apart the identification of poxvirus gene products with multiple functions will become the norm, rather than the exception.

Most poxvirus genomes are relatively AT rich. The entomopoxviruses represent the extreme at 81% AT, while the opposite end is represented by the parapoxviruses and the molluscipoxviruses at only 34% AT rich. The remaining members are around 65% AT rich. As well, codon usage across the genera is quite divergent, although it tends to be more conserved within members of the genera (Barrett *et al.*, 2006).

Genus Avipoxvirus

Any poxvirus identified from a skin lesion of a wild or domestic bird is generally grouped within the *Avipoxvirus* genus. This is therefore a large and divergent group of poxviruses with infection currently described in 232 species of birds (Bolte *et al.*, 1999). Virus transmission is thought to be primarily by biting arthropods, particularly the mosquito. Avipoxviruses generally have the largest genomes of the vertebrate poxviruses, ranging up to 360 kb. They share the common features exhibited by chordopoxviruses, including replication within the cytoplasm and characteristic poxvirion morphology. With the completion of the genomic sequences of two members, it was clear that avipoxviruses exhibited large-scale genome rearrangements,

TABLE 19.2 Evolution of poxviral immune evasion molecules with multiple functions

Gene	Species	Primary function	Secondary function	References
M-T2	Myxoma virus	Secreted TNFR homolog	Intracellular inhibitor of apoptosis	Sedger *et al.*, 2006; Schreiber *et al.*, 1997
M-T5	Myxoma virus	Binds Cullin-1 to prevent cell cycle arrest	Binds and activates Akt	Johnston *et al.*, 2005; Wang *et al.*, 2006
M-T7	Myxoma virus	IFN-R homolog	Chemokine binding protein	Upton *et al.*, 1992; Lalani *et al.*, 1997
M11L/F1L	Myxoma virus/vaccinia virus	Binds to mt PTP	Binds Bak and prevents Bax conformational change	Wang *et al.*, 2004; Su *et al.*, 2006; Wasilenko *et al.*, 2005; Everett *et al.*, 2002
G2R	Variola virus	TNFR homologue (CrmB)	Chemokine binding protein	Alejo *et al.*, 2006
GIF	Orf virus	Binds GM-CSF	Binds IL-2	Deane *et al.*, 2000
gp38	Tanapox virus	Secreted inhibitor of TNF	Binds multiple cytokines	Essani *et al.*, 1994; Brunetti *et al.*, 2003b; Paulose *et al.*, 1998

extensive host-acquired gene families, and uniquely distinctive host range genes. For example, 31 genes were identified within the ankyrin repeat superfamily, 10 genes comprise the N1R-p28 virulence family, and 6 genes were members of the B22R gene family and together these represented about 32% of the total fowlpox genome (Afonso *et al.*, 2000). In addition, genes encoding normally cellular functions, such as steroid biogenesis, antioxidant functions and vesicle trafficking were also identified, suggesting an evolved role in controlling the cellular environment during infection. Genetic evidence suggests that the avipoxviruses have co-evolved with their avian and proto-avian hosts and have progressively acquired host genes that controlled the innate immune response over at least 150 million years. Therefore, based on two known complete sequences, there is more variation between members within the avipoxviruses then there is between the other seven genera within the subfamily *Chordopoxvirinae*. Phylogenetic analysis also suggests that the avipoxviruses are the most genetically divergent (Figure 19.2). A substantial amount of interest in the avipoxviruses is directed at their use as non-replicating vaccine platforms for the delivery of foreign antigens (Taylor and Paoletti, 1988; Taylor *et al.*, 1995).

Genus Capripoxvirus

There are three recognized members of the genus *Capripoxvirus* that are important pathogens of ungulates (Table 19.1). Lumpy skin disease virus, sheeppox virus, and goatpox virus are isolated from African cattle, sheep, and goats of Asia and Africa, respectively. Lumpy skin disease is found in most African countries south of 10°N latitude and is thought to be vector borne (Losos, 1986). All capripoxviruses can be spread by direct contact within a herd. The genomic sequence of these three members are available and they average about 150 kb and encode between 147 and 156 putative genes (Tulman *et al.*, 2001, 2002). The genomes are highly conserved, being 96% identical at the nucleotide level. The capripoxviruses also have the highest AT content among the chordopoxviruses at 73–75%. Comparative genomic analysis suggests that, based on the close genetic similarity among the members of the capripoxviruses and the observation that several lumpy skin

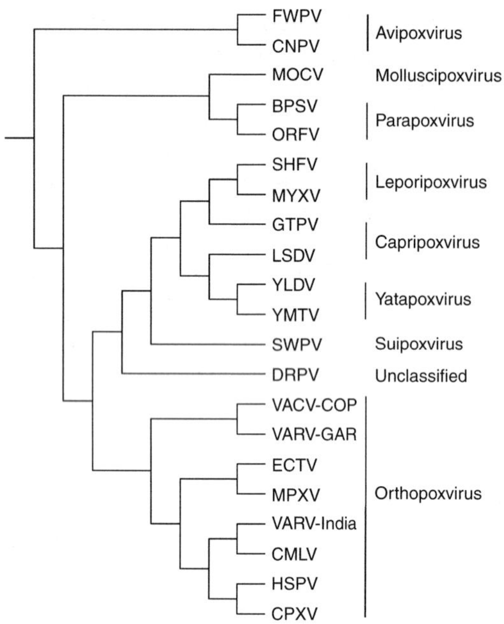

FIGURE 19.2 Poxvirus organization based on the presence of immunomodulatory genes. Poxvirus genome sequences were screened for the presence of immunomodulatory genes identified in Seet *et al.* (2003) and these characters were used to generate the shortest tree using MacClade 3.08a (Maddison and Maddison, 1992). BPSV, bovine papular stomatitis virus; CMLV, camelpox virus; CNPV, canarypox virus; CPXV, cowpox virus; DRPV, deerpox virus; ECTV, ectromelia virus; FWPV, fowlpox virus; GTPV, goatpox virus; HSPV, horsepox virus; LSDV, lumpy skin disease virus; MOCV, molluscum contagiosum; MPXV, monkeypox virus; MYXV, myxoma virus; ORFV, orf virus; SHFV, Shope fibroma virus; SWPV, swinepox virus; VACV-COP, vaccinia virus strain Copenhagen; VARV-GAR, variola virus strain Garcia; VARV-India, variola virus strain India; YLDV, yaba-like disease virus; YMTV, yaba monkey tumor virus.

disease virus genes are fragmented within the sheeppox and goatpox viruses, these members all likely were derived from lumpy skin disease virus-like ancestor (Tulman *et al.*, 2002). However, historical evidence on sheeppox and goatpox virus occurrence, and the wider distribution of these two viruses across Asia and Africa suggest an older evolutionary history. Lumpy skin disease virus was only identified in 1929, sheeppox was originally described in the first century AD, and goatpox was reported in Northern Africa and Southern Europe in the 1880s (Diallo and Viljoen, 2007). This suggests that lumpy skin disease is in fact an old disease that was only recently identified, but the true evolutionary history and spread of the capripoxviruses is still an unresolved issue.

Genus Leporipoxvirus

The leporipoxviruses are restricted to rabbit and squirrel species of North and South America (Table 19.1). There is only one member, hare fibroma virus, which is naturally found in Europe. The best characterized leporipoxviruses are myxoma virus, which has co-evolved with *Sylvilagus* species in the southwestern USA, Central America, and South America, and shope fibroma virus, which has evolved with the cottontail rabbit (*Sylvilagus floridanus*) of eastern North America (Barrett and McFadden, 2007). The leporipoxviruses cause small localized fibromas in their evolutionary host and the lesions are clinically similar (Fenner and Ratcliffe, 1965). The viruses are transmitted by biting insects, primarily mosquitoes, and secondary infection does not normally proceed from the primary lesion when the virus is in the evolutionary host. Interest in the leporipoxviruses stems from the production of myxomatosis, a lethal disease caused specifically by myxoma virus infection of European rabbits (*Oryctolagus cuniculus*).

Two members of the genus *Leporipoxvirus* have been sequenced and they are 86% identical at the nucleotide level (Cameron *et al.*, 1999; Willer *et al.*, 1999). They both encode approximately 165 genes, ten of which are found as duplicated copies within the terminal inverted repeats. Pathogenic studies of myxoma virus infection of laboratory rabbits has been revealing about the complement of immune evasion molecules carried by poxviruses (Johnston and McFadden, 2004; Stanford *et al.*, 2007). The use of myxoma virus as a biological control agent for rabbit populations in Australia has revealed the ability of the virus to respond rapidly to environmental conditions to ensure its spread within susceptible rabbit populations (discussed later).

Genus Molluscipoxvirus

Molluscum contagiosum virus is the only member of the genus *Molluscipoxvirus*. It is restricted to human infections of the skin, primarily in children and immuno-compromised adults. The genome is 190 kb and is predicted to encode 163 genes (Senkevich *et al.*, 1996). The molluscum genome is GC rich (63% GC), which is similar to the members of the genus *Parapoxvirus*. The molluscum contagiosum virion has a typical pox-like morphology and the viral genome is organized in a standard fashion. However, there are several unique features of molluscum contagiosum, including its restriction to human keratinocytes, lack of an available tissue culture system, and a genome that is more divergent from the other mammalian chordopoxviruses. Most of the central core genes show strong similarity and genomic organization to the housekeeping and structural genes of other poxviruses. In contrast, many of the genes of the variable regions (Figure 19.1) are totally unique to this virus (Senkevich *et al.*, 1997). Phylogenetic analysis suggests that molluscum contagiosum radiated from a common pox ancestor following the divergence of the avipoxviruses (Figure 19.2) (Senkevich *et al.*, 1997).

Genus Orthopoxvirus

An enormous amount of information is known about members of the genus *Orthopoxvirus*, primarily because of their historical significance. All the major orthopoxviruses have been sequenced except for those species identified as North American specific (raccoonpox, volepox, and skunkpox). Members of this genus infect a wide range of reservoir species (Table 19.1) and exhibit a broad spectrum of host species tropisms. Cowpox virus is considered the most closely related to the ancestral orthopoxviruses because it has the largest genome and possesses the broadest complement of intact ORFs (Gubser *et al.*, 2004). Other orthopoxvirus members have fragmented or deleted versions of many of these ORFs in the variable regions of the cowpox genome (Figure 19.1). Cowpox virus exhibits a broad host range, while several of the other orthopoxvirus members are much more host species restricted (e.g. variola, camelpox, and ectromelia) and can be highly lethal in their reservoir host (McFadden, 2005; Bray, 2006). There are two strains of cowpox virus presently sequenced. The strains (Brighton and GRI-90) were isolated 50 years apart and geographically separated, one from Great Britain, the other from Russia. Genomic analysis suggests that they are quite divergent and their identification as strains of the same species may have to be re-evaluated in the future (Gubser *et al.*, 2004). Phylogenetic analysis suggests that the orthopoxviruses have evolved as a distinct clade within the subfamily *Chordopoxvirinae* (Figure 19.1) for some time (Gubser *et al.*, 2004; Esposito *et al.*, 2006).

Genus Parapoxvirus

Members of the genus *Parapoxvirus* (Table 19.1) cause highly contagious pustular skin infections of sheep, goats, and cattle and the virus can be transmitted zoonotically into humans. Proliferative pustular lesions observed in camels, seals, red deers of New Zealand, and reindeer have all been attributed to parapoxviruses. A parapoxvirus of historical note is pseudocowpox virus, which infects cattle and humans and which Jenner identified as being different from the virus ("true" cowpox) that would elicit protection from smallpox (Baxby, 1981). Genomic sequences are available for orf virus and bovine papular stomatitis virus (Delhon *et al.*, 2004). The genomes are GC rich (64% GC) and include the shortest of the known poxvirus sequences at about 135 kb (Delhon *et al.*, 2004). Parapoxviruses have evolved for replication in the highly specialized tissue environment of the epidermis. Approximately 70% of the genes encoded by the sequenced parapoxviruses are found in common with other members of the subfamily *Chordopoxvirinae*. However, the GC composition of the genomes, their predicted codon usage (Barrett *et al.*, 2006) and genetic make-up of their genomes suggest that members of this genus are most similar to molluscum contagiosum by phylogenetic analysis (Figure 19.2).

In common with molluscum contagiosum, the parapoxviruses do not encode many genes that are otherwise conserved in the mammalian chordopoxviruses, including genes for nucleotide metabolism, serine proteinase (serpin) inhibitors and kelch-like proteins (Delhon *et al.*, 2004). Unique features of the parapoxviruses include a distinctive virion with a "ball of yarn" appearance and genomes that encode for a vascular endothelial growth factor (VEGF), a homologue of a mammalian IL-10 and a granulocyte–macrophage colony stimulating factor/IL-2-binding protein (Deane *et al.*, 2000; Delhon *et al.*, 2004). As for the capripoxviruses, infection by the parapoxviruses does not induce long-lasting immune protection, which may explain why parapoxvirus infection of herd animals is worldwide.

Genus Suipoxvirus

Swinepox virus is the single member of the genus *Suipoxvirus*. Productive infection of swinepox virus is restricted to swine worldwide. The virus is introduced via skin abrasions and replicates within the epidermal keratinocytes. Infection causes a mild, self-limiting disease and offers protective immunity. The genome has been sequenced (146 kb) and is one of the shorter poxvirus genomes (Afonso *et al.*, 2002). Phylogenetic analysis indicates that swinepox virus groups in a clade along with the capripox-, leporipox-, and yatapoxviruses (Figure 19.2) (Afonso *et al.*, 2002).

Genus Yatapoxvirus

This genus contains two members, yaba monkey tumor virus (YMTV) and tanapox virus (TPV), that infect primates, including humans (Niven *et al.*, 1961; Grace and Mirand, 1963). A yatapoxvirus infection of primates in US primate centers and their handlers exhibited similar disease symptoms to tanapox virus, and this virus was called yaba-like disease virus (YLDV). Sequencing and molecular analysis has demonstrated that YLDV is best considered a strain of tanapoxvirus (Downie and Espana, 1972; Lee *et al.*, 2001; Nazarian *et al.*, 2007). Yaba monkey tumor virus infects old world monkeys and produces large tumors (histiocytomas) on the extremities in primates. No natural infection of humans has occurred, however accidental infections and infection of volunteers has demonstrated that yaba monkey tumor virus produces similar histiocytomas in humans. Tanapox virus has been identified in outbreaks in native populations and travelers of equatorial Africa (Dhar *et al.*, 2004). Tanapox virus infection results in the development of a few nodular skin lesions and a brief febrile illness. Genomic sequencing of yaba monkey tumor virus (Brunetti *et al.*, 2003a), yaba-like disease virus (Lee *et al.*, 2001), and tanapox virus (Nazarian *et al.*, 2007) demonstrate several unique aspects of the genus *Yatapoxvirus*. The genomes of tanapox virus and yaba-like disease virus are 98% identical confirming that YLDV is a strain of tanapox virus (Downie and Espana, 1973; Nazarian *et al.*, 2007). The genomes are AT rich (70% for YMTV and 73% for TPV) and among the shortest poxvirus genomes sequenced (135 kb YMTV, 144 kb TPV). Phylogenetic analysis places the yatapoxvirus closer to the capripox- and leporipoxviruses (Figure 19.2).

Unclassified Chordopoxviruses

Two recently published poxvirus genomic sequences that have become available are for members that have not been assigned a place within the recognized chordopoxvirus genera. Two strains of pathogenic deerpox virus have been fully sequenced (Afonso *et al.*, 2005). The strains are 95% identical, about 167 kb and encode approximately 167 genes. Phylogenetic analysis indicates that deerpox is genetically distinct and diverged before the separation of the suipox, yatapox, capripox, and leporipoxviruses (Figure 19.2).

The genome sequence of the Nile crocodile is 190 kb, 62% GC rich and predicted to encode 173 genes (Afonso *et al.*, 2006). The central core ORFs are generally conserved and collinear with those of other chordopoxviruses but the variable regions are unique.

This genome lacks recognizable homologues of most virulence and host range genes found in other chordopoxviruses. The high GC content and presence of genomic similarities with molluscum contagiosum suggest that crocodilepox diverged following the separation of the avipoxviruses from the mammalian chordopoxviruses (Afonso *et al.*, 2006).

It is interesting to note that poxvirus infection of marsupials indigenous to Australia (kangaroo and quokka) have been reported but no genomic sequence data are available. It might be expected that these viruses would exhibit intermediate features between crocodilepox virus and molluscum contagiosum indicating divergence from the mammalian branch before the rise of placental mammals.

SURVIVAL STRATEGIES OF THE POXVIRUSES

The success of the poxviruses is the result of expression of viral genes that allow for manipulation and modulation of the host pathogen defense responses. The mammalian innate and acquired immune systems have evolved to respond to foreign pathogens in a regulated and regimented manner. Upon detecting the initial infection, using a range of cell surface and intracellular sensors, the innate host response becomes activated and early proinflammatory cytokines are generated by sentinel cells of the immune system, particularly neutrophils, natural killer cells, macrophages, and dendritic cells. It is this panel of host immune cells which first respond to microbial infection and synthesize the first wave of key antiviral cytokines, including tumor necrosis factors (TNFs), interferons (IFNs), chemokines, and interleukins (IL) such as IL-1β and IL-18 (Table 19.3). These cytokines initiate leukocyte activation and infiltration to sites of infection and are responsible for stimulating a cellular-based response to initiate viral clearance (Smith and Kotwal, 2002). Poxviruses have evolved multiple strategies to manipulate and manage these sentinel molecules of the immune system, but the specifics of how any one poxvirus accomplishes this goal can be quite varied (Seet *et al.*, 2003). In some cases, there is evidence that the immunomodulatory genes encoded by poxviruses have been adapted from the sequential capture of cellular genes from ancestrally infected hosts. Selection pressure exerted on the virus by the host immune system over the millennia have shaped these viral molecules into highly effective modulators necessary to perturb the host immune response in a fashion that favors virus survival. A second strategy has been the evolution of poxviral molecules that have multiple activities (Table 19.2). For example, myxoma virus M-T5, which is a cytoplasmic ankyrin repeat-containing host range protein, binds to cellular cullin-1, likely through the M-T5 C-terminal F-box domain and prevents cell cycle arrest (Johnston *et al.*, 2005). The same viral molecule (M-T5) has been shown to also bind cellular Akt in human cancer cells and activates this multifunctional host cell signaling molecule (Wang *et al.*, 2006).

EVOLUTION OF IMMUNE EVASION

Phylogenetic analysis of the poxviruses indicate that the chordopoxviruses comprise a monophyletic clade and have likely evolved from a common ancestoral virus-like entity (McLysaght *et al.*, 2003). The gene order and gene spacing is highly conserved among the central conserved genes of all mammalian poxvirus genomes. This gene order has undergone some rearrangement in the avipoxvirus genomes (Afonso *et al.*, 2000; McLysaght *et al.*, 2003; Tulman *et al.*, 2004) and is more dramatic when compared with the entomopoxviruses (Afonso *et al.*, 1999; Bawden *et al.*, 2000). However, the pattern of maintaining essential housekeeping and structural genes within the central core of the genome, and arranging the host interactive functions distal to this core gene set, is highly conserved. The acquisition and accumulation of the various immune evasion genes (MHC class 1, IL-10, IL-24, IL-18, IFNγ receptor, TNF receptor, etc.) is evidence of more recent horizontal gene transfer (HGT) from host to poxvirus genome (Hughes and

TABLE 19.3 Immunomodulators of the family *Poxviridae*

Cellular homologue	Poxvirus response	Function	Poxvirus protein
Type I interferons	Receptor homologue	Blocks type I IFN	VV B19R
	IFN inhibition	dsRNA binding	VV E3L
	IFN inhibition	eIF2α homologue	VV K3L
Type II interferons	Receptor homologue	Blocks type II IFN	M-T7, VV B8R
IL-1β	Receptor homologue	Blocks IL-1β	VV B15R
IL-18	Binding protein	Inhibits IL-18	MOCV 54L, EV p13
Tumor necrosis factor	Receptor homologue	Blocks TNF from binding to receptor	M-T2, CPV CrmB/C/D/E
	Binding protein	Binds hTNF	TPV 2L
		Binds monkey TNF	YMTV 2L
		Binds pig TNF	SPV 2L
Chemokines	MIP-1β homologue		MOCV 148R
	Binding proteins	CC	MV M-T1, VV CCI
		C, CC, CXC	M-T7
	Receptor homologue	CCR8	YLDV 7L, 145R
Multi-cytokine	Binding protein	GMCSF/IL-2	Orf virus GIF
Serpins	Serine proteases	Blocking inflammation response	MV Serp-1
Complement	Complement binding	C3 convertase	VV VCP
Cytokine inhibition	Toll-like receptors	Intracellular signaling inhibition	VV A52R, A46R
		Inflammation	VV SPI-3
		IL-10 inflammation	Orf virus IL-10
Semaphorin	Semaphorin homologue		VV A39R
Apoptosis inhibitors		Ankyrin repeats ensures ER egress	M-T5
	ER		M-T4
	Mitochondria	Binding Bax and Bak	MV M11L, VV F1L

Friedman, 2005) and represents a selective advantage for the virus within the host. In contrast, gene families that were identified as being related but distinct between vertebrate host animals and poxviruses were probably not recent HGT events. These more ancient viral genes include RNA polymerase subunits and proteins involved in nucleic acid biosynthesis and metabolism (Hughes and Friedman, 2005). Thus, genetic variability within poxviral genomes occurs more towards the terminal regions of the genomes and defines the specific species requirements for any one particular poxvirus maintaining a sustained transmission relationship with its host.

POXVIRUS TROPISM AND HOST RANGE

The poxviruses are widespread and found on all continents except Antarctica. The co-evolution of the poxviruses with their hosts, likely over hundreds of millions of years, has led to a situation where many poxviruses have become so host restricted that they no longer are able to cause disease in other species (Bray, 2006). There is a range of possible outcomes when one specific poxvirus is transmitted from its reservoir host to another animal species. These outcomes can range from the complete absence of any visible evidence of infection, to the formation of a localized self-limiting skin lesion, to the development of a systemic lethal disease (Bray, 2006). The ability to detect a poxvirus infection in a non-reservoir host is normally the appearance of a unique disease or novel gross pathology. In general terms, poxvirus tropism is generally controlled at three levels (McFadden, 2005). These levels of control that mediate host range can be divided into cellular, tissue, and organismal tropisms. Most poxviruses are capable of cell binding and entry into most mammalian cells *in vitro*. This promiscuous ability to bind and enter a wide spectrum of cells from many species does not seem to be dependent on whether the infection ultimately results in the production of infectious progeny or not. This capacity of poxviruses to at least initiate infection in most cells, whether the particular cell supports a permissive, semi-permissive, or abortive infection, implies that any block to productive poxvirus infections has evolved not at the level of cellular receptors for entry but rather at the level of intracellular mechanisms within the individually infected cells.

STUDIES IN EVOLUTION AT THE HOST SPECIES LEVEL

The poxviruses provide some fascinating examples of DNA virus evolution. For example, the rate of poxvirus evolution has recently been estimated (Babkin and Shchelkunov, 2006). These authors examined a stretch of conserved central DNA (102 kb) among the orthopoxviruses and the viral DNA polymerase gene (3 kb) from members of the other genera and using the known time of variola virus introduction from West Africa to South America, and have determined that poxviruses have been undergoing mutations at $0.9–1.2 \times 10^{-6}$ nucleotide substitutions per site per year. At this rate of accumulation of change, it is estimated that the chordopoxviruses diverged from a common ancestor 500 000 years ago. Using this rate of mutational drift, the orthopoxviruses are postulated to have emerged about 300 000 years ago and modern members arose approximately 14 000 years ago (Babkin and Shchelkunov, 2006). Other studies of specific viruses in three different genera can be used as examples of the speed and amount of genetic variation observed in poxviruses (Fenner, 1995; Esposito *et al.*, 2006; Nazarian *et al.*, 2007).

Variola Virus (Orthopoxvirus)

Scientists at the Centers for Disease Control, Atlanta, performed comparative genomic analysis of 45 strains of variola separated by geographic location (South America, Africa, Asia), time (1944–1977), and disease severity (1–30% fatality rates) and found low sequence diversity and good conservation of ORFs. In addition, phylogenetic analysis demonstrated that the strains clustered based on their geographic location and virus severity (Esposito *et al.*, 2006). Most gene variation occurred in the ORFs towards the genome termini and selection pressure of the variola ancestor apparently favored the evolution into a devastating human pathogen (Esposito *et al.*, 2006). The most closely related poxvirus to variola virus sequence is the taterapox virus isolated from a gerbil of West Africa (Benin) and camelpox virus (Gubser and Smith, 2002), but neither of these viruses infect humans. The basis for this inability to leap into humans and ability to cause zoonotic disease is unknown.

Tanapox Virus (Yatapoxvirus)

Two clinical isolates of tanapox virus, isolated 50 years apart and from different locations in Africa have now been sequenced completely (Nazarian *et al.*, 2007). An isolate from the original outbreak of tanapox virus in the Tana River valley of Kenya in 1957 (TPV-Kenya) was compared with a sample isolated from a visitor to Equatorial Africa in 2004 (TPV-RoC). Although isolated 50 years apart, the two genomes were 99.98% identical and differed at only 35 of 144 565 nucleotide positions, with 31 of the 35 differences occurring within coding sequences. There were only two changes to the ORF organization between the two isolates. One ORF (11L) of TPV-RoC had a transversion that causes premature termination of the full-length 11L gene in TPV-Kenya, resulting in the division into two ORFs (11.1L and 11.2L). A second putative ORF identified in the related yatapoxvirus, yaba monkey tumor virus, was truncated in each isolate so that TPV-RoC encoded the N-terminal half (first 38 amino acids) and TPV-Kenya encoded the C-terminal half (last 50 amino acids). The sequencing results indicated a slow evolutionary rate of change that suggests a stable, confined evolutionary niche, which is surprising considering the amount of environmental changes in Equatorial Africa, such as deforestation and widespread human relocations (Nazarian *et al.*, 2007).

Myxoma Virus (Leporipoxvirus)

In contrast to tanapox virus, where selection pressures seem to have remained relatively constant over the last 50 years, there is dramatic evidence for rapid evolution in the virulence of myxoma virus in a natural setting. In 1950 scientists in Australia introduced the virulent but rabbit-specific South American strain of myxoma virus into a naïve environment of Australia in an effort to control the introduced European rabbit (now feral) populations. *Oryctolagus cuniculus* had been introduced into Australia over 150 years ago, but without any natural predators the rabbits had reproduced and expanded to the point where they caused extensive ecological damage throughout the continent. Thus, feral rabbits became a major agriculture and ranching pest. During the summer of 1950–1951, the introduced myxoma virus spread, largely by mosquito vectors, and estimates suggested that there was an initially high mortality rate in infected rabbits of over 99%. But by the next year, less virulent myxoma virus strains emerged and eventually replaced the more virulent original strain (called SLS). Australian scientists concluded that transmission by mosquito favored less virulent but still highly pathogenic virus because the original highly virulent virus killed rabbits within 3–4 days of their lesions becoming infectious, whereas the highly attenuated virus variants produced too little virus in the skin for transmission and the rabbits recovered too quickly. This led to recovery rates that evolved from 1% to 10% of infected animals and furthermore allowed for the selection of resistance among rabbits. This co-evolution led to selection for virus variants with greater disease involvement because they were now more effectively transmitted from rabbit to rabbit (Myers *et al.*, 1954; Fenner and Ratcliffe, 1965; Fenner, 1995).

CONCLUSIONS

At least one strain of most of the major viral members of the chordopoxviruses has been sequenced (Table 19.1). Comparative sequence analysis indicates that the genome structure of the chordopoxviruses is highly conserved (Smith and McFadden, 2002; McLysaght *et al.*, 2003; Upton *et al.*, 2003). There is limited sequence information regarding the entomopoxviruses but what is available suggests that the invertebrate poxviruses diverged before the separation of the chordopoxvirus genera (Gubser *et al.*, 2004). A website (www.poxvirus.org) is dedicated to the genomic and sequencing information of all poxviruses in the public domain.

Genetic information has been both lost and acquired as part of poxvirus genome evolution, and continues today. The rate of loss has varied among gene families and the gene acquistion has apparently increased, at least in the orthopoxviruses (McLysaght *et al.*, 2003; Hughes and Friedman, 2005). There are 90 poxviral genes that are highly conserved in all chordopoxviruses. And if one includes all poxviruses, this number of completely conserved genes is reduced to 45 (Upton *et al.*, 2003; Gubser *et al.*, 2004). This core complement of conserved poxvirus genes is found within the central domain of all the genomes, and no single gene is maintained in all sequenced members from the variable region of the genome. This is clear evidence of a common ancestral precursor of all poxviruses, and extensive evolutionary divergence in poxviruses found today.

Host cellular genes are acquired primarily by horizontal gene transfer (McLysaght *et al.*, 2003; Hughes and Friedman, 2005). Although gene acquisition from the host has never been observed in the lab, our best guess is that host genes in the form of cDNAs (e.g. in cells co-infected with a retrovirus) recombine into replicating poxvirus genomes. Many of these cellular acquisitions are directed at manipulation of the host immune system (Seet *et al.*, 2003) and the unique evolutionary histories of each individual poxvirus member probably contributes to the diversity of the various poxvirus family members that are observed today.

REFERENCES

Afonso, C.L., Tulman, E.R., Lu, Z., Oma, E., Kutish, G.F. and Rock, D.L. (1999) The genome of *Melanoplus sanguinipes* entomopoxvirus. *J. Virol.* **73**, 533–552.

Afonso, C.L., Tulman, E.R., Lu, Z., Zsak, L., Kutish, G.F. and Rock, D.L. (2000) The genome of fowlpox virus. *J. Virol.* **74**, 3815–3831.

Afonso, C.L., Tulman, E.R., Lu, Z., Zsak, L., Osorio, F. A., Balinsky, C. *et al.* (2002) The genome of swinepox virus. *J. Virol.* **76**, 783–790.

Afonso, C.L., Delhon, G., Tulman, E.R., Lu, Z., Zsak, A., Becerra, V.M. *et al.* (2005) Genome of deerpox virus. *J. Virol.* **79**, 966–977.

Afonso, C.L., Tulman, E.R., Delhon, G., Lu, Z., Viljoen, G. J., Wallace, D.B. *et al.* (2006) Genome of crocodilepox virus. *J Virol.* **80**, 4978–4991.

Alejo, A., Ruiz-Arguello, M.B., Ho, Y., Smith, V.P., Saraiva, M. and Alcami, A. (2006) A chemokine-binding domain in the tumor necrosis factor receptor from variola (smallpox) virus. *Proc. Natl Acad. Sci. USA* **103**, 5995–6000.

Babkin, I.V. and Shchelkunov, S.N. (2006) [The time scale in poxvirus evolution]. *Mol. Biol. (Mosk.)* **40**, 20–24.

Barrett, J.W. and McFadden, G. (2007) Genus Leporipoxviruses. In: *Poxviruses* (A.A. Mercer, A. Schmidt and O. Weber, eds). Heidelberg: Birkhauser.

Barrett, J.W., Sun, Y., Nazarian, S.H., Belsito, T.A., Brunetti, C.R. and McFadden, G. (2006) Optimization of codon usage of poxvirus genes allows for improved transient expression in mammalian cells. *Virus Genes* **33**, 15–26.

Bawden, A.L., Glassberg, K.J., Diggans, J., Shaw, R., Farmerie, W. and Moyer, R.W. (2000) Complete genomic sequence of the *Amsacta moorei* entomopoxvirus: Analysis and comparison with other poxviruses. *Virology* **274**, 120–139.

Baxby, D. (1981) *Jenner's Smallpox Vaccine: The Riddle of Vaccinia Virus and its Origin*. London: Heinemann Educational.

Becker, M. and Moyer, R.W. (2007) Subfamily Entomopoxvirinae. In: *Poxviruse* (A.A. Mercer, A. Schmidt and O. Weber, eds). Berlin: Birkhauser Verlag.

Bolte, A.L., Meurer, J. and Kaleta, E.F. (1999) Avian host spectrum of avipoxviruses. *Avian Pathol.* **28**, 415–432.

Bray, M. (2006) Cross-species transmission of poxviruses. In: *Emerging Pathogens of the 21st Century* (C. Fong and K. Alibeck, eds). Berlin: Springer.

Brunetti, C.R., Li, X., Barrett, J., Amano, H., Yoshiaki, U., Miyamura, T. *et al.* (2003a) The complete genome sequence and comparative analysis of the tumorigenic poxvirus Yaba monkey tumor virus. *J. Virol.* **77**, 13335–13347.

Brunetti, C.R., Paulose-Murphy, M., Singh, R., Qin, J., Barrett, J.W., Tardivel, A. *et al.* (2003b) A secreted high-affinity inhibitor of human TNF from Tanapox virus. *Proc. Natl Acad. Sci. USA* **100**, 4831–4836.

Cameron, C., Hota-Mitchell, S., C hen, L., Barrett, J., Cao, J.-X., Macaulay, C. *et al.* (1999) The complete DNA sequence of myxoma virus. *Virology* **264**, 298–318.

Deane, D., McInnes, C.J., Percival, A., Wood, A., Thomson, J., Lear, A. *et al.* (2000) Orf virus encodes a novel secreted protein inhibitor of granulocyte-macrophage colony-stimulating factor and interleukin-2. *J. Virol.* **74**, 1313–1320.

Delhon, G., Tulman, E.R., Afonso, C.L., Lu, Z., De La Concha-Bermejillo, A., Lehmkuhl, H.D. *et al.* (2004) Genomes of the parapoxviruses ORF virus and bovine papular stomatitis virus. *J. Virol.* **78**, 168–177.

Dhar, A.D., Werchniak, A.E., Li, Y., Brennick, J.B., Goldsmith, C.S., Kline, R. *et al.* (2004) Tanapox infection in a college student. *New Engl. J. Med.* **350**, 361–366.

Diallo, A. and Viljoen, G.J. (2007) Genus Capripoxvirus. In: *Poxviruses* (A.A. Mercer, A. Schmidt and O. Weber, eds). Boston, Birkhauser Verlag.

Downie, A.W. and Espana, C. (1972) Comparison of Tanapox virus and Yaba-like viruses causing epidemic disease in monkeys. *J. Hyg. (Lond.)* **70**, 23–32.

Downie, A.W. and Espana, C.A. (1973) A comparative study of tanapox and yaba viruses. *J. Gen. Virol.* **19**, 37–49.

Esposito, J.J., Sammons, S.A., Frace, A.M., Osborne, J.D., Olsen-Rasmussen, M., Zhang, M. *et al.* (2006) Genome sequence diversity and clues to the evolution of variola (smallpox) virus. *Science* **313**, 807–812.

Essani, K., Chalasani, S., Eversole, R., Beuving, L. and Birmingham, L. (1994) Multiple anti-cytokine activities secreted from tanapox virus-infected cells. *Microb. Pathog.* **17**, 347–353.

Everett, H., Barry, M., Sun, X., Lee, S.F., Frantz, C., Berthiaume, L.G. *et al.* (2002) The myxoma poxvirus protein, M11L, prevents apoptosis by direct interaction with the mitochondrial permeability transition pore. *J. Exp. Med.* **196**, 1127–1139.

Fenner, F. (1995) Classical studies of virus evolution. In: *Molecular Basis of Viral Evolution* (A. Gibbs, C. Calisher and F. Garcia-Arenal, eds). New York: University Cambridge Press.

Fenner, F. and Ratcliffe, F.N. (1965) *Myxomatosis*. Cambridge: Cambridge University Press.

Grace, J.T.J. and Mirand, E.A. (1963) Human susceptibility to a simian tumor virus. *Annu. NY Acad. Sci.* **108**, 1123–1128.

Gubser, C. and Smith, G.L. (2002) The sequence of camelpox virus shows it is most closely related to variola virus, the cause of smallpox. *J. Gen. Virol.* **83**, 855–872.

Gubser, C., Hue, S., Kellam, P. and Smith, G.L. (2004) Poxvirus genomes: a phylogenetic analysis. *J. Gen. Virol.* **85**, 105–117.

Hughes, A.L. and Friedman, R. (2005) Poxvirus genome evolution by gene gain and loss. *Mol. Phylogenet. Evol.* **35**, 186–195.

Johnston, J.B. and McFadden, G. (2004) Technical knockout: understanding poxvirus pathogenesis by selectively deleting viral immunomodulatory genes. *Cell. Microbiol.* **9**, 695–705.

Johnston, J.B., Wang, G., Barrett, J.W., Nazarian, S.H., Colvill, K., Moran, M. and McFadden, G. (2005) Myxoma virus M-T5 protects infected cells from the stress of cell cycle arrest through its interaction with host cell cullin-1. *J. Virol.* **79**, 10750–70763.

Lalani, A.S., Graham, K., Mossman, K., Rajarathnam, K., Clark-Lewis, I., Kelvin, D. and McFadden, G. (1997) The purified myxoma virus gamma interferon receptor homolog M-T7 interacts with the heparin-binding domains of chemokines. *J. Virol.* **71**, 4356–4363.

Lee, H.-J., Essani, K. and Smith, G.L. (2001) The genome sequence of yaba-like disease virus, a yatapoxvirus. *Virology* **281**, 170–192.

Losos, G.J. (1986) *Infectious Tropical Diseases of Domestic Animals*. Essex, UK: Longman Scientific and Technical.

McFadden, G. (2005) Poxvirus tropism. *Nat. Rev. Microbiol.* **3**, 201–213.

McLysaght, A., Baldi, P.F. and Gaut, B.S. (2003) Extensive gene gain associated with adaptive evoluton of poxviruses. *Proc. Natl Acad. Sci. USA* **100**, 15655–15660.

Myers, K., Marshall, I.D. and Fenner, F. (1954) Studies in the epidemiology of infectious myxomatosis of rabits. III. Observations on two succeeding epizootics in Australian wild rabbits on the Riverine plain of south-eastern Australia. *J. Hyg.* **52**, 337–340.

Nazarian, S.H., Barrett, J.W., Frace, A.M., Olsen-Rasmussen, M., Khristova, M., Shaban, M. *et al.* (2007) Comparative genetic analysis of genomic DNA sequences of two human isolates of Tanapox virus. *Virus Res.* **129**, 11–25.

Niven, J.S., Armstrong, J.A., Andrewes, C.H., Pereira, H. G. and Valentine, R.C. (1961) Subcutaneous "growths" in monkeys produced by a poxvirus. *J. Pathol. Bacteriol.* **81**, 1–14.

Paulose, M., Bennet, B.L., Manning, A.M. and Essani, K. (1998) Selective inhibition of TNF-alpha induced cell adhesion molecular gene expression by tanapox virus. *Microb. Pathogen.* **25**, 33–41.

Schreiber, M., Sedger, L. and McFadden, G. (1997) Distinct domains of M-T2, the myxoma virus TNF receptor homolog, mediate extracellular TNF binding and intracellular apoptosis inhibition. *J. Virol.* **71**, 2171–2181.

Sedger, L.M., Osvath, S.R., Xu, X.M., Li, G., Chan, F.K., Barrett, J.W. and McFadden, G. (2006) Poxvirus tumor necrosis factor receptor (TNFR)-like T2 proteins contain a conserved preligand assembly domain that inhibits cellular TNFR1-induced cell death. *J. Virol.* **80**, 9300–9309.

Seet, B.T., Johnston, J.B., Brunetti, C.R., Barrett, J.W., Everett, H., Cameron, C. *et al.* (2003) Poxviruses and immune evasion. *Annu. Rev. Immunol.* **21**, 377–423.

Senkevich, T.G., Bugert, J.J., Sisler, J.R., Koonin, E.V., Darai, G. and Moss, B. (1996) Genome sequence of a human tumorigenic poxvirus: Prediction of specific host response-evasion genes. *Science* **273**, 813–816.

Senkevich, T.G., Koonin, E.V., Bugert, J.J., Darai, G. and Moss, B. (1997) The genome of molluscum contagiosum virus: analysis and comparison with other poxviruses. *Virology* **233**, 19–42.

Smith, G.L. and McFadden, G. (2002) Smallpox: Anything to declare?. *Nat. Rev. Immunol.* **2**, 521–528.

Smith, S.A. and Kotwal, G.J. (2002) Immune response to poxvirus infections in various animals. *Crit. Rev. Microbiol.* **28**, 149–185.

Stanford, M.M., Werden, S.J. and McFadden, G. (2007) Myxoma virus in the European rabbit: interactions between the virus and its susceptible host. *Vet. Res.* **38**, 299–318.

Su, J., Wang, G., Barrett, J.W., Irvine, T.S., Gao, X. and McFadden, G. (2006) Myxoma virus M11L blocks apoptosis through inhibition of conformational activation of Bax at the mitochondria. *J. Virol.* **80**, 1140–1151.

Taylor, J. and Paoletti, E. (1988) Fowlpox virus as a vector in non-avian species. *Vaccine* **6**, 466–468.

Taylor, J., Meignier, B., Tartaglia, J., Languet, B., Vanderhoeven, J., Franchini, G. *et al.* (1995) Biological and immunogenic properties of a canarypox-rabies recombinant, ALVAC RG (vCP65) in non-avian species. *Vaccine* **13**, 539–549.

Tulman, E.R., Afonso, C.L., Lu, Z., Zsak, L., Kutish, G.F. and Rock, D.L. (2001) Genome of lumpy skin disease virus. *J. Virol.* **75**, 7122–7130.

Tulman, E.R., Afonso, C.L., Lu, Z., Zsak, L., Sur, J.H., Sandybaev, N.T. *et al.* (2002) The genomes of sheeppox and goatpox viruses. *J. Virol.* **76**, 6054–6061.

Tulman, E.R., Afonso, C.L., Lu, Z., Zsak, L., Kutish, G.F. and Rock, D.L. (2004) The genome of canarypox virus. *J. Virol.* **78**, 353–366.

Upton, C., Mossman, K. and McFadden, G. (1992) Encoding of a homolog of the IFN-gamma receptor by myxoma virus. *Science* **258**, 1369–1372.

Upton, C., Slack, S., Hunter, A.L., Ehlers, A. and Roper, R. L. (2003) Poxvirus orthologous clusters: toward defining the minimum essential poxvirus genome. *J. Virol.* **77**, 7590–7600.

Wang, G., Barrett, J.W., Nazarian, S.H., Everett, H., Gao, X., Bleackley, C. *et al.* (2004) Myxoma virus M11L prevents apoptosis through constitutive interaction with Bak. *J. Virol.* **78**, 7097–7111.

Wang, G., Barrett, J.W., Stanford, M., Werden, S.J., Johnston, J.B., Gao, X. *et al.* (2006) Infection of human cancer cells with myxoma virus requires Akt activation via interaction with a viral ankyrin-repeat host range factor. *Proc. Natl Acad. Sci. USA* **103**, 4640–4645.

Wasilenko, S.T., Banadyga, L., Bond, D. and Barry, M. (2005) The vaccinia virus F1L protein interacts with the proapoptotic protein Bak and inhibits Bak activation. *J. Virol.* **79**, 14031–14043.

Willer, D.O., McFadden, G. and Evans, D.H. (1999) The complete genome sequence of shope (rabbit) fibroma virus. *Virology* **264**, 319–343.

CHAPTER

20

Molecular Evolution of the *Herpesvirales*

Duncan J. McGeoch, Andrew J. Davison, Aidan Dolan, Derek Gatherer, and Edgar E. Sevilla-Reyes

ABSTRACT

The herpesviruses are a group of large DNA viruses, originally defined by their characteristic virion structure. On the basis of genome sequences they have been assigned to an order, *Herpesvirales*, containing three families: the *Herpesviridae*, infecting mammals, birds and reptiles; the *Alloherpesviridae*, infecting amphibians and fish; and the *Malacoherpesviridae*, populated only by an oyster virus. Viruses in the *Herpesviridae* are descended from a common ancestor, as are those in the *Alloherpesviridae*, but connections between the families are tenuous.

The *Herpesviridae* include eight human viruses. Three widely diverged subfamilies, the *Alpha-*, *Beta-*, and *Gammaherpesvirinae*, are defined. A robust phylogenetic tree has been constructed for this family, based on amino acid sequences from conserved genes. Within subfamilies, aspects of branching patterns resemble those of mammalian host lineages, indicating long-term co-evolution of virus and host lines and thus enabling inference of a timeframe for the tree. On this basis the tree is estimated to be about 400 million years in depth.

Some 40 genes are conserved across the *Herpesviridae*, with recognized roles mainly in capsid structure and DNA replication machinery. Functions of non-conserved genes include roles in immune modulation and latency. Some herpesvirus genes appear to have originated by capture from cellular genomes, and others by genesis *de novo*. Multigene families are common, notably in the *Betaherpesvirinae*. Aspects of DNA replication systems have diverged among subfamilies, with complex arrangements for initiation of DNA synthesis in the *Gammaherpesvirinae* and part of the *Betaherpesvirinae*, and disabling in the *Betaherpesvirinae* of genes for nucleotide anabolism. Comparative genomic sequencing of herpesvirus isolates is revealing novel aspects of recent evolution. Recombination among strains has emerged as a general phenomenon. Also, certain latent cycle genes of gammaherpesviruses uniquely evince signs of widespread diversifying selection.

Genomic organizations in the *Alloherpesviridae* and *Malacoherpesviridae* look generally similar to those of the *Herpesviridae*, and the *Alloherpesviridae* are also widely diverse. It appears likely that capsid components of

Origin and Evolution of Viruses
ISBN 978-0-12-374153-0

the three families possess a shared, distant ancestry, but there may be little else of common origin in the families' genetic complements. A yet more distant relationship is suggested by similarities between elements of *Herpesviridae* capsids and those of tailed DNA bacteriophages.

INTRODUCTION

The herpesviruses are a numerous group of large DNA viruses, almost all with a vertebrate species as host. They are widely distributed in nature and the range of disease associations and biological characteristics found across the group is substantial. Historically, assignment as a herpesvirus was defined by virion architecture. Early images of herpes simplex virus particles, in electron microscopy studies of the 1950s, showed an icosahedral structure with 162 capsomeric spikes, of about 100nm diameter and enclosed in a membranous bag. This characteristic appearance was also observed with some other viruses, and formed the basis for defining the herpesvirus group. In modern terms, the herpes virion has four components: a densely packed DNA core in an icosahedral capsid ($T = 16$), surrounded by an amorphous proteinaceous layer (the tegument) and enveloped in a protein–lipid membrane. By 1971 around 30 members of the group were recognized, from many different host species (Wildy, 1971), and it was known that herpesviruses possessed DNA genomes, large by virus standards. The taxonomic status of family was applied by the International Committee on Taxonomy of Viruses (ICTV) in 1976, initially as the *Herpetoviridae* (Fenner, 1976). The name *Herpesviridae* made its debut in 1979, together with the three subfamilies, the *Alpha-*, *Beta-*, and *Gammaherpesvirinae*, which were defined largely on biological attributes (Matthews, 1979); this nomenclature has persisted. By 1981, 63 herpesviruses with species of mammals as hosts were recognized, plus 12 avian, six reptilian, two amphibian, and six piscine herpesviruses (Roizman *et al.*, 1981). The number of herpesvirus species known in 2007 easily exceeds 100, and if viruses detected only as fragments of genomic sequence are included, the total rises to over 200.

DNA sequence analyses of herpesvirus genes and genomes was under way by the 1980s, and comparisons among DNA sequences and encoded amino acid sequences had by the mid-1990s greatly clarified overall relationships involving mammalian and avian herpesviruses, and the views obtained continue to the present day to increase in scope and resolution. Analyses based on genomic sequences form most of the content of this chapter, and at this stage we make only some general points. First, sequence comparisons have shown that herpesviruses of mammals, birds and reptiles are all descended from a common ancestor. Second, the extent of evolutionary divergence across these viruses is large. Third, their previous classification into three subfamilies has remained valid, indeed is a fundamental feature. Fourth, herpesviruses of amphibians and fish are also among themselves clearly related by descent, but by genomic criteria this group shows only tenuous signs of being related to the mammalian, avian, and reptilian herpesviruses. And last, the single characterized herpesvirus of an invertebrate (oyster) forms a third genomic group, extremely distant from the other two.

Sequence comparisons have thus replaced virion morphology as the main route to identifying and classifying herpesviruses, but have revealed a greater complexity of evolutionary relationships than previously suspected, and this has necessitated an expansion of the taxonomic framework. A new top level taxon has recently been established by ICTV, the order *Herpesvirales*, which encompasses all the entities identified as herpesviruses through having the characteristic virion architecture or through genomic sequence relationships to characterized herpesviruses (Eighth Report of ICTV plus subsequent developments; Davison *et al.*, 2005a). The order *Herpesvirales* contains three families, the *Herpesviridae*, *Alloherpesviridae*, and *Malacoherpesviridae*. The family *Herpesviridae* as now defined comprises the three previously recognized subfamilies, so includes all known mammalian, avian, and reptilian herpesviruses; this means that the taxonomic arrangements for the great majority of species have been left undisturbed. The amphibian and piscine herpesviruses have been assigned to the family *Alloherpesviridae* and the oyster herpesvirus to the family *Malacoherpesviridae*, presently as sole member. By far the greatest amount of effort in studying the *Herpesvirales* has concerned members of the *Herpesviridae, sensu novo.* This chapter is therefore primarily concerned with that group, and treats aspects of the two other families only late in the text. In all further use in this chapter, the term *Herpesviridae* is to be taken in the revised sense. As of early 2007, more than 80 complete or near complete sequences are known, which represent 39 species in the *Herpesviridae.* In addition, many other partial sequences are available, for single genes or larger genomic sections. Table 20.1 gives an outline of current taxonomical arrangements and of sequenced genomes in the *Herpesviridae*, with abbreviations (not repeated in text) of virus names for the species listed.

In order to illustrate the diversity of diseases caused by herpesviruses and also the diversity in their underlying biological properties, we employ the device of enumerating the herpesviruses presently known to have the human species as their natural host, and in effect allowing the list of their properties, in Table 20.2, to speak for itself. Eight human herpesviruses are known, with representatives in all three subfamilies. Genome sizes range from 125 to 236 kbp, genome base compositions from 36% to 70% G+C, and estimates for complements of protein coding genes from 70 to 165. Most of these viruses occur at high prevalence in human populations, while one (HSV-2) is characteristically sexually transmitted. All establish a lifetime presence in a host, involving diverse mechanisms to achieve a latent state. In their latent existence, three are neurotropic (HSV-1, HSV-2, VZV), four are lymphotropic (EBV, HHV-6, HHV-7, HHV-8) and one is involved with the monocyte lineage (HCMV). Two are associated with human cancers (EBV and HHV-8). Infections can range from inapparent through disabling to lethal. All this for the viruses of one host species, albeit the most extensively studied.

We aim in this chapter to provide an overview of the evolutionary history of the herpesviruses. Our treatment is in essence comparative: we use genome sequences and encoded protein sequences to discern processes that have occurred during evolution, and to construct a phylogenetic description of the mammalian, avian, and reptilian viruses. We next move to a finer level, treating aspects of specific genes and gene subsystems. The topics in this part of the chapter were selected on the basis of the varied contributions they could make to illuminating the overall picture; they do not constitute a general tour of herpesvirus genes. We then turn to herpesviruses of fish and amphibians to recapitulate our descriptions in terms of those very different entities. Finally, we attempt to summarize implications of the current situation in our understanding of herpesvirus evolution.

COMPARATIVE HERPESVIRUS GENOMICS

Structures of Herpesvirus DNAs

A herpesvirus genome is constituted of a single molecule of double-stranded DNA. The genomes fall into a substantial range of sizes: among the sequenced genomes in the *Herpesviridae*, the smallest are SVV and VZV at 125 kbp and the largest is CCMV at 241 kbp. Most of the genome consists of so-called unique sequences, and the majority of genomes also contain large repeated elements (of order 10^3–10^4 bp in size). Distinct copies of these major repeats are usually considered to possess indistinguishable sequences, with homogeneity being maintained by recombination. Different patterns of placement of unique and repeat sequences define at least six genome types, as shown in Figure 20.1. In most cases, viruses in a sublineage are of the same genome

TABLE 20.1 Taxonomy and sequenced genomes in the *Herpesviridae*

	Virus[a]	Name abbreviation	Genome size (kbp)[b]
Subfamily			
Alphaherpesvirinae	Herpes simplex virus type 1	HSV-1	152
Simplexvirus	Herpes simplex virus type 2	HSV-2	155
	Simian agent 8	SA-8	151
	Herpesvirus papio 2	HVP-2	156
	Simian B virus	SBV	157
Varicellovirus	Bovine herpesvirus 1	BHV-1	135
	Bovine herpesvirus 5	BHV-5	138
	Pseudorabies virus	PRV	143
	Equid herpesvirus 1	EHV-1	150
	Equid herpesvirus 4	EHV-4	146
	Varicella-zoster virus	VZV	125 (20)
	Simian varicella virus	SVV	125
	Marek's disease virus type 1	MDV-1	171–178 (4)
	Marek's disease virus type 2	MDV-2	164
	Herpesvirus of turkeys	HVT	159, 161 (2)
	Infectious laryngotracheitis virus	ILTV	149
	Psittacid herpesvirus 1	PsHV-1	163
Subfamily			
Betaherpesvirinae	Human cytomegalovirus	HCMV	236[c]
Cytomegalovirus	Chimpanzee cytomegalovirus	CCMV	241
	Rhesus cytomegalovirus	RhCMV	216, 221 (2)
	Simian cytomegalovirus	SCMV	225[d]
	Aotine cytomegalovirus	AoCMV	218[d]
(no genus)	Tupaiid herpesvirus	THV	196
Muromegalovirus	Murine cytomegalovirus	MCMV	230
	Rat cytomegalovirus	RCMV	230
Roseolovirus	Human herpesvirus 6	HHV-6	159–162 (3)
	Human herpesvirus 7	HHV-7	145, 153
Subfamily			
Gammaherpesvirinae			
Lymphocryptovirus	Epstein-Barr virus	EBV	172–173 (3)
	Rhesus lymphocryptovirus	RLV	171
	Callitrichine herpesvirus 3	CalHV-3	150
Macavirus	Alcelaphine herpesvirus 1	AlHV-1	138[e]
	Ovine herpesvirus 2	OHV-2	135[e]
Percavirus	Equid herpesvirus 2	EHV-2	184
Rhadinovirus	Herpesvirus ateles	HVA	108[e]
	Herpesvirus saimiri	HVS	113[e] (2)
	Human herpesvirus 8	HHV-8	138[e, f]
	Rhesus rhadinovirus	RRV	131–134[e] (3)
	Bovine herpesvirus 4	BHV-4	109[e]
	Murid herpesvirus 4	MHV-4	119, 120[e] (2)

[a] Sequence data were obtained from GenBank except where noted. Annotations of completely sequenced herpesvirus genomes, including the locations of genes and the sequences of encoded proteins, are available from the webpages of the National Center for Biotechnology Information, under Genomic Biology, Viral Genomes.

[b] Genome sizes are rounded to nearest kbp. Where more than one sequence is available, the number is given in parentheses. Where two sequences are available, both are listed if sizes differ. For multiple sequences the range is given.

[c] Eight almost complete HCMV sequences are also available.

[d] Unpublished data of A. Dolan.

[e] Genome sizes are given exclusive of extensive terminal repeats.

[f] Two almost complete HHV-8 sequences are also available.

TABLE 20.2 Human herpesviruses

Virus	Abbrev	Genus	Genome Size (kbp)	G+C Content (%)	Genome Type[a]	Encoded Proteins	Pathology
Herpes simplex virus type 1	HSV-1	*Simplexvirus*	152	68	E	74	Recurrent epithelial lesions; rarer, serious neural disease
Herpes simplex virus type 2	HSV-2	*Simplexvirus*	155	70	E	74	As for HSV-1
Varicella-zoster virus	VZV	*Varicellovirus*	125	46	D	70	Primary infection chickenpox; recurrence as shingles
Human cytomegalovirus	HCMV	*Cytomegalovirus*	236	57	E	165	Cause of congenital abnormalities; severe infections inimmuno-compromised
Human herpesvirus 6	HHV-6	*Roseolovirus*	159–162	43	A	85	Primary infection exanthem subitum
Human herpesvirus 7	HHV-7	*Roseolovirus*	153, 145	36	A	84	No definitive disease association
Epstein-Barr virus	EBV	*Lymphocryptovirus*	172–173	60	C	83	Primary infection mononucleosis; associated with Burkitt's lymphoma and other neoplasias
Human herpesvirus 8	HHV-8	*Rhadinovirus*	138	53[b]	B	86	Associated with Kaposi's sarcoma and other neoplasias

[a] See Figure 20.1.
[b] G+C content for HHV-8 excludes terminal reiterations.

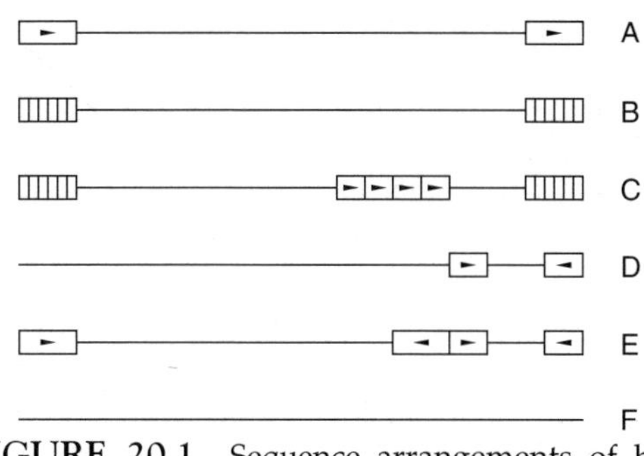

FIGURE 20.1 Sequence arrangements of herpesvirus DNAs. A schematic representation of the layouts in herpesvirus DNAs of unique sequences (shown as single lines) and repeat elements (shown as open boxes, with relative orientations indicated for large repeats by arrowheads). In types B and C the number of terminal reiterations, and in type C the number of internal repeats, are considered variable. In type E the sequences of the large and small repeat elements are distinct. Type F is intended to encompass genomes with very small or absent terminal repeat sequences. Adapted from Roizman *et al.* (1992).

type, but this classification does not necessarily correlate with evolutionary relatedness. For instance, HSV-1 in the *Alphaherpesvirinae* has a type E genome, and so does HCMV in the *Betaherpesvirinae*, but the R_L, R_S, and U_S elements of HCMV are considered to be evolutionarily distinct from the corresponding elements of HSV-1. Major repeat elements in herpesvirus genomes are dynamic entities on an evolutionary time-scale, as evidenced by comparison of the S regions of *Simplexvirus* and *Varicellovirus* genomes. These all consist of an unique sequence (U_S) flanked by a pair of repeats (IR_S and TR_S) in opposing orientations. Comparisons of the gene contents of S regions show that gene orders are variable although clearly related. Certain genes that in some genomes are in the U_S segment are in others duplicated, with one copy in each flanking repeat; and in other comparisons, certain genes that lie in U_S are near either one extremity or the other relative to the overall layout of genes.

These movements can be accounted for as the result of recombination events between two copies of the S region, with one end of a double crossover located legitimately in a flanking repeat, and the other illegitimately between heterologous points in the two copies of U_S (Davison and McGeoch, 1986).

In addition to these large features of genomic structure, other finer scale aspects bear describing in an evolutionary context. The most prominent concerns base compositions of herpesvirus DNAs, which at the whole genome level cover an impressively wide range, from 32% to 75% G+C content (Roizman *et al.*, 1992). There is no general correlation between virus lineage and base composition; the *Varicellovirus* genus, for instance, contains both canine herpesvirus (32% G+C) and pseudorabies virus (74% G+C). Wide differences in base composition among regions in one genome are also commonplace. Generally, large repeat elements have a higher G+C content than unique sequences in the same genome. The unique sequence DNA of HVS is 34% G+C, while the flanking repeats are 71% G+C (Albrecht *et al.*, 1992). In HSV-1 and HSV-2, with overall G+C contents of 68% and 70% respectively, the region with the highest G+C content is R_S (6600 bp) at close to 80% G+C, and this is inclusive of 3900 bp of coding sequence for one protein (McGeoch *et al.*, 1986; Dolan *et al.*, 1998). At such extreme base compositions, changes relative to orthologous genes with lesser compositional bias cannot be accommodated only through synonymous changes in the coding sequence, so that the overall amino acid composition of encoded protein also changes.

Although compositionally biased DNA sequences have been widely observed in other contexts, including genomes of eukaryotes and prokaryotes, the effects seen in herpesviruses are very striking by any standards. The extremity of base composition to which even some protein coding sequences in genomes of herpesviruses have evolved argues for the action of powerful forces, but the nature of these remains obscure. We can discern three functional classes of components, that would (1) introduce base substitutions, (2) impart directionality to the substitutions fixed, and (3) impose a differential effect between unique and repeat sequences (McGeoch *et al.*, 1986). The first of these roles and also the second could be ascribed to the herpesvirus DNA polymerase and to imbalances in dNTP pools in infected cells. In order to account for the differential effect on repeat sequences we have to invoke recombination processes between repeat copies in an intracellular pool of genome sequences, to accelerate changes relative to those in unique sequences. A recombinational mechanism could also be responsible for imposing directionality, in a form of biased gene conversion; evidence for operation of such a system in mammalian genomes has been developing recently (for instance, Galtier, 2003).

Examination of patterns of dinucleotide frequencies in genomic DNAs has revealed marked differences among herpesviruses (Honess *et al.*, 1989). In the alphaherpesviruses the dinucleotide frequencies exhibit only small deviations from random expectation. In gammaherpesvirus genomes there is a markedly low occurrence of 5′-CG. In betaherpesvirus genomes this restriction of 5′-CG occurs only in the major immediate-early gene regions. This phenomenon, often referred to as the "CpG shortage," has been known for many years as a global feature of vertebrate nuclear DNA, and is considered to result from methylation of cytosine residues in DNA, in the sequence 5′-CG, to give 5-methylcytosine; any hydrolytic deamination of the 5-methylcytosine then transforms the sequence to 5′-TG, which is presumably not subject to a repair process. A 5′-CG deficit in the herpesvirus context thus appears to indicate that over an evolutionary period of time the virus DNA has existed in a cellular environment where methylation of DNA cytosine was active, and in a form that made it susceptible to this modification. In the EBV genome, certain promoter sequences become methylated while in an episomal latent state in B cells, as an element in controlling viral gene expression and maintaining latency (reviews by Ambinder *et al.*, 1999; Minarovits, 2006).

The last fine-structure feature of herpesvirus genomes to mention is the occurrence of families of short, tandemly reiterated sequences. Such tandem repeats are found in all herpesvirus groups, with single copy sizes in the range, very roughly, of 5–100 bp. Simple arrays of identical elements and also more complicated mixtures of distinct repeat motifs occur. Copy numbers of repeat units range up to many tens, even hundreds. Some arrays clearly lie within protein coding sequences. Where comparative data are available for different strains or clones of the same virus, it is seen that at least some repeat families can exhibit variation among isolates both in copy number and in sequence of the repeat unit. These effects are presumed to be mediated by recombination or by slippage during DNA replication. Lastly, it is our impression that length variations in homopolymeric tracts within herpesvirus DNAs arise with some facility, as evidenced by their occurrence in virus isolates, molecular clones and mutants, and these could be regarded as the lower, unit-length limit case of tandem reiteration.

Gene Complements of Herpesviruses

The first wave of complete herpesvirus genome sequencing projects now lies two decades in the past (Baer *et al.*, 1984; Davison and Scott, 1986; McGeoch *et al.*, 1988; Chee *et al.*, 1990). We consider that most herpesvirus genome sequence determinations have been carried out to a very good standard of accuracy. Nonetheless, in almost all cases a low level of corrections to the sequences has come to light over time. There has also been a parallel ongoing process of refinement to catalogues of proposed genes, and we believe that herpesvirus gene listings have by now reached a state of some maturity and accuracy—but not necessarily of unchangeable finality (discussed by McGeoch *et al.*, 2006). A notably radical revision has been with HCMV, whose genome is large and challenging, and prone to mutating during virus isolation in cell culture (Cha *et al.*, 1996). Critical evaluation of an essentially wild-type sequence and comparisons with the closely related CCMV sequence have reduced estimates of gene numbers from well over 200 to 165 (Davison *et al.*, 2003; Dolan *et al.*, 2004).

Unique regions in herpesvirus genomes are generally closely packed with protein coding genes, but there are exceptions—notably, betaherpesviruses and also some gammaherpesviruses contain substantial "voids" of heteropolymeric but evidently non-coding DNA, of presently uncertain functionalities; some of these regions are known to be transcribed. Protein coding sequences also occur within major repeat elements but are typically arranged sparsely. As summarized in Table 20.2 for the human viruses, our estimates of the number of genes encoding distinct proteins lie in the range 70–86 for alpha-, beta-, and gammaherpesviruses of the *Roseolovirus* genus, but the estimate for viruses in the betaherpesvirus *Cytomegalovirus* genus is much higher—165 for HCMV. Because herpesvirus genome sequences are widely diverged, it is generally most useful to compare genes between viruses by way of their encoded amino acid sequences. Systematic comparisons of genes from pairs of herpesviruses have shown that some 30 clear orthologues are present in fully sequenced mammalian herpesviruses and avian herpesviruses. To these can be added another 10, which present less definitive cases (in terms of marginal amino acid sequence similarity, or patterns of hydrophilic and hydrophobic residues, or equivalent genomic locations) but whose ubiquity we regard as overall reasonably convincing. The present view is thus that there are about 40 genes that have detectable counterparts across the alpha-, beta-, and gammaherpesvirus subfamilies; these conserved genes have been termed the "core" set (Davison and Taylor, 1987; McGeoch, 1989). There are three more genes, to be discussed later, that we believe were present in the common progenitor of the three subfamilies but have been lost from all or part of the betaherpesvirus lineage. We thus recognize 43 genes that were present in an ancestral herpesvirus genome. In summary, present day herpesvirus gene complements comprise about 40 core members plus (except in the case of the cytomegaloviruses (CMVs)) around another 40 which are specific to subfamily, genus, sublineage or individual virus. With CMVs, the genus and lineage specific set is much expanded, with many multigene families (see below). Examination of the genomic locations of the core set of genes has shown that within each subfamily their layout is near constant, with the same orders and relative orientations. Among the three subfamilies, short range ordering of core genes is mostly maintained, but on the whole genome scale there are substantial relative differences in the arrangements of several blocks of genes, with seven major blocks recognized (Gompels *et al.*, 1995). In all cases the core genes occupy roughly the central part of the U_L region. Genes that do not belong to the core set are found mainly towards the extremities of U_L and in other genomic regions (repeat elements and U_S), and to a lesser extent interspersed among core genes. We emphasize that this description of comparative gene layout presents only a digest of global features from a quite complex picture.

We cannot attempt here a comprehensive account of herpesvirus gene functions. Instead, we have summarized functions known for the best studied herpesvirus, HSV-1, into several groupings, and in Table 20.3 have listed numbers of genes (total and core) in each of these. This presentation shows that genes for replication, processing, and repair of DNA are universal, with one exception. Those for peripheral enzymes of nucleotide metabolism are not so, but assignments here are complicated by the occurrence of two enzymatically inactive proteins in the betaherpesviruses (see later). The non-conserved HSV-1 DNA replication protein is concerned with initiation of DNA synthesis from origins of replication; this part of the DNA replicative machinery turns out to differ substantially among different lineages of herpesviruses, while components concerned with subsequent chain elongation are common to all lineages; this topic is raised again in a later section. Genes for control functions are almost all not universal, and indeed the two registered in the core set were assigned

TABLE 20.3 Functional and structural classes of HSV-1 proteins

Class of protein[a]	Number of proteins		Core	Non-core
1. Concerned with genomic DNA	20			
(a) Central DNA synthetic machinery		7	6	1
(b) Peripheral enzymes		4	2[b]	2
(c) Processing, packaging and repair of DNA		9	9	0
2. Control and modulation	10		2	8
3. Virion	30			
(a) Capsid assembly and structure		6	6	0
(b) Tegument		10	3	7
(c) Surface and membrane		14	5	9
4. Other and unknown	14		7	7
TOTALS	74		40	34

[a] Proteins were assigned to one category only, that judged most characteristic of the function. Thus, certain proteins that are found in the tegument are listed under "Control and modulation."

[b] Two proteins are listed as in the core set, but they are not enzymatically active in the betaherpesvirus lineage versions; see text.

as "control" only on relaxed criteria—they specify a protein kinase of unknown function and a post-transcriptional mRNA-processing protein. Genes for capsid assembly and capsid structural proteins, unsurprisingly, belong to the core set, but many of those for tegument proteins and virion surface proteins do not. The numbers for the set of tegument proteins lack precision because this group remains incompletely defined experimentally and because certain tegument proteins were pre-emptively assigned to the "control" class. Of the universal surface proteins, three (glycoprotein B and the glycoprotein H:glycoprotein L complex) are known to be essential in HSV-1, and are involved in cell entry and cell to cell spread (reviewed by Spear, 1993), while the remaining two (the glycoprotein M:glycoprotein N complex) are non-essential in HSV-1 and are considered to act in trafficking of other membrane proteins (MacLean *et al.*, 1991; Pyles *et al.*, 1992; Crump *et al.*, 2004). In summary, most of the core gene complement of herpesviruses is concerned with specifying the icosahedral capsid structure, with synthesizing and packaging progeny DNA, and with cell entry and exit mechanisms, while genes supplying all well-characterized control functions and also those for most virion tegument and surface components are specific to subfamilies or to lineages within subfamilies.

Prominent among non-core gene systems are those that effect some form of interaction with the host. In this category we include: interactions with the host's immune system; aspects bearing on the establishment, maintenance, and reactivation from latent states; and lastly oncogenesis. Regarding the first of these classes, virus and host can be seen as engaged in an ongoing arms race, with the host attempting to suppress the invader and the virus to elude destruction. Lineage-specific genes for evasion or modulation of host immune responses to the infecting virus are widespread, and probably occur in all herpesviruses; their mechanisms of action cover a great range of immune components. HHV-8 is quoted as an instance in which such viral capabilities are notably extensive: 22 of the virus's 86 genes are considered to have the potential for activities in immune modulation (Rezaee *et al.*, 2006).

For the second class of virus–host interaction, persistence in the host lifelong is unvaryingly stated to be a primary attribute of the *Herpesviridae*, but no such universality prevails in the underlying mechanisms. In the case of neuronal alphaherpesvirus latency, the specific genetic basis remains quite elusive (reviewed by Efstathiou and Preston, 2005). In HSV-1 latency in neurons, the only transcribed portion of the genome is of a locus in the R_L region, but after two decades of extensive study the roles played by the latency associated transcript (LAT), an untranslated, stable species, remain tentative. The most developed analyses of a latent condition are those of EBV's existence in B lymphocytes (reviewed by Kieff and Rickinson, 2001); persistent states of a virus genome in non-dividing neurons and in dividing lymphocytes predicate quite different mechanisms, and EBV possesses a suite of genes specifically for deployment in latency. These encode six nuclear proteins (EBNA1, EBNA2, EBNA3A, EBNA3B, EBNA3C, and EBNALP), three membrane proteins (LMP1, LMP2A, and LMP2B) and two small RNA species (EBER1 and EBER2). The EBNA genes are scattered across about 100 kbp of the EBV genome, and are all expressed by way of a huge transcription unit that traverses most of the genome, with differential splicing to give an assortment of mRNA species from which the individual proteins are translated. The LMP1 and LMP2 proteins are encoded in separate, highly spliced transcription units; LMP2A and 2B proteins are the closely similar results of differential splicing of one gene, notable in that it runs across the genomic termini and so is only assembled intracellularly by circularization of the genome. Present understanding is that the most important species for transforming cells to indefinite growth are EBNA2 and EBNA3C (both modulators of gene expression) and LMP1 (membrane signaling). EBNA1 acts on a latent origin of DNA replication (distinct from the lytic cycle origins) to enable maintenance replication of the virus genome by host cell machinery. The latent genes thus comprise an elaborate superposed subsystem, which has evidently evolved solely within the *Lymphocryptovirus* lineage.

On the last class of interaction listed above, the subject of possible oncogenicity of herpesviruses has a long history, but at present the only mammalian herpesviruses usually regarded as tumorigenic are in the *Gammaherpesvirinae*; the two human viruses in this subfamily, EBV and HHV-8, are both associated with human malignancies. Their oncogenic propensities are in effect incidental consequences of arrangements for latent existence. Separately, members of the alphaherpesvirus *Mardivirus* genus are well known as oncogenic in birds, and reptilian herpesviruses are also associated with tumors (for instance, green turtle herpesvirus (GTHV); Greenblatt *et al.*, 2005).

In addition to mRNAs from protein coding genes, the mammalian herpesviruses also express other RNA species that do not encode proteins; none of these are of core status. These must definitely be treated as parts of the gene complement of the viruses, but not much can be said of functions at present. They form a heterogeneous collection which includes, among others: the LATs of HSV-1 and HSV-2 (Efstathiou and Preston, 2005); the EBER RNA polymerase III transcripts of EBV seen in the latent state (Rosa *et al.*, 1981); RNA polymerase III transcripts of MHV-4, of tRNA-like sequences (Bowden *et al.*, 1997); and proposed functional microRNA species from several viruses (for instance, Pfeffer *et al.*, 2004, 2005).

PHYLOGENY OF HERPESVIRUSES

The Phylogenetic Tree of the *Herpesviridae*

Analyses of phylogenetic relationships among members of the *Herpesviridae* have been quite extensively pursued using genomic sequence data; this text presents primarily the contributions of our own group. We have employed substitutional differences in amino acid sequence alignments for members of the core set of genes, to give quantitative estimates of relationships across the *Herpesviridae* through building phylogenetic trees. The high levels of divergence, both of substitutions and addition/deletion changes, constitute an important limitation, which focused our analyses on using encoded amino acid sequences in preference to DNA sequences, and to concentrating on eight genes that were judged most suitable in terms of the quality of alignment achievable for a set of homologous sequences. Our tree building work initially used the straightforward neighbor-joining method and latterly more sophisticated and computationally demanding modeling by maximum-likelihood and Bayesian approaches (McGeoch and Cook, 1994; McGeoch *et al.*, 1995, 2000, 2005, 2006; McGeoch, 2001; McGeoch and Gatherer, 2005). The version of the *Herpesviridae* tree shown in Figure 20.2 is based on six genes (of the eight used in most published work) from 43 viruses, and was chosen to maximize jointly numbers of input genes and of virus species. The position of the root was estimated as the midpoint between the mean of branch tips in the *Alphaherpesvirinae* and those in the *Beta-* and *Gammaherpesvirinae*. The tree is robust and is mostly well resolved, although there are two loci whose branching patterns have been refractory to complete definition. Consistent trees can be obtained with fewer input genes, and data from such trees can with appropriate caution be interpolated into the primary tree to give a version that has expanded membership but is still robust. Other aspects of genome sequences than alignments of core gene amino acid sequences can be used to examine phylogeny: relationships among herpesviruses have been examined on the basis of differing arrangements among conserved genomic blocks (Hannenhalli *et al.*, 1995), and by analysing differences in total sets of genes (Montague and Hutchison, 2000). Results from these approaches are overall compatible with the tree in Figure 20.2 but of lower resolution.

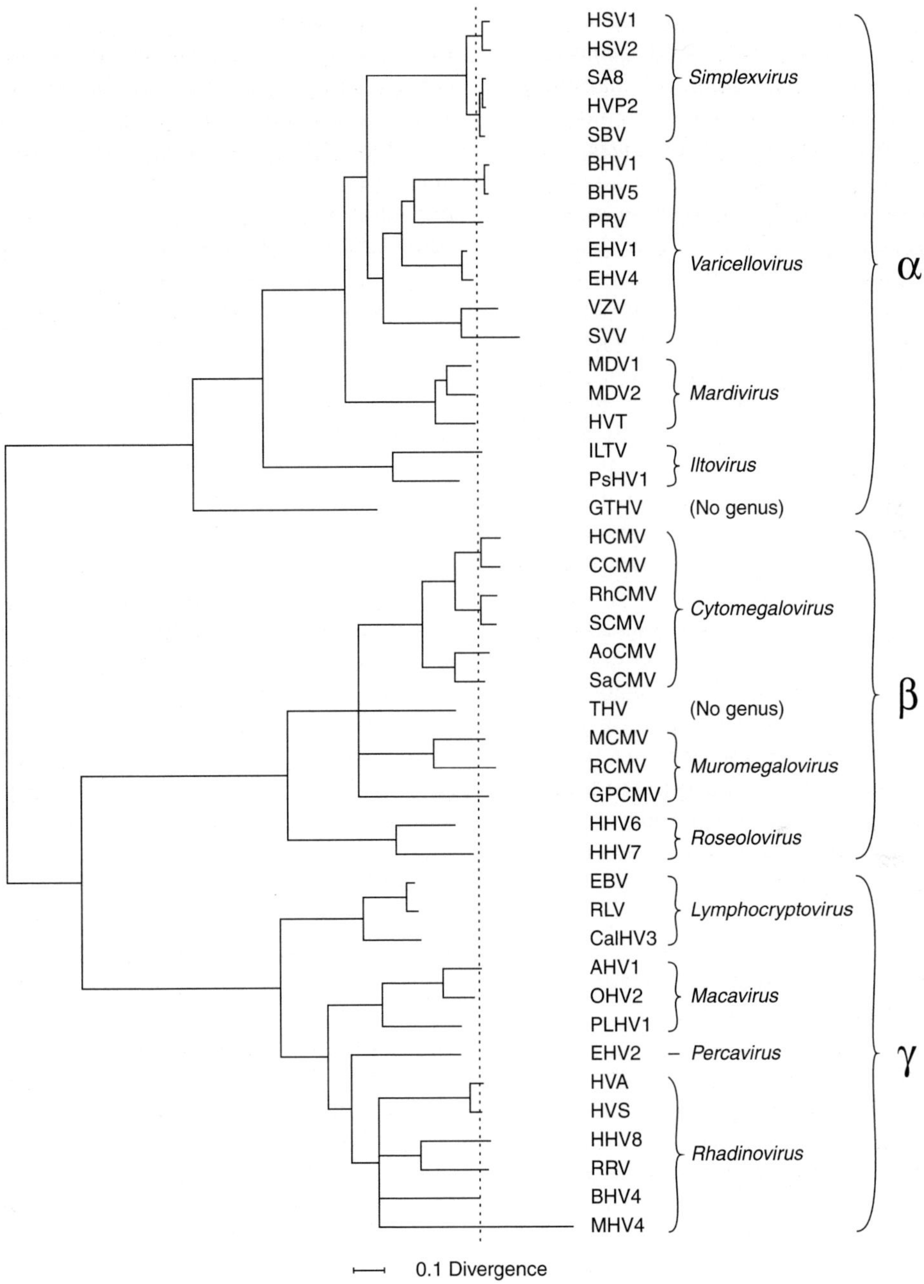

FIGURE 20.2 Forty-three-species phylogenetic tree for the *Herpesviridae*. The tree was based on amino acid sequence alignments for six genes from 43 species; genes used were the orthologues of HSV-1 genes UL15, UL19, UL27, UL28, UL29, and UL30. The tree was constructed by Bayesian Markov chain and maximum likelihood methods, as described by McGeoch *et al.* (2006). The mean position of branch tips is marked by a dashed line. The divergence scale at the foot indicates substitutions per site. Abbreviations for virus names are given in Table 20.1, with addition of GTHV (green turtle herpesvirus), SaCMV (saimiriine cytomegalovirus), GPCMV (guinea pig cytomegalovirus), and PLHV1 (porcine lymphocryptovirus 1). Genera are indicated for all species except GTHV and THV. Assignments to the *Alpha-*, *Beta-*, and *Gammaherpesvirinae* are indicated by the respective Greek letters.

Figure 20.2 demonstrates that the three subfamilies, the *Alpha-*, *Beta-*, and *Gammaherpesvirinae*, represent three primary, unambiguously resolved lineages in the *Herpesviridae*. As presented, the alphaherpesvirus lineage was the earliest to become distinct, but given the limitations of the midpoint rooting method this interpretation must be treated with a little caution. Forty-one of the virus species in the tree belong to defined genera as shown, with GTHV and THV not presently assigned to genera. Mammalian herpesviruses are spread across all three subfamilies, while reptilian and avian herpesviruses appear only in the *Alphaherpesvirinae*, with one reptilian virus lineage and two avian virus lineages. The reptilian and avian lineages all originate from points deep in the tree, outside the clade of mammalian alphaherpesviruses. From examination of the positions of branch tips across the tree, it is evident that in general (and under the assumption of midpoint rooting) lineages have experienced similar secular rates of amino acid sequence change in the gene set utilized for deriving the tree; striking outliers are the low rates apparent for lineages of the reptilian GTHV and the lymphocryptoviruses (LCVs), and the high rate for the MHV-4 line.

Co-speciation of Herpesvirus and Host Lineages

It has become clear that many parts of the branching patterns within trees for herpesvirus subfamilies are congruent with branching patterns for corresponding lineages of mammalian host species. This observation was initially made for the *Simplexvirus* and *Varicellovirus* genera, using glycoprotein B sequences (McGeoch and Cook, 1994), and has since been extended to other parts of the tree. Such correspondence between virus and host lines of descent points strongly to these virus lineages having arisen through long-term association with the host lineage to give co-speciation of virus and host. In turn, if this hypothesis is valid, then dates from vertebrate paleontology for divergences of host lineages can be transferred to events in the herpesvirus tree, so that a timeframe and rates of sequence change can be estimated for this virus family. With accumulation of sequence data for large numbers of herpesviruses, various exceptions and complications to co-speciational interpretation have become visible. Overall, we consider the hypothesis of co-evolution of lineages of mammalian hosts and their herpesviruses to be widely applicable, but by no means does co-speciation represent the sole mode by which mammalian herpesvirus species have arisen. In the following paragraphs we take a look at some features in current trees for each of the subfamilies. Figure 20.3 presents aspects of vertebrate phylogenetic trees for comparison with the herpesvirus trees to be discussed.

A tree was constructed based on glycoprotein B sequences for 25 mammalian alphaherpesvirus species (Figure 20.4A). Branching patterns in the *Varicellovirus* clade show extensive correspondences with the host tree: in the overall arrangement for lineages of herpesviruses of Primates, Artiodactyla, Perissodactyla and Carnivora; within the artiodactyl herpesvirus clade, in separation of the Suidae virus PRV from herpesviruses of Bovidae and Cervidae; and within the carnivore herpesvirus clade, with the arrangement of feline, canine, and seal virus lineages. Within the close grouping of viruses from bovines, goats, and deer, co-speciational ordering is not universal. In the *Simplexvirus* clade, there is overall co-speciational correspondence for herpesvirus lines of New World monkeys, Old World monkeys, and Hominidae. We leave unresolved the nature of the relationship between the two primate virus lineages, in the *Simplexvirus* and *Varicellovirus* clades. The locations of bovine herpesvirus 2 and two marsupial herpesviruses within the *Simplexvirus* clade represent striking examples of non-cospeciational origins. Interestingly, the recently characterized chimpanzee alphaherpesvirus is closer to HSV-2 than to HSV-1 (Luebcke *et al.*, 2006). This is consistent with the previous estimate, from a co-speciational timeframe, that HSV-1 and HSV-2 diverged around 8.5 million years ago

(MA) (McGeoch *et al.*, 1995), i.e. before separation of the human and chimpanzee lineages at 6 MA. These two well-known human viruses, usually viewed as closely related, have thus had substantial separate histories with respect to the human evolutionary timeframe. The HSV-1 and HSV-2 pair represents a rather common phenomenon, of two or more similar but definitely distinct herpesvirus species occurring in the same host species. Other examples in the alphaherpesviruses include BHV-1 plus BHV-5, and EHV-1 plus EHV-4, and instances can be seen in the other subfamilies. At least some of these can be rationalized as correlating with distinct host tissue tropisms.

Turning to the *Betaherpesvirinae*, Figure 20.4B shows a tree based on DNA polymerase sequences for 14 viruses in the subfamily. The primate viruses of genus *Cytomegalovirus* exhibit a straightforward correspondence with the host tree in Figure 20.3B. The three rodent CMVs (of mouse, rat, and guinea pig) also look to have a co-speciational arrangement, although the tree's resolution is insufficient to ascertain whether guinea pig cytomegalovirus (GPCMV) truly forms a clade with MCMV, and RCMV. Next, branching patterns for the primate CMVs, the rodent CMVs, and THV (whose host belongs to the order Scandentia) are consistent with a co-speciational grouping with hosts in the Euarchontoglires; to which can be added porcine CMV representing the Laurasiatheria and elephant herpesvirus 1 for the Afrotheria. HHV-6 and HHV-7 have no place in this grand synthesis; we speculate that they could have arisen by transfer into the primate lineage of an unknown virus or viruses from an ungulate or carnivore host line.

The *Gammaherpesvirinae* have been the most refractory of the subfamilies for both resolving and interpreting the phylogenetic tree (McGeoch and Gatherer, 2005), and recent work on characterization of novel gammaherpesviruses has increased complexities yet further (Ehlers *et al.*, 2007; Wibbelt *et al.*, 2007). Figure 20.4C shows a summary tree of currently defined major lineages in the *Gammaherpesvirinae*, constructed from partial glycoprotein B and DNA polymerase sequences (B. Ehlers, G. Dural, N. Yasmum, T. Lembo, B. deThoisy, M.-P. Ryser-Degiorgis, R.G. Ulrich and D.J. McGeoch, unpublished). Of the 11 major lineages depicted, five contain viruses whose host species come from two or more distant taxonomic orders, so that widespread transfer of viruses between host species must have occurred. In parts of the tree, definite indications of co-speciational behavior can be seen, notably in the *Macavirus* and *Lymphocryptovirus* clades, but no overall consistent, higher level framework of co-speciational history can be discerned.

We consider that for purposes of applying a timeframe to the herpesvirus tree, the most reliable instances of co-speciational divergence are those intermediate in depth, of herpesviruses within one host order or between sister orders, in preference either to events very deep in the tree or to recent divergences of herpesviruses within a single host family. The latter exclusion is also influenced by the fact that cross-species infections between related species of hosts are known to occur at readily observable levels, for instance with humans and other primates or between captive animals. In our recent estimates we have favored using separation of the ruminant and suid lineages, and separation of artiodactyl and perissodactyl lineages, to best effect in dating the alphaherpesvirus tree. Taking these two dates as 63.8 and 82.1 MA respectively (Springer *et al.*, 2003), we estimated the base of the herpesvirus tree to be equivalent to about 400 MA (McGeoch and Gatherer, 2005). Any such extrapolation to the distant past must be treated with considerable caution, and indeed we note that the first such estimate by our group in 1995 yielded a date of near 200 MA for the base of the herpesvirus tree (McGeoch *et al.*, 1995); doubling to the current figure has resulted from application of larger datasets, improved phylogenetic modeling and revisions in paleontological dating. Nonetheless, we consider our overall view of evolution of the *Herpesviridae* to account very well for many aspects of the phylogenetic tree, and we have good confidence in the general validity of the derived timeframe. The family

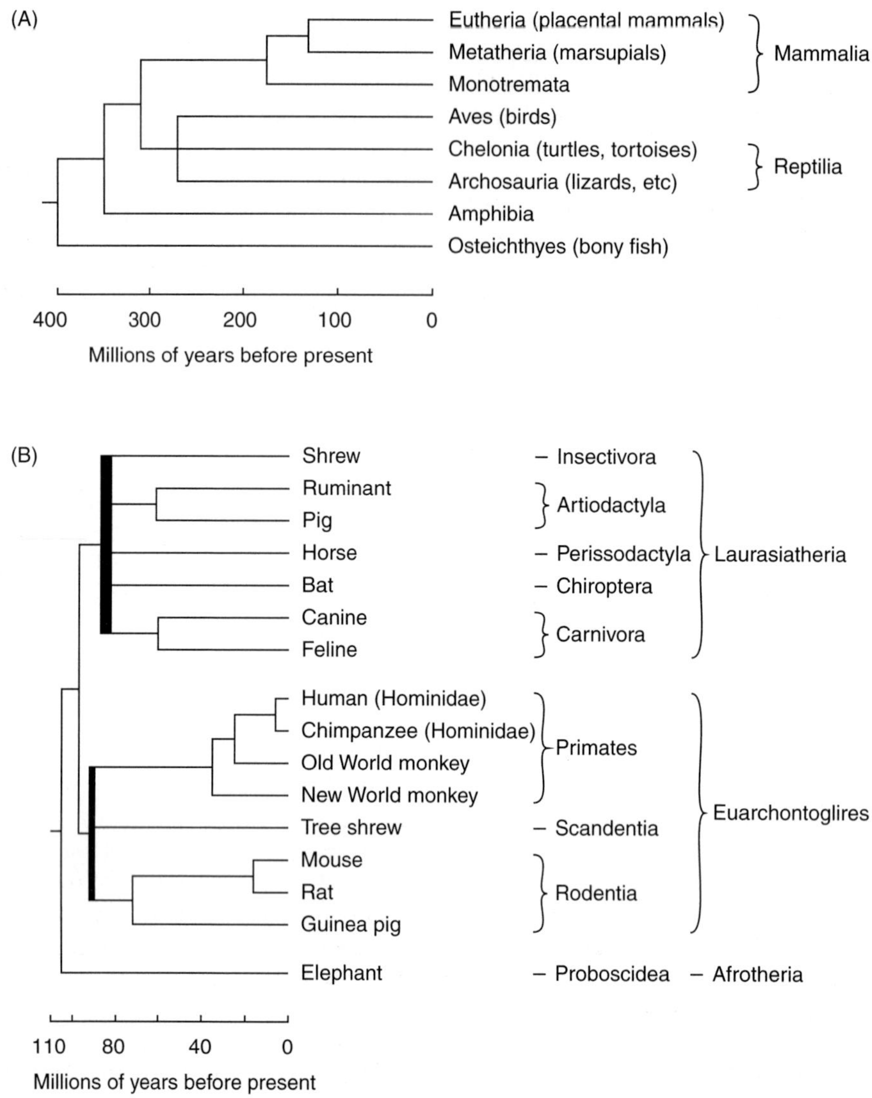

FIGURE 20.3 Phylogenetic trees for vertebrate lineages. Panels A and B sketch phylogenetic trees for the Vertebrata and for the placental mammals respectively, emphasizing lineages to which species of studied herpesviruses belong, with time-scales. Loci of unresolved or contentious branching order are shown as multifurcated. The trees are based mainly on Springer *et al.* (2003), Benton and Donoghue (2007), Murphy *et al.* (2007), and Nikolaev *et al.* (2007).

Herpesviridae, we think, is of ancient origin, far preceding the advent of mammals.

Considerations of origins of lineages have so far not treated the avian and reptilian alphaherpesviruses. While sequence data on reptilian herpesviruses are limited, it seems that several other viruses of both chelonians and lizards also belong to the lineage defined by GTHV in the alphaherpesvirus tree (Figure 20.2), and thus that it represents a major grouping of reptilian herpesviruses (Herbst *et al.*, 2004; Wellehan *et al.*, 2004). From a molecular clock version of the tree in Figure 20.2, we estimate the base of the GTHV lineage to be 239 MA, and those of the *Iltovirus* and *Mardivirus* lineages to be 202 and

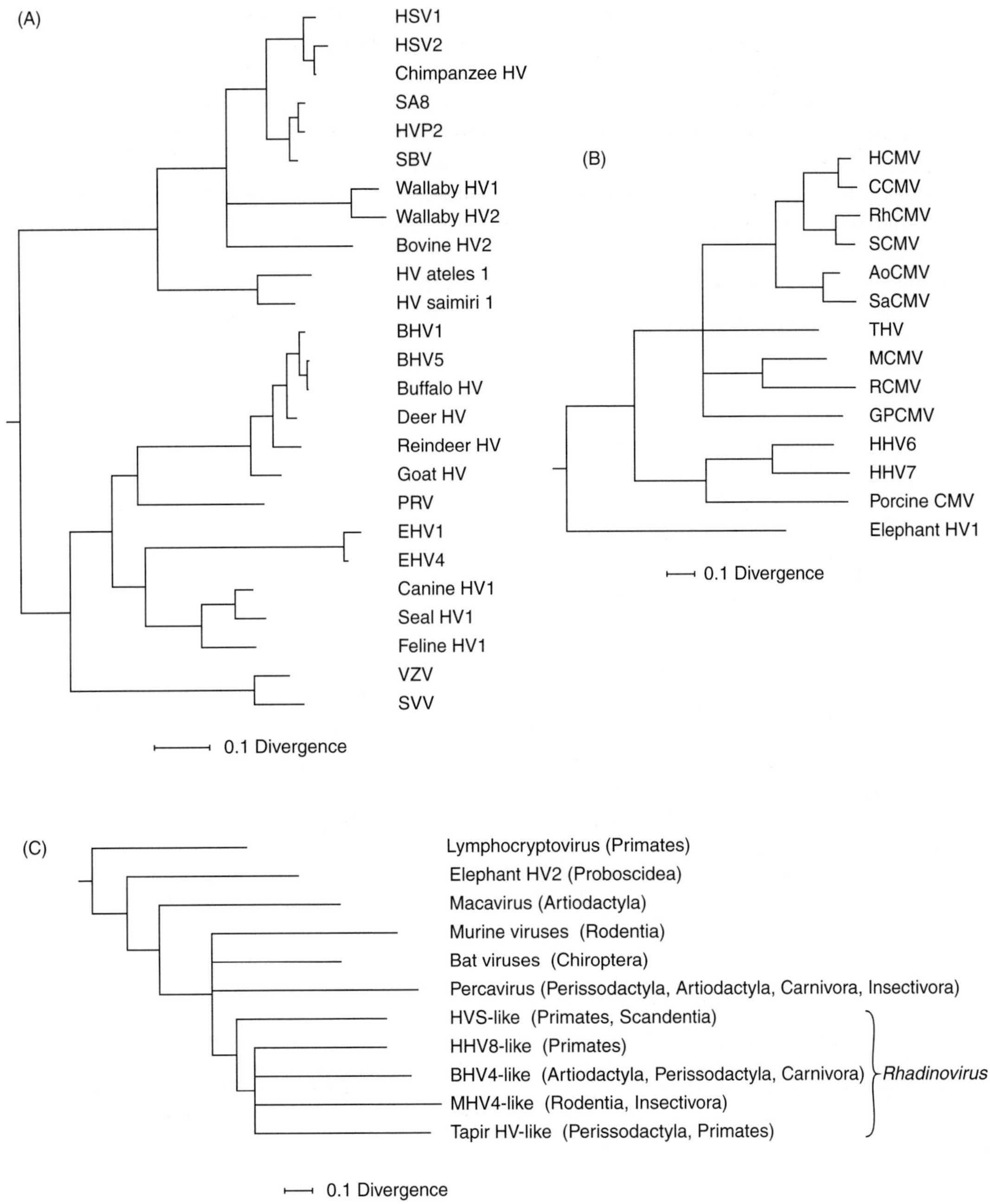

FIGURE 20.4 Phylogenetic trees for subfamilies of the *Herpesviridae*. Panel A shows a tree based on a 738-residue alignment of glycoprotein B sequences for 25 alphaherpesvirus species. The tree in Panel B is based on a 929-residue alignment of DNA polymerase sequences for 14 betaherpesvirus species. Panel C shows a tree for 45 gammaherpesvirus species, based on alignments of partial sequences for glycoprotein B and DNA polymerase, of concatenated length 949 residues; in panel C individual virus species are not shown but instead 11 major lineages (approximately equivalent to genera), with taxonomic orders of the host species listed for each. Trees were constructed by Bayesian Markov chain and maximum likelihood methods. Each was rooted by reference to trees containing additional herpesvirus species as outgroups. Virus names are abbreviated (as in Table 20.1 and Figure 20.2), except for species that do not feature elsewhere. HV, herpesvirus; CMV, cytomegalovirus. The divergence scales at the foot indicate substitutions per site, and are specific to each panel.

132 MA respectively. As can be seen by comparisons of lineages in Figures 20.2 and 20.3A, the lines of descent of reptilian, avian and mammalian herpesviruses are not congruent with those of the host classes Reptilia, Aves, and Mammalia. We can discern two, alternative scenarios to rationalize deep features in the alphaherpesvirus tree with those of the hosts (McGeoch and Gatherer, 2005). The first is that the base of the GTHV lineage corresponds to the divergence at 310 MA of diapsid and synapsid reptiles (from the latter of which mammals eventually evolved); in this case, the origin of mammalian alphaherpesviruses could be co-speciational but that of the two avian herpesvirus lineages could not. The second scenario is that the base of the GTHV lineage corresponds to radiations at 270–285 MA in the diapsid reptile lineage which gave rise to separate lines leading to chelonians, birds, and lizards; in this case, the avian herpesvirus lineages could have arisen co-speciationally but the mammalian alphaherpesvirus lineage could not. Both schemes emphasize the major involvement of reptilian hosts early in the development of the alphaherpesviruses. By extension, such reptilian involvement might possibly also have occurred in the beta-, and gammaherpesviruses; in principle there may exist as yet undetected beta-, and gammaherpesviruses in reptiles or birds.

EVOLUTION OF HERPESVIRUS GENE SYSTEMS

Origins of Herpesvirus Genes

Among the 43 genes inferred to have been present in the genome of the common ancestor of present-day mammalian herpesviruses, nine are clearly related by similarities in encoded amino sequences to cellular genes, to the extent that they and the cellular genes must have common evolutionary origins (see the list in Table 20.4). In addition, another gene, encoding the alkaline exonuclease (HSV-1 UL12), has been reported to be distantly related to non-herpesvirus nucleases (Bujnicki and Rychlewski, 2001); the similarities seen are at a much lower level and their import less clear than for the nine genes in Table 20.4, and we put aside the case of the nuclease for the remainder of this discussion. All the genes listed in Table 20.4 encode enzymes, of which all but one are involved in DNA metabolism and replication. We hold as almost axiomatic that these genes must have been acquired from some other genetic element, with the primary source a cellular genome. In contrast, for none of the proteins assigned as capsid, tegument, or virion surface components is there any substantive indication of relatedness with a cellular protein. Gene capture remains, nonetheless, an obviously attractive origin for members of this class of herpesvirus genes. In all of the completely sequenced herpesvirus genomes there are among non-core genes further examples that have clear non-herpesviral homologues. Some of these are restricted to a subfamily, some to a sublineage, some to a virus species; evidently gene capture has proceeded throughout herpesvirus evolution. The encoded products are diverse, including nucleotide anabolic enzymes, other enzymes, immune system proteins, transcriptional regulators, cell-state regulatory proteins, and a parvovirus DNA replication protein. All of these are taken to have been captured from

TABLE 20.4 ***Herpesviridae*** **ancestral genes with cellular homologues**

HSV Gene	Function of encoded protein
UL2	Uracil-DNA glycosylase
UL5	Component of DNA helicase
UL9	DNA helicase active at origin of DNA replication
UL13	Protein kinase
UL23	Deoxynucleoside kinase (thymidine kinase)
UL30	Catalytic subunit of DNA polymerase
UL39	Large subunit of ribonucleotide reductase
UL40	Small subunit of ribonucleotide reductase
UL50	Deoxyuridine triphosphatase

another genome; in most cases the host cell's is presumed, but a cogent exception is seen with the presence in HHV-6 of a gene closely related to the *rep* gene of adeno-associated virus 2 (AAV-2), a helper dependent species in the *Parvoviridae* (Thomson *et al.*, 1991, 1994). This gene was likely acquired from AAV-2 via the non-homologous recombination process by which its replicative form genome integrates into the host genome. Attempts to determine origins and dates for gene acquisitions by analyzing jointly the phylogeny of host and virus homologues have in some cases given clear results. For instance, it was demonstrated that the occurrence of interleukin 10 genes in certain gammaherpesviruses (EBV and a related baboon virus, and EHV-2) resulted from two independent capture events, both around 10 MA, from primate and equine host lineages to the corresponding virus genomes (McGeoch, 2001). The most recent gene transfer documented is that of a gene for β-1,6-*N*-acetylglucosaminyltransferase-mucin, acquired by a virus of the BHV-4 lineage from a progenitor of the African buffalo around 1.5 MA (Markine-Goriaynoff *et al.*, 2003).

In their forms in herpesvirus genomes, genes thought to have been gained from other genetic elements are generally devoid of introns, suggesting that a mechanism involving reverse transcription of mRNA species may have been commonly involved in the process of capture. Experimental studies have been made of interactions between herpesvirus and retrovirus genomes in mixed infections of the avian alphaherpesvirus MDV and a chicken retrovirus, and MDV isolates with retroviral sequences inserted into their genome were obtained; in most cases the stably inserted retrovirus sequences were single LTRs (long terminal repeats) (reviewed by Brunovskis and Kung, 1996). One serotype of MDV was found to contain sequences resembling LTRs, which may well represent the mutated remains of natural retroviral insertion events (Isfort *et al.*, 1992). In principle, then, retroviruses or other retrotransposing entities may have been involved in gene capture by herpesviruses or, by LTR insertion, in disrupting coding sequences or altering patterns of transcription. Such processes could thus have been of widespread significance in herpesvirus evolution, although direct evidence is likely to remain unobtainable.

Examples of multigene families are visible in all three herpesvirus subfamilies, and are presumed to have arisen through successive gene duplication events. These are most developed in the CMVs, and account for a substantial part of the larger size of CMV genomes relative to those of other herpesviruses (Chee *et al.*, 1990; Rawlinson *et al.*, 1996). For an essentially wild-type HCMV genome, Dolan *et al.* (2004) listed 13 gene families, with memberships ranging from 2 to 14 and accounting for 63 genes in all. Members of families occur both in tandem arrays and distributed across the genome. The encoded proteins are considered mostly to be membrane-translated species, that then are secreted or become membrane-associated. Some genes in families are seen to be highly variable among virus strains, typically with several diverged lineages. For instance, 14 distinct lineages of the HCMV UL146 gene have been found, and the encoded protein (a CXC chemokine) shows a level of diversity in the HCMV isolates comparable to that seen with the glycoprotein B gene across the entire *Herpesviridae* (Dolan *et al.*, 2004). Some of these families occur also in HHV-6 and HHV-7 (Gompels *et al.*, 1995; Nicholas, 1996; Megaw *et al.*, 1998), emphasizing that their presence has had very long-term stability, and thus that the members of the family must be presumed to have distinct functional significances. The precise natures of such distinct roles remain rather obscure, but are presumed to involve interactions with the host's immune system. One of the multigene families in betaherpesviruses has recently been shown to be derived from the deoxyuridine triphosphatase (dUTPase) gene (see below).

It is interesting to consider whether genes may have arisen *de novo* in herpesvirus evolution (McGeoch and Davison, 1995). Our position, considering the large size of the genomes and the extent and complexity of genomic changes during the evolution of the family, is that gene development in this mode might well have been not uncommon; the challenge, however, is to identify convincing candidates for such genes. One such class would be authenticated gene pairs with extensively overlapping reading frames, where we presume that one of the pair developed *in situ*, and some such cases occur in herpesvirus genomes. As an example of a gene whose coding region does not overlap another and that may have arisen *de novo* we discuss gene US12 of HSV-1 and HSV-2 (McGeoch *et al.*, 1985; Dolan *et al.*, 1998).

US12, expressed in the immediate early class, appears unique to the *Simplexvirus* genus, and is dispensable for virus growth. Its product is active in interdicting antigen presentation in infected cells, by binding to the TAP peptide transporter (Früh *et al.*, 1995; Hill *et al.*, 1995)—that is, it should provide a selectable advantage for the virus, but is nonetheless basically a "luxury" function. The encoded HSV-1 polypeptide is only 88 amino acids in length and the active part is contained in the *N*-terminal 35 residues (Galocha *et al.*, 1997). When purified and in solution it has no detectable stable secondary structure, so that it is more appropriately viewed as a TAP-binding oligopeptide than as a globular protein (P.N. Barlow, H.W.M. Moss, and D.J. McGeoch, unpublished). The US12 gene lies at one end of the U_S region, with its coding region overlying the transcription initiation sequences of the downstream US11 gene. US12 is transcribed from an immediate early promoter located in the immediately adjacent TR_S element (Figure 20.5), and the twin copy of this promoter in the IR_S element drives expression of gene US1 at the opposite extremity of U_S (Rixon and McGeoch, 1985). Since US1 encodes a substantial protein species (420 amino acid residues), it is attractive to regard the US1/US12 promoter sequence as having evolved primarily for expressing gene US1. In this view the association with the US12 region then occurred as a by-product, perhaps in a genomic rearrangement that moved the R_S/U_S boundaries (as discussed above), so that the resulting immediate early transcription of the undeveloped US12 locus then provided fertile conditions for genesis of a new function. These aspects, of lineage specificity, a function that is both non-essential and selectable, small size and lack of ordered structure of the product, and characteristics of the genomic location, together build a persuasive case for a history of generation of US12 *de novo*.

Herpesviral DNA Replication Systems

This section addresses aspects of the genome replication machinery of herpesviruses, which clearly constitutes an ancient set of functions in the virus family. The theme of our treatment

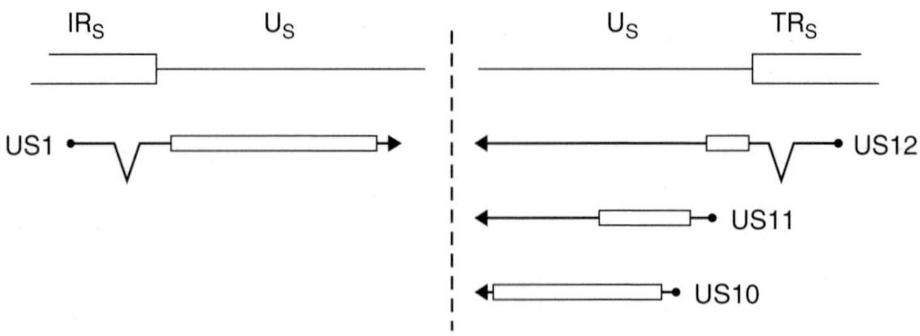

FIGURE 20.5 Genomic organization around the HSV-1 US12 gene. The layouts of genes at each end of the HSV-1 U_S region are shown. At the top of the figure, the left and right extremities of U_S are indicated by solid lines and the abutting parts of repeats IR_S and TR_S by open boxes. In the lower part of the figure, transcripts for genes US1, US10, US11, and US12 are shown as heavy lines with coding regions as open boxes; 5′ termini of mRNAs are marked by solid circles and 3′ termini by arrowheads. After McGeoch *et al.* (1985) and Rixon and McGeoch (1985).

concerns how such a central system has been modulated and changed during the evolutionary development of the virus subfamilies. Across the family of mammalian herpesviruses, certain aspects of arrangements for replication of the genome are universal and others are variable. All viruses have an equivalent set of six genes that encode the proteins needed for replicative elongation of DNA chains during lytic infection, and the set of genes for subsequent processing and packaging of nascent DNA appears to be similarly invariant (reviewed by Challberg, 1996). However, there are distinct ways in which initiation of DNA synthesis is organized, and there are differences between virus subfamilies in the manner in which supply of dNTP precursors is ensured. In addition, in at least one sublineage, the LCVs in the *Gammaherpesvirinae*, there is a separate DNA replication system for maintenance of genome copies when the virus is in a latent state in dividing cells (reviewed by Yates, 1996). In this condition the genome exists as a circular nuclear episome and its replication requires only one virus gene product, EBNA1, as mentioned above.

In the alphaherpesviruses, there is a gene (HSV-1 UL9) whose product is a DNA helicase that binds to origins of DNA replication, and in association with other proteins acts to initiate synthesis of progeny DNA (Challberg, 1996; Boehmer and Lehman, 1997). The origins of replication (three in the HSV-1 genome) are short, around 140 bp, with dyad symmetry and a central A+T rich segment. An equivalent situation exists in betaherpesvirus species HHV-6 and HHV-7, with a UL9 homologue and a similar origin structure (Gompels *et al.*, 1995; Nicholas, 1996). The UL9 homologues occur in an equivalent location in alphaherpesvirus and in HHV-6/HHV-7 genomes, in terms of the identities and relative orientations of neighboring genes. In betaherpesvirus CMVs and in gammaherpesviruses, however, there is no UL9 homologue, and the mechanism of DNA chain initiation is more complex. In the gammaherpesvirus LCV and betaherpesvirus CMV lineages, origins of lytic replication are larger, bipartite entities containing complex sets of repeats (Challberg, 1996; Yates, 1996). Initiation of replication involves both a virus encoded transcriptional regulator acting on one of the segments of the origin and a host protein or proteins interacting with the other. The mechanism involving UL9 is likely to represent the ancestral state, with adoption of a new mode of chain initiation occurring independently in branches leading to the LCVs and CMVs. We thus regard UL9 as an ancestral gene, one of the three not in the core set. The significance of this mechanistic change remains uninvestigated: we suggest that the more elaborate mechanism may allow superior control over the onset of DNA synthesis, by criteria of whether both virus and cellular gene expression are in appropriate states.

Alpha- and gammaherpesviruses encode three enzymes active in pathways of synthesis of dNTPs for DNA replication, namely thymidine kinase (TK), dUTPase, and ribonucleotide reductase (RR). The last of these has two subunits (RR1 and RR2) encoded by separate viral genes. Some viruses encode additional enzymes, namely thymidylate synthase (VZV and certain gammaherpesviruses) and dihydrofolate reductase (certain gammaherpesviruses) (Davison and Scott, 1986; Albrecht *et al.*, 1992; Russo *et al.*, 1996). The standard rationale for occurrence of these herpesvirus genes is that in some aspect of its life cycle the virus requires to replicate DNA in a cellular environment that has inadequate pools of DNA precursors, and that direct supply of nucleotide synthesizing capability alleviates this situation. In contrast, viruses of the betaherpesvirus subfamily do not have genes for TK and RR2, and parsimony criteria indicate that these genes are ancestral but were lost in the lineage leading to the betaherpesviruses. Additionally, inspection of the putative RR1 and dUTPase sequences of betaherpesviruses indicates that these proteins are actually unlikely to possess the relevant enzymatic activity, since they both lack sequence features otherwise generally conserved.

This argument is most clear with dUTPase: the betaherpesvirus candidates lack five local conserved motifs that are directly involved in the dUTPase active site (Cedergren-Zeppezauer *et al.*, 1992; McGeehan *et al.*, 2001); their assignment as dUTPase homologues is based on the presence of an additional motif that is unique to herpesvirus dUTPases (and is not considered to be involved in the enzymatic mechanism) and on the genomic location of the gene; and, finally, the HCMV protein has been found to be enzymatically inactive (Caposio *et al.*, 2004). We therefore come to the conclusion that the strategy of the virus genome encoding enzymes of nucleotide synthesis was completely abandoned near the start of the betaherpesvirus lineage. This correlates neatly with observations that CMVs switch on systems of host DNA synthesis at an early stage in the infectious cycle (St Jeor *et al.*, 1974; Mocarski, 1996); presumably host enzymes then provide adequate dNTP synthesis capacity to support virus DNA replication. An intriguing aspect is that in betaherpesvirus genomes the reading frames for the RR1 and dUTPase equivalents have remained open through the extensive evolutionary changes that have occurred in the subfamily's existence, rather than being lost as with the TK and RR2 genes. This definitely suggests that each encoded protein continues to have some function, distinct from the known enzymatic activity. This account explains the dUTPase and RR1 listings in Table 20.3.

There are more elaborations in the tale of herpesvirus dUTPase's evolutionary adventures. Comparisons of amino acid sequences of alpha- and gammaherpesvirus dUTPases with those from non-herpesvirus sources indicated that the herpesvirus gene has experienced an intragenic duplication, which has resulted in a longer polypeptide chain with conserved motifs differentially preserved in one or the other of the two portions (McGeoch, 1990), and it is known that herpesvirus dUTPase is active as a single polypeptide chain whereas other dUTPases form active trimers (Cedergren-Zeppezauer *et al.*, 1992; Tarbouriech *et al.*, 2005). The development of the novel herpesvirus-specific sequence mentioned above is enabled by this radical reorganization of the protein, replacing a redundant active site component, and it is suggested that the herpesvirus-specific locus is associated with the unknown proposed second function (McGeehan *et al.*, 2001). In a further twist, Davison and Stow (2005) demonstrated that one of the betaherpesvirus multi-gene families is related to the "ex-dUTPase" gene, and also that there are two such genes in gammaherpesviruses. The gene that was captured by an ancestral virus of the *Herpesviridae*, presumably for its dUTPase activity, has evidently been utilized by herpesvirus evolution on several occasions subsequently as amenable material for building new functionalities.

FRONTIERS IN HERPESVIRUS GENOMIC EVOLUTION

This section presents some topics of particular current activity in herpesvirus genomics and evolution, with the unifying feature that all depend on access to sets of whole genome or single gene sequences for multiple isolates of one virus. A notable current feature in genomic analyses of herpesviruses is the burgeoning number of complete genome sequences for multiple strains of a single virus, with the formidable capabilities of modern DNA sequencing technologies being brought to bear (in such re-sequencing, knowing almost all of the answer in advance might also be helpful). There is thus a developing focus on in-depth analyses of single species genomic diversity. The most sequenced herpesvirus genome as of early 2007 is VZV, with 20 complete sequences available, followed by HCMV with sequences of seven strains published. As it happens, in comparisons of levels of diversity among genomes of human herpesviruses, VZV appears as much the quietest and HCMV the most variable and complicated.

Homologous recombination between distinguishable strains of herpes simplex virus cultured *in vitro* has been known for many years (Wildy, 1955; Brown *et al.*, 1973), and HSV-1/HSV-2 recombinants were described in the 1970s (Timbury and Subak-Sharpe,

1973). On the other hand, experimental demonstration of recombination with certain other herpesviruses has proved difficult; recombination between HCMV strains in cultured cells has been reported (Haberland *et al.*, 1999), but it is the present authors' experience with such experiments that genuine recombination events may occur at such a low level that they might scarcely be distinguishable from PCR artifacts in the detection assays used (Sevilla-Reyes, 2006). Measurable incidence of recombination between two strains of a herpesvirus in natural infection was widely discounted for many years, given that the two parental types would both have to gain access to the same cell, perhaps within the same short timeframe; and phenomena of superinfection resistance were also regarded as likely. However, recent DNA sequence studies of multiple natural strains of a single herpesvirus, either of several selected regions or of complete genomes, have shown unambiguously that crossover events are readily discernible involving strains of a herpesvirus where several sequences are available, by criteria of changing associations of polymorphisms and strains along a sequence alignment.

Viruses for which crossovers have been detected from sequence data include HSV-1 (Bowden *et al.*, 2004), VZV (Norberg *et al.*, 2006; Peters *et al.*, 2006), HCMV (Dolan *et al.*, 2004), EBV (Midgley *et al.*, 2000; McGeoch and Gatherer, 2007), and HHV-8 (Poole *et al.*, 1999). In the regions of the HCMV genome containing genes of high diversities (near the extremities of U_L), there is a high frequency of crossovers between genes, so that effectively each strain possesses an unique combination of alleles of variable genes (Dolan *et al.*, 2004; E.E. Sevilla-Reyes, unpublished). We note that Pagamjav *et al.* (2005) have recently described an apparently naturally occurring virus that is a recombinant between EHV-1 and EHV-4. In all of these systems, the "unlikely" circumstance of genomes from two strains of a virus colliding in the same cell in a host to yield recombinant progeny does actually occur. Regarding the frequency at which such events might take place, it must be sufficiently low relative to the rate of accumulation of polymorphisms for distinguishable strains to develop in the first instance, but high enough for crosses between such strains to appear in small samples of isolates. In analysis of HSV-1 in three separate areas of the genome, Bowden *et al.* (2004) found that the recombination rate was of the same order as the nucleotide substitution rate. Complementary to findings of natural recombinants are reports of the common appearance of multiple strains in clinical samples, for instance of HCMV (Pignatelli *et al.*, 2004). We register here (but do not develop) that modern capabilities in computer-based modeling can allow detailed analyses of virus populations, as was done by Bowden *et al.* (2006) in studying geographical correlates in a wide collection of sequences from loci in the genome of HSV-1.

Work from our group (Dolan *et al.*, 2006; McGeoch and Gatherer, 2007) has clarified the relationship between the two so-called types of EBV (EBV-1 and EBV-2), which has presented a puzzle of some years' standing. The types differ in their distribution: EBV-2 is common in Africa and New Guinea, less so in Caucasian populations (Zimber *et al.*, 1986). There is no clear distinction between the two types in pathology, but there are differences in biology of cell transformation—EBV-1 appears more efficient in transforming B cells, and transformed lines grow more vigorously (Rickinson *et al.*, 1987). The complete sequence for a type 2 strain and comparison with two available type 1 sequences has now shown that: first, the types characteristically differ solely in that they possess well diverged alleles of four latent cycle genes, EBNA2, EBNA3A, EBNA3B, and EBNA3C (as was previously known; Dambaugh *et al.*, 1984; Sample *et al.*, 1990); and second, that in the remainder of the genomes all three strains each display a distinct pattern of low diversity haplotype regions separated by crossovers, with the type 2 strain not distinguished from the other two in terms of divergence of haplotype regions and levels of crossovers (Dolan *et al.*, 2006; McGeoch and Gatherer, 2007). The following scenario is suggested to account for these findings: that at some ancient point two host populations became isolated, so that their EBV populations diverged, with the latent genes diverging most markedly; and that subsequently the populations regained contact and recombinational mixing of the two diverged EBV strains proceeded, to give the patterns now seen.

A currently active area in molecular evolution in general concerns detection of loci in genes that are under diversifying, or positive, selection, and ambitiously complex software has been developed to detect positively selected sites (e.g. Yang, 2007). The approach consists of evaluating synonymous (d_S) and non-synonymous (d_N) rates of substitution in an aligned set of coding sequences showing some diversity, with positive selection having a signature of d_N greater than d_S. With availability of sequences for multiple strains of many herpesvirus genes and genomes, we have searched across the family for examples of positive selection. Cogent examples of this evolutionary mode turn out to be quite rare in herpesvirus systems. In some cases, isolated codons or groups of codons register as positively selected; these may correlate with mutation to avoid selection by some particular host MHC alleles (Midgley *et al.*, 2003; Burrows *et al.*, 2004) or often may be of limited apparent interpretive value. We have, however, found that two latent cycle genes of EBV, encoding the LMP1 and EBNA1 proteins, show strong signs of widespread positive selection, visible simply in pairwise plots of overall d_S versus d_N. The latent cycle K1 gene of HHV-8 gives similar results. With LMP1 the sites responsible are distributed across the whole coding sequence (alignment length of 323 codons), and with EBNA1 across the coding sequence of the DNA-binding domain (203 codons) for which multiple sequences are available. Figure 20.6 illustrates these results: panel A shows data for a large set of HSV-1 TK gene sequences (376 codons) as a control; and panel B shows data for a major clade (one of three) of EBV LMP1 gene sequences. The picture seen with the TK gene is typical of the great majority of genes: d_S is greater than d_N in almost all pairwise comparisons, with the two classes of mutation accumulating in a definite proportion under some stochastic noise. With the LMP1 gene, however, d_N is greater than d_S for most pairwise relationships; this analysis can be taken further in favorable cases by deriving a tree for

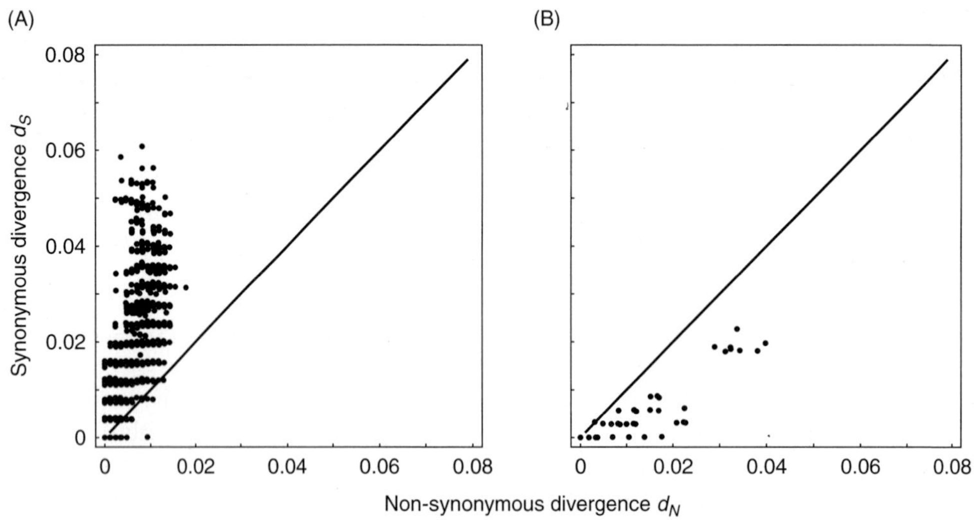

FIGURE 20.6 Divergence characteristics of coding regions in the HSV-1 TK gene and EBV LMP1 gene. Panels A and B show plots of pairwise synonymous (d_S) versus non-synonymous (d_N) divergence rates for alignments of coding sequences for isolates of the HSV-1 TK gene (72 sequences) and for one of three major clades in the LMP1 gene (10 sequences), respectively. The solid line in each panel is the locus of d_S=d_N. Divergences were calculated by the maximum likelihood method of Yang (2007).

the variants and examining changes from the most recent ancestor directly (not shown here). We take these results to indicate that EBV genes which are expressed over a prolonged period in the latent state can have an uniquely high exposure to surveillance by the immune system, so that variants which evade MHC restriction are selected by enabling survival of the latently infected cell.

EVOLUTIONARY RELATIONSHIPS OF THE ORDER *HERPESVIRALES*

We turn to considering the nature of evolutionary relationships among the three families of the *Herpesvirales*, and their possible deeper antecedents. Details on taxonomy and sequenced genomes for the *Alloherpesviridae* and *Malacoherpesviridae* are listed in Table 20.5. In the *Alloherpesviridae*, six complete genome sequences are known, representing four virus species—two with an amphibian host (Davison *et al.*, 2006) and two with teleost fish hosts, channel catfish and koi (the latter are ornamental varieties of carp) (Davison, 1992; Aoki *et al.*, 2007). Partial genomic data are available for other fish herpesviruses of carp and salmon (Bernard and Mercier, 1993; Davison, 1996; Waltzek *et al.*, 2005). The genomes and their contents present a parallel universe to that of the *Herpesviridae*: many similar features but little suggestion of any common ancestry. Similarities include: the general appearance of densely arranged coding sequences; core and non-core gene sets; multigene families; genes evidently captured from host genomes; and genome blocks rearranged between genomes. The smallest of the sequenced genomes is that of channel catfish virus (CCV) (135 kbp) and the largest koi herpesvirus, which at 295 kbp is the largest known in any family of the *Herpesvirales*. Base compositions all fall in the range 53–59% G + C. Comparisons of the gene complements demonstrate clear descent from a common ancestor, but none of the sequenced viruses are closely related; indeed they form a widely diverged group, perhaps of greater diversity than that described above for the *Herpesviridae*. Formal phylogenetic investigation of the family has not as yet been reported. Both frog viruses exhibit a genomic peculiarity not seen in any other member of the *Herpesvirales*: their DNAs are extensively methylated, apparently as 5-methylcytosine; and both encode a homologue of DNA (cytosine-5-)-methyltransferase, presumed to be responsible for the genomic methylation (Davison *et al.*, 2006). These genomes do not evince any marked deficit of 5′-CG content (see above).

In the *Malacoherpesviridae*, the genome of OsHV1 also exhibits familiar themes—densely arrayed coding sequences, multigene families and captured genes—but is distinct from

TABLE 20.5 Sequenced members of the *Alloherpesviridae* and *Malacoherpesviridae*

	Virus[a]	Name abbreviation	Genome size (kbp)
Family *Alloherpesviridae*			
Ictalurivirus	Channel catfish virus	CCV	134
(no genus)	Koi herpesvirus	KHV	295 (3)
(no genus)	Lucké tumor herpesvirus	LTHV	221
(no genus)	Frog virus 4	FV4	232
Family *Malacoherpesviridae*			
(no genus)	Ostreid herpesvirus 1	OsHV1	207

[a] Annotations are as for Table 20.1.

members of both other families (Davison *et al.*, 2005b). Comparisons among the three herpesvirus families of encoded amino acid sequences of gene sets have revealed only one instance of inter-family sequence similarity that could be regarded as characteristic of herpesviruses, namely in the putative ATP-utilizing terminase subunit that is known in experimentally studied members of the *Herpesviridae* to be involved in packaging of nascent DNA molecules into capsids. Some other cases of similar sequences exist, mostly for enzymes of nucleotide metabolism, but they are not specific to herpesviruses. For instance, DNA polymerase of CCV was assigned by Knopf (1998) to a subset of DNA polymerases distinct from the group related to cellular δ polymerase to which DNA polymerases of *Herpesvirales* members belong; and the dUTPases of fish, amphibian, and oyster herpesviruses are not the long-chain type described above as characteristic of the alpha- and gammaherpesvirus subfamilies, but belong to the otherwise universal short-chain class.

Thus, if the three families do indeed possess some element of common evolutionary antecedent, it is too distant to study by criteria of genomics. In this circumstance it is reasonable to re-visit the original classification criterion, of virion structure. Capsids of species from all three families have been studied by cryoelectron microscopy and computer-based three-dimensional reconstruction to a resolution (in the worst case) of 3 nm (Booy *et al.*, 1996; Davison *et al.*, 2005b), giving a much superior view to negative-stain images. The capsids from the three families are seen to be closely similar overall and in structural details, although not identical; judgment remains that they should be regarded as evolutionarily related (Davison *et al.*, 2005b). Given the distinct phylogenetic association of CCV DNA polymerase and the lack of detectable counterparts of other proteins comprising the DNA replicative machinery of mammalian herpesviruses, it is possible that genes for DNA replication were acquired independently in the three families, either by way of introduction of a new capability to the virus or by substitution. This interpretation then raises the possibility that the last common ancestor of herpesviruses in the three families might have been a simpler entity, that possessed genes encoding the capsid but lacked some other parts of the genetic capabilities of today's herpesviruses.

There is another possible ancient evolutionary connection of herpesviruses, with tailed DNA bacteriophages. A number of similarities between aspects of capsid assembly and structure in the *Herpesviridae* and in the phage order *Caudovirales* have emerged over the years. These include: common amino acid sequence motifs in ATP-utilizing terminase subunits; overall similarity of portal complexes for packaging DNA into capsids (Trus *et al.*, 2004); and detailed structural similarity between protein domains, in the floors of capsids (Baker *et al.*, 2005). The herpesvirus and phage capsids may thus have a common element in their ancestry. We regard this as an important extension for considering the evolutionary origins of herpesviruses. Nonetheless, it is important to recognize a potential fallacy, or at least a narrowness of vision, concerning this long view: given that the assignments as herpesviruses were defined by possession of a characteristic particle, it ought not to be surprising that at the greatest distances, aspects of this structure should be all that is visible. Other lineages of genes than those specifying the capsid are also essential and substantial components of modern herpesviruses' genomic makeup, but our account has not been aimed specifically at tracing their antecedents.

CONCLUSIONS

The main concern of this chapter has been with aspects of the macroevolution of the *Herpesviridae*. By comparing genomes, gene sequences, and gene sets we have been able to outline various processes that have taken place in the evolution of these viruses and to discern major events in their evolutionary history. The overall scheme of herpesvirus evolution that we have elaborated places the development of herpesviruses in much the same timeframe as that of the vertebrates. The basis for this assignment is the extensive congruence of branching patterns of the viruses within sublineages of the herpesviruses of placental mammals and the branching patterns of their host lineages, that strongly supports a basis of co-speciation. Extrapolation of the resulting time-scale then suggests that a common ancestor of today's mammalian, avian, and reptilian herpesviruses existed around 400 MA, in the Devonian Period. This ancestral species was already a developed herpesvirus and thus must also have had an extensive history. The *Alloherpesviridae* appear to represent an evolutionary development of complexity at least equivalent to that of the *Herpesviridae*, and presumably thus having a comparable time dimension. The singleton species of the *Malacoherpesviridae* presents a narrow base for evolutionary inference, but here too a deep history is plausible. The nature of earlier events, and in particular of the relationships among the three families of the *Herpesvirales*, quickly becomes wholly speculative. A possible glimpse of that earlier history is afforded by comparisons of present day families in the *Herpesvirales*: the common capsid architecture argues for a common descent, while the gene complements appear deeply distinct. A resolution of this puzzle envisages descent from a prototypic entity that was in effect a pre-herpesvirus, possessing genes that specified capsid structure but lacking significant other parts of modern gene complements. A well-grounded understanding of the origins and evolutionary rationale of such a hypothetical creature may always be beyond our grasp.

The evident widespread occurrence of very long-term associations of herpesvirus and host species implies directly that extensive functional co-evolution must have occurred, molding the genomes of both host and virus. We presume that we see the outcome of this process in the present day phenotypes of the viruses—these are very successful parasites, many infecting a high proportion of the host population and maintaining that infection for the host's lifetime, while in terms of effects on host population they are not severe pathogens. Conversely, the high virulence seen in certain herpesviral infections of a host of a species close to but distinct from the natural host (for instance, the lethal disease caused by SBV in humans) also serves as an indicator of the moderating outcome of co-evolution. An important point is that the extended nature of the co-evolutionary association may well have acted to incorporate adaptations so fundamentally into the genetic organization and capabilities of the contemporary species, both host and virus, that it could be very difficult to recognize and dissect out specific adaptations.

Finally, we are all too aware that our account has been heavily phenomenological—we have outlined the effects of evolutionary processes on herpesvirus genomes and given a time-scale to certain aspects, but we have not achieved any very detailed insight into the evolutionary forces that have operated on these genomes, or the opportunities that were associated with new genomic elaborations. Given that strategic mechanisms in herpesvirus infections are highly complicated and still only partially understood, and given the complex histories sketched above, it is rather likely that such a functional account of the evolution of these viruses will only ever be fragmentary. Conversely, we also believe that evolutionary analysis must comprise an essential part of the effort toward understanding herpesvirus strategies.

REFERENCES

Albrecht, J.-C., Nicholas, J., Biller, D., Cameron, K.R., Biesinger, B., Newman, C. *et al.* (1992) Primary structure of the herpesvirus saimiri genome. *J. Virol.* **66**, 5047–5058.

Ambinder, R.F., Robertson, K.D. and Tao, Q. (1999) DNA methylation and the Epstein-Barr virus. *Cancer Biol.* **9**, 369–375.

Aoki, T., Hirono, I., Kurokawa, K., Fukuda, H., Nahary, R., Eldar, A. *et al.* (2007) Genome sequences of three koi herpesvirus isolates representing the expanding distribution of an emerging disease threatening koi and common carp worldwide. *J. Virol.* **81**, 5058–5065.

Baer, R., Bankier, A.T., Biggin, M.D., Deininger, P.L., Farrell, P.J., Gibson, T.J. *et al.* (1984) DNA sequence and expression of the B95-8 Epstein-Barr virus genome. *Nature (Lond.)* **310**, 207–211.

Baker, M.L., Jiang, W., Rixon, F.J. and Chiu, W. (2005) Common ancestry of herpesviruses and tailed DNA bacteriophages. *J. Virol.* **79**, 14967–14970.

Benton, M.J. and Donoghue, P.C.J. (2007) Paleontological evidence to date the tree of life. *Mol. Biol. Evol.* **24**, 26–53.

Bernard, J. and Mercier, A. (1993) Sequence of two EcoRI fragments from salmonis herpesvirus 2 and comparison with ictalurid herpesvirus 1. *Arch. Virol.* **132**, 437–442.

Boehmer, P.E. and Lehman, I.R. (1997) Herpes simplex virus DNA replication. *Annu. Rev. Biochem.* **66**, 347–384.

Booy, F.P., Trus, B.L., Davison, A.J. and Steven, A.C. (1996) The capsid architecture of channel catfish virus, an evolutionarily distant herpesvirus, is largely conserved in the absence of discernible sequence homology with herpes simplex virus. *Virology* **215**, 134–141.

Bowden, R.J., Simas, J.P., Davis, A.J. and Efstathiou, S. (1997) Murine gammaherpesvirus 68 encodes tRNA-like sequences which are expressed during latency. *J. Gen. Virol.* **78**, 1675–1687.

Bowden, R., Sakaoka, H., Donnelly, P. and Ward, R. (2004) High recombination rate in herpes simplex virus type-1 natural population suggests significant co-infection. *Infect. Genet. Evol.* **4**, 115–123.

Bowden, R., Sakaoka, H., Ward, R. and Donnelly, P. (2006) Patterns of Eurasian HSV-1 molecular diversity and inferences of human migrations. *Infect. Genet. Evol.* **6**, 63–74.

Brown, S.M., Ritchie, D.A. and Subak-Sharpe, J.H. (1973) Genetic studies with herpes simplex virus type 1. The isolation of temperature-sensitive mutants, their arrangement into complementation groups and recombination analysis leading to a linkage map. *J. Gen. Virol.* **18**, 329–346.

Brunovskis, P. and Kung, H.-J. (1996) Retrotransposition and herpesvirus evolution. *Virus Genes* **11**, 259–270.

Bujnicki, J.M. and Rychlewski, L. (2001) The herpesvirus alkaline exonuclease belongs to the restriction endonuclease PD-(D/E)XK superfamily: insight from molecular modeling and phylogenetic analysis. *Virus Genes* **22**, 219–230.

Burrows, J.M., Broomham, L., Woolfit, M., Piganeau, G., Tellam, J., Connolly, G. *et al.* (2004) Selection pressure-driven evolution of the Epstein-Barr virus-encoded oncogene LMP1 in virus isolates from Southeast Asia. *J. Virol.* **78**, 7131–7137.

Caposio, P., Riera, L., Hahn, G., Landolfo, S. and Gribaudo, G. (2004) Evidence that the human cytomegalovirus 46-kDA UL2 protein is not an active dUTPase but a late protein dispensable for replication in fibroblasts. *Virology* **325**, 264–276.

Cedergren-Zeppezauer, E.S., Larsson, G., Nyman, P.O., Dauter, Z. and Wilson, K.S. (1992) Crystal structure of a dUTPase. *Nature (Lond.)* **355**, 740–743.

Cha, T.-A., Tom, E., Kemble, G.W., Duke, G.M., Mocarski, E.S. and Spaete, R.R. (1996) Human cytomegalovirus clinical isolates carry at least 19 genes not found in laboratory strains. *J. Virol.* **70**, 78–83.

Challberg, M. (1996) Herpesvirus DNA replication. In: *DNA Replication in Eukaryotic Cells* (M.L. De Pamphilis, ed.), pp. 721–750. Cold Spring Harbor: Cold Spring Harbor Laboratory Press.

Chee, M.S., Bankier, A.T., Beck, S., Bohni, R., Brown, C.M., Cerny, R. *et al.* (1990) Analysis of the protein coding content of the sequence of human cytomegalovirus strain AD169. *Curr. Top. Microbiol. Immunol.* **154**, 125–169.

Crump, C.M., Bruun, B., Bell, S., Pomeranz, L.E., Minson, T. and Browne, H.M. (2004) Alphaherpesvirus glycoprotein M causes the relocalization of plasma membrane proteins. *J. Gen. Virol.* **85**, 3517–3527.

Dambaugh, T., Hennessy, K., Chamnankit, L. and Kieff, E. (1984) U2 region of Epstein-Barr virus DNA may encode Epstein-Barr nuclear antigen 2. *Proc. Natl Acad. Sci. USA* **81**, 7632–7636.

Davison, A.J. (1992) Channel catfish virus: a new type of herpesvirus. *Virology* **186**, 9–14.

Davison, A.J. (1998) The genome of salmonid herpesvirus 1. *J. Virol.* **72**, 1974–1982.

Davison, A.J. and McGeoch, D.J. (1986) Evolutionary comparisons of the S segments in the genomes of herpes simplex virus type 1 and varicella-zoster virus. *J. Gen. Virol.* **67**, 597–611.

Davison, A.J. and Scott, J.E. (1986) The complete DNA sequence of varicella-zoster virus. *J. Gen. Virol.* **67**, 1759–1816.

Davison, A.J. and Stow, N.D. (2005) New genes from old: redeployment of dUTPase by herpesviruses. *J. Virol.* **79**, 12880–12892.

Davison, A.J. and Taylor, P. (1987) Genetic relations between varicella-zoster virus and Epstein-Barr virus. *J. Gen. Virol.* **68**, 1067–1079.

Davison, A.J., Dolan, A., Akter, P., Addison, C., Dargan, D.J., Alcendor, D.J. *et al.* (2003) The human cytomegalovirus genome revisited: comparison with the chimpanzee cytomegalovirus genome. *J. Gen. Virol.* **84**, 17–28.

Davison, A.J., Eberle, R., Hayward, G.S., McGeoch, D.J., Minson, A.C., Pellett, P.E. *et al.* (2005a) *Herpesviridae.* In: *Virus Taxonomy. Classification and Nomenclature of Viruses. Eighth Report of the International Committee on Taxonomy of Viruses* (C.M. Fauquet, M.A. Mayo, J. Maniloff, U. Desselberger and L.A. Ball, eds), pp. 193–212. San Diego and London: Elsevier Academic Press.

Davison, A.J., Trus, B.L., Cheng, N., Steven, A.C., Watson, M.S., Cunningham, C. *et al.* (2005b) A novel class of herpesvirus with bivalve hosts. *J. Gen. Virol.* **86**, 41–53.

Davison, A.J., Cunningham, C., Sauerbier, W. and McKinnell, R.G. (2006) Genome sequences of two frog herpesviruses. *J. Gen. Virol.* **87**, 3509–3514.

Dolan, A., Jamieson, F.E., Cunningham, C., Barnett, B.C. and McGeoch, D.J. (1998) The genome sequence of herpes simplex virus type 2. *J. Virol.* **72**, 2010–2021.

Dolan, A., Cunningham, C., Hector, R.D., Hassan-Walker, A.F., Lee, L., Addison, C. *et al.* (2004) Genetic content of wild-type human cytomegalovirus. *J. Virol.* **85**, 1301–1312.

Dolan, A., Addison, C., Gatherer, D., Davison, A.J. and McGeoch, D.J. (2006) The genome of Epstein-Barr virus type 2 strain AG876. *Virology* **350**, 164–170.

Efstathiou, S. and Preston, C.M. (2005) Towards an understanding of the molecular basis of herpes simplex virus latency. *Virus Res.* **111**, 108–119.

Ehlers, B., Küchler, J., Yasmum, N., Dural, G., Voigt, S., Schmidt-Chanasit, J. *et al.* (2007) Identification of novel rodent herpesviruses, including the first gammaherpesvirus of *Mus musculus*. *J. Virol.* **81**, 8091–8100.

Fenner, F. (1976) Classification and nomenclature of viruses. Second Report of the International Committee on Taxonomy of Viruses. *Intervirology* **7**, 1–115.

Früh, K., Ahn, K., Djaballah, H., Sempe, P., van Endert, P.M., Tampé, R. *et al.* (1995) A viral inhibitor of peptide transporters for antigen presentation. *Nature (Lond.)* **375**, 415–418.

Galocha, B., Hill, A., Barnett, B.C., Dolan, A., Raimondi, A., Cook, R.F. *et al.* (1997) The active site of ICP47, a herpes simplex virus-encoded inhibitor of the major histocompatibility complex (MHC)-encoded peptide transporter associated with antigen processing (TAP), maps to the NH_2-terminal 35 residues. *J. Exp. Med.* **185**, 1565–1572.

Galtier, N. (2003) Gene conversion drives GC evolution in mammalian histones. *Trends Genet.* **19**, 65–68.

Gompels, U.A., Nicholas, J., Lawrence, G., Jones, M., Thomson, B.J., Martin, M.E.D. *et al.* (1995) The DNA sequence of human herpesvirus-6: structure, coding content, and genome evolution. *Virology* **209**, 29–51.

Greenblatt, R.J., Quackenbush, S.L., Casey, R.N., Rovnak, J., Balazs, G.H., Work, T.M. *et al.* (2005) Genomic variation of the fibropapilloma-associated marine turtle herpesvirus across seven geographic areas and three host species. *J. Virol.* **79**, 1125–1132.

Haberland, M., Meyer-König, U. and Hufert, F.T. (1999) Variation within the glycoprotein B gene of human cytomegalovirus is due to homologous recombination. *J. Gen. Virol.* **80**, 1495–1500.

Hannenhalli, S., Chappey, C., Koonin, V. and Pevzner, P.A. (1995) Genome sequence comparison and scenarios for gene rearrangements: a test case. *Genomics* **30**, 299–311.

Herbst, L., Ene, A., Su, M., Desalle, R. and Lenz, J. (2004) Tumor outbreaks in marine turtles are not due to recent herpesvirus mutations. *Curr. Biol.* **14**, R697–R699.

Hill, A., Jugovic, P., York, I., Russ, G., Bennink, J., Yewdell, J. *et al.* (1995) Herpes simplex virus turns off the TAP to evade host immunity. *Nature (Lond.)* **375**, 411–415.

Honess, R.W., Gompels, U.A., Barrell, B.G., Craxton, M., Cameron, K.R., Staden, R. *et al.* (1989) Deviations from expected frequencies of CpG dinucleotides in herpesvirus DNAs may be diagnostic of differences in the states of their latent genomes. *J. Gen. Virol.* **70**, 837–855.

Isfort, R., Jones, D., Kost, R., Witter, R. and Kung, H.-J. (1992) Retrovirus insertion into herpesvirus *in vitro* and *in vivo*. *Proc. Natl Acad. Sci. USA* **89**, 991–995.

Kieff, E. and Rickinson, A.B. (2001) Epstein-Barr virus and its replication. In: *Fields Virology* (D.M. Knipe and P.M. Howley, eds), Vol. 2, pp. 2511–2573. Philadelphia: Lippincott Williams & Wilkins.

Knopf, C.W. (1998) Evolution of viral DNA-dependent DNA polymerases. *Virus Genes* **16**, 47–58.

Luebcke, E., Dubovi, E., Black, D., Ohsawa, K. and Eberle, R. (2006) Isolation and characterization of a chimpanzee alphaherpesvirus. *J. Gen. Virol.* **87**, 11–19.

MacLean, C.A., Efstathiou, S., Elliott, M.L., Jamieson, F.E. and McGeoch, D.J. (1991) Investigation of herpes simplex virus type 1 genes encoding multiply inserted membrane proteins. *J. Gen. Virol.* **72**, 897–906.

Markine-Goriaynoff, N., Georgin, J.-P., Goltz, M., Zimmermann, W., Broll, H., Wamwayi, H.M. *et al.* (2003) The core 2 β-1,6-*N*-acetylglucosaminyltransferase-mucin encoded by bovine herpesvirus 4 was acquired from an ancestor of the African buffalo. *J. Virol.* **77**, 1784–1792.

Matthews, R.E.F. (1979) Classification and nomenclature of viruses. Third Report of the International Committee on Taxonomy of Viruses. *Intervirology* **12**, 132–296.

McGeehan, J.E., Depledge, N.W. and McGeoch, D.J. (2001) Evolution of the dUTPase gene of mammalian and avian herpesviruses. *Curr. Prot. Pept. Sci.* **2**, 325–333.

McGeoch, D.J. (1989) The genomes of the human herpesviruses: contents, relationships and evolution. *Annu. Rev. Microbiol.* **43**, 235–265.

McGeoch, D.J. (1990) Protein sequence comparisons show that the "pseudoproteases" encoded by poxviruses and certain retroviruses belong to the deoxyuridine triphosphatase family. *Nucleic Acids Res.* **18**, 4105–4110.

McGeoch, D.J. (2001) Molecular evolution of the γ-*Herpesvirinae*. *Phil. Trans. R. Soc. Lond. B Biol. Sci.* **356**, 421–435.

McGeoch, D.J. and Cook, S. (1994) Molecular phylogeny of the Alphaherpesvirinae subfamily and a proposed evolutionary timescale. *J. Mol. Biol.* **238**, 9–22.

McGeoch, D.J. and Davison, A.J. (1995) Origins of DNA viruses. In: *Molecular Basis of Virus Evolution* (A.J. Gibbs, C.H. Calisher and F. Garcia-Arenal, eds), pp. 67–75. Cambridge: Cambridge University Press.

McGeoch, D. and Gatherer, D. (2005) Integrating reptilian herpesviruses into the family *Herpesviridae*. *J. Virol.* **79**, 725–731.

McGeoch, D.J. and Gatherer, D. (2007) Lineage structures in the genome sequences of three Epstein-Barr virus strains. *Virology* **359**, 1–5.

McGeoch, D.J., Dolan, A., Donald, S. and Rixon, F.J. (1985) Sequence determination and genetic content of the short unique region in the genome of herpes simplex virus type 1. *J. Mol. Biol.* **181**, 1–13.

McGeoch, D.J., Dolan, A., Donald, S. and Brauer, D.T.K. (1986) Complete DNA sequence of the short repeat region in the genome of herpes simplex virus type 1. *Nucleic Acids Res.* **14**, 1727–1745.

McGeoch, D.J., Dalrymple, M.A., Davison, A.J., Dolan, A., Frame, M.C., McNab, D. *et al.* (1988) The complete

DNA sequence of the long unique region in the genome of herpes simplex virus type 1. *J. Gen. Virol.* **69**, 1531–1574.

McGeoch, D.J., Cook, S., Dolan, A., Jamieson, F.E. and Telford, E.A.R. (1995) Molecular phylogeny and evolutionary timescale for the family of mammalian herpesviruses. *J. Mol. Biol.* **247**, 443–458.

McGeoch, D.J., Dolan, A. and Ralph, A.C. (2000) Toward a comprehensive phylogeny for mammalian and avian herpesviruses. *J. Virol.* **74**, 10401–10406.

McGeoch, D.J., Gatherer, D. and Dolan, A. (2005) On phylogenetic relationships among major lineages of the *Gammaherpesvirinae*. *J. Gen. Virol.* **86**, 307–316.

McGeoch, D.J., Rixon, F.J. and Davison, A.J. (2006) Topics in herpesvirus genomics and evolution. *Virus Res.* **117**, 90–104.

Megaw, A.M., Rapaport, D., Avidor, B., Frenkel, N. and Davison, A.J. (1998) The DNA sequence of the RK strain of human herpesvirus 7. *Virology* **244**, 119–132.

Midgley, R.S., Blake, N.W., Yao, Q.Y., Croom-Carter, D., Cheung, S.T., Leung, S.F. *et al.* (2000) Novel intertypic recombinants of Epstein-Barr virus in the Chinese population. *J. Virol.* **74**, 1544–1548.

Midgley, R.S., Bell, A.I., McGeoch, D.J. and Rickinson, A.B. (2003) Latent gene sequencing reveals familial relationships among Chinese Epstein-Barr virus strains and evidence for positive selection of A11 epitope changes. *J. Virol.* **77**, 11517–11530.

Minarovits, J. (2006) Epigenotypes of latent herpesvirus genomes. *Curr. Top. Microbiol. Immunol.* **310**, 61–80.

Mocarski, E.S., Jr. (1996) Cytomegaloviruses and their replication. In: *Fields Virology* (B.N. Fields, D.M. Knipe and P.M. Howley, eds), 3rd edn, pp. 2447–2492. Philadelphia and New York: Lippencott-Raven Publishers.

Montague, M.G. and Hutchison, C.A. (2000) Gene content phylogeny of herpesviruses. *Proc. Natl Acad. Sci. USA* **97**, 5334–5339.

Murphy, W.J., Pringle, T.H., Crider, T.A., Springer, M.S. and Miller, W. (2007) Using genomic data to unravel the root of the placental mammal phylogeny. *Genome Res.* **17**, 413–421.

Nicholas, J. (1996) Determination and analysis of the complete nucleotide sequence of human herpesvirus 7. *J. Virol.* **70**, 5975–5989.

Nikolaev, S., Montoya-Burgos, J.I., Margulies, E.H., NISC Comparative Sequencing Program, Rougemont, J., Nyffeler, B. and Antonarakis, S.E. (2007) Early history of mammals is elucidated with the ENCODE multiple species sequencing data. *PLoS Genet.* **3**, e2. doi:10.1371/journal.pgen.0030002.

Norberg, P., Liljeqvist, J.A., Bergström, T., Sammons, S., Schmid, D.S. and Loparev, V.N. (2006) Complete-genome phylogenetic approach to varicella-zoster virus evolution: genetic divergence and evidence for recombination. *J. Virol.* **80**, 9569–9576.

Pagamjav, O., Sakata, T., Matsumura, T., Yamaguchi, T. and Fukushi, H. (2005) Natural recombinant between equine herpesviruses 1 and 4 in the ICP4 gene. *Microbiol. Immunol.* **49**, 167–179.

Peters, G.A., Tyler, S.D., Grose, C., Severini, A., Gray, M.J., Upton, C. and Tipples, G.A. (2006) A full-genome phylogenetic analysis of varicella-zoster virus reveals a novel origin of replication-based genotyping scheme and evidence of recombination between major circulating clades. *J. Virol.* **80**, 9850–9860.

Pfeffer, S., Zavolan, M., Grasser, F.A., Chien, M., Russo, J.J., Ju, J. *et al.* (2004) Identification of virus-encoded microRNAs. *Science* **304**, 734–736.

Pfeffer, S., Sewer, A., Lagos-Quintana, M., Sheridan, R., Sander, C., Grasser, F.A. *et al.* (2005) Identification of microRNAs of the herpesvirus family. *Nature Methods* **2**, 269–276.

Pignatelli, S., Dal Monte, P., Rossini, G. and Landini, M.P. (2004) Genetic polymorphisms among human cytomegalovirus (HCMV) wild-type strains. *Rev. Med. Virol.* **14**, 383–410.

Poole, L.J., Zong, J.C., Ciufo, D.M., Alcendor, D.J., Cannon, J.S., Ambinder, R. *et al.* (1999) Comparison of genetic variability at multiple loci across the genomes of the major subtypes of Kaposi's sarcoma-associated herpesvirus reveals evidence for recombination and for two distinct types of open reading frame K15 alleles at the right-hand end. *J. Virol.* **73**, 6646–6660.

Pyles, B.R., Sawtell, N.M. and Thompson, R.L. (1992) Herpes simplex virus type 1 dUTPase mutants are attenuated for neurovirulence, neuroinvasiveness, and reactivation from latency. *J. Virol.* **66**, 6706–6713.

Rawlinson, W.D., Farrell, H.E. and Barrell, B.G. (1996) Analysis of the complete DNA sequence of murine cytomegalovirus. *J. Virol.* **70**, 8833–8849.

Rezaee, S.A.R., Cunningham, C., Davison, A.J. and Blackbourn, D.J. (2006) Kaposi's sarcoma-associated herpesvirus immune modulation: an overview. *J. Gen. Virol.* **87**, 1781–1804.

Rickinson, A.B., Young, L.S. and Rowe, M. (1987) Influence of the Epstein-Barr virus nuclear antigen EBNA2 on the growth phenotype of virus-transformed B cells. *J. Virol.* **61**, 1310–1317.

Rixon, F.J. and McGeoch, D.J. (1985) Detailed analysis of the mRNAs mapping in the short unique region of herpes simplex type 1. *Nucleic Acids Res.* **13**, 953–973.

Roizman, B., Carmichael, L.E., Deinhardt, F., de-The, G., Nahmias, A.J., Plowright, W. *et al.* (1981) Herpesviridae. Definition, provisional nomenclature, and taxonomy. *Intervirology* **16**, 201–217.

Roizman, B., Desrosiers, R.C., Fleckenstein, B., Lopez, C., Minson, A.C. and Studdert, M.J. (1992) The family *Herpesviridae:* an update. *Arch. Virol.* **123**, 425–449.

Rosa, M.D., Gottlieb, E., Lerner, M.R. and Steitz, J.A. (1981) Striking similarities are exhibited by two small Epstein-Barr virus-encoded ribonucleic acids and the adenovirus-associated ribonucleic acids VAI and VAII. *Mol. Cell. Biol.* **1**, 785–796.

Russo, J.J., Bohenzky, R.A., Chien, M.-C., Chen, J., Yan, M., Maddalena, D. *et al.* (1996) Nucleotide sequence of the Kaposi sarcoma-associated herpesvirus (HHV8). *Proc. Natl Acad. Sci. USA* **93**, 14862–14867.

Sample, J., Young, L., Martin, B., Chatman, T., Kieff, E., Rickinson, A. and Kieff, E. (1990) Epstein-Barr virus types 1 and 2 differ in their EBNA-3A, EBNA-3B, and EBNA-3C genes. *J. Virol.* **64**, 4084–4092.

Sevilla-Reyes, E.E. (2006) Recombination in human cytomegalovirus. Ph.D. thesis, University of Glasgow.

Spear, P.G. (1993) Entry of alphaherpesviruses into cells. *Semin. Virol.* **4**, 167–180.

Springer, M.S., Murphy, W.J., Eizirik, E. and O'Brien, S. J. (2003) Placental mammal diversification and the Cretaceous-Tertiary boundary. *Proc. Natl Acad. Sci. USA* **100**, 1056–1061.

St Jeor, S.C., Albrecht, T.B., Funk, F.D. and Rapp, F. (1974) Stimulation of cellular DNA synthesis by human cytomegalovirus. *J. Virol.* **13**, 353–362.

Tarbouriech, N., Buisson, M., Seigneurin, J.M., Cusack, S. and Burmeister, W.P. (2005) The monomeric dUTPase from Epstein-Barr virus mimics trimeric dUTPases. *Structure* **13**, 1299–1310.

Thomson, B.J., Efstathiou, S. and Honess, R.W. (1991) Acquisition of the human adeno-associated virus type-2 *rep* gene by human herpesvirus type-6. *Nature (Lond.)* **351**, 78–80.

Thomson, B.J., Weindler, F.W., Gray, D., Schwaab, V. and Heilbronn, R. (1994) Human herpesvirus 6 (HHV-6) is a helper virus for adeno-associated virus type 2 (AAV-2) and the AAV-2 *rep* gene homologue in HHV-6 can mediate AAV-2 DNA replication and regulate gene expression. *Virology* **204**, 304–311.

Timbury, M.C. and Subak-Sharpe, J.H. (1973) Genetic interactions between temperature-sensitive mutants of types 1 and 2 herpes simplex viruses. *J. Gen. Virol.* **18**, 347–357.

Trus, B.L., Chen, N.Q., Newcomb, W.W., Homa, F.L., Brown, J.C. and Steven, A.C. (2004) Structure and polymorphism of the UL6 portal protein of herpes simplex virus type 1. *J. Virol.* **78**, 12668–12671.

Waltzek, T.B., Kelley, G.O., Stone, D.M., Way, K., Hanson, L., Fukuda, H. *et al.* (2005) Koi herpesvirus represents a third cyprinid herpesvirus (CyHV-3) in the family. *Herpesviridae. J. Gen. Virol.* **86**, 1659–1667.

Wellehan, J.F., Nichols, D.K., Li, L.L. and Kapur, V. (2004) Three novel herpesviruses associated with stomatitis in Sudan plated lizards (*Gerrhosaurus major*) and a black-lined plated lizard (*Gerrhosaurus nigrolineatus*). *J. Zool. Wildlife Med.* **35**, 50–54.

Wibbelt, G., Kurth, A., Yasmum, N., Bannert, M., Nagel, S., Nitsche, A. and Ehlers, B. (2007) Discovery of herpesviruses in bats. *J. Gen. Virol.* **88**, 2651–2655.

Wildy, P. (1955) Recombination with herpes simplex virus. *J. Gen. Microbiol.* **13**, 346–360.

Wildy, P. (1971) Classification and nomenclature of viruses. First Report of the International Committee on Nomenclature of Viruses. In: *Monographs in Virology* (J.L. Melnick, ed.), Vol. 5, pp. 1–81. Basel: Karger.

Yang, Z. (2007) PAML 4: phylogenetic analysis by maximum likelihood. *Mol. Biol. Evol.* **24**, 1586–1591.

Yates, J.L. (1996) Epstein-Barr virus DNA replication. In: *DNA Replication in Eukaryotic Cells* (M.L. De Pamphilis, ed.), pp. 751–773. Cold Spring Harbor: Cold Spring Harbor Laboratory Press.

Zimber, U., Adldinger, H.K., Lenoir, G.M., Vuillaume, M., Knebel-Doeberitz, M.V., Laux, G. *et al.* (1986) Geographical prevalence of two Epstein-Barr virus types. *Virology* **154**, 56–66.

CHAPTER

21

The Widespread Evolutionary Significance of Viruses

Luis P. Villarreal

ABSTRACT

In the last 30 years, the study of virus evolution has undergone a transformation. Originally concerned with disease and its emergence, virus evolution had not been well integrated into the general study of evolution. This chapter reviews the developments that have brought us to this new appreciation for the general significance of virus evolution to all life. We now know that viruses numerically dominate all habitats of life, especially the oceans. Theoretical developments in the 1970s regarding quasispecies, error rates, and error thresholds have yielded many practical insights into virus–host dynamics. The human diseases of HIV-1 and hepatitis C virus cannot be understood without this evolutionary framework. Yet recent developments with poliovirus demonstrate that viral fitness can be the result of a consortia, not one fittest type, a basic Darwinian concept in evolutionary biology. Darwinian principles do apply to viruses, such as with Fisher population genetics, but other features, such as reticulated and quasispecies-based evolution distinguish virus evolution from classical studies. The available phylogenetic tools have greatly aided our analysis of virus evolution, but these methods struggle to characterize the role of virus populations. Missing from many of these considerations has been the major role played by persisting viruses in stable virus evolution and disease emergence. In many cases, extreme stability is seen with persisting RNA viruses. Indeed, examples are known in which it is the persistently infected host that has better survival. We have also recently come to appreciate the vast diversity of phage (DNA viruses) of prokaryotes as a system that evolves by genetic exchanges across vast populations (Chapter 10). This has been proposed to be the "big bang" of biological evolution. In the large DNA viruses of aquatic microbes we see surprisingly large, complex and diverse viruses. With both prokaryotic and eukaryotic DNA viruses, recombination is the main engine of virus evolution, and virus host co-evolution is common, although not uniform. Viral emergence appears to be an unending phenomenon and we can currently witness a selective sweep by retroviruses that infect and become endogenized in koala bears.

Origin and Evolution of Viruses
ISBN 978-0-12-374153-0

INTRODUCTION

Our understanding of virus evolution has reached a threshold in that it now appears to provide a much broader vista regarding its general significance and influence on host evolution. Several developments have brought us to this point. One has been the realization that viruses often evolve by processes involving the collective action of a consortia, or quasispecies. And the resulting adaptability and power of such evolution is unmatched by any other genetic entity. Much of this volume is dedicated to this issue. In such consortia, the concept of a "wild-type" virus is no longer considered to be the fittest type, as the quasispecies itself provides fitness (see Chapters 3 and 4). The quasispecies model resembles population genetics in some ways, but it has led to some significant departures from population genetics, and these departures are very well supported by experiments. Another development that has recalibrated our view of the overall significance of viruses is information on the scale and diversity of viruses. Viruses are present at a previously unappreciated global level and appear to have affected the evolution of all life on Earth. Much of this realization has been brought about by the development of metagenomic methods as applied to various habitats. Measurements of major habitats (the oceans, soil, extreme environments) have established that our biological world is predominantly viral, in terms of both numbers and diversity (Paul *et al.*, 2002; Breitbart *et al.*, 2003; Breitbart and Rohwer, 2005a, 2005b; Edwards and Rohwer, 2005; Comeau *et al.*, 2006).

These two developments would seem reason enough to consider virus evolution in a new light. However, there have also been numerous theoretical proposals suggesting viral involvement in some of the very earliest events and major transitions in the evolution of life. We no longer think of viruses as recent agents that escaped from the host chromosomes as run away replicons. Viruses now appear very old to us and they relate to and trace all branches of life. The last 30 years have been very active regarding virus evolution. Major developments in theory, technology, medicine, and the study of human disease with respect to virus evolution have all occurred. And as we seek to grow and manage various life forms for human use, virus evolution has also had major impact on such efforts.

As a science, virus evolution has benefitted greatly from traditional evolutionary biology. However, since viruses are molecular genetic parasites that are inscrutable by casual observation, our understanding of virus evolution has been dependent upon measuring sequence variation and sequence diversity in a large number of virus genomes. Because of this, these small genetic parasites have been the last domain of life to yield their secrets of evolution. And viruses harbor some clearly distinct evolutionary abilities. For one, they are polyphyletic. All major viral lineages have their own distinct origins. They are also difficult if not impossible to define as species and are able to exchange genetic information across normal boundaries. Even "dead" and defective viruses can participate in such exchanges, which confuses the definitions of fitness. We now know that viruses can evolve by a consortia process and also exchange information by recombination across vast genetic pools to assemble new mosaic combinations of genes. We thus no longer think of a specific genetic lineage in understanding virus evolution, but instead think of a cloud, matrix, or a population as the basis of virus evolution. Viruses are inherently fuzzy entities that can differ from their relatives in any specific feature. Yet even with such fuzziness, it is clear that common themes also link them. Patterns of evolution have become clear. Diversity and variation are often (but not always) observed. Stability and host congruence can also be observed. Nevertheless, the evolutionary power of viruses has been learned at a human cost.

The application of numerous analytical and phylogenetic tools have provided crucial insights into virus origin and evolution. Yet these methods struggle to incorporate the fuzzy nature of viruses and have clear limits, especially regarding quasispecies and high recombination rates. Structural biology now also adds tools that extend our vision of virus evolution beyond what can be seen in the genetic sequence. For example, common structural motifs from phage to eukaryotic DNA viruses (T4 and herpesvirus) suggest very ancient links in virus evolution that span all domains of life (see below). Nevertheless, our analytical methods are currently lacking as we struggle to understand complex genetic mixtures that provide fitness, reticulated relationships, polyphyletic origins, and virus–host congruence. Virus evolution has, for the most part, been considered to be a specific, esoteric part of broader evolutionary biology and has been given limited attention in reference works on evolutionary biology (see Pagel, 2002). Historically, the focus has been on various RNA viruses and some DNA viruses that cause disease in humans and domesticated animals and plants (Domingo *et al.*, 1999). However, I have asserted that all life forms must be examined from the perspective of virus evolution (Villarreal, 2005), not just those pathogens that impact on us. How viruses evolve in a more general sense informs us of evolutionary paradigms that have not been previously well understood (especially the evolution of consortia or the dynamics of vast reticulated gene pools). This volume now extends these traditional topics of virus evolution to include the vast virology of the prokaryotic world. In so doing, it illuminates the global consequences viruses have had on all life forms.

Prior to the 1970s we saw some stunning successes in vaccine control for major viral human diseases, such as polio, measles, mumps, influenza, and especially smallpox virus. Due to this historic success, American health agencies and educators considered virus disease as a thing of the past, no longer a serious threat. The era of infectious disease that they represented was now one for the historical record, an unfortunate part of human history, or so we thought. In what now seems to have been a clear case of hubris and naivety, we have been humbled by the evolutionary power of viruses, which was woefully underappreciated, even by most virologists. By the end of that decade, the evolution and emergence of HIV-1 permanently changed our views (see Chapters 13 and 14). This has also been followed by a seemingly never-ending series of viral threats as newly emerging viral diseases have come to our attention. HIV-1 provides the only example of a public health situation that has reversed centuries of progress for extending human health and lifespan. It now limits human life expectancy in many parts of the world, especially sub-Saharian Africa. This development could not have been imagined in the 1970s. We are much less confident now about predicting the future of virus evolution and its potential impact on human health.

Diseases of domesticated microbiological, plant, and animal species have also experienced the trauma of the consequences of emerging viral diseases along with huge losses. However, the human HIV-1 story may not be a fluke of virology but may be telling us something basic about human and primate evolution. As we sought to understand the origin of this new virus, we have come to appreciate a much broader virus–host story which involves simian immunodeficiency virus (SIV), foamy viruses, and the speciation of Old World primates. We have also come to learn that the genomes of these primates show much evidence of past viral interaction and ongoing endogenous retrovirus colonization. The evolution of retroviral endogenization has taken on a much greater significance in basic evolutionary biology. Thus it is with great interest that we now study the ongoing endogenization of retroviruses in the koala bear genome (see below).

Historically, we are compelled to study viruses because they can cause serious disease. New viruses come to our attention also mainly because of disease. It is therefore understandable that most evolutionary biologists mainly think of viruses strictly as agents of disease. These are the products of run away replicons that provide negative selection to host survival. In this light, the application of predator–prey based mathematical models has seemed most appropriate. With such viral disease, variation has long been observed and was initially used for the generation of most vaccines. However, this disease-centric view has also occluded another more prevalent virus–host relationship. For example the emergence of HIV-1 has led us conclude that it likely evolved from various versions of SIV. But SIV is not pathogenic in its native African primate host. Nor does it show the genetic diversification of HIV-1 in these native primate hosts. It is a silent, asymptomatic infection. Genomic and metagenomic analysis now allows us to identify many more silent, asymptomatic viruses that would not have previously been observed. We now know of many such viruses that are prevalent in a specific host. Evolutionary biology must escape the confines of disease-centric thinking and seek to understand these relationships as well.

A Role for Persistence in Viral Evolution

In the last ten years I have attempted to provide another view concerning virus–host evolution. I have argued that viruses often attain evolutionary stability by species-specific persistence and that such states apply to all domains of life, including prokaryotes. On an evolutionary time-scale, the majority of viral lineages tend to exist as species-specific persistent (aka temperate, latent, and chronic) infections in which individual hosts will be colonized by mostly silent (asymptomatic) viruses for the duration of their life (Villarreal *et al.*, 2000). Such persistence can have major consequences to the evolution of both virus and host, which also leads us to more directly link virus evolution to broader issues of host evolution. It is from this perspective that we start to clearly see that viruses indeed belong on the tree of life as major participants (Villarreal, 2005, 2006). Persisting viruses are not simply agents responsible for destruction of life, but are also agents that create genetic novelty on a vast scale that influences all life and promotes symbiosis (Marquez *et al.*, 2007; Ryan, 2007; Villarreal, 2007). The persistent lifestyle of such symbiotic virus–host relationships is not simply a less efficient, acute infection; nor is it simply a "reservoir" for acute virus (as epidemiologists are prone to assert). Neither can it be attributable to concepts of selfish DNA. Persistence represents a major virus life strategy that is both fundamental and highly adapted. It has distinct genetic, fitness, and evolutionary characteristics that require intimate, host (tissue)-specific viral strategies and precise gene functions to attain stable maintenance in the presence of immunity and to allow biologically controlled reactivation. Persistence also must resist displacement by similar viruses and competitors. It is virus–host persistence that provides the thread that allows us to link these polyphyletic viral lineages (and their clouds) with the entire tree of life. In turn, this link identifies a much more fundamental role for viruses in the evolution of host, visible from the very earliest to the most recent events in host evolution. It is from such species-specific persistent states that the large majority of acute diseases evolve and emerge by various mechanisms.

We know much about virus replication and disease. However, our understanding of the specific mechanisms of persistence is generally poor. Persistence is a generally silent and inscrutable state, it does not lend itself to *in vitro* or cell culture experimental models. We are left with but a few examples from which to attempt to extrapolate the possible existence of general relationships. The study of virus evolution thus struggles to incorporate concepts of persistence.

Viruses Mediating Innovation

Another recent and major development in virus evolution is the arrival of various proposals suggesting that viruses have been involved in some major innovation and transition in the evolution of life. In all these proposals, however, it is necessary that the virus in question has attained a stable genomic persistence with its host. These evolutionary events thus seem to be the products of viral-mediated symbiogenesis of host (Ryan, 2007; Villarreal, 2007). Proposals include the possibility that viruses may have originated the DNA replication system of all three cellular domains (archaea, bacteria, eukarya) of life (Forterre, 1999, 2005, 2006a, 2006b; Filee *et al.*, 2003; Forterre *et al.*, 2007). The discovery and analysis of the largest DNA virus (1200 ORF mimivirus), a lytic cytoplasmic virus of amebae (a distant relative of phycodnavirus and poxviruses), has also led to proposals that this virus lineage may represent an ancient fourth domain of life (Raoult *et al.*, 2004; Desjardins *et al.*, 2005; Claverie *et al.*, 2006). It is interesting that in an initial structural analysis, the large complex replication centers for mimivirus were confused with the host nucleus (Suzan-Monti *et al.*, 2007). Thus, it seems relevant that others have proposed that a distant relative of phycodnaviruses and poxviruses may have originated the eukaryotic nucleus (Villarreal, 1999; Villarreal and DeFilippis, 2000; Bell, 2001, 2006; Takemura, 2001).

Such proposals, although consistent with various observations, however, remain outside of the consensus of most evolutionary biologists. Nevertheless, numerous other observations continue to suggest viral involvement in other major host innovations, such as a viral origin of RAG1/2 of the adaptive immune system (Dreyfus *et al.*, 1999 Kapitonov and Jurka, 2005; Fugmann *et al.*, 2006) or the role of endogenous retroviruses (ERVs) in the evolution of the placenta (Villarreal and Villareal, 1997; Harris, 1998; Blond *et al.*, 2000; Mi *et al.*, 2000; Dupressoir *et al.*, 2005; Caceres and Thomas, 2006). Such possible roles for viruses in host evolution are at odds with accepted views of virus–host relationships, but might be the products of viral symbiogenesis.

The Virosphere: a New Evolutionary Reality

The metagenomic viral measurements mentioned above for prokaryotic DNA viruses, along with the increasing realization that viruses and host can co-evolve, has led to various calls that a viral tree of life needs to be considered and developed (Forterre, 2003; Villarreal, 2006; Filee *et al.*, 2006b). A virosphere clearly exists but its nature and boundaries are not so clear. Multiple viral origins, their diversity and numerical dominance in distinct and sometimes harsh environments as well as their presence in host genomes suggest that any viral tree of life will be huge, multidimensional, and connected to the host tree of life. As discussed below regarding double-stranded (ds)DNA viruses of prokaryotes, they have all the above characteristics. Such viruses may represent the big bang of biological novelty. With their unmatched capacity to generate diversity they can function as the mass creators of biological novelty as well as destroyers of most species. Surely, such capacity must have had big influences on the evolution of life.

Symbiosis, simply defined, is the stable coexistence of two previously separate lineages of organisms. There can be little doubt that many temperate phage can stably colonize a bacterial cell resulting in a stable descendent from two lineages. This is clearly symbiotic. Endogenous retroviruses can similarly be found to persist in vertebrate genomes and also appear symbiotic. Yet studies of symbiosis seldom consider a role for virus (Villarreal, 2007).

How important are viruses in general to evolutionary biology? The core concepts of evolutionary biology were developed well before we had a modern understanding and definition of viruses (Luria, 1950). After all, the basic lysogenic model of phage integration was only clarified in 1962 when developed by Campbell (see Campbell, 2007). That cryptic and defective phage are ubiquitous in the genomes of all prokaryotes is generally considered uninteresting by many in the field of evolutionary biology. I suggest that the seemingly applicable concepts of selfish DNA effectively derailed any thinking that persisting genetic parasites might have a more germinal role in the evolution of life (Doolittle and Sapienza, 1980; Orgel and Crick, 1980). Yet as outlined above, virus footprints in major evolutionary transitions are clear and a direct role in such events now seems much more plausible. We therefore must seek to defining the nature of the virosphere and how its evolution relates to the tree of life.

EXEMPLARS OF VIRUS EVOLUTION

This book represents the first integration of the entire field of virus evolution, including both prokaryotic and eukaryotic life forms. However, because our understanding of virus–host relationships remains uneven, the chapters necessarily focus on well-studied models (exemplars). These exemplars also tend to reflect a historic disease focus (i.e., *E. coli*, flowering plants, mouse, humans). It is unfortunate that the silent species-specific viruses that tend to exist in stable states with long evolutionary histories seldom provide our examplars. We understand these infections poorly and lack basic definitions concerning fitness or selective advantages. Only metagenomics tools now seem able to inform us of their presence, but not their biology.

Errors at the Start of Life: Quasispecies

Self-organization and the evolution of RNA molecules as an origin of biological information are discussed in Chapter 1. Autocatalytic chemical reactions, such as replication of RNA, presents issues such as how to optimize a rugged fitness landscape yet allow the study of evolution *in vitro* with RNA. RNA *in vitro* reduces genotype–phenotype issues to RNA secondary structure and minimal free energy states. This allows both continuity and discontinuity to be measured. These same issues are crucial for the study of RNA viruses, whose sites of secondary structure often define replicator identity. These models currently offer the best system to evaluate an early and simple biological world for evolutionary principles (see Chapter 3). Viruses and viroids with their RNA genomes may be the only extant survivors of this pre-DNA world. Since it was the consideration of error-prone replication that led to the development of the concept of quasispecies (see Chapter 2), such models have provided a conceptual foundation which led to several basic concepts. Chapter 4 discussed the foundations and various aspects of quasispecies theory. Viruses appear to operate close to the error threshold, thus allowing maximum evolutionary exploration (Biebricher and Eigen, 2006). However, as presented below, the loss of the "fittest" type concept has also led to clear experimental evaluations of consortia-based evolutionary behaviors. Such behaviors were not predicted by classical Darwinian models.

Although virologists were initially attracted to quasispecies models, many evolutionary biologists were initially hostile to the application of the quasispecies concept to evolution. It was thought that the classical mathematical models of population biology, as originally developed by Wright and Fisher, and later applied by Kimura and Maruyama to asexual haploid populations at the mutation–selection balance, had already fully developed the needed models and precluded additional need for the quasispecies concept. The classical models were thus argued to provide adequate mathematical coverage for viruses, including quasispecies and error threshold (Wilke, 2005). However, these two approaches differ fundamentally with regard to the significance of error-prone replication and it was the quasispecies approach that led to the clear experimental establishment that quasispecies selection, *per se*, is important for viral pathology and fitness (see below).

The development of quasispecies theory to virology does indeed demonstrate distinct differences with population genetics. Various phenomena, such as complementation, cooperation, competition, and even defective mediated extinction (Domingo *et al.*, 2001, Domingo 2006; Grande-Perez *et al.*, 2005) have been observed, all of which fall outside of the parameters of classical population genetics. Viral fitness has indeed been shown to be due to interaction within a diverse population, and not to the fittest or master type. And with RNA viruses, error threshold has become a central issue (Biebricher and Eigen, 2005).

The collective experience has thus made clear the value of theory to biology. Many working biologists understand that life seems overly complex and defies most generalizations. Thus they do not always appreciate attempts at general theory. Although in reality, biology may indeed often be too complex for accurate theoretical predictions, these theories have nevertheless clearly stimulated crucial concepts and experimental evaluation, and biologists should be encouraged by them. By providing new ways of thinking, entirely new experimental approaches can be developed.

The existence of error-prone replication and quasispecies also raises the issue of the conservation of information. How is information stability and higher fitness attained with such errors? How can genetic complexity be created in such a circumstance? Can cooperation (or consortia behavior) result from any of these models? These issues have yet to be resolved. Some interesting suggestions, however, have been proposed. One involved cooperative evolution that results from ligated genomes. A model was proposed in 2000 by Stadler and Schuster in which they considered the dynamics of replicator networks resulting from higher order reactions involving the templated ligation of smaller genomes (Stadler *et al.*, 2000). Although this was based on concepts such as triple-stranded nucleic acids, this clearly has some elements that also resemble the ligation of recombinational processes for DNA phage (presented below).

A most interesting outcome of these models is that, depending on initial replicator concentrations, permanent coexistence of replicators could result in a cooperative network. Such cooperation is a rare outcome for most models and given the conclusion that early DNA based life was a "horizontal" consortium, such models are of special interest. The issue of consortia will come up often. Consortia selection directly implies cooperation, but cooperation of selfish replicators presents dilemmas. Replicator networks with interaction functions that give highly non-linear dynamics can result in complex mixtures, with behaviors ranging from survival of the fittest to also including attainment of globally stable equilibrium tantamount to permanent coexistence. The fitness of populations, however, is inherent to current quasispecies concepts in virology. There may also be other ways to explain the genetic origin of cooperation (such as stable persistence involving addiction strategies). The stable persistence of a genetic parasite can compel cooperation and promote the conservation of information (see below).

FITNESS, CONSORTIA, AND PERSISTENCE

The definitions of fitness with respect to a virus in a natural habitat are far from clear. Although the concept of relative replicative fitness is often applied to lab experiments of virus growth, we know many situations in which virus replication is not maximized in natural settings and many viruses can exist in relatively non-replicative states for long periods. Even in the context of an acutely replicating virus in a host organism, the concept of fitness is clearly conditional, as the virus must replicate through various *in vivo* habitats that can have opposing selection. As presented below, *in vivo* models that study fitness and viral diversity have clearly indicated that diversity *per se* is important and fitness is the result of consortia. How do we define such fitness since the mixture clearly matters? Also, how do we define information content or integrity of a consortia? Currently, we cannot. In the lab, the viability of a virus is usually measured by the ability to produce plaques. This has been a crucial and main assay for many experimental systems that study virus population. Here, the definition of fitness seems direct: plaque formation equals fitness. Various highly useful models have thus been defined and developed that depend on these plaque assays (see Chapter 4). With this, populations and population growth are defined as relative growth of plaque-forming units. However, the concept has always been problematic when considered from a natural virology perspective. Plaque formation is clearly not equal to fitness in natural habitats. There are many examples of highly successful viruses that either plaque poorly or not at all. Consider the roughly 100 types of human papillomavirus (HPV), a simple small circular DNA virus of epithelia; this does not form plaques in any known system (Chapter 18). HPV is clearly fit, well adapted to, and stable in its human host. In addition, HPV evolution is phylogenetically congruent with their primate host, as are most persistent viral infections. We have yet to understand the definition of fitness in this situation.

In some cases, it seems selection for plaque propagation has clearly resulted in loss of highly conserved genes; such as with the plaque-adapted laboratory strains of cytomegalovirus (CMV). The problem posed by viruses with inefficient plaque formation is not limited to DNA viruses. Many persisting RNA viruses also do not plaque well or at all, such as most RNA viruses of plants or many insect picorna-like viruses, such as those found in *Drosophila* and bees (which also conserve an extra ORF). Nor do most persistent infections make lots of virus. Low-level persistence, such as hantavirus in rodents, for example, is common (Hart and Bennett, 1999).

Clearly, our simplifying assumptions of viral fitness and population dynamics cannot apply to these stable evolutionary states. However, if we limit our definition of viral fitness to relative replication or plaque formation we can perform some clear and quantitative evaluations.

Fitness Theory for Persistence

Experimental evaluation forces us to study fitness by only those definitions that we can currently measure. As fitness appears to be a relativistic and transient concept, depending very much on the tissue, time, place, extant adaptive and innate immunity, and competition, it is likely that we can only measure with any accuracy one aspect of fitness at any one time. HIV infection of humans shows evidence of this in that the R5 virus is more fit for transmission and early disease whereas the X4 virus is fit later during the AIDS disease phase. Clearly conditional, time-dependent issues relate to fitness definitions. However, much more problematic is that we have no theory for viral persistence or its fitness. We lack specific or measurable parameters other than the simple maintenance of genetic material. Yet it seems clear that some distinguishing features of persistence can already be recognized. For example, the possible participation of viral defectives (normally considered unfit), which in numerous circumstances can modify or mediate persistence, would need to be included. Clearly, a defective role in persistence would also preclude them from being considered as genetic "junk," or selfish elements, since they would then matter in measurable ways to the biological outcome of virus persistent infection.

Persistence also requires an extended duration of infection, not simply maximized replication. In fact, persistence generally requires mechanisms to limit the replication of at least the same virus for at least some time. Thus, limited replication must be an essential element for this life strategy. In my judgment, and much like the quasispecies concept, the concept of persistence will eventually be recognized for the fundamental (symbiotic) force it represents in virus evolution.

The experimental work of Domingo and Holland spans the modern assessment of quasispecies theory that occurred in the 1980s and 1990s. These investigators were the chief proponents of this theory, bringing it to the attention of the broader virology community (Chapter 4). This work has transformed our thinking and laid the experimental foundations that we now build upon. This current volume is an extension from an earlier book on quasispecies (Domingo *et al.*, 1999) and now encompasses both prokaryotic and eukaryotic viruses. Since early experimental phage studies provided the foundations for quasispecies theory (Eigen *et al.*, 1988), using mathematical descriptions (differential equations) of mutation rates in T-even phage (Luria and Delbruck, 1943), this inclusion is appropriate.

Interestingly, a second early paper measuring replication rates by these same authors also noted the problems of viral interference and defectives (Delbruck, 1945). Other early experiments of phage RNA polymerase (Batschelet *et al.*, 1976), especially with RNA phage Qβ (Domingo *et al.*, 1978), helped set the stage for the subsequent experiments of the 1980s and 1990s. From the test tube to mouse models to the study of human disease, the work of Domingo and colleagues has spanned the entire history of viral quasispecies (Domingo and Gomez, 2007; Chapter 4).

Quasispecies deals with the products of error-prone replication. However, it is worth repeating that products of error-prone replication are not behaving in a simple "selfish DNA" capacity and are not devoid of biological relevance and phenotype. In their complex populations, they create clear and varied affects on viral adaptability, competition, and fitness. Since quasispecies necessarily involve defective and mutant virus, it is easy (and common) to think of these entities simply as genetic junk (Villarreal, 2005, 2006). Defective and even lethal or interfering variation in viral genomes can contribute to adaptability. Thus, viruses can clearly adapt as a cloud with a mutant spectra. In addition, unending competition and exclusion, consistent with the Red Queen hypothesis, has also been observed (Clarke *et al.*, 1994).

The poliovirus–mouse model (see below) in particular has provided a solid experimental system for evaluating the adaptive consequence of quasispecies. It is thus ironic that these same experiments have also made clear that the original simplifying assumptions of the quasispecies ordinary differential as presented by Eigen are violated by the resulting quasispecies. These products of error-prone replication do indeed strongly interact with each other in both positive and negative ways and such interactions contribute significantly to the observed fitness of the population.

Errors and interaction are important for fitness. For example, defectives have been reported to mediate extermination of a competing wild-type virus (Grande-Perez *et al.*, 2005). Complementation has also been observed (Garcia-Arriaza *et al.*, 2004), as has *trans*-dominant inhibition (Crowder and Kirkegaard, 2005). Genetic memory of past selection has been shown to be maintained in a minority of the population (Ruiz-Jarabo *et al.*, 2000; Briones *et al.*, 2006). Such cooperative (consortia) behavior, which can also depend on unfit or defective members, is at odds with classical Darwinian notions regarding survival of the fittest. Consider, for example, the fitness of a defective or mutant outside of its role in a quasispecies. Such a consideration ignores the very nature of a quasispecies yet it is an issue that has often been posed and experimentally evaluated. We should refrain from thinking of viruses simply as fit or non-fit individual types since they clearly exist in populations that provide population-based adaptability.

The selection of viral consortia or population raises some fundamental issues for evolutionary biology. This is essentially group selection in which a population, not the fittest individual, is selected. This view makes cooperation or interaction of individual genomes a significant component of selection, which is not commonly thought to be a general or accepted process in evolution. Yet population selection is no longer a contestable issue in RNA virus evolution (see below). I expect that many classical evolutionary biologists might interpret this as evidence that viruses really are an oddity in this feature and are not representative of broader processes of evolutionary biology. Furthermore, viral-based group selection may not be limited to quasispecies-based evolution. As presented below, persistent viral infections may also provide population-based selective advantage (see below for the P1 and mouse hepatitis virus (MHV) persistence exemplars; Villarreal *et al.*, 2000; Villarreal, 2006, 2007).

Since viruses are ancient, numerically dominant, and the most diverse biological entities on Earth, no life form can escape exposure to them. All extant life forms have evolved in a viral habitat. Thus we should expect that the viral footprints (including defectives) that we now find in all genomes have likely played an active role in their evolution; a role, I would argue, that is fundamental, dynamic, and unending. If we can accept this assertion, we may start to see and appreciate the vast evolutionary power that viruses can bring to bear onto host evolution. We can start to attain a global perspective and appreciation for their ability to assemble genetic function from enormous, complex mixtures of genomes, and select gene sets needed to solve multivariant, temporally dynamic evolutionary problems. We can then seek evidence for the role of viral elements in fundamental host innovation and be open to evaluating the occurrence of viral entities from a constructive perspective and not instinctively dismiss such observations as due to coincidence, "junk," or selfish DNA.

The advantage of such a perspective is that it will promote the specific experimental evaluations that can better assess any constructive role genetic parasites might have played in host evolution. For example, there is much reason to think ERVs have played an active role in human evolution (for references see Ryan, 2007). The quasispecies concept has provided the foundation for us to understand virus evolution and informed us of the evolutionary power viruses possess. If that power also links to host evolution, then the tree of life becomes enriched by virus, much larger and more dynamic.

The Poliovirus–Mouse Exemplar: "Quasispecies *per se* Rather than the Selection of Individual Adaptive Mutations Correlates with Enhanced Pathogenesis"

The recent experimental studies from Andino and colleagues using poliovirus in the mouse model should, in my judgment, provide the keystone exemplar regarding the *in vivo* fitness of quasispecies (see Chapters 6 and 7; Vignuzzi *et al.*, 2006). These studies make clear the importance of quasispecies and error-prone replication. Such detailed *in vivo* experiments were made possible by a long and detailed history of poliovirus studies that has identified the nature of RNA polymerase fidelity as well as developed mouse models for the study of pathogenesis. Few other virus–host systems could have provided such potential for high resolution. These results also provide the experimental observations that distinguish quasispecies-based evolution from the classical Fisher-based population genetics. The general importance of this story for understanding virus evolution thus deserves special emphasis.

The very origins of modern animal virology stem from poliovirus studies with the need to develop *in vitro* cell culture technology in order to grow and evaluate poliovirus and generate variants. The live poliovirus vaccine is of special interest with regards to virus evolution and adaptability. The "live" oral Sabin vaccine can be considered to have been a miracle of the practical approach to virology developed in the 1950s (Horaud, 1993) in that it was used well before our understanding of the relevant evolutionary theory. The Sabin vaccine strain was the result of rodent-adapted virus and differs from the neurovirulent Mahoney strain by 56 point mutations (in the consensus sequence), although only a small number of these mutations were needed for neurovirulence (Christodoulou *et al.*, 1990). One of the important neurovirulent mutations was within the RNA polymerase gene (Tardy-Panit *et al.*, 1993). However, the significance of this observation took many years to unravel and exploit. In time it became apparent that 3Dpol mutants could affect replication fidelity. One poliovirus point mutant, 3DG64S, was shown to have enhanced high-fidelity replication and that selective pressure could be designed to increase fidelity in RNA polymerase (Pfeiffer and Kirkegaard, 2005).

Another major development was the molecular identification of the poliovirus receptor and the subsequent creation of transgenic mice expressing this receptor, making them susceptible to poliovirus infection. One of these transgenic lines allowed mouse brain infections with neurovirulent versions of poliovirus (Crotty *et al.*, 2002), and has provided a very useful animal model that allowed the evaluation of viral fitness in the context of *in vivo* pathogenesis. Although 3DG64S replicates well in culture (with lowered error rate), it was less pathogenic in this mouse model and competed poorly with 3D wild-type virus. It seemed that the decreased viral diversity was less able to generate the variation needed to get past bottlenecks due to multiple selective differences presented *in vivo* in tissues in the host, such as brain infection (Pfeiffer and Kirkegaard, 2006). This experimental system also makes clear the greater complexity of fitness *in vivo* relative to that typically measured in culture.

Thus it seems that *in vivo* there may not be one fitness but several that cannot be distinguished or individually measured. It is likely that various *in vivo* barriers require distinct fitness solutions that tend to create bottlenecks and that the diversity *per se* is essential to get past such bottlenecks. A population, not a clone or a consensus, appeared more fit as higher titer infections of 3DG64S also failed to be pathogenic. Thus, higher levels of a consensus virus are not equivalent to higher diversity.

The relationship between RNA polymerase structure, error rates, and ribavirin action is discussed by Cameron in Chapter 6 and has been the subject of numerous studies (Crotty *et al.*, 2001; Crotty and Andino, 2002; Vignuzzi *et al.*, 2005). Knowledge of the structure and catalytic mechanism of RNA polymerase function has allowed a greatly enhanced level of detail to be considered into what affects error rate (see Castro *et al.*, 2007; Korneeva and Cameron, 2007; Marcotte *et al.*, 2007). This has provided insight into the likely action of ribavarin on product fidelity (Harki *et al.*, 2002). Thus, it appeared that even a mutant of RNA polymerase with increased fidelity could still generate elevated diversity by various methods. Such control of fidelity allowed for the design of control experiments in which the same consensus virus genome could be forced to generate either less or more diverse progeny populations. In no other virus–host system have we attained such detailed insight into issues of error rate as those that were put to such excellent use in the poliovirus–mouse system.

How generally important is this poliovirus *in vivo* quasispecies result? Although the poliovirus–mouse system provides us with a firm experimental result, it seems likely that the generality of this relationship will be questioned by evolutionary biologists for several reasons. For one, this was observed in a lab constructed model system, which, it could be argued, is not an accurate representation of *in vivo* virus–host fitness. Also, as mentioned above, group selection is a process that will not readily be accepted as representative by the broader community. Is there any evidence that this result with poliovirus indeed represents a general virus–host evolutionary relationship in natural settings? As presented in Chapters 13–15, retroviruses and also human hepatitis virus C clearly exist as quasispecies populations that affect disease outcome. In the case of the retroviruses, viral populations show diversity that far exceeds that seen for other RNA viruses. In both HIV-1 and HCV there is clear circumstantial evidence for the importance of quasispecies for *in vivo* disease outcome, drug resistance, and fitness. In addition, with HCV, CNS infection may sometimes result, and such brain infections appear to be mediated by distinct quasispecies (Forton *et al.*, 2004; Forton *et al.*, 2006), reminiscent of the poliovirus mouse model.

Quasispecies memory, as mentioned above, also seems to be an important issue with regard to failure of antiretroviral therapy (Kijak *et al.*, 2002) and it appears that pol gene mutations could also be involved in this (Carobene *et al.*, 2004). Measurements of HIV quasispecies in individual patients indicates that multiple evolutionary patterns can be found in typical individual patients (Casado *et al.*, 2001), thus mixtures of HIV exist in patients (Bello *et al.*, 2004, 2005). And HIV-1 recombination is clearly contributing to diversity (Kijak and McCutchan, 2005). Thus, with both HIV-1 and HCV, their capacity to cause human disease is clearly associated with quasispecies compositions that affect fitness in complex ways. The poliovirus mouse system therefore appears to reflect quasispecies issues as observed in natural virus–host situations.

Consideration of retrovirus–host evolution introduces another large issue in evolution: genomic viruses. Unlike poliovirus and most

RNA viruses, retroviruses (e.g. non-lentivirus) have colonized the genomes of animal species in large numbers and represent a large fraction of these genomes. Genomic retroviruses are present in vast numbers, most of which are defective and mutant copies. In this genomic colonization they resemble the dsDNA viruses of prokaryotes (discussed below) that also colonize all prokaryotes although at a much lower numbers. The human genome has fewer than 26 000 genes, but appears to have 500 000 retroviral-related LTR elements. Some of these elements are intact and conserved (human ERVs (HERVs)) and this genomic population has some clear characteristics of a viral quasispecies. Such large amounts of genetic material have previously been dismissed simply as selfish or junk DNA of no fitness consequences to the host. However, given the importance of quasispecies mutant genomes for viral fitness and persistence, we might need to re-evaluate this dismissal. Retroviruses are clearly part of the human ancestry thus we should seek to understand, not dismiss their role in human evolution.

Evolution of High Fidelity

In contrast to the story above in which polio infection of mouse brain was dependent on the quasispecies resulting from lowered fidelity replication, a different relationship has been proposed for the nidoviruses. These are also positive single-stranded polycistronic RNA viruses (Gorbalenya *et al.*, 2006). This group of virus includes the coronaviruses (e.g. mouse hepatitis virus and SARS-associated coronavirus), which are the largest RNA viruses known (26–32 kb). It has been proposed that such large genomes have required the adaptation of a high-fidelity RNA polymerase in order to increase the error threshold and accommodate large RNA genomes. Based on the phylogenetics of this polymerase and other RNA-processing enzymes, this group of viruses appears to be monophyletic and it is thought that the acquisition of a high-fidelity RNA replicase was central to the origin of this lineage. This type of replicase is unique to RNA viruses. The monophyletic view stems from an analysis of a small set of conserved genes. Overall, however, these larger genomes have many other genes that show no similarities to related viruses. The origins and evolution of these more diverse and numerous genes cannot be currently traced.

This is an inherent problem in the analysis of virus evolution: a small selected set of hallmark genes with some similarity are assumed to trace an apparently linear (tree-based) viral lineage whereas the larger number of genes are not included and cannot be traced. If most of RNA virus evolution is indeed mediated by a mixed cloud of genomes, any role for mutant mixtures thus becomes obscure. But perhaps there is little else we can currently do given the lack of information. How might we explain the increased fidelity and genome size of the nidoviruses? Was there some change in viral adaptation in which quasispecies and generation of mixtures was no longer as important for adaptation? Did the need and selection for a larger genome override the use of error to generate adaptability as seen in poliovirus and HIV-1? If so, what selective pressures might have changed this seemingly basic feature? What do we know about the natural biology of these viruses, which might provide some insight into this?

Unfortunately, the natural distribution and gene functions of the nidoviruses are generally poorly understood. In terms of coronaviruses, numerous mammal and avian species can be infected and the virus will cause acute disease. In several of these acute infections, the virus involved seems to have recently been adapted to the new host from other, often unknown sources. With the recent emergence of the SARS virus and human infections, however, much greater attention has been focussed on trying to understand the origin and evolution of this virus. It has recently become clear that there indeed appears to exist an evolutionary stable source of this virus from which adaptation to humans was possible.

Various bat species have been found to support persistent asymptomatic infections by specific versions of SARS viruses (Tang *et al.*, 2006; Wang *et al.*, 2006; Vijaykrishna *et al.*,

2007). These studies also indicate that there appear to be three different and independent groups of SARS viruses in bats. In fact six novel coronaviruses were isolated from six different bat species showing an astonishing diversity in bats. Furthermore, phylogenetic analysis indicates that all bat coronaviruses appear to have descended from a common ancestor. Only one of these bat groups includes SARS and SARS-like coronaviruses that adapted to acute human infections. Thus, a prevalent and species-specific persistence of SARS viruses is found in particular geographical populations. Why is this relationship stable? Could the adaptation to a host-specific persistence-based basal life strategy provide some explanations for the evolution of the higher fidelity RNA replicase of these coronaviruses? As I have argued, persistent viral infection represents the majority of evolutionary stable viral lineages (Villarreal, 2006). However, we have almost no knowledge regarding how these bat SARS viruses persist and escape elimination by innate and adaptive immunity and what, if any, role the high-fidelity replicase (or other genes) have in this life strategy.

MHV—Mouse Exemplar (A Case for Persistence and Virus Addiction)

Although we cannot yet evaluate natural SARS virus persistence in native bat hosts, another coronavirus may be more informative regarding the effects of persistence on host populations. Mouse hepatitis virus (MHV) may provide our best exemplar of virus–host relationships and show how the concept of virus addiction relates to population persistence. MHV is the best-studied coronavirus. As a natural and prevalent virus of rodents, MHV is our best natural model of persistent RNA virus–host relationships for any mammal. In general, rodents are the most studied non-domestic mammals with regard to natural virus distribution.

Overall, we know that wild-caught rodents seldom show signs of acute virus infection (Kashuba *et al.*, 2005). However, asymptomatic virus persistence is ubiquitous in wild rodents (Descoteaux *et al.*, 1977; Gannon and Carthew, 1980; Schoondermark-van de Ven *et al.*, 2006), including voles (Descoteaux and Mihok, 1986). Some field studies have evaluated broader patterns of virus persistence in mice (Singleton *et al.*, 1993; Becker *et al.*, 2007) which indicated that wild house mice are highly colonized with MHV (80–100% prevalence). In addition to MHV, mouse cytomegalovirus, mouse parvovirus, mouse thymic virus, and mouse adenovirus are also prevalent. Other well-studied mouse viruses, such as lymphocytic choriomeningitis virus (LCMV) and polyomavirus (PyV), were at low natural prevalence. Interestingly, some non-native house mice that have colonized isolated islands may lack MHV (Moro *et al.*, 1999), although most other isolated island populations retain MHV (Moro *et al.*, 2003).

Other small mammals have yet to show any viral disease whatsoever (hedgehogs, chinchillas, prairie dogs, gerbils, sugar gliders) (Kashuba *et al.*, 2005). Thus, asymptomatic persistent viral infection is clearly the norm in rodents. Yet, in spite of this usual asymptomatic viral persistence, historically, some zoonotic viral disease outbreaks have occasionally been documented in natural populations. One such early outbreak was an epizootic diarrhea that occurred in infant mice (Adams and Kraft, 1963). Later, it was established that one such infection was due to mouse hepatitis virus (Carthew, 1977; Ishida *et al.*, 1978). In spite of this disease outbreak, with MHV, it has since become clear that asymptomatic persistent infections are the norm and are highly stable. Yet MHV disease outbreaks, especially in virus-free mouse facilities, are also common and severe. How does MHV attain such stable and prevalent persistence in natural population yet retain the ability to cause disease in naive populations? What maintains the MHV fitness of natural persistence?

It is well known that once MHV is established in a mouse or rat colony it can be very difficult to eliminate (Gannon and Carthew, 1980; Lussier and Descoteaux, 1986), clearly indicating that stability is rapidly attained and likely genetically programmed by the virus. I propose that these stable evolutionary states of viral persistence are due to a strategy we can call virus addiction (Villarreal, 2005) and that MHV can provide the exemplar of such a state. With MHV, only persistently infected mice colonies are protected from the disease that is otherwise caused by the virus. In wild asymptomatic mice, MHV is found mostly as an enteric infection. The CNS demyelinated disease that MHV can induce is most observed in newborn pups (Homberger, 1997; Nash *et al.*, 2001) and once in the brain, MHV can persist in CNS with recurring disease (Marten *et al.*, 2001). This recurring CNS disease is also associated with quasispecies (in the S gene) and recombination (Rowe *et al.*, 1998). The most serious CNS disease is in S-gene variant of MHV-4 (JHM), thus as with the polio–mouse model, pathogenic fitness with MHV is also associated with quasispecies.

Such MHV disease is the bane of all mouse colonies (Knobler *et al.*, 1982). However, once MHV persistence is attained, the problem to a mouse facility is not due to acute disease, but because immunological measurements are significantly affected by MHV persistence. Thus MHV alters mouse molecular identity regarding immunological (T-cell) reactions (Wilberz *et al.*, 1991). To establish stable asymptomatic persistence, however, MHV needs to infect newborns (Weir *et al.*, 1987), in which acute disease is prevented due to maternal passive immune antibody transfer (Gustafsson *et al.*, 1996). Being born to immune mothers thus protects against CNS disease and promotes enteric (not brain) virus colonization. In addition, it appears that persistence also promotes cross-species transfer (Baric *et al.*, 1999).

MHV persistence may involve genome stability and result in a distinct evolutionary dynamic. Asymptomatic persisting infections in a Lewis rat, for example, showed no variation in MHV S gene sequence, and no quasispecies as seen in brain infections (Stuhler *et al.*, 1997). The need to establish stable persistence could then be providing a strong selection for increased genome complexity and stability and might better explain the selection for the enhanced RNA polymerase fidelity in nidoviruses. How might such selection operate in natural populations? Evolutionary biologists often consider what might differentiate one group from another very similar group in a way that leads to two isolated and distinct populations.

Consider two hypothetical adjacent hay stacks harboring two *Mus musculus* colonies, one of which is persistently infected with MHV the other which is not. What is the fitness consequence to the colony harboring MHV relative to its uninfected neighbor? Our experience with MHV in mouse breeding colony provides a clear answer. The colony that is persistently infected with MHV will have a distinct advantage over its neighbor as MHV introduced into this uninfected colony will have severe effects on the offspring. Eventually, we can expect only the MHV-harboring colony will prevail in both hay stacks. This is a state I have called virus addiction. Only mice harboring persistent MHV are protected against the potential pathogenic consequence of acute MHV (or related virus) infection. The population is addicted to the virus.

Such a state, however, is clearly affecting colonies (or groups) of host, not individuals. An individual either quickly succumbs to the virus infection or, if infected, transmits it to others in the colony. A colony is thus under selection by MHV. To generalize this state, we expect that the persistence of SARS in specific bat populations would be expected to also affect the fitness of the corresponding specific bat populations. Persistence is a more demanding phenotype than acute replication. It requires greater gene complexity to counter host immunity and also to promote self-regulation. Thus the enhanced fidelity of RNA replication is selected in order to conserve this greater genetic complexity and stability.

We know that the high-fidelity RNA replication system (including RNA pol, helicase, endoribonuclease, and other activities) is also present in an ancient nidovirus relative of coronaviruses, such as fish-isolated white bream virus (26 kb RNA). I suggest there will also likely be species-specific persistent infections with this virus that require this enhanced replication fidelity and maintain this virus in its natural habitat. Thus, I suggest, an ancient persistent life strategy could more easily explain the monophyletic character of the nidovirus virus lineage. It is particularly interesting that one of these unique and conserved replication proteins (ADP ribose-1-monophosphate) is dispensable for culture growth (Putics *et al.*, 2005). I suggest it will not be dispensable for persistence.

THE REAL WORLD OF VIRAL RNA IN HUMAN DISEASE

HIV-1

The HIV-1 pandemic is an unfinished story. HIV-1 represents a real-time biological event in human evolution that confirms for us the importance of quasispecies and retroviruses to human biology. However, even though its human toll is huge, modern medicine and culture has responded rapidly enough to limit the impact of HIV-1 to the point at which it will not likely be the cause of a selective evolutionary sweep that could have altered human genetic makeup (in contrast to the koala bear endogenization presented below). As described earlier, its amazing adaptability via quasispecies along with extensive recombination contribute directly to HIV-1's diversity (Charpentier *et al.*, 2006) and makes it the most dynamic genetic entity ever studied. Many studies track the dominant HIV population and fail to examine minority populations. Yet it is precisely these minority populations, which evolve independently of the majority population, that can determine drug resistance phenotype and biological outcome (Charpentier *et al.*, 2004; Briones *et al.*, 2006; Morand-Joubert *et al.*, 2006).

Clearly, the specific makeup of a complex HIV population matters. Furthermore, HIV defectives and variants can also have major consequences. In some cases, long-term non-progressors of HIV-1 have shown mixed populations and unusual polymorphism in the early phase of HIV infection, sometimes contributing to long-term non-progression (LTNP) (Alexander *et al.*, 2000). One population of LTNPs was reported to have been colonized by an HIV variant that showed low virus replication and slow or arrested evolution (Bello *et al.*, 2005). In another case, a stable non-progressor was colonized by a replication incompetent version of HIV-1 (Wang *et al.*, 2003). Some of these non-progressors also appear to resist super-infection (Zhu *et al.*, 2003). It seems clear that at least in these exceptional situations, non-majority HIVs are crucial to the outcome.

There is also reason to think that other retroviruses have had a major influence on recent primate and human evolution, such as apathogenic persisting foamy virus in primates (Switzer *et al.*, 2005; Murray and Linial, 2006). Human antiretroviral genes seem to have undergone recent adaptations, such as APOBEC3, which can interfere with exogenous retroviruses (such as MLV and SIV) and underwent an expansion in the hominid lineage (Esnault *et al.*, 2005). It thus seems clear that human and primate evolution has been significantly affected by earlier, prevalent primate retroviruses.

HCV

Another important human–virus quasispecies story that has long been recognized is with hepatitis C virus (HCV), (see Chapter 15; Domingo and Gomez, 2007). HCV seems to have adapted to humans in the recent past, possibly from asymptomatic enteric primate viruses currently found in Africa (Smith *et al.*, 1997). As HCV remains an infection predominantly transmitted by blood, it does not appear to have fully adapted to the tissues of and transmission within its human host. However, like HIV-1, HCV has long been recognized to generate quasispecies in chronically infected people (Martell *et al.*, 1992) and it soon became apparent that the viral quasispecies are affected by and affects the outcome of antiviral therapy (Enomoto *et al.*, 1994; Hohne *et al.*, 1994; Kurosaki *et al.*, 1994; Okamoto and Mishiro, 1994).

Thus, successful antiviral therapy is directly correlated with an initial dramatic reduction in genetic diversity. Unfortunately, it has become clear that only a minority of HCV-infected individuals will respond favorably to a combination of interferon and ribivarin. Thus it seems to be diversity *per se* and the resulting structure of an HCV quasispecies that has a direct consequence to human health. However, since HCV is less well-adapted to humans compared with HIV-1, it does not pose the same threat to potentially provoke an evolutionary event in human evolution.

VSV

VSV is a negative-stranded RNA virus that has been a very important experimental model and has provided many laboratory measurements regarding quasispecies theory (see Chapter 4). Using VSV, evidence supporting the Red Queen hypothesis, involving unending adaptation to greater competition and Mueller's ratchet has been presented (Clarke *et al.*, 1994; Novella *et al.*, 1995; Elena *et al.*, 1996). When VSV was evaluated as an arbovirus, requiring adaptation to alternating and opposing fitness of insect and mammalian host, it was also apparent that minority quasispecies populations were responsible for maintaining the apparently antagonistic phenotypes (Novella *et al.*, 1999). Thus here too, the consortia character of a quasispecies is clear. Yet in natural settings several very different virus–host relationships can be seen with rhabdoviruses.

A distant relative of VSV (VHSV) is also known to be responsible for mass die-off of commercially important fish (Marty *et al.*, 2003). This virus infects many teleost species and has shown 100% mortality in many experiments (i.e. with i.p. inoculation). In natural outbreaks, however, it has also shown surprising genetic stability (Einer-Jensen *et al.*, 2006). Clearly error-prone rhabdovirus replication must be kept in check by purifying selection in this situation. In contrast, another rhabdovirus, sigma virus of *Drosophila*, is associated with no mortality but is a vertically transmitted persisting virus in specific *Drosophila* populations (Fleuriet, 1996). Yet in some recent population measurements, sigma virus infected *Drosophila* are expanding for unknown reasons (Fleuriet, 1994). Clearly this particular virus–host persistent relationship has some undefined selective advantage that operates beyond the lab-based concepts as measured above. Other rhabdoviruses also have peculiar host-specific relationships, such as bats that tend to support many persistent infections (Badrane and Tordo, 2001; Li *et al.*, 2005), or birds that seem to be free of almost all rhabdoviruses. Clearly, although VSV lab results have been highly informative, we still have much to learn regarding natural settings that affect rhabdovirus adaptation and evolution.

Another major paradigm for the high rates of negative-strand virus evolution is found with influenza virus. Due to its history and potential for initiating great human epidemics, it has long held the special interest of evolutionary virologists (see Chapter 5; Nelson and Holmes, 2007). However, this research has not much emphasized the quasispecies character of influenza virus evolution. Instead, it concentrates on the evolution of the master template or clades of template for the purposes of vaccine development (Webster and Govorkova, 2006). The views stemming from this type of evolution have lent themselves well to master template-based phylogenetic analysis and have dominated how many researchers think of virus–host evolution. Thus it is curious, given the above emphasis, that the quasispecies character of influenza populations often seems of low relevance to issues of acute disease and vaccination, other then to provide a source of diversity. In some situations, viral competitive interference may contribute to drift variation and displacement in antigenic epitopes (Levin *et al.*, 2004). Yet outcomes of individual human and bird infections do not seem much affected by specific quasispecies structures, as we saw with HIV-1 and HCV.

With influenza, we are mainly concerned with epidemic human disease. However, by shear numbers of infections and deaths worldwide, it must be admitted that influenza virus is really a virus that affects mostly birds. For example, during the 2005 outbreak in China, only 251 humans died whereas 230 million domestic birds died (Smith *et al.*, 2006). Although our concern on the large potential for human disease is understandable, these numbers should inform us of a more basic virus–host biology. In this case, influenza shows a high affinity for various birds; migratory water birds in particular can have high prevalence (Wallensten *et al.*, 2006). Some waterfowl, such as wild mallard ducks, have been called the stealth (asymptomatic) carriers of influenza H5N1 and free grazing ducks seem to introduce virus into domestic bird populations (Gilbert *et al.*, 2006). Thus waterfowl represent the well-accepted epidemiological concept of a reservoir species (Louz *et al.*, 2005). But these wild waterfowl, shorebirds, and gulls that are a natural host for avian influenza also seem to show a much slower rate of evolution (Spackman *et al.*, 2005). In contrast, the much higher rate of evolution as seen in chickens and turkeys indicates that these hosts should not be considered as natural reservoirs (Suarez, 2000). In waterfowl, influenza infections show several distinctions, such as virus co-infection or virus interference (Sharp *et al.*, 1997) as well as phylogenetically distinguishable waterfowl dendograms, including specific M lineages (Makarova *et al.*, 1998; Widjaja *et al.*, 2004). The diverse and stable avian pool of influenza virus appears to be ancestral to the influenza viruses that infected human populations.

THE ANALYTICAL PROBLEM OF QUASISPECIES: GROUP SELECTION, RETICULATION, AND RECOMBINATION

The Good

The phylogenetic methods that have been adapted from evolutionary biology have been tremendously helpful and have allowed us to trace the seemingly untraceable, virus evolution (see Chapter 5). Thus, we have often been able to make informed judgments concerning broader patterns of virus evolution and this has become the major tool for the current study of virus evolution, such as influenza virus (Nelson and Holmes, 2007). Influenza A, for example can be seen to show extended periods of stasis followed by periods of rapid adaptation that necessitates adaptations in vaccine strategy (Wolf *et al.*, 2006). However, the evolutionary variation between seemingly similar viruses can be surprisingly large (see above VSV section). For example, the very different phylogenetic behaviors between influenza A and measles virus, both acute human respiratory infections due to membrane-bound negative-stranded RNA viruses, are striking. The reasons for the maintained genetic stability of measles virus remain poorly understood, but may well involve more complex fitness associated with systemic infections.

Phylogenetic methods can also be highly informative regarding the likely origins of viral lineages and possible sources of emergence. For example, the studies of dengue virus by Holmes and colleagues suggest that this virus first entered its human host about 1000 years ago, and that sylvatic (African jungle) asymptomatic infection of primates may have provided the origin of this virus that later became a human pathogen (Holmes and Twiddy, 2003; Holmes, 2006). Such insight provides valuable clues concerning the likely selective pressures that may lead to the emergence of dengue virus. Phylogenetic methods are also highly informative regarding classification and taxonomy relationships and have allowed us to understand viral relationships across broad species definitions (Zanotto *et al.*, 1996).

The Bad

However, phylogenetic approaches necessarily assume the master template is the fittest type and that mutations or variants in the RNA populations are a source of genetic load that are deleterious and limiting to virus adaptation (Pybus *et al.*, 2007). Such variation is mostly due to "unfit" mutations, which indicates that a viral cloud is mostly and unfit consortia. It would seem that such conclusions go against the concept of quasispecies as being fit *per se* as described above. In this consideration we see a major weakness of extant phylogenetic methods. They were not developed to access the evolutionary relationship and fitness of interacting mixtures. Nor were they designed to follow the evolution of systems with high rates of recombination between numerous parental templates. We currently lack the analytical tools for such a population analysis. Without such tools, however, it seems we can only evaluate those parameters we can define and will remain confused by those we cannot. Evolution of a consortia thus provides a new directions for theoretical and laboratory research. We should seek to investigate the mixture, not just its average.

Plants

Another major virus–host system that has been highly studied is the viruses of agricultural plants. Our understanding of plant viruses has also been highly influenced by disease associated with agricultural domestic species, thus natural virus–plant relationships are much less understood, although some recent field studies are starting to change this situation (see Chapter 12). We currently have a rather uneven understanding of broader virus–host relationships and evolution in plants. For example, viruses of the more ancient ferns, if they exist, are essentially unknown.

The prevalence and diversity of positive-stranded RNA viruses in plants is striking. In addition, we are starting to appreciate that virus–virus interactions are also frequently involved, although this issue remains poorly studied. One well-studied family of plant virus are the tobamoviruses of angiosperms (see Chapter 11; Gibbs, 1999). Progenitors of this virus family appear to also be found in algae and fungi consistent with a very long evolutionary history. Both high transmission between host and virus–host congruence are observed with these viruses. Virus–virus interactions also seem to be important. For example, tobacco mosaic virus (TMV) and tomato golden mosaic virus (TMGV) appear to have shown interactions in Australia which have apparently led to the extinction of TMV, but the retention of TMGV with no increase in genetic diversity (Fraile *et al.*, 1997).

Plant viruses have also been seen as quasispecies in some but not all settings (see Chapter 12; Roossinck, 2003; Roossinck and Schneider, 2006). Besides the interactions expected for typical viral quasispecies, plants often show evidence of more extensive mixed virus infections. There are, for instance, many examples of satellite viruses that must necessarily interact with other RNA viruses of plants. It is also clear that the subviral elements of even a single viral lineage can greatly affect the virus–host relationship. Such subviral elements (DIs) have been observed to both reduce and intensify disease, and also interact with satellite viruses (Qiu and Scholthof, 2001), thus virus–virus interactions are clearly crucial in many situations (Simon *et al.*, 2004) and viral interactions and synergism appear to have led to significant events in plant virus emergence (Fargette *et al.*, 2006).

Virus–virus interactions are not limited to plant RNA viruses. The ssDNA plant geminiviruses also display complex interactions with satellites as well as high diversity in field isolates of East Africa (Ndunguru *et al.*, 2005). Thus, plant viruses seem particularly prone to interactions. More recently, virus-mediated symbiosis with respect to host survival has been reported (Roossinck, 2005) (discussed below).

Phylogenetic methods also struggle to address the occurrence of high rates of recombination in viral lineages. Such a situation complicates the analysis, creating hard-to-define, reticulated trees, although these limitations can be partially overcome by using sliding windows for the analysis. Such approaches have allowed surveys of recombination in some viral lineages, such as with the plant potyviruses (Chare and Holmes, 2006). However, the rampant recombination and quasispecies generation of HIV-1 makes a quantitative assessment of the virus population problematic. One proposed solution is to use a composition vector method (Gao and Qi, 2007). The issue of measuring recombination and tracing evolution in large populations is especially a problem that applies to the DNA viruses (phage) of prokaryotes (see below).

THE BIG BANG OF BIOLOGY

Prokaryotic DNA Viruses, Mosaic Swarms, the Origin of DNA-Based Cells.

Our perception regarding the overall importance of DNA viruses of prokaryotes to the evolution of life on Earth has undergone a major shift in recent years. The main realization is that DNA phage are the numerically dominant genetic entity in most habitats on Earth (mentioned above). In addition, as discussed in Chapter 10, it is now clear that some of these viruses are surprisingly complex and that essentially the entire pool of dsDNA viruses of prokaryotes may be exchanging DNA via recombination at high rates. This would constitute by far the largest common gene pool on Earth. Historically, the evolution of the DNA viruses of prokaryotes has seldom been considered in the broader context of virus evolution or evolutionary biology. Although it has long been realized that there are many basic similarities between viruses of bacteria and eukaryotes (Luria *et al.*, 1959), not until structural studies solved the capsid genes of prokaryotic and eukaryotic viruses did the evolutionary relationships between these viruses become clear. In addition, there have been a number of striking proposals that suggest that DNA viruses of prokaryotes may be involved in the origin of several major systems used by cells and that viruses appear to be involved in several major transitions during host evolution. Thus we now consider the possibility that these DNA phage were fundamental to the origin and evolution of life on Earth.

The Big Bang of Phage and Cellular DNA

It now seems likely that some large DNA viruses infecting eubacteria, archaea, and eukaryotes share some common evolutionary histories. It also seems clear that such viruses can link all three domains of life. This realization was not apparent based on phylogenetic sequence conservation, which is absent. It stems from the structure and assembly of virion capsids in which T4 phage, halophage, and the herpesviruses all show clear similarity as well as similarity in replication strategies. In addition, phage PRD1 and adenoviruses show similar broad structural and strategic conservation. Some biochemical (DNA pol family) and genetic similarities (gene order, gene programming) are also apparent, which taken together supports the common origins of these viruses (Hendrix, 1999, 2002; Hendrix *et al.*, 1999, 2003). T4-like viruses in particular seem to represent a major source of global genetic diversity.

This giant genetic pool represents a huge potential to affect life (Filee *et al.*, 2005) and the viral genetic creativity represented by this pool would also be vast (Nolan *et al.*, 2006). Since T4-like phage that infect cyanobacteria also encode virus-specific type II photosynthetic core genes, viruses appear able to create the most complex of genes as well (Clokie *et al.*, 2006; Sullivan *et al.*, 2006).

As presented in Chapter 10, phage are now thought to evolve by distinct and highly mosaic "horizontal" processes of rampant recombination (Hendrix, 2002, 2003). Large DNA phage appear to be ancient, present before the split of the three main branches of cellular life: bacteria, archaea, and eukarya (Benson *et al.*, 2004). LUCA, the Last Universal Common Ancestor, would represent the putative cell ancestor prior to this split. However, phyogenetic analysis of common or conserved genes of LUCA identifies only about 325 or fewer genes in extant cellular genomes (Mushegian, 1999; Koonin *et al.*, 2001; Mirkin *et al.*, 2003). Ironically, the genes needed for DNA replication are not part of this conserved set, calling into question the nature of the first DNA-based cell. Large-scale "horizontal" transfer seems to have clearly prevailed early in the evolution of DNA-based cellular life and it has recently been asserted that LUCA existed in a highly horizontal "consortia" of cooperative genes that developed the common genetic code (Vetsigian *et al.*, 2006).

Since the DNA replication proteins in the extant three domains of life have distinct compositions, it has been proposed by Forterre that DNA viruses and retroviruses were directly involved in the invention of the three extant cellular DNA replication systems (Forterre *et al.*, 2005). According to this view, early cellular life was completely entangled with viral (phage) lineages; hence cells must have evolved from an ancestral "virus"-mediated population not a single genetic lineage. Thus the evolution of early life would have clear similarity to the quasispecies (consortia) state of genetic information as seen in RNA viruses above. Thus the huge creative and adaptive potential of virus would have been directly involved in the very earliest evolution of life. Clearly, such conjectures regarding the most ancient events in the evolution of life are hard to substantiate. But, these theories are as viable as any other and deserve serious consideration.

In spite of this seemingly unending mosaic exchange in dsDNA phage, some phage isolates show surprisingly stable genetic makeup. We now accept that T4-related phage are an important source of the larger global phage genetic diversity (Liu *et al.*, 2006) and that most such viral genes are novel (Filee *et al.*, 2006b; Nolan *et al.*, 2006). Yet even with T4-like viruses, there can be clear barriers to horizontal gene transfer which promote the evolution of stable viral lineages (Filee *et al.*, 2006a). In T4-type phage, 24 similar core genes could be seen in all genomes, which seem to be inherited in gene blocks that preclude recombination. However, these blocks were not seen in the broader T-even and pseudo T-even genomes. Other phage also show surprising genetic stability when repeatedly isolated from similar habitats, such as soil phages of *Burkholderia* (Summer *et al.*, 2006) and Bam35 (Saren *et al.*, 2005) as well as some hot spring isolates (Khayat *et al.*, 2005). This Bam35 capsid also identifies another structural motif mentioned above that is broadly conserved in evolution and shows clear similarity to that capsids found in PRD1 and PBCV-1 (discussed below). SH1 also has a clear PRD1-related capsid, membrane, and genome; thus this halophilic euryarchaeon virus, although showing no sequence similarity to PRD1 or any other bacterial phage, is clearly structurally related (Bamford *et al.*, 2005a).

It is interesting that overall the viruses of hyperthermophilic Crenarchaeota generally show no sequence relationship to phage of bacteria. In addition, the use of the term phage for these viruses can also be questioned as most establish non-lytic chronic infections. Many of these Crenarchaeota viruses have unique morphologies not found in any other domain of life (Prangishvili *et al.*, 2006a, 2006b; Ortmann *et al.*, 2006). Some, however, have clear structural and genetic similarity to specific phage (i.e. T4).

GENOMIC STABILITY: PERSISTENCE AND TEMPERATE LIFESTYLE

Considerations of phage evolution and rampant recombination (especially with T4 and T-even phage) often emphasize the viral lytic lifestyle and host death. In fact this lytic relationship was argued by many early phage researchers to be the fundamental and only character of phage–host relationships in general. We now know, however, that persisting (temperate) phage are also common, some of which have no independent lytic phase. The fundamental model of phage persistence by unique integration into host chromosomes (temperate lysogeny) marks a major development in our understanding of molecular virology and virus–host relationships which was first clarified by Campbell in 1962 (see Campbell, 2007). All free-living prokaryotes show the presence of colonized phage in their genomes. Both complete and defective genomes of dsDNA viruses have been observed in the sequenced DNA of all free living prokaryotic genomes (Gelfand and Koonin, 1997) (exceptions are some intracellular parasites and plastids). Thus, the massive genetic diversity and novelty of phage evolution as presented above has a direct conduit into the genetic composition of all prokaryotes via lysogeny.

The fitness and evolutionary consequences of such colonization to the evolution of the host and its virus should be considerable but is in need of theoretical development. Fitness of temperate phage, however, is more complicated then that of a lytic virus and, like fitness of persistence discussed above, cannot be simply described by relative replication or efficient virion production. Here too, successful phage colonization must inherently limit the replication of the same virus. Thus, a temperate lifestyle also requires an autoinhibitory capacity. This generally involves an immunity gene set that not only limits self-replication but can also affect replication of other temperate and lytic viruses, i.e. lambda (even as a defective) precludes T4 and other T-even phage. Uncolonized hosts are thus susceptible to lysis by highly prevalent acute tailed phage. Host fitness is thus strongly affected by a temperate phage due to its ability to preclude and survive other competing phage. I suggest this situation is similar to the MHV–mouse exemplar above, in that virus-colonized hosts are in a state of "virus addiction" in which persistence is needed to provide protection from the same or similar virus (Villarreal, 2005, 2006).

It is well established that most natural populations of bacteria have specific patterns of phage colonization, hence the utility of phage typing for strain identification. From this, we can infer that virus–virus competition is a prevalent and major issue regarding the prokaryotic fitness resulting from a symbiotic temperate phage–host combination. In addition, such virus–host symbiosis can also affect competition with other bacteria. This would be very much like the virus addiction concept outlined above for the MHV examplar.

The original observation of a lysogenic process and coining of this term occurred in the 1920s when two pure cultures of bacteria were grown together. It was observed that in some combinations, one strain would lyse the other strain (was lysogenic). Later, it became clear that such lysis was mediated by reactivation of temperate phage present in the lysogenic strain, but absent from the non-lysogenic susceptible strain. In this relationship, we see another example of group selection operating on bacterial populations harboring a persistent virus. Thus, what host is fit depends very much on the prevalent viruses it will encounter as well as the viruses that colonize it. Bacterial populations that are colonized by the same or similar phage express the appropriate immunity functions and are protected from lysis by the same or similar phage.

Such a situation has significant implication for the evolution of immunity and group identity for cells. Host stability becomes a major fitness issue for a persistent virus life strategy. It is generally thought that a temperate virus attains a stable colonization of its host by simply integrating into and become one with the host genome. However, there are also clear examples of stable phage persistence that does not integrate and uses other strategies to attain host stability (similar to eukaryotic DNA viruses; see below for the P1 phage exemplar of this). Like a temperate phage, a host that is colonized by episomal persisting viruses has also been much affected in its evolutionary potential.

EPISOMAL STABILITY: THE P1 EXEMPLAR OF PERSISTENCE

It is clear that phage can have complex effects on host populations, but these phage themselves often exist in complex and mixed states that can be difficult to unravel (Harcombe and Bull, 2005). It has been known for some time that the presence of otherwise silent phage can greatly affect the growth of other virus and susceptibility of host. One such silent and common phage that has long been studied is P1. P1 was initially discovered due to its effect on T4 and lambda. However, P1 has been a very interesting model, not because it causes disease or offers potential therapy against bacterial pathogens, but simply because it persists efficiently as an episome and competes effectively with many other phage (Yarmolinsky, 2004). Since it does so without integrating, P1 provides us with one of the only well-studied models that can inform us regarding the molecular strategies and details of how stability in non-genomic persistence is attained.

Curiously, a main strategy by which P1 attains this stability was inapparent and not suspected after several decades of study. It became apparent only after replication mutations were made that induced self-destruction and uncovered the existence of what came to be called "addiction modules" (Lehnherr *et al.*, 1993). P1 encodes several gene pairs (toxins/antitoxins, such as the PhD/DOC pair) that protect bacteria harboring P1, but kill daughter bacteria that have lost the P1 genome (Gazit and Sauer, 1999). This strategy compels colonized *E. coli* to maintain P1 or die (DOC, death on curing). However, these very same addiction systems are also involved in protecting a P1-colonized colony from T4 and lambda infection and will also induce self-destruction when cells are infected by those viruses, protecting the colony (population). P1 also provides an exquisite level of molecular self-identification in that it will recognize a single second copy of its own genome (Yarmolinsky, 2000).

What then is the fitness and evolutionary consequence to *E. coli* harboring P1? Clearly it is major, but mostly host fitness is affected relative to other viruses. Accordingly, when contemplating the amazing complexity of the P1 immunity and how it evolved, Yarmolinsky posed the question; "Could the byzantine complexity of the controls at ImmI be the outcome, not of successive host–parasite accommodations, but of competition among related phages?" (Yarmolinsky, 2004). If we answer yes to this question, then we would also conclude that virus–virus interactions and competition in general are major forces in the adaptability and evolution of persisting phage and surviving colonized host. In this light, viral persistence takes on a major role in virus and host evolution. The P1 exemplar has thus provided us the concept of viral addiction that also promotes host group selection.

PROKARYOTES AND THEIR VIRUSES AS ONE EVOLUTIONARY POOL

Historically, we are biased to think of viruses (and phage) as agents that simply kill their host. Some have proposed that the prokaryotic global biomass is phage partitioned into those populations that live and those that die due to viral lysis. From such a perspective, viral novelity would seen of little relevance to host evolution. Metagenomic projects as noted above, have sequenced nearly 2 million phage genomes and report that most of these phage genes are unique, not in the database, and likely not derived from host (Edwards and Rohwer, 2005). The protein repertoire of sequenced phage indicates that 80% of conserved phage genes are specific to phage and show an evolutionary independence from genes of host (Liu *et al.*, 2006). This identifies a massive genetic novelty from virus, which is especially apparent in large DNA phage. As just discussed above, however, those hosts that live are also products of phage selection, and persisting temperate phage play a major role in this. Such phage colonization allows this massive phage novelty to find its way into host genomes, which allows viral complex gene sets to be applied to novel problems of host adaptation. Host novelty can thus be introduced by phage (Comeau and Krisch, 2005).

That persistence is a major life strategy of phage is confirmed by the large numbers of genes associated with persistence (i.e. integrases, immunity) observed in metagenomic screens. There is also much practical experience that supports the crucial role of prophage in host evolution. One particularly well-studied system that has been studied for over 50 years is the ongoing evaluation of phage evolution as observed in the dairy industry (Canchaya *et al.*, 2003, 2004; Brussow *et al.*, 2004). The temperate phage analysis of these bacteria follows a long tradition of lambda and *E. coli* studies (Campbell *et al.*, 1992; Canchaya *et al.*, 2003, 2004). Since lytic phage can severely disrupt dairy fermentation, it was of particular interest to understand and trace their evolution. These studies have led Brussow to conclude that much of the more recent dairy bacteria evolution can be considered to have resulted from the action of temperate phage.

A similar view applies to *E. coli* and cyanobacteria. In addition, the ECOR collection of 72 sequenced *E. coli* genomes of medical interest shows that they differ from each other mainly due to patterns of genetic colonization, mostly by prophage, but they also show the presence tRNA-adjacent defective prophage and plasmid elements that differentiate these strains (Hurtado and Rodriguez-Valera, 1999; Mazel *et al.*, 2000; Nilsson *et al.*, 2004). Cyanobacteria (*Prochlorococcus*) is major model for the study of the origin of the type II (plant-like) photosynthetic system. Since such genes show much evidence of recent and massive horizontal movement, it seem quite likely that prophage are mediators of such transfers, especially as these phage encode their own version of these photosynthetic genes (Lindell *et al.*, 2004; Sullivan *et al.*, 2006).

Very similar *Prochlorococcus* strains exist in distinct oceanic populations in various habitats known as ecotypes. Some think that such ecotypes represent the initial type of genetic variation that leads to speciation. The sequencing of six ecotypes has shown that they are 99% similar to one another, but the genetic variation that distinguishes them is mostly due to patterns of prophage colonization (called phage islands) (Bouman *et al.*, 2006; Coleman *et al.*, 2006).

Thus in all these prokaryotic models, persisting viruses play a fundamental role in host evolution and host genetic novelty is mostly phage derived. Such observations have led some to propose that "war is peace" regarding virus–host evolution (Comeau and Krisch, 2005). Massive and complex innovation by phage appears to be a major force in the prokaryotic world.

Prokaryotes are the most adaptable of all cells. If we can accept the above conclusion concerning the role for viruses in the evolution of prokaryotes, we must then ask why such a successful evolutionary strategy was not apparently maintained in eukaryotes? In eukaryotes we see little evidence that large-scale integration by DNA viruses is an important evolutionary process (although the story with retroviruses is different). Why should prokaryotes and eukaryotes differ is such a fundamental way? Nevertheless, as noted at the start of this section, we do see good evidence that links the evolution of large DNA viruses of prokaryotes to the large DNA viruses of eukaryotes.

DNA Quasispecies

In case we were becoming comfortable with the apparently clear distinctions between RNA and DNA virus evolution as outlined above (quasispecies vs. domain recombination respectively), the evolution of the parvoviruses informs us that DNA viruses can also evolve by a quasispecies process. Parvovirus evolution (see Chapter 17) can show a sharp contrast to the evolutionary pattern displayed by other small dsDNA viruses above (HPV, Py). With the emergence of an acute pandemic in domestic dogs and cats (as well as other wild carnivore species), we see what is essentially evolution driven by single point mutations, mostly affecting the capsid genes and host cell receptor binding. This system provides us with one of the better studied examples of the evolutionary dynamics of an emergent viral disease. In addition, *in vivo* mouse studies with minute virus of mouse (MVM) now make it clear that parvoviruses can behave much like RNA viruses, generating quasispecies of diverse progeny that allow a high adaptability for the generation of fitness and disease *in vivo* (Lopez-Bueno *et al.*, 2006). This story is very reminiscent of the study of poliovirus in mice mentioned above. Human studies with B19 parvovirus are also consistent with high mutation rates (Parsyan *et al.*, 2007; Shackelton and Holmes, 2006).

THE TRANSITION TO EUKARYOTIC DNA VIRUSES

The Phycodnavirus Exemplar

Although not specifically addressed in this volume, the viruses of eukaryotic unicellular green algae are of special interest from the perspective of DNA virus evolution. These large, complex dsDNA membrane-containing icosahedral viruses are abundant in some water habitats (Van Etten, 2003; Ghedin and Claverie,

2005). The reason they deserve special attention is that they clearly have many features that are characteristic of both prokaryotic and eukaryotic viruses. They resemble prokaryotic viruses in that their life cycle is clearly phage-like, such as external virion attachment, injection of DNA and no pinocytosis. In addition, they also encode many phage-like genes, such as restriction-modification enzymes and homing endonucleases (Filee *et al.*, 2006c). They also resemble eukaryotic viruses in that they have eukaryotic DNA replication proteins (DNA polymerase beta and PCNA; Chen and Suttle, 1996; Nagasaki *et al.*, 2005; Villarreal and DeFilippis, 2000) as well as many genes associated with eukaryotic signal transduction (Van Etten *et al.*, 2002). Thus they represent a clear link between prokaryotic and eukaryotic DNA viruses. For example, the DNA polymerase of paramecium bursaria chlorella virus (PBCV-1) is the most conserved gene and most closely resembles that found in human herpesvirus and is distantly related to the similar family DNA pol encoded by T4. This polymerase is distinct from that of the poxviruses or PRD1/adenoviruses (associated with protein-primed DNA replication). However, numerous other genes of the phycodnaviruses are similar to some genes found in the mimiviruses (giant DNA virus of ameba), including the presence of conserved intenes in the DNA pol gene (Ogata *et al.*, 2005).

In view of this it is most curious that in structural similarity, polydnavirus capsids clearly resemble PRD1 capsid (Khayat *et al.*, 2005; Nandhagopal *et al.*, 2002). PRD1 contains the double-barrel trimer capsid structure that was first observed in adenovirus (for references see Saren *et al.*, 2005). Adenovirus also closely resembles PRD1 in DNA replication strategy (i.e. linear DNA with covalently closed ends (Benson *et al.*, 2004; Khayat *et al.*, 2005). The lineage of adenovirus-like DNA viruses, however, is thought to be distinct from that herpes and poxviruses and its DNA polymerase is clearly distinct from polyndavirus. It is clear that related elements of all these viruses can be found in phycodnaviruses. Overall, the phycodnaviruses, like phage, also appear to be creating genes in large numbers and they encode many genes unrelated to their host. What then is the evolutionary relationship that links all of these seemingly distinct viruses?

As outlined above, the pattern of evolution of dsDNA phage involves lots of exchange by recombination from a vast gene pool. This pool resembles a cloud from which various mosaic subelements and substrategies are assembled to allow viral gene acquisition and novelty (Blum *et al.*, 2001; Benson *et al.*, 2004). Does such a distributed pattern of evolution and gene novelty also apply to the phycodnaviruses?

Recently, another distinct phycodnavirus has been sequenced: coccolithovirus (EhV-86) (Allen *et al.*, 2006a, 2006b) conserves only 24 core genes in common with PBCV-1 and is unique to the phycodnaviruses in that it has acquired six DNAdep RNA polymerase subunit genes, which are absent in all other phycodnaviruses.

As RNA polymerase is considered a core viral gene function, it is clear that phycodnaviruses can alter some very basic molecular functions during their evolution. Oceanic phycodnaviruses are thought to have large influence on the free-living populations of eukaryotic algae, such as the termination of algal blooms reported for Emilian Huxley virus (Martinez *et al.*, 2007; Schroeder *et al.*, 2003). However, not all phycodnaviruses are lytic. Another lineage of phycodnaviruses is represented by two viruses of filamentous brown algae, EsV-1 and FirrV-1 (Delaroque *et al.*, 2003). Unlike the lytic phycodnaviruses noted above, these two viruses are "temperate phage" like. That is they exist as silent viruses whose DNA is integrated into the germlines of their host. In this, they are unique to all known eukaryotic DNA viruses; host chromosome integration is a normal part of their persistent life strategy. EsV-1 has a 335 593-bp genome and encodes 231 likely genes (Delaroque *et al.*, 2001). These genes are mostly unique and only 28 are clearly related to PBCV-1 genes. The gene differences include many replication genes and their gene order is completely different. Like the temperate phage–host evolutionary relationship outlined above, it would be most interesting to understand how the integration of these large DNA viruses has affected host evolution.

Thus, the phycodnaviruses appear to represent a basal but diverse viral lineage that has both acute and persistent lifestyle and have some clear relationships to most large eukaryotic DNA viruses and many phage.

A Comment on the Proposed Monophylogeny of Nucleo-Cytoplasmic Large DNA Viruses (NCLDVs)

The phycodnavirus exemplar above should leave us with several impressions regarding the nature and evolution of these large and ubiquitous DNA viruses of algae, an early eukaryotic host. They show clear linkages by structure and function to both phage and various eukaryotic DNA viruses. They also show major variation and novelty in their own genetic composition, including their core genes. In addition, they show clear relationships to distinct and seemingly separate viral lineages (adenoviruses, herpesviruses, poxviruses, iridoviruses). The picture we are left with is that they seem to resemble phage evolution in that they appear to have evolved from a diverse pool that has exchanged many basic viral features and created many new genes.

This view, however, contrasts sharply with the work of Iyer *et al.* (2001, 2006). By considering the small number of conserved genes in four families of eukaryotic DNA viruses (poxviruses, asfarviruses, iridoviruses, phycodnaviruses), they suggest that these viruses are monophyletic, evolving from a common nucleo-cytoplasmic large DNA virus (NCLDV) with an icosahedral capsid.

Given the above information, I find this view unhelpful and possibly confusing. It has numerous problems. The main problem is that it fails to acknowledge the clear link between prokaryotic and eukaryotic viruses. Furthermore, by focussing on a small set of related genes, it represents a traditional perspective as found in evolutionary biology that assumes a common (fittest) linear lineage, not a cloud, cooperative, or mosaic pool as the main source of novelty resulting in the matrix pattern of virus evolution. The virosphere is clearly not disconnected from itself, but it is also clearly not a linear or tree-like evolutionary system as suggested above. We must learn to think of virus evolution in its own terms; fuzzy, mixed, reticulated, and cloud-like.

HERPESVIRUS; MOSTLY PERSISTING AND CO-SPECIATION

As mentioned in the phage section, there have been various publications that suggest a deep evolutionary relationship between the herpesviruses and dsDNA viruses of prokaryotes (Rice *et al.*, 2004; Khayat *et al.*, 2005; Duda *et al.*, 2006; Akita *et al.*, 2007). Such enormously distant relationships, however, cannot now be measured by any reliable metric. Although herpes-like viruses are found in invertebrates (such as ostreid herpesvirus 1 (OsHV-1)) in both lytic and asymptomatic states (Barbosa-Solomieu *et al.*, 2005), our interest in their evolution has been mainly focussed on the vertebrate herpesviruses. Vertebrate herpesvirus do tend to show clear sequence conservation that suggests broad patterns of evolution. One interesting feature of this evolution is the apparent link between the biology of the virus and its evolution. A common, but not universal pattern is that of virus and host co-evolution (McGeoch *et al.*, 2000, 2006; McGeoch and Gatherer, 2005). This trend has maintained several biological characteristics, such as highly species host- and tissue-specific persistence (i.e. neuronal and lymphoid persistence).

The discovery of HHV-8 has further stimulated studies of herpesvirus evolution in that HHV-8 appears to have undergone much recombination with herpesviruses of related primate lineages (McGeoch and Davison, 1999). Thus recombination seems prevalent in herpesviruses. The apparent link between herpesvirus evolution and recent human evolution, as well as an apparent link to primate retroviral evolution, is fascinating, but of unknown significance (Kung and Wood, 1994; Lacoste *et al.*, 2000).

Herpesvirus Gene Acquisition

The herpesviruses lineages will often show the presence of lineage-specific genes. Many of these genes affect innate and adaptive host functions, whereas others affect host metabolism. When the source of such genes has been contemplated, in contrast to phage, phycodnaviruses, or baculoviruses (Herniou *et al.*, 2001), it is often proposed that most such herpes genes originate from the host. It is well accepted that the three major lineages of herpesviruses descended from a common ancestor in vertebrates (McGeoch *et al.*, 2006). There have been numerous proposals that most new lineage-specific herpesvirus genes have originated from host (see Becker and Darai, 2000). This includes herpesvirus dUTPase (Davison and Stow, 2005), and viral chemokines and viral Bcl-2 (Nicholas *et al.*, 1998). In my evaluation of such claims, however, it seems clear that the possibility that there was an ancient viral source of such genes was not considered and cannot now be dismissed.

We currently believe that ancient herpesvirus ancestors can be traced to tailed phage (Hendrix, 1999; Bamford, 2003; Baker *et al.*, 2005; Duda *et al.*, 2006; McGeoch *et al.*, 2006). Other phage lineages also appear to trace to eukaryotic viruses (Bamford *et al.*, 2005b). Within the herpesviruses, the same T-16 icosahedral structure, as well as invertable DNA regions are also present in the very distant but much more recognizable oceanic ostreid herpesvirus 1 (Davison *et al.*, 2005).

Given the highly diverse and mosaic nature of large DNA virus evolution in prokaryotes and lower eukaryotes described above, it seem quite possible that many other viral genes might also trace far back in virus evolution. Consider the example of dUTPase in avian and mammalian herpesvirus (Davison and Stow, 2005; McGeehan *et al.*, 2001). The current view requires very complicated gene rearrangements to account for the viral source of this gene from its host. Yet we know that diverse dUTPases are found in many ancient viral lineages. For example, the ERVs present in all vertebrate genomes also conserve dUTPase (Jern *et al.*, 2005), as do exogenous retroviruses (i.e. lentiviruses) (McIntosh and Haynes, 1996). In fact, since the herpesviruses genes are especially poor in introns, it would seem likely that any herpesviral gene acquisition would necessarily involve a retrovirus via a cDNA.

The oceans are especially filled with large complex DNA viruses (such as mimivirus and phycodnavirus, plus numerous relatives of OsHV-1) thought to be ancient ancestors of herpesvirus. The phycodnavirus (chlorella virus, PBCV-1) provides a clear bridge between phage and eukaryotic DNA viruses. PBCV-1 also encodes a dUTPase that has the highly conserved motif III (Zhang *et al.*, 2005). Many phage are also known to encode dUTPases of diverse types, such as *B. subtilis* (SPbeta) (Persson *et al.*, 2005), and a phage of *Thermus thermophilus* (Naryshkina *et al.*, 2006). This *Thermus* phage (phiYS40) is of special note since its dUTPase gene is clearly related to the dUTPases of eukaryotic viruses and has a version that has undergone multiple events of recombination from apparently distinct phage, exactly as expected for mosaic phage genes.

Thus, the origin of new herpesvirus genes might not be so different than that seen in other large DNA viruses and a potential ancient source of new genes from these ancestral viruses remains plausible.

Similar considerations apply to other possible examples of herpesvirus gene capture. For example, the herpes thymidylate synthase (TS) has also been considered to have originated by host gene capture (Chen *et al.*, 2001). Yet distinct versions of these genes are also found in different herpesviral lineages, which would necessitate multiple independent "capture" events of different version of host TS genes. TS genes are present in ancient virus sources. For example, *Bacillus* phage beta 22 encodes TS, which also has a self-splicing intron (Bechhofer *et al.*, 1994). Also, phage phiKZ has a highly conserved TS (Mesyanzhinov *et al.*, 2002), yet this virus lacks a DNA polymerase or other replication proteins, clearly indicating that the viral TS genes has a basic viral role. Similarly, the cytokines-like genes (such as IL-10) as found in poxviruses and herpesviruses appear to have originated in at least three independent events prior to the divergence of mammalian eutherian orders. Yet it is still presupposed that they are necessarily the products of host gene capture (Hughes, 2002).

Comparative genomics supports the idea that the herpesviruse lineages are originating viral genes. A broader phylogenetic analysis of all herpesvirus genomes identified only 17 genes in common to all 30 taxa of herpesvirus (Wang *et al.*, 2006). Thus only 30 genes appear to be in common to all the herpesviruses. In this analysis, only a few genes of recent origin could be identified as possibly having been transferred between virus and host (e.g. new genes found at tips of phylogenetic dendograms). Thus, gene gain in the herpesviruses (as in DNA phage and phycodnavirus) is prevalent but the origination of such genes from the host is not prevalent. I suggest that our tendency to assume that new viral genes are usually "stolen" from the host should be revised (Moreira and Lopez-Garcia, 2005).

Poxvirus as Mostly Acute, with Frequent Species Shifts

In contrast to the herpesviruses, the poxviruses evolution tend to have little congruence to host evolution (see Chapter 19). Yet, they too show evidence of ancient linkages to other viruses. The replication of poxvirus DNA is distinct in that it involves a linear genome with inverted ends that have covalently closed "snapback" DNA. The resulting replication structures involve head-to-tail and tail-to-tail intermediates. This replication strategy is very different from that used by the host (and most other DNA viruses), but is clearly related to that found in other eukaryotic and prokaryotic viruses. Similar replication mechanisms are seen in all poxviruses, as well as African swine fever virus and phycodnaviruses (PBCV-1). This exact replication strategy is also present in archaeal lipothrixviruses (SIRV1 and SIRV2) which has been proposed to be ancestral to phycodnaviruses and poxviruses (Persson *et al.*, 2005). A similar replication strategy is also seen with N15 (Lobocka *et al.*, 1996), an unusual phage of *E. coli* that persists as a linear DNA (Casjens *et al.*, 2004).

Conservation of such replication similarities clearly suggests ancestral relationships, but no sequence similarity can be seen between these viruses. The similarity between poxvirus and PBCV-1 DNA replication deserves some additional comment. PBVC-1 and herpesvirus have very similar DNA polymerase genes, yet differ fundamentally in replication strategy. Furthermore, the poxviral DNA polymerase gene is very different from that found in the herpesviruses. Yet, the PBCV-1 capsid was clearly similar to that of adenoviruses and PRD1 phage (and iridovirus capsids). How then do we link poxvirus evolution to other more ancient DNA viruses, such as PBCV-1 which has the same DNA replication mechanism, but distinct replication proteins? Such observations might seem confusing, but they are clearly consistent with mosaic, reticulated evolution of DNA viruses. Various distinct phage lineages can link in multiple ways to various distinct eukaryotic DNA viruses. The concept of a net or matrix rather than a tree is thus a better way to describe the broad topology of DNA virus evolution.

The issue of gene gain and gene loss is also of central interest to orthopoxvirus evolution. Typically, we seek to understand poxviruses evolution from the perspective of pathogenesis, such as the origin of human-specific smallpox virus. With the comparative genomics of several orthopoxviruses now possible, we see curious overall patterns of gene loss in their evolution (Randall *et al.*, 2004). For example, comparing human smallpox to cowpox DNA (a rodent virus that is phylogenetically basal to smallpox), we observe an overall diminution of gene content in smallpox virus. Several poxviruses seem to have also lost genes relative to cowpoxvirus, especially genes that appear to affect immunity (Hughes and Friedman, 2005).

I suggest that this evolutionary tendency for gene reduction is associated with a switch from a more demanding species-specific persistent life strategy to a less demanding, acute life strategy in a new host. Cowpox is a naturally persistent infection in rodents (bank voles) (Feore *et al.*, 1997; Chantrey *et al.*, 1999), which has been called a natural virus reservoir (Hazel *et al.*, 2000). Smallpox is a strictly acute and human-specific disease. Such gene loss in association with lost persistence could be a general situation and might also explain why clinical isolates of human cytomegalovirus isolates show a strong tendency to delete genes with passage in culture (Davison *et al.*, 2003).

Most orthopoxviruses are not phylogenetically congruent with their vertebrate host. Host switching and acute replication seem to be relatively common but recent occurrences in their evolution (Babkin and Shchelkunov, 2006). The avian poxviruses are not as well studied in this context, but curiously have significantly more complex genomes than the orthopoxviruses (Jarmin *et al.*, 2006). The entomopoxviruses are even less well understood from both a biological and molecular perspective, although they do conserve 49 genes found in all poxvirus family members (Gubser *et al.*, 2004). Clearly these poxviruses share some degree of evolutionary history.

It is most curious that entomopoxviruses have even larger, more diverse and complex genomes than the other poxviruses. Why? As insects lack an adaptive immune system (the target of many orthopoxvirus genes), they would seem to present a simpler host for virus adaptation. This group appears to be the most basal phylogenetically, but evolutionary relationships between entomopoxvirus and insect evolution have not been studied. The enotomopoxviruses are particularly prevalent in grasshopper and locust species, often in unapparent states. Interestingly, within these viruses we can find examples of major shifts in core replication genes, such as the family of DNA pol gene that is used (a shift from DNA pol X to DNA pol B in two entomopoxvirus lineages). We can recall that the DNA pol B gene closely resembles that found in phycodnaviruses (and herpesvirus), but is distinct from that in orthopoxvirus (Zhu, 2003). We also see in the entomopoxviruses some clear links to phage genes, such as T4-like RNA ligase found in all entomopoxviruses (Ho and Shuman, 2002) as well as a lambda-like integrase seen in D1EPV (Hashimoto and Lawrence, 2005). This integrase in D1EPV implies possible integration and persistence, thus it is most significant that D1EPV also shows a clear persistent host infection as well as symbiosis and apparent phylogenetic congruence between virus and host.

This virus is symbiotic in its parasitoid wasp host in that virus is injected into larval host along with the wasp egg (and also along with a second D1RhV rhabdovirus) and virus is needed for successful host parasitization. This symbiosis is clearly very reminiscent of the genomic polydnaviruses of other parasitoid wasp species. DIEPV is also expressed in the male poison gland. However, it is unknown if DIEPV is integrating into the host DNA. Clearly, D1EPV it is part of a complex virus–virus–host symbiotic interaction.

Small DNA Viruses

The overall evolution of orthopoxviruses contrasts sharply with that of the papillomaviruses as presented by Bernard in Chapter 18. Here, highly species-specific and tissue-specific host infection are the norm and the viral evolution is typically highly congruent with the host (with some exceptions). The resolution between virus and host can be high, in that human racial and geographical populations, for example, can often be differentiated based on the type of HPV they harbor. Yet here too there is evidence of significant shifts in core gene usage early during papillomavirus evolution. In the human and rodent viruses, a highly conserved gene function associated with replication and cell control are the E6 and E7 early genes. In particular, the pRB-binding domain of the E7 gene is thought to be central to the biological strategy of the virus. Thus, it is most curious that the papillomaviruses of lagomorphs, such as bovine and reindeer papillomavirus, lack an E7 Rb-binding domain and instead appear to use E5 or E9 genes for this regulatory function (Narechania *et al.*, 2004). It seems an early but significant and bifurcating shift occurred in the molecular strategy during the virus–host evolution of this group of viruses for unknown reasons.

Other small DNA viruses (JCV, BKV, Py) can also show similar high-resolution host congruence (Shadan and Villarreal, 1995). As well as similar curious shifts in basic molecular strategies. For example, the presence of a middle T-antigen in mouse virus (a third early gene), but its absence from primate viruses (Gottlieb and Villarreal, 2001), clearly differentiates these viral lineages. Although the origins of these entire small DNA viruses are obscure, and any links to prokaryotic viruses are unknown, it does appear they have tended to retain their overall biological strategy and show a strong tendency for tissue-specific (especially kidney) persistence and virus–host congruence.

Persistence as Symbiosis, another Foundation for Virus–Host Evolution

Since persistence requires the stable coexistence of a virus and its host, it also fits the simple definition of symbiosis (the stable living together of two distinct lineages of organisms). Viral involvement in symbiosis is a foreign idea to many and possibly presents a fundamentally different view of the role viruses may have in host evolution. A major role for persisting (temperate, cryptic) viruses in the evolution of prokaryotes is no longer a controversial idea. Thus, at least in the prokaryotic world, virus persistence can be accepted as adaptive. In eukaryotes, however, viral persistence is seldom considered adaptive. The MHV–mouse exemplar as presented above has suggested how persistence can directly affect host survival. Can this be considered an example of symbiosis in the accepted sense?

A crowning achievement in the field of symbiosis has been to explain the origin of plastids (chloroplasts, mitochondria) from symbiotic prokaryotes in eukaryotic cytoplasm (Margulis and Bermudes, 1985). This idea involves the high adaptability of prokaryotes to provide innovation but would seem not to involve virus in any way. Yet here too we can find viral footprints that suggest some involvement. For example, various plastid-specific RNA and DNA polymerases clearly resemble polymerases from T3/T7-like phage (Cermakian *et al.*, 1996; Shutt and Gray, 2006). Other models of symbiosis also show evidence of a viral role, such as the sexual isolation of *Buchnera* (Moran *et al.*, 2005).

Another very popular topic in the field of symbiosis is the symbiotic origin of the photosynthetic sea slug, *Elysa chlorotica*. What could be more fascinating than a green sea slug—an animal that can use light for photosynthesis? *E. chlorotica* eats photosynthetic eukaryotic algae (*Vaucheria litorea*) and retains the functional chloroplast from algae for months. Here too, however, there lies a viral footprint. This slug harbors an unusual endogenous retrovirus which is expressed in large numbers during sexual reproduction, following which all slugs die via synchronized apoptosis and in which the chloroplasts have accumulated numerous viral particles (Pierce *et al.*, 1999; Mondy and Pierce, 2003). Since there is reason to think gene movement from the algae to the slug genome is involved in this symbiosis, the presence of this retrovirus is a strong candidate to also be involved in symbiogenesis.

Clearly we should thus investigate retroviral elements as possible symbiotic participants and not dismiss them beforehand as irrelevant or "junk DNA" (as is automatically done in many database screens).

If viral persistence is a kind of symbiosis, viruses may also mediate the establishment of other symbiotic relationships (Villarreal, 2007). The recent studies by Roossinck and colleagues (see Chapter 12), in which a persisting virus, a plant, and a fungus were all symbiotically involved in altering the thermal tolerance of the plant could be an example of this (Marquez *et al.*, 2007). Many other virus–host relationships should also be examined for possible symbiosis. For example, placental vertebrate evolution has involved various endogenous retroviruses (i.e. HERV-W, HERV-FRD). Intact HERV genomes, including *env* ORFs, are important for placental trophoblast fusion (for references see Ryan, 2007).

Some will dismiss this situation as the quirky usurping of a viral gene for host function which is of little general significance. The specific ERV involved is simply selfish and mostly defective genetic material of no general consequence. If so, why is it that in sheep a distinctly different lineage of retrovirus (enJSRV) was also selected to provide a related placental function to a another mammal with significantly different placental reproductive biology? It has been experimentally well established the enJSRV *env* is essential for sheep embryo implantation (Dunlap *et al.*, 2006a, 2006b). enJSRV is the endogenous version of JSRV, a problematic sheep-specific retrovirus that induces lung tumors (responsible for the death of Dolly, the famous first cloned sheep). The endogenous virus (enJSRV) is present in 20 copies in the sheep genome and all sheep have this virus. Sheep genomes also encoded a *trans*-dominant enJRSV *gag* that is inhibitory to exogenous JSRV (Mura *et al.*, 2004; Oliveira *et al.*, 2006; Murcia *et al.*, 2007). It seems clear that this situation can also be considered from the perspective of viral symbiosis and/or virus addiction in host evolution. We should thus seek to understand why colonization by an ERV population might generally provide a good solution to the evolutionary demands of placental biology.

Open Questions Regarding Virus in Human Evolution

There are many other opportunities to examine the potential role of persistent and symbiotic viruses in the evolution of viruses, animals, primates, and humans. For example, as we seek to understand the origins of the adaptive immune system we should pay attention to viral footprints. We can ask, for example, why the major histocompatibility complex (MHC) locus, the most polymorphic, diverse, and rapidly evolving gene set in our chromosome, is so densely colonized with retroviral elements (Andersson *et al.*, 1998). Why is a retrovirus also the basic element of the duplication unit that was thought to have been the progenitor for the expansion of the MHC class I (and II) genes (Gaudieri *et al.*, 1997; Kulski *et al.*, 1998, 2005)? Why do similar HERV element (L and 16) also differentiate between human and chimpanzee MHC I (Watkins, 1995; Kulski *et al.*, 1999)? What was the role for SIV in the evolution of primate MHC (Vogel *et al.*, 1999)?

Humans and primates appear to have undergone some significant and relatively recent evolution with regard to their endogenous and exogeneous retroviruses. Along these lines, APOBEC-like genes are basic component of the adaptive immune response but they are also antiretroviral genes that act on retroviral cDNA and *gag* (OhAinle *et al.*, 2006). The APOBEC3 antiviral system has expanded recently in humans, but not chimpanzees (Sawyer *et al.*, 2004; OhAinle *et al.*, 2006). Why? All African primates support unapparent foamy viruses (and also SIV co-infection), but not humans (Murray and Linial, 2006). APOBEC3C is active against foamy viruses (Delebecque *et al.*, 2006). Old World primates also underwent an expansion of ERVL colonization (a clear relative of foamy virus) (Sawyer *et al.*, 2004). Was this ERVL colonization of relevance to the ancient co-speciation of simian foamy virus and their primate host (Switzer *et al.*, 2005)? What exactly was the relevance of HERV endogenization to human survival and adaptations?

Curiously, human brain (neocortex) specifically expresses many of these more recent ERVs as transcripts (Nakamura *et al.*, 2003; Yi *et al.*, 2004). If we consider these situations as possible examples of virus-mediated symbiosis in human evolution, perhaps they may make more sense of the otherwise confusing role or HERVs.

REAL-TIME VIRUS–HOST EVOLUTION: KOALA BEAR EXEMPLAR

As noted, all primates, but especially humans show much evidence of recent endogenization by retroviruses. But these events mostly occurred in our extinct ancestors and we do not see ongoing evidence that any HERVs remain active. However, we are currently witnessing a related virus–host evolutionary event of considerable interest. Koala bears, native marsupials of Australia, are currently experiencing a major epidemic caused by a leukemia-inducing retrovirus. As a consequence, they are undergoing massive endogenization by a gammaretrovirus (MLV-related). This virus is similar to Gibbon ape leukemia virus, but most likely originated from rodent ancestry (Tarlinton *et al.*, 2006; Fiebig *et al.*, 2006). The expectation is that extinction awaits those koalas that do not adapt or endogenize the retrovirus successfully (Stoye, 2006). This event has the appearances of a retroviral-driven addiction that will result in a genetic variant of koala bear that has acquired a new antiretroviral state. This seems equivalent to the expansion of human APOBEC3; or perhaps a closer analogy is the endogenization of a suppressive *gag* as occurred with enJSRV. The surviving koala bears will likely tolerate or be persistently infected with this retrovirus pool. The genome of the species will have undergone considerable (but unpredictable) genetic perturbations and likely contain a large pool of variant and defective retrovirus. However, in so doing, the descendent koalas will likely present a biological hazard to any koala species that remain virus-free (as in virus addiction). Currently, one island colony of koalas is sufficiently isolated to have remained virus-free. This population will henceforth be under persistent threat from populations of endogenized koalas, now favored by group selection.

From the very earliest events in evolution of prebiotic replicators to very recent events in human evolution, including the emergence of human-specific HIV, we expect viral evolution to show profound effects on the evolution of all life. Unlike accepted host evolution, viruses also employ consortia and mixed populations to evolve, sometimes at unprecedented rates. Thus viruses have informed us of quasispecies, group dynamics, and group selection in evolution. Virus evolution should now be considered as basic science, not just a medical concern. We must acknowledge that the tree of life cannot be properly understood without virus evolution. This book helps to lay the foundation for such understanding.

REFERENCES

Adams, W.R. and Kraft, L.M. (1963) Epizootic diarrhea of infant mice: indentification of the etiologic agent. *Science* **141**, 359–360.

Akita, F., Chong, K.T., Tanaka, H., Yamashita, E., Miyazaki, N., Nakaishi, Y. *et al.* (2007) The crystal structure of a virus-like particle from the hyperthermophilic archaeon *Pyrococcus furiosus* provides insight into the evolution of viruses. *J. Mol. Biol.* **368**, 1469–1483.

Alexander, L., Weiskopf, E., Greenough, T.C., Gaddis, N.C., Auerbach, M.R., Malim, M.H. *et al.* (2000) Unusual polymorphisms in human immunodeficiency virus type 1 associated with nonprogressive infection. *J. Virol.* **74**, 4361–4376.

Allen, M.J., Schroeder, D.C., Donkin, A., Crawfurd, K.J. and Wilson, W.H. (2006a) Genome comparison of two Coccolithoviruses. *Virol. J.* **3**, 15.

Allen, M.J., Schroeder, D.C., Holden, M.T. and Wilson, W.H. (2006b) Evolutionary history of the Coccolithoviridae. *Mol. Biol. Evol.* **23**, 86–92.

Andersson, G., Svensson, A.C., Setterblad, N. and Rask, L. (1998) Retroelements in the human MHC class II region. *Trends Genet.* **14**, 109–114.

Babkin, I.V. and Shchelkunov, S.N. (2006) [The time scale in poxvirus evolution]. *Mol. Biol. (Mosk.)* **40**, 20–24.

Badrane, H. and Tordo, N. (2001) Host switching in Lyssavirus history from the Chiroptera to the Carnivora orders. *J. Virol.* **75**, 8096–8104.

Baker, M.L., Jiang, W., Rixon, F.J. and Chiu, W. (2005) Common ancestry of herpesviruses and tailed DNA bacteriophages. *J. Virol.* **79**, 14967–14970.

Bamford, D.H. (2003) Do viruses form lineages across different domains of life?. *Res. Microbiol.* **154**, 231–236.

Bamford, D.H., Grimes, J.M. and Stuart, D.I. (2005a) What does structure tell us about virus evolution?. *Curr. Opin. Struct. Biol.* **15**, 655–663.

Bamford, D.H., Ravantti, J.J., Ronnholm, G., Laurinavicius, S., Kukkaro, P., Dyall-Smith, M. *et al.* (2005b) Constituents of SH1, a novel lipid-containing virus infecting the halophilic euryarchaeon *Haloarcula hispanica*. *J. Virol.* **79**, 1107–9097.

Barbosa-Solomieu, V., Degremont, L., Vazquez-Juarez, R., Ascencio-Valle, F., Boudry, P. and Renault, T. (2005) Ostreid herpesvirus 1 (OsHV-1) detection among three successive generations of Pacific oysters (*Crassostrea gigas*). *Virus Res.*. **107**, 47–56.

Baric, R.S., Sullivan, E., Hensley, L., Yount, B. and Chen, W. (1999) Persistent infection promotes cross-species transmissibility of mouse hepatitis virus. *J. Virol.* **73**, 638–649.

Batschelet, E., Domingo, E. and Weissmann, C. (1976) The proportion of revertant and mutant phage in a growing population, as a function of mutation and growth rate. *Gene* **1**, 27–32.

Bechhofer, D.H., Hue, K.K. and Shub, D.A. (1994) An intron in the thymidylate synthase gene of *Bacillus* bacteriophage beta 22: evidence for independent evolution of a gene, its group I intron, and the intron open reading frame. *Proc. Natl Acad. Sci. USA* **91**, 11669–11673.

Becker, S.D., Bennett, M., Stewart, J.P. and Hurst, J.L. (2007) Serological survey of virus infection among wild house mice (*Mus domesticus*) in the UK. *Lab. Anim.* **41**, 229–238.

Becker, Y. and Darai, G. (2000) *Molecular Evolution of Viruses, Past and Present: Evolution of Viruses by Acquisition of Cellular RNA and DNA*. Boston: Kluwer Academic.

Bell, P.J. (2001) Viral eukaryogenesis: was the ancestor of the nucleus a complex DNA virus? *J. Mol. Evol.* **53**, 251–256.

Bell, P.J. (2006) Sex and the eukaryotic cell cycle is consistent with a viral ancestry for the eukaryotic nucleus. *J. Theor. Biol.* **243**, 54–63.

Bello, G., Casado, C., Garcia, S., Rodriguez, C., del Romero, J. and Lopez-Galindez, C. (2004) Co-existence of recent and ancestral nucleotide sequences in viral quasispecies of human immunodeficiency virus type 1 patients. *J. Gen. Virol.* **85**, 399–407.

Bello, G., Casado, C., Sandonis, V., Alonso-Nieto, M., Vicario, J.L., Garcia, S. *et al.* (2005) A subset of human immunodeficiency virus type 1 long-term non-progressors is characterized by the unique presence of ancestral sequences in the viral population. *J. Gen. Virol.* **86**, 355–364.

Benson, S.D., Bamford, J.K., Bamford, D.H. and Burnett, R.M. (2004) Does common architecture reveal a viral lineage spanning all three domains of life?. *Mol. Cell.* **16**, 673–685.

Biebricher, C.K. and Eigen, M. (2005) The error threshold. *Virus Res.* **107**, 117–127.

Biebricher, C.K. and Eigen, M. (2006) What is a quasispecies?. *Quasispecies: Concept and Implications for Virology* **299**, 1–31.

Blond, J.L., Lavillette, D., Cheynet, V., Bouton, O., Oriol, G., Chapel-Fernandes, S. *et al.* (2000) An envelope glycoprotein of the human endogenous retrovirus HERV-W is expressed in the human placenta and fuses cells expressing the type D mammalian retrovirus receptor. *J. Virol.* **74**, 3321–3329.

Blum, H., Zillig, W., Mallok, S., Domdey, H. and Prangishvili, D. (2001) The genome of the archaeal virus SIRV1 has features in common with genomes of eukaryal viruses. *Virology* **281**, 6–9.

Bouman, H.A., Ulloa, O., Scanlan, D.J., Zwirglmaier, K., Li, W.K., Platt, T. *et al.* (2006) Oceanographic basis of the global surface distribution of Prochlorococcus ecotypes. *Science* **312**, 918–921.

Breitbart, M. and Rohwer, F. (2005a) Here a virus, there a virus, everywhere the same virus?. *Trends Microbiol.* **13**, 278–284.

Breitbart, M. and Rohwer, F. (2005b) Method for discovering novel DNA viruses in blood using viral particle selection and shotgun sequencing. *Biotechniques* **39**, 729–736.

Breitbart, M., Hewson, I., Felts, B., Mahaffy, J.M., Nulton, J., Salamon, P. and Rohwer, F. (2003) Metagenomic analyses of an uncultured viral community from human feces. *J. Bacteriol.* **185**, 6220–6223.

Briones, C., de Vicente, A., Molina-Paris, C. and Domingo, E. (2006) Minority memory genomes can influence the evolution of HIV-1 quasispecies in vivo. *Gene* **384**, 129–138.

Brussow, H., Canchaya, C. and Hardt, W.D. (2004) Phages and the evolution of bacterial pathogens: from genomic rearrangements to lysogenic conversion. *Microbiol. Mol. Biol. Rev.* **68**, 560–602.

Caceres, M. and Thomas, J.W. (2006) The gene of retroviral origin Syncytin 1 is specific to hominoids and is inactive in Old World monkeys. *J. Hered.* **97**, 100–106.

Campbell, A. (2007) Phage integration and chromosome structure. A personal history. *Annu. Rev. Genet.*

Campbell, A., Schneider, S.J. and Song, B. (1992) Lambdoid phages as elements of bacterial genomes. *Genetica* **86**, 259–267.

Canchaya, C., Fournous, G., Chibani-Chennoufi, S., Dillmann, M.L. and Brussow, H. (2003) Phage as agents of lateral gene transfer. *Curr. Opin. Microbiol.* **6**, 417–424.

Canchaya, C., Fournous, G. and Brussow, H. (2004) The impact of prophages on bacterial chromosomes. *Mol. Microbiol.* **53**(1), 9–18.

Carobene, M.G., Rubio, A.E., Carrillo, M.G., Maligne, G.E., Kijak, G.H., Quarleri, J.F. and Salomon, H. (2004) Differences in frequencies of drug resistance-associated mutations in the HIV-1 pol gene of B subtype and BF intersubtype recombinant samples. *J. Acquir. Immune Defic. Syndr.* **35**, 207–209.

Carthew, P. (1977) Lethal intestinal virus of infant mice is mouse hepatitis virus. *Vet. Rec.* **101**, 465.

Casado, C., Garcia, S., Rodriguez, C., del Romero, J., Bello, G. and Lopez-Galindez, C. (2001) Different evolutionary patterns are found within human immunodeficiency virus type 1-infected patients. *J. Gen. Virol.* **82**, 2495–2508.

Casjens, S.R., Gilcrease, E.B., Huang, W.M., Bunny, K.L., Pedulla, M.L. *et al.* (2004) The pK02 linear plasmid prophage of *Klebsiella oxytoca*. *J. Bacteriol.* **186**, 1818–1832.

Castro, C., Smidansky, E., Maksimchuk, K.R., Arnold, J.J., Korneeva, V.S., Gotte, M. *et al.* (2007) Two proton transfers in the transition state for nucleotidyl transfer catalyzed by RNA- and DNA-dependent RNA and DNA polymerases. *Proc. Natl Acad. Sci. USA* **104**, 4267–4272.

Cermakian, N., Ikeda, T.M., Cedergren, R. and Gray, M.W. (1996) Sequences homologous to yeast mitochondrial and bacteriophage T3 and T7 RNA polymerases are widespread throughout the eukaryotic lineage. *Nucleic Acids Res.* **24**, 648–654.

Chantrey, J., Meyer, H., Baxby, D., Begon, M., Bown, K.J., Hazel, S.M. *et al.* (1999) Cowpox: reservoir hosts and geographic range. *Epidemiol. Infect.* **122**, 455–460.

Chare, E.R. and Holmes, E.C. (2006) A phylogenetic survey of recombination frequency in plant RNA viruses. *Arch. Virol.* **151**, 933–946.

Charpentier, C., Dwyer, D.E., Mammano, F., Lecossier, D., Clavel, F. and Hance, A.J. (2004) Role of minority populations of human immunodeficiency virus type 1 in the evolution of viral resistance to protease inhibitors. *J. Virol.* **78**, 4234–4247.

Charpentier, C., Nora, T., Tenaillon, O., Clavel, F. and Hance, A.J. (2006) Extensive recombination among human immunodeficiency virus type 1 quasispecies makes an important contribution to viral diversity in individual patients. *J. Virol.* **80**, 2472–2482.

Chen, F. and Suttle, C.A. (1996) Evolutionary relationships among large double-stranded DNA viruses that infect microalgae and other organisms as inferred from DNA polymerase genes. *Virology* **219**, 170–178.

Chen, H.H., Tso, D.J., Yeh, W.B., Cheng, H.J. and Wu, T.F. (2001) The thymidylate synthase gene of Hz-1 virus: a gene captured from its lepidopteran host. *Insect. Mol. Biol.* **10**, 495–503.

Christodoulou, C., Colbere-Garapin, F., Macadam, A., Taffs, L.F., Marsden, S. *et al.* (1990) Mapping of mutations associated with neurovirulence in monkeys infected with Sabin 1 poliovirus revertants selected at high temperature. *J. Virol.* **64**, 4922–4929.

Clarke, D.K., Duarte, E.A., Elena, S.F., Moya, A., Domingo, E. and Holland, J. (1994) The red queen reigns in the kingdom of RNA viruses. *Proc. Natl Acad. Sci. USA* **91**, 4821–4824.

Claverie, J.M., Ogata, H., Audic, S., Abergel, C., Suhre, K. and Fournier, P.E. (2006) Mimivirus and the emerging concept of "giant" virus. *Virus Res.* **117**, 133–144.

Clokie, M.R., Shan, J., Bailey, S., Jia, Y., Krisch, H.M., West, S. and Mann, N.H. (2006) Transcription of a 'photosynthetic' T4-type phage during infection of a marine cyanobacterium. *Environ. Microbiol.* **8**, 827–835.

Coleman, M.L., Sullivan, M.B., Martiny, A.C., Steglich, C., Barry, K., Delong, E.F. and Chisholm, S.W. (2006) Genomic islands and the ecology and evolution of Prochlorococcus. *Science* **311**, 1768–1770.

Comeau, A.M. and Krisch, H.M. (2005) War is peace—dispatches from the bacterial and phage killing fields. *Curr. Opin. Microbiol.* **8**, 488–494.

Comeau, A.M., Chan, A.M. and Suttle, C.A. (2006) Genetic richness of vibriophages isolated in a coastal environment. *Environ. Microbiol.* **8**, 1164–1176.

Crotty, S. and Andino, R. (2002) Implications of high RNA virus mutation rates: lethal mutagenesis and the antiviral drug ribavirin. *Microb. Infect.* **4**, 1301–1307.

Crotty, S., Cameron, C.E. and Andino, R. (2001) RNA virus error catastrophe: direct molecular test by using ribavirin. *Proc. Natl Acad. Sci. USA* **98**, 6895–6900.

Crotty, S., Hix, L., Sigal, L.J. and Andino, R. (2002) Poliovirus pathogenesis in a new poliovirus receptor transgenic mouse model: age-dependent paralysis and a mucosal route of infection. *J. Gen. Virol.* **83**, 1707–1720.

Crowder, S. and Kirkegaard, K. (2005) Trans-dominant inhibition of RNA viral replication can slow growth of drug-resistant viruses. *Nat. Genet.* **37**, 701–709.

Davison, A.J. and Stow, N.D. (2005) New genes from old: redeployment of dUTPase by herpesviruses. *J. Virol.* **79**, 12880–12892.

Davison, A.J., Dolan, A., Akter, P., Addison, C., Dargan, D.J., Alcendor, D.J. *et al.* (2003) The human cytomegalovirus genome revisited: comparison with the chimpanzee cytomegalovirus genome. *J. Gen. Virol.* **84**, 17–28.

Davison, A.J., Trus, B.L., Cheng, N., Steven, A.C., Watson, M.S., Cunningham, C. *et al.* (2005) A novel class of herpesvirus with bivalve hosts. *J. Gen. Virol.* **86**, 41–53.

Delaroque, N., Muller, D.G., Bothe, G., Pohl, T., Knippers, R. and Boland, W. (2001) The complete DNA sequence of the *Ectocarpus siliculosus* Virus EsV-1 genome. *Virology* **287**, 112–132.

Delaroque, N., Boland, W., Gerhard Müller, D. and Knippers, R. (2003) Comparisons of two large phaeoviral genomes and evolutionary implications. *J. Mol. Evol.* **57**, 613–622.

Delbruck, M. (1945) Interference between bacterial viruses: III. The mutual exclusion effect and the depressor effect. *J. Bacteriol.* **50**, 151–170.

Delebecque, F., Suspene, R., Calattini, S., Casartelli, N., Saib, A., Froment, A. *et al.* (2006) Restriction of foamy viruses by APOBEC cytidine deaminases. *J. Virol.* **80**, 605–614.

Descoteaux, J.P. and Mihok, S. (1986) Serologic study on the prevalence of murine viruses in a population of wild meadow voles (*Microtus pennsylvanicus*). *J. Wildl. Dis.* **22**, 314–319.

Descoteaux, J.P., Grignon-Archambault, D. and Lussier, G. (1977) Serologic study on the prevalence of murine viruses in five Canadian mouse colonies. *Lab. Anim. Sci.* **27**, 621–626.

Desjardins, C., Eisen, J.A. and Nene, V. (2005) New evolutionary frontiers from unusual virus genomes. *Genome Biol.* **6**, 212.

Domingo, E., Biebricher, C., Eigen, M. and Holland J.J. (2001) *Quasispecies and RNA Virus Evolution: Principles and Consequences.* Austin, TX: Landes Bioscience.

Domingo, E. (2006) Quasispecies: concept and implications for virology. In: *Curr. Top. Microbiol. Immunol,* Vol. 299. Berlin, New York: Springer.

Domingo, E. and Gomez, J. (2007) Quasispecies and its impact on viral hepatitis. *Virus Res.* **127**, 131–150.

Domingo, E., Sabo, D., Taniguchi, T. and Weissmann, C. (1978) Nucleotide sequence heterogeneity of an RNA phage population. *Cell* **13**, 735–744.

Domingo, E., Webster, R.G. and Holland, J.J. (eds) (1999) *Origin and Evolution of Viruses.* San Diego: Academic Press.

Doolittle, W.F. and Sapienza, C. (1980) Selfish genes, the phenotype paradigm and genome evolution. *Nature* **284**, 601–603.

Dreyfus, D.H., Jones, J.F. and Gelfand, E.W. (1999) Asymmetric DDE (D35E)-like sequences in the RAG proteins: implications for V(D)J recombination and retroviral pathogenesis. *Med. Hypoth.* **52**, 545–549.

Duda, R.L., Hendrix, R.W., Huang, W.M. and Conway, J.F. (2006) Shared architecture of bacteriophage SP01 and herpesvirus capsids. *Curr. Biol.* **16**, R11–R13.

Dunlap, K.A., Palmarini, M. and Spencer, T.E. (2006a) Ovine endogenous betaretroviruses (enJSRVs) and placental morphogenesis. *Placenta* **27**(Suppl A), S135–S140.

Dunlap, K.A., Palmarini, M., Varela, M., Burghardt, R.C., Hayashi, K., Farmer, J.L. and Spencer, T.E. (2006b) Endogenous retroviruses regulate periimplantation placental growth and differentiation. *Proc. Natl Acad. Sci. USA* **103**, 14390–14395.

Dupressoir, A., Marceau, G., Vernochet, C., Benit, L., Kanellopoulos, C., Sapin, V. and Heidmann, T. (2005) Syncytin-A and syncytin-B, two fusogenic placenta-specific murine envelope genes of retroviral origin conserved in Muridae. *Proc. Natl Acad. Sci. USA* **102**, 725–730.

Edwards, R.A. and Rohwer, F. (2005) Viral metagenomics. *Nat. Rev. Microbiol.* **3**, 504–510.

Eigen, M., McCaskill, J. and Schuster, P. (1988) Molecular quasi-species. *J. Phys. Chem.* **92**, 6881–6891.

Einer-Jensen, K., Ahrens, P. and Lorenzen, N. (2006) Genetic stability of the VHSV consensus sequence of G-gene in diagnostic samples from an acute outbreak. *Bull. Eur. Assoc. Fish Pathol.* **26**, 62–67.

Elena, S.F., Gonzalez-Candelas, F., Novella, I.S., Duarte, E.A., Clarke, D.K., Domingo, E. *et al.* (1996) Evolution of fitness in experimental populations of vesicular stomatitis virus. *Genetics* **142**, 673–679.

Enomoto, N., Kurosaki, M., Tanaka, Y., Marumo, F. and Sato, C. (1994) Fluctuation of hepatitis C virus quasispecies in persistent infection and interferon treatment revealed by single-strand conformation polymorphism analysis. *J. Gen. Virol.* **75**, 1361–1369.

Esnault, C., Heidmann, O., Delebecque, F., Dewannieux, M., Ribet, D., Hance, A.J. *et al.* (2005) APOBEC3G cytidine deaminase inhibits retrotransposition of endogenous retroviruses. *Nature* **433**, 430–433.

Fargette, D., Konate, G., Fauquet, C., Muller, E., Peterschmitt, M. and Thresh, J.M. (2006) Molecular ecology and emergence of tropical plant viruses. *Annu. Rev. Phytopathol.* **44**, 235–260.

Feore, S.M., Bennett, M., Chantrey, J., Jones, T., Baxby, D. and Begon, M. (1997) The effect of cowpox virus infection on fecundity in bank voles and wood mice. *Proc. R. Soc. Lond. B Biol. Sci.* **264**, 1457–1461.

Fiebig, U., Hartmann, M.G., Bannert, N., Kurth, R. and Denner, J. (2006) Transspecies transmission of the endogenous koala retrovirus. *J. Virol.* **80**, 5651–5654.

Filee, J., Forterre, P. and Laurent, J. (2003) The role played by viruses in the evolution of their hosts: a view based on informational protein phylogenies. *Res. Microbiol.* **154**, 237–243.

Filee, J., Tetart, F., Suttle, C.A. and Krisch, H.M. (2005) Marine T4-type bacteriophages, a ubiquitous component of the dark matter of the biosphere. *Proc. Natl Acad. Sci. USA* **102**(35), 12471–12476.

Filee, J., Bapteste, E., Susko, E. and Krisch, H.M. (2006a) A selective barrier to horizontal gene transfer in the T4-type bacteriophages that has preserved a core genome with the viral replication and structural genes. *Mol. Biol. Evol.* **23**, 1688–1696.

Filee, J., Comeau, A.M., Suttle, C.A. and Krisch, H.M. (2006b) T4-type bacteriophages: ubiquitous components of the "dark matter" of the biosphere. *Med. Sci. (Paris)* **22**, 111–112.

Filee, J., Siguier, P. and Chandler, M. (2006c) I am what I eat and I eat what I am: acquisition of bacterial genes by giant viruses. *Trends Genet.* (in press).

Fleuriet, A. (1994) Female characteristics in the *Drosophila melanogaster* sigma-virus system in natural-populations from Languedoc (Southern France). *Arch. Virol.* **135**, 29–42.

Fleuriet, A. (1996) Polymorphism of the *Drosophila melanogaster* Sigma virus system. *J. Evol. Biol.* **9**, 471–484.

Forterre, P. (1999) Displacement of cellular proteins by functional analogues from plasmids or viruses could explain puzzling phylogenies of many DNA informational proteins. *Mol. Microbiol.* **33**, 457–465.

Forterre, P. (2003) The great virus comeback—from an evolutionary perspective. *Res. Microbiol.* **154**, 223–225.

Forterre, P. (2005) The two ages of the RNA world and the transition to the DNA world: a story of viruses and cells. *Biochimie* **87**, 793–803.

Forterre, P. (2006a) The origin of viruses and their possible roles in major evolutionary transitions. *Virus Res.* **117**, 5–16.

Forterre, P. (2006b) Three RNA cells for ribosomal lineages and three DNA viruses to replicate their genomes: A hypothesis for the origin of cellular domain. *Proc. Natl Acad. Sci. USA* **103**, 3669–3674.

Forterre, P., Gribaldo, S. and Brochier, C. (2005) Luca: the last universal common ancestor. *Med. Sci. (Paris)* **21**, 860–865.

Forterre, P., Gribaldo, S., Gadelle, D. and Serre, M.C. (2007) Origin and evolution of DNA topoisomerases. *Biochimie* **89**, 427–446.

Forton, D.M., Karayiannis, P., Mahmud, N., Taylor-Robinson, S.D. and Thomas, H.C. (2004) Identification of unique hepatitis C virus quasispecies in the central nervous system and comparative analysis of internal translational efficiency of brain, liver and serum variants. *J. Virol.* **78**, 5170–5183.

Forton, D.M., Taylor-Robinson, S.D. and Thomas, H.C. (2006) Central nervous system changes in hepatitis C virus infection. *Eur. J. Gastroenterol. Hepatol.* **18**, 333–338.

Fraile, A., Escriu, F., Aranda, M.A., Malpica, J.M., Gibbs, A.J. and Garcia-Arenal, F. (1997) A century of tobamovirus evolution in an Australian population of *Nicotiana glauca*. *J. Virol.* **71**, 8316–8320.

Fugmann, S.D., Messier, C., Novack, L.A., Cameron, R.A. and Rast, J.P. (2006) An ancient evolutionary origin of the Rag1/2 gene locus. *Proc. Natl Acad. Sci. USA* **103**, 3728–3733.

Gannon, J. and Carthew, P. (1980) Prevalence of indigenous viruses in laboratory animal colonies in the United Kingdom 1978–1979. *Lab. Anim.* **14**, 309–311.

Gao, L. and Qi, J. (2007) Whole genome molecular phylogeny of large dsDNA viruses using composition vector method. *BMC Evol. Biol.* **7**, 41.

Garcia-Arriaza, J., Manrubia, S.C., Toja, M., Domingo, E. and Escarmis, C. (2004) Evolutionary transition toward defective RNAs that are infectious by complementation. *J. Virol.* **78**, 11678–11685.

Gaudieri, S., Kulski, J.K., Balmer, L., Giles, K.M., Inoko, H. and Dawkins, R.L. (1997) Retroelements and segmental duplications in the generation of diversity within the MHC. *DNA Seq.* **8**, 137–141.

Gazit, E. and Sauer, R.T. (1999) The Doc toxin and Phd antidote proteins of the bacteriophage P1 plasmid addiction system form a heterotrimeric complex. *J. Biol. Chem.* **274**, 16813–16818.

Gelfand, M.S. and Koonin, E.V. (1997) Avoidance of palindromic words in bacterial and archaeal genomes: a close connection with restriction enzymes. *Nucleic Acids Res.* **25**, 2430–2439.

Ghedin, E. and Claverie, J.M. (2005) Mimivirus relatives in the Sargasso sea. *Virol. J.* **2**, 62.

Gibbs, A. (1999) Evolution and origins of tobamoviruses. *Philos. Trans. R. Soc. Lond. B Biol. Sci.* **354**, 593–602.

Gilbert, M., Chaitaweesub, P., Parakamawongsa, T., Premashthira, S., Tiensin, T., Kalpravidh, W. *et al.* (2006) Free-grazing ducks and highly pathogenic avian influenza, Thailand. *Emerg. Infect. Dis.* **12**, 227–234.

Gorbalenya, A.E., Enjuanes, L., Ziebuhr, J. and Snijder, E.J. (2006) Nidovirales: evolving the largest RNA virus genome. *Virus Res.* **117**, 17–37.

Gottlieb, K.A. and Villarreal, L.P. (2001) Natural biology of polyomavirus middle T antigen. *Microbiol. Mol. Biol. Rev.* **65**, 288–318.

Grande-Perez, A., Lazaro, E., Lowenstein, P., Domingo, E. and Manrubia, S.C. (2005) Suppression of viral infectivity through lethal defection. *Proc. Natl Acad. Sci. USA* **102**, 4448–4452.

Gubser, C., Hue, S., Kellam, P. and Smith, G.L. (2004) Poxvirus genomes: a phylogenetic analysis. *J. Gen. Virol.* **85**, 105–117.

Gustafsson, E., Blomqvist, G., Bellman, A., Holmdahl, R., Mattsson, A. and Mattsson, R. (1996) Maternal antibodies protect immunoglobulin deficient neonatal mice from mouse hepatitis virus (MHV)-associated wasting syndrome. *Am. J. Reprod. Immunol.* **36**, 33–39.

Hambly, E. and Suttle, C.A. (2005) The viriosphere, diversity and genetic exchange within phage communities. *Curr. Opin. Microbiol.* **8**, 444–450.

Harcombe, W.R. and Bull, J.J. (2005) Impact of phages on two-species bacterial communities. *Appl. Environ. Microbiol.* **71**, 5254–5259.

Harki, D.A., Graci, J.D., Korneeva, V.S., Ghosh, S.K., Hong, Z., Cameron, C.E. and Peterson, B.R. (2002) Synthesis and antiviral evaluation of a mutagenic and non-hydrogen bonding ribonucleoside analogue: 1-beta-D-Ribofuranosyl-3-nitropyrrole. *Biochemistry* **41**, 9026–9033.

Harris, J.R. (1998) Placental endogenous retrovirus (ERV): structural, functional and evolutionary significance. *Bioessays* **20**, 307–316.

Hart, C.A. and Bennett, M. (1999) Hantavirus infections: epidemiology and pathogenesis. *Microbes Infect.* **1**, 1229–1237.

Hashimoto, Y. and Lawrence, P.O. (2005) Comparative analysis of selected genes from *Diachasmimorpha longicaudata* entomopoxvirus and other poxviruses. *J. Insect Physiol.* **51**, 207–220.

Hazel, S.M., Bennett, M., Chantrey, J., Bown, K., Cavanagh, R., Jones, T.R. *et al.* (2000) A longitudinal study of an endemic disease in its wildlife reservoir: cowpox and wild rodents. *Epidemiol. Infect.* **124**, 551–562.

Hendrix, R.W. (1999) Evolution: the long evolutionary reach of viruses. *Curr. Biol.* **9**, R914–R917.

Hendrix, R.W. (2002) Bacteriophages: evolution of the majority. *Theor. Popul. Biol.* **61**, 471–480.

Hendrix, R.W. (2003) Bacteriophage genomics. *Curr. Opin. Microbiol.* **6**, 506–511.

Hendrix, R.W., Hatfull, G.F. and Smith, M.C. (2003) Bacteriophages with tails: chasing their origins and evolution. *Res. Microbiol.* **154**, 253–257.

Hendrix, R.W., Smith, M.C., Burns, R.N., Ford, M.E. and Hatfull, G.F. (1999) Evolutionary relationships among diverse bacteriophages and prophages: all the world's a phage. *Proc. Natl Acad. Sci. USA* **96**, 2192–2197.

Herniou, E.A., Luque, T., Chen, X., Vlak, J.M., Winstanley, D., Cory, J.S. and O'Reilly, D.R. (2001) Use of whole genome sequence data to infer baculovirus phylogeny. *J. Virol.* **75**, 8117–8126.

Ho, C.K. and Shuman, S. (2002) Bacteriophage T4 RNA ligase 2 (gp24.1) exemplifies a family of RNA ligases found in all phylogenetic domains. *Proc. Natl Acad. Sci. USA* **99**, 12709–12714.

Hohne, M., Schreier, E. and Roggendorf, M. (1994) Sequence variability in the env-coding region of hepatitis C virus isolated from patients infected during a single source outbreak. *Arch. Virol.* **137**, 25–34.

Holmes, E.C. (2006) The evolutionary biology of dengue virus. *Novartis Found. Symp.* **277**, 177–187. discussion 187–192, 251–253.

Holmes, E.C. and Twiddy, S.S. (2003) The origin, emergence and evolutionary genetics of dengue virus. *Infect. Genet. Evol.* **3**, 19–28.

Homberger, F.R. (1997) Enterotropic mouse hepatitis virus. *Lab. Anim.* **31**, 97–115.

Horaud, F. (1993) Albert B. Sabin and the development of oral poliovaccine. *Biologicals* **21**, 311–316.

Hughes, A.L. (2002) Origin and evolution of viral interleukin-10 and other DNA virus genes with vertebrate homologues. *J. Mol. Evol.* **54**, 90–101.

Hughes, A.L. and Friedman, R. (2005) Poxvirus genome evolution by gene gain and loss. *Mol. Phylogenet. Evol.* **35**, 186–195.

Hurtado, A. and Rodriguez-Valera, F. (1999) Accessory DNA in the genomes of representatives of the *Escherichia coli* reference collection. *J. Bacteriol.* **181**, 2548–2554.

Ishida, T., Taguchi, F., Lee, Y.S., Yamada, A., Tamura, T. and Fujiwara, K. (1978) Isolation of mouse hepatitis virus from infant mice with fatal diarrhea. *Lab. Anim. Sci.* **28**, 269–276.

Iyer, L.M., Aravind, L. and Koonin, E.V. (2001) Common origin of four diverse families of large eukaryotic DNA viruses. *J. Virol.* **75**, 11720–11734.

Iyer, L.M., Balaji, S., Koonin, E.V. and Aravind, L. (2006) Evolutionary genomics of nucleo-cytoplasmic large DNA viruses. *Virus Res.* **117**, 156–184.

Jarmin, S., Manvell, R., Gough, R.E., Laidlaw, S.M. and Skinner, M.A. (2006) Avipoxvirus phylogenetics: identification of a PCR length polymorphism that discriminates between the two major clades. *J. Gen. Virol.* **87**, 2191–2201.

Jern, P., Sperber, G.O. and Blomberg, J. (2005) Use of endogenous retroviral sequences (ERVs) and structural markers for retroviral phylogenetic inference and taxonomy. *Retrovirology* **2**, 50.

Kapitonov, V.V. and Jurka, J. (2005) RAG1 core and V(D)J recombination signal sequences were derived from Transib transposons. *PLoS Biol.* **3**, e181.

Kashuba, C., Hsu, C., Krogstad, A. and Franklin, C. (2005) Small mammal virology. *Vet. Clin. North Am. Exot. Anim. Pract.* **8**, 107–122.

Khayat, R., Tang, L., Larson, E.T., Lawrence, C.M., Young, M. and Johnson, J.E. (2005) Structure of an archaeal virus capsid protein reveals a common ancestry to eukaryotic and bacterial viruses. *Proc. Natl Acad. Sci. USA* **102**, 18944–18949.

Kijak, G.H. and McCutchan, F.E. (2005) HIV diversity, molecular epidemiology and the role of recombination. *Curr. Infect. Dis. Rep.* **7**, 480–488.

Kijak, G.H., Simon, V., Balfe, P., Vanderhoeven, J., Pampuro, S.E., Zala, C. *et al.* (2002) Origin of human immunodeficiency virus type 1 quasispecies emerging after antiretroviral treatment interruption in patients with therapeutic failure. *J. Virol.* **76**, 7000–7009.

Knobler, R.L., Lampert, P.W. and Oldstone, M.B. (1982) Virus persistence and recurring demyelination produced by a temperature-sensitive mutant of MHV-4. *Nature* **298**, 279–280.

Koonin, E.V., Makarova, K.S. and Aravind, L. (2001) Horizontal gene transfer in prokaryotes: quantification and classification. *Annu. Rev. Microbiol.* **55**, 709–742.

Korneeva, V.S. and Cameron, C.E. (2007) Structure-function relationships of the viral RNA-dependent RNA polymerase: Fidelity, replication speed and initiation mechanism determined by a residue in the ribose-binding pocket. *J. Biol. Chem.* **282**, 16135–16145.

Kulski, J.K., Gaudieri, S., Bellgard, M., Balmer, L., Giles, K., Inoko, H. and Dawkins, R.L. (1998) The evolution of MHC diversity by segmental duplication and transposition of retroelements. *J. Mol. Evol.* **46**, 734.

Kulski, J.K., Gaudieri, S., Inoko, H. and Dawkins, R.L. (1999) Comparison between two human endogenous retrovirus (HERV)-rich regions within the major histocompatibility complex. *J. Mol. Evol.* **48**, 675–683.

Kulski, J.K., Anzai, T. and Inoko, H. (2005) ERVK9, transposons and the evolution of MHC class I duplicons within the alpha-block of the human and chimpanzee. *Cytogenet. Genome Res.* **110**, 181–192.

Kung, H.J. and Wood, C. (1994) *Interactions between Retroviruses and Herpesviruses*. Singapore; River Edge, NJ: World Scientific.

Kurosaki, M., Enomoto, N., Marumo, F. and Sato, C. (1994) Evolution and selection of hepatitis C virus variants in patients with chronic hepatitis C. *Virology* **205**, 161–169.

Lacoste, V., Mauclere, P., Dubreuil, G., Lewis, J., Georges-Courbot, M.C. and Gessain, A. (2000) KSHV-like herpesviruses in chimps and gorillas. *Nature* **407**, 151–152.

Lehnherr, H., Maguin, E., Jafri, S. and Yarmolinsky, M.B. (1993) Plasmid addiction genes of bacteriophage P1: doc, which causes cell death on curing of prophage and phd, which prevents host death when prophage is retained. *J. Mol. Biol.* **233**, 414–428.

Levin, S.A., Dushoff, J. and Plotkin, J.B. (2004) Evolution and persistence of influenza A and other diseases. *Math. Biosci.* **188**, 17–28.

Li, W., Shi, Z., Yu, M., Ren, W., Smith, C., Epstein, J.H. *et al.* (2005) Bats are natural reservoirs of SARS-like coronaviruses. *Science* **310**, 676–679.

Lindell, D., Sullivan, M.B., Johnson, Z.I., Tolonen, A. C., Rohwer, F. and Chisholm, S.W. (2004) Transfer of photosynthesis genes to and from Prochlorococcus viruses. *Proc. Natl Acad. Sci. USA* **101**, 11013–11018.

Liu, J., Glazko, G. and Mushegian, A. (2006) Protein repertoire of double-stranded DNA bacteriophages. *Virus Res.* **117**, 68–80.

Lobocka, M.B., Svarchevsky, A.N., Rybchin, V.N. and Yarmolinsky, M.B. (1996) Characterization of the primary immunity region of the *Escherichia coli* linear plasmid prophage N15. *J. Bacteriol.* **178**, 2902–2910.

Lopez-Bueno, A., Villarreal, L.P. and Almendral, J.M. (2006) Parvovirus variation for disease: a difference with RNA viruses?. *Curr. Top. Microbiol. Immunol.* **299**, 349–370.

Louz, D., Bergmans, H.E., Loos, B.P. and Hoeben, R.C. (2005) Cross-species transfer of viruses: implications for the use of viral vectors in biomedical research, gene therapy and as live-virus vaccines. *J. Gene Med.* **7**, 1263–1274.

Luria, S.E. (1950) Bacteriophage: an essay on virus reproduction. *Science* **111**, 507–511.

Luria, S.E. and Delbruck, M. (1943) Mutations of bacteria from virus sensitivity to virus resistance. *Genetics* **28**, 491–511.

Luria, S.E., Kellenberger, E., Harrison, B.D., Schafer, W., Hirst, G.K., Isaacs, A. *et al.* (1959) *Virus growth and variation: Ninth symposium of the society for general microbiology*. London: Cambridge University Press.

Lussier, G. and Descoteaux, J.P. (1986) Prevalence of natural virus infections in laboratory mice and rats used in Canada. *Lab. Anim. Sci.* **36**, 145–148.

Makarova, K.S., Wolf, Y.I., Tereza, E.P. and Ratner, V.A. (1998) Different patterns of molecular evolution of influenza A viruses in avian and human populations. *Genetika* **34**, 890–896.

Marcotte, L.L., Wass, A.B., Gohara, D.W., Pathak, H.B., Arnold, J.J., Filman, D.J. *et al.* (2007) Crystal structure of poliovirus 3CD protein: virally encoded protease and precursor to the RNA-dependent RNA polymerase. *J. Virol.* **81**, 3583–3596.

Margulis, L. and Bermudes, D. (1985) Symbiosis as a mechanism of evolution: status of cell symbiosis theory. *Symbiosis* **1**, 101–124.

Marquez, L.M., Redman, R.S., Rodriguez, R.J. and Roossinck, M.J. (2007) A virus in a fungus in a plant: three-way symbiosis required for thermal tolerance. *Science* **315**, 513–515.

Martell, M., Esteban, J.I., Quer, J., Genesca, J., Weiner, A., Esteban, R. *et al.* (1992) Hepatitis C virus (HCV) circulates as a population of different but closely related genomes: quasispecies nature of HCV genome distribution. *J. Virol.* **66**, 3225–3229.

Marten, N.W., Stohlman, S.A. and Bergmann, C.C. (2001) MHV infection of the CNS: mechanisms of immune-mediated control. *Viral Immunol.* **14**, 1–18.

Martinez, J.M., Schroeder, D.C., Larsen, A., Bratbak, G. and Wilson, W.H. (2007) Molecular dynamics of *Emiliania huxleyi* and cooccurring viruses during two separate mesocosm studies. *Appl. Environ. Microbiol.* **73**, 554–562.

Marty, G.D., Quinn, T.J., Carpenter, G., Meyers, T.R. and Willits, N.H. (2003) Role of disease in abundance of a Pacific herring (*Clupea pallasi*) population. *Can. J. Fisheries Aquat. Sci.* **60**, 1258–1265.

Mazel, D., Dychinco, B., Webb, V.A. and Davies, J. (2000) Antibiotic resistance in the ECOR collection: integrons and identification of a novel aad gene. *Antimicrob. Agents Chemother.* **44**, 1568–1574.

McGeehan, J.E., Depledge, N.W. and McGeoch, D.J. (2001) Evolution of the dUTPase gene of mammalian and avian herpesviruses. *Curr. Protein Pept. Sci.* **2**, 325–333.

McGeoch, D.J. and Davison, A.J. (1999) The descent of human herpesvirus 8. *Semin. Cancer Biol.* **9**, 201–209.

McGeoch, D.J. and Gatherer, D. (2005) Integrating reptilian herpesviruses into the family herpesviridae. *J. Virol.* **79**, 725–731.

McGeoch, D.J., Dolan, A. and Ralph, A.C. (2000) Toward a comprehensive phylogeny for mammalian and avian herpesviruses. *J. Virol.* **74**, 10401–10406.

McGeoch, D.J., Rixon, F.J. and Davison, A.J. (2006) Topics in herpesvirus genomics and evolution. *Virus Res.* **117**, 90–104.

McIntosh, E.M. and Haynes, R.H. (1996) HIV and human endogenous retroviruses: an hypothesis with therapeutic implications. *Acta Biochim. Pol.* **43**, 583–592.

Mesyanzhinov, V.V., Robben, J., Grymonprez, B., Kostyuchenko, V.A., Bourkaltseva, M.V., Sykilinda, N.N. *et al.* (2002) The genome of bacteriophage phiKZ of *Pseudomonas aeruginosa*. *J. Mol. Biol.* **317**, 1–19.

Mi, S., Lee, X., Li, X., Veldman, G.M., Finnerty, H., Racie, L. *et al.* (2000) Syncytin is a captive retroviral envelope protein involved in human placental morphogenesis. *Nature* **403**, 785–789.

Mirkin, B.G., Fenner, T.I., Galperin, M.Y. and Koonin, E.V. (2003) Algorithms for computing parsimonious evolutionary scenarios for genome evolution, the last universal common ancestor and dominance of horizontal gene transfer in the evolution of prokaryotes. *BMC Evol. Biol.* **3**, 2.

Mondy, W.L. and Pierce, S.K. (2003) Apoptotic-like morphology is associated with annual synchronized death in kleptoplastic sea slugs (*Elysia chlorotica*). *Invertebrate Biol.* **122**, 126–137.

Moran, N.A., Degnan, P.H., Santos, S.R., Dunbar, H.E. and Ochman, H. (2005) The players in a mutualistic symbiosis: insects, bacteria, viruses and virulence genes. *Proc. Natl Acad. Sci. USA* **102**, 16919–16926.

Morand-Joubert, L., Charpentier, C., Poizat, G., Chene, G., Dam, E., Raguin, G. *et al.* (2006) Low genetic barrier to large increases in HIV-1 cross-resistance to protease inhibitors during salvage therapy. *Antivir. Ther.* **11**, 143–154.

Moreira, D. and Lopez-Garcia, P. (2005) Comment on 'The 1.2-megabase genome sequence of Mimivirus.' *Science* **308**, 1114.

Moro, D., Lloyd, M.L., Smith, A.L., Shellam, G.R. and Lawson, M.A. (1999) Murine viruses in an island population of introduced house mice and endemic short-tailed mice in Western Australia. *J. Wildl. Dis.* **35**, 301–310.

Moro, D., Lawson, M.A., Hobbs, R.P. and Thompson, R.C. (2003) Pathogens of house mice on arid Boullanger Island and subantarctic Macquarie Island, Australia. *J. Wildl. Dis.* **39**, 762–771.

Mura, M., Murcia, P., Caporale, M., Spencer, T.E., Nagashima, K., Rein, A. and Palmarini, M. (2004) Late viral interference induced by transdominant Gag of an endogenous retrovirus. *Proc. Natl Acad. Sci. USA* **101**, 11117–11122.

Murcia, P.R., Arnaud, F. and Palmarini, M. (2007) The transdominant endogenous retrovirus enJS56A1 associates with and blocks intracellular trafficking of Jaagsiekte sheep retrovirus Gag. *J. Virol.* **81**, 1762–1772.

Murray, S.M. and Linial, M.L. (2006) Foamy virus infection in primates. *J. Med. Primatol.* **35**, 225–235.

Mushegian, A. (1999) The minimal genome concept. *Curr. Opin. Genet. Dev.* **9**, 709–714.

Nagasaki, K., Shirai, Y., Tomaru, Y., Nishida, K. and Pietrokovski, S. (2005) Algal viruses with distinct intraspecies host specificities include identical intein elements. *Appl. Environ. Microbiol.* **71**, 3599–3607.

Nakamura, A., Okazaki, Y., Sugimoto, J., Oda, T. and Jinno, Y. (2003) Human endogenous retroviruses with transcriptional potential in the brain. *J. Hum. Genet.* **48**, 575–581.

Nandhagopal, N., Simpson, A.A., Gurnon, J.R., Yan, X., Baker, T.S., Graves, M.V. *et al.* (2002) The structure and evolution of the major capsid protein of a large, lipid-containing DNA virus. *Proc. Natl Acad. Sci. USA* **99**, 14758–14763.

Narechania, A., Terai, M., Chen, Z., DeSalle, R. and Burk, R.D. (2004) Lack of the canonical pRB-binding domain in the E7 ORF of artiodactyl papillomaviruses is associated with the development of fibropapillomas. *J. Gen. Virol.* **85**, 1243–1250.

Naryshkina, T., Liu, J., Florens, L., Swanson, S.K., Pavlov, A.R., Pavlova, N.V. *et al.* (2006) *Thermus thermophilus* bacteriophage phiYS40 genome and proteomic characterization of virions. *J. Mol. Biol.* **364**, 667–677.

Nash, A.A., Dutia, B.M., Stewart, J.P. and Davison, A.J. (2001) Natural history of murine gamma-herpesvirus infection. *Philos. Trans. R. Soc. Lond. B Biol. Sci.* **356**, 569–579.

Ndunguru, J., Legg, J.P., Aveling, T.A., Thompson, G. and Fauquet, C.M. (2005) Molecular biodiversity of cassava begomoviruses in Tanzania: evolution of cassava geminiviruses in Africa and evidence for East Africa being a center of diversity of cassava geminiviruses. *Virol. J.* **2**, 21.

Nelson, M.I. and Holmes, E.C. (2007) The evolution of epidemic influenza. *Nat. Rev. Genet.* **8**, 196–205.

Nicholas, J., Zong, J.C., Alcendor, D.J., Ciufo, D.M., Poole, L.J., Sarisky, R.T. *et al.* (1998) Novel organizational features, captured cellular genes and strain variability within the genome of KSHV/HHV8. *J. Natl Cancer Inst. Monogr.* **23**, 79–88.

Nilsson, A.S., Karlsson, J.L. and Haggard-Ljungquist, E. (2004) Site-specific recombination links the evolution of P2-like coliphages and pathogenic enterobacteria. *Mol. Biol. Evol.* **21**, 1–13.

Nolan, J.M., Petrov, V., Bertrand, C., Krisch, H.M. and Karam, J.D. (2006) Genetic diversity among five T4-like bacteriophages. *Virol. J.* **3**, 30.

Novella, I.S., Duarte, E.A., Elena, S.F., Moya, A., Domingo, E. and Holland, J.J. (1995) Exponential increases of RNA virus fitness during large population transmissions. *Proc. Natl Acad. Sci. USA* **92**, 5841–5844.

Novella, I.S., Hershey, C.L., Escarmis, C., Domingo, E. and Holland, J.J. (1999) Lack of evolutionary stasis during alternating replication of an arbovirus in insect and mammalian cells. *J. Mol. Biol.* **287**, 459–465.

Ogata, H., Raoult, D. and Claverie, J.M. (2005) A new example of viral intein in Mimivirus. *Virol. J.* **2**, 8.

OhAinle, M., Kerns, J.A., Malik, H.S. and Emerman, M. (2006) Adaptive evolution and antiviral activity of the conserved mammalian cytidine deaminase APOBEC3H. *J. Virol.* **80**, 3853–3862.

Okamoto, H. and Mishiro, S. (1994) Genetic heterogeneity of hepatitis C virus. *Intervirology* **37**, 68–76.

Oliveira, N.M., Farrell, K.B. and Eiden, M.V. (2006) In vitro characterization of a koala retrovirus. *J. Virol.* **80**, 3104–3107.

Orgel, L.E. and Crick, F.H. (1980) Selfish DNA: the ultimate parasite. *Nature* **284**, 604–607.

Ortmann, A.C., Wiedenheft, B., Douglas, T. and Young, M. (2006) Hot crenarchaeal viruses reveal deep evolutionary connections. *Nat. Rev. Microbiol.* **4**, 520–528.

Pagel, M.D. (2002) *Encyclopedia of Evolution*. Oxford and New York: Oxford University Press.

Parsyan, A., Szmaragd, C., Allain, J.P. and Candotti, D. (2007) Identification and genetic diversity of two human parvovirus B19 genotype 3 subtypes. *J. Gen. Virol.* **88**, 428–431.

Paul, J.H., Sullivan, M.B., Segall, A.M. and Rohwer, F. (2002) Marine phage genomics. *Comp. Biochem. Physiol. B Biochem. Mol. Biol.* **133**, 463–476.

Persson, R., McGeehan, J. and Wilson, K.S. (2005) Cloning, expression, purification and characterisation of the dUTPase encoded by the integrated *Bacillus subtilis* temperate bacteriophage SPbeta. *Protein Expr. Purif.* **42**, 92–99.

Pfeiffer, J.K. and Kirkegaard, K. (2005) Increased fidelity reduces poliovirus fitness and virulence under selective pressure in mice. *PLoS Pathog.* **1**, e11.

Pfeiffer, J.K. and Kirkegaard, K. (2006) Bottleneck-mediated quasispecies restriction during spread of an RNA virus from inoculation site to brain. *Proc. Natl Acad. Sci. USA* **103**, 5520–5525.

Pierce, S.K., Maugel, T.K., Rumpho, M.E., Hanten, J.J. and Mondy, W.L. (1999) Annual viral expression in a sea slug population: Life cycle control and symbiotic chloroplast maintenance. *Biol. Bull.* **197**, 1–6.

Prangishvili, D., Forterre, P. and Garrett, R.A. (2006a) Viruses of the Archaea: a unifying view. *Nat. Rev. Microbiol.* **4**, 837–848.

Prangishvili, D., Garrett, R.A. and Koonin, E.V. (2006b) Evolutionary genomics of archaeal viruses: unique viral genomes in the third domain of life. *Virus Res.* **117**, 52–67.

Putics, A., Filipowicz, W., Hall, J., Gorbalenya, A.E. and Ziebuhr, J. (2005) ADP-ribose-1″-monophosphatase: a conserved coronavirus enzyme that is dispensable for viral replication in tissue culture. *J. Virol.* **79**, 12721–12731.

Pybus, O.G., Rambaut, A., Belshaw, R., Freckleton, R.P., Drummond, A.J. and Holmes, E.C. (2007) Phylogenetic evidence for deleterious mutation load in RNA viruses and its contribution to viral evolution. *Mol. Biol. Evol.* **24**, 845–852.

Qiu, W. and Scholthof, K.B. (2001) Defective interfering RNAs of a satellite virus. *J. Virol.* **75**, 5429–5432.

Randall, A.Z., Baldi, P. and Villarreal, L.P. (2004) Structural proteomics of the poxvirus family. *Artif. Intell. Med.* **31**, 105–115.

Raoult, D., Audic, S., Robert, C., Abergel, C., Renesto, P., Ogata, H. *et al.* (2004) The 1.2-megabase genome sequence of Mimivirus. *Science* **306**, 1344–1350.

Rice, G., Tang, L., Stedman, K., Roberto, F., Spuhler, J., Gillitzer, E. *et al.* (2004) The structure of a thermophilic archaeal virus shows a double-stranded DNA viral capsid type that spans all domains of life. *Proc. Natl Acad. Sci. USA* **101**, 7716–7720.

Roossinck, M.J. (2003) Plant RNA virus evolution. *Curr. Opin. Microbiol.* **6**, 406–409.

Roossinck, M.J. (2005) Symbiosis versus competition in plant virus evolution. *Nat. Rev. Microbiol.* **3**, 917–924.

Roossinck, M.J. and Schneider, W.L. (2006) Mutant clouds and occupation of sequence space in plant RNA viruses. *Curr. Top. Microbiol. Immunol.* **299**, 337–348.

Rowe, C.L., Baker, S.C., Nathan, M.J., Sgro, J.Y., Palmenberg, A.C. and Fleming, J.O. (1998) Quasispecies development by high frequency RNA recombination during MHV persistence. *Adv. Exp. Med. Biol.* **440**, 759–765.

Ruiz-Jarabo, C.M., Arias, A., Baranowski, E., Escarmis, C. and Domingo, E. (2000) Memory in viral quasispecies. *J. Virol.* **74**, 3543–3547.

Ryan, F.P. (2007) Viruses as symbionts. *Symbiosis* **44**. (in press)

Saren, A.M., Ravantti, J.J., Benson, S.D., Burnett, R.M., Paulin, L., Bamford, D.H. and Bamford, J.K. (2005) A snapshot of viral evolution from genome analysis of the tectiviridae family. *J. Mol. Biol.* **350**, 427–440.

Sawyer, S.L., Emerman, M. and Malik, H.S. (2004) Ancient adaptive evolution of the primate antiviral DNA-editing enzyme APOBEC3G. *PLoS Biol.* **2**, E275.

Schoondermark-van de Ven, E.M., Philipse-Bergmann, I.M. and van der Logt, J.T. (2006) Prevalence of naturally occurring viral infections, *Mycoplasma pulmonis* and *Clostridium piliforme* in laboratory rodents in Western Europe screened from 2000 to 2003. *Lab. Anim.* **40**, 137–143.

Schroeder, D.C., Oke, J., Hall, M., Malin, G. and Wilson, W.H. (2003) Virus succession observed during an *Emiliania huxleyi* bloom. *Appl. Environ. Microbiol.* **69**, 2484–2490.

Shackelton, L.A. and Holmes, E.C. (2006) Phylogenetic evidence for the rapid evolution of human B19 erythrovirus. *J. Virol.* **80**, 3666–3669.

Shadan, F.F. and Villarreal, L.P. (1995) The evolution of small DNA viruses of eukaryotes: Past and present considerations. *Virus Genes* **11**, 239–257.

Sharp, G.B., Kawaoka, Y., Jones, D.J., Bean, W.J., Pryor, S.P., Hinshaw, V. and Webster, R.G. (1997) Coinfection of wild ducks by influenza a viruses: Distribution patterns and biological significance. *J. Virol.* **71**, 6128–6135.

Shutt, T.E. and Gray, M.W. (2006) Bacteriophage origins of mitochondrial replication and transcription proteins. *Trends Genet.* **22**, 90–95.

Simon, A.E., Roossinck, M.J. and Havelda, Z. (2004) Plant virus satellite and defective interfering RNAs: new paradigms for a new century. *Annu. Rev. Phytopathol.* **42**, 415–437.

Singleton, G.R., Smith, A.L., Shellam, G.R., Fitzgerald, N. and Muller, W.J. (1993) Prevalence of viral antibodies and helminths in field populations of house mice (Mus domesticus) in southeastern Australia. *Epidemiol. Infect.* **110**, 399–417.

Smith, D.B., Pathirana, S., Davidson, F., Lawlor, E., Power, J., Yap, P.L. and Simmonds, P. (1997) The origin of hepatitis C virus genotypes. *J. Gen. Virol.* **78**, 321–328.

Smith, G.J., Fan, X.H., Wang, J., Li, K.S., Qin, K., Zhang, J.X. *et al.* (2006) Emergence and predominance of an H5N1 influenza variant in China. *Proc. Natl Acad. Sci. USA*, **103**, 16936–16941.

Spackman, E., Stallknecht, D.E., Slemons, R.D., Winker, K., Suarez, D.L., Scott, M. and Swayne, D.E. (2005) Phylogenetic analyses of type A influenza genes in natural reservoir species in North America reveals genetic variation. *Virus Res.* **114**, 89–100.

Stadler, B.M.R., Stadler, P.F. and Schuster, P. (2000) Dynamics of autocatalytic replicator networks based on higher-order ligation reactions. *Bull. Math. Biol.* **62**, 1061–1086.

Stoye, J.P. (2006) Koala retrovirus: a genome invasion in real time. *Genome Biol.* **7**, 241.

Stuhler, A., Flory, E., Wege, H., Lassmann, H. and Wege, H. (1997) No evidence for quasispecies populations during persistence of the coronavirus mouse hepatitis virus JHM: sequence conservation within the surface glycoprotein gene S in Lewis rats. *J. Gen. Virol.* **78**, 747–756.

Suarez, D.L. (2000) Evolution of avian influenza viruses. *Vet. Microbiol.* **74**, 15–27.

Sullivan, M.B., Lindell, D., Lee, J.A., Thompson, L.R., Bielawski, J.P. and Chisholm, S.W. (2006) Prevalence and evolution of core photosystem II genes in marine cyanobacterial viruses and their hosts. *PLoS Biol.* **4**.

Summer, E.J., Gonzalez, C.F., Bomer, M., Carlile, T., Embry, A., Kucherka, A.M. *et al.* (2006) Divergence and mosaicism among virulent soil phages of the *Burkholderia cepacia* complex. *J. Bacteriol.* **188**, 255–268.

Suzan-Monti, M., Scola, B.L., Barrassi, L., Espinosa, L. and Raoult, D. (2007) Ultrastructural characterization of the giant volcano-like virus factory of *Acanthamoeba polyphaga* mimivirus. *PLoS ONE* **2**, e328.

Switzer, W.M., Salemi, M., Shanmugam, V., Gao, F., Cong, M.E., Kuiken, C. *et al.* (2005) Ancient co-speciation of simian foamy viruses and primates. *Nature* **434**, 376–380.

Takemura, M. (2001) Poxviruses and the origin of the eukaryotic nucleus. *J. Mol. Evol.* **52**, 419–425.

Tang, X.C., Zhang, J.X., Zhang, S.Y., Wang, P., Fan, X.H., Li, L.F. *et al.* (2006) Prevalence and genetic diversity of coronaviruses in bats from China. *J. Virol.* **80**, 7481–7490.

Tardy-Panit, M., Blondel, B., Martin, A., Tekaia, F., Horaud, F. and Delpeyroux, F. (1993) A mutation in the RNA polymerase of poliovirus type 1 contributes to attenuation in mice. *J. Virol.* **67**, 4630–4638.

Tarlinton, R.E., Meers, J. and Young, P.R. (2006) Retroviral invasion of the koala genome. *Nature* **442**, 79–81.

Van Etten, J.L. (2003) Unusual life style of giant chlorella viruses. *Annu. Rev. Genet.* **37**, 153–195.

Van Etten, J.L., Graves, M.V., Muller, D.G., Boland, W. and Delaroque, N. (2002) Phycodnaviridae—large DNA algal viruses. *Arch. Virol.* **147**, 1479–1516.

Vetsigian, K., Woese, C. and Goldenfeld, N. (2006) Collective evolution and the genetic code. *Proc. Natl Acad. Sci. USA* **103**, 10696–10701.

Vignuzzi, M., Stone, J.K. and Andino, R. (2005) Ribavirin and lethal mutagenesis of poliovirus: molecular mechanisms, resistance and biological implications. *Virus Res.* **107**, 173–181.

Vignuzzi, M., Stone, J.K., Arnold, J.J., Cameron, C.E. and Andino, R. (2006) Quasispecies diversity determines pathogenesis through cooperative interactions in a viral population. *Nature* **439**, 344–348.

Vijaykrishna, D., Smith, G.J., Zhang, J.X., Peiris, J.S., Chen, H. and Guan, Y. (2007) Evolutionary insights into the ecology of coronaviruses. *J. Virol.* **81**, 4012–4020.

Villarreal, L.P. (1999) DNA virus contribution to host evolution. In: *Origin and Evolution of Viruses* (E. Domingo, R.G. Webster and J.J. Holland, eds), pp. 391–420. San Diego: Academic Press.

Villarreal, L.P. (2005) *Viruses and the Evolution of Life*. Washington, DC: ASM Press.

Villarreal, L.P. (2006) How viruses shape the tree of life. *Future Virol.* **1**, 587–595.

Villarreal, L.P. (2007) Virus-host symbiosis mediated by persistence. *Symbiosis* **44**, 1–9.

Villarreal, L.P. and DeFilippis, V.R. (2000) A hypothesis for DNA viruses as the origin of eukaryotic replication proteins. *J. Virol.* **74**, 7079–7084.

Villarreal, L.P. and Villareal, L.P. (1997) On viruses, sex and motherhood. *J. Virol.* **71**, 859–865.

Villarreal, L.P., Defilippis, V.R. and Gottlieb, K.A. (2000) Acute and persistent viral life strategies and their relationship to emerging diseases. *Virology* **272**, 1–6.

Vogel, T.U., Evans, D.T., Urvater, J.A., O'Connor, D.H., Hughes, A.L. and Watkins, D.I. (1999) Major histocompatibility complex class I genes in primates: co-evolution with pathogens. *Immunol. Rev.* **167**, 327–337.

Wallensten, A., Munster, V.J., Karlsson, M., Lundkvist, A., Brytting, M., Stervander, M. *et al.* (2006) High prevalence of influenza A virus in ducks caught during spring migration through Sweden. *Vaccine*. .

Wang, B., Mikhail, M., Dyer, W.B., Zaunders, J.J., Kelleher, A.D. and Saksena, N.K. (2003) First demonstration of a lack of viral sequence evolution in a nonprogressor, defining replication-incompetent HIV-1 infection. *Virology* **312**, 135–150.

Wang, L.F., Shi, Z., Zhang, S., Field, H., Daszak, P. and Eaton, B.T. (2006) Review of bats and SARS. *Emerg. Infect. Dis.* **12**, 1834–1840.

Wang, N., Baldi, P.F. and Gaut, B.S. (2006) Phylogenetic analysis, genome evolution and the rate of gene gain in the Herpesviridae. *Mol. Phylogenet. Evol.*

Watkins, D.I. (1995) The evolution of major histocompatibility class I genes in primates. *Crit. Rev. Immunol.* **15**, 1–29.

Webster, R.G. and Govorkova, E.A. (2006) H5N1 influenza—continuing evolution and spread. *N. Engl. J. Med.* **355**, 2174–2177.

Weir, E.C., Bhatt, P.N., Barthold, S.W., Cameron, G.A. and Simack, P.A. (1987) Elimination of mouse hepatitis virus from a breeding colony by temporary cessation of breeding. *Lab. Anim. Sci.* **37**, 455–458.

Widjaja, L., Krauss, S.L., Webby, R.J., Xie, T. and Webster, R.G. (2004) Matrix gene of influenza a viruses isolated from wild aquatic birds: Ecology and emergence of influenza A viruses. *J. Virol.* **78**, 8771–8779.

Wilberz, S., Partke, H.J., Dagnaes-Hansen, F. and Herberg, L. (1991) Persistent MHV (mouse hepatitis virus) infection reduces the incidence of diabetes mellitus in non-obese diabetic mice. *Diabetologia* **34**, 2–5.

Wilke, C.O. (2005) Quasispecies theory in the context of population genetics. *BMC Evol. Biol.* **5**, 44.

Wolf, Y.I., Viboud, C., Holmes, E.C., Koonin, E.V. and Lipman, D.J. (2006) Long intervals of stasis punctuated by bursts of positive selection in the seasonal evolution of influenza A virus. *Biol. Direct*, **1**, 34.

Yarmolinsky, M.B. (2000) A pot-pourri of plasmid paradoxes: effects of a second copy. *Mol. Microbiol.* **38**, 1–7.

Yarmolinsky, M.B. (2004) Bacteriophage P1 in retrospect and in prospect. *J. Bacteriol.* **186**, 7025–7028.

Yi, J.M., Kim, T.H., Huh, J.W., Park, K.S., Jang, S.B., Kim, H.M. and Kim, H.S. (2004) Human endogenous retroviral elements belonging to the HERV-S family from human tissues, cancer cells and primates: expression, structure, phylogeny and evolution. *Gene* **342**, 283–292.

Zanotto, P.M., Gibbs, M.J., Gould, E.A. and Holmes, E.C. (1996) A reevaluation of the higher taxonomy of viruses based on RNA polymerases. *J. Virol.* **70**, 6083–6096.

Zhang, Y., Moriyama, H., Homma, K. and Van Etten, J.L. (2005) Chlorella virus-encoded deoxyuridine triphosphatases exhibit different temperature optima. *J. Virol.* **79**, 9945–9953.

Zhu, T., Corey, L., Hwangbo, Y., Lee, J.M., Learn, G.H., Mullins, J.I. and McElrath, M.J. (2003) Persistence of extraordinarily low levels of genetically homogeneous human immunodeficiency virus type 1 in exposed seronegative individuals. *J. Virol.* **77**, 6108–6116.

Zhu, X.Y. (2003) [Phylogenetic reconstruction of DNA polymerase X family]. *Yi Chuan Xue Bao* **30**, 867–872.

Index

Origin and Evolution of Viruses
ISBN 978-0-12-374153-0